Yard & Garden Tractor

SERVICE MANUAL ■ 1ST EDITION

INTERTEC PUBLISHING
P.O. Box 12901 ■ Overland Park, KS 66282-2901
Phone: 800-262-1954 Fax: 800-633-6219
www.intertecbooks.com

Library of Congress Catalog Card Number 91-77842

Cover photo courtesy of Deere & Co.

Yard & Garden Tractor

SERVICE MANUAL ■ 1ST EDITION

Tractor Manufacturers:

- Allis-Chalmers
- Ariens
- Bolens
- Case/Ingersoll
- Cub Cadet
- John Deere
- Deutz-Allis
- Engineering Products Co.
- Ford
- Gilson
- Gravely
- International Harvester
- Kubota
- MTD
- Murray
- Simplicity
- Snapper
- Speedex
- Wheel Horse

Engine Manufacturers:

- Briggs & Stratton
- Kawasaki
- Kohler
- Kubota
- Onan
- Yanmar

Covers Multi-cylinder Models

INTERTEC PUBLISHING CORP.

President and CEO Raymond E. Maloney

Group Vice President Dan Fink

EDITORIAL

Editorial Director
Randy Stephens

Editors
Mike Hall
Mark Jacobs

Technical Writers
Robert Mills
Michael Morlan
Edward Scott
Ron Wright

Product Planning and Support Manager
Terry Distin

Lead Editorial Assistant
Shirley Renicker

Editorial Assistants
Veronica Bollin
Elizabeth Couzens

Technical Illustrators
Steve Amos
Robert Caldwell
Mitzi McCarthy
Diana Kirkland

MARKETING

Director of Marketing
Chris Charlton

Advertising and Promotions Manager
Katherine Hughes

Advertising Assistant
Hilary Lindsey

Art Director
Anita Blattner

SALES AND ADMINISTRATION

Director of Sales
Dutch Sadler

Accounts Manager
Ted Metzger

Sales Coordinator
Lynn Reynolds

Customer Service and Administration Manager
Joan Dickey

The following books and guides are published by Intertec Publishing Corp.

CLYMER SHOP MANUALS
Boat Motors and Drives
Motorcycles and ATVs
Snowmobiles
Personal Watercraft

ABOS/INTERTEC BLUE BOOKS AND TRADE-IN GUIDES
Recreational Vehicles
Outdoor Power Equipment
Agricultural Tractors
Lawn and Garden Tractors
Motorcycles and ATVs
Snowmobiles
Boats and Motors
Personal Watercraft

AIRCRAFT BLUEBOOK-PRICE DIGEST
Airplanes
Helicopters

I&T SHOP MANUALS
Tractors

INTERTEC SERVICE MANUALS
Snowmobiles
Outdoor Power Equipment
Personal Watercraft
Gasoline and Diesel Engines
Recreational Vehicles
Boat Motors and Drives
Motorcycles
Lawn and Garden Tractors

CONTENTS

GENERAL

TRACTOR SERVICE SECTION

ENGINE SECTION

Dual Dimensions

This service manual provides specifications in both the U.S. Customary and Metric (SI) systems of measurement. The first specification is given in the measuring system perceived by us to be the preferred system when servicing a particular component. The second specification (given in parentheses) is the converted measurement. For instance, the specification "0.011 inch (0.28 mm)" indicates that we feel the preferred measurement, in this instance, is the U.S. measurement system and the metric equivalent of 0.011 inch is 0.28 mm.

FUNDAMENTALS SECTION

ENGINE FUNDAMENTALS

OPERATING PRINCIPLES

The one, two or four cylinder engines used to power riding lawn mowers, garden tractors, pumps, generators, welders, mixers, windrowers, hay balers and many other items of power equipment in use today are basically similar. All are technically known as "Internal Combustion Reciprocating Engines."

The source of power is heat formed by burning a combustible mixture of petroleum products and air. In a reciprocating engine, this burning takes place in a closed cylinder containing a piston. Expansion resulting from the heat of combustion applies pressure on the piston to turn a shaft by means of a crank and connecting rod.

The fuel-air mixture may be ignited by means of an electric spark (Otto Cycle Engine) or by heat formed from compression of air in the engine cylinder (Diesel Cycle Engine). The complete series of events which must take place in order for the engine to run occurs in two revolutions of the crankshaft (four strokes of the piston in cylinder) and is referred to as a "Four-Stroke Cycle Engine."

OTTO CYCLE. In a spark ignited engine, a series of five events is required in order for the engine to provide power. This series of events is called the "Cycle" (for "Work Cycle") and is repeated in each cylinder of the engine as long as work is being done. This series of events which comprise the "Cycle" is as follows:

1. The mixture of fuel and air is pushed into the cylinder by atmospheric pressure when the pressure within the engine cylinder is reduced by the piston moving downward in the cylinder.
2. The mixture of fuel and air is compressed by the piston moving upward in the cylinder.
3. The compressed fuel-air mixture is ignited by a timed electric spark.
4. The burning fuel-air mixture expands, forcing the piston downward in the cylinder thus converting the chemical energy generated by combustion into mechanical power.
5. The gaseous products formed by the burned fuel-air mixture are exhausted from the cylinder so a new "Cycle" can begin.

The above described five events which comprise the work cycle of an engine are commonly referred to as (1), INTAKE; (2), COMPRESSION; (3), IGNITION; (4), EXPANSION (POWER); and (5), EXHAUST.

DIESEL CYCLE. The Diesel Cycle differs from the Otto Cycle in that air alone is drawn into the cylinder during the intake period. The air is heated from being compressed by the piston moving upward in the cylinder then a finely atomized charge of fuel is injected into the cylinder where it mixes with the air and is ignited by the heat of the compressed air. In order to create sufficient heat to ignite the injected fuel, an engine operating on the Diesel Cycle must compress the air to a much greater degree than an engine operating on the Otto Cycle where the fuel-air mixture is ignited by an electric spark. The power and exhaust events of the Diesel Cycle are similar to the power and exhaust events of the Otto Cycle.

FOUR-STROKE CYCLE. In a four-stroke cycle engine operating on the Otto Cycle (spark ignition), the five events of the cycle take place in four strokes of the piston, or in two revolutions of the engine crankshaft. Thus, a power stroke occurs only on alternate downward strokes of the piston.

In view "A" of Fig. 1-1, the piston is on the first downward stroke of the cycle. The mechanically operated intake valve has opened the intake port and, as the downward movement of the piston has reduced the air pressure in the cylinder to below atmospheric pressure, air is forced through the carburetor, where fuel is mixed with the air, and into the cylinder through the open intake port. The intake valve remains open and the fuel-air mixture continues to flow into the cylinder until the piston reaches the bottom of its downward stroke. As the piston starts on its first upward stroke, the mechanically operated intake valve closes and, since the exhaust valve is closed, the fuel-air mixture is compressed as in view "B".

Just before the piston reaches the top of its first upward stroke, a spark at the spark plug electrodes ignites the compressed fuel-air mixture. As the engine crankshaft turns past top center, the burning fuel-air mixture expands rapidly and forces the piston downward on its power stroke as shown in view "C". As the piston reaches the bottom of the power stroke, the mechanically operated exhaust valve starts to open and as the pressure of the burned fuel-air mixture is higher than atmospheric pressure, it starts to flow out the open exhaust port. As the engine crankshaft turns past bottom center, the exhaust valve is almost completely open and remains open dur-

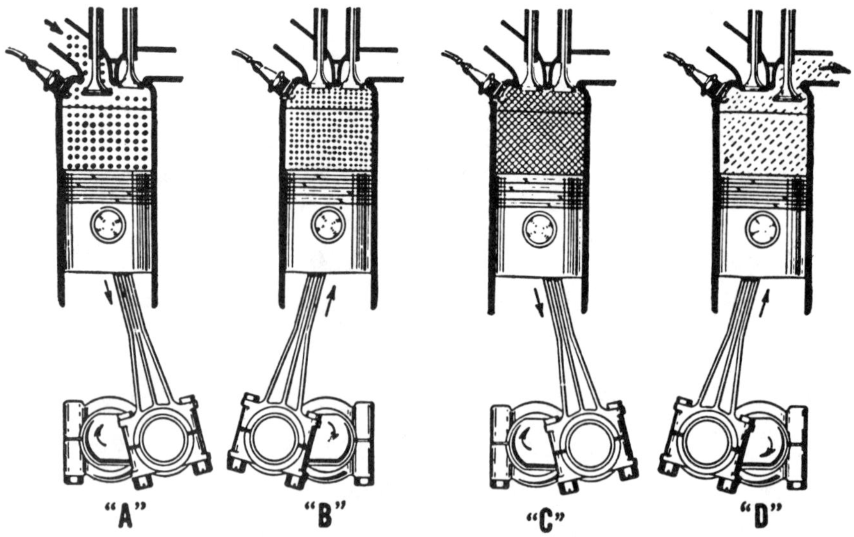

Fig. 1-1 – Schematic diagram of four-stroke cycle engine operating on the Otto (spark ignition) cycle. In view "A", piston is on first downward (intake) stroke and atmospheric pressure is forcing fuel-air mixture from carburetor into cylinder through the open intake valve. In view "B", both valves are closed and piston is on its first upward stroke compressing the fuel-air mixture in cylinder. In view "C", spark across electrodes of spark plug has ignited fuel-air mixture and heat of combustion rapidly expands the burning gaseous mixture forcing the piston on its second downward (expansion or power) stroke. In view "D", exhaust valve is open and piston on its second upward (exhaust) stroke forces the burned mixture from cylinder. A new cycle then starts as in view "A".

ing the upward stroke of the piston as shown in view "D". Upward movement of the piston pushes the remaining burned fuel-air mixture out of the exhaust port. Just before the piston reaches the top of its second upward or exhaust stroke, the intake valve opens and the exhaust valve closes. The cycle is completed as the crankshaft turns past top center and a new cycle begins as the piston starts downward as shown in view "A".

In a four-stroke cycle engine operating on the Diesel Cycle, the sequence of events of the cycle is similar to that described for operation on the Otto Cycle, but with the following exceptions: On the intake stroke, air only is taken into the cylinder. On the compression stroke, the air is highly compressed which raises the temperature of the air. Just before the piston reaches top dead center, fuel is injected into the cylinder and is ignited by the heated, compressed air. The remainder of the cycle is similar to that of the Otto Cycle.

CARBURETOR FUNDAMENTALS

OPERATING PRINCIPLES

Function of the carburetor on a spark-ignition engine is to atomize the fuel and mix the atomized fuel in proper proportions with air flowing to the engine intake port or intake manifold. Carburetors used on engines that are to be operated at constant speeds and under even loads are of simple design since they only have to mix fuel and air in a relatively constant ratio. On engines operating at varying speeds and loads, the carburetors must be more complex because different fuel-air mixtures are required to meet the varying demands of the engine.

FUEL-AIR MIXTURE RATIO REQUIREMENTS. To meet the demands of an engine being operated at varying speeds and loads, the carburetor must mix fuel and air at different mixture ratios. Fuel-air mixture ratios required for different operating conditions are approximately as follows:

	Fuel	Air
Starting, cold weather	1 lb.	7 lbs.
Accelerating	1 lb.	9 lbs.
Idling (no load)	1 lb.	11 lbs.
Part open throttle	1 lb.	15 lbs.
Full load, open throttle	1 lb.	13 lbs.

BASIC DESIGN. Carburetor design is based on the venturi principle which simply means that a gas or liquid flowing through a necked-down section (venturi) in a passage undergoes an increase in velocity (speed) and a decrease in pressure as compared to the velocity and pressure in full size sections of the passage. The principle is illustrated in Fig. 2-1, which shows air passing through a carburetor venturi. The figures given for air speeds and vacuum are approximate for a typical wide-open throttle operating condition. Due to low pressure (high vacuum) in the venturi, fuel is forced out through the fuel nozzle by the atmospheric pressure (0 vacuum) on the fuel; as fuel is emitted from the nozzle, it is atomized by the high velocity air flow and mixes with the air.

In Fig. 2-2, the carburetor choke plate and throttle plate are shown in relation to the venturi. Downward pointing arrows indicate air flow through the carburetor.

At cranking speeds, air flows through the carburetor venturi at a slow speed; thus, the pressure in the venturi does not usually decrease to the extent that atmospheric pressure on the fuel will force fuel from the nozzle. If the choke plate is closed as shown by dotted line in Fig. 2-2, air cannot enter into the carburetor and pressure in the carburetor decreases greatly as the engine is turned at cranking speed. Fuel can then flow from the fuel nozzle. In manufacturing the carburetor choke plate or disc, a small hole or notch is cut in the plate so that some air can flow through the plate when it is in closed position to provide air for the starting fuel-air mixture. In some instances, after starting a cold engine, it is advantageous to leave the choke plate in a partly closed position as the restriction of air flow will decrease the air pressure in carburetor venturi, thus causing more fuel to flow from the nozzle, resulting in a richer fuel-air mixture. The choke plate or disc should be in fully open position for normal engine operation.

If, after the engine has been started, the throttle plate is in wide-open position as shown by the solid line in Fig. 2-2, the engine can obtain enough fuel and air to run at dangerously high speeds. Thus, the throttle plate or disc must be partly closed as shown by the dotted lines to control engine speed. At no load, the engine requires very little air and fuel to run at its rated speed and the throttle must be moved on toward the closed position as shown by the dash lines. As more load is placed on the engine, more fuel and air are required for the engine to operate at its rated speed and throttle must be moved closer to the wide open position as shown by the solid line. When the engine is required to develop maximum power or speed the throttle must be in the wide open position.

Although some carburetors may be as simple as the basic design just described, most engines require more complex design features to provide variable fuel-air mixture ratios for different operating conditions. These design features will be described in the following paragraph.

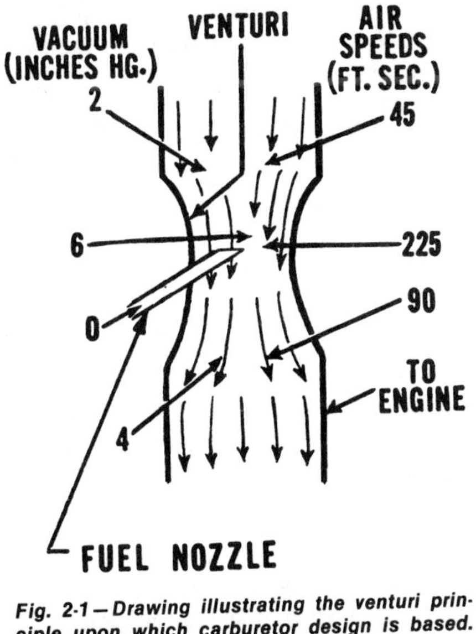

Fig. 2-1 – Drawing illustrating the venturi principle upon which carburetor design is based. Figures at left are inches of mercury vacuum and those at right are air speeds in feet per second that are typical of conditions found in a carburetor operating at wide open throttle. Zero vacuum in fuel nozzle corresponds to atmospheric pressure.

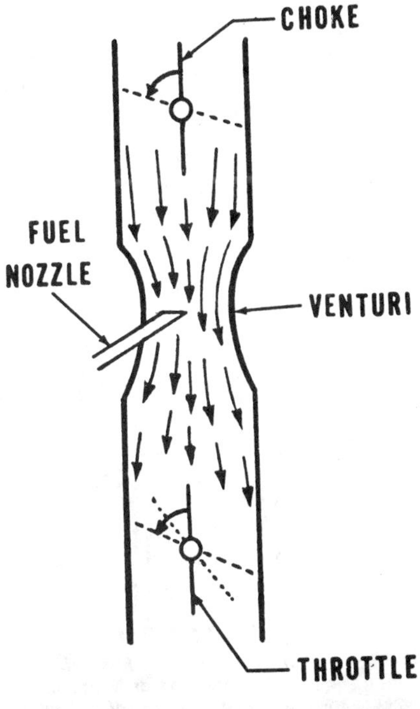

Fig. 2-2 – Drawing showing basic carburetor design. Text explains operation of the choke and throttle valves. In some carburetors, a primer pump may be used instead of the choke valve to provide fuel for the starting fuel-air mixture.

CARBURETOR TYPE

Carburetors are the float type and are either of the downshaft or side draft design. The following paragraph describes the features and operating principles of the float type carburetor.

FLOAT TYPE CARBURETOR. The principle of float type carburetor operation is illustrated in Fig. 2-3. Fuel is delivered at inlet (I) by gravity with fuel tank placed above carburetor, or by a fuel lift pump when tank is located below carburetor inlet. Fuel flows into the open inlet valve (V) until fuel level (L) in bowl lifts float against fuel valve needle and closes the valve. As fuel is emitted from the nozzle (N) when engine is running, fuel level will drop, lowering the float and allowing valve to open so fuel will enter the carburetor to meet the requirements of the engine.

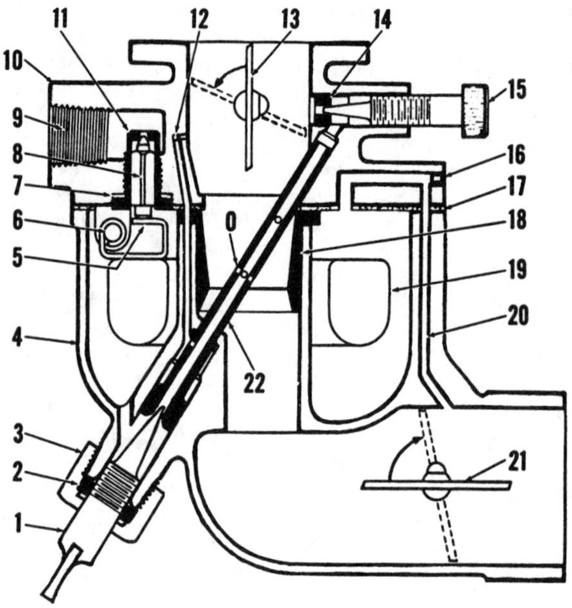

Fig. 2-4—Cross-sectional drawing of float type carburetor used on some engines.

0. Orifice
1. Main fuel needle
2. Packing
3. Packing nut
4. Carburetor bowl
5. Float tang
6. Float hinge pin
7. Gasket
8. Inlet valve
9. Fuel inlet
10. Carburetor body
11. Inlet valve seat
12. Vent
13. Throttle plate
14. Idle orifice
15. Idle fuel needle
16. Plug
17. Gasket
18. Venturi
19. Float
20. Fuel bowl vent
21. Choke
22. Fuel nozzle

In Fig. 2-4, a cut-away view of a well known make of float type carburetor is shown. Atmospheric pressure is maintained in fuel bowl through passage (20) which opens into carburetor air horn ahead of the choke plate (21). Fuel level is maintained at just below level of opening (O) in nozzle (22) by float (19) actuating inlet valve needle (8). Float height can be adjusted by bending float tang (5).

When starting a cold engine, it is necessary to close the choke plate (21) as shown by dotted lines so as to lower the air pressure in carburetor venturi (18) as engine is cranked. Then, fuel will flow up through nozzle (22) and will be emitted from openings (O) in nozzle. When an engine is hot, it will start on a leaner fuel-air mixture than when cold and may start without the choke plate being closed.

When engine is running at slow idle speed (throttle plate nearly closed as indicated by dotted lines in Fig. 2-4), air pressure above the throttle plate is low and atmospheric pressure in fuel bowl forces fuel up through the nozzle and out through orifice in seat (14) where it mixes with air passing the throttle plate. The idle fuel mixture is adjustable by turning needle (15) in or out as required. Idle speed is adjustable by turning the throttle stop screw (not shown) in or out to control amount of air passing the throttle plate.

When throttle plate is opened to increase engine speed, velocity of air flow through venturi (18) increases, air pressure at venturi decreases and fuel will flow from openings (O) in nozzle instead of through orifice in idle seat (14). When engine is running at high speed, pressure in nozzle (22) is less than at vent (12) opening in carburetor throat above venturi. Thus, air will enter vent and travel down the vent into the nozzle and mix with the fuel in the nozzle. This is referred to as air bleeding and is illustrated in Fig. 2-5.

Many different designs of float type carburetors will be found when servicing the different makes and models of engines. Reference should be made to the engine repair section of this manual for adjustment and overhaul specifications. Refer to carburetor servicing paragraphs in fundamentals sections for service hints.

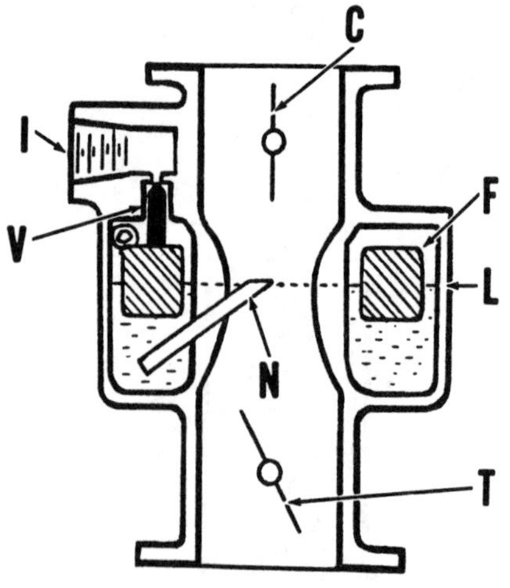

Fig. 2-3—Drawing showing basic float type carburetor design. Fuel must be delivered under pressure either by gravity or by use of fuel pump, to the carburetor fuel inlet (I). Fuel level (L) operates float (F) to open and close inlet valve (V) to control amount of fuel entering carburetor. Also shown are the fuel nozzle (N), throttle (T) and choke (C).

Fig. 2-5—Illustration of air bleed principle explained in text.

IGNITION SYSTEM FUNDAMENTALS

The ignition system provides a properly timed surge of extremely high voltage electrical energy which flows across the spark plug electrode gap to create the ignition spark. Engines may be equipped with either a magneto or battery ignition system. A magneto ignition system generates electrical energy, intensifies (transforms) this electrical energy to the extremely high voltage required and delivers this electrical energy at the proper time for the ignition spark. In a battery ignition system, a storage battery is used as a source of electrical energy and the system transforms the relatively low electrical voltage from the battery into the high voltage required and delivers the high voltage at proper time for the gnition spark. Thus, the function of the two systems is somewhat similar except for the basic source of electrical energy. The fundamental operating principles of ignition systems are explained in the following paragraphs.

MAGNETISM AND ELECTRICITY

The fundamental principles upon which ignition systems are designed are presented in this section. As the study of magnetism and electricity is an entire scientific field, it is beyond the scope of this manual to fully explore these subjects. However, the following information will impart a working knowledge of basic principles which should be of value in servicing engines.

MAGNETISM. The effects of magnetism can be shown easily while the theory of magnetism is too complex to be presented here. The effects of magnetism were discovered many years ago when fragments of iron ore were found to attract each other and also attract other pieces of iron. Further, it was found that when suspended in air, one end of the iron ore fragment would always point in the direction of the North Star. The end of the iron ore fragment pointing north was called the "north pole" and the opposite end the "south pole." By stroking a piece of steel with a "natural magnet," as these iron ore fragments were called, it was found that the magnetic properties of the natural magnet could be transferred or "induced" into the steel.

Steel which will retain magnetic properties for an extended period of time after being subjected to a strong magnetic field are called "permanent magnets;" iron or steel that loses such magnetic properties soon after being subjected to a magnetic field are called "temporary magnets." Soft iron will lose magnetic properties almost immediately after being removed from a magnetic field, and so is used where this property is desirable.

The area affected by a magnet is called a "field of force." The extent of this field of force is related to the strength of the magnet and can be determined by use of a compass. In practice, it is common to illustrate the field of force surrounding a magnet by lines as shown in Fig. 3-1 and the field of force is usually called "lines of force" or "flux." Actually, there are no "lines;" however, this is a convenient method of illustrating the presence of the invisible magnetic forces and if a certain magnetic force is defined as a "line of force," then all magnetic forces may be measured by comparison. The number of "lines of force" making up a strong magnetic field is enormous.

Most materials when placed in a magnetic field are not attracted by the magnet, do not change the magnitude or direction of the magnetic field, and so are called "non-magnetic materials." Materials such as iron, cobalt, nickel or their alloys, when placed in a magnetic field will concentrate the field of force and hence are magnetic conductors or "magnetic materials." There are no materials known in which magnetic fields will not penetrate and magnetic lines of force can be deflected only by magnetic materials or by another magnetic field.

Alnico, an alloy containing aluminum, nickel and cobalt, retains magnetic properties for a very long period of time after being subjected to a strong magnetic field and is extensively used as a permanent magnet. Soft iron, which loses magnetic properties quickly, is used to concentrate magnetic fields as in Fig. 3-1.

ELECTRICITY. Electricity, like magnetism, is an invisible physical force whose effects may be more readily explained than the theory of what electricity consists of. All of us are familiar with the property of electricity to produce light, heat and mechanical power. What must be explained for the purpose of understanding ignition system operation is the inter-relationship of magnetism and electricity and how the ignition spark is produced.

Electrical current may be defined as a flow of energy in a conductor which, in some ways, may be compared to flow of water in a pipe. For electricity to flow, there must be a pressure (voltage) and a complete circuit (closed path) through which the electrical energy may return, a comparison being a water pump and a pipe that receives water from the outlet (pressure) side of the pump and returns the water to the inlet side of the pump. An electrical circuit may be completed by electricity flowing through the earth (ground), or through the metal framework of an engine or other equipment ("grounded" or "ground" connections). Usually, air is an insulator through which electrical energy will not flow. However if the force (voltage) becomes great, the resistance of air to the flow of electricity is broken down and a current will flow, releasing energy in the form of a spark. By high voltage electricity breaking down the resistance of the air gap between the spark plug electrodes, the ignition spark is formed.

ELECTRO-MAGNETIC INDUCTION. The principle of electro-magnetic induction is as follows: When a wire (conductor) is moved through a field of magnetic force so as to cut across the lines of force (flux), a potential voltage or electromotive force (emf) is induced in the wire. If the wire is a part of a completed electrical circuit, current will flow through the circuit as illustrated in Fig. 3-2. It should be noted that the movement of the wire through the lines of

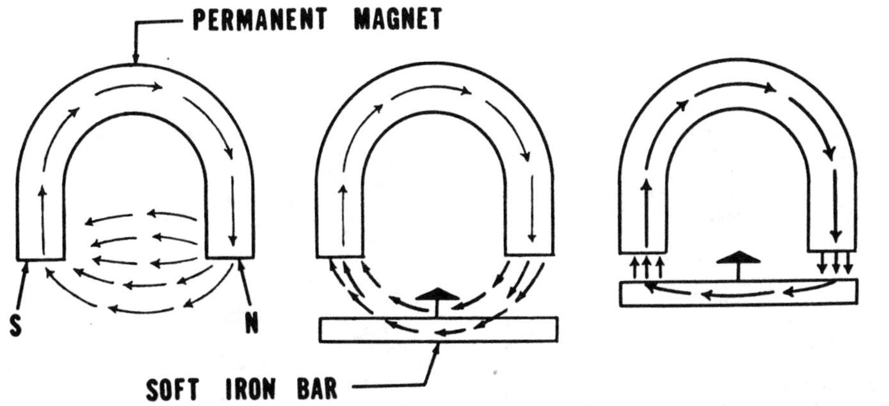

Fig. 3-1 — In left view, field of force of permanent magnet is illustrated by arrows showing direction of magnetic force from north pole (N) to south pole (S). In center view, lines of magnetic force are being attracted by soft iron bar that is being moved into the magnetic field. In right view, the soft iron bar has been moved close to the magnet and the field of magnetic force is concentrated within the bar.

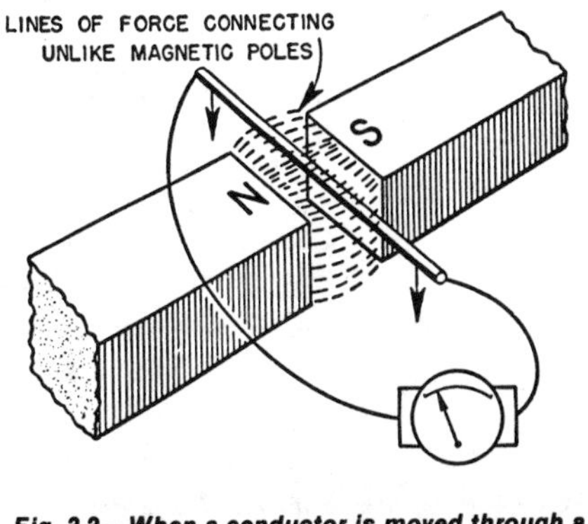

Fig. 3-2 — When a conductor is moved through a magnetic field so as to cut across lines of force, a potential voltage will be induced in the conductor. If the conductor is a part of a completed electrical circuit, current will flow through the circuit indicated by the gage.

magnetic force is a relative motion; that is, if the lines of force of a moving magnetic field cut across a wire, this will also induce an emf to the wire.

The direction of an induced current is related to the direction of magnetic force and also to the direction of movement of the wire through the lines of force, or flux. The voltage of an induced current is related to the strength, or concentration of lines of force, of the magnetic field and to the rate of speed at which the wire is moved through the flux. If a length of wire is wound into a coil and a section of the coil is moved through magnetic lines of force, the voltage induced will be proportional to the number of turns of wire in the coil.

ELECTRICAL MAGNETIC FIELDS. When a current is flowing in a wire, a magnetic field is present around the wire as illustrated in Fig. 3-3. The direction of lines of force of this magnetic field is related to the direction of current in the wire. This is known as the left hand rule . and is stated as follows: If a wire carrying a current is grasped in the left hand with thumb pointing in direction electrons are moving, the curved fingers will point the direction of lines of magnetic force (flux) encircling the wire.

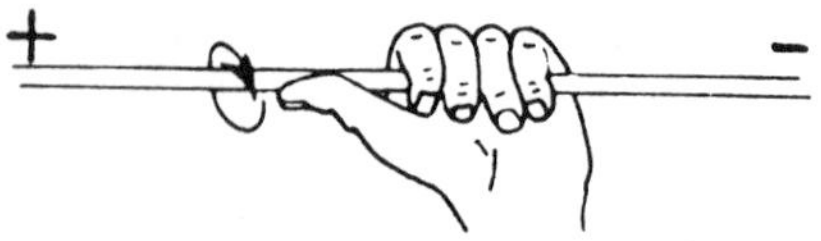

Fig. 3-3—A magnetic field surrounds a wire carrying an electrical current. The direction of magnetic force is indicated by the left hand rule; that is, if thumb of left hand points in direction that electrical current is flowing in conductor, fingers of the left hand will indicate direction of magnetic force.

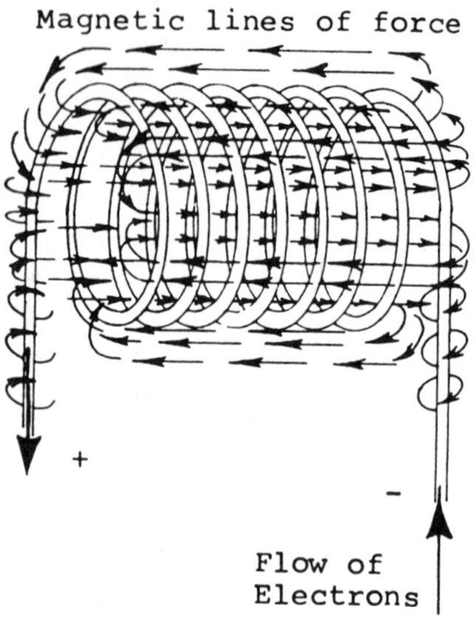

Fig. 3-4—When a wire is wound in a coil, the magnetic force created by a current in the wire will tend to converge in a single strong magnetic field as illustrated. If the loops of the coil are wound closely together, there is little tendency for lines of force to surround individual loops of the coil.

NOTE: The currently used electron theory explains the movement of electrons from negative to positive. Be sure to use the LEFT HAND RULE with the thumb pointing the direction electrons are moving (toward positive end of a conductor).

If a current is flowing in a wire that is wound into a coil, the magnetic flux surrounding the wire converge to form a stronger magnetic field as shown in Fig. 3-4. If the coils of wire are very close together, there is little tendency for magnetic flux to surround individual loops of the coil and a strong magnetic field will surround the entire coil. The strength of this field will vary with the current flowing through the coil.

STEP-UP TRANSFORMERS (IGNITION COILS). In both battery and magneto ignition systems, it is necessary to step-up, or transform, a relatively low primary voltage to the 15,000 to 20,000 volts required for the ignition spark. This is done by means of an ignition coil which utilizes the interrelationship of magnetism and electricity as explained in preceding paragraphs.

Basic ignition coil design is shown in Fig. 3-5. The coil consists of two separate coils of wire which are called the primary coil winding and the secondary coil winding, or simply the primary winding and secondary winding. The primary winding as indicated by the heavy, black line is of larger diameter wire and has a smaller number of turns when compared to the secondary winding indicated by the light line.

A current passing through the primary winding creates a magnetic field (as indicated by the "lines of force") and this field, concentrated by the soft iron core, surrounds both the primary and secondary windings. If the primary winding current is suddenly interrupted, the magnetic field will collapse and the lines of force will cut through the coil windings. The resulting induced voltage in the secondary winding is greater than the voltage of the current that was flowing in the primary winding and is related to the number of turns of wire in each winding. Thus:

$$\text{Induced secondary voltage} = \text{primary voltage} \times \frac{\text{No. of turns in secondary winding}}{\text{No. of turns in primary winding}}$$

For example, if the primary winding of an ignition coil contained 100 turns of wire and the secondary winding contained 10,000 turns of wire, a current having an emf of 200 volts flowing in the primary winding, when suddenly interrupted, would result in an emf of:

$$200 \text{ Volts} \times \frac{10{,}000 \text{ turns of wire}}{100 \text{ turns of wire}} = 20{,}000 \text{ volts}$$

SELF-INDUCTANCE. It should be noted the collapsing magnetic field resulting from interrupted current in the primary winding will also induce a current in the primary winding. This effect is termed "self-inductance." This self-induced current is such as to oppose any interruption of current in the primary winding, slowing the collapse of the magnetic field and reducing the efficiency of the coil. The self-induced primary current flowing across the slightly open breaker switch, or contact

Fig. 3-5—Drawing showing principles of ignition coil operation. A current in primary winding will establish a magentic field surrounding both the primary and secondary windings and the field will be concentrated by the iron core. When primary current is interrupted, the magnetic field will "collapse" and the lines of force wil cut the coil windings inducing a very high voltage in the secondary winding.

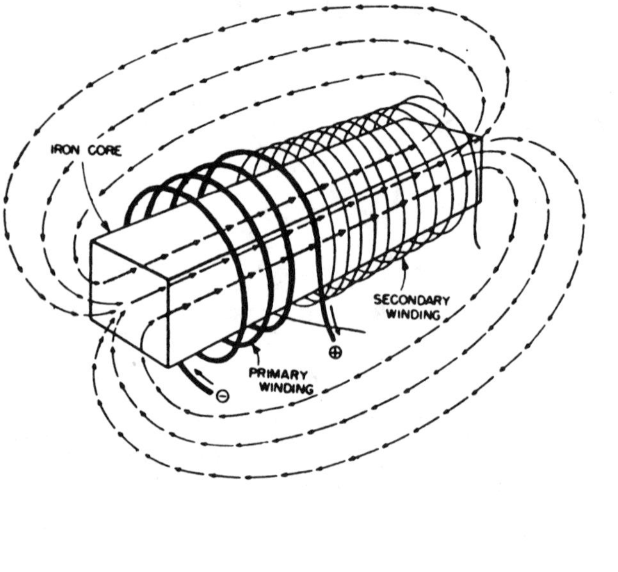

oints, will damage the contact surfaces ue to the resulting spark.

To momentarily absorb, then stop the ow of current across the contact oints, a capacitor or, as commonly alled, a condenser is connected in arallel with the contact points. A simle condenser is shown in Fig. 3-6; owever, the capacity of such a condenser to absorb current (capacitance) is limited by the small surface area of the plates. To increase capacity to absorb current, the condenser used in ignition systems is constructed as shown in Fig. 3-9.

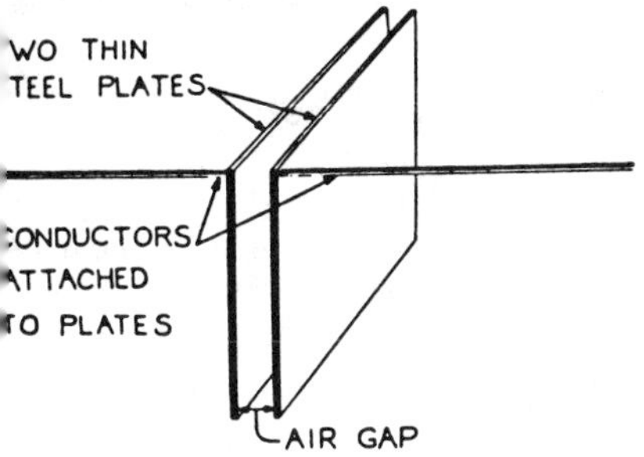

Fig. 3-6—Drawing showing construction of a simple condenser. Capacity of such a condenser to absorb current is limited due to the relatively small surface area. Also, there is a tendency for current to arc across the air gap. Refer to Fig. 3-9 for construction of typical ignition system condenser.

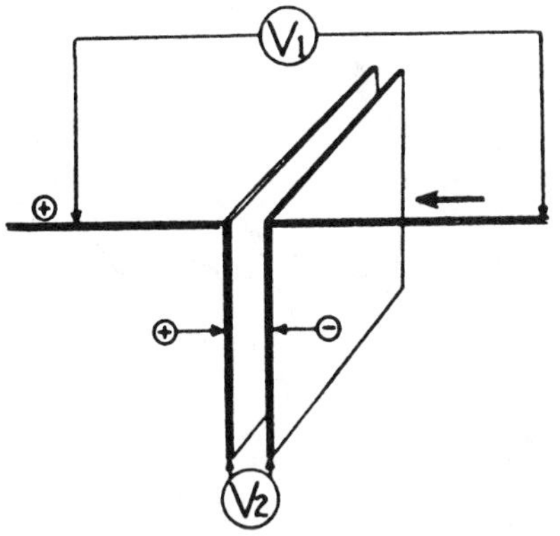

Fig. 3-7—A condenser in an electrical circuit will absorb electrons until opposing voltage (V2) is built up across condenser plates which is equal to the voltage (V1) of the electrical current.

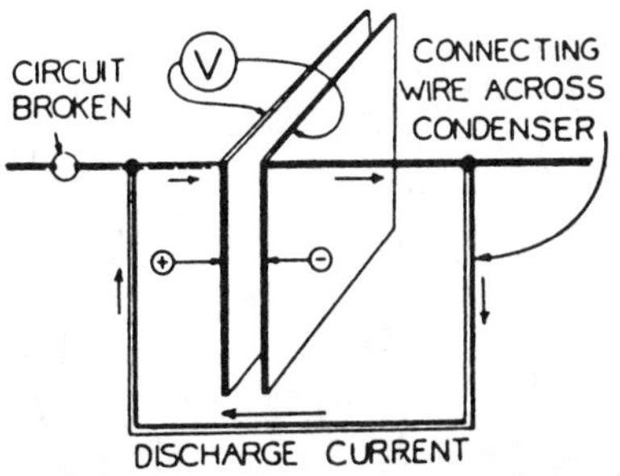

Fig. 3-8—When a circuit containing a condenser is interrupted (circuit broken), the condenser will retain a potential voltage (V). If a wire is connected across the condenser, a current will flow in reverse direction of charging current until condenser is discharged (voltage across condenser plates is zero).

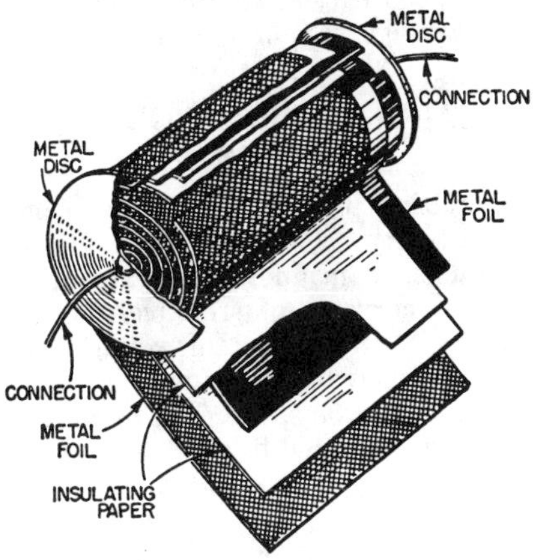

Fig. 3-9—Drawing showing construction of typical ignition system condenser. Two layers of metal foil, insulated from each other with paper, are rolled tightly together and a metal disc contacts each layer, or strip of foil. Usually, one disc is grounded through the condenser shell.

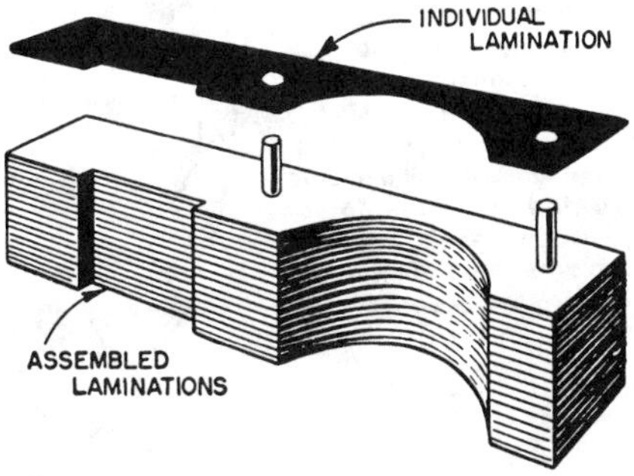

Fig. 3-10—To prevent formation of "eddy currents" within soft iron cores used to concentrate magnetic fields, core is assembled of plates or "laminations" that are insulated from each other. In a solid iron core, there is a tendency for counteracting magnetic forces to build up from stray currents induced in the core.

EDDY CURRENTS. It has been found that when a solid soft iron bar is used as a core for an ignition coil, stray electrical currents are formed in the core. These stray, or "eddy currents," create opposing magnetic forces causing the core to become hot and also decrease efficiency of the coil. As a means of preventing excessive formation of eddy currents within the core, or other magnetic field carrying parts of a magneto, a laminated plate construction as shown in Fig. 3-10 is used instead of solid material. The plates, or laminations, are insulated from each other by a natural oxide coating formed on the plate surfaces or by coating the plates with varnish. The cores of some ignition coils are constructed of soft iron wire instead of plates and each wire is insulated by a varnish coating. This type construction serves the same purpose as laminated plates.

BATTERY IGNITION SYSTEMS

Some engines are equipped with a battery ignition system. A schematic diagram of a typical battery ignition system for a single cylinder engine is shown in Fig. 3-11. Designs of battery ignition systems may vary, especially as to location of breaker points and method for actuating the points; however, all operate on the same basic principles.

BATTERY IGNITION SYSTEM PRINCIPLES. Refer to the schematic diagram in Fig. 3-11. When the timer cam is turned so the contact points are closed, a current is established in the primary circuit by the emf of the bat-

Fig. 3-11—Schematic diagram of typical battery ignition system used on single cylinder engine. On unit shown, breaker points are actuated by timer cam; on some units, points may be actuated by cam on engine camshaft. Refer to Fig. 3-12 for cut-away view of typical battery ignition coil. In this Fig., primary coil winding is shown as heavy black line (outside coil loops) and secondary winding is shown by lighter line (inside coil loops).

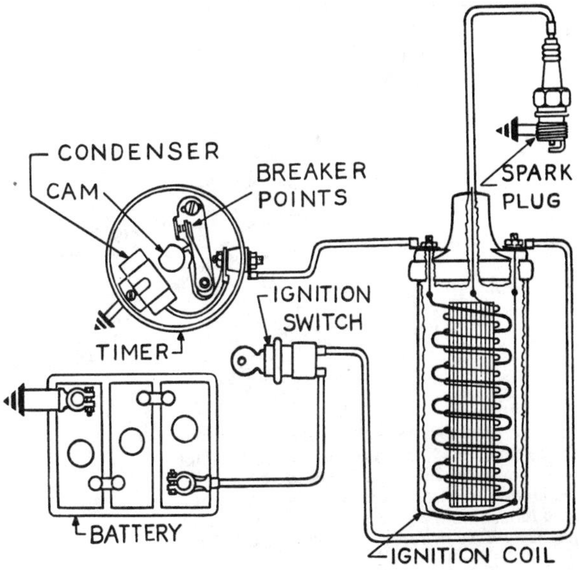

tery. This current flowing through the primary winding of the ignition coil establishes a magnetic field concentrated in the core laminations and surrounding the windings. A cut-away view of a typical ignition coil is shown in Fig. 3-12. At the proper time for the ignition spark, contact points are opened by the timer cam and primary ignition circuit is interrupted. The condenser, wired in parallel with breaker contact points between timer terminal and ground, absorbs self-induced current in the primary circuit for an instant and brings the flow of current to a quick, controlled stop. The magnetic field surrounding the coil rapidly cuts the primary and secondary windings creating an emf as high as 250 volts in the primary winding and up to 25,000 volts in the secondary winding. Current absorbed by the condenser is discharged as breaker points close, grounding the condenser lead wire.

Due to resistance of the primary winding, a certain period of time is required for maximum primary current flow after breaker contact points are closed. At high engine speeds, points remain closed for smaller interval of time, hence primary current does not build up to maximum and secondary voltage is somewhat less than at low engine speed. However, coil design is such that the minimum voltage available at high engine speed exceeds the normal maximum voltage required for ignition spark.

MAGNETO IGNITION SYSTEMS

By utilizing principles of magnetism and electricity as outlined in previous paragraphs, a magneto generates an electrical current of relatively low voltage, then transforms this voltage into the extremely high voltage necessary to produce ignition spark. This surge of high voltage is timed to create the ignition spark and ignite the compressed fuel-air mixture in the engine cylinder at the proper time in the Otto cycle as described in the paragraphs on fundamentals of engine operation principles.

Two different types of magnetos are used on air-cooled engines and, for discussion in this section of the manual, will be classified as "flywheel type magnetos" and "self-contained unit type magnetos."

Flywheel Type Magnetos

The term "flywheel type magneto" is derived from the fact that the engine flywheel carries the permanent magnets and is the magneto rotor. In some

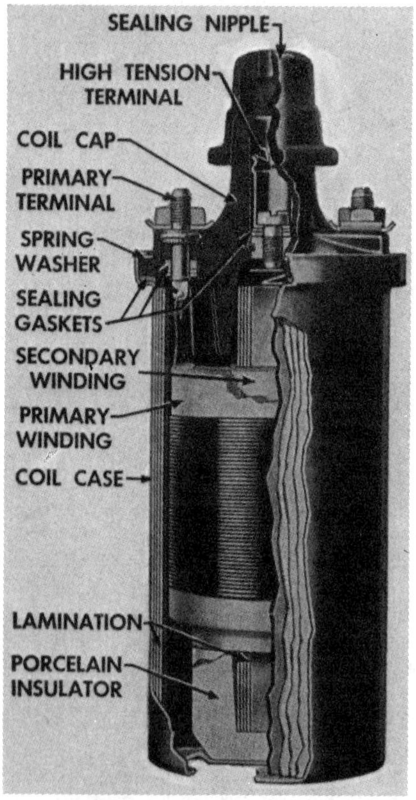
SEALING NIPPLE
HIGH TENSION TERMINAL
COIL CAP
PRIMARY TERMINAL
SPRING WASHER
SEALING GASKETS
SECONDARY WINDING
PRIMARY WINDING
COIL CASE
LAMINATION
PORCELAIN INSULATOR

Fig. 3-12—Cut-away view of typical battery ignition system coil. Primary winding consists of approximately 200-250 turns (loops) of heavier wire; secondary winding consists of several thousand turns of fine wire. Laminations concentrate magnetic lines of force and increase efficiency of the coil.

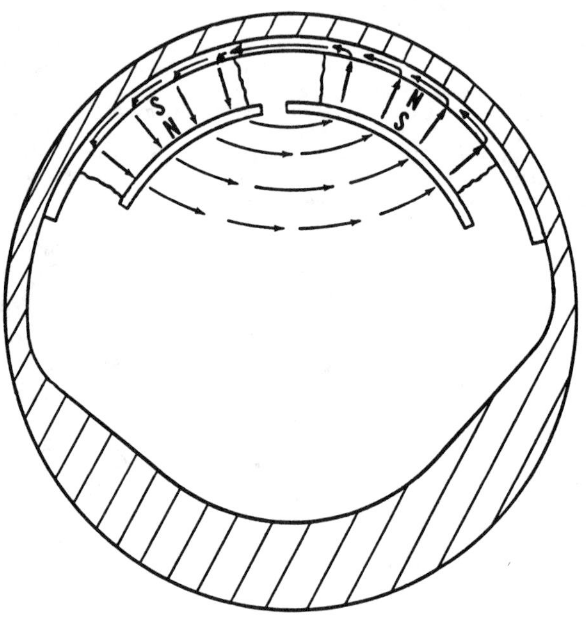

Fig. 3-13—Cut-away view of typical engine flywheel used with flywheel magneto type ignition system. The permanent magnets are usually cast into the flywheel. For flywheel type magnetos having the ignition coil and core mounted to outside of flywheel, magnets would be flush with outer diameter of flywheel.

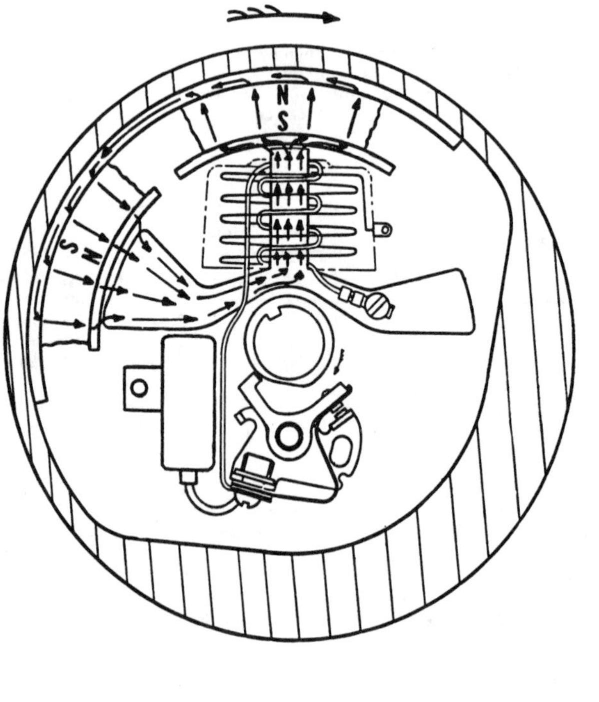

Fig. 3-14—View showing flywheel turned to a position so lines of force of the permanent magnets are concentrated in the left and center core legs and are interlocking the coil windings.

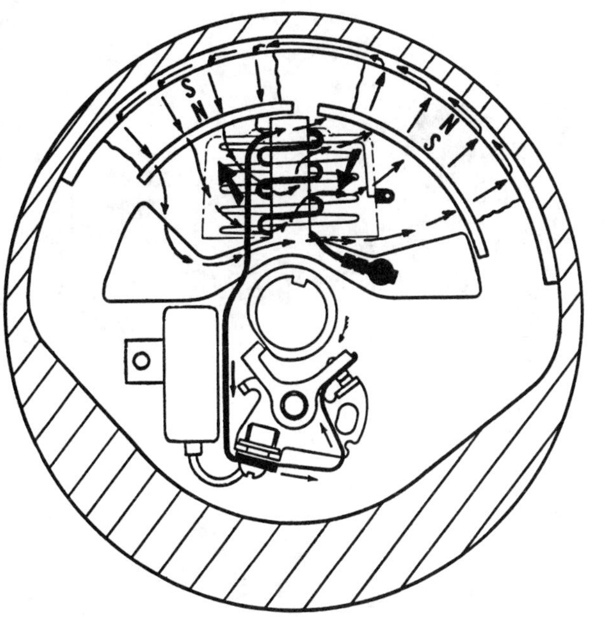

Fig. 3-15—View showing flywheel turned to a position so lines of force of the permanent magnets are being withdrawn from the left and center core legs and are being attracted by the center and right core legs. While this event is happening, the lines of force are cutting up through the coil windings section between the left and center legs and are cutting down through the section between the right and center legs as indicated by the heavy black arrows. As the breaker points are now closed by the cam, a current is induced in the primary ignition circuit as the lines of force cut through the coil windings.

similar systems, magneto rotor is mounted on engine crankshaft as is the flywheel, but is a part separate from flywheel.

FLYWHEEL MAGNETO OPERATING PRINCIPLES. In Fig. 3-13, a cross-sectional view of a typical engine flywheel (magneto rotor) is shown. The arrows indicate lines of force (flux) of the permanent magnets carried by the flywheel. As indicated by arrows, direction of force of magnetic field is from north pole (N) of left magnet to south pole (S) of right magnet.

Figs. 3-14, 3-15, 3-16 and 3-17 illustrate operational cycle of flywheel type magneto. In Fig. 3-14, flywheel magnets have moved to a position over left and center legs of armature (ignition coil) core. As magnets moved into this position, their magnetic field was attracted by armature core as illustrated in Fig. 3-1 and a potential voltage (emf) was induced in coil windings. However, this emf was not sufficient to cause current to flow across spark plug electrode gap in high tension circuit and points were open in primary circuit.

In Fig. 3-15, flywheel magnets have moved to a new position to where their magnetic field is being attracted by center and right legs of armature core, and is being withdrawn from left and center legs. As indicated by heavy black arrows, lines of force are cutting up through the section of coil windings between left and center legs of armature and are cutting down through coil windings section between center and right legs. If the left hand rule, as explained in a previous paragraph, is applied to the lines of force cutting through the coil sections, it is seen that the resulting emf induced in the primary circuit will cause a current to flow through primary windings and breaker points which have now been closed by action of the cam.

At the instant movement of lines of force cutting through coil winding sections is at maximum rate, maximum flow of current is obtained in primary circuit. At this time, cam opens breaker points interrupting primary circuit and, for an instant, flow of current is absorbed by condenser as illustrated in Fig. 3-16. An emf is also induced in secondary coil windings, but voltage is not sufficient to cause current to flow across spark plug gap.

Flow of current in primary windings creates a strong electromagnetic field surrounding coil windings and up through center leg of armature core as shown in Fig. 3-17. As breaker points were opened by the cam, interrupting primary circuit, magnetic field starts to collapse cutting coil windings as indicated by heavy black arrows. The emf induced in primary circuit would be sufficient to cause a flow of current across opening breaker points were it not for condenser absorbing flow of current and bringing it to a controlled stop. This allows electromagnetic field to collapse at such a rapid rate to induce a very high voltage in coil high tension or secondary windings. This voltage, in order of

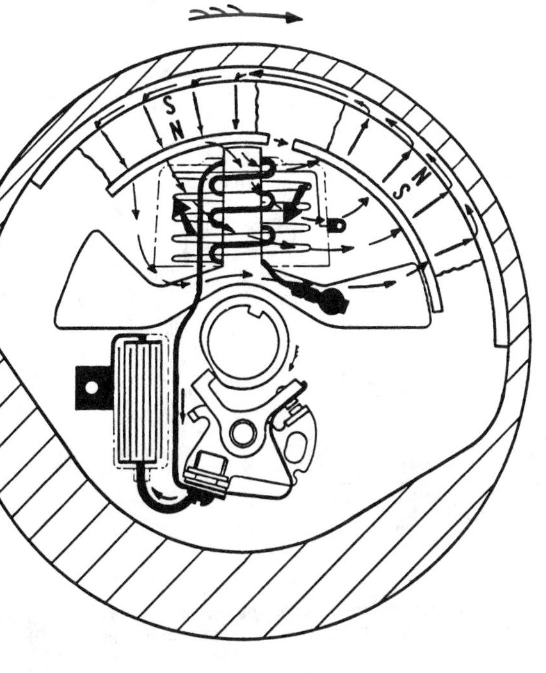

Fig. 3-16—The flywheel magnets have now turned slightly past the position shown in Fig. 3-15 and the rate of movement of lines of magnetic force cutting through the coil windings is at the maximum. At this instant, the breaker points are opened by the cam and flow of current in the primary circuit is being absorbed by the condenser, bringing the flow of current to a quick, controlled stop. Refer now to Fig. 3-17.

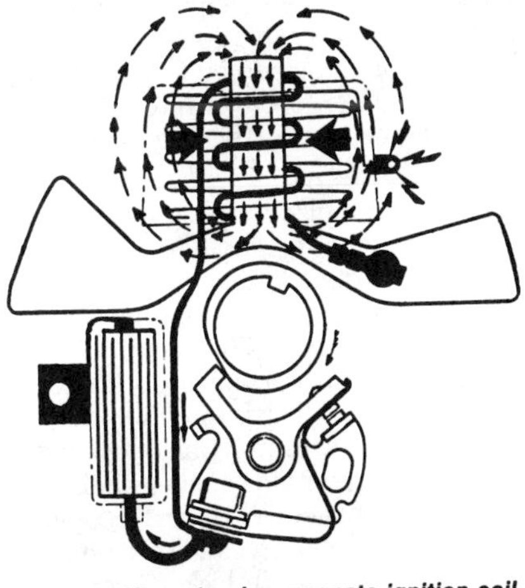

Fig. 3-17—View showing magneto ignition coil, condenser and breaker points at same instant as illustrated in Fig. 3-16; however, arrows shown above illustrate lines of force of the electromagnetic field established by current in primary coil windings rather than the lines of force of the permanent magnets. As the current in the primary circuit ceases to flow, the electromagnetic field collapses rapidly, cutting the coil windings as indicated by heavy arrows and inducing a very high voltage in the secondary coil winding resulting in the ignition spark.

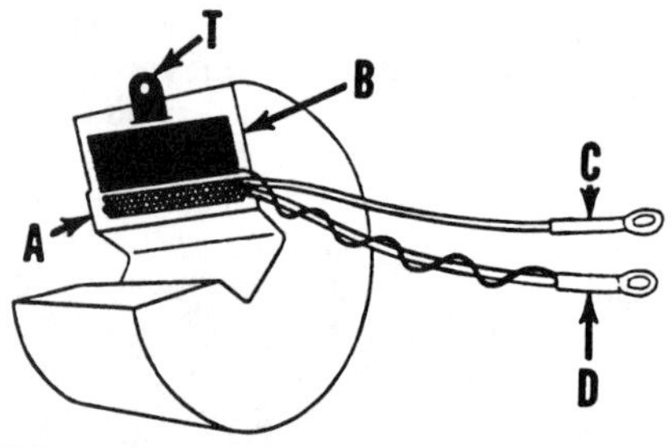

Fig. 3-18—Drawing showing construction of a typical flywheel magneto ignition coil. Primary winding (A) consists of about 200 turns of wire. Secondary winding (B) consists of several thousand turns of fine wire. Coil primary and secondary ground connection is (D); primary connection to breaker point and condenser terminal is (C); and coil secondary (high tension) terminal is (T).

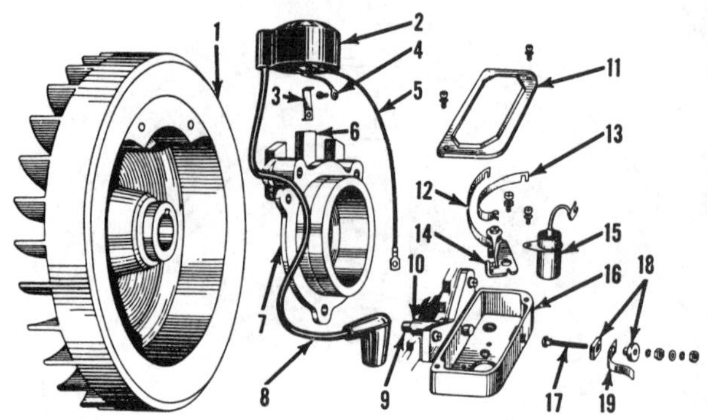

Fig. 3-19—Exploded view of a typical flywheel type magneto used on single cylinder engines in which the breaker points (14) are actuated by a cam on engine camshaft. Push rod (9) rides against cam to open and close points. In this type unit an ignition spark is produced only on alternate revolutions of the flywheel as the camshaft turns at one-half engine speed.

1. Flywheel
2. Ignition coil
3. Coil clamps
4. Coil ground lead
5. Breaker point lead
6. Armature core laminations
7. Crankshaft bearing RETAINER
8. High tension lead
9. Push rod
10. Bushing
11. Breaker box cover
12. Point lead strap
13. Breaker point spring
14. Breaker point assy.
15. Condenser
16. Breaker box
17. Terminal bolt
18. Insulators
19. Grounding (stop) spring

15,000 to 25,000 volts, is sufficient to break down resistance of air gap between spark plug electrodes and a current will flow across gap. This creates ignition spark which ignites compressed fuel-air mixture in engine cylinder.

Self-Contained Unit Type Magnetos

Some four-stroke cycle engines are equipped with a magneto which is a self-contained unit as shown in Fig. 3-20. This type magneto is driven from engine timing gears via a gear or coupling. All components of the magneto are enclosed in one housing and magneto can be removed from engine as a unit.

UNIT TYPE MAGNETO OPERATING PRINCIPLES. In Fig. 3-21, a schematic diagram of a unit type magneto is shown. Magneto rotor is driven through an impulse coupling (shown at right side of illustration). Function of impulse coupling is to increase rotating speed of rotor, thereby increasing magneto efficiency, at engine cranking speeds.

A typical impulse coupling for a single cylinder engine magneto is shown in Fig. 3-22. When engine is turned at cranking speed, coupling hub pawl engages a stop pin in magneto housing as engine piston is coming up on compression stroke. This stops rotation of coupling hub assembly and magneto rotor. A spring within coupling shell (see Fig. 3-23) connects shell and coupling hub; as engine continues to turn, spring winds up until pawl kickoff contacts pawl and disengages it from stop pin. This occurs at the time an ignition spark is required to ignite compressed fuel-air mixture in engine cylinder. As pawl is released, spring connecting coupling shell and hub unwinds and rapidly spins magneto rotor.

Magneto rotor (see Fig. 3-21) carries permanent magnets. As rotor turns, alternating position of magnets, lines of force of magnets are attracted, then withdrawn from laminations. In Fig. 3-21, arrows show magnetic field concentrated within laminations, or armature core. Slightly further rotation of magnetic rotor will place magnets to where laminations will have greater attraction for opposite poles of magnets. At this instant, lines of force as indicated by arrows will suddenly be withdrawn and an opposing field of force will be established in laminations. Due to this rapid movement of lines of force, a current will be induced in primary magneto circuit as coil windings are cut by lines of force. At instant maximum current is induced in primary windings, breaker points are opened by a cam on magnetic rotor shaft interrupting primary circuit. The lines of magnetic force established by primary current (refer to Fig. 3-5) will cut through secondary windings at such a rapid rate to induce a very high voltage in secondary (or high tension) circuit. This voltage will break down resistance

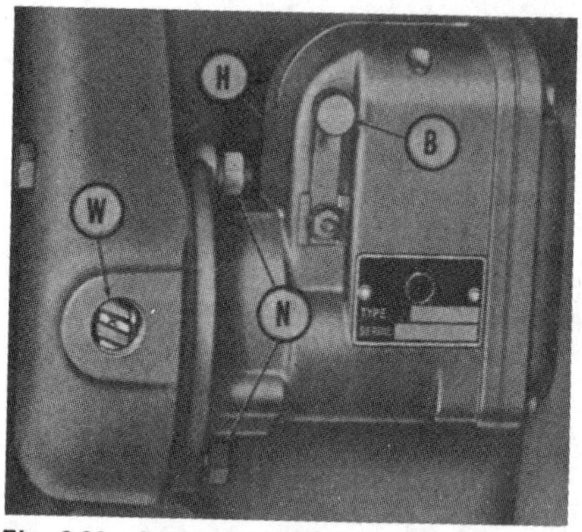

Fig. 3-20—Some engines are equipped with a unit type magneto having all components enclosed in a single housing (H). Magneto is removable as a unit after removing retaining nuts (N). Stop button (B) grounds out primary magneto circuit to stop engine. Timing window is (W).

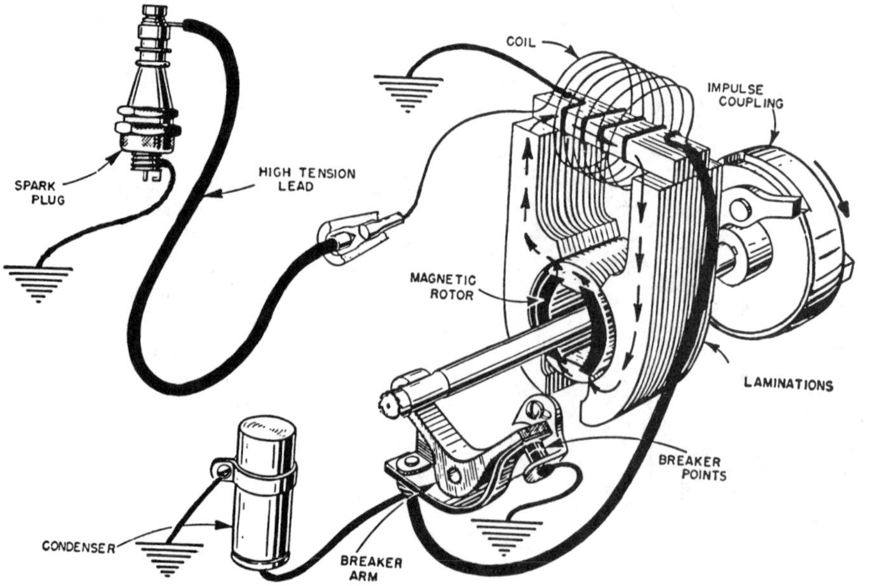

Fig. 3-21—Schematic diagram of typical unit type magneto for single cylinder engine. Refer to Figs. 3-22, 3-23 and 3-24 for views showing construction of impulse couplings.

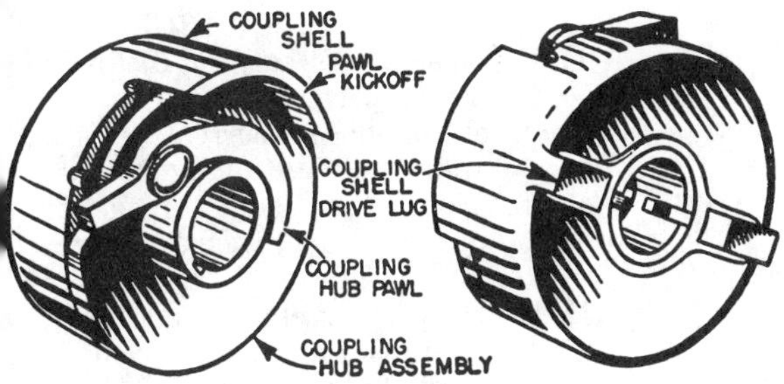

Fig. 3-22—Views of typical impulse coupling for magneto driven by engine shaft with slotted drive connection. Coupling drive spring is shown in Fig. 3-23. Refer to Fig. 3-24 for view of combination magneto drive gear and impulse coupling used on some magnetos.

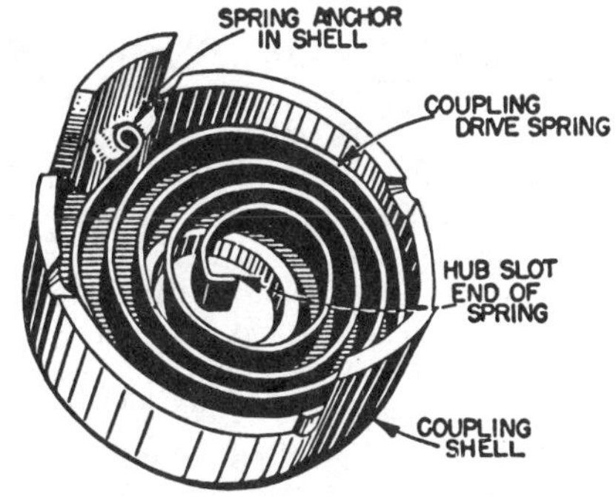

Fig. 3-23—View showing impulse coupling shell and drive spring removed from coupling hub assembly. Refer to Fig. 3-22 for views of assembled unit.

of spark plug electrode gap and a spark across electrodes will result.

At engine operating speeds, centrifugal force will hold impulse coupling hub pawl (See Fig. 3-22) in a position so it cannot engage stop pin in magneto housing and magnetic rotor will be driven through spring (Fig. 3-23) connecting coupling shell to coupling hub. The impulse coupling retards ignition spark, at cranking speeds, as engine piston travels closer to top dead center while magnetic rotor is held stationary by pawl and stop pin. The difference in degrees of impulse coupling shell rotation between position of retarded spark and normal running spark is known as impulse coupling lag angle.

SOLID STATE IGNITION SYSTEM

BREAKERLESS MAGNETO SYSTEM. Solid state (breakerless) magneto ignition sytem operates somewhat on the same basic principles as conventional type flywheel magneto previously described. The main difference is breaker contact points are replaced by a solid state electronic Gate Controlled Switch (GCS) which has no moving parts. Since, in a conventional system breaker points are closed over a longer period of crankshaft rotation than is the "GCS", a diode has been added to the circuit to provide the same characteristics as closed breaker points.

BREAKERLESS MAGNETO OPERATING PRINCIPLES. The same basic principles for electro-magnetic induction of electricity and formation of magnetic fields by electrical current as outlined for conventional flywheel type magneto also apply to the solid state magneto. Therefore principles of different components (diode and GCS) will complete operating principles of the solid state magneto.

The diode is represented in wiring diagrams by the symbol shown in Fig. 3-25. The diode is an electronic device that will permit passage of electrical current in one direction only. In electrical schematic diagrams, current flow is opposite the direction the arrow part of the symbol is pointing.

The symbol shown in Fig. 3-26 is used to represent gate controlled switch (GCS) in wiring diagrams. The GCS acts as a switch to permit passage of current from cathode (C) terminal to anode (A) terminal when in "ON" state and will not permit electric current to flow when in "OFF" state. The GCS can be turned "ON" by a positive surge of electricity at gage (G) terminal and will remain "ON" as long as current remains positive at gate terminal or as long as current is flowing through GCS from cathode (C) terminal to anode (A) terminal.

The basic components and wiring diagram for solid state breakerless magneto are shown schematically in Fig. 3-27. In Fig. 3-28, magneto rotor (flywheel) is turning and ignition coil magnets have just moved into position so their lines of force are cutting ignition coil windings and producing a negative surge of current in primary windings. The diode allows current to flow opposite to the direction of diode symbol arrow and action is same as conventional magneto with breaker points closed. As rotor (flywheel) continues to turn as shown in Fig. 3-29, direction of magnetic flux lines will reverse in armature center leg. Direction of current will change in primary coil circuit and previously conducting diode will be shut off. At this point, neither diode is conducting. As voltage begins to build up as rotor continues to turn, condenser acts

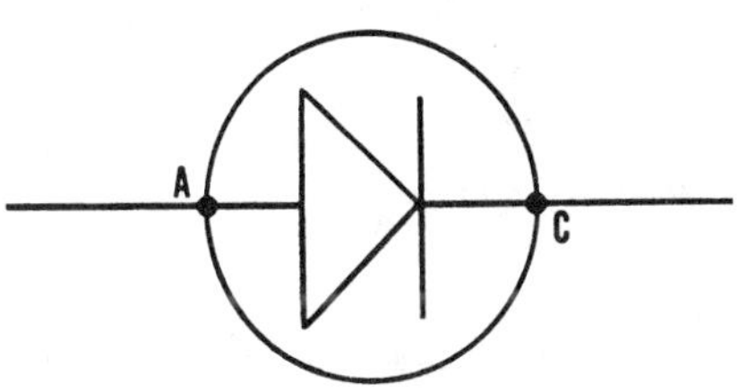

Fig. 3-25—In a diagram of an electrical circuit, the diode is represented by the symbol shown above. The diode will allow current to flow in one direction only, from cathode (C) to anode (A).

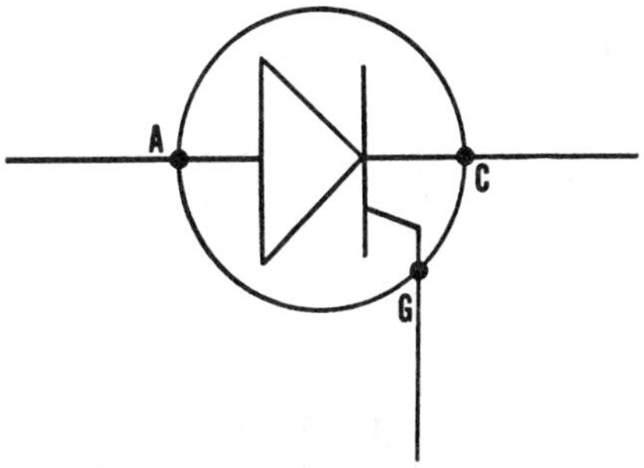

Fig. 3-26—The symbol used for a Gate Controlled Switch (GCS) in an electrical diagram is shown above. The GCS will permit current to flow from cathode (C) to anode (A) when "turned on" by a positive electrical charge at gate (G) terminal.

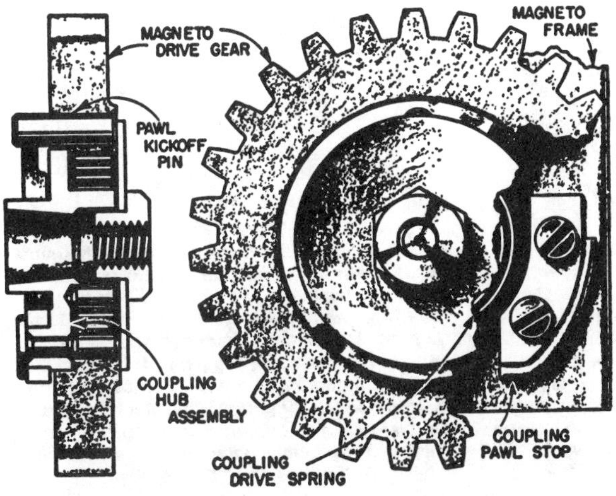

Fig. 3-24—Views of combination magneto drive gear and impulse coupling used on some magnetos.

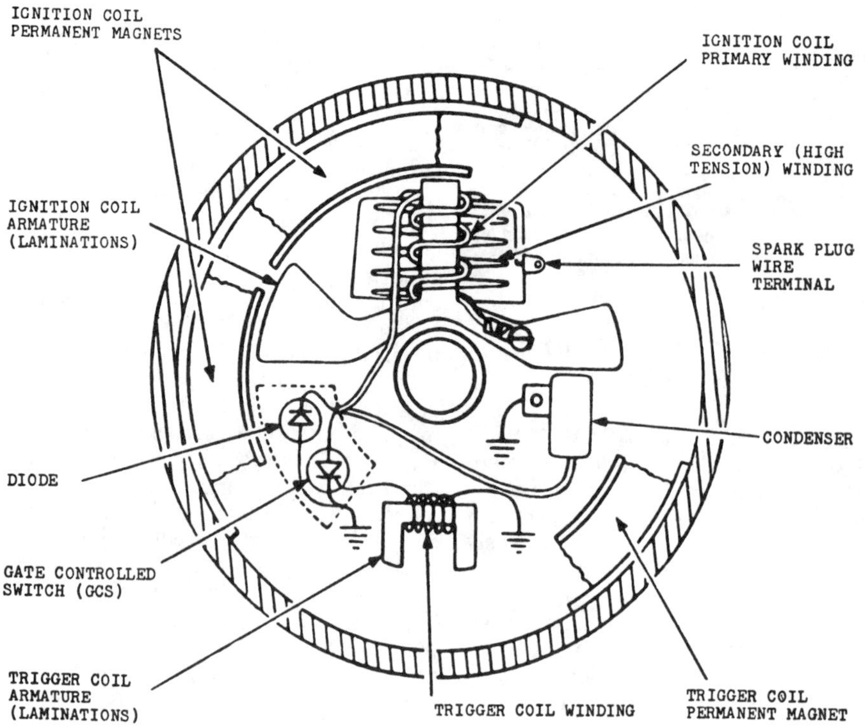
IGNITION COIL PERMANENT MAGNETS
IGNITION COIL PRIMARY WINDING
SECONDARY (HIGH TENSION) WINDING
IGNITION COIL ARMATURE (LAMINATIONS)
SPARK PLUG WIRE TERMINAL
CONDENSER
DIODE
GATE CONTROLLED SWITCH (GCS)
TRIGGER COIL ARMATURE (LAMINATIONS)
TRIGGER COIL WINDING
TRIGGER COIL PERMANENT MAGNET

Fig. 3-27—Schematic diagram of typical breakerless magneto ignition system. Refer to Figs. 3-28, 3-29 and 3-30 for schematic views of operating cycle.

as a buffer to prevent excessive voltage build up at GCS before it is triggered.

When rotor reaches approximate position shown in Fig. 3-30, maximum flux density has been achieved in center leg of armature. At this time the GCS is triggered. Triggering is accomplished by triggering coil armature moving into field of a permanent magnet which induces a positive voltage on the gate of GCS. Primary coil current flow results in the formation of an electromagnetic field around primary coil which induces a voltage of sufficient potentential in secondary coil windings to "fire" spark plug.

When rotor (flywheel) has moved magnets past armature, GCS will cease to conduct and revert to "OFF" state until it is triggered. The condenser will discharge during time GCS was conducting.

CAPACITOR DISCHARGE SYSTEM. Capacitor discharge (CD) ignition system uses a permanent magnet rotor (flywheel) to induce a current in a coil, but unlike conventional flywheel magneto and solid state breakerless magneto described previously, current is stored in a capacitor (condenser). Then, stored current is discharged through a transformer coil to create ignition spark. Refer to Fig. 3-31 for a schematic of a typical capacitor discharge ignition system.

CAPACITOR DISCHARGE OPERATING PRINCIPLES. As permanent flywheel magnets pass by input generating coil (1–Fig. 3-31), current produced charges capacitor (6). Only half of the generated current passes through diode (3) to charge capacitor. Reverse current is blocked by diode (3) but passes through Zener diode (2) to complete reverse circuit. Zener diode (2) also limits maximum voltage of forward current. As flywheel continues to turn and magnets pass trigger coil (4), a small amount of electrical current is generated. This current opens gate controlled switch (5) allowing capacitor to discharge through pulse transformer (7). The rapid voltage rise in transformer primary coil induces a high voltage secondary current which forms ignition spark when it jumps spark plug gap.

THE SPARK PLUG

In any spark ignition engine, the spark plug (See Fig. 3-32) provides means for igniting compressed fuel-air mixture in

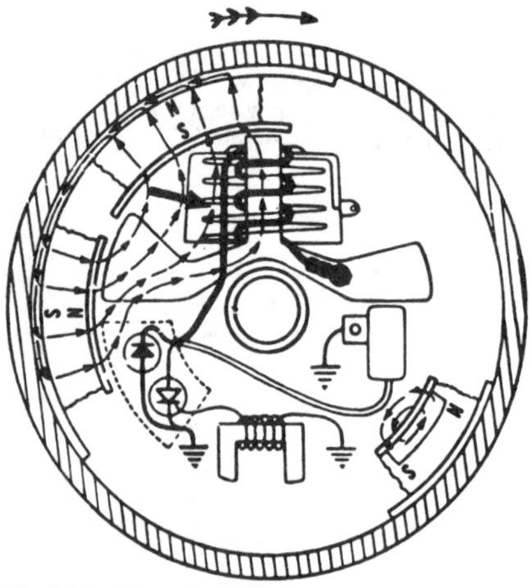

Fig. 3-28—View showing flywheel of breakerless magneto system at instant of rotation where lines of force of ignition coil magnets are being drawn into left and center legs of magneto armature. The diode (see Fig. 3-25) acts as a closed set of breaker points in completing the primary ignition circuit at this time.

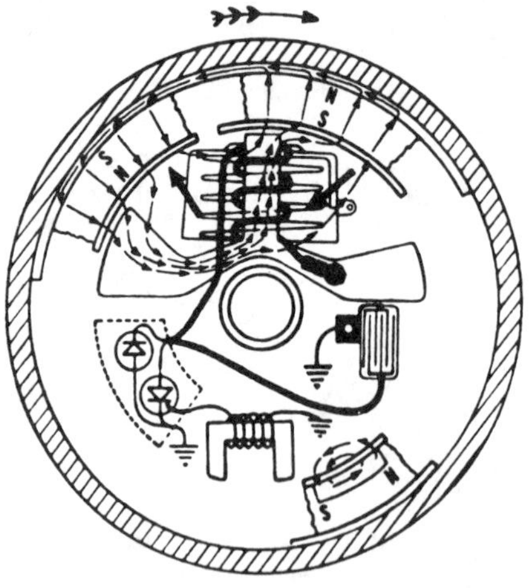

Fig. 3-29—Flywheel is turning to point where magnetic flux lines through armature center leg will reverse direction and current through primary coil circuit will reverse. As current reverses, diode which was previously conducting will shut off and there will be no current. When magnetic flux lines have reversed in armature center leg, voltage potential will again build up, but since GCS is in "OFF" state, no current will flow. To prevent excessive voltage build up, the condenser acts as a buffer.

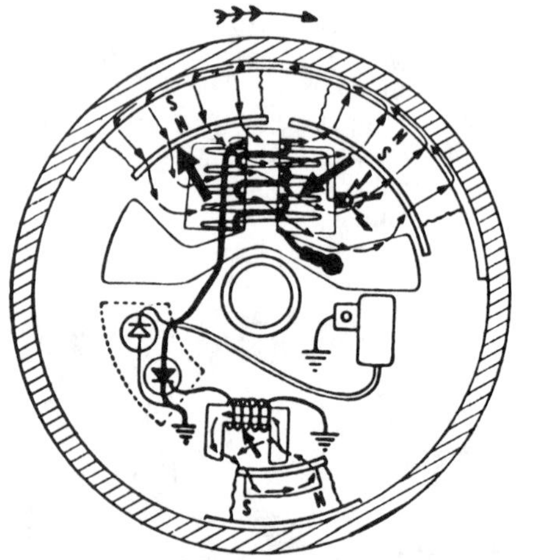

Fig. 3-30—With flywheel in the approximate position shown, maximum voltage potential is present in windings of primary coil. At this time the triggering coil armature has moved into the field of a permanent magnet and a positive voltage is induced on the gate of the GCS. The GCS is triggered and primary coil current flows resulting in the formation of an electromagnetic field around the primary coil which induces a voltage of sufficient potential in the secondary windings to "fire" the spark plug.

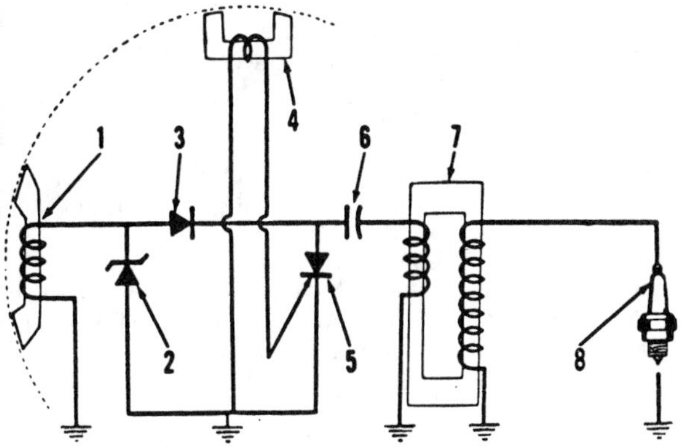
1
2
3
4
5
6
7
8

Fig. 3-31—Schematic diagram of a typical capacitor discharge ignition system.

1. Generating coil
2. Zener diode
3. Diode
4. Trigger coil
5. Gate controlled switch
6. Capacitor
7. Pulse transformer (coil)
8. Spark plug

cylinder. Before an electric charge can move across an air gap, intervening air must be charged with electricity, or ionized. If spark plug is properly gapped and system is not shorted, not more than 7,000 volts may be required to initiate a spark. Higher voltage is required as spark plug warms up, or if compression pressures or distance of air gap is increased. Compression pressures are highest at full throttle and relatively slow engine speeds, therefore, high voltage requirements or a lack of available secondary voltage most often shows up as a miss during maximum acceleration from a slow engine speed.

There are many different types and sizes of spark plugs which are designed for a number of specific requirements.

THREAD SIZE. The threaded, shell portion of the spark plug and the attaching hole in cylinder are manufactured to meet certain industry established standards. The diameter is referred to as "Thread Size." Those commonly used are: 10 mm, 14 mm, 18 mm, 7/8-inch and 1/2-inch pipe.

REACH. The length of thread, and thread depth in cylinder head or wall are also standardized throughout the industry. This dimension is measured from gasket seat of plug to cylinder end of thread. See Fig. 3-33. Four different reach plugs commonly used are: 3/8-inch, 7/16-inch, 1/2-inch and 3/4-inch.

HEAT RANGE. During engine operation, part of the heat generated during combustion is transferred to the spark plug, and from plug to cylinder through steel threads and gasket. The operating temperature of spark plug plays an important part in engine operation. If too much heat is retained by plug, fuel-air mixture may be ignited by contact with heated surface before ignition spark occurs. If not enough heat is retained, partially burned combustion products (soot, carbon and oil) may build up on plug tip resulting in "fouling" or shorting out of plug. If this happens, secondary current is dissipated uselessly as it is generated instead of bridging plug gap as a useful spark, and engine will misfire.

Operating temperature of plug tip can be controlled, within limits, by altering length of the path heat must follow to reach threads and gasket of plug. Thus, a plug with a short, stubby insulator around center electrode will run cooler than one with a long, slim insulator. Refer to Fig. 3-34. Most plugs in more popular sizes are available in a number of heat ranges which are interchangeable within the group. The proper heat range is determined by engine design and type of service. Refer to SPARK PLUG SERVICING FUNDAMENTALS for additional information on spark plug selection.

SPECIAL TYPES. Sometimes, engine design features or operating conditions call for special plug types designed for a particular purpose.

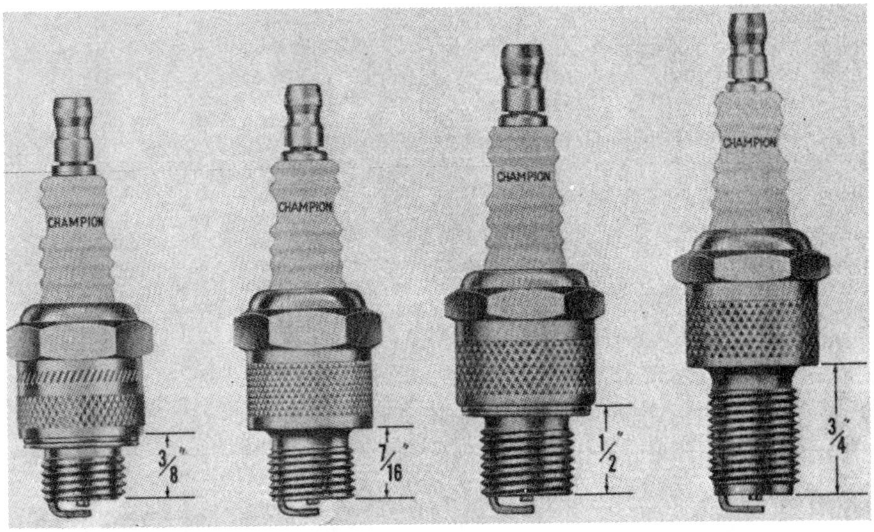
CHAMPION
CHAMPION
CHAMPION
CHAMPION
3/8"
7/16"
1/2"
3/4"

Fig. 3-33—Views showing spark plugs with various "reaches" available. A 3/8-inch reach spark plug measures 3/8-inch from firing end of shell to gasket surface of shell.

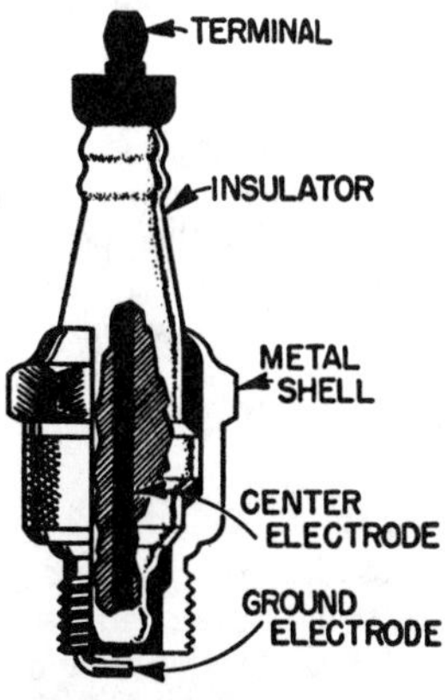
TERMINAL
INSULATOR
METAL SHELL
CENTER ELECTRODE
GROUND ELECTRODE

Fig. 3-32—Cross-sectional drawing of spark plug showing construction and nomenclature.

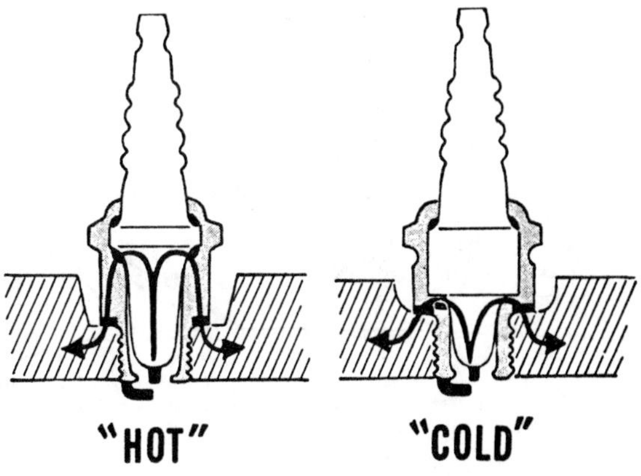
"HOT"
"COLD"

Fig. 3-34—Spark plug tip temperature is controlled by the length of the path heat must travel to reach cooling surface of the engine cylinder head.

ENGINE POWER AND TORQUE RATINGS

The following paragraphs discuss terms used in expressing engine horsepower and torque ratings and explains methods for determining different ratings. Some engine repair shops are now equipped with a dynamometer for measuring engine torque and/or horsepower and the mechanic should be familiar with terms, methods of measurement and how actual power developed by an engine can vary under different conditions.

GLOSSARY OF TERMS

FORCE. Force is an action against an object that tends to move the object from a state of rest, or to accelerate movement of an object. For use in calculating torque or horsepower, force is measured in pounds.

WORK. When a force moves an object from a state of rest, or accelerates movement of an object, work is done. Work is measured by multiplying force applied by distance force moves object, or:

$$\text{work} = \text{force} \times \text{distance}.$$

Thus, if a force of 50 pounds moved an object 50 feet, work done would equal 50 pounds times 50 feet, or 2500 pounds-feet (or as it is usually expressed, 2500 foot-pounds).

POWER. Power is the rate at which work is done; thus, if:

$$\text{work} = \text{force} \times \text{distance},$$

then:

$$\text{power} = \frac{\text{force} \times \text{distance}}{\text{time}}$$

From the above formula, it is seen that power must increase if time in which work is done decreases.

HORSEPOWER: Horsepower is a unit of measurement of power. Many years ago, James Watt, a Scotsman noted as inventor of the steam engine, evaluated one horsepower as being equal to doing 33,000 foot-pounds of work in one minute. This evaluation has been universally accepted since that time. Thus, the formula for determining horsepower is:

$$\text{horsepower} = \frac{\text{pounds} \times \text{feet}}{33{,}000 \times \text{minutes}}$$

Horsepower (hp.) ratings are sometimes converted to kilowatt (kW) ratings by using the following formula:

$$\text{kW} = \text{hp.} \times 0.745\ 699\ 9$$

When referring to engine horsepower ratings, one usually finds the rating expressed as brake horsepower or rated horsepower, or sometimes as both.

BRAKE HORSEPOWER. Brake horsepower is the maximum horsepower available from an engine as determined by use of a dynamometer, and is usually stated as maximum observed brake horsepower or as corrected brake horsepower. As will be noted in a later paragraph, observed brake horsepower of a specific engine will vary under different conditions of temperature and atmospheric pressure. Corrected brake horsepower is a rating calculated from observed brake horsepower and is a means of comparing engines tested at varying conditions. The method for calculating corrected brake horsepower will be explained in a later paragraph.

RATED HORSEPOWER. An engine being operated under a load equal to the maximum horsepower available (brake horsepower) will not have reserve power for overloads and is subject to damage from overheating and rapid wear. Therefore, when an engine is being selected for a particular load, the engine's brake horsepower rating should be in excess of the expected normal operating load. Usually, it is recommended that the engine not be operated in excess of 80% of the engine maximum brake horsepower rating; thus, the "rated horsepower" of an engine is usually equal to 80% of maximum horsepower the engine will develop.

TORQUE. In many engine specifications, a "torque rating" is given. Engine torque can be defined simply as the turning effort exerted by the engine output shaft when under load.

Torque ratings are sometimes converted to newton-meters (N·m) by using the following formula:

$$\text{N·m} = \text{foot pounds of torque} \times 1.355\ 818$$

It is possible to calculate engine horsepower being developed by measuring torque being developed and engine output speed. Refer to the following paragraphs.

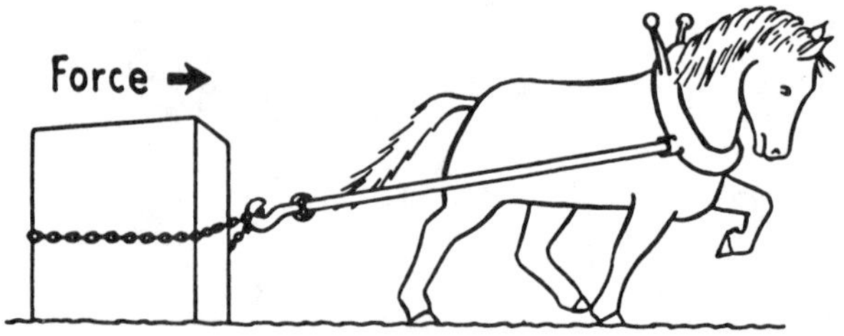

Fig. 4-1—A force, measured in pounds, is defined as an action tending to move an object or to accelerate movement of an object.

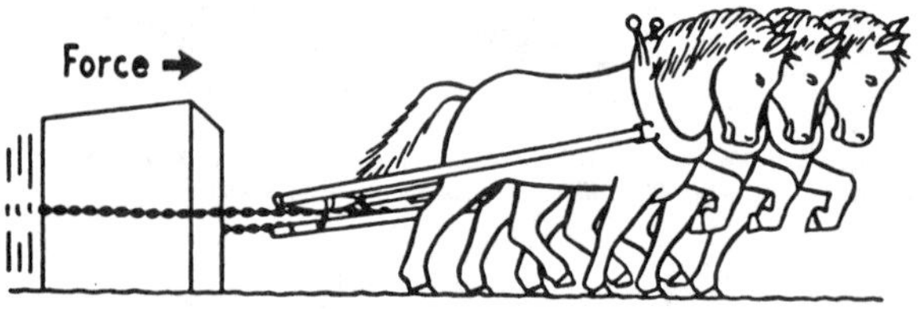

Fig. 4-2—If a force moves an object from a state of rest or accelerates movement of an object, then work is done.

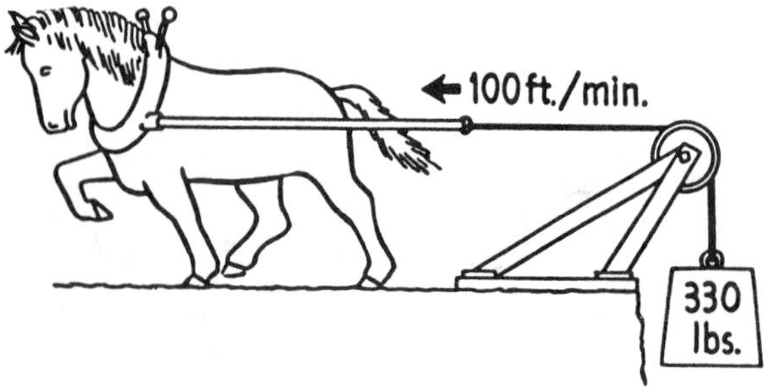

Fig. 4-3—This horse is doing 33,000 foot-pounds of work in one minute, or one horsepower.

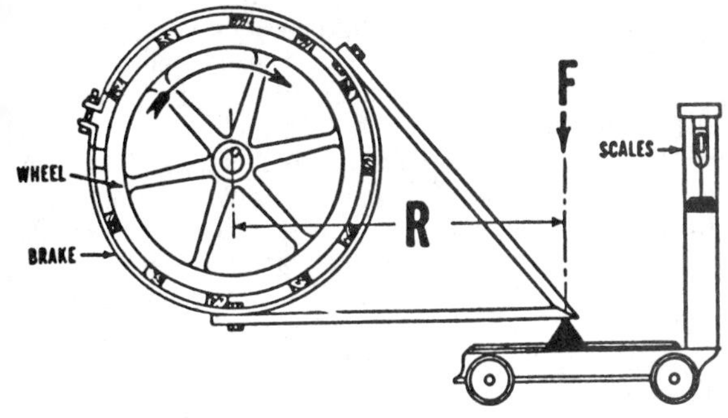

Fig. 4-4—Diagram showing a prony brake on which the torque being developed by an engine can be measured. By also knowing the rpm of the engine output shaft, engine horsepower can be calculated.

MEASURING ENGINE TORQUE AND HORSEPOWER

PRONY BRAKE. The prony brake is the most simple means of testing engine performance. Refer to diagram in Fig. 4-4. A torque arm is attached to a brake on wheel mounted on engine output shaft. The torque arm, as the brake is applied, exerts a force (F) on scales. Engine torque is computed by multiplying force (F) times length of torque arm radius (R), or:

$$\text{engine torque} = F \times R.$$

If, for example, torque arm radius (R) is 2 feet and force (F) being exerted by torque arm on scales is 6 pounds, engine torque would be 2 feet × 6 pounds, or 12 foot-pounds.

To calculate engine horsepower being developed by use of the prony brake, we must also count revolutions of engine output shaft for a specific length of time. In formula for calculating horsepower:

$$\text{horsepower} = \frac{\text{feet} \times \text{pounds}}{33{,}000 \times \text{minutes}}$$

Feet in formula will equal circumference transcribed by torque arm radius multiplied by number of engine output shaft revolutions. Thus:

$$\text{feet} = 2 \times 3.14 \times \text{radius} \times \text{revoultions}.$$

Pounds in formula will equal force (F) of torque arm. If, for example, force (F) is 6 pounds, torque arm radius is 2 feet and engine output shaft speed is 3300 revolutions per minute, then:

$$\text{horsepower} = \frac{2 \times 3.14 \times 2 \times 3300 \times 6}{33{,}000 \times 1}$$

or,

$$\text{horsepower} \times 7.54$$

DYNAMOMETERS. Some commercial dynamometers for testing small engines are now available, although the cost may be prohibitive for all but larger repair shops. Usually, these dynamometers have a hydraulic loading device and scales indicating engine speed and load; horsepower is then calculated by use of a slide rule type instrument.

HOW ENGINE HORSEPOWER OUTPUT VARIES

Engine efficiency will vary with the amount of air taken into the cylinder on each intake stroke. Thus, air density has a considerable effect on horsepower output of a specific engine. As air density varies with both temperature and atmospheric pressure, any change in air temperature, barometric pressure, or elevation will cause a variance in observed engine horsepower. As a general rule, engine horsepower will:

A. Decrease approximately 3% for each 1000 foot increase above 1000 ft. elevation;

B. Decrease approximately 3% for each 1 inch drop in barometric pressure; or,

C. Decrease approximately 1% for each 10° rise in temperature (Fahrenheit).

Thus, to fairly compare observed horsepower readings, observed readings should be corrected to standard temperature and atmospheric pressure conditions of 60°F., and 29.92 inches of mercury. The correction formula specified by the Society of Automotive Engineers is somewhat involved; however for practical purposes, the general rules stated above can be used to approximate corrected brake horsepower of an engine when observed maximum brake horsepower is known.

For example, suppose the engine horsepower of 7.54 as found by use of the prony brake was observed at an altitude of 3000 feet and at a temperature of 100°F. At standard atmospheric pressure and temperature conditions, we could expect an increase of 4% due to temperature (100° − 60° x 1% per 10°) and an increase of 6% due to altitude (3000 ft.-1000 ft.x3% per 1000 ft.) or a total increase of 10%. Thus, corrected maximum horsepower from this engine would be approximately 7.54 + .75, or approximately 8.25 horsepower.

TROUBLESHOOTING

When servicing an engine to correct a specific complaint, such as engine will not start, is hard to start, etc., a logical step-by-step procedure should be followed to determine cause of trouble before performing any service work. This procedure is "TROUBLESHOOTING."

Of course, if an engine is received in your shop for a normal tune up or specific repair work is requested, troubleshooting procedure is not required and work should be performed as requested. It is wise, however, to fully check the engine before repairs are made and recommend any additional repairs or adjustments necessary to ensure proper engine performance.

The following procedures, as related to a specific complaint or trouble, have proven to be a satisfactory method for quickly determining cause of trouble in a number of engine repair shops.

NOTE: It is not suggested the troubleshooting procedure as outlined in following paragraphs be strictly adhered to at all times. In many instances, customer's comments on when trouble was encountered will indicate cause of trouble. Also, the mechanic will soon develop a diagnostic technique that can only come with experience. In addition to the general troubleshooting procedure, reader should also refer to special notes following this section and to the information included in engine, carburetor and magneto servicing fundamentals sections.

If Engine Will Not Start—Or Is Hard To Start

1. If engine is equipped with a rope or crank starter, turn engine slowly. As engine piston is coming up on compression stroke, a definite resistance to turning should be felt on rope or crank. This resistance should be noted every other crankshaft revolution on a single cylinder engine, on every revolution of a two cylinder engine and on every ½-revolution of a four cylinder engine. If correct cranking resistance is noted, engine compression can be considered as not the cause of trouble at this time.

NOTE: Compression gages for gasoline engines are available and are of value in troubleshooting engine service problems.

Where available from engine manufacturer, specifications will be given for engine compression pressure in engine service sections of this manual. On engines having electric starters, remove spark plug and check engine compression with gage; if gage is not available, hold thumb so spark plug hole is partly covered. An alternating blowing and suction action should be noted as engine is cranked.

If very little or no compression is noted, refer to appropriate engine repair section for repair of engine. If check indicates engine is developing compression, proceed to step 2.

2. Remove spark plug wire and hold wire terminal about 1/8-inch (3.18 mm) away from cylinder (on wires having rubber spark plug boot, insert a small screw or bolt in terminal).

NOTE: A test plug with 1/8-inch (3.18 mm) gap is available or a test plug can be made by adjusting the electrode gap of a new spark plug to 0.125 inch (3.18 mm).

While cranking engine, a bright blue spark should snap across the 1/8-inch (3.18 mm) gap. If spark is weak or yellow, or if no spark occurs while cranking engine, refer to IGNITION SYSTEM SERVICE FUNDAMENTALS for information on appropriate type system.

If spark is satisfactory, remove and inspect spark plug. Refer to SPARK PLUG SERVICE FUNDAMENTALS. If in doubt about spark plug condition, install a new plug.

NOTE: Before installing plug, make certain electrode gap is set to proper dimension shown in engine repair section of this manual. Refer also to Fig. 5-1.

If ignition spark is satisfactory and engine will not start with new plug, proceed with step 3.

3. If engine compression and ignition spark are adequate, trouble within the fuel system should be suspected.

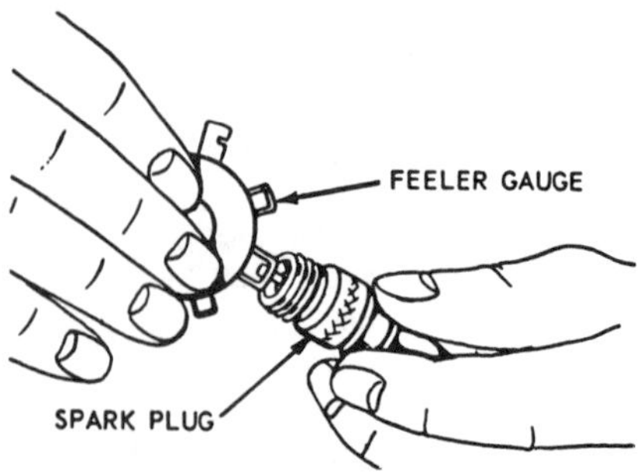

Fig. 5-1 — Be sure to check spark plug electrode gap with proper size feeler gage and adjust gap to specification recommended by manufacturer.

Remove and clean or renew air cleaner or cleaner element. Check fuel tank (Fig. 5-2) and make certain it is full of fresh fuel as prescribed by engine manufacturer. If equipped with a fuel shut-off valve, make certain valve is open.

If engine is equipped with remote throttle controls that also operate carburetor choke plate, check to be certain that when controls are placed in choke position, carburetor choke plate is fully closed. If not, adjust control linkage so choke will fully close; then, try to start engine. If engine does not start after several turns, remove air cleaner assembly; carburetor throat should be wet with gasoline. If not, check for reason fuel is not getting to carburetor. On models with gravity feed from fuel tank to carburetor (fuel tank above carburetor), disconnect fuel line at carburetor to see that fuel is flowing through the line. If no fuel is flowing, remove and clean fuel tank, fuel line and any fuel filters or shut-off valve.

On models having a fuel pump separate from carburetor, remove fuel line at carburetor and crank engine through several turns; fuel should spurt from open line. If not, disconnect fuel line from tank to fuel pump at pump connection. If fuel will not run from open line, remove and clean fuel tank, line and if so equipped, fuel filter and/or shut-off valve. If fuel runs from open line, remove and overhaul or renew the fuel pump.

After making sure clean, fresh fuel is available at carburetor, again try to start engine. If engine will not start, refer to recommended initial adjustments for carburetor in appropriate engine repair section of this manual and adjust carburetor idle and/or main fuel needles.

If engine will not start when compression and ignition test within specifications and clean, fresh fuel is available to carburetor, remove and clean or overhaul carburetor as outlined in CARBURETOR SERVICING FUNDAMENTALS section of this manual.

4. The preceding troubleshooting techniques are based on the fact that to run, an engine must develop compression, have an ignition spark and receive proper fuel-air mixture. In some instances, there are other factors involved. Refer to special notes following this section for service hints on finding common causes of engine trouble that may not be discovered in normal troubleshooting procedure.

If Engine Starts, Then Stops

This complaint is usually due to fuel starvation, but may be caused by a faulty ignition system. Recommended troubleshooting procedure is as follows:

1. Remove and inspect fuel tank cap; on all engines, fuel tank is vented through breather in fuel tank cap so air can enter tank as fuel is used. If engine stops after running several minutes, a clogged breather should be suspected. On some engines, it is possible to let engine run with fuel tank cap removed and if this permits engine to run without stopping, clean or renew cap.

CAUTION: Be sure to observe safety precautions before attempting to run engine without fuel tank cap in place. If there is any danger of fuel being spilled on engine or spark entering open tank, do not attempt to run engine without fuel tank cap in place. If in doubt, try a new cap.

2. If clogged breather in fuel tank cap is eliminated as cause of trouble, a partially clogged fuel filter or fuel line should be suspected. Remove and clean fuel tank and line and if so equipped, clean fuel shut-off valve and/or fuel tank filter. On some engines, a screen or felt type fuel filter is located in carburetor fuel inlet; refer to engine repair section for appropriate engine make and model for carburetor construction.

3. After cleaning fuel tank, line, filters, etc., if trouble is still encountered, a sticking or faulty carburetor inlet needle valve or float may be cause of trouble. Remove, disassemble and clean carburetor using data in engine repair section and in CARBURETOR SERVICE FUNDAMENTALS as a guide.

4. If fuel system is eliminated as cause of trouble by performing procedure outlined in steps 1, 2 and 3, check magneto or battery ignition coil on tester if such equipment is available. If not, check for ignition spark immediately after engine stops.

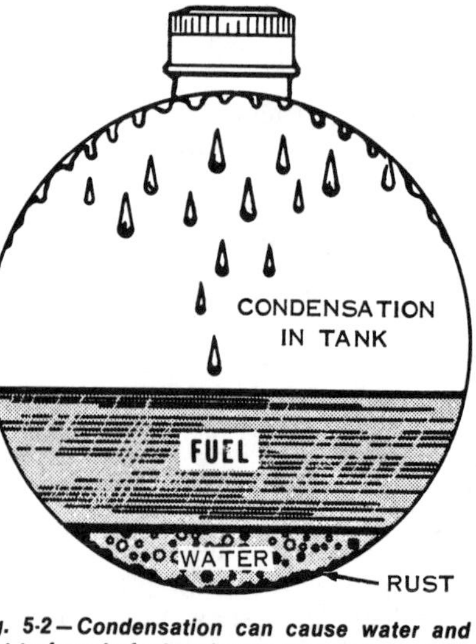

Fig. 5-2 — Condensation can cause water and rust to form in fuel tank even though only clean fuel has been poured into tank.

Renew coil, condenser and breaker points if no spark is noted. Also, check for engine compression immediately after engine stops; trouble may be caused by sticking intake or exhaust valve or cam followers (tappets). If no or little compression is noted immediately after engine stops, refer to ENGINE SERVICE FUNDAMENTALS section and to engine repair data in appropriate engine repair section of this manual.

Engine Overheats

When air cooled engines overheat, check for:

1. Air inlet screen in blower housing plugged with grass, leaves, dirt or other debris.
2. Remove blower housing and shields and check for dirt or debris accumulated on or between cooling fins on cylinder.
3. Missing or bent shields on blower housing. (Never attempt to operate an air cooled engine without all shields and blower housing in place.)
4. A too lean main fuel-air adjustment of carburetor.
5. Improper ignition spark timing. Check breaker point gap, and on engine with unit type magneto, check magneto to engine timing. On battery ignition units with timer or distributor, check for breaker points opening at proper time.
6. Engines being operated under loads in excess of rated engine horsepower or at extremely high ambient (surrounding) air temperatures may overheat.

Engine Surges When Running

Trouble with an engine surging is usually caused by improper carburetor adjustment or improper governor adjustment.

1. Refer to CARBURETOR paragraphs in appropriate engine repair section and adjust carburetor as outlined.
2. If adjusting carburetor did not correct surging condition, refer to GOVERNOR paragraph and adjust governor linkage.
3. If any wear is noted in governor linkage and adjusting linkage did not correct problem, renew worn linkage parts.
4. If trouble is still not corrected, remove and clean or overhaul carburetor as necessary. Also check for any possible air leaks between the carburetor to engine gaskets or air inlet elbow gaskets.

Special Notes on Engine Troubleshooting

ENGINES WITH COMPRESSION RELEASE. Several different makes of four-stroke cycle engines now have a compression release that reduces compression pressure at cranking speeds, thus making it easier to crank the engine. Most models having this feature will develop full compression when turned in a reverse direction. Refer to the appropriate engine repair section in this manual for detailed information concerning compression release used on different makes and models.

IGNITION SYSTEM SERVICE

The fundamentals of servicing ignition systems are outlined in the following paragraphs. Refer to appropriate heading for type ignition system being inspected or overhauled.

BATTERY IGNITION SERVICE FUNDAMENTALS

Usually all components are readily accessible and while use of test instruments is sometimes desirable, condition of the system can be determined by simple checks. Refer to following paragraphs.

GENERAL CONDITION CHECK. Remove spark plug wire and if terminal is rubber covered, insert small screw or bolt in terminal. Hold uncovered end of terminal or bolt inserted in terminal about 1/8-inch (3.18 mm) away from engine or connect spark plug wire to test plug. Crank engine while observing gap between spark plug wire terminal and engine; if a bright blue spark snaps across gap, condition of system can be considered satisfactory. However, ignition timing may have to be adjusted. Refer to timing procedure in appropriate engine repair section.

VOLTAGE, WIRING AND SWITCH CHECK. If no spark, or a weak yellow-orange spark occurred when checking system as outlined in preceding paragraph, proceed with following checks:

Test battery condition with hydrometer or voltmeter. If check indicates a dead cell, renew battery; recharge battery if a discharged condition is indicated.

NOTE: On models with electric starter or starter-generator unit, battery can be assumed in satisfactory condition if the starter cranks the engine freely.

If battery checks within specifications, but starter unit will not turn engine, a faulty starter unit is indicated and ignition trouble may be caused by excessive current draw of such a unit. If battery and starting unit, if so equipped, are in satisfactory condition, proceed as follows:

Remove battery lead wire from ignition coil and connect a test light of same voltage as battery between disconnected lead wire and engine ground. Light should go on when ignition switch is in "off" position. If not, renew switch and/or wiring and recheck for satisfactory spark. If switch and wiring are functioning properly, but no spark is obtained, proceed as follows:

BREAKER POINTS AND CONDENSER. Remove breaker box cover and, using small screwdriver, separate and inspect breaker points. If burned or deeply pitted, renew breaker points and condenser. If point contacts are clean to grayish in color and are only slightly pitted, proceed as follows: Disconnect condenser and ignition coil lead wires from breaker point terminal and connect a test light and battery between terminal and engine ground. Light should go on when points are closed and should go out when points are open. If light fails to go out when points are open, breaker arm insulation is defective and breaker points must be renewed. If light does not go on when points are in closed position, clean or renew breaker points. In some instances, new breaker point contact surfaces may have an oily or wax coating or have foreign material between the surfaces so proper contact is prevented. Check ignition timing and breaker point gap as outlined in appropriate engine repair section of this manual.

Connect test light and battery between condenser lead and engine ground; if light goes on, condenser is shorted out and should be renewed. Capacity of condenser can be checked if test instrument is available. It is usually good practice to renew condenser whenever new breaker points are being installed if tester is not available.

IGNITION COIL. If a coil tester is available, condition of coil can be checked. However, if tester is not available, a reasonably satisfactory performance test can be made as follows:

Disconnect high tension wire from spark plug. Turn engine so cam has allowed breaker points to close. With ignition switch on, open and close points with small screwdriver while holding high tension lead about 1/8 to 1/4-inch (3.18 to 6.35 mm) away from engine ground. A bright blue spark should snap across gap between spark plug wire and ground each time points are opened. If no spark occurs, or spark is weak and yellow-orange, renewal of ignition coil is indicated.

Sometimes, an ignition coil may perform satisfactorily when cold, but fail after engine has run for some time and coil is hot. Check coil when hot if this condition is indicated.

FLYWHEEL MAGNETO SERVICE FUNDAMENTALS

In servicing a flywheel magneto ignition system, the mechanic is concerned with troubleshooting, service adjustments and testing magneto components. The following paragraphs outline basic steps in servicing a flywheel type magneto. Refer to appropriate engine section for adjustment and test specifications for a particular engine.

Troubleshooting

If engine will not start and malfunction of ignition system is suspected, make the following checks to find cause of trouble.

Check to be sure ignition switch (if so equipped) is in "On" or "Run" position and the insulation on wire leading to ignition switch is in good condition. Switch can be checked with timing and test light as shown in Fig. 5-3. Disconnect lead from switch and attach one clip of test light to switch terminal and remaining clip to engine. Light should go on when switch is in "Off" or "Stop" position, and should go off when switch is in "On" or "Run" position.

Inspect high tension (spark plug) wire for worn spots in insulation or breaks in wire. Frayed or worn insulation can be repaired temporarily with plastic electrician's tape.

If no defects are noted in ignition switch or ignition wires, remove and inspect spark plug as outlined in SPARK PLUG SERVICING section. If spark plug is fouled or is in questionable condition, connect a spark plug of known quality to high tension wire, ground base of spark plug to engine and turn engine rapidly with starter. If spark across electrode gap of spark plug is a bright blue, magneto can be considered in satisfactory condition.

NOTE: Some engine manufacturers specify a certain type spark plug and a specific test gap. Refer to appropriate engine service section; if no specific spark plug type or electrode gap is recommended for test purposes, use spark plug type and electrode gap recommended for engine make and model.

If spark across gap of test plug is weak or orange colored, or no spark occurs as engine is cranked, magneto should be serviced as outlined in the following paragraphs.

Magneto Adjustments

BREAKER CONTACT POINTS. Adjustment of breaker contact points affects both ignition timing and magneto edge gap. Therefore, breaker contact point gap should be carefully adjusted according to engine manufacturer's specifications. Before adjusting breaker contact gap, inspect contact points and renew if condition of contact surfaces is questionable. It is sometimes desirable to check condition of points as follows: Disconnect condenser and primary coil leads from breaker point terminal. Attach one clip of test light (See Fig. 5-3) to breaker point terminal and remaining clip of test light to magneto ground.

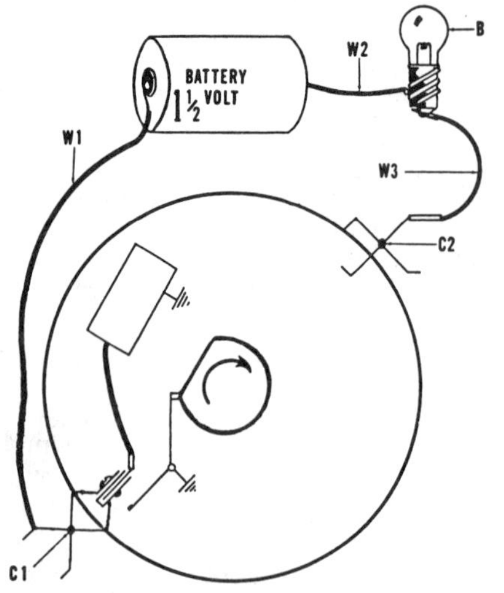

Fig. 5-3—Drawing showing a simple test lamp for checking ignition timing and/or breaker point opening.

B. 1½ volt bulb
C1. Spring clamp
C2. Spring clamp
W1. Wire
W2. Wire
W3. Wire

Light should be out when contact points are open and should go on when engine is turned to close breaker contact points. If light stays on when points are open, insulation of breaker contact arm is defective. If light does not go on when points are closed, contact surfaces are dirty, oily or are burned.

Adjust breaker point gap as follows unless manufacturer specifies adjusting breaker gap to obtain correct ignition timing. First, turn engine so points are closed to be sure contact surfaces are in alignment and seat squarely. Then, turn engine so breaker point opening is maximum and adjust breaker gap to manufacturer's specification. A wire type feeler gage is recommended for checking and adjusting the breaker contact gap. Be sure to recheck gap after tightening breaker point base retaining screws.

IGNITION TIMING. On some engines, ignition timing is nonadjustable and a certain breaker point gap is specified. On other engines, timing is adjustable by changing position of magneto stator plate with a specified breaker point gap or by simply varying breaker point gap to obtain correct timing. Ignition timing is usually specified either in degrees of engine (crankshaft) rotation or in piston travel before piston reaches top dead center position. In some instances, a specification is given for ignition timing even though timing may be nonadjustable; if a check reveals timing is incorrect on these engines, it is an indication of incorrect breaker point adjustment or excessive wear of breaker cam. Also, on some engines, it may indicate a wrong breaker cam has been installed or cam has been installed in a reversed position on engine crankshaft.

Some engines may have a timing mark or flywheel locating pin to locate flywheel at proper position for ignition spark to occur (breaker points begin to

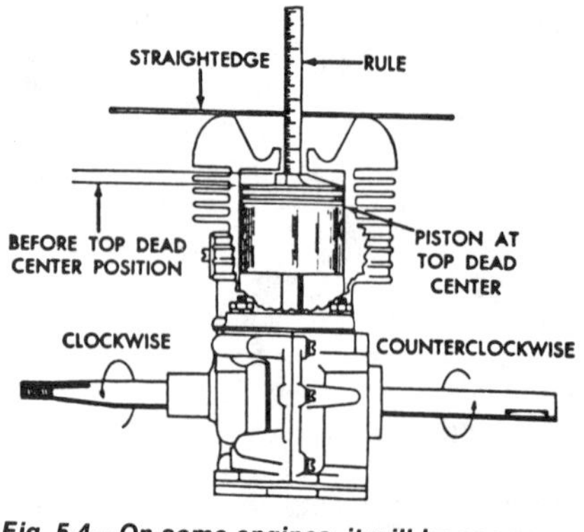

Fig. 5-4—On some engines, it will be necessary to measure piston travel with rule, dial indicator or special timing gage when adjusting or checking ignition timing.

open). If not, it will be necessary to measure piston travel as illustrated in Fig. 5-4 or install a degree indicating device on engine crankshaft.

A timing light as shown in Fig. 5-3 is a valuable aid in checking or adjusting engine timing. After disconnecting ignition coil lead from breaker point terminal, connect leads of timing light as shown. If timing is adjustable by moving magneto stator plate, be sure breaker point gap is adjusted as specified. Then, to check timing, slowly turn engine in normal direction of rotation past point at which ignition spark should occur. Timing light should be on, then go out (breaker points open) just as correct timing location is passed. If not, turn engine to proper timing location and adjust timing by relocating magneto stator plate or varying breaker contact gap as specified by engine manufacturer. Loosen screws retaining stator plate or breaker points and adjust position of stator plate or points so points are closed (timing light is on). Then, slowly move adjustment until timing light goes out (points open) and tighten retaining screws. Recheck timing to be sure adjustment is correct.

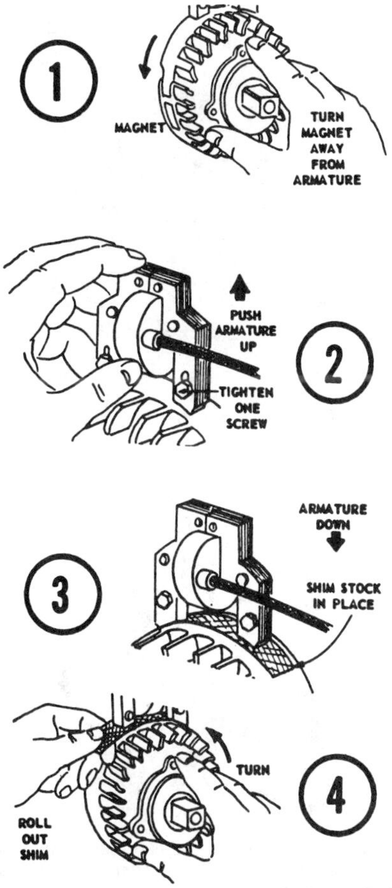

Fig. 5-5 — Views showing adjustment of armature air gap when armature is located outside flywheel. Refer to Fig. 5-6 for engines having armature located inside flywheel.

ARMATURE AIR GAP. To fully concentrate magnetic field of flywheel magnets within armature core, it is necessary that flywheel magnets pass as closely to armature core as possible without danger of metal to metal contact. Clearance between flywheel magnets and legs of armature core is called armature air gap.

On magnetos where armature and high tension coil are located outside of the flywheel rim, adjustment of armature air gap is made as follows: Turn engine so flywheel magnets are located directly under legs of armature core and check clearance between armature core and flywheel magnets. If measured clearance is not within manufacturers specifications, loosen armature core mounting screws and place shims of thickness equal to minimum air gap specification between magnets and armature core (Fig. 5-5). The magnets will pull armature core against shim stocks. Tighten armature core mounting screws, remove shim stock and turn engine through several revolutions to be sure flywheel does not contact armature core.

Where armature core is located under or behind flywheel, the following methods may be used to check and adjust armature air gap: On some engines, slots or openings are provided in flywheel through which armature air gap can be checked. Some engine manufacturers provide a cutaway flywheel that can be installed temporarily for checking armature air gap. A test flywheel can be made out of a discarded flywheel (See Fig. 5-6), or out of a new flywheel if service volume on a particular engine warrants such expenditure. Another method of checking armature air gap is to remove flywheel and place a layer of plastic tape equal to minimum specified air gap over legs of armature core. Reinstall flywheel and turn engine through several revolutions and remove flywheel; no evidence of contact between flywheel magnets and plastic tape should be noticed. Then cover legs of armature core with a layer of tape of thickness equal to maximum specified air gap; then, reinstall flywheel and turn engine through several revolutions. Indication of flywheel magnets contacting plastic tape should be noticed after flywheel is again removed. If magnets contact first thin layer of tape applied to armature core legs, or if they do not contact second thicker layer of tape, armature air gap is not within specifications and should be adjusted.

NOTE: Before loosening armature core mounting screws, scribe a mark on mounting plate against edge of armature core so that adjustment of air gap can be gaged.

In some instances it may be necessary to slightly enlarge armature core mounting holes before proper air gap adjustment can be made.

MAGNETO EDGE GAP. The point of maximum acceleration of movement of flywheel magnetic field through high tension coil (and therefore, the point of maximum current induced in primary coil windings) occurs when trailing edge of flywheel magnet is slightly past left hand leg of armature core. The exact point of maximum primary current is determined by using electrical measuring devices, distance between trailing edge of flywheel magnet and leg of armature core at this point is measured and becomes a service specification. This distance, which is stated either in thousandths of an inch or in degrees of flywheel rotation, is called the Edge Gap or "E" Gap.

For maximum strength of ignition spark, breaker points should just start to open when flywheel magnets are at specified edge gap position. Usually, edge gap is nonadjustable and will be maintained at proper dimension if contact breaker points are adjusted to recommended gap and correct breaker cam is installed. However, magneto edge gap can change (and spark intensity thereby reduced) due to the following:

a. Flywheel drive key sheared
b. Flywheel drive key worn (loose)
c. Keyway in flywheel or crankshaft worn (oversized)
d. Loose flywheel retaining nut which can also cause any above listed difficulty
e. Excessive wear on breaker cam
f. Breaker cam loose on crankshaft
g. Excessive wear on breaker point rubbing block or push rod so points cannot be properly adjusted.

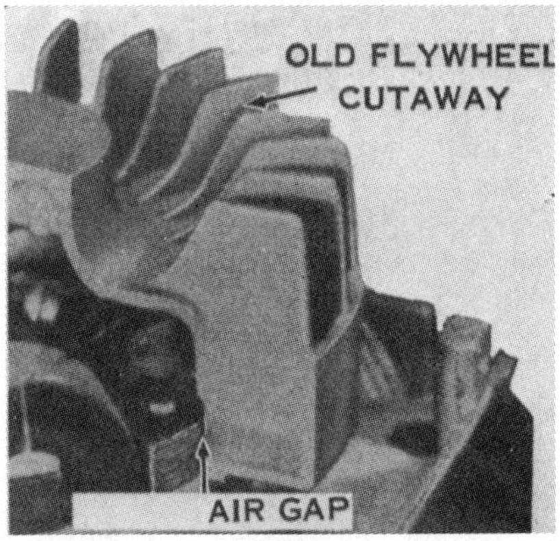

Fig. 5-6 — Where armature core is located inside flywheel, check armature gap by using a cutaway flywheel unless other method is provided by manufacturer; refer to appropriate engine repair section. Where possible, an old discarded flywheel should be used to cut-away section for checking armature gap.

Unit Type Magneto Service Fundamentals

Improper functioning of carburetor, spark plug or other components often causes difficulties that are thought to be an improperly functioning magneto. Since a brief inspection will often locate other causes for engine malfunction, it is recommended one be certain magneto is at fault before opening magneto housing. Magneto malfunction can easily be determined by simple tests as outlined in following paragraph.

Troubleshooting

With a properly adjusted spark plug in good condition, ignition spark should be strong enough to bridge a short gap in addition to actual spark plug gap. With engine running, hold end of spark plug wire not more than 1/16-inch (1.59 mm) away from spark plug terminal. Engine should not misfire.

To test magneto spark if engine will not start, remove ignition wire from magneto end cap socket. Bend a short piece of wire so when it is inserted in end cap socket, other end is about 1/8-inch (3.18 mm) from engine casting. Crank engine slowly and observe gap between wire and engine; a strong blue spark should jump gap the instant impulse coupling trips. If a strong spark is observed, it is recommended magneto be eliminated as source of engine difficulty and spark plug, ignition wire and terminals be thoroughly inspected.

If, when cranking engine, impulse coupling does not trip, magneto must be removed from engine and coupling overhauled or renewed. It should be noted that if impulse coupling will not trip, a weak spark will occur.

Magneto Adjustments and Service

BREAKER POINTS. Breaker points are accessible for service after removing magneto housing end cap. Examine point contact surfaces for pitting or pyramiding (transfer of metal from one surface to the other); a small tungsten file or fine stone may be used to resurface points. Badly worn of badly pitted points should be renewed. After points are resurfaced or renewed, check breaker point gap with rotor turned so points are opened maximum distance. Refer to MAGNETO paragraph in appropriate engine repair section for point gap specifications.

When installing magneto end cap, both end cap and housing mating surfaces should be thoroughly cleaned and a new gasket be installed.

CONDENSER. Condenser used in unit type magneto is similar to that used in other ignition systems. Refer to MAGNETO paragraph in appropriate engine repair section for condenser test specifications. Usually, a new condenser should be installed whenever breaker points are being renewed.

COIL. The ignition coil can be tested without removing the coil from housing. Instructions provided with coil tester should have coil test specifications listed.

ROTOR. Usually, service on magneto rotor is limited to renewal of bushings or bearings, if damaged. Check to be sure rotor turns freely and does not drag or have excessive end play.

MAGNETO INSTALLATION. When installing a unit type magneto on an engine, refer to MAGNETO paragraph in appropriate engine repair section for magneto to engine timing information.

SOLID STATE IGNITION SERVICE FUNDAMENTALS

Because of differences in solid state ignition construction, it is impractical to outline a general procedure for solid state ignition service. Refer to specific engine section for testing, overhaul notes and timing of solid state ignition systems.

SPARK PLUG SERVICING

ELECTRODE GAP. Spark plug electrode gap should be adjusted by bending the ground electrode. Refer to Fig. 5-7. Recommended gap is listed in SPARK PLUG paragraph in appropriate engine repair section of this manual.

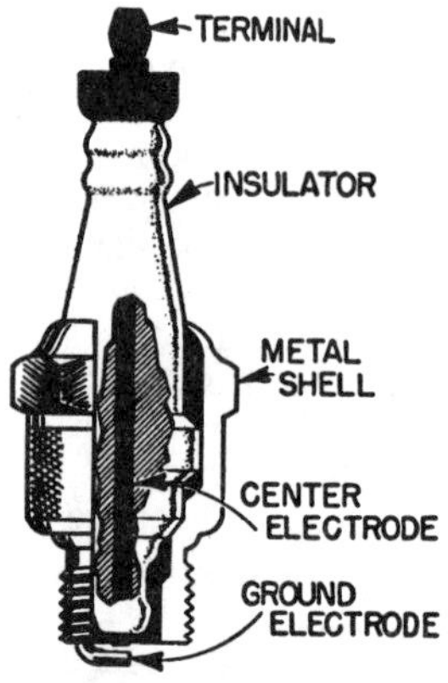
TERMINAL
INSULATOR
METAL SHELL
CENTER ELECTRODE
GROUND ELECTRODE

Fig. 5-7—Cross-sectional drawing of spark plug showing construction and nomenclature.

Fig. 5-8—Normal plug appearance in four-stroke cycle engine. Insulator is light tan to gray in color and electrodes are not burned. Renew plug at regular intervals as recommended by engine manufacturer.

Fig. 5-9—Appearance of spark plug indicating cold fouling. Causing of cold fouling may be use of a too-cold plug, excessive idling or light loads, carburetor choke out of adjustment, defective spark plug wire or boot, carburetor adjusted too "rich" or low engine compression.

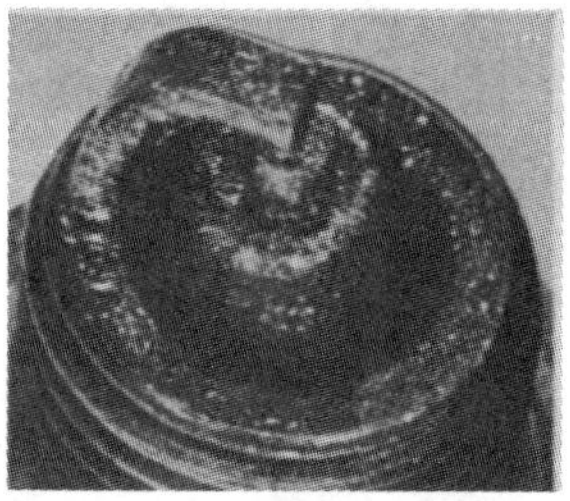

Fig. 5-10—Appearance of spark plug indicating wet fouling; a wet, black oily film is over entire firing end of plug. Cause may be oil getting by worn valve guides, worn oil rings or plugged breather or breather valve in tappet chamber.

Fig. 5-11—Appearance of spark plug indicating overheating. Check for plugged cooling fins, bent or damaged blower housing, engine being operated without all shields in place or other causes of engine overheating. Also can be caused by too lean a fuel-air mixture or spark plug not tightened properly.

CLEANING AND ELECTRODE CONDITIONING. Spark plugs are most usually cleaned by abrasive action commonly referred to as "sand blasting." Actually, ordinary sand is not used, but a special abrasive which is non-conductive to electricity even when melted, thus the abrasive cannot short out plug current. Extreme care should be used in cleaning plugs after sand blasting, however, as any particles of abrasive left on plug may cause damage to piston rings, piston or cylinder walls. Some engine manufacturers recommend spark plug be renewed rather than cleaned because of possible engine damage from cleaning abrasives.

After plug is cleaned by abrasive, and before gap is set, electrode surfaces between grounded and insulated electrodes should be cleaned and returned as nearly as possible to original shape by filing with a point file. Failure to properly dress electrodes can result in high secondary voltage requirements, and misfire of the plug.

PLUG APPEARANCE DIAGNOSIS. Appearance of a spark plug will be altered by use, and an examination of plug tip can contribute useful information which may assist in obtaining better spark plug life. Figs. 5-8 through 5-11 are provided by Champion Spark Plug Company to illustrate typical observed conditions. Listed in captions are probable causes and suggested corrective measures.

CARBURETOR SERVICING FUNDAMENTALS

The bulk of carburetor service consists of cleaning, inspection and adjustment. After considerable service it may become necessary to overhaul the carburetor and renew worn parts to restore original operating efficiency. Although carburetor condition affects engine operating economy and power, ignition and engine compression must also be considered to determine and correct causes of poor performance.

Before dismantling carburetor for cleaning or overhaul, clean all external surfaces and remove accumulated dirt and grease. Refer to appropriate engine repair section for carburetor exploded or cross-sectional views. Dismantle carburetor and note any discrepancies to assure correction during overhaul. Thoroughly clean all parts and inspect for damage or wear. Wash jets and passages and blow clear with clean, dry compressed air. Do not use a drill or wire to clean jets as possible enlargement of calibrated holes will disturb operating balance. Measurement of jets to determine extent of wear is difficult and new parts are usually installed to assure satisfactory results.

Carburetor manufacturers provide for many of their models an assortment of gaskets and other parts usually needed to do a correct job of cleaning and overhaul. These assortments are usually catalogued as Gasket Kits and Overhaul Kits respectively.

On float type carburetors, inspect float pin and needle valve for wear and renew if necessary. Check metal floats for leaks and where a dual type float is installed, check alignment of float sections. Check cork floats for loss of protective coating and absorption of fuel.

NOTE: Do not attempt to recoat cork floats with shellac or varnish or to resolder leaky metal floats. Renew part if defective.

Check fit of throttle and choke valve shafts. Excessive clearance will cause improper valve plate seating and will permit dust or grit to be drawn into engine. Air leaks at throttle shaft bores due to wear will upset carburetor calibration and contribute to uneven engine operation. Rebush valve shaft holes where necessary and renew dust seals. If rebushing is not possible, renew body part supporting shaft. Inspect throttle and choke valve plates for proper installation and condition.

Power or idle adjustment needles must not be worn or grooved. Check condition of needle seal packing or "O" ring and renew packing or "O" ring if necessary.

Reinstall or renew jets, using correct size listed for specific model. Adjust power and idle settings as described for specific carburetors in engine service section of manual.

It is important that carburetor bore at idle discharge ports and in vicinity of throttle valve be free of deposits. A partially restricted idle port will produce a "flat spot" between idle and mid-range rpm. This is because the restriction makes it necessary to open throttle wider than the designed opening to obtain proper idle speed. Opening throttle wider than the design specified amount will uncover more of the port than was intended in calibration of carburetor. As a result an insufficient amount of the port will be available as a reserve to cover transition period (idle to the mid-range rpm) when the high speed system begins to function.

When reassembling float type carburetors, be sure float position is properly adjusted. Refer to CARBURETOR paragraph in appropriate engine repair section for float level adjustment specifications.

ENGINE SERVICE

DISASSEMBLY AND ASSEMBLY

Special techniques must be developed in repair of engines of aluminum alloy or magnesium alloy construction. Soft threads in aluminum or magnesium castings are often damaged by carelessness in overtightening fasteners or in attempting to loosen or remove seized fasteners. Manufacturer's recommended torque values for tightening screw fasteners should be followed closely.

NOTE: If damaged threads are encountered, refer to following paragraph, "REPAIRING DAMAGED THREADS."

A given amount of heat applied to aluminum or magnesium will cause it to expand a greater amount than will steel under similar conditions. Because of different expansion characteristics, heat is usually recommended for easy installation of bearings, pins, etc., in aluminum or magnesium castings. Sometimes, heat can be used to free parts that are seized or where an interference fit is used. Heat, therefore, becomes a service tool and the application of heat one of the required service techniques. An open flame is not usually advised because it destroys paint and other protective coatings and because a uniform and controlled temperature with open flame is difficult to obtain. Methods commonly used are heating in oil or water, with a heat lamp, electric hot plate, electric hot air gun, or in an oven or kiln. The use of water or oil gives a fairly accurate temperature control but is somewhat limited as to the size and type of part that can be handled. Thermal crayons are available which can be used to determine temperature of a heated part. These crayons melt when the part reaches a specified temperature, and a number of crayons for different temperatures are available. Temperature indicating crayons are usually available at welding equipment supply houses.

Use only specified gaskets when reassembling, and use an approved gasket cement or sealing compound unless otherwise stated. Seal all exposed threads and repaint or retouch with an approved paint.

REPAIRING DAMAGED THREADS

Damaged threads in castings can be renewed by use of thread repair kits which are recommended by a number of equipment and engine manufacturers. Use of thread repair kits is not difficult, but instructions must be carefully followed. Refer to Figs. 5-12 through 5-15 which illustrate the use of Heli-Coil thread repair kits that are manufactured by the Heli-coil Corporation, Danbury, Connecticut.

Heli-Coil thread repair kits are available through parts departments of most engine and equipment manufacturers; thread inserts are available in all National Coarse (USS) sizes from #4 to 1½ inch and National Fine (SAE) sizes from #6 to 1½ inch. Also, sizes for repairing 14 mm and 18 mm spark plug ports are available.

Fig. 5-12—Damaged threads in casting before repair. Refer to Figs. 5-13, 5-14 and 5-15 for steps in installing thread insert. (Series of photos provided by Heli-Coil Corp., Danbury, Conn.)

Fig. 5-13—First step in repairing damaged threads is to drill out old threads using exact size drill recommended in instructions provided with thread repair kit. Drill all the way through an open hole or all the way to bottom of blind hole, making sure hole is straight and that centerline of hole is not moved in drilling process.

VALVE SERVICE FUNDAMENTALS

When overhauling engines, obtaining proper valve sealing is of primary importance. The following paragraphs cover fundamentals of servicing intake and exhaust valves, valve seats and valve guides.

REMOVING AND INSTALLING VALVES. A valve spring compressor, one type of which is shown in Fig. 5-16, is a valuable aid in removing and installing intake and exhaust valves. This tool is used to hold spring compressed while removing or installing pin, collars or retainer from valve stem. Refer to Fig. 5-17 for views showing some of the different methods of retaining valve spring to valve stem.

Fig. 5-14—Special drill taps are provided in thread repair kit for threading drilled hole to correct size for outside of thread insert. A standard tap cannot be used.

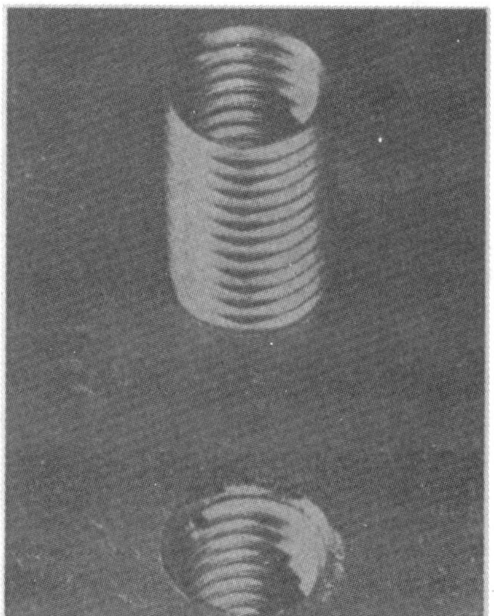

Fig. 5-15—A thread insert and a completed repair are shown above. Special tools are provided in thread repair kit for installation of thread insert.

VALVE REFACING. If valve face (See Fig. 5-18) is slightly worn, burned or pitted, valve can usually be refaced providing proper equipment is available. Many shops will usually renew valves, however, rather than invest in somewhat costly valve refacing tools.

Before attempting to reface a valve, refer to specifications in appropriate engine repair section for valve face angle. On some engines, manufacturer recommends grinding the valve face to an angle of ½° to 1° less than that of the valve seat. Refer to Fig. 5-19. Also, nominal valve face angle may be either 30° or 45°.

After valve is refaced, check thickness of valve "margin" (See Fig. 5-18). If margin is less than manufacturer's minimum specification (refer to specifications in appropriate engine repair section), or is less than one-half the margin of a new valve, renew valve. Valves having excessive material removed in refacing operation will not give satisfactory service.

When refacing or renewing a valve, the seat should also be reconditioned, or in engines where valve seat is renewable, a new seat should be installed. Refer to following paragraph "RESEATING OR RENEWING VALVE SEATS." Then, the seating surfaces should be lapped in using a fine valve grinding compound.

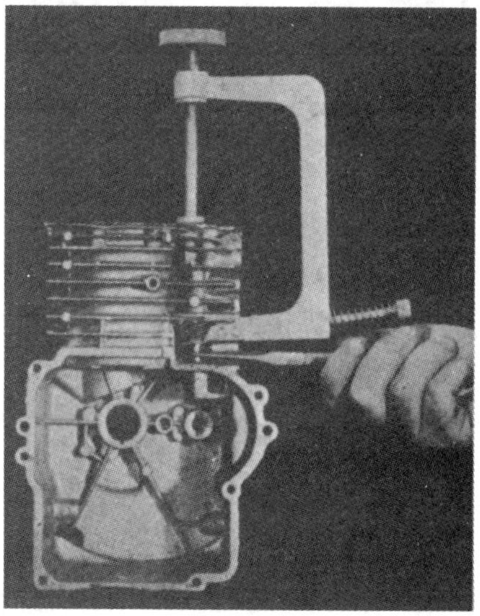

Fig. 5-16—View showing one type of valve spring compressor being used to remove keeper. (Block is cut-away to show valve spring.)

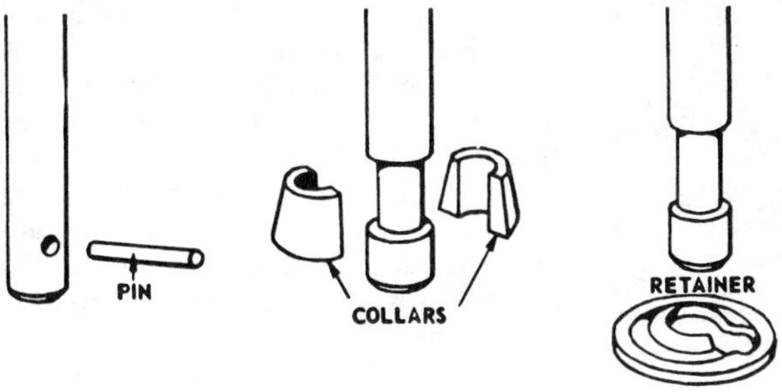

Fig. 5-17—Drawing showing three types of valve spring keepers used.

RESEATING OR RENEWING VALVE SEATS. On engines having valve seat machined in cylinder block casting, seat can be reconditioned by using a correct angle seat grinding stone or valve seat cutter. When reconditioning valve seat, care should be taken that only enough material is removed to provide a good seating on valve contact surface. The width of seat should then be measured (See Fig. 5-20) and if width exceeds manufacturer's maximum specifications, seat should be narrowed by using one stone or cutter with an angle 15° greater than valve seat angle and a second stone or cutter with an angle 15° less than seat angle. When narrowing seat, coat seat lightly with Prussian blue and check where seat contacts valve face by inserting valve in guide and rotating valve lightly against seat. Seat should contact approximate center of valve face. By using only narrow angle seat narrowing stone or cutter, seat contact will be moved towards outer edge of valve face.

On engines having renewable valve seats, refer to appropriate engine repair section in this manual for recommended method of removing old seat and installing new seat. Refer to Fig. 5-21 for one method of installing new valve seats. Seats are retained in cylinder block bore by an interference fit; that is, seat is slightly larger than bore in block. It sometimes occurs that valve seat will become loose in bore, especially on engines with aluminum crankcase. Some manufacturers provide oversize valve seat inserts (insert O.D. larger than standard part) so that if standard size insert fits loosely, bore can be cut oversize and a new insert be tightly installed. After installing valve seat insert in engines of aluminum construction, metal around seat should be peened as shown in Fig. 5-22. Where a loose insert is encountered and an oversize insert is not available, loose insert can usually be tightened by centerpunching cylinder block material at three equally spaced points around insert, then peening completely around insert as shown in Fig. 5-22.

For some engines with cast iron cylinder blocks, a service valve seat insert is available for reconditioning valve

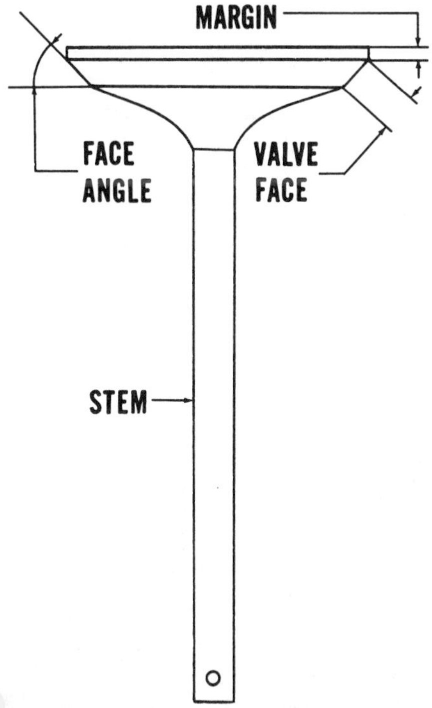

Fig. 5-18—Drawing showing typical four-stroke cycle engine valve. Face angle is usually 30° or 45°. On some engines, valve face is ground to an angle of ½ or 1 degree less than seat angle.

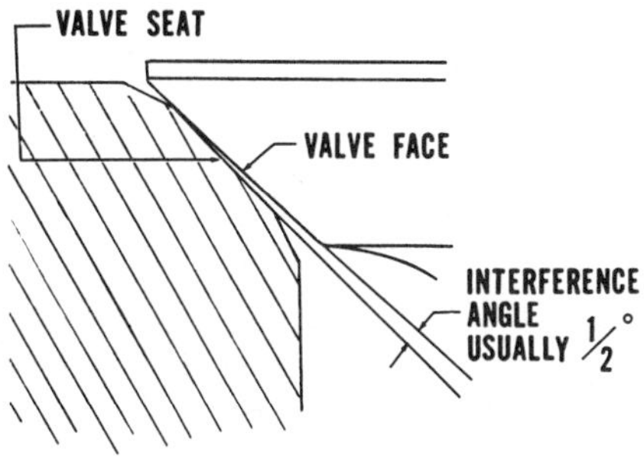

Fig. 5-19—Drawing showing line contact of valve face with valve seat when valve face is ground at smaller angle than valve seat; this is specified on some engines.

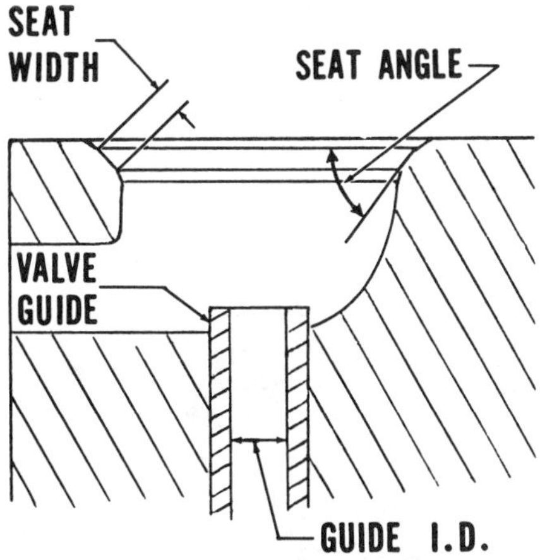

Fig. 5-20—Cross-sectional drawing of typical valve seat and valve guide as used on some engines. Valve guide may be integral part of cylinder block; on some models so constructed, valve guide I.D. may be reamed out and an oversize valve stem installed. On other models, a service guide may be installed after counterboring cylinder block.

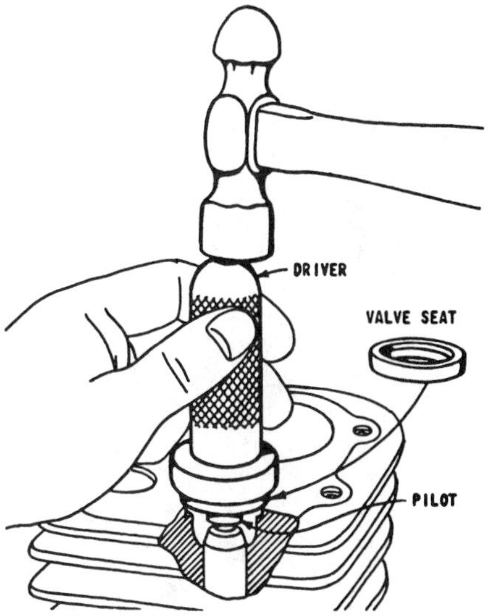

Fig. 5-21—View showing one method used to install valve seat insert. Refer to appropriate engine repair section for manufacturer's recommended method.

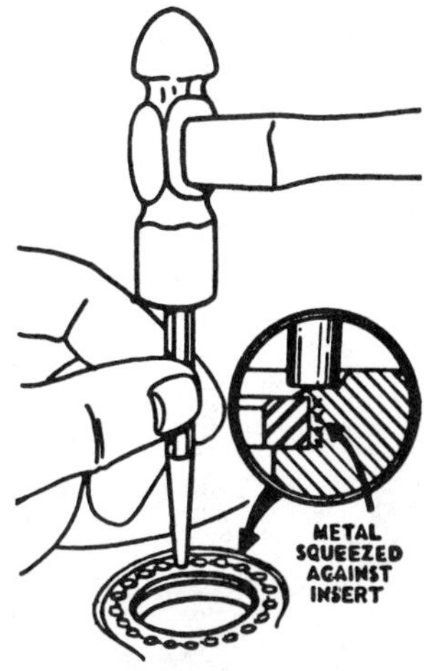

Fig. 5-22—It is usually recommended that on aluminum block engines, metal be peened around valve seat insert after insert is installed.

seat, and is installed by counterboring cylinder block to specified dimensions, then driving insert into place. Refer to appropriate engine repair section in this manual for information on availability and installation of service valve seat inserts for cast iron engines.

INSTALLING OVERSIZE PISTON AND RINGS

Some engine manufacturers have oversize piston and ring sets available for use in repairing engines in which cylinder bore is excessively worn and standard size piston and rings cannot be used. If care and approved procedure are used in boring the cylinder oversize, installation of an oversize piston and ring set should result in a highly satisfactory overhaul.

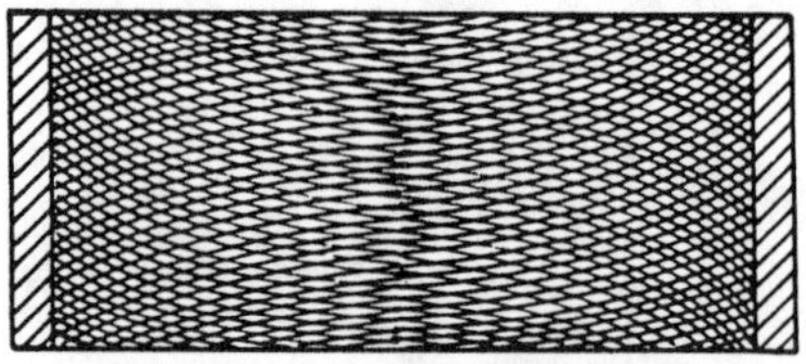

Fig. 5-23—A cross-hatch pattern as shown should be obtained when honing cylinder. Pattern is obtained by moving hone up and down cylinder bore as it is being turned by slow speed electric drill.

Cylinder bore may be oversized by using either a boring bar or a hone; however, if a boring bar is used, it is usually recommended cylinder bore be finished with a hone. Refer to Fig. 5-23.

Where oversize piston and rings are available, it will be so noted in appropriate engine repair section of this manual. Also, the standard bore diameter will also be given. Before attempting to rebore or hone the cylinder to oversize, carefully measure the cylinder bore to be sure standard size piston and rings will not fit within tolerance. Also, it may be possible cylinder is excessively worn or damaged and reboring or honing to largest oversize will not clean up worn or scored surface.

DIESEL ENGINE FUNDAMENTALS

PRINCIPLE OF OPERATION

The diesel engines covered in this manual are all technically known as "Internal Combustion Reciprocating Engines." The source of power is heat formed by the burning of a combustible mixture of petroleum products and air. This burning takes place in a closed cylinder containing a piston. Expansion resulting from the heat of combustion applies downward pressure on the piston to produce reciprocating (up-and-down) motion of the piston. The reciprocating motion is transmitted to the crankshaft and converted to rotary motion by means of a connecting rod and the offset "crank" on the crankshaft.

Diesel engines may be classified as "compression ignition." Their operation depends upon the air temperature in the engine cylinder being increased above the combustion point of the fuel to ignite the fuel-air mixture, rather than using an electrical spark to ignite the fuel-air mixture as in "spark ignition" engines. In diesel engines, there is no premixing of the fuel and air outside the cylinder; air only is taken into the cylinder through the intake manifold. The pressure created by the piston moving upward on the compression stroke causes a rapid temperature increase of the air in the cylinder. When the piston nears the top of the cylinder, finely atomized fuel is injected into the cylinder where it mixes with the air and is ignited by the heat of the compressed air. Expansion resulting from the heat of combustion then forces the piston downward on the power stroke and turns the crankshaft.

As with spark ignition engines, diesel engines are produced in both 2-stroke cycle and 4-stroke cycle design. However, since all of the engines covered in this manual are of 4-stroke cycle design, only this type will be discussed.

With the 4-stroke cycle design, two revolutions of the engine crankshaft (four piston strokes) are required to provide one power stroke of the piston. See Fig. FN1. The four-stroke cycle begins with the engine intake valve open and the piston being pulled to the bottom of the cylinder by crankshaft rotation (intake stroke). Atmospheric pressure (see also TURBOCHARGING paragraph) pushes air into the cylinder past the open intake valve. As the piston reaches the bottom of the stroke, the intake valve closes and crankshaft rotation forces the piston upward compressing the air trapped in the cylinder (compression stroke). As the piston nears the end of the compression stroke, the compressed air in the cylinder becomes very hot. At this point, a timed injection of fuel enters the cylinder and is ignited by the temperature of the air. The combustion process begins and as the piston reaches top dead center (TDC) and starts downward, the burning fuel-air mixture expands rapidly and forces the piston down (power stroke), imparting a turning motion to the engine crankshaft. As the piston reaches the bottom of the cylinder, the exhaust valve opens. The crankshaft continues to turn from inertia of the rotating engine flywheel, moving the piston up and expelling the burned fuel-air mixture out through the open exhaust valve (exhaust stroke). The exhaust valve closes and the intake valve opens as piston reaches the top of cylinder. The piston then is pulled back down (intake stroke) by the crankshaft and connecting rod, beginning another 4-stroke engine operating cycle.

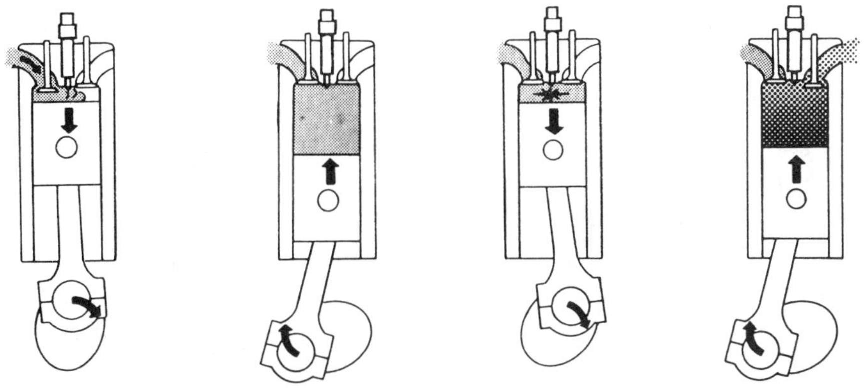
INTAKE
COMPRESSION
POWER
EXHAUST

Fig. FN1—Schematic diagram of four-stroke cycle diesel engine. The "compression ignition" diesel engine depends on the heat produced from compressing the air in the cylinder to ignite the fuel as it is sprayed into the cylinder.

The above is a simplified explanation of how a four-stroke cycle diesel engine operates for a single cylinder engine. In multi-cylinder engines, the crankshaft operates each piston in a sequence referred to as the "firing order" of the engine. This provides a balance between the piston power strokes of each cylinder. For example, the single cylinder engine has one power stroke in two revolutions of the engine crankshaft. A two-cylinder engine will have a power stroke on every revolution of the crankshaft and a four-cylinder engine will have two power strokes for each crankshaft revolution.

DIESEL FUEL SYSTEM

The basic diesel fuel system consists of a fuel supply tank (1—Fig. FN2), an injection pump (3) capable of creating a high pressure flow of fuel, and a fuel injector (4) for each engine cylinder which receives the fuel from the pump and sprays it into the engine cylinder. The diesel fuel system may include one or more fuel filters (2) to protect the system from contamination, and a low pressure pump that feeds or lifts fuel from the supply tank to the fuel injection pump. On some diesel engines, the fuel feed pump is mounted on the injection pump and is actuated by a cam lobe on the fuel injection pump camshaft, or it may be an integral part of the fuel injection pump.

In order for the diesel engine to operate properly, the fuel injection system must supply the correct quantity of fuel, time the fuel delivery, atomize the fuel and distribute the fuel evenly through the cylinder.

FUEL INJECTION SYSTEM TYPES

There are two basic types of diesel fuel injection systems: direct injection and indirect injection. The cylinder head configuration is different with each type of fuel injection system.

DIRECT FUEL INJECTION. Direct fuel injection utilizes an injector with the nozzle tip extending directly into the engine combustion chamber; see Fig. FN3. Usually, the combustion chamber is located in the piston crown and the cylinder head surface is flat. The injector nozzle protrudes from the head surface directly into the combustion chamber. The injector nozzle tip has three to five orifices (holes) of approximately 0.25-0.30 mm (0.010-0.012 inch) diameter. The spray holes are spaced equidistantly around the nozzle tip. The spray from each orifice should form a cone and the spray from all orifices should be alike. The fuel spray is finely atomized so that it will ignite readily from the heat of the compressed air and will burn steadily through the engine power stroke.

INDIRECT FUEL INJECTION. The injector nozzle for indirect injection systems does not extend into the engine combustion chamber. The indirect injection nozzle sprays a single spray cone into a precombustion chamber (3—Fig. FN4), or to an air cell (3—Fig. FN5) directly across the cylinder head combustion chamber from the injector nozzle. The fuel is concentrated in the center of the spray cone, but the outer circumference of the cone is finely atomized so as to provide instant combustion. Some indirect system injectors have two spray orifices, one for the major volume of fuel that enters the air cell and another small orifice that atomizes the fuel that directly enters the combustion chamber.

With the precombustion chamber design, fuel is sprayed into the precombustion chamber and ignites there, but cannot completely burn due to lack of air. What combustion does occur heats the fuel and forces it into the cylinder where complete combustion occurs on the power stroke of the piston.

With the air cell design, the fine atomization surrounding the core of the fuel spray ignites in the engine combustion chamber, increasing the temperature and pressure in the combustion chamber and air cell. The major portion of the fuel enters the air cell, where partial combustion of the fuel occurs. This partial combustion heats the remainder of the fuel and forces it out of the air cell into the cylinder where complete combustion occurs on the power stroke.

DIESEL FUEL SYSTEM COMPONENTS

FUEL INJECTION PUMPS. There are a number of different fuel injection pump designs, but basically all incorporate a cam actuated plunger operating in a cylinder with extremely close tolerances and capable of pumping fuel under high pressure to the fuel injector. A clean supply of fuel free of water is very important to the precision parts of the diesel injection system.

A single cylinder diesel engine requires a fuel injection pump with only one pumping element. See Fig FN6. Some multi-cylinder engines are fitted with a separate pump, comparable to the single cylinder engine pump, for each cylinder of the engine. Many multi-cylinder engines utilize a common pump housing containing a pumping element

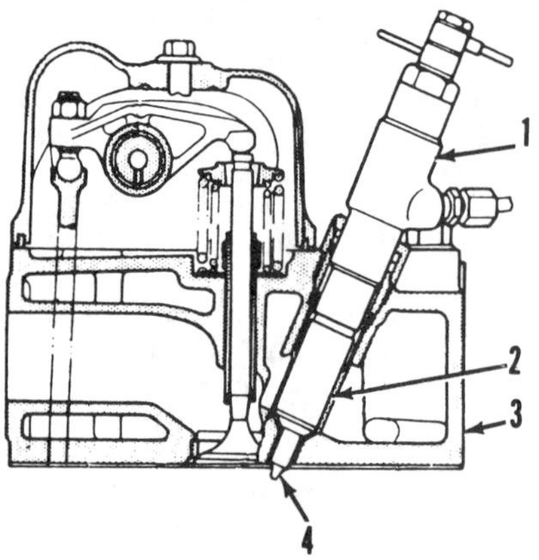

Fig. FN3—Cross-sectional view of cylinder head and injector typical of direct injection engines. Note that nozzle tip extends into the combustion chamber.

1. Injector
2. Nozzle sleeve
3. Cylinder head
4. Nozzle tip

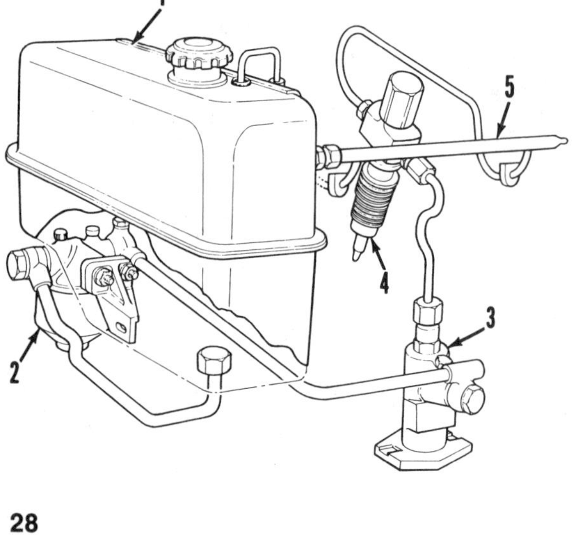

Fig. FN2—Drawing of basic fuel injection system components.

1. Fuel tank
2. Fuel filter
3. Fuel injection pump
4. Fuel injector
5. Fuel return line

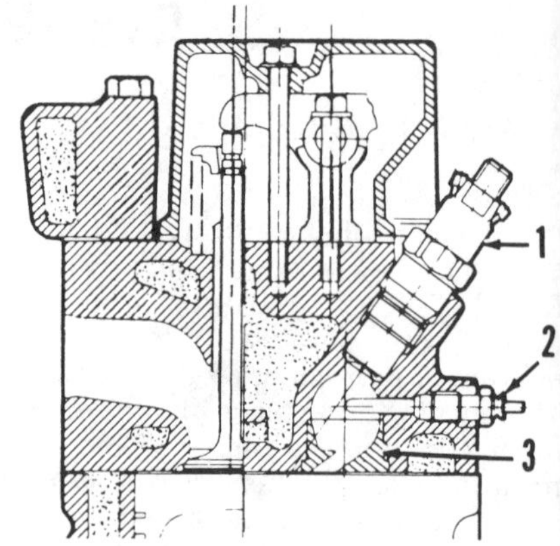

Fig. FN4—Cross-sectional view of cylinder head, injector (1) and precombustion chamber (3) typical of indirect injection engines. A glow plug (2) may be used as a starting aid on indirect injection engines.

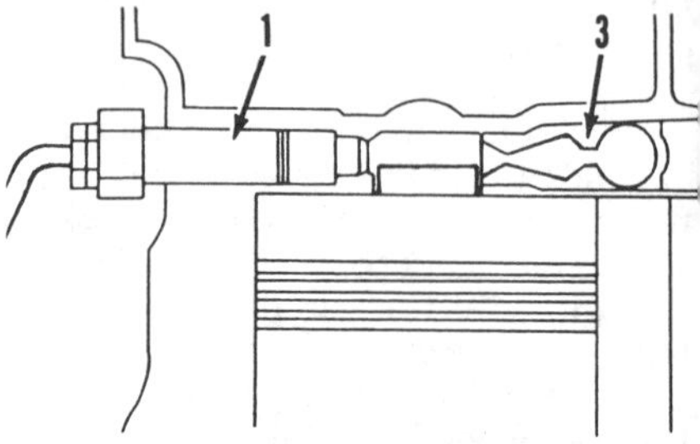

Fig. FN5—Some indirect injection engines may be equipped with an air cell (3) rather than a precombustion chamber. Injector (1) sprays fuel across the combustion chamber into the air cell where partial combustion of the fuel occurs.

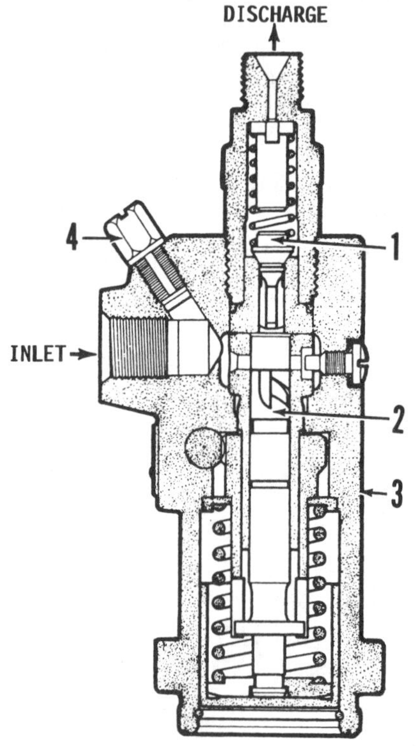

Fig. FN6—Cross-sectional view of typical single plunger injection pump used on some engines.

1. Delivery valve
2. Pump plunger
3. Pump body
4. Air bleed screw

(in-line type) for each cylinder of the engine. A different arrangement for multi-cylinder engines is a pump with only one pumping element that uses a distributor type arrangement within the pump to direct a timed pressure pulse to the injector of each cylinder in the firing order of the engine. See Fig. FN9.

SINGLE OR MULTIPLE ELEMENT PUMPS. On engines with a separate pump for each cylinder or with an in-line type pump having a pumping element for each cylinder, a camshaft is used to actuate the pump plunger(s). When the engine is running, there is a constant supply of fuel provided by a lift pump or gravity available at the inlet port and spill port of the pumping element (Fig. FN7). With the plunger at its lowest point (A) in the cylinder barrel, fuel is free to flow into the barrel through the inlet port. As the plunger is forced upward (B) by the camshaft, the top of the plunger closes the inlet port and spill port, trapping the fuel in the barrel. Further movement of the plunger causes the pressure of the trapped fuel to increase until it overcomes the delivery valve spring pressure and flows out the top of the barrel to the injector. The injector nozzle valve is forced open by the pressurized fuel and sprays a quantity of fuel into the cylinder corresponding to the amount of fuel delivered by the pump plunger. The injection of fuel into the cylinder continues until pump plunger has risen to position (C), where a helical land on the plunger (which is connected to the top of the plunger by a vertical slot) uncovers the spill port in the barrel and allows pressurized fuel in the barrel to escape back to inlet side of the pumping element. This causes an immediate drop in pressure, resulting in the delivery valve and injector valve closing under spring pressure.

The total length of the plunger stroke is always constant, but the effective pumping stroke can be controlled by altering the position of the bottom of the plunger helix in relation to the spill port. The position of the helix is changed by rotating the plunger in the barrel, and the point of fuel cut-off can then be made to occur earlier (less fuel delivery) or later (more fuel delivery) in the stroke.

With the engine running under load, the plunger is rotated to the left (Fig. FN8) providing a long effective pumping stroke before the helix uncovers the spill port. At idle or very light load, plunger is rotated to the right so that the helix uncovers the spill port early in the stroke, reducing the amount of fuel pumped to the injector. To stop the engine, plunger is moved further to the right so that vertical slot in plunger is aligned with the spill port during the complete plunger stroke, preventing any build-up of pressure in the barrel.

The rotation of the helix is accomplished by a toothed quadrant on the pumping element mating with a control rack, which is controlled by the engine governor. On a multi-cylinder engine, all pumps are controlled by a common control rack.

DISTRIBUTOR TYPE PUMP. On distributor type injection pumps, a drive shaft rotates a fuel transfer pump, distributor rotor and pump element in a hydraulic head to deliver fuel to each injector. See Fig. FN9. The operating principles for typical distributor type pump are outlined below:

Fuel entering the injection pump is pressurized by the positive displacement transfer pump (2). The transfer pump pressure is controlled by pump speed and a pressure regulating valve in the pump end plate (1). The pressurized fuel flows through the pump housing, providing cooling and lubrication of injection pump cam, rollers, governor and other parts. Part of the excess fuel not required for injection is bypassed back to the inlet side of the pump and the remainder to the fuel tank.

The pressurized fuel for charging the injection pumping element flows through passages in the hydraulic head to a metering valve (8), which is controlled by governor action. The position of the metering valve and transfer pump pressure determines the quantity of fuel which is available to charge the pumping element. As the rotor turns, its charging channel comes into register with the inlet port to the pumping elements (Fig. FN10). As fuel enters the pumping cylinder, the two plungers are forced outward a distance proportionate to the quantity of fuel to be injected. The quantity of fuel entering the pumping cylinder is controlled by fuel pressure at the charging passage, by the amount of time available for charging, and by the maximum outward travel of the plungers which is limited by the roller shoes contacting an adjustable stop.

As the rotor continues to turn, the inlet port is closed and the single distributor port (Fig. FN10) in the rotor aligns with an outlet port in the hydraulic head. An internal cam ring (7—Fig. FN9) mounted in the pump housing, which has as many lobes as there are engine cylinders, operates the opposed pump plungers through cam rollers carried in shoes sliding in the rotor body. The rollers (4) simultaneously contact the diametrically opposed cam lobes and force the plungers (6) toward each other, discharging the fuel through the hydraulic head discharge port, into the fuel injector line (3) and opening the injector nozzle valve. The cam lobes are contoured to allow a very slight outward movement of the plungers at the end of the injection cycle. This gives a sharp cut-off of fuel to the injector and prevents "dribble" at injector nozzle tip.

INJECTOR. The injector must atomize the fuel to provide good combustion and must spray the fuel into the cylinder in such a manner that it mixes thoroughly with the air. Most injector nozzles used on modern diesel engines

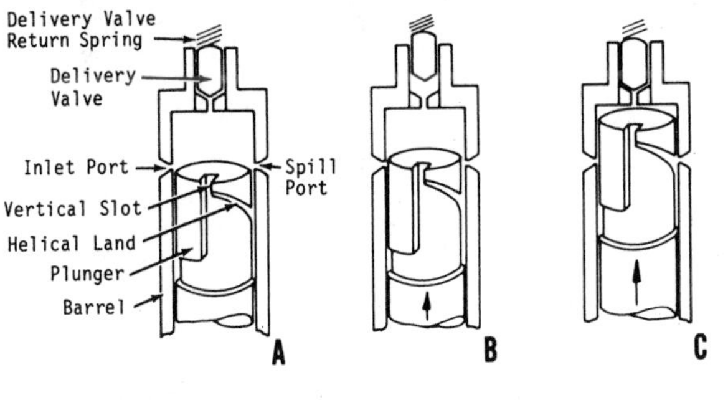
Delivery Valve
Return Spring
Delivery
Valve
Inlet Port
Spill
Port
Vertical Slot
Helical Land
Plunger
Barrel
A
B
C

Fig. FN7—Injection pump plunger is forced upward by a camshaft. Beginning of injection occurs when top of plunger closes inlet and spill ports in pump barrel. End of injection occurs when helical land on plunger uncovers the spill port.

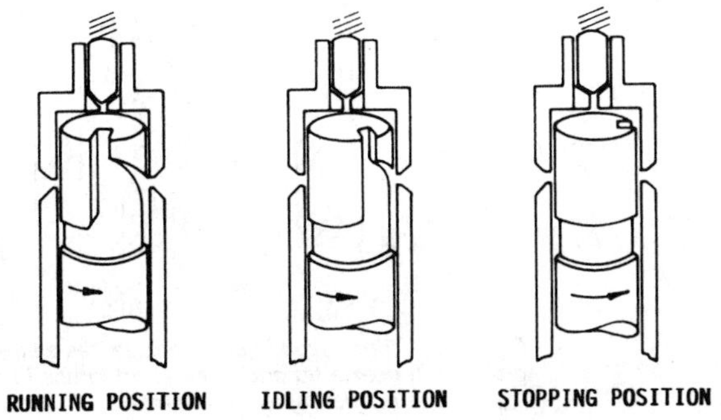
RUNNING POSITION
IDLING POSITION
STOPPING POSITION

Fig. FN8—Injection pump plunger is rotated to alter the position of the plunger helix in relation to the spill port, which varies the effective length of the pumping stoke and controls the amount of fuel delivery.

are closed, inward opening type. See Figs. FN11 and FN12. They use a spring (S) to hold the injector nozzle valve (V) on its seat, and are operated hydraulically when fuel pressure acting on one side of the nozzle valve causes the valve to lift off its seat against spring pressure. The fuel pressure required to open the nozzle valve is determined by the spring pressure against the top end of the valve and is normally adjustable.

Hole type nozzles (Fig. FN11) have more than one spray hole in the nozzle tip. The size and angle of the holes can be varied by the manufacturer to achieve the spray pattern within the cylinder needed for proper combustion. This type nozzle is normally used in engines with open combustion chambers and direct injection.

Pintle type nozzles (Fig. FN12) have a single orifice hole that sprays fuel in a direct line from the injector. This type injector is normally used on indirect injection engines using a precombustion chamber or air cell.

GOVERNOR

Engine speed and power output are determined by the amount of fuel injected into the cylinder. To protect the engine from overspeeding or exceeding its designed power output limit, a means of controlling the amount of fuel delivered to the engine must be used. Depending upon the type of fuel injection pump, various methods are used to meter the amount of fuel that is delivered per each stroke of the pumping element, and a governor is then used to control the fuel metering system. A flyweight type governor is most commonly used, and may be contained in the injection pump housing or may be part of the engine.

The governor maintains a nearly constant engine speed by varying the amount of fuel supplied to the engine to satisfy varying load demands placed on the engine. In operation, governor main spring tension (which increases or decreases relative to the position of throttle control lever on variable speed governors) tends to move the fuel metering linkage to increase the amount of fuel being delivered to the engine, while centrifugal force created by the spinning governor flyweights tends to move the fuel metering linkage to decrease the amount of fuel delivery. A constant engine speed is maintained when the centrifugal force created by the spinning flyweights counterbalances the tension of governor main spring.

An increase in the load being applied to the engine causes a decrease in engine speed which results in a corresponding decrease in flyweight centrifugal force, allowing the flyweights to

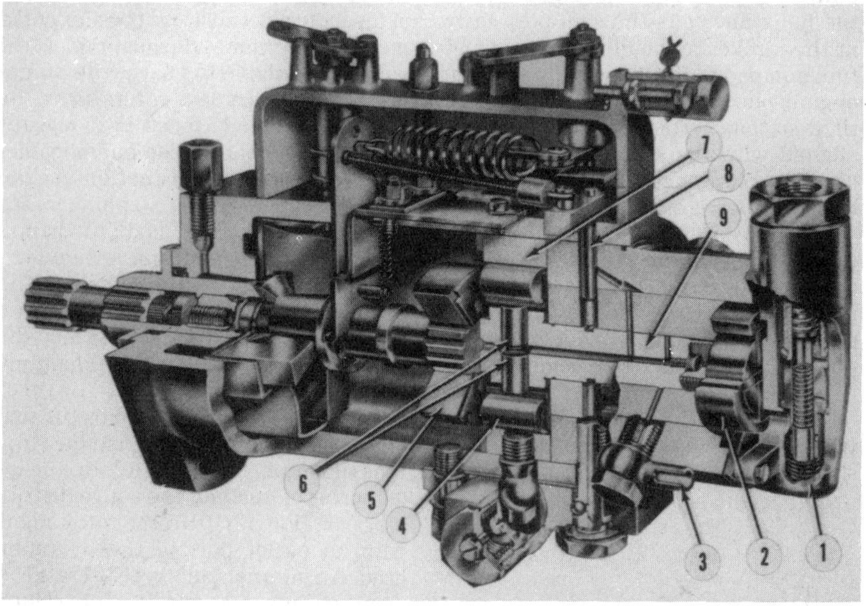

Fig. FN9—Cutaway view of a rotary distributor type fuel injection pump.

1. End plate & pressure regulating valve
2. Fuel transfer pump
3. Injector line
4. Cam roller
5. Rotor
6. Pump plungers
7. Cam ring
8. Metering valve
9. Pump rotor

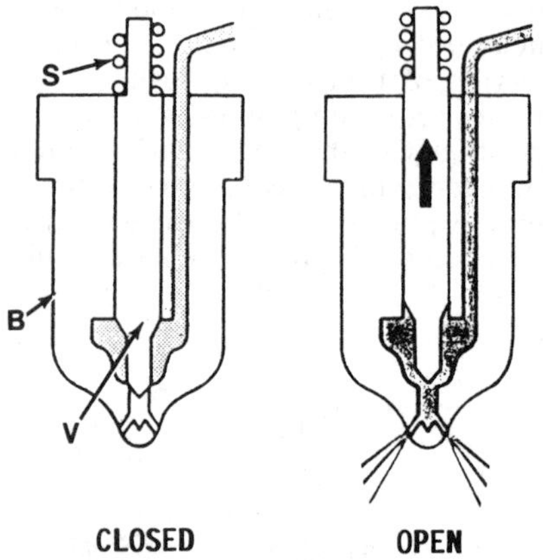

Fig. FN11—The hole type injector nozzle has more than one spray hole in nozzle tip and is most often used in direct injection engines.

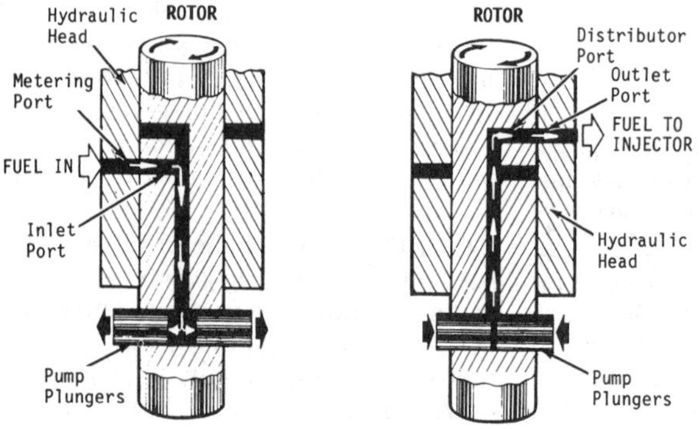

Fig. FN10—On the charging cycle, inlet port in pump rotor is in register with fuel metering port in hydraulic head. On pumping cycle, distributor port in rotor is in register with one of the outlet ports in hydraulic head.

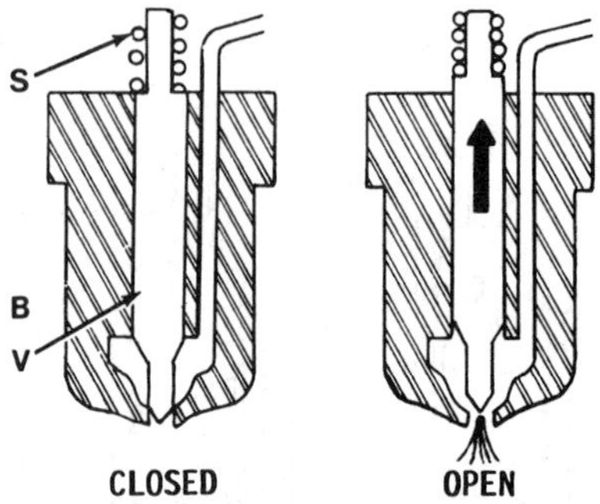

Fig. FN12—Pintle type injector nozzle has a single hole in nozzle tip and is most often used in indirect injection engines.

move inward. Movement of the flyweights is transmitted through the governor linkage to move the injection pump fuel metering control to obtain increased fuel delivery to the engine, resulting in increased power and speed. The engine speed increases until the centrifugal force of the flyweights and governor spring tension are again balanced.

When the load on the engine is reduced, engine speed increases and the opposite reactions take place in the governor. As the governor flyweight centrifugal force exceeds the governor spring force, the governor linkage moves the fuel metering controls to reduce the fuel delivery rate to the engine and maintain desired speed setting.

TURBOCHARGING

Power obtained from an internal combustion engine is limited by the amount of air, thus oxygen, in the cylinder for combustion of the fuel. Engines with normal air intake through air cleaner and intake ports, called "naturally aspirated" engines, are limited by atmospheric pressure to the amount of air that can be taken in on the intake stroke. This limit is reduced at higher elevations because of decrease in atmospheric pressure. Increasing the amount of fuel above that which will be burned causes excessive exhaust smoke and loss of engine power. If the fuel delivery system is adjusted for proper air/fuel ratio at sea level or lower elevations, fuel delivery must be decreased for the engine to operate efficiently at higher elevations.

The turbocharger utilizes a turbine blade (rotor) powered by the engine exhaust to drive a turbine blade in the engine air intake to compress the air entering the engine cylinder, thus increasing the volume of air going into the engine cylinder on the intake stroke. As the volume of air in the cylinder is increased, more fuel can be injected and burned to provide more power per unit of engine displacement. The turbocharger is especially beneficial for high altitude engine operation, but will increase power output at any elevation.

COLD WEATHER STARTING

As diesel engines rely solely on the temperature of air compressed in the cylinder to ignite the fuel when it is injected into the cylinder, cold temperatures have an adverse effect on engine starting. To improve diesel engine starting characteristics in cold ambient temperatures, engine starting aids are used. Several different methods used to aid in cold weather starting are:

a. Increasing cranking speed with more powerful starting motor and battery. An optional, more powerful electric starter and higher capacity battery are available on some engines. The higher cranking speeds increases compression temperature for easier starting.
b. Heating the air entering the engine through the intake manifold by an electric heater. If the air at beginning of compression stroke is warmer, the temperature at end of compression stroke will be relatively higher and result in easier ignition of fuel injected into the cylinder. This method is usually utilized in direct injection engines.
c. Use of electric "glow plugs" in the engine cylinder to cause ignition of the fuel injected into the cylinder. This method is most generally used in indirect injection engines.
d. Adding a small amount of engine oil in the engine intake to increase sealing of the piston rings and increase compression pressure, thereby raising air temperature to aid ignition of fuel injected in cylinder.
e. Using ether sprayed into the air intake to provide a more volatile and easily ignited fuel-air mixture. This method is not recommended on glow plug equipped engines as extreme engine damage can result.

ENGINE COOLING

To remove heat transferred from the combustion process to the engine cylinder and cylinder head, a cooling system must be provided. Two different systems are used which are air cooling and liquid cooling.

AIR COOLING. Engines may be cooled either by air flow around cooling fins cast as a part of the engine cylinder, and on some engines, cooling fins on the cylinder head. The air flow can be created by fan blades attached to the engine flywheel, or by a belt driven cooling fan. The air cooled engine is fitted with a cover, or shroud, to concentrate air flow from the cooling fan to the engine parts requiring cooling.

On air cooled engines, maintenance of the cooling system is extremely important for engines operating in dirty or trashy conditions. In time, the air intake screen or air passage between the engine cooling fins and cover or shrouding may be blocked with trash or dirt. If this happens, the air cooled engine will overheat and severe engine damage will occur.

LIQUID COOLING. Some engines are cooled by a liquid flowing through passages within the cylinder block and cylinder head. A system for removing heat from the cooling liquid must be used and several different methods are utilized.

The simplest method of removing heat from engine coolant is fitting a large jacket or hopper to the engine cylinder; heat is transferred from the coolant liquid through the water hopper material to surrounding air.

On many engines used for industrial purposes, the coolant, after picking up heat from the engine, is passed through a radiator. Many small tubes attached to two radiator manifold tanks carry liquid through the cooling fin area. The radiator has a fan pulling or pushing air through the cooling fins, and heat from coolant passing through the radiator tubes is transmitted to the air. On most engines, a pump is used to circulate the cooling liquid through the engine, to the radiator where heat is removed, and back to the engine. Some engines utilize "thermosiphon" cooling where circulation through the radiator and engine is created by percolation of the heated coolant.

Some engines for marine use are cooled from water pumped up through an inlet in the bottom of the marine vessel directly through the engine liquid cooling system and then the water is discharged back to source, normally some or all with the engine exhaust. On other marine liquid cooled engines, the water pumped through the bottom of the vessel passes through a heat exchanger where heat is removed from the engine's fresh water cooling liquid. The heat exchanger method is very well suited to uses where the marine vessel is operated in salt, or sea water. Water pumps used for this purpose have a rubber impeller which should be inspected and renewed, if necessary, on a periodic basis.

Where salt or sea water is used for engine cooling, the salt will tend to corrode metal parts quite rapidly. To reduce the corrosive effect, zinc anodes are installed in the sea water flow so that the chemical corrosion will be concentrated on the anodes rather than other engine or heat exchanger parts. The anodes should be inspected periodically and renewed whenever they are reduced to about 25 percent of their original size.

ENGINE TEMPERATURE CONTROL. Air cooled engines do not normally need any method of engine temperature control except by circulating fan speed changing with engine speed.

In liquid cooled engines, the engine coolant must be kept within a desired temperature range for efficient operation. Therefore, except those having a water hopper or thermosiphon system, most systems have a thermostat valve that stops or reduces flow of coolant through the engine until engine temperature reaches a level predetermined by design of the thermostat. If the thermostat fails, flow of engine coolant may either be blocked causing the engine to overheat, or flow may not be controlled at all causing the engine to operate at coolant temperature too low. In either case, engine will be damaged if operated with a defective thermostat. Some engine manufacturers recommend periodic replacement of the cooling system thermostat.

DIESEL FUEL

There are two universal diesel fuel classifications: API Number 1 and Number 2. The engine manufacturer will specify which fuel should be used in their engines. Ambient temperature in which the engine is operating may affect choice of fuel type. Number 1 diesel fuel will flow at a much lower temperature than Number 2 diesel fuel and will be recommended for some engines operating in very low ambient temperatures. However, in higher ambient and engine operating temperatures, Number 2 diesel fuel is needed for its lubricating quality and to reduce leakage past the fuel injection pump and injector elements. Distillate, heating oils or kerosene should never be substituted for diesel fuel.

Extreme care must be taken in the handling and storage of diesel fuel. Diesel fuel should never be stored in a container previously used to store gasoline. Diesel fuel storage tanks should be of a design to inhibit condensation within the tank; water is very damaging to the diesel engine fuel system. Sediment will remain in suspension in diesel fuel for a number of hours and may cause early clogging of the fuel filter element. The diesel engine fuel tank should be refilled at end of operation period and before storage so that there is no room for air to enter the tank and cause condensation.

TROUBLE-SHOOTING

The following are problems which may occur during operation of the engine and their possible causes.

1. Engine hard to start or will not start.
   a. Low cranking speed due to faulty starting system components.
   b. Incorrect oil viscosity (too heavy) for ambient temperature.
   c. No fuel in tank or incorrect grade of fuel.
   d. Fuel filter plugged with water or dirt.
   e. Air in fuel system.
   f. Injection pump timing incorrect.
   g. Injection pump control rack stuck or governor linkage broken.
   h. Faulty injector nozzle operation.
   i. Faulty injection pump or pump drive.
   j. Low engine compression.

2. Engine misses.
   a. Water in fuel system.
   b. Air in fuel system.
   c. Faulty injector nozzle.
   d. Faulty injection pump.

3. Engine lacks power.
   a. Fuel filter plugged or other restriction in fuel lines.
   b. Dirty air cleaner or other restriction in intake system.
   c. Restriction in exhaust system.
   d. Wrong grade of fuel.
   e. Injection pump timing retarded.
   f. Faulty injector pump or injector.
   g. Incorrect valve tappet clearance.
   h. High idle no-load speed too low.
   i. Low compression.

4. Engine knocks.
   a. Faulty injection timing.
   b. Foreign material in cylinder.
   c. Valve sticking and contacting piston.
   d. Worn engine bearings.
   e. Excessive crankshaft end play.
   f. Excessive clearance between piston and cylinder.

5. Blue exhaust smoke.
   a. Piston rings worn, stuck or broken.
   b. Cylinder bore worn.

6. White exhaust smoke.
   a. Water in fuel.
   b. Low cetane fuel.
   c. Incomplete combustion due to low compression or incorrect injection timing.

7. Black exhaust smoke.
   a. Engine overloaded.
   b. Air filter plugged.
   c. Faulty injection pump delivering excessive amount of fuel.
   d. Faulty injector nozzle.

8. Engine uses too much oil.
   a. Incorrect oil viscosity (too light) for ambient temperature.
   b. Restricted air intake.
   c. Restricted oil return passage in cylinder head.
   d. Worn pistons and rings.
   e. Worn valve guides or valve stem seals.

9. Engine overheating — water cooled.
   a. Faulty radiator cap.
   b. Radiator fins plugged.
   c. Faulty thermostat.
   d. Coolant level too low.
   e. Engine lubricating oil level too low.
   f. Water pump belt slipping.
   g. Engine overloaded.
   h. Incorrect injection timing.

10. Engine overheating — air cooled.
    a. Cooling air being recirculated.
    b. Fins of cylinder head or cylinder plugged.
    c. Cooling air inlet or outlet blocked.
    d. Lubricating oil level too low.
    e. Engine overloaded.
    f. Incorrect injection timing.

11. Lubricating oil pressure too low.
    a. Incorrect oil viscosity (too light) for ambient temperature.
    b. Low oil level.
    c. Oil diluted with fuel.
    d. Engine overheating.
    e. Engine oil cooler (if used) plugged.
    f. Insufficient output from oil pump or leaking connections.
    g. Relief valve not seating.
    h. Worn engine bearings.

OFF-SEASON STORAGE

A proper storage procedure can extend the life of equipment by preventing damage when the equipment is not used. Exact procedures for storing depend on the type of equipment, length of storage, time of year stored and storage location. To obtain satisfactory results, equipment storage must be coordinated with a regular maintenance program. The following outline lists procedures applicable for extended storage of most powered equipment.

ENTERING INTO STORAGE

If equipped with a liquid cooling system, do not attempt to drain water in an effort to prevent freezing. Water trapped in the cooling system can freeze and result in extensive damage. A good quality antifreeze should be added per manufacturer's instructions to provide protection against freezing, rust and corrosion damage. Run engine until normal operating temperature is attained to circulate the antifreeze throughout system.

Drain old oil from all regularly serviced compartments, such as engine crankcase, gearboxes, chain cases, etc., while oil is warm. Refill with new approved oil to the level recommended by the equipment manufacturer.

Clean and dry all exterior surfaces. Remove all accumulated dirt and repair any damaged surface. Paint exposed surfaces to prevent rust.

Clean all cooling air passages and straighten, repair or renew any part that would interfere with normal air flow. On liquid-cooled systems, be sure radiator and grilles are clean. On air-cooled systems, remove shrouds and deflectors necessary to inspect and clean all cooling air passages. Clean all grease, oil, dirt, leaves, etc., from all cooling system components.

Lubricate all normally serviced surfaces with approved oil or grease. If equipped with grease fittings, attempt to purge old grease from joint.

Inspect for worn or broken parts. Make necessary adjustments and repair all damage. Tighten all loose hardware.

Fuel should be either drained or treated with an approved stabilizer. All fuel should be drained from tank, filters, lines, pumps and carburetor unless specifically discouraged by manufacturer. Do not add fuel stabilizer to any fuel containing alcohol. Fuel containing alcohol will separate if permitted to sit for long and internal parts may be extensively damaged by corrosion. Some manufacturers recommend coating inside of tank with a small amount of oil to deter rusting in tank. Filters should be serviced initially and water traps should be serviced regularly while in storage.

Loosen all drive belts and remove pressure from friction drive components. Inspect and note condition of drive belts. If condition is questionable, a new belt should be installed when removing equipment from storage.

Install new filter elements. Some filters can be cleaned and serviced, but most should be installed new at this time.

Pour a small amount (usually 1 tablespoon) of oil into each cylinder of engine through spark plug hole. Crank engine with starter about 12 revolutions to distribute oil, then install spark plugs and reconnect spark plug wires.

Support all mounted assemblies, removing weight from lift arms and hydraulic systems (of models so equipped).

Retract all hydraulic cylinders, if possible. Coat exposed parts of cylinder rods and other exposed machined surfaces with grease to prevent rust.

Install protective caps at ends of all disconnected lines. Seal openings of exhaust, air intake, engine dipstick and crankcase breather tube.

Remove battery and store in a cool, dry place. Do not permit battery to freeze. Maintain battery fully charged, checking approximately every 30 days.

Block suspension to remove all weight from tires and suspension components. If unit is stored outside, wheels and tires should be stored in a cool, dark, dry location away from operating electrical equipment.

Store the equipment in a dry protected place. If necessary to store outside, cover equipment to prevent entrance of water, but do not seal tightly. Sealing may cause condensation and accelerate rusting.

REMOVING FROM STORAGE

Check for obvious damage to covering and equipment and repair as necessary.

Install tires and wheels. Inflate tires with correct amount of air. Remove any blocks used during storage.

Charge battery, then install in equipment making sure that battery is properly retained. Clean battery cables and battery posts, then attach cables to battery terminals.

Remove covers from exhaust, air intake, engine dipstick and crankcase breather tube. Remove any protective caps from lines disconnected during disassembly. Be sure ends are clean, then reconnect lines.

Check all filters. Install new filters, or clean and service existing filters as required.

Adjust all drive belts and friction drive components to correct tension as recommended by the manufacturer. New belts should be installed if condition is questionable.

Fill fuel tank with correct type of fuel. Check for leaks. Gaskets may dry up or carburetor needle valve may stick during storage. Repair any problems before attempting to start. Drain water traps and check condition of fuel filters.

Lubricate all surfaces normally lubricated with oil or grease. Check cooling passages for restrictions, such as insect, bird or animal nests. Check oil in all compartments — engine crankcase, gearboxes, chain cases, etc. — for proper level. Evidence of too much oil may indicate water settled below oil. Drain oil if contamination is suspected or if time of storage exceeds recommended time change interval. Fill to proper level with correct type of oil.

Fill cooling system to correct level on liquid-cooled models. Antifreeze solutions, which improve summer cooling ability and provide rust and corrosion protection, should be used. Check for leaks if fluid has leaked from system during storage.

SERVICE SHOP TOOL BUYER'S GUIDE

This listing of service shop tools is solely for the convenience of manual users and does not imply endorsement or approval by Intertec Publishing Corporation of the tools and equipment listed. The listing is in response to many requests for information on sources for purchasing special tools and equipment. Every attempt has been made to make the listing as complete as possible at time of publication and each entry is made from the latest material available.

Special engine service tools, such as drivers, pullers, gages, etc., which are available from the engine manufacturer, are not listed in this section of the manual. Where a special service tool is listed in the engine service section of this manual, the tool is available from the central parts or service distributors listed at the end of most engine service sections, or from the manufacturer.

NOTE TO MANUFACTURERS AND NATIONAL SALES DISTRIBUTORS OF ENGINE SERVICE TOOLS AND RELATED SERVICE EQUIPMENT. To obtain either a new listing for your products, or to change or add to an existing listing, write to Intertec Publishing Corporation, Book Division, P.O. Box 12901, Overland Park, KS 66212.

ENGINE SERVICE TOOLS

Ammco Tools, Inc.
Wacker Park
North Chicago, Illinois 60064
Valve spring compressor, torque wrenches, cylinder hones, ridge reamers, piston ring compressors, piston ring expanders.

Bloom, Inc.
Highway 939 West
Independence, Iowa 50644
Engine repair stand with crankshaft straightening attachment.

Brush Research Mfg. Co., Inc.
4642 East Floral Drive
Los Angeles, California 90022
Cylinder hones.

E-Z Lok
P.O. Box 2069
Gardena, California 90247
Thread repair insert kits.

Fairchild Fastener Group
3000 W. Lomita Blvd.
Torrance, California 90505
Thread repair insert kits (Keenserts) and installation tools.

Foley-Belsaw Company
Outdoor Power Equipment Parts Div.
6301 Equitable Road
P.O. Box 419593
Kansas City, Missouri 64141
Crankshaft straightener and repair stand, valve refacer, valve seat grinder, parts washer, cylinder hone, ridge reamer, piston ring expander and compressor, flywheel puller, torque wrench.

Frederick Manufacturing Corp.
4840 E. 12th Street
Kansas City, Missouri 64127
Piston groove cleaner, compression tester, piston ring expander and compressor, valve spring compressor, valve seat grinder, valve refacer, cylinder hone, flywheel puller, flywheel wrench, flywheel holder, starter spring rewinder, condenser pliers.

Heli-Coil
Shelter Rock Lane
Danbury, Connecticut 06810
Thread repair kits, thread inserts, installation tools.

K-D Tools
3575 Hempland Road
Lancaster, Pennsylvania 17604
Thread repair kits, valve spring compressors, reamers, micrometers, dial indicators, calipers.

Keystone Reamer & Tool Co.
South Front Street
P.O. Box 308
Millersburg, Pennsylvania 17061
Valve seat cutter and pilots, reamers, screw extractors, taps and dies.

Ki-Sol Corporation
100 Larkin Williams Ind. Ct.
Fenton, Missouri 63026
Cylinder hone, ridge reamer, ring compressor, ring expander, ring groove cleaner, torque wrenches, valve spring compressor, valve refacing equipment.

K-Line Industries, Inc.
315 Garden Avenue
Holland, Michigan 49424
Cylinder hone, ridge reamer, ring compressor, valve guide tools, valve spring compressor, reamers.

K.O. Lee Company
200 South Harrison
P.O. Box 1416
Aberdeen, South Dakota 57402

Kwik-Way Mfg. Co.
500 57th Street
Marion, Iowa 52302
Cylinder boring equipment, valve facing equipment, valve seat grinding equipment.

Lisle Corporation
807 East Main
Clarinda, Iowa 51632
Cylinder hones, ridge reamers, ring compressors, valve spring compressors.

Microdot, Inc.
P.O. Box 3001
Fullerton, California 92634
Thread repair insert kits.

Mighty Midget Mfg. Co.,
Div. of Kansas City Screw Thread Co.
2908 E. Truman Road
Kansas City, Missouri 64127
Crankshaft straightener.

Neway Manufacturing, Inc.
1013 N. Shiawassee
Corunna, Michigan 48817
Valve seat cutters.

OTC
655 Eisenhower Drive
Owatonna, Minnesota 55060
Valve tools, spark plug tools, piston ring tools, cylinder hones.

Power Lawnmower Parts, Inc.
1920 Lyell Avenue
P.O. Box 60860
Rochester, New York 14606-0860
Flywheel pullers, starter wrench, flywheel holder, gasket cutter tool, gasket scraper tool, crankshaft

cleaning tool, ridge reamer, valve spring compressor, valve seat cutters, thread repair kits, valve lifter, piston ring expander.

Precision Manufacturing & Sales Co., Inc.
2140 Range Road
Clearwater, Florida 34625
Cylinder boring equipment, measuring instruments, valve equipment, hones, hand tools, test equipment, threading tools, presses, parts washers, milling machines, lathes, drill presses, glass beading machines, dynos, safety equipment.

Sioux Tools, Inc.
2901 Floyd Blvd.
P.O. Box 507
Sioux City, Iowa 51102
Valve refacing and seat grinding equipment.

Sunnen Product Company
7910 Manchester Avenue
St. Louis, Missouri 63143
Cylinder hones, rod reconditioning, valve guide reconditioning.

TEST EQUIPMENT AND GAGES

AW Dynamometer, Inc.
P.O. Box 428
Colfax, Illinois 61728
Engine test dynamometer.

B.C. Ames Company
131 Lexington
Waltham, Massachusetts 02254
Micrometers, dial gages, calipers.

Dixson, Inc.
287 27 Road
Grand Junction, Colorado 81503
Tachometer, compression gage, timing light.

Foley-Belsaw Company
Outdoor Power Equipment Parts Div.
6301 Equitable Road
P.O. Box 419593
Kansas City, Missouri 64141
Cylinder gage, amp/volt testers, condenser and coil tester, magneto tester, ignition testers, tachometers, spark testers, compression gages, timing lights and gages, micrometers and calipers, carburetor testers, vacuum gages.

Frederick Manufacturing Corp.
4840 E. 12th Street
Kansas City, Missouri 64127
Ignition tester, tachometer, compression gage.

Graham-Lee Electronics, Inc.
4220 Central Avenue N.E.
Minneapolis, Minnesota 55421
Coil and condenser tester.

K-D Tools
3575 Hempland Road
Lancaster, Pennsylvania 17604
Diode tester and installation tools, compression gage, timing light, timing gages.

Ki-Sol Corporation
100 Larkin Williams Ind. Ct.
Fenton, Missouri 63026
Micrometers, telescoping gages, compression gages, cylinder gages.

K-Line Industries, Inc.
315 Garden Avenue
Holland, Michigan 49424
Compression gage, leakdown tester, micrometers, dial gages.

Merc-O-Tronic Instruments Corporation
215 Branch Street
Almont, Michigan 48003
Igition analyzers for conventional, solid-state and magneto systems, electric tachometers, electronic tachometer and dwell meter, power timing lights, ohmmeters, compression gages, mechanical timing devices.

OTC
655 Eisenhower Drive
Owatonna, Minnesota 55060
Feeler gages, hydraulic test gages.

Power Lawnmower Parts, Inc.
1920 Lyell Avenue
P.O. Box 60860
Rochester, New York 14606-0860
Compression gage, cylinder gage, magneto tester, compression gage, condenser and coil tester.

Prestolite Electronic Div.
An Allied Company
4 Seagate
Toledo, Ohio 43691
Magneto test plug.

Simpson Electric Company
853 Dundee Avenue
Elgin, Illinois 60120
Electrical and electronic test equipment.

L.S. Starrett Company
121 Crescent St.
Athol, Massachusetts 01331
Micrometers, dial gages, bore gages, feeler gages.

Stevens Instrument Company
P.O. Box 193
Waukegan, Illinois 60079
Igition analyzers, timing lights, volt-ohm meter, tachometer, spark checkers, CD ignition testers.

Stewart-Warner Corporation
580 Slawin Ct.
Mt. Prospect, Illinois 60056
Compression gage, ignition tachometer, timing light ignition analyzer.

SHOP TOOLS AND EQUIPMENT

AC Delco Division General Motors Corp.
3031 W. Grand Blvd.
P.O. Box 33115
Detroit, Michigan 48232
Spark plug tools.

Black & Decker Mfg. Co.
626 Hanover Pike
Hampstead, Maryland 21074
Electric power tools.

Champion Pneumatic Machinery Co.
1301 N. Euclid Avenue
Princeton, Illinois 61356
Air compressors.

Champion Spark Plug Company
P.O. Box 910
Toledo, Ohio 43661
Spark plug cleaning and testing equipment, gap tools and wrenches.

Chicago Pneumatic Tool Co.
2200 Bleecker St.
Utica, New York 13501
Air impact wrenches, air hammers, air drills and grinders, nut runners, speed ratchets.

Cooper Tools
P.O. Box 728
Apex, North Carolina 27502
Chain-type engine hoists and utility slings.

E-Z Lok
P.O. Box 2069
Gardena, California 90247
Thread repair insert kits.

Fairchild Fastener Group
3000 W. Lomita Blvd.
Torrance, California 90505
Thread repair insert kits (Keenserts) and installation tools.

Foley-Belsaw Company
Outdoor Power Equipment Parts

Div.
6301 Equitable Road
P.O. Box 419593
Kansas City, Missouri 64141
Torque wrenches, parts washers, micrometers and calipers.

Frederick Manufacturing Corp.
4840 E. 12th Street
Kansas City, Missouri 64127
Torque wrenches, gear pullers.

G&H Products, Inc.
P.O. Box 770
St. Paris, Ohio 43027
Equipment lifts.

General Scientific Equipment Co.
525 Spring Garden St.
Philadelphia, Pennsylvania 19122
Safety equipment.

Graymills Corporation
3705 N. Lincoln Avenue
Chicago, Illinois 60613
Parts washing equipment.

Heli-Coil
Shelter Rock Lane
Danbury, Connecticut 06810
Thread repair kits, thread inserts, installation tools.

Ingersoll-Rand
253 E. Washington Avenue
Washington, New Jersey 07882
Air and electric impact wrenches, electric drills and screwdrivers.

Jaw Manufacturing Co.
39 Mulberry St.
P.O. Box 213
Reading, Pennsylvania 19603
Files for repair or renewal of damaged threads, rethreader dies, flexible shaft drivers and extensions, screw extractors, impact drivers.

Jenny Division of Homestead Ind., Inc.
700 Second Avenue
Coraopolis, Pennsylvania 15108
Steam cleaning equipment, pressure washing equipment.

K-Line Industries, Inc.
315 Garden Avenue
Holland, Michigan 49424
Pullers, crack detectors, gloves, aprons, eyewear.

Keystone Reamer & Tool Co.
South Front Street
P.O. Box 308
Millersburg, Pennsylvania 17061
Adjustable reamers, twist drills, taps, dies.

Microdot, Inc.
P.O. Box 3001
Fullerton, California 92634
Thread repair insert kits.

OTC
655 Eisenhower Drive
Owatonna, Minnesota 55060
Bearing and gear pullers, hydraulic shop presses.

Power Lawnmower Parts, Inc.
1920 Lyell Avenue
P.O. Box 60860
Rochester, New York 14606-0860
Flywheel pullers, starter wrench.

Shure Manufacturing Corp.
1601 S. Hanley Road
St. Louis, Missour 63144
Steel shop benches, desks, engine overhaul stand.

Sioux Tools, Inc.
2901 Floyd Blvd.
P.O. Box 507
Sioux City, Iowa 51102
Air and electric impact tools, drills, grinders.

Sturtevant Richmont
3203 N. Wolf Rd.
Franklin Park, Illinois 60131
Torque wrenches, torque multipliers, torque analyzers.

MECHANIC'S HAND TOOLS

Channellock, Inc.
1306 South Main St.
P.O. Box 519
Meadville, Pennsylvania 16335

John H. Graham & Company
P.O. Box 739
Oradell, New Jersey 07649

Jaw Manufacturing Co.
39 Mulberry St.
P.O. Box 213
Reading, Pennsylvania 19603

K-D Tools
3575 Hempland Road
Lancaster, Pennsylvania 17604

K-Line Industries, Inc.
315 Garden Avenue
Holland, Michigan 49424

OTC
655 Eisenhower Drive
Owatonna, Minnesota 55060

Snap-On Tools
2801 80th Street
Kenosha, Wisconsin 53140

Triangle Corporation-Tool Division
P.O. Box 1807
Orangeburg, South Carolina 29115

SHOP SUPPLIES (chemicals, metallurgy products, seals, sealers, common parts items, etc.)

Clayton Manufacturing Co.
4213 N. Temple City Blvd.
El Monte, California 91731
Steam cleaning compounds and solvents.

E-Z Lok
P.O. Box 2069
Gardena, California 90247
Thread repair insert kits.

Eutectic+Castolin
40-40 172nd Street
Flushing, New York 11358
Specialized repair and maintenance welding alloys.

Fairchild Fastener Group
3000 W. Lomita Blvd.
Torrance, California 90505
Thread repair insert kits (Keenserts) and installation tools.

Foley-Belsaw Company
Outdoor Power Equipment Parts Div.
6301 Equitable Road
P.O. Box 419593
Kansas City, Missouri 64141
Parts washers, cylinder head rethreaders, micrometers, calipers, gasket material, nylon rope, oil and grease products, and bolt, nut, washer and spring assortments.

Frederick Manufacturing Corp.
4840 E. 12th Street
Kansas City, Missouri 64127
Thread repair kits.

Graymills Corporation
3705 N. Lincoln Avenue
Chicago, Illinois 60613
Parts cleaning fluids.

Heli-Coil
Shelter Rock Lane
Danbury, Connecticut 06810
Thread repair kits, thread inserts, installation tools.

K-Line Industries, Inc.
315 Garden Avenue
Holland, Michigan 49424
Machining lubricants, solvents, cleaning solutions.

Loctite Corporation
705 North Mountain Road
Newington, Connecticut 06111
Compounds for locking threads, retaining bearings and securing machine parts; sealants and adhesives.

Microdot, Inc.
P.O. Box 3001
Fullerton, California 92634
Thread repair insert kits.

Permatex Industrial
30 Tower Lane
Avon Park South
Avon, Connecticut 06001
Cleaning chemicals, gasket sealers, pipe sealants, adhesives, lubricants, thread locking and forming compounds.

Power Lawnmower Parts, Inc.
1920 Lyell Avenue
P.O. Box 60860
Rochester, New York 14606-0860
Grinding compound paste, gas tank sealer stick, shop aprons, eyewear.

Radiator Specialty Co.
1900 Wilkinson Blvd.
Charlotte, North Carolina 28208
Cleaning chemicals (Gunk) and solder seal.

ALLIS-CHALMERS

CONDENSED SPECIFICATIONS

	MODELS			
	616	620	718	720
Engine Make	Onan	Onan	Kohler	Onan
Model	CCKA	CCKB	K361QS	CCKB
Bore	3.25 in. (82.5 mm)	3.25 in. (82.5 mm)	3.75 in. (95.3 mm)	3.25 in. (82.5 mm)
Stroke	3 in. (76.2 mm)	3 in. (76.2 mm)	3.25 in. (82.5 mm)	3 in. (76.2 mm)
Displacement	49.8 cu. in. (816 cc)	49.8 cu. in. (816 cc)	35.9 cu. in. (589 cc)	49.8 cu. in. (816 cc)
Power Rating	16.5 hp. (12.3 kW)	19.5 hp. (14.5 kW)	18 hp. (13.4 kW)	19.5 hp. (14.5 kW)
Slow Idle	1200 rpm	1200 rpm	1800 rpm	1200 rpm
High Speed (No-Load)	3850 rpm	3850 rpm	3800 rpm	3850 rpm
Capacities –				
Crankcase	3.5 qts. (3.3 L)	3.5 qts. (3.3 L)	2 qts. (1.9 L)	3.5 qts. (3.3 L)
Hydraulic System	5.5 qts. (5.2 L)	5.5 qts. (5.2 L)	3 qts. (2.8 L)	5.5 qts. (5.2 L)
Transmission	See Hyd.	See Hyd.	See Hyd.	See Hyd.
Final Drive	0.5 qts. (0.47 L)	0.5 qts. (0.47 L)	....	0.5 qts. (0.47 L)
Bevel Gear Box	....	....	*	....
Fuel Tank	3.8 gal. (14.3 L)	3.8 gal. (14.3 L)	3 gal. (11.4 L)	3.8 gal. (14.3 L)

*Refer to **MAINTENANCE** paragraph in **BEVEL GEAR BOX** section.

	MODELS			
	816	818	917	919
Engine Make	B&S	B&S	Kohler	Kohler
Model	402707	422707	KT17	KT19
Bore	3.44 in. (87.4 mm)	3.44 in. (87.4 mm)	3.125 in. (79.4 mm)	3.125 in. (79.4 mm)
Stroke	2.16 in. (54.9 mm)	2.28 in. (57.9 mm)	2.75 in. (69.8 mm)	3.063 in. (78 mm)
Displacement	40 cu. in. (659 cc)	42.3 cu. in. (695 cc)	42.18 cu. in. (691.4 cc)	47 cu. in. (770 cc)
Power Rating	16 hp. (11.9 kW)	18 hp. (13.5 kW)	17 hp. (12.7 kW)	19 hp. (14.2 kW)
Slow Idle	1350 rpm	1350 rpm	1200 rpm	1200 rpm
High Speed (No-Load)	3600 rpm	3600 rpm	3600 rpm	3600 rpm
Capacities –				
Crankcase	3 pts. (1.4 L)	3 pts. (1.4 L)	3 pts. (1.4 L)	3 pts. (1.4 L)
Hydraulic System	....	....	3 qts. (2.8 L)	3 qts. (2.8 L)
Transmission	3.5 pts. (1.6 L)	3.5 pts. (1.6 L)	See Hyd.	See Hyd.
Bevel Gear Box	....	....	*	*
Fuel Tank	2.2 gal. (8.3 L)	2.2 gal. (8.3 L)	3 gal. (11 L)	3 gal. (11 L)

*Refer to **MAINTENANCE** paragraph in **BEVEL GEAR BOX** section.

FRONT AXLE AND STEERING SYSTEM

MAINTENANCE

All Models Except 720

It is recommended steering spindles, axle pivot and steering gear be lubricated at 25 hour intervals of normal operation and front wheel bearings should be cleaned and repacked at 100 hour intervals. Use multi-purpose, lithium base grease. Clean all fittings before and after lubrication. Check for any excessive looseness of parts, bearings, gears or tie rods and repair as necessary.

Model 720

It is recommended steering spindles, axle pivot and steering gear be lubricated at 50 hour intervals of normal operation and front wheel bearings should be cleaned and repacked at 400 hour intervals. Use multi-purpose, lithium base grease. Clean all fittings before and after lubrication. Check for any excessive looseness of parts, bearings, gears or tie rods and repair as necessary.

AXLE MAIN MEMBER

Models 616-620-720

To remove axle main member (6 – Fig. AC1) assembly from tractor, support front of tractor and disconnect drag link (21) from pitman arm at steering gear box. Remove pivot bolts (1 and 10). Lower axle assembly and roll out from under tractor. Renew bushings (3, 5 and 16) and bolts (1 and 10) as necessary.

Reinstall by reversing removal procedure.

Models 816-818

To remove axle main member (6 – Fig. AC2) assembly from tractor, support front of tractor and disconnect drag link at left spindle arm. Remove pivot bolt (2), pull axle assembly forward out of center pivot and remove axle. Renew bushing (11) and bolt (2) as necessary.

Reinstall by reversing removal procedure.

Models 718-917-919

To remove axle main member (1 – Fig. AC3) assembly from tractor, support front of tractor and disconnect drag link (12) at left steering arm (14). Remove pivot bolt (4), pull axle assembly forward out of center pivot and remove axle. Renew bushing (2) and bolt (4) as necessary.

Reinstall by reversing removal procedure.

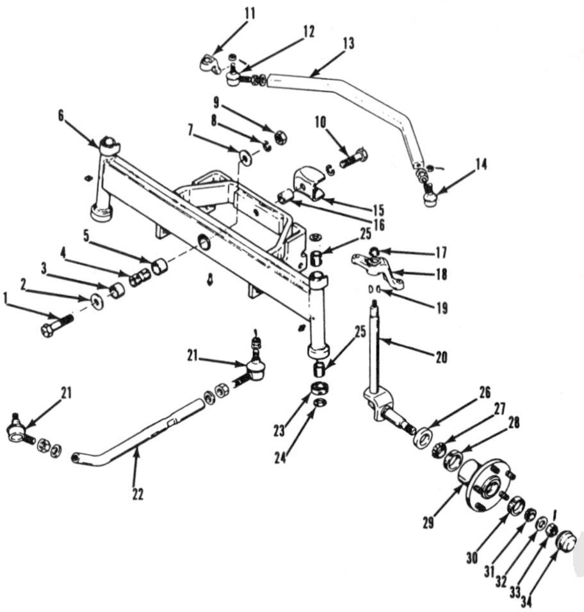

Fig. AC1 – Exploded view of front axle used on 616, 620 and 720 models.

1. Front pivot belt
2. Washer
3. Pivot bushing
4. Spacer
5. Pivot bushing
6. Main member
7. Washer
8. Lockwasher
9. Nut
10. Rear pivot bolt
11. Right steering arm
12. Tie rod end
13. Tie rod
14. Tie rod end
15. Clevis
16. Spacer
17. Nut
18. Left steering arm
19. Keys
20. Spindle
21. Drag link ends
22. Drag link
23. Thrust bearing
24. Spacer
25. Bushings
26. Seal
27. Bearing
28. Bearing cup
29. Hub
30. Bearing cup
31. Bearing
32. Washer
33. Nut
34. Dust cap

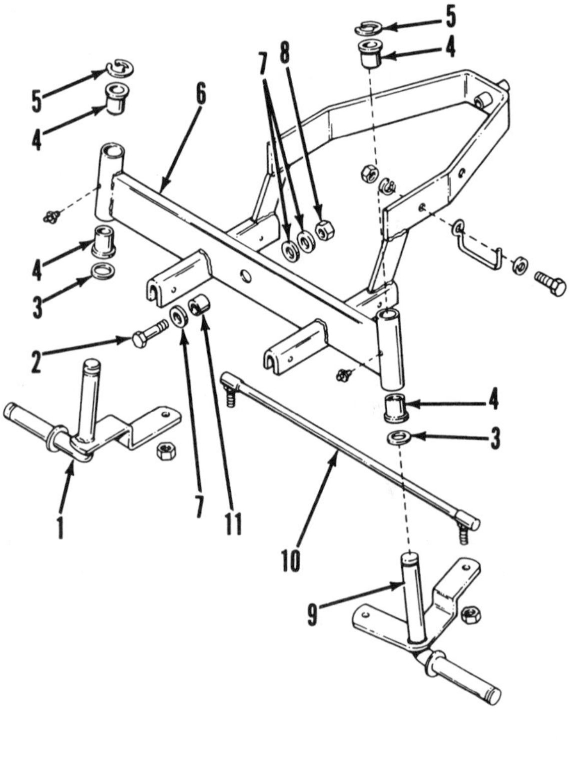

Fig. AC2 – Exploded view of front axle used on 816 and 818 models.

1. Spindle
2. Front pivot bolt
3. Washer
4. Bushings
5. Retaining ring
6. Main member
7. Washers
8. Nut
9. Spindle
10. Tie rod
11. Spacer

TIE ROD AND TOE-IN

All Models

On models equipped with renewabl tie rod ends, toe-in should be adjusted b lengthening or shortening tie rod unti

1/16 to 1/8-inch (1.588 to 3.175 mm) toe-in is obtained.

STEERING SPINDLES

Models 616-620-720

To remove spindles (20–Fig. AC1), raise and support front of tractor. Remove dust cap (34), cotter key and nut (33). Remove wheel, hub assembly and bearings. Remove nut (17), steering arms (11 or 18), keys (19) and spindle (20). Renew bushings (25) and spindle (20) as necessary.

Reinstall by reversing removal procedure.

Models 816-818

To remove spindle (1 or 9–Fig. AC2), raise and support front of tractor. Remove front wheel and tire. Disconnect tie rod (10) from both sides and drag link from left side. Remove snap ring (5) and spindle. Renew bushings (4), washers (3) and spindles (1 or 9) as necessary.

Models 718-917-919

To remove spindles (5 and 16–Fig. AC3), raise and support front of tractor. Remove dust cap (26), loosen setscrew (24), then remove collar (25) and washer (23). Remove wheel, hub, bearings and tie rod (11). On left side, disconnect drag link (12), remove the two square head setscrews (15) and use a suitable puller to remove steering arm (14). Remove the two keys (13) and left spindle. On right side, remove snap ring (9) and spindle. Renew washers (6 and 8), bushings (7) and spindles as necessary.

Reinstall by reversing removal procedure.

FRONT WHEEL BEARINGS

Models 616-620-720

To remove front wheel bearings, raise and support side to be serviced. Remove dust cap (34–Fig. AC1), cotter key and nut (33). Remove washer (32), bearing (31) and hub (29). Remove seal (26) and inner bearing (27) from hub. Renew bearing cups (28 and 30) in hub as necessary.

Clean all parts thoroughly and inspect for wear or damage. Repack bearings using a good quality multi-purpose, lithium base grease. Apply coating of grease on spindle and in hub.

When reinstalling, bearing cups (28 and 30) must be squarely seated in hub. Place inner bearing (27) in cup (28), then install seal (26). Install hub assembly on spindle. Install outer bearing (31), washer (32) and nut (33). Tighten until a slight preload is noted on bearing as wheel is being rotated. Install cotter key to lock nut in this position. Install dust cap (34).

Models 816-818

To renew wheel bearing bushings, raise and support front of tractor. Remove wheel cover. Remove snap ring and washer from end of spindle. Remove wheel. Inner and outer bushings may be renewed as necessary. If bushings are loose in wheel, repair or renew wheel. Apply coating of good quality multi-purpose, lithium base grease to spindle and bushings.

Reinstall by reversing removal procedure.

Models 718-917-919

To remove front wheel bearings, raise and support front of tractor. Remove dust cap (26–Fig. AC3). Loosen setscrew (24), then remove collar (25) and washer (23). Remove outer bearing (22) and wheel and hub assembly. Remove seal (18) and inner bearing (19) from wheel hub. Remove spacer (17) from spindle. Renew bearing cups (20 or 21), located in wheel hub, as necessary.

Clean all parts thoroughly and inspect for wear or damage. Repack bearings using a good quality multi-purpose, lithium base grease. Apply coating of grease to spindle and in hub.

When reinstalling, seat bearing cups squarely in hub. Place inner bearing (19) in cup in hub. Install seal (18). Install spacer (17) on spindle, then install wheel, hub and bearing assembly. Install outer bearing (22) and washer (23). Install collar (25). Push collar against washer until all end play is just removed from bearings. Tighten setscrew (24). Install dust cap (26).

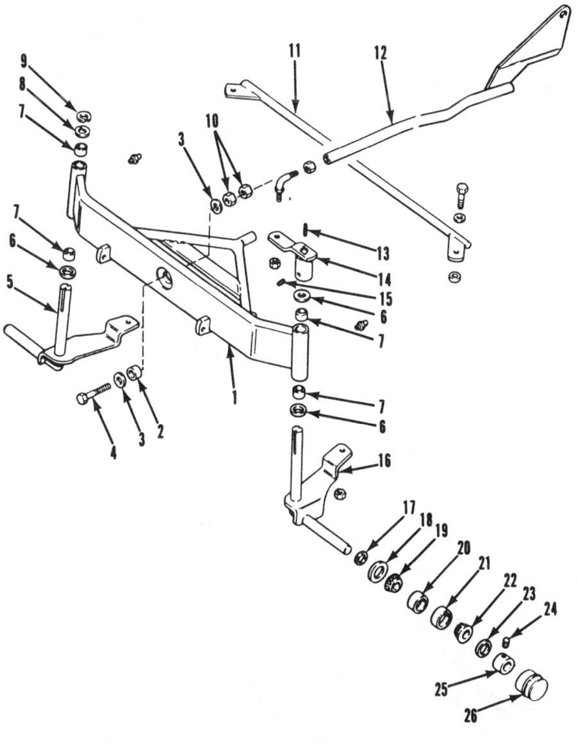

Fig. AC3—Exploded view of front axle used on 718, 917 and 919 models.

1. Main member
2. Spacer
3. Washer
4. Pivot bolt
5. Spindle
6. Washer
7. Bushings
8. Washer
9. Retaining ring
10. Nut
11. Tie rod
12. Drag link
13. Key
14. Steering arm
15. Setscrew
16. Spindle
17. Spacer
18. Seal
19. Bearing
20. Bearing cup
21. Bearing cup
22. Bearing
23. Washer
24. Setscrew
25. Retaining collar
26. Dust cap

STEERING GEAR

Models 616-620-720

R&R AND OVERHAUL. To remove the rotating ball nut and sector type steering gear box, disconnect battery, remove oil cooler, rear engine shroud and fuel tank. Disconnect clutch rod. Unbolt pto clutch and driveshaft assembly. Remove steering wheel. Disconnect choke and throttle cables. Remove instrument panel. Remove roll pin in steering shaft "U" joint and pull shaft up into support tube. Loosen clutch pivot arm and tilt it away from frame. Remove shims from pivot support assembly.

NOTE: When reinstalling pivot, shims must be placed in their original positions.

Remove nut from pitman shaft and pitman arm. Unbolt steering gear case from tractor frame. Remove "U" joint from steering gear. Lift assembly from frame.

To disassemble steering gear assembly, clamp gear housing in a vise with worm shaft horizontal. Determine center of travel on worm shaft by rotating shaft from stop to stop. Place a container in position to catch lubricant from case as cover is removed. Remove side cover plate cap screws. Tap pitman shaft and cover out of gear case. Remove locknut (17 – Fig. AC4) and adjuster plug (16) with lower bearing (15) and bearing retainer (14) from gear case. Pull worm shaft (12) with ball nut (13) from case through lower end. Remove upper bearing (11) from case. Remove ball guides (18) from nut and allow the 48 steel balls to drop into a container. Remove ball nut from worm shaft.

Clean all parts thoroughly and inspect all bearings, cups, worm shaft and ball nut for wear or damage. Bearing cup for bearing (15) and adjusting plug (16) are serviced as an assembly only. Renew pitman shaft needle bearing (9) and seal (8) as needed. Use seal protectors when installing shafts through seals and lubricate all bearings prior to installation.

To reassemble ball nut, place nut as near center of worm shaft as possible with nut holes aligned with races in worm shaft. Install ball guides (18) in ball nut. Drop 24 steel balls into each circuit of ball nut while oscillating the nut to ensure proper seating of balls in worm shaft races. Apply heavy coating of multi-purpose, lithium base grease to worm shaft races and rotate ball nut assembly on worm shaft until steel balls are well lubricated. Install upper shaft seal (7) in case (6). Install worm shaft and ball nut assembly into case. Install bearing and retainer in adjusting plug and install assembly into case. Preload bearings until 3 to 6 in.-lbs. (0.339 to 0.678 N·m) force is required to rotate worm shaft. Install locknut and tighten while holding adjustment plug from turning. Pack 9 ounces (266 mL) of multi-purpose, lithium base grease into gear case. Place seal protector on splines of pitman shaft and install shaft in case so center tooth of sector gear enters center groove of ball nut. Install side cover (3) and gasket (4). Rotate lash adjustment screw counter-clockwise to allow cover to seat properly, then install side cover cap screws and tighten. Center ball nut on worm shaft and turn lash adjustment screw clockwise to remove all sector to ball nut lash without binding. Tighten adjuster screw locknut while holding screw in this location. Ball nut should travel through entire cycle with no binding or roughness.

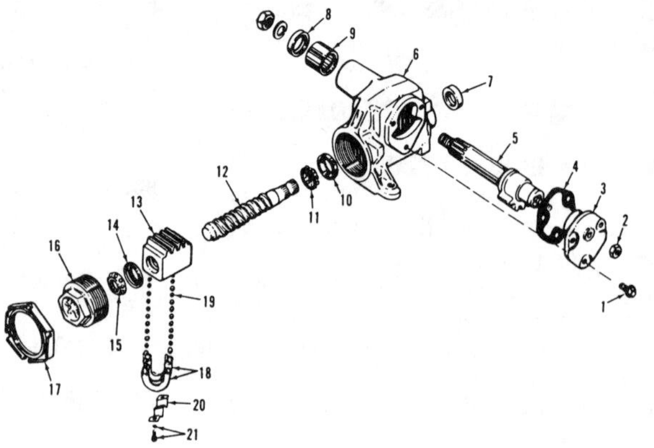

Fig. AC4 – Exploded view of steering gear assembly used on 616, 620 and 720 models.

1. Side cover bolts
2. Lash adjuster locknut
3. Side cover & needle bearing assy.
4. Side cover gasket
5. Pitman shaft & lash adjuster
6. Steering gear housing
7. Worm shaft seal
8. Pitman shaft seal
9. Pitman shaft needle bearing
10. Bearing race
11. Bearing (upper)
12. Worm shaft
13. Ball nut
14. Bearing race
15. Bearing (lower)
16. Adjuster plug
17. Adjuster plug locknut
18. Ball guides
19. Balls
20. Ball guide clamp
21. Clamp screw & washer

Reinstall by reversing removal procedure.

Models 816-818

R&R AND OVERHAUL. To remove steering gear, remove steering wheel, upper dash, fuel tank and lower dash. Disconnect drag link (11 – Fig. AC5) from steering rod (10). Unbolt and remove upper steering plate (6) and steering shaft (8). Remove lower plate (9) and steering rod (10) as an assembly. Remove the two set screws (13) and use a suitable puller or press to remove steering gear (12) from steering rod (10).

Clean and inspect all parts. Renew bushings (5) as necessary.

To reassemble, install key (14) with rounded end toward bent end of steering rod (10) until top edge of gear is flush with end of key (14). Steering rod should extend approximately 1/4-inch (6 mm) above steering gear. Install assembly on lower plate (9). Place spacer (7) on steering shaft (8) with bevel towards steering gear. Center steering gear on lower plate and install steering shaft so hole in shaft faces front and back.

Reassemble by reversing disassembly procedure.

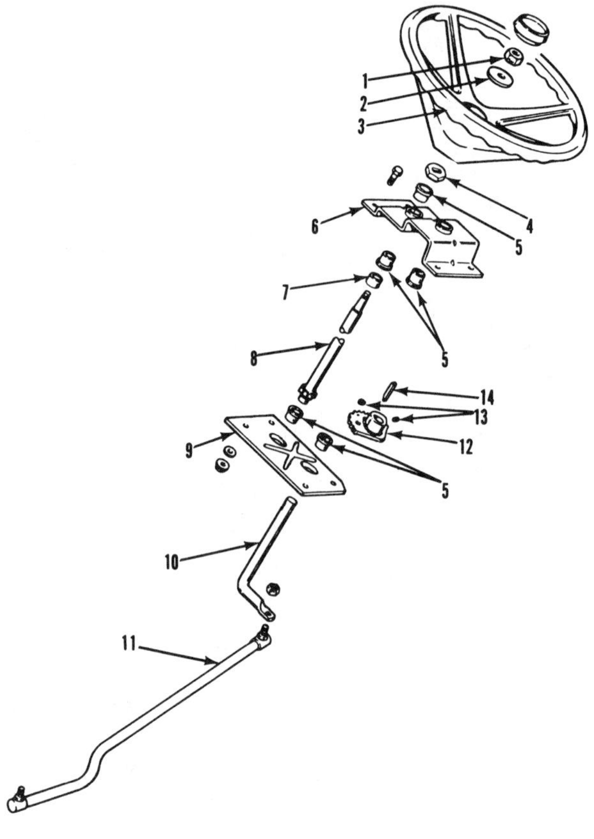

Fig. AC5 – Exploded view of steering shaft and mechanism used on 816 and 818 models.

1. Locknut
2. Washer
3. Steering wheel
4. Hex washer
5. Bushings
6. Upper steering plate
7. Spacer
8. Steering shaft
9. Lower steering plate
10. Steering rod
11. Drag link
12. Steering gear
13. Setscrews
14. Key

Models 718-917-919

R&R AND OVERHAUL. To remove steering gear, remove steering wheel and battery. Disconnect drag link from steering arm (13–Fig. AC6) and turn steering arm to allow access to mounting bolts. Remove mounting bolts and move steering gear assembly forward until casting lug clears edge of frame opening and lower entire assembly.

To disassemble steering gear, clamp support casting in a vise and remove locknut and washer from steering arm (13). Use a plastic mallet to separate steering arm (13) from bevel gear (11) and pull steering arm out of casting (8). Position steering shaft (3) in a vise. Remove retaining ring (10) and use a suitable puller to remove pinion gear (9). Remove key from shaft and slide shaft out of bushing. Remove adjusting plate (6) and bushing (7). Use a bearing puller to remove needle bearings from casting.

Clean and inspect all parts for wear or damage. During reassembly, press needle bearings into each end of bore with end of bearing with manufacturers name facing outward. Bearing at gear end must be 1/8-inch (3 mm) below surface of casting and bearing at steering arm end must fit flush with casting. Install bushing (7) and adjusting plate assembly, then tighten adjusting plate retaining cap screws until approximately 1/64-inch (0.4 mm) of bushing is above casting surface.

Reinstall by reversing removal procedure.

ENGINE

MAINTENANCE

All Models

Regular engine maintenance is required to maintain peak performance and long engine life.

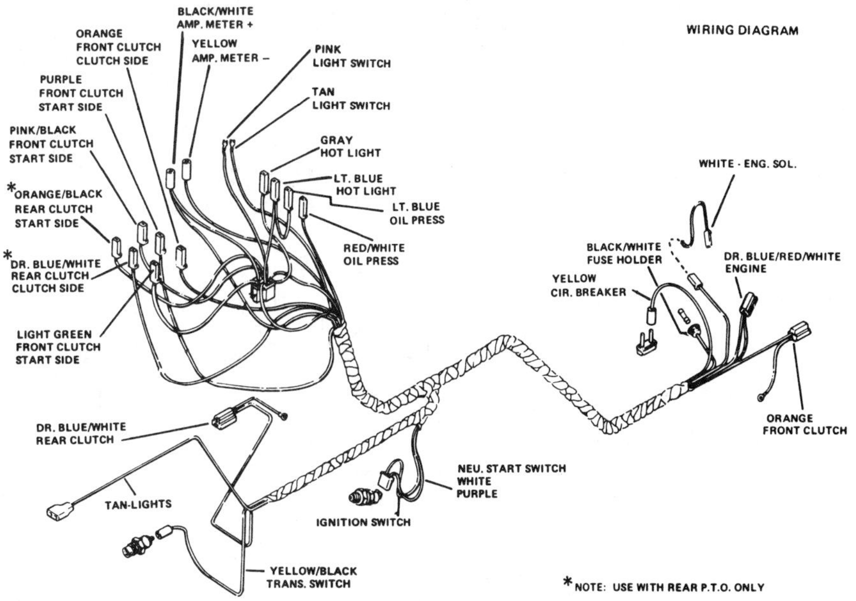

Fig. AC7 — View of typical 616 models wiring harness. Wiring may vary slightly when electric options are added.

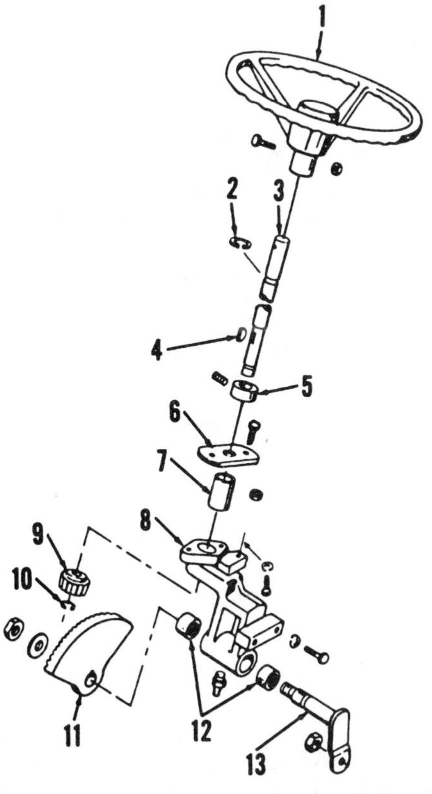

Fig. AC6 — Exploded view of steering shaft and mechanism used on 718, 917 and 919 models.

1. Steering wheel
2. Retaining ring
3. Steering shaft
4. Key
5. Retaining collar
6. Steering plate
7. Bushing
8. Casting
9. Pinion
10. "E" ring
11. Bevel gear
12. Needle bearings
13. Steering arm assy.

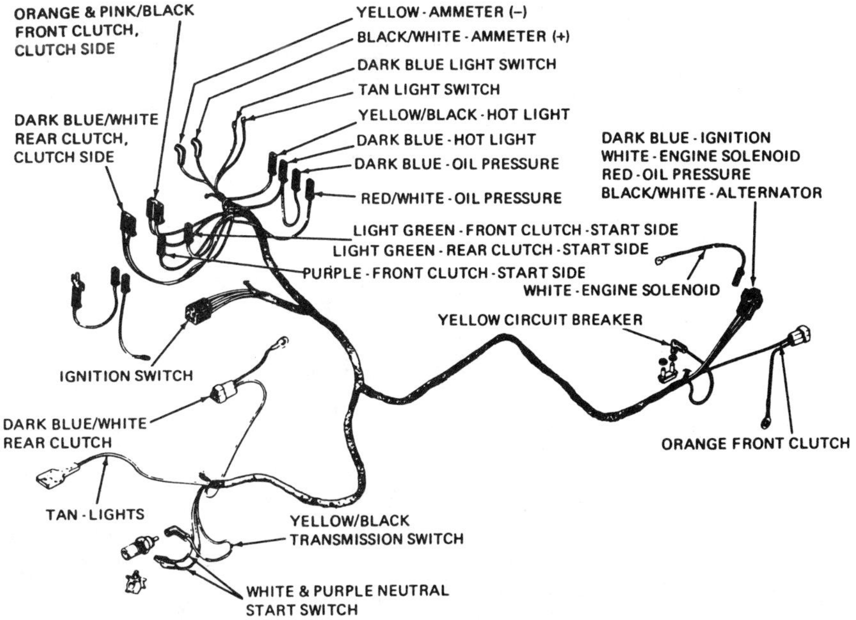

Fig. AC8 — View of typical 620 models wiring harness. Wiring may vary slightly when electric options are added.

Check oil level and clean air intake screen at 5 hour intervals under normal operating conditions.

Clean engine air filter at 25 hour intervals and clean crankcase breather, governor linkage and engine cooling fins at 200 hour intervals.

Change engine oil and filter, perform tune-up, valve adjustment and clean carbon from cylinder heads as recommended in appropriate engine section in this manual.

REMOVE AND REINSTALL

Models 616-620-720

Remove hood and battery, then disconnect throttle and choke cables.

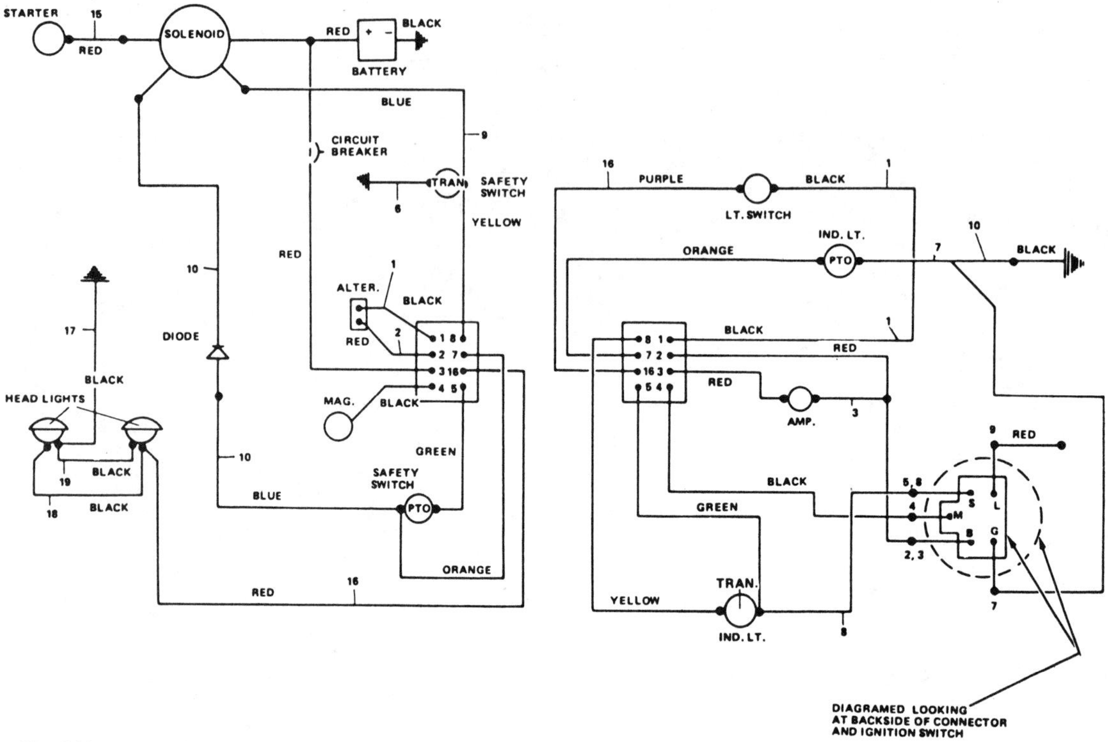

Fig. AC9—View of typical 816 and 818 models wiring diagram. Wiring may vary slightly when electric options are added.

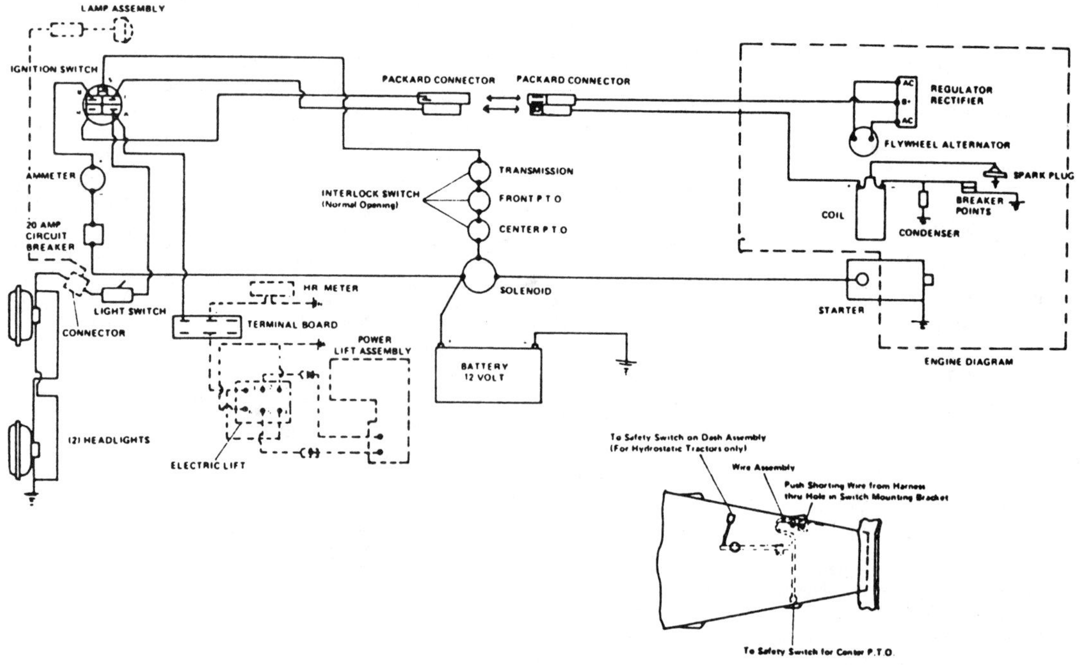

Fig. AC10—View of typical 718, 917 and 919 models wiring diagram. Wiring may vary slightly when electric options are added.

Disconnect electrical connections and fuel lines. Remove oil cooler and shrouds, as equipped. If equipped with pto, remove pto yoke from engine flywheel. Depress clutch pedal and remove clutch belts from engine pulley. Disconnect clutch rod and lower clutch assembly. Remove bottom cover attached to tractor frame and remove engine mounting bolts. Remove engine.

Reinstall by reversing removal procedure.

Models 816-818

Disconnect battery ground cable. Disconnect fuel line, throttle and choke cables and all necessary electrical wiring. Remove drive belts and engine mounting bolts. Remove engine.

Reinstall by reversing removal procedure.

Models 718-917-919

Disconnect fuel line from fuel pump and cap opening. Disconnect battery ground cable, all necessary electrical wires and choke and throttle cables. Disconnect driveshaft from engine flywheel. Remove engine mounting bolts and remove engine.

Reinstall by reversing removal procedure.

OVERHAUL

All Models

Engine make and model are listed at the beginning of this section. To overhaul engine or accessories, refer to appropriate engine section in this manual.

ELECTRICAL SYSTEM

MAINTENANCE AND SERVICE

All Models

Battery electrolyte level should be checked at 25 hour intervals of normal operation. If necessary, add distilled water until level is just below base of vent well. **DO NOT** overfill. Keep battery posts clean and cable ends tight.

For alternator or starter service, refer to appropriate engine section in this manual.

Refer to Fig. AC7, AC8, AC9 or AC10 for a view of typical wiring diagram. Wiring may vary slightly according to optional equipment installed.

BRAKE

ADJUSTMENT

Models 616-620-720

Unbolt and remove frame top cover. Loosen jam nuts (6 and 8 – Fig. AC12) at each end of turnbuckles (7) and adjust turnbuckles until brake pedal free travel is 1-1/4 inches (31.75 mm) for each pedal. Brakes must be adjusted equally so both brakes are activated at the same time when pedals are locked together. Tighten jam nuts and reinstall frame cover.

Models 816-818

Pull forward on brake band located to the inside of left rear tire to remove all slack. Make certain brake pedal is in full "up" position and measure gap (E – Fig. AC13) between spacer (C) and brake band (B). Gap (E) should be 5/8 to 3/4-inch (16 to 19 mm). Adjust by tightening or loosening brake rod nut (D) as necessary. A minimum of two full threads must extend beyond nut to properly secure nut on rod.

Models 718-917-919

Loosen jam nut on front end of parking brake rod, then turn handle and rod end until parking brake is tight when brake handle is against fender in brake lock position. With parking brake engaged and foot pedal released, adjust

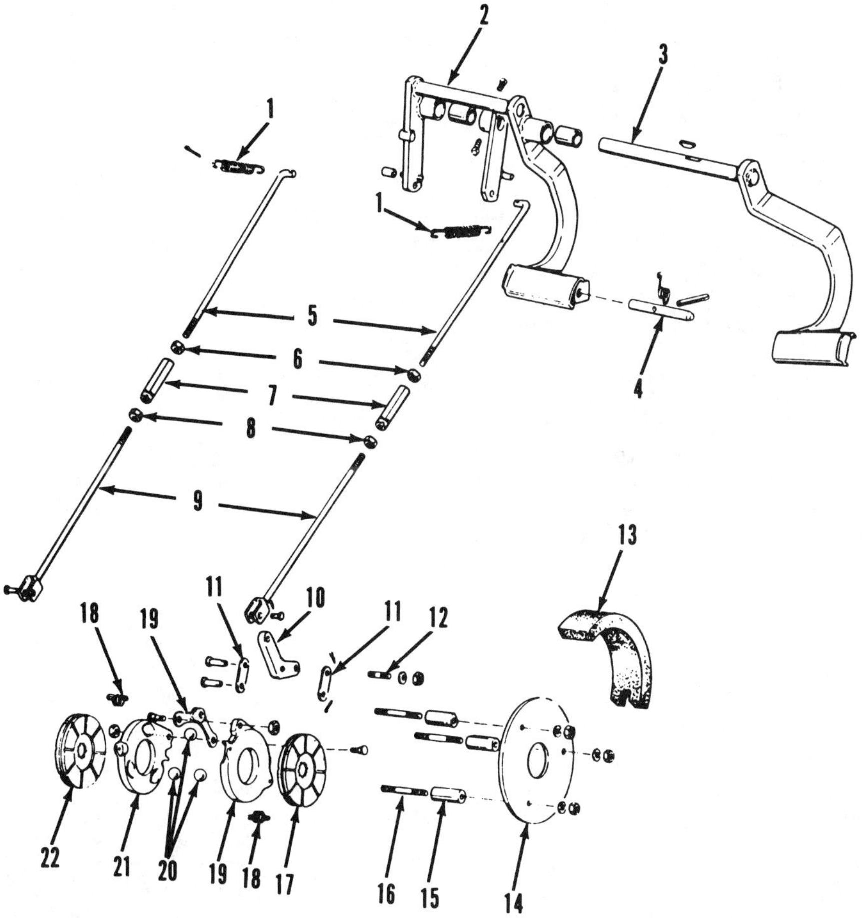

Fig. AC12 – Exploded view of brake mechanism used on 616, 620 and 720 models.

1. Return springs
2. Left lever assy.
3. Right lever assy.
4. Pedal lock rod
5. Front brake rod
6. Jam nut
7. Turnbuckle
8. Jam nut
9. Rear brake rod
10. Brake lever
11. Link plates
12. Stud
13. Brake pad
14. Disc assy.
15. Spacer
16. Stud
17. Brake disc
18. Spring
19. Actuating disc
20. Steel balls
21. Actuating disc
22. Brake disc

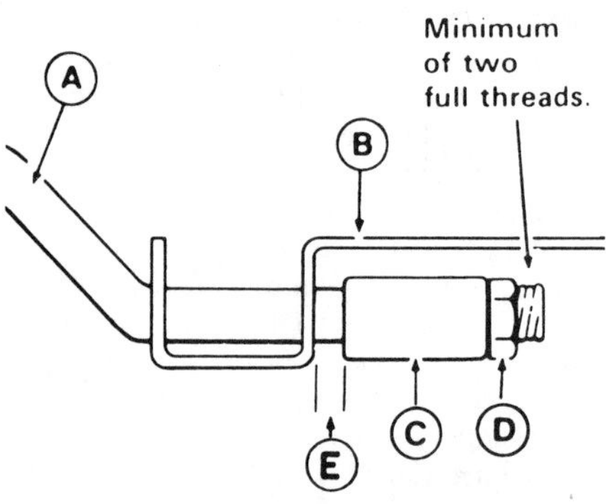

Fig. AC13—Adjust brake on 816 and 818 models by loosening or tightening locknut (D) on brake rod (A) until gap (E) between spacer (C) and band (B) is ⅝ to ¾-inch (16 to 19 mm). Refer to text.

jam nuts on front of brake rod to obtain ¾-inch (19 mm) clearance between jam nut and rod guide (Fig. AC14).

R&R BRAKES

Models 616-620-720

Raise and support rear of tractor. Remove rear tires, wheels and wheel guards. Remove cap screws retaining each drop housing to axle extension, slide drop housing away from axle extension until pinion shaft clears coupling and remove housing. Disconnect hydraulic lift cylinder from left axle extension. Remove lift shaft assembly, drawbar assembly and transmission filter. Remove the four cap screws retaining axle extensions to transmission case and the hex nuts and lockwashers from stationary brake disc studs. Remove axle extension and coupling. Remove outer stationary disc (14–Fig. AC12), differential output shaft with outer rotating disc (17) and retaining ring, spacers (15) and pin attaching link plate to yoke link assembly. Remove disc actuating assembly and inner disc (22). Remove springs (18) from actuating assembly. Remove oil seals from differential supports.

Clean and inspect all parts. Steel balls (20) should be smooth and round and ramp area of actuating plates (19 and 21) should be smooth with no grooves, pits or worn spots. Renew discs (17 and 22) and actuating plates (19 and 21) if worn, cracked or scored.

Install new oil seals in differential supports and install supports in transmission case. Install inner disc (22). Install the two spacers (15) with flat edges toward yoke link assembly on studs (16). Install spacer (15) which is round, on remaining stud. Reassemble acuating plates and install assembly in brake housing. Connect yoke link assembly and actuating link to link plate using a pin and cotter pin. Install retaining ring and outer rotating disc on differential output shaft, lubricate splines and carefully install shaft. Do not damage oil seal. Install outer stationary disc (14) and axle extension.

NOTE: If axle extension does not have a drain hole in bottom area, drill a ⅜-inch hole to allow water and any oil leakage to drain.

Install coupling on differential shaft, spring on pinion shaft and install drop housing. Reinstall by reversing removal procedure. Adjust brakes as previously outlined.

Models 718-816-818-917-919

Brake band and drum are located on left side of transmission and renewal procedure is obvious after examination. Adjust brakes as previously outlined.

CLUTCH

MAINTENANCE

All Models

Clutch belts should be checked and adjusted at 100 hour intervals or whenever it is suspected drive belts may be slipping. Periodically check all belts, stops, idler pulley, drive pulley and linkage for wear, looseness or damage.

ADJUSTMENT

Models 616-620-720

Unbolt hydraulic oil cooler and raise left end of cooler to expose clutch belt tension adjuster. Make certain clutch pedal is in engaged (up) position. Refer to Fig. AC15 and adjust plug at top of spring until spring length is 8 inches (203 mm). Adjust belt stop on right side of tractor so there is a clearance of 1/16 to ⅛-inch (1.588 to 3.175 mm) between belt stop and belts when clutch is engaged. Unbolt and remove cover from bottom of frame. Adjust locknut on clutch rod until clutch pedal free travel is 1½ to 1¾ inches (38 to 45 mm). Fully depress clutch pedal and check clearance between engine pulley (2) and driveshaft pulley (1). Clearance should be 1/16-inch (1.588 mm). If not, remove plug and belt

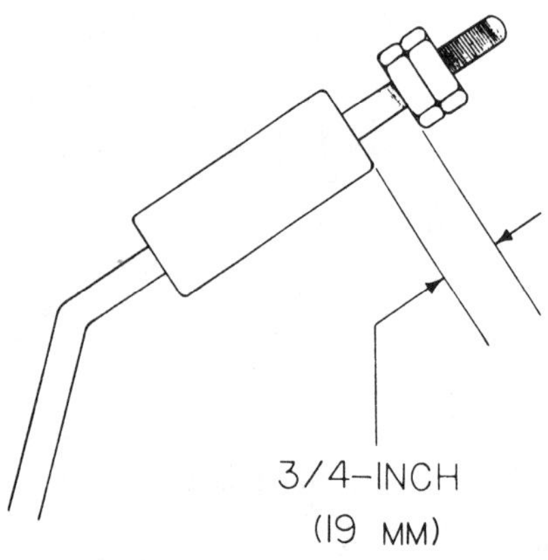

Fig. AC14—When adjusting brake on 718, 917 and 919 models, gap between brake rod guide and adjustment nuts should be ¾-inch (19 mm). Refer to text.

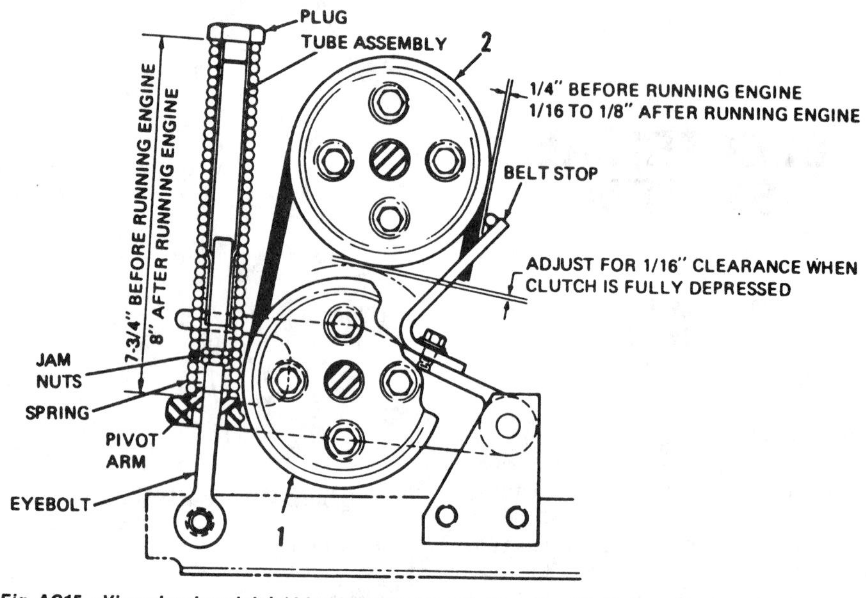

Fig. AC15—View showing clutch (drive) belt tension adjustment and belt stop adjustment on 616, 620 and 720 models. With clutch pedal depressed, a clearance of 1/16-inch (1.6 mm) should exist between engine crankshaft pulley (2) and drive shaft pulley (1) Refer to text for adjustment procedure.

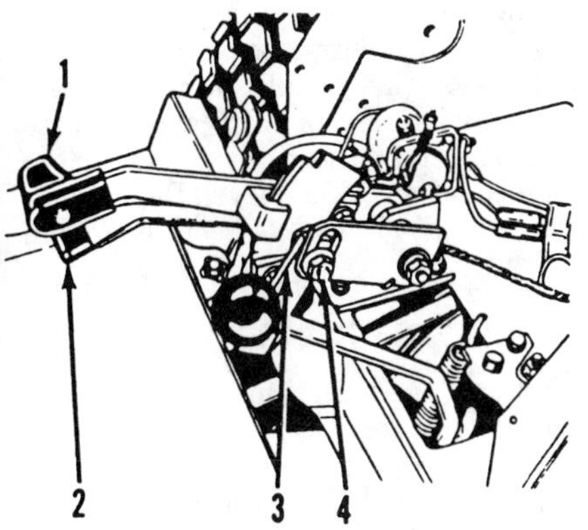

Fig. AC16—View of variable speed control linkage used on 816 and 818 models. Refer to text.

1. Control lever
2. Handle
3. Bar
4. Shoulder bolt

tension spring and adjust jam nuts on eyebolt until correct clearance is obtained. Readjust belt tension and clutch pedal free travel. Bolt oil cooler in place and install bottom frame cover.

Models 816-818

Variable speed control and clutch are adjusted together. Place speed control lever up in high speed position, then loosen shoulder bolt (4–Fig. AC16). With transmission in neutral, start engine, depress clutch pedal, set parking brake and stop engine. Unlatch parking brake and allow pedal to come up slowly, then measure distance from pedal shaft to forward edge of foot rest. Distance should be 5½ inches (140 mm); if not, adjust nut (2–Fig. AC17) towards spring to increase measurement. Place speed control lever down in low speed position, then pull upper handle (1–Fig. AC16) upward and hold to lock lever in position. Push bar (3) down and tighten shoulder bolt.

Models 718-917-919

Jam nuts on clutch rod should be adjusted so spring length is ½-inch (12.7 mm) between washers when clutch-brake pedal is in "up" position.

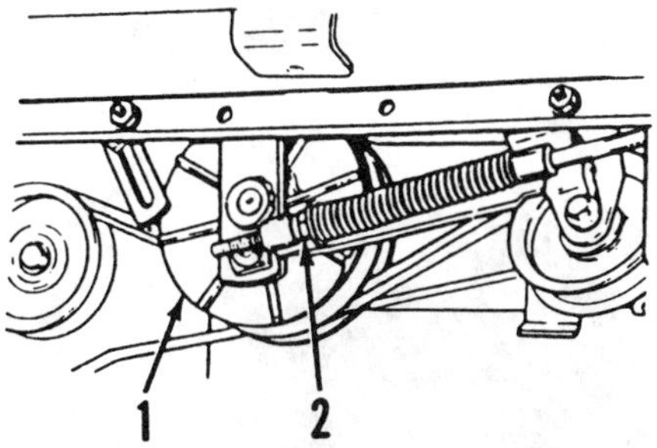

Fig. AC17—Variable speed pulley (1) clutch rod used on 816 and 818 models. Adjusting nut (2) is used to adjust clutch. Refer to text.

HYDROSTATIC TRANSMISSION

MAINTENANCE

Models 616-620-720

Hydrostatic transmission, tractor hydraulic system and three speed gear transmission all use oil from the three speed gear type transmisison housing. To check fluid level on early 616 models, open check cock (Fig. AC18) counterclockwise two turns or until fluid runs out. If no fluid runs out of check cock, remove filler plug (Fig. AC18) and add Dexron Automatic Transmission Fluid, or equivalent, until oil drips from check cock. Install filler plug and tighten. Close check cock. To check fluid level on all other models, a dipstick is located as shown in Fig. AC19. Fluid should be maintained at the "FULL" mark on dipstick.

On all models, transmission oil should be checked and oil cooler cleaned at five hour intervals. Change fluid and filter at 400 hour intervals. System oil capacity is 5.5 quarts (5.2 L) when filter is changed.

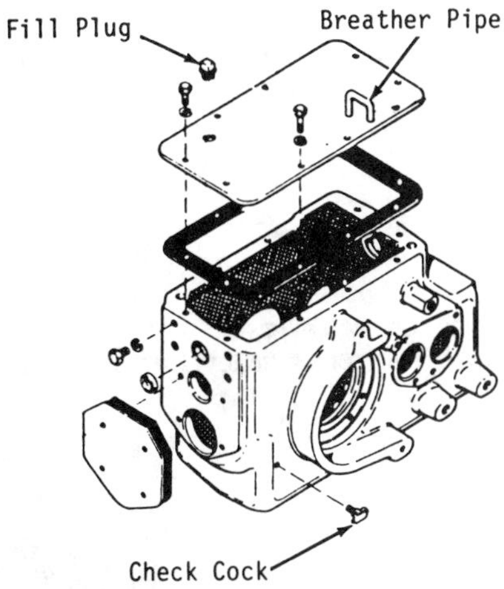

Fig. AC18—View of filler plug and check cock on early 616 models. Refer to text.

Models 718-917-919

Hydrostatic transmission and tractor hydraulic system use fluid from gear transmission (reduction gear) housing. Dexron Automatic Transmission Fluid, or equivalent, is recommended and fluid level should be maintained ⅛-inch (3.175 mm) below top rear edge of fill tube (Fig. AC20) when upper, forward edge of tube is ½-inch (12.7 mm) lower than rolled edge of pump support plate (Fig. AC20). Fluid should be checked at 25 hour intervals and fluid and filter should be changed at 400 hour intervals. Fluid

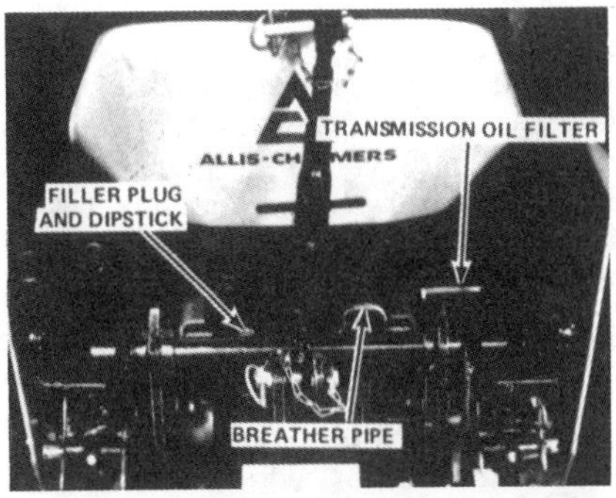

Fig. AC19—View showing location of dipstick, transmission oil filter and breather pipe on late 616 and all 620 and 720 models.

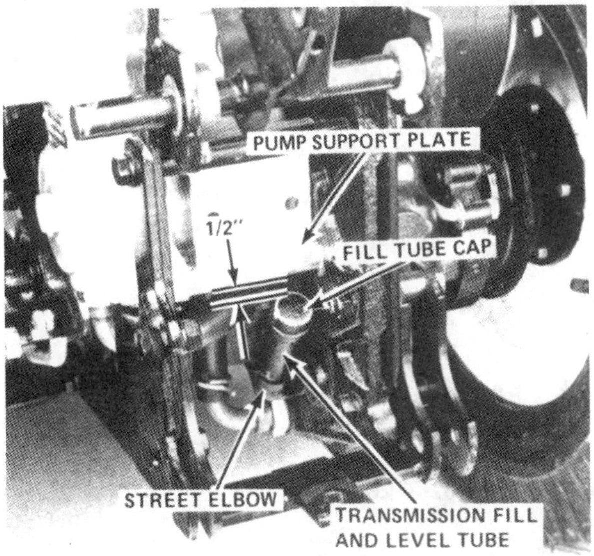

Fig. AC20—View showing location of fill and level tube on 718, 917 and 919 models. Refer to text.

capacity is 3 quarts (2.8 L) when filter is changed.

ADJUSTMENTS

Models 616-620-720

To adjust hydrostatic transmission for neutral position, remove top frame cover. Raise and support tracxtor so rear wheels are off the ground. Loosen locknuts (A–Fig. AC21). Start engine and operate at high idle speed. Move hydrostatic control lever to forward position, then into neutral position. Rear wheels should not rotate when control lever is in neutral position. Turn turnbuckle (B) as required. Tighten locknuts and recheck adjustment.

Models 718-917-919

Raise and support tractor so rear wheels are off the ground. Start engine and run at high idle. Move hydrostatic control lever to forward position, then neutral position. If wheels rotate when lever is in neutral position, stop engine. Raise seat deck and check if pump control arm (E–Fig. AC22) is exactly centered with centering mark (D). If not, loosen bolt (C) and move control cam (B) until centering mark (D) is centered on roller (E). Tighten bolt (C). If control arm is centered with centering mark, but rear wheels still rotate with lever in neutral position, loosen jam nut (H) on end of cam pivot shaft (G). If wheel rotation is in reverse, turn adjusting nut (I) 1/8 to 1/4-turn clockwise as viewed from right side of tractor. If rotation is forward, turn adjusting nut (I) 1/8 to 1/4-turn counter-clockwise. Lock jam nuts and recheck adjustment.

PRESSURE CHECK AND ADJUST

All Models

To check and adjust hydrostatic transmission operating pressure, refer to appropriate paragraph in HYDRAULIC SYSTEM section in this manual.

R&R HYDROSTATIC TRANSMISSION

Models 616-620-720

Remove frame top and bottom covers. Disconnect lower end of control linkage

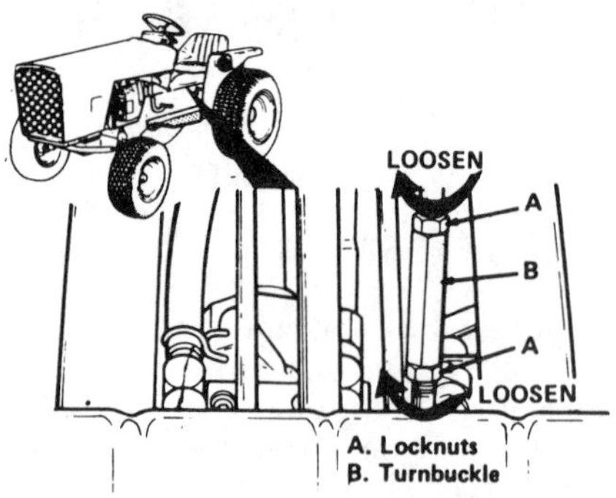
LOOSEN
A
B
A
LOOSEN
A. Locknuts
B. Turnbuckle

Fig. AC21—View showing adjustment location for hydrostatic neutral on 616, 620 and 720 models. Refer to text.

and remove control arm from hydrostatic unit. Remove wire from transmission oil temperature sending unit. Remove sending unit and drain transmission. Disconnect oil filter hose from charge pump elbow. Remove oil cooler return hose and hydraulic lift pressure line from top of pump housing. Cap all openings. Unbolt driveshaft rear coupling. Remove flexible disc, slide yoke from driveshaft; then loosen clamp screws and remove yoke and key from pump input shaft. Remove access hole cover in left side of tractor frame. Remove the two flange nuts securing hydrostatic drive unit to the three speed transmission and remove unit through opening in bottom of frame.

Reinst al by reversing removal procedure. Fill to proper level with recommended fluid. Position hydrostatic control lever in neutral position and start and run engine at idle speed. Operate transmission in forward and reverse direction for short distances. Stop engine and check fluid level. Repeat procedure until all air has been bled from system.

Models 718-917-919

Remove seat and fender assembly. Support left rear side of tractor and remove the wheel. Remove transmission fan, shroud and deflector. Loosen and remove drive belt, input pulley and fan. Disconnect pump control arm spring. Remove bolt, control arm roller, washer and nut. Remove oil filter and drain lubricant from reduction gear housing. Disconnect hydraulic hoses at oil filter assembly. Remove mounting bolts and slide transmission out of gear case.

Reinstall by reversing removal procedure. Lift pin in relief valve (Fig. AC23) and fill system with recommended fluid until fluid is visible in filler tube (Fig. AC20). Raise rear wheels off the ground and disconnect spark plug cables. Set speed control lever half-way forward and crank engine for short intervals. When wheels start to rotate, connect spark plugs. Start and run tractor for 1-2 minutes. Stop engine and lift pin in relief valve. Add fluid until at proper level. Recheck fluid level after five hours of operation.

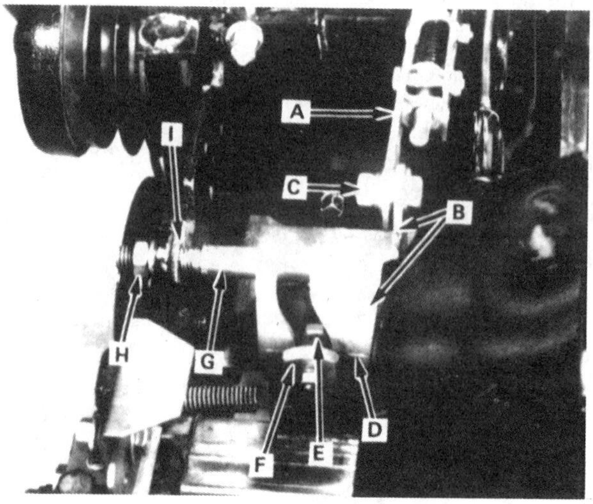
A
I
C
B
H
G
F
E
D

Fig. AC22—View showing location of hydrostatic transmission neutral adjustment points on 718, 917 and 919 models. Refer to text.

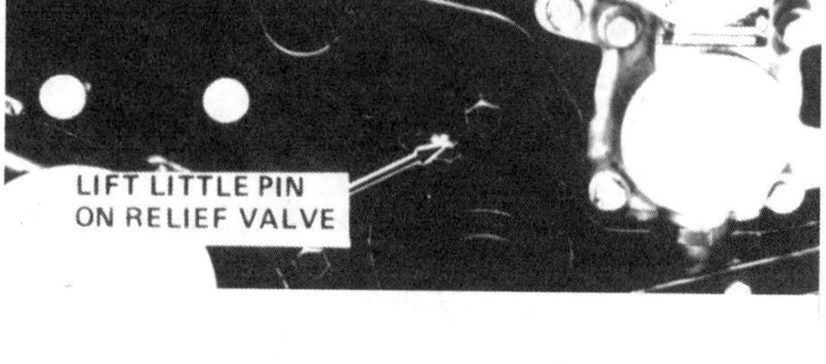
LIFT LITTLE PIN
ON RELIEF VALVE

Fig. AC23—View showing location of relief valve pin on 718, 917 and 919 models. Refer to text.

OVERHAUL HYDROSTATIC TRANSMISSION

Models 616-620-720

Scribe marks across charge pump cover (60–Fig. AC24) and pump housing (45), pump housing and center section (16) and center section and motor housing (3) for aid in reassembly. Remove all rust, paint or burrs from end of pump shaft (42). Unbolt and remove charge pump cover (60), then remove charge pump (58) and withdraw drive pin (59) from pump shaft. Remove oil seal (61) from cover. Remove hydraulic lift relief valve assembly (52 and 56) and charge pressure relief valve (67 through 71).

CAUTION: Keep valve assemblies separated. Do not mix component parts.

Remove the four through-bolts and separate pump assembly, center section and motor assembly. Valve plates (12 and 22) may stick to cylinder blocks (7 and 36). Be careful not to let them drop. Remove pump cylinder block and piston assembly (32 through 38) and lay aside for later disassembly. Remove thrust plate (39), drive out roll pins (40), then withdraw control shaft (62) with snap ring (66) and washer (65) and stub shaft (47) with snap ring (51) and washer (50). Lift out swashplate (41). Using a hot air gun or by dipping bearing end of housing in oil heated to approximately 250° F (121° C), heat housing and drive out input shaft (42) with bearing (44). Press bearing from input shaft. Oil seals (49 and 64) and needle bearings (48 and 63) can now be removed from housing (45).

If valve plates (12 and 22) stayed on center section (16), identify plates and lay them aside. Remove check valve assemblies (18 through 21), then remove forward acceleration valve assembly (24 through 30) and spring (31). Remove reverse acceleration valve from opposite side. Acceleration valves are not interchangeable. Reverse acceleration valve can be identified by small orifice hole in side of valve body (15). Needle bearings (14 and 23) can now be removed.

Remove motor cylinder block and piston assembly (5 through 11) and thrust plate (4) from motor housing. Heat bearing end of housing (3), using a hot air gun or by dipping in oil heated to approximately 250° F (121° C), then drive output shaft and bearing assembly from housing. Remove output bevel pinion and shims, then press bearing (1) from shaft (2).

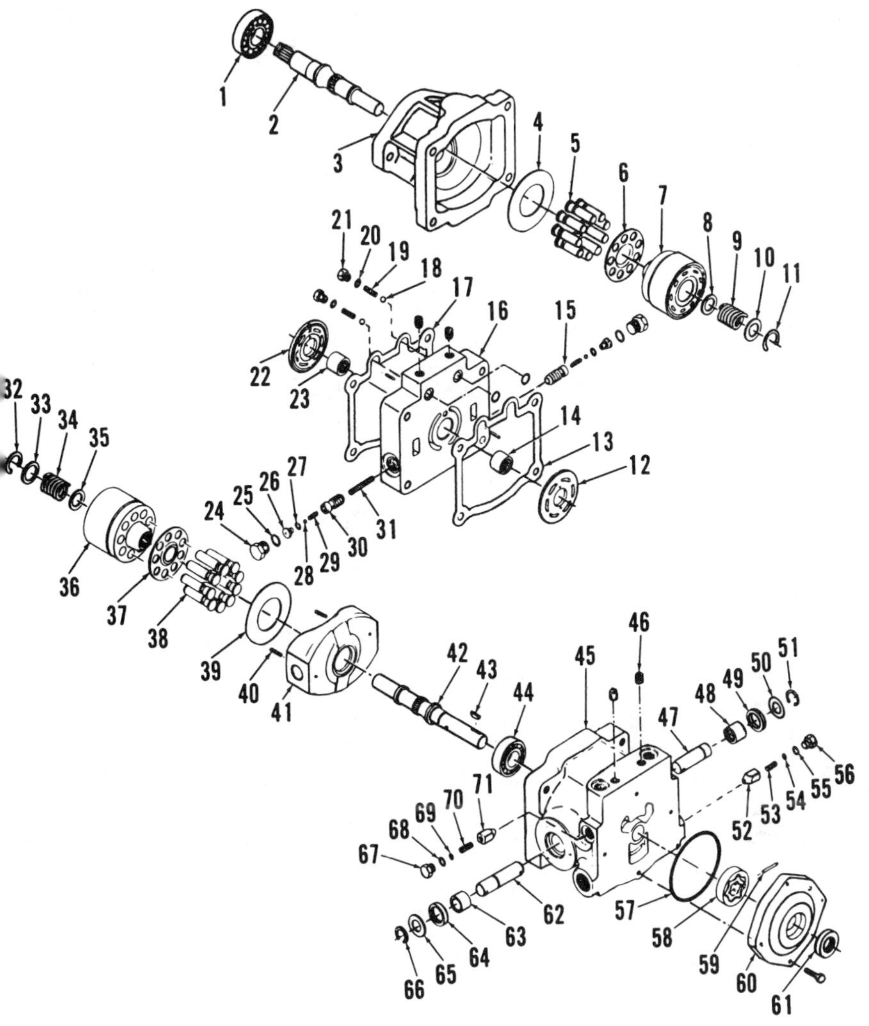

Fig. AC24—Exploded view of typical Sundstrand hydrostatic drive unit used on 616, 620 and 720 models.

1. Ball bearing
2. Motor (output) shaft
3. Motor housing
4. Thrust plate
5. Pistons (9)
6. Shoe plate
7. Cylinder block
8. Washer
9. Spring
10. Washer
11. Snap ring
12. Pump valve plate
13. Gasket
14. Needle bearing
15. Reverse acceleration valve
16. Center section
17. Gasket
18. Check valve ball (2)
19. Spring (2)
20. "O" ring (2)
21. Plug (2)
22. Motor valve plate
23. Needle bearing
24. Plug (2)
25. "O" ring (2)
26. Cap (2)
27. "O" ring (2)
28. Ball (2)
29. Spring (2)
30. Forward acceleration valve
31. Acceleration spring
32. Snap ring
33. Washer
34. Spring
35. Washer
36. Cylinder block
37. Shoe plate
38. Pistons (9)
39. Thrust plate
40. Roll pin (2)
41. Swashplate
42. Pump (input) shaft
43. Key
44. Ball bearing
45. Pump housing
46. Pressure check port plug
47. Stub shaft
48. Needle bearing
49. Oil seal
50. Washer
51. Snap ring
52. Hydraulic lift relief valve
53. Spring
54. Shim
55. "O" ring
56. Plug
57. "O" ring
58. Charge pump assy.
59. Drive pin
60. Charge pump cover
61. Oil seal
62. Control shaft
63. Needle bearing
64. Oil seal
65. Washer
66. Snap ring
67. Plug
68. "O" ring
69. "O" ring
70. Spring
71. Charge pressure relief valve

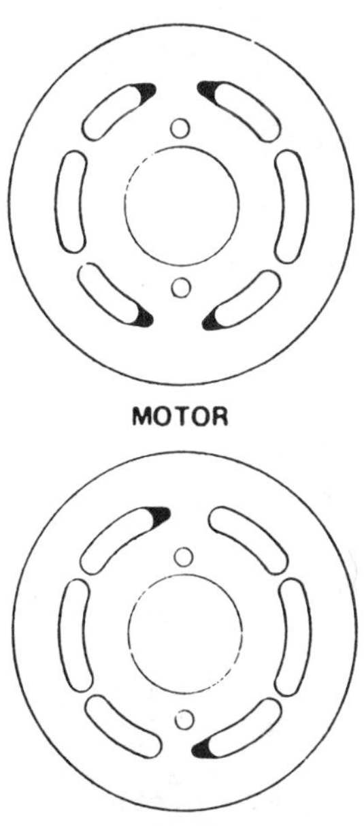

Fig. AC25—Motor plate has four notches (dark areas) and pump plate has two notches. Refer to text.

Carefully remove shoe plate and pistons (5 and 6) and (37 and 38) from cylinder blocks (7 and 36). Place each cylinder block on wood blocks in a press, compress spring (9 or 34) and remove snap ring (11 or 32). Release press and remove spring and washers.

Inspect pistons and bores in cylinder blocks for excessive wear or scoring. Light scratches on piston shoes can be removed by lapping. Inspect valve plates (12 and 22) and valve plate contacting surfaces of cylinder blocks (7 and 36) for excessive wear or scoring and renew as necessary. Check thrust plates (4 and 39) for excessive wear or other damage. Inspect charge pump (58) for wear, pitting or scoring.

Renew all oil seals, "O" rings and gaskets, lubricate all internal parts with new transmission fluid. Reassemble by reversing disassembly procedure. Heat bearing end of housings (3 and 45) before installing shaft and bearing assemblies. Install needle bearings (14 and 23) in center section (16) until they are 0.100 inch (2.54 mm) above machined surfaces. Install pump valve plate (12) and motor valve plate (22) in original positions with retaining pins in grooves. Pump valve plate (12) has two relief notches and motor valve plate (22) has four relief notches. See Fig. AC25. Check valve assemblies are interchangeable and are installed at rear side of center section. When installing acceleration valves, reverse acceleration valve (with small orifice hole in the side) must be installed in left side of center section. Install charge pressure relief valve assembly (67 through 71–Fig. AC24) using original shim (69). Install

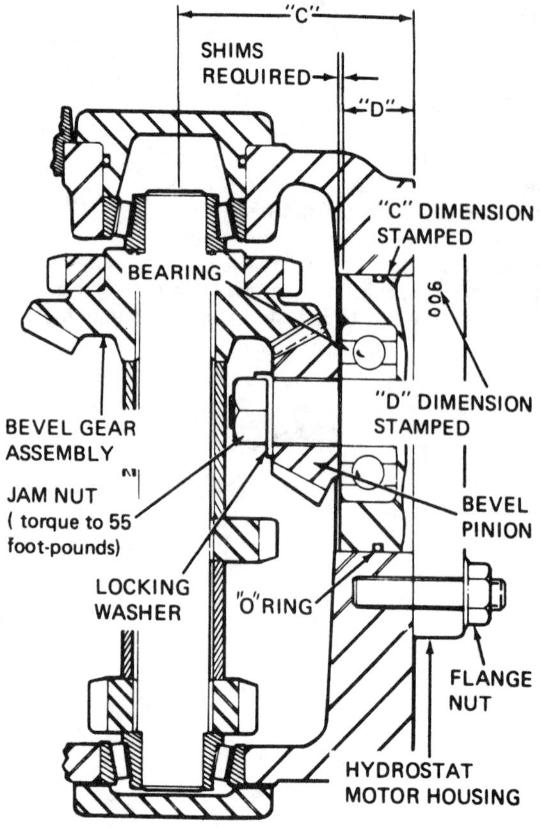

Fig. AC26—View showing location of "C" and "D" dimension code numbers stamped on three speed transmission case and hydrostatic motor housing. Stamped code numbers are used with chart in Fig. AC26A to determine shim pack thickness to be installed between bearing and output bevel pinion. Refer to text.

DIMENSION "D"
HYDROSTATIC PUMP-MOTOR (On Motor Housing)

DIMENSION "C" TRANSMISSION NUMBER (On Case)

	900	901	902	903	904	905	906	907	908	909	910	911	912	913	914	915
3000	.038	.037	.036	.035	.034	.033	.032	.031	.030	.029	.028	.027	.026	.025	.024	.023
3001	.039	.038	.037	.036	.035	.034	.033	.032	.031	.030	.029	.028	.027	.026	.025	.024
3002	.040	.039	.038	.037	.036	.035	.034	.033	.032	.031	.030	.029	.028	.027	.026	.025
3003	.041	.040	.039	.038	.037	.036	.035	.034	.033	.032	.031	.030	.029	.028	.027	.026
3004	.042	.041	.040	.039	.038	.037	.036	.035	.034	.033	.032	.031	.030	.029	.028	.027
3005	.043	.042	.041	.040	.039	.038	.037	.036	.035	.034	.033	.032	.031	.030	.029	.028
3006	.044	.043	.042	.041	.040	.039	.038	.037	.036	.035	.034	.033	.032	.031	.030	.029
3007	.045	.044	.043	.042	.041	.040	.039	.038	.037	.036	.035	.034	.033	.032	.031	.030
3008	.046	.045	.044	.043	.042	.041	.040	.039	.038	.037	.036	.035	.034	.033	.032	.031
3009	.047	.046	.045	.044	.043	.042	.041	.040	.039	.038	.037	.036	.035	.034	.033	.032
3010	.048	.047	.046	.045	.044	.043	.042	.041	.040	.039	.038	.037	.036	.035	.034	.033
3011	.049	.048	.047	.046	.045	.044	.043	.042	.041	.040	.039	.038	.037	.036	.035	.034
3012	.050	.049	.048	.047	.046	.045	.044	.043	.042	.041	.040	.039	.038	.037	.036	.035
3013	.051	.050	.049	.048	.047	.046	.045	.044	.043	.042	.041	.040	.039	.038	.037	.036
3014	.052	.051	.050	.049	.048	.047	.046	.045	.044	.043	.042	.041	.040	.039	.038	.037
3015	.053	.052	.051	.050	.049	.048	.047	.046	.045	.044	.043	.042	.041	.040	.039	.038
3016	.054	.053	.052	.051	.050	.049	.048	.047	.046	.045	.044	.043	.042	.041	.040	.039
3017	.055	.054	.053	.052	.051	.050	.049	.048	.047	.046	.045	.044	.043	.042	.041	.040
3018	.056	.055	.054	.053	.052	.051	.050	.049	.048	.047	.046	.045	.044	.043	.042	.041
3019	.057	.056	.055	.054	.053	.052	.051	.050	.049	.048	.047	.046	.045	.044	.043	.042
3020	.058	.057	.056	.055	.054	.053	.052	.051	.050	.049	.048	.047	.046	.045	.044	.043
3021	.059	.058	.057	.056	.055	.054	.053	.052	.051	.050	.049	.048	.047	.046	.045	.044
3022	.060	.059	.058	.057	.056	.055	.054	.053	.052	.051	.050	.049	.048	.047	.046	.045
3023	.061	.060	.059	.058	.057	.056	.055	.054	.053	.052	.051	.050	.049	.048	.047	.046
3024	.062	.061	.060	.059	.058	.057	.056	.055	.054	.053	.052	.051	.050	.049	.048	.047

Fig. AC26A—Chart used in determining shim pack thickness to be installed between bearing and output bevel pinion. Refer to text for procedure.

lift relief valve assembly (52 through 56) with original shim (54). Tighten the four through-bolts to 35 ft.-lbs. (48 N•m). Install drive pin (59), charge pump (58) and cover (60). Tighten cover retaining cap screws to 20 ft.-lbs. (27 N•m).

If original motor housing (3), output shaft (2), bearing (1) and output bevel pinion are being reused, install output bevel gear with original shim pack and tighten retaining nut to 55 ft.-lbs. (75 N•m).

If new output bevel pinion, output shaft, bearing, motor housing or complete new hydrostatic assembly is being installed, correct shim pack must be installed between output shaft bearing (1) and output bevel pinion. To determine correct thickness shim pack, refer to Fig. AC26 and note dimension (D) code number stamped on motor housing flange and dimension (C) code number stamped on three speed transmission case. Refer to chart in Fig. AC26A and using dimension (D and C–Fig. AC26) code numbers, determine correct thickness shim pack from chart. For example: If dimension (D) number on motor housing is 905 and dimension (C) number is 3012, correct thickness shim pack would be 0.045 inch. Install shim pack between output shaft bearing (1–Fig. AC24) and output bevel pinion and tighten nut to 55 ft.-lbs. (75 N•m).

Before installing hydrostatic drive unit, place unit on a bench with charge pump inlet port upward. Rotate pump shaft counter-clockwise while pouring new fluid into charge pump inlet port. When resistance is felt on rotation of input (pump) shaft, enough fluid has been installed to prime system and lubricate unit during initial operation. Plug all openings until unit is installed.

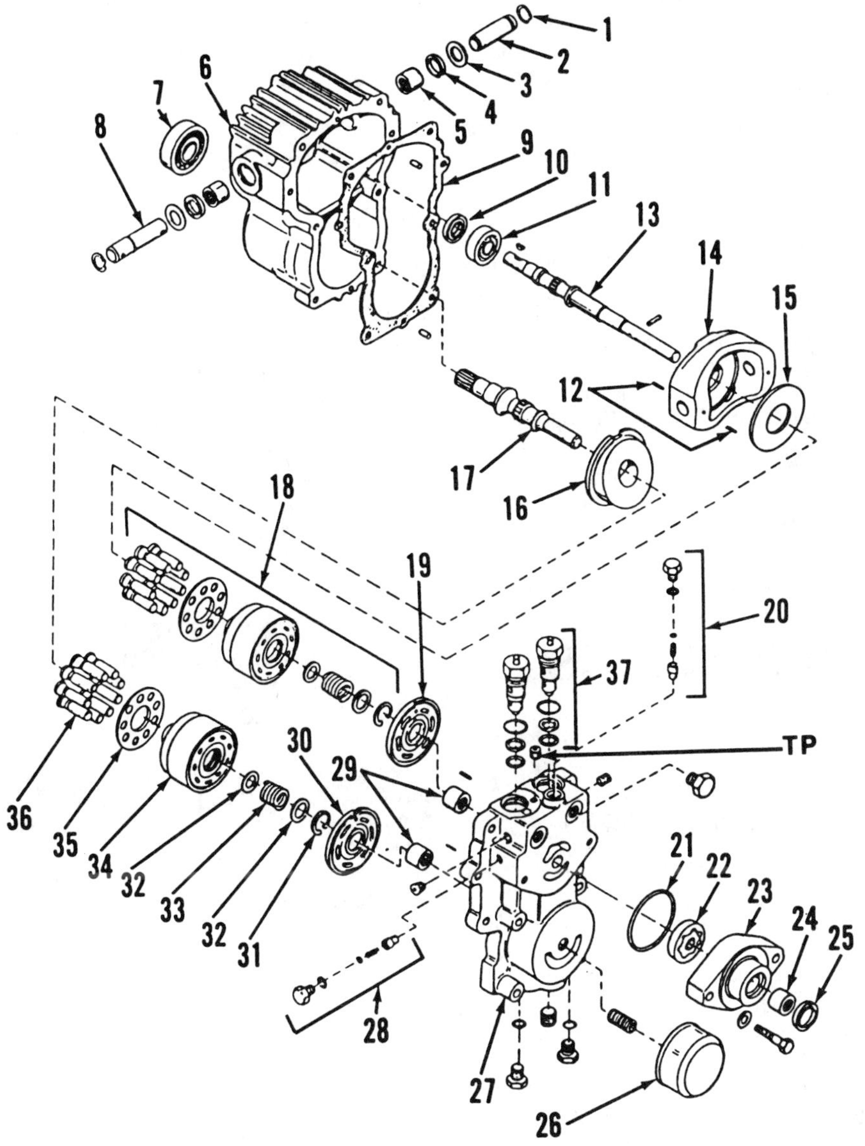

Fig. AC27–Exploded view of typical Sundstrand hydrostatic transmission used on 718, 917 and 919 models. Early models use a remote mounted oil filter in place of filter (26) shown. On tractors not equipped with hydraulic lift, relief valve (28) is not used and valve (20) is charge pressure relief valve.

TP. Test port plug
1. Retaining ring
2. Stub shaft
3. Washer
4. Seal
5. Needle bearing
6. Housing
7. Ball bearing
8. Control shaft
9. Gasket
10. Seal
11. Ball bearing
12. Roll pins (3)
13. Pump shaft
14. Swashplate (pump)
15. Thrust plate
16. Swashplate (motor)
17. Motor shaft
18. Pump assy.
19. Valve plate (pump)
20. Relief valve (implement)
21. "O" ring
22. Gerotor assy.
23. Charge pump housing
24. Needle bearing
25. Seal
26. Oil filter
27. Center section
28. Relief valve (charge)
29. Needle bearings
30. Valve plate (motor)
31. Retaining ring
32. Washers
33. Spring
34. Cylinder block (motor)
35. Slipper retainer
36. Pistons (motor)
37. Check valves

Models 718-917-919

Clean exterior of unit and remove hydraulic hoses, fan and control cam assembly from unit. Scribe a mark on control side of housing, center section and charge pump as an assembly aid. Remove charge pump housing (23–Fig. AC27). Remove gerotor assembly (22) and drive pin. Pry oil seal (25) from housing and press needle bearing (24) out front of housing. Remove both check valves (37), back-up washers and "O" rings. Remove relief valves (20 and 28) from bores in center section (27). Unbolt and remove center section.

CAUTION: Valve plates (19 and 30) may stick to center section. Taking care not to drop them, remove and identify valve plates.

Lift out both cylinder block and piston assemblies. Drive roll pins (12) into swashplate (14) and remove stub shaft (2), control shaft (8), swashplate (14) and thrust plate (15). Withdraw pump shaft (13) and bearing (11). Remove cap screws securing motor swashplate (16) to housing, then lift out swashplate and motor shaft (17). Bearings and oil seals can now be removed from housing as required.

To disassemble cylinder block and piston assemblies, carefully withdraw slipper retainer (35) with pistons (36) from cylinder block (34). Place cylinder block on a wood block in a press; compress spring (33) and remove retaining ring (31). Spring (33) should have a free length of 1-3/64 to 1-1/16 inches (26.6 to 27.9 mm) and should test 63-75 pounds (280-334 N) force when compressed to a length of 19/32-inch (15 mm). Check cylinder blocks for scratches or other damage and renew as necessary. Inspect

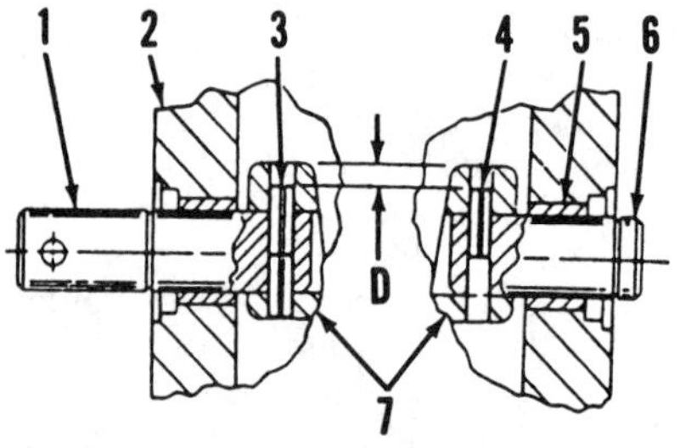

Fig. AC28—When reinstalling pump swashplate (7), drive roll pins in until ¼-inch (6 mm) "D" below swashplate surface. Two pins are used on control shaft (1) side.

1. Control shaft
2. Housing
3. Roll pins (2)
4. Roll pin
5. Bearing
6. Stub shaft
7. Swashplate

pistons and bores for excessive wear or scoring. Piston slippers can be lapped to remove light scratches. Minimum slipper thickness is 0.021 inch (0.533 mm) for both pump and motor. Slipper thickness must not vary more than 0.002 inch (0.051 mm) for all nine pistons in each block.

Check pump and motor valve plates for excessive wear and note pump valve plate has two notches and motor valve plate has four notches. See Fig. AC25. Valve plates are not interchangeable. Inspect motor swashplate and pump swashplate thrust plate for wear, embedded material or scoring. Check charge pump housing (23–Fig. AC27) and rotor assembly (22) for wear or scoring and renew as necessary.

Renew all "O" rings, gaskets and seals. Reassemble by reversing disassembly procedure. Lubricate all parts with clean oil when reassembling. When installing new bearings (29), press bearings into center section (27) until 1/16 to ⅛-inch (1.5 to 3 mm) of bearing protrudes; bearings are pilots for valve plates when unit is reassembled. Check valves (37) are interchangeable and are serviced only as an assembly. Pump swashplate (14) must be installed with thin pad towards top of transmission. Press pins (3 and 4–Fig. AC28) into swashplate (7) to dimension (D), which should be ¼-inch (6 mm) below swashplate surface. Two pins (3) are used in control shaft (1) side of swashplate. Install motor swashplate (16–Fig. AC27) with notch at top and high point of cam angle towards bottom. Flat end of charge pump housing (23) must be installed towards right side of unit. Position screws, which hold center section (27) and housing (6) together, in their respective holes and tighten evenly to 20-30 ft.-lbs. (27-41 N·m). Tighten charge pump mounting bolts to 50-55 ft.-lbs. (68-75 N·m).

Models 816-818

Remove seat and fender assembly. Disconnect brake rod from brake band and clutch rod from idler bracket. Remove drive belt and transmission pulley. Unbolt and remove shift lever assembly. Remove "U" bolt clamp from right axle and remove frame support and transmission support. Lower transmission and slide it clear of tractor.

Reinstall by reversing removal procedure. Adjust clutch and brake and check transmission fluid level.

Models 718-917-919

Raise and support rear of tractor. Drain transmission fluid and remove hydrostatic drive unit from gear transmission housing as outlined in appropriate paragraph in HYDROSTATIC TRANSMISSION section in this manual. Remove brake linkage and brake drum. Disconnect oil suction line from bottom of transmisson housing. Place jack under transmission housing and remove rear wheels, rear axles and differential assembly. Remove bolts securing transmission and lower transmission.

Reinstall by reversing removal procedure. Fill system to correct level with recommended fluid.

GEAR TYPE TRANSMISSIONS

MAINTENANCE

Models 616-620-720

Refer to appropriate MAINTENANCE paragraph in HYDROSTATIC TRANSMISSION section in this manual.

Models 816-818

Transmission fluid level should be checked at 25 hour intervals or if excessive leakage becomes apparent. To check fluid level, remove oil level check cap screw (A–Fig. AC29). Oil level should be at lower edge of cap screw opening. Recommended fluid is SAE 90 EP gear lubricant and fluid capacity is 3.5 pints (1.6 L).

All belts, guides, stops and pulleys should be inspected for wear, looseness or damage at 25 hour intervals and repaired as necessary.

Models 718-917-919

Refer to appropriate MAINTENANCE paragraph in HYDROSTATIC TRANSMISSION section in this manual.

R&R GEAR TYPE TRANSMISSIONS

Models 616-620-720

Remove seat supports and frame top and bottom covers. Disconnect lower end of hydrostatic control linkage and remove wire from transmission oil temperature sending unit. Remove sending unit and drain transmission. Disconnect oil filter hose from charge pump inlet elbow and remove oil cooler return hose and hydraulic lift pressure line from top of pump housing. Cap all openings. Disconnect driveshaft rear coupling and, if so equipped, unbolt rear pto shaft hub from electric clutch. Place a floor jack under transmission housing, support tractor under frame and remove rear wheels, drop housings, axle extensions and brakes. Remove lift cylinder, lift shaft and drawbar. Remove transmission retaining cap screws and remove transmission assembly.

Reinstall by reversing removal procedure.

OVERHAUL GEAR TYPE TRANSMISSIONS

Models 616-620-720

Remove top cover. Remove center cap screw and washers, then remove hub from front of top pto shaft. Pry out oil seal, remove bearing retaining ring and move top pto shaft forward. Remove rear snap ring and gear from shaft and withdraw pto shaft from front of transmission housing. Remove cap screws securing differential supports (39 and 56–Fig. AC30) to transmission

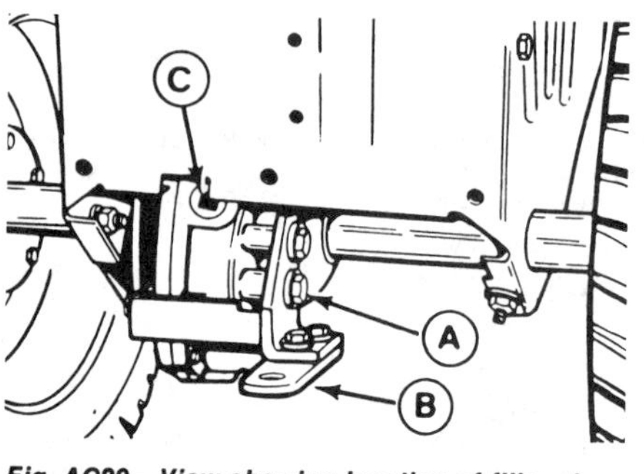

Fig. AC29—View showing location of filler plug (C), level check cap screw (A) and drain plug (B) on 816 and 818 models.

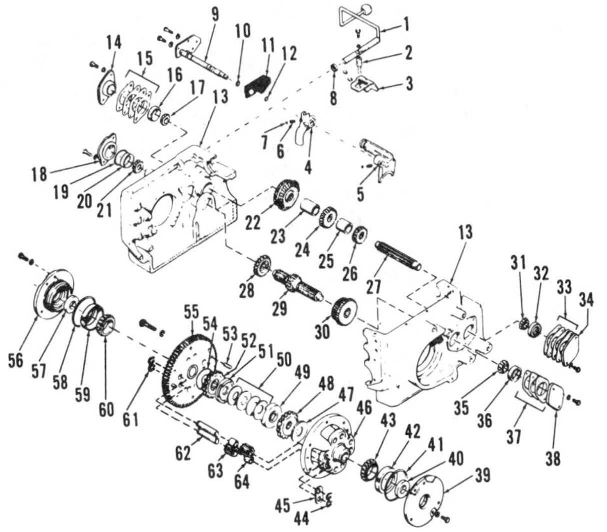

Fig. AC30—Exploded view of three speed gear type transmission and differential used on 616, 620 and 720 models.

1. Shift lever & shaft
2. Shifter stem
3. Shift guide
4. Third shift lock
5. First & second shift fork
6. Detent spring
7. Detent ball
8. Oil seal
9. Shifter shaft
10. "O" ring
11. Gasket
12. "O" ring
13. Transmission housing
14. Bearing cap
15. Shims
16. Bearing cup
17. Bearing
18. Bearing cap
19. "O" ring
20. Bearing cup
21. Bearing
22. Bevel & third gear
23. Spacer
24. Second gear
25. Spacer
26. First gear
27. Shaft
28. Third sliding gear
29. Sliding gear shaft
30. First & second sliding gear
31. Bearing
32. Bearing cup
33. Shims
34. Bearing cap
35. Bearing
36. Bearing cup
37. Shims
38. Bearing cap
39. Differential support
40. Oil seal
41. "O" ring
42. Bearing cup
43. Bearing
44. Snap rings
45. Lock plate (4)
46. Differential case
47. Thrust washer
48. Differential side gear
49. Clutch plate
50. Belleville washers
51. Clutch plate
52. Differential side gear
53. Dowel pin
54. Thrust washer
55. Output gear
56. Differential support
57. Oil seal
58. "O" ring
59. Bearing cup
60. Bearing
61. Snap rings
62. Pinion shafts (8)
63. Differential pinions (8)
64. Spacers (8)

housing. Hold up on differential assembly, remove differential support assemblies (39 through 42 and 56 through 59); then remove differential assembly from transmission housing.

Unbolt and remove shifter stem (2) and withdraw shift lever and shaft (1). Remove oil seal (8). Unbolt and remove shift guide (3) with two spacers. Remove cap screws securing shifter shaft (9) to housing and withdraw shifter shaft with "O" rings (10 and 12) and gasket (11).

CAUTION: Cover holes in forks (4 and 5) to prevent detent balls and springs from flying out when shifter shaft is removed.

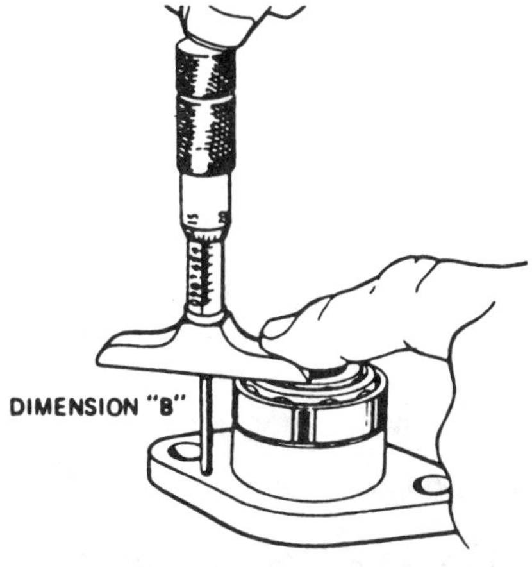

Fig. AC31—Use a depth gage to measure dimension "B" when determining thickness of shim pack (15—Fig. AC30) to be used with new bearing assembly (16 and 17—Fig. AC30).

Remove shift forks. Unbolt and remove cap (18) with "O" ring (19) and bearing cup (20). Remove bearing cone (21) and keep with cap (18) and bearing cup (20). Remove sliding gear (28), then move shaft (29) to the left and remove bearing cone (35) and sliding gear (30). Lift out sliding gear shaft (29). Unbolt and remove cap (38) with shims (37) and bearing cup (36). Remove cap screws securing cap (14) to housing, then withdraw cap with shims (15) and "O" ring, if so equipped. Remove bearing cup (16) from housing and place it with cap (14) and shims (15). Move shaft (27) to the left as far as possible, remove bearing cone (31), gears (24 and 26) and spacers (23 and 25) from shaft. Lift out shaft (27) with bevel gear (22) and bearing cone (17). Unbolt and remove cap (34), shims (33) and bearing cup (32).

Clean and inspect all parts and renew any showing excessive wear or other damage.

NOTE: Do not remove the press fit bearing cone (17) from shaft (27) unless bearing is to be renewed.

If bearing cup (16) and cone (17) are to be renewed, place new bearing assembly on cap (14) as shown in Fig. AC31. Using a depth gage, measure distance from cap flange to top of cone inner race as shown. This dimension will be 1.25 + inches. A four digit number (3.11 +) is stamped on the transmission case. Refer to Fig. AC32 and use chart to determine correct thickness shim pack (15–Fig. AC30) to be used with new bearing assembly (16 and 17). For example: If dimension "B" in Fig. AC31 measures 1.265 inches and dimension "A" stamped on transmission case is 3.122 inches, correct shim pack thickness to be installed is 0.018 inch. Shims are available in a variety of thicknesses.

When reassembling, place bevel gear (22–Fig. AC30) on shaft (27), then press bearing cone (17) on end of shaft. Place assembly in housing (13), move shaft to the left as far as possible and install long spacer (23), second speed gear (24) with bevel edge of teeth toward right side, short spacer (25), first speed gear (26) with bevel edge of teeth toward left side and bearing cone (31). Install bearing cup (16), coat shim pack (15) with a light coat of gasket sealer; then install shims and cap (14) with "O" ring, if so equipped. Apply a suitable sealer to cap screw threads, then install and tighten cap screws to 20 ft.-lbs. (27 N·m). Install bearing cup (32), make up shim pack (33) to a thickness of 0.040 inch and install shim pack and bearing cap (34). Tighten cap screws evenly to 20 ft.-lbs. (27 N·m). Check input shaft (27) for end play. Add or remove shims as required until shaft will rotate freely with zero end play. Do not preload bearings. When adjustment is correct, remove cap and shims. Apply a light coat of sealer to shims and cap screw threads, reinstall cap and shims, then tighten cap screws to 20 ft.-lbs. (27 N·m). Place first and second sliding gear (30) on sliding gear shaft (29), then position assembly in transmission housing. Install third speed sliding gear (28) with bevel edge of teeth to right side and install bearing

DIMENSION "A" TRANSMISSION NUMBER (On Case)

DIMENSION "B"	3.118	3.119	3.120	3.121	3.122	3.123	3.124	3.125	3.126	3.127	3.128
1.255	.012	.011	.010	.009	.008	.007	.006	.005	.004	.003	.002
1.257	.014	.013	.012	.011	.010	.009	.008	.007	.006	.005	.004
1.259	.016	.015	.014	.013	.012	.011	.010	.009	.008	.007	.006
1.261	.018	.017	.016	.015	.014	.013	.012	.011	.010	.009	.008
1.263	.020	.019	.018	.017	.016	.015	.014	.013	.012	.011	.010
1.265	.022	.021	.020	.019	.018	.017	.016	.015	.014	.013	.012
1.267	.024	.023	.022	.021	.020	.019	.018	.017	.016	.015	.014
1.269	.026	.025	.024	.023	.022	.021	.020	.019	.018	.017	.016
1.271	.028	.027	.026	.025	.024	.023	.022	.021	.020	.019	.018
1.272	.030	.029	.028	.027	.026	.025	.024	.023	.022	.021	.020

Fig. AC32—Chart used in determining thickness of shim pack (15—Fig. AC30) when installing new bearing assembly (16 and 17—Fig. AC30). Refer to text.

cone (21) on shaft. Install bearing cap (18) with "O" ring (19) and bearing cup (20). Use sealer on cap screw threads and tighten cap screws to 20 ft.-lbs. (27 N•m). Install bearing cone (35) and bearing cup (36). Make up a 0.040 inch thick shim pack (37) and install shim pack and bearing cap (38). Tighten cap screws evenly to 20 ft.-lbs. (27 N•m). Check sliding gear shaft for end play. Add or remove shims as required until the shaft will rotate freely with zero end play. Do not preload bearings. When adjustment is correct, remove bearing cap (38) and shim pack (37). Apply a light coat of sealer to shims and cap screw threads, reinstall shim pack and cap; then tighten cap screws to 20 ft.-lbs. (27 N•m).

Using new oil seal (8), gasket (11) and "O" rings (10 and 12), install shifter components (1 through 9). Reinstall differential assembly, top pto shaft and hydrostatic drive unit.

Reassemble by reversing disassembly procedure.

Models 816-816

Remove tires, wheels and hubs from both axles. Clean ends of axle shafts sufficiently so that axles can be withdrawn through bearings (61 and 64) and seals (62 and 63). Remove retaining ring (32–Fig. AC33), brake drum (30) and key (31). Remove cap screws and separate case halves (4 and 6). Lift out axles and differential assembly. Remove shaft and gear assembly (25). Remove nuts from shift rails (11 and 33) and reverse gear shaft (44). Remove shift rail (33), shift fork (37), shaft (44) and gear (46). Remove shift rail (11), shift fork (15), shaft (10) and gear (16). The balance of disassembly is evident after examination.

Clean and inspect all parts for wear or damage. Renew as necessary.

Reassemble by reversing disassembly procedure. Note differential bolts are tightened to 20 ft.-lbs. (27 N•m). Nuts securing shift rails and reverse shaft are tightened to 50 ft.-lbs. (68 N•m). Refill transmission with SAE 90 EP gear lubricant.

Models 718-917-919

Remove all paint, rust or burrs from brake drum shaft (1–Fig. AC34) and axle tube (37) to prevent bearing or seal damage during removal or installation. Make certain all grease fittings and keys have been removed from axle tube and remove cover (33), intermediate shaft (26), gear (27) and washers (25 and 28). Remove drive gear (17), axle tube (37) and thrust washers. Remove all seals from case halves. Continue disassembly as necessary.

Reassemble by reversing disassembly procedure.

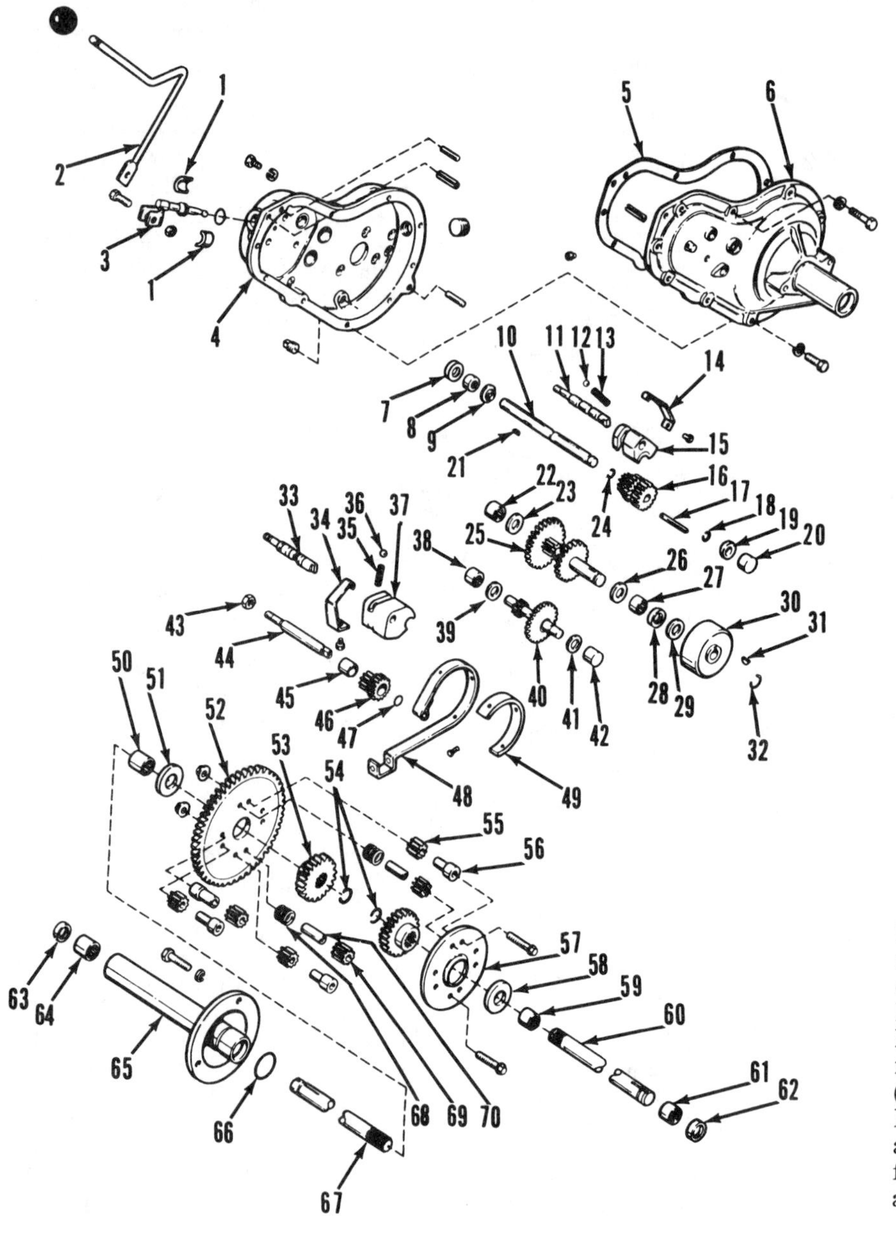

Fig. AC33—Exploded view of three-speed transmission used for 816 and 818 models.

1. Bushings
2. Shift lever
3. Pivot rod
4. Case
5. Gasket
6. Case
7. Seal
8. Bearing
9. Washer
10. Input shaft
11. Shift rail (2nd & 3rd)
12. Detent ball
13. Detent spring
14. Spring
15. Shift fork (2nd & 3rd)
16. Pinion gear assy.
17. Key
18. Retaining ring
19. Washer
20. Bearing
21. Key
22. Bearing
23. Washer
24. Retaining ring
25. Shaft & gear assy.
26. Washer
27. Bearing
28. Seal
29. Washer
30. Brake drum
31. Key
32. Retaining ring
33. Shift rail (1st & Rev.)
34. Spring
35. Detent spring
36. Detent ball
37. Shift fork (1st & Rev.)
38. Bearing
39. Washer
40. Gear assy. (1st)
41. Washer
42. Bearing
43. Locknut
44. Reverse gear shaft
45. Bushing
46. Gear (1st & Rev.)
47. "O" ring
48. Brake band
49. Brake lining
50. Bearing
51. Washer
52. Drive gear
53. Differential gears
54. Snap ring
55. Differential pinion gears
56. Differential studs
57. Differential plate
58. Washer
59. Bearing
60. Axle shaft
61. Bearing
62. Seal
63. Seal
64. Bearing
65. Axle tube
66. "O" ring
67. Axle shaft
68. Spring
69. Differential pinion gear
70. Spacer

DIFFERENTIAL

MAINTENANCE

All Models

Differential assembly on all models except 718, 917 and 919 models is an integral part of transmission assemblies and fluid level is maintained according to appropriate transmission maintenance section in this manual.

Models 718, 917 and 919 use a differential assembly which is a sealed unit, packed at the factory with multipurpose, lithium base grease. Differential assembly is located on right axle tube.

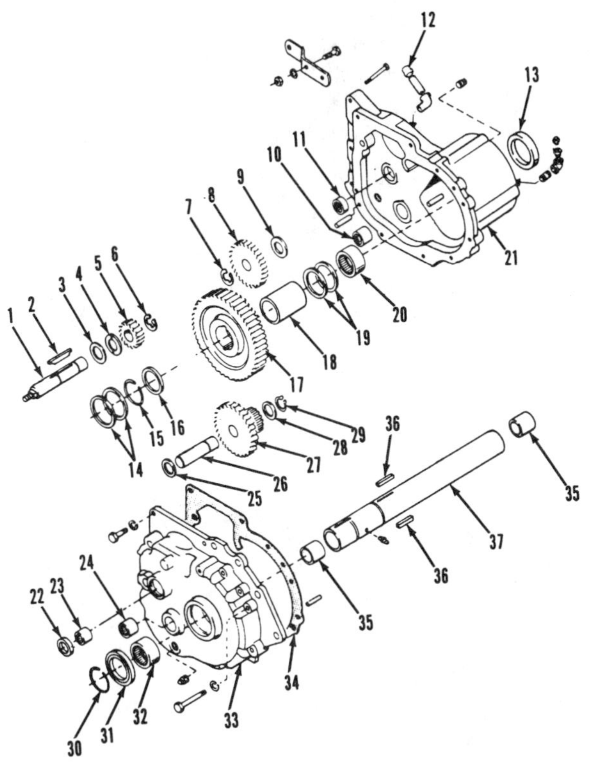

Fig. AC34—Exploded view of gear case used for 718, 917 and 919 models in conjunction with the hydrostatic transmission.

1. Brake shaft
2. Key
3. Washer
4. Washer
5. Gear
6. Retaining ring
7. Retaining ring
8. Gear (1st)
9. Washer
10. Bearing
11. Bearing
12. Cap
13. Seal
14. Washers
15. Retaining ring
16. Washer
17. Drive gear
18. Spacer
19. Washers
20. Bearing
21. Case
22. Seal
23. Bearing
24. Bearing
25. Washer
26. Intermediate shaft
27. Gear assy.
28. Washer
29. Retaining ring
30. Retaining ring
31. Seal
32. Bearing
33. Cover
34. Gasket
35. Bushings
36. Keys
37. Axle tube

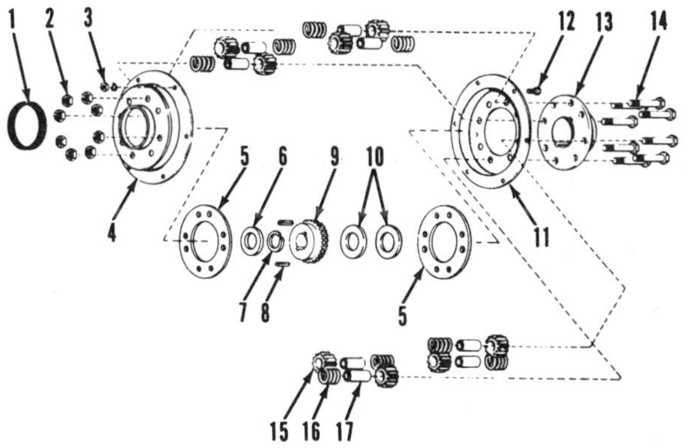

Fig. AC36—Exploded view of differential assembly used on 718, 917 and 919 models.

1. Seal
2. Nuts
3. Nuts
4. Differential cover
5. Spacer plate
6. Axle washer
7. Snap ring
8. Keys
9. Differential gear
10. Axle washers
11. Differential cover
12. Cap screws
13. Carrier
14. Cap screws
15. Pinion gears
16. Springs
17. Shafts

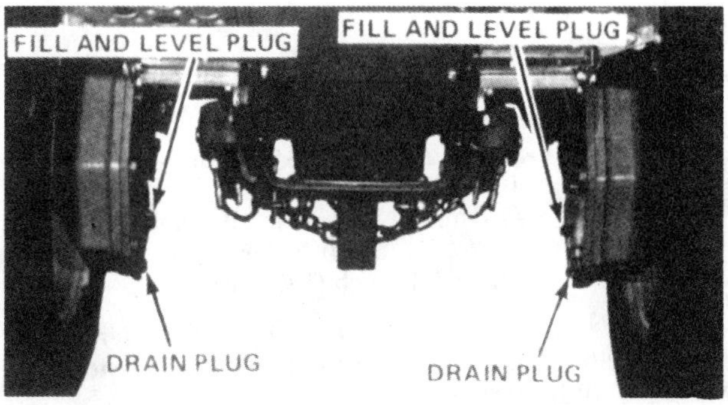

Fig. AC37—View showing fill and drain plugs on final drive units of 616, 620 and 720 models. Refer to text.

R&R DIFFERENTIAL

Models 616-620-720-816-818

Differential is an integral part of the transmission. Refer to appropriate TRANSMISSION section for service information.

Models 718-917-919

Raise and support rear of tractor. Remove both rear tires and wheels. Loosen setscrew on hub retaining collar for right wheel hub and remove collar and hub. Loosen left collar setscrew and slide axle to the right until collar is against left hub. Remove recessed washer, retaining ring and the two keys from right end of axle shaft. Slide differential assembly off and remove axle shaft.

Reinstall by reversing removal procedure.

OVERHAUL DIFFERENTIAL

Models 718-917-919

Remove the six bolts (12–Fig. AC36) and nuts (3) from around outer rim of differential covers. Slowly and evenly loosen, then remove the eight locknuts (2). Remove inner differential cover (4) and all internal parts.

Clean and inspect all parts for wear or damage. Renew parts as necessary.

Reassemble by reversing disassembly procedure. Pack housing with multipurpose, lithium base grease.

FINAL DRIVE

Models 616, 620 and 720 are equipped with final drive assemblies.

MAINTENANCE

Left and right final drive fluid levels should be checked at 50 hour intervals and changed at 400 hour intervals. Fluid level should be maintained at lower edge of fill plug (Fig. AC37). Recommended fluid is SAE 90 EP gear lubricant and capacity of each final drive housing is 0.5 quarts (0.47 L).

R&R FINAL DRIVE

Raise and support rear of tractor, then remove wheel and tire from side to be serviced. Remove fender assembly, then unbolt and remove final drive assembly. Retaining ring and coupling (2 and 3–Fig. AC38) will remain in axle housing.

OVERHAUL FINAL DRIVE

Drain fluid from final drive unit. Remove spring (4–Fig. AC38), then unbolt and remove housing plate (1) with bearing (5), slinger (6), bull pinion (7) and thrust washer (13). Press bull pinion and bearing assembly from plate (1), then press bull pinion from bearing and remove slinger. Remove locknut (8), washer (9) and lift out bull gear (10). Withdraw axle (20), then remove bearing cones (11 and 18) and oil seal (19) from housing. If necessary, press roller bearing (16) and bearing cups (12 an 17) from housing.

Clean and inspect all parts and renew any showing excessive wear or other damage.

To reassemble, press roller bearing (16) and bearing cups (12 and 17) in housing, if removed. Lubricate bearing cone (18) with SAE 90 EP gear lube and install in bearing cup (17). Install oil seal (19), with lip inward, in housing (15). Insert axle (20) through seal and bearing. Lubricate bearing cone (11) with SAE 90 EP gear lube and install it over axle until it is seated in bearing cup (12). Install bull gear (10), washer (9) and locknut (8). Tighten locknut until a force of 12 in.-lbs. (1.5 N•m) is required to rotate axle.

Reassemble by reversing disassembly procedure.

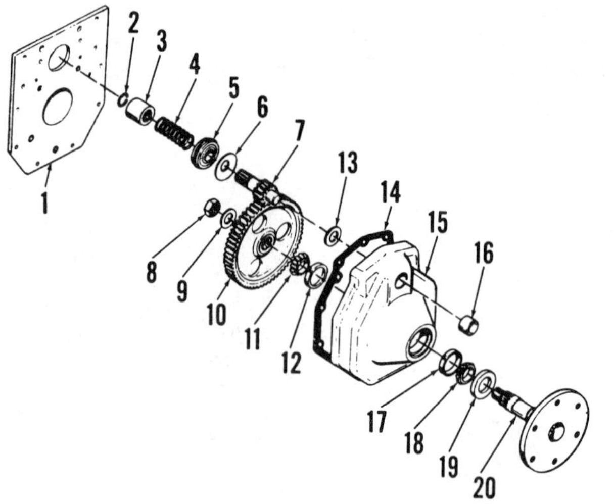

Fig. AC38—Exploded view of final drive unit used on 616, 620 and 720 models.

1. Housing plate
2. Retaining ring
3. Coupling
4. Spring
5. Bearing
6. Oil slinger
7. Bull pinion
8. Locknut
9. Washer
10. Bull gear
11. Bearing
12. Bearing cup
13. Thrust washer
14. Gasket
15. Housing
16. Roller bearing
17. Bearing cup
18. Bearing
19. Oil seal
20. Axle

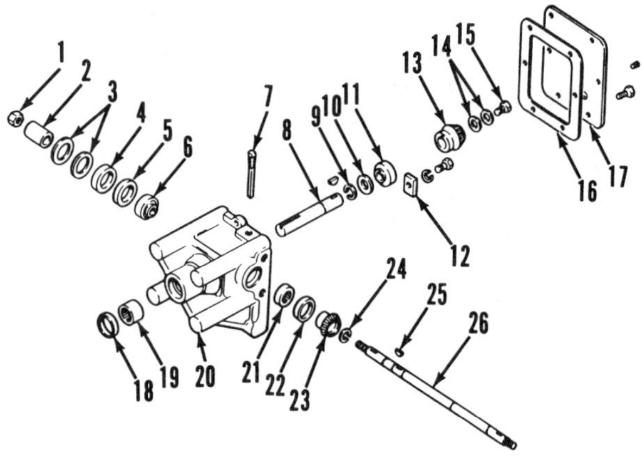

Fig. AC39—Exploded view of bevel gear box used on 718, 917 and 919 models.

1. Nut
2. Spacer
3. Shims
4. Seal
5. Spacer
6. Bearing
7. Dipstick
8. Input shaft
9. Retaining ring
10. Washer
11. Bearing
12. Clamp plate
13. Bevel gear (drive)
14. Washers
15. Cap screw
16. Gasket
17. Cover
18. Seal
19. Bearing
20. Housing
21. Bearing
22. Seal
23. Bevel gear (driven)
24. Retaining ring
25. Key
26. Output shaft

BEVEL GEAR BOX

On 718, 917 and 919 models, main drive from engine is connected to a bevel gear box located under the operator's seat.

MAINTENANCE

Fluid level in bevel gear box should be checked at 25 hour intervals. Fluid level should be maintained at a level which just touches end of roll pin in check plug when plug is just contacting gear box housing. **DO NOT** screw plug in to check level. Recommended fluid is SAE 90 EP gear lube. Fluid should be changed at 400 hour intervals.

R&R OVERHAUL

To remove bevel gear unit, complete drive unit, pulleys and belts must first be removed. Refer to appropriate TRANSMISSION section for removal procedure. Refer to POWER TAKE-OFF section and remove pto clutch assembly and idler pulley assembly. Remove pulley from right side of bevel gear unit. Disconnect drive shaft from bevel gear input shaft. Remove cap screws securing gear box to side plates. Pull side plates apart and withdraw bevel gear unit.

Drain lubricant and remove case cover (17–Fig. AC39). Remove nut (1) and key from shaft (26). Support gear (23) and drive shaft (26) through gear and case to expose key (25). Remove key. Remove retaining ring (24) from groove and slide it toward gear. Remove shaft (26) from case and retaining ring and gear from inside case. Remove cap screw and clamp plate (12). Remove key from external portion of shaft (8) and drive shaft, gear and ball bearing assembly out of case. Remove cap screw (15) and press shaft from gear (13). Remove key and press shaft from bearing.

Reassemble and reinstall by reversing disassembly and removal procedures.

POWER TAKE-OFF

ADJUSTMENTS

Models 616-620-720

FRONT CLUTCH ADJUSTMENT. To check front electric clutch adjustment, insert a 0.010 inch (0.254 mm) feeler gage (Fig. AC40) through each of the openings approximately 1/16-inch (1.6 mm). Nuts (A) should each be turned 1/12-turn until feeler gage just does enter 1/16-inch (1.6 mm).

CAUTION: Do not insert feeler gage more than 1/16-inch (1.6 mm) into any opening.

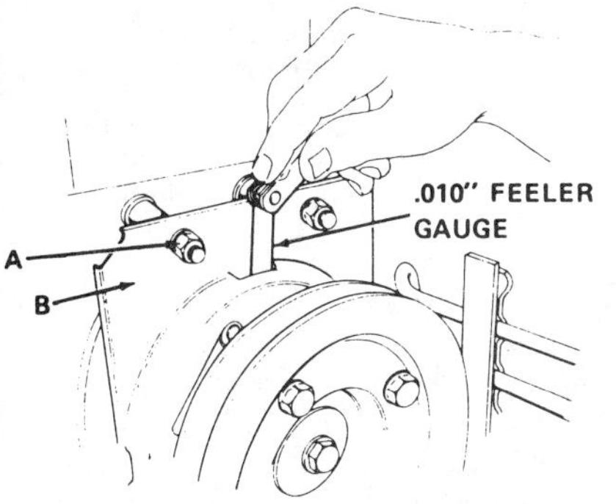

Fig. AC40—View showing front electric clutch adjustment on 616, 620 and 720 models. Refer to text.

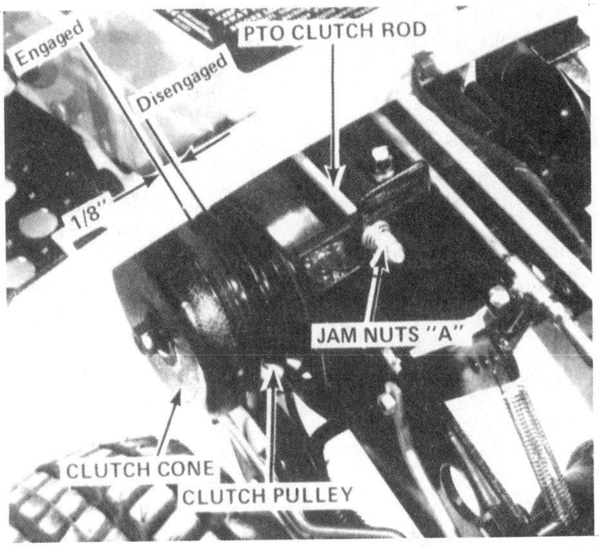

Fig. AC42—View showing clutch (pto) adjustment location on 718, 917 and 919 models. Refer to text.

MODELS 816-818

BELT STOP ADJUSTMENT. Belt stops (A–Fig. AC41) are adjusted by loosening belt stop mounting bolts and positioning them so there is 1/16-inch (1.6 mm) clearance between each stop and belt. Retighten mounting bolts.

Models 718-917-919

PTO CLUTCH ADJUSTMENT. When pto clutch is properly adjusted, clutch pulley (Fig. AC42) will move exactly 1/8-inch (3.2 mm) toward clutch cone when pto lever is moved from fully disengaged to fully engaged position. Adjust jam nuts (A) on end of pto clutch rod to obtain correct movement.

R&R AND OVERHAUL

Models 616-620-720

FRONT PTO CLUTCH. Remove belt, cap screws and pulley from pto clutch (Fig. AC43). Remove cap screw and washers from end of crankshaft, then withdraw clutch-armature assembly and remove key. Disconnect clutch wires at connector, then unbolt and remove clutch field unit from engine.

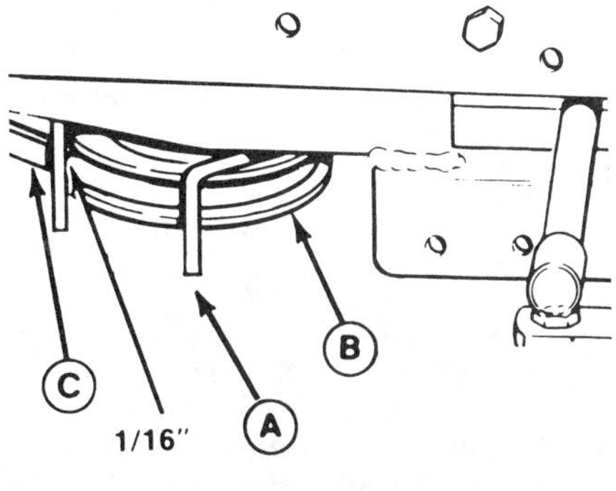

Fig. AC41—View showing pto belt stop adjustment on 816 and 818 models. Refer to text.

Disassemble clutch-armature assembly by pressing armature from ball bearing, then remove front retaining ring and press bearing from clutch.

Clean and inspect all parts. Normal resistance reading for clutch field winding is 2.75 to 3.60 ohms. If reading is higher or lower, field must be renewed.

Reassemble by reversing disassembly procedure. Tighten field retaining screws to 30 in.-lbs. (3.4 N•m). Tighten clutch-armature retaining cap screw in crankshaft to 20 ft.-lbs. (27 N•m) and pulley retaining cap screws to 10 ft.-lbs. (14 N•m).

REAR PTO AND CLUTCH. Remove seat support and top and bottom frame covers. Clean top of transmission and surrounding area. Unbolt and remove flexible coupling (1–Fig. AC44), yoke (2), felt ring (17) and rubber washer (16). Disconnect electric wires and link (9). Unbolt electric clutch assembly (7) from hub (3). Lower front end of pto drive shaft (15) and remove clutch and shaft assembly through frame top opening. Remove cap screw (5) with lockwasher and collar (6). Remove clutch-armature assembly and Woodruff key. Press armature from bearing, remove bearing retaining ring and press clutch from bearing. Remove the four socket head screws securing field to bearing housing (12) and withdraw field and washer (10). Remove retaining ring (14) and washer (13), then remove bearing (11) and housing (12) from driveshaft (15).

Unbolt and remove transmission top cover and pto shield (34). Remove cap screw (4) and washers, then remove hub (3). Pry out oil seal (18), remove bearing retaining ring (20) and move shaft (22) forward. Remove snap ring (24) and gear (23) from shaft, then withdraw shaft from front of transmission housing. Spacer (19) and bearing (21) can now be removed from shaft (22) and needle bearing (25) can be removed from rear of transmission housing.

Remove the two studs and withdraw idler shaft (33) with "O" ring (32 and spring (31). Remove idler gear (29), bearing (28), thrust washers (30 and 27) and spacer (26).

Spread retaining ring in collar (35) and remove collar, retainers (36 and 39), spring (38) and balls (37). Remove snap ring (40), then unbolt and remove bearing cap (42) with oil seal (41), "O" ring (43) and shims (44). Withdraw output shaft (47) with bearing cup (45) and cone (46). Remove output gear (48) and bearing cone (52) from above. If necessary, remove bearing cup (53) from housing. Remove snap ring (51), plug (50) and "O" ring (49) from output shaft (47).

Clean and inspect all parts and renew any showing excessive wear or other

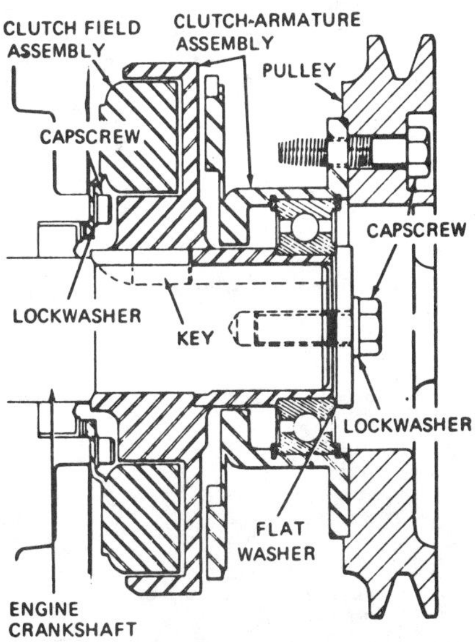

Fig. AC43—Cross-sectional view of electric pto clutch used on some models. Refer to text.

damage. To check clutch field, connect leads from an ohmmeter across field connector terminals. Normal resistance reading should be between 2.75 and 3.60 ohms. If reading is higher or lower, coil winding is defective and field must be renewed.

When reassembling, renew all oil seals, "O" rings and gaskets. Reassemble by reversing disassembly procedure. Output shaft (47) should have an end play of 0.002-0.005 inch (0.0508-0.1270 mm). Use a dial indicator to check end play and, if necessary, add or remove shims (44) to correct end play. When installing cap screws (4 and 5), use "Loctite", or equivalent, on threads and tighten to 45 ft.-lbs. (61 N·m).

Models 816-818

A spring loaded idler is used to apply tension to pto belt. Correct belt tension is maintained by adjusting idler tension spring length. Tension of idler pulley against drive belt should be sufficient to drive implement without belt slippage. Excessive tension will cause premature failure of belts and pulley bearings. Service procedures are apparent after examination.

Models 718-917-919

Remove drive belt from pto pulley. Remove retaining nut (15–Fig. AC45) and withdraw clutch plate (14). Protect threads on pto shaft and pry key out of shaft keyway.

CAUTION: Pto pulley is spring loaded. Install C-clamp to compress spring and hold pulley before attempting to remove pulley retaining ring.

Compress pulley internal spring with a C-clamp, then remove retaining ring. Remove pivot assembly retaining screws. Slowly release clamp and remove pto clutch assembly. Remove cotter pin retaining idler pivot to right side plate, then withdraw idler assembly.

Check pto shaft runout at outer retaining ring groove with a dial indicator. If runout exceeds 0.010 inch (0.254 mm), shaft should be renewed or straightened.

Inspect all parts for excessive wear or damage and renew as necessary.

Reassemble by reversing disassembly procedure. Make certain bearing (11) is installed with locking groove facing outward. If clutch plate retaining nut (15) is ¾-inch, tighten to 70 ft.-lbs. (95 N·m); if ½-inch nut is used, tighten to 50 ft.-lbs. (68 N·m). Adjust clutch as previously outlined.

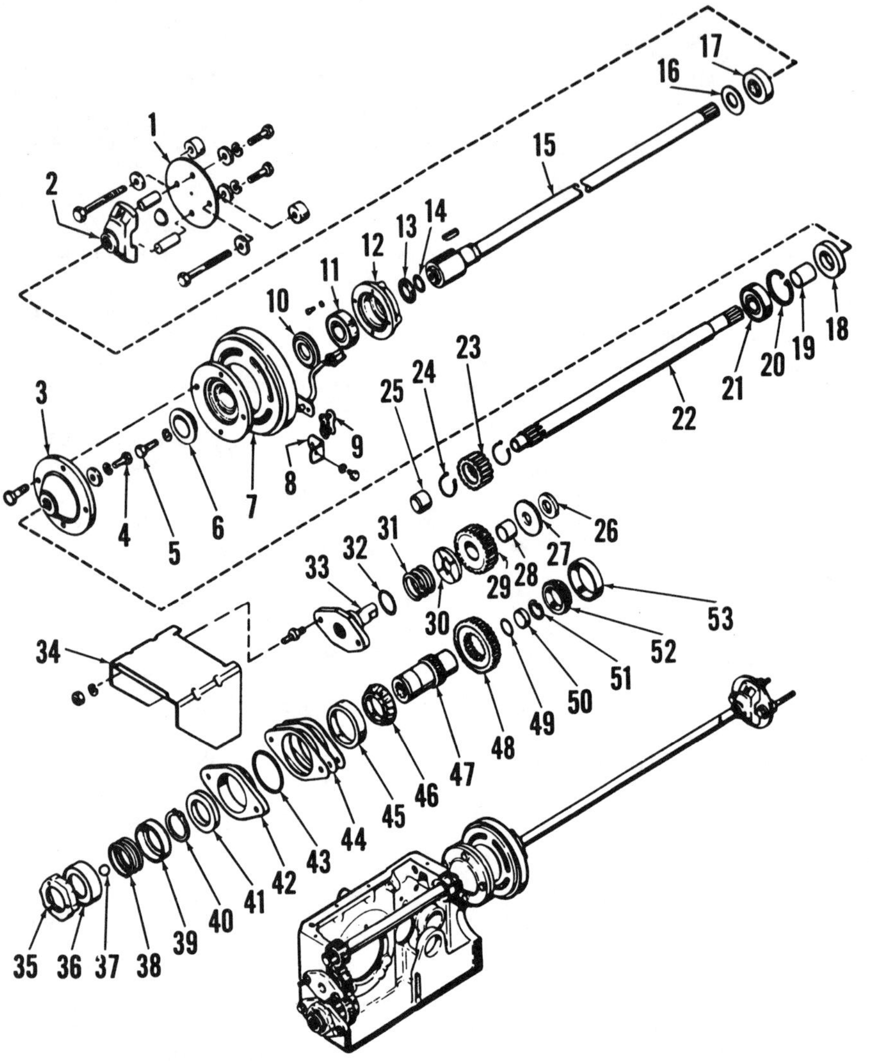

Fig. AC44–Exploded view of rear pto shafts and gears used on 620 and 720 models. Model 616 is similar.

1. Flexible coupling
2. Yoke
3. Hub
4. Cap screw
5. Cap screw
6. Collar
7. Pto clutch assy.
8. Bracket
9. Connecting link
10. Washer
11. Bearing
12. Bearing housing
13. Washer
14. Retaining ring
15. Pto drive shaft
16. Rubber washer
17. Felt ring
18. Seal
19. Spacer
20. Retaining ring
21. Bearing
22. Top pto shaft
23. Gear
24. Snap ring
25. Needle bearing
26. Spacer
27. Thrust washer
28. Needle bearing
29. Idler gear
30. Thrust washer
31. Spring
32. "O" ring
33. Idler shaft
34. Pto shield
35. Collar
36. Spring retainer
37. Ball
38. Spring
39. Retainer
40. Snap ring
41. Seal
42. Bearing cap
43. "O" ring
44. Shims
45. Bearing cup
46. Bearing
47. Output shaft
48. Output gear
49. "O" ring
50. Plug
51. Snap ring
52. Bearing
53. Bearing cup

HYDRAULIC SYSTEM

MAINTENANCE

Hydraulic system utilizes pressurized oil from charge pump of hydrostatic drive unit. Refer to HYDROSTATIC DRIVE paragraphs in TRANSMISSION section for maintenance and service procedures.

TROUBLESHOOTING

All Models So Equipped

Some problems which may occur during operation of hydraulic lift system and their possible causes are as follows:

1. System lifts load slowly. Could be caused by:

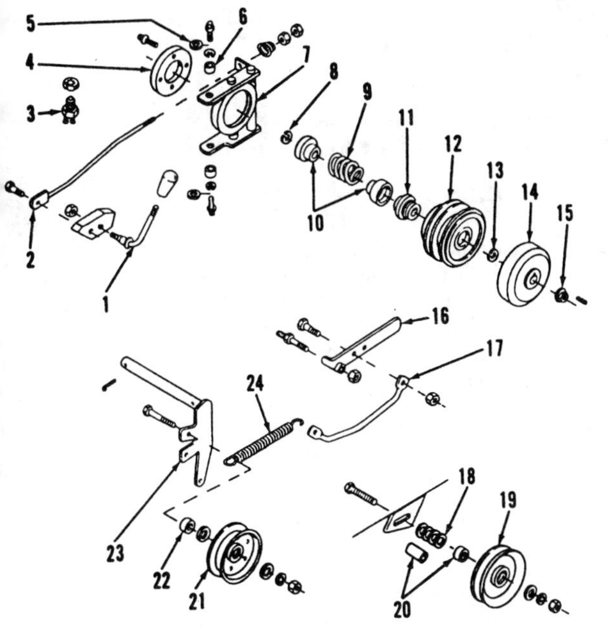

Fig. AC45—Exploded view of pto clutch and idler pulley assembly used on 718, 917 and 919 models.

1. Clutch control handle
2. Control rod
3. Interlock switch
4. Brake disc
5. Spring washer (2)
6. Spacer (2)
7. Pivot assy.
8. Retaining ring
9. Spring
10. Spring guides
11. Bearing
12. Pulley
13. Retaining ring
14. Clutch plate
15. Retaining nut
16. Tensioning lever
17. Tension rod
18. Washers (4)
19. Rear idler pulley
20. Spacers
21. Front idler pulley
22. Spacer
23. Pivot arm
24. Tension spring

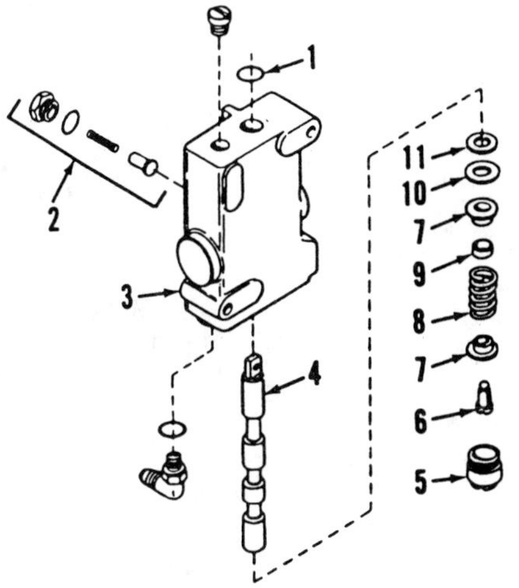

Fig. AC47—Exploded view of AICO lift control valve used on some models.

1. "O" ring
2. Check valve assy.
3. Valve body
4. Valve spool
5. End cap
6. Spool stem
7. Spring seats
8. Spring
9. Spool stop
10. Washer
11. Quad ring

a. Excessively worn charge pump.
b. Faulty lift relief valve.
c. Relief valve pressure adjusted low.
d. Damaged lift cylinder or lines.

2. Load lowers with control valve in raise position. Could be caused by:
a. Faulty lift check plunger or seat in control valve.

3. Load lowers with control valve in neutral position. Could be caused by:
a. External leakage (lines or fittings).
b. Internal leakage (worn spool or valve body).

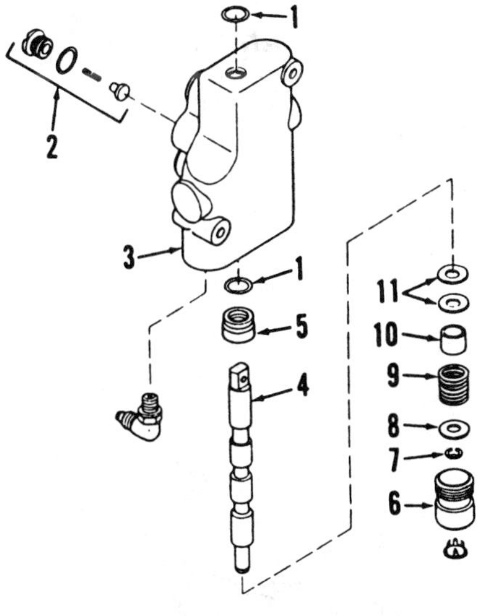

Fig. AC46—Exploded view of CESSNA lift control valve used on some models.

1. "O" rings
2. Check valve assy.
3. Valve body
4. Valve spool
5. Bushing
6. End cap
7. Retaining ring
8. Washer
9. Spring
10. Spool
11. Washers

PRESSURE CHECK AND ADJUST

Models 616-620-720

Remove pressure check test port plug (46–Fig. AC24) and install a 0-1000 psi (0-7000 kPa) test gage. Start engine and operate tractor until fluid reaches normal operating temperature. Test gage reading should be 70-150 psi (483-1034 kPa). Hold lift control valve with lift cylinder at end of stroke. Test gage should read 550-700 psi (3792-4826 kPa). If not, stop engine and add or remove shims (54) behind plug (56) to obtain correct lift pressure relief setting. Remove test gage and reinstall plug.

NOTE: If lift relief pressure 550-700 psi (3792-4826 kPa) cannot be reached, remove and inspect charge pump as outlined in HYDROSTATIC DRIVE UNIT paragraph in TRANSMISSION section.

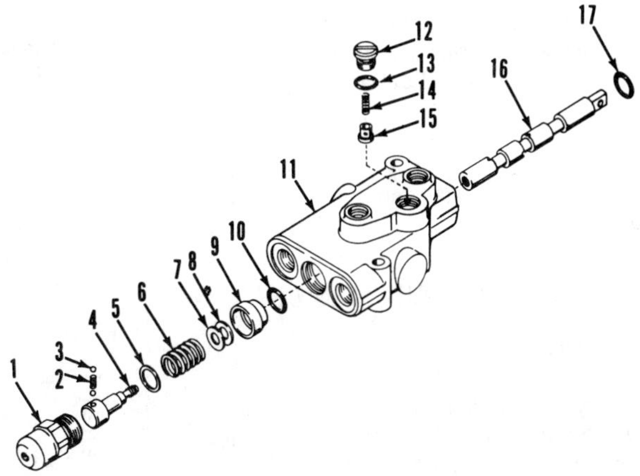

Fig. AC48—Exploded view of single spool control valve with a spring and ball type detent mechanism used on late models. Two spool valve is similar.

1. End cap
2. Detent spring
3. Detent balls
4. Detent adapter
5. Washer
6. Spring
7. Washer
8. Washer
9. Bushing
10. "O" ring
11. Valve body
12. Plug
13. "O" ring
14. Spring
15. Plunger
16. Spool
17. "O" ring

Models 718-917-919

Remove test port plug (TP–Fig. AC27) and install a 0-1000 psi (0-7000 kPa) test gage. Start engine and run at near full speed. Gage reading should be 70-150 psi (482-1034 kPa). Renew relief valve (28) or vary shims, as necessary, to obtain correct charge pressure.

CONTROL VALVE

All Models So Equipped

A variety of hydraulic control valves have been used according to model or date of manufacture. Overhaul of valves is apparent after inspection and reference to Fig. AC46, AC47 or AC48 according to model being serviced. Renew all "O" rings and lubricate all parts during reasembly. Tighten caps to 20 ft.-lbs. (27 N•m).

ARIENS

CONDENSED SPECIFICATIONS

	MODELS		
	GT-17	GT-18	GT-19
Engine Make	Kohler	Kohler	Kohler
Model	KT-17	K-361	KT-19
Bore	3.12 in. (79.3 mm)	3.75 in. (95.25 mm)	3.12 in. (79.3 mm)
Stroke	2.75 in. (69.9 mm)	3.25 in. (82.55 mm)	3.12 in. (78 mm)
Displacement	42.18 cu. in. (690.5 cc)	35.89 cu. in. (588.2 cc)	46.98 cu. in. (770.5 cc)
Power Rating	17 hp. (12.7 kW)	18 hp. (13.4 kW)	19 hp. (14.2 kW)
Slow Idle	1200 rpm	1800 rpm	1200 rpm
High Speed (No-Load)	3600 rpm	3600 rpm	3600 rpm
Capacities –			
Crankcase	3 pts. (1.4 L)	3.5 pts. (1.7 L)	3 pts. (1.4 L)
Hydraulic System	5 qts. (4.7 L)	5 qts. (4.7 L)	5 qts. (4.7 L)
Transaxle	See Hyd.	See Hyd.	See Hyd.
Fuel Tank	4.5 gal. (17 L)	4.5 gal. (17 L)	4.5 gal. (17 L)

FRONT AXLE AND STEERING SYSTEM

MAINTENANCE

All Models

It is recommended that steering system be lubricated at 50 hour intervals. Lubricate left and right wheel bearings, spindles, axle pivot and steering gear box.

NOTE: Position wheels at full right turn position and lubricate steering gear box until grease is forced out around steering pin.

Use a good quality multi-purpose, lithium base grease and clean each fitting before and after lubrication.

R&R AXLE MAIN MEMBER

All Models

To remove axle main member (7–Fig. A1), remove any pto driven accessories from pto drive shaft. Disconnect drag link from steering arm (4). Support front of tractor, unbolt pto housing and remove from frame bracket. Withdraw pto drive from front of tractor until axle is clear of shaft. Raise front of tractor and roll front axle assembly forward from tractor.

Reinstall by reversing removal procedure.

TIE ROD AND TOE-IN

All Models

Remove tie rod (14–Fig. A1) by disconnecting tie rod ends (12) at steering spindles (11 and 15).

Reinstall by reversing removal pro-

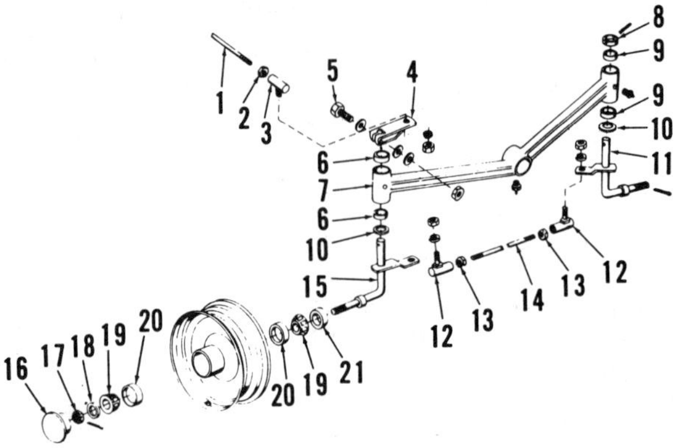

Fig. A1 – Exploded view of typical front axle assembly.

1. Drag link
2. Nut
3. Ball joint
4. Steering arm
5. Bolt
6. Bushing
7. Axle main member
8. Collar
9. Bushing
10. Washer
11. Steering spindle
12. Tie rod end
13. Jam nut
14. Tie rod
15. Steering spindle
16. Grease cap
17. Castle nut
18. Washer
19. Bearing cone
20. Bearing cup
21. Seal

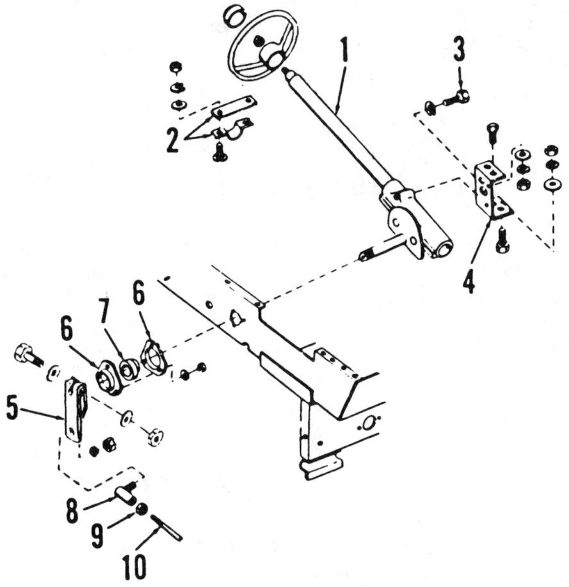

Fig. A2—Exploded view of steering mechanism.

1. Steering gear & shaft
2. Clamp
3. Cap screw
4. Bracket
5. Steering arm
6. Bearing flange
7. Bearing & collar
8. Drag link end
9. Jam nut
10. Drag link

cedure. Adjust toe-in by loosening jam nuts (13) and turning tie rod as required to obtain 1/8-inch (3.175 mm) toe-in. Tighten jam nuts.

R&R STEERING SPINDLES

All Models

To remove steering spindles (11 and 15–Fig. A1), support front of tractor and remove front wheels. Disconnect and remove tie rod from steering spindles, then disconnect drag link from steering arm (4). Loosen clamp bolt (5), remove steering arm and lower steering spindle (15) from axle main member. Drive out roll pin, remove collar (8) and lower steering spindle (11) from axle main member. Bushings (6 and 9) can now be inspected and renewed as necessary.

Reinstall by reversing removal procedure.

R&R STEERING GEAR

All Models

To remove steering column and gear assembly (1–Fig. A2), raise hood. Remove floor plate and battery. Remove air cleaner assembly.

CAUTION: Carefully cover carburetor opening to prevent accidental entrance of dirt or foreign objects.

Disconnect coil wire, blue (charging) wire and engine ground wire. Remove starter cable from solenoid. Disconnect choke and throttle cables. Remove steering wheel. Remove the four side mounting bolts and support clamp bolts and remove console/dash support. Remove steering arm (5), bearing flanges (6) and driveshaft. Remove the three steering column support bracket retaining bolts. Tilt steering column towards steering arm and remove assembly from tractor.

Reinstall by reversing removal procedure.

STEERING GEAR OVERHAUL

All Models

To disassemble steering gear, remove jam nuts (5–Fig. A3). Remove cotter pin (11), adjustment plug (10) and slide steering shaft assembly (8) out of housing. Remove Belleville washer (9), retainer cups (6) and bearings (7).

Inspect parts for wear, scoring or other damage and renew as necessary.

To reassemble, apply heavy grease to bearings (7) and retainer cups (6) and install on steering shaft (8). Apply heavy grease to steering gear and install assembly into housing (3). Install Belleville washer (9) and plug (10). Tighten plug to 4-6 ft.-lbs. (5-8 N•m) and install cotter pin (11). Shaft should turn freely in housing. Install new seal (16) and retainer (15) on housing. Remove steering pin (13) from steering arm (14) and install steering arm. Place a 5/32-in. shim between steering arm plate and seal retainer, then install washer (4) and tighten one jam nut (5) until just snug. Remove shim stock, install second jam nut and while keeping first nut in position on shaft, tighten second jam nut against first nut. Tighten to 22-25 ft.-lbs. (30-34 N•m). Install steering pin with steering arm in center of travel until pin is finger tight. Hold pin in this position and tighten jam nut (12) to 35-45 ft.-lbs. (48-61 N•m). Steering wheel should turn freely.

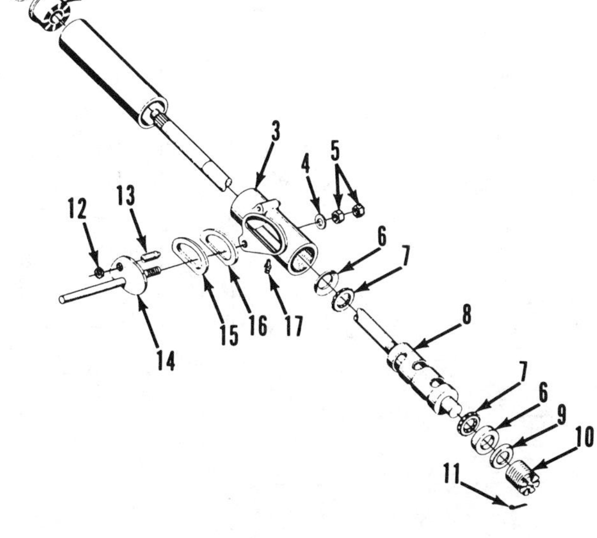

Fig. A3—Exploded view of steering gear assembly.

1. Dust cover
2. Bushing
3. Housing
4. Washer
5. Jam nuts
6. Retainer cups
7. Bearings
8. Steering shaft & worm gear assy.
9. Belleville washer
10. Adjustment plug
11. Cotter pin
12. Jam nut
13. Steering pin
14. Steering arm
15. Retainer
16. Seal
17. Grease fitting

STEERING GEAR ADJUSTMENT (IN TRACTOR)

All Models

To adjust steering gear, raise front of tractor so tires clear the ground and are in a straight ahead position. Remove cotter pin (11–Fig. A3) and turn adjustment plug (10) counter-clockwise 1½ turns, then tighten adjusting plug to 4-6 ft.-lbs. (5-8 N•m) and install cotter pin. Loosen jam nut (12) and tighten steering pin (13) until steering wheel free play is removed. **DO NOT OVERTIGHTEN STEERING PIN.** Hold pin in this position and tighten jam nut to 35-45 ft.-lbs. (48-61 N·m).

STEERING STOP ADJUSTMENT

All Models

To adjust steering stops, set toe-in and position front wheels at full right turn position. If right spindle arm does not contact axle stop, remove drag link rod end from steering arm and adjust drag link length until rod end is ½-hole to rear of hole in steering arm with spindle contacting axle stop. Reinstall drag link and turn wheels to full left turn position. Left spindle arm should contact axle stop. If not, shorten drag link slightly until both spindle arms contact their respective axle stops. Steering gear must not bottom before spindles contact axle stops.

ENGINE

MAINTENANCE

All Models

Regular engine maintenance is required to maintain peak performance and long engine life.

Check oil level at five hour intervals under normal operating conditions.

Wash engine air cleaner pre-cleaner at 25 hour intervals. Clean or renew dry filter element at 100 hour intervals.

REMOVE AND REINSTALL

All Models

To remove engine, disconnect battery. Remove front cowling. Remove front pto belts and disconnect wiring to pto clutch, rectifier and positive wire from coil. Disconnect choke and throttle cables. Disconnect fuel line and remove driveshaft guard and floor plate. Disconnect lead to starter and engine retaining bolts. Remove engine.

Reinstall by reversing removal procedure.

OVERHAUL

All Models

Engine make and model are listed at the beginning of this section. For tune-up specifications, engine overhaul procedures and engine maintenance, refer to appropriate engine section in this manual.

ELECTRICAL SYSTEM

MAINTENANCE

All Models

Battery electrolyte level should be checked at 50 hour intervals of normal

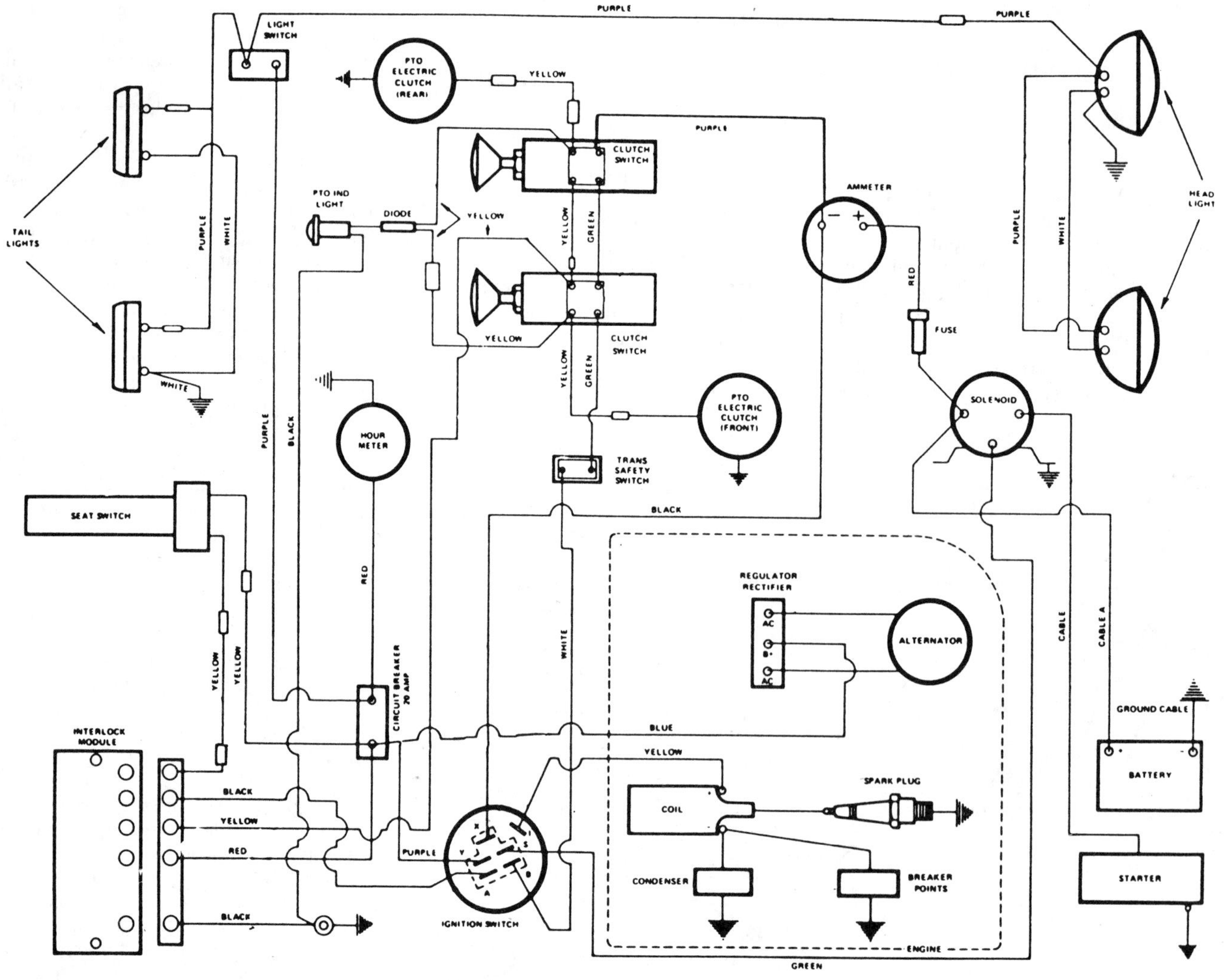

Fig. A4—Wiring diagram for all models.

operation. If necessary, add distilled water until level is just below base of vent well. **DO NOT** overfill. Keep battery posts clean and cable ends tight.

For alternator or starter service, refer to appropriate engine section in this manual.

Refer to Fig. A4 for wiring diagram for all models.

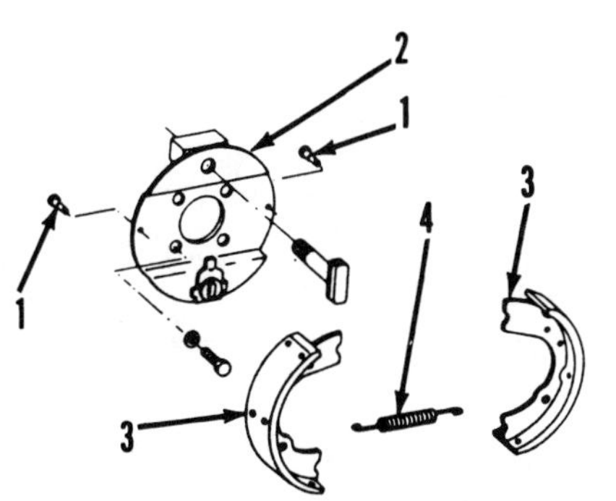

Fig. A7—Exploded view of drum brake system. Refer to text.

1. Hold-down springs
2. Backing plate
3. Brake shoes
4. Spring

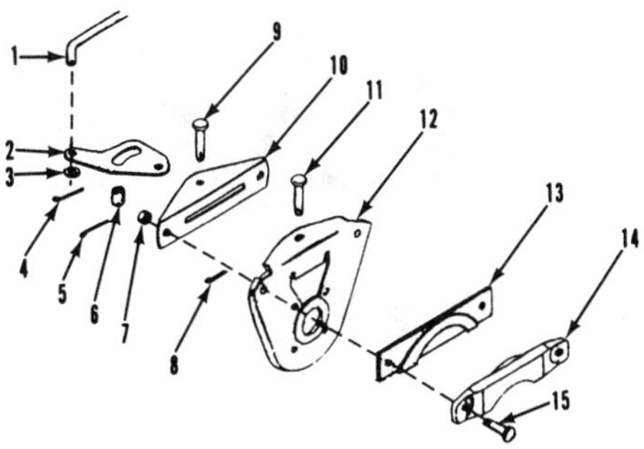

Fig. A8—Exploded view of disc brake system. Refer to text.

1. Brake rod
2. Plate
3. Washer
4. Cotter pin
5. Cotter pin
6. Roller
7. Nut
8. Cotter pin
9. Pin
10. Pressure plate
11. Pin
12. Caliper mounting plate
13. Friction plate &pad
14. Caliper plate & pad

BRAKES

ADJUSTMENT

Drum Brakes

If brake pedal can be depressed over two inches (50.8 mm), brakes must be adjusted. Raise and support rear of tractor with wheels off the ground. Remove rear wheels and tires. Use a flat blade screwdriver inserted through slot in brake drum (Fig. A5) to tighten star adjuster until drum cannot be turned. Back star adjuster off until brake drum turns with no drag (approximately one turn). Repeat procedure for remaining side. Reinstall tires and wheels and check brake operation.

Disc Brakes

If brake pedal can be depressed over two inches (50.8 mm), brakes must be adjusted. Raise and support rear of tractor with wheels off the ground. Open free-wheeling valve on hydrostatic transmission so wheels may be rotated. Tighten the two adjusting nuts (Fig. A6) evenly while rotating wheel until slight drag is felt. Back each nut off 1/8-turn. Repeat procedure for remaining side.

R&R BRAKE SHOES/PADS

Drum Brakes

Raise and support rear of tractor. Remove tire and wheel and hub and drum assemblies. Disconnect hold-down springs (1–Fig. A7) and remove shoes (3). Remove spring (4).

Reinstall by reversing removal procedure. Adjust brakes.

Disc Brakes

Raise and support rear of tractor. Remove tire and wheel. Disconnect brake rod (1–Fig. A8), remove pin (9) and roller (6). Remove pin (11) and plate (2). Remove nuts (7) and bolts (15). Remove pads (13 and 14).

Reinstall by reversing removal procedure. Adjust brakes.

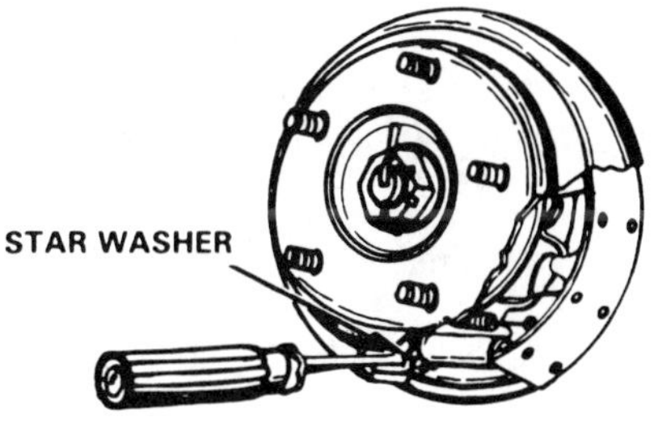

Fig. A5—View showing location of brake adjustment opening. Refer to text.

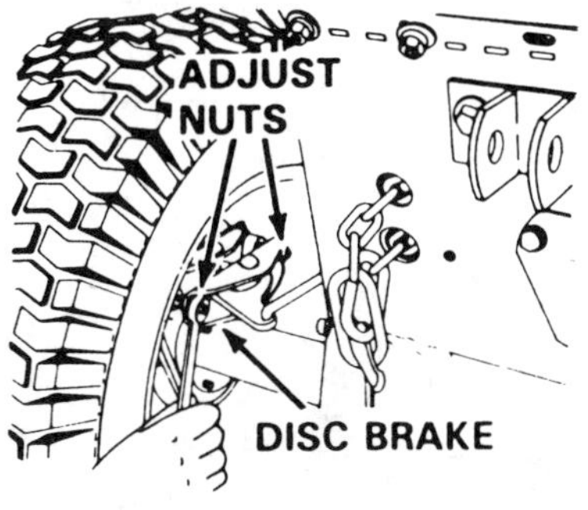

Fig. A6—View showing location of adjustment nuts for models equipped with disc brakes. Refer to text.

HYDROSTATIC TRANSMISSION

MAINTENANCE

All Models

Hydraulic system, hydrostatic drive unit and reduction gear and differential assembly have a common fluid reservoir in the reduction gear housing. Check fluid level at 25 hour intervals and maintain fluid level at transmission filler plug opening (Fig. A9).

Fluid and filter should be changed at 500 hour intervals. Recommended fluid is SAE 5W-30 SE motor oil and approximate capacity is 3.5 quarts (3.3 L). Do not overfill.

ADJUSTMENTS

All Models

CONTROL LEVER FRICTION ADJUSTMENT. If hydrostatic control lever (1–Fig. A11) will not maintain set position, adjust friction plate spring pressure. Raise rear deck and insert screwdriver (3) through access holes (4) and tighten screws (A and B) evenly. If tractor tends to gain speed, tighten screw (A) more than (B). If tractor loses speed, tighten screw (B) more than (A).

NEUTRAL POSITION ADJUSTMENT (MODELS PRIOR TO 1982). Raise and support rear of tractor so wheels are free to turn.

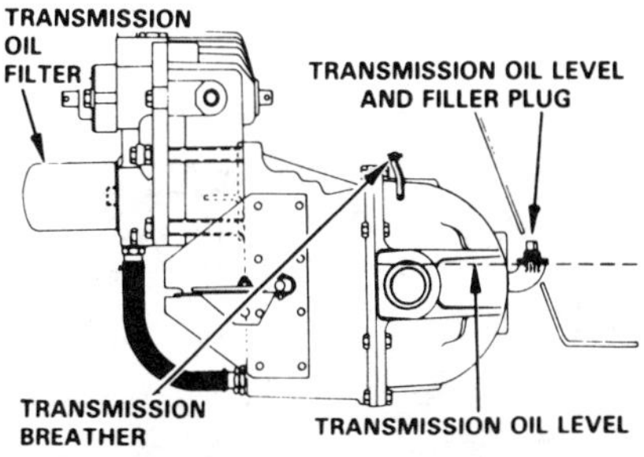

Fig. A9—Hydraulic system, hydrostatic drive unit and reduction gear and differential assembly share a common fluid reservoir located in reduction gear and differential housing. Fluid should be level with filler plug opening.

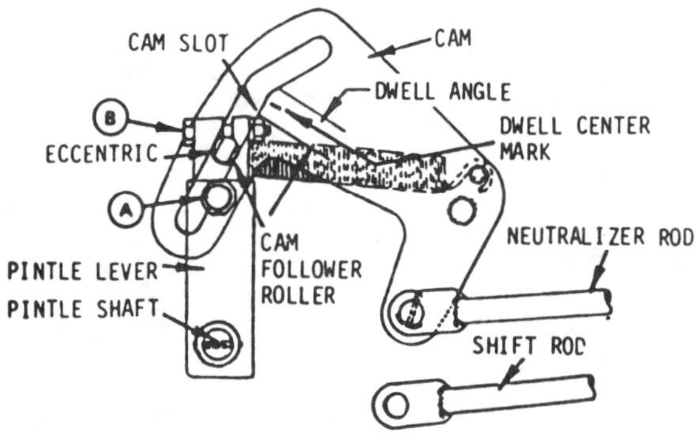

Fig. A10—View showing locations for hydrostatic drive control linkage adjustments. Refer to text.

NOTE: Use care when working around rotating drive shaft and tires.

Start engine and place control level in neutral position. With engine speed at full throttle, loosen clamp bolt (B–Fig. A10) and turn eccentric as required until rear wheels stop turning. Retighten clamp bolt (B). Check adjustment by moving lever into forward and reverse positions and back to neutral. Wheels should not turn with shift lever in neutral.

If fine adjustment of cam follower eccentric does not stop rear wheel rotation in neutral, or if shift linkage has been disassembled, use the following procedure for neutral adjustment.

Raise and support rear of tractor so wheels are free to turn. Loosen clamp bolt (B–Fig. A10) and turn eccentric until flats on eccentric are parallel with cam slot as shown, then retighten clamp bolt. Move cam to align cam follower roller with cam dwell center mark. Loosen pintle lever clamp bolt (A) so pintle shaft can rotate freely in pintle lever. Start engine and place control lever in neutral position. Move pintle lever as required until wheels stop turning. Be sure cam follower roller is still aligned with cam dwell center mark, then shut off engine. Carefully tighten bolt (A) making sure cam or pintle lever position is not disturbed. Loosen adjusting nuts on shift rod and neutralizer rod, then depress neutralizer pedal. Make sure cam follower roller is aligned with cam dwell center mark and adjust nuts snug against neutralizer plate. Loosen shift control fork bolt (D–Fig. A11). Hold cam to prevent movement of cam follower roller in cam slot and move control fork until it is aligned with neutral slot in control console. Tighten nut (D).

If wheels still creep when in neutral, loosen clamp bolt (B–Fig. A10) and adjust eccentric roller until wheels stop. Tighten clamp bolt (B).

NEUTRAL POSITION ADJUSTMENT. (1982 MODELS AND AFTER). Raise and support rear of tractor so wheels are free to turn. Remove floor plate, then disconnect shift rod yoke (2–Fig. A12) from neutralizer rod (3). If shift lever does not return to neutral position when neutral return pedal is depressed, adjust length of neutralizer rod using adjusting nuts (4). Loosen jam nut (1–Fig. A13) so eccentric (2) can be turned with a wrench. Start engine and move shift lever to neutral position.

NOTE: Use care when working around rotating drive shaft and tires.

With engine running at full throttle, adjust eccentric until wheels stop turning and tighten jam nut. Shut off engine and adjust shift rod yoke until shift rod (7) is centered in slot in transmission lever (6).

REMOVE AND REINSTALL

All Models

Disconnect battery and drain lubricant. Remove rear deck and fuel tank. Disconnect drive shaft rear coupler and remove coupler half from transmission input shaft. Disconnect hydraulic hoses and cap all openings to prevent entrance of dirt. Disconnect control linkage from transmission. Unbolt transmission, then remove from reduction gear and differential assembly.

Reinstall by reversing removal procedure. Fill reservoir with recommended fluid. Prime transmission by removing implement relief valve, located on top of unit between directional valves, and pour in one pint (0.5 L) of fluid. Reinstall relief valve but do not tighten valve plug. Start engine and run until fluid runs past threads of plug, then tighten plug.

OVERHAUL

All Models

Remove hydrostatic drive unit previously outlined and thoroughly clean exterior of unit. Place unit in a holding fixture with charge pump facing upward. Scribe a mark across charge pump housing and center section to ensure correct reassembly.

Unbolt and remove charge pump assembly. Pry oil seal (29–Fig. A15) from housing (28), then press needle

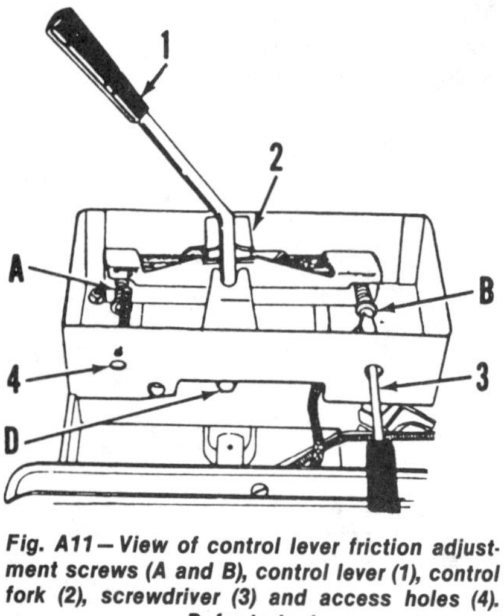

Fig. A11—View of control lever friction adjustment screws (A and B), control lever (1), control fork (2), screwdriver (3) and access holes (4). Refer to text.

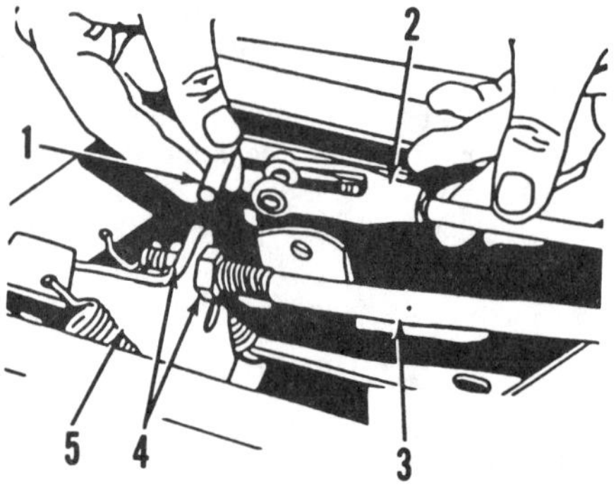

Fig. A12—To adjust hydrostatic shift linkage on 1982 and after models, shift rod (2) must be disconnected from neutralizer rod (3). Refer to text.

1. Pin
2. Shift rod
3. Neutralizer rod
4. Adjusting nuts
5. Neutralizer plate

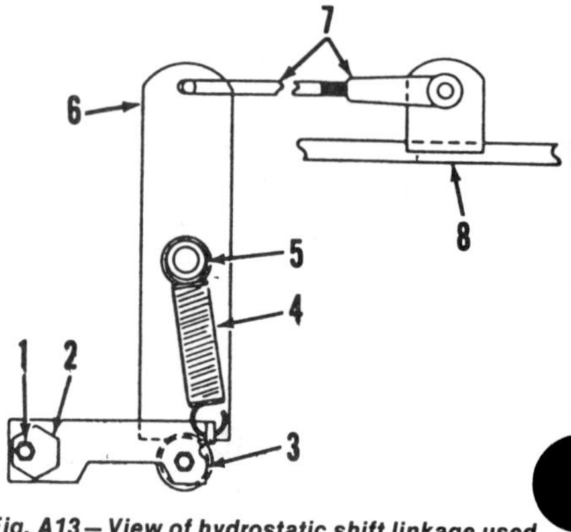

Fig. A13—View of hydrostatic shift linkage used on 1982 and later models. Refer to text for adjustment procedure.

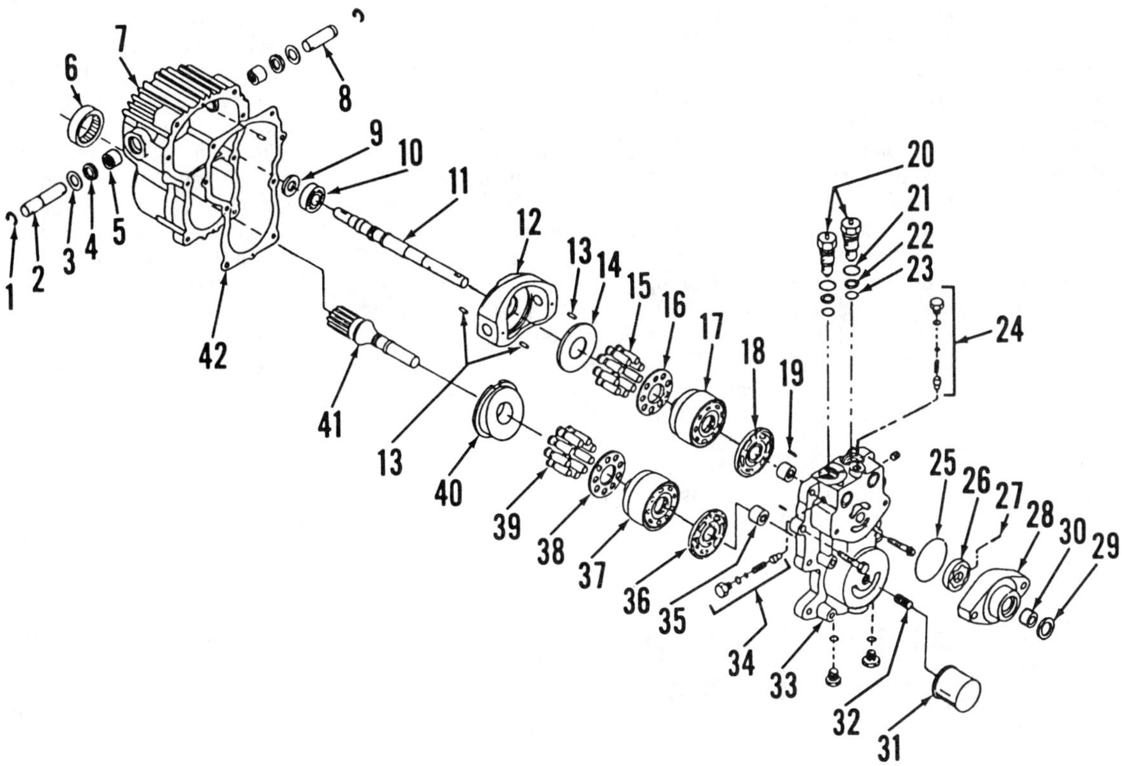

Fig. A15—Exploded view of hydrostatic drive unit.

1. Retaining ring (2)
2. Control shaft
3. Washer (2)
4. Seal (2)
5. Bearing (2)
6. Bearing
7. Housing
8. Trunnion shaft
9. Seal
10. Bearing
11. Pump shaft
12. Swashplate (pump)
13. Spring pins
14. Thrust plate
15. Pump pistons
16. Slipper retainer
17. Cylinder block (pump)
18. Valve plate
19. Dowel pin (2)
20. Check valves
21. Back-up washer
22. "O" ring
23. "O" ring
24. Implement lift relief valve
25. "O" ring
26. Rotor assy.
27. Drive pin
28. Charge pump housing
29. Seal
30. Bearing
31. Oil filter
32. Filter fitting
33. Center section
34. Charge relief valve
35. Bearing (2)
36. Valve plate
37. Cylinder block (motor)
38. Slipper retainer
39. Motor pistons
40. Swashplate (motor)
41. Motor shaft
42. Gasket

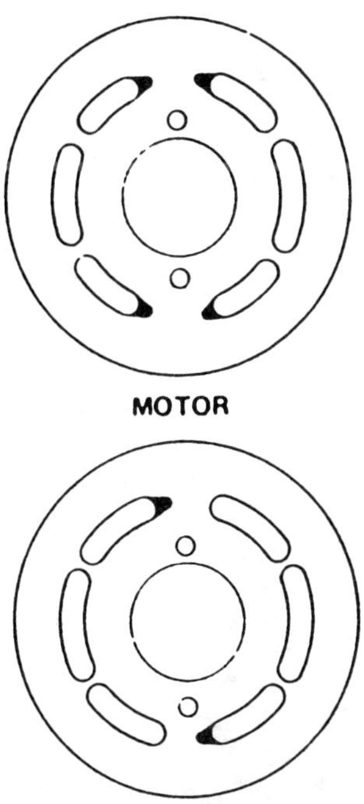

Fig. A16—Motor valve plate has four notches (dark areas) and pump valve plate has two notches (dark areas). Refer to text.

bearing (30) out front of housing. Remove check valves (20) and relief valves (24 and 34). Unbolt and lift off center section (33).

CAUTION: Valve plates (18 and 36) may stick to center housing. Be careful not to drop them. Remove and identify valve plates. Pump valve plate has two relief notches and motor valve plate has four notches. See Fig. A16.

Tilt housing (7–Fig. A15) on its side, identify and remove cylinder block and piston assemblies from pump and motor shafts. Lay cylinder block assemblies aside for later disassembly.

To remove pump swashplate (12), tilt and hold swashplate in full forward position while driving pins (13) out of shafts (2 and 8). Be careful not to damage housing when driving out pins. When pins are free of shafts, withdraw shafts and swashplate. Withdraw pump shaft (11) and bearing (10). Remove cap screws securing motor swashplate (40) to housing, then lift out swashplate and motor shaft (41). Bearings (5 and 6) and oil seals (4 and 9) can now be removed from housing. Remove needle bearings (35) from center section (33).

NOTE: Pump cylinder block and pistons are identical to motor cylinder block and pistons and complete assemblies are interchangeable. However, since pistons or cylinder blocks are not serviced separately, it is advisable to keep each piston set with its original cylinder block.

Carefully remove slipper retainer (16) with pistons (15) from pump cylinder block (17). Check cylinder block valve face and piston bores for scratches or other damage. Inspect pistons for excessive wear or scoring. Check piston slippers for excessive wear, scratches or embedded material. Make certain center oil passage is open in pistons. If excessive wear or other damage is noted on cylinder block or pistons, install a new cylinder block kit which includes pistons, slipper retainer and new cylinder block assembly. If original parts are serviceable, thoroughly clean parts, reassemble and wrap cylinder block assembly in clean paper. Repeat operation on motor cylinder block assembly (37 through 39–Fig. A15) using same checks as used on pump cylinder block assembly.

Check pump valve plate (18) and motor valve plate (36) for excessive wear or other damage and renew as necessary. Inspect motor swashplate (40) and pump swashplate thrust plate (14) for wear, embedded material or scoring and renew as required.

Check charge pump housing (28) and rotor assembly (26) for excessive wear, scoring or other damage and renew as necessary. Charge pump relief valve cone should be free of nicks and scratches.

Check valves (20) are interchangeable and are serviced only as assemblies. Wash check valves in clean solvent and air dry. Thoroughly lubricate check valve assemblies with clean oil before installation.

Renew all "O" rings, gaskets and seals and lubricate all parts with clean fluid. When installing new bearings (35), press bearings in until they are 0.100 inch (2.54 mm) above machined surface of the center housing. Pump swashplate (12) must be installed with thin stop pad towards top of transmission. Be sure control shaft (2) is installed on correct side. Drive new pins (13) into pump swashplate and shafts, using two pins on control shaft (2). Pins should be driven in until ¼-inch (6 mm) below surface of swashplate. Tighten motor swashplate cap screws to 67 in.-lbs. (7.5 N•m). Be sure pump and motor valve plates (18 and 36) are installed correctly and located on needle bearing (35) and pin (19). Tighten center section to housing cap screws to 30 ft.-lbs. (41 N•m). Rotate pump and motor shafts while tightening these screws to check for proper internal assembly.

Reinstall unit and prime with oil as previously outlined.

REDUCTION GEARS AND DIFFERENTIAL

MAINTENANCE

All Models

Refer to **MAINTENANCE** paragraph in **TRANSMISSION** section for maintenance schedule and fluid requirements.

R&R AND OVERHAUL

All Models

To remove gear reduction and differential unit, remove hydrostatic transmission as outlined in previous paragraph. Detach parking pawl assembly from housing and disconnect brake control rods. Support rear of tractor and unbolt axle brackets from frame. Roll unit forward until oil filter pipe is clear of rear hitch plate, raise tractor and roll assembly rearward from tractor.

To disassemble unit, drain lubricant from housing, then remove wheels and hubs (1–Fig. A17) from axles (6). Unbolt axle bearing retainer and withdraw axle and bearing assemblies from rear housing (7). Unbolt and separate front and rear housings (23 and 7). Be sure differential bearing caps are marked for correct reassembly as they are matched to front housing. Unbolt and remove bearing caps, then pry differential assembly out of housing. Drive a pointed punch through pinion shaft expansion plug and out of housing. Remove snap ring (19) and shim (18) from end of pinion shaft. Remove side cover (25) and place a screwdriver blade under edge of reduction gear (24) to prevent gear from binding. Press pinion gear (13) out of housing and remove outer bearing (21), spacer (17) and reduction gear (24). Press pinion shaft out of pinion bearing (14). Press bearing cups (15 and 22) out of housing.

To disassemble differential, drive lock pin (2) out of pinion shaft (3) as shown in Fig. A18. Drive pinion shaft (3) from housing. Rotate pinion gears (27–Fig. A17) 90° to openings in differential case and remove pinion gears (27), side gears (28) and thrust washers (26 and 29). To remove differential case bearings (9), use a suitable puller inserting puller jaws into indentations provided in differential case. Remove cap screws securing ring gear (12) to case, then drive ring gear off case using a hammer and wood block.

Clean and inspect all parts and renew any parts damaged or excessively worn.

Reassemble by reversing disassembly procedure. Reuse original shim pack (16), when reinstalling inner bearing cup (15) in original housing (23). If housing (23) is being renewed, determine proper shim pack by installing bearing cup (15) in housing without shims. Press bearing cone (14) on pinion shaft and position shaft and bearing in housing. Using a

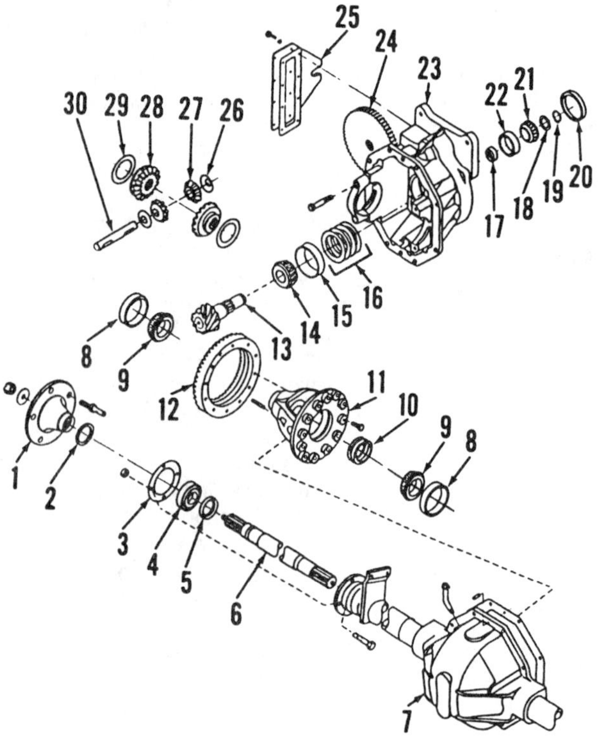

Fig. A17—Exploded view of gear reduction, differential and rear axle assembly.

1. Hub
2. Felt seal
3. Retainer
4. Axle bearing
5. Oil seal
6. Axle
7. Rear housing
8. Bearing cup
9. Bearing cone
10. Shims
11. Differential case
12. Ring gear
13. Pinion gear
14. Bearing cone (inner)
15. Bearing cup (inner)
16. Shims
17. Spacer
18. Shim
19. Snap ring
20. Expansion plug
21. Bearing cone (outer)
22. Bearing cup (outer)
23. Front housing
24. Reduction gear
25. Cover
26. Thrust washer (2)
27. Pinion gear (2)
28. Side gear (2)
29. Thrust washer (2)
30. Pinion shaft

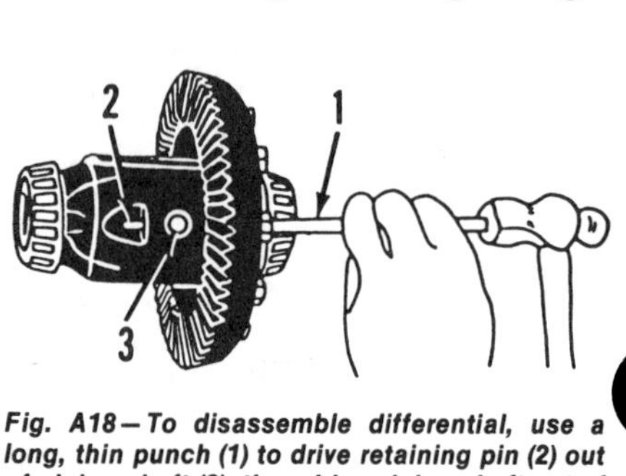

Fig. A18—To disassemble differential, use a long, thin punch (1) to drive retaining pin (2) out of pinion shaft (3), then drive pinion shaft out of differential case.

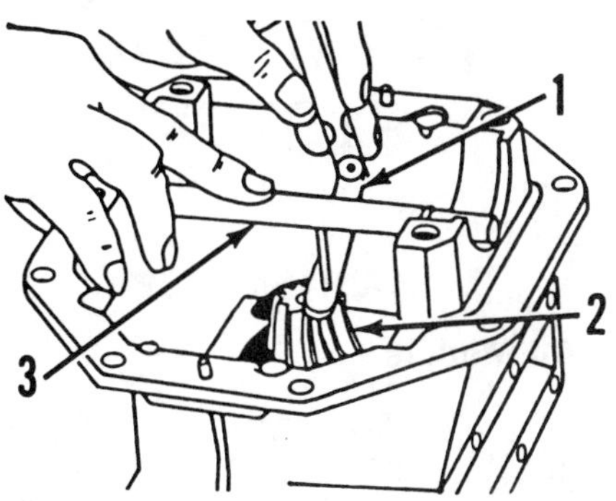

Fig. A19—If front housing is renewed, measure from bottom of bearing cradles to pinion gear surface as shown to determine pinion shaft shim pack thickness. Refer to text.

depth measuring tool similar to one shown in Fig. A19, measure distance from bottom of bearing cradles to pinion gear surface. Subtract measured dimension (in inches) from 1.2097 and difference will be required shim pack thickness (in inches). Remove inner bearing cup, install required shim pack and reinstall cup in housing.

Position pinion gear, reduction gear (chamfered side of splines towards pinion gear) and spacer in housing. Press outer bearing cone onto pinion shaft until a slight drag is felt when gear is turned by hand. Then, install thickest shim (18–Fig. A17) possible which will still allow installation of snap ring (19). Use sealer when installing expansion plug (20) and side cover (25).

Position ring gear (12) on differential case (11) and pull gear into place by tightening retaining cap screws evenly. Final torque on cap screws should be 50-55 ft.-lbs. (68-75 N·m).

Press bearing cones (9) on original differential case (11) using original shims (10). If differential case is being renewed, install 0.020 inch shim pack under each bearing. Position differential assembly in front housing with ring gear facing same side as reduction gear cover. Install bearing caps in their original positions and tighten cap screws to 40-45 ft.-lbs. (54-61 N·m).

Using a dial indicator, check for proper ring gear to pinion backlash of 0.003-0.007 inch (0.076-0.178 mm). If necessary, adjust backlash by moving shims (10) from one side to the other until correct backlash is obtained. To check gear teeth contact pattern, paint teeth with gear pattern compound, then rotate pinion while applying light load to ring gear. Compare contact area on teeth with patterns illustrated in Fig. A20. Correct as necessary. Desired tooth contact pattern on ring gear is shown at "A". To move toe pattern "B" towards heel, shim ring gear away from pinion, within 0.003-0.007 inch (0.076-0.178 mm) backlash limits. To move heel pattern "C" towards toe, shim ring gear closer to pinion, within backlash limits. If pattern is low "D", remove shims located under pinion inner bearing cup. If pattern is high "E", increase shim pack under pinion inner bearing cup.

Assemble front housing to rear housing and tighten retaining screws to 18-23 ft.-lbs. (24-32 N·m).

POWER TAKE-OFF

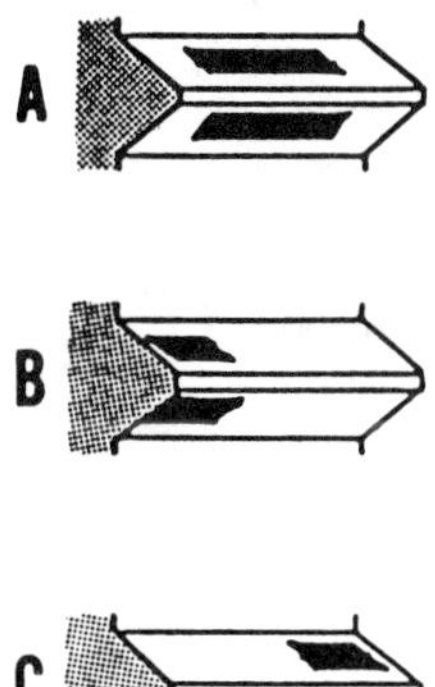

Fig. A20—Illustrations of typical gear teeth contact patterns encountered when checking ring gear and pinion. "A" pattern desired; "B" too close to toe; "C" too close to heel; "D" contact too low; "E" contact too high. Refer to text and correct as necessary.

ADJUSTMENT

All Models

With electric pto clutch disengaged, check clearance (C–Fig. A21) using a feeler gage inserted through each of the four slots (1) in brake flange. Adjust flange mounting nuts (5) until measured gap at each of the four slots (1) is 0.010-0.015 inch (0.254-0.381 mm).

PTO CLUTCH

All Models

R&R FRONT PTO CLUTCH. Front electric pto clutch is mounted on engine crankshaft and pto shaft (12–Fig. A22) rides in bearings and a tube (18) which supports front axle assembly. Belt tension is maintained by spring loaded idler pulley (5).

To remove front electric pto clutch, unbolt and remove brake flange (7). Remove cap screw (9) and spacer (8). Use a suitable puller to remove clutch pulley (6) and rotor (3). Unbolt and remove field assembly (1).

When reinstalling clutch, add or remove shims (4) as required to obtain 0.060-0.125 inch (1.524-3.175 mm) clearance (A) between rotor and armature. Tighten retaining cap screw (8) to 25 ft.-lbs. (34 N·m). Adjust clutch.

R&R REAR PTO CLUTCH. Remove rear pto shield and disconnect rear pto clutch electric lead. Loosen setscrew in

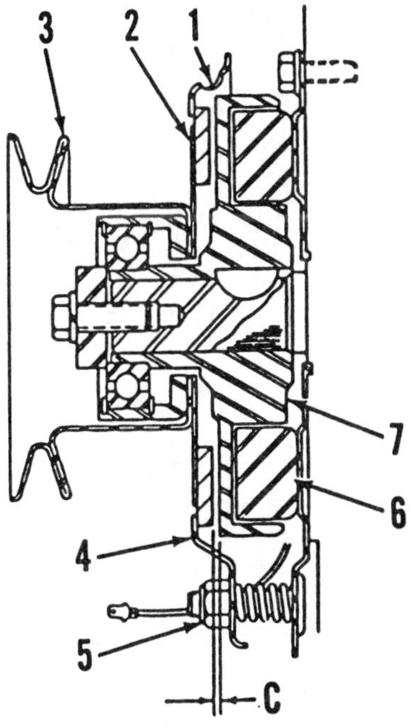

Fig. A21—Cross-sectional view of typical electric pto clutch. Refer to text for adjustment procedure.

1. Adjustment slot (4)
2. Armature-brake assy.
3. Pulley
4. Brake flange
5. Locknut
6. Field assy.
7. Rotor assy.

bearing lock collar (6 – Fig. A23) and use a drift punch to rotate and free collar. Disassemble bearing flange plates (2 and 4). Remove pto shaft (8) and clutch assembly (19). Continue disassembly as necessary.

Reinstall by reversing removal procedure. Note there must be clearance between snap ring (7) and collar (6). Failure to leave clearance will result in damage to charge pump, rear seal or both.

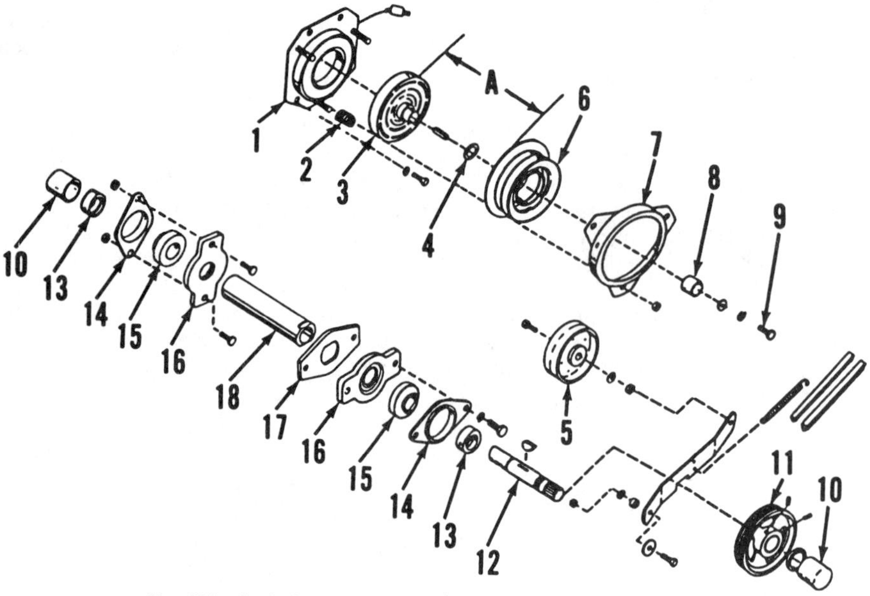

Fig. A22—Exploded view of front electric clutch and pto assembly.

1. Field assy.
2. Spring
3. Rotor assy.
4. Shim
5. Idler pulley
6. Drive pulley
7. Brake flange
8. Spacer
9. Retainer screw
10. Rubber cover
11. Driven pulley
12. Pto shaft
13. Locking collar
14. Flange plates
15. Bearing
16. Flange
17. Anti-pivot plate
18. Tube

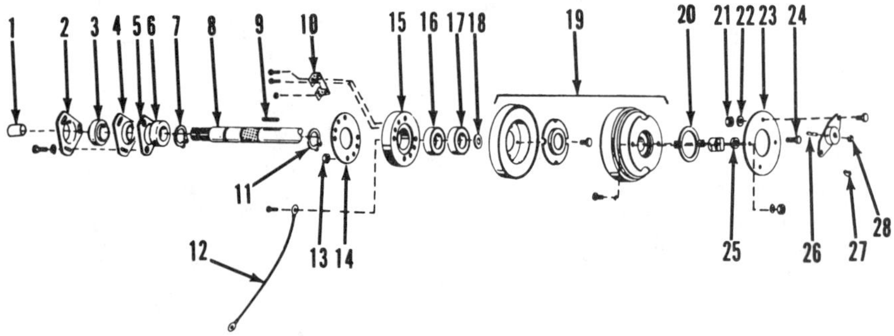

Fig. A23—Exploded view of rear electric clutch and pto assembly.

1. Cap
2. Flange plate
3. Bearing
4. Flange plate
5. Spacer flange
6. Collar
7. Snap ring
8. Rear pto shaft
9. Key
10. Link assy.
11. Snap ring
12. Ground wire
13. Locknut
14. Ring
15. Bearing hub
16. Bearing
17. Bearing
18. Washer
19. Clutch assy.
20. Pilot plate
21. Nut
22. Washer
23. Coupling
24. Cap screw
25. Retainer
26. Stud
27. Key
28. Locknut

HYDRAULIC SYSTEM

MAINTENANCE

All Models

Refer to **MAINTENANCE** paragraph in **TRANSMISSION** section for maintenance schedule and fluid requirements.

PRESSURE CHECK

All Models

To check hydrostatic drive transmission charge pressure and system (lift) operating pressure, install a 0 to 1000 psi (0 to 7000 kPa) test gage as shown in Fig. A24. Raise and support rear of tractor so tires are off the ground. Start and run engine at ¾ throttle. Move transmission speed control lever to full forward or reverse position.

CAUTION: Use care when working near rotating parts and rear tires.

Transmission charge pressure should be 70 to 150 psi (483 to 1034 kPa).

To test lift system pressure, place transmission speed control lever in neutral position. With engine running at ¾ throttle, work control valve lever so lift cylinder reaches end of stroke. Gage pressure reading at this time should be 550 to 700 psi (3792 to 4826 kPa).

If pressures are low, refer to **HYDROSTATIC TRANSMISSION** section for service procedures.

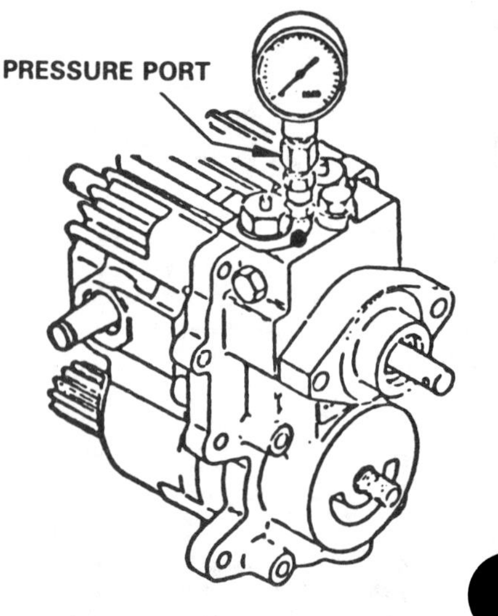

Fig. A24 – View showing location of test port used to check hydrostatic unit charge pump and lift system pressure. Refer to text.

BOLENS

CONDENSED SPECIFICATIONS

	MODELS			
	QT-16	1600	1700	1900
Engine Make	Onan	B&S	Kohler	Kohler
Model	BF	402437	KT-17	KT-19
Bore	3.125 in. (79.38 mm)	3.44 in. (87.3 mm)	3.12 in. (79.3 mm)	3.12 in. (79.3 mm)
Stroke	2.625 in. (66.68 mm)	2.16 in. (54.8 mm)	2.75 in. (69.9 mm)	3.06 in. (78 mm)
Displacement	40.3 cu. in. (660 cc)	40 cu. in. (656 cc)	42.18 cu. in. (691 cc)	46.98 cu. in. (771 cc)
Idle Speed	1200 rpm	1200 rpm	1200 rpm	1200 rpm
High Speed (No-Load)	3600 rpm	3600 rpm	3600 rpm	3600 rpm
Capacities–				
Crankcase	4 pts. (1.9 L)	3 pts.* (1.4 L)	3 pts. (1.4 L)	3 pts. (1.4 L)
Hydraulic system	10 qts. (9.5 L)	....	....	10 qts. (9.5 L)
Transaxle	....	4 pts. (1.9 L)	4 pts. (1.9 L)	....
Right Angle Gear Box	....	4 oz. (118 mL)	4 oz. (118 mL)	....
Fuel Tank	3.5 gal. (13.2 L)	3.5 gal. (13.2 L)	3.5 gal. (13.2 L)	3.5 gal. (13.2 L)

*Early production engine crankcase capacity is 3.5 pints (1.6 L).

	MODELS		
	HT-18	HT-20	HT-23
Engine Make	Kohler	Kohler	Kohler
Model	K-482	K-532	K-582
Bore	3.25 in. (82.6 mm)	3.38 in. (85.7 mm)	3.5 in. (88.9 mm)
Stroke	2.88 in. (73 mm)	3 in. (76.2 mm)	3 in. (76.2 mm)
Displacement	47.8 cu. in. (782 cc)	53.7 cu. in. (880 cc)	57.7 cu. in. (946 cc)
Idle Speed	1000 rpm	1000 rpm	1000 rpm
High Speed (No-Load)	3600 rpm	3600 rpm	3600 rpm
Capacities–			
Crankcase	3 qts.** (2.8 L)	3 qts.** (2.8 L)	3 qts.** (2.8 L)
Hydraulic System & Reduction Gear	10 qts. (9.5 L)	10 qts. (9.5 L)	10 qts. (9.5 L)
Fuel Tank	5 gal. (18.9 L)	5 gal. (18.9 L)	5 gal. (18.9 L)

**Add 0.5 qt. (0.47 L) if filter is also changed.

FRONT AXLE AND STEERING SYSTEM

MAINTENANCE

All Models

Lubricate left and right spindles, axle pivots, steering shaft and steering arm, tie rods and drag link at 10 hour intervals. Use multi-purpose grease where grease fittings are present and engine oil at remaining locations.

Repack front wheel bearings on all models annually. Use a good quality wheel bearing grease.

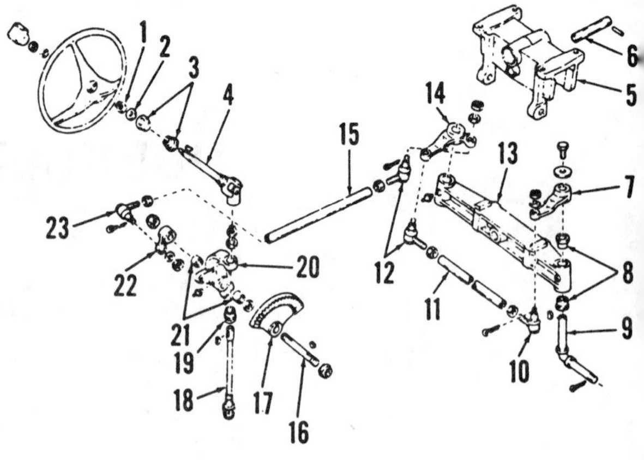

Fig. BO1 – Exploded view of front axle and steering system used on QT-16 model.

1. Retainer ring
2. Washer
3. Shaft bearings
4. Upper shaft assy.
5. Axle support
6. Pivot pin
7. Steering arm
8. Bearing set
9. Spindle
10. Ball joint
11. Tie rod
12. Ball joints
13. Axle main member
14. Steering arm
15. Drag link
16. Sector shaft
17. Steering sector
18. Lower shaft & pinion
19. Flange bushing
20. Support casting
21. Bushings
22. Pitman arm
23. Ball joint

R&R AXLE MAIN MEMBER

Model QT-16

Raise and support front of tractor. Disconnect drag link (15–Fig. BO1) at steering arm (14). Support axle main member (13), remove roll pin from pivot pin (6) and push pin out to rear of axle support (5). Lower axle and roll assembly away from tractor.

Reinstall by reversing removal procedure.

Model 1600

Raise and support front of tractor. Disconnect drag link (17–Fig. BO2) at steering arm (20). Support axle main member (25) and remove pivot bolt (27) and axle support mounting bolts. Remove axle support and roll axle assembly away from tractor.

Reinstall by reversing removal procedure.

Models 1700-1900

Raise and support front of tractor. Disconnect drag link (3–Fig. BO3) at steering arm (8). Support axle main member (12), remove pin (11) and pull pivot tube (9) out of main member and axle support (10). Lower axle main member and roll assembly away from tractor.

Reinstall by reversing removal procedure.

Models HT-18 – HT-20 – HT-23

Raise and support front of tractor. Remove pto drive belts. Disconnect drag link end (12–Fig. BO4) at steering arm (16). Support axle main member (15) and remove pivot pin (17). Lower main member assembly and roll axle assembly forward. Front pto shaft should slide out of axle.

Reinstall by reversing removal procedure.

TIE ROD AND TOE-IN

All Models

Drag link and tie rod each have renewable ends. Toe-in is adjusted by turning tie rod ends to lengthen or shorten tie rod to provide 1/8 to 3/8-inch (3.18 to 9.53 mm) toe-in. Tighten tie rod end jam nuts to lock ends in correct position.

R&R STEERING SPINDLES

All Models

Raise and support front of tractor. Disconnect drag link at steering arm and tie rod ends at each steering arm or spindle. Remove tire, wheel and hub assemblies. Remove steering arms and keys and pull spindles down out of axle main member.

Renew all bushings, bearings and thrust washers as necessary. Reinstall by reversing removal procedure.

R&R AND OVERHAUL STEERING GEAR

Model QT-16

Remove logo cap from hub of steering wheel, back off retainer nut and remove wheel. Remove Woodruff key from tapered portion of upper shaft assembly (4–Fig. BO1). Disconnect ball joint (23) from pitman arm (22) to separate drag link. Remove nut from inboard end of sector shaft (16) then bump shaft out of steering sector (17) and while holding sector in one hand, grasp pitman arm (22) and pull shaft clear of steering sup-

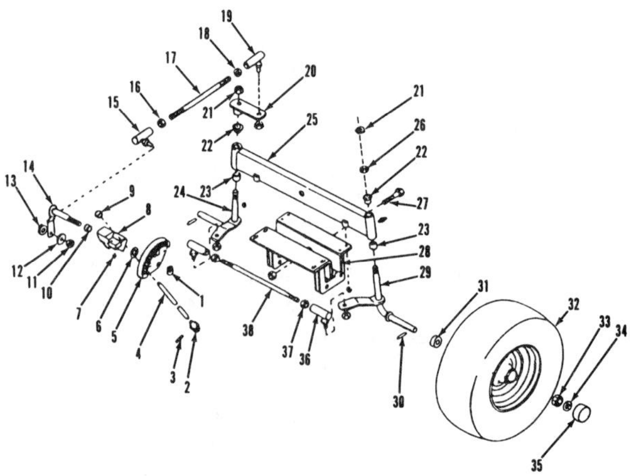

Fig. BO2 – Exploded view of front axle and steering system used on 1600 model.

1. Retaining ring
2. Pinion gear
3. Pin
4. Steering shaft
5. Gear
6. Shim
7. Grease fitting
8. Steering support
9. Bushing
10. Spacer
11. Locknut
12. Spacer
13. Spacer
14. Steering arm
15. Drag link end
16. Jam nut
17. Drag link
18. Jam nut
19. Drag link end
20. Steering arm
21. Locknut
22. Bushing
23. Bushing
24. Spindle
25. Axle main member
26. Washer
27. Pivot bolt
28. Axle support
29. Spindle
30. Pin
31. Bearing
32. Tire & hub
33. Bearing
34. Washer
35. Dust cap
36. Tie rod end
37. Jam nut
38. Tie rod

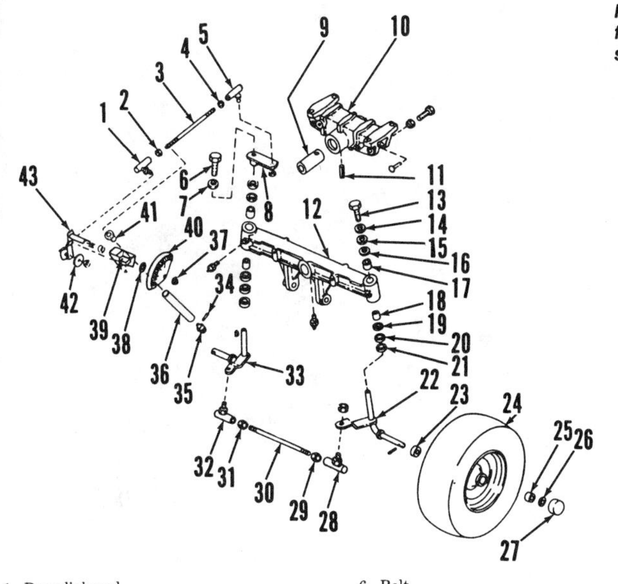

Fig. BO3—Exploded view of front axle and steering system used on 1700 and 1900 models.

1. Drag link end
2. Jam nut
3. Drag link
4. Jam nut
5. Drag link end
6. Bolt
7. Washer
8. Steering arm
9. Pivot tube
10. Axle pivot support
11. Machine screw
12. Axle main member
13. Bolt
14. Washer
15. "O" ring
16. Thrust washer
17. Bushing
18. Bushing
19. Thrust washer
20. Thrust race
21. Thrust race
22. Spindle
23. Bearing
24. Tire & hub
25. Bearing
26. Washer
27. Dust cap
28. Tie rod end
29. Jam nut
30. Tie rod
31. Jam nut
32. Tie rod end
33. Spindle
34. Pin
35. Pinion gear
36. Steering shaft
37. Retainer ring
38. Shim washer
39. Support bracket
40. Gear
41. Bushing
42. Spacer
43. Pitman arm

port casting (20) with pitman arm attached. If shaft (16) is to be renewed, remove pitman arm (22).

To remove upper and lower steering shafts (2 and 18), remove retaining rings (1) from each shaft so shafts can be separated at swivel coupling at lower end of upper shaft. Take care not to lose Woodruff key. Remove lower shaft (18) with pinion from under tractor.

Lubricate during reassembly. Set front wheels straight ahead, center quadrant gear on pinion and adjust length of drag link by its threaded ends to equalize turning radius each way from centered position of steering wheel.

Model 1600

Disconnect drag link from steering arm (14–Fig. BO2). Remove snap ring (1) from steering arm (14). Keep track of all shims and their locations during disassembly. Remove steering arm shaft, shims and quadrant gear. Remove roll pin from steering wheel and withdraw steering shaft from bottom side of tractor. It may be necessary to tap steering shaft down until flange bushing is pushed out of steering support (8).

Check bushings and gears for wear and renew as necessary. Reassemble steering gear by reversing disassembly procedure. If pinion and quadrant gears do not engage fully, adjust shims between quadrant gear and steering support. Adjust drag link as required to have equal turning radius on right or left turns.

Models 1700-1900

Disconnect drag link end (1–Fig. BO3) at pitman arm (43). Remove retainer ring (37) and pull pitman arm from support (39) and gear (40). Remove shims (38) and gear (40). Remove pin (34) and pinion gear (35).

Reassemble by reversing removal procedure. Adjust gear mesh by varying number and thickness of shims (38). Gear backlash should be minimal, however, steering wheel should turn through entire left/right cycle with no binding.

Models HT-18 – HT-20 – HT-23

Disconnect drag link (13–Fig. BO4) from lever portion of quadrant gear (8). Remove logo plate from center of steering wheel followed by retaining nut and washers so steering wheel can be pulled off steering shaft (5). Take care not to lose Woodruff key from tapered portion of shaft. Remove cotter key from lower end of steering shaft (5) below support plate (9) so shaft can be lifted up to disengage its pinion from quadrant (8) and clear of support plate. Tip shaft and column so shaft can be lowered out of tube down past rear edge of support plate (9) and out from under tractor. Take care not to damage self-aligning bearing (2) and protect bronze bushing (4) from dirt. Back out pivot bolt and remove quadrant gear (8).

Thoroughly clean and inspect gears and bushings for undue wear and damage. All parts are serviced for individual renewal as needed.

When reinstalling, set steering wheel and tractor front wheels in straight ahead position and reinstall quadrant gear (8) properly centered on steering pinion of shaft (5). Engine oil is specified as correct lubricant for steering shaft, bushings and drag link ends.

Steering radius stops which limit right and left movement of quadrant gear are fitted in right side frame rail. Be sure these adjusters (square-head setscrews) are set to limit turning radius so as to prevent wheel interference with undertractor attachments and that locknuts are secure.

POWER STEERING CONTROL VALVE AND CYLINDER ASSEMBLY

All Models So Equipped

Power steering cylinder and control valve are a single assembly. Refer to Fig. BO5 exploded view to aid disassembly.

Renew all "O" rings and seals and inspect cylinder body (27) for scratches,

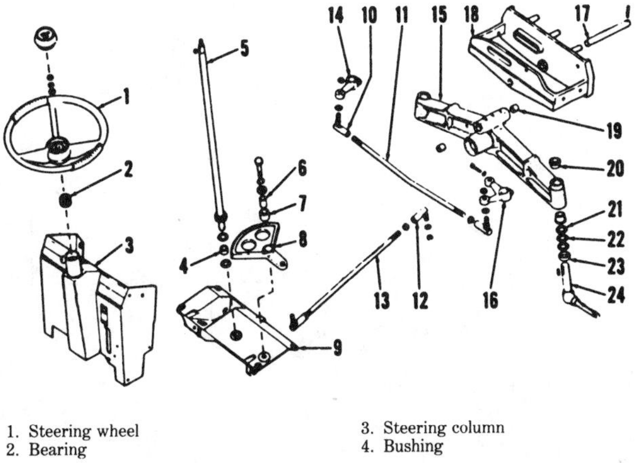

Fig. BO4—Exploded view of front axle and steering system used on HT-18, HT-20 and HT-23 models.

1. Steering wheel
2. Bearing
3. Steering column
4. Bushing
5. Steering shaft
6. Spacer
7. Bushing
8. Quadrant gear
9. Steering support
10. Tie rod end
11. Tie rod
12. Drag link end
13. Drag link
14. Steering arm
15. Axle main member
16. Steering arm
17. Pivot pin
18. Axle support
19. Pivot bushing
20. Flange bearing
21. Thrust race
22. Thrust bearing
23. Dust seal
24. Steering knuckle

scores or other damage. Lubricate all parts during reassembly. After installation, adjust tie rod (3) so it is 20.75 inches (527 mm) in length and piston rod is extended 2.25 inches (57 mm) when front wheels are in straight ahead position. See Fig. BO6. Adjust stops (1 and 2) so cylinder does not bottom out during full left or right turns.

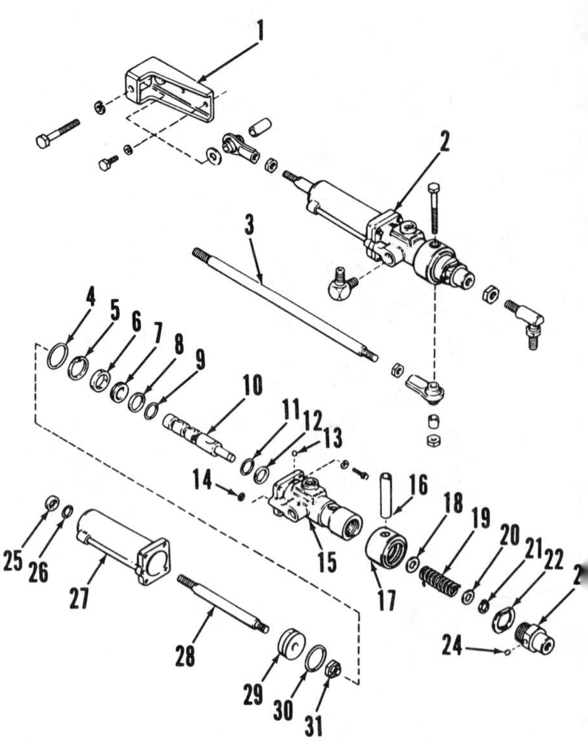

Fig. BO5—View showing assembled power steering control valve and cylinder (2) and exploded view of control valve and cylinder (4 through 31).

1. Bracket
2. Control valve/cylinder assy.
3. Drag link
4. "O" ring
5. Retaining ring
6. Spacer
7. Plug
8. "O" ring
9. "O" ring
10. Spool
11. "O" ring
12. Back-up washer
13. Filter
14. "O" ring
15. Valve body
16. Sleeve
17. Rubber seal
18. Washer
19. Spring
20. Washer
21. Retaining ring
22. Lock ring
23. Cap & breather assy.
24. Filter
25. Seal
26. "O" ring
27. Cylinder body
28. Piston rod
29. Piston
30. "O" ring
31. Locknut

ENGINE

MAINTENANCE

All Models

Regular engine maintenance is required to maintain peak performance and long engine life.

Check oil level and clean air intake screen at five hour intervals under normal operating conditions.

Change engine oil and filter, perform tune-up, valve adjustment and clean carbon from cylinder heads as recommended in appropriate ENGINE section of this manual.

REMOVE AND REINSTALL

Model QT-16

Raise hood and loosen the three screws at bottom of each side panel and release the two quarter-turn fasteners at top of each panel for removal. Disconnect all necessary electrical wiring and remove battery. Disconnect fuel line at carburetor and cap openings. Disconnect throttle and choke cables. Remove pto drive belts. Separate rear of engine crankshaft from driveshaft by removing through-bolt and Woodruff key (as equipped). Remove engine mounting bolts, raise engine and shift it forward while lifting it from engine compartment.

Reinstall by reversing removal procedure.

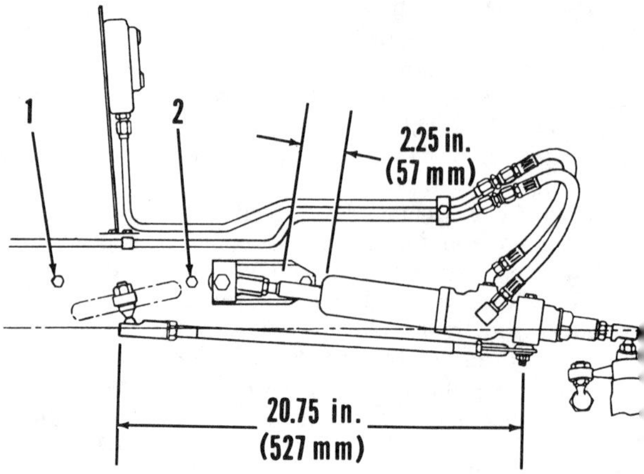

Fig. BO6—View showing location of cylinder stop adjustment bolts (1 and 2) and correct measurements for power steering system adjustments. Refer to text.

All Other Models

Disconnect battery cables and headlight wiring. Remove hood and stop arm assembly. Remove bolt and setscrew at forward end of driveshaft. Disconnect fuel line and cap opening. Remove fuel tank and battery platform as an assembly. Disconnect throttle and choke cables at carburetor. Remove pto drive belts and engine mounting bolts. Slide engine forward to disengage driveshaft and remove engine.

Reinstall by reversing removal procedure.

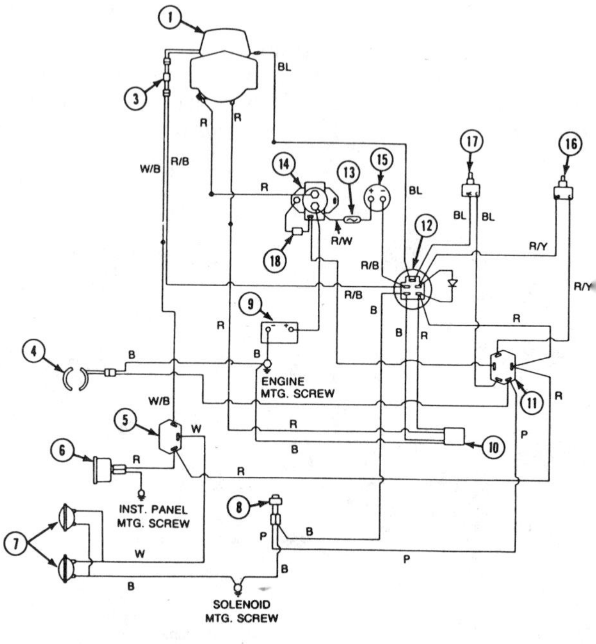

Fig. BO7—Wiring diagram for 1600 models. Color code is as follows: R-Red, B-Black, W-White, P-Purple, BL-Blue, R/W-Red/White, R/B-Red/Black, R/Y-Red/Yellow, W/B-White/Black.

1. Engine
3. Diode
4. Electric clutch
5. Light switch
6. Hourmeter
7. Headlights
8. Attachment drive indicator light
9. Battery
10. Electric plug
11. Attachment drive switch
12. Ignition switch
13. Fuse (30 amp)
14. Solenoid
15. Ammeter
16. Brake interlock switch
17. Seat interlock switch
18. Capacitor

OVERHAUL

All Models

Engine make and model are listed at the beginning of this section. To overhaul engine or accessories, refer to appropriate engine section of this manual.

ELECTRICAL SYSTEM

MAINTENANCE

All Models

Battery electrolyte level should be checked at 50 hour intervals of normal operation. If necessary, add distilled water until level is just below base of vent well. **DO NOT** overfill. Keep battery posts clean and cable ends tight.

For alternator or starter service, refer to appropriate engine accessory section in this manual.

Refer to Fig. BO7 through BO10 for appropriate wiring diagram for model being serviced.

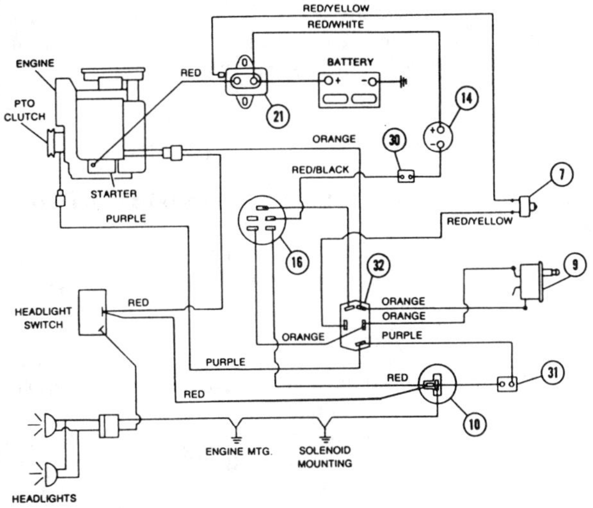

Fig. BO8—Wiring diagram for 1700 models.

7. Switch
9. Switch
10. Hourmeter
14. Ammeter
16. Ignition switch
21. Solenoid
30. Circuit breaker
31. Clutch indicator lamp
32. Switch (pto)

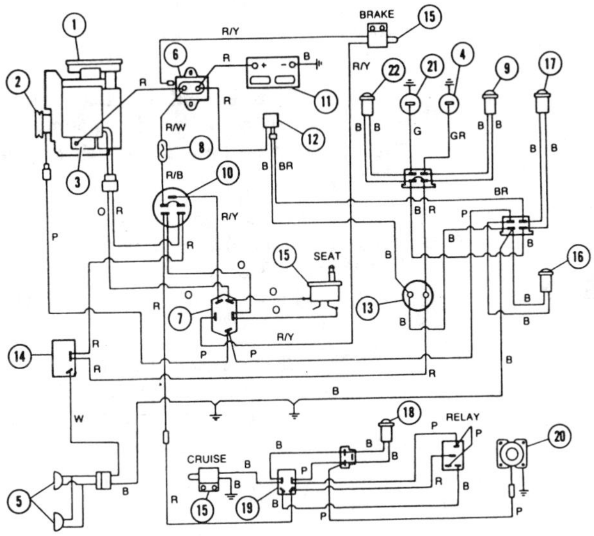

Fig. BO9—Wiring diagram for 1900 models. Color code as follows: R-Red, B-Black, W-White G-Green, O-Orange, P-Purple, BR-Brown, GR-Gray, R/Y-Red/Yellow, R/W-Red/White, R/B-Red/Black.

1. Engine
2. Electric clutch
3. Starter
4. Oil pressure switch
5. Headlights
6. Solenoid
7. Attachment drive switch
8. Fuse (25 amp)
9. Oil pressure light
10. Ignition switch
11. Battery
12. Voltage sensor
13. Hourmeter
14. Light switch
15. Interlock switch
16. Pto light
17. Battery light
18. Cruise light
19. Cruise switch
20. Cruise clutch
21. Oil temp. switch
22. Oil temp. light

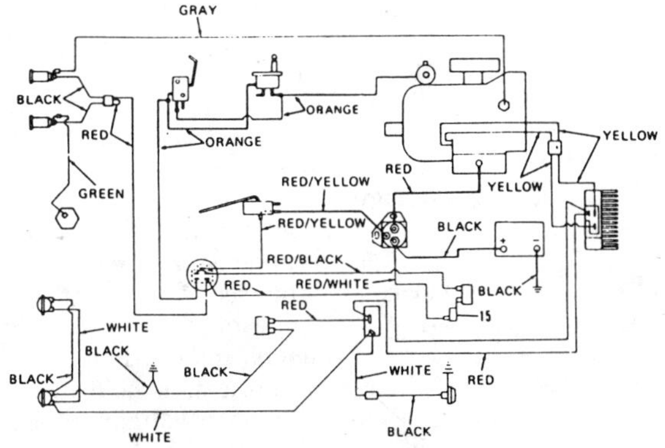

Fig. BO10—Typical wiring diagram similar to HT-18, HT-20 and HT-23. Models with varying serial numbers may be slightly different.

BRAKES

ADJUSTMENT

All Models

Adjust nut (B–Fig. BO11) on brake rod until 0.010 inch (0.254 mm) clearance between brake pad and brake disc (A) is obtained.

R&R BRAKE PADS

All Models

Remove brake rods (17–Fig. BO12) from brake arms (19 and 20) and remove caliper assembly from tractor. Remove bolt (6) and pads (5).

Reinstall by reversing removal procedure. Adjust brakes.

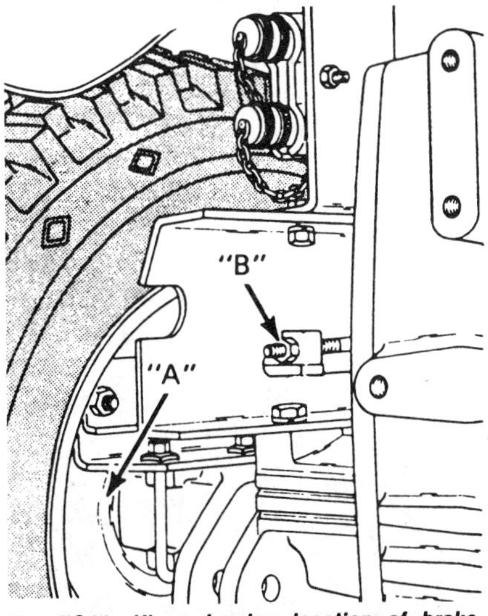

Fig. BO11—View showing location of brake caliper and disc (A) and brake adjustment nut (B). Refer to text.

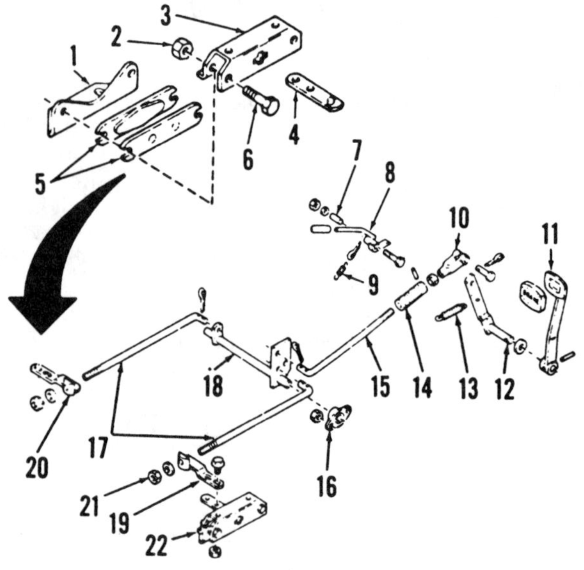

Fig. BO12 – Exploded view of brake system showing component parts and their locations.

1. Brake support
2. Locknut
3. Bracket
4. Lever
5. Pad plates
6. Bolt
7. Spacer
8. Latch
9. Extension spring
10. Clevis
11. Pedal
12. Brake arm
13. Return spring
14. Ratchet
15. Adjusting rod
16. Flange bearing
17. Brake rods (2)
18. Transfer shaft
19. Arm extension
20. Arm extension
21. Locknut
22. Brake assy. R.H.

HYDROSTATIC DRIVE SYSTEM

MAINTENANCE

All Hydrostatic Drive Models

Check hydrostatic drive fluid level at 30 hour intervals and maintain fluid level in operating zone marked on dipstick. Change hydrostatic drive unit oil filter at 100 hour intervals and change fluid and filter at 500 hour intervals or when fluid becomes discolored. Recommended fluid is Bolens oil number 1738157, or equivalent. Approximate fluid capacity is 10 quarts (9.5 L).

ADJUSTMENT

All Hydrostatic Drive Models

Remove seat and fender assemblies. Be sure free wheeling valve on left side of unit is closed (as equipped), and fluid level is correct. Raise and support rear of tractor so rear wheels are off the ground. With travel pedal in neutral, start engine and release parking brake. If rear wheels are rotating, loosen nuts (A – Fig. BO13) at rear end of rod (B) as necessary to stop wheel rotation. Tighten nuts (A) to maintain adjustment.

R&R HYDROSTATIC DRIVE UNIT

Models QT-16 – HT-18 – HT-20

Remove seat and fender assembly. Drain hydrostatic fluid. Loosen setscrew and remove bolt which secures driveshaft to engine crankshaft and slide driveshaft forward off of hydrostatic drive unit. Disconnect and remove all hydraulic lines. Cap all openings. Remove fluid filter. Disconnect control linkage at control shaft and remove unit mounting screws. Lift hydrostatic drive unit out of tractor.

Reinstall by reversing removal procedure.

Models 1900 – HT-23

Eaton hydrostatic drive unit must be removed with the gear reduction drive unit. Refer to REDUCTION GEAR section in this manual.

OVERHAUL

Models QT-16 – HT-18 – HT-20

Remove hydrostatic drive unit as previously outlined and clean exterior of unit.

Remove all hydraulic fittings. Drive roll pin out of control lever and shaft and remove control lever. Remove snap ring and pinion gear from motor shaft. Remove the four cap screws securing motor housing (3 – Fig. BO14) to pump case (53), then lift motor housing off pump case and center section. Remove snap ring (1), motor shaft (5) and bearing (2). Remove thrust plate (4), cylinder block assembly (6) and motor valve plate (14).

NOTE: Care should be take· when removing cylinder block assemblies. Pistons (7) must be kept in their original bores.

Remove center section (21) and pump valve plate (65). Discard all gaskets and "O" rings. Carefully remove pump cylinder block assembly (57) and thrust plate (26).

NOTE: Care should be taken when removing cylinder block assemblies. Pistons (58) must be kept in their original bores.

Drive spring pins (54) out of swashplate (27). Remove trunnion shaft (52), washers (48) and swashplate.

Note position of charge pump housing (38), remove the four cap screws and housing. Do not misplace serial number plate. Remove outer rotor gear (37) and rotor (41). Remove pin (42), snap ring (28), pump shaft (56) and bearing (29).

Clean and inspect all parts. Renew all parts showing excessive wear or damage.

Piston retainers (8 and 59) must be flat. Pistons and piston bores in cylinder blocks must be free of scores or damage and outer edge of piston slipper must not be rounded more than 1/32-inch (0.79 mm). Lubrication hole in slipper face must be open. Minimum slipper thickness is 0.121 inch (3.073 mm) for

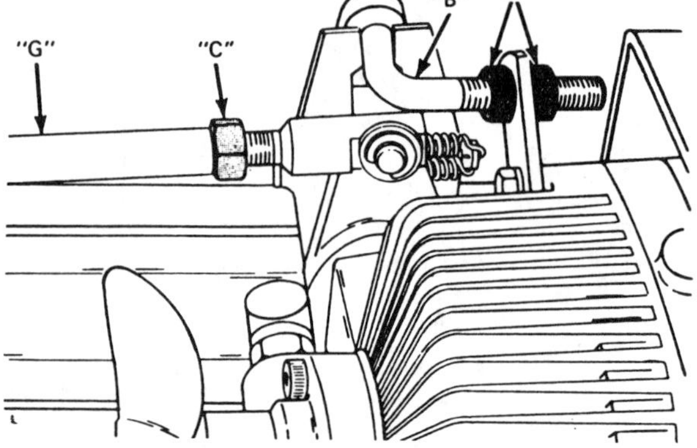

Fig. BO13 – View showing hydrostatic drive linkage neutral adjustment points. Refer to text.

A. Jam nuts
B. Control rod
C. Jam nut
G. Pedal rod

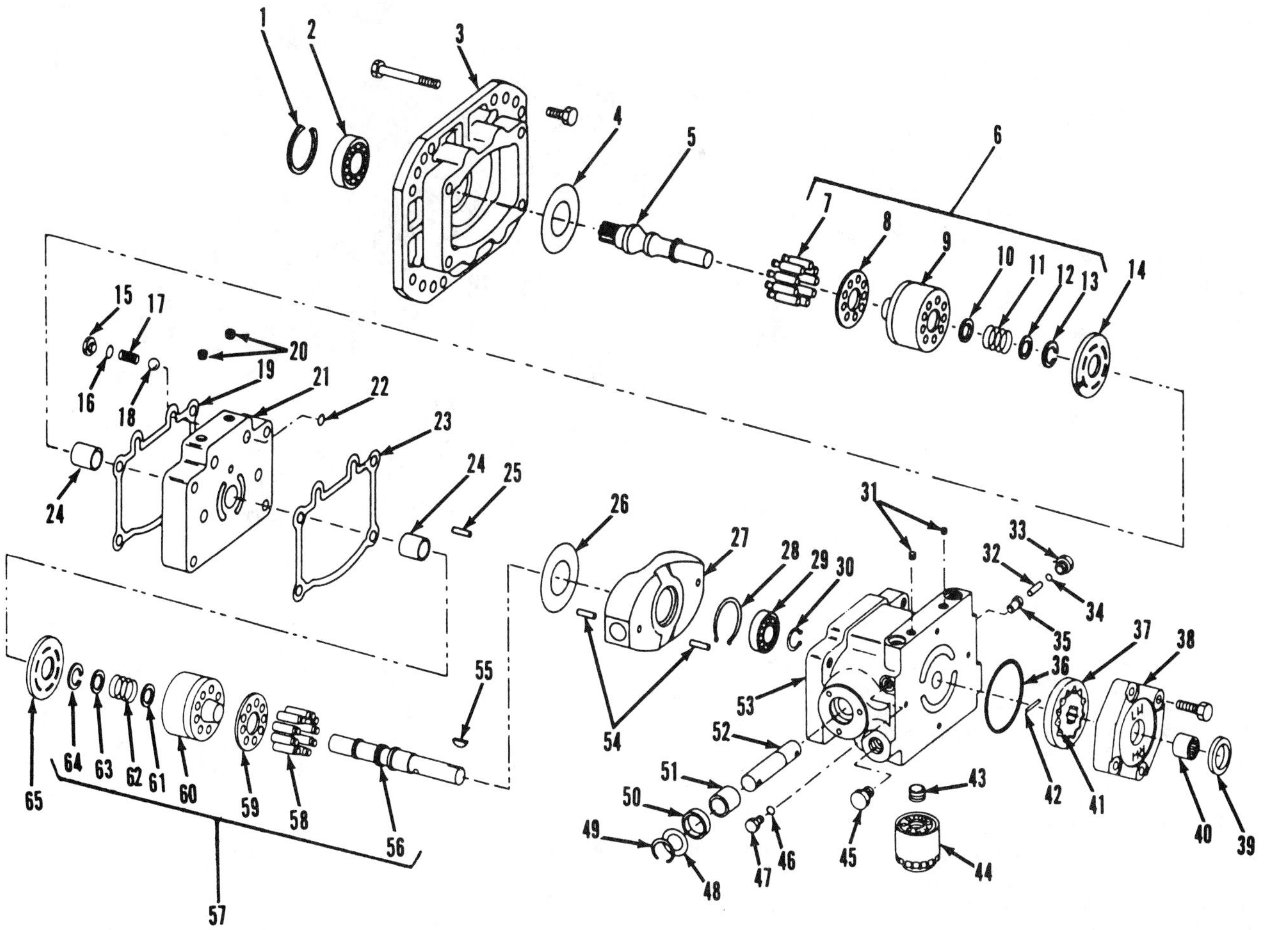

Fig. BO14—Exploded view of hydrostatic drive unit used on QT-16, HT-18 and HT-20 models.

1. Retaining ring
2. Bearing
3. Motor housing
4. Thrust plate
5. Motor shaft
6. Motor cylinder block assy.
7. Pistons (9 used)
8. Slipper retainer
9. Cylinder block
10. Washer
11. Spring
12. Washer
13. Snap ring
14. Valve plate (motor)
15. Check valve plug (2)
16. "O" ring
17. Spring (2 used)
18. Ball (2 used)
19. Gasket
20. Pipe plugs
21. Center section
22. "O" ring (2)
23. Gasket
24. Bearing (2)
25. Pin (2)
26. Thrust plate
27. Swashplate
28. Retaining ring
29. Bearing
30. Retaining ring
31. Pipe plugs
32. Plug (service only)
33. Plug
34. "O" ring
35. Cone
36. "O" ring
37. Stator
38. Charge pump housing
39. Seal
40. Bearing
41. Rotor
42. Pin
43. Oil filter fitting
44. Oil filter
45. Plug
46. "O" ring
47. Plug
48. Washer (2)
49. Retaining ring (2)
50. Seal
51. Bearing
52. Trunnion shaft (2)
53. Pump housing
54. Pins (2)
55. Key
56. Pump shaft
57. Pump cylinder block assy.
58. Pistons (9)
59. Slipper retainer
60. Cylinder block (pump)
61. Washer
62. Spring
63. Washer
64. Retaining ring
65. Valve plate (pump)

both motor and pump. Slipper thickness must not vary more than 0.002 inch (0.051 mm) for all pistons. Piston and cylinder block assemblies are serviced as assemblies only.

Check pump valve plate (65) and motor valve plate (14) for excessive wear or damage. If wear or scratches can be felt by running fingernail across face of plates, renew plates. Pump valve plate has two notches and motor plate has four notches. See Fig. BO15.

Reinstall by reversing removal procedure.

Models 1900 – HT-23

Remove hydrostatic drive unit as previously outlined and drain unit.

Place unit in holding fixture with input shaft pointing up. Remove dust shield (1–Fig. BO16) and snap ring (3). Remove cap screws from charge pump body (7). One cap screw is ½-inch (12.7 mm) longer than the others and must be installed in original position. Remove charge pump body (7) with ball bearing (4). Ball bearing and oil seal (6) can be removed after removing retaining ring (2). Remove snap rings (5 and 8) and charge pump rotor assembly. Remove "O" rings (10) and pump plate (11). Turn hydrostatic unit over in fixture and remove output gear. Unscrew the two cap screws until two threads are engaged. Raise body (42) until it contacts cap screw heads. Insert a special fork tool (Fig. BO17) between motor rotor (39–Fig. BO16) and pintle (28). Remove cap screws, lift off body and motor assembly with fork tool and place assembly on a bench or in a holding fixture with output shaft pointing down. Remove fork and place a wide rubber band around motor rotor to hold ball pistons (38) in their bores. Carefully remove motor rotor assembly and lay aside for later disassembly. Remove motor race (41) and output shaft (40). Remove retainer (45), bearing (44) and oil seal (43). With housing assembly (12) resting in holding fixture, remove pintle assembly (28).

CAUTION: Do not allow pump to raise with pintle as ball pistons (22) may fall out of rotor (21). Hold pump in position by inserting a finger through hole in pintle.

Remove plug (37), spring (36) and charge relief ball (35). To remove direc-

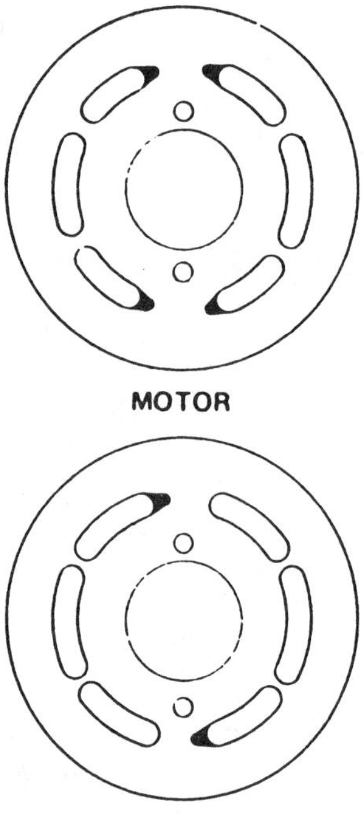

Fig. BO15—Motor valve plate has four slots (dark areas) and pump valve plate has two slots (dark areas).

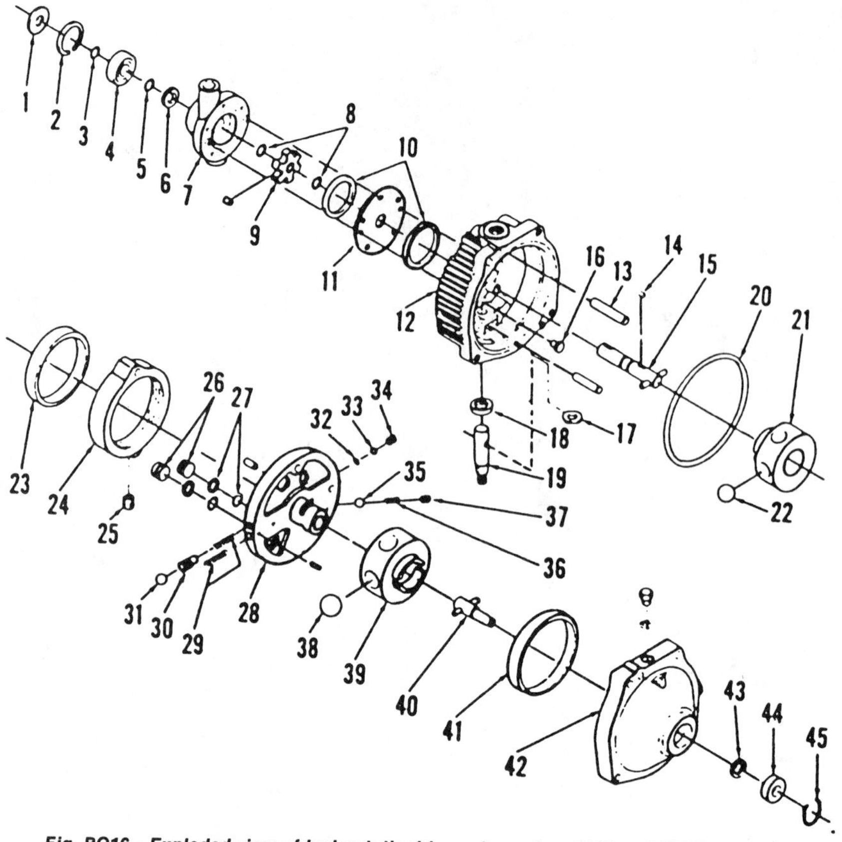

Fig. BO16—Exploded view of hydrostatic drive unit used on 1900 and HT-23 models.

1. Dust shield
2. Retaining ring
3. Snap ring
4. Bearing
5. Snap ring
6. Seal
7. Charge pump body
8. Snap rings
9. Charge pump rotor
10. "O" rings
11. Pump plate
12. Housing
13. Cam pivot pin
14. Key
15. Input shaft
16. Cap
17. Washer
18. Seal
19. Control shaft
20. "O" ring
21. Pump rotor
22. Bell pistons
23. Pump race
24. Pump cam ring
25. Insert
26. Dampening pistons
27. "O" rings
28. Pintle
29. Spring
30. Acceleration valve body
31. Acceleration valve ball
32. Retaining ring
33. Check valve ball
34. Check valve body
35. Charge relief ball
36. Spring
37. Relief valve plug
38. Motor ball piston
39. Motor rotor
40. Output shaft
41. Motor race
42. Body
43. Oil seal
44. Ball bearing
45. Retainer

tional check valves, it may be necessary to drill through pintle with a drill bit that will pass freely through roll pins. Redrill holes from opposite side with a ¼-inch drill bit. Press roll pin from pintle. Newer units are drilled at factory. Using a 5/16-18 tap, thread inside of valve bodies (34), then remove valve bodies using a draw bolt or slide hammer puller. Remove check valve balls (33) and retaining ring (32). To remove acceleration valves, remove retaining pin, insert a 3/16-inch (5 mm) rod 8 inches (203 mm) long through passage in pintle and carefully drive out spring (29), body (30) and ball (31). To remove dampening pistons (26), carefully tap outside edge of pintle on work bench to jar pistons free.

CAUTION: If pintle journal is damaged, pintle must be renewed. Use care when removing dampening pistons.

Remove pump cam ring (24) and pump race (23). Place a wide rubber band around pump rotor to prevent ball pistons (22) from falling out. Carefully remove pump assembly and input shaft (15).

To remove control shaft (19), drill a 11/32-inch hole through aluminum housing (12) directly in line with center line of dowel pin. Press dowel pin from control shaft, then withdraw control shaft. Remove oil seal (18). Thread drilled hole in housing with ⅛-inch pipe tap. Apply a light coat of "Loctite" grade 35 to a ⅛-inch pipe plug, install plug and tighten until snug. Do not overtighten.

Number piston bores (1 through 5) on pump rotor and on motor rotor. Use a plastic ice cube tray or equivalent and mark cavities 1P through 5P for pump ball pistons and 1M through 5M for motor ball pistons. Remove ball pistons (22) one at a time, from pump rotor and place each ball in the correct cavity in tray. Remove ball pistons (38) and springs from motor rotor in the same manner.

Clean and inspect all parts and renew any showing excessive wear or other damage. Renew all gaskets, seals and "O" rings. Ball pistons are a select fit to 0.0002-0.0006 inch (0.0050-0.0152 mm) clearance in rotor bores and must be reinstalled in their original bores. If rotor bushing to pintle journal clearance is 0.002 inch (0.051 mm) or more, bushing wear or scoring is excessive and pump rotor or motor rotor must be renewed. Check clearance between input shaft (15) and housing bushing. Normal clearance is 0.0013-0.0033 inch (0.033-0.0838 mm). If clearance is excessive, renew input shaft and/or housing assembly.

Install ball pistons (22) in pump rotor (21) and ball pistons (38) and springs in motor rotor (39), then use wide rubber bands to hold pistons in their bores.

Install charge relief valve ball (35) and spring (36) in pintle. Screw plug (37) into

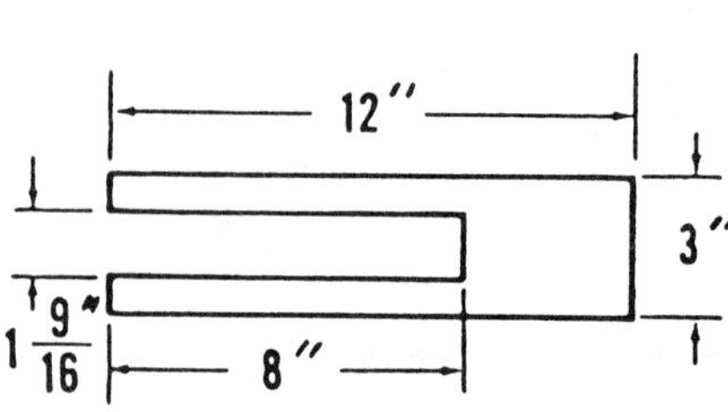

Fig. BO17—Special fork tool fabricated from a piece of ⅛-inch (3 mm) flat stock, used in disassembly and reassembly of hydrostatic drive unit.

pintle until just below outer surface of pintle. Install acceleration valve springs (29) and bodies (30) making sure valves move freely. Tap balls (31) into pintle until roll pins will go into place. Install snap rings (32), check valve balls (33) and valve bodies (34) in pintle and secure with new roll pins.

NOTE: When installing oil seals (6, 18 or 43), apply a light coat of "Loctite" grade 35 to seal outer diameter.

Renew oil seal (18) and install special washer (17), then press dowel pin through control shaft until 1¼ inches (32 mm) of pin extends from control shaft. Renew oil seal (43) and reinstall output shaft (40), bearing (44), retainer (45), output gear and snap ring.

Insert input shaft (15) in housing (12). Install snap ring (8) in its groove on input shaft. Place "O" ring (10), pump plate (11) and "O" ring in housing, then install charge pump drive key (14), charge pump rotor (9) and snap ring (8). Apply light grease to pump rollers and place rollers in rotor slots. Install oil seal (6) and pump race in charge pump body (7), then install body assembly. Secure with the five cap screws, making certain long cap screw is installed in its original location (in heavy section of pump body). Tighten cap screws to 28-30 ft.-lbs. (38-41 N·m). Install snap ring (5), bearing (4), retaining ring (2), snap ring (3) and dust shield (1).

Place charge pump and housing assembly in a holding fixture with input shaft pointing downward. Install pump race (23) and insert (25) in cam ring (24), then install cam ring assembly over cam pivot pin (13) and control shaft dowel pin. Turn control shaft (19) back and forth and check movement of cam ring. Cam ring must move freely from stop to stop. If not, check installation of insert (25) in cam ring.

Install pump rotor assembly and remove rubber band used to retain pistons. Install pintle assembly (28) over cam pivot pin (13) and into pump rotor. Place "O" ring (20) in position on housing.

Place body assembly (42) in a holding fixture with output gear down. Install motor race (41) in body, then install motor rotor assembly and remove rubber band used to retain pistons in rotor.

Using special fork tool (Fig. BO17) to retain motor assembly in body, carefully install body and motor assembly over pintle journal. Remove fork tool, align bolt holes and install the two cap screws. Tighten cap screws to 15 ft.-lbs. (20 N·m).

Place hydrostatic unit on holding fixture with reservoir adapter opening and venting plug opening facing upward. Fill unit with recommended fluid until fluid flows from fitting hole in body. Plug all openings to prevent dirt or other foreign material from entering hydrostatic unit.

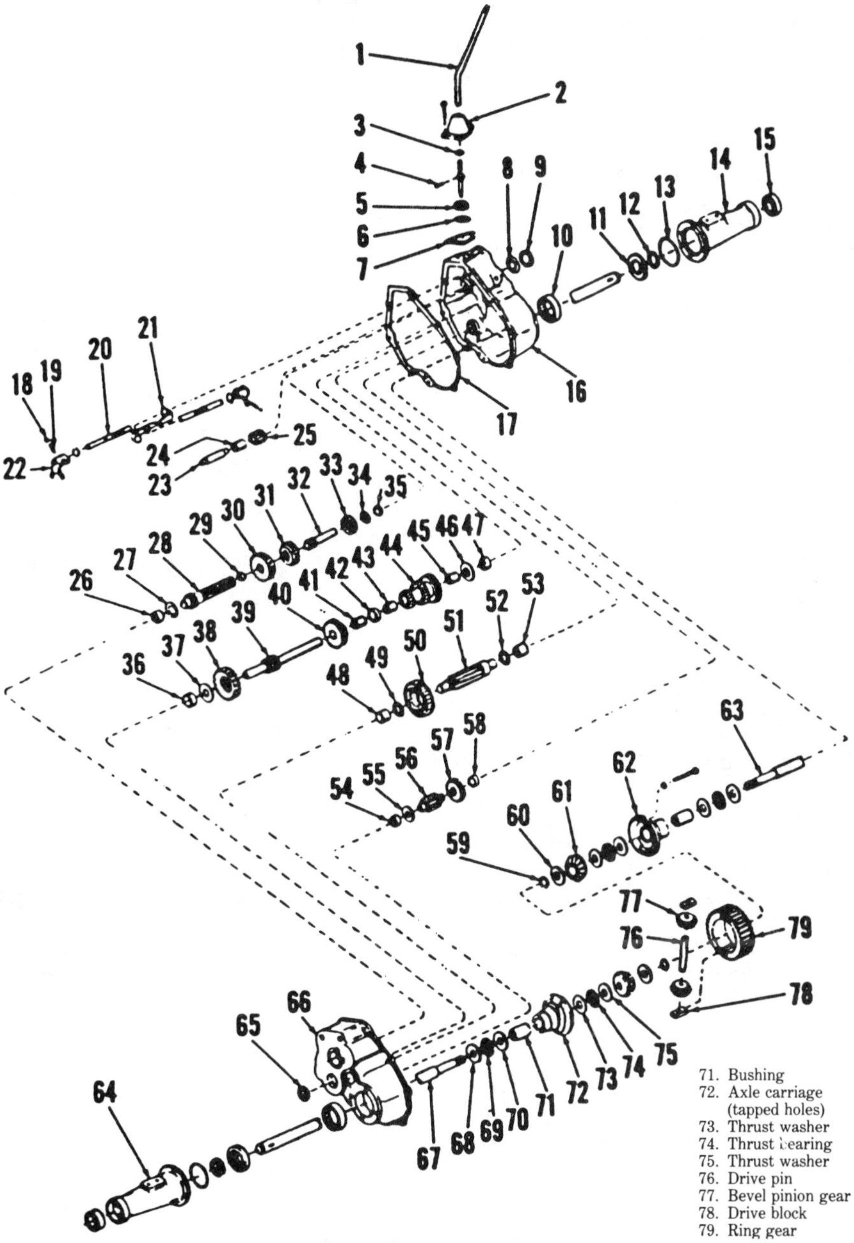

Fig. BO19—Exploded view of four speed transaxle.

1. Shift lever
2. Shift lever housing
3. Seal ring
4. Roll pin
5. Retainer
6. Snap ring
7. Gasket
8. Ball bearing
9. Oil seal
10. Carrier bearing
11. Seal retainer
12. Oil seal
13. "O" ring
14. Axle housing
15. Axle outer bearing
16. Transaxle case
17. Gasket
18. Detent ball
19. Spring
20. Shifter rod
21. Shifter stop
22. Shifter fork
23. Reverse idler shaft
24. Spacer
25. Reverse idler gear
26. Needle bearing
27. Thrust washer
28. Shifter shaft
29. Needle bearing
30. First, second & reverse gear
31. Third & fourth gear
32. Input shaft
33. Input gear
34. Thrust washer
35. Needle bearing
36. Needle bearing
37. Thrust washer
38. Idler gear
39. Brake & cluster shaft
40. Two-cluster gear
41. Bushing
42. Spacer
43. Bushing
44. Three cluster gear
45. Bushing
46. Thrust washer
47. Needle bearing
48. Needle bearing
49. Thrust washer
50. Output gear
51. Output shaft
52. Thrust washer
53. Needle bearing
54. Needle bearing
55. Thrust washer
56. Low reduction shaft
57. Low reduction gear
58. Needle bearing
59. Snap ring
60. Thrust washer
61. Axle gear
62. Axle carriage (plain holes)
63. Axle shaft
64. Axle housing
65. Oil seal
66. Transaxle cover
67. Axle shaft
68. Thrust washer
69. Thrust bearing
70. Thrust washer
71. Bushing
72. Axle carriage (tapped holes)
73. Thrust washer
74. Thrust bearing
75. Thrust washer
76. Drive pin
77. Bevel pinion gear
78. Drive block
79. Ring gear

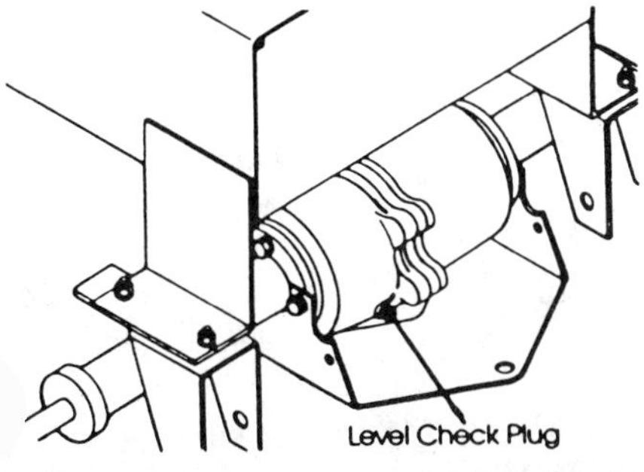

Fig. BO18—View showing location of transaxle fluid level check plug on models equipped with four speed transaxle.

TRANSAXLE

MAINTENANCE

Models 1600–1700

Transaxle type transmission fluid level should be checked at 100 hour intervals. Maintain fluid level at level check plug (Fig. BO18). Fluid capacity is 4 pints (1.9 L) and recommended fluid is SAE 90 EP gear lubricant.

R&R TRANSAXLE

Models 1600-1700

Raise and support rear of tractor. Disconnect brake rod and remove drive belt. Remove the bolts retaining shift lever and remove lever assembly. Support transaxle, then remove bolts retaining transaxle and lower and remove transaxle assembly.

Reinstall by reversing removal procedure.

OVERHAUL TRANSAXLE

Models 1600–1700

Remove drain plug and drain lubricant. Remove rear wheel and hub assemblies. Remove brake caliper and brake disc. Place shift lever (1–Fig. BO19) in neutral position, then unbolt and remove shift lever assembly. Remove axle housings (14 and 64) and remove seal retainers (11) with oil seals (12) and "O" rings (13) by pulling each axle shaft out of case and cover as far as possible. Place transaxle unit on the edge of a bench with left axle shaft pointing downward. Remove cap screws securing case (16) to cover (66) and drive aligning dowel pins out of case. Lift case (16) up 1½ to 2 inches (40 to 50 mm), tilt case about 45°, rotate case clockwise and remove it from the assembly. Input shaft (32) and input gear (33) will be removed with case. Withdraw differential and axle shaft assembly and lay aside for later disassembly. Remove the three-cluster gear (44) with its thrust washer (46) and spacer (42). Lift out reverse idler gear (25), spacer (24) and shaft (23). Hold upper ends of shifter rods together and lift out shifter rods, forks, shifter stop (21), sliding gears (30 and 31) and shaft (28) as an assembly. Remove low reduction gear (57), reduction shaft (56) and thrust washer (55), then remove the two-cluster gear (40) from brake shaft. Lift out the output gear (50), shaft (51) and thrust washers (49 and 52). To remove brake shaft (39) and gear (38) from cover (66), block up under gear (38) and press shaft out of gear.

CAUTION: Do not allow cover or low reduction gear bearing boss to support any part of pressure required to press brake shaft from gear.

Remove input shaft (32) with input gear (33) and thrust washer (34) from case (16).

To disassemble differential, remove the four cap screws and separate axle shaft and carriage assemblies from ring gear (79). Drive blocks (78), bevel pinion gears (77) and drive pin (76) can now be removed from ring gear. Remove snap rings (59) and withdraw axle shafts (63 and 67) from axle gears (61) and carriages (62 and 72).

Clean and inspect all parts and renew any showing excessive wear or other damage. When installing new needle bearings, press bearing (29) in spline shaft (28) to a depth of 0.010 inch (0.254 mm) below end of shaft and low reduction shaft bearings (54 and 58) 0.010 inch (0.254 mm) below thrust surfaces of bearing bases. Carrier bearings (10) should be pressed in from inside of case and cover until bearings are 0.290 inch (7.366 mm) below face of axle housing mounting surface. All other needle bearings are to be pressed in from inside of case and cover to a depth of 0.015-0.020 inch (0.381-0.508 mm) below thrust surfaces.

Renew all seals and gaskets. Reassemble by reversing disassembly procedure. When installing brake shaft (39) and idler gear (38), beveled edge of gear teeth must be up away from cover. Install reverse idler shaft (23), spacer (24) and reverse idler gear (25) with rounded end of gear teeth facing spacer. Install input gear (33) and shaft (32) so chamfered side of input gear is facing case (16).

Tighten transaxle cap screws to the following torque:

Differential cap screws	7 ft.-lbs. (9 N·m)
Case to cover cap screws	10 ft.-lbs. (14 N·m)
Axle housing cap screws	13 ft.-lbs. (18 N·m)
Shift lever housing cap screws	10 ft.-lbs. (14 N·m)

Fill transaxle, after unit is installed to correct operating level with recommended fluid.

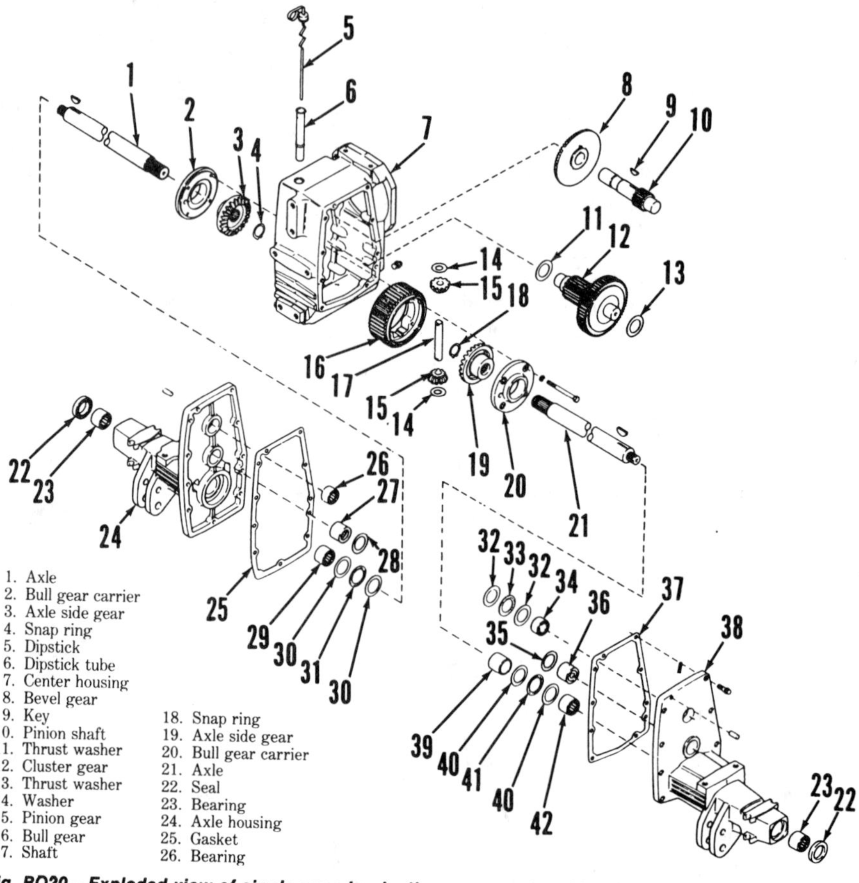

1. Axle
2. Bull gear carrier
3. Axle side gear
4. Snap ring
5. Dipstick
6. Dipstick tube
7. Center housing
8. Bevel gear
9. Key
10. Pinion shaft
11. Thrust washer
12. Cluster gear
13. Thrust washer
14. Washer
15. Pinion gear
16. Bull gear
17. Shaft
18. Snap ring
19. Axle side gear
20. Bull gear carrier
21. Axle
22. Seal
23. Bearing
24. Axle housing
25. Gasket
26. Bearing

Fig. BO20—Exploded view of single speed reduction gear used on QT-16, HT-18 and HT-20 models.

27. Bearing
28. Shims
29. Bearing
30. Thrust washers
31. Thrust bearing
32. Thrust washers
33. Thrust bearing
34. Bearing
35. Shims
36. Bearing
37. Gasket
38. Axle housing
39. Spacer
40. Thrust washer
41. Thrust bearing
42. Bearing

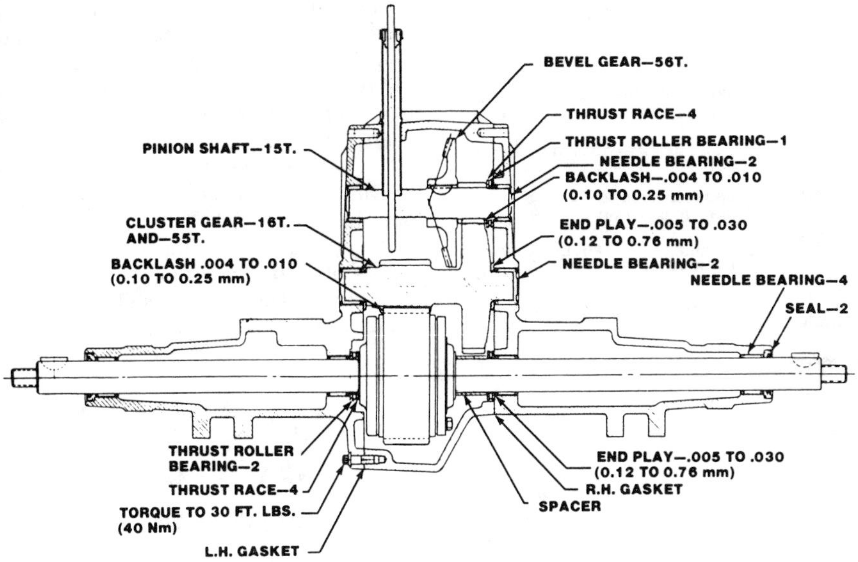

Fig. BO21—View showing correct end play for pinion shaft, cluster gear and axle shafts on single speed reduction gear assembly. Refer to text.

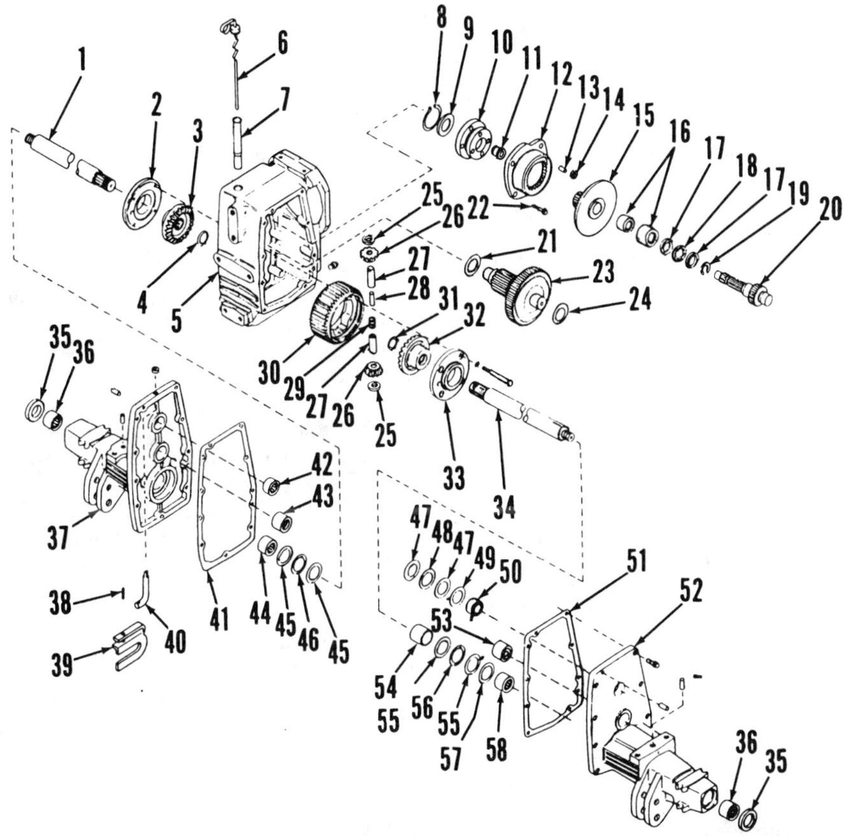

Fig. BO22—Exploded view of two speed reduction gear assembly used on 1900 and HT-23 models.

1. Axle
2. Bull gear carrier
3. Axle side gear
4. Snap ring
5. Center housing
6. Dipstick
7. Dipstick tube
8. Retaining ring
9. Thrust washer
10. Planet carrier
11. Sliding gear
12. Internal ring gear
13. Pin (4)
14. Planet gears (4)
15. Bevel gear
16. Bearings
17. Thrust washers
18. Thrust bearing
19. Positioning ring
20. Pinion shaft
21. Thrust washer
22. Cap screws (4)
23. Cluster gear
24. Thrust washer
25. Washers
26. Pinion gears
27. Shaft halves
28. Pin
29. Spring
30. Bull gear
31. Snap ring
32. Axle side gear
33. Bull gear carrier
34. Axle
35. Seal
36. Bearing
37. Axle housing
38. Pin
39. Shift fork
40. Shift shaft
41. Gasket
42. Bearing
43. Bearing
44. Bearing
45. Thrust washers
46. Thrust bearing
47. Thrust washers
48. Thrust bearing
49. Shim
50. Bearing
51. Gasket
52. Axle housing
53. Bearing
54. Spacer
55. Thrust washers
56. Thrust bearing
57. Shim
58. Bearing

REDUCTION GEAR

MAINTENANCE

All Hydrostatic Drive Models

Reduction gear fluid and hydrostatic drive unit fluid reservoir is located in reduction gear housing. Refer to HYDROSTATIC DRIVE SYSTEM section for appropriate model being serviced.

R&R REDUCTION GEAR

All Hydrostatic Drive Models

Drain fluid and remove seat and fender assemblies. Raise tractor and support at mid-frame. Disconnect brake arms at cross shaft. Remove brake arm extensions. Disconnect all hydraulic lines and control linkage at hydrostatic drive unit. Cap all openings. Remove the four bolts securing transaxle to frame and the two top mounting bolts. Balance transaxle assembly and roll assembly away from tractor. Separate hydrostatic drive unit from transaxle housing.

Reinstall by reversing removal procedure. Fill to correct operating level with recommended fluid.

OVERHAUL REDUCTION GEAR

Models QT-16 – HT-18 – HT-20

Separate right axle housing (38 – Fig. BO20) from center housing (7). Entire reduction gear set, cluster gear (12), pinion shaft (10) and input bevel gear (8) can be removed from housing. Remove all thrust washers and bearings. Separate left axle housing (24) from center housing (7). Remove differential and axle assemblies. Separate carriers (2 and 20) from bull gear (16), remove snap rings (4 and 18) and axle side gears (3 and 19). Remove pinion gears (15), washers (14) (as equipped) and shaft (17).

Inspect all gears and bearings and renew all parts showing excessive wear or damage. Renew all seals and gaskets.

Reassemble by reversing disassembly procedure. Vary thickness of thrust washers and shims to obtain correct pinion shaft, cluster gear and axle shaft end play as shown in Fig. BO21. Tighten bull gear carrier (2 and 20) retaining bolts to 25-30 ft.-lbs. (34-40 N·m). Tighten axle housing to center housing retaining bolts to 30 ft.-lbs. (40 N·m).

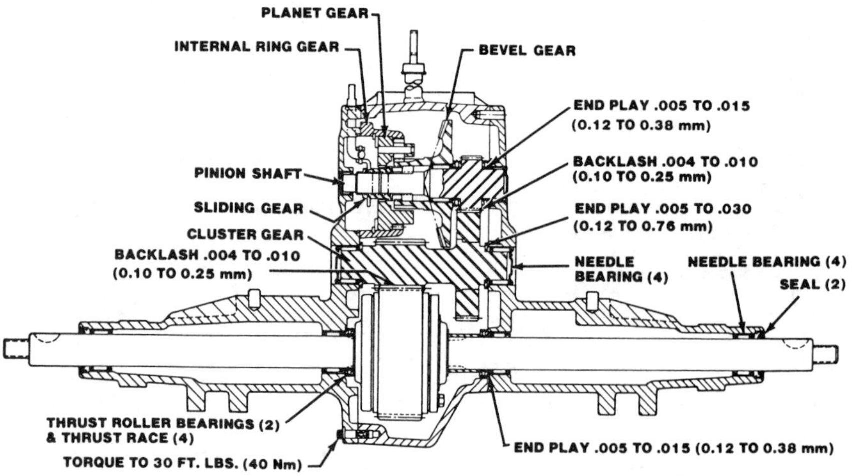

Fig. BO23 – View showing correct end play for pinion shaft, cluster gear and axle shafts on 1900 and HT-23 models. Refer to text.

Models 1900 – HT-23

Separate left axle housing (37 – Fig. BO22) from center housing (5). Remove all shims, thrust washers and thrust bearings. Remove the four screws retaining planetary carrier and remove planetary carrier-pinion shaft assembly. Remove sliding gear (11), roll pin (38), shift fork (39) and shift shaft (40) as necessary. Separate right axle housing (52) from center housing (5). Remove cluster gear (23) and all thrust washers. Remove differential and axle assemblies. Separate carriers (2 and 33) from bull gear (30), remove snap rings (4 and 31) and axle side gears (3 and 32). Remove pinion gears (26) and washers (25). Separate shaft halves (27) and remove pin (28) and spring (29).

Inspect all gears and bearings and renew all parts showing excessive wear or damage. Renew all seals and gaskets.

Reassemble by reversing disassembly procedure. Tighten bull gear carrier (2 and 33) cap screws to 25-30 ft.-lbs. (34-40 N·m) and the four planetary carrier cap screws (22) to 18 ft.-lbs. (24 N·m). Vary thickness of thrust washers and shims to obtain correct pinion shaft, cluster gear and axle shaft end play as shown in Fig. BO23. Tighten axle housing to center housing retaining cap screws to 30 ft.-lbs. (40 N·m).

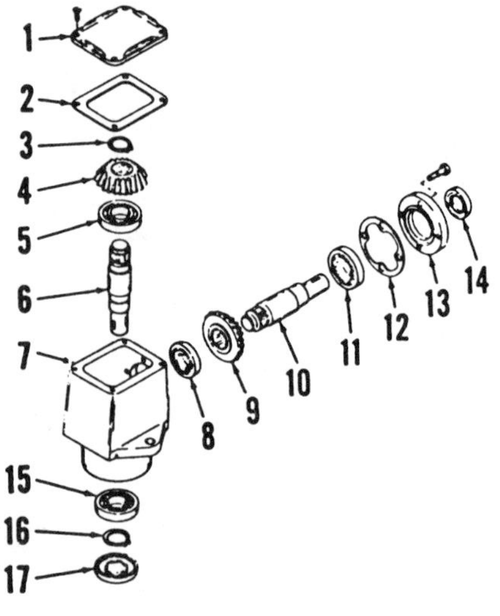

Fig. BO23A – Exploded view of right angle drive gear box used on 1600 and 1700 models.

1. Cover
2. Gasket
3. Snap ring
4. Drive Gear
5. Bearing
6. Input shaft
7. Housing
8. Bearing
9. Driven gear
10. Output shaft
11. Bearing
12. Gasket
13. Retainer
14. Oil seal
15. Bearing
16. Snap ring
17. Oil seal

RIGHT ANGLE DRIVE GEAR BOX

R&R AND OVERHAUL

Models 1600 – 1700

To remove right angle drive gear box, unbolt driveshaft rear coupling. Remove drive belt from output pulley, then remove pulley assembly. Unbolt and remove right angle drive gear box from tractor.

To disassemble unit, remove cover (1 – Fig. BO23A) and gasket (2). Unbolt and remove retainer (13) with oil seal (14) and gasket (12). Withdraw output shaft (10) and bearing (11), then remove driven gear (9) through cover opening. Remove oil seal (17) and snap ring (16). Push input shaft (6), bearing (5) and drive gear (4) out through cover opening. Gear and bearing can be removed from input shaft after first removing snap ring (3). To remove bearing (8), either tap outside of housing (7) behind bearing with a mallet or apply heat to housing.

Renew all seals and gaskets and reassemble by reversing disassembly procedure. Pack 4 ounces (118 mL) of multi-purpose grease in housing before installing cover (1).

HYDRAULIC SYSTEM

MAINTENANCE

All Models So Equipped

Hydraulic system utilizes pressurized oil from the hydrostatic drive unit charge pump. Refer to HYDROSTATIC DRIVE SYSTEM section for maintenance information.

SYSTEM PRESSURE TEST

All Models So Equipped

To test hydraulic system operating pressure, install a 0-1000 psi (0-7000 kPa) test gage in line with the hydraulic cylinder. Start engine and allow hydrostatic drive fluid to reach operating temperature. Set engine throttle lever at full speed and hold control lever until cylinder is at end of stroke and pressure gage shows a reading. Correct operating pressure is 450-550 psi (3100-3790 kPa).

On QT-16, HT-18 and HT-20 models, shims located under implement relief valve plug in hydrostatic drive unit are changed to vary pressure. See Fig. BO27.

On 1900 and HT-23 models, screw implement relief valve plug (8 – Fig. BO26) in or out until correct pressure is obtained.

HYDRAULIC CYLINDER

All Models So Equipped

To remove piston rod assembly from cylinder, elbow hose fittings must first be removed from cylinder tube. Refer to appropriate Fig. BO24 or BO25. Then, remove cylinder gland (8) from cylinder tube and withdraw piston rod assembly from cylinder.

Inspect piston and cylinder for wear, scoring or other damage and renew as necessary. Renew all "O" rings and back-

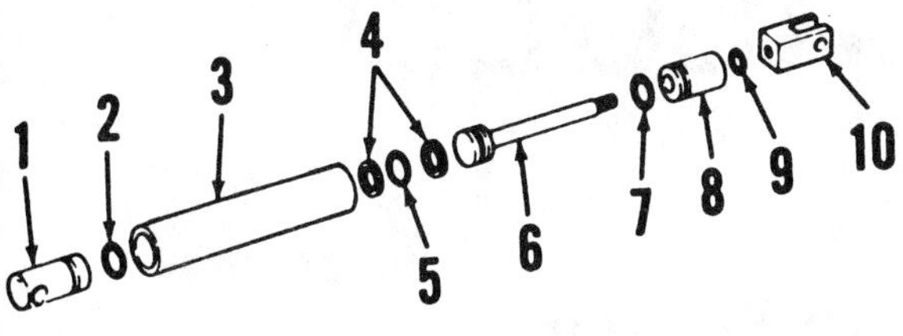

Fig. BO24—Exploded view of early style hydraulic cylinder used on some models.

1. Cylinder head
2. "O" ring
3. Cylinder tube
4. Washers
5. "O" ring
6. Piston & rod
7. "O" ring
8. Gland
9. "O" ring
10. Clevis

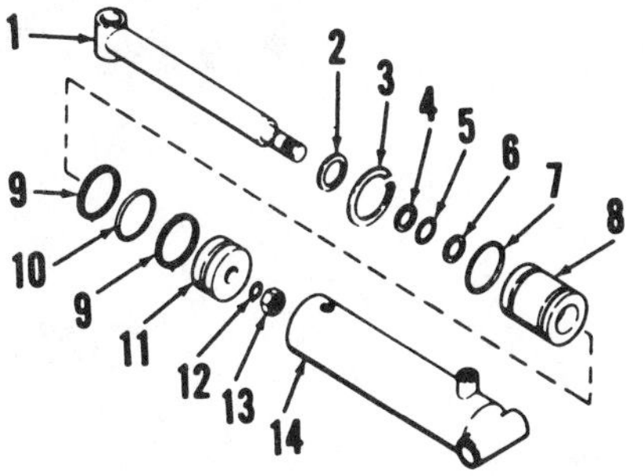

Fig. BO25—Exploded view of late style hydraulic cylinder used on some models.

1. Piston rod
2. Seal
3. Snap ring
4. Washer
5. "O" ring
6. Washer
7. "O" ring
8. Gland
9. Washers
10. "O" ring
11. Piston
12. Washer
13. Nut
14. Cylinder

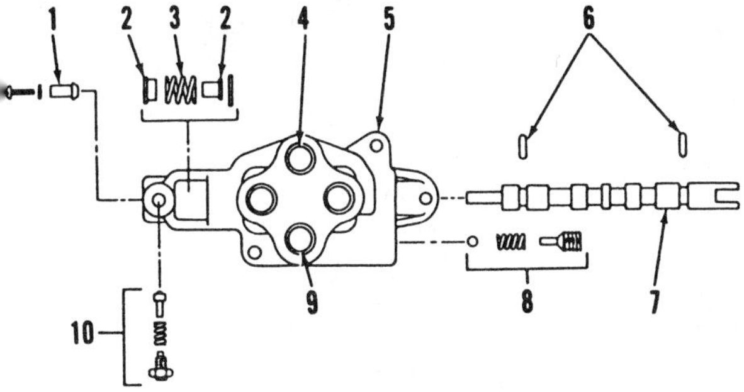

Fig. BO26—Exploded view of typical hydraulic lift control valve. Some models are equipped with a two spool control valve, however, service procedure is similar. Refer to text.

1. Detent ramp
2. Spring guides
3. Centering spring
4. Outlet
5. Valve body
6. Lip seals
7. Valve spool
8. Relief valve assy.
9. Inlet
10. Detent assy.

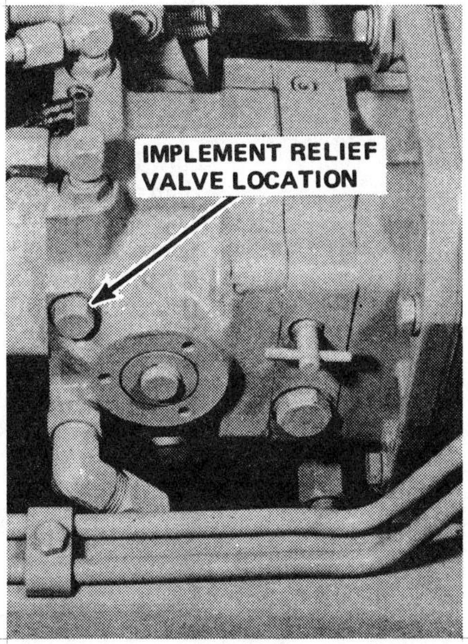

Fig. BO27—View showing location of implement relief valve on QT-16, HT-18 and HT-20 models. Refer to text.

lip washers. Coat all "O" rings with grease to prevent damage during reassembly.

CONTROL VALVE

All Models So Equipped

To remove spool (7–Fig. BO26) from valve body (5), remove screw from end of spool and remove detent assembly, centering spring and guides. Withdraw spool from lever end of valve body.

If spool or valve body is damaged, renew complete valve assembly. Install new lip seal rings (6) onto spool with lip openings towards inside of spool. Coat seal rings with grease and reinstall spool into valve body from lever end.

CASE/INGERSOLL

CONDENSED SPECIFICATIONS

	Models 446	448
Engine Make	Onan	Onan
Model	B43M	B48M
Bore	3.25 in. (82.5 mm)	3.25 in. (82.5 mm)
Stroke	2.62 in. (66.6 mm)	2.875 in. (73 mm)
Displacement	43.7 cu. in. (775 cc)	47.7 cu. in. (782 cc)
Power Rating	16 hp. (11.9 kW)	18 hp. (13.4 kW)
Slow Idle	1200 rpm	1200 rpm
High Speed (No-Load)	3600 rpm	3600 rpm
Capacities–		
Crankcase (W/O Filter)	1.75 qts (1.6 L)	1.75 qts. (1.6 L)
Crankcase (W/Filter)	....	2 qts. (1.8 L)
Hydraulic System	6.5 qts. (6 L)	6.5 qts. (6 L)
Transaxle	3 qts. (2.8 L)	3 qts. (2.8 L)
Fuel Tank	3 gal. (11.4 L)	3 gal. (11.4L)

FRONT AXLE AND STEERING SYSTEM

MAINTENANCE

All Models

It is recommended that steering knuckles, axle pivot pin, front wheel bearings and steering gear be lubricated at 25 hour intervals of normal operation. Use multi-purpose, lithium base grease and clean all grease fittings before and after lubrication. Check for any excessive looseness of parts, bearings, gears or tie rods at this time and repair as necessary.

AXLE MAIN MEMBER

All Models

Axle main member (4–Fig. C1) is mounted to main frame and pivots on pin (6). To remove front axle assembly, disconnect drag link from right steering knuckle. Using a suitable jack under main frame, raise front of tractor until weight is removed from front wheels. Unbolt and remove pivot pin (6), then raise front of tractor to clear axle main

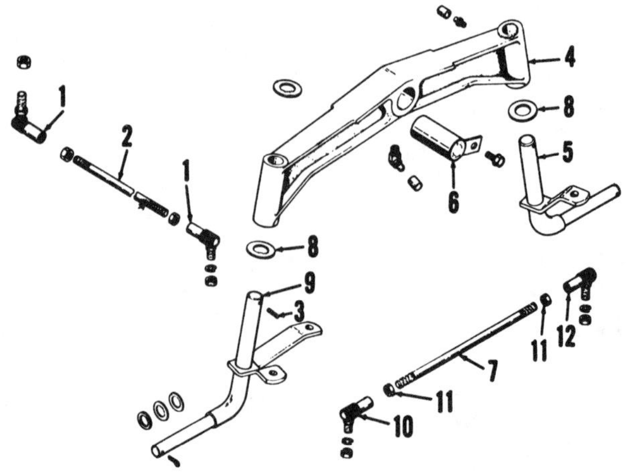

Fig. C1–Exploded view of typical front axle assembly.
1. Drag link ends
2. Drag link
3. Roll pin
4. Axle main member
5. Steering knuckle (left)
6. Pivot pin
7. Tie rod
8. Thrust washers
9. Steering knuckle (right)
10. Tie rod end
11. Locknuts
12. Tie rod end

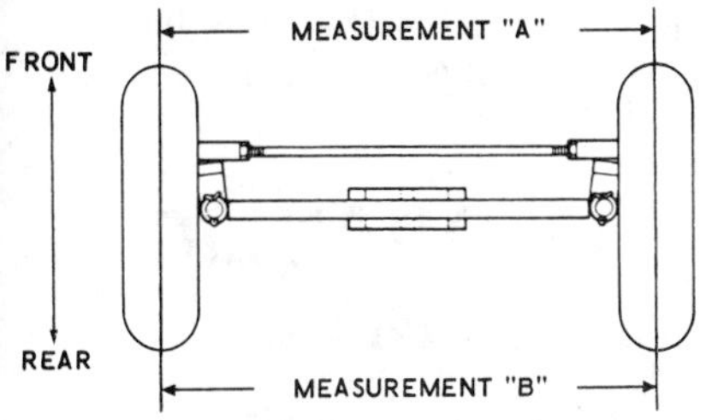

Fig. C2—Toe-in measurement "A" should be 1/8 to 3/8-inch (3.175 to 9.525 mm) less than measurement "B" with both measurements taken at hub height.

member. Roll front axle assembly forward from tractor.

Reinstall by reversing removal procedure.

TIE ROD AND TOE-IN

All Models

Tie rod used on all models is adjustable and front wheel toe-in is adjusted by loosening locknuts (11–Fig. C1) and rotating tie rod (7) until front wheel toe-in is 1/8 to 3/8-inch (3.175 to 9.525 mm) measured as shown in Fig. C2. Retighten locknuts.

STEERING KNUCKLES

All Models

To remove steering knuckles (5 and 9–Fig. C1), block up under axle main member and remove front wheels. Disconnect drag link from right steering knuckle and remove tie rod. Drive out roll pins (3) and lower steering knuckles from axle main member.

STEERING GEAR

All Models

R&R AND OVERHAUL. Refer to Fig. C3 and remove steering wheel retaining nut and steering wheel (1). Remove key (17) and tube (16). Raise front of tractor and support on suitable stands. Disconnect drag link from quadrant gear (11). Remove cap screw (15), lockwasher (14), washer (13), quadrant gear (11) and shims (10 and 12). Shims (10 and 12) should be kept in the order they are removed to aid in reassembly. Withdraw steering shaft and pinion (3–Fig. C3) from steering support (7) and from underside of tractor. Remove nylon bushing (2) from steering column. Unbolt and remove steering support (7). Remove locknut (6), then unscrew stub shaft (9) from support.

Clean and inspect all parts and renew any showing excessive wear or other damage. To reassemble, reverse disassembly procedure. Tighten steering wheel retaining nut to remove excessive end play of steering shaft (3). Shims (10–Fig. C3) are used to adjust gear backlash between quadrant gear (11) and steering shaft pinion. Adjust gear backlash by repositioning shims (10) to upper side of quadrant gear to increase backlash. Backlash should be minimal without causing binding of gears. Gears should be lubricated with a good quality multi-purpose, lithium base grease at 25 hour intervals of normal operation.

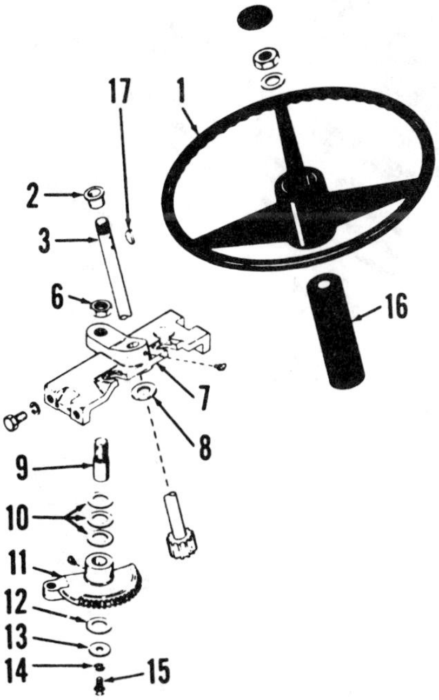

Fig. C3—Exploded view of steering gear assembly used on all models.

1. Steering wheel
2. Nylon bushing
3. Steering shaft & pinion
6. Locknut
7. Steering support
8. Washer
9. Stub shaft
10. Shims
11. Quadrant gear
12. Shim
13. Washer
14. Lockwasher
15. Cap screw
16. Tube
17. Key

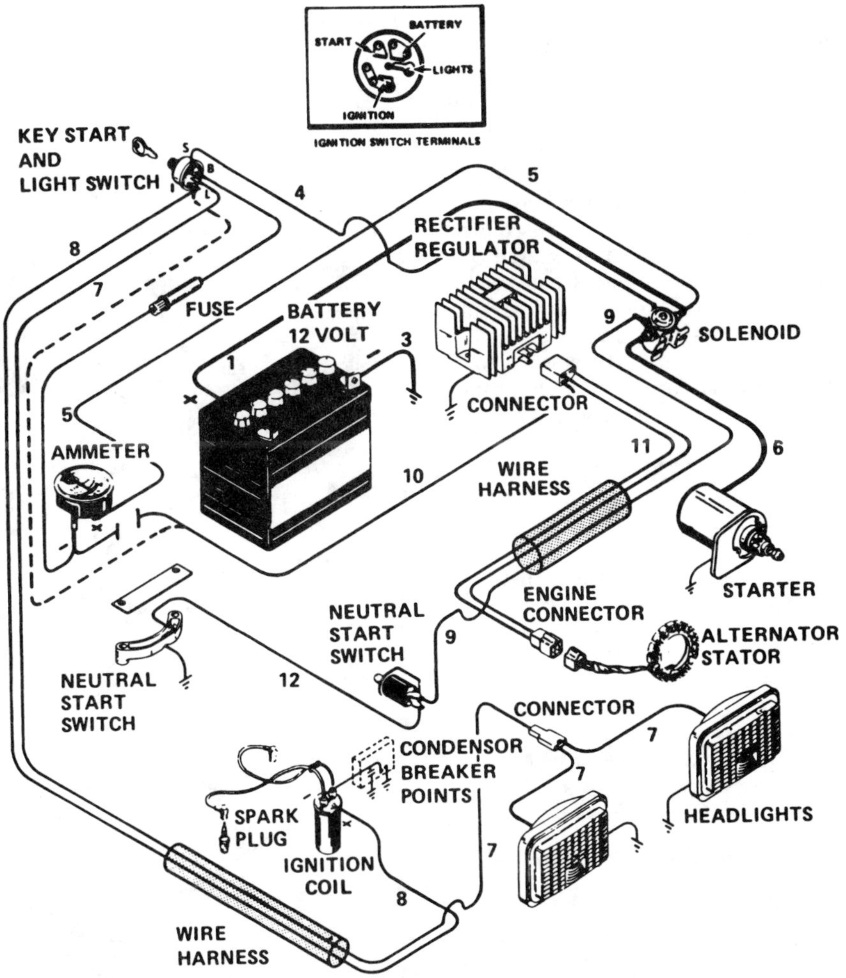

Fig. C4—Wiring diagram for 446 models with serial number prior to 9742953. Wiring color is as listed.

1. Red
2. Orange
3. Black
4. Yellow
5. Black
6. Black
7. Pink
8. Black/white
9. Black
10. Red
11. White

ENGINE

MAINTENANCE

All Models

Regular engine maintenance is required to maintain peak performance and long engine life.

Check oil level, clean air intake screen and oil cooler daily under normal operating conditions.

Wash engine air filter pre-cleaner at 25 hour intervals and clean or renew dry element after each 50 hours of normal operation.

Change engine oil, perform tune-up, valve adjustment and clean carbon from cylinder heads as recommended by engine manufacturer.

REMOVE AND REINSTALL

All Models

To remove engine, tilt hood and grille forward and disconnect battery cables. Disconnect cable from starter motor and unplug charging circuit wires. Disconnect headlight wire and ignition coil wire. Disconnect choke and throttle control cables and the fuel line at fuel pump. Disconnect hydraulic line at lower right side of oil cooler and drain hydraulic oil. Disconnect pump suction line at reservoir, then unbolt and remove reservoir and oil cooler assembly. Disconnect pump pressure line and pto clutch control rod. Unbolt engine from tractor frame and lift engine and hydraulic pump assembly from tractor.

To reinstall engine, reverse removal procedure. Make certain hydraulic reservoir is refilled to proper level.

OVERHAUL

All Models

Engine make and model are listed at the beginning of this section. To overhaul engine or accessories, refer to appropriate engine section in this manual.

ELECTRICAL SYSTEM

MAINTENANCE AND SERVICE

All Models

Battery electrolyte level should be checked at 50 hour intervals of normal operation. If necessary, add distilled water until level is just below base of vent well. **DO NOT** overfill. Keep battery posts clean and cable ends tight.

For alternator or starter service, refer to appropriate engine section in this manual.

Refer to Fig. C4 for wiring diagram for 446 model (prior to serial number 9742953) or to Fig. C5 for wiring diagram for 446 model (serial number 9742953 and after) and all 448 models.

BRAKE

ADJUSTMENT

All Models

Refer to Fig. C6 and adjust brake linkage mounting nuts to obtain a clearance of 0.010-0.015 inch (0.254-0.381 mm) between washers and brake arms. Disconnect brake rod clevis from vertical link. Place range transmission in neutral position. Tighten brake band adjusting nut until tractor cannot be moved manually, then back nut off 1½ turns. Hold brake vertical link in vertical position and push it rearward until all slack is removed from linkage. With brake pedal in full release (up) position, adjust clevis on linkage rod until holes in clevis are aligned with hole in vertical link. Install clevis pin.

HYDRAULIC DRIVE

MAINTENANCE

All Models

Hydraulic oil should be changed at 500 hour intervals. Oil approved by manufacturer must meet API service classification SE or CC. Use SAE 20W40 oil if temperature is above 32° F (0° C) and SAE 5W20 oil if below 32° F (0° C). Maintain fluid level 5 to 6 inches (127 to 152 mm) below top of filler neck. Approximate capacity is 6.5 quarts (6 L).

OPERATION

All Models

The three main components of the hydraulic drive are the hydraulic pump,

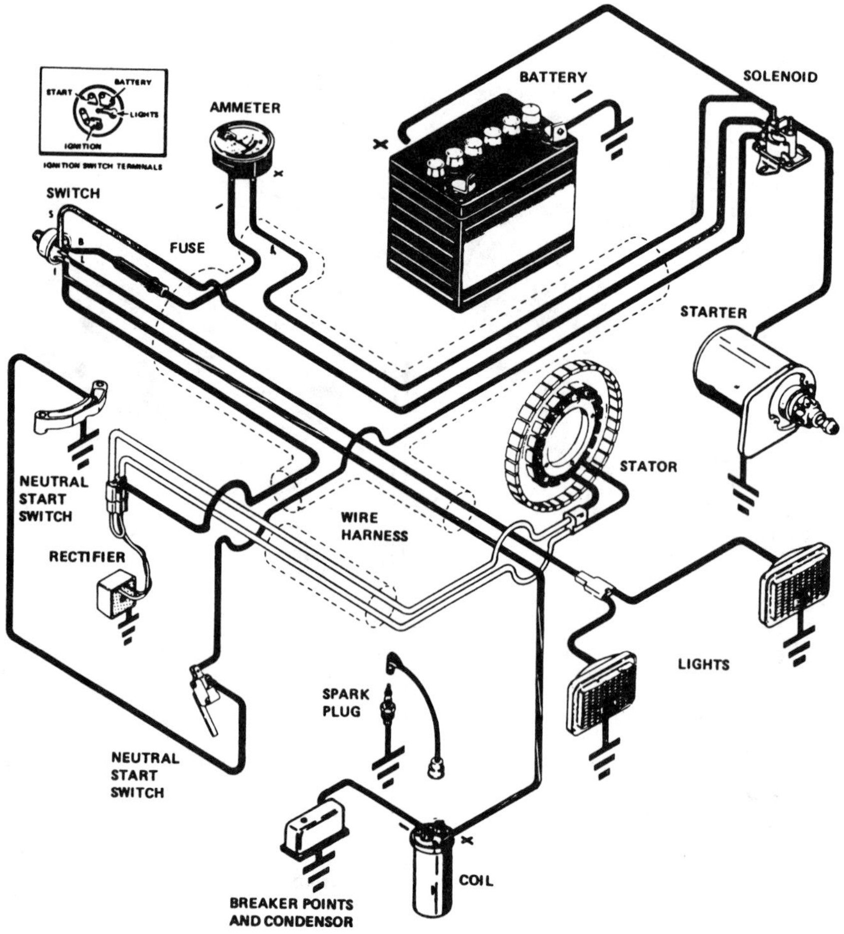

Fig. C5 – Wiring diagram for 446 models with serial number 9742953 and after and all 448 models.

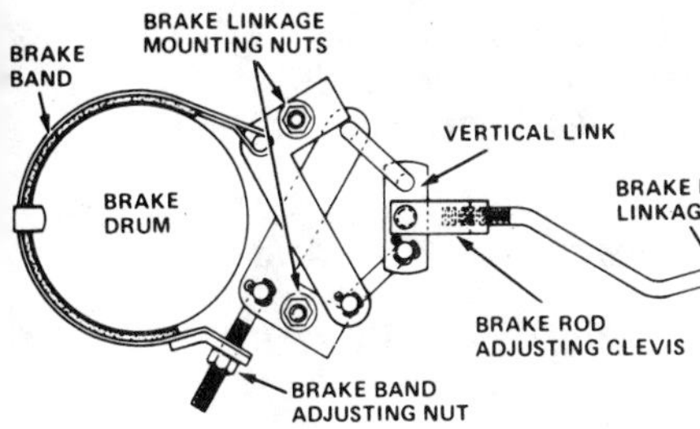

Fig. C6—View showing brake linkage used on all models.

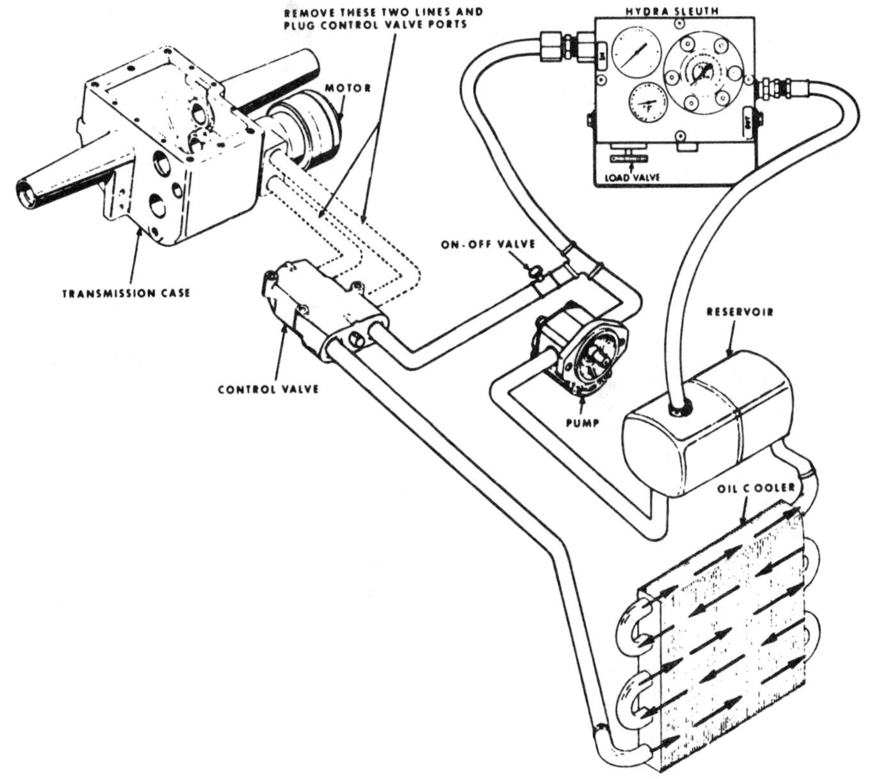

Fig. C7—View showing "Hydra Sleuth" tester installed to check flow and pressure of hydraulic drive system.

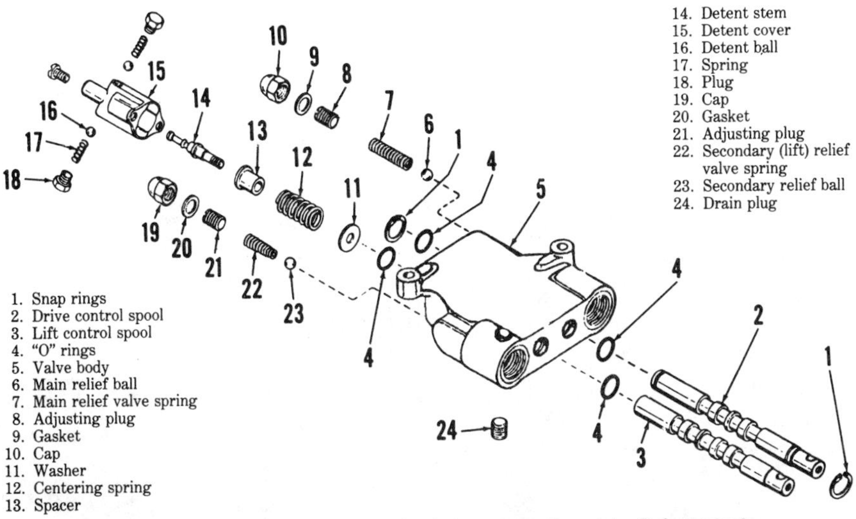

1. Snap rings
2. Drive control spool
3. Lift control spool
4. "O" rings
5. Valve body
6. Main relief ball
7. Main relief valve spring
8. Adjusting plug
9. Gasket
10. Cap
11. Washer
12. Centering spring
13. Spacer
14. Detent stem
15. Detent cover
16. Detent ball
17. Spring
18. Plug
19. Cap
20. Gasket
21. Adjusting plug
22. Secondary (lift) relief valve spring
23. Secondary relief ball
24. Drain plug

Fig. C9—Exploded view of control valve used on all models. Refer to text.

control valve and hydraulic motor. Hydraulic pump draws oil from reservoir located at front of tractor behind oil cooler. Oil is pumped to drive control valve and when control valve is in neutral, oil passes through valve and to oil cooler. After flowing through oil cooler, oil is returned to reservoir. If control valve is in forward or reverse position, oil is directed to the hydraulic motor. This causes motor shaft to rotate which in turn drives range transmission input gear. Oil returning from motor, flows through valve, through oil cooler and to reservoir. When control level is returned to neutral position, oil flow between valve and motor stops and this stops rotation of motor shaft. Dynamic braking action occurs and tractor is halted.

OIL FLOW AND PRESSURE CHECK

All Models

To check oil flow and system pressure, install a Hydra Sleuth, Flo-Rater or equivalent tester as shown in Fig. C7. Remove hydraulic lines from control valve to hydraulic motor and plug valve ports. Close shut-off valve in line and fully open load valve on tester. Start and operate engine at 3600 rpm until hydraulic oil temperature is approximately 120° F (50° C). Check and record flow at 0 pressure. Slowly close tester load valve until pressure gage reading is 1000 psi (6895 kPa) and note flow reading. This flow reading must not be more than 25% less than previously recorded 0 pressure flow reading. If pump output flow drops more than 25%, pump is worn and must be overhauled or renewed.

If pump output flow drop is less than 25%, record flow at 1000 psi (6895 kPa) and continue tests. Fully open shut-off valve in line and load valve on tester. With engine still operating at 3600 rpm, move control lever in full forward or reverse position. Slowly close load valve on tester until pressure gage reading is 1000 psi (6895 kPa) and note flow reading. This reading must not exceed ½-gpm (1.9 L/min.) less than previously recorded pump gpm flow at 1000 psi (6895 kPa). If control valve internal leakage is more than ½-gpm (1.9 L/min.), check and adjust relief valve pressure.

To check relief valve pressure, close load valve on tester and note pressure gage reading. Relief valve should open between 2050-2150 psi (14135-14825 kPa). Turn adjusting plug (8-Fig. C9) in to increase pressure or out to decrease pressure as required. Recheck control valve for internal leakage.

If control valve leakage still exceeds

½-gpm (1.9 L/min.), valve body is excessively worn or cracked and valve must be renewed.

Remove test equipment and install original lines. If pump and control valve check out good and adequate tractor performance cannot be obtained, remove hydraulic motor and overhaul or renew motor.

CONTROL VALVE

All Models

R&R CONTROL VALVE. To remove control valve, first drain hydraulic reservoir. Disconnect control lever linkage and hydraulic lines. Plug or cap openings to prevent dirt from entering system. Unbolt and remove control valve assembly. To reinstall control valve, reverse removal procedure.

CONTROL VALVE OVERHAUL. To disassemble control valve, refer to Fig. C9. Remove snap rings (1) and withdraw drive control spool (2). Remove plugs (18), springs (17) and detent balls (16), then unbolt and remove detent cover (15). Withdraw lift spool assembly from valve body (5). Detent stem (14), spacer (13), centering spring (12) and washer (11) can now be removed. Unscrew cap (10), then remove gasket (9), adjusting plug (8), spring (7) and main relief ball (6). Place body (5) on a bench so drain plug (24) is pointing upward. Remove cap (19), gasket (20), adjusting plug (21), spring (22) and secondary (lift) relief ball (23). Remove "O" rings (4) from valve body.

Clean and inspect all parts and renew any showing excessive wear or other damage. When installing relief valves, turn about ½ of adjusting screw threads into the valve. Install new "O" ring seals (4) in grooves in front end of valve body. Lubricate seals and insert spools (2 and 3) into correct bores at rear (relief valve) end of valve body. Push spools forward through front seals until rear seal grooves in bores are exposed. Install and lubricate rear seal rings, then move spools back to normal centered position. Install snap rings on travel spool (2) and using "Loctite" on detent stem (14) install stem on lift spool (3). Install detent cover (15), balls (16), springs (17) and plugs (18).

After installing control valve on tractor, adjust main relief valve pressure as previously outlined and secondary relief valve pressure to 575 psi (3965 kPa).

HYDRAULIC PUMP

A variety of hydraulic pumps by several manufacturers (Parker Hannifin, Borg Warner and Cessna) have been used.

Model 446, prior to serial number 9728158, is equipped with Parker Hannifin C15481 or C19045 hydraulic pumps.

Model 446, between serial number 9728158 and 9742953, is equipped with Borg Warner C20757, C19743 or C22771 hydraulic pump.

Model 446, serial number 9742953 and after and all 448 models, is equipped with Cessna C25179 hydraulic pump, however, this pump has been replaced in service with Parker Hannifin C24828, C24868 or C25053 hydraulic pumps.

Refer to appropriate paragraph for model being serviced.

Model 446 Prior To Serial Number 9646800

R&R HYDRAULIC PUMP. Drain hydraulic system and remove panel from left side of control tower. Disconnect inlet tube at pump and disconnect and remove outlet tube. Remove left foot rest. Disconnect brake rod and spring at brake pedal. Hold brake pedal down and slide pedal of shaft to remove. Remove pump support from engine, then pull pump and support through left side of control tower. Mark coupling position on pump shaft, loosen set screw and remove coupling. Separate pump from support.

Reinstall by reversing disassembly procedure. Fill hydraulic system to correct operating level with specified fluid. Bleed air from system as outlined.

Model 446 With Serial Numbers Between 9646800 And 9728158

R&R HYDRAULIC PUMP. Drain hydraulic system and remove battery and battery mounting plate. Disconnect lines at pump and cap openings. Remove the four bolts retaining pump support to engine and pull pump and pump support up through opening created by removal of battery and support. Mark coupling position on pump shaft, loosen set screw and remove coupling. Separate pump from support.

Reinstall by reversing removal procedure. Fill hydraulic system to correct operating level with specified fluid. Bleed air from system as outlined.

Models 446 With Serial Numbers Between 9728158 And 9742953

R&R HYDRAULIC PUMP. Drain hydraulic system and remove battery. Remove voltage regulator and starter solenoid mounting screws and move regulator and solenoid out of the way. Remove battery tray. Disconnect suction line and return line from reservoir, then remove reservoir. Disconnect lines at pump. Remove the four bolts retaining pump support to engine and pull pump and support out through opening

Fig. C10—Exploded view of Parker Hannifin hydraulic pump. Note pump numbers C24828, C24868 and C25053 are service replacement pumps for Cessna pumps and drive gear (12) is held in position with retaining rings which are not shown. Refer to text.

1. Key
2. Section (body)
3. Wear plate
4. Heat shield
5. Gasket
6. Seal
7. "O" ring
8. Front cover
9. Seal
10. Rear cover
11. "O" ring
12. Drive gear
13. Driven gear
14. Dowel

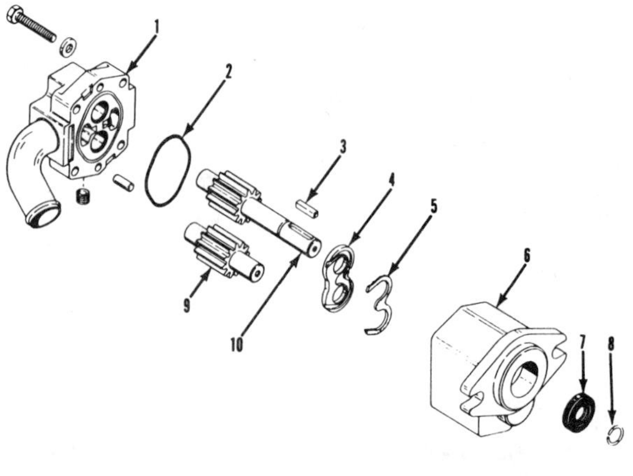

Fig. C11—Exploded view of Borg Warner hydraulic pump used on some models.

1. Rear housing
2. "O" ring
3. Key
4. Wear plate
5. Pressure seal
6. Front housing
7. Seal
8. Snap ring
9. Driven gear
10. Drive gear

created by battery and reservoir removal. Mark coupling position on pump shaft, loosen set screw and remove coupling. Separate pump from support.

Reinstall by reversing removal procedure. Fill hydraulic system to correct operating level with specified fluid. Bleed air from system as outlined.

Models 446 (After Serial Number 9742953) And All 448 Models

R&R HYDRAULIC PUMP. Drain hydraulic system and remove battery. Remove access cover from control tower and lines at hydraulic pump. Remove right and left heat exchanger brackets. Remove the four bolts retaining pump to engine and remove pump.

Reinstall by reversing removal procedure. Fill hydraulic system to correct operating level with specified fluid. Bleed air from system as outlined.

All Models

BLEED HYDRAULIC SYSTEM. With reservoir filled to correct level with specified fluid, start and run engine at half throttle for 30 seconds while operating all hydraulic controls. Stop engine. Check hydraulic fluid level and fill as necessary. Start and run engine at half throttle and apply short intervals of load for 3 minutes. Increase engine speed to full throttle and apply short intervals of load for 3 minutes. Stop engine and check for leaks. Check fluid level and fill as necessary.

HYDRAULIC PUMP OVERHAUL (PARKER HANNIFIN). Clean exterior of pump and remove key (1–Fig. C10). Scribe mark across pump sections and end covers to aid reassembly, then remove the four bolts which hold pump sections together. Carefully separate pump sections.

NOTE: DO NOT pry sections apart as pump damage may occur. If necessary, lightly tap sides of back cover (10) with plastic hammer.

On pump numbers C24828, C24868 and C25053, remove retaining ring from drive gear shaft. Place a thin steel bar between drive gear and front cover on each side of gear, place assembly in press and press gear from shaft. Remove key and second retaining ring, then remove shaft.

On all models, remove wear plates (3). Mark front cover next to pressure hole in heat shield (4) as an aid for reassembly. Use a hooked wire in pressure hole to remove heat shield (4), gasket (5) and seal (6). Remove "O" rings (11 and 7) from front and rear covers. To renew seal (9), heat cover to approximately 250° F (121° C) and use a suitable puller to remove seal. New seal should be pressed in until flush with cover.

Inspect all parts for damage or wear. Section (2) must be 0.0002 to 0.0018 inch (0.005 to 0.045 mm) wider than width of gears. Minimum gear shaft diameter is 0.4993 inch (12.682 mm) and minimum gear diameter is 1.2390 inch (31.468 mm). Minimum gear width is 1.0557 inch (26.815 mm). Maximum bearing bore inside diameter in covers is 0.5025 inch 12.764 mm). Maximum gear wear ridge in rear cover and wear plate is 0.0005 inch (0.0127 mm). Maximum gear wear ridge in rear cover and wear plate is 0.0005 inch (0.0127 mm). Maximum gear bore diameter in center section (2) is 1.243 inch (31.572 mm). Split in bearing bores in covers must not be more than 1/16-inch (1.58 mm) from a vertical center line through both bearings. Front cover bearing must be even with face of bearing bore and rear cover bearing must be recessed below face of bearing bore.

Reassemble by reversing disassembly procedure. Make certain all alignment marks are in register. Lubricate parts during assembly with clean fluid specified for hydraulic system. Tighten pump bolts to 24-26 ft.-lbs. (33-35 N·m) torque for C15481 and C19045 pumps and to 17 ft.-lbs. (23 N·m) for remaining pumps. Reinstall pump, fill system with fluid specified for hydraulic system and bleed air from system as outlined.

OVERHAUL HYDRAULIC PUMP (BORG WARNER). Clean exterior of pump and remove key (3–Fig. C11). Scribe mark across pump housing to aid reassembly. Remove the four bolts which hold pump housings together. Use a screwdriver in each slot on opposite sides of housings and carefully separate housings. Remove "O" ring (2) from rear housing (1). Remove driven gear (9) from front housing (6). Remove snap ring (8) and drive gear (10). Remove wear plate (4) and pressure seal (5). To renew seal (7), heat front housing to 250°F (121°C) and use a suitable puller to remove seal. New seal is pressed into bore until 0.188 inch (4.78 mm) below rim.

Inspect all parts for damage or wear. Split in bearing bores in housings must not be more than 1/16-inch (1.58 mm) from a vertical center line through both bearings.

Reassemble by reversing disassembly procedure. Make certain all alignment marks are in register. Lubricate parts during assembly with clean fluid specified for hydraulic system. Tighten pump bolts to 28-32 ft.-lbs. (38-43 N·m). Reinstall pump, fill system with fluid specified for hydraulic system and bleed air from system as outlined.

OVERHAUL HYDRAULIC PUMP (CESSNA). Clean exterior of pump and remove key (1–Fig. C12). Scribe mark across pump covers and center section to aid reassembly, then remove the four bolts which hold pump sections and covers together. Remove wear plate (2), the two springs (6) and the two 7/32-inch (5.55 mm) steel balls (7). Note locations of springs and balls. Remove gaskets (3 and 4) and seal (5). Remove shaft seal (9).

Inspect all parts for damage or wear Oil grooves in bearings in each cover must align with dowel pin holes and grooves must be 180° apart. Minimum gear shaft diameter is 0.5605 inch (14.237 mm) and minimum gear width is 0.803 inch (20.40 mm). Maximum gear bore inside diameter in pump section (13) is 1.404 inches (35.61 mm). Maximum bearing bore inside diameter for front and rear covers is 0.5655 inch (14.364 mm) and maximum wear ridge in rear cover (10) is 0.0015 inch (0.038 mm).

To reassemble, push seal (5) into grooves in front cover using a dull tool. Open part of "V" section must be to the inside. Make certain inner lip of seal does not turn out during assembly. Push protector gasket (4) and back-up gasket (3) into seal (5). Install the steel balls (7) and springs (6) on their seats. Install wear plate (2), bronze face up, in inside

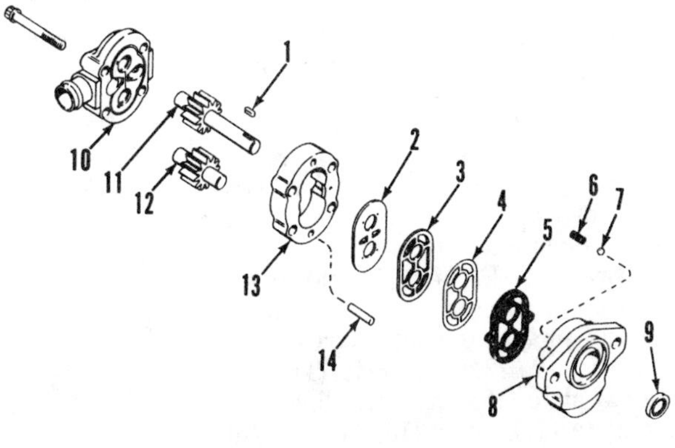

Fig. C12—Exploded view of Cessna C25179 hydraulic pump used on some models.

1. Key
2. Wear plate
3. Gasket
4. Gasket
5. Seal
6. Spring
7. Ball
8. Front cover
9. Seal
10. Rear cover
11. Drive gear
12. Driven gear
13. Center section
14. Dowel

rim of seal (5). Install gear assemblies in front cover. Install dowel pins (14) in section (13). Apply thin layer of grease to both faces of section (13), align scribe marks and install section (13) on front cover (8). Align scribe marks and install rear cover (10). Tighten pump bolts to 23-25 ft.-lbs. (31-34 N•m). Carefully install shaft seal (9) over shaft using a rotating motion. Use plastic hammer to drive seal squarely into seal bore until fully seated. Fill pump with specified hydraulic fluid. Pump should turn freely. Reinstall pump, fill system with fluid specified for hydraulic system and bleed air from system as outlined.

HYDRAULIC MOTOR

All Models

R&R HYDRAULIC MOTOR. Place range transmission shift lever in neutral position and block tractor up under tractor frame. Place a rolling floor jack under range transmission and differential housing. Unbolt and remove fenders, seat and seat support. Drain fuel tank, disconnect fuel line, loosen mounting straps and remove tank. Disconnect hydraulic lines from hydraulic drive motor and cap all openings. Remove cap screws securing range transmission and differential assembly to tractor frame and roll assembly rearward from tractor. Drain lubricant from transmission and differential housing, then unbolt and remove top cover. Remove left rear wheel. Unbolt hydraulic drive motor, hold range transmission sliding gear and withdraw drive motor. Let sliding gear rest in bottom of transmission housing.

When reinstalling hydraulic drive motor, use a new "O" ring on motor housing. Hold range transmission sliding gear in position (shift fork between gears) and insert motor output shaft through housing and gear. Apply "Loctite" to cap screw threads and tighten motor retaining cap screws to 110-125 ft.-lbs. (149-169 N•m). The balance of installation is the reverse of removal procedure. Fill transmission and differential housing to level plug opening with SAE 80 EP gear oil. Capacity is approximately 3 quarts (2.8 L).

OVERHAUL HYDRAULIC MOTOR. To disassemble the hydraulic drive motor, clamp motor body port boss in a padded jaw vise with output shaft pointing downward. Remove the seven cap screws (24–Fig. C13), then remove end cover (23), seal ring (22), commutator (19) and commutator ring (18). Remove sleeve (21), manifold (17) and manifold plate (16). Lift drive link (11), wear plate (12), rotor (13), rollers (14) and stator (15) off body (5). Remove output shaft (10), then remove snap ring (1), spacer (2), shim (3) and oil seal (4). Remove seal ring (9). Do not remove needle bearing (8), thrust bearing (7) or thrust washer (6) from body (5) as these parts are not serviced separately.

Clean and inspect all parts for excessive wear or other damage and renew as necessary. A seal ring and seal kit (items 2, 3, 4, 9, 20 and 22) is available for resealing motor. To reassemble motor, clamp body port boss in a padded vise with the seven tapped holes upward. Insert shaft (10) and drive link (11). Install new seal ring (9) in groove on body (5). Place stator (15) on wear plate (12) and install rotor and rollers (13 and 14) with counterbore in rotor facing upward. Place wear plate and rotor assembly over drive link and on body.

NOTE: Two ⅜-inch NF by 4½-inch guide studs can be used to align bolt holes in body (5) with holes in wear plate (12), stator (15), manifold plate (16), manifold (17), commutator plate (18) and end cover (23).

Install manifold plate (16) with slots toward rotor. Install manifold (17) with swirl grooves toward rotor and diamond shaped holes upward. Place commutator ring (18) and commutator (19) on manifold with bronze ring groove facing upward. Place bronze seal ring (20) into groove with rubber side downward. Lubricate seal ring (9) and install sleeve (21) over assembled components. Install new seal ring (22) on end cover (23), lubricate seal ring and install end cover. Remove alignment studs and install the seven cap screws (24). Tighten cap screws evenly to 50 ft.-lbs. (68 N•m). Pack seal spacer (2) with multi-purpose, lithium base grease and install seal (4), shim (3), spacer (2) and snap ring (1). Fill motor with specified fluid for initial lubrication.

RANGE TRANSMISSION AND DIFFERENTIAL

MAINTENANCE

All Models

It is recommended that range transmission and differential oil be changed at 500 hour intervals of normal operation. Oil approved by manufacturer should meet API service classification SE or CC. Use either SAE 20W40 or SAE 80 EP Gear Lube for all temperatures. Transmission and differential oil capacity is approximately 3 quarts (2.8 L).

REMOVE AND REINSTALL

All Models

Place range transmission shift lever in neutral position and block tractor up under tractor frame. Place a rolling

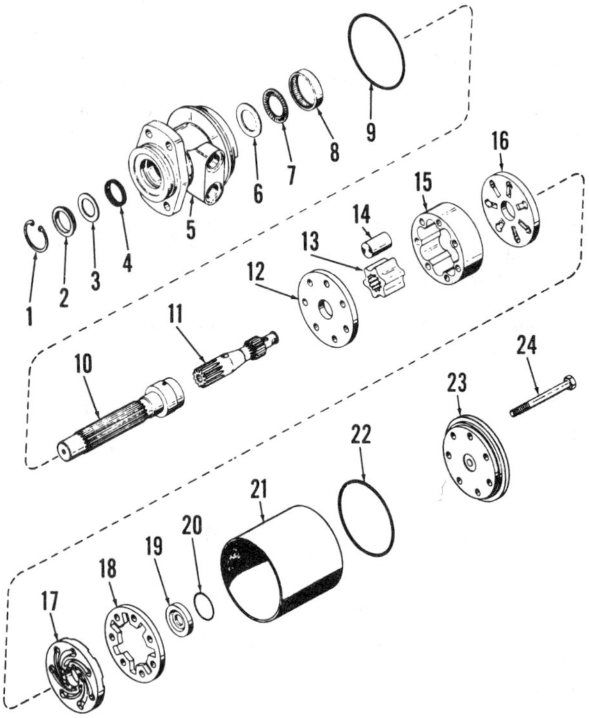

Fig. C13—Exploded view of hydraulic drive motor.

1. Snap ring
2. Spacer
3. Shim washer
4. Oil seal
5. Body
6. Thrust washer
7. Thrust bearing
8. Needle bearing
9. Seal ring
10. Output shaft
11. Drive link
12. Wear plate
13. Rotor
14. Roller (6)
15. Stator
16. Manifold plate
17. Manifold
18. Commutator ring
19. Commutator
20. Seal ring
21. Sleeve
22. Seal ring
23. End cover
24. Cap screw (7)

floor jack under range transmission and differential housing. Unbolt and remove fenders, seat and seat support. Drain fuel tank, disconnect fuel line, loosen mounting straps and remove tank. Disconnect hydraulic lines from hydraulic drive motor and cap all openings. Remove cap screws securing range transmission and differential assembly to tractor frame and roll assembly rearward from tractor.

Reinstall by reversing removal procedure.

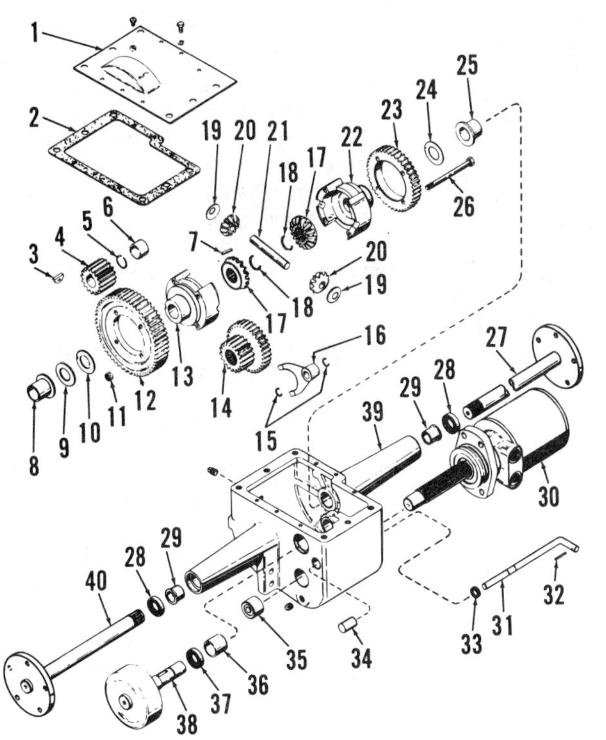

Fig. C14—Exploded view of range transmission and differential assembly.

1. Cover
2. Gasket
3. Key
4. Brake idler gear
5. Retaining ring
6. Bushing
7. Roll pin
8. Flanged bushing
9. Spacer
10. Thrust washer
11. Locknut (4)
12. Low speed ring gear
13. Differential case half
14. Hi-Lo sliding gear
15. Retaining rings
16. Shift fork
17. Axle gears
18. "C" rings
19. Thrust washers
20. Bevel pinion gears
21. Drive pin
22. Differential case half
23. High speed ring gear
24. Thrust washer
25. Flanged bushing
26. Bolt (4)
27. Axle shaft
28. Oil seal
29. Flanged bushing (2)
30. Drive motor assy.
31. Shift rod
32. Roll pin
33. Oil seal
34. Plug
35. Needle bearing
36. Bushing
37. Oil seal
38. Brake drum & shaft
39. Transmission & differential housing
40. Axle shaft

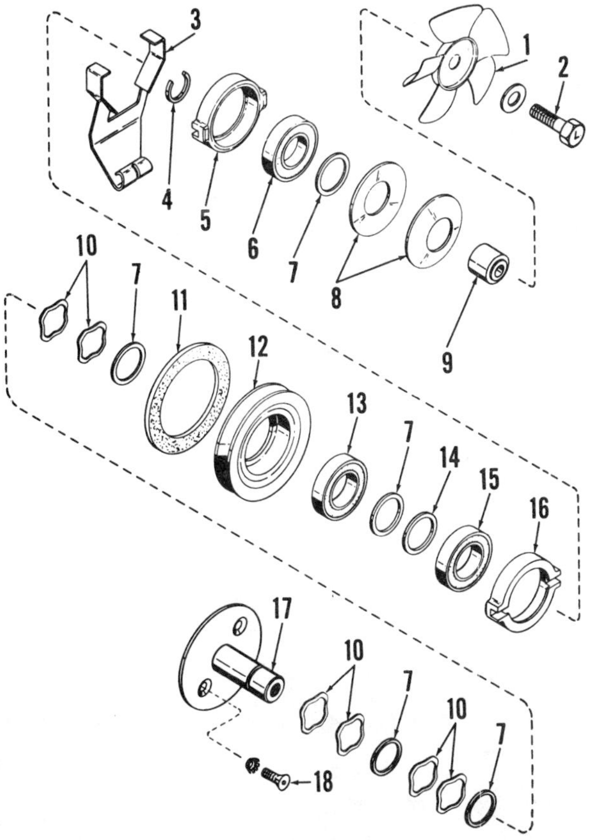

Fig. C16—Exploded view of typical pto clutch assembly used on all models.

1. Fan
2. Cap screw (L.H. thread)
3. Cam actuator lever
4. Retaining ring
5. Outer cam
6. Bearing
7. Spacer
8. Belleville springs
9. Hub spacer
10. Springs
11. Friction disc
12. Clutch pulley
13. Bearing
14. Shim
15. Bearing
16. Inner cam
17. Drive hub & clutch plate
18. Machine screw (2)

OVERHAUL

All Models

To disassemble range transmission and differential assembly, drain fluid, remove top cover and rear wheels. Place a large drive punch against inside end of brake shaft (38–Fig. C14) and hit drift sharply with a hammer. This will dislodge retaining ring (5) from its groove in brake shaft. Remove retaining ring and brake idler gear (4) as brake shaft is withdrawn. Unbolt hydraulic drive motor (30), hold sliding gear (14) and remove drive motor and sliding gear. Unseat retaining rings (15), withdraw shift rod (31) and remove shift fork (16). Remove "C" ring (18) from inner end of axle shafts (27 and 40), withdraw axle shafts and lift out differential assembly with spacer (9) and thrust washers (10 and 24). Remove locknuts (11) and bolts (26), then remove low speed ring gear (12) and high speed ring gear (23). Separate differential case halves (13 and 22) and remove drive pin (21) with roll pin (7), thrust washers (19), pinion gears (20) and axle gears (17). Remove oil seals (28, 33 and 37). Remove plug (34) and needle bearing (35). If necessary, remove brake shaft bushings (6 and 36) and axle shaft flanged bushings (8, 25 and 29).

Clean and inspect all parts and renew any showing excessive wear or other damage. If brake shaft bushings were removed, press new bushings into position, then ream inner bushing (6) to 1.004-1.005 inches (25.50-25.53 mm) and outer bushing (36) to 1.192-1.193 inches (30.28-30.30 mm). If axle shaft bushings were removed, install inner bushings (8 and 25) with oil groove downward, then install outer bushings (29). Ream all four axle bushings to 1.876-1.877 inches (30.165-30.167 mm). Reassemble differential and ring gears. Use new locknuts (11) and tighten them to 50 ft.-lbs. (68 N·m). Use new oil seals (28, 33 and 37), new gasket (2) and new "O" ring on drive motor flange. Use 0.015 inch thick spacers (9) as required to adjust differential unit side play until side play is between 0.005 inch (0.127 mm) minimum and 0.030 inch (0.762 mm) maximum. When installing drive motor, apply "Loctite" to cap screw threads and tighten cap screws to 110-125 ft.-lbs. (149-170 N·m). Fill transmission and differential housing to level plug opening.

PTO CLUTCH

R&R AND OVERHAUL

All Models

To remove pto clutch, remove tractor hood and unbolt oil cooler supports from flange. Move oil cooler ahead for access to clutch. Disconnect pto control rod from cam actuator lever (3–Fig. C16). Remove left hand thread cap screw (2), fan (1), hub spacer (7), outer cam (5) with bearing (6) and cam actuator (3) as an assembly. Remove retaining ring (4), inner cam (16) with bearing (15), spacer and shim pack (7 and 14), pulley (12) with bearing (13) and friction disc (11). Spacers (7) and springs (10) can now be removed.

Clean and inspect all parts and renew any showing excessive wear or other damage. Check friction disc for glaze and wear. If friction surface is glazed or if friction disc thickness measures less than 1/8-inch (3.175 mm), renew friction disc. Sealed bearings (6, 13 and 15) must rotate freely and quietly.

When reassembling, install spacers (7) and springs in same sequence as original assembly. Place friction disc (11) over shoulder of clutch pulley (12), then place both on drive hub. Install original spacer and shim pack (7 and 14). Measure diameter of inner cam (16) at each side of lever notches. One side will measure about 3 inches (76.2 mm) and the other about 2⅞ inches (73.0 mm). Install cam and bearing assembly, with shorter side of cam downward, on drive hub. Install retaining ring (4) on drive hub. Place washer, fan (1), hub spacer (9), Belleville spring (8) and spacer (7) on left hand thread cap screw (2). Position outer cam (5) with bearing (6) on hub spacer (9) so longer diameter of cam between notches is downward. Install cam actuating lever (3) in notches of inner cam (16), then install outer cam assembly. Tighten cap screw (2) to 35-40 ft.-lbs. (48-54 N·m).

With clutch disengaged, measure friction disc clearance using two feeler gages 180° apart. Clearance should measure 0.015-0.025 inch (0.381-0.635 mm). To increase clearance, remove shims (14) or to decrease clearance, add shims (14) as required.

NOTE: If clutch will not disengage, check for incorrect assembly of cams (5 and 16). When correctly installed, facing notches on inner and outer cams are out of alignment when clutch is either engaged or disengaged.

Reinstall by reversing removal procedure.

HYDRAULIC LIFT

Hydrostatic pump supplies fluid and pressure for hydraulic lift mechanism and control valve is an integral part of hydrostatic drive control valve. For pump or control valve service, refer to appropriate paragraphs in **HYDRAULIC DRIVE** section.

CUB CADET

CONDENSED SPECIFICATIONS

	MODELS			
	580	582	682 782 784	982 984 986
Engine Make	B&S	B&S	Kohler	Onan
Model	401417	401417	KT-17	B48G
Bore	3.44 in. (87.3 mm)	3.44 in. (87.3 mm)	3.12 in. (79.3 mm)	3.25 in. (82.55 mm)
Stroke	2.16 in. (54.8 mm)	2.16 in. (54.8 mm)	2.75 in. (69.9 mm)	2.88 in. (76.03 mm)
Displacement	40 cu. in. (656 cc)	40 cu. in. (656 cc)	42.18 cu. in. (690.5cc)	47.7 cu. in. 781.7cc)
Power Rating	16 hp., (11.9 kW)	16 hp. 11.9 kW)	17 hp. 12.7 kW)	20 hp. (14.9 kW)
Idle Speed	1200 rpm	1200 rpm	1200 rpm	1200 rpm
High Speed (No Load)	3600 rpm	3600 rpm	3600 rpm	3600 rpm
Capacities –				
Crankcase	3 pts.* (1.4 L)	3 pts.* (1.4 L)	3 pts. (1.4 L)	3.5 pts. (1.6 L)
Transmission Or Transaxle	4 pts. (1.9L)	8 pts. (3.8L)	7 qts. (6.6L)	7 qts. (6.6L)
Creeper Drive	.....	½-pt. (0.24 L)	.....	.....
Right Angle Gear Box	4 oz. (118 mL)	.....	.....	.....
Fuel Tank	4 gal. (15.1 L)	4 gal. (15.1 L)	4 gal. (15.1 L)	4 gal. (15.1 L)

*Early production engine crankcase capacity is 3.5 pints (1.6 L).

FRONT AXLE AND STEERING SYSTEM

MAINTENANCE

All Models

Lubricate steering spindles and steering arm at 10 hour intervals. Lubricate axle pivot bolt at 30 hour intervals and steering gear box at 100 hour intervals.

NOTE: Do not over lubricate steering gear box. Four strokes of a hand grease gun is sufficient. Do not use a high volume or high pressure grease gun.

Use multi-purpose, lithium base grease and clean all fittings before and after lubrication.

R&R AXLE MAIN MEMBER

Models 580 – 582 – 682 – 782 – 784

To remove front axle main member, disconnect drag link from left hand steering spindle. Support front of tractor and remove axle pivot bolt (3 – Fig. CC1). Raise tractor and roll axle out.

Reinstall by reversing removal procedure.

Models 982 – 984 – 986

To remove front axle main member (2 – Fig. CC2), disconnect drag link (7 – Fig. CC3) from steering lever (5). Support front of tractor and remove axle pivot bolt (4 – Fig. CC2). Raise front of tractor and roll axle out.

Reinstall by reversing removal procedure.

TIE ROD, STEERING LEVER AND TOE-IN

All Models

A single tie rod with renewable ends and a single drag link with renewable ends are used on 580, 582, 682, 782 and 784 models.

Two tie rods with renewable ends, a single drag link with renewable ends and a steering lever mounted on axle main member are used on 982, 984 and 986 models.

Steering lever pivot pin (4–Fig. CC3) for 982, 984 and 986 models is retained by a snap ring (6). Pin is threaded for slide hammer to aid removal.

Toe-in for all models should be 1/32 to 1/8-inch (0.794 to 3.175 mm) and is adjusted on 580, 582, 682, 782 and 784 models by lengthening or shortening tie rod (17–Fig. CC1) as necessary. Toe-in for 982, 984 and 986 models is adjusted by lengthening and shortening each of the two tie rods (6 and 7–Fig. CC2) as necessary. Tie rods should be kept as nearly equal in length as possible.

R&R STEERING SPINDLES

Models 580 – 582 – 682 – 782 – 784

Raise and support front of tractor. Remove front tires, wheels and hubs, disconnect tie rod (17–Fig. CC1) at steering arms (1 and 9) and drag link at steering arm (9). Remove bolts (2 and 8) and pull spindles (1 and 9) from axle main member (6).

Reinstall by reversing removal procedure. Renew bushings (11 and 13) and/or bolts (2 and 8) as necessary. Tighten nuts (14) on bolts to 80 ft.-lbs. (108 N·m).

Models 982 – 984 – 986

Raise and support front of tractor. Remove front tires, wheels and hubs. Disconnect tie rods (6 and 7–Fig. CC2) at spindles (5 and 8). Remove cap screws and flat washers from upper ends of spindles (5 and 8) and pull spindles from axle main member (2).

Reinstall by reversing removal procedure. Renew bushings (1 and 3) and/or spindles (5 and 8) as necessary.

FRONT WHEEL BEARINGS

All Models

Front wheel bearings are a press fit in wheel hub and a slip fit on spindle. Bearings should be removed and cleaned at 100 hour intervals. Repack bearings with multi-purpose, lithium base grease.

To remove bearings, raise and support front of tractor. Remove wheel and tire assembly. Bearings can be driven from wheel hub using a long drift punch and hammer. Bearings should be driven from inside wheel hub toward outside of hub.

If bearings are a loose fit in wheel hub, renew or repair hub.

Reinstall by reversing removal procedure.

R&R STEERING GEAR

Models 580 – 582 – 682 – 782 – 784

Remove steering wheel cover and a steering wheel retaining nut. Use a suitable puller and remove steering wheel.

On models with hydraulic control valve, disconnect hydraulic lines from transmission at control valve. Disconnect linkage at control valve and remove valve assembly. Mark location of control valve mounting plate on steering column and remove plate.

On all models, disconnect drag link from steering lever. Remove mounting bolts and pull steering assembly down through control panel and remove.

Reinstall by reversing removal procedure.

Models 982 – 984 – 986

Remove steering wheel cover and steering wheel retaining nut. Use a suitable puller and remove steering wheel. Shut off fuel at tank, disconnect fuel line. Remove fire wall and fuel tank as an assembly. Disconnect drag link at steering lever. Disconnect hydraulic lines from transmission at control valve and remove valve assembly. Mark location of control valve mounting plate on steering column and remove plate. Remove the three steering column retaining bolts and pull steering assembly down through control panel and remove.

Reinstall by reversing removal procedure.

STEERING GEAR OVERHAUL

Models 580 – 582 – 682 – 782 – 784

Remove steering column and gear assembly as previously outlined. Remove steering lever jam nut (17–Fig. CC4), adjusting nut (16) and washer (15). Remove steering lever (11), seal retainer (12) and seal (13). Remove cotter pin (23) and adjusting plug (22). Pull steering gear and bearing assembly from housing. Remove bearing race retainer snap rings (18). Remove bearing retainers (19), bearing balls (20) and retainers (21). Remove jam nut (7) and steering pin (6) from steering lever (11).

Inspect all parts for wear, cracks or damage and renew as necessary.

To reassemble, coat steering gear with multi-purpose, lithium base grease.

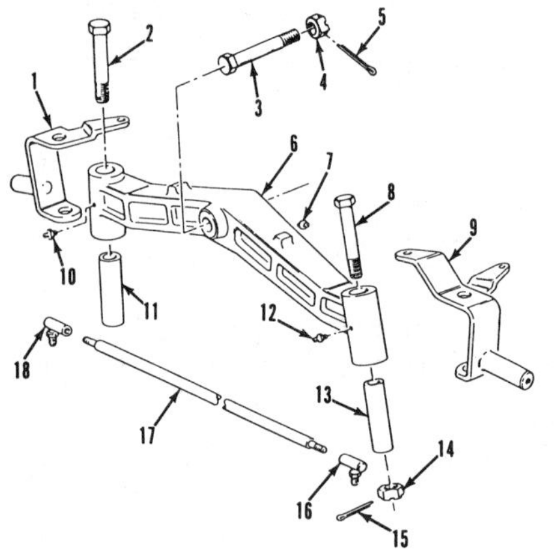

Fig. CC1 – Exploded view of front axle assembly used on 580, 582, 682, 782 and 784 models.

1. Right steering spindle
2. Spindle bolt
3. Axle pivot bolt
4. Nut
5. Cotter pin
6. Axle main member
7. Grease fitting
8. Spindle bolt
9. Left steering spindle
10. Grease fitting
11. Bushing
12. Grease fitting
13. Bushing
14. Nut
15. Cotter pin
16. Tie rod end
17. Tie rod
18. Tie rod end

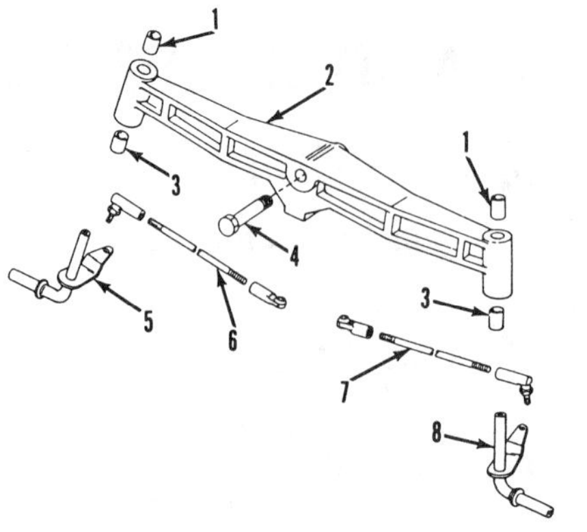

Fig. CC2 – Exploded view of front axle assembly used on 982, 984 and 986 models.

1. Upper bushing
2. Axle main member
3. Lower bushing
4. Axle pivot bolt
5. Right wheel spindle
6. Tie rod
7. Tie rod
8. Left wheel spindle

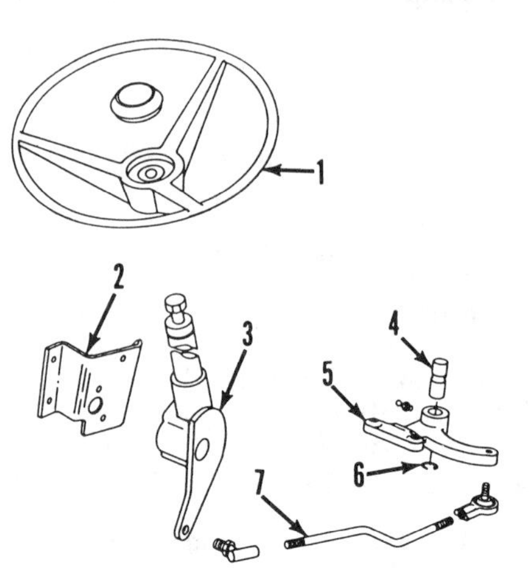

Fig. CC3 – View of steering column assembly and steering lever.

1. Steering wheel
2. Bracket
3. Steering gear & column assy.
4. Steering lever pivot pin
5. Steering lever
6. Snap ring
7. Drag link

Illustrations courtesy Cub Cadet Corp.

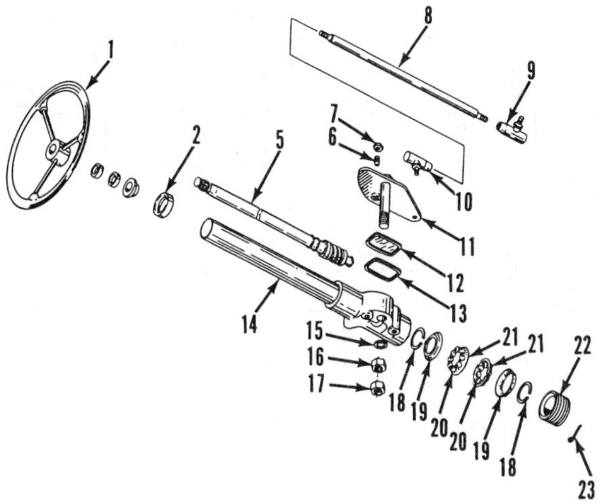

Fig. CC4—Exploded view of 582, 682, 782 and 784 models steering gear and column assembly.

1. Steering wheel
2. Bearing
5. Steering gear & shaft assy.
6. Steering pin
7. Jam nut
8. Drag link
9. Ball joint end
10. Ball joint end
11. Steering lever plate
12. Seal retainer
13. Seal
14. Housing
15. Washer
16. Adjusting nut
17. Jam nut
18. Snap ring
19. Bearing retainer
20. Bearing balls
21. Ball retainers
22. Adjusting plug
23. Cotter pin

Install retainers (21), bearing balls (20) and retainers (19) on steering gear (5). Install snap rings (18). Install steering gear and bearing assembly in housing (14). Make certain bearing retainers (19) enter bore of housing squarely and do not bind. Install adjusting plug (22) and tighten until all end play is removed from shaft and gear. Shaft and gear assembly should turn freely. Install cotter pin (23). Pack housing with multipurpose, lithium base grease. Install seal (13), retainer (12) and steering lever plate (11). Place a 3/32-inch (2.4 mm) thick shim stock between steering lever plate (11) and seal retainer (12). Install washer (15) and adjusting nut (16). Tighten adjusting nut (16) snug and remove the 3/32-inch (2.4 mm) shim stock. Install jam nut (17) and tighten to 40 ft.-lbs. (54 N·m) while holding adjusting nut in place. Install steering pin (6) in steering lever plate until it just engages steering gear. Center steering gear by rotating shaft to position halfway between full left and full right turn. Adjust steering pin (6) inward until all backlash is removed but steering gear turns through full left/right cycle with no binding. Install jam nut (7) and tighten to 40 ft.-lbs. (54 N·m) while holding steering pin in proper position.

Models 982–984–986

To disassemble steering gear, remove jam nuts (5–Fig. CC5). Remove steering lever plate (14). Remove cotter pin (11), adjustment plug (10) and slide steering shaft assembly (8) out of housing. Remove Belleville washer (9), retainer cups (6) and bearings (7).

Inspect parts for wear, scoring or damage and renew as necessary.

To reassemble, apply multi-purpose, lithium base grease to bearings (7) and retainer cups (6) and install on steering shaft (8). Apply grease to steering gear and install assembly into housing (3). Install Belleville washer (9) and plug (10). Tighten plug to 4-6 ft.-lbs. (5-8 N·m) and install cotter pin (11). Shaft should turn freely in housing. Pack housing with grease and install new seal (16) and retainer (15) on housing. Remove steering pin (13) from steering lever plate (14) and install steering lever plate on housing. Place a 5/32-inch (4 mm) shim between steering lever plate and seal retainer, install washer (4) and tighten one jam nut (5) until snug. Install second jam nut and tighten against first nut to 22-25 ft.-lbs. (30-34 N·m) while holding first nut in position on stud. Remove shim stock. Install steering pin in steering lever plate until it just engages steering gear. Center steering gear by rotating shaft to position halfway between full left and full right turn. Adjust steering pin (13) inward until all backlash is removed but steering gear turns through full left/right cycle with no binding. Tighten jam nut (12) to 35-45 ft.-lbs. (48-61 N·m).

ENGINE

MAINTENANCE

All Models

Regular engine maintenance is required to maintain peak performance and long engine life.

Check oil level and clean air intake screen at five hour intervals under normal operating conditions.

Clean engine air filter and cooling fins at 25 hour intervals and check for any loose nuts, bolts or linkage. Repair as necessary.

Change engine oil and filter, perform tune-up, valve adjustment and clean carbon from cylinder heads as recommended in appropriate ENGINE section in this manual.

REMOVE AND REINSTALL

Models 580–582

Disconnect battery ground cable. Raise hood and remove engine side panels which are secured with wing nuts and a spring. Disconnect headlight wiring and remove hood and grille as an assembly. Disconnect alternator-regulator wire, starter cables and pto clutch wire. Remove air cleaner assembly and disconnect choke and throttle cables. Disconnect engine shut-off wire. Shut off fuel and disconnect fuel line at carburetor. Remove engine mounting bolts, slide engine forward and remove.

Reinstall by reversing removal procedure.

Models 682–782–784

Disconnect battery ground cable. Raise hood and remove engine side

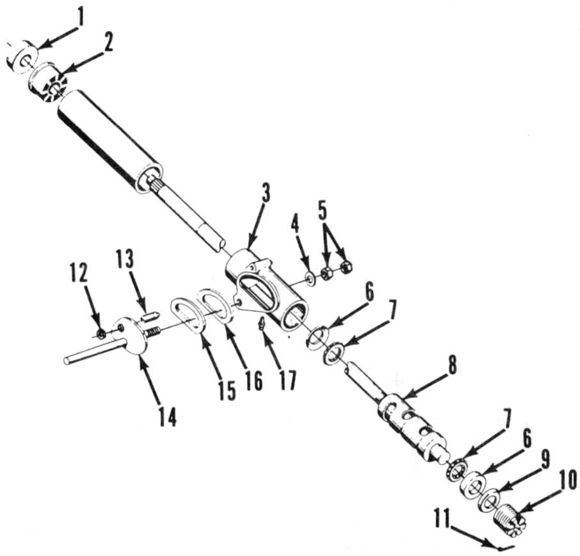

Fig. CC5—Exploded view of 982, 984 and 986 models steering gear and column assembly.

1. Dust cover
2. Upper bearing
3. Housing
4. Washer
5. Nuts
6. Cup
7. Bearing
8. Steering gear
9. Belleville washer
10. Adjustment plug
11. Cotter pin
12. Nut
13. Steering pin
14. Steering lever plate
15. Seal retainer
16. Seal
17. Grease fitting

Illustrations courtesy Cub Cadet Corp.

panels which are secured with wing nuts and a spring. Disconnect headlight wiring and remove hood and grille as an assembly. Remove air cleaner assembly and disconnect choke cable, throttle cable and wiring harness. Disconnect pto clutch wire and starter cable. Shut off fuel and disconnect fuel line at tank. Remove front flex coupler to flywheel flange retaining nuts. Remove engine mounting bolts and remove engine.

Reinstall by reversing removal procedure.

Models 982–984–986

Disconnect battery ground cable. Raise hood and remove engine side panels which are secured with wing nuts and a spring. Disconnect pto clutch wire and starter cable. Remove air cleaner assembly and disconnect choke and throttle cables. Shut off fuel and disconnect fuel line from fuel pump. Disconnect positive terminal coil wire and wire at rectifier. Remove engine mounting bolts and remove engine.

Reinstall by reversing removal procedure.

OVERHAUL

All Models

Engine make and model are listed at the beginning of this section. To overhaul engine or accessories, refer to appropriate engine section of this manual.

ELECTRICAL SYSTEM

MAINTENANCE AND SERVICE

All Models

Battery electrolyte should be checked at 50 hour intervals of normal operation. If necessary, add distilled water until level is just below base of vent well. **DO NOT** overfill. Keep battery posts clean and cable ends tight.

For alternator or starter service, refer to appropriate engine accessory section in this manual.

Refer to Fig. CC6 through CC8 for appropriate wiring diagram for model being serviced.

BRAKES

ADJUSTMENT

Model 580

To adjust disc type brake on 580 model, place brake pedal in up position (Fig. CC9). Remove all slack from brake

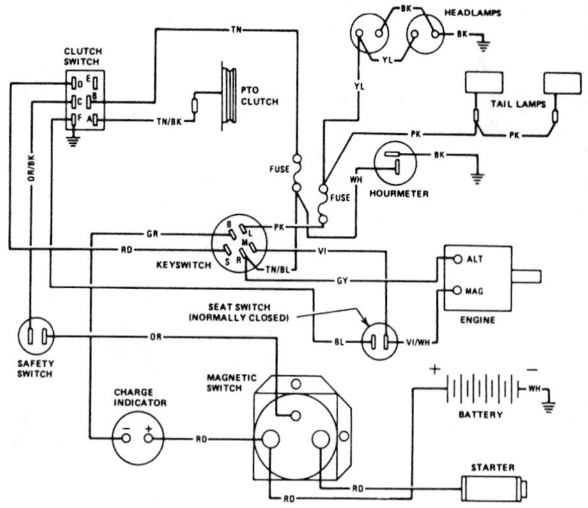

Fig. CC6—Wiring diagram for 580 and 582 models.

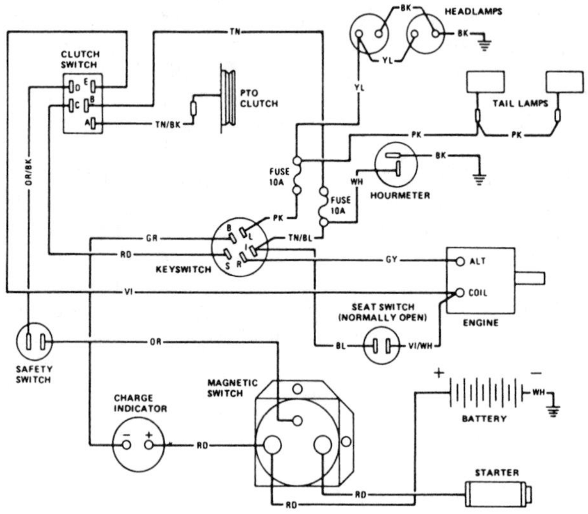

Fig. CC7—Wiring diagram for 682, 782 and 784 models.

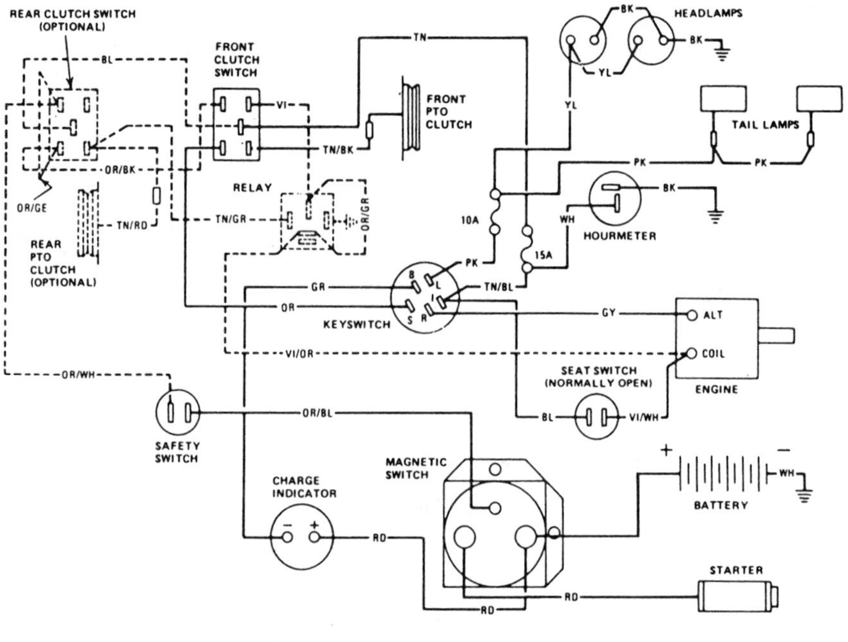

Fig. CC8—Wiring diagram for 982, 984 and 986 models.

Illustrations courtesy Cub Cadet Corp.

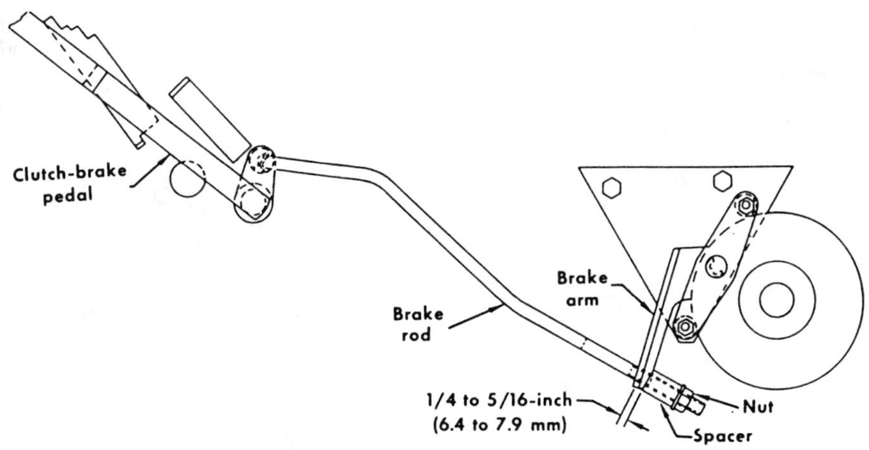

Fig. CC9—View of disc brake system used on 580 model. Refer to text for adjustment procedure.

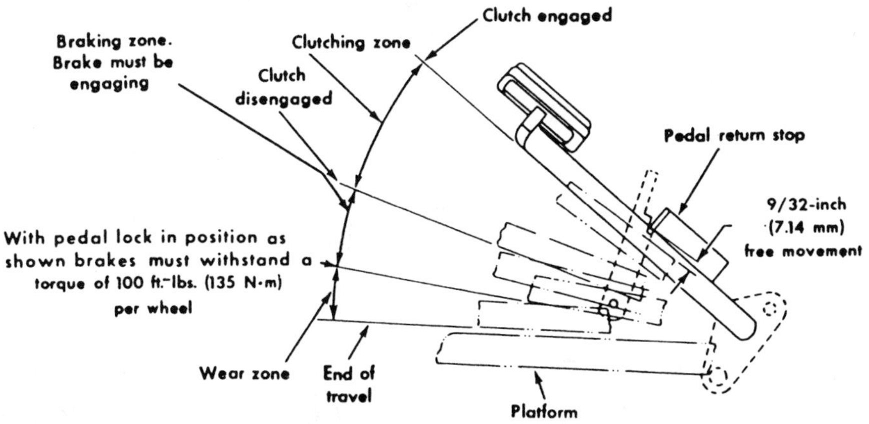

Fig. CC10—Brake should engage when pedal is pushed down to within a maximum of 1-5/16 inches (33 mm) and a minimum of ¾-inch (19 mm) above top of platform.

linkage by moving brake arm up and pulling brake rod down. Adjust nut to allow ¼ to 5/16-inch (6.4 to 7.9 mm) clearance between spacer and brake arm. Check brake operation.

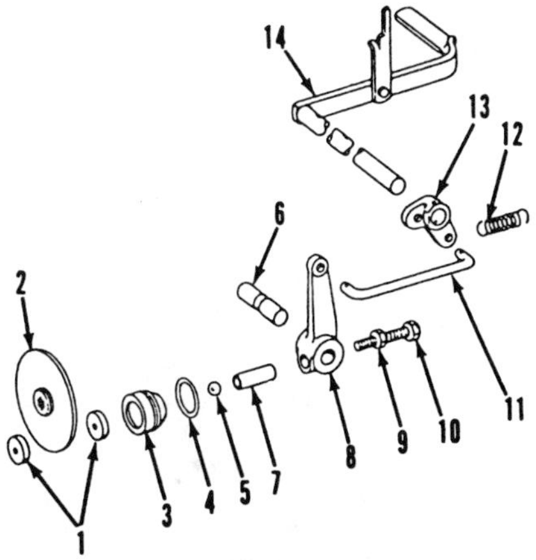

Fig. CC11—Exploded view of brake system used on 582 model prior to S/N 720000.

1. Brake pads
2. Brake disc
3. Retainer
4. "O" ring
5. Ball
6. Pivot pin
7. Push rod
8. Brake lever
9. Jam nut
10. Adjustment bolt
11. Brake rod
12. Return spring
13. Arm
14. Clutch & brake pedal

Model 582 (Prior To S/N 720000)

Internal type brake on 582 model (prior to S/N 720000) should engage when pedal arm is pressed down to within a maximum of 1-5/16 inches (33 mm) and a minimum of ¾-inch (19 mm) distance above top of platform (Fig. CC10).

To adjust, loosen locknut (9–Fig. CC11) and turn adjusting bolt (10) in or out as required. Brake pedal is spring loaded and pedal may be pushed down against platform even if correctly adjusted.

Models 682–782–784 (Prior To S/N 720000)

To adjust internal type brakes on 682, 782 and 784 models (prior to S/N 720000), raise and support rear wheels off the ground. With brake pedal in full up position (Fig. CC12), loosen jam nut (14–Fig. CC13) and tighten adjustment bolt (13) to 8-10 in.-lbs. (0.9-1.1 N·m). Operate brake pedal through several full strokes and readjust. If brake drags with pedal in full up position after adjustment, loosen adjustment screw slightly. Tighten jam nut while holding adjustment screw in correct position.

With brake properly adjusted, it should require 100 ft.-lbs. (135 N·m) torque to turn wheel with brake pedal locked in down position. See Fig. CC12.

Models 582–682–782–784 (After S/N 720000) And All 982–984–986 Models

Raise and support rear wheels off the ground. Adjust left brake first. Disconnect brake rod (13–Fig. CC14) at clevis and shorten or lengthen rod by turning clevis (14) until gap between inner brake pad and brake disc is 0.030-0.035 inch (0.762-0.889 mm).

Repeat procedure for right side.

R&R BRAKE PADS/DISCS

Model 580

Remove nut and spacer from brake rod (Fig. CC9) and disconnect brake rod from actuator cam (5–Fig. CC15). Unbolt brake assembly from frame. Remove retaining ring from brake pad and disassemble brake.

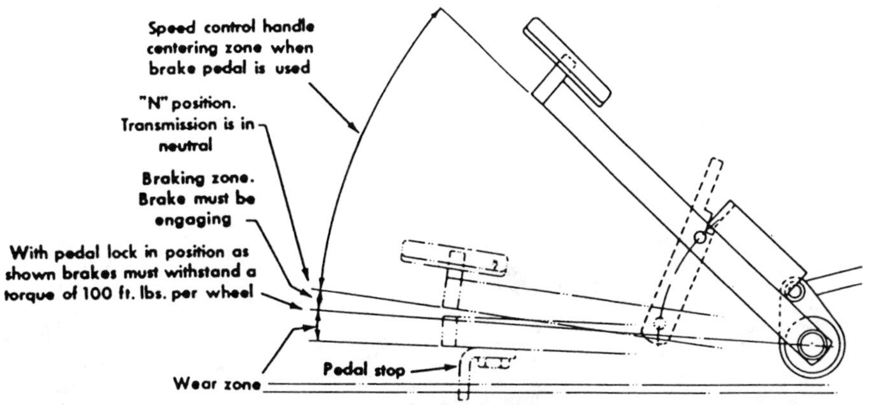

Fig. CC12—View showing pedal positions for 682, 782 and 784 models prior to S/N 720000. Refer to text for adjustment procedure.

Reinstall by reversing removal procedure.

Model 582 (Prior to S/N 720000)

Drain transmission lubricant. Remove brake adjustment bolt (10–Fig. CC11) and jam nut (9). Remove brake lever (8), pivot pin (6) and push rod (7). Remove creeper drive unit, if so equipped. Remove reduction housing front cover plate (4–Fig. CC16). Remove bolt (3) which secures reduction gear (2) to mainshaft (1). Remove reduction gear. Move brake disc (8) along countershaft (5) to remove brake pads (6). Both pads and disc may be removed without removing front retainer (3–Fig. CC11), however, "O" ring (4) should be inspected and renewed as necessary.

Reinstall by reversing removal procedure. Bolt (3–Fig. CC16) should be tightened to 55 ft.-lbs. (75 N·m). Make certain ball (5–Fig. CC11) is in place before installing push rod (7). Fill transmission to proper level with recommended fluid. Adjust brake as previously outlined.

Models 682–782–784 (Prior To S/N 720000)

To renew internal type brake pads on 682, 782 and 784 models prior to S/N 720000, first drain transmission lubricant. Remove transmission as outlined in HYDROSTATIC DRIVE SYSTEM section in this manual. Remove differential assembly. Remove bevel pinion shaft

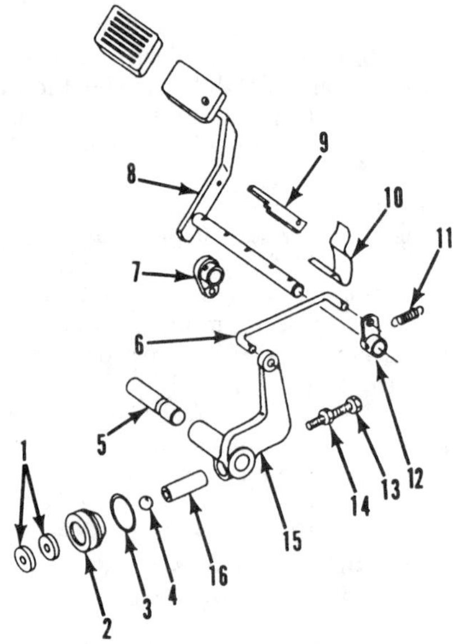

Fig. CC13—View showing component parts of 682, 782 and 784 models prior to S/N 720000 brake system. Refer to text.

1. Brake pads
2. Disc retainer
3. "O" ring
4. Ball
5. Pivot shaft
6. Brake rod
7. Neutral return lever
8. Pedal assy.
9. Brake lock
10. Spring
11. Brake return spring
12. Arm
13. Adjustment bolt
14. Jam nut
15. Brake lever
16. Push rod

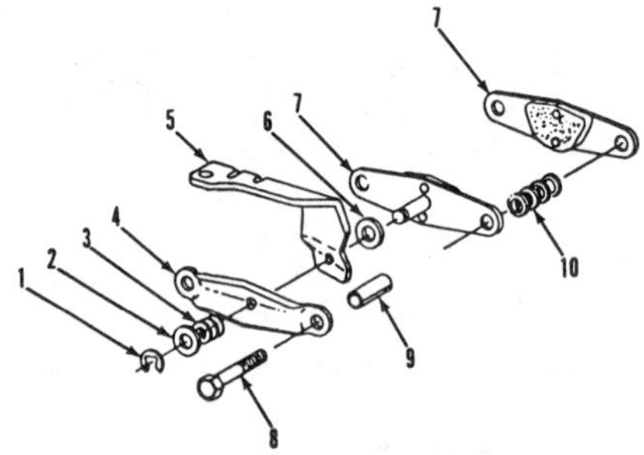

Fig. CC15—View of component parts of 580 model brake system.

1. Retaining ring
2. Thrust washer
3. Spring
4. Actuator plate
5. Cam
6. Hardened washer
7. Brake pads
8. Bolt
9. Spacer
10. Spring

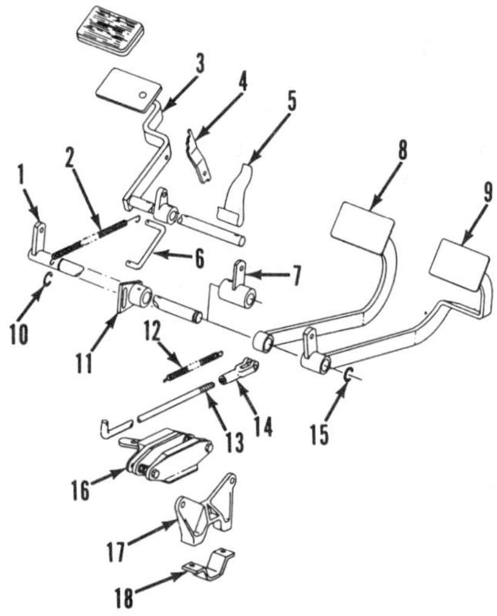

Fig. CC14—View showing component parts of 582, 682, 782, 784 (after 720000) and all 982, 984 and 986 models. Single pedal system is standard, dual pedal system is optional.

1. Brake shaft
2. Return spring
3. Pedal assy.
4. Brake lock
5. Spring
6. Center brake rod
7. Right brake lever
8. Left brake pedal
9. Right brake pedal
10. Retaining ring
11. Arm assy.
12. Return spring
13. Brake rod
14. Clevis
15. Retaining ring
16. Brake assy.
17. Bracket
18. Flange

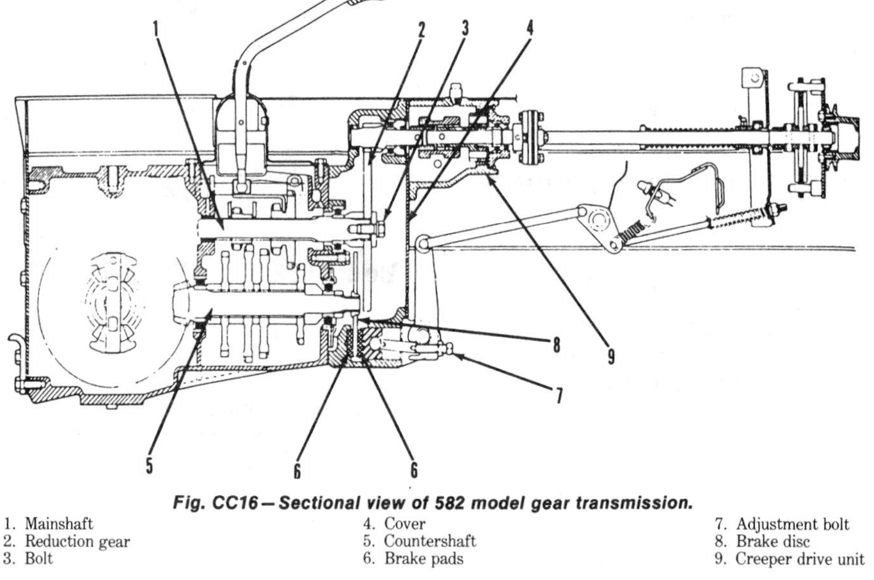

Fig. CC16—Sectional view of 582 model gear transmission.

1. Mainshaft
2. Reduction gear
3. Bolt
4. Cover
5. Countershaft
6. Brake pads
7. Adjustment bolt
8. Brake disc
9. Creeper drive unit

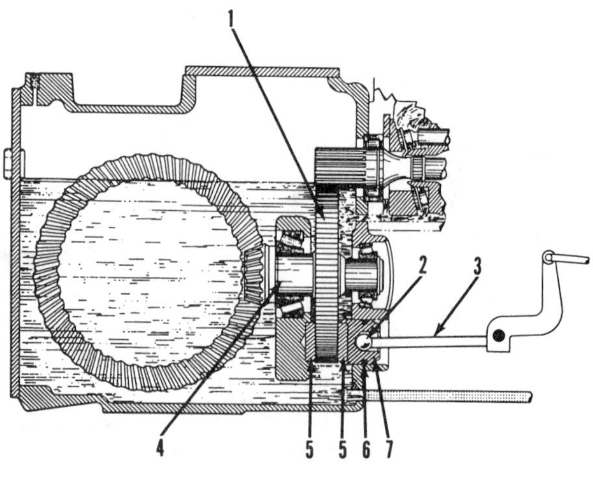

Fig. CC17—Sectional view of 682, 782 and 784 models (prior to S/N 720000) reduction drive showing component parts of internal type brake system.

1. Constant mesh gear
2. Ball
3. Push rod
4. Pinion shaft
5. Brake pads
6. "O" ring
7. Retainer

(4–Fig. CC17) and constant mesh gear (1). Remove brake pads (5) and retainer (7). Remove "O" ring (6).

Reinstall by reversing removal procedure. Fill transmission to proper level with recommended fluid and adjust brake as previously outlined.

Models 582–682–782–784 (After S/N 720000) And All 982–984–986 Models

Remove cotter pin and washer and disconnect brake rod from brake lever (1–Fig. CC18). Remove cap screws from mounting flange and remove brake assembly from axle carrier. Remove cap screws (7) and disassemble caliper assembly.

Reassemble by reversing disassembly procedure. Make certain spacers (6) and springs (5) are correctly installed.

CLUTCH

ADJUSTMENT

Model 580

Transaxle used on 580 models is belt driven. Clutch linkage which releases clutch belt tension is in conjunction with brake linkage. If belt is in good condition, proper adjustment is obtained by adjusting brakes as previously outlined. See Fig. CC9.

Model 582

A clearance of 0.050 inch (1.27 mm) must be maintained between clutch release lever (10–Fig. CC18A) and clutch release bearing (9). Maintain 9/32-inch (7 mm) pedal free movement which is measured at pedal arm point of contact with front edge of pedal return stop. See Fig. CC18A.

To adjust clearance, turn adjusting nut on clutch release rod (14) as necessary.

Models 682–782–784–982–984–986

Models 682, 782, 784, 982, 984 and 986 have hydrostatic transmissions. Brake pedal, when fully depressed, should move speed control lever to neutral position. Refer to HYDROSTATIC DRIVE SYSTEM section of this manual for adjustment procedure.

R&R CLUTCH

Model 580

Model 580 is equipped with a transaxle which is belt driven. Combination brake/clutch pedal, when depressed, releases drive belt tension. To renew drive belt, disconnect battery ground cable. Remove drawbar assembly and center frame cover. Depress clutch/brake pedal and lock in lowest position. Loosen drive belt guides and remove idler pulley. Remove the two mounting bolts securing right angle drive gear box to cross support and rotate gear box to bring pulley down. Remove drive belt.

Install new drive belt on drive pulley and transaxle pulley. Rotate gear box back into position and reinstall bolts. Install idler pulley. Release clutch/brake pedal and secure drive belt guides so there is 1/8 to 3/16-inch (3 to 5 mm) clearance between belt and guides. Install center frame cover, drawbar assembly and reconnect battery.

Model 582

To remove clutch assembly, remove engine side panels and frame cover. Disconnect battery ground cable. Remove pivot pin and hanger assembly (11–Fig. CC18A). Remove bolts from flex coupling, drive roll pins out of driveshaft coupling and coupling arm. Slide couplings forward on clutch shaft and remove clutch shaft assembly including pressure plates, drive plate and clutch release lever. Remove drive plate, and pressure plate and release lever from shaft. Clamp shaft in vise jaws slowly. Remove spring and remaining parts.

Inspect all parts for wear or damage. Clutch loading spring (6–Fig. CC18) free length should be 6.7 inches (170 mm) and should require 235-240 pounds (1053-1067 N) pressure to compress to 5.2 inches (132 mm). Teaser spring (17) free length should be a 0.442 inch (11 mm) and should require 50 pounds (22 N) pressure to compress to 0.37 inch (9 mm).

Reinstall by reversing removal procedure. Adjust clutch as previously outlined.

HYDROSTATIC DRIVE SYSTEM

MAINTENANCE

Models 682–782–784–982–984–986

Check fluid level and clean hydrostatic drive unit cooling fins at 30 hour intervals. Maintain fluid level at "FULL" mark on dipstick which is located just in front of tractor seat. Change transmission filter at 100 hour intervals and change fluid and filter at 500 hour intervals. Recommended lubricant is Cub Cadet Hydraulic/Transmission Fluid or equivalent lubricant. Approximate fluid capacity is 7 quarts (6.6 L).

ADJUSTMENT

Models 682–782–784–982–984–986

SPEED CONTROL LEVER ADJUSTMENT. Speed control lever should require a pulling force of 7 to 8 pounds (3.2 to 3.6 kg) to maintain speed control lever position. Adjust by removing left side engine panel and tighten or loosen friction control nut as necessary.

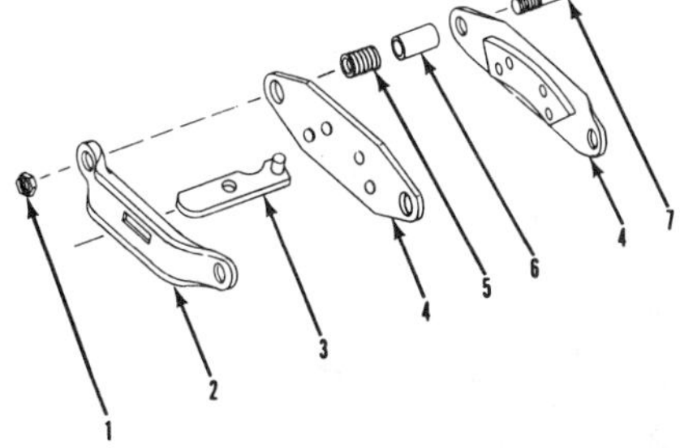

Fig. CC18—Exploded view of brake caliper and component parts used on 582, 682, 782, 784 models (after S/N 720000) and all 982, 984 and 986 models.

1. Nut
2. Cam bracket
3. Cam
4. Pads
5. Spring
6. Spacer
7. Bolt

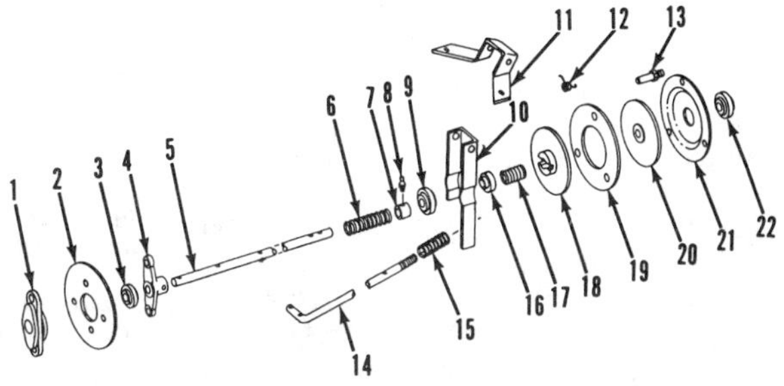

Fig. CC18A—Exploded view of clutch system used on 582 models with gear transmissions. Bushing (7) and clutch release bearing (9) are a single unit on some models.

1. Coupler
2. Flex coupler disc
3. Bushing
4. Coupling arm
5. Shaft
6. Spring
7. Bushing
8. Grease fitting
9. Clutch release bearing
10. Release lever
11. Hanger
12. Spring
13. Drive stud
14. Release rod
15. Release rod spring
16. Spacer
17. Teaser spring
18. Hub assy.
19. Clutch disc
20. Hub assy.
21. Disc
22. Bushing

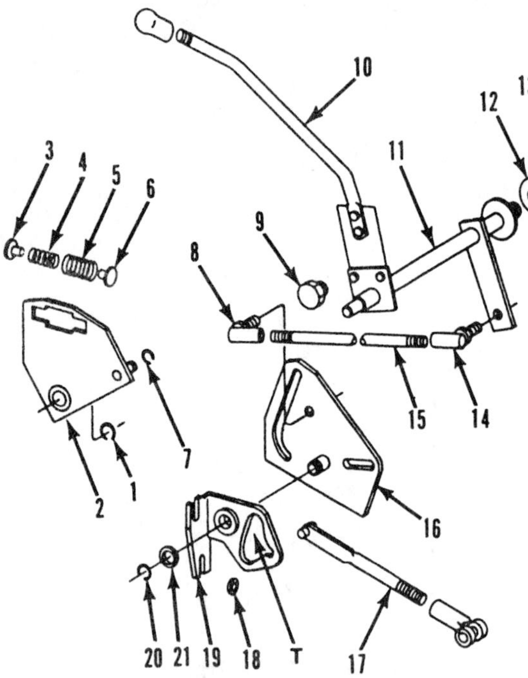
1
2
3
4
5
6
7
8
9
10
11
12
13
14
15
16
17
18
19
20
21
T

Fig. CC19—Exploded view of hydrostatic transmission speed and direction control linkage.
1. Retaining ring
2. Damper spring plate
3. Damper spring guide pin
4. Damper spring (light)
5. Damper spring (heavy)
6. Damper spring guide pin
7. Retaining ring
8. Ball joint
9. Cross-shaft bearing
10. Speed control lever
11. Cross-shaft
12. Friction disc
13. Belleville washers
14. Ball joint
15. Linkage shaft
16. Cam bracket assy.
17. Rod
18. Washer
19. Bracket
20. Retaining ring
21. Washer
T. "T" slot

CAM BRACKET ADJUSTMENT. If tractor creeps when speed control lever is in neutral position or if linkage has been removed and is being reinstalled, adjust brake pedal and speed control lever, then raise and support tractor so rear tires are off the ground. Remove frame cover and lubricate "T" slot (T–Fig. CC19). Move speed control lever to fast forward position. Loosen cam bracket mounting bolts and move cam bracket to its highest position in slotted holes and tighten bolts slightly to retain in this position. Start engine and use punch and hammer to adjust cam bracket downward until wheels stop turning.

CAUTION: Use care when working near rotating tires or parts.

Move speed and directional control lever to forward position. Depress brake pedal and lock in this position. If there is excessive transmission noise or vibration with brake pedal depressed, adjust cam bracket to position where noise or vibration stops. Release brake and shut off engine. Move speed control lever to fast forward position and tighten cam bracket retaining bolts. Start engine, move speed control lever to fast foward position, depress brake pedal fully and release. Wheels should stop turning and speed control lever should return to neutral position. If speed control lever is not in neutral position, loosen jam nut on neutral return rod and adjust until lever is in neutral position. Tighten jam nut. Speed control rod should not touch end of slot when brake pedal is fully depressed. If rod touches, disconnect clevis from brake cross shaft, loosen jam nut and lengthen rod until clearance is obtained. Tighten jam nut and connect clevis to brake cross shaft.

R&R HYDROSTATIC DRIVE UNIT

Models 682–782–784

Remove frame cover and an engine side panel. Remove hydraulic lines from center section to lift control valve, as equipped. On all models, disconnect flex coupling. Remove driveshaft coupling roll pin and slide coupling forward on shaft. Remove coupling arm roll pin and slide coupling arm rearward on transmission input shaft. Disconnect front flex coupling at flywheel flange and remove driveshaft. Remove retaining ring securing control cam assembly to damper spring plate. Remove locknut connecting stud ball joint and linkage rod to control cam. Remove cam bracket mounting bolts and move bracket and linkage up out of the way. Disconnect brake rod from brake lever and remove brake adjusting screw. Disconnect suction line and cap openings. Remove hydrostatic unit mounting bolts from rear frame housing, bring unit forward, tilting top of unit downward and bring it up and out.

Reinstall by reversing removal procedure. Tighten all hydrostatic unit to frame mounting bolts except those which hold cam bracket, to 30 ft.-lbs. (41 N·m). Adjust brakes and cam bracket assembly. Fill unit to proper level with recommended fluid.

Models 982–984–986

Raise hood and remove engine side panels. Remove frame cover. Disconnect headlight wiring and remove hood and grille as an assembly. Remove hydraulic lines from center section to lift control valve. Remove engine oil filter. Remove coupling arm roll pin. Remove engine mounting bolts and slide engine forward until coupling arm is free of input shaft. Remove cam bracket mounting bolts and move bracket and linkage up out of the way. Remove brake rod and return spring. Disconnect suction line from hydrostatic unit. Remove hydrostatic mounting bolts from rear frame housing and bring unit forward, up and out.

Reinstall by reversing removal procedure. Tighten all hydrostatic unit to frame mounting bolts except those which hold cam bracket, to 30 ft.-lbs. (41 N·m). Fill unit to proper level with Cub Cadet Hydraulic/Transmission Fluid or equivalent lubricant.

OVERHAUL HYDROSTATIC DRIVE UNIT

Models 682–782–784–982–984–986

Thoroughly clean exterior of unit. A holding fixture, similar to one shown in Fig. CC20, makes servicing unit more

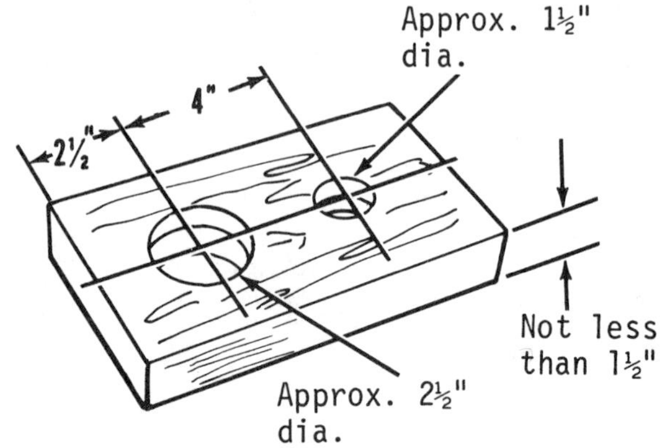
Approx. 1½" dia.
4"
2½"
Not less than 1½"
Approx. 2½" dia.

Fig. CC20—View showing dimensions of wooden fixture which may be used to hold hydrostatic transmission while servicing unit.

Illustrations courtesy Cub Cadet Corp

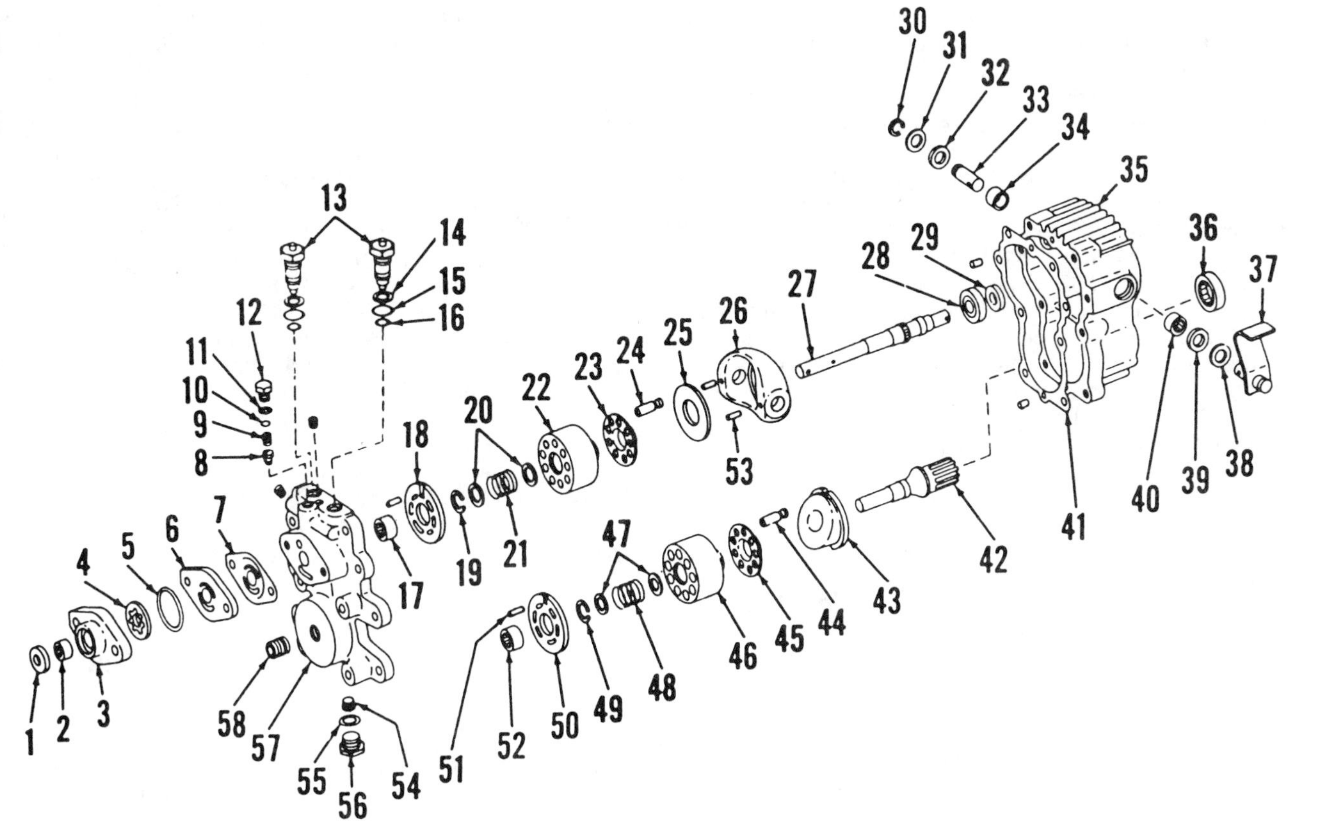

Fig. CC21 — Exploded view of hydrostatic drive unit. Later models do not use port plate (6) and gasket (7). See Fig. CC22.

1. Oil seal
2. Needle bearing
3. Charge pump housing
4. Rotor assy.
5. "O" ring
6. Port plate
7. Gasket
8. Relief valve cone
9. Spring
10. Shim
11. "O" ring
12. Plug
13. Check valves
14. Back-up washer
15. "O" ring
16. "O" ring
17. Needle bearing
18. Valve plate
19. Retaining ring
20. Washers
21. Spring
22. Cylinder block
23. Slipper retainer
24. Piston (9 used)
25. Thrust plate
26. Swashplate
27. Pump shaft
28. Ball bearing
29. Oil seal
30. Snap ring
31. Washer
32. Oil seal
33. Stub shaft
34. Bushing
35. Housing
36. Roller bearing
37. Trunnion shaft
38. Washer
39. Oil seal
40. Needle bearing
41. Gasket
42. Motor shaft
43. Swashplate (motor)
44. Piston (9 used)
45. Slipper retainer
46. Cylinder block
47. Washer
48. Spring
49. Retaining ring
50. Valve plate
51. Pin
52. Needle bearing
53. Roll pin
54. Plug
55. "O" ring
56. Plug
57. Center housing
58. Oil filter fitting

convenient. Mark charge pump housing and center section before disassembly as it is possible to install pump incorrectly.

Remove charge pump housing (3 – Fig. CC21). Remove rotor assembly (4) and rotor drive pin, then remove port plate (6) and gasket (7).

NOTE: On later models, charge pump port plate and gasket are not used. See Fig. CC22.

Pry oil seal (1 – Fig. CC21) from charge pump housing and press needle bearing (1) out the front of housing. Remove both check valves (13), back-up washers (14) and "O" ring (11), shim (10), spring (9) and relief valve cone (8) from relief bore in center housing. Unbolt and remove center housing (57).

CAUTION: Valve plates (18 and 50) may stick to center housing. Be careful not to drop or damage them.

Remove and identify valve plates. Pump valve plate has two relief notches. Tilt housing (35) on its side and remove cylinder block and piston assemblies from pump and motor shafts. Lay cylinder block assemblies aside for later disassembly.

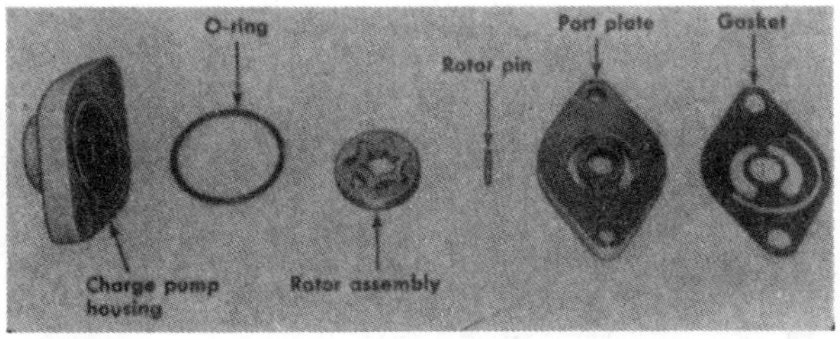

Fig. CC22 — Exploded view of charge pump assembly used on early models. Gasket is installed with circular groove towards top of unit and flat end to right side of unit.

CAUTION: Some swashplates (26) have spring pin holes drilled through both walls. When removing stub shaft (33) and trunnion shaft (37), do not drive spring pins through the shaft and into holes in of bottom swashplate, as removal is then very difficult.

Mark or tape a punch exactly 15/32-inch (12 mm) from end. Carefully drive roll pins (53) into swashplate until mark on punch is even with top surface of swashplate; at this point spring pins should be centered in trunnion shaft. Remove swashplate and thrust plate (25), then withdraw pump shaft and bearing (27 and 28). Remove the two socket head screws securing motor swashplate (43) to housing, then lift out swashplate and motor shaft (42). Bearing (36 and 40), bushing (34) and oil seals (29, 32 and 39) can now be removed from housing (35). Remove needle bearings (17 and 52) from center housing (57).

Carefully remove slipper retainer (23) with nine pistons (24) from pump cylinder block (22), then remove slipper retainer (45) with nine pistons (44) from motor cylinder block. Place each cylinder block on wood block in a press, compress spring (21 and 48) and remove retaining ring. Release press and remove spring and washers. Springs (21 and 48) should have a free length measurement of 1-3/64 to 1-1/16 inches (26.67 to 27 mm) and should test 63-75 pounds (280.24-333.62 N) when compressed to a length of 19/32-inch (15 mm). Check cylinder blocks (22 and 46) on valve face end and renew if deep scratches or other damage is evident. Inspect pistons and bores in cylinder blocks for excessive wear or scoring. Piston slippers can be lapped to remove light scratches. Minimum slipper thickness is 0.121 inch (3.073 mm) for both pump and motor. Slipper thickness must not vary more than 0.002 inch (0.051 mm) for all nine pistons in each block.

Check pump valve plate and motor valve plate (Fig. CC23) for excessive wear and renew as necessary. Inspect motor swashplate (43–Fig. CC21) and pump swashplate thrust plate (25) for wear, embedded material or scoring and renew as indicated.

Check charge pump housing (3), rotor assembly (4) and port plate (6), on models so equipped, for wear or scoring and renew as necessary. Relief valve cone (8) should be free of nicks and scratches. Relief valve sprng (9) should have a free length of 1.057 inches (26.848 mm) and should test 7.0-7.6 pounds (31.14-33.81 N) when compressed to a length of 0.525 inch (13.33 mm).

Renew all "O" rings, gaskets and seals and reassemble by reversing removal procedure. Thoroughly lubricate all parts with Hy-Tran, or equivalent. When installing new bearings (17), press bearings in until 0.100 inch (2.54 mm) above machined surface of center housing. Check valves (13–Fig. CC21) are interchangeable and are serviced only as an assembly. Pump swashplate (26) must be installed with thin pad towards top of transmission. When installing charge pump port plate gasket, on models so equipped, circular groove in gasket must be towards top of unit. Flat end of gasket and port plate on early models and charge pump housing (3) on all must be towards top of unit. Flat end of gasket must be towards right side of unit. Motor swashplate socket head screws should be tightened to 67 ft-lbs. (91 N·m). Charge pump cap screws to 35 ft.-lbs. (47 N·m).

Fill transmission to proper operating level with recommended fluid and adjust linkage as necessary.

GEAR TRANSMISSION

MAINTENANCE

Model 580

Check transmission fluid level at 100 hour intervals. Maintain fluid level at lower edge of fill plug opening (Fig. CC24). Recommended fluid is IH-135H EP Gear Lubricant, or equivalent. Approximate fluid capacity is 4 pints (1.9 L).

Model 582

Check transmission fluid level at 100 hour intervals. Maintain fluid level at lower edge of fill plug opening (Fig. CC25). Recommended lubricant is Cub Cadet Hydraulic/Transmission Fluid or equivalent lubricant. Approximate capacity is 8 pints (3.8 L).

R&R GEAR TRANSMISSION

Model 580

Disconnect and remove battery. Remove the four bolts in battery box which secure fender assembly to rear frame. Remove drawbar assembly. Remove foot platform mounting screws and remove fender assembly. Remove center frame cover. Place wedges be tween front axle and frame at each side to prevent tractor from tipping. Depress clutch pedal and lock in lowest position.

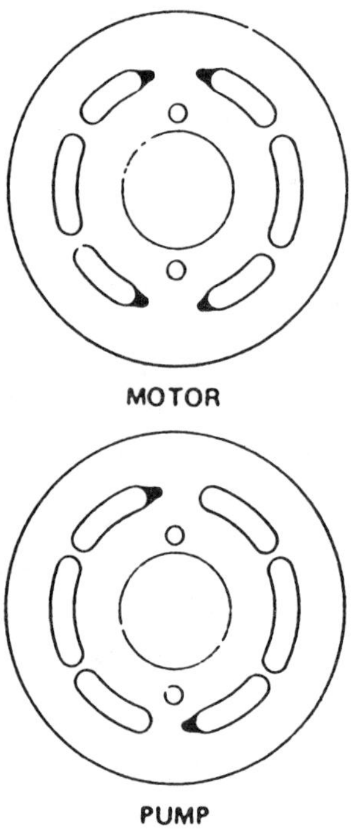
MOTOR
PUMP

Fig. CC23—Check pump and motor valve plates for wear or other damage. Note motor valve plate has four notches (dark areas) and pump valve plate has two notches.

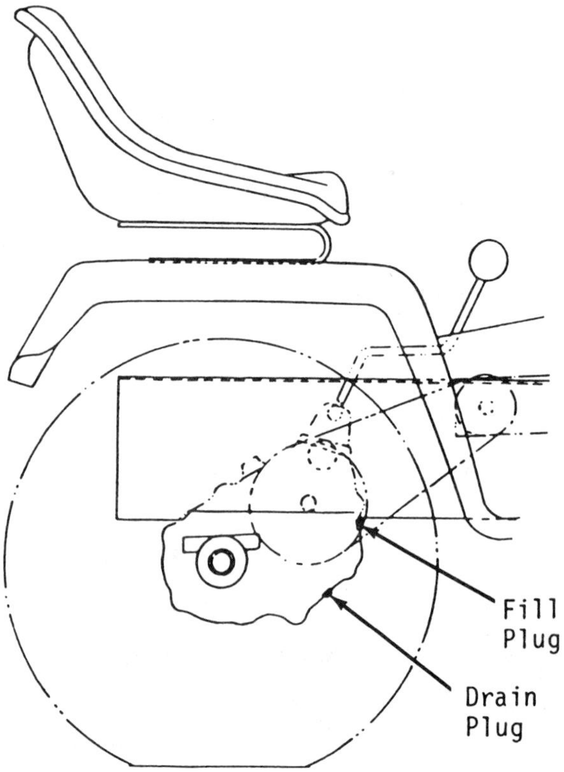
Fill
Plug
Drain
Plug

Fig. CC24—View showing location of fill plug and drain plug on 580 model.

Illustrations courtesy Cub Cadet Corp.

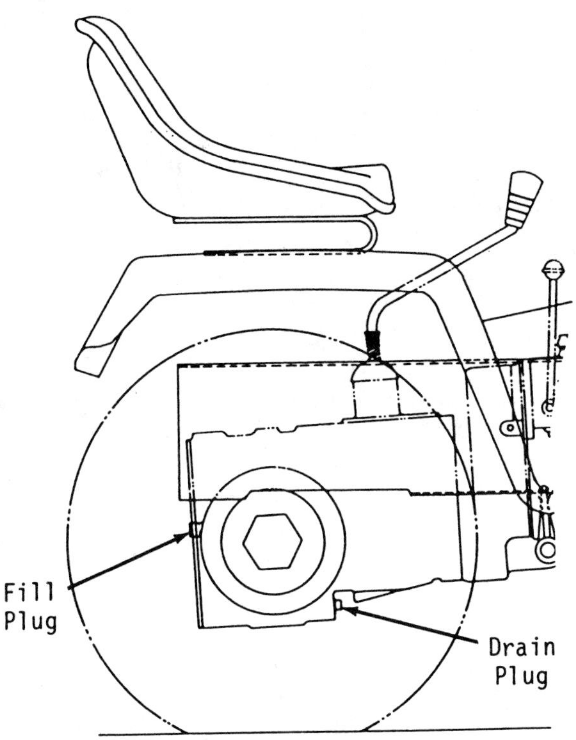
Fill
Plug
Drain
Plug

Fig. CC25—View showing location of fill plug and drain plug on 582 model.

Loosen the two belt guides and work belt off transaxle input pulley. Remove bolts securing transaxle to frame and roll assembly out of frame.

Reinstall by reversing removal procedure. Adjust belt guides so there is 1/8 to 3/16-inch (3 to 5 mm) clearance between belt and guides.

Model 582

Remove battery and disconnect electrical leads from solenoid, tail lights and seat safety switch. Remove rear fender to frame bolts and battery ground wire. Remove foot platform mounting screws and remove fender assembly. Remove frame cover. Disconnect brake rod and rear flex coupling. Remove the three point hitch lift lever if so equipped. On all models, support frame, remove frame mounting bolts and roll differential and transmission assembly out of frame.

Reinstall by reversing removal procedure.

OVERHAUL GEAR TRANSMISSION

Model 580

Drain lubricant remove transaxle assembly from tractor. Remove tires and wheels. Remove snap ring securing wheel hub to axle and remove hub and Woodruff key. Remove brake disc from brake shaft. Loosen setscrew and remove input pulley. Place shift lever in neutral position and remove shift lever housing and gasket.

NOTE: If disassembly of shift lever is necessary, scribe match marks to make certain shift lever is not reassembled 180° out of line.

Scribe match marks on axle housings and case and cover and remove axle housings. Remove "O" rings (54–Fig. CC26), oil seals (53 and 71) and retainers (52 and 72). Axle support bearings (56) are removed by driving them out from the inside of axle housing. Support transaxle assembly on bench with left axle downward. Drive alignment dowels into cover, remove the eight cap screws and separate case from cover. Remove differential assembly. Remove thrust washer (33), three gear cluster (32) from brake shaft (28). Remove reverse idler gear (8), spacer (9) and shaft (10). Hold upper ends of shifter rods together and lift out shifter rods, forks, shifter stop (5), sliding gears (15 and 16) and shaft (18) as an assembly. Remove low reduction gear (48), reduction shaft (49) and thrust washer (50), then remove the two-cluster gear (29) from brake shaft. Lift out output gear (46), shaft (45) and

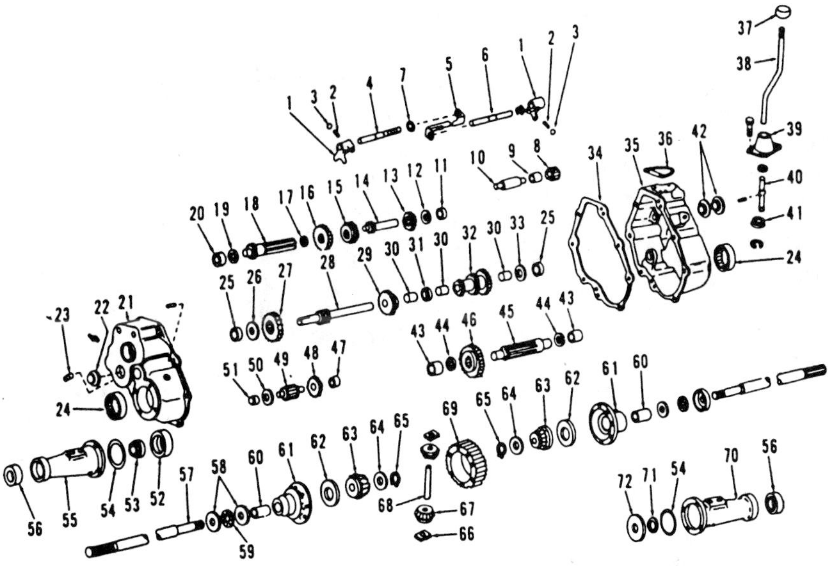

Fig. CC26—Exploded view of transaxle assembly used on 580 model.

1. Shift fork
2. Spring
3. Ball
4. Shift rod
5. Shifter stop
6. Shift rod
7. Snap ring
8. Reverse idler
9. Spacer
10. Reverse idler shaft
11. Bearing
12. Thrust washer
13. Input gear
14. Input shaft
15. Third & fourth gear
16. First, second & reverse gear
17. Bearing
18. Shifter shaft
19. Thrust washer
20. Needle bearing
21. Cover
22. Oil seal
23. Drain plug
24. Needle bearing
25. Bearing
26. Thrust washer
27. Idler gear
28. Brake shaft
29. Two-cluster gear
30. Bushing
31. Spacer
32. Cluster gear assy.
33. Thrust washer
34. Gasket
35. Case
36. Gasket
37. Knob
38. Shift lever
39. Shift lever housing
40. Shifter
41. Keeper
42. Bearing & seal
43. Needle bearing
44. Thrust washer
45. Shaft
46. Output gear
47. Bearing
48. Low reduction gear
49. Low reduction shaft
50. Washer
51. Bearing
52. Seal retainer
53. Oil seal
54. "O" ring
55. Axle housing
56. Ball bearing
57. Axle
58. Thrust race
59. Thrust bearing
60. Bushing
61. Carrier assy.
62. Washer
63. Axle gear
64. Thrust washer
65. Snap ring
66. Drive block
67. Bevel pinion gear
68. Drive pin
69. Ring gear
70. Axle housing
71. Oil seal
72. Seal retainer

thrust washers (44). To remove brake shaft (28) and gear (27) from cover (21), block up under gear (27) and press shaft out of gear.

CAUTION: Do not allow cover or low reduction gear bearing boss to support any part of pressure required to press brake shaft from gear.

Remove input shaft (14) with input gear (13) and thrust washer (12) from case (35).

To disassemble differential, remove four cap screws and separate axle shaft and carrier assemblies from ring gear (69). Drive blocks (66), bevel pinion gears (67) and drive pin (68) can now be removed from ring gear. Remove snap rings (65) and withdraw axle shaft (57) from axle gears (63) and carriers (61).

Clean and inspect all parts and renew any showing excessive wear or other damage. When installing new needle bearings, press bearing (17) in spline shaft (18) to a depth of 0.010 inch (0.254 mm) below end of shaft and low reduction shaft bearings (51 and 47) 0.010inch (0.254 mm) below thrust surfaces of bearing bosses. Carrier bearings (24) should be pressed in from inside of case and cover until bearings are 0.290 inch (7.366 mm) below face of axle housing mounting surface. All other needle bearings are to be pressed in from inside of case and cover to a depth of 0.015-0.020 inch (0.381-0.508 mm) below thrust surfaces.

Renew all seals and gaskets and reassemble by reversing disassembly procedure. Note when installing brake shaft (28) and idler gear (27), beveled edge of gear teeth must be up away from cover. Install reverse idler shaft (10), spacer (9) and reverse idler gear (8) with rounded end of teeth facing spacer. Install input gear (13) and shaft (14) with chamfered side of input gear facing case (35).

Tighten transaxle cap screws to the following torque:

Differential cap screws . . . . . 7-10 ft.-lbs. (9-14 N·m)
Axle housing cap screws . . . . 8-10 ft.-lbs. (11-14 N·m)
Shift lever housing cap screws . . . . 8-10 ft.-lbs. (11-14 N·m)

Reinstall transaxle and fill to proper level with specified fluid. Adjust brake and drive belt as necessary.

Model 582

Drain lubricant and remove transmission and differential assembly from tractor. Disassemble reduction gear unit and if prior to S/N 720000, remove internal brake disc. On all models, remove reduction housing mounting bolts and remove housing. Remove differential assembly. Remove gear shift lever and cover assembly. Shift transmission into two speeds to lock transmission and remove nut (26–Fig. CC27) from countershaft (13). Remove shifter fork setscrews. Cover gearshift detent ball and spring holes and drive shift rods forward out of transmission. Remove detent balls and springs. Remove bolts from mainshaft front bearing retainer and pull mainshaft forward out of transmission as gears are removed. Push countershaft rearward out of transmission noting location of gears and spacers as they are removed. Pull mainshaft needle bearings from housing. Remove reverse idler shaft and gear. Remove countershaft front bearing, retainer and shims.

Reassemble by reversing disassembly procedure. Note beveled edge of spacer (24) is toward bearing (28). Rounded edge of gear teeth on first gear (19) are toward bearing (15), rounded edge of gear teeth on second gear (21) are toward bearing (28) and rounded edge of gear teeth on third gear (23) are toward bearing (15). Nut (26) should be tightened to 85 ft.-lbs. (115 N·m). Make certain pinion gear depth and backlash are properly adjusted.

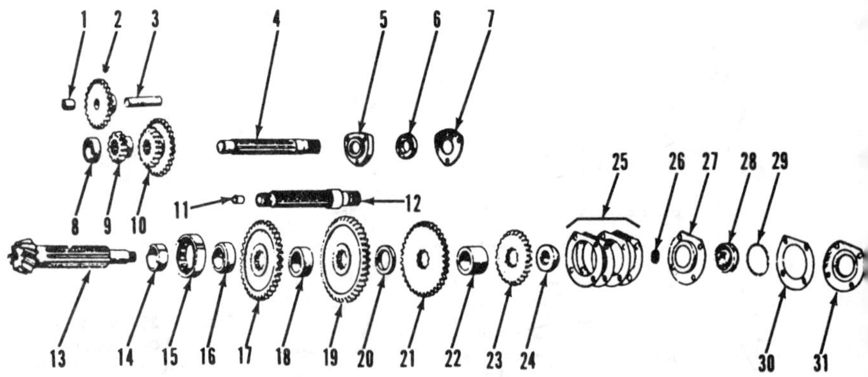

Fig. CC27 – Exploded view of transmission used on 582 models (S/N 719999 and before). Models after S/N 720000 are similar but do not have internal brake mechanism.

1. Bushing
2. Reverse gear
3. Reverse idler shaft
4. Mainshaft
5. Bearing
6. Bearing
7. Bearing retainer
8. Bearing
9. Sliding gear (1st & Rev.)
10. Sliding gear (2nd & 3rd)
11. Bushing
12. Input shaft
13. Countershaft
14. Bearing race
15. Bearing
16. Spacer
17. Reverse gear
18. Spacer
19. First gear
20. Spacer
21. Second gear
22. Spacer
23. Third gear
24. Spacer
25. Shims
26. Nut
27. Retainer
28. Bearing
29. Snap ring
30. Gasket
31. Retainer

REDUCTION GEAR

R&R REDUCTION GEAR

Model 582

Reduction gear assembly is removed with the gear type transmission. Refer to appropriate GEAR TRANSMISSION section for removal procedure.

Models 682–782–784–982–984–986

Reduction gear assembly is removed with the hydrostatic drive and differential assembly as a unit. Refer to the HYDROSTATIC DRIVE SYSTEM section for removal procedure.

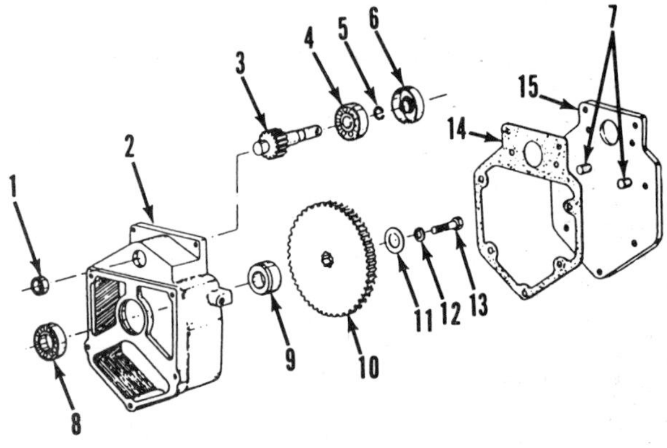

Fig. CC28 – Exploded view of reduction gear assembly used on 582 models with S/N 720000 or after. Refer to Fig.CC16 for 582 models below S/N 720000 with internal type brake.

1. Bearing
2. Housing
3. Drive gear & shaft
4. Bearing
5. Snap ring
6. Seal
7. Dowels
8. Bearing
9. Spacer
10. Gear
11. Flat washer
12. Lockwasher
13. Bolt
14. Gasket
15. Cover

Illustrations courtesy Cub Cadet Corp.

REDUCTION GEAR OVERHAUL

Model 582

Drain lubricant and remove creeper drive unit, if so equipped.

On models having internal type transmission brakes (S/N 719999 or below) remove brake adjusting screw, pivot pin, brake lever and push rod.

On all models, remove reduction housing cover (15–Fig. CC28). Remove bolt (13) and washers (12 and 11). Remove gear (10). Use a suitable puller to remove reduction gear, shaft and bearing assembly from housing.

On models with creeper drive, press spring pin out of gear and shaft assembly and remove splined coupling.

On all models, press bearings off of gear and shaft assembly. Press needle bearing from rear of housing.

To reassemble, press needle bearing in until flush with housing. Continue to reverse removal procedure noting copper sealing washers are installed on the two lower mounting bolts. Bolts should be tightened to 80 ft.-lbs. (108 N·m). Bolt (13) is tightened to 55 ft.-lbs. (75 N·m). Fill to proper level with recommended fluid and adjust brakes as necessary.

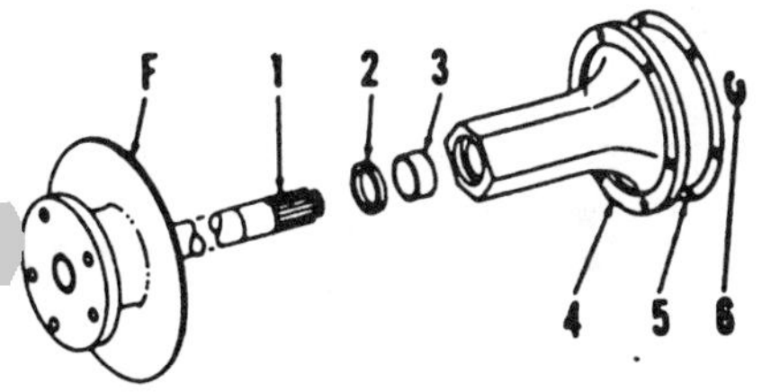

Fig. CC29–Exploded view of rear axle assembly.

F. Brake disc
1. Axle
2. Oil seal
3. Needle bearing
4. Axle carrier
5. Gasket
6. "C" ring retainer

Models 682–782–784–982–984–986

Reduction gear is serviced with the differential assembly. Refer to DIFFERENTIAL section. Note late production models (S/N 720000 or after) do not have an internal brake mechanism.

DIFFERENTIAL

R&R AND OVERHAUL

Model 580

Differential assembly is an integral part of the transaxle assembly. Refer to appropriate GEAR TRANSMISSION section for service procedure.

Model 582

Remove rear cover and remove "C" ring retainers (6–Fig. CC29) from inner ends of axle shafts (1). Remove bolts securing axle carriers (4) to differential housing and remove axle and carrier assemblies. Remove differential carrier bearing cages (2–Fig. CC30) noting location and thickness of shims (B). Angle ring gear carrier assembly (8) and remove from housing. Remove pin (12) and shaft (11). Rotate side gears 90° and remove spider gears (9) and side gears (10).

NOTE: Models with S/N 720000 or after use a ring and pinion and spider gears as shown in Fig. CC31.

Remove bearing cones (5–Fig. CC30) as necessary. Ring gear is riveted to carrier and is available separately for some models. Always check part availability before removing. To remove pinion gear and shaft assembly, refer to GEAR TRANSMISSION section.

Check all bearings for looseness, or damage and check all gears for wear or damage.

To reassemble, install ring gear and carrier assembly in differential housing and install bearing cages (2–Fig. CC30) and shims (B). Adjust bearing preload by adding or removing shims (B) until a preload of 1 to 8 pounds (0.5 to 3.6 kg) pull on a spring scale measured as shown in Fig. CC32. Remove ring gear and carrier assembly noting thicknesses of shim packs (B–Fig. CC30). Install pinion gear and shaft assembly as outlined in GEAR TRANSMISSION section and tighten nut on pinion shaft to 85 ft.-lbs. (115 N·m). Install ring gear and carrier assembly and adjust gear backlash by moving shims (B) from side to side as necessary to obtain 0.003-0.005 inch (0.076-0.127 mm) backlash. Paint ring gear teeth with Prussian Blue or red lead and rotate ring gear. Observe contact pattern on tooth surfaces. Refer to Fig. CC33. If pattern is too low (pinion depth to great) as "B", remove shims (25–Fig. CC27) as necessary to correct pattern. If pattern is too high (pinion depth to shallow) as "C", add shims (25) to correct pattern. Make certain ring

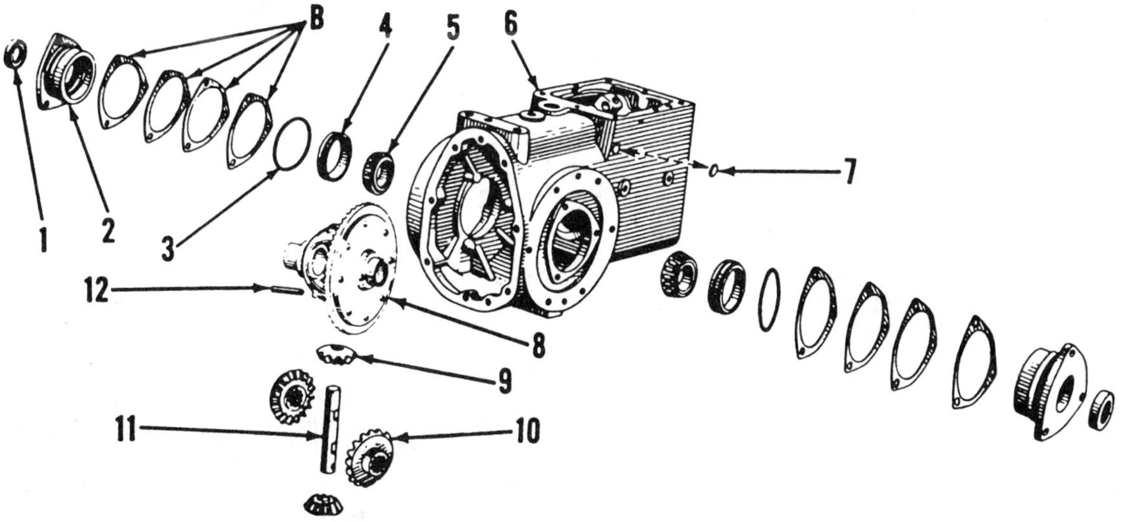

Fig. CC30–Exploded view of differential assembly used for all models with S/N 719999 or below except 580 models.

B. Shims
1. Oil seal
2. Bearing retainer
3. "O" ring
4. Bearing cup
5. Bearing cones
6. Case
7. Expansion plug
8. Ring gear & carrier
9. Spider gear
10. Side gear
11. Cross-shaft
12. Pin

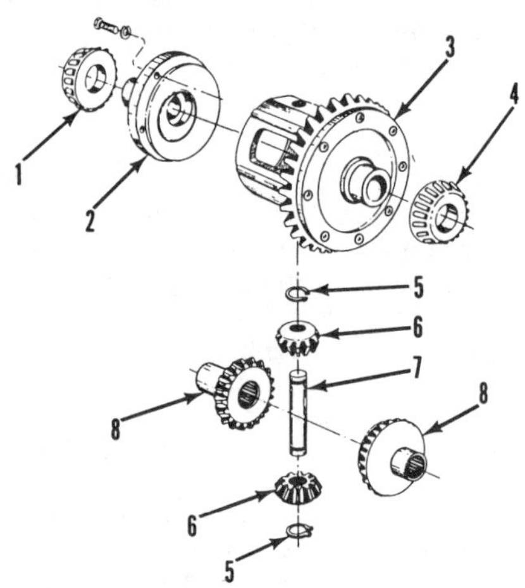

Fig. CC31—Exploded view of differential and spider gear assemby used on all models after S/N 720000 except 580 model.

1. Bearing
2. Flange
3. Ring gear & carrier
4. Bearing
5. Snap rings
6. Spider gears
7. Cross-shaft
8. Axle gears

gear backlash is correct after each shim change. Adjust as necessary. Carrier bearing cage retaining bolts should be tightened to 20 ft.-lbs. (27 N·m) during final assembly. Right side axle carrier mounting bolt at 9 o'clock position should be coated with thread sealer to prevent fluid loss. Continue to reverse disassembly procedure and refill to correct level with recommended fluid.

Models 682—782—784—982—984—986

Remove transmission and differential assembly as outlined in HYDRO-

Fig. CC32—Differential bearing preload is correct when a steady pull of 1 to 8 pounds (0.5 to 3.6 kg) on a spring scale is necessary to rotate differential as shown.

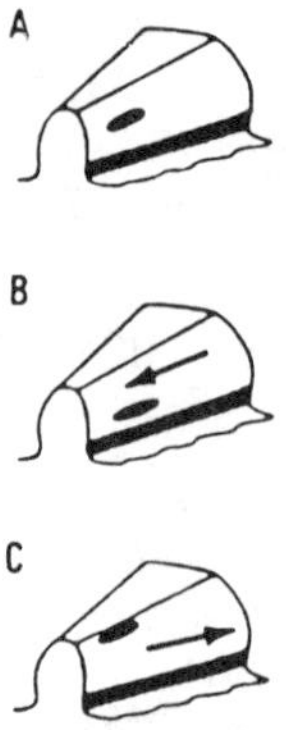

Fig. CC33—Bevel pinion tooth contact pattern is correct at "A", too low at "B" and too high at "C".

STATIC DRIVE SYSTEM section.

Drain fluid from differential and remove the axles and their carriers. Remove left and right bearing cages (29–Fig. CC34) and note thickness of each shim pack (31). Turn differential assembly in housing until it can be removed.

NOTE: IT may be necessary on some models to remove one of the bearing cones (4 or 11) before differential can be removed.

Remove differential flange (5), snap rings (7), shaft (9) and spider gears (8). Remove axle gears (10).

NOTE: On models with S/N 719999 c before, differential and spider gea assembly is similar to ones shown in Fi CC30. Shaft (11–Fig. CC30) is retained b pin (12) in carrier and ring gear assemb (8).

Ring gear and case assembly (6–Fig CC34) are serviced as an assembly only

Remove hydrostatic drive unit and o models with S/N 719999 or before remove expansion plug (28) and sna ring (27). Press bevel pinion shaft out c front bearing and constant mesh reduc tion gear. Remove rear bearing cup (14 and shim pack (15).

On models with S/N 720000 and after remove expansion plug (28) and sna ring (22), spacer (21), gear (20) and sna ring (19). Remove bevel pinion shaft (12 from housing (18). Remove bearing cu (14) and shim pack (15).

To reassemble, install differentia assembly in housing and adjust preloa on bearings (4 and 11) by varyin thickness of shim packs (31) until steady pull of 1 to 8 pounds (0.5 to 3. kg) is required to rotate differentia assembly. See Fig. CC32. Remove dif ferential assembly keeping shim pack (31–Fig. CC34) with their respectiv bearing cage.

If a new bevel pinion shaft, transmis sion case or rear bearing and cup are t be installed, take the number stampe

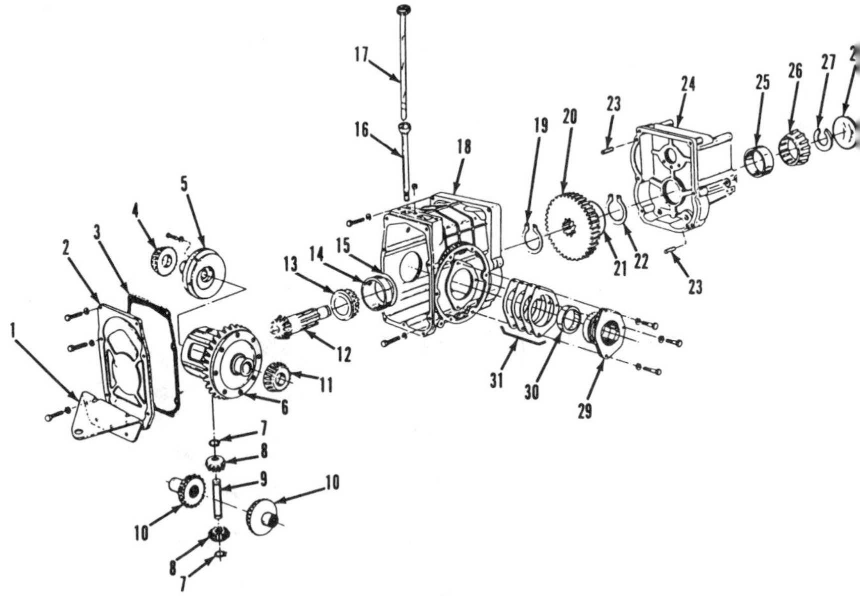

Fig. CC34—Exploded view of reduction gear and differential assembly used on 682, 782, 784, 982, 984 and 986 models after S/N 72000.

1. Drawbar
2. Cover
3. Gasket
4. Bearing
5. Differential flange
6. Ring gear & carrier
7. Snap rings
8. Spider gears
9. Cross-shaft
10. Axle gears
11. Bearing
12. Pinion shaft
13. Bearing
14. Bearing cup
15. Shims
16. Dipstick tube
17. Dipstick
18. Case
19. Snap ring
20. Reduction gear
21. Spacer
22. Snap ring
23. Dowel pin
24. Cover
25. Bearing cup
26. Bearing
27. Snap ring
28. Expansion plug
29. Cage
30. Bearing cup
31. Shim pack

on the top of case just ahead of rear cover (2) and the number stamped on the end of the bevel pinion shaft (12), add these two figures together and add 0.015 inch to their total. Place shim pack of a thickness equal to this total under bearing cup (15).

On models with S/N 719999 or before, install pinion shaft and bearing assembly in housing (18) making certain it goes through reduction gear (20). Install front bearing (26).

On models with S/N 720000 or after, install pinion shaft and bearing assembly in housing. Install snap ring (19), gear (20), spacer (21) and snap ring (22). Install cover (24), bearing cup (25) and bearing (26).

On all models, install a snap ring (27) which will allow 0.003 inch (0.0762 mm) pinion shaft end play. Snap rings are available in a variety of thicknesses. Install a new expansion plug (28).

Install differential assembly and install a dial indicator to measure ring gear backlash. Move shims (31) from side to side until 0.003-0.005 inch (0.0762-0.1270 mm) backlash is obtained.

RIGHT ANGLE DRIVE GEAR BOX

MAINTENANCE

Model 580

Right angle drive gear box is a sealed unit, packed by the manufacturer with 4 ounces (118 mL) of IH-251 EP, or equivalent grease.

R&R RIGHT ANGLE DRIVE GEAR BOX

Model 580

Disconnect negative battery cable and remove center frame cover. Lock clutch/brake pedal in depressed position. Drive roll pin from rear half of flex coupler. Remove the two mounting bolts retaining right angle drive gear box to cross support and remove gear box.

Reinstall by reversing removal procedure.

OVERHAUL RIGHT ANGLE GEAR BOX

Model 580

Remove cover (2–Fig. CC35) and clean lubricant from inside housing. Remove snap ring retaining output pulley, loosen pulley setscrew and remove pulley and key. Remove cap (17).

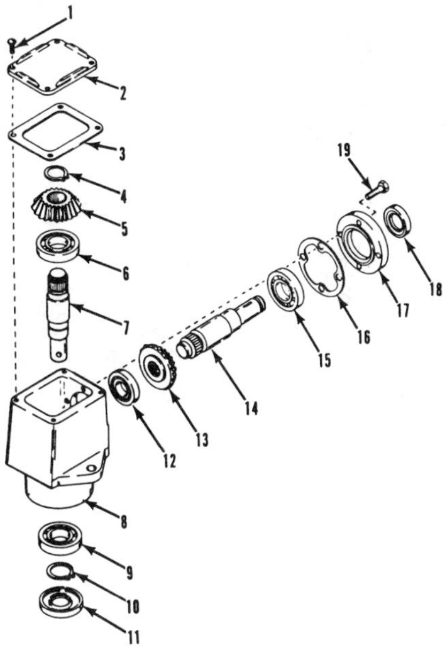

Fig. CC35–Exploded view of right angle drive gear box used on 580 model.

1. Screw
2. Cover
3. Gasket
4. Snap ring
5. Gear
6. Bearing
7. Shaft
8. Housing
9. Bearing
10. Snap ring
11. Seal
12. Bearing
13. Gear
14. Shaft
15. Bearing
16. Gasket
17. Cap
18. Seal
19. Bolt

Remove seal (18) from cap. Remove bearing (15) and use a wood dowel or brass rod to drive gear (13) off shaft (14). Remove shaft, gear and bearing (12). Remove seal (11) and snap ring (10). Press shaft (7) out of bearing (9) and housing (8). Remove snap ring (4), gear (5) and bearing (6).

Lubricate parts with IH-251 EP, or equivalent grease and reassemble by reversing disassembly procedure.

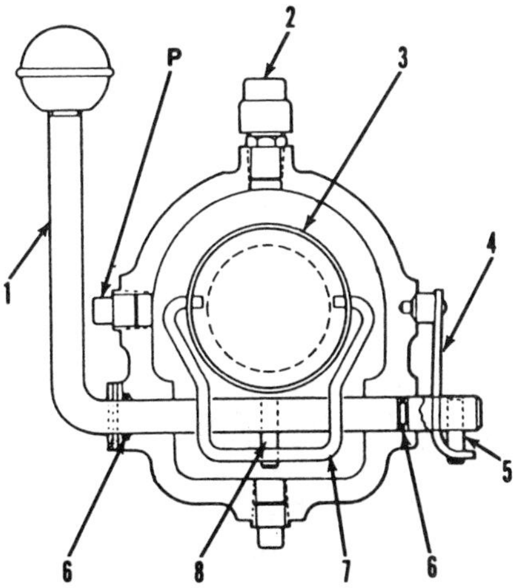

Fig. CC36–Sectional view of creeper drive unit.

1. Shift lever
2. Breather
3. Collar
4. Detent
5. Pin
6. "O" rings
7. Yoke
8. Pin
P. Plug

Before installing cover (2), fill housing (8) with 4 ounces (118 mL) of IH-251 EP, or equivalent grease. Seal (11) should be 0.045 inch (1.1 mm) below edge of cap (17).

CREEPER DRIVE

MAINTENANCE

Model 582

Model 582 may be equipped with an optional creeper drive unit. Check fluid level at 100 hour intervals. Maintain fluid level at lower edge of check plug opening (P–Fig. CC36). Recommended fluid is Cub Cadet Hydraulic/Transmission Fluid or equivalent lubricant. Approximate capacity is 8 oz. (0.24 L).

R&R CREEPER DRIVE

Model 582

Remove frame cover and drive roll pins from flex couplings and slide couplers forward on driveshaft. Place drain pan under tractor and remove unit mounting bolts. Bump unit to free from dowels and remove.

Reinstall by reversing removal procedure. Fill unit to proper level through breather opening (2–Fig. CC36) until level with lower edge of plug opening (P) in side of housing with recommended fluid. Reinstall breather (2) and plug (P).

CREEPER DRIVE OVERHAUL

Model 582

Remove snap ring (11–Fig. CC37) and pull input shaft (8), bearing (6),

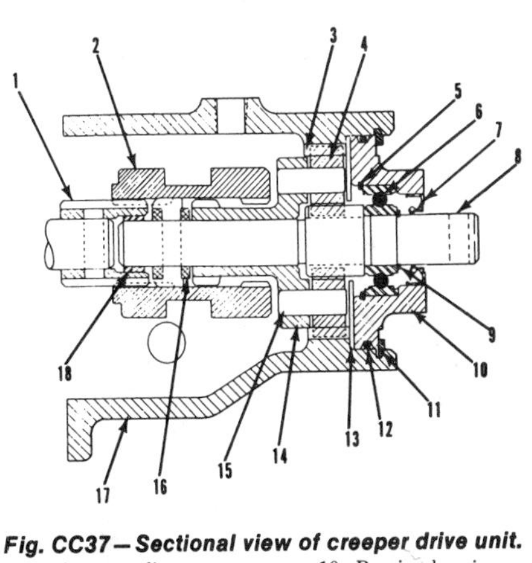

Fig. CC37–Sectional view of creeper drive unit.

1. Driven coupling
2. Shift collar
3. Ring gear
4. Planet gear
5. Snap ring
6. Bearing
7. Seal
8. Input shaft
9. Snap ring
10. Bearing housing
11. Snap ring
12. "O" ring
13. Thrust washer
14. Planet carrier
15. Carrier pin
16. Direct drive coupler
17. Housing
18. Pilot bushing

Illustrations courtesy Cub Cadet Corp.

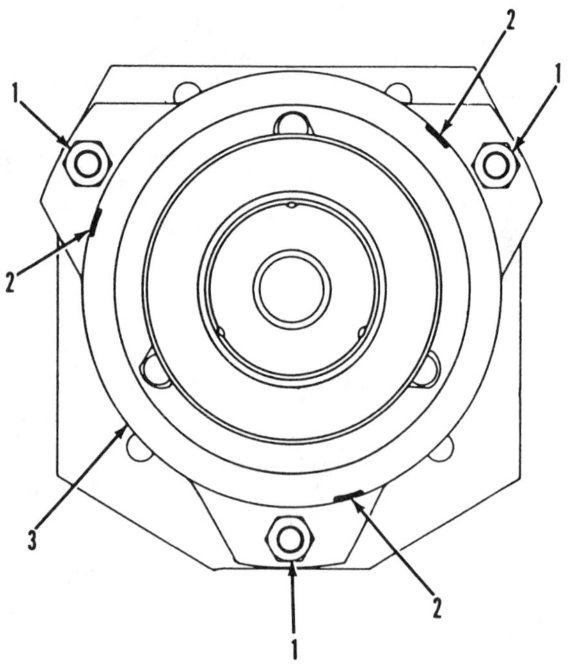

Fig. CC38—Front view of electric pto clutch.
1. Locknuts
2. Slots
3. Drum

planetary assembly and direct drive coupler (16) from housing (17).

NOTE: It may be necessary to rotate input shaft to align spline grooves of direct drive coupling with shift collar for removal.

Remove spring pin from drive coupling and remove coupling from shaft. Slide planet carrier (14) off input shaft (8) and planet gears (4) off carrier pins (15). Remove thrust washer (13). Remove snap ring (5) and press bearing and shaft assembly out of bearing housing (10). Remove oil seal (7) from bearing housing (10). Remove snap ring (9) and press input shaft (8) out of bearing (6). Drive pin (5–Fig. CC36) from lever (1) and remove shift detent (4). Shift lever and shift collar toward rear of housing while lifting shift collar up to disengage shift collar (3) from yoke (7). Remove pin (8) and slide lever (1) out of housing and yoke. Remove yoke. Remove "O" rings (6).

Clean all parts and inspect for wear or damage. Ring gear (3–Fig. CC37) is an integral part of housing (17) and if damaged, complete assembly must be renewed.

Reassemble by reversing disassembly procedure. Fill to proper level with recommended fluid.

POWER TAKE-OFF

ADJUSTMENT

All Models

Insert a 0.120 inch (3.05 mm) feeler gage between aramature assembly and rotor through access holes (2–Fig. CC38) and adjust locknuts (1) until feeler gage can just be removed. Repeat procedure at all three access hole (2) locations.

R&R ELECTRIC PTO CLUTCH

All Models

Location and construction of electric pto clutch may vary from model to model, however, removal procedure is similar.

Remove flange (6–Fig. CC39), springs (4) and retaining bolt, washer and armature assembly (5). Remove rotor (3). Remove field assembly (2).

Reinstall by reversing removal procedure. Adjust clutch as previously outlined.

HYDRAULIC SYSTEM

MAINTENANCE

Models 682–782–784–982–984–986

Pressurized oil is utilized from the charge pump in hydrostatic drive unit to provide hydraulic lift system and supply remote outlets. Refer to HYDROSTATIC DRIVE SYSTEM section for maintenance information.

PRESSURE CHECK AND ADJUST

Models 682–782–784–982–984–986

Install a 0-1000 psi (0-7000 kPa) pressure gage in test port (3–Fig. CC40). Start engine and allow transmission fluid to reach operating temperature. Run engine at full speed. With hydraulic control valve (as equipped) in neutral position, gage will

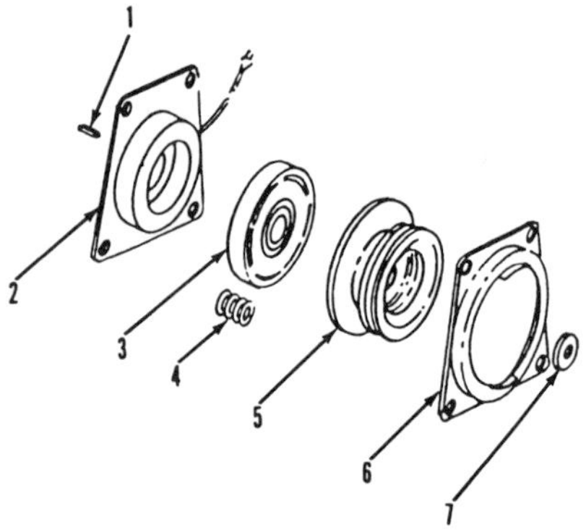

Fig. CC39—Exploded view of electric pto clutch.
1. Key
2. Field assy.
3. Rotor
4. Spring
5. Armature
6. Drum
7. Washer

Fig. Fig. CC40—View of hydrostatic drive unit showing locations of check valves (1), implement relief valve (2), test port (3) and charge pump relief valve (4).

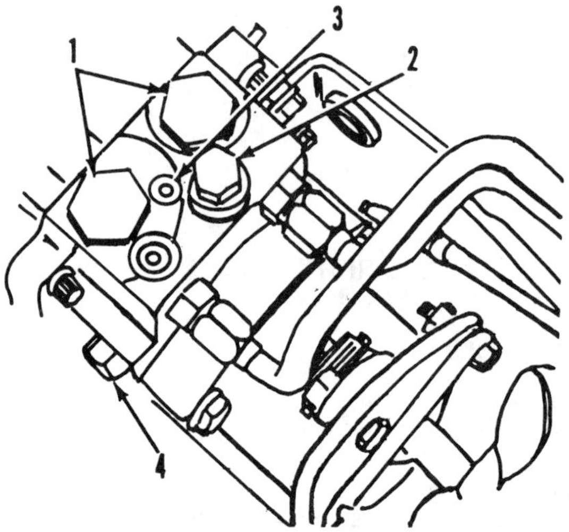

Illustrations courtesy Cub Cadet Corp.

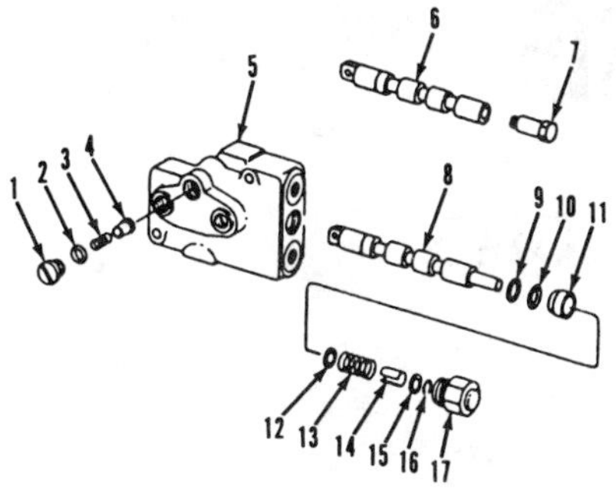

Fig. CC41—Exploded view of double spool valve. Single spool valve is similar.

1. Plug
2. "O" ring
3. Spring
4. Plunger
5. Body
6. Spool (single)
7. Bolt
8. Spool (double)
9. "O" ring
10. "O" ring
11. Spacer
12. Washer
13. Spring
14. Spacer
15. Washer
16. Retainer ring
17. Cap

indicate charge pump pressure as follows:

Model 682 . . . . 90-165psi (620-1138kPa)

Models 782, 784, 982, 984 & 986 . . . . 90-200 psi (620-1379 kPa)

With control valve in raise position and the lift cylinder at end of stroke, gage will indicate lift system pressure as follows:

Model 782 & 784 . . . . 500-625 psi (3447-4309 kPa)

Model 982, 984 & 986 . . . . . . 700-900 psi (4826-6205 kPa)

If charge pressure is not as specified, remove charge pump relief valve (4) and add or remove shims as necessary.

If lift system pressure is not as specified, remove implement relief valve (2) and add or remove shims as necessary.

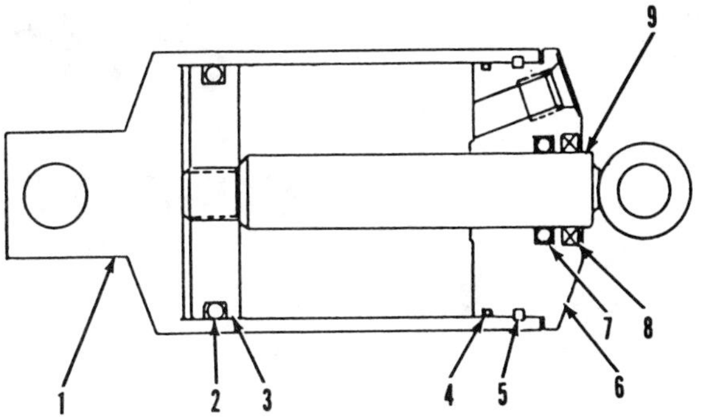

Fig. CC42—Cutaway view of hydraulic lift cylinder. Refer to text.

1. Cylinder body
2. "O" ring
3. Piston
4. "O" ring
5. Wire retainer
6. Head
7. "O" ring
8. Seal
9. Rod

CONTROL VALVE

Models 782—784—982—984—986

Models 782, 982 and 984 are equipped with a single spool hydraulic control valve. Models 784 and 986 are equipped with a double spool control valve. Service procedure for each valve is similar.

To remove control valve, raise hood and remove an engine side panel. Mark locations of hydraulic lines at control valve and disconnect all lines and cap openings. Remove connecting links at valve spools, remove valve mounting bolts and remove valve.

To disassemble control valve, remove cap (17–Fig. CC41) and pull spool (8 or 6) from body (5). Disassemble spool only if necessary to renew springs.

Renew all internal "O" rings and seals. Lubricate all parts during assembly and reassemble by reversing removal procedure.

LIFT CYLINDER

Models 782—784—982—984—986

Two different lift cylinders which are similar in appearance have been used. Make certain cylinder is correctly identified before ordering service parts.

To remove cylinder, remove frame cover, mark locations of hydraulic hoses and disconnect hoses and cap openings. Remove pin securing cylinder to lift bracket through access hole in right hand side of frame. Remove cylinder mounting bolt and remove cylinder.

To disassemble cylinder, note location of 45° and 90° fittings and remove fittings. Place cylinder in a holding fixture and use a suitable spanner wrench to turn cylinder head (6–Fig.CC42) until end of wire retainer (5) appears in access hole in cylinder body. Pry end of wire retainer out of hole and rotate cylinder head to remove wire retainer. Pull cylinder head out of cylinder body (1).

NOTE: Cylinder head "O" ring (4) may catch in wire retainer groove in cylinder body. If so, cut "O" ring through access hole and use wire retainer removal procedure to remove "O" ring.

Turn piston and rod assembly as it is pulled from cylinder body.

Lubricate all parts and renew all seals and "O" rings during reassembly. Reassemble by reversing disassembly procedure.

JOHN DEERE

CONDENSED SPECIFICATIONS

	MODELS			
	316	317	318	322
Engine Make	Onan	Kohler	Onan	Yanmar
Model	B43M	KT-17	B43G	3TG66UJ
Bore	3.25 in. (82.55 mm)	3.12 in. (79.3 mm)	3.25 in. (82.55 mm)	66 mm (2.6 in.)
Stroke	2.62 in. (66.6 mm)	2.75 in. (69.9 mm)	2.62 in. (66.6 mm)	64.2 mm (2.53 in.)
Displacement	43.3 cu. in. (712.4 cc)	42.18 cu. in. (690.5 cc)	43.3 cu. in. (712.4 cc)	658 cc (40.2 cu. in.)
Power Rating	16 hp (11.9 kW)	17 hp (12.7 kW)	18 hp (13.4 kW)	18 hp (13.4 kW)
Slow Idle	1350 rpm	1200 rpm	1250 rpm	1350 rpm
High Speed (No-Load)	3600 rpm	3400 rpm	3500 rpm	3425 rpm
Capacities—				
Crankcase	1.5 qts. (1.4 L)	1.5 qts. (1.4 L)	1.5 qts. (1.4 L)	2.6 qts. (2.5 L)
w/filter	2 qts. (1.9 L)	—	2 qts. (1.4 L)	2.9 qts. (2.8 L)
Hydraulic System	5 qts. (4.7 L)	5 qts. (4.7 L)	4.7 qts. (4.5 L)	6.5 qts. (6.2 L)
Transaxle or Transmission	See Hyd.	See Hyd.	See Hyd.	See Hyd.
Cooling System	—	—	—	2.9 qts. (2.8 L)
Fuel Tank	4.5 gals. (17 L)	4.5 gals. (17 L)	4.5 gals. (17 L)	4.5 gals. (17 L)

	MODELS			
	330, 332	400	420	430
Engine Make	Yanmar	Kohler	Onan	Yanmar
Model	3TN66UJ	K-532	B48G	3TNA72UJ
Bore	66 mm (2.6 in.)	3.38 in. (85.7 mm)	3.25 in. (82.6 mm)	72 mm (2.83 in.)
Stroke	2.62 in. (66.6 mm)	3 in. (76.2 mm)	2.88 in. (73 mm)	72 mm (2.83 in.)
Displacement	43.3 cu. in. (712.4 cc)	53.67 cu. in. (879.7 cc)	47.7 cu. in. (781.7 cc)	879 cc (54 cu. in.)
Power Rating	16 hp (11.9 kW)	20 hp (15 kW)	20 hp (15 kW)	20 hp. (15 kW)
Slow Idle	450 rpm	1800 rpm	1250 rpm	1350 rpm
High Speed (No-Load)	3635 rpm	3500 rpm	3500 rpm	3425 rpm
Capacities—				
Crankcase	2.6 qts. (2.5 L)	3 qts. (2.8 L)	1.5 qts. (1.4 L)	2.6 qts. (2.5 L)
w/filter	3.1 qts. (2.9 L)	—	—	3.1 qts. (2.9 L)
Hydraulic System	5 qts. (4.7 L)	5.5 qts. (5.2 L)	4.6 qts. (4.4 L)	7.5 qts. (7.1 L)
Transaxle or Transmission	See Hyd.	See Hyd.	See Hyd.	See Hyd.
Cooling System	2.9 qts. (2.8 L)	—	—	4 qts. (3.8 L)
Fuel Tank	4.5 gals. (17 L)	4.5 gals. (17 L)	6.5 gals. (24.6 L)	6.5 gals. (24.6 L)

FRONT AXLE AND STEERING SYSTEM

MAINTENANCE

All Models Equipped With Manual Steering

It is recommended that front wheel hubs, axle spindles, steering pivot and steering gear be lubricated at 25 hour intervals.

NOTE: Do not overlubricate steering gear. Two strokes of a hand grease gun are usually sufficient.

Use multipurpose, lithium base grease and clean all grease fittings before and after lubrication. Check for any excessive looseness of parts at this time and repair as necessary.

All Models Equipped With Power Steering

It is recommended that steering pivot and axle spindles be lubricated at 25 hour intervals. Use multipurpose, lithium base grease and clean all fittings before and after lubrication.

Power steering system utilizes pressurized oil from the hydrostatic drive unit. Refer to appropriate TRANSMISSION section for maintenance information.

Check for any excessive looseness of parts, bearings or tie rods and repair as necessary.

R&R AXLE MAIN MEMBER

Model 316

Raise and support front of tractor. Disconnect tie rod ends (5 and 11—Fig. JD1) at spindles (1). Support axle main member (17), then remove cotter key (12) and nut (14). Remove pivot bolt (13) and lower axle from tractor.

Inspect all parts for excessive wear or damage. Bushing (15) and pivot bolt (13) are renewable.

Reinstall axle by reversing removal procedure.

Model 317

Raise and support front of tractor. Disconnect drag link (5—Fig. JD2) at steering arm (6). Support axle main member (16) and remove pivot bolt (13). Lower axle assembly and roll away from tractor.

Inspect all parts for wear or damage. Renew bushings (14 and 15) and pivot bolt (13) as necessary.

Reinstall axle by reversing removal procedure.

Models 318-322-332

Raise and support front of tractor. Disconnect drag link at steering arm (12—Fig. JD3). Disconnect and remove power steering cylinder if equipped. Support axle main member and loosen jam nuts (3) at left and right side, then turn axle deflector adjustment bolts (6) in to provide maximum axle movement. Remove pivot bolt (15). Lower axle assembly and roll away from tractor.

Inspect all parts for wear or damage. Renew bushings (16, 17, 18 and 19) and pivot bolt (15) as necessary.

Reinstall axle by reversing removal procedure. Adjust left and right axle deflector adjustment bolts outward until axle just pivots freely and tighten jam nuts (3).

Model 400

Raise and support front of tractor. Dis-

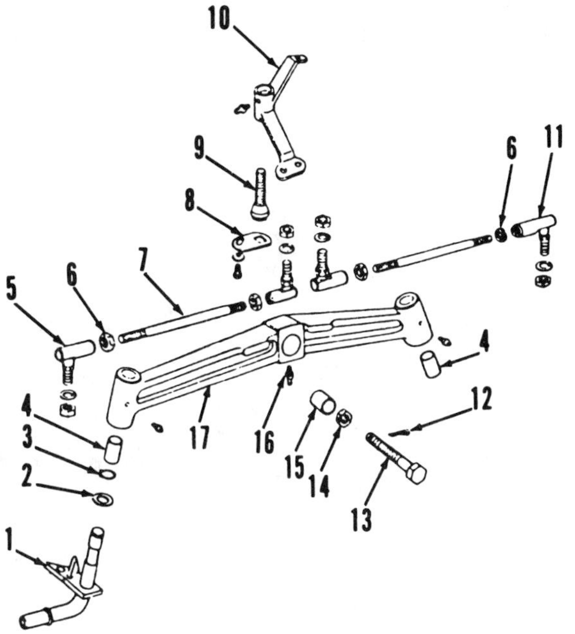

Fig. JD1—Exploded view of front axle assembly used on 316 model.

1. Spindle
2. Washer
3. Snap ring
4. Bushing
5. Tie rod end
6. Jam nut
7. Tie rod
8. Lock plate
9. Conical bolt
10. Steering pivot arm
11. Tie rod end
12. Cotter key
13. Pivot bolt
14. Nut
15. Bushing
16. Grease fitting
17. Axle main member

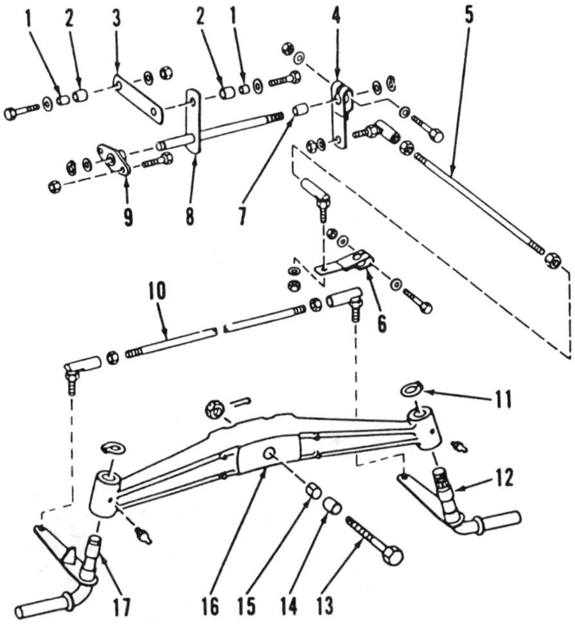

Fig. JD2—Exploded view of front axle assembly used on 317 model.

1. Bushing
2. Bearing
3. Link
4. Steering arm (rear)
5. Drag link
6. Steering arm (front)
7. Bushing
8. Pivot shaft
9. Bearing
10. Tie rod
11. Snap ring
12. Steering spindle
13. Pivot bolt
14. Bushing
15. Spacer
16. Axle main member
17. Steering spindle

connect drag link at steering arm (9—Fig. JD4). Disconnect and remove power steering cylinder. Refer to POWER TAKE-OFF section and remove front pto driven pulley and shaft. Support axle main member and remove axle pivot (16). Lower axle assembly from tractor.

Reinstall axle by reversing removal procedure.

Models 420-430

Raise and support front of tractor. Disconnect drag link at steering arm (12—Fig. JD5). Disconnect and remove power steering cylinder. Refer to POWER TAKE-OFF section and remove front pto driven pulley and shaft. Support axle main member, loosen jam nuts (3) at left and right side, then turn axle deflector adjustment bolts (6) in to provide maximum axle movement. Remove axle pivot (15) and lower axle assembly from tractor.

Reinstall axle by reversing removal procedure. Adjust left and right axle deflector adjustment bolts outward until axle just pivots freely and tighten jam nuts (3).

TIE ROD AND TOE-IN

All Models

Two tie rods, each with renewable ends, are used on 316 model and a single tie rod with renewable ends is used on all other models.

Front wheel toe-in should be $^{3}/_{16}$ inch (5 mm) and is adjusted by lengthening or shortening tie rod length.

NOTE: Each tie rod on 316 model must be adjusted equally to provide correct turning radius in each direction.

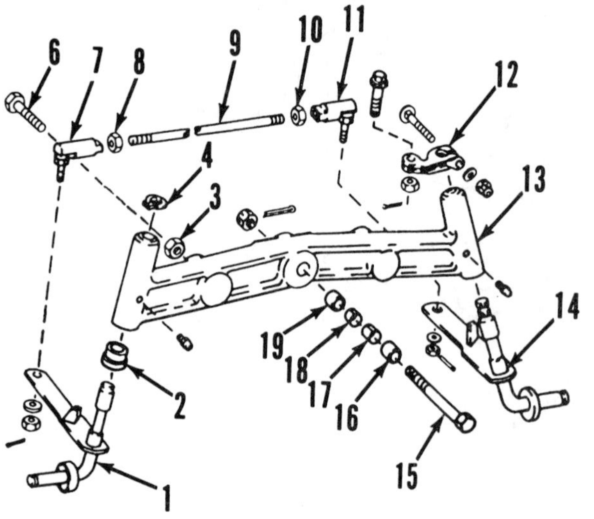

Fig. JD3—Exploded view of front axle assembly used on Model 318. Models 322, 330 and 332 are similar.

1. Spindle
2. Bushing
3. Jam nut
4. Snap ring
6. Bolt
7. Tie rod end
8. Jam nut
9. Tie rod
10. Jam nut
11. Tie rod end
12. Steering arm
13. Axle main member
14. Spindle
15. Pivot bolt
16. Bushing
17. Bushing
18. Bushing
19. Bushing

STEERING SPINDLES

All Models

Raise and support side to be serviced. Remove wheel and tire. Disconnect tie rod ends, drag link or steering cylinder as necessary. Remove snap ring securing spindle or steering arm, then remove spindle from axle main member.

Inspect spindle and bushing (if equipped) for wear or damage and renew as necessary.

Reinstall spindle by reversing removal procedure.

FRONT WHEEL BEARINGS

All Models

To remove front wheel bearings/bushings, raise and support side to be serviced. On 318, 322, 332, 420 and 430 models, remove dust cap from hub and snap ring from spindle end. On all other models, remove cap screw (1—Fig. JD6), spindle cap (2) and spring washer (3). On all models, slide wheel and hub assembly off spindle. Wheel bearings/bushings are a press fit in wheel hub. If bearings/bushings are loose fit in hub, renew or repair hub as necessary.

Pack bearings/bushings and hub with multipurpose, lithium base grease. Reinstall by reversing removal procedure. Tighten cap screw (1) to 35 ft.-lbs. (46 N·m).

STEERING GEAR

All Models Equipped With Manual Steering Except Late 316-330.

R&R AND OVERHAUL. To remove steering gear, use a suitable puller and remove steering wheel. Remove battery and battery box. Disconnect drag link

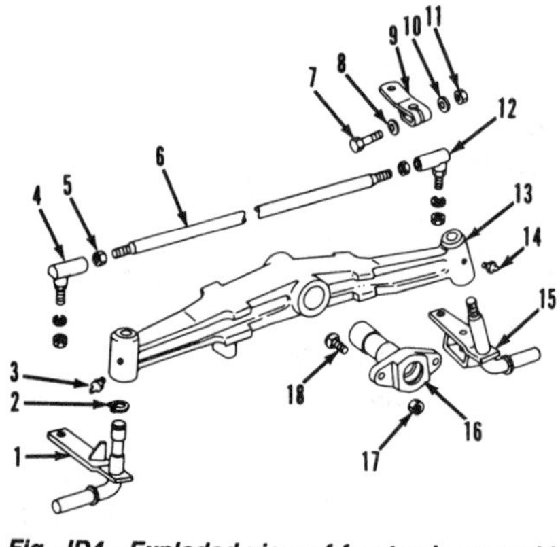

Fig. JD4—Exploded view of front axle assembly used on 400 model.

1. Spindle
2. Snap ring
3. Grease fitting
4. Tie rod end
5. Jam nut
6. Tie rod
7. Bolt
8. Washer
9. Steering arm
10. Washer
11. Nut
12. Tie rod end
13. Axle main member
14. Grease fitting
15. Spindle
16. Pivot
17. Nut
18. Bolt

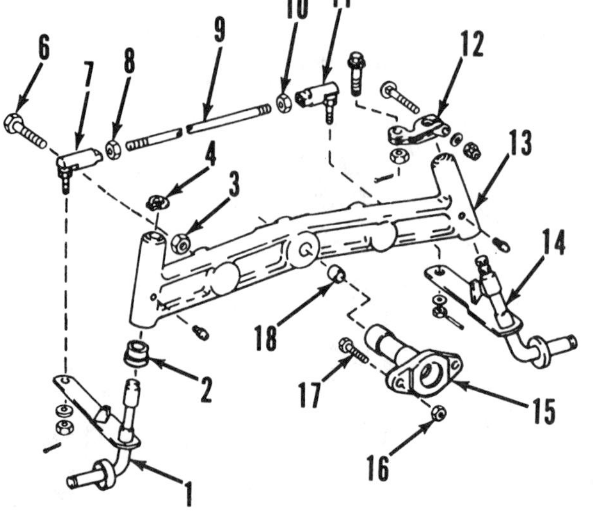

Fig. JD5—Exploded view of front axle assembly used on Models 420 and 430.

1. Spindle
2. Bushing
3. Jam nut
4. Snap ring
6. Bolt
7. Tie rod end
8. Jam nut
9. Tie rod
10. Jam nut
11. Tie rod end
12. Steering arm
13. Axle main member
14. Spindle
15. Axle pivot
16. Nut
17. Bolt
18. Bushing

at steering lever plate (14—Fig. JD7). Remove cap screws securing steering gear to frame and disconnect necessary wiring to permit steering gear and column to be lowered through dash and removed.

To disassemble steering gear, remove jam nuts (5). Remove steering lever plate (14). Remove cotter pin (11) and adjustment plug (10), then slide steering shaft assembly (8) out of housing. Remove Belleville washer (9), retainer cups (6) and bearings (7).

Inspect parts for wear, scoring or damage and renew as necessary.

To reassemble, apply multipurpose grease to bearings (7) and retainer cups (6) and install on steering shaft (8). Apply grease to steering gear and install assembly into housing (3). Install Belleville washer (9) and plug (10). Tighten plug to 12 ft.-lbs. (16 N·m) and install cotter pin (11). Shaft should turn freely in housing. Pack housing with multipurpose grease and install new seal (16) and retainer (15) on housing. Remove steering pin (13) from steering lever plate (14) and install steering lever plate on housing. Place a $^{5}/_{32}$-inch (4 mm) shim between steering lever plate and seal retainer, install washer (4) and tighten one jam nut (5) until just snug. Install second jam nut and tighten against first nut to 22-25 ft.-lbs. (30-34 N·m) while holding first nut in position on stud. Remove shim stock. Install steering pin in steering lever plate until it just does engage steering gear. Center steering gear by rotating shaft to a position halfway between full left and full right turn. Adjust steering pin (13) inward until all backlash is removed but steering gear turns through full left/right cycle with no binding. Tighten jam nut (12) to 35-45 ft.-lbs. (48-61 N·m).

Reinstall steering gear by reversing removal procedure.

Models Late 316-330

R&R AND OVERHAUL. To remove steering gear, proceed as follows: Remove steering wheel (do not strike end of steering shaft). On Model 316, remove battery, side panels and battery base. On Model 330, disconnect negative battery cable and remove side panels. On all models, disconnect seat safety switch and push switch through hole in deck. Remove rear deck/fender assembly and screen under frame. On Model 330, detach right and left pedestals and remove radiator. On all models, detach and plug interfering hydraulic lines. Detach brake neutral return linkage. Detach pitman arm. Detach brake arm from left side of frame. On 316 models, remove left pedestal panel. On all models, disconnect drive shaft and remove. On all models, detach brake arm components on right side of frame and withdraw brake crossshaft. Detach interfering rockshaft link, remove steering gear mounting screws, and remove steering gear with shaft. Detach steering shaft from steering gear.

Position steering gear in a vise and remove cover (15—Fig. JD8) screws. Loosen nut (13) and remove pitman shaft (18) and cover assembly. Unscrew wormshaft nut (1) and adjuster plug (2). Withdraw wormshaft assembly while being careful not to rotate shaft to the end of its travel as ball guides may be damaged.

Inspect components for damage and excessive wear. Detach ball guide clamp (6) to remove balls so wormshaft (10) and ball nut (9) can be separated. There are 50 balls, which are available only as a set. Balls must be renewed if flat or otherwise damaged. Wormshaft and ball nut are available only as a set. Tips of ball guides (8) must not be bent.

Lubricate wormshaft with multipurpose grease, then assemble wormshaft and ball nut so narrow end of ball nut

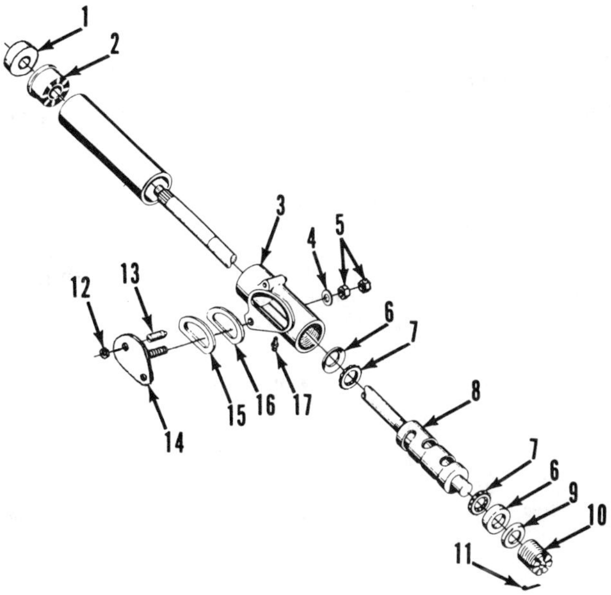

Fig. JD7—Exploded view of steering gear assembly used on early 316 and all 317 and 400 models.

1. Seal
2. Bearing
3. Housing
4. Washer
5. Jam nuts
6. Retainer cups
7. Bearings
8. Steering shaft & gear assy.
9. Belleville washer
10. Adjustment plug
11. Cotter key
12. Jam nut
13. Steering pin
14. Steering lever plate
15. Retainer
16. Seal
17. Grease fitting

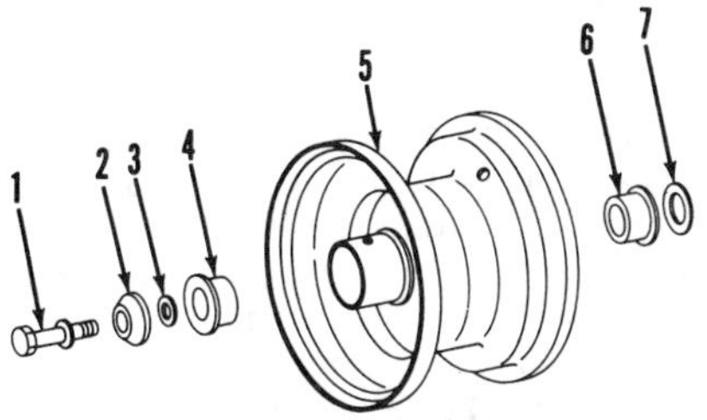

Fig. JD6—View showing typical wheel bearing assembly used on all models. Some models may use a snap ring to retain the wheel in place of bolt (1) and cap (2) shown.

1. Bolt
2. Cap
3. Washer
4. Bearing
5. Wheel & hub
6. Bearing
7. Washer

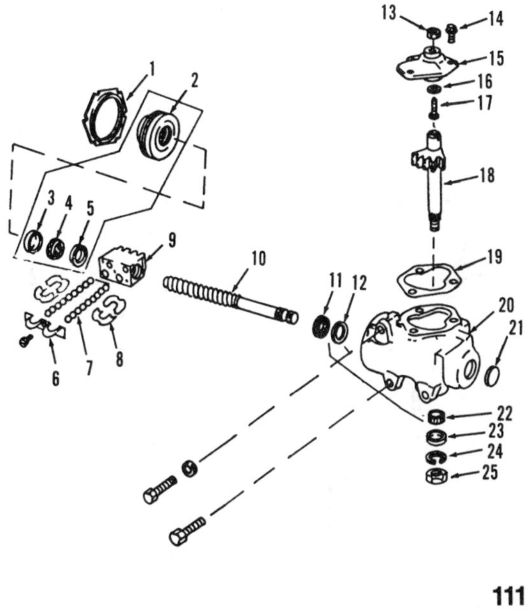

Fig. JD8—Exploded view of steering gear used on later Model 316 and Model 330.

1. Locknut
2. Adjuster
3. Thrust bearing
4. Bearing
5. Retainer
6. Clamp
7. Balls (50)
8. Ball guides
9. Ball nut
10. Wormshaft
11. Bearing
12. Thrust bearing
13. Nut
14. Screw
15. Cover
16. Shim
17. Adjuster
18. Pitman shaft
19. Gasket
20. Gearbox
21. Seal
22. Bearing
23. Seal
24. Lockwasher
25. Nut

teeth face upward and splined end of wormshaft is located on left side of ball nut. Install one half of ball guide (8), install half of balls (7) and other half of ball guide. Install remaining ball guides and balls and secure ball guides with clamp (6). When installing wormshaft and ball nut in housing, be sure ball nut is centered in housing so middle of nut is aligned with pitman shaft bore. Fill housing with 312 g (11 oz.) of multipurpose grease.

End play of adjuster (17) must be less than 0.05 mm (0.002 in.). Install shims (16) as needed to obtain desired end play. Install pitman shaft in cover and turn adjuster counterclockwise until bottomed, then turn adjuster ½ turn clockwise. Install pitman shaft assembly in housing so pitman gear teeth are centered on ball nut teeth. Tighten side cover screws to 40 N·m (30 ft.-lbs.).

To adjust wormshaft bearing preload, tighten adjuster plug (2) until bottomed, then back out ¼ turn. Rotate worm shaft until at end of travel—do not force—then rotate in opposite direction ½ turn. Tighten adjuster plug so a rolling torque of 0.6-1.0 N·m (5-9 in.-lbs.) is required to rotate wormshaft. Do not rotate wormshaft to end of travel when reading torque. Install nut (1).

Back out adjuster (17) and turn back in one turn. Turn wormshaft so gear is centered and check rolling torque required to turn wormshaft. Rotate adjuster in so an additional 0.5-1.2 N·m (4-11 in.-lbs.) torque is required to rotate wormshaft. Tighten adjuster locknut (13) to 34 N·m (25 ft.-lbs.) while preventing rotation of adjuster.

Tighten steering gear retaining screws to 95 N·m (70 ft.-lbs.). Tighten pitman arm retaining nut to 224 N·m (165 ft.-lbs.).

POWER STEERING SYSTEM

Pressurized oil from hydrostatic drive unit is supplied to a steering control valve and cylinder to provide power steering on some 318, 322, 332, 400, 420 and 430 models. Refer to appropriate TRANSMISSION section for maintenance and service information for hydrostatic drive unit.

STEERING CONTROL VALVE

Model 400

R&R AND OVERHAUL. Power steering control valve is attached to drag link. To remove, note locations of hydraulic hoses and remove hoses. Remove ball joint from power steering valve, loosen clamp bolt and screw valve off drag link.

To disassemble valve, remove retaining ring (2—Fig. JD10) and press dowel pin (10) out of valve. Remove sleeve (9) and carefully pull spool (8) out of bore in control valve body (13).

NOTE: Do not remove snap ring (3), washer (11), spring (4) or washer (5) from spool (8) unless spring (4) is to be renewed.

Renew all "O" rings (12 and 6) and back-up ring (7). Lubricate all parts with clean hydraulic fluid.

Reassemble by reversing disassembly procedure.

Models 318-322-332-420-430

REMOVE AND REINSTALL. Model 430. Remove grille and side engine covers. Remove battery and battery base. Disconnect seat safety switch and taillight wire connectors. Detach interfering handle knobs and remove rear fender and deck. Remove bottom frame guard. Detach battery cable clamp from firewall. Disconnect wire leads from circuit breaker on right pedestal panel and detach fuse block from panel. Remove right pedestal panel. Detach shock absorber attached to hydrostatic control mechanism. Mark location of hydraulic lines on steering valve and note steering valve location in relation to mounting bracket. Disconnect hydraulic lines and cap openings. Remove steering wheel; do not strike end of steering shaft. Unscrew mounting nuts and remove steering valve.

Models 318, 322, 332 and 420. Power steering control valve is attached to lower end of steering shaft. To remove, use a suitable puller and remove steering wheel. Remove battery and battery tray. Remove right side panel and remove protective screen/shield from lower side of frame. Disconnect drive shaft. Mark location of hydraulic lines on steering valve and note steering valve location in reference to mounting bracket. Disconnect hydraulic lines and cap openings. Remove nuts securing steering unit and lower unit through dash and remove.

OVERHAUL. Observe the following before proceeding with service on steering control valve.

NOTE: Do not clamp steering control valve in a vise for service work. Fabricate a holding fixture as shown in Fig. JD11.

Before disassembling steering valve, note alignment grooves machined in side of unit. Place steering valve in holding fixture with steering shaft down and secure with 5/16-24 NF nuts. Remove the four nuts securing port cover (1—Fig. JD12) and remove cover. Remove sealing ring (5) and four or five "O" rings (6). Remove plug (4) and ball (2). Remove "O" ring (3) from plug. Remove port manifold (7) and the three springs (9). Remove alignment pins (8). Remove valve ring (11) and discard sealing rings (10 and 12). Remove valve plate (13) and the three springs (14) from isolation manifold pockets. Remove hex drive and pin assembly (16), then remove isolation manifold (17). Remove drive link (18), seal (19), metering ring (20) and seal (21). Lift off metering assembly (22) and remove seal (24). Remove the 11 Allen head screws (23) and lift commutator cover (25) off. Remove commutator ring (26), then carefully remove commutator (27) and the five alignment pins (28). Remove drive link spacer (29) and separate rotor (30) from stator (31). Remove drive plate (32), thrust bearing (33), seal

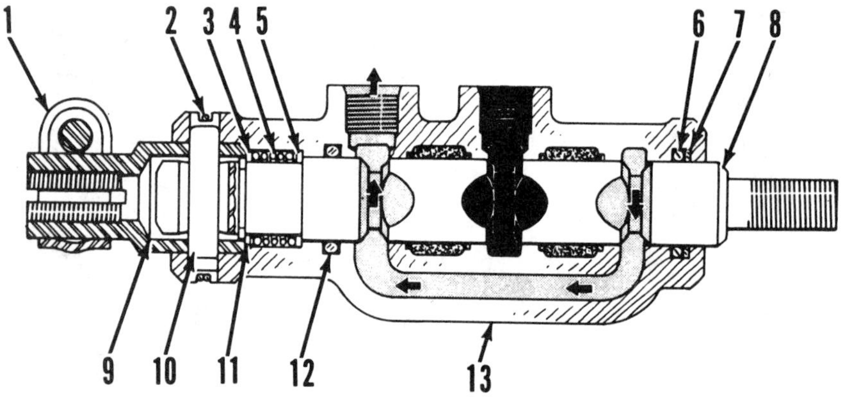

Fig. JD10—Sectional view of power steering control valve used on Model 400.

1. Clamp
2. Retaining ring
3. Snap ring
4. Spring
5. Washer
6. "O" ring
7. Back-up ring
8. Spool
9. Sleeve
10. Pin
11. Washer
12. "O" ring
13. Valve body

pacer (34), face seal (35) and face seal ack-up ring (36). Remove upper cover late (37). Remove steering shaft (39) nd service upper bearing seat, washer nd retaining ring as necessary.

Clean and inspect all parts and renew ny showing signs of wear, scoring or amage. Springs (9) should have a free ength of 0.75 inch (19 mm) and springs 14) should have a free length of 0.50 nch (12.7 mm). Springs must be re- ewed as sets only. Use a feeler gage to neasure clearance between tip of rotor 30) and stator body (31). If clearance xceeds 0.003 inch (0.076 mm) renew ro- or and stator assembly. Renew all "O" ings and lubricate all parts with clean ydraulic fluid prior to installation.

Reassemble by reversing disassembly rocedure and referral to the following otes: Be sure alignment grooves on up- er cover plate (37—Fig. JD12), isolation nanifold (17) and port manifold (7) are n same side. See Fig. JD13. Install ro- or (30—Fig. JD12) so pin holes face out- vard. Install commutator so long rooves face outward as shown in Fig. D14. When installed, pins (28—Fig. D12) must not protrude above face of ommutator. Apply low strength thread- ocking solution to threads of screws 23) and install screws finger tight. In- ert metering valve assembly in meter- ng ring (20). Make six shims that are .007 inch thick, $^1/_2$ inch wide and $1^1/_2$ nches long. Insert two shims at three lifferent locations (120 degrees apart) etween drive plate and metering ring o center metering valve assembly. Turn assembly over and push metering valve down until bottomed. Tighten screws (23) to 11-13 in.-lbs. (1.24-1.47 N·m) using sequence shown in Fig. JD15. Commutator ring (26) should now be concentric within 0.005 inch (0.127 mm) of drive plate (32). Remove the shims. Install commutator seal (24) so yellow mark is toward commutator cover (25). Install valve plate (13) so surface marked "shaft side" is down over hex drive (16) and so smaller cavity and "PORT SIDE" marking are positioned opposite from alignment grooves on isolation manifold (17) as shown in Fig. JD16. Install port manifold (7—Fig. JD12) and port cover (1) with new "O" rings and tighten valve assembly nuts in a crossing pattern to 20-24 ft.-lbs. (27-33 N·m). Tighten plug (4) to 97-151 in.-lbs. (11-17 N·m).

STEERING CYLINDER

All Models So Equipped

R&R AND OVERHAUL. Mark hydraulic hose locations and disconnect hoses. Disconnect steering cylinder at each end. Steering cylinder used on all models is a welded assembly and must be renewed as an entire unit.

ENGINE

MAINTENANCE

All Models

Regular engine maintenance is required to maintain peak performance and long engine life.

Check engine oil level and clean air intake screen at five-hour intervals. Clean engine air filter precleaner at 25-hour intervals.

On models equipped with Kohler or Onan engine, manufacturer recommends changing engine oil and filter after every 100 hours of operation. Check spark plug electrode gap and breaker point gap after every 100 hours of operation. Renew air cleaner element after every 100 hours of operation. Renew spark plug and breaker points after every 250 hours of operation or as necessary.

On models equipped with Yanmar engine, manufacturer recommends changing engine oil and filter after every 200

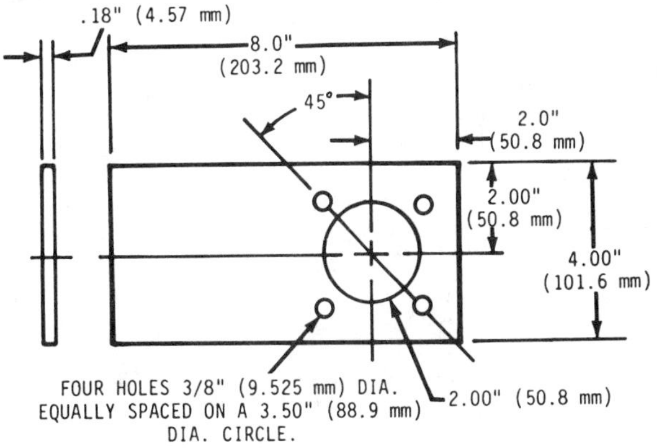

Fig. JD11—Fabricate a holding fixture as shown to aid service work on power steering control valve.

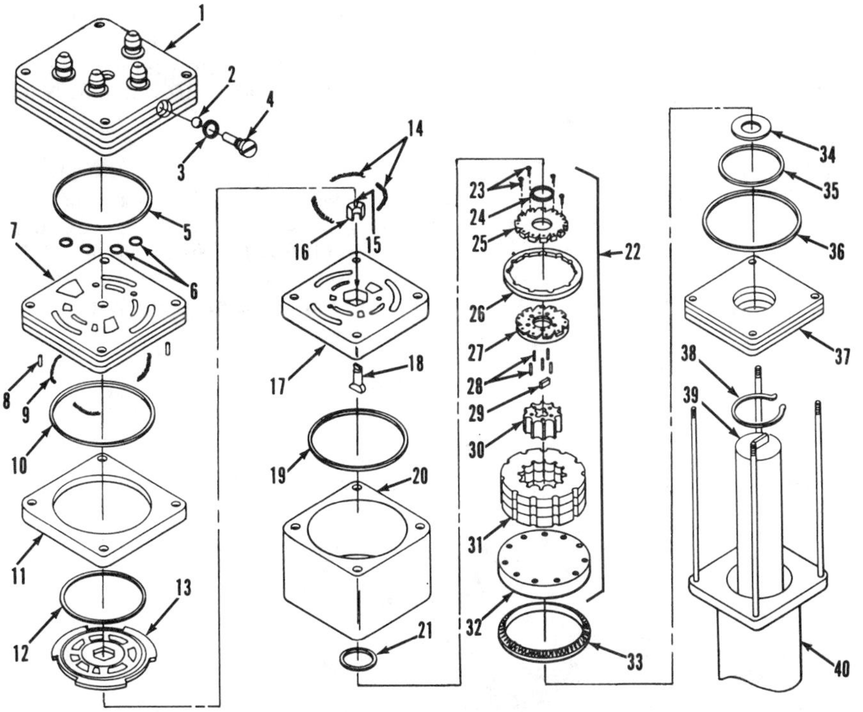

Fig. JD12—Exploded view of power steering control valve used on Models 318, 322, 332, 420 and 430.

1. Port cover
2. Ball
3. "O" ring
4. Plug
5. Seal ring
6. "O" rings
7. Port manifold
8. Alignment pins (2)
9. Springs
10. Seal ring
11. Valve ring
12. Seal ring
13. Valve plate
14. Springs
15. Pin
16. Hex drive assy.
17. Isolation manifold
18. Drive link
19. Seal
20. Metering ring
21. Seal
22. Metering assy.
23. Screws (11)
24. Commutator seal
25. Commutator cover
26. Commutator ring
27. Commutator
28. Alignment pins (5)
29. Drive link spacer
30. Rotor
31. Stator
32. Drive plate
33. Thrust bearing
34. Seal spacer
35. Face seal
36. Face seal back-up ring
37. Upper cover plate
38. Snap ring
39. Steering shaft
40. Tube assy.

hours of operation. After every 600 hours of operation, adjust engine valve clearance and check engine idle speeds. Once a year or as necessary, renew fuel filter, spark plug (gasoline engine), air filter and thermostat. Drain, flush and refill cooling system annually. Manufacturer recommends low silicate antifreeze that does not contain stop leak additive.

REMOVE AND REINSTALL

Models 322 and 332

Remove grille, side panels, battery, battery base and panel on tractor underside. Disconnect leads for headlights and hourmeter. Remove hood, hood support and support for battery base. Drain cooling system. Disconnect all interfering wires, cables and hoses. Detach rubber drive shaft isolator at engine. Loosen screws securing drive shaft universal joint at transmission and slide drive shaft toward transmission. Detach motor mounts and lift engine out of tractor.

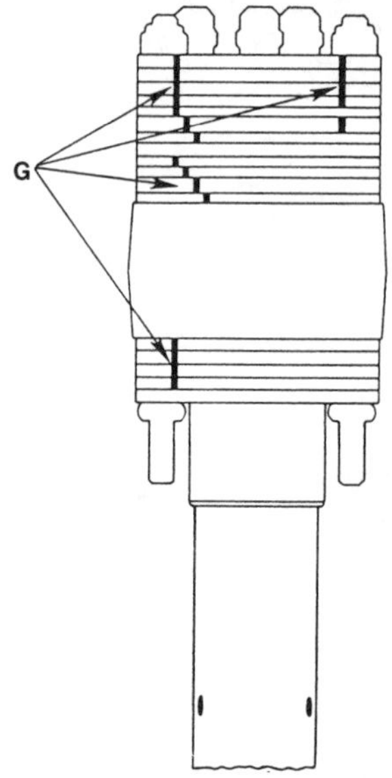

Fig. JD13—Components of steering control valve having alignment grooves (G) must be positioned as shown, or unit will not operate correctly.

When installing engine, tighten engine mounting nuts to 27 ft.-lbs. (37 N·m). On Model 332, bleed fuel injection system as outlined in Yanmar engine section.

Model 330

Remove grille and engine side panels. Remove battery, disconnect wiring for lights and hourmeter, and remove hood. Drain cooling system. Remove air cleaner and muffler. Detach all interfering hoses and wiring. Detach drive shaft. Unscrew engine mounting nuts and lift engine out of tractor.

When installing engine, tighten mounting nuts to 32-40 ft.-lbs. (44-54 N·m). Tighten drive shaft screws to 17-21 ft.-lbs. (23-29 N·m). Bleed fuel injection system as outlined in Yanmar engine section.

Model 430

Remove grill and side panels. Remove battery and base. Drain cooling system. Remove fuel/water separator. Disconnect fuel injection pump inlet hose and fuel return hose. Cap fuel lines to prevent contamination. Remove coolant tank and disconnect water pump inlet hose. Disconnect headlight wire. Remove hood. Disconnect any interfering wiring and cables. Remove muffler. Detach belts from pto pulley and disconnect drive shaft. Remove fan shroud. Attach lift to engine, detach engine from mounts, and lift engine from tractor.

When installing engine, tighten engine mounting nuts to 33-40 ft.-lbs. (44-54 N·m). Tighten drive shaft screws to 17-21 ft.-lbs. (23-29 N·m). Tighten pto spring retaining nut so spring length is 1.38 inches (35 mm). Bleed fuel injection system as outlined in Yanmar engine section.

All Other Models

Remove grille, side panels, hood, hood support and, if necessary, air cleaner and muffler. Disconnect all interfering wiring, cables and hoses. Remove pto belts and disconnect drive shaft at engine. Detach motor mounts and lift engine out of tractor.

After installing engine on 420 models, tighten pto belt tension spring so spring length is 1.38 inches (35 mm) if spring is vertical, or so spring length is 1.6 inches (41 mm) if spring is horizontal.

OVERHAUL

All Models

Engine make and model are listed at the beginning of this section. To overhaul engine or accessories, refer to appropriate engine section of this manual.

ELECTRICAL SYSTEM

MAINTENANCE AND SERVICE

All Models

Battery electrolyte level should be checked at 50-hour intervals of normal operation. If necessary, add distilled water until level is just below base of vent well. DO NOT overfill. Keep battery posts clean and cable ends tight.

For alternator or starter service, refer to appropriate engine service section in this manual.

Refer to appropriate wiring diagram for model being serviced (Fig. JD17 through JD27).

Fig. JD14—Install commutator (27) so long grooves (G) are out.

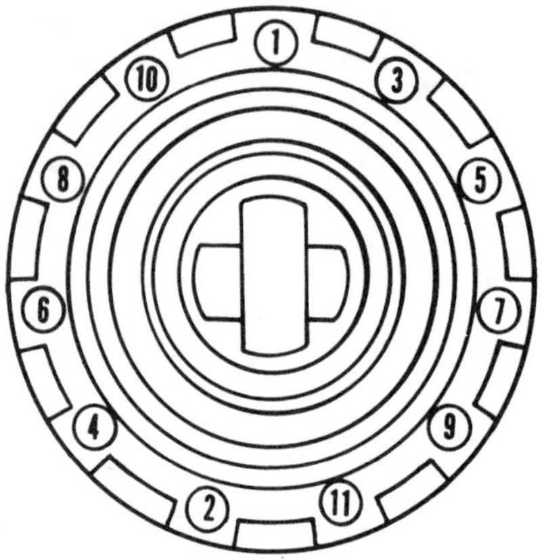

Fig. JD15—Tighten the 11 Allen head screws to 11-13 in.-lbs. (1.25-1.45 N·m) in sequence shown. Refer to text.

Fig. JD16—Install valve plate (13) so smaller cavity and "PORT SIDE" marking are positioned opposite from alignment grooves on isolation manifold (17).

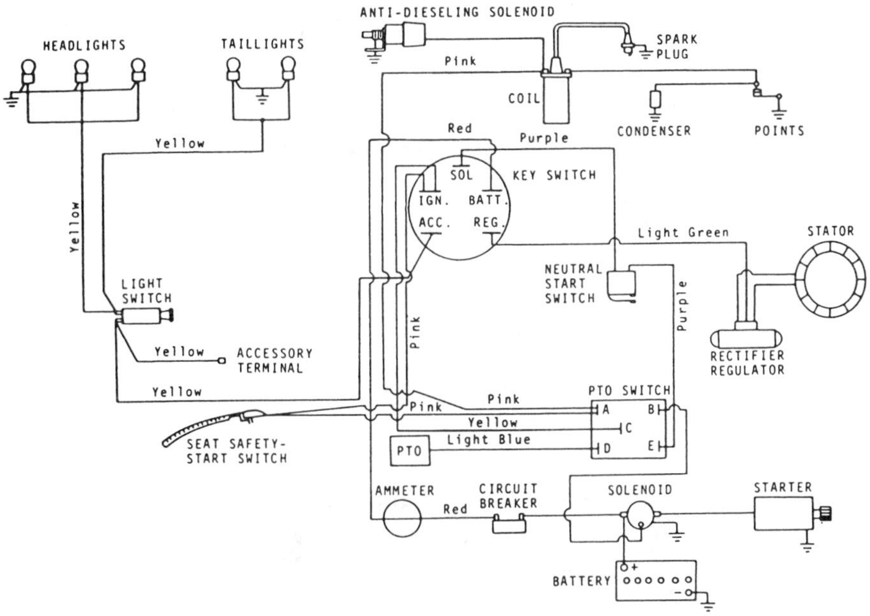

Fig. JD17—Wiring diagram for Model 316 prior to serial number 70,000.

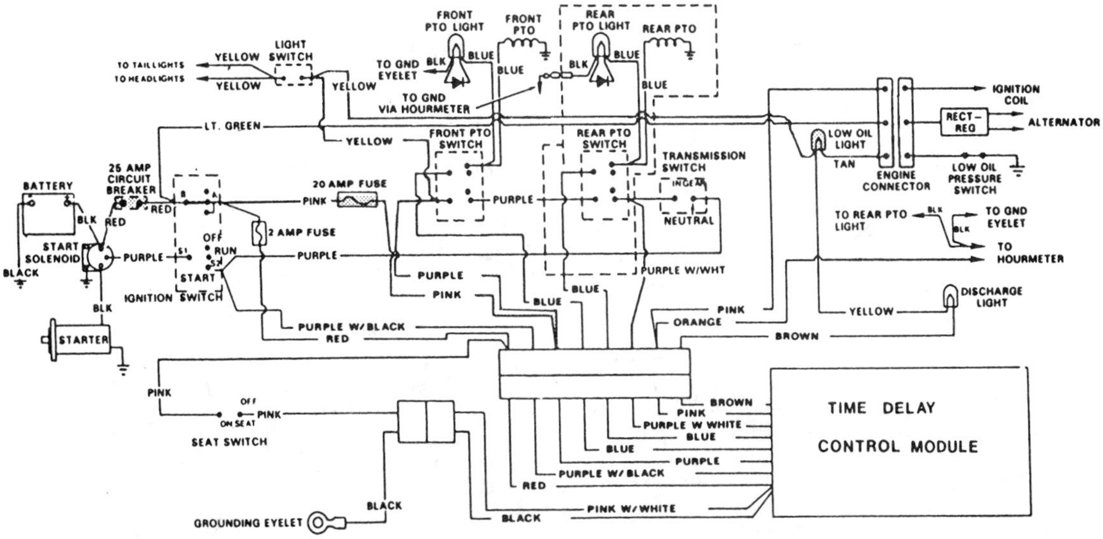

Fig. JD18—Wiring diagram for Model 316 serial number 70,001-285,000 and 318 and 420 models prior to 475,000.

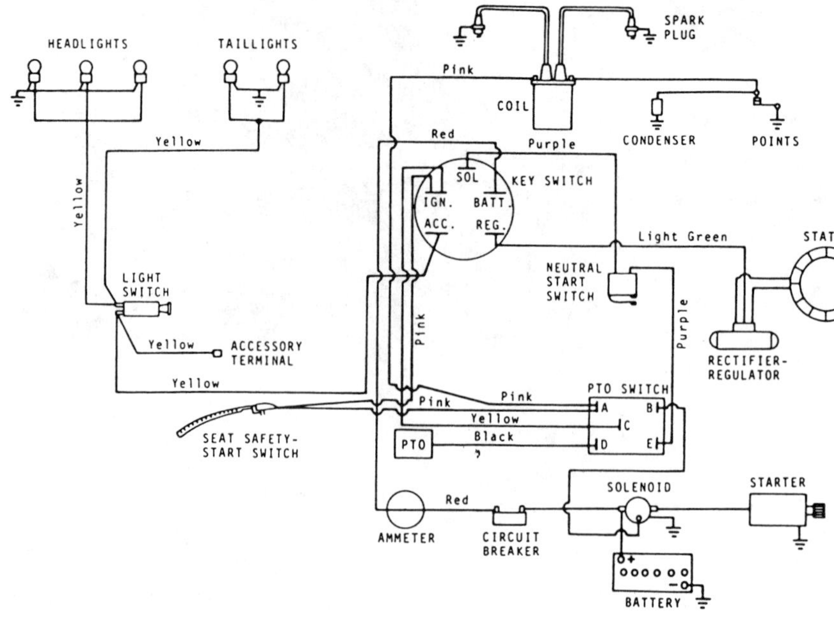

Fig. JD19—Wiring diagram for Model 317.

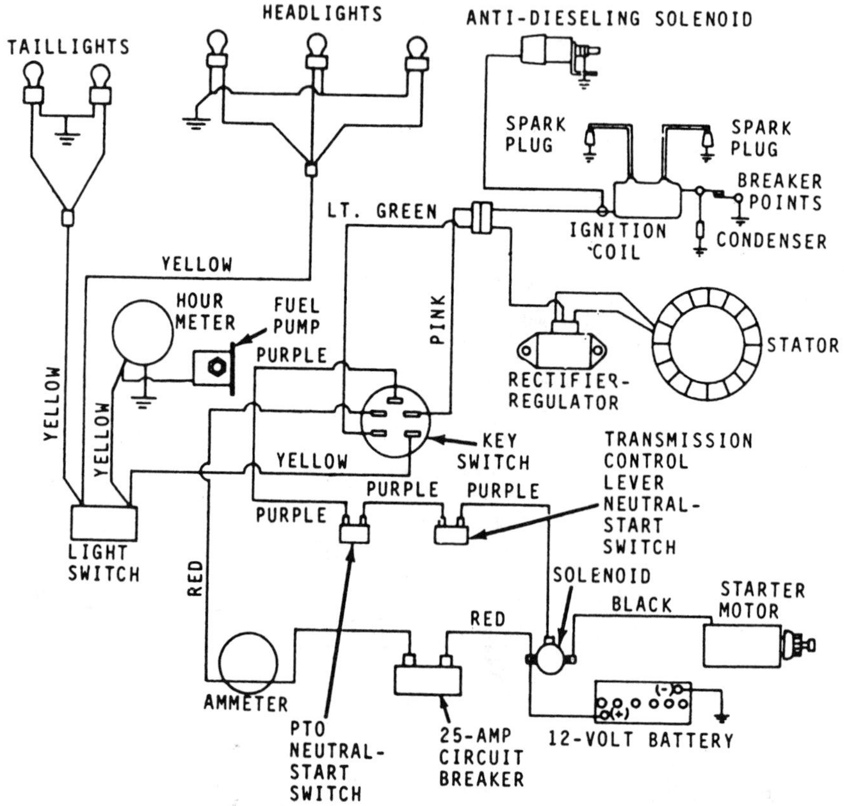

Fig. JD20—Wiring diagram for Model 400 prior to serial number 70,000.

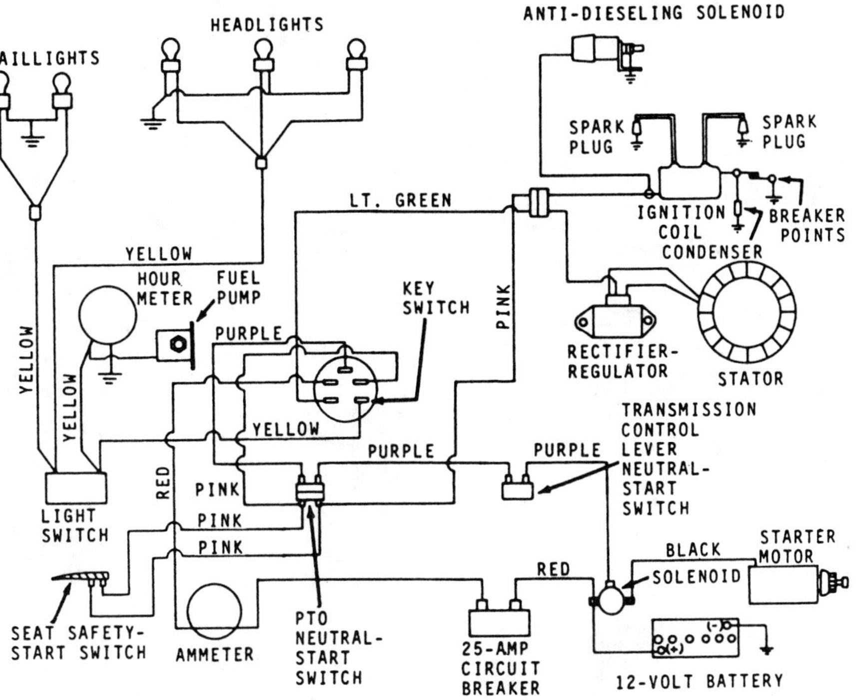

Fig. JD21—Wiring diagram for Model 400 after serial number 70,001.

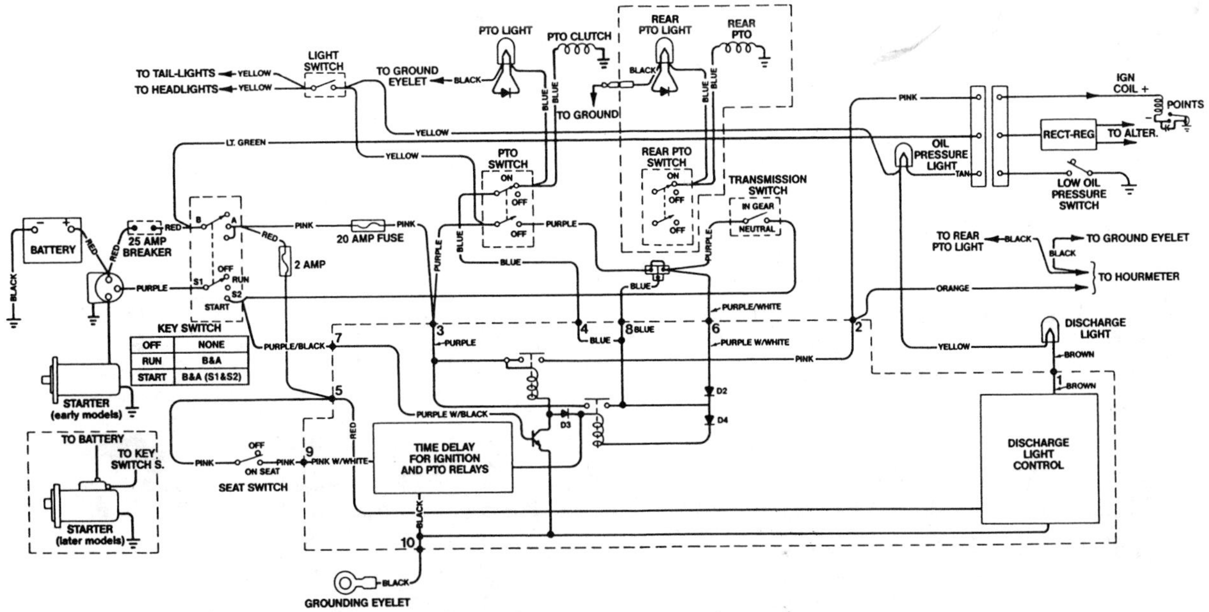

Fig. JD22—Wiring diagram for Model 316 after serial number 285,001.

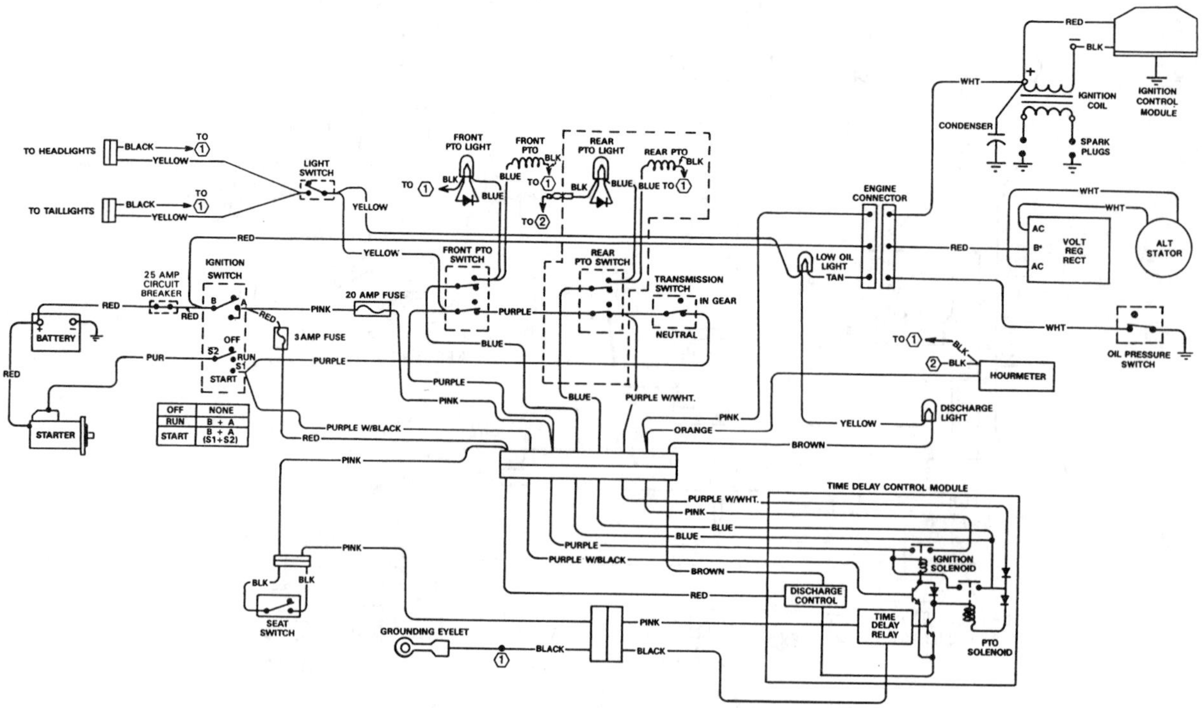

Fig. JD23—Wiring diagram for Models 318 and 420 after serial number 475,001.

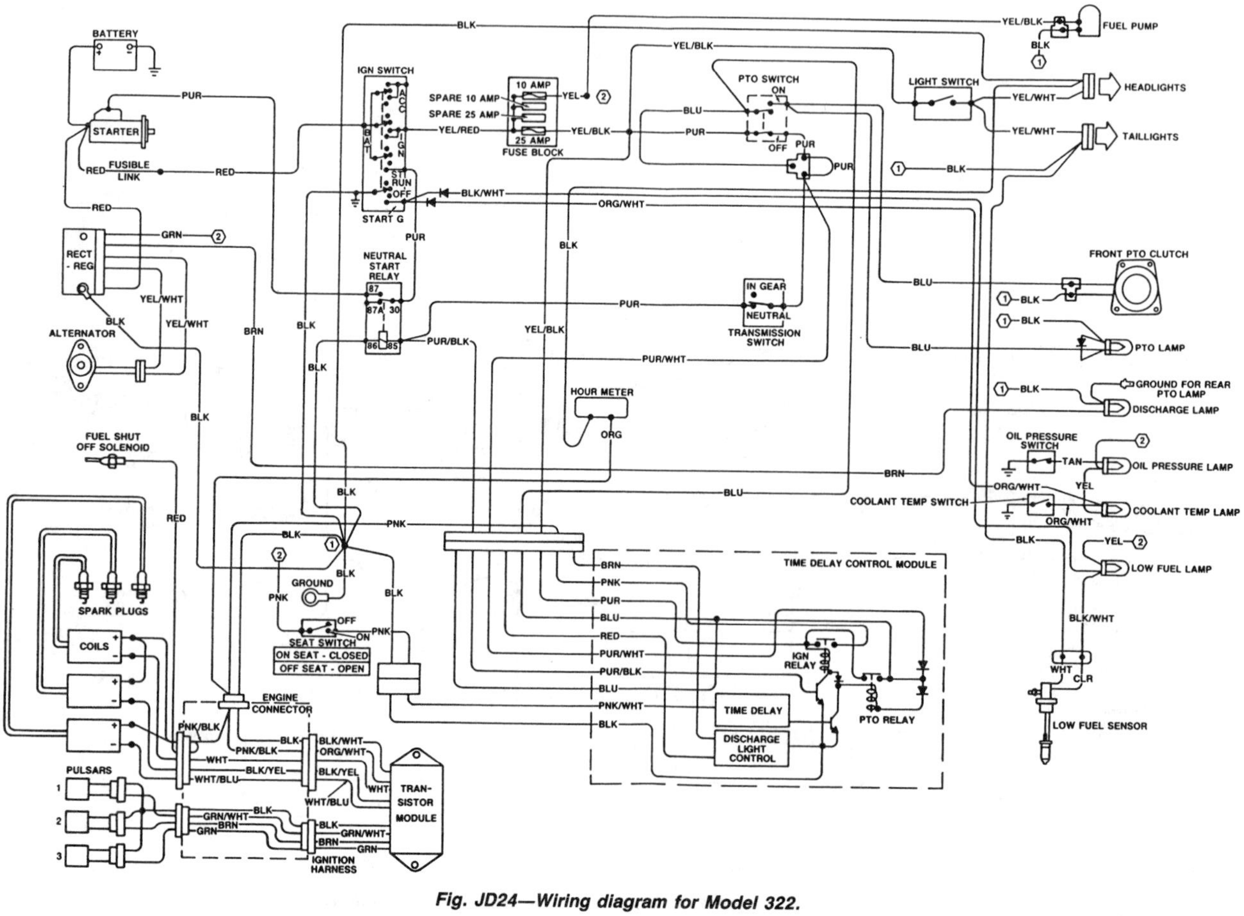

Fig. JD24—Wiring diagram for Model 322.

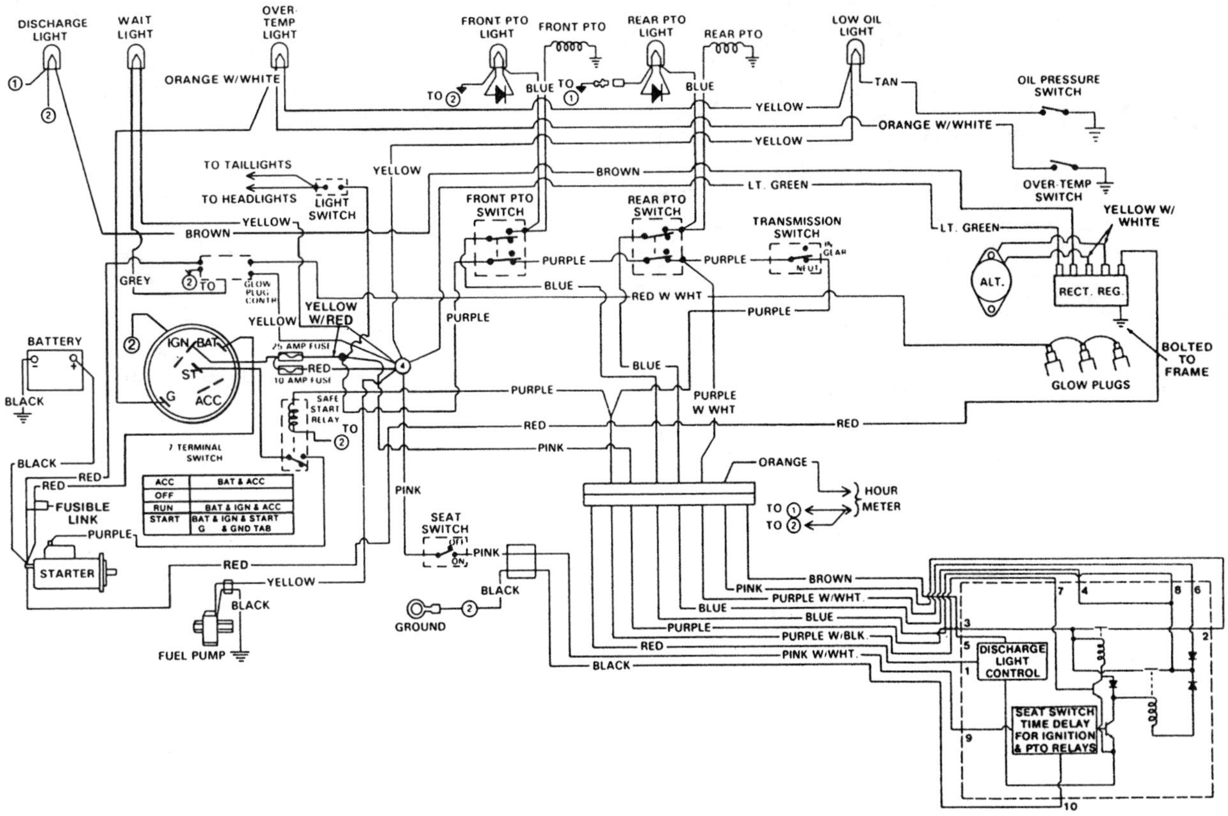

Fig. JD25—Wiring diagram for Model 330.

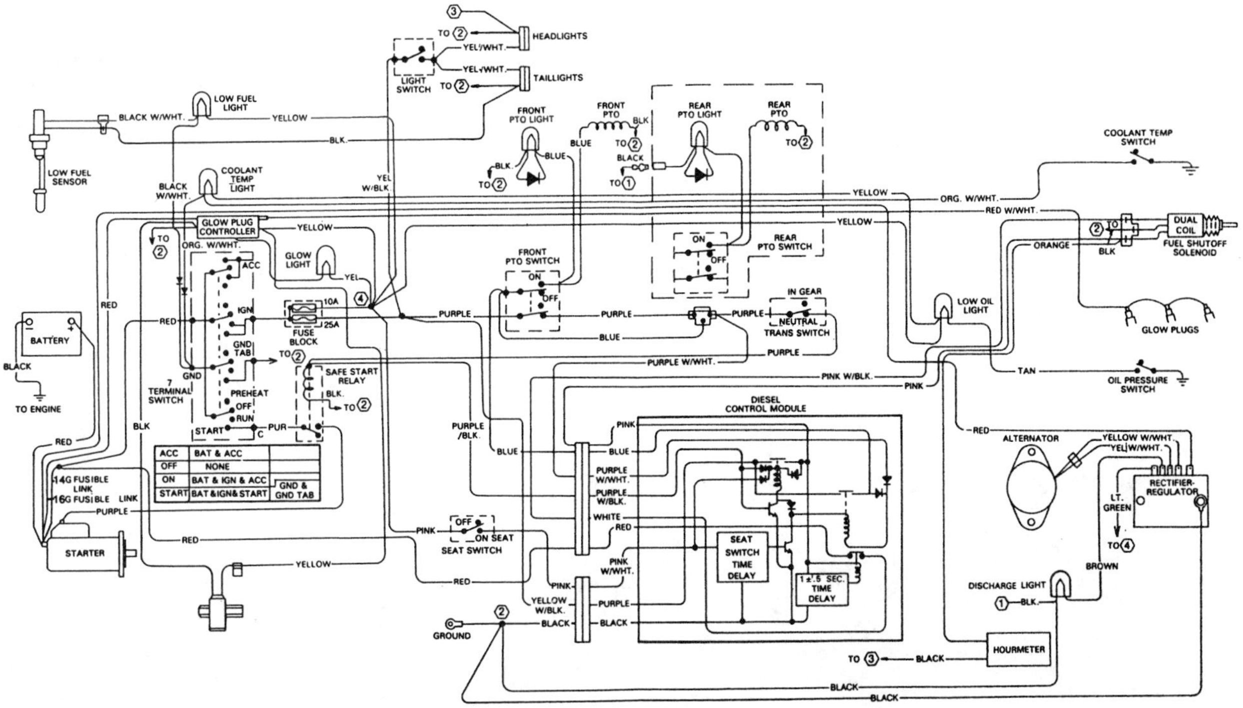

Fig. JD26—Wiring diagram for Model 332. Early models used external grounds.

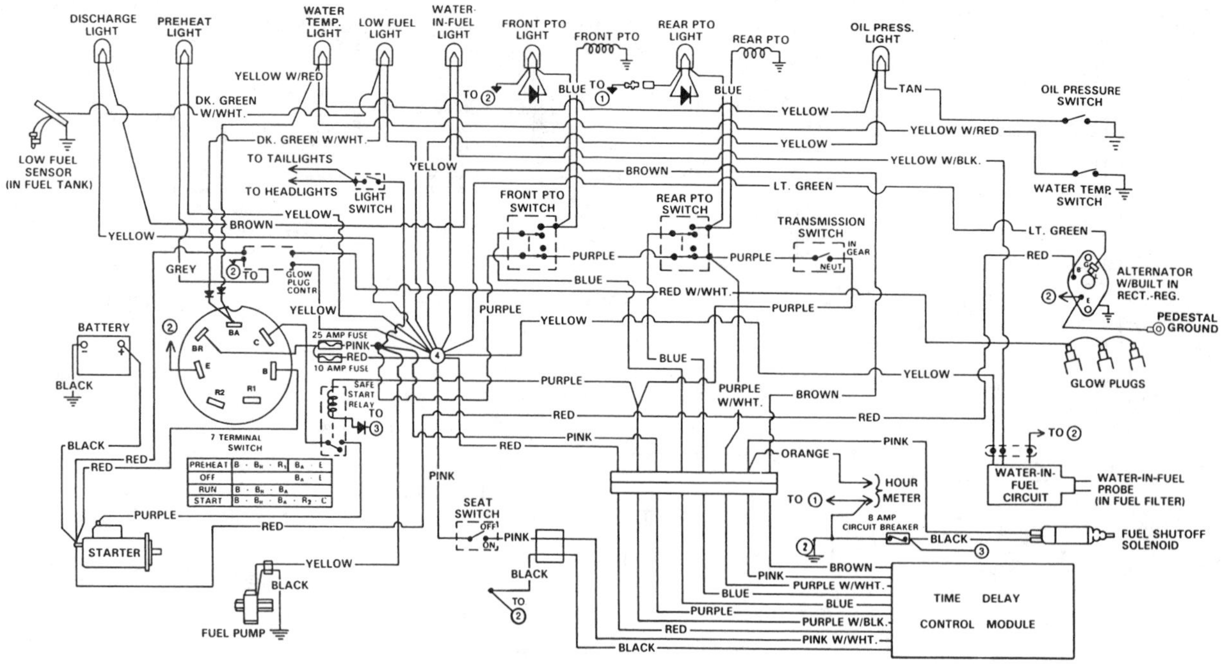

Fig. JD27—Wiring diagram for Model 430.

BRAKES

MAINTENANCE

All Models

Lubricate brake pedal shaft on early models after every 25 hours of operation and adjust brake shoes periodically to maintain safe brake performance. Systems having individual rear wheel brakes must be adjusted so pedal height is equal for each side.

ADJUSTMENT

Models Early 316-317-400

Remove pin (15—Fig. JD30) and turn yoke (3) as necessary to obtain good braking action. Repeat procedure for opposite side making certain brake pedal height for each side is equal.

Models Later 316-318-322-330-332-420-430

Raise and support rear of tractor. Insert brake adjusting tool or screwdriver through brake adjustment slot in backing plate (1—Fig. JD31) and turn adjustment wheel (6) until a slight drag is felt as tire and wheel are rotated. Depress brake pedal firmly to seat shoes. Back adjuster wheel off until slight drag is just removed. Adjust opposite side in the same manner making certain left and right pedal height is equal. Apply both brakes and set parking brake so tractor will not move. Turn adjusting nut (8—Fig. JD32) until spring (5) compressed length is 1.93-1.96 inches (49-50 mm) on Models 316, 317, 400 and 420, or 1.65 inches (42 mm) on all other models. Turn nut (2) until nut and washer (3) just contact brake lever (4).

R&R BRAKES

Models Early 316-317-400

Raise and support rear of tractor. Remove tires and wheels. Remove cotter key (10—Fig. JD30), nut (11) and wash-

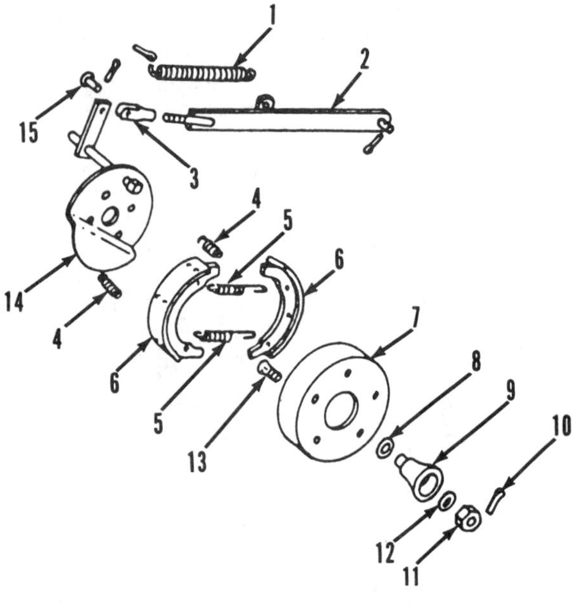

Fig. JD30—Exploded view of brake assembly used on early Model 316 and all 317 and 400 models.

1. Spring
2. Brake rod
3. Yoke
4. Hold-down spring
5. Return springs
6. Brake shoes
7. Brake drum
8. Felt seal
9. Hub
10. Cotter key
11. Nut
12. Washer
13. Stud
14. Backing plate
15. Pin

er (12). Use a suitable puller to remove brake drum (7). Disconnect hold-down springs (4) and return springs (5). Remove brake shoes (6).

Reinstall by reversing removal procedure. Tighten nut (11) to 35-40 ft.-lbs. (47-54 N·m).

Models Later 316-318-322-330-332-420-430

Raise and support rear of tractor. Remove tires and wheels. Bend tabs of lock plate from axle nut and remove nut. Remove brake drum (7—Fig. JD31). Disconnect hold-down springs (9) and return spring (5). Remove brake shoes (4).

Reinstall by reversing removal procedure. Tighten axle nut to 50-80 ft.-lbs. (68-108 N·m) and secure nut with lock plate.

TRANSMISSION

MAINTENANCE

All Models

Hydrostatic transmission fluid level should be checked before operating tractor. Transmission must be cold and tractor must be on level ground. Maintain fluid level at midpoint in sight tube at rear of tractor (Fig. JD33). Transmission oil and filter should be changed after every 200 hours of operation. Recommended fluid is John Deere All-Weather Hydrostatic Fluid or type "F" automatic transmission fluid. Refer to specifications table at beginning of section for fluid capacity. Do not overfill.

ADJUSTMENTS

Models Early 316-317

NEUTRAL POSITION ADJUSTMENT. Raise and support rear of tractor so one tire is off the ground. Remove cotter pin (21—Fig. JD34). Remove pin (19) from clevis (20). Position control arm (23) so roller (25) is seated in control arm detent notch. Start engine and run at ¾ throttle.

CAUTION: Make certain tractor is blocked securely and use care when working near rotating tire or moving parts.

If raised wheel is rotating, loosen locknut (16) and turn eccentric (17) until wheel rotation stops. Tighten locknut (16). Loosen jam nut (15) and turn clevis (20) until pin (19) can be installed easily. Install cotter key (21) and tighten jam nut (15). Move motion control lever (4) forward and backward, then recheck neutral position adjustment.

LINKAGE BRAKE ADJUSTMENT. If motion control lever (4—Fig. JD34) creeps from set position or is difficult to move, turn locknut (13) as needed.

Model 400

NEUTRAL POSITION ADJUSTMENT. Raise and support rear of tractor so one tire is off the ground. Place motion control lever (8—Fig. JD35) in neutral position and remove cotter key, washer and pin (2) from yoke (3). Position control arm (1) so roller (18) is seated in control arm notch. Start engine and run at ¾ throttle.

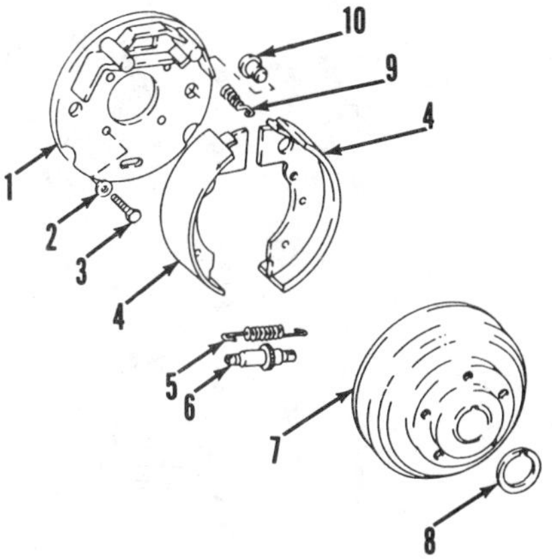

Fig. JD31—Exploded view of brake assembly used on later Model 316 and all 318, 322, 330, 332, 420 and 430 models.

1. Backing plate
2. Lockwasher
3. Bolt
4. Brake shoes
5. Return spring
6. Brake adjustment wheel
7. Brake drum
8. Lock plate
9. Hold-down spring
10. Plug

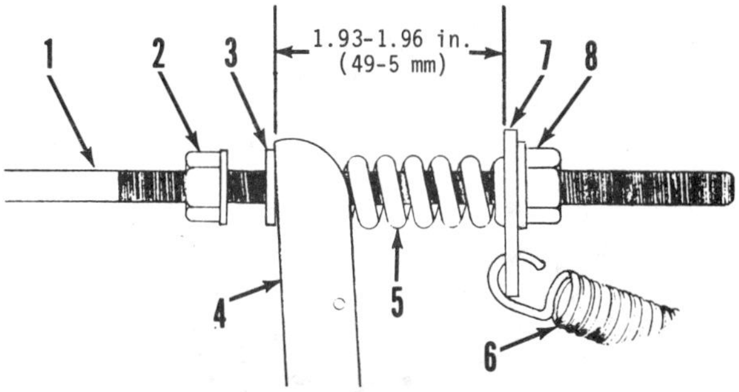

Fig. JD32—View showing correct adjustment specification for brake adjustment on later Model 316 and all 318, 322, 330, 332, 420 and 430 models.

1. Brake rod
2. Nut
3. Washer
4. Brake lever
5. Spring
6. Return spring
7. Spring link
8. Adjustment nut

Fig. JD33—View showing location of fill tube (A), sight tube (B) and accumulator-vent (C). Refer to text.

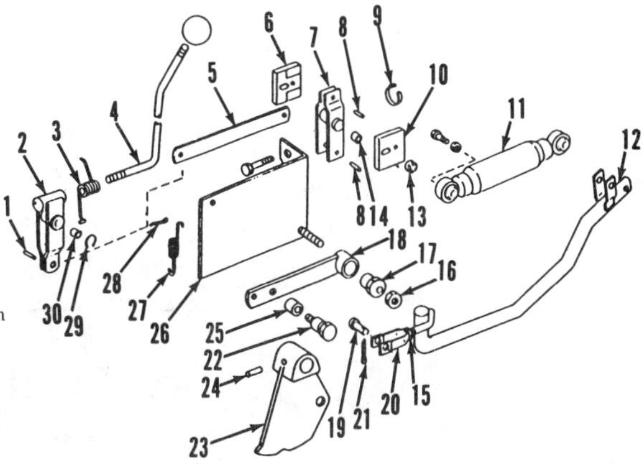

Fig. JD34—Exploded view of hydrostatic drive motion control linkage used on early Model 316 and Model 317.

1. Pin
2. Crank
3. Spring
4. Motion control lever
5. Link
6. Pad
7. Pivot
8. Pin
9. Toe band
10. Pad
11. Shock absorbing link
12. Tube
13. Locknut
14. Bushing
15. Jam nut
16. Nut
17. Eccentric
18. Arm
19. Pin
20. Clevis
21. Cotter pin
22. Bolt
23. Control arm
24. Pin
25. Roller
26. Bracket
27. Spring
28. Cotter pin
29. Snap ring
30. Bushing

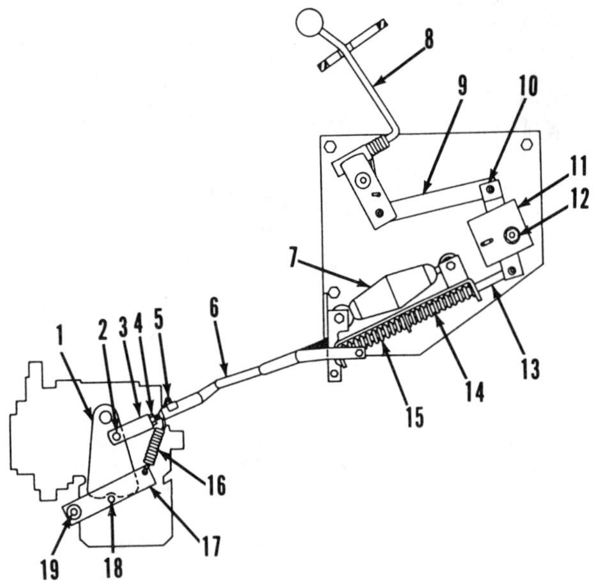

Fig. JD35—View showing hydrostatic drive motion control linkage used on 400 models with serial numbers 30,001 to 55,000. Models with serial number greater than 55,001 are similar, except parts 6, 15, 14 and 13 were replaced with a solid control rod.

1. Control arm
2. Pin
3. Clevis
4. Jam nut
5. "S" link
6. Rod
7. Shock absorbing link
8. Motion control lever
9. Upper link
10. Lower link
11. Linkage friction brake
12. Locknut
13. Slider tube
14. Spring
15. Spring
16. Spring
17. Roller arm
18. Roller
19. Eccentric

CAUTION: Make certain tractor is blocked securely and use care when working near rotating tire or moving parts.

If raised wheel is rotating, loosen locknut on eccentric (19) and turn eccentric until wheel rotation stops. Tighten locknut. Disconnect "S" link (5) from spring (16), loosen jam nut (4) and turn clevis (3) until pin (2) can be inserted easily. Install pin (2), washer and cotter key. Tighten jam nut (4). Connect "S" link (5) to spring (16). Move motion control lever forward and backward and recheck neutral position adjustment.

LINKAGE BRAKE ADJUSTMENT. If motion control lever (8—Fig. JD35) creeps from set position or is difficult to move, turn locknut (12) as necessary.

Models Later 316-318-322-332-420-430

NEUTRAL POSITION ADJUSTMENT. On all models, block up rear of tractor so rear wheels can rotate and remove bottom frame cover. Three different linkage configurations have been used. Inspect shift linkage and determine if tractor has a "J" bolt (J—Fig. JD36), turnbuckle (T—Fig. JD37) or detent spring (D—Fig. JD38).

To adjust linkage on models with a "J" bolt (J—Fig. JD36) and turnbuckle, proceed as follows: Movement of hydrostatic control lever should require 7-10 pounds (31-44.5 N). If not, remove side panel and turn nut (A) as needed. Apply both brakes and allow hydrostatic control lever to return to neutral position. Adjust jam nuts (M) on "J" bolt (J) so roller (R) just enters neutral slot (S) in cam plate (C), but does not contact end of neutral slot. Move lever forward, apply brakes and make certain lever correctly returns to neutral position. If necessary, loosen jam nut (N) and adjust eccentric nut (E) so lever goes squarely into STOP notch of slot in tractor dash. See following section for TRANSMISSION NEUTRAL ADJUSTMENT.

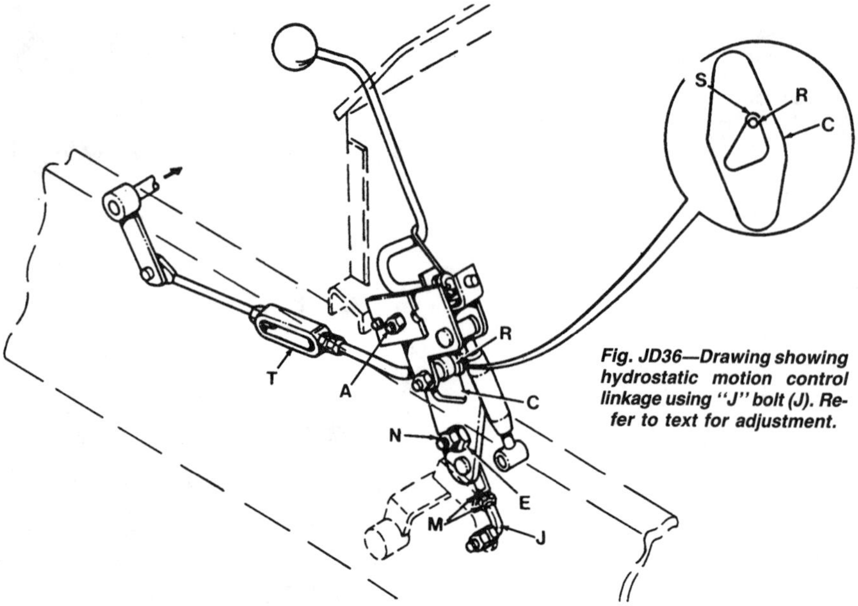

Fig. JD36—Drawing showing hydrostatic motion control linkage using "J" bolt (J). Refer to text for adjustment.

To adjust linkage on models with a turnbuckle (T—Fig. JD37) but no "J" bolt, proceed as follows: Movement of hydrostatic control lever should require 7-10 pounds (31-44.5 N). If not, turn locknut attached to screw (A) as needed. Apply both brakes and allow hydrostatic control lever to return to neutral position. Loosen nuts (N) and position cam plate (C) so neutral roller (R) enters slot (S); roller must not touch top of slot. Tighten nuts (N). If control lever is not in neutral notch in tractor dash, loosen locknut and turn eccentric nut (E). Retighten locknut. See following section for TRANSMISSION NEUTRAL ADJUSTMENT.

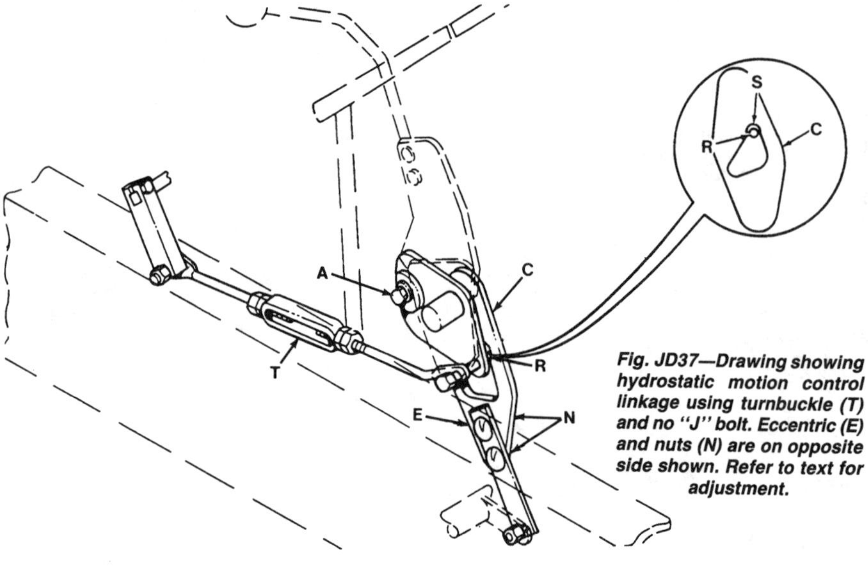

Fig. JD37—Drawing showing hydrostatic motion control linkage using turnbuckle (T) and no "J" bolt. Eccentric (E) and nuts (N) are on opposite side shown. Refer to text for adjustment.

To adjust linkage on models with detent spring (D—Fig. JD38), proceed as follows: Remove right pedestal panel and fender deck. Block up rear of tractor so it is securely supported and rear wheels can rotate. Start and run engine at fast idle with hydrostatic control lever in neutral. Wheels should not rotate. If wheels rotate, stop engine and loosen bolts (B1 and B2). Move transmission control arm (T) so roller (R1) is centered in detent notch. Rotate nut adjacent to detent spring (D) so spring length is 1.97 inch (50 mm).

CAUTION: Exercise extreme caution during next steps as rear wheel and drive components will be spinning near work area.

Start and run engine at fast idle. Rotate eccentric (E1) so rear wheels do not rotate. Tighten eccentric and stop engine. Turn eccentric (E2) so bolt (B2) is centered in hole; roller (R2) must be centered in cam plate notch. Tighten eccentric (E2) and bolt (B2). Engage parking brake and loosen bolt (B3); some models have two bolts. Move plate (P) up until plate is against roller (R2) and tighten bolt (B3). Position hydrostatic control lever in neutral notch in dash panel. Tighten bolts (B1) and disengage parking brake. Check force required to move hydrostatic control lever. If more or less than 7-10 pounds (31-44.5 N), turn locknut (L) as needed.

TRANSMISSION NEUTRAL ADJUSTMENT. On models with a turnbuckle (T—Fig. JD36 or JD37), movement of the shift mechanism and transmission shift arm must be synchronized. Remove fender deck and block up rear of tractor so it is securely supported and rear wheels can rotate. On Model 420, shift two-speed axle to low position. Start and run engine at fast idle with hydrostatic control lever in neutral. Wheels should not rotate. If wheel rotates, stop engine and loosen jam nuts on turnbuckle.

CAUTION: Exercise extreme caution during next step as rear wheels and drive components will be spinning near work area.

Start and run engine at fast idle. Rotate turnbuckle so rear wheels do not rotate and tighten jam nuts.

R&R HYDROSTATIC DRIVE UNIT

Models Early 316-317

Drain transmission fluid by removing transmission to axle cooling tube. Remove fender deck and fuel tank. Disconnect transmission control linkage and remove from hydrostatic drive unit. Disconnect drive shaft at flex coupling. Disconnect hydraulic lines at couplings located at front of hydrostatic drive unit and cap openings. Remove the six mounting bolts and remove transmission through bottom of frame.

Reinstall by reversing removal procedure. Adjust transmission linkage.

Model 400

Drain transmission by removing drain plug at rear, in bottom of differential housing. Remove fender and deck assembly and fuel tank. Remove upper fan guard. Disconnect supply tube and return tube from hydrostatic drive unit. Remove bottom frame screen and lower fan guard. Remove all hydraulic lines and filter. Cap openings. Disconnect all control linkage. Drive spring pin out of drive shaft coupler and slip coupling back. Remove key. Remove the four cap screws securing hydrostatic drive unit. Move unit forward and lift up to remove.

Reinstall by reversing removal procedure. Adjust transmission linkage.

Models Later 316-318-322-330-332-420-430

Disconnect necessary electrical wiring and two-speed shift knob, if so equipped, then remove seat and deck assembly. Note location of fuel lines, disconnect fuel lines, then drain and remove fuel tank. Remove drain plug at bottom of differential housing and drain fluid. Disconnect all hydraulic lines, filler hose and sight tube. Cap all openings and disconnect transmission control linkage and brake rods. Disconnect transmission shift linkage and differential lock linkage, if so equipped. Raise and support frame of tractor. Support transmission assembly, remove mounting bolts and lower assembly. Roll assembly away from tractor.

Separate hydrostatic drive unit from reduction gear/range transmission and differential assembly as needed, making certain spacers between hydrostatic drive unit and range transmission are marked for correct reinstallation.

Reinstall by reversing removal procedure. Adjust transmission linkage and brakes.

OVERHAUL

All Except Model 400

Remove hydrostatic drive unit as previously outlined and clean exterior of unit.

To disassemble charge pump, first scribe a line across center section and charge pump body to aid in correct reassembly, then unbolt and remove charge pump housing (3—Fig. JD40). Remove rotor assembly (4) and rotor drive pin. Pry oil seal (2) from housing and press needle bearing out front of housing. Remove both check valves (13), back-up washer (14) and "O" ring (15). Remove plugs (8), "O" ring (9), shim (10), spring (11) and relief valve cone (12) from relief bore in center housing. Unbolt and remove center housing (57).

CAUTION: Valve plates (18 and 50) may stick to center housing. Be careful not to drop them.

Remove and identify valve plates. Pump valve plate has two relief notches and motor valve plate has four notches. Refer to Fig. JD41. Tilt housing (35—Fig. JD40) on its side and remove cylinder block and piston assemblies from pump and motor shafts. Lay cylinder block assemblies aside for later disassembly. Drive roll pins (53) into swashplate (26) and remove stub shaft (33) and trunnion shaft (40) assemblies. Be careful not to damage housing when driving roll pins out. Remove swashplate (26) and thrust plate (25), then withdraw pump shaft (27) and bearing (28). Remove two socket head cap screws securing motor swashplate (43) to housing, then lift out swashplate and

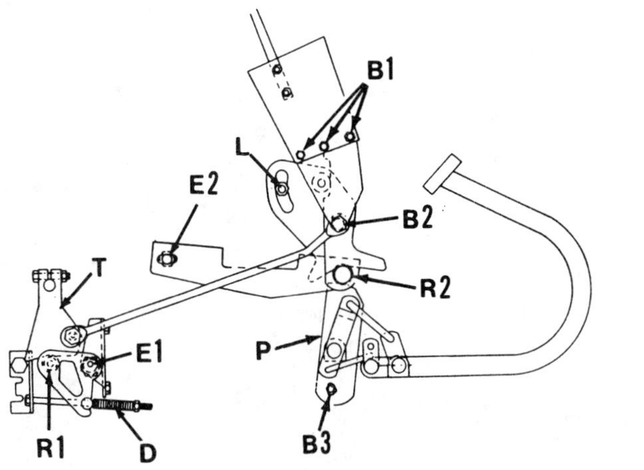

Fig. JD38—Drawing showing hydrostatic motion control linkage using detent spring (D). Refer to text for adjustment.

motor shaft (42). Bearings (36 and 41), bushing (34) and oil seals (29, 32 and 37) can now be removed from housing (35). Remove needle bearing (17) from center housing (57).

Carefully remove slipper retainer (23) with nine pistons (24) from pump cylinder block (22), then remove slipper retainer (45) with nine pistons (44) from motor cylinder block. Place each cylinder block on wood blocks in a press, compress spring (21 or 48) and remove retaining ring. Release pressure and remove spring and washers.

Check cylinder blocks (22 and 46) on valve face end and renew if scratches or other damage is evident. Inspect pistons and bores in cylinder blocks for excessive wear or scoring. Piston slippers can be lapped to remove light scratches. Outer edge of piston slipper must not be rounded more than 1/32 inch (0.79 mm) and lubricant hole in slipper face must be open. Minimum slipper thickness is 0.121 inch (3.073 mm) for both pump and motor. Slipper thickness must not vary more than 0.002 inch (0.051 mm) for all nine pistons in each block.

Check pump valve plate (18) and motor valve plate (50) for excessive wear and damage. If wear or scratches can be felt by running a fingernail across face of plates, renew plates. Inspect motor swashplate (43) and pump swashplate thrust plate (25) for wear, embedded material and scoring, and renew if necessary.

Check charge pump housing (3) and rotor assembly (4) for wear and scoring, and renew if necessary. Relief valve cones (12) should be free of nicks and scratches. Note implement relief valve spring (11) is heavier than charge pump relief valve spring (7). Shims (10) are available to increase relief valve pressure.

Renew all "O" rings, gaskets and seals. Thoroughly lubricate all parts with clean hydraulic fluid during reassembly. When installing new bearings (17 and 52), press bearings in until they are 7/64 inch (2.8 mm) above machined surface of center housing. Check valves (13) are interchangeable and are serviced only as an assembly. Pump swashplate (26) must be installed with thin pad toward top of transmission. Flat end of charge pump housing (3) must be installed on same side as charge relief valve. On models so equipped, motor swashplate socket head cap screws should be tightened to 67 ft.-lbs. (91 N·m). Tighten housing screws to 25-30 ft.-lbs. (34-40 N·m). Tighten charge pump cap screws to 52 ft.-lbs. (71 N·m).

With unit reassembled, reinstall assembly on differential housing and tighten mounting cap screws to 31 ft.-lbs. (42 N·m). Install oil intake tube and oil filter.

Reinstall by reversing removal procedure.

Model 400

Remove hydrostatic drive unit as previously outlined and clean exterior of unit.

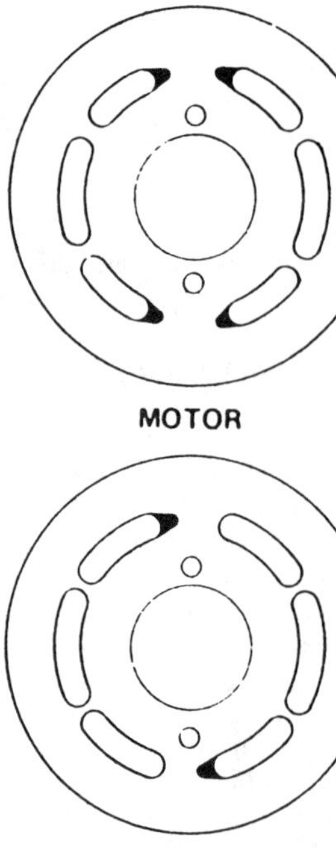

Fig. JD41—Motor valve plate has four notches (dark areas) and pump valve plate has two notches (dark areas). Refer to text.

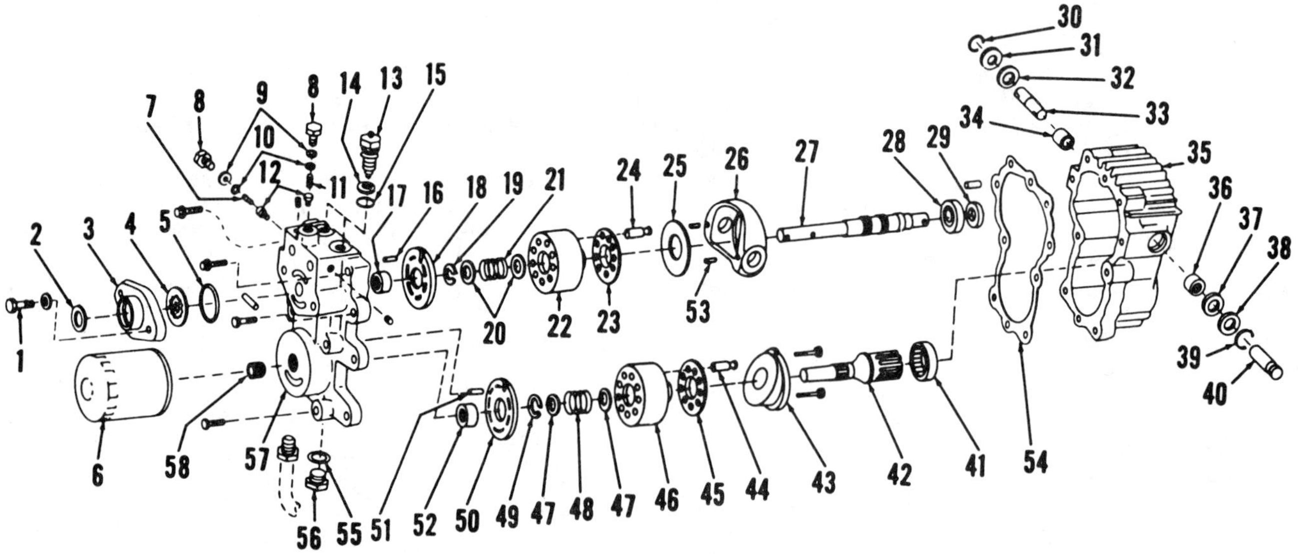

Fig. JD40—Exploded view of hydrostatic transmission used on all models except Model 400.

1. Cap screw
2. Oil seal
3. Charge pump housing
4. Rotor assy.
5. "O" ring
6. Filter
7. Charge pump relief valve spring
8. Plug
9. "O" ring
10. Shim
11. Implement relief valve spring
12. Relief valve cone
13. Check valve
14. Back-up washer
15. "O" ring
16. Pin
17. Needle bearing
18. Valve plate
19. Retaining ring
20. Washers
21. Spring
22. Cylinder block
23. Slipper retainer
24. Piston (9 used)
25. Thrust plate
26. Swashplate
27. Pump shaft
28. Ball bearing
29. Oil seal
30. Snap ring
31. Washer
32. Oil seal
33. Stub shaft
34. Bushing
35. Housing
36. Needle bearing
37. Oil seal
38. Washer
39. Retaining ring
40. Trunnion shaft
41. Roller bearing
42. Motor shaft
43. Swashplate (motor)
44. Piston (9 used)
45. Slipper retainer
46. Cylinder block
47. Washers
48. Spring
49. Retaining ring
50. Valve plate
51. Pin
52. Needle bearing
53. Roll pin
54. Gasket
55. "O" ring
56. Plug
57. Center housing
58. Oil filter fitting

Remove all hydraulic fittings. Drive roll pin out of control lever and shaft and remove control lever. Remove snap ring and pinion gear from motor shaft. Remove the four cap screws securing motor housing (3—Fig. JD42) to pump case (53), then lift motor housing off pump case and center section. Remove snap ring (1), motor shaft (5) and bearing (2). Remove thrust plate (4), cylinder block assembly (6) and motor valve plate (14).

NOTE: Care should be taken when removing cylinder block assemblies. Pistons (7) must be kept in their original bores.

Remove center section (21) and pump valve plate (65). Discard all gaskets and "O" rings. Carefully remove pump cylinder block assembly (57) and thrust plate (26).

NOTE: Care should be taken when removing cylinder block assemblies. Pistons (58) must be kept in their original bores.

Drive spring pins (54) out of swashplate (27). Remove trunnion shaft (52), washers (48) and swashplate.

Note position of charge pump housing (38), remove the four cap screws and housing. Do not misplace serial number plate. Remove outer rotor gear (37) and rotor (41). Remove pin (42), snap ring (28), pump shaft (56) and bearing (29).

Clean and inspect all parts. Renew all parts showing excessive wear or damage. Piston retainers (8 and 59) must be flat. Pistons and piston bores in cylinder blocks must be free of scores or damage and outer edge of piston slipper must not be rounded more than 1/32 inch (0.79 mm). Lubrication hole in slipper face must be open. Minimum slipper thickness is 0.121 inch (3.073 mm) for both motor and pump. Slipper thickness must not vary more than 0.002 inch (0.051 mm) for all pistons. Piston and cylinder block assemblies are serviced as assemblies only.

Check pump valve plate (65) and motor valve plate (14) for excessive wear and damage. If wear or scratches can be felt by running fingernail across face of plates, renew plates. Pump valve plate has two notches and motor plate has four notches. See Fig. JD41.

Reassemble by reversing disassembly procedure. Lubricate all parts with clean hydraulic fluid during reassembly. Renew all gaskets, seals and "O" rings.

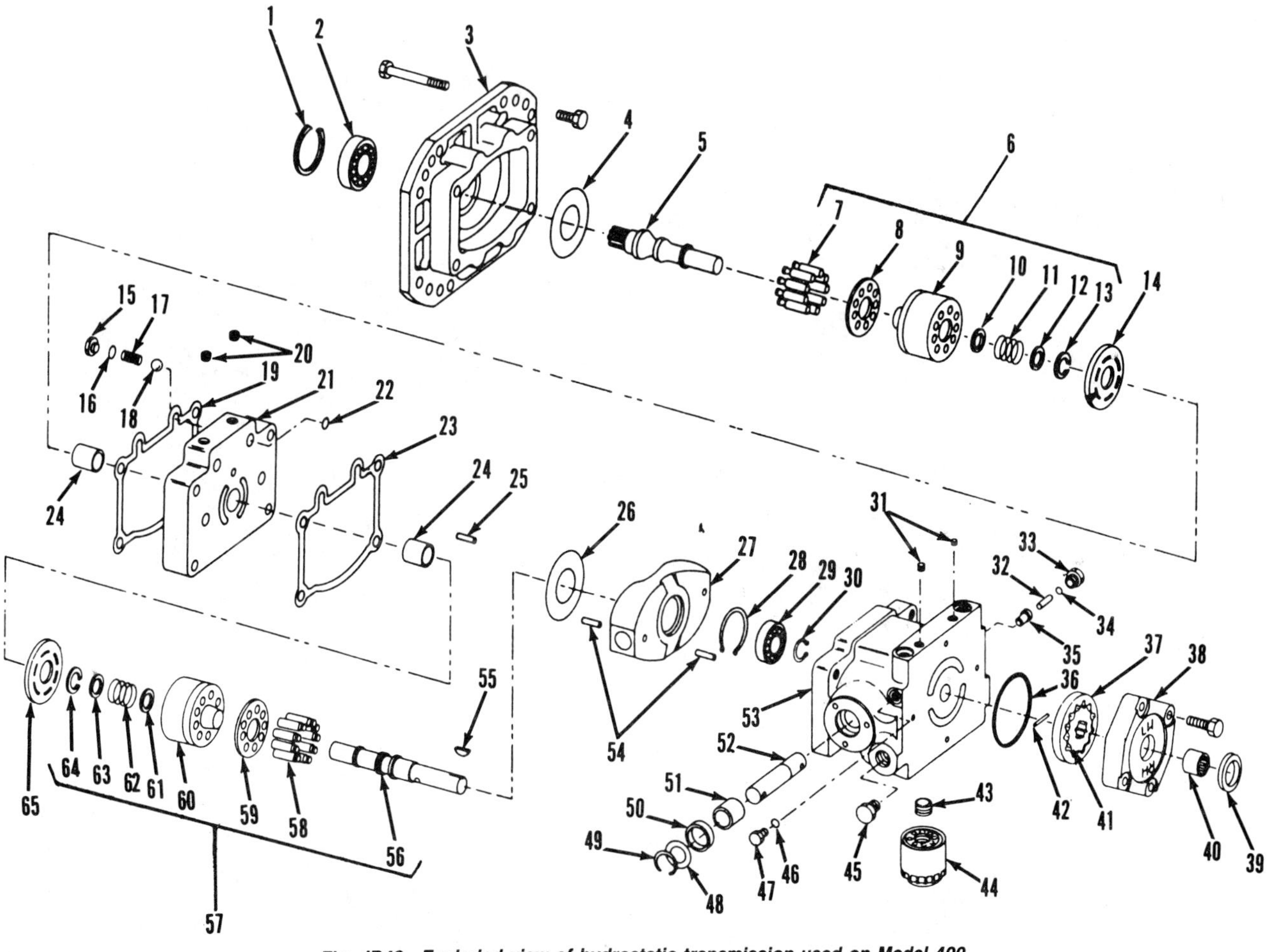

Fig. JD42—Exploded view of hydrostatic transmission used on Model 400.

1. Retaining ring
2. Bearing
3. Motor housing
4. Thrust plate
5. Motor shaft
6. Motor cylinder block assy.
7. Pistons (9)
8. Slipper retainer
9. Cylinder block
10. Washer
11. Spring
12. Washer
13. Snap ring
14. Valve plate (motor)
15. Check valve plug (2)
16. "O" ring
17. Spring (2)
18. Ball (2)
19. Gasket
20. Pipe plug
21. Center section
22. "O" ring (2)
23. Gasket
24. Bearing (2)
25. Pin (2)
26. Thrust plate
27. Swashplate
28. Retaining ring
29. Bearing
30. Retaining ring
31. Pipe plugs
32. Plug (service only)
33. Plug
34. "O" ring
35. Cone
36. "O" ring
37. Stator
38. Charge pump housing
39. Seal
40. Bearing
41. Rotor
42. Pin
43. Oil filter fitting
44. Oil filter
45. Plug
46. "O" ring
47. Plug
48. Washer (2)
49. Retaining ring (2)
50. Seal (2)
51. Bearing (2)
52. Trunnion shaft (2)
53. Pump housing
54. Pins (2)
55. Key
56. Pump shaft
57. Pump cylinder block assy.
58. Pistons (9)
59. Slipper retainer
60. Cylinder block (pump)
61. Washer
62. Spring
63. Washer
64. Retaining ring
65. Valve plate (pump)

DIFFERENTIALS, REDUCTION GEARS AND RANGE TRANSMISSIONS

R&R AND OVERHAUL

Models Early 316-317

To remove reduction gear and differential assembly, raise and support rear of tractor. Remove fender and deck assembly and disconnect drive shaft at rear flex coupling. Remove transmission fan shield and fan. Disconnect pto shaft extension, if so equipped. Disconnect transmission control linkage at control cam and brake rods at brake lever. Drain fluid from differential housing, disconnect all hydraulic lines and cap all openings. Support reduction gear and differential assembly, remove all mounting bolts and lower assembly and roll it away from tractor. Separate hydrostatic drive unit from reduction gear and differential assembly as needed.

To disassemble differential and reduction gear assembly, remove tires and wheels. Remove brake drums, brake shoes and backing plates. Remove seal retainer (6—Fig. JD43), gaskets (7) and bearing retainer (8). Withdraw axle (11) and bearing (9).

NOTE: A ring is epoxyed to bearing (10) to prevent bearing cone (9) from separating from bearing cup. See Fig. JD44. If ring separates from bearing cup when bearing cone is removed, bearing cup must be removed separately. Press bearing cone off axle shaft.

Remove cap screws and separate front housing (28—Fig. JD43) from rear housing (2). Mark differential case bearing caps to aid in reassembly and remove caps. Remove differential case (15) from housing. It may be necessary to pry case from housing using two wooden handles. Drive pinion pin (16) out of pinion shaft (19) and drive pinion shaft from differential case (15). Remove pinion gears (18) and thrust washers (17). Remove side gears (20) and thrust washers (21). If bearing renewal is required, pull case bearings (13) from case. Unscrew cap screws and drive ring gear (22) off case.

To remove pinion and spur gears, remove expansion plug (34) from front differential housing. Remove snap ring (33) and shim (32). Press pinion gear (23) out of housing. Before applying pressure to pinion shaft, remove side cover of housing and insert 1/8-inch (3 mm) steel

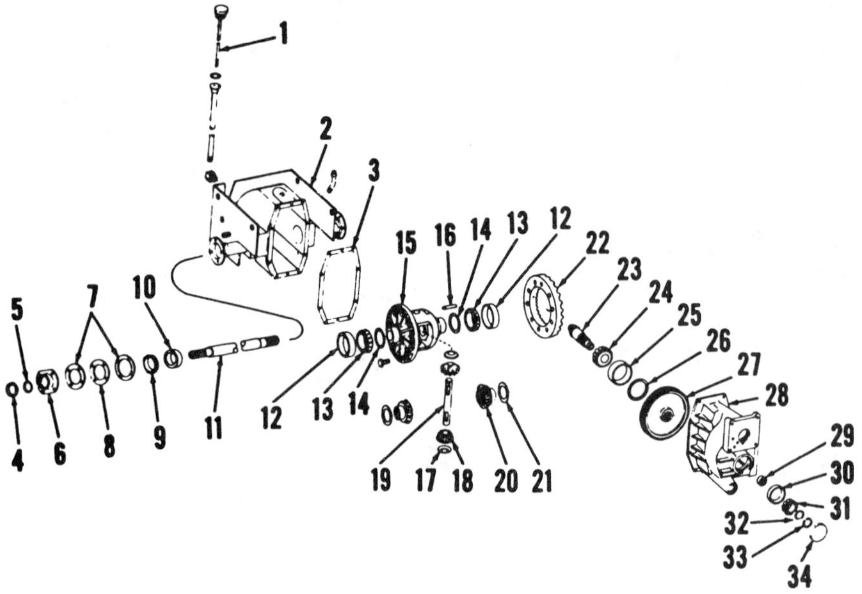

Fig. JD43—Exploded view of differential and reduction gear assembly used on early Model 316 and Model 317.

1. Dipstick
2. Rear housing & mounting bracket
3. Gasket
4. Felt seal
5. "O" ring
6. Seal retainer
7. Gaskets
8. Bearing retainer
9. Bearing cone
10. Bearing cup
11. Axle shaft
12. Bearing cup
13. Bearing cone
14. Shims
15. Case
16. Pin
17. Thrust washer
18. Pinion gear
19. Pinion shaft
20. Side gear
21. Thrust washer
22. Ring gear
23. Drive pinion gear & shaft
24. Bearing cone
25. Bearing cup
26. Shims
27. Spur gear
28. Front housing
29. Spacer
30. Bearing cup
31. Bearing cone
32. Shim
33. Snap ring
34. Expansion plug

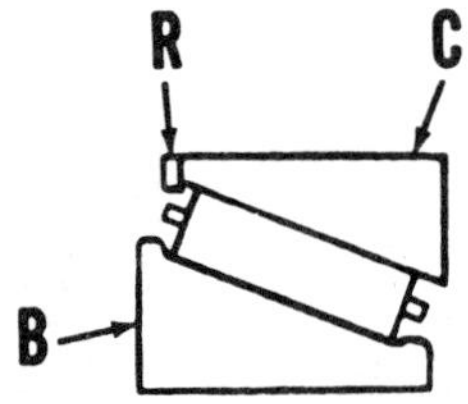

Fig. JD44—Axle bearing cone (B) is retained in cup (C) by ring (R) that is epoxyed to edge of cup.

Fig. JD45—Spur gear should be supported as shown to prevent housing damage when pressing pinion gear shaft out.

spacer or suitable screwdriver blade under edge of spur gear (27) as shown in Fig. JD45. This will prevent spur gear from cocking and possibly cracking differential housing. Catch pinion gear after it is pressed free. Spur gear (27—Fig. JD43), spacer (29) and outer pinion bearing may now be removed. Press pinion gear out of bearing cone (24). Remove bearing cups (25 and 30) and shim (26) from housing.

Clean and inspect all parts for excessive wear and damage and renew as needed.

If new pinion gear (23), front housing (28) or bearing cups are installed, select correct shim pack (26) as follows: Install pinion gear with bearing cup and cone in housing without shims. Position tool JDST-10 in bearing cradles as shown in Fig. JD46 and measure distance between tool depth pin and face of pinion gear. Measured distance will be thickness of shim pack (26—Fig. JD43) to be installed. Shims are available in a variety of sizes. Remove pinion gear, bearing cone and cup and install shims. Press outer pinion bearing (31) on pinion gear shaft (23) until there is a slight drag felt when pinion is turned by hand. Install thickest shim (32) that will allow installation of snap ring (33).

If differential case (15) is renewed, install a 0.020 inch shim pack (14) under each differential bearing (13). If differential case is not renewed, reinstall shims that were removed during disassembly. Tighten ring gear cap screws to 50-55 ft.-lbs. (68-75 N·m). Be sure to install bearing caps in their original positions. Tighten bearing cap retaining bolts to 40-45 ft.-lbs. (54-61 N·m).

Gear backlash between ring and pinion gears should be 0.003-0.007 inch (0.076-0.178 mm). Transfer shims (14) from one side of differential case to the other until correct backlash is obtained. Paint ring and pinion gears with a suitable gear pattern compound and check mesh position of gears. Adjust thickness or number of shims (14 or 26) to obtain correct mesh position shown in Fig. JD47.

Assemble bearing cone, bearing cup and ring on axle before installing axle. Cement ring to bearing cup as shown in Fig. JD44 using an epoxy adhesive. If bearing protrudes slightly, use more than one gasket between end of axle tube and bearing retainer. Make sure hub seal is aligned with axle, then tighten backing plate retaining nuts to 15 ft.-lbs. (20 N·m). Reinstall brake assemblies. Reassemble by reversing disassembly procedure. Fill reduction gear and differential assembly to correct level with specified fluid and adjust transmission linkage and brakes as previously outlined.

Models Later 316-318-322-330-332

Refer to R&R HYDROSTATIC DRIVE UNIT paragraph in TRANSMISSION section for removal and installation procedures.

To disassemble reduction gear and differential assembly, remove hitch plate and the six bolts securing each axle housing to differential housing. Separate differential housing and axle housings. Remove cap screws and separate case cover (12—Fig. JD48) from

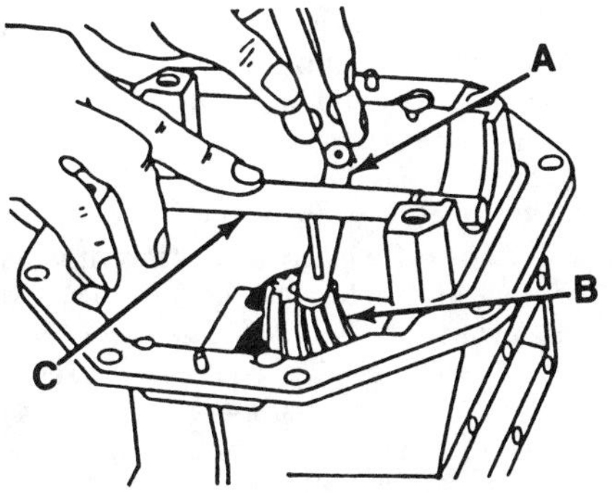

Fig. JD46—To determine thickness of pinion gear shim pack, measure as shown and refer to text.

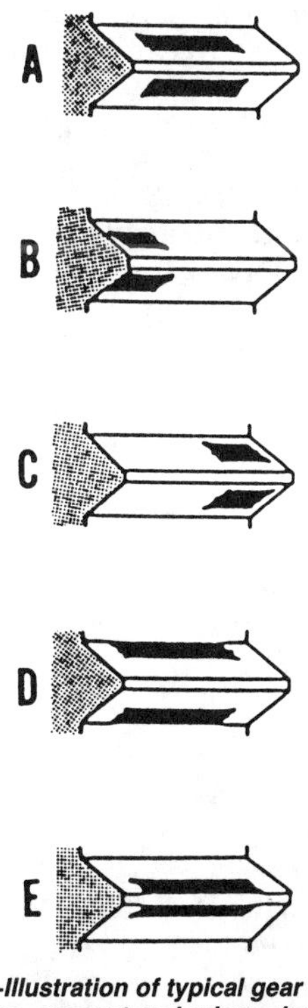

Fig. JD47—Illustration of typical gear tooth contact patterns encountered when checking ring gear and pinion. "A" pattern desired; "B" too close to toe; "C" too close to heel; "D" contact too low; "E" contact too high.

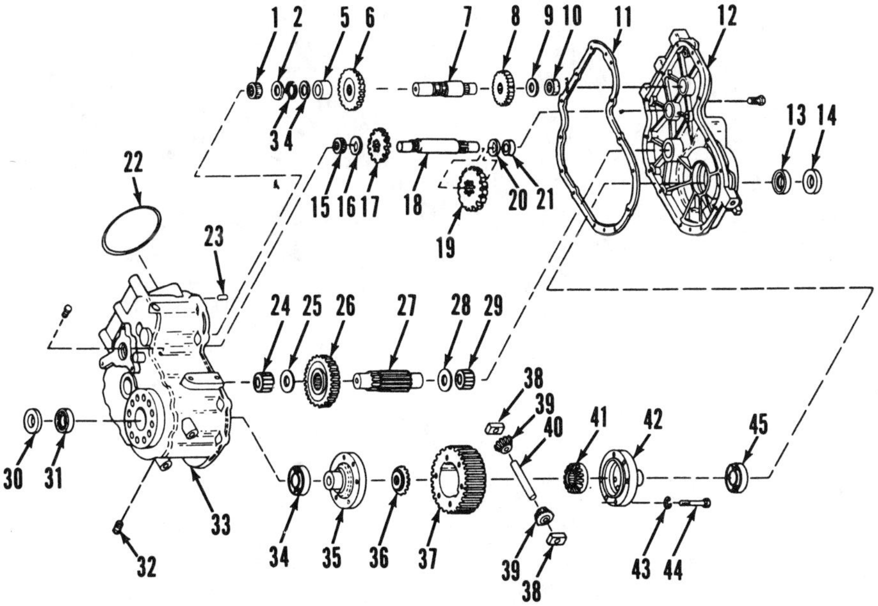

Fig. JD48—Exploded view of reduction gear and differential assembly used on later Model 316 and Models 318, 322, 330 and 332.

1. Bearing
2. Thrust washer
3. Thrust bearing
4. Thrust washer
5. Spacer
6. Bevel gear
7. Countershaft
8. Spur gear
9. Thrust washer
10. Bearing
11. Gasket
12. Cover
13. Seal
14. Washer
15. Bearing
16. Thrust bearing
17. Spur gear
18. Intermediate shaft
19. Intermediate gear
20. Thrust washer
21. Bearing
22. "O" ring
23. Dowel pin
24. Bearing
25. Thrust washer
26. Output gear
27. Output shaft
28. Thrust washer
29. Bearing
30. Washer
31. Seal
32. Drain plug
33. Case
34. Bearing
35. Differential carrier
36. Axle gear
37. Ring gear
38. Pinion blocks
39. Pinion gears
40. Cross shaft
41. Axle gear
42. Differential gear
43. Washer
44. Bolt
45. Bearing

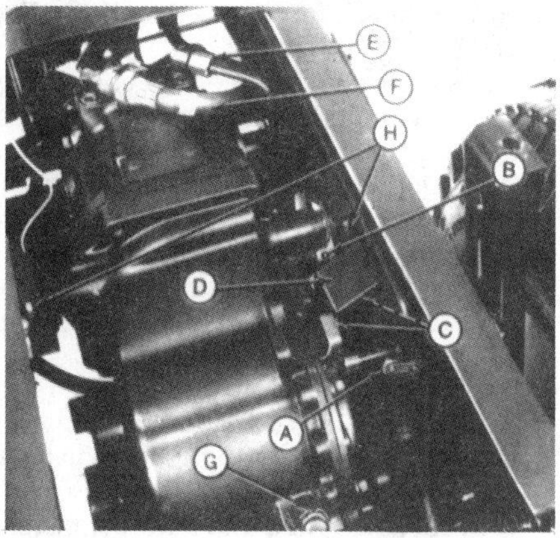

Fig. JD49—View of hydrostatic drive unit and range transmission and differential assembly with fender/deck assembly removed.

A. Coupler
B. Pin
C. Cap screws
D. Control rod bracket
E. Supply tube
F. Return hose
G. Vent
H. Cap screws

case (33). Remove countershaft (7) and gear assemblies, ring gear (37) and differential assembly and output shaft (27) and gear assemblies. Remove cap screws (44) and disassemble differential as needed.

Inspect all parts for wear, looseness and damage and renew as needed.

Reassemble by reversing disassembly procedure. Renew all gaskets and seals. Apply Loctite to cap screws (44) and tighten to 35-40 ft.-lbs. (47-54 N·m).

Model 400

Remove drain plug and drain fluid. Remove depth control knob and jam nut and two-speed range transmission shift knob. Remove fuel tank cap. Disconnect all interfering electrical wiring, remove all attaching bolts and remove fender/deck assembly. Disconnect fuel lines, drain fuel tank and remove tank. Disconnect coupler (A—Fig. JD49) and remove pin (B). Remove cap screws (C and control rod bracket (D). Disconnect hydraulic supply tube (E) and remove return hose (F). Remove vent tube (G), if so equipped, and remove cap screws (H). Remove bottom screen and fan guard. Disconnect all interfering hydraulic lines and cap openings. Disconnect transmission control linkage and brake rods. Place jackstands under each side of frame and support front of transmission. Remove rockshaft cylinder. Remove all transmission retaining bolts and roll transmission away from tractor.

Remove hitch plate, wheels and tires, and axle housing. Separate hydrostatic drive unit from range transmission and differential assembly as needed.

To disassemble range transmission and differential assembly, remove brake caliper assembly and brake disc. Thoroughly clean exterior of unit.

Remove axle housings (15—Fig. JD50). Position unit with cover up, then remove cover (13). Lift out differential assembly and axles (40 through 54). Remove output shaft (36), gear (35) and thrust washers (37). Unscrew setscrew (2) and remove spring (3) and ball (4). Remove brake shaft (32), sliding gear (33), shift fork (10) and rod (8). Remove input shaft and gear components (18 through 25).

To disassemble differential assembly, remove cap screws (53) and remove differential carriers (41 and 52) and axles from ring gear assembly. Remove snap ring to separate axle gear, carrier and axle. Remove pinions (47) and separate body cores (45 and 48) from ring gear (46).

Inspect components for damage and excessive wear. To reassemble unit, reverse disassembly procedure. Check movement of shift rod when tightening setscrew (2). Install gears (22 and 25) so bevels of gears face together as shown in Fig. JD51. Install carrier cap screws

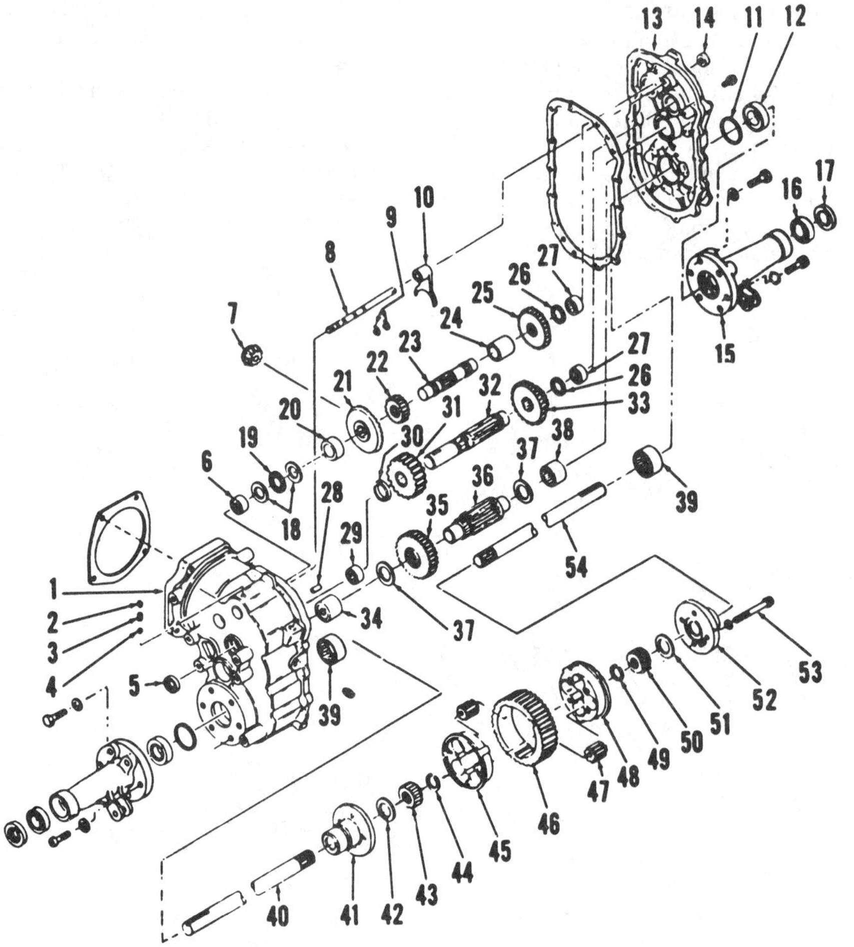

Fig. JD50—Exploded view of two-speed range transmission and differential unit used on Model 400.

1. Case
2. Setscrew
3. Spring
4. Ball
5. Seal
6. Needle bearing
7. Transmission output gear
8. Shift rail
9. Snap rings
10. Shift fork
11. Quad ring
12. Tapered roller bearing
13. Cover
14. Seal
15. Axle housing
16. Ball bearing
17. Seal
18. Thrust washers
19. Thrust bearing
20. Spacer
21. Bevel gear
22. Gear (low range)
23. Shaft
24. Spacer
25. Gear (high range)
26. Thrust washer
27. Needle bearing
28. Dowel pin
29. Needle bearing
30. Spacer
31. Gear
32. Brake shaft
33. Sliding gears
34. Needle bearing
35. Gear
36. Output shaft
37. Thrust washer
38. Needle bearing
39. Needle bearing
40. Axle shaft
41. Differential carrier
42. Thrust washer
43. Axle gear
44. Snap ring
45. Body core
46. Ring gear
47. Pinion gears
48. Body core
49. Snap ring
50. Axle gear
51. Thrust washer
52. Differential carrier
53. Cap screw
54. Axle shaft

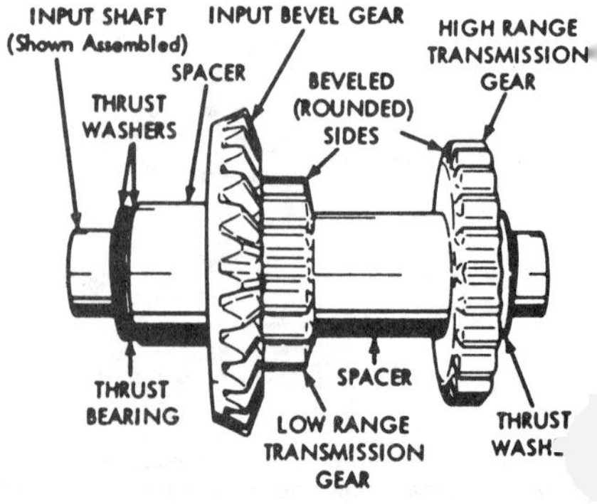

Fig. JD51—View showing correct installation positions for gear, spacer, thrust washers and thrust bearing on input shaft for 400 model. Refer to text.

(53—Fig. JD50) so head of cap screw is on side of shorter carrier (52). Do not rotate axle housings after housing has been pressed tight against seal (11) as seal may be cut.

Reinstall by reversing removal procedure.

Models 420-430

Refer to R&R HYDROSTATIC DRIVE UNIT paragraph in TRANSMISSION section for removal and installation procedure.

To disassemble reduction gear and differential assembly, remove hitch plate and remove the six bolts securing each axle housing to differential housing. Separate differential housing and axle housings. Remove cap screws and separate cover (13—Fig. JD52) from case (39). Remove differential, locking shifter rod, thrust washers and high range driven gear as an assembly. Remove bolt (40), spring (41) and ball (42). Remove shift rod (19), shift fork (20), sliding gear (22) and shaft (21) as an assembly. Remove output shaft (49), thrust washers (47 and 50) and gear (48) as an assembly. Disassemble gear and shaft assemblies and differential unit as needed.

Inspect all parts for wear, looseness or damage and renew as needed.

Reassemble by reversing disassembly procedure. Renew all gaskets and seals. Apply Loctite to carrier (60) cap screws and tighten to 35-40 ft.-lbs. (47-54 N·m).

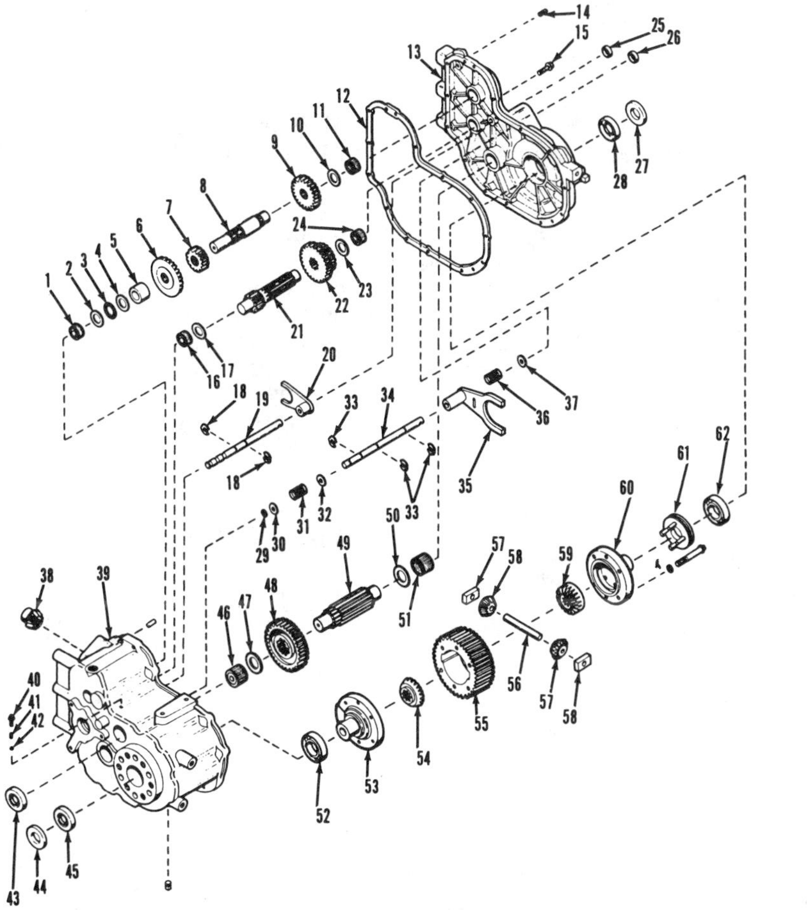

Fig. JD52—Exploded view of range transmission and differential assembly with locking differential used on Models 420 and 430.

1. Bearing
2. Thrust washer
3. Thrust bearing
4. Thrust washer
5. Spacer
6. Bevel gear
7. Spur gear
8. Countershaft
9. Spur gear
10. Thrust washer
11. Bearing
12. Gasket
13. Cover
14. Plug
15. Cap screw
16. Bearing
17. Thrust washer
18. Snap rings
19. Shift rod (range transmission)
20. Shift fork
21. Shaft & gear
22. Sliding gear
23. Thrust washer
24. Bearing
25. Seal (as equipped)
26. Seal (as equipped)
27. Spacer
28. Seal
29. Retaining ring
30. Thrust washer
31. Spring
32. Thrust washer
33. Snap rings
34. Shift rod (differential lock)
35. Shift fork
36. Spring
37. Thrust washer
38. Pinion gear (drive)
39. Case
40. Bolt
41. Detent spring
42. Detent ball
43. Seal (as equipped)
44. Spacer
45. Seal
46. Bearing
47. Thrust washer
48. Output gear
49. Output shaft
50. Thrust washer
51. Bearing
52. Bearing
53. Carrier
54. Axle gear
55. Ring gear
56. Cross shaft
57. Pinion gears
58. Pinion blocks
59. Axle gear
60. Carrier
61. Locking collar
62. Bearing

POWER TAKE-OFF

ADJUSTMENT

Models Early 316-317

Adjust electric pto clutch by turning each of the four adjustment nuts (D—Fig. JD53) until there is 0.012-0.014 inch (0.31-0.35 mm) clearance between each adjustment nut and brake plate (F).

Models Later 316-318-322-330-332-420-430

Adjust electric pto clutch by inserting a 0.018 inch (0.46 mm) feeler gage between rotor and armature through the slots in brake plate (Fig. JD54) and turning each adjusting nut (four adjusting nuts on early models and three adjusting nuts on late models) until feeler gage can just be removed. Adjust all nuts equally.

Model 400

Engage pto and measure total length of spring (23—Fig. JD55). Spring should

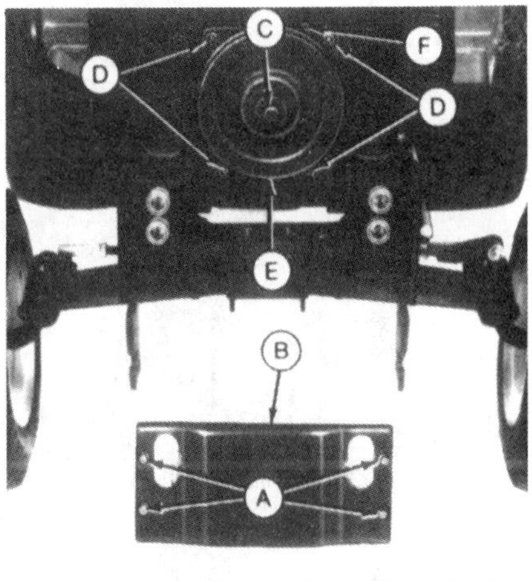

Fig. JD53—View of electric pto clutch installed on early Model 316 and Model 317.

A. Cap screws
B. Front bumper
C. Cap screw
D. Adjusting nuts
E. Pulley & rotor
F. Brake plate

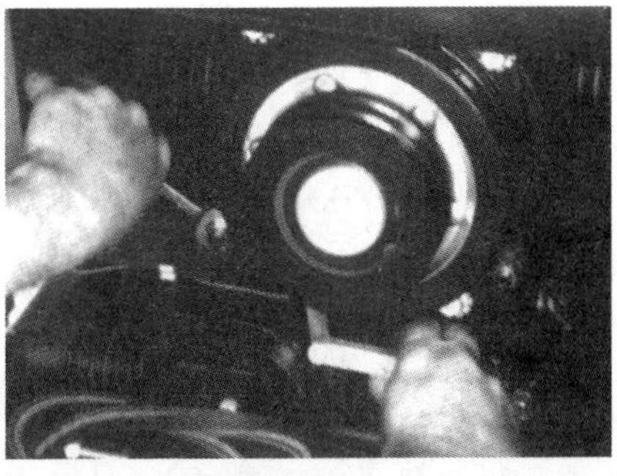

Fig. JD54—Use a 0.018 inch (0.46 mm) feeler gage to check electric pto clutch adjustment on later Model 316 and Models 318, 322, 330, 332, 420 and 430. Refer to text.

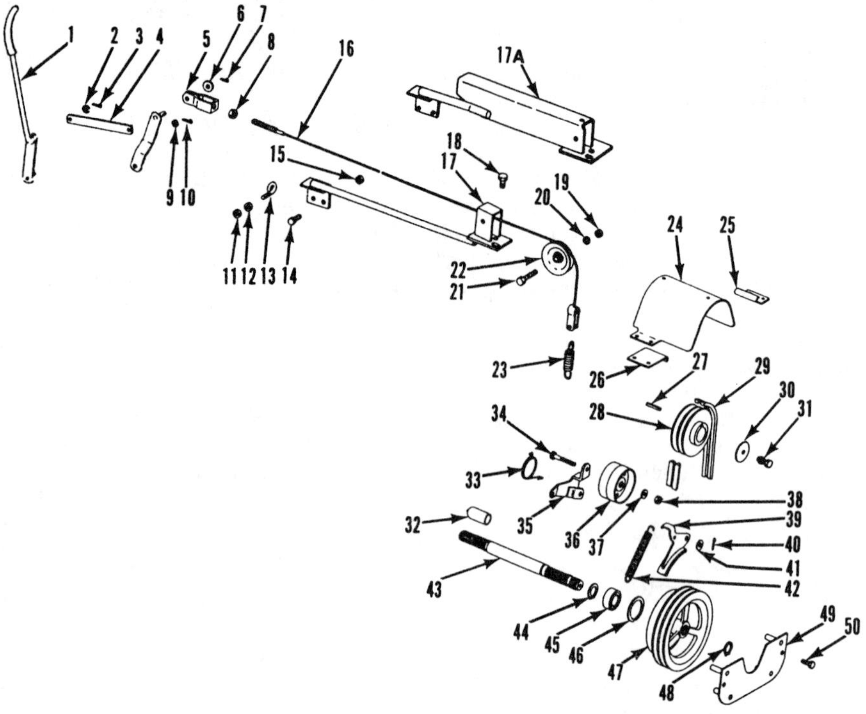

Fig. JD55—Exploded view of mechanical pto assembly used on Model 400.

1. Lever
2. Washer
3. Cotter pin
4. Link
5. Clevis
6. Washer
7. Cotter pin
8. Jam nut
9. Washer
10. Cotter pin
11. Jam nut
12. Jam nut
13. Cable guide
14. Cap screw
15. Nut
16. Cable
17. Idler support (prior to serial number 55,001)

17A. Idler support (after serial number 55,000)

18. Cap screw
19. Nut
20. Washer
21. Bolt
22. Cable pulley
23. Cable spring
24. Upper pto shield
25. Left belt guide
26. Right belt guide
27. Key
28. Drive sheave
29. Drive belts (pto)
30. Washer
31. Cap screw
32. Cover
33. Return spring
34. Bolt
35. Tension arm
36. Idler pulley
37. Washer
38. Nut
39. Brake (pto)
40. Cotter pin
41. Washer
42. Brake spring (pto)
43. Shaft
44. Snap ring
45. Bearing
46. Snap ring
47. Driven sheave
48. Snap ring
49. Shield
50. Cap screw

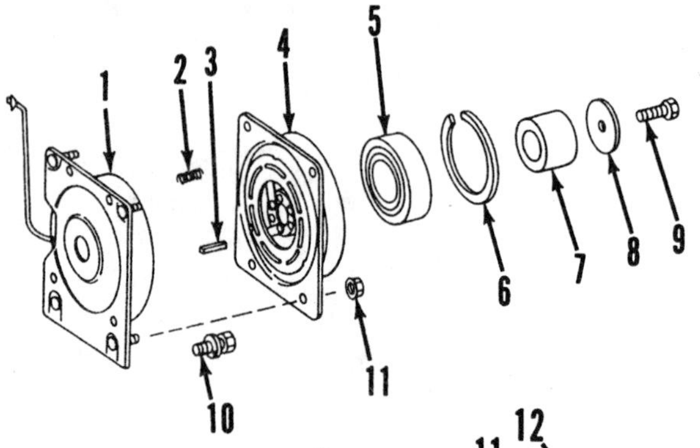

Fig. JD56—Exploded view of Warner electric pto clutch used on some models.

1. Coil
2. Spring
3. Key
4. Rotor & pulley
5. Bearing
6. Snap ring
7. Bushing
8. Washer
9. Cap screw
10. Cap screw
11. Adjustment nut

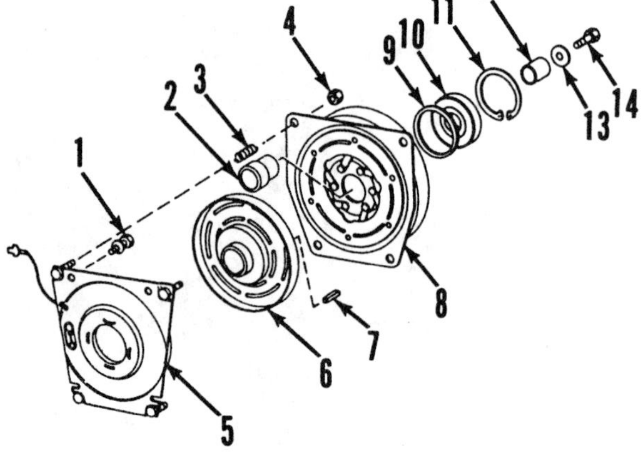

Fig. JD57—Exploded view of Ogura electric pto clutch used on some models.

1. Cap screw
2. Bearing hub
3. Spring
4. Adjustment nut
5. Coil
6. Rotor
7. Key
8. Armature & pulley
9. Spacer
10. Bearing
11. Snap ring
12. Bushing
13. Washer
14. Cap screw

measure $3^{3}/_{4}$ to 4 inches (92.25 to 107.95 mm). If not, disengage pto, remove cotter pin (7) and washer (6). Disconnect clevis (5), loosen locknut (8) and thread clevis onto pto cable to increase spring length or off pto cable to decrease spring length. Reassemble cable and clevis , engage pto and recheck spring (23) length.

R&R PTO ASSEMBLY

Models Early 316-317

To remove electric pto clutch, first remove pto drive belt. Remove the four cap screws (A—Fig. JD53) and remove front bumper (B). Disconnect pto clutch electrical lead, Remove cap screw (C), adjusting nuts (D) and slide assembly off crankshaft. Remove key.

Reinstall by reversing removal procedure.

Models Later 316-318-322-330-332-420-430

Remove side panels, grille, cover and hood. Remove pto drive belt. Remove adjustment nuts (4—Fig. JD57). remove cap screw (14) and washer (13). Disconnect clutch electric lead and remove pulley assembly (8) from crankshaft. Remove springs (3), rotor (6) and key (7). Remove cap screws (1) and coil (5).

Reinstall by reversing removal procedure.

Model 400

Remove grille and pto shield (24—Fig. JD55), left and right belt guides (25 and 26) and lower shield (49). Remove pto drive belts (29), cap screw (31) and washer (30), then use a suitable puller to remove drive sheave (28). Engage pto lever to relieve pto brake tension, then remove snap ring (48) and driven sheave (47). Remove snap ring (46) and drive pto shaft (43) out front of tractor. Bearing (45) is a press fit on shaft (43). Remove spring (23) and brake spring (42). Remove cotter pin (40), washer (41) and brake (39). Remove belt tension arm (35) and return spring (33). Remove nut (38), washer (37) and idler pulley (36) from tension arm (35).

Inspect all parts for wear or damage. Renew loose or rough bearings. Brake spring (43) free length is 3.315 to 3.565 inches (84.2 to 90.6 mm) and pto cable spring (23) free length is 2.65 to 2.85 inches (67.3 to 72.4 mm).

Reassemble by reversing disassembly procedure.

OVERHAUL ELECTRIC PTO CLUTCH

All Models

A variety of electric pto clutches have been used, however, overhaul procedure is similar for all models.

Remove snap ring (11—Fig. JD57) and carefully press bearing (10) out of bearing bore. Remove shims (9). Press hub (2) from bearing (10).

Windings in coil assembly (5) may be checked by connecting lead wire to positive (+) battery post and touching metal flange of coil assembly to negative (−) battery post. A strong electromagnetic field should be present at face of coil.

Reassemble by reversing disassembly procedure.

CONTROL VALVE

All Models

A variety of multi-spool control valves from various manufacturers have been used according to model, optional equipment or serial number. Refer to Fig. JD60 for an exploded view of a typical three-spool control valve.

Before removing control valve from tractor, mark all hydraulic lines to assure correct installation during reassembly. Cap all openings.

To disassemble control valve, remove button plugs (1—Fig. JD60). Remove spool cap (5) and carefully remove detent balls (6) and spring (7). Remove spool (8), spool caps (2) and spools (24).

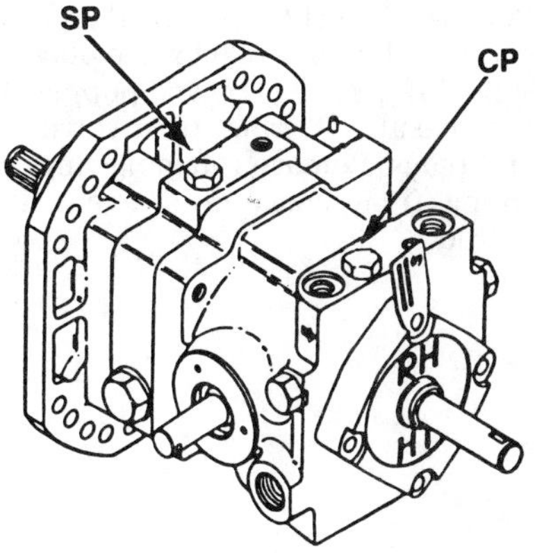

Fig. JD58—View showing location of test ports in hydrostatic drive unit used on 400 model.

CP. Charge pump pressure test port
SP. System pressure test port

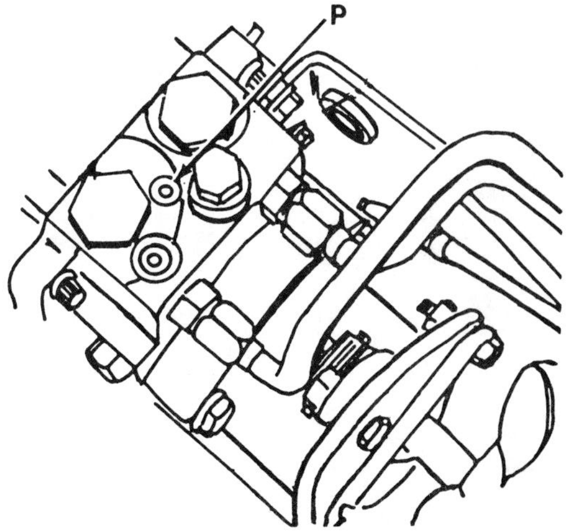

Fig. JD59—View showing location of test port (P) on hydrostatic drive unit used on all models except Model 400.

HYDRAULIC SYSTEM

MAINTENANCE

All Models

The hydrostatic drive unit supplies pressurized oil for hydraulic system functions. Refer to appropriate TRANSMISSION section for maintenance and service information for hydrostatic drive unit.

PRESSURE TESTS

CHARGE PUMP PRESSURE TEST. Install a 0-1000 psi (0-7000 kPa) pressure gage in port (CP—Fig. JD58 for Model 400, or P—Fig. JD59 on all other models). Start and run engine at slow idle. Charge pump pressure must be 90-180 psi (620-1240 kPa). If low, be sure transmission filter and charge pump relief valve are in good condition. On some models, relief valve pressure setting is adjustable using shims (10—Fig. JD40).

LIFT SYSTEM PRESSURE TEST. Install a 0-3000 psi (0-20,000 kPa) pressure gage in test port (SP—Fig. JD58 for Model 400, or P—Fig. JD59 on all other models). Start and run engine at fast idle. Operate hydraulic lever to obtain pressure reading. System pressure should be 650-750 psi (4485-5175 kPa) on early models and 850-950 psi (5865-6555 kPa) on later models. If low, be sure transmission filter and implement relief valve are in good condition. On some models, relief valve pressure setting is adjustable using shims (10—Fig. JD40).

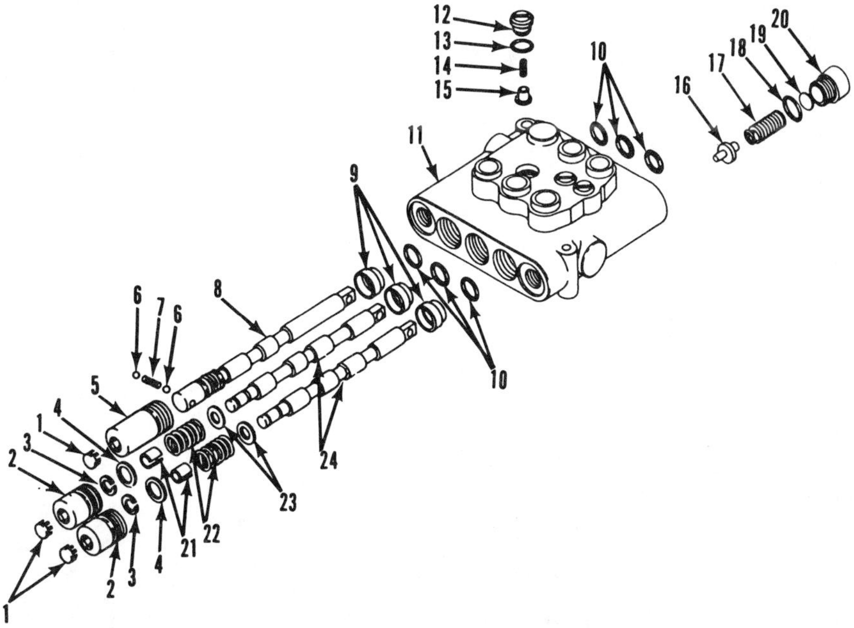

Fig. JD60—Exploded view of typical three-spool valve.

1. Button plugs
2. Spool caps
3. Retaining rings
4. Washers
5. Spool cap
6. Detent balls
7. Detent spring
8. Spool
9. Bushings
10. "O" rings
11. Valve body
12. Plug
13. "O" ring
14. Spring
15. Plunger
16. Poppet
17. Spring
18. "O" ring
19. Shim
20. Plug
21. Spacers
22. Springs
23. Washers
24. Spools

Remove plugs (12), spring (14) and plungers (15). Remove relief valve plug (20), shims (19), spring (17) and poppet (16). Remove all "O" rings from valve body and plugs. Remove retaining rings and springs from spools only if renewal is required.

Inspect all parts for wear and damage. Renew all "O" rings.

Reassemble by reversing disassembly procedure.

HYDRAULIC CYLINDER

All Models

Hydraulic cylinder used on all models is a welded assembly and no service parts are available. Renew cylinder if internal or external leakage is apparent.

JOHN DEERE

CONDENSED SPECIFICATIONS

MODEL 285

Engine Make . . . . Kawasaki
Model FD590V
Bore . . . . 74 mm (2.91 in.)
Stroke . . . . 68 mm (2.68 in.)
Displacement . . . . 585 cc (35.68 cu. in.)
Power Rating . . . . 13.4 kW (18 hp.)
Slow Idle . . . . 1400 rpm
High Speed (No-Load) . . . . 3350 rpm
Capacities—
Crankcase . . . . 1.9 L (2 qts.)
Hydrostatic Transmission . . . . 2.5 L (2.6 qts.)
Cooling System . . . . 3.1 L (3.3 qts.)
Fuel Tank . . . . 11.4 L (3.0 gals.)

FRONT AXLE AND STEERING SYSTEM

MAINTENANCE

The front wheel spindles should be lubricated after every 50 hours of operation. Jack up front axle so wheels are suspended and apply grease to spindles through grease fitting at each end of axle. One or two shots from a hand-held grease gun should be sufficient. Use a good quality, lithium based grease. Check for looseness and binding in front axle components.

R&R AXLE MAIN MEMBER

Raise and support front of tractor. Disconnect drag link end (14—Fig. JD101) from steering arm (13). Using a suitable jack, support axle assembly. Unscrew pivot pin nut (1). Unscrew pivot pin mounting screws and lower axle assembly out of tractor.

Inspect components for excessive wear and damage. Bushings in axle are renewable.

When reinstalling axle, tighten pivot pin nut (1) until axle end play is removed but axle still pivots freely.

TIE ROD AND TOE-IN

The tie rod is equipped with renewable ends that are not interchangeable.

Front wheel toe-in should be 3-6 mm ($^1/_8$-$^1/_4$ in.). Adjust toe-in by altering tie rod length.

STEERING SPINDLES

Raise and support side to be serviced. Remove wheel and tire. Disconnect tie rod end and, if needed, drag link end. If removing right spindle (7—Fig. JD101), detach snap ring securing top of spindle and withdraw spindle from axle. If removing left spindle (17), mark end of spindle and steering arm (13) so arm can be returned to original position. Remove clamp bolt, remove steering arm and withdraw spindle from axle.

Inspect components for excessive wear and damage. Renewable bushings (5) are located in axle.

DRAG LINK

A drag link with adjustable ends is located between steering arm and pitman arm. Adjust length of drag link so equal turns of steering wheel are required to reach full left and right turn from neutral position.

Steering effort and response can be adjusted by changing the connecting point of drag link in pitman arm. Connecting drag link to lower hole in pitman arm will provide quick steering response when tractor is not heavily loaded. Connecting drag link to upper hole in pitman arm will provide reduced steering effort when tractor is equipped with heavy loads on front axle, but steering response will be slower because more turns of steering wheel will be required to turn the front wheels.

FRONT WHEEL BEARINGS

Front wheel bearings are sealed and should not require lubrication. Wheel bearings are a press fit in wheel hub.

STEERING GEAR

R&R AND OVERHAUL. Detach drag link from pitman arm. Scribe match marks on pitman arm and end of pitman shaft, then remove pitman arm. Raise hood and remove lower pedestal panel. Disconnect steering column shaft from gearbox shaft. Unscrew fasteners securing gearbox and remove gearbox.

The steering gear is a recirculating ball type. Refer to Fig. JD102 for an exploded view of gearbox. To disassemble, remove side cover (3) and end cover (18) and withdraw pitman shaft (7) and wormshaft assembly (15). The wormshaft is available only as an assembly

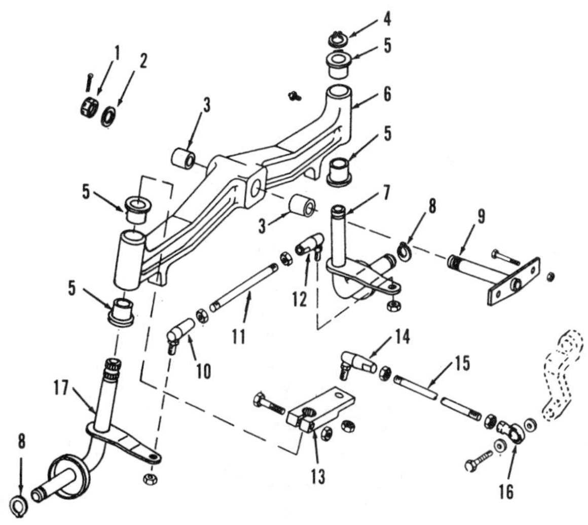

Fig. JD101—Exploded view of front axle assembly. Note that rod ends (12 and 14) have left-hand threads.

1. Slotted nut
2. Washer
3. Bushing
4. Snap ring
5. Bushing
6. Axle main member
7. Spindle (R.H.)
8. Snap ring
9. Pivot
10. Tie rod end
11. Tie rod
12. Tie rod end
13. Steering arm
14. Drag link end
15. Drag link
16. Drag link end
17. Spindle (L.H.)

and no attempt should be made to disassemble the unit.

When assembling gearbox, perform the following adjustments: Back out adjusting screw (6) until it stops. Turn wormshaft (15) in (clockwise) until stopped, then turn back one-half turn. Install needed shims (17) so rolling torque when turning steering shaft is 0.4 N·m (3.5 in.-lbs.). Install thickest shim (5) allowed when assembling unit and make sure pitman shaft (7) is centered on ball nut when nut is at midpoint of travel. Turn adjusting screw (6) so steering shaft rolling torque increases to 1.1 N·m (9.5 in.-lbs.), then tighten locknut (1) while holding adjusting screw in place.

Gearbox oil capacity is 170 mL (6 oz.). Recommended oil is SAE 90 gear oil rated GL-5. After installation, check oil level by removing plug from oil fill hole. Oil level should be 25 mm (1 in.) below top of plug hole.

ENGINE

MAINTENANCE

LUBRICATION. Check oil level before operation and after every four hours of operation. Determine oil level by resting dipstick on top of dipstick tube; do not screw in.

Manufacturer recommends oil with API classification SE or SF. Use SAE 30 oil for temperatures above 32° F (0° C); use SAE 10W-30 oil for temperatures between −4° F (−20° C) and 86° F (30° C); SAE 5W30 may be used between 50° F (10° C) and −31° F (−35° C).

Change oil and filter after first five hours of operation and then after every 100 hours of operation, or more frequently if operation is severe. Crankcase capacity is 1.9 L (2 qts.).

AIR FILTER. The air filter consists of a foam precleaner element and a dry paper filter element. Both elements should be cleaned after every 25 hours of operation or more frequently if operating conditions are severe.

Clean dry paper element with low pressure compressed air blown from the inside toward the outside. Clean foam precleaner element by washing in a solution of mild detergent and water. Allow to dry completely, then apply approximately 30 mL (1 oz.) of clean engine oil. Squeeze out excess oil and install.

COOLING SYSTEM. Drain, flush and refill cooling system annually. Manufacturer recommends low silicate antifreeze that does not contain stop leak additive. Cooling system capacity is 3.1 L (3.3 qts.).

A thermostat is used to maintain cooling system temperature. The thermostat should begin to open at 82° C (180° F) and open fully at 95° C (203° F).

SPARK PLUG. Recommended spark plug is NGK BMR6A. Spark plug gap should be 0.7 mm (0.028 in.). Tighten spark plug to 20 N·m (180 in.-lbs.).

FUEL FILTER. The fuel filter should be changed periodically based on cleanliness of fuel and operating conditions. Pay particular attention to clean fuel on fuel injected models due to problems that may occur if contaminated fuel is used. The fuel filter on carbureted engines is located in fuel hose adjacent to engine. The fuel filter on fuel injected engines is located at rear of tractor on a panel attached to rear of tractor frame.

CARBURETOR. Carbureted engines are equipped with the float type carburetor shown in Fig. JD103.

Adjust carburetor control cables as follows: Place engine throttle lever in fast position. Using a 15/64-inch drill bit, insert bit D—Fig. JD104) through hole in throttle control lever (A) and plate (B). Loosen throttle cable clamp screw, pull cable housing tight and retighten cable clamp screw. Rotate adjusting screw so there is a gap between screw and choke control lever (A), then turn back in until screw just touches lever. Remove drill bit. With throttle control lever in choke position, carburetor choke plate should be closed, if not, repeat procedure.

Initial setting of idle mixture screw (8—Fig. JD103) is 1 1/8 turns out from lightly seated position. To adjust idle mixture screw, turn idle speed screw (10) so engine runs at 1400 rpm. Run engine until it reaches normal operating temperature. Turn idle mixture screw (8) in until engine runs rough, then turn screw out until engine runs rough again. Turn screw to midpoint position (should be highest engine speed) then back out 1/4 turn. Adjust idle stop screw so engine idles at 1400 rpm.

High speed mixture is controlled by main jet (19). Standard main jet is 112.5 and should provide satisfactory operation up to 3000 feet. Main jet 110.0 is available for use from 3000 to 7000 feet, while main jet 107.5 is designed for operation above 7000 feet.

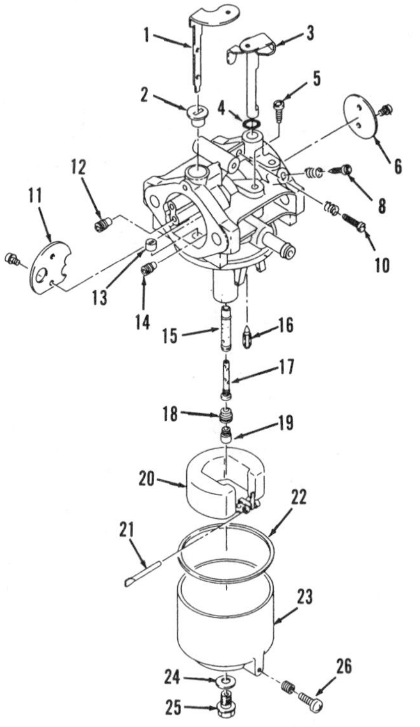

Fig. JD103—Exploded view of carburetor.

1. Choke shaft
2. Bushing
3. Throttle shaft
4. Seal
5. Idle jet
6. Throttle plate
8. Idle mixture screw
10. Idle speed screw
11. Choke plate
12. Air jet
13. Bushing
14. Air jet
15. Nozzle
16. Fuel inlet valve
17. Nozzle
18. Jet holder
19. Main jet
20. Float
21. Pin
22. Gasket
23. Fuel bowl
24. Gasket
25. Screw
27. Drain screw

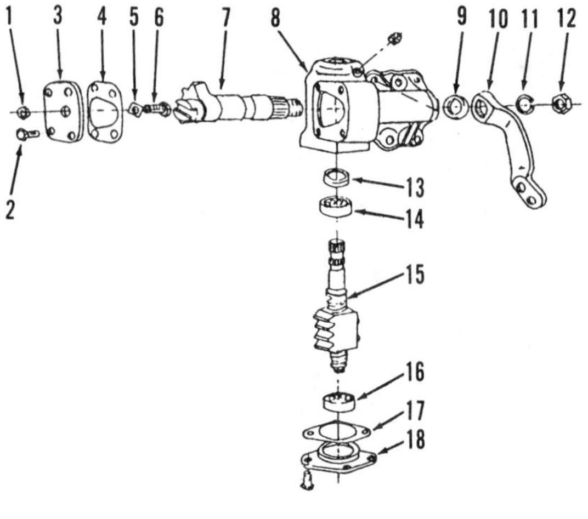

Fig. JD102—Exploded view of steering gearbox.

1. Nut
2. Screw
3. Cover
4. Gasket
5. Shim
6. Adjuster
7. Pitman shaft
8. Gearbox
9. Seal
10. Pitman arm
11. Lockwasher
12. Nut
13. Seal
14. Bearing
15. Shaft & nut assy.
16. Bearing
17. Shim
18. Cover

Overhaul of carburetor is evident after inspection and referral to Fig. JD103. Do not use drill bits or wire to clean carburetor jets or fuel passages as calibration of carburetor will be affected if these openings are enlarged. Float should be parallel with carburetor float bowl mating surface when carburetor is inverted. Bend float tab to adjust float level, being careful not to push on inlet needle valve.

GOVERNOR. To adjust governor linkage, first check throttle cable adjustment as outlined in CARBURETOR section. Place throttle control lever in fast position. Loosen governor lever clamp nut (N—Fig. JD104), hold lever and turn governor shaft (S) as far as possible counterclockwise. Retighten nut.

Maximum no-load engine speed should be 3250-3450 rpm on carbureted engines and is adjusted as follows: Run engine until normal operating temperature is reached. Align holes in throttle lever (A) and plate (B) and insert a 15/64 inch drill bit (D). Run engine with transmission in neutral and determine engine speed using an accurate tachometer. If engine speed is not 3250-3450 rpm, loosen plate retaining screws and reposition plate (B) to obtain desired engine speed. Retighten screws and recheck engine speed. Check choke operation as outlined in CARBURETOR section.

IGNITION SYSTEM. A breakerless, solid-state ignition system is used. Ignition timing is not adjustable. If malfunction is suspected, refer to REPAIRS section for service procedure.

VALVE ADJUSTMENT. The engine must be cold when checking valve stem end clearance. Remove rocker arm cover and unscrew spark plug for cylinder being checked. Rotate crankshaft so piston is at top dead center on compression stroke. Clearance between valve stem and rocker arm should be 0.25 mm (0.010 in.). Loosen jam nut (N—Fig. JD105) and turn adjusting screw (W) to obtain desired clearance. Tighten jam nut to 9 N·m (79 in.-lbs.) and recheck clearance.

REPAIRS

TIGHTENING TORQUES. Recommended tightening torque specifications are as follows:

Connecting rod	21 N·m (186 in.-lbs.)
Crankcase cover	21 N·m (186 in.-lbs.)
Cylinder head	21 N·m (186 in.-lbs.)
Flywheel	108 N·m (79 ft.-lbs.)
Rocker arm	9 N·m (79 in.-lbs.)
Spark plug	20 N·m (180 in.-lbs.)

IGNITION SYSTEM. The ignition system consists of an ignition module, two pulsar coils and two ignition coils. Voltage is provided by the 12 volt battery circuit. A safety switch operated by weight on the tractor seat disconnects current to the ignition system if the operator leaves the seat while the engine is running.

Refer to Fig. JD106 for a wiring dia-

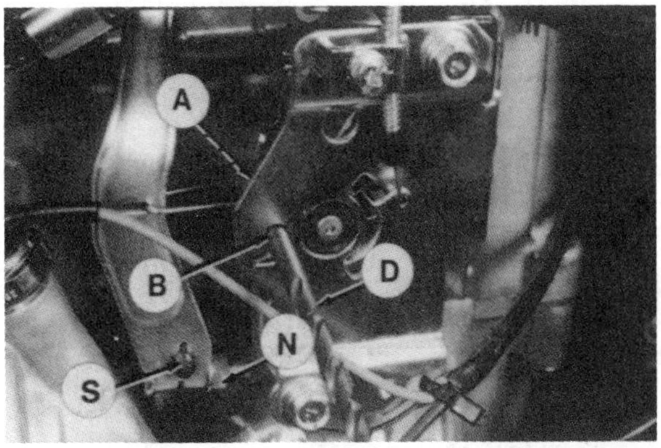

Fig. JD104—To adjust governed speed, place throttle control lever in fast position. Loosen governor lever clamp nut (N), hold lever and turn governor shaft (S) as far as possible counterclockwise. Insert drill bit (D) in holes of throttle arm (A) and plate (B). Refer to text.

Fig. JD105—Loosen locknut (N) and turn screw (W) to adjust valve clearance.

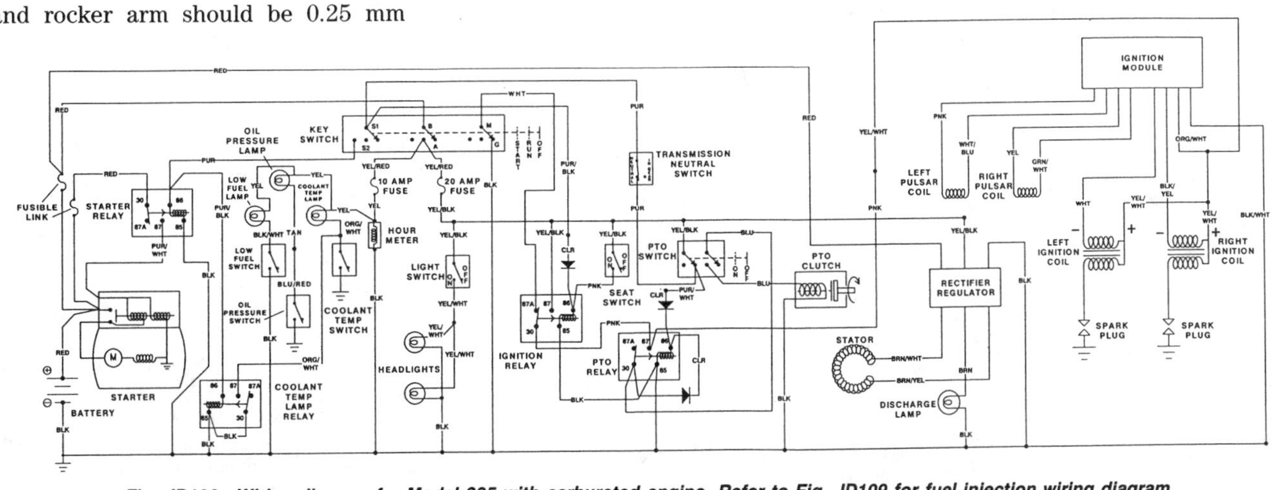

Fig. JD106—Wiring diagram for Model 285 with carbureted engine. Refer to Fig. JD109 for fuel injection wiring diagram.

gram that includes ignition system components. Before testing ignition system, be sure safety circuit components are not faulty and affecting performance of ignition system. Note the following test specifications when trouble-shooting ignition system:

Pulsar coil resistance . . . 85-270 ohms

Ignition coil resistance:
Primary . . . . 3.4-4.6 ohms
Secondary . . . . 10.4-15.5k ohms

To check ignition module, detach lead connectors and, using an ohmmeter set on K ohm scale, perform resistance checks as indicated in Fig. JD107. Note that values in table are approximate and

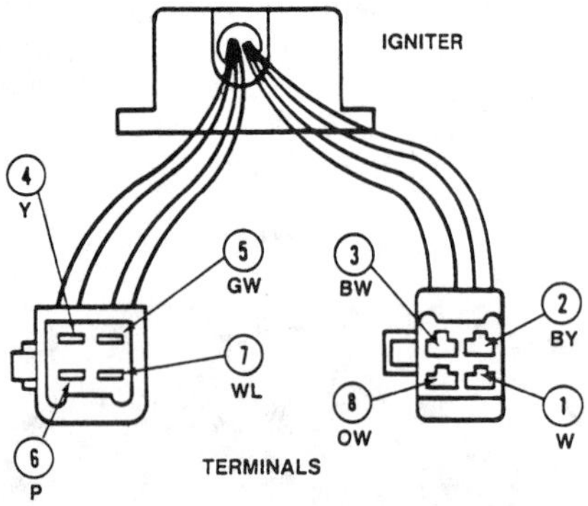

KΩ		TESTER ⊕ TERMINAL (RED)							
	TERMINAL	1	2	3	4	5	6	7	8
TESTER (−) TERMINAL (BLACK)	1		∞	1~8KΩ	4~16	2~8	4~16	2~8	2~10
	2	∞		1~8	4~16	2~8	4~16	2~8	2~10
	3	∞	∞		1~6	0	1~6	0	0.5~2
	4	∞	∞	1~6		0	3~12	1~6	2~8
	5	∞	∞	0	1~6		1~6	0	0.5~2
	6	∞	∞	1~6	3~15	0.5~2		1~6	1~6
	7	∞	∞	0	1~6	2~8	1~6		0.5~2
	8	∞	∞	0.5~2	2~8	0.5~2	2~8	0.5~2	

Fig. JD107— To check ignition module, detach lead connectors and, using an ohmmeter set on K ohm scale, perform resistance checks as indicated in table.

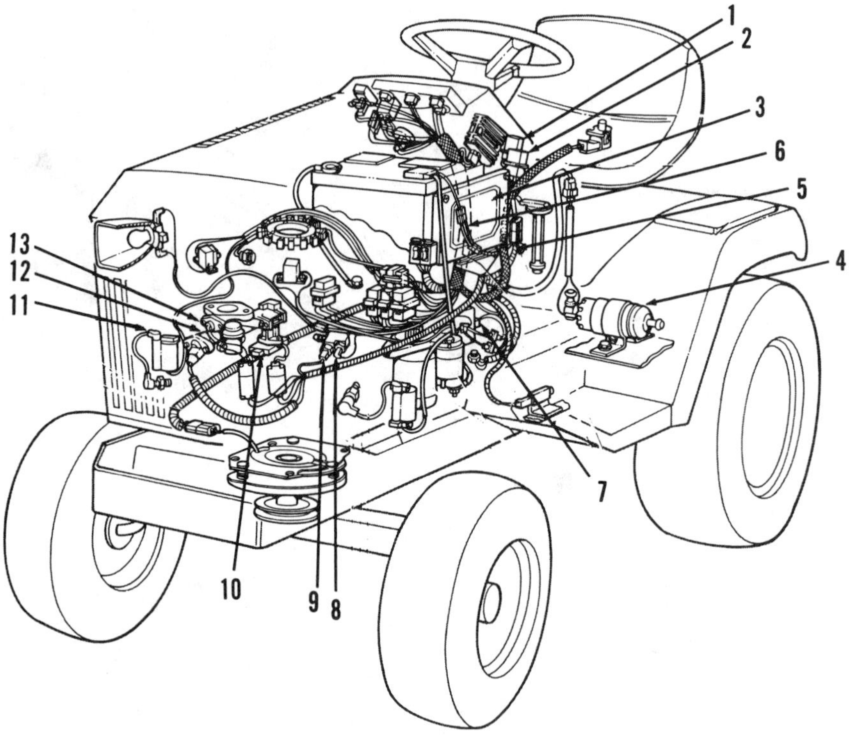

Fig. JD108—Drawing showing location of fuel injection system components.

1. Fuel injection module relay (green)
2. Fuel pump relay (blue)
3. Fuel injection module
4. Fuel pump
5. Altitude compensation switch
6. Fusible links
7. Ignition module
8. Fuel injection coolant temperature sensor
9. Coolant temperature sensor
10. Throttle sensor
11. Ignition coil
12. Fuel injector
13. Air temperature sensor

test values are determined by accuracy and condition of ohmmeter.

FUEL PUMP. The fuel pump on carbureted engines is a mechanical type pump driven by a lobe on the camshaft. The pump should maintain a minimum fuel pressure of 19.6 kPa (2.8 psi) at full throttle. Fuel flow at full throttle should be 160 ml (5.4 ozs.) in 15 seconds. Fuel pump must be serviced as a unit assembly.

The fuel pump on fuel injected engines is driven electrically. Check fuel pressure at inlet line to fuel injector. Pressure should be 274-284 kPa (40-41 psi). Check for restrictions and leaks in filter, hoses and pressure regulator if pressure is insufficient. Renew pump if pressure remains low. Check return line, pressure regulator vacuum hose and pressure regulator if pressure is high.

FUEL PRESSURE REGULATOR. Fuel Injected Engines. A vacuum-controlled fuel pressure regulator is located on the throttle body of fuel injected engines. The regulator varies fuel pressure according to air pressure in the intake manifold through a vacuum hose. When manifold pressure decreases, fuel pressure is increased, and vice versa when manifold pressure increases. The fuel pressure regulator must be serviced as a unit assembly.

FUEL INJECTION SYSTEM. The engine may be equipped with a fuel injection system. Refer to Fig. JD108 for a diagram that indicates the location of the fuel injection components. Refer to Fig. JD109 for wiring diagram of fuel injection electrical circuit.

Operation. The fuel injection system is an electronically controlled, throttle body type system. The system is divided into two subsystems, fuel delivery and control.

The fuel delivery system consists of the fuel filter, fuel pump, fuel injector and pressure regulator. Fuel from the fuel tank passes through the filter to the fuel pump. Pressurized fuel from the pump is directed to the fuel injector (15), which is located in the throttle body (6) attached to the intake manifold (see Fig. JD110). Excess fuel is routed from the injector to the pressure regulator (11), which maintains system pressure (see FUEL PRESSURE REGULATOR). Leak-off fuel from the regulator is directed back to the fuel tank.

The control system consists of the fuel injection control module, module relay, fuel pump relay and several sensors. The sensors provide the control module with voltage readings related to throttle position, coolant temperature and air temperature. The control module also senses engine speed, starter circuit voltage and battery voltage. The control module relay directs current to the control module when the ignition key switch is in "RUN." The fuel pump relay is energized by the control module when output from the fuel pump is required. Time and duration of fuel injector output are controlled by the control module. Activation of the injector by the control module is determined by input from sensors and related circuits. Note that two fusible links are located in the system to protect system components.

Trouble-shooting. Refer to Fig. JD109 for voltage readings that will help determine a faulty component. Normal trouble-shooting techniques will serve to isolate the faulty component. Voltage readings are taken with connectors connected.

The following fuel system circuits may prevent engine from starting if a malfunction occurs (including starter and ignition circuits):

- Fuel injection module
- Fuel pump circuit
- Fuel injector circuit
- Engine speed circuit
- Starter sensing circuit

A rough running engine may be caused by a malfunction in the following:

- Air temperature sensor circuit
- Coolant temperature sensor circuit
- Throttle sensor circuit
- Engine speed circuit

Note the following when performing voltage checks:

Throttle Sensor—voltage reading of 0.3-6.5 volts at pin 11 of fuel injection module is obtained when throttle plate is closed and throttle lever is moved from slow idle to fast idle.

Starter Voltage Sensing—voltage reading of 10-13.8 volts at blue/red wire terminal on solenoid and pin 19 of fuel injection module is obtained with ignition key switch held in START position (ground spark plug wires to prevent starting).

Fuel Pump Circuit—when ignition key switch is turned ON, there should be battery voltage for 5 seconds at fuel injection module pin 21. If not, renew module. Battery voltage should be present at fuel pump relay terminal 2 for 5 seconds when key switch is turned to ON position. With spark plugs grounded, turn key switch to ON until fuel

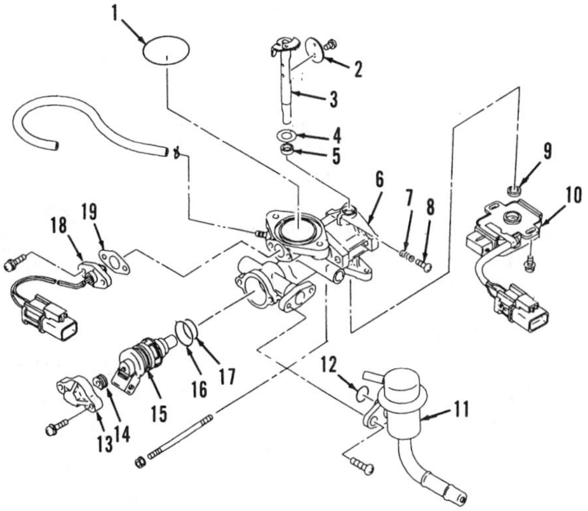

Fig. JD110—Exploded view of throttle body injection components.

1. "O" ring
2. Throttle plate
3. Throttle shaft
4. Washer
5. Seal
6. Body
7. Spring
8. Idle speed screw
9. Seal
10. Throttle sensor
11. Fuel regulator
12. "O" ring
13. Cap
14. Damper
15. Injector
16. "O" ring
17. "O" ring
18. Air temperature sensor
19. Gasket

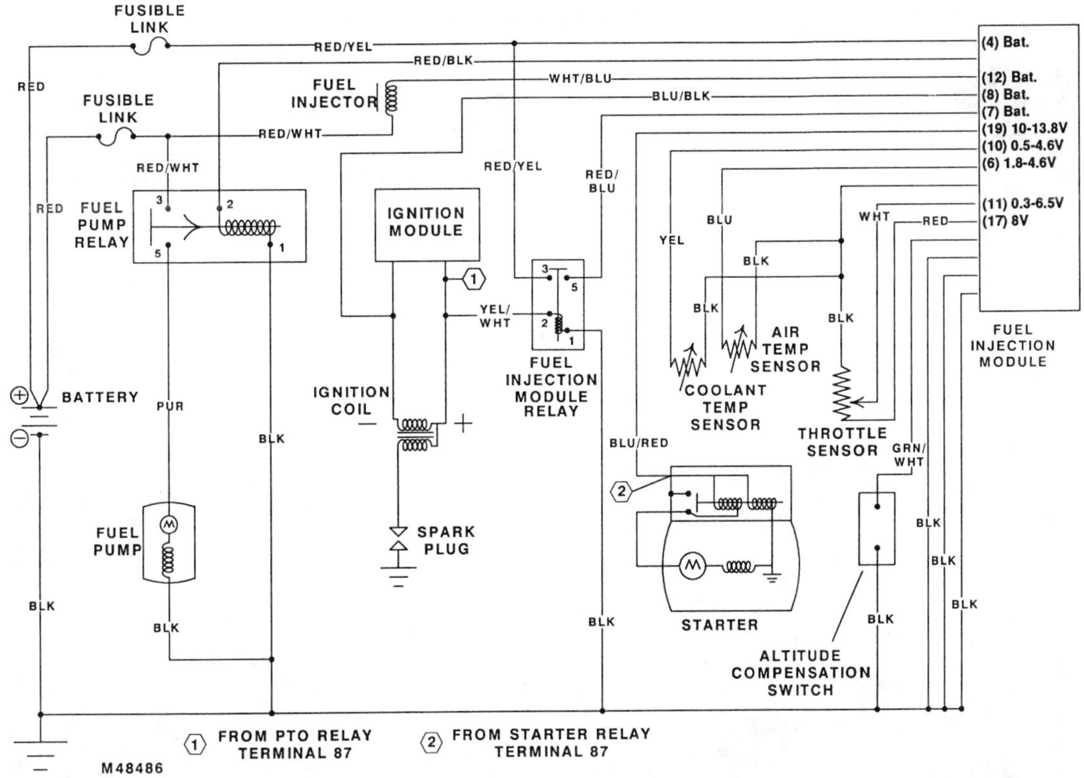

Fig. JD109—Wiring diagram of fuel injection system on models so equipped. Note voltage readings at fuel injection module and connector. Reading is taken at module or connector, depending on current direction.

pump stops, then turn to START and measure voltage at fuel pump relay terminal 2; there should be 10-13.8 volts. If not, check circuit and renew fuel injection module if no other cause is found. Turn key switch to ON until fuel pump stops, then connect a jumper wire to battery negative terminal and to blue/black connector between ignition coil and ignition module and measure voltage at fuel pump relay terminal 2; reading should indicate battery voltage. If not, check circuit and renew fuel injection module if no other cause is found.

Component Testing. The following components may be tested using an ohmmeter. Ignition key switch must be OFF.

THROTTLE SENSOR. Detach throttle sensor connector. Connect ohmmeter leads to black wire pin and red wire pin. Move throttle lever while noting meter reading. Resistance should decrease when throttle lever is moved from slow to fast. Resistance readings should fall in range from 3.3k to 6.8k ohms. Connect ohmmeter leads to white wire pin and black wire pin. Resistance should be 0.2-0.5k ohms when throttle lever is in idle position.

AIR TEMPERATURE SENSOR. Detach air temperature sensor connector. Resistance reading between sensor connector terminals at a temperature range of 0°-20°C (32°-86°F) should be in range of 1.3-7.4k ohms.

Fig. JD111—View of fuel pump relay pins. Refer to text for resistance tests.

COOLANT TEMPERATURE SENSOR. Specified temperatures and resistance readings are same as for air temperature sensor.

FUEL PUMP RELAY. Refer to Fig. JD111 for connector pin identification. Continuity should exist between pins 1 and 2. There should be infinite resistance between pins 3 and 5. Connect a jumper wire to battery positive terminal and to pin 1 and connect a jumper wire to battery negative terminal and to pin 2. Continuity should now exist between pins 3 and 5.

FUEL INJECTION MODULE RELAY. Follow testing instructions for fuel pump relay.

FUEL INJECTOR. Detach injector wire connector. Resistance between injector pins should be 11.1-12.9 ohms. Connect a jumper wire from battery negative terminal to one injector pin and tap a wire from positive battery terminal to other injector pin. Injector should "click" each time connection is made. If not, renew injector.

CYLINDER HEAD. To remove cylinder head, remove intake manifold and detach exhaust pipe. Rotate crankshaft so piston is at top dead center on compression stroke. Remove rocker arm cover, detach snap rings on rocker shaft, withdraw rocker shaft and remove rocker arms. Remove push rods and mark them so they can be reinstalled in their original position. Unscrew cylinder head screws and remove cylinder head.

Clean cylinder head and inspect for cracks or other damage. Use a feeler gage and straightedge or surface plate to check cylinder head gasket surface for distortion. Renew cylinder head if a 0.05 mm (0.002 in.) feeler gage can be inserted between the head and flat surface.

Install cylinder head with new gasket. Lightly oil threads of cylinder head screws, then tighten in sequence shown in Fig. JD112. Tighten screws in four steps until a final torque of 21 N·m (186 in.-lbs.) is attained.

VALVE SYSTEM. The valves are located in the cylinder head. Press down on top spring retainer to disengage retainer from valve stem. Removal of bottom spring retainer will result in damage to valve stem seal. If retainer and seal are removed, install new stem seal.

Valve seats and guides are not renewable. Valve face and seat angles should be 45 degrees. Manufacturer does not recommend grinding face of exhaust valve as life of valve will be shortened. Renew exhaust valve if it is burned, pitted or distorted.

Valve seat width should be 0.80 mm (0.031 in.) for intake and exhaust. Use a 30 degree stone or cutter to narrow

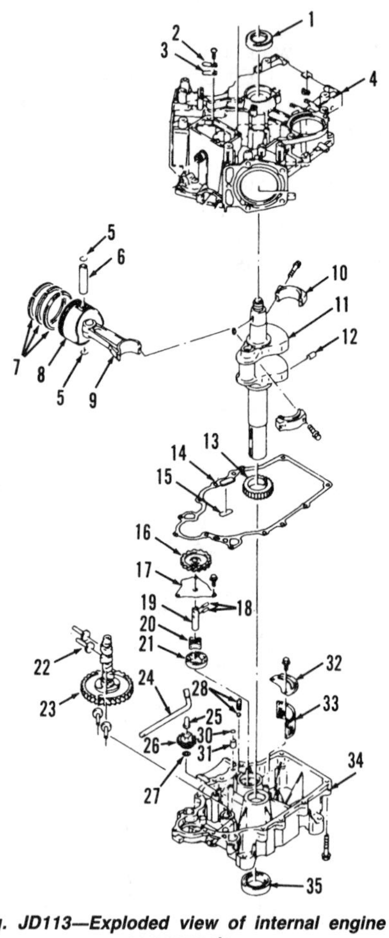

Fig. JD113—Exploded view of internal engine components.

1. Seal
2. Plate
3. Breather valve
4. Cylinder block
5. Snap ring
6. Piston pin
7. Piston rings
8. Piston
9. Connecting rod
10. Rod cap
11. Crankshaft
12. Pin
13. Gear
14. Gasket
15. "O" ring
16. Oil pump gear
17. Cover
18. Pins
19. Shaft
20. Inner rotor
21. Outer rotor
22. Tappets
23. Camshaft
24. Governor shaft
25. Sleeve
26. Governor
27. Washer
28. Relief valve
30. "O" ring
31. Dowel pin
32. Plate
33. Strainer
34. Cover
35. Seal

Fig. JD112—Tighten cylinder head screws in sequence shown above.

valve seat if necessary. Minimum valve margin is 0.60 mm (0.024 in.). Maximum allowable valve guide inside diameter is 6.05 mm (0.238 in.). Minimum valve spring length is 29.70 mm (1.170 in.). Maximum valve stem runout, measured with valve supported at ends, is 0.03 mm (0.001 in.). Maximum push rod runout, measured with push rod supported at ends, is 0.30 mm (0.012 in.). Minimum rocker shaft diameter is 11.95 mm (0.470 in.). Maximum rocker arm inner diameter is 12.07 mm (0.475 in.).

PISTON, PIN, RINGS AND ROD. Remove cylinder head. Remove crankcase cover for access to piston and rod assemblies. Unscrew rod cap screws, detach rod cap and remove piston and rod assembly through cylinder bore.

Maximum allowable pin bore in piston is 17.04 mm (0.671 in.). Minimum allowable piston pin diameter is 16.98 mm (0.668 in.). Maximum allowable piston ring end gap is 1.00 mm (0.039 in.). Maximum allowable piston ring side clearance is 0.10 mm (0.004 in.) for compression rings. Pistons and rings are available in standard and 0.25, 0.50 and 0.75 mm oversizes. Refer to CYLINDER BLOCK section if oversize pistons are required.

The connecting rod rides directly on the crankpin. Maximum allowable inside diameter at big end of rod is 34.06 mm (1.341 in.). Minimum allowable crankpin journal diameter is 33.93 mm (1.336 in.). Maximum allowable inside diameter at pin end of rod is 17.05 mm (0.671 in.). Connecting rod is available with standard size big end diameter as well as with 0.50 mm undersize big end diameter.

Assemble piston and rod so arrow (A) on piston crown will be toward flywheel, as shown in Fig. JD114, and chamfered side (C) of rod big end is next to crankshaft web. Install piston rings with lettered side toward piston crown. See Fig. JD114A. Position piston rings so ring gaps are 180 degrees apart. Install piston and rod assembly so arrow on piston crown points toward flywheel end of crankshaft. Make sure rod cap matches rod and tighten rod screws to 21 N·m (186 in.-lbs.). Tighten crankcase cover screws to 21 N·m (186 in.-lbs.) in sequence shown in Fig. JD115. Install cylinder head as previously outlined.

CAMSHAFT. The camshaft is accessible after removing the crankcase cover. The camshaft rides in nonrenewable bushings in the cylinder block and crankcase cover. Renew cylinder block or cover if bushing diameter is greater than 16.07 mm (0.633 in.).

Minimum allowable camshaft bearing journal diameter is 15.91 mm (0.626 in.). Minimum allowable camshaft lobe height is 25.23 mm (0.993 in.). Minimum fuel pump lobe height is 19.50 mm (0.768 in.).

When installing camshaft, be sure to align timing marks (M—Fig. JD116) on camshaft and crankshaft gears. Tighten crankcase cover screws to 21 N·m (186 in.-lbs.) in sequence shown in Fig. JD115.

CRANKSHAFT AND MAIN BEARINGS. The crankshaft rides in nonrenewable bushings in the cylinder block and crankcase cover. Maximum allowable bushing diameter in block and cover is 34.07 mm (1.341 in.). Minimum allowable crankshaft main bearing journal diameter is 33.91 mm (1.335 in.). Minimum allowable crankpin diameter is 33.93 mm (1.336 in.). The crankpin may be reground to accept a connecting rod with a 0.50 mm undersize big end. Maximum allowable crankshaft runout is 0.05 mm (0.002 in.) measured at main bearing journals with crankshaft supported at ends.

Be sure timing marks (M—Fig. JD116) on crankshaft and camshaft gears are aligned. Tighten crankcase cover screws to 21 N·m (186 in.-lbs.) in sequence shown in Fig. JD115.

CYLINDER BLOCK. The cylinders may be bored to accept oversize pistons. Standard bore diameter is 73.98-74.00 mm (2.9126-2.9134 in.) and maximum wear limit is 74.07 mm (2.916 in.). Refer to following table for oversize dimensions:

Piston Diameter	Bore Diameter
0.25 mm	74.21-74.23 mm
(0.010 in.)	(2.9216-2.9224 in.)
0.50 mm	74.46-74.48 mm
(0.020 in.)	(2.9315-2.9322 in.)
0.75 mm	74.71-74.73 mm
(0.030 in.)	(2.9413-2.9421 in.)

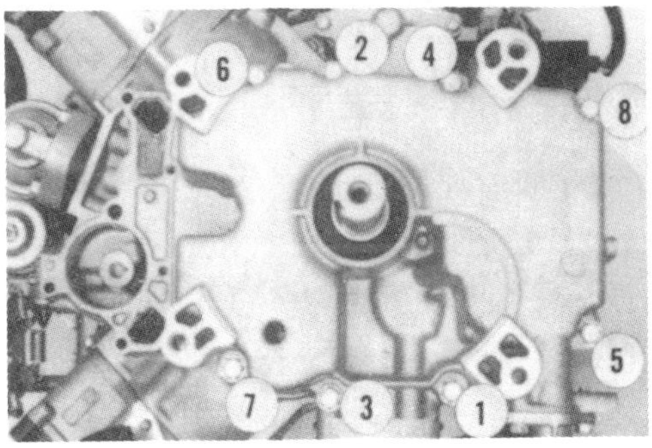

Fig. JD115—Tighten crankcase cover screws in sequence shown above.

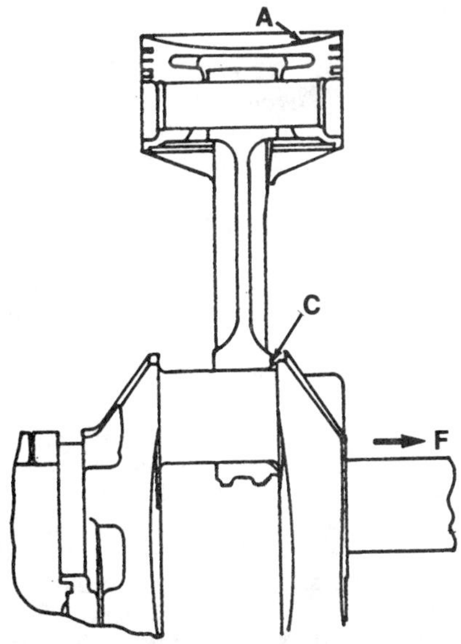

Fig. JD114—Assemble piston and rod so arrow (A) on piston crown is toward flywheel (F) and chamfer (C) is toward adjacent crankshaft web. Note that when both rods are installed on crankshaft, the non-chamfered sides will be together.

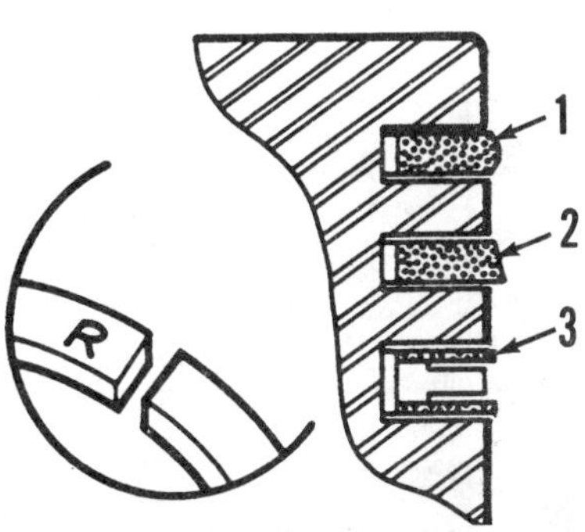

Fig. JD114A—Cross section of piston and rings showing correct installation. Be sure that "R" or "NPR" mark on end of top ring (1) and second ring (2) faces up. Oil ring (3) can be installed either way.

Fig. JD116—Align timing marks (M) on crankshaft and camshaft gears when installing camshaft.

GOVERNOR. A flyweight type governor is located in the crankcase cover. The governor gear is driven by the camshaft gear. Remove crankcase cover for access to governor assembly. Governor components must move freely without binding.

NOTE: Removing governor assembly from shaft will damage governor and require installation of a new governor assembly.

Remove governor assembly by prying or pulling off shaft. Shaft must protrude above boss 32.2-32.8 mm (1.27-1.29 in.). Install governor by pushing down until it snaps onto the locating groove. Tighten crankcase cover screws to 21 N·m (186 in.-lbs.) in sequence shown in Fig. JD115.

OIL PUMP. A gerotor type oil pump is located in the crankcase cover. The pump provides pressurized oil that is directed to the crankshaft main bearing journals and crankpin. Drilled passages in the connecting rods squirt oil on the underside of the pistons.

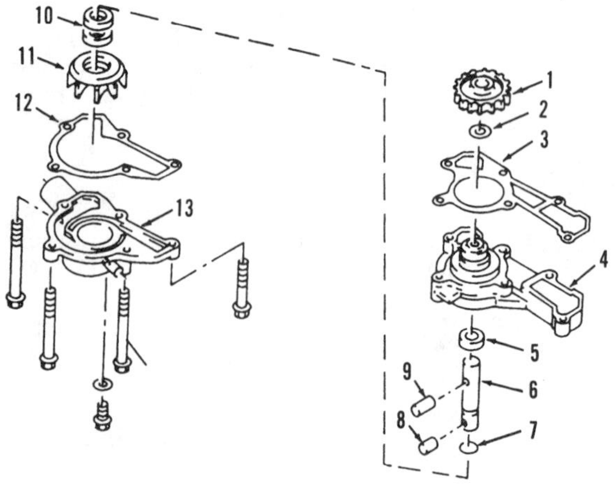

Fig. JD117—Exploded view of water pump.
1. Gear
2. Washer
3. Gasket
4. Housing
5. Seal
6. Shaft
7. "O" ring
8. Pin
9. Pin
10. Seals
11. Impeller
12. Gasket
13. Cover

To check oil pressure, remove oil pressure switch in side of cylinder block adjacent to oil filter and connect an oil pressure gage. With engine at operating temperature and running at fast throttle setting, oil pressure should be no less than 240 kPa (35 psi).

Remove crankcase cover for access to pump. Minimum allowable shaft diameter is 10.92 mm (0.430 in.). Maximum allowable bore diameter in crankcase cover and pump cover is 11.07 mm (0.436 in.). Maximum rotor bore diameter in crankcase cover is 40.80 mm (1.606 in.). Renew relief valve spring (28—Fig. JD113) if length is less than 19.50 mm (0.768 in.).

WATER PUMP. Refer to Fig. JD117 for an exploded view of water pump. Impeller (11), shaft (6) and pin (8) are available as a unit assembly only. Minimum allowable diameter of shaft is 9.94 mm (0.391 in.). Maximum allowable diameter of housing bore is 10.09 mm (0.397 in.). Install seal (10) using driver provided in seal kit. Tighten 6 mm mounting screws to 8 N·m (71 in.-lbs.) and 8 mm mounting screw to 21 N·m (186 in.-lbs.).

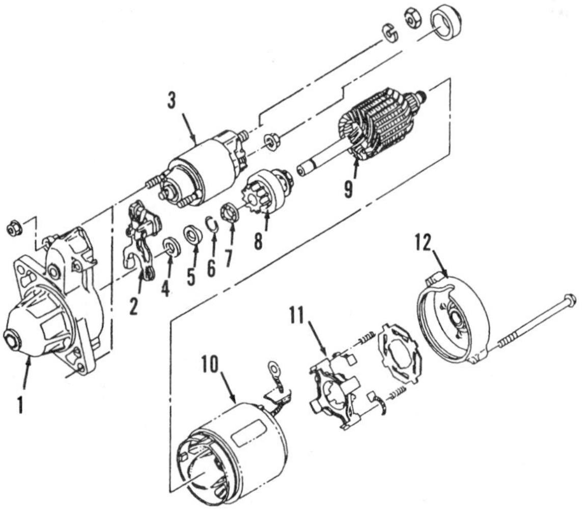

Fig. JD118—Exploded view of starter.
1. Drive housing
2. Fork
3. Solenoid
4. Washer
5. Pinion stop half
6. Snap ring
7. Pinion stop half
8. Clutch
9. Armature
10. Frame
11. Brush plate
12. End cap

ELECTRICAL SYSTEM

The tractor is equipped with a 12 volt, group 22F battery. Cold cranking amp rating is 341 amps with reserve capacity of 68 minutes.

Refer to Fig. JD106 for wiring diagram.

ELECTRIC STARTER. Refer to Fig. JD118 for an exploded view of electric starter. Place alignment marks on pinion housing, frame and end cover before disassembly so they can be reinstalled in original position. Renew brushes if length is less than 10.5 mm (0.413 in.). Starter no-load current draw should be 50 amps at starter speed of 6000 rpm. Starter current draw under load should be 72 amps at starter speed of 500 rpm.

ALTERNATOR. The alternator stator is located under the flywheel. To check alternator, disconnect stator from regulator/rectifier. Refer to wiring diagram in Fig. JD106 and check voltage reading at stator leads. With engine running at full speed, the minimum voltage reading is 26 volts ac. If voltage is insufficient, check resistance of stator (engine stopped) by attaching ohmmeter leads to stator wires. Renew stator if resistance is not 0.12-0.19 ohms.

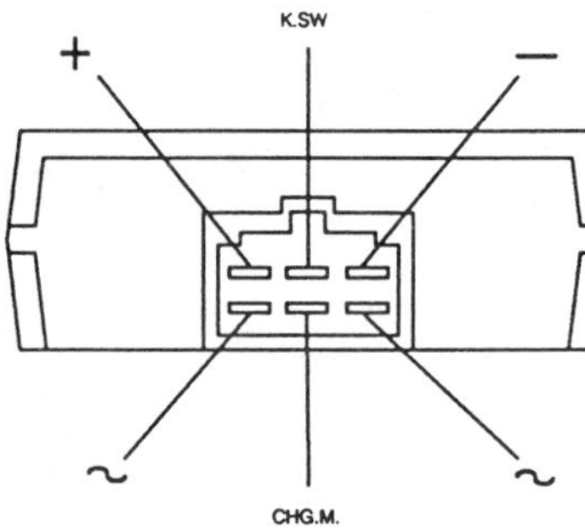

RANGE: RX1KΩ

TESTER (+) / (−)	+	K.SW	—	~	~	CHG.M.
+		∞	∞	∞	∞	∞
K.SW	4 KΩ ~ 20KΩ		200 Ω 1 KΩ	1 KΩ 5 KΩ	1 KΩ ~ 5 KΩ	200K Ω ~ ∞
—	3 KΩ ~ 15KΩ	200 Ω ~ 1 KΩ		1 KΩ 5 KΩ	1 KΩ 5 KΩ	200K Ω ~ ∞
~	1 KΩ ~ 5KΩ	∞	∞		∞	∞
~	1 KΩ ~ 5KΩ	∞	∞	∞		∞
CHG.M.	10 KΩ ~ 50KΩ	1 KΩ ~ 5K Ω	1.5 K ~ 7.5KΩ	4 KΩ ~ 20K Ω	4 KΩ ~ 20KΩ	

NOTE: Resistance value may vary with individual meters.

Fig. JD119—Refer to table and diagram shown above to perform resistance readings on regulator/rectifier.

REGULATOR/RECTIFIER. The regulator/rectifier should provide 15 amps direct current at 12.2-13.8 volts at full engine speed. If regulator/rectifier is suspected faulty, refer to Fig. JD119 and perform resistance checks. Renew regulator/rectifier if desired readings are not obtained.

BRAKES

Model 285 tractors are equipped with a drum type brake located on the differential housing. The brake drum is attached to the differential shaft. Depressing the brake pedal operates the brake and shifts the transmission into neutral (see transmission section).

ADJUSTMENT

To adjust brake, place parking brake in locked position. Measure gap between washer (W—Fig. JD120) and spring retainer (R). Gap should be 8-14 mm (0.31-0.55 in.). If not, loosen nut (N) and turn nut (T). Retighten nut (N).

R&R BRAKES

Refer to Fig. JD121 for an exploded view of brake assembly. To disassemble, detach rod from brake arm and remove screws securing cover (6). Inspect components for excessive wear and damage. Brake arm shaft diameter should be 17.96-18.00 mm (0.707-0.709 in.). Shaft bore in cover should be 18.02-18.04 mm (0.709-0.710 in.). Maximum allowable shaft clearance is 0.5 mm (0.020 in.). Install spring (4) so open portion is toward brake arm. Apply grease to brake arm shaft.

While tightening cover screws, push brake arm forward so brake shoes will contact drum thereby centering cover. Tighten cover screws to 24 N·m (18 ft.-lbs.) if holes are threaded, or to 29 N·m (22 ft.-lbs.) on new cases with untapped holes. Adjust brake as outlined in previous section.

TRANSMISSION

MAINTENANCE

Check transmission fluid level after every 25 hours of operation. Fluid level should be maintained at top mark on dipstick attached to fill plug. Operate tractor with engine running and shift several times into forward and reverse before checking level. Be sure tractor is on level surface and engine is stopped when checking.

Transmission fluid and filter should be changed after first 50 hours of operation and then after every 250 hours of operation thereafter. Partially fill filter with oil before installing.

Recommended fluid is John Deere All-Weather Hydrostatic Fluid or good quality type "F" automatic transmission fluid.

Oil capacity is 2.5 L (2.6 qts.).

ADJUSTMENTS

BELT GUIDES. Adjust location of belt guides so guide (1—Fig. JD123) is 1-4 mm (0.04-0.16 in.) from belt, guides (2) are 1-5 mm (0.04-0.20 in.) from belt, and guide (3) is centered over belt.

NEUTRAL START SWITCH. The tractor is equipped with a neutral start switch that prevents engine starting unless brake pedal is depressed and transmission is in neutral. Plunger on switch (22—Fig. JD125) should be forced in when brake pedal is depressed. Adjust position of switch by loosening switch mounting bolts and moving switch.

CONTROL LEVER FRICTION. Force required to move control lever should be 22-36 N (5-8 lbs.). Adjust control lever friction by rotating nut (N—Fig. JD124).

NEUTRAL POSITION. Securely support rear of tractor so rear wheels are off ground. Run engine at fast throttle setting and move transmission control lever so wheels stop turning (this may not be neutral position indicated on shift panel). Stop engine.

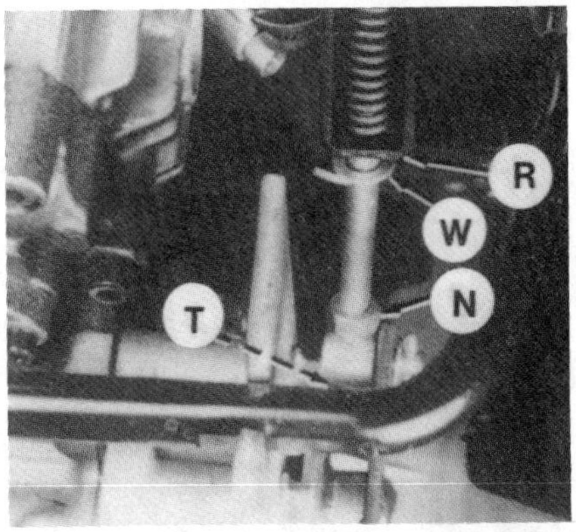

Fig. JD120—Gap between washer (W) and spring retainer (R) on brake rod should be 8-14 mm (0.31-0.55 in.). If not, loosen nut (N) and turn nut (T). Retighten nut (N).

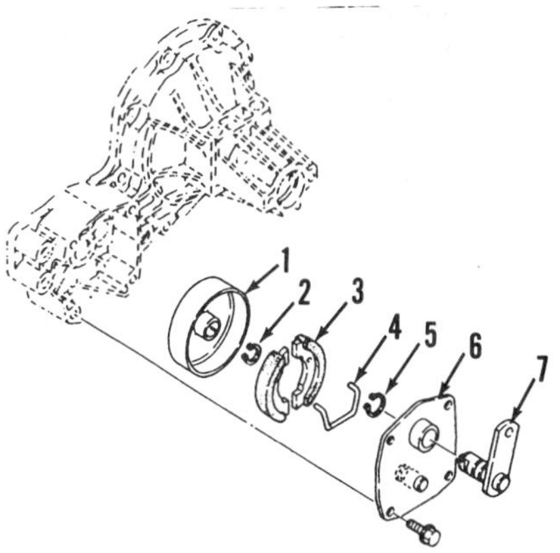

Fig. JD121—Exploded view of brake assembly.

1. Drum
2. Snap ring
3. Brake shoes
4. Spring
5. Snap ring
6. Cover
7. Arm

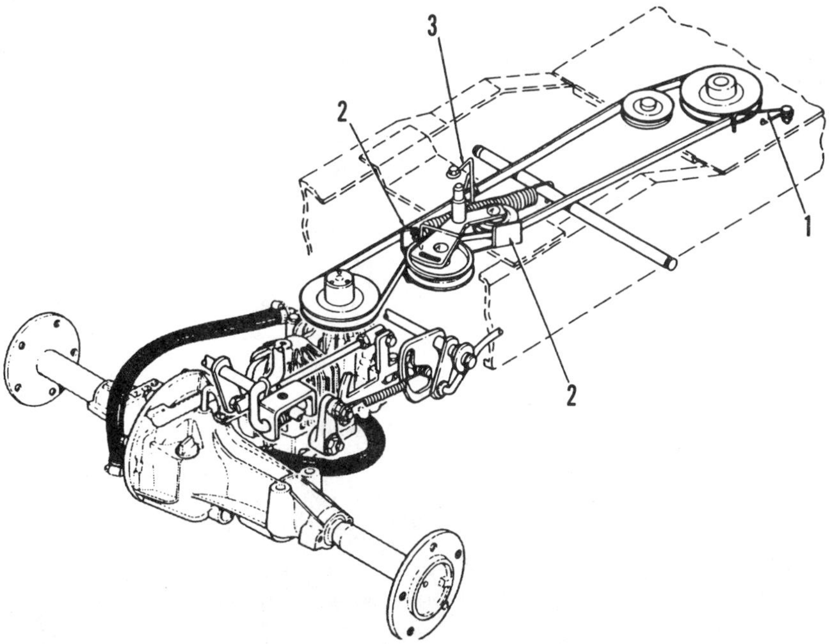

Fig. JD123—Drawing of drive belt. Position guides (1, 2 and 3) as specified in text.

Loosen eccentric retaining screw (18—Fig. JD125) and rotate eccentric (14—Fig. JD125 and JD126) so roller (12) is centered in notch of control arm (11). Loosen nut (19—Fig. JD125). Engage parking brake. Turn eccentric (7) so roller (20) is centered in slot of cam (6), then retighten nut (19). Being certain roller (12) is centered in notch of control arm (11), turn nuts (21) so shift control lever is centered in neutral notch of shift panel. Run engine and check adjustments. Shift lever should return to neutral when brake is depressed and rear wheels should not turn when shift lever is in neutral.

Fig. JD124—Adjust control lever friction by rotating nut (N) so force required to move control lever is 22-36 N (5-8 lbs.).

PRESSURE TEST

CHARGE PUMP PRESSURE TEST. To check charge pump pressure, first start and run engine at fast idle until transmission oil is warm. Stop engine and install a 0-150 psi (0-1000 kPa) pressure gage in either port (P—Fig. JD127) located in front of transmission. Move transmission control lever to neutral, run engine at slow idle and check pressure.

NOTE: Transmission must be in neutral and wheels must not rotate. Do not move transmission control lever out of neutral as pressure will increase and may damage pressure gage.

Charge pump pressure must be 193-490 kPa (28-71 psi). If low, be sure transmission filter and charge pump relief valve are in good condition. Relief valve pressure is not adjustable.

OVERHAUL

Drain fluid from transmission and remove differential and transmission as a unit assembly from the tractor. Clean transmission and surrounding area. Remove fan, pulley, hoses, shift control arm, eccentric arm and mounting bracket from transmission. Unscrew retaining screws and separate transmission from differential.

To disassemble transmission, remove charge pump housing (1—Fig. JD128). Remove pump rotors (3), charge relief valve (4) and drive pin in shaft. Remove screws securing cover (54) and separate cover from housing (48). Remove valves and plugs as shown in Fig. JD128 being careful not to interchange parts or lose shims.

Position housing (48) on its side. Remove pump cylinder block (17) and piston assembly. Remove motor cylinder block (31) and piston assembly and shaft. Lay cylinder block assemblies aside for later disassembly.

NOTE: Pistons (21) must be identified prior to removal so they can be reinstalled in original cylinder bore. Do not lose ball (20) in each piston.

Remove thrust washer (24), thrust bearing (25), thrust washer (26), swashplate (28) and bearings (29). Remove snap ring (45) and extract input shaft (40) assembly. Remove control arm (49).

Inspect components for damage and excessive wear. Inspect pistons and cylinder bores for scoring and scratches. Inspect valve cones for nicks and scratches. Maximum allowable clearance between tip of charge pump inner rotor and lobe of outer rotor is 0.13 mm (0.005 in.). See Fig. JD129. Use straightedge and feeler gage to check end clearance between rotors and pump housing. Maximum allowable rotor end clearance is 0.05 mm (0.002 in.). Inspect charge pump rotors and face of cover (54—Fig. JD128) for scoring and scratches.

Coat parts with transmission oil during assembly. Be sure pistons are returned to original bore in cylinder. Note that thrust washer (26) is thinner than thrust washer (24). Press bearing (16) into cover (54) so bearing is bottomed. Press bearing (30) into cover until flush with seal seat. Bottom seal (5) in housing. Install charge pump cover so

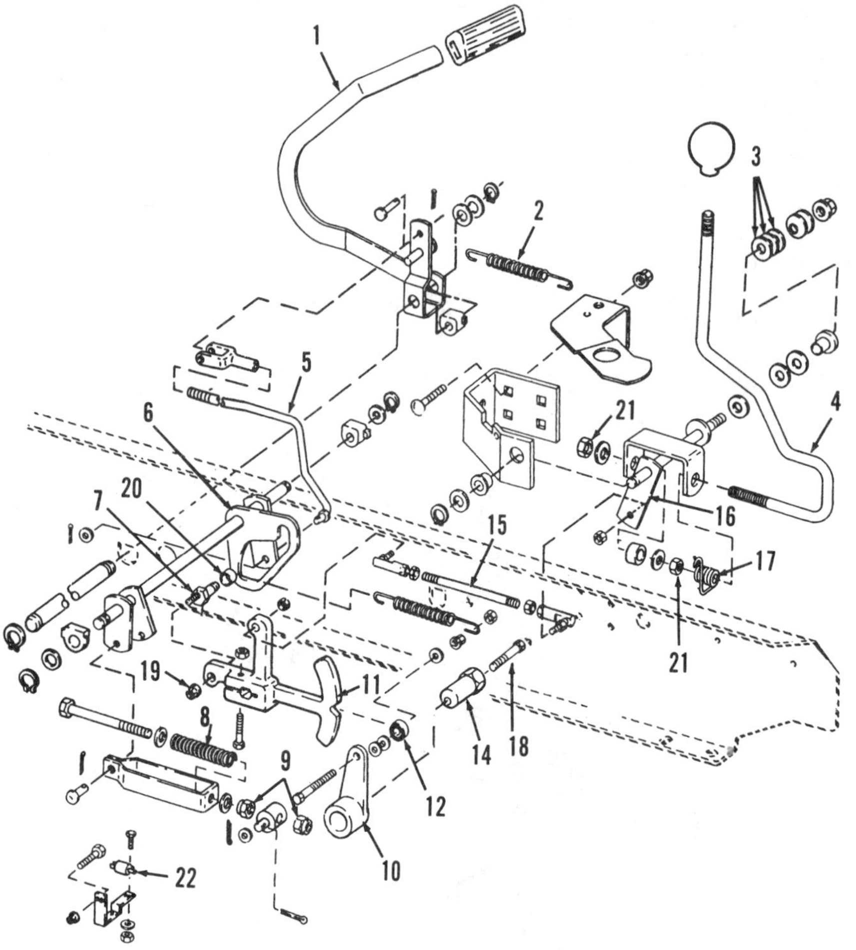

Fig. JD125—Exploded view of transmission control linkage.

1. Pedal
2. Spring
3. Spring washers
4. Shift lever
5. Control rod
6. Cam plate
7. Eccentric
8. Spring
9. Nuts
10. Control lever
11. Control arm
12. Roller
13. Spring
14. Eccentric
15. Rod
16. Lever
17. Spring
18. Screw
19. Nut
20. Roller
21. Nut
22. Neutral switch

flat edge (F—Fig. JD129) of cover is toward control arm side of transmission. Tighten charge pump housing screws to 11 N·m (97 in.-lbs.). Top of drive pulley should be 8 mm (0.31 in.) from top of pump shaft. Tighten screws securing hose fittings to 10 N·m (88 in.-lbs.).

DIFFERENTIAL

R&R AND OVERHAUL

After removing differential and transmission assembly from tractor, remove wheels then remove transmission from differential as outlined in TRANSMISSION section. Remove brake components.

Refer to Fig. JD130 for an exploded view of reduction gear and differential assembly. Remove screws securing housings together. Separate housings while noting location of all washers and shims. Use a suitable press to remove axles. Refer to Fig. JD130 and disassemble unit.

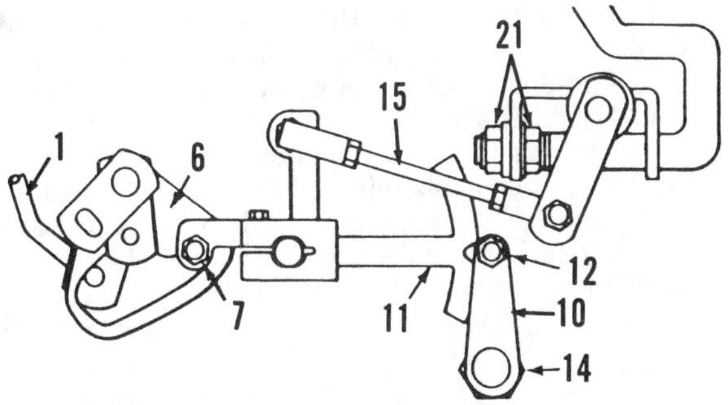

Fig. JD126—Drawing of transmission control mechanism. Refer to text for adjustment.

Fig. JD127—Connect a suitable pressure gage to either port (P) to check hydraulic pressure. See text.

1. Charge pump housing
2. "O" ring
3. Pump rotors
4. Charge relief valve
5. Seal
6. "O" ring
7. Plug
8. Bypass valve
9. "O" ring
10. "O" ring
11. Plug
12. Plug
13. "O" ring
14. Spring
15. Check valve
16. Bearing
17. Cylinder block
18. Spring
19. Spring seat
20. Ball
21. Piston
22. Spring
23. Washer
24. Thrust washer (thick)
25. Thrust bearing
26. Thrust washer (thin)
27. Sleeve
28. Swashplate
29. Bearing
30. Bearing
31. Cylinder block
32. Spring
33. Washer
34. Motor shaft
35. Snap ring
36. Bearing
37. Thrust washer
38. Bearing
39. Pin
40. Pump shaft
41. Snap rings
42. Bearing
43. Spacer
44. Seal
45. Snap ring
46. Dowel
47. Gasket
48. Housing
49. Control arm
50. Bushing
51. Seal
52. Plug
53. "O" ring
54. Cover

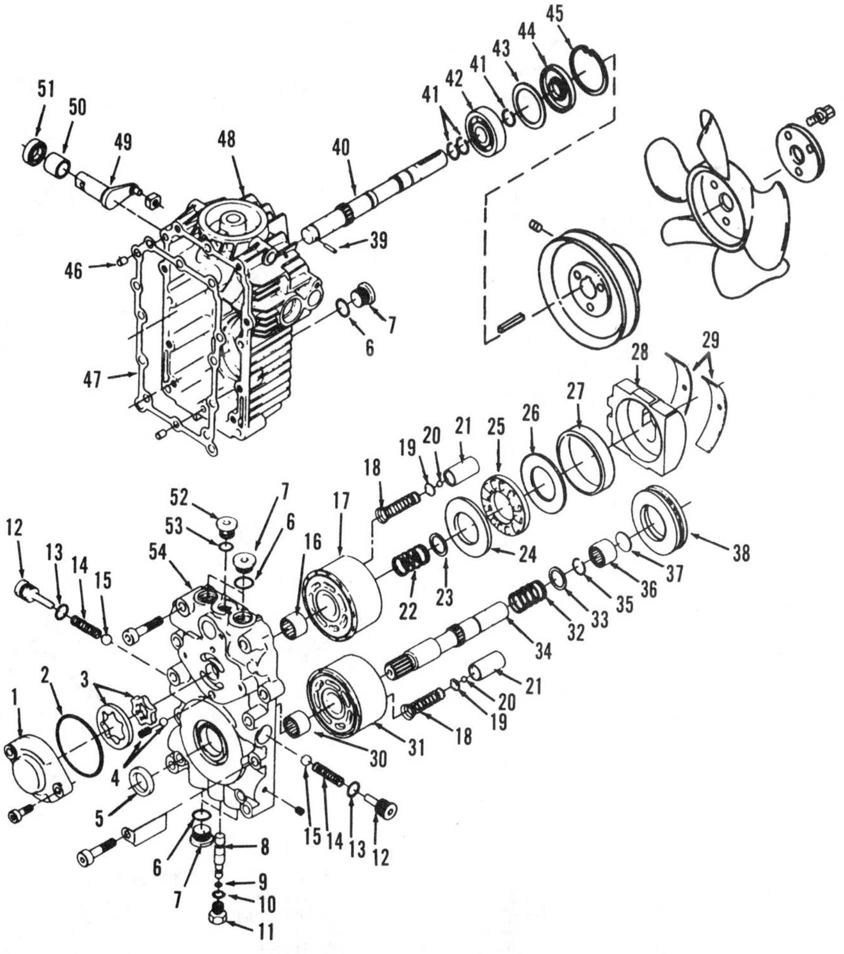

Fig. JD128—Exploded view of hydrostatic transmission.

Minimum allowable thickness of washers (14) is 0.5 mm (0.020 in.). Minimum allowable thickness of washers (28 and 31) is 2.16 mm (0.085 in.). Pinion gear (18) inside diameter should be 14.02-14.03 mm (0.5520-0.5524 in.) and pinion shaft diameter should be 13.97-13.98 mm (0.5500-0.5504 in.). Maximum allowable clearance between pinion gears and shaft is 0.4 mm (0.016 in.). Minimum allowable thickness of washers (11) is 1.5 mm (0.060 in.).

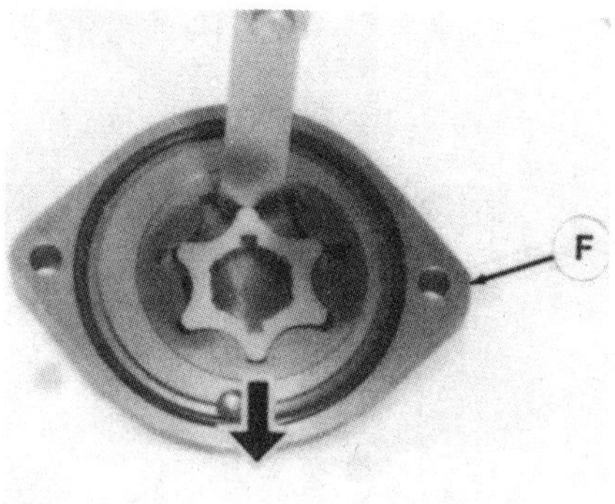

Fig. JD129—Check clearance between tip of inner rotor and lobe of outer rotor, and between surface of pump housing and face of rotors. Refer to text for specifications. Install charge pump housing so flat edge (F) is toward control arm side of transmission.

Press bearings (8) into housings so printed side is out. Insert axles into housings and press bearings (10) onto axles so bearing is 4-5 mm (0.16-0.20 in.) below snap ring groove on axle. Place washer and snap ring on axle then tap inner (splined) end of axle until snap ring forces washer against bearing. Bearing (3) must be pressed onto pinion (4). Bearings (23 and 33) must be pressed onto gear shaft (32). Install gear (26) on shaft (25) and press bearings (24) onto shaft. Washers (28 and 31) must be installed on shaft (27) with oil grooves toward bearing (29).

Be sure pinion shaft (17) is properly centered within ring gear (16) when assembling differential housing components. Apply Loctite to differential housing bolts and tighten to 27 N·m (20 ft.-lbs.). Assemble components in right housing (9) and tap idler shaft (27) to seat in housing. Install shims and washers in left housing using grease to hold them in place. Apply sealer to housing mating surface. Tighten housing screws to 26 N·m (19 ft.-lbs.) if original housing is being reused. If new housing with untapped holes is being installed, tighten screws to 30 N·m (22 ft.-lbs.).

Measure backlash of pinion (4) while pulling outward on splined, brake end of shaft (32). Backlash should be 0.06-0.24 mm (0.002-0.009 in.). Add or delete shims (34) as needed to obtain desired backlash.

POWER TAKE-OFF

ADJUSTMENT

Adjust electric pto clutch by turning each of the three adjusting nuts (N—Fig. JD131) until clearance between rotor and armature is 0.38-0.64 mm (0.015-0.025 in.) on Warner clutch, or 0.30-0.51 mm (0.012-0.020 in.) on Ogura clutch. Do not overtighten nuts. Operate clutch several times and recheck clearance.

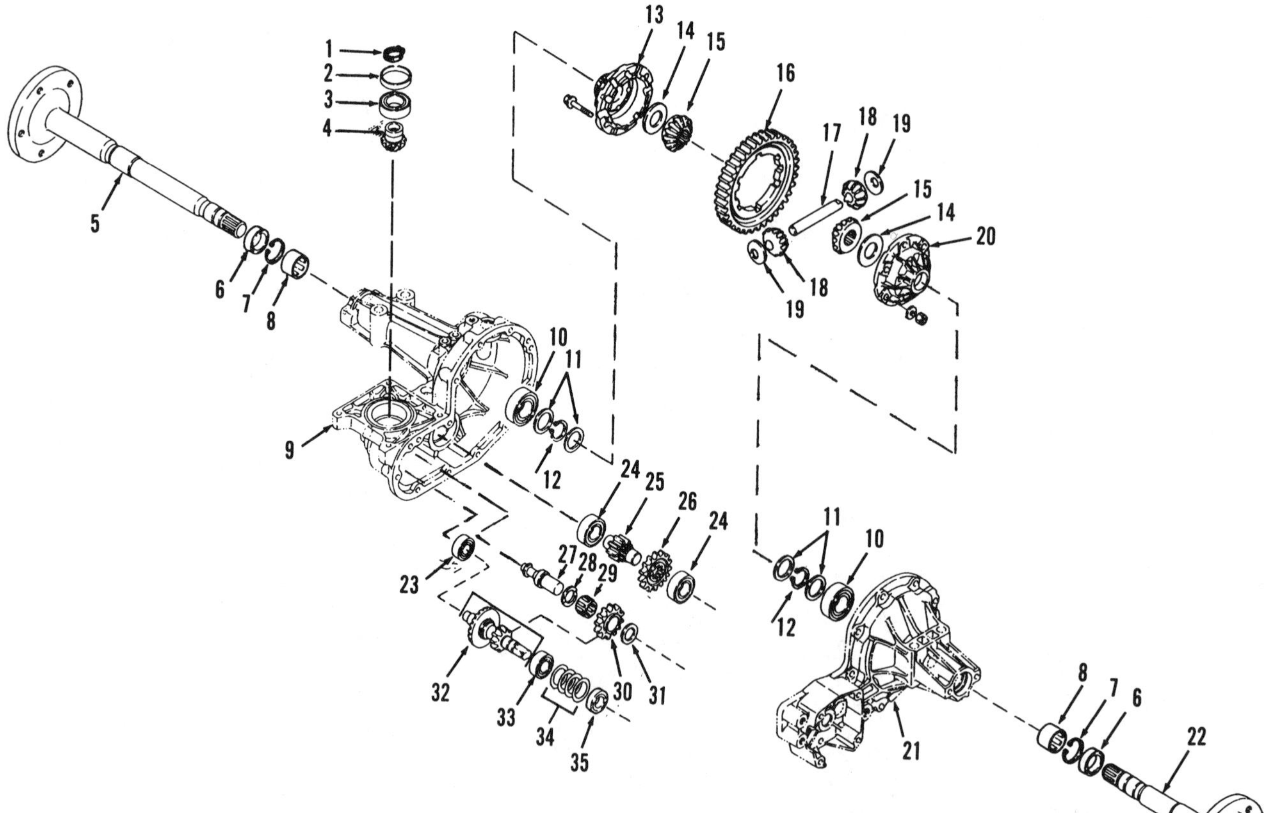

Fig. JD130—Exploded view of differential and gear reduction.

1. Snap ring
2. Bushing
3. Bearing
4. Pinion
5. Axle
6. Seal
7. Snap ring
8. Bearing
9. Housing
10. Bearing
11. Washers
12. Snap ring
13. Carrier half
14. Washer
15. Side gear
16. Ring gear
17. Pinion shaft
18. Pinion
19. Washer
20. Carrier half
21. Housing
22. Axle
23. Bearing
24. Bearing
25. Gear
26. Gear
27. Shaft
28. Washer
29. Bearing
30. Gear
31. Washer
32. Brake shaft
33. Bearing
34. Shims
35. Seal

OVERHAUL

Two types of electric pto clutches have been used. Refer to Fig. JD132 for an exploded view of Ogura pto clutch or to Fig. JD133 for an exploded view of Warner pto clutch. Either clutch may be removed after unscrewing center screw. The bearing of the Ogura clutch may be removed by pressing out bearing after removing the snap ring. The Warner clutch is available only as a unit assembly.

Fig. JD131—Adjust pto clutch by turning each of the three adjusting nuts (N) until clearance between rotor and armature is 0.38-0.64 mm (0.015-0.025 in.) on Warner clutch or 0.30-0.51 mm (0.012-0.020 in.) on Ogura clutch.

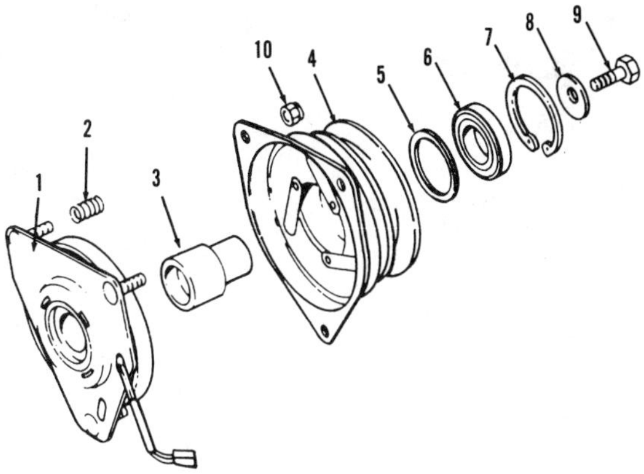

Fig. JD132—Exploded view of Ogura electric pto clutch.

1. Coil
2. Spring
3. Bearing hub
4. Armature & pulley
5. Washer
6. Bearing
7. Snap ring
8. Belleville washer
9. Screw
10. Nut

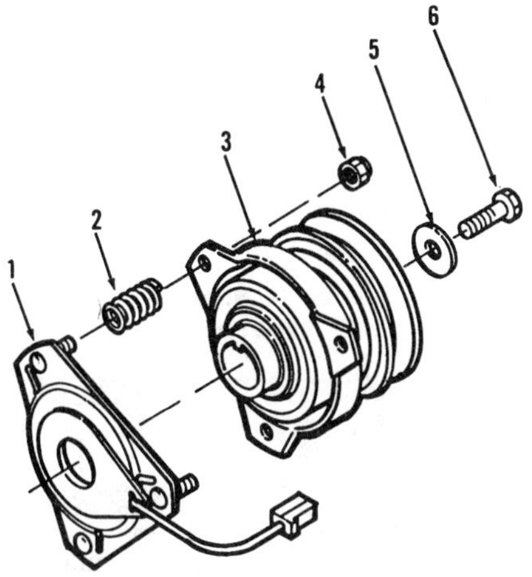

Fig. JD133—Exploded view of Warner electric pto clutch.

1. Coil
2. Spring
3. Armature & pulley
4. Nut
5. Belleville washer
6. Screw

DEUTZ-ALLIS

CONDENSED SPECIFICATIONS

	MODELS		
	613 Special	613H	616 Special
Engine Make	B&S	B&S	B&S
Model	404707	404707	402707
Bore	3.44 in. (87.3 mm)	3.44 in. (87.3 mm)	3.44 in. (87.3 mm)
Stroke	2.16 in. (54.8 mm)	2.16 in. (54.8 mm)	2.16 in. (54.8 mm)
Displacement	40 cu. in. (656 cc)	40 cu. in. (656 cc)	40 cu. in. (656 cc)
Power Rating	12.5 hp (9.3 kW)	12.5 hp (9.3 kW)	16 hp (11.9 kW)
Slow Idle	1350 rpm	1350 rpm	1350 rpm
High Speed (No-Load)	3400 rpm	3400 rpm	3400 rpm
Capacities—			
Crankcase	3 pts.* (1.4 L)	3 pts.* (1.4 L)	3 pts.* (1.4 L)
Transmission	24 ozs. (0.7 L)	**	24 ozs. (0.7 L)
Differential	.....	2.75 pts. (1.3 L)	.....

*Early production engine crankcase capacity is 3.5 pints (1.6 L).
**See HYDROSTATIC TRANSMISSION section.

	MODELS		
	616H	1814	1816
Engine Make	B&S	B&S	B&S
Model	402707	400700	402707
Bore	3.44 in. (87.3 mm)	3.44 in. (87.3 mm)	3.44 in. (87.3 mm)
Stroke	2.16 in. (54.8 mm)	2.16 in. (54.8 mm)	2.16 in. (54.8 mm)
Displacement	40 cu. in. (656 cc)	40 cu. in. (656 cc)	40 cu. in. (656 cc)
Power Rating	16 hp (11.9 kW)	14 hp (10.4 kW)	16 hp (11.9 kW)
Slow Idle	1350 rpm	1350 rpm	1350 rpm
High Speed (No-load)	3400 rpm	3400 rpm	3400 rpm
Capacities—			
Crankcase	3 pts.* (1.4 L)	3 pts.* (1.4 L)	3 pts.* (1.4 L)
Transmission	**	12.5 pts. (5.8 L)	12.5 pts. (5.8 L)
Differential	2.75 pts. (1.3 L)	See Trans.	See Trans.

*Early production engine crankcase capacity is 3.5 pints (1.6 L).
**See HYDROSTATIC TRANSMISSION section.

	MODELS 1817	1918	1920
Engine Make	Kohler	Kohler	Kohler
Model	KT-17	MV-18	MV-20
Bore	3.12 in. (79.3 mm)	3.12 in. (79.3 mm)	3.12 in. (79.3 mm)
Stroke	2.75 in. (69.9 mm)	2.75 in. (69.9 mm)	3.06 in. (78 mm)
Displacement	42.18 cu. in. (690.5 cc)	42.18 cu. in. (690.5 cc)	46.98 cu. in. (818.9 cc)
Power Rating	17 hp (12.7 kW)	18 hp (13.4 kW)	20 hp (14.9 kW)
Slow Idle	1350 rpm	1200 rpm	1200 rpm
High Speed (No-Load)	3400 rpm	3600 rpm	3600 rpm
Capacities—			
Crankcase	3.5 pts. (1.6 L)	3.5 pts. (1.6 L)	3.5 pts. (1.6 L)
w/filter	4 pts. (1.9 L)	4 pts. (1.9 L)	4 pts. (1.9 L)
Transmission	12.5 pts. (5.8 L)	12.5 pts. (5.8 L)	12.5 pts. (5.8 L)
Differential	See Trans.	See Trans.	See Trans.

FRONT AXLE AND STEERING SYSTEM

MAINTENANCE

Models 613 Special-613H-616 Special-616H-1814-1816-1817

Steering spindles, axle pivot and steering gear should be lubricated after every 25 hours of operation. Front wheel bearings should be cleaned and repacked after every 100 hours of operation. Recommended lubricant is lithium based, multipurpose grease. Periodically check for loose or damaged components.

Models 1918-1920

Use a good quality oil and lubricate pivot points after every 25 hours of operation. Inject lithium based, multipurpose grease into all grease fittings after every 100 hours of operation. Periodically check for loose or damaged components.

AXLE MAIN MEMBER

Models 613 Special-613H-616 Special-616H-1814-1816-1817

To remove axle main member (6—Fig. DA1) assembly from tractor, support front of tractor and disconnect drag link at left spindle arm. Remove pivot bolt (2), pull axle assembly forward out of center pivot and remove axle. Renew bushing (11) and bolt (2) as needed.

Reinstall by reversing removal procedure.

Models 1918-1920

To remove axle main member (15—Fig. DA2), support front of tractor and remove hood. Detach drag link from steering arm (3). Remove guard for front pto. Push against belt idler and detach pto belts. Unscrew bolts (B—Fig. DA3) retaining pto bearing support and remove pto shaft assembly. Unscrew bolts securing pivot pins (23—Fig. DA2) and drive out pins toward front of tractor. Support axle assembly, remove bolt (10) and lower axle away from tractor.

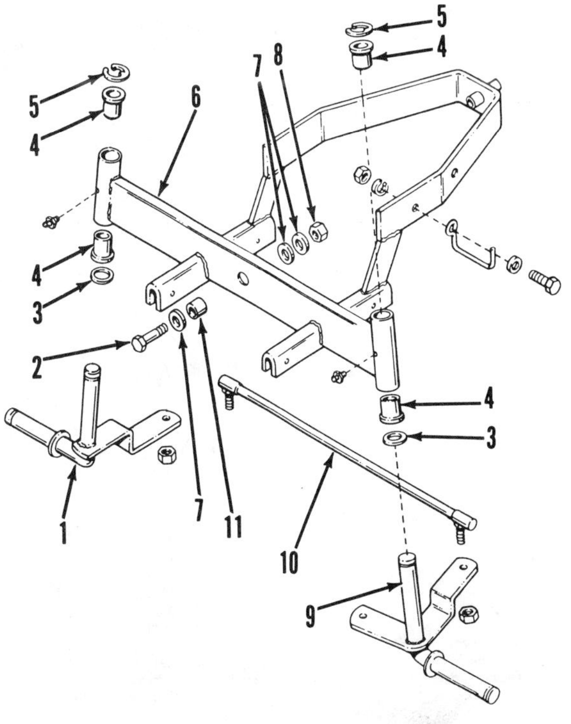

Fig. DA1—Exploded view of front axle used on Models 613 Special, 613H, 616 Special, 616H, 1814, 1816 and 1817.

1. Spindle, R.H.
2. Front pivot bolt
3. Washer
4. Bushings
5. Retaining ring
6. Main member
7. Washers
8. Nut
9. Spindle, L.H.
10. Tie rod
11. Spacer

Inspect components for excessive wear and damage. Reinstall by reversing removal procedure. Tighten bolt (10) to 90-120 ft.-lbs. (122-163 N·m). Lubricate pivot pins by injecting grease through grease fittings.

TIE ROD AND TOE-IN

All Models

Front wheel toe-in should be 1/2-3/4 inch (13-19 mm) on Models 1918 and 1920, and 1/8-3/8 inch (3.2-9.5 mm) on all other models. Toe-in is adjustable on Models 1918 and 1920 by rotating tie rod ends to alter length of tie rod. Toe-in is not adjustable on other models.

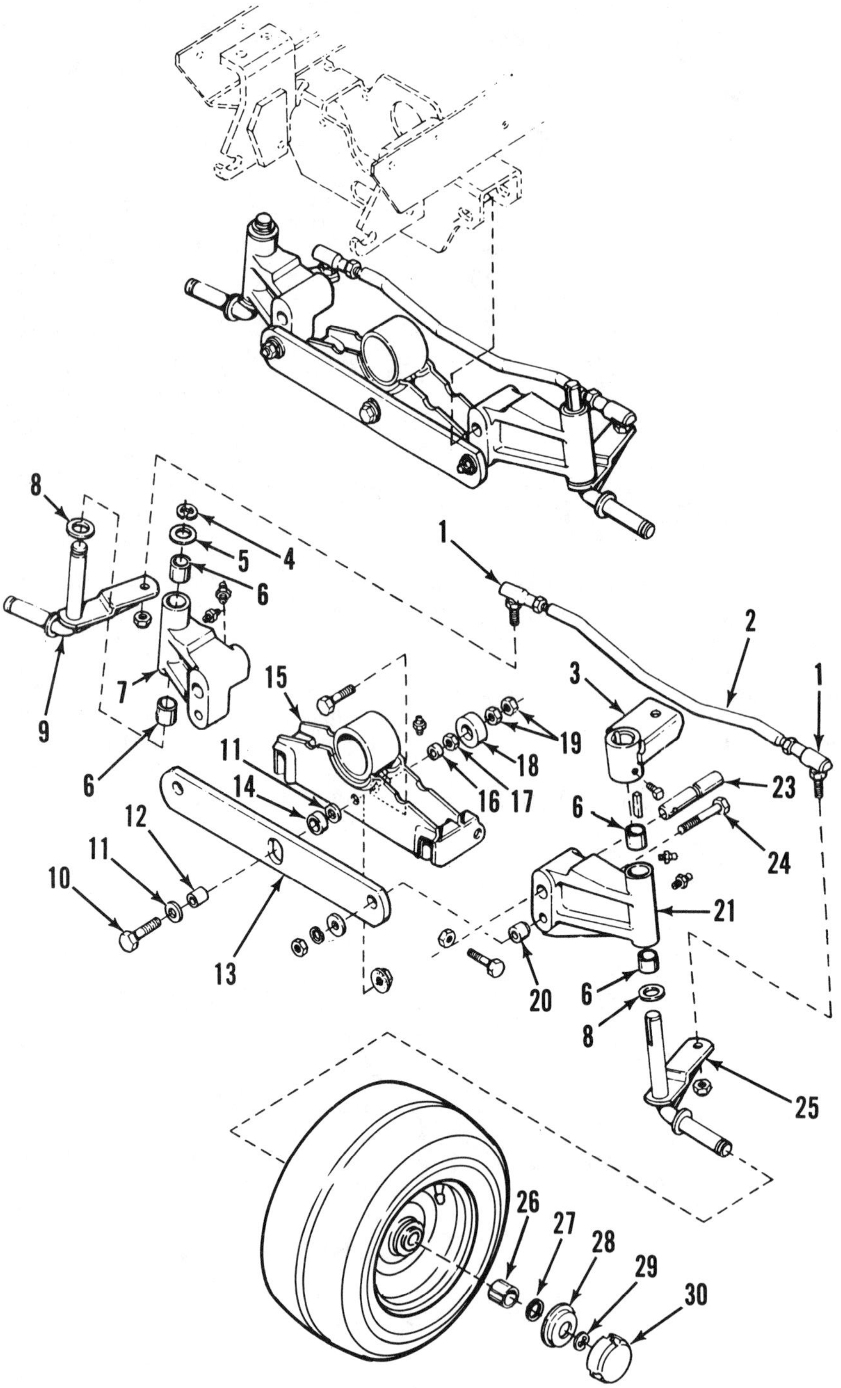

Fig. DA2—Exploded view of front axle assembly used on Models 1918 and 1920.

1. Tie rod end
2. Tie rod
3. Steering arm
4. "E" ring
5. Washer
6. Bushing
7. Axle pivot, R.H.
8. Washer
9. Spindle, R.H.
10. Cap screw
11. Washer
12. Spacer
13. Axle link
14. Spacer
15. Axle main member
16. Spacer
17. Nut
18. Retainer
19. Jam nuts
20. Spacer
21. Axle pivot, L.H.
23. Pivot pin
24. Cap screw
25. Spindle, L.H.
26. Bushing
27. Shim washer
28. Retainer
29. "E" ring
30. Hub cap

STEERING SPINDLES

Models 613 Special-613H-616 Special-616H-1814-1816-1817

To remove spindle (1 or 9—Fig. DA1), raise and support front of tractor. Remove front wheel and tire. Disconnect tie rod (10) from both sides and drag link from left side. Remove snap ring (5) and spindle. Renew bushings (4), washers (3) and spindles (1 or 9) as needed.

Models 1918-1920

To remove spindle (9 or 25—Fig. DA2), remove wheel then detach tie rod from spindle. On right spindle, remove "E" ring (4) and slide spindle out of axle end. To remove left spindle, loosen set screws and pull steering arm (3) off spindle. Remove key and slide spindle out of axle end.

Bushings (6) are renewable. Tighten tie rod ends to 45-60 ft.-lbs. (61-81 N·m). Tighten steering arm set screws to 13-19 ft.-lbs. (18-25 N·m).

FRONT WHEEL BEARINGS

All Models

To renew wheel bushings, raise and support front of tractor. Remove wheel cover. Remove snap ring and washer from spindle end. Remove wheel. Inner and outer bushings may be renewed as needed. If bushings are loose in wheel, repair or renew wheel. Apply a coating of lithium based, multipurpose grease to spindle and bushings.

Reinstall by reversing removal procedure. On Models 1918 and 1920, inject grease into fitting after installing wheel. Adjust wheel side play on Models 1918 and 1920 by installing shims (27—Fig. DA2) so side play is 1/32 inch (0.8 mm).

DRAG LINK

Model 1918 With Manual Steering

ADJUSTMENT. After adjusting toe-in as previously outlined, adjust drag link length as follows: Turn steering wheel until hole in steering gear (11—Fig. DA4) aligns with hole in plate (14). Steering arm (29) should be vertical, if not, relocate arm. Apply Loctite to nut securing steering arm (29) and tighten to 35-40 ft.-lbs. (47-54 N·m). Adjust length of drag link (32) so center-to-center distance between ends is 19 5/16

Illustrations courtesy Deutz-Allis.

inches (490 mm). Be sure there is equal thread engagement at both ends. Install drag link. Turn steering wheel counterclockwise until stopped. Adjust length of drag link so steering arm on left spindle contacts axle just before spacer (12) contacts plate (14). Tighten all fasteners and check steering operation.

STEERING GEAR

Model 1918 With Manual Steering

R&R AND OVERHAUL. To service steering gear assembly, remove hood, air plenum and battery. Remove dash side panels and tunnel cover. Remove steering wheel nut (1—Fig. DA4) and pull steering wheel from steering shaft (7) using a suitable puller. Remove snap ring (10) from lower end of steering shaft, raise shaft until clear of support plate (14) and withdraw shaft from bottom of tractor. Note that some early model tractors use a two-piece steering shaft. Disconnect drag link end (31) from steering arm (29). Unbolt and remove steering gear assembly from tractor.

Remove bevel gear (19) and steering arm (29) from steering arm shaft (24). Remove snap rings (27) and withdraw steering arm shaft from gear housing (21). Remove snap ring (18), bevel pinion (17), thrust washer (16) and steering gear and shaft (11) from housing.

Bushings (4, 15 and 22) are renewable. Renew thrust washers (8 and 16) as necessary if steering shaft end play is excessive.

When assembling, note the following: Pinion gear (17) must be centered on bevel gear (19) when front wheels are pointing straight ahead. Apply Loctite to nut securing bevel gear (19) to shaft and tighten to 35-40 ft.-lbs. (47-54 N·m). Apply Loctite to screws securing clamp (20) and tighten to 25-30 ft.-lbs. (34-40 N·m). Apply Loctite to nut securing steering arm (29) to shaft and tighten to 35-40 ft.-lbs. (47-54 N·m). Tighten bolt securing drag link end (31) to 45-60 ft.-lbs. (62-81 N·m).

Models 613 Special-613H-616 Special-616H-1814-1816-1817

R&R AND OVERHAUL. To remove steering gear, remove steering wheel, upper dash, fuel tank and lower dash. Disconnect drag link (11—Fig. DA5) from steering rod (10). Unbolt and remove upper steering plate (6) and steering shaft (8). Remove lower plate (9) and steering rod (10) as an assembly. Remove two set screws (13) and use a suitable puller or press to remove steering gear (12) from steering rod (10).

Clean and inspect all parts. Renew bushings (5) as needed.

To reassemble, install key (14) with rounded end toward bent end of steering rod (10). Press gear (12) onto steering rod until top edge of gear is flush with end of key (14). Steering rod should extend approximately 1/4 inch (6 mm) above steering gear. Install assembly on lower plate (9). Place spacer (7) on steering shaft (8) with bevel toward steering gear. Center steering gear on lower plate and install steering shaft so hole in shaft faces front and back. Complete reassembly by reversing disassembly procedure.

POWER STEERING SYSTEM

Pressurized oil from the hydrostatic drive unit is supplied to a steering control valve and cylinder to provide power steering for all 1920 models and as an

Fig. DA3—Remove bolts (B) so pto shaft and support assembly can be removed on Models 1918, 1920.

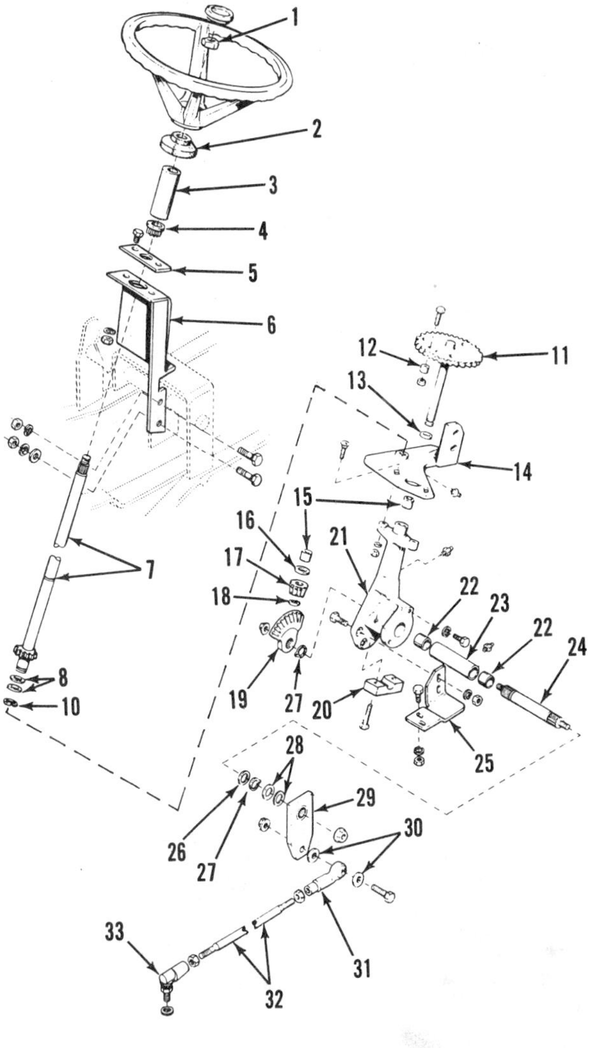

Fig. DA4—Exploded view of steering gear assembly used on Models 1918 with manual steering.

1. Nut
2. Boot
3. Tube
4. Bushing
5. Retainer plate
6. Support bracket
7. Steering shaft
8. Thrust washers
10. Retaining ring
11. Gear & shaft
12. Spacer
13. Curved washer
14. Support plate
15. Bushings
16. Thrust washer
17. Pinion gear
18. Retaining ring
19. Sector gear
20. Clamp
21. Steering gear housing
22. Bushings
23. Tube
24. Sector shaft
25. Bracket
26. Washer
27. Snap ring
28. Washers
29. Steering arm
30. Washers
31. Ball joint
32. Drag link
33. Ball joint

Illustrations courtesy Deutz-Allis.

option on 1918 models. Refer to section on HYDROSTATIC TRANSMISSION for maintenance and service information on hydrostatic unit.

STEERING CONTROL VALVE

All Models So Equipped

R&R AND OVERHAUL. To remove control valve, remove the hood. Remove pto guard assembly at front of tractor. Detach pto drive belts. Loosen fasteners securing engine plate and slide engine forward so drive shaft is separated from transmission input shaft. Position drive shaft out of way. Remove cleanout panel adjacent to oil cooler. Remove air induction plenum. Remove side dash panels, tunnel cover and left lower panel. Remove battery. Remove steering wheel and boot using a suitable puller (do not strike steering shaft). Mark location of hydraulic lines on steering valve and note steering valve location in reference to mounting bracket. Disconnect hydraulic lines from bottom of control valve and cap openings. Hydraulic lines connected to "B" ports (Fig. DA6) should be removed from tractor to provide clearance for removal of valve. Remove nuts securing steering unit and lower unit through dash and remove.

NOTE: Do not clamp steering control valve in a vise for service work. Fabricate a holding fixture as shown in Fig. DA7.

Before disassembling steering valve, note alignment grooves machined in side of unit. All marked components must be reinstalled in their original positions.

Place steering valve in holding fixture with steering shaft down and secure with 5/16-24 NF nuts. Remove the four nuts securing port cover (4—Fig. DA8) and remove cover. Remove sealing ring (6) and five "O" rings (5). Remove plug (1) and ball (3). Remove port manifold (7) and the three springs (8). Remove valve ring (9) and discard sealing rings (6). Remove valve plate (11) and the three springs (10) from isolation manifold pockets. Remove hex drive and pin assembly (12), then remove isolation manifold (14). Remove drive link (15), metering ring (16) and seal rings (6). Lift off metering assembly and remove commutator seal (18). Remove the 11 Allen head screws (17) and lift commutator cover (19) off. Remove commutator ring (20), then carefully remove commutator (21) and the five alignment pins (22). Remove drive link spacer (23) and separate rotor (24) from stator (25). Remove drive plate (26), spacer (27), thrust bearing (28), seal spacer (31), face seal backup ring (30) and face seal (29). Remove upper cover plate (32). Remove steering shaft (36) and service retainer plate (35), upper bushing (38) and seal (39) as necessary.

Clean and inspect all parts and renew any showing signs of wear, scratches, scoring or other damage. Springs (8) should have a free length of 0.75 inch (19 mm) and springs (10) should have a free length of 0.50 inch (12.7 mm). Springs must be renewed as sets only. The commutator (21) and commutator ring (20) are a matched set and must be renewed as an assembly. Check rotor to stator clearance using a feeler gage as shown in Fig. DA9. If clearance exceeds 0.005 inch (0.13 mm), renew rotor and stator assembly. Renew all seals and "O" rings. Lubricate all parts with clean hydraulic fluid prior to installation.

Reassemble by reversing disassembly procedure and referral to the following notes: Be sure alignment grooves on upper cover plate (32), isolation manifold (14) and port manifold (7) are on same side. Install rotor (24) so pin holes are out. Install commutator so long grooves (G) are out as shown in Fig. DA10. When installed, pins (22) must not protrude above face of commutator (21). Apply low strength thread-locking solution to threads of screws (17) and install screws finger tight. Insert metering valve assembly in metering ring (16—Fig. DA8) with commutator cover (19) facing downward. Place a suitable size wood block under metering valve assembly so that drive plate (26) is partially out of metering ring. Make six shims that are 0.007 inch thick, 1/2-inch wide and 1 1/2 inches long. Insert shims between drive plate and metering ring so two shims

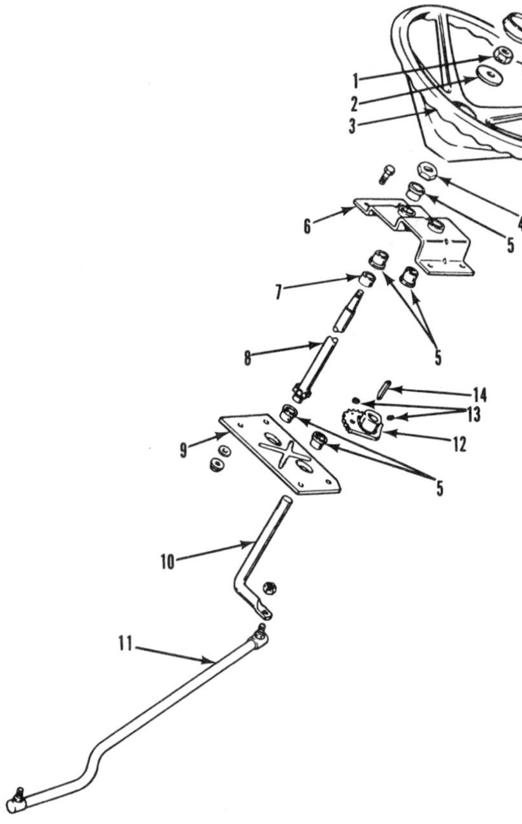

Fig. DA5—Exploded view of steering shaft and mechanism used on Models 613 Special, 613H, 616 Special, 616H, 1814, 1816 and 1817.

1. Locknut
2. Washer
3. Steering wheel
4. Hex washer
5. Bushings
6. Upper steering plate
7. Spacer
8. Steering shaft
9. Lower steering plate
10. Steering rod
11. Drag link
12. Steering gear
13. Set screws
14. Key

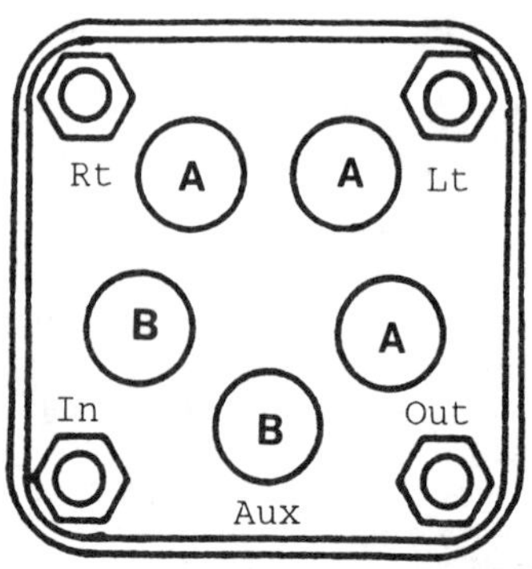

Fig. DA6—Drawing of bottom of power steering control valve showing hydraulic hose ports.

Illustrations courtesy Deutz-Allis.

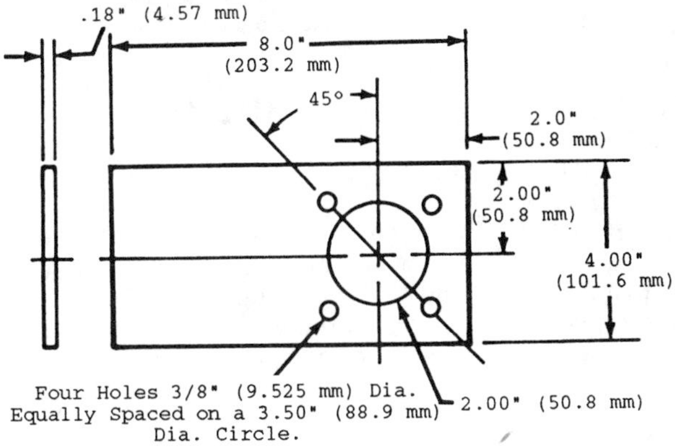
.18" (4.57 mm)
8.0"
(203.2 mm)
45°
2.0"
(50.8 mm)
2.00"
(50.8 mm)
4.00"
(101.6 mm)
Four Holes 3/8" (9.525 mm) Dia.
Equally Spaced on a 3.50" (88.9 mm)
Dia. Circle.
2.00" (50.8 mm)

Fig. DA7—Fabricate a holding fixture as shown to aid service work on power steering control valve used on models with power steering.

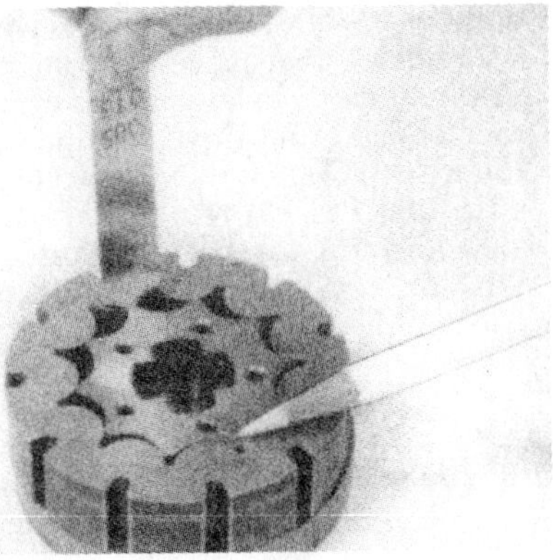

Fig. DA9—Use a feeler gage to check rotor to stator clearance. Refer to text.

1. Plug
2. "O" ring
3. Ball
4. Port cover
5. "O" rings
6. Seal rings
7. Port manifold
8. Springs
9. Valve ring
10. Springs
11. Valve plate
12. Hex drive assy.
13. Alignment pins (2)
14. Isolation manifold
15. Drive link
16. Metering ring
17. Screws (11)
18. Commutator seal
19. Commutator cover
20. Commutator ring
21. Commutator
22. Alignment pins (5)
23. Drive link spacer
24. Rotor
25. Stator
26. Drive plate
27. Spacer
28. Thrust bearing
29. Face seal
30. Back-up ring
31. Seal spacer
32. Upper cover plate
33. Snap ring
34. Washer
35. Retainer
36. Steering shaft
37. Steering tube
38. Bearing
39. Dust seal
40. Special bolts
41. Nut

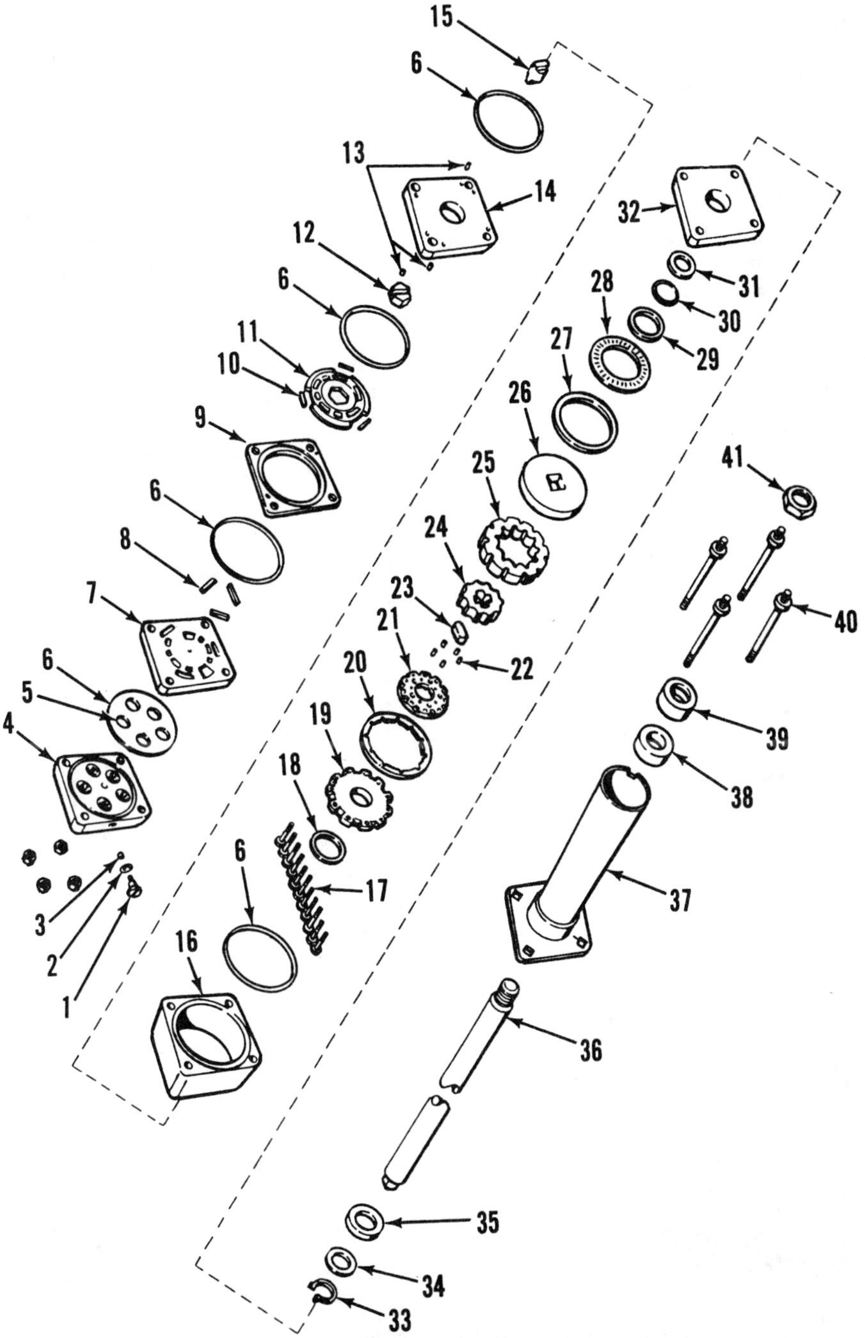

Fig. DA8—Exploded view of power steering control valve used on models equipped with power steering.

Illustrations courtesy Deutz-Allis.

abut every 120 degrees. Remove the wood block and push metering valve down until drive plate is flush with surface of metering ring. Tighten screws (17) in several steps to final torque of 11-13 in.-lbs. (1.24-1.47 N·m) using sequence shown in Fig. DA11. Commutator ring (20—Fig. DA8) should now be concentric within 0.005 inch (0.127 mm) of

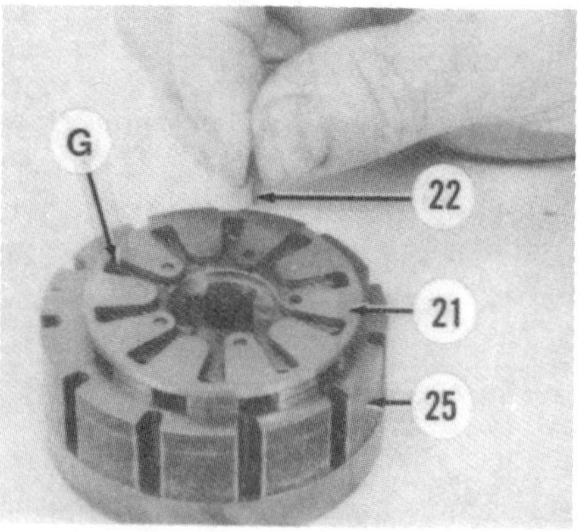

Fig. DA10—Install commutator (21) so long grooves (G) are out.

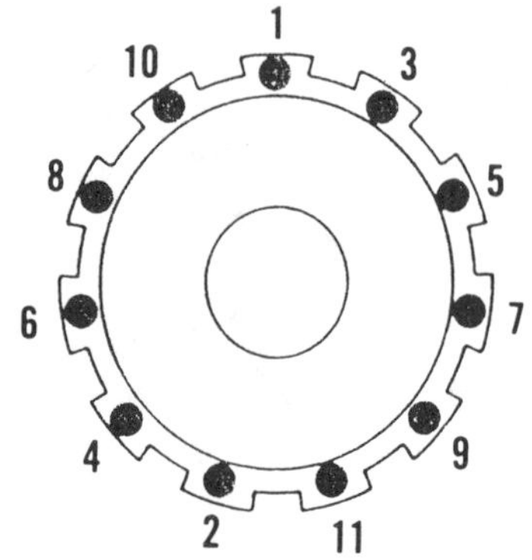

Fig. DA11—Tighten the 11 Allen head screws in metering assembly to 11-13 in.-lbs. (1.24-1.47 N·m) in sequence shown. Refer to text.

Fig. DA12—Install valve plate (11) so "PORT SIDE" on plate is opposite from alignment grooves (A) on isolation manifold.

drive plate (26). Insert large tang end of drive link (15) into slot in rotor (24) and rotate metering assembly to make certain parts do not bind. Install commutator seal (18) so yellow mark on seal is toward commutator cover (19). Be sure ½ inch (13 mm) springs (10) are installed in isolation manifold (14) and ¾ inch (19 mm) springs (8) are installed in port manifold (7). Install valve plate (11) so that spring slot with smallest cavity and the words "PORT SIDE" are positioned opposite from alignment grooves (A) on isolation manifold as shown in Fig. DA12. Tighten valve assembly nuts in a crossing pattern to 20-24 ft.-lbs. (27-33 N·m). Tighten plug (1—Fig. DA8) to 97-151 in.-lbs. (11-17 N·m). Make sure that all parts with alignment grooves are aligned as shown in Fig. DA12A, otherwise unit will not operate correctly.

To reinstall steering control valve, reverse the removal procedure. Start engine and turn steering wheel from full left to full right several times to bleed air from system.

STEERING CYLINDER

All Models So Equipped

R&R AND OVERHAUL. Mark hydraulic hose locations and disconnect hoses. Disconnect steering cylinder at each end. Steering cylinder used on all models is a welded assembly and must be renewed as an entire unit. Tighten mounting bolts at both ends of cylinder to 200-225 ft.-lbs. (272-306 N·m).

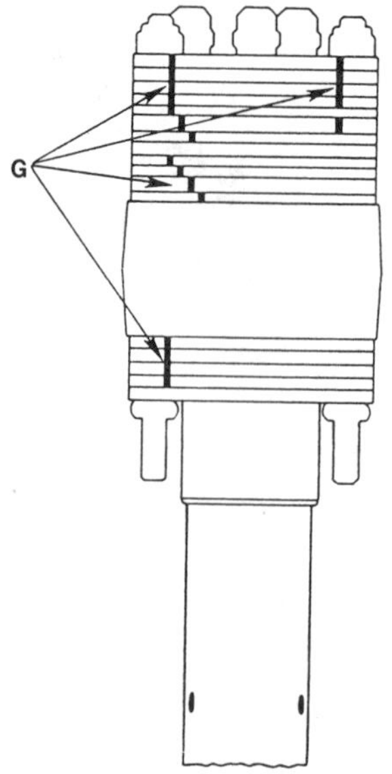

Fig. DA12A—Components of steering control valve having alignment grooves (G) must be positioned as shown, or unit will not operate correctly.

ENGINE

MAINTENANCE AND REPAIR

All Models

Engine make and model are listed at the beginning of this section. Refer to appropriate engine section for information concerning recommended maintenance and repair procedures.

REMOVE AND REINSTALL

Models 613 Special-613H-616 Special-616H-1814-1816-1817

Disconnect battery ground cable and remove hood. Disconnect fuel line, throttle and choke cables, and all interfering electrical wires. Remove drive belts and engine mounting bolts. Remove engine.

Reinstall by reversing removal procedure.

Models 1918-1920

To remove engine, disconnect battery ground cable and remove hood. Remove dash side panels. Disconnect all interfering electrical wires. Disconnect fuel line and remove air cleaner. Detach throttle and choke cables. Remove pto guard and detach pto drive belts. Loosen screws securing engine to engine plate and slide engine forward so drive shaft separates from transmission input shaft. Remove engine mounting bolts, lift engine toward front of tractor, detach drive shaft and remove engine.

Reinstall by reversing removal procedure.

ELECTRICAL SYSTEM

MAINTENANCE AND SERVICE

All Models

Battery electrolyte level should be checked at 25-hour intervals of normal operation. If necessary, add distilled water until level is just below base of vent well. DO NOT overfill. Keep battery posts clean and cable ends tight.

For alternator or starter service, refer to appropriate engine service section in this manual.

Refer to appropriate wiring diagram for model being serviced (Fig. DA13 through DA18).

Illustrations courtesy Deutz-Allis.

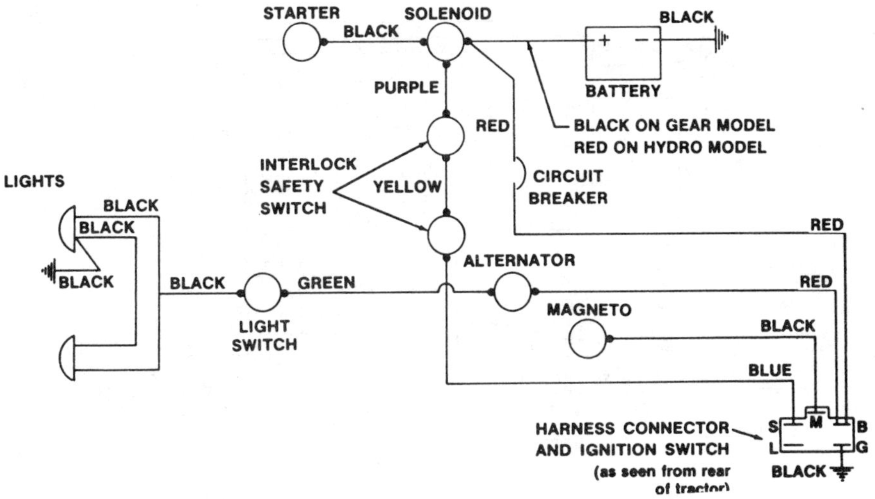

Fig. DA13—Typical wiring diagram for Models 613 Special, 613H, 616 Special and 616H.

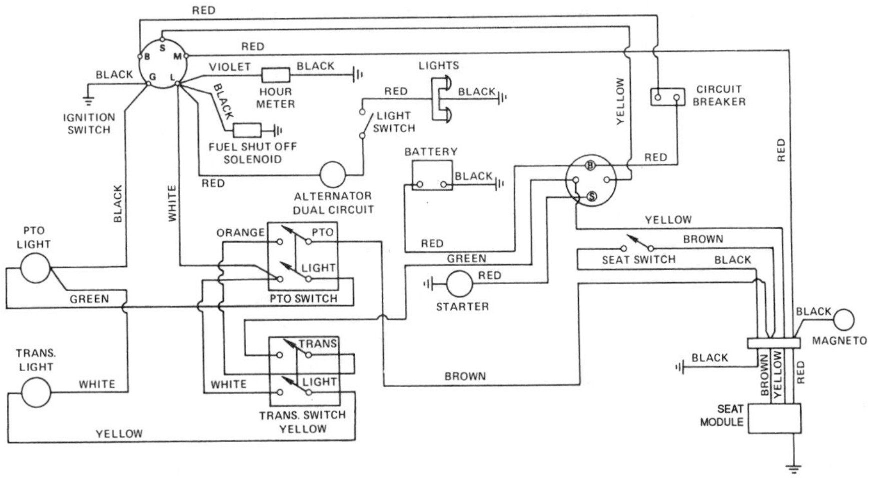

Fig. DA14—Typical wiring diagram for early Models 1814 and 1816 (prior to serial number 9066).

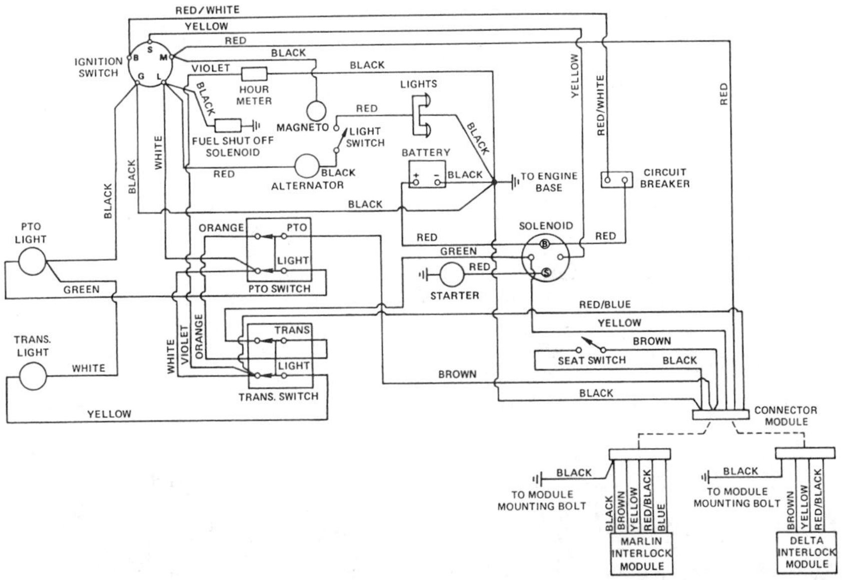

Fig. DA15—Typical wiring diagram for some later production Models 1814 and 1816.

Illustrations courtesy Deutz-Allis.

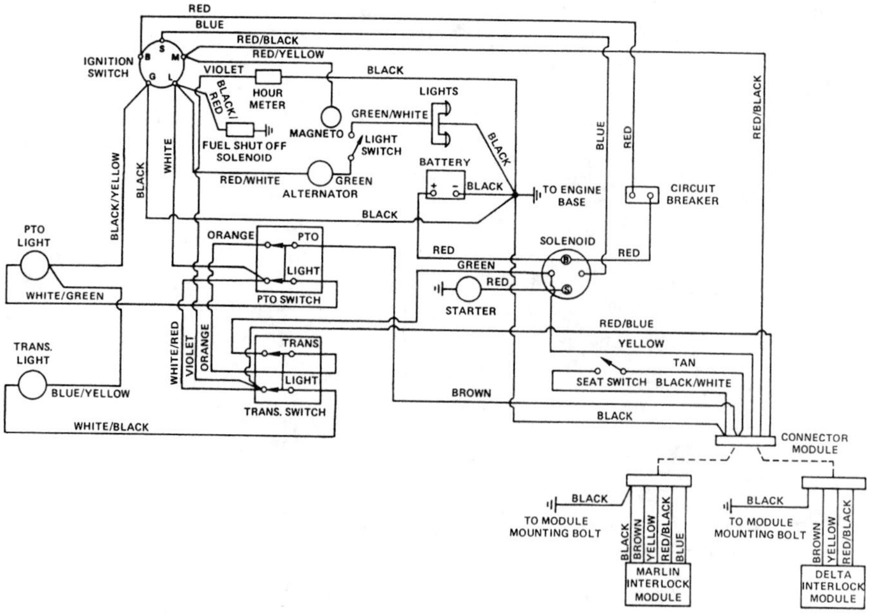

Fig. DA16—Typical wiring diagram for latest production Models 1814 and 1816.

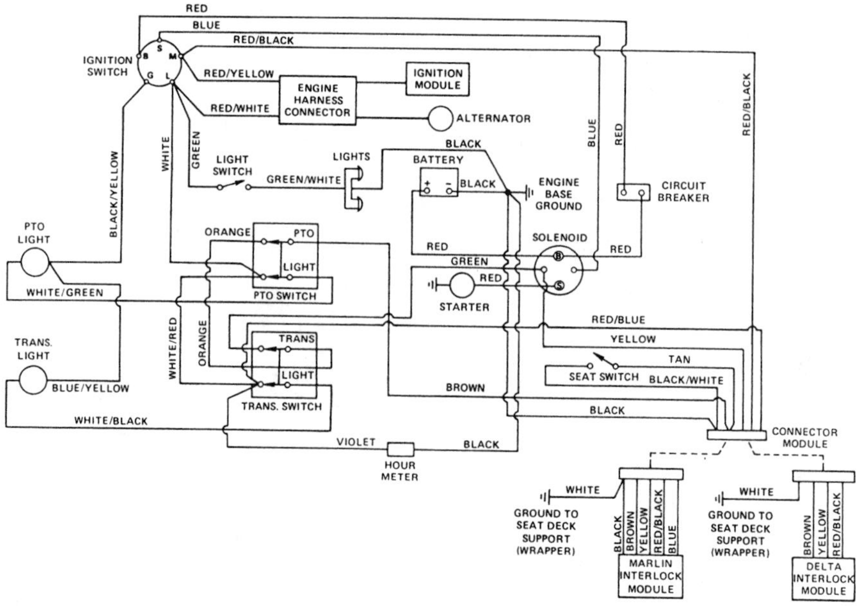

Fig. DA17—Typical wiring diagram for Model 1817.

Illustrations courtesy Deutz-Allis.

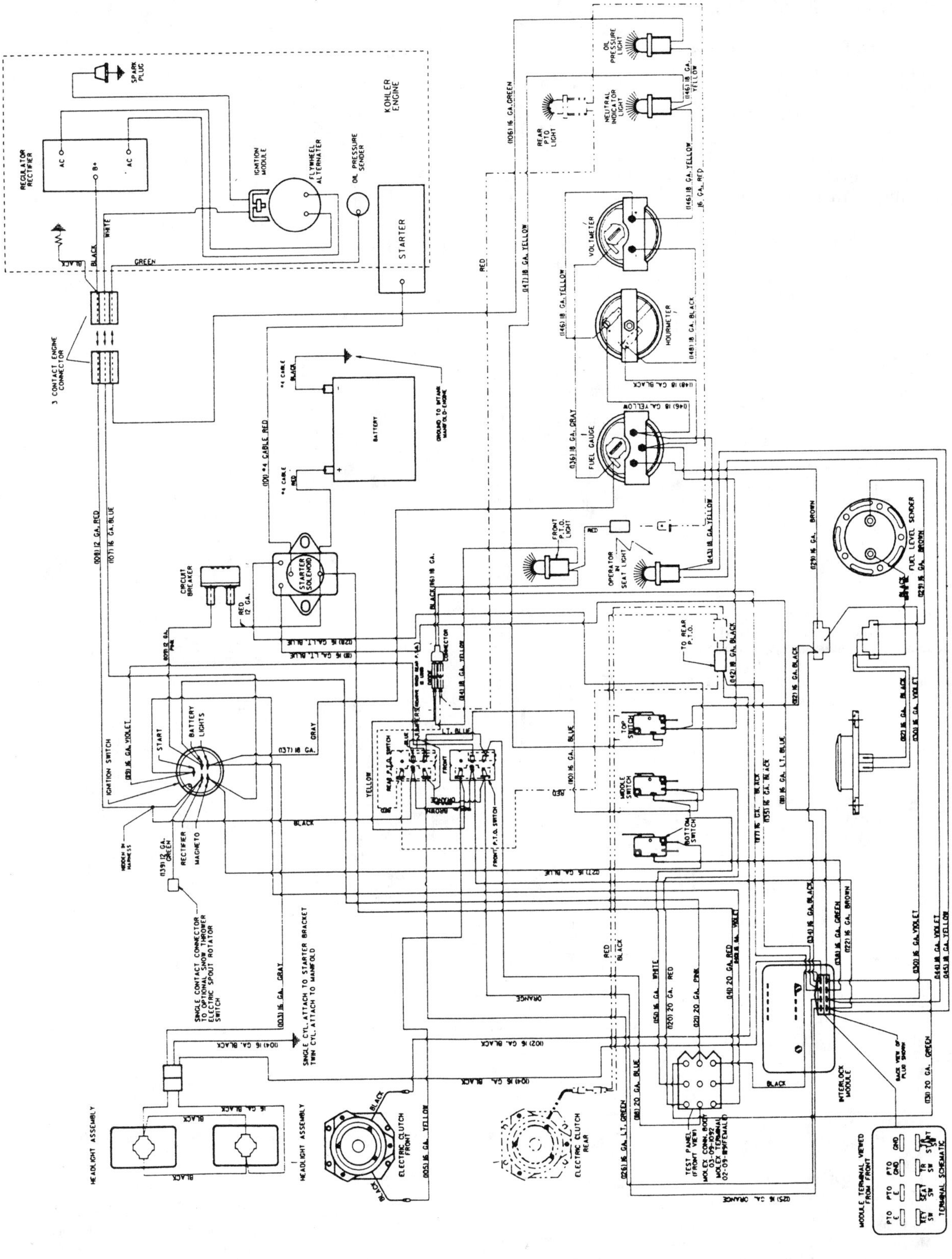

Fig. DA18—Typical wiring diagram for Models 1918 and 1920.

Illustrations courtesy Deutz-Allis.

BRAKE

ADJUSTMENT

Models 613 Special-616 Special

Place gear shift lever so a gear is engaged. Parking brake must be released. Move brake rod back and forth to be sure brake is not engaged. Push brake lever (R—Fig. DA19) forward to remove slack and measure gap between lever and stop (S). Gap should be $^1/_8$ inch (3.2 mm). To adjust gap, loosen outer nut (N) then turn inner nut until desired gap is obtained. Retighten outer nut and recheck gap.

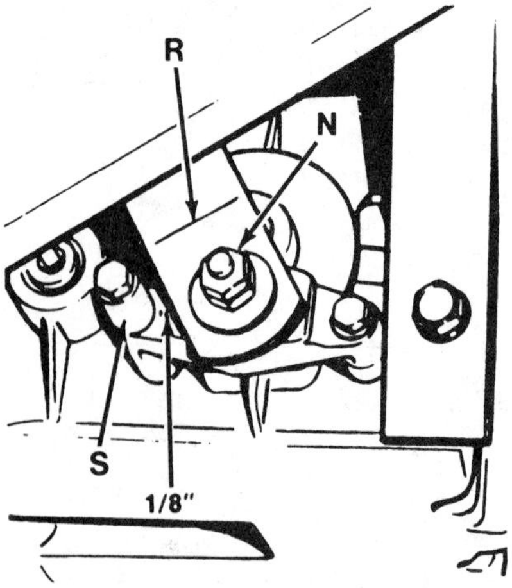

Fig. DA19—To adjust brake on Models 613 Special and 616 Special, push brake lever (R) forward to remove slack and measure gap between lever and stop (S). Gap should be $^1/_8$ inch (3.2 mm). Loosen outer nut (N) and turn inner nut to obtain correct gap.

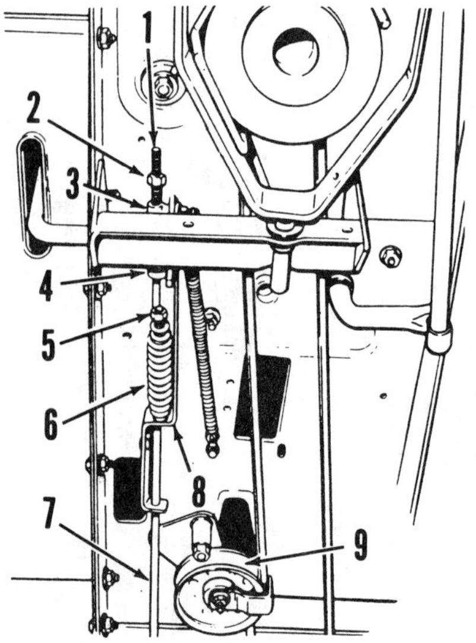

Fig. DA20—View of brake and clutch mechanism used on Models 613 Special and 616 Special. Refer to text for adjustment.

1. Clutch rod
2. Nut
3. Clutch rod guide
4. Set collar
5. Nut
6. Spring
7. Brake rod
8. Brake rod guide
9. Idler pulley

Push brake rod guide (8—Fig. DA20) forward as far as possible while pulling brake rod (7) back. Turn nut (5) so there is a clearance of $^1/_4$ inch (6.4 mm) between nut and end of spring (6).

Models 613H-616H

Parking brake must be released and brake linkage must be free. Clearance between brake disc (D—Fig. DA21) and pad should be 0.010 inch (0.25 mm). Adjust clearance by loosening outer nut (N) and rotating inner nut. Retighten outer nut.

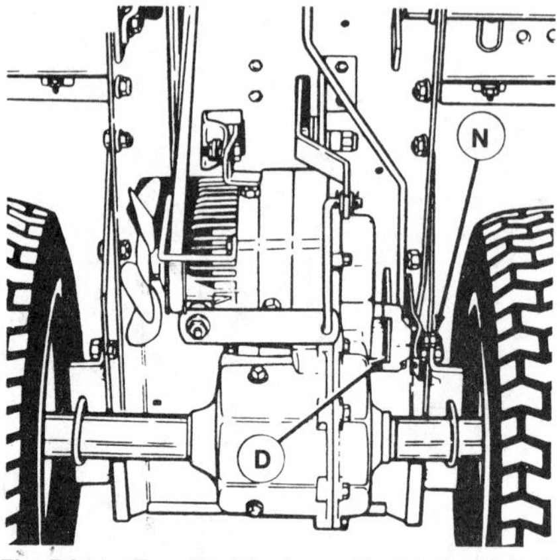

Fig. DA21—To adjust brake on Models 613H and 616H, loosen outer nut (N) and rotate inner nut so clearance between brake disc (D) and pad is 0.010 inch (0.25 mm).

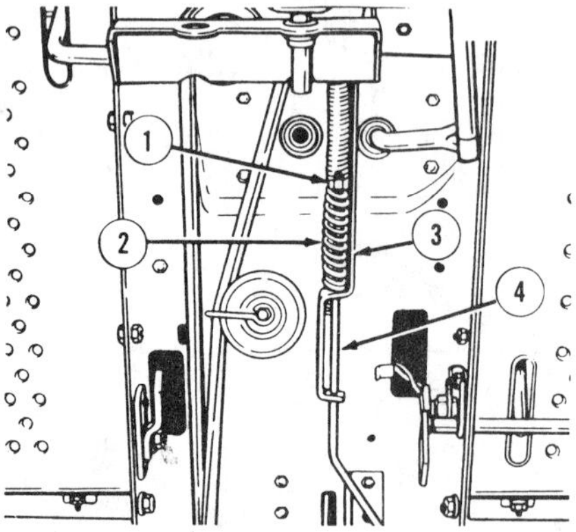

Fig. DA22—On Models 613H and 616H, push rod guide (3) forward while pulling brake rod (4) rearward. Rotate nut (1) so spring is snug but will still turn.

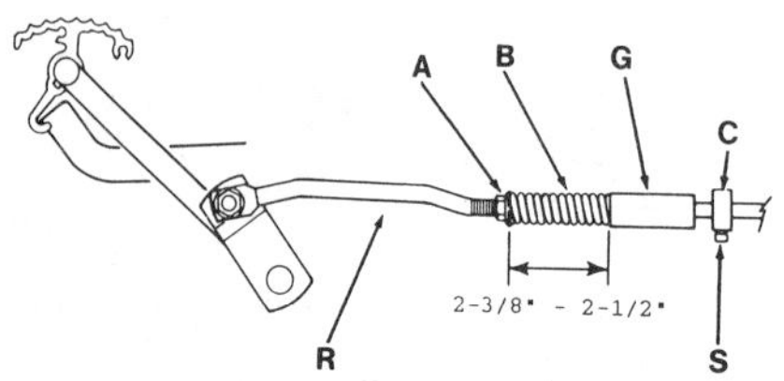

Fig. DA23—On Models 1814, 1816 and 1817, brake spring (B) length should be $2^3/_8$ to $2^1/_2$ inches (60.3-63.5 mm) and is adjusted by turning nut (A). Release parking brake, pull brake rod (R) to rear and position collar (C) against rod guide (G).

Push brake rod guide (3—Fig. DA22) forward while pulling brake rod (4) rearward. Rotate nut (1) so spring (2) is snug but will still turn.

Models 1814-1816-1817

Fully depress brake pedal and engage parking brake latch, then measure length of spring (B—Fig. DA23). Spring length should be $2^3/_8$ to $2^1/_2$ inches (60.3-63.5 mm). If adjustment is needed, loosen collar setscrew (S) and turn nut (A) to compress or release spring tension. Spring length must not be less than $2^3/_8$ inches (60.3 mm). Release parking brake, pull brake rod (R) to rear and position collar (C) against rod guide (G). Retighten collar setscrew.

Models 1918-1920

Raise and support rear of tractor. Remove rubber plug (12—Fig. DA24) from brake backing plate (14). Insert brake adjusting tool or screwdriver through brake adjustment slot in backing plate and turn adjustment wheel (16) until it is difficult to rotate tire and wheel by hand. Depress brake pedal firmly to seat shoes, then release pedal. Back adjuster wheel off until just a slight drag is felt when rotating tire and wheel by hand. Adjust opposite side in the same manner making certain left and right pedal height is equal.

Apply both brakes and set parking brake so tractor will not move. Loosen outer nut and turn inner nut (11) until brake rod spring (10) compressed length is $1^5/_8$ inches (41.3 mm). Tighten outer nut against inner nut. Repeat spring tension adjustment on opposite brake rod.

R&R BRAKES

Models 613 Special-613H-616 Special-616H

Remove adjustment nut (1—Fig. DA25) and brake lever (11). Remove bolts (3) and spacer (4). Remove brake caliper (5), outer brake pad (6), metal backing plate (9), brake disc (8) and inner brake pad (7).

Make certain the two push pins (10) and metal backing plate (9) remain in place in brake pad holder and metal spacer (4) is in place on rear mounting bolt during reassembly.

Adjust brake as previously outlined.

Illustrations courtesy Deutz-Allis.

Models 1814-1816-1817

The brake is located inside the transmission unit. Refer to transmission section for brake service.

Models 1918-1920

Raise and support rear of tractor. Remove tires and wheels. Bend tabs of lock plate from axle nut and remove nut. Remove brake drum. Disconnect hold-down springs (13—Fig. DA24) and spring (15). Remove brake shoes (17).

Reinstall by reversing removal procedure. Tighten axle nut to 75-100 ft.-lbs. (102-136 N·m) and secure nut with lock plate.

If originally not equipped with a nut retainer, install retainer.

CLUTCH

ADJUSTMENT

Models 613 Special-616 Special

Clutch/brake pedal must be up. Move idler pulley (9—Fig. DA20) tightly against drive belt. Rotate nut (2) so gap between nut and rod guide (3) is 7/8-1 inch (22-25 mm). Push down and lock clutch/brake pedal. Loosen set screw in collar (4), then move collar against rod guide so spring is compressed 1/4-3/8 inch (6.4-9.5 mm). Retighten set screw.

Models 1814-1816-1817

Clutch/brake pedal must be up. Rotate nut (N—Fig. DA26) so spring length (L) is 7/8-1 inch (22-25 mm). Recheck brake adjustment.

Models 1919-1920

A mechanical clutch is not used on these tractors. Refer to HYDROSTATIC TRANSMISSION section for transmission neutral adjustment procedure.

PTO

MAINTENANCE

Models 1918-1920

Belt idler (7—Fig. DA27) is equipped with a grease fitting. Inject a lithium based, multipurpose grease into fitting after every 100 hours of operation.

PTO SHAFT ASSEMBLY

Models 1918-1920

Models 1918 and 1920 are equipped with a pto shaft assembly (Fig. DA27) located in the front frame support. Maximum allowable shaft end play is 1/8 inch (3.2 mm). If end play is excessive, inspect shaft (18) and bearings (17 and 20) and renew as necessary.

Bearings are a press fit on shaft and in housing (19). Bearings must contact inner snap rings (15). Install rear bearing (20) on shaft, pressing against bearing inner race. Then press shaft and bearing assembly into housing, pressing against bearing outer race. Install remaining bearing (17) onto shaft and into housing, pressing against bearing inner and outer races while supporting rear end of shaft and outer race of rear bearing. Bearings are sealed and no periodic lubrication is required.

PTO CLUTCH

Models 1918-1920

To adjust electric pto clutch, first remove the hood. Check clearance be-

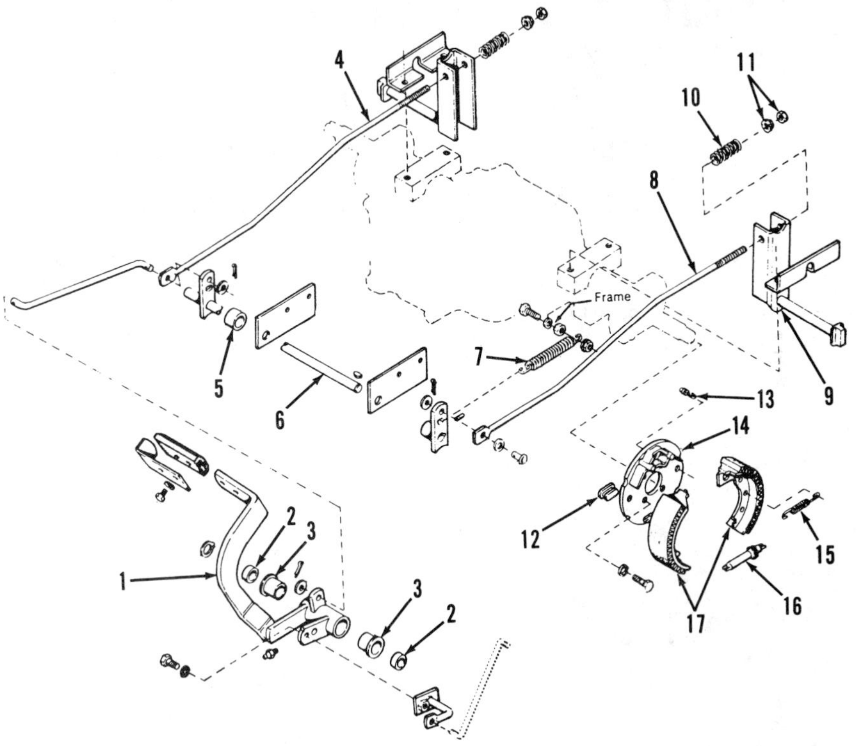

Fig. DA24—Exploded view of brake assembly and control linkage used on Models 1918 and 1920.

1. Brake pedal
2. Plugs
3. Bushings
4. Brake rod, R.H.
5. Special spacer
6. Brake pivot shaft
7. Return spring
8. Brake rod, L.H.
9. Brake actuator assy.
10. Compression spring
11. Jam nuts
12. Plug
13. Retainer springs
14. Brake backing plate
15. Spring
16. Brake adjusting wheel
17. Brake shoes

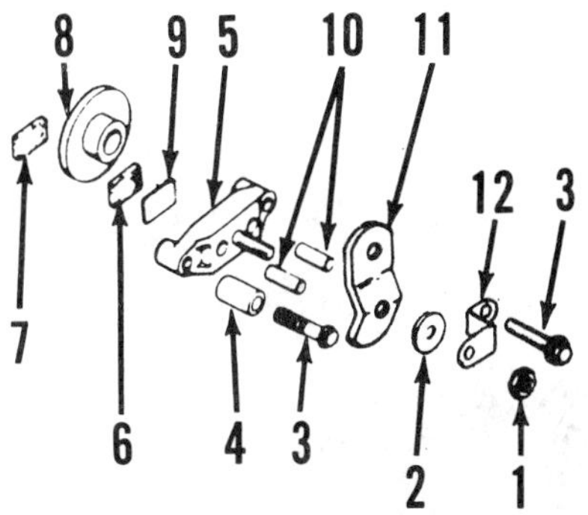

Fig. DA25—Exploded view of brake assembly used on Models 613 Special, 613H, 616 Special and 616H.

1. Adjusting nut
2. Washer
3. Bolts
4. Spacer
5. Brake caliper
6. Outer brake pad
7. Inner brake pad
8. Brake disc
9. Metal backing plate
10. Pins
11. Brake lever
12. Bracket

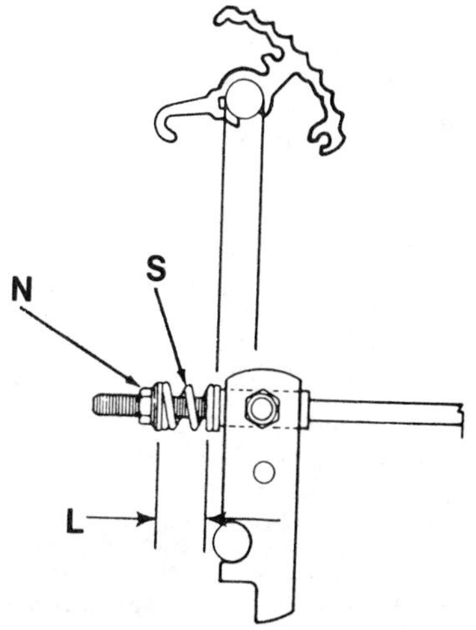

Fig. DA26—To adjust clutch on Models 1814, 1816 and 1817, rotate nut (N) so spring length is 7/8-1 inch (22-25 mm).

tween armature (1—Fig. DA27) and rotor (2) using a feeler gage at the three slots provided in the armature. Turn each of the adjusting nuts (N) as necessary until clearance between armature and rotor is 0.015-0.020 inch (0.4-0.5 mm) at each of the three slots. Do not overtighten nuts. Operate clutch several times and recheck clearance.

GEAR TYPE TRANSAXLE

MAINTENANCE

Models 613 Special-616 Special

The five-speed transaxle is packed with 36 ounces (1065 mL) of multipurpose, lithium base grease during manufacture and does not require periodic maintenance.

REMOVE AND REINSTALL

Remove mower deck. Drain fuel tank and remove the battery. Position 4 × 4 wood block behind tractor, then use a hoist to raise front of tractor so that rear frame support rests on wood block. Disconnect transmission interlock switch. Disconnect brake rod and return spring. Disconnect shift linkage and remove drive belt from transmission pulley. Remove wheels and support transaxle, then remove bolts retaining transaxle to front and rear supports. Tilt transmission forward and slide to left to remove transaxle assembly from tractor.

Reinstall by reversing removal procedure.

OVERHAUL TRANSAXLE

Models 613 Special-616 Special

Remove transaxle assembly as previously outlined. Place shift lever in neutral, then unbolt and remove shift lever. Remove disc brake assembly. Remove setscrew (1—Fig. DA28), spring (2) and index ball (3). Unbolt cover (4) and push shift fork assembly (13) in while removing cover. Before removing gear and shaft assemblies, shifter fork (13) should be removed. Note position of parts before removal. Remove gear and shaft assemblies (Fig. DA29) from case taking care not to disturb drive chain (39—Fig. DA28). Remove needle bearings (34), flat washer (32), square cut seals (33), output gear (35) and output pinion (36) from countershaft. Angle the two shafts together, then mark position of chain on sprocket collars and remove chain. Remove sprocket (38), bevel gear (41), spur gears (42 through 46), thrust washer (47) and flange bushing (48). All gears are splined to the countershaft. Disassembly of shifter-brake shaft (14 through 30) is apparent after inspection. Remove snap ring (12) and input bevel gear (11), then pull input shaft (8) through cover.

Disassemble differential by driving out the roll pin (53) securing pinion shaft (60) and remove shaft from differential housing (59). Remove pinion gears (62) by rotating gears in opposite directions. Remove snap rings (56) and slide axles out of side gears (55) and thrust washers (54). Note axle shafts (50 and 57) are different lengths.

Clean and inspect all parts and renew any showing excessive wear or other damage. When installing new input shaft needle bearings, press lower bearing in to a depth of 0.135-0.150 inch (3.43-3.81 mm) below flush and upper bearing flush with housing as shown in Fig. DA30. When installing new needle bearings (34—Fig. DA28) in output pinion (36), press bearings in until each bearing is 0.135-0.150 inch (3.43-3.81 mm) below flush. Pack all needle bearings with grease during assembly.

Install shift collar (20—Fig. DA28) so thick segment is toward shoulder as shown in Fig. DA31. Note that replacement shafts have four keyways and keys. When installing thrust washers (23—Fig. DA28) and shifting gears on brake shaft of early units, the 45 degree chamfer on inside diameter of thrust washers must face shoulder on brake shaft. On later units and replacement parts, washers (23) are curved and rounded side must face shaft shoulder. See Fig. DA32. Flat side of gears must face shoulder on shaft. Install drive chain (39—Fig. DA28) on sprockets, then install shafts and gears as an assembly in housing. Note that "V" protrusion on flanged bushings must index in "V" notch in housing (bushings on some models do not have protrusion).

Complete reassembly and pack case with approximately 36 ounces (1065 mL) of multipurpose, lithium base grease. Tighten case cover cap screws to 90-100 in.-lbs. (10-11 N·m).

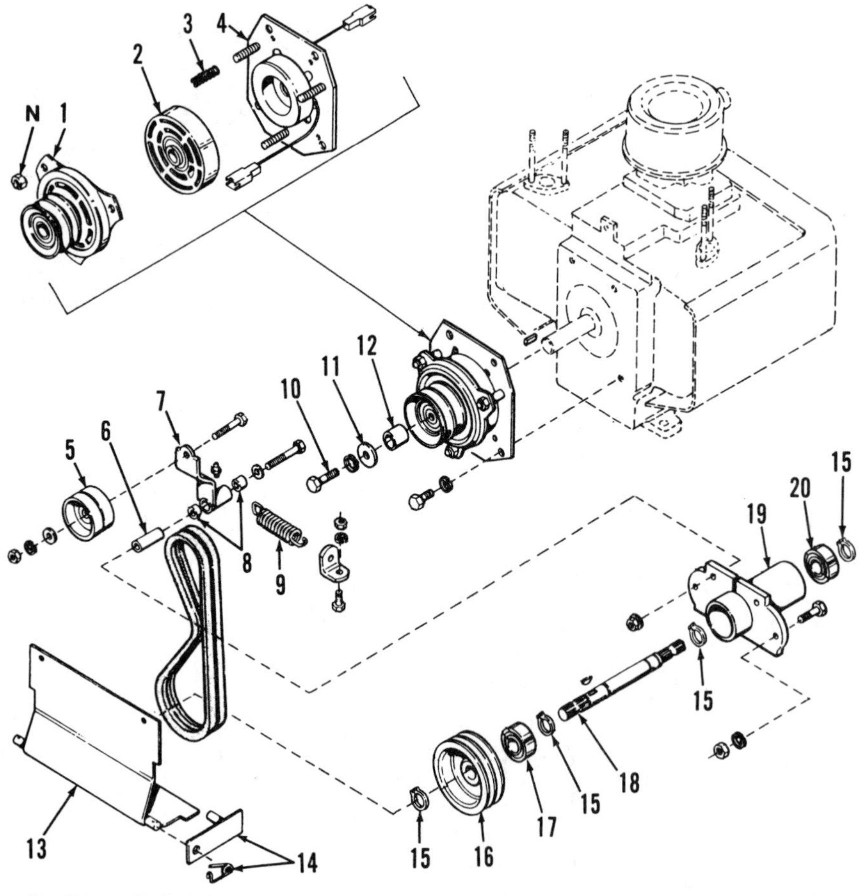

Fig. DA27—Exploded view of pto clutch and shaft assembly used on Models 1918 and 1920.

1. Armature
2. Rotor
3. Springs
4. Field
5. Pulley
6. Spacer
7. Idler arm
8. Bushings
9. Spring
10. Cap screw
11. Washer
12. Spacer
13. Pto guard
14. Clip pin & link
15. Snap ring
16. Pto drive pulley
17. Bearing
18. Pto shaft
19. Bearing support
20. Bearing

Illustrations courtesy Deutz-Allis.

HYDROSTATIC TRANSMISSION

MAINTENANCE

Models 613H-616H

The hydrostatic transmission is equipped with an integral reservoir. Recommended transmission fluid is Power Fluid 821.

If contamination is observed in reservoir tank or color of fluid has changed (black or milky), fluid should be drained and renewed. To drain fluid, remove transmission from reduction gear housing, remove reservoir and vent plug from body. Turn unit upside down and allow to drain.

NOTE: Reservoir tank has left-hand threads.

When refilling transmission, add oil until fluid overflows out vent hole in body. Then reinstall vent plug and fill reservoir tank to COLD mark on tank.

Models 1814-1816-1817

Oil level in transmission oil reservoir should be at "FULL" mark when oil is cold. Recommended oil is high quality SAE 30W oil.

The transmission oil filter should be changed after the first 10 hours of operation and after every 400 hours of operation thereafter.

Models 1918-1920

Hydrostatic oil level should be maintained at a level that is 1 inch (25 mm) above fitting (F—Fig. DA33). Check oil level when cold. Recommended oil is Type "F" Automatic Transmission Fluid. Add oil through fill tube located under hood adjacent to battery.

LINKAGE ADJUSTMENT

Models 613H-616H

If tractor moves when control lever is in neutral position, adjust as follows: With engine running, move control lever (L—Fig. DA34) until transmission is in neutral. Shut off engine. Raise the seat, loosen attaching screws of quadrant panel (P) and move panel so control lever is positioned in neutral slot. Tighten screws and check operation.

Models 1814-1816-1817

Prior to adjusting linkage, perform clutch adjustment as previously out-

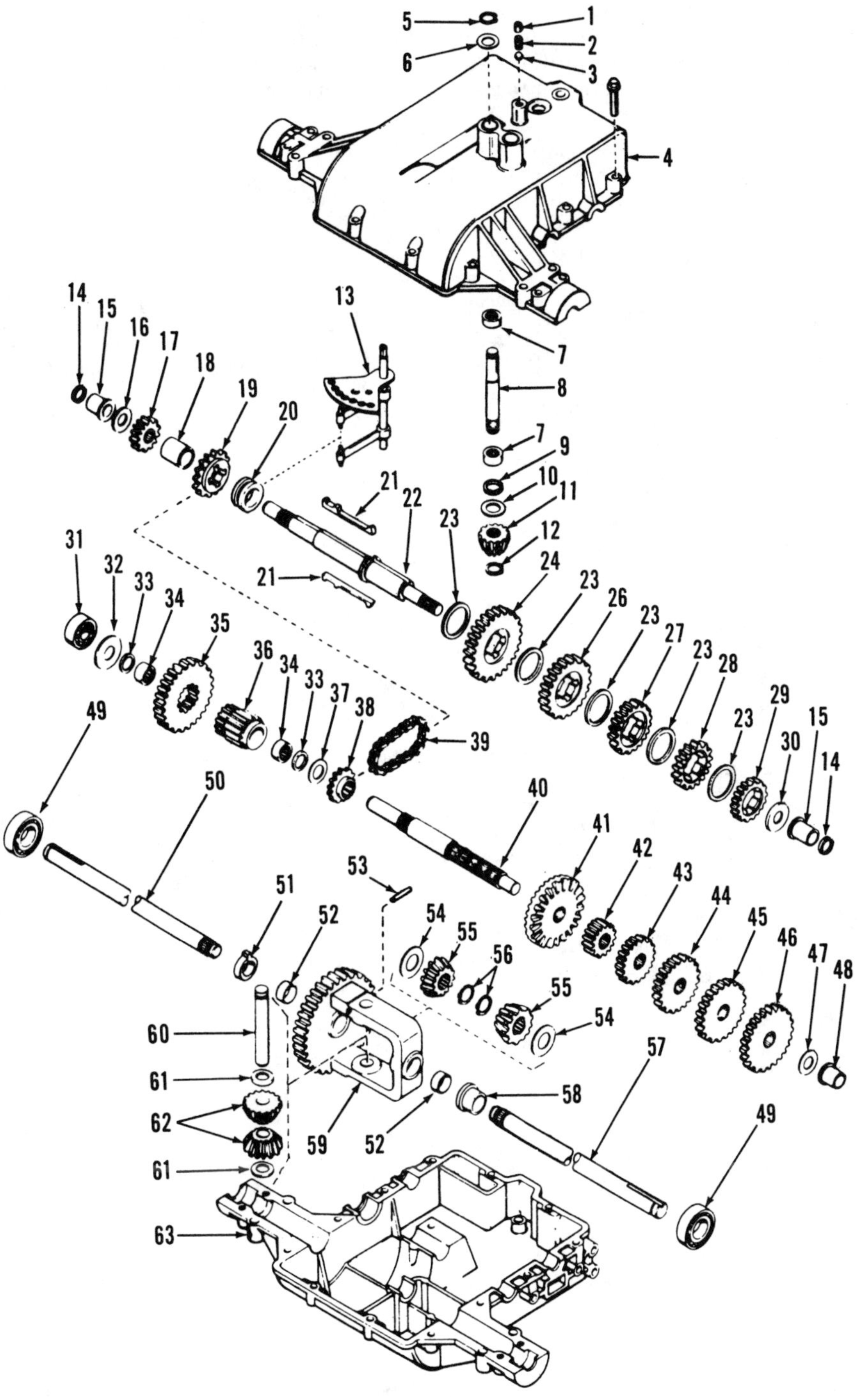

Fig. DA28—Exploded view of transaxle used on Models 613 Special and 616 Special.

1. Setscrew
2. Spring
3. Ball
4. Cover
5. Snap ring
6. Thrust washer
7. Needle bearings
8. Input shaft
9. Square cut ring
10. Thrust washer
11. Input pinion
12. Snap ring
13. Shift fork assy.
14. Square cut ring
15. Flanged bushing
16. Thrust washer
17. Spur gear (15 T)
18. Spacer
19. Sprocket (18 T)
20. Shift collar
21. Key
22. Brake shaft
23. Thrust washers
24. Spur gear (37 T)
26. Spur gear (30 T)
27. Spur gear (25 T)
28. Spur gear (22 T)
29. Spur gear (20 T)
30. Thrust washer
31. Ball bearing
32. Flat washer
33. Square cut seal
34. Needle bearings
35. Output gear
36. Output pinion
37. Flat washer
38. Sprocket
39. Roller chain
40. Countershaft
41. Bevel gear
42. Spur gear (12 T)
43. Spur gear (20 T)
44. Spur gear (25 T)
45. Spur gear (28 T)
46. Spur gear (30 T)
47. Thrust washer
48. Flanged bushing
49. Ball bearing
50. Axle shaft
51. Bushing
52. Bushing
53. Roll pin
54. Thrust washers
55. Bevel side gears
56. Snap rings
57. Axle shaft
58. Flanged bushing
59. Differential carrier & gear assy.
60. Pinion shaft
61. Thrust washers
62. Pinion gears
63. Housing

Illustrations courtesy Deutz-Allis.

lined. Raise and securely support rear of tractor so rear wheels are off ground. Run engine at half throttle and move speed control lever to forward and reverse positions, then return lever to center of neutral slot. Wheels should not rotate. If wheels rotate, loosen screws securing quadrant panel. Move control lever (C—Fig. DA35) to position where wheels do not rotate, then relocate panel (P) so lever is in center of neutral slot. If there is not sufficient slot length so lever will be positioned in neutral, then the length of control link (L) must be adjusted. Position panel so top of slots align with top of washers and tighten mounting screws. Run engine and move control lever to forward then return to center of neutral slot.

CAUTION: Make certain tractor is blocked securely and use care when working near rotating tire or moving parts.

Loosen jam nut on control link (L) and turn nut (N) on link to adjust length of link so wheels do not rotate. Tighten jam nut and recheck operation.

If control lever does not maintain its position when in forward or reverse, tighten nut (M) to increase tension of compression spring (S).

Models 1918-1920

If tractor moves when control lever is in neutral position, remove right side dash panel, then raise and securely support rear of tractor so wheels are off ground. Engage parking brake. Loosen nuts (1, 2 and 3—Fig. DA36). Position speed control lever in full rear location and tighten nut (1). Disengage parking brake and move speed control lever to neutral. Run engine at full throttle, move speed control lever forward so wheels rotate then move lever rearward until wheels stop turning. Stop engine and engage parking brake (nut 3 must be loose). As brake is applied, cam plate (5) will drop down. Align spacer (4) on nut (3), then tighten nut (3) without moving the linkage. Loosen nut (1) and move speed control lever to neutral. Tighten nuts (1 and 2). Check operation.

TROUBLE-SHOOTING

The following problems and possible causes should be used as an aid in locating and correcting transmission problems.

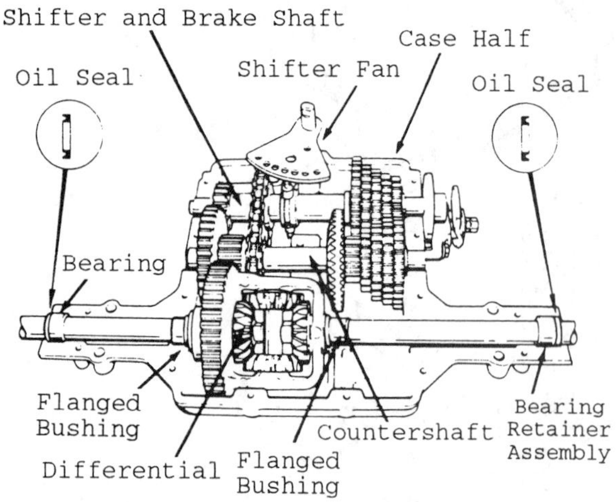

Fig. DA29—When disassembling transaxle, remove gear and shaft assemblies from case by lifting both shafts at once. Be careful not to disturb chain.

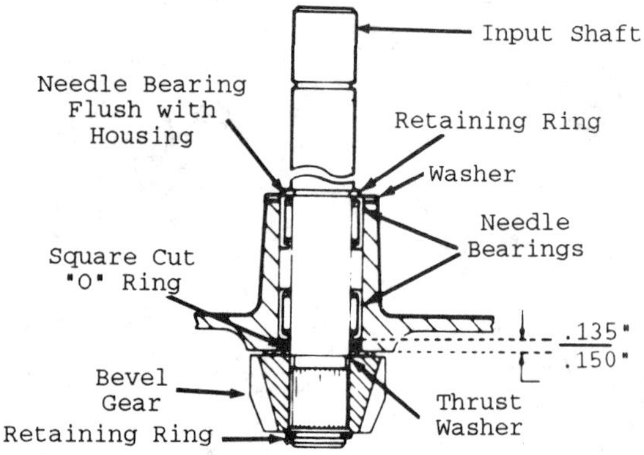

Fig. DA30—Cross-sectional view of transaxle input shaft used on Models 613 Special and 616 Special showing correct installation of needle bearings.

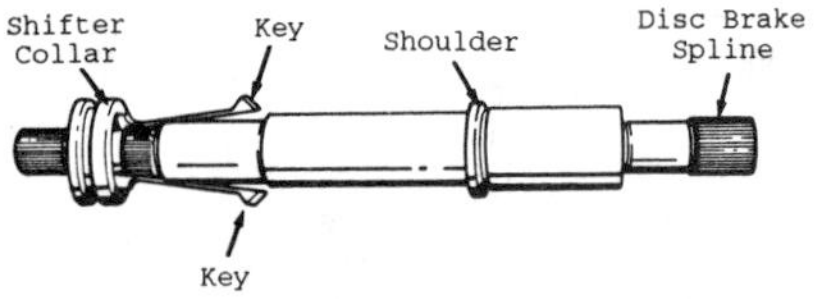

Fig. DA31—Install shift collar on shifter/brake shaft so thick segment is nearer shaft shoulder.

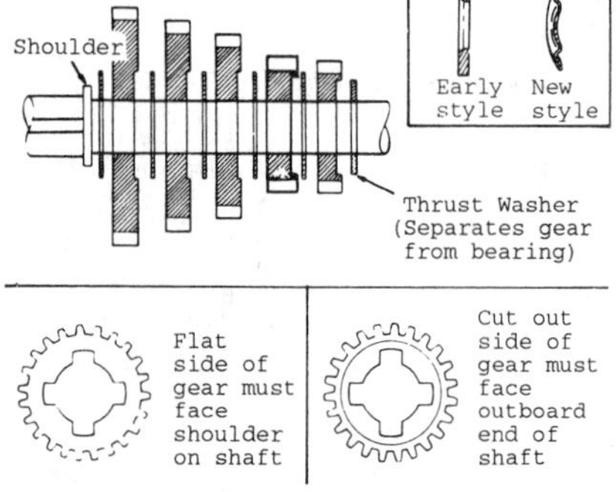

Fig. DA32—When assembling shift shaft, install thrust washers with 45 degree chamfer toward shoulder on output shaft. Install curved washers so rounded side is toward shaft shoulder. Flat side of gears also must face shoulder on output shaft.

Fig. DA33—On Models 1918 and 1920, hydrostatic oil level should be maintained at a level that is one inch (25 mm) above fitting (F). Check oil level when cold.

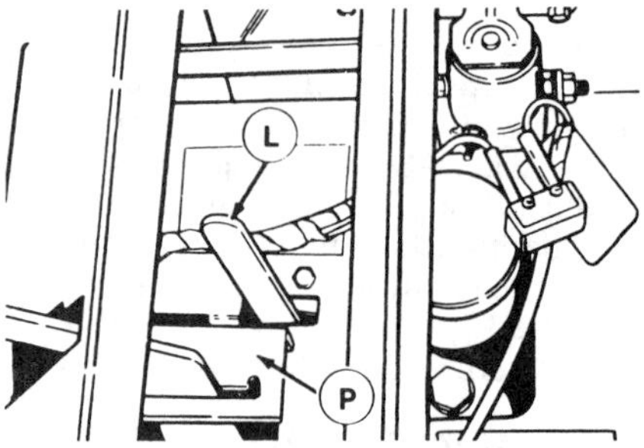

Fig. DA34—To adjust shift control linkage on Models 613H and 616H, loosen attaching screws of quadrant panel (P) and move panel as needed. Refer to text.

Illustrations courtesy Deutz-Allis.

Models 613H-616H

1. Loss of power or transmission will not operate in either direction. Could be caused by:
   a. Drive belt slipping or broken.
   b. Speed control linkage broken.
   c. Transmission oil level low.
   d. Wrong type of transmission oil or oil contaminated with water.
   e. Transmission oil temperature too hot.
   f. Transmission dump valve defective.
   g. Transmission pump and/or motor worn or damaged.
   h. Internal damage to reduction gear unit.
2. Transmission operating too hot. Could be caused by:
   a. Transmission oil level low.
   b. Wrong type of oil in transmission.
   c. Cooling fan defective.
   d. Transmission cooling fins blocked.
   e. Dump valve faulty.
   f. Transmission pump and/or motor worn.
3. Transmission jerks when starting or operates in one direction only. Could be caused by:
   a. Faulty control linkage.
   b. Check valve in pintle assembly faulty.
   c. Acceleration valve in pintle assembly faulty.
4. Tractor creeps when in neutral. Could be caused by:
   a. Worn or misadjusted control linkage.

Models 1814-1816-1817

1. Loss of power or transmission will not operate in either direction. Could be caused by:
   a. Drive belt slipping or broken.
   b. Speed control linkage broken.
   c. Transmission oil level low.
   d. Wrong type of transmission oil or oil contaminated with water.
   e. Transmission oil temperature too hot.
   f. Transmission roll release valve in wrong position.
   g. Oil filter plugged.
   h. Charge pump or charge pressure relief valve faulty.
   i. Transmission pump and/or motor worn or damaged.
   j. Drive pulley slipping on transmission input shaft.
   k. Internal damage to reduction gear assembly.
2. Transmission operating too hot. Could be caused by:
   a. Transmission oil level low.
   b. Wrong type of oil in transmission.
   c. Cooling fan defective.
   d. Transmission cooling fins blocked.
   e. Roll release valve faulty.
   f. Tractor is being overloaded.
   g. Transmission pump and/or motor worn.
3. Transmission jerks when starting or operates in one direction only. Could be caused by:
   a. Faulty speed control linkage.
   b. Faulty charge check valve.
4. Tractor creeps when in neutral. Could be caused by:
   a. Worn or misadjusted control linkage.

Models 1918-1920

1. Loss of power or transmission will not operate in either direction. Could be caused by:
   a. Drive belt slipping or broken.
   b. Transmission oil level low.
   c. Transmission oil temperature too hot.
   f. Oil filter plugged.
   g. Speed control linkage broken.
   h. Charge pump or charge pressure relief valve faulty.
   i. Transmission pump and/or motor worn or damaged.
   k. Internal damage to reduction gear assembly.
2. Transmission operating too hot. Could be caused by:
   a. Transmission oil level low.
   b. Wrong type of oil in transmission.
   c. Oil filter plugged.
   d. Transmission oil cooler fins blocked.
   e. Tractor is being overloaded.
   f. Charge pump or charge pressure relief valve faulty.
   g. Transmission pump and/or motor worn.
3. Transmission jerks when starting or operates in one direction only. Could be caused by:
   a. Faulty speed control linkage.
   b. Faulty charge check valve.
4. Tractor creeps when in neutral. Could be caused by:
   a. Worn or misadjusted control linkage.

REMOVE AND REINSTALL

Models 613H-616H

To remove hydrostatic transmission from reduction drive unit, remove mower deck. Remove battery, drain fuel tank and remove seat deck. Remove transmission drain plug and drain oil. Remove transmission reservoir.

NOTE: Reservoir tank has left-hand threads.

Cover opening to prevent entry of foreign material. Raise and support rear of tractor, or position 4 × 4 wooden block behind tractor and use a suitable hoist to raise front of tractor so rear frame rests on wood block. Remove transmission cooling fan. Remove belt stop and disengage drive belt from transmission pulley. Unscrew transmission mounting screws and separate transmission from gear reduction housing.

Reverse removal procedure to install transmission. Tighten transmission mounting screws to 20 ft.-lbs. (27 N·m). When refilling transmission, add oil until fluid overflows vent hole in body. Then reinstall vent plug and fill reservoir tank to COLD mark on tank. If transmission is installed without oil, rotate shafts to purge trapped air, then refill. Check oil level after operating tractor.

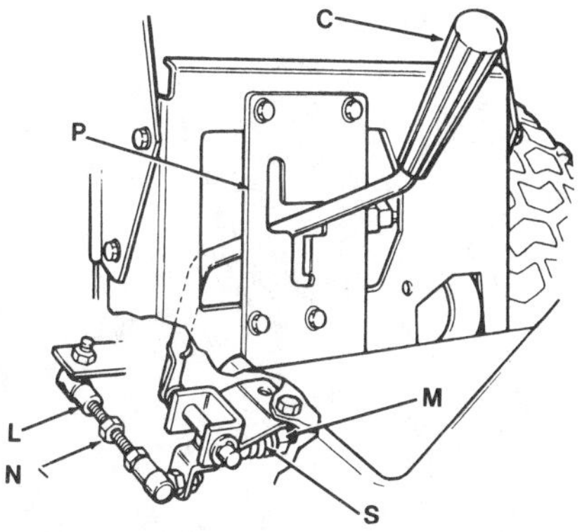

Fig. DA35—The rear wheels on Models 1814, 1816 and 1817 should not rotate when the shift lever (C) is centered in neutral slot of panel (P). Refer to text for adjustment procedure.

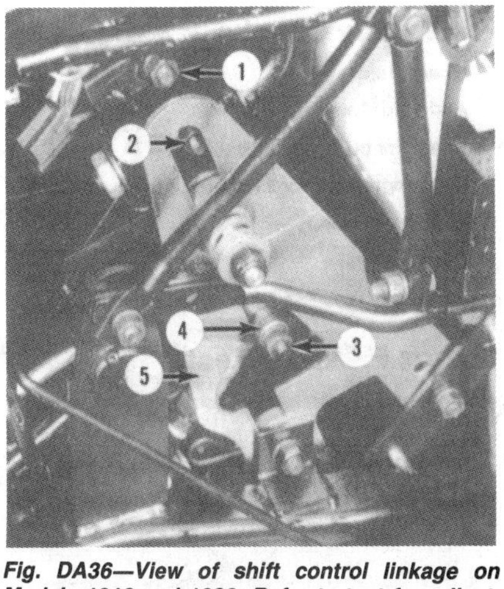

Fig. DA36—View of shift control linkage on Models 1918 and 1920. Refer to text for adjustment procedure.

Models 1814-1816-1817

To remove transmission unit, remove battery and drain fuel tank. Unbolt and remove drawbar and seat deck from seat support. Remove transmission belt stop. Unscrew drain plugs under each axle housing and drain oil. Disconnect oil hoses and tubing. Raise and support rear of tractor, or stand tractor on end using a suitable hoist attached to front end of tractor. Remove rear wheels. Detach hydrostatic release valve lever spring. Detach speed control link. Disconnect brake linkage. Engage parking brake and remove drive belt from transmission pulley. Remove screws securing braces to transmission. Remove screws securing axle housings and remove transmission unit.

To reinstall unit, reverse removal procedure. Tighten screws securing axle housings to 23-27 ft.-lbs. (31-37 N·m). Apply Loctite to screws securing right front brace. Belt stops should be positioned $^{1}/_{16}$-$^{1}/_{8}$ inch (1.6-3.2 mm) from transmission pulley. Check brake/clutch adjustment and hydrostatic control linkage adjustment as previously outlined.

Models 1918-1920

To remove hydrostatic transmission, remove pto guard and detach pto drive belts. Loosen screws securing engine to engine plate and slide engine forward so drive shaft separates from transmission input shaft. Remove seat, fuel tank and tunnel cover. Disconnect lower transmission hose and drain transmission oil. Detach hydraulic lines from transmission. Remove control damper and disconnect control linkage. Remove screws securing transmission to transaxle and remove transmission.

To reinstall transmission, reverse removal procedure. Tighten transmission mounting screws to 25-30 ft.-lbs. (34-41 N·m). Refill transmission with Type F automatic transmission fluid, then prime transmission as follows: Raise and support rear of tractor so rear wheels are off the ground. Remove spark plugs, reconnect wires to spark plugs and ground plugs to engine. Move speed control lever to full forward position, then crank engine with starter until rear wheels begin to move. System is now primed. Check fluid level and reinstall spark plugs.

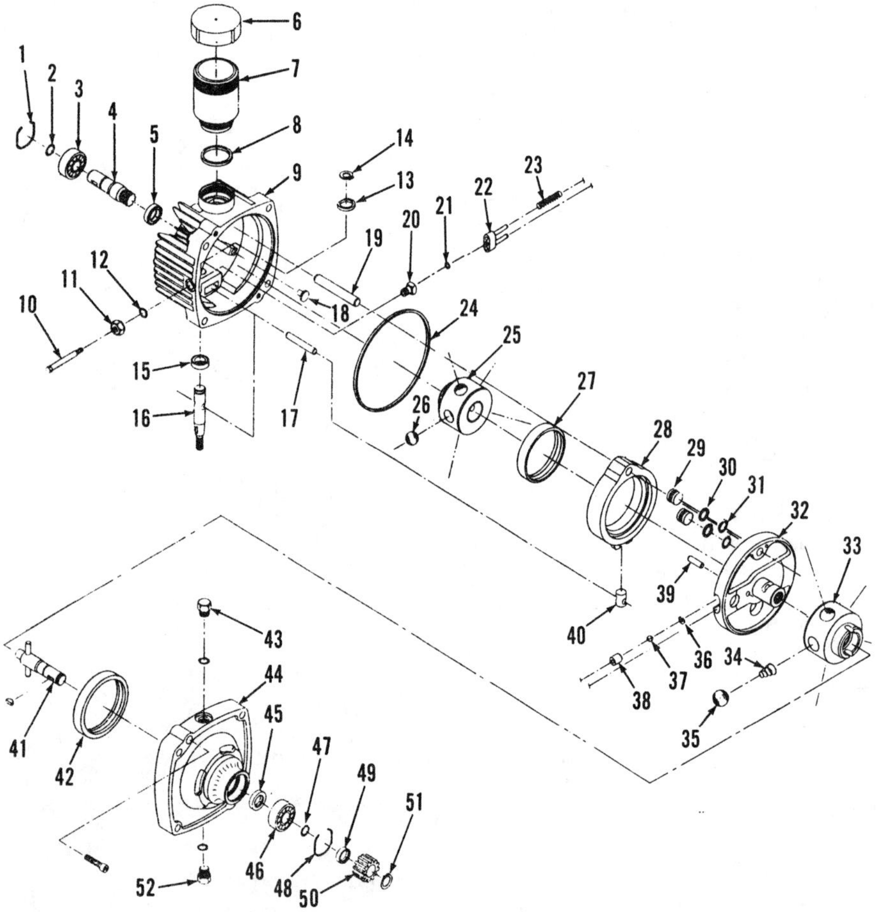

Fig. DA38—Exploded view of Eaton Model 7 hydrostatic transmission used on Models 613H and 616H.

1. Retaining ring
2. Snap ring
3. Ball bearing
4. Input shaft
5. Oil seal
6. Cap
7. Reservoir
8. Seal ring
9. Cover
10. Dump valve shaft
11. Nut
12. "O" ring
13. Washer
14. Snap ring
15. Oil seal
16. Control shaft
17. Dowel pin
18. Button
19. Pivot pin
20. Adapter
21. "O" ring
22. Dump valve bracket
23. Spring
24. Seal ring
25. Pump rotor
26. Ball piston
27. Pump race
28. Cam ring
29. Dampening pistons
30. Back-up rings
31. "O" rings
32. Pintle assy.
33. Motor rotor
34. Spring
35. Ball piston
36. Snap ring
37. Ball
38. Check valve body
39. Coil pin
40. Cam ring insert
41. Output shaft
42. Motor race
43. Vent plug
44. Body
45. Oil seal
46. Bearing
47. Snap ring
48. Retaining ring
49. Oil seal
50. Bearing
51. Snap ring
52. Drain plug

OVERHAUL

Models 613H-616H

These models are equipped with an Eaton Model 7 hydrostatic transmission. To overhaul transmission, first thoroughly clean outside of unit and proceed as follows:

Scribe a line across cover (9—Fig. DA38) and body to ensure correct reassembly. Place transmission in a holding fixture with input shaft pointing up. Unscrew dump valve stem (10) from dump valve bracket (22). Remove cap screws retaining cover to body. Tap cover with a plastic hammer and lift cover from body being careful not to allow pump rotor (25), cam ring (28) or pintle (32) to lift with the cover. Remove button (18) from cover. Remove nut (11) and dump valve guide (20) and discard "O" rings (12 and 21). Remove snap ring (1) and pull input shaft (4) and bearing (3) from cover. Drive oil seal (5) out of cover bore. Pierce control shaft oil seal (15) with a sharp tool and pry seal from cover, being careful not to damage cover bore or control shaft. Cover, control shaft (16) and dowel (17) are available only as an assembly.

CAUTION: The ball pistons (26 and 35) are a select fit in pump rotor (25) and motor rotor (32) bores and must be reinstalled in original bores if removed from rotors. Use a wide rubber band, piston ring compressor or other suitable means to retain balls in the rotors when servicing pump and motor assemblies.

Remove pump cam ring (28). Remove dump valve bracket (22) and springs (23) from pintle (32). While holding pintle in position in body, remove pump rotor assembly. Lift pintle from body, then carefully lift out motor rotor assembly. Note that motor balls are spring loaded in rotor bores. Remove snap ring (51), drive gear (50), key and spacer (49) from output shaft (41). Remove small snap ring (47) and tap output shaft inward. Re-

Illustrations courtesy Deutz-Allis.

move large snap ring (48) and drive bearing (46) and seal (45) outward from body.

Inspect pump ball race (27) and motor ball race (42) for wear or damage and renew as necessary. Number piston bores on pump rotor and motor rotor. Use a plastic ice cube tray or egg carton with cavities marked to correspond to rotor bores as a container for ball pistons. Remove ball pistons one at a time and place in correct cavity in tray, then inspect each ball and rotor bore. Pump and motor ball pistons and rotor bores must be smooth and free of any irregularities. Balls and rotors must be renewed as assemblies. Be sure that ball pistons are returned to their original bores after inspection. Disassembly of pintle assembly (32) for cleaning is not recommended. If pintle is worn or damaged, renew pintle assembly. Renew all "O" rings and seals.

To reassemble, press new oil seal (45—Fig. DA38) into body with lip of seal facing inward. Position output shaft (41) in the body. Support inner end of shaft so cross pin in shaft is held against body. Install output shaft bearing (46), pressing against outer race of bearing. Make sure output shaft rotates freely, then install snap rings (47 and 48). Install spacer, key, drive gear and snap ring on output shaft. Use suitable retaining device to hold ball pistons and springs in motor rotor, then align slot in rotor with output shaft drive pin and install motor in body. Remove ball retaining device.

Install new control shaft seal (15) and input shaft seal (5) in cover (9). Install input shaft assembly (4) and retaining ring (1). Install button (18), dump valve guide (20) and nut (11) in cover. Do not install dump valve shaft at this time. Install cam ring (28) with flush side of ball race (27) facing the cover and with notch in cam ring aligned with control shaft pin and pivot (40). Push cam ring into cover until bottomed. Be sure that cam ring moves freely. Use suitable retaining device to hold ball pistons in pump rotor (25), align internal spline in pump rotor with external spline on input shaft and install pump rotor assembly in cover. Remove ball retaining device. Install springs (23) and dump valve bracket (22) in pintle (32). Compress and hold dump valve bracket in the pintle to clear dump valve guide (20) and install pintle through pump rotor. Pintle should be a slip fit through pump rotor and pump rotor should turn freely on pintle. Thread dump valve stem (10) into dump valve bracket.

Position new seal ring (24) in cover. While holding motor rotor assembly in position in body, align previously scribed line on cover and body and install motor rotor on pintle. Install two of cover retaining screws and tighten to 15 ft.-lbs. (20 N·m). (The other two retaining screws are installed after unit is installed on gear reduction housing.) Install reservoir. Install pulley and fan and tighten retaining cap screw to 70 in.-lbs. (7.9 N·m).

To fill transmission with fluid, remove plug (43) from top of body and fill with fluid through reservoir until fluid flows from plug opening. Rotate input and output shafts to purge air from transmission. Refill until fluid flows from plug opening, then reinstall plug. Continue to fill reservoir to COLD oil level mark.

Models 1814-1816-1817

Models 1814, 1816 and 1817 are equipped with an Eaton 850 hydrostatic transmission. Prior to disassembly, remove fan, pulley, and all external brackets, levers and fittings. Thoroughly clean outside of unit. Drain oil from unit, remove oil filter and detach oil filter base. During disassembly, mark components so they may be returned to their original position.

Service procedure is the same for either axle housing assembly. Remove cap screws securing axle housing (6—Fig. DA39) and withdraw axle housing and planetary assembly (components 1 through 21) from hydrostatic housing (29).

CAUTION: Be very careful not to dislodge hydrostatic motor assembly. The spring-loaded balls (22) are a selective fit in motor rotor (24) and must be installed in matching bore in rotor if removed.

To disassemble axle assembly, remove wear plate (21), first planetary assembly (16, 17 and 18), ring gear (19), brake gear (11) and second planetary assembly (13, 14 and 15). Remove large thrust washer (9), snap ring (8) and small thrust washer (7) from axle shaft. Remove outer snap ring (1), then remove

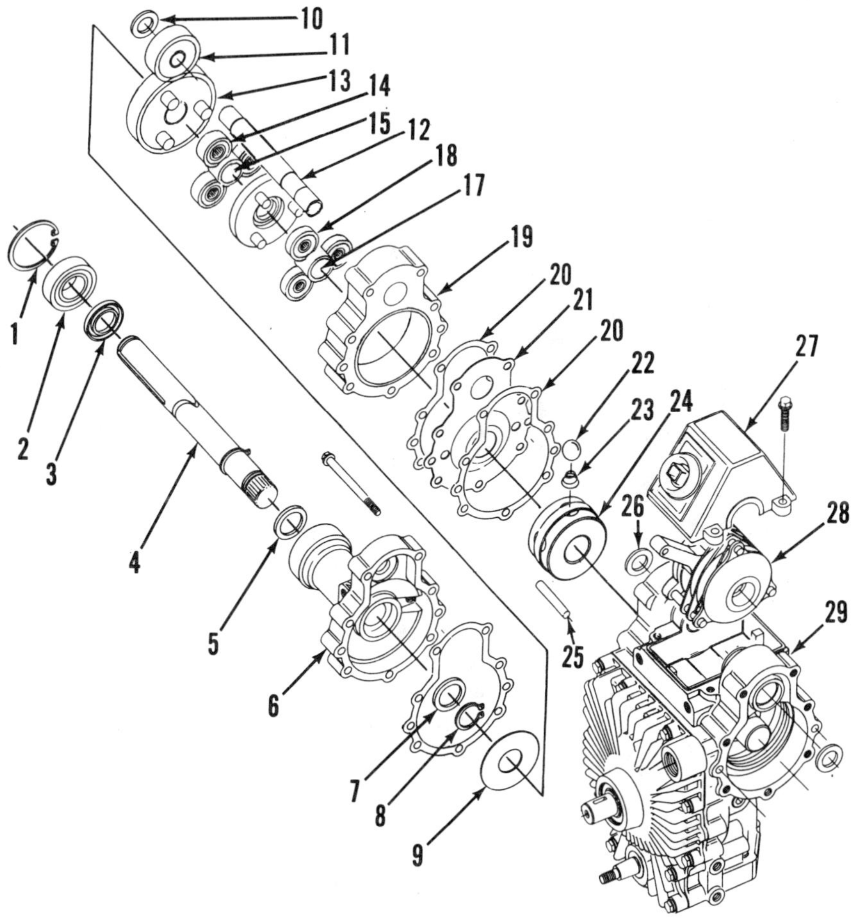

Fig. DA39—Exploded view of Eaton 850 transmission right axle housing assembly used on Models 1814, 1816 and 1817.

1. Snap ring
2. Bearing
3. Seal
4. Axle shaft
5. Thrust washer
6. Axle housing
7. Thrust washer
8. Snap ring
9. Thrust washer
10. Spacer
11. Brake gear
12. Brake shaft
13. Planetary carrier
14. Planetary gear (second)
15. Sun gear (second)
16. Planetary carrier
17. Sun gear (first)
18. Planetary gear (first)
19. Ring gear
20. Gaskets
21. Wear plate
22. Ball piston
23. Spring
24. Motor rotor
25. Dowel
26. Seal
27. Brake cover
28. Parking brake assy.
29. Transmission housing

Illustrations courtesy Deutz-Allis.

axle shaft (4), oil seal (3) and bearing (2) from axle housing (6) by tapping on splined end of shaft with a soft hammer.

Detach brake housing (27) and remove parking brake assembly (28) being careful not to move brake lever. The brake is self-adjusting and moving lever will expand brake assembly making installation more difficult. Brake assembly is available only as a unit and should not be disassembled.

To remove hydrostatic motors, place a large rubber band or similar device around the motor rotor (24) to hold spring-loaded ball pistons (22) in their bores. Carefully withdraw motor assembly. Due to selective fit, if balls are removed, they must be returned to original rotor bore.

To disassemble hydrostatic pump, unbolt and remove oil filter base from housing. Unscrew fasteners retaining pump cover (35—Fig. DA40). Note that there is a screw located in the drain hole. Separate cover from housing and remove snap ring (30). Tap or press input shaft (32) and bearing (31) out of cover by forcing shaft toward outside of cover. Shaft and bearing assembly is available only as a unit assembly. Drive input shaft seal (33) out of cover bore. Remove charge pressure relief valve spring (45) and ball (46). Remove buttons (36), control shaft (37) and cam ring insert (38) from cam ring (40). Remove pump rotor drive ring (41). Place a wide rubber band or similar device around pump rotor (42) to hold the ball pistons (43) in their bores. The balls are a selective fit in rotor and must be installed in their matching bore if removed from rotor. Carefully lift pump rotor from housing.

Dump valve bracket (51) can be removed from housing after removing nut (48). Check valves (Fig. DA41) should be flushed with suitable solvent, but not removed, and are available only with housing. Removal of dampening pistons (39—Fig. DA40) is also not recommended.

Charge pump assembly (1—Fig. DA42) may be removed from housing for inspection after removing pump plate (2). Charge pump components are available only with the housing (47) as an assembly.

Inspect components for excessive wear and damage. When removing ball pistons from pump and motor rotors, number rotor bores and place balls in a plastic ice cube tray or egg carton with cavities numbered to correspond to rotor bores. Piston balls and rotor bores must be smooth and free of any irregularities. Pump and motor rotors and balls are unit assemblies as balls are selectively fitted to rotor bores.

Assemble by referring to appropriate exploded view and reversing disassembly procedure while noting the following: Lubricate all assemblies with clean transmission oil. Install dump valve spring (52—Fig. DA40) so right angle bend points toward "O" ring (50). Tighten dump valve nut (48) to 150 in.-lbs. (17 N·m). Install charge pump cover so arrow (A—Fig. DA42) points toward "CW" on housing. Do not tighten cover screws until pump rotor is installed. Note that pins on bottom of pump rotor (42—Fig. DA40) must engage holes in charge pump inner rotor. Rotate pump rotor clockwise to center pump cover then tighten cover screws in a crossing pattern to 44 in.-lbs. (5.0 N·m). Install cam ring (40) so flush side is out. Install seals (33 and 34) so lip is to inside of cover. Be sure snap ring (30) is firmly seated in cover after installing input shaft assembly (32). Hold buttons (36) in place with petroleum jelly. Be sure charge relief valve ball (46) is not dislodged when installing cover assembly on housing. Splines on input shaft must engage splines of drive ring (41); do not force cover onto housing. Tighten cover screws to 105 in.-lbs. (11.9 N·m) using pattern shown in Fig. DA43. Carefully install motor rotors in housing and remove rubber band used to retain ball pistons in rotors.

Install brake assembly so lever is toward cover side of housing. Note that axle housings and sides of transaxle housing are identified with an "A" or "B" as shown in Fig. DA44. The "B"

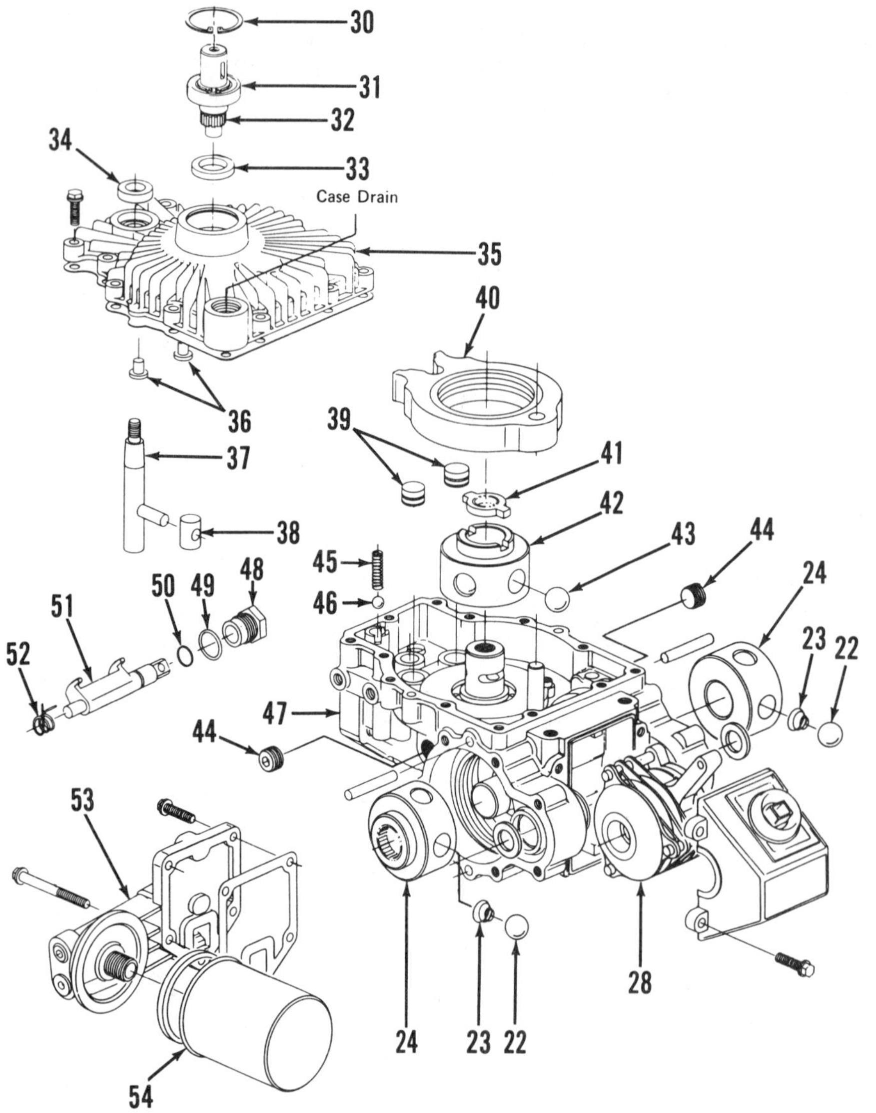

Fig. DA40—Exploded view of pump and housing components on Eaton 850 hydrostatic transmission.

22. Ball piston
23. Spring
24. Motor rotor
28. Brake assy.
30. Snap ring
31. Bearing
32. Input shaft assy.
33. Oil seal
34. Oil seal
35. Cover
36. Buttons
37. Control shaft
38. Insert
39. Dampening pistons
40. Cam ring
41. Drive ring
42. Pump rotor
43. Ball piston
44. Plugs
45. Charge relief valve spring
46. Charge relief valve ball
47. Housing
48. Plug
49. "O" ring
50. "O" ring
51. Dump valve bracket
52. Spring
53. Oil filter base
54. Oil filter

Illustrations courtesy Deutz-Allis.

side of transaxle should be assembled first. Note that "B" side brake shaft (12—Fig. DA39) is longer than the "A" side shaft. Insert the "B" side brake shaft into the brake assembly and rotate so the splines are engaged. Install "A" side brake shaft. When both shafts are properly installed, the shafts will rotate independently.

Install axle components while noting the following: Inner sun gear (17) is taller than outer sun gear (15). Install ring gear (19) so side with needle bearing mates with axle housing (6). Install outer sun gear (15) so convex (pointed) end is toward inner sun gear (17). Install axle housing seal (3) so lip points inward. Install wear plate (21) so bowed side is facing motor rotor. Tighten axle housing screws to 125 in.-lbs. (14 N·m). Tighten oil filter base screws to 125 in.-lbs. (14 N·m).

Models 1918-1920

Remove hydrostatic drive unit as previously outlined and clean exterior of unit.

To disassemble charge pump, first scribe a line across center section (7—Fig. DA45) and charge pump body (3) to aid in correct reassembly, then unbolt and remove charge pump housing. Remove rotor assembly (5) and rotor drive pin (4). Pry oil seal (1) from housing and press needle bearing (2) out front of housing.

Remove both check valves (26), "O" rings (23 and 25) and back-up washer (24) from center section (7). Remove plug (9), "O" ring (10), shims (11), spring (12) and charge relief valve cone (13). Remove plug (22), "O" ring (21), shims (20), spring (19) and implement relief valve cone (18). Unbolt and remove center housing (7).

CAUTION: Valve plates (15 and 16) may stick to center housing. Be careful not to drop them.

Remove and identify valve plates. Pump valve plate (16) has two relief notches (Fig. DA46) and motor valve plate has four notches. Tilt housing (46—Fig. DA45) on its side and remove cylinder block and piston assemblies from pump and motor shafts. Lay cylinder block assemblies aside for later disassembly. Tilt pump swashplate (37) to full forward position, then tap single roll pin (36) out of trunnion shaft (45) until it hits the housing. Tap double roll pins (36) out of control shaft (41) until first pin contacts the housing. Move swashplate back toward neutral position, trunnion shaft roll pin and first control shaft roll pin should fall into housing. Tilt swashplate back to full forward and tap second roll pin out of control shaft. Be careful not to damage housing when driving roll pins out. Remove trunnion shaft and control shaft assemblies. Remove swashplate (37) and thrust plate (35), then remove pump shaft (38), bearing (39) and seal (47) from housing. Tap motor shaft (31) out of housing and remove bearing (48). Bearings (44) and oil seals (43) can now be removed from housing (46). Inspect needle bearings (14) and remove from center housing (7) if necessary.

Carefully remove slipper retainer (33) with nine pistons (34) from pump cylinder block (32), then remove slipper retainer (28) with nine pistons (29) from motor cylinder block (27). Manufacturer does not recommend removal of snap rings and springs from cylinder blocks

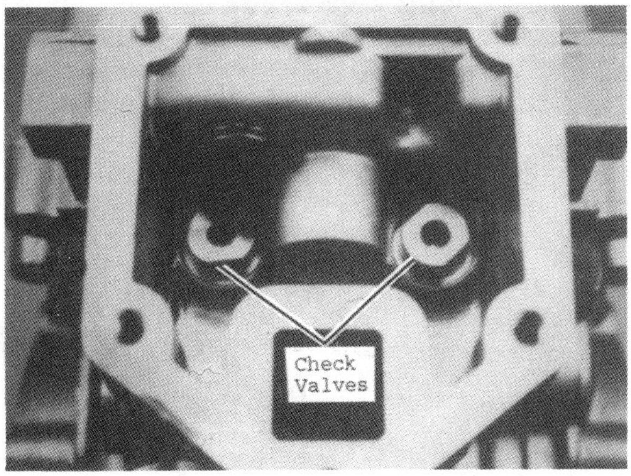

Fig. DA41—Check valves are available only with hydrostatic housing and should be flushed, but not removed.

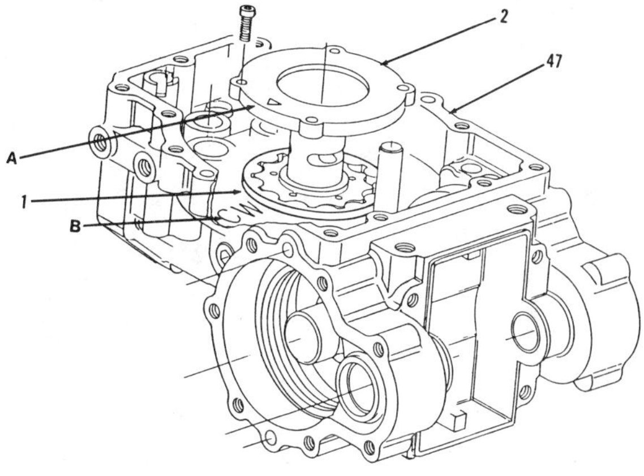

Fig. DA42—Install charge pump cover (2) so arrow (A) points towards "CW" on housing.

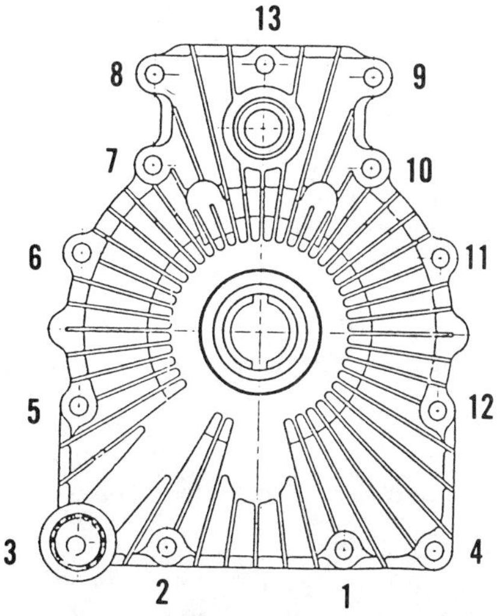

Fig. DA43—Tighten cover screws to 105 in.-lbs. (11.9 N·m) using sequence shown above.

Fig. DA44—Note that axle housings and sides of transaxle housing are identified with an "A" or "B" as shown above. Refer to text.

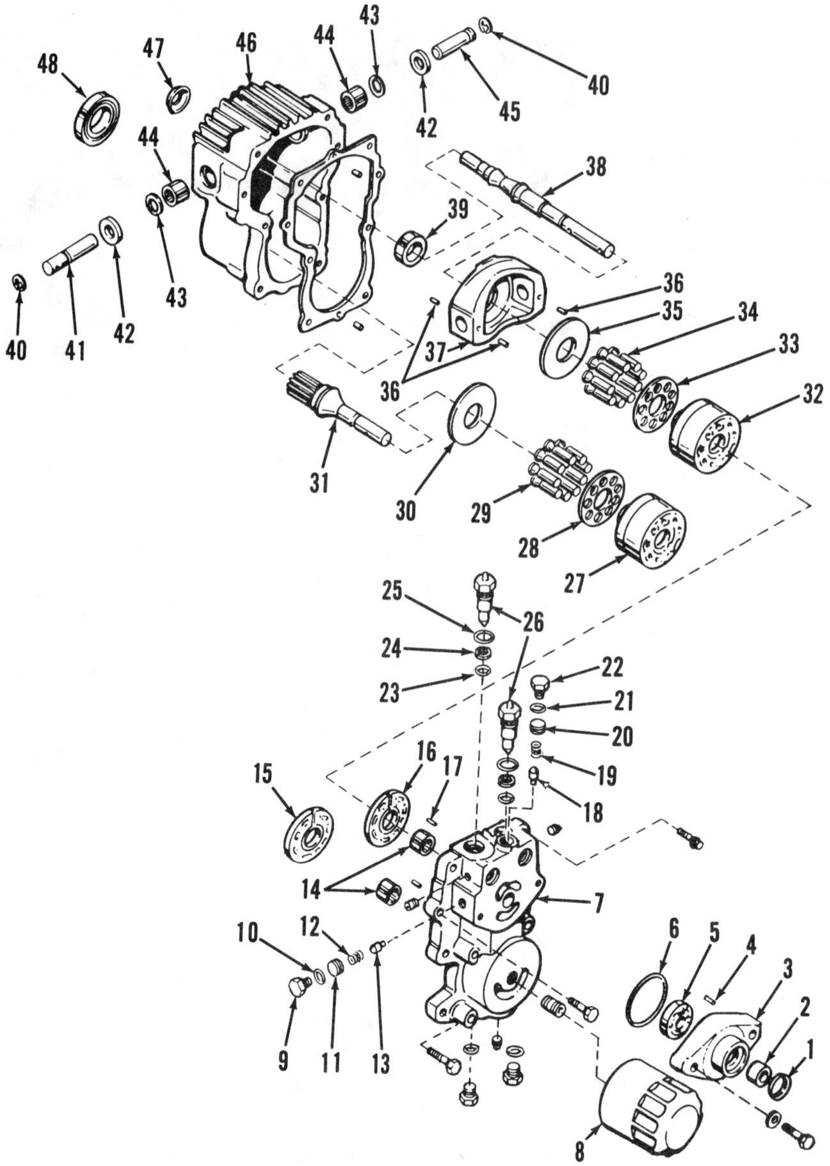

Fig. DA45—Exploded view of hydrostatic transmission used on Models 1918 and 1920.

1. Oil seal
2. Needle bearing
3. Charge pump housing
4. Pin
5. Rotor assy.
6. "O" ring
7. Center section
8. Filter
9. Plug
10. "O" ring
11. Shim
12. Charge relief valve spring
13. Relief valve cone
14. Needle bearings
15. Motor valve plate
16. Pump valve plate
17. Alignment pin
18. Relief valve cone
19. Implement relief valve spring
20. Shim
21. "O" ring
22. Plug
23. "O" ring
24. Back-up ring
25. "O" ring
26. Check valve assy.
27. Motor cylinder block
28. Slipper retainer
29. Piston (9)
30. Thrust plate
31. Motor shaft
32. Pump cylinder block
33. Slipper retainer
34. Piston (9)
35. Thrust plate
36. Spring pins
37. Swashplate
38. Pump shaft
39. Ball bearing
40. Snap ring
41. Control shaft
42. Washer
43. Oil seal
44. Needle bearings
45. Trunnion shaft
46. Cover
47. Seal
48. Bearing

as cylinder blocks are available only as unit assemblies.

Check cylinder blocks (27 and 32) on valve face end and renew if scratches or other damage is evident. Inspect piston slippers, pistons and bores in cylinder blocks for excessive wear or scoring.

Check pump valve plate (16) and motor valve plate (15) for scratches, excessive wear or other damage. If wear or scratches can be felt by running a fingernail across face of plates, renew plates. Inspect motor and pump swashplate thrust plates (30 and 35) for wear, embedded material and scoring. Renew if necessary.

Check charge pump housing (3) and rotor assembly (5) for wear and scoring. Renew if necessary. Relief valve cones (13 and 18) should be free of nicks and scratches. Note that implement relief valve spring (19) is heavier than charge pressure relief valve spring (12). Shims (11 and 20) are available to adjust relief valve pressure.

Renew all "O" rings, gaskets and seals. Thoroughly lubricate all parts with clean hydraulic fluid during reassembly. When installing new bearings (14), press bearings in until they are $^1/_{16}$-$^1/_8$ inch (1.6-3.2 mm) above machined surface of center housing. Check valves (26) are interchangeable and are serviced only as an assembly. Pump swashplate (37) must be installed with thin pad toward top of transmission. Drive top of roll pins (36) ¼ inch (6.35 mm) below swashplate surface. Flat end of charge pump housing (3) must be installed on same side as charge relief valve (13). Tighten housing screws to 25-30 ft.-lbs. (34-40 N·m). Tighten charge pump cap screws to 52 ft.-lbs. (71 N·m). Be sure that input shaft can be turned by hand. If shaft cannot be turned, disassemble and check for incorrect assembly of valve plates on locating pins (17).

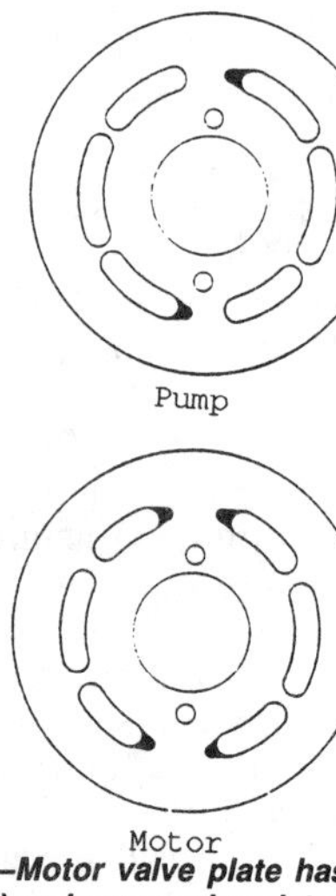

Fig. DA46—Motor valve plate has four notches (dark areas) and pump valve plate has two notches (dark areas). Refer to text.

Illustrations courtesy Deutz-Allis.

REDUCTION GEAR AND RANGE TRANSMISSION

REMOVE AND REINSTALL

Models 613H-616H

Hydrostatic transmission and gear reduction are removed as a unit. Support rear of tractor and remove rear wheels. Remove drive belt from transmission pulley. Detach control rod from transmission control arm. Detach brake rod and return spring. Unscrew gear reduction mounting screws in floor pan. Support unit, remove axle mounting "U" bolts and lower unit out of tractor. Drain oil from gear reduction housing. Unscrew transmission mounting screws and separate transmission from gear reduction housing.

To reinstall unit, reverse removal procedure. Tighten transmission mounting screws to 20 ft.-lbs. (27 N·m). Tighten gear reduction housing-to-floor pan mounting screws to 20 ft.-lbs. (27 N·m).

Models 1918-1920

To remove transmission and gear reduction or range transmission as a unit, proceed as follows: Remove tunnel cover and upper and lower side panels on left side. Disconnect rear wiring harness. Disconnect fuel line and detach line from oil cooler bulkhead. Detach fender supports and footrest supports from frame. Remove seat, fender and fuel tank assembly. Drain oil from gear reduction and remove fill tube. Remove 90 degree elbow on bottom of transmission. Detach brake and control linkage. Disconnect hoses and tubing from front of transmission. Support rear of tractor and gear reduction, unscrew fasteners securing gear reduction to frame and lower drive unit. Roll drive unit away from tractor.

Separate hydrostatic transmission from gear reduction while noting location of spacers. Remove wheels and brackets.

Reverse removal procedure for installation. Tighten rear axle mounting screws to 40-50 ft.-lbs. (54-68 N·m). Tighten drawbar mounting screws to 100 ft.-lbs. (136 N·m). Tighten axle nuts to 75-110 ft.-lbs. (102-150 N·m). Tighten screws securing front of gear reduction to crossmember to 40-50 ft.-lbs. (54-68 N·m). Tighten transmission to gear reduction screws to 25-30 ft.-lbs. (34-41 N·m).

OVERHAUL

Models 613H-616H

If not removed previously, remove wheel and hub assemblies from axles. Remove park brake disc from brake shaft. Clean axle shaft and remove any burrs from shafts. Unscrew cap screws and drive out dowel pins in cover (29—Fig. DA47). Lift cover off case and axle shaft. Withdraw brake shaft (5), idler gear (4) and thrust washers (3 and 6) from case. Remove output shaft (11), output gear (10), spacer (9), thrust washer (8) and differential assembly from case. Axle shaft housings (20 and 22) must be pressed from case and cover.

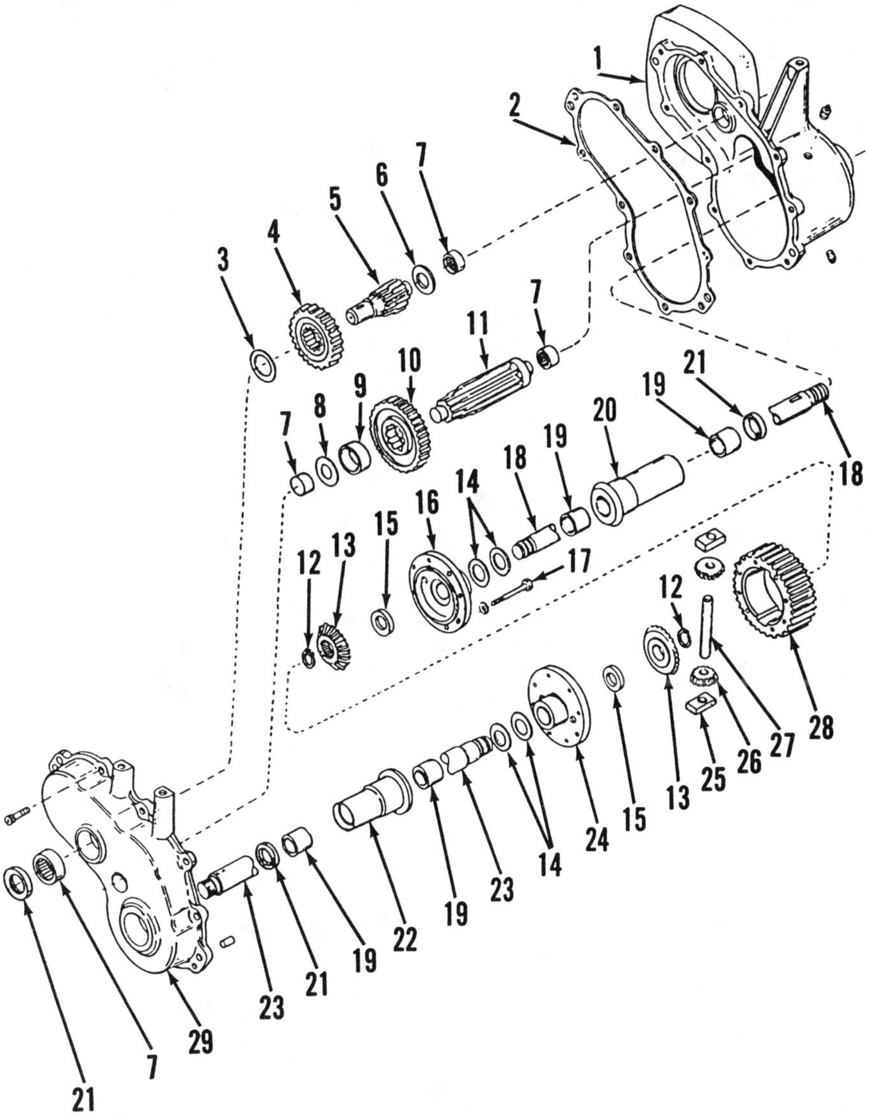

Fig. DA47—Exploded view of Peerless reduction gear and differential unit used on Models 613H and 616H.

1. Case
2. Gasket
3. Thrust washer
4. Idler gear
5. Brake shaft
6. Thrust washer
7. Bearing
8. Thrust washer
9. Spacer
10. Output gear
11. Output shaft
12. Snap ring
13. Side gears
14. Thrust washers
15. Thrust washer
16. Differential carrier
17. Bolt
18. Axle shaft, R.H.
19. Bushing
20. Axle housing
21. Oil seal
22. Axle housing
23. Axle shaft, L.H.
24. Differential carrier
25. Drive block
26. Drive pinion
27. Drive pin
28. Ring gear
29. Cover

Illustrations courtesy Deutz-Allis.

To disassemble differential, unscrew cap screws (17) and separate axle shaft and carrier assemblies from ring gear (28). Drive blocks (25), bevel pinion gears (26) and drive pin (27) can now be removed from ring gear. Remove snap rings (12) and slide axle shaft (18 and 23) from axle gears (13) and carriers (16 and 24).

Clean and inspect all parts and renew any parts damaged or excessively worn. When installing needle bearings, press bearings in from inside of case or cover until bearings are 0.015-0.020 inch (0.38-0.51 mm) below thrust surfaces. Be sure heads of differential cap screws (17) and right axle shaft (18) are installed in right carrier housing (16). Right axle shaft is installed through case (1). Tighten differential cap screws to 7 ft.-lbs. (9 N·m) and cover cap screws to 10 ft.-lbs. (14 N·m). Differential assembly and output shaft (11) must be installed in case at same time. Remainder of assembly is reverse of disassembly procedure. Refill differential unit with $2^3/_4$ pints (1.3 L) of SAE 90 EP gear oil.

Models 1918

To disassemble reduction gear and differential assembly, remove hitch plate and the six bolts securing each axle housing to differential housing. Separate differential housing and axle housings. Remove cap screws and separate case cover (12—Fig. DA48) from case (33). Remove countershaft (7) and gear assemblies, intermediate shaft (18) and gear assemblies, output shaft (27) and gear (26), and ring gear (37) and differential assembly. Remove cap screw (44) and disassemble differential as needed.

Inspect all parts for wear, looseness and damage and renew as needed. When renewing needle bearings in cover (12) or case (33), drive bearings in until they are 0.015-0.020 inch (0.38-0.50 mm) below thrust surfaces of cover and case.

Reassemble by reversing disassembly procedure. Renew all gaskets and seals. Apply Loctite to cap screws (44) and tighten to 7-10 ft.-lbs. (10-13 N·m).

Models 1920

To disassemble range transmission and differential assembly, remove hitch plate and remove the six bolts securing each axle housing to differential housing. Separate differential housing and axle housings. Remove cap screws and separate cover (13—Fig. DA49) from case (39). Remove differential assembly and locking shifter fork (35). Remove shift detent bolt (40), spring (41) and ball (42). Withdraw thrust washers, high range gear (9), sliding gear (22) and shift rod (19) and fork (20). Remove input shaft (8), countershaft (21) and output shaft (49) along with remaining gears. Disassemble gear and shaft assemblies and differential unit as needed.

Inspect all parts for wear, looseness or damage and renew as needed. When renewing needle bearings in cover (13) or case (39), drive bearings in until they are 0.015-0.020 inch (0.38-0.50 mm) below thrust surfaces of cover and case.

Reassemble by reversing disassembly procedure. Renew all gaskets and seals. Apply Loctite to carrier (60) cap screws and tighten to 7-10 ft.-lbs. (10-13 N·m). Tighten cover cap screws to 20 ft.-lbs. (27 N·m).

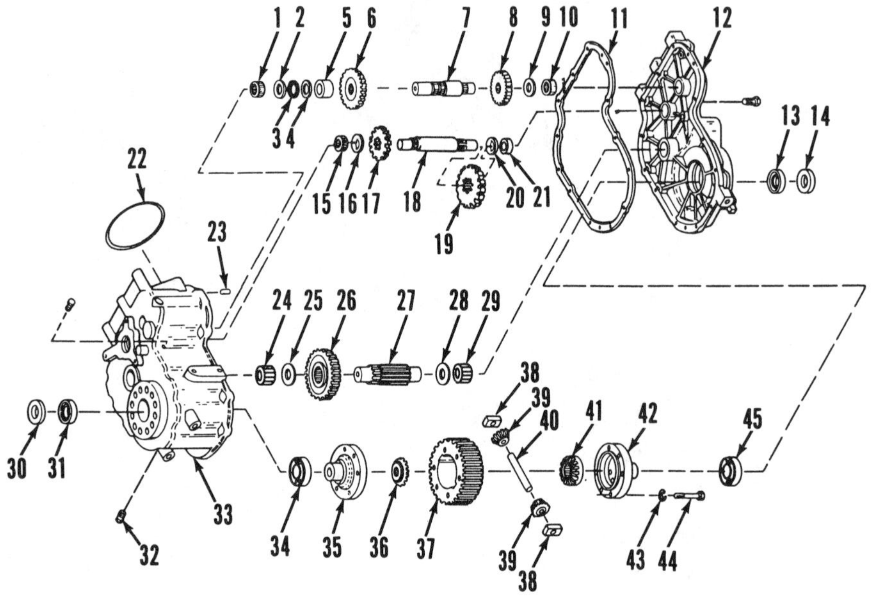

Fig. DA48—Exploded view of gear reduction used on Model 1918.

1. Bearing
2. Thrust washer
3. Thrust bearing
4. Thrust washer
5. Spacer
6. Bevel gear
7. Countershaft
8. Spur gear
9. Thrust washer
10. Bearing
11. Gasket
12. Cover
13. Seal
14. Washer
15. Bearing
16. Thrust bearing
17. Spur gear (17 T)
18. Intermediate shaft
19. Intermediate gear (27 T)
20. Thrust washer
21. Bearing
22. "O" ring
23. Dowel pin
24. Bearing
25. Thrust washer
26. Output gear (33 T)
27. Output shaft
28. Thrust washer
29. Bearing
30. Washer
31. Seal
32. Drain plug
33. Case
34. Bearing
35. Differential carrier
36. Axle gear
37. Ring gear
38. Pinion blocks
39. Pinion gears
40. Cross shaft
41. Axle gear
42. Differential carrier
43. Washer
44. Bolt
45. Bearing

HYDRAULIC SYSTEM

MAINTENANCE

Models 1918-1920

The hydrostatic drive unit supplies pressurized oil for the power steering system (models so equipped) and the hydraulic lift system. Refer to appropriate TRANSMISSION section for maintenance and service information for hydrostatic drive unit.

PRESSURE TESTS

Models 1918-1920

CHARGE PUMP PRESSURE TEST. To check charge pressure, remove tunnel cover. Remove pipe plug from top of hydrostatic housing and install a 0-300 psi (0-2000 kPa) dampened pressure gage or a gage with snubber in port (P—Fig. DA50). Start and run engine at fast idle. Charge pump pressure should be 90-180 psi (620-1240 kPa). If low, be sure transmission filter and charge pump relief valve are in good condition. Relief valve pressure setting is adjustable using shims (11—Fig. DA45).

LIFT SYSTEM PRESSURE TEST. To check lift system pressure, install a 0-1000 psi (0-7000 kPa) pressure gage in hydrostatic housing test port (P—Fig. DA50) as outlined in CHARGE PUMP PRESSURE TEST. Start and run engine at fast idle. Operate lift control lever to fully extend lift cylinder and obtain system relief pressure reading. System pressure should be 700-900 psi (4830-6205 kPa). If low, be sure transmission filter and implement relief valve are in good condition. Relief valve pressure setting is adjustable using shims (20—Fig. DA45).

CONTROL VALVE

Models 1918-1920

Before removing control valve from tractor, mark all hydraulic lines to assure correct installation during reassembly. Remove upper dash panel from left side of tractor. Disconnect hydraulic hoses from valve and cap all openings. Disconnect control linkage from valve spool. Remove retaining bolts and remove valve assembly.

Refer to Fig. DA51 for an exploded view of control valve. Prior to disassembly, thoroughly clean outside of valve housing. Remove hydraulic fittings from housing. Remove valve cap (1) and withdraw valve spool (6) and spring (7) from valve body. Remove and discard "O" rings (8 and 10) and wiper (11). Remove plug (15), spring (13) and lift check plunger (12).

Inspect all parts for wear and damage. Renew all "O" rings. Valve spool and body are a select fit and must be renewed as an assembly if either is excessively worn or damaged.

Reassemble by reversing disassembly procedure. Lubricate all internal components with hydraulic oil prior to assembly.

HYDRAULIC CYLINDER

Models 1918-1920

Hydraulic cylinder is a welded assembly and no service parts are available. Renew cylinder if internal or external leakage is apparent.

To check cylinder for internal leakage, start engine and fully extend cylinder. Stop engine and disconnect front hose of cylinder from control valve. Place end of hose in a container. Start engine and move lift control lever to extend cylinder. A small amount of oil may initially flow from open end of hose, then oil

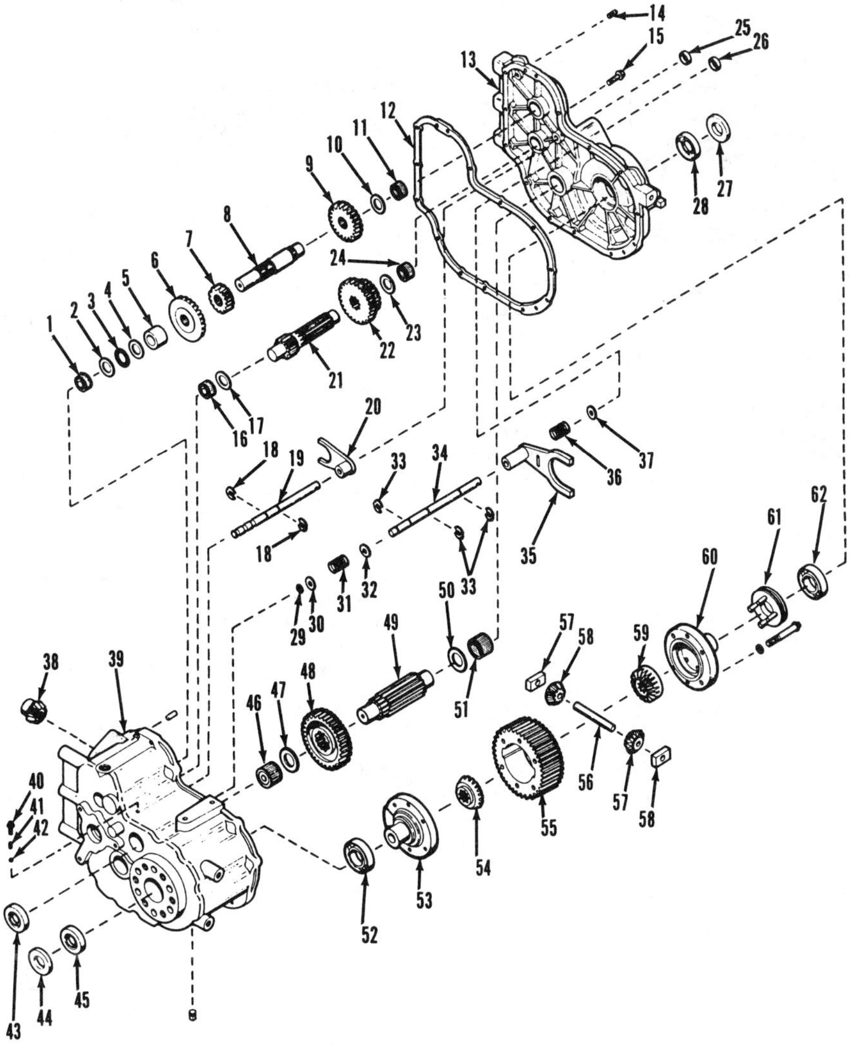

Fig. DA49—Exploded view of two-speed range transmission and differential used on Model 1920.

1. Bearing
2. Thrust washer
3. Thrust bearing
4. Thrust washer
5. Spacer
6. Bevel gear
7. Spur gear (16 T)
8. Input shaft
9. Spur gear (20 T)
10. Thrust washer
11. Bearing
12. Gasket
13. Cover
14. Plug
15. Cap screw
16. Bearing
17. Thrust washer
18. Snap rings
19. Shift rod (range transmission)
20. Shift fork
21. Countershaft
22. Sliding gear cluster
23. Thrust washer
24. Bearing
25. Oil seal
26. Oil seal
27. Spacer
28. Oil seal
29. Retaining ring
30. Thrust washer
31. Spring
32. Thrust washer
33. Snap rings
34. Shift rod (differential lock)
35. Shift fork
36. Spring
37. Thrust washer
38. Pinion gear (drive)
39. Case
40. Bolt
41. Detent spring
42. Detent ball
43. Seal (as equipped)
44. Spacer
45. Oil seal
46. Bearing
47. Thrust washer
48. Output gear (37 T)
49. Output shaft
50. Thrust washer
51. Bearing
52. Bearing
53. Carrier
54. Axle gear
55. Ring gear
56. Cross shaft
57. Pinion gears
58. Pinion blocks
59. Axle gear
60. Carrier
61. Locking collar
62. Bearing

flow should stop. If oil flow continues, renew cylinder.

To remove cylinder, remove tunnel cover and upper dash panel from right side of tractor. Clean outside of cylinder and hydraulic hoses, then disconnect hose from front of cylinder. Cap openings to prevent entry of dirt in system. Remove retainer clips and pins retaining cylinder. Disconnect hose from rear of cylinder and remove cylinder from tractor.

To install cylinder, reverse the removal procedure. Fully extend and retract cylinder several times to purge air from system.

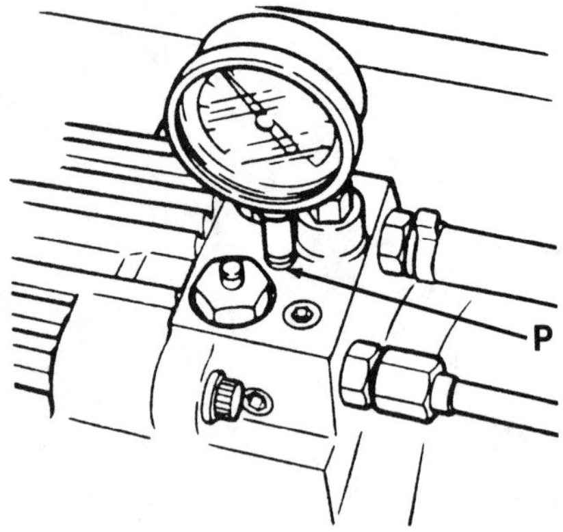

Fig. DA50—To check charge pressure or implement lift pressure on Models 1918 and 1920, remove plug from top of hydrostatic housing and install a pressure gage in test port. Refer to text.

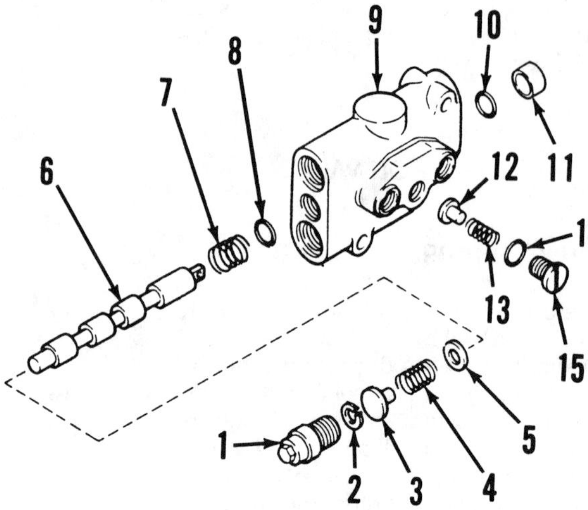

Fig. DA51—Exploded view of lift control valve used on Models 1918 and 1920.

1. Plug
2. Snap ring
3. Spacer
4. Spring
5. Washer
6. Valve spool
7. Spring
8. "O" ring
9. Body
10. "O" ring
11. Wiper
12. Lift check valve
13. Spring
14. "O" ring
15. Plug

Illustrations courtesy Deutz-Allis.

ENGINEERING PRODUCTS CO.

CONDENSED SPECIFICATIONS

	POWER KING MODELS				
	1217	1617	1618	2417	2418
Engine Make	Kohler	Kohler	Kohler	Kohler	Kohler
Model	KT-17	KT-17	K-361	KT-17	K-361
Bore	3.12 in. (79.3 mm)	3.12 in. (79.3 mm)	3.75 in. (95.3 mm)	3.12 in. (79.3 mm)	3.75 in. (95.3 mm)
Stroke	2.75 in. (69.9 mm)	2.75 in. (69.9 mm)	3.25 in. (82.5 mm)	2.75 in. (69.9 mm)	3.25 in. (82.5 mm)
Displacement	42.18 cu. in. (690.5 cc)	42.18 cu. in. (690.5 cc)	35.9 cu. in. (589 cc)	42.18 cu. in. (690.5 cc)	35.9 cu. in. (589 cc)
Power Rating	17 hp. (12.7 kW)	17 hp. (12.7 kW)	18 hp. (13.4 kW)	17 hp. (12.7 kW)	18 hp. (13.4 kW)
Slow Idle	1200 rpm	1200 rpm	1800 rpm	1200 rpm	1800 rpm
High Speed (No-Load)	3600 rpm	3600 rpm	3800 rpm	3600 rpm	3800 rpm
Capacity–					
Crankcase	3 pts. (1.4 L)	3 pts. (1.4 L)	2 qts. (1.9 L)	3 pts. (1.4 L)	2 qts. (1.9 L)
Hydraulic Systems	.....	2 gal. (7.6 L)	5 qts. (4.7 L)	2 gal. (7.6 L)	5 qts. (4.7 L)
Transmission	1.5 pts. (0.7 L)	0.5 pt.* (0.2 L)	0.5 pt. (0.2 L)	0.5 pt.* (0.2 L)	0.5 pt. (0.2 L)
Range Transmission	3.5 qts. (3.3 L)	.....	0.5 pt. (0.2 L)	.....	0.5 pt. (0.2 L)
Final Drives	.....	1.5 pts. (0.7 L)	1. pts. (0.7 L)	1.5 pts. (0.7 L)	1.5 pts. (0.7 L)
Differential	.....	2 pts. (0.9 L)	2 pts. (0.9 L)	2 pts. (0.9 L)	2 pts. (0.9 L)
Fuel Tank	5 gal. (18.9 L)	8.25 gal. (31.2 L)	3 gal. (11.4 L)	8.25 gal. (31.2 L)	3 gal. (11.4 L)

*Transmission fluid capacity for 1617 and 2417 models after S/N 62750 is 1³/₈ pints (0.65 L).

FRONT AXLE AND STEERING SYSTEM

MAINTENANCE

All Models

Lubricate steering spindles, tie rod ends, drag link ends, axle pivot and front wheel bearings at 50 hour intervals. Use multi-purpose, lithium base grease. Clean all grease fittings before and after lubrication.

R&R AXLE MAIN MEMBER

Model 1217

Raise and support front of tractor. Disconnect drag link end (18–Fig. E1) at steering lever (19). Support axle main member (14), remove cotter pin (10), nut (11) and washer (12). Pull main member forward off pivot pin and lower axle to the ground. Roll axle assembly away from tractor.

Renew bushing (13) as necessary. Reinstall by reversing removal procedure.

Models 1617-1618-2417-2418

Raise and support front of tractor. Disconnect drag link end (9–Fig. E3) at steering lever (11). Support axle main member (3), remove cotter pin (6), nut (5) and washer (4). Slide axle assembly forward off of pivot pin, lower axle to the ground and roll assembly away from tractor.

Renew bushings (2) as necessary. Reinstall by reversing removal procedure.

TIE ROD AND TOE-IN

All Models

To remove tie rod, disconnect tie rod ends at each steering arm.

Renew ends as necessary. Reinstall by

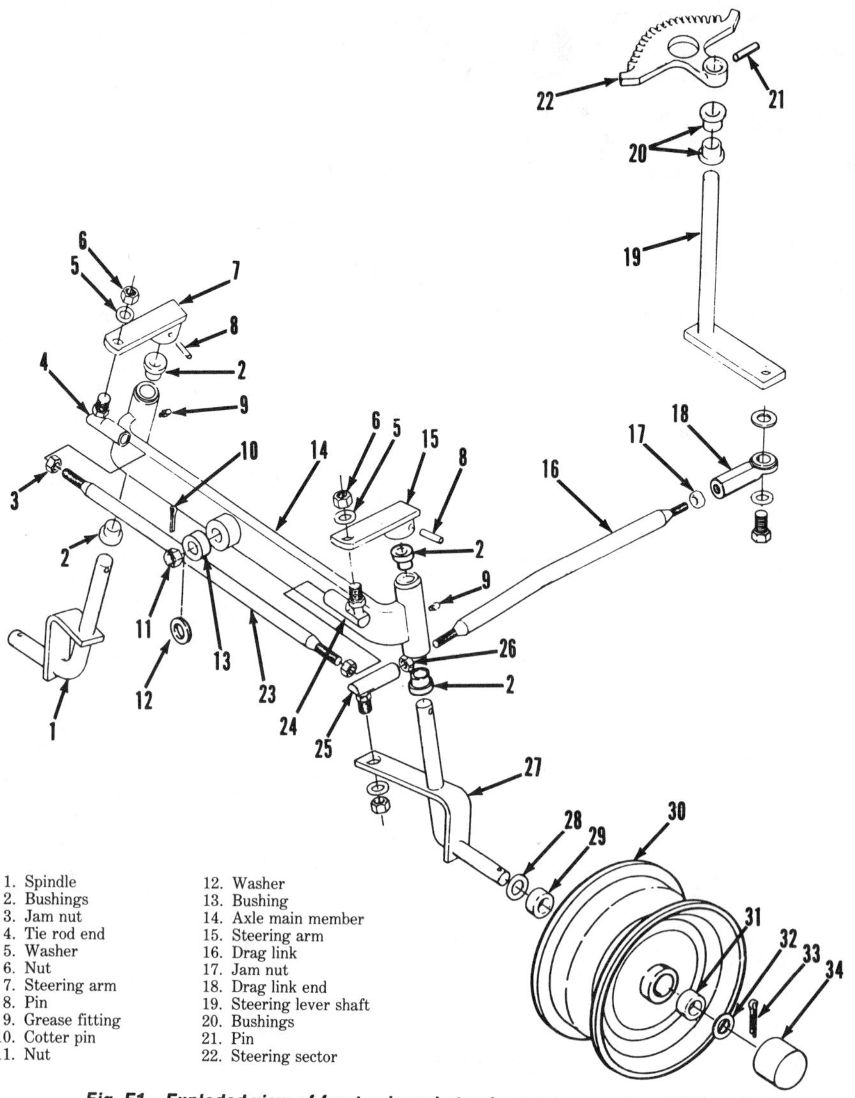

1. Spindle
2. Bushings
3. Jam nut
4. Tie rod end
5. Washer
6. Nut
7. Steering arm
8. Pin
9. Grease fitting
10. Cotter pin
11. Nut
12. Washer
13. Bushing
14. Axle main member
15. Steering arm
16. Drag link
17. Jam nut
18. Drag link end
19. Steering lever shaft
20. Bushings
21. Pin
22. Steering sector

Fig. E1 – Exploded view of front axle and steering system used on 1217 models.

23. Tie rod
24. Tie rod end
25. Drag link end
26. Jam nut
27. Spindle
28. Spacer
29. Bearing
30. Wheel & hub assy.
31. Bearing
32. Washer
33. Cotter pin
34. Dust cap

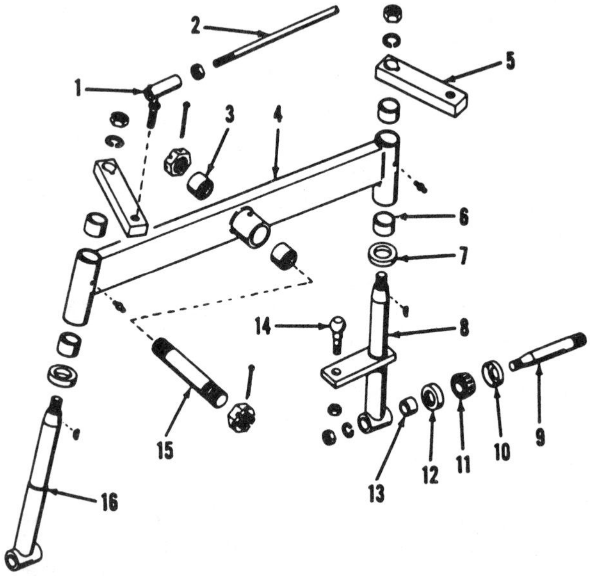

Fig. E2 – Exploded view of front axle assembly used on 1618, 2418 models and 1617, 2417 models prior to S/N 62750.

1. Tie rod end
2. Tie rod
3. Pivot bearing
4. Axle main member
5. Steering arm
6. Bushing
7. Load bearing
8. L.H. steering spindle shaft
9. Steering spindle
10. Bearing cup
11. Bearing cone
12. Oil seal
13. Spacer
14. Ball stud
15. Axle pivot stud
16. R.H. steering spindle shaft

reversing removal procedure. Adjust length of tie rod to obtain 1/8-inch (3.175 mm) toe-in. Tighten jam nuts against tie rod ends.

R&R STEERING SPINDLES

Model 1217

Raise and support front of tractor. Remove dust cap (34–Fig. E1), cotter pin (33) and tire and wheel assembly from each side. Disconnect drag link end (25) at steering spindle (27). Remove tie rod (23). Drive out pins (8) and remove steering arms (7 and 15). Remove spindles (1 and 27).

Renew bushings (2) and/or spindles (1 and 27) as necessary. Reinstall by reversing removal procedure and adjust toe-in.

Models 1617-1618-2417-2418

Raise and support front of tractor, then remove tire and wheel. Remove dust cap (22–Fig. E3), nut (21), bearing (20) and hub (17) from each side. Disconnect drag link end (9) at steering lever (11). Remove tie rod. Drive pins (10) from left and right steering arms, then pull spindles out of steering arms and axle main member.

Renew bushings (12) and/or spindles as necessary. Reinstall by reversing removal procedure and adjust toe-in.

R&R FRONT WHEEL BEARINGS

Model 1217

Bearings (29 and 31–Fig. E1) are a press fit in wheel hub (30). To remove bearings, first remove hub from spindle; then use a hammer and punch to drive bearings outward from inside of hub.

Renew or repair hub (30) if bearings are a loose fit in hub.

Models 1617-1618-2417-2418

Raise and support side to be serviced. Remove tire and wheel. Remove dust cap (22–Fig. E3), cotter pin, nut (21) and bearing (20). Remove hub (17) and use a punch to carefully drive bearing cup and bearing (16), seal (15) and spacer (19) out of hub. Drive outer bearing cup out of hub as necessary.

Clean and repack or renew bearings. Reinstall by reversing removal procedure. Adjust nut (21) so bearings turn freely with a slight amount of end play. Install cotter pin.

R&R STEERING GEAR

Model 1217

Disconnect seat safety switch, then

Illustrations courtesy Engineering Products Co.

remove seat and seat mounting bracket. Remove all bolts and nuts securing rear fender/body assembly. Remove both shift lever knobs and place transmission shift lever in reverse position. Remove fuel tank gage and cap assembly. Tilt entire fender/body assembly forward far enough to clear shift lever and lift assembly from tractor. Reinstall fuel gage and cap assembly. Raise hood and remove the six nuts, retaining clips and screws and lift firewall off tractor. Remove battery and battery support shelf. Remove steering wheel, disconnect all electrical leads, remove grommet (10–Fig. E4), bushing (9) and snap ring (2), as equipped. Remove console and steering support bracket (3). Turn the front wheels all the way to the left and drive pin (21–Fig. E1) out of steering sector (22). Lift sector off of steering lever shaft (19).

Renew bushings as required. Reassemble by reversing disassembly procedure.

Models 1618-2418-Models 1617-2417 Prior To S/N 62750

Remove nut (12–Fig. E5) and use a suitable puller to remove steering arm (10). Remove bolts securing steering bracket to frame and remove complete steering gear and bracket assembly.

Reinstall by reversing removal procedure.

Models 1617-2417 After S/N 62750

Raise and remove hood. Disconnect choke, throttle cable and all necessary electrical connections. Disconnect battery and remove battery and battery support shelf. Remove both heat shields from between engine and console. Remove hydraulic pump and drive belt, as equipped. Remove the four cap screws securing driveshaft to engine and slide driveshaft back on clutch input shaft. Remove pin (20–Fig. E6) and use a suitable puller to remove steering arm (21). Remove mounting bracket (6) and slide steering gear assembly to the left, out of bushing (22) and frame bracket. Raise column and slide forward out of console sheet metal.

Reinstall by reversing removal procedure.

OVERHAUL STEERING GEAR

Models 1618-2418-Models 1617-2417 Prior To S/N 62750

Lubricate all parts during reassembly. Install bearing cup (13–Fig. E5) in housing (8). Place bearings (14 and 16)

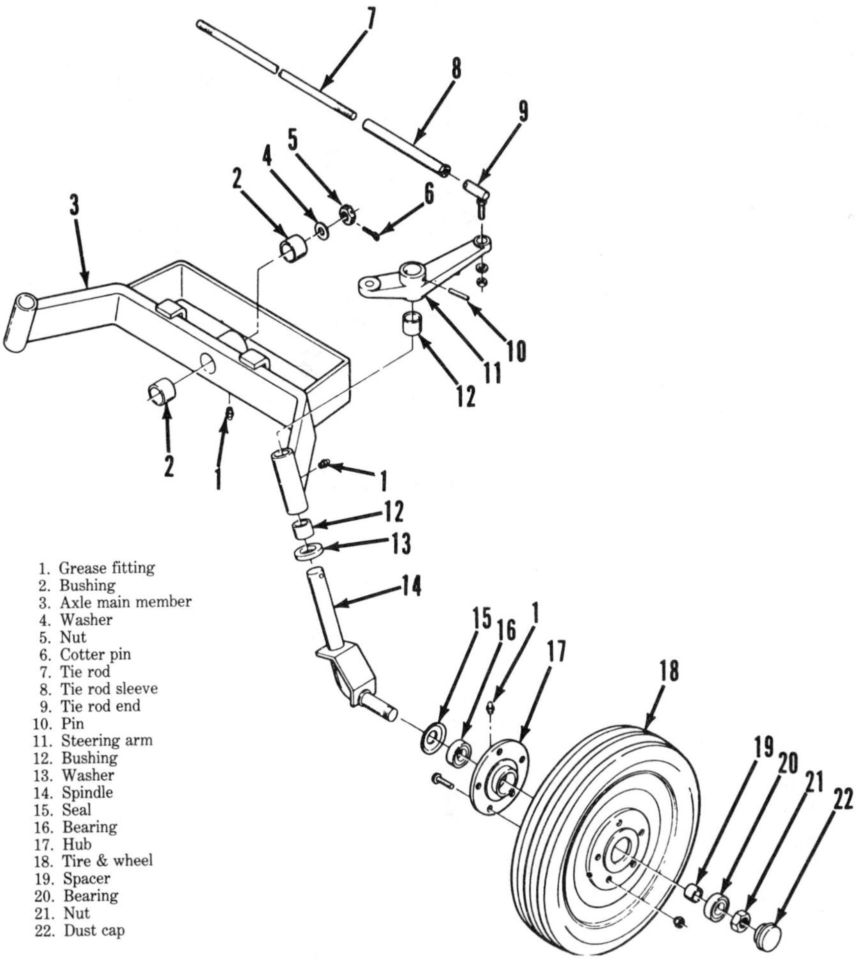

1. Grease fitting
2. Bushing
3. Axle main member
4. Washer
5. Nut
6. Cotter pin
7. Tie rod
8. Tie rod sleeve
9. Tie rod end
10. Pin
11. Steering arm
12. Bushing
13. Washer
14. Spindle
15. Seal
16. Bearing
17. Hub
18. Tire & wheel
19. Spacer
20. Bearing
21. Nut
22. Dust cap

Fig. E3—Exploded view of front axle asembly used on 1617 and 2417 models after S/N 62750.

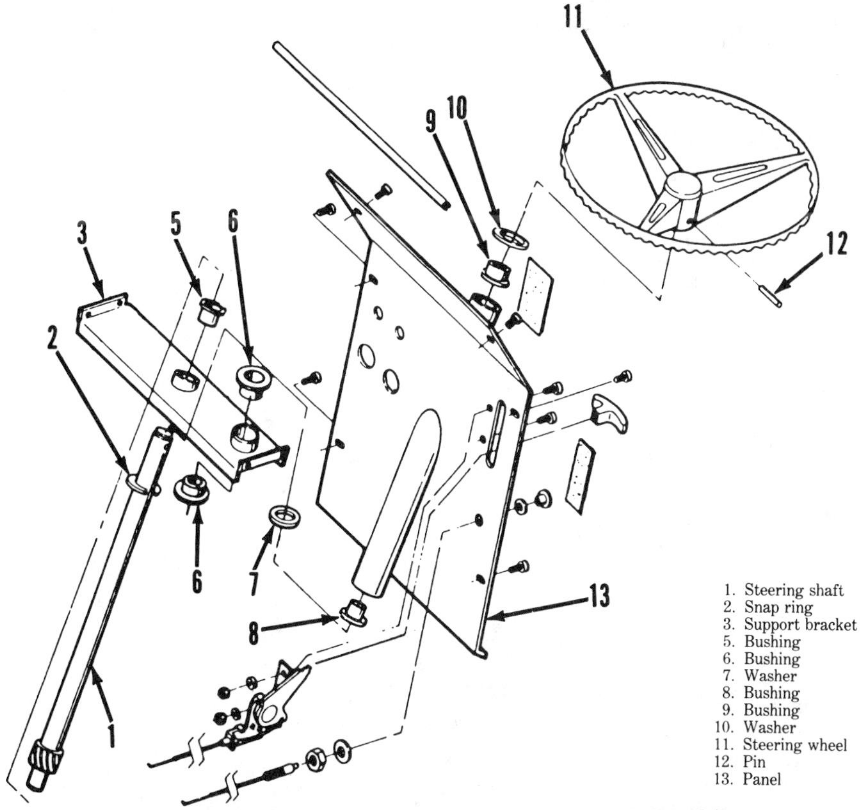

1. Steering shaft
2. Snap ring
3. Support bracket
5. Bushing
6. Bushing
7. Washer
8. Bushing
9. Bushing
10. Washer
11. Steering wheel
12. Pin
13. Panel

Fig. E4—Exploded view of steering system used on 1217 models.

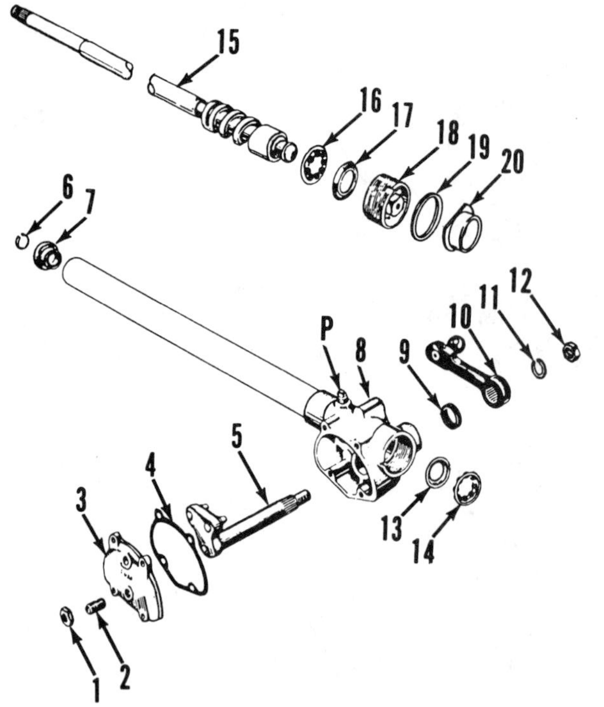

Fig. E5—Exploded view of steering gear assembly used on 1618, 2418 models and 1617, 2417 models prior to S/N 62750.

P. Plug
1. Jam nut
2. Stud
3. Cover
4. Gasket
5. Shaft
6. Spring seat
7. Bushing
8. Housing
9. Seal
10. Steering arm
11. Washer
12. Nut
13. Bearing cup
14. Bearing
15. Shaft
16. Bearing
17. Bearing cup
18. Adjustment plug
19. Seal
20. Lockring

on steering gear and shaft (15), then install shaft assembly in housing (8). Install bearing cup (17) in housing (8) making certain it enters bore squarely. Install adjustment plug (18) and tighten until a slight drag is felt as steering shaft (15) is turned. Install seal (19) and lockring (20). Stake lockring in position. Install lever shaft (5) and cover (3) and tighten adjustment stud (2) until a slight drag is felt as steering shaft (15) is turned past mid-position. Secure with jam nut (1). Fill housing (8) with SAE 90 EP gear lubricant and install plug (P).

Models 1617-2417 After S/N 62750

Remove jam nut (12 – Fig. E6) adjustment nut (11), washer (10) and pull steering lever (23) from housing (5). Remove cotter pin (15), adjustment plug (14), washer (13) and press steering shaft (7) and bearing assemblies out of housing (5). Remove bearings (9) from shaft (7) and bearing cup (8) from housing (5) as necessary.

Lubricate all parts during reassembly. Install cup (8) in housing (5) making certain it seats squarely in bore. Install bearings (9) on shaft (7) and install assembly in housing (5). Install bearing cup (8) squarely in housing bore, install washer (13) and adjustment plug (14). Tighten adjustment plug to 10-14 ft.-lbs. (13-19 N·m) and install cotter pin (15). Pack housing with multi-purpose lithium base grease. Loosen adjustment stud (19) two full turns and install steering plate (23) on housing (5). Tighten adjustment nut (12) until a 0.100 inch (2.54 mm) feeler gage can be inserted between steering plate (23) and housing (5). Install jam nut (12) and tighten to 40 ft.-lbs. (54 N·m). Tighten adjustment stud (19) until a slight drag is felt as steering shaft (7) is turned past mid-position. Secure with jam nut (16).

ENGINE

MAINTENANCE

All Models

Regular engine maintenance is required to maintain peak performance and long engine life.

Check oil level at five hour intervals under normal operating conditions.

Clean and re-oil foam precleaner at 25 hour intervals and clean or renew dry filter element at 100 hour intervals.

REMOVE AND REINSTALL

Model 1217

Remove hood and disconnect negative battery cable. Remove the six cap screws, retaining washers and locknuts and remove firewall. Disconnect choke cable, throttle linkage, fuel line and all necessary electrical connections. Remove front footrest rod and the grass screen located beneath tractor. Remove cap screws securing flexible coupler to crankshaft adapter. Remove the four cap screws securing engine, slide engine forward until flexible coupler is free and lift engine from tractor.

Reinstall by reversing removal procedure.

Models 1618-2418-Models 1617-2417 Prior To S/N 62750

Remove hood, front shell, tie rod, pto clutch, muffler and dash assemblies. Remove clutch cover and flywheel housing bolts. Remove the four cap screws securing engine to mounting plate. If equipped with hydraulic pump, remove heat shield and hydraulic pump (pump may be moved to one side, leaving hoses attached). On all models, slide engine forward until transmission input shaft

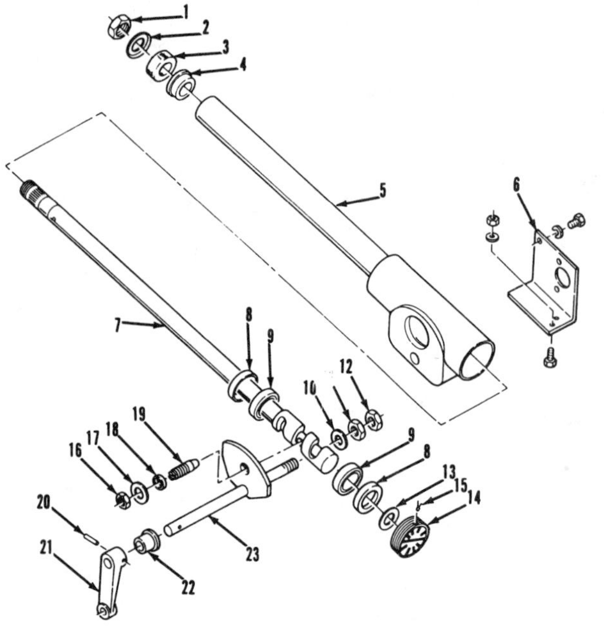

Fig. E6—Exploded view of steering gear assembly used on 1617 and 2417 models after S/N 62750.

1. Nut
2. Washer
3. Seal
4. Bushing
5. Housing
6. Bracket
7. Shaft
8. Bearing cup
9. Bearing
10. Washer
11. Adjustment nut
12. Jam nut
13. Washer
14. Adjustment plug
15. Cotter pin
16. Jam nut
17. Washer
18. Seal
19. Stud
20. Pin
21. Steering arm
22. Bushing
23. Steering lever

Illustrations courtesy Engineering Products Co.

disengages and lift engine out.

Reinstall by reversing removal procedure.

Models 1617-2417 After S/N 62750

Remove hood and disconnect battery cables. Disconnect all necessary electrical leads and choke and throttle cables. Close fuel valve and disconnect fuel line at fuel pump. Remove heat shrouds between engine and dash console. Remove hydraulic pump belt, as equipped. Disconnect driveshaft at engine, remove engine retaining bolts and lift engine out of tractor.

Reinstall by reversing removal procedure. Note driveshaft must be installed so keyway in rear universal joint yoke is 90 degrees to the right of engine flywheel key position.

OVERHAUL

All Models

Engine make and model are listed at the beginning of this section. For tune-up specifications, engine overhaul procedures and engine maintenance, refer to appropriate engine section in this manual.

ELECTRICAL SYSTEM

MAINTENANCE AND SERVICE

All Models

Battery electrolyte level should be checked at 50 hour intervals of normal operation. If necessary, add distilled water until level is just below base of vent well. **DO NOT** overfill. Keep battery posts clean and cable ends tight.

For alternator or starter service, refer to appropriate engine in this manual.

Refer to approximate wiring diagram shown in Figs. E7 through E7B for model being serviced.

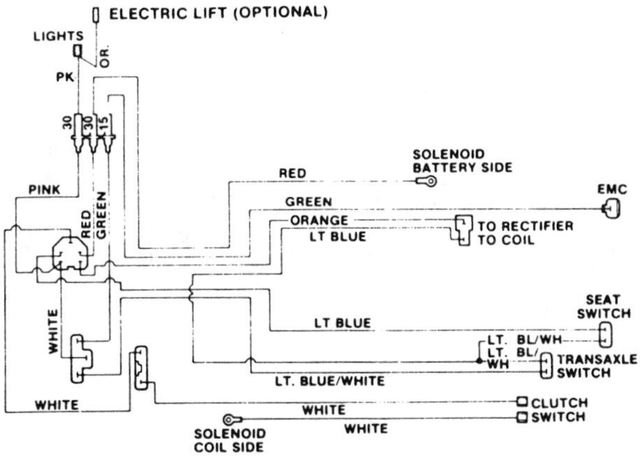

Fig. E7—Typical wiring diagram for 1217 model.

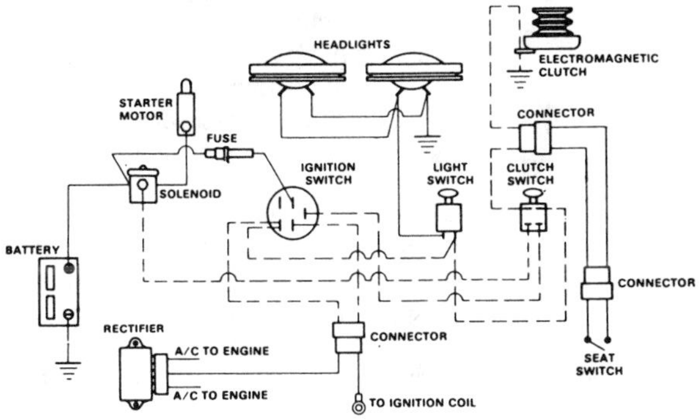

Fig. E7A—Wiring diagram for 1618 and 2418 models and 1617 and 2417 models prior to S/N 62750.

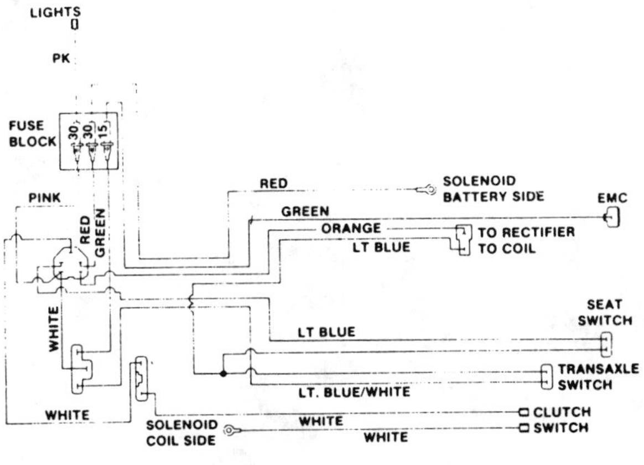

Fig. E7B—Typical wiring diagram for 1617 and 2417 models after S/N 62750.

BRAKES

ADJUSTMENT

Model 1217

With brake pedal in fully raised position, adjust nut (8–Fig. E8) so a 0.020 inch (0.508 mm) feeler gage can be inserted between outer friction pad (2) and disc (9).

Models 1617 and 2417

Adjust each wheel brake individually. Loosen nut (11–Fig. E9) and tighten nut (7) until brake band (1) is tight on brake drum. Tighten nut (11) until spring (9) measures 2¼ inches (57.15 mm) in length. Brake pedals must have ½-inch (12.7 mm) free travel before spring (9) begins to compress.

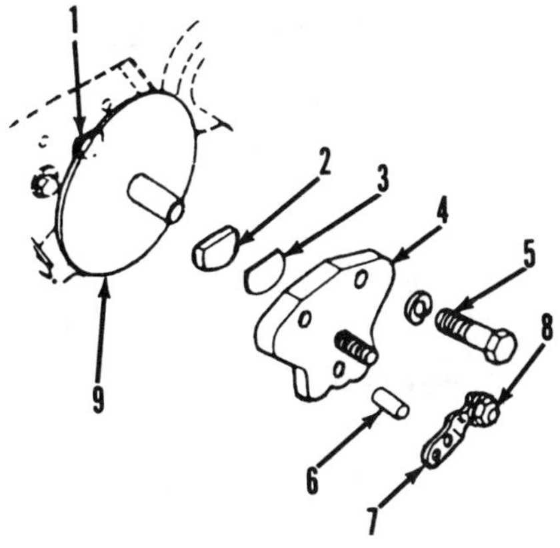

Fig. E8—Exploded view of disc brake system used on 1217 models.

1. Inner brake pad
2. Outer brake pad
3. Metal backing plate
4. Caliper
5. Cap screw
6. Pin
7. Brake lever
8. Adjustment nut
9. Brake disc

Illustrations courtesy Engineering Products Co.

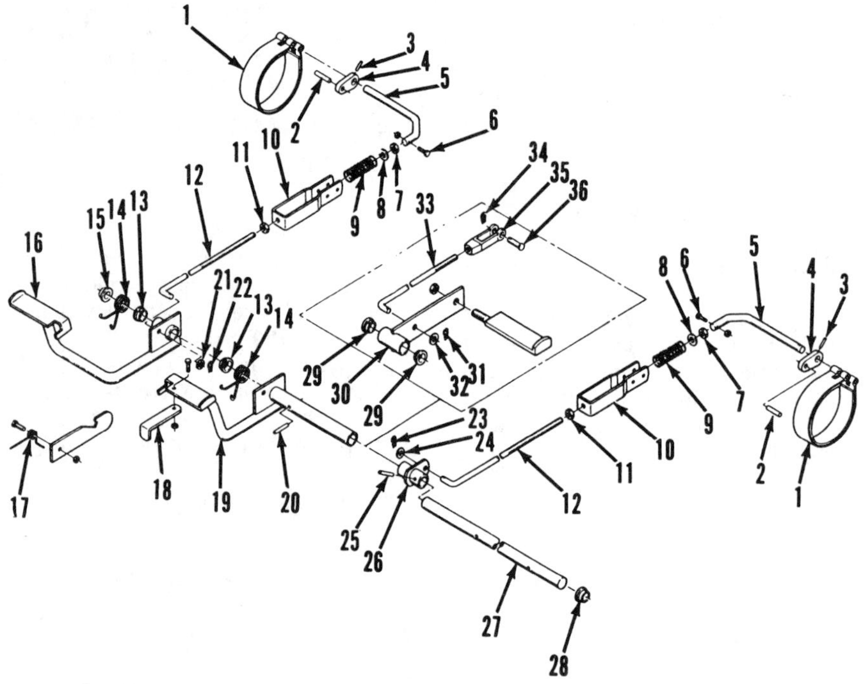

Fig. E9—Exploded view of brake system used on all models, except 1217 model.

1. Brake band
2. Pin
3. Pin
4. Brake arm
5. Shaft
6. Cap screw
7. Nut
8. Washer
9. Spring
10. Clevis
11. Nut
12. Brake rod
13. Bushing
14. Spring
15. Bushing
16. Pedal
17. Spring
18. Latch
19. Pedal
20. Cotter pin
21. Washer
22. Cotter pin
23. Cotter pin
24. Washer
25. Pin
26. Brake arm
27. Shaft
28. Bushing
29. Bushings
30. Brake pedal lever
31. Cotter pin
32. Washer
33. Brake rod
34. Cotter pin
35. Clevis
36. Pin

R&R BRAKE PADS/BANDS

Model 1217

Disconnect brake rod at brake lever (7–Fig. E8). Remove cap screws (5), caliper (4), pin (6), outer brake pad (2) and metal plate (3). Remove brake disc (9) and inner pad (1).

Reinstall by reversing removal procedure.

Models 1617-1618-2417-2418

Raise and support rear of tractor, then remove tire and wheel assemblies. Remove fenders as necessary to obtain additional working space. Use a hammer and punch to drive pins (3–Fig. E9) from cams (4). Remove cam and brake band assemblies. Drive pins (2) out of bands.

Reinstall by reversing removal procedure and adjust brakes.

CLUTCH

ADJUSTMENT

Model 1217

Forward or reverse tractor motion is stopped by returning hydrostatic drive unit to neutral position. If tractor creeps forward or backward when in neutral, adjust motion control linkage by raising and supporting rear of tractor off the ground. Remove left rear wheel and tire. Start and run engine at 3/4 throttle. Loosen both nuts on eccentric adjusting screw located under the tractor in pivot bracket (Fig. E10) and adjust screw to a position which stops wheel rotation when motion control linkage is in neutral position.

CAUTION: Use care when working around rotating parts.

Tighten both nuts on eccentric adjusting screw to lock screw in this position. Install tire and lower tractor to the ground.

Models 1618-2418-Models 1617-2417 Prior To S/N 62750

Slight changes in clutch pedal free play can be obtained by lengthening or shortening bolts (14–Fig. E11) to adjust lever (15) height. If clutch slips even with adequate pedal free play, clutch disc facings (5 and 7) are worn and must be renewed.

Models 1617-2417 After S/N 62750

Remove keeper (22–Fig. E12), pin (23) and clevis (21) from clutch arm (20). Move clutch arm (20) forward until it contacts throwout bearing and stops. Holding clutch arm in this position, adjust clevis (21) so pin (23) can just be inserted through clevis and clutch arm, then turn clevis out exactly 1½ additional turns. Reassemble linkage.

R&R CLUTCH

Models 1618-2418- Models 1617 2417 Prior To S/N 62750

Remove engine as outlined in appropriate paragraph in ENGINE section in this manual. Unbolt and remove clutch assembly from flywheel (29–Fig. E11). Remove flywheel as necessary making certain all setscrews (4), located at bottom of pulley groove, are removed first.

Clutch disc lining consists of 8 segments (four thick and four thin). When relining clutch disc (6), the 4 thick segments (5) must be installed on flywheel side of disc. Ball type clutch release bearing (24) can be inspected and renewed, if necessary.

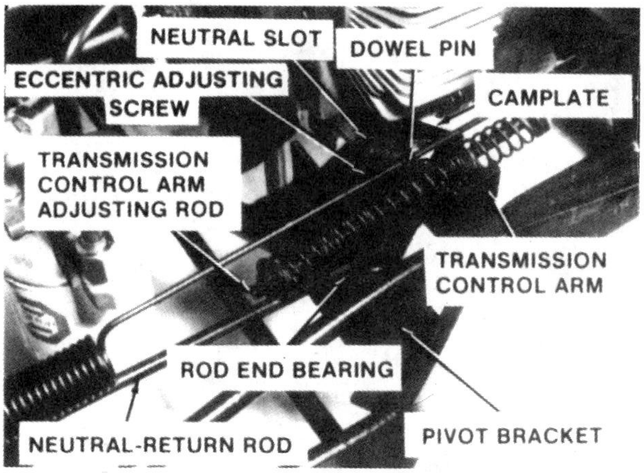

Fig. E10—View showing neutral adjustment for 1217 model. Refer to text.

Illustrations courtesy Engineering Products Co.

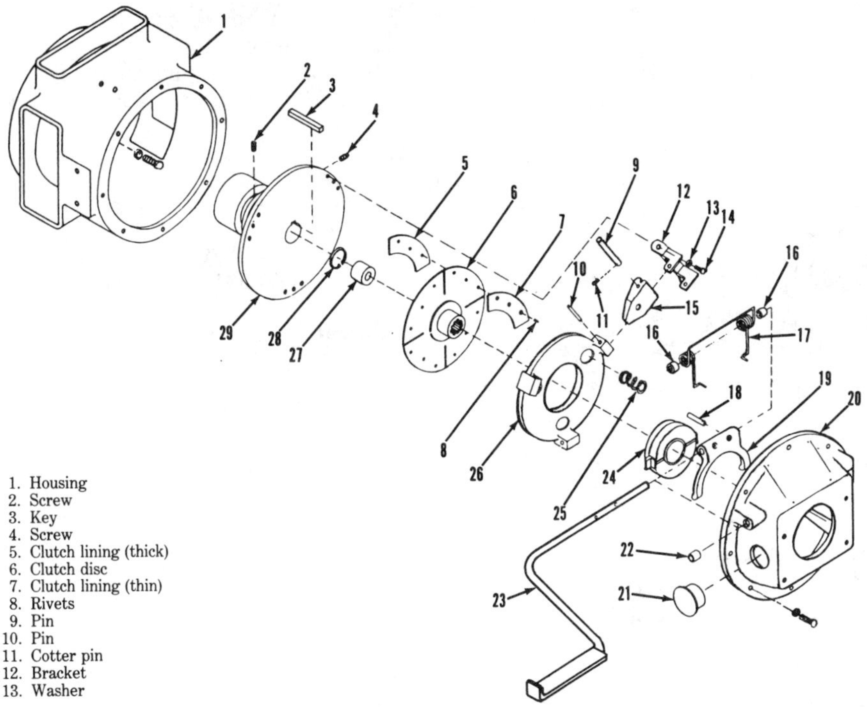

1. Housing
2. Screw
3. Key
4. Screw
5. Clutch lining (thick)
6. Clutch disc
7. Clutch lining (thin)
8. Rivets
9. Pin
10. Pin
11. Cotter pin
12. Bracket
13. Washer

Fig. E11—Exploded view of clutch assembly used on 1618, 2418 models and 1617, 2417 models prior to S/N 62750.

14. Cap screw
15. Lever
16. Spacers
17. Spring
18. Pins
19. Fork
20. Housing
21. Plug
22. Bushing
23. Pedal
24. Throwout bearing
25. Springs
26. Clutch plate
27. Bushing
28. Snap ring
29. Flywheel

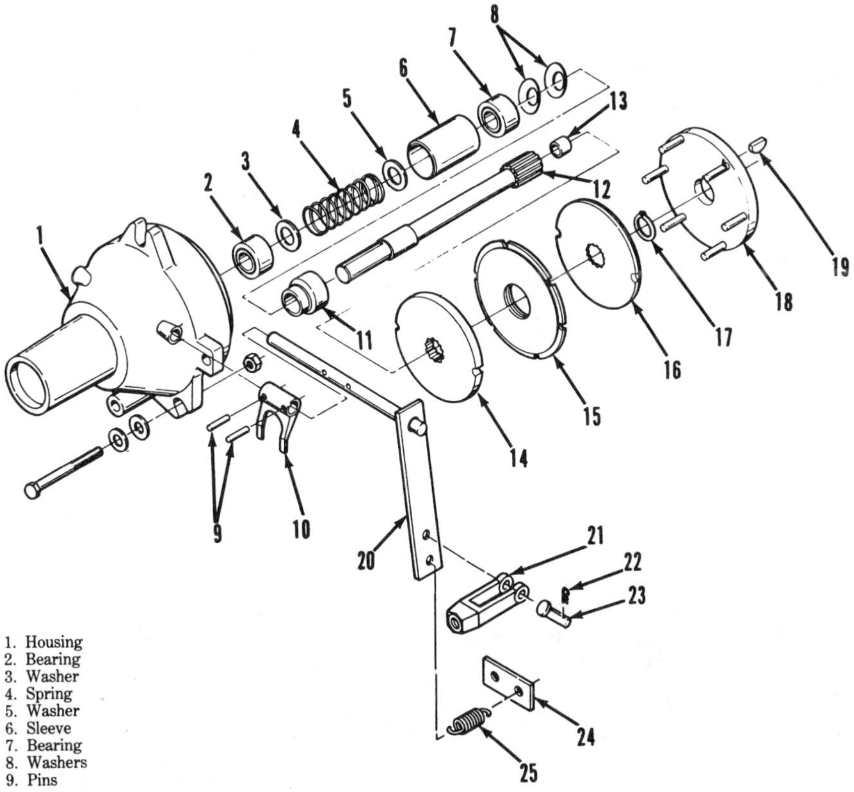

1. Housing
2. Bearing
3. Washer
4. Spring
5. Washer
6. Sleeve
7. Bearing
8. Washers
9. Pins

Fig. E12—Exploded view of clutch assembly used on 1617 and 2417 models after S/N 62750.

10. Fork
11. Collar
12. Shaft
13. Bushing
14. Plate
15. Plate
16. Disc
17. Snap ring
18. Plate
19. Key
20. Clutch arm shaft
21. Clevis
22. Cotter pin
23. Pin
24. Bracket
25. Spring

Before reinstalling clutch assembly on flywheel, lubricate clutch shaft pilot bearing (27). Align clutch and clutch shaft splines by turning engine front pulley. Reassemble by reversing removal procedure.

Models 1617-2417 After S/N 62750

Raise hood and remove rear body section. Remove heat shrouds at rear of engine. Loosen hydraulic pump drive belt, as equipped. Mark position of driveshaft, remove the four cap screws securing driveshaft to engine and slide driveshaft forward off the clutch input shaft. Remove drive belt and engine pulley. Remove rear driveshaft section of rear pto, as equipped. Disconnect clutch arm return spring (25–Fig. E12) and clevis (21) from clutch arm (20). Install a ½ by 1½ inch NF cap screw in lug (A–Fig. E13) and tighten against clutch arm to relieve pressure on clutch plates. Remove the three cap screws securing clutch housing. Remove clutch plates. Remove the ½ by 1½ inch NF cap screw from lug (A–Fig. E13). Use a large punch and hammer to drive clutch arm shaft out of housing, shearing pins (9–Fig. E12) retaining throwout bearing fork (10). Remove fork. Lift out input shaft (12) including bearings, washers, spring and sleeve. Place shaft in press with splined end down, resting on throwout collar (11). Press shaft through upper bearing.

CAUTION: When pressing shaft through bearing, spring (4) will be compressed and when bearing is free on shaft, spring pressure will thrust bearing upward.

Remove bearing (2), washer (3), spring (4), washer (5) and sleeve (6). Place shaft back in press, then remove bearing (7) and washers (8 and 9). Remove bushing (13) as necessary.

To reassemble, install bushing (13) and ream to 0.516 inch (13.16 mm). Slide throwout bearing collar (11) over for-

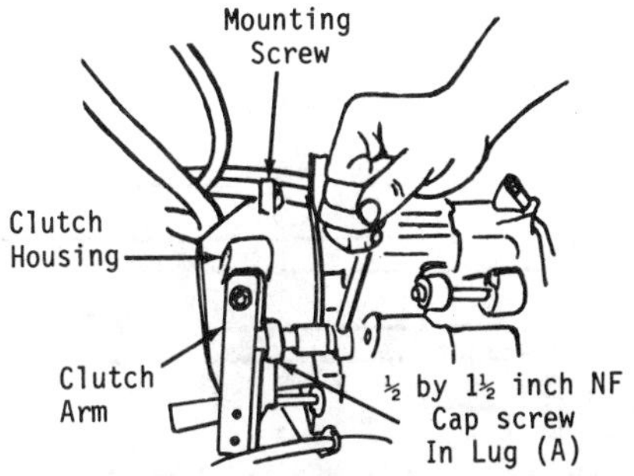

Fig. E13—Install a ½ x 1½-inch NF cap screw in lug (A) to relieve clutch pressure during disassembly. Refer to text.

Illustrations courtesy Engineering Products Co.

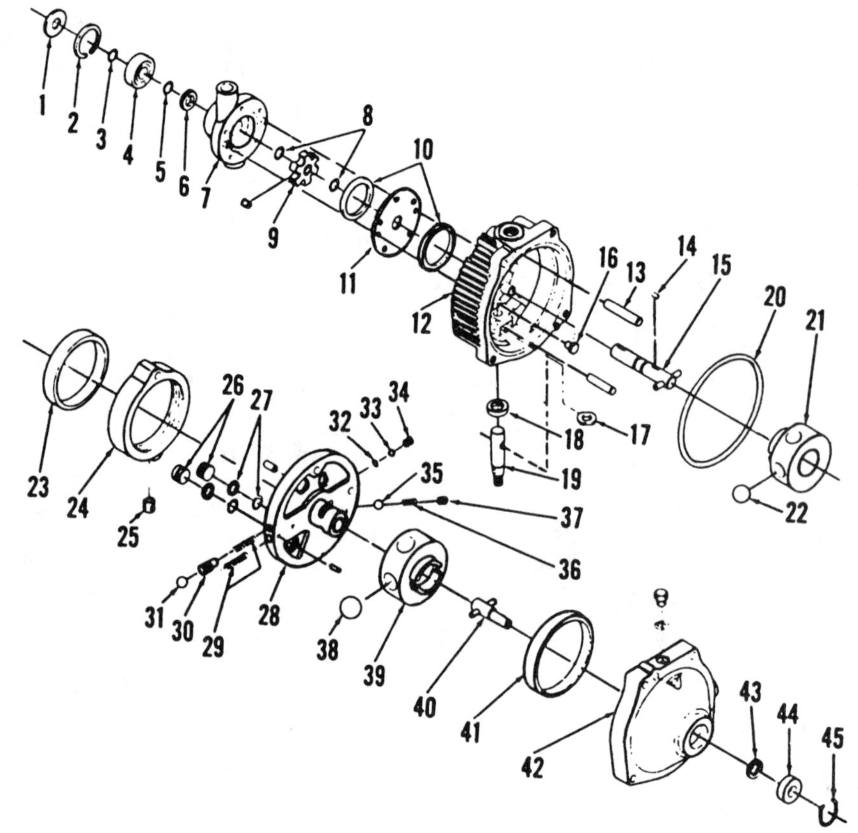

Fig. E16—Exploded view of hydrostatic drive unit used on 1217 models.

1. Dust shield
2. Retaining ring
3. Snap ring
4. Ball bearing
5. Snap ring
6. Oil seal
7. Charge pump body
8. Snap rings
9. Charge pump rotor assy.
10. Square cut seals
11. Pump plate
12. Housing
13. Cam pivot pin
14. Key
15. Input shaft
16. Neutral spring cap
17. Washer
18. Oil seal
19. Control shaft
20. "O" ring
21. Pump rotor
22. Pump ball pistons
23. Pump race
24. Pump cam ring
25. Cam ring insert
26. Dampening pistons
27. "O" rings
28. Pintle
29. Spring
30. Acceleration valve body
31. Acceleration valve ball
32. Retaining ring
33. Check valve ball
34. Check valve body
35. Charge relief ball
36. Spring
37. Relief valve plug
38. Motor ball piston
39. Motor rotor
40. Output shaft
41. Motor race
42. Body
43. Oil seal
44. Ball bearing
45. Retainer

ward end of shaft (12), install wave washers (8), bearing (7) and sleeve (6) and press into position. Install washer (5), spring (4), washer (3) and bearing (2). Press bearing (2) onto shaft until it is 2-11/16 inches (68.26 mm) from forward end of shaft to forward face of bearing. Install input shaft assembly in clutch housing. Install clutch arm shaft (20) and fork (10). Secure fork with new pins (9). Install clutch plates as shown in Fig. E12. Install the ½ by 1½ inch NF cap screw in lug to relieve pressure. Line up the six pin plate (18) with slotted plate (15) in clutch housing and install housing on transmission. Tighten the clutch housing retaining cap screws to 22 ft.-lbs. (30 N·m). Complete reassembly and adjust clutch as previously outlined.

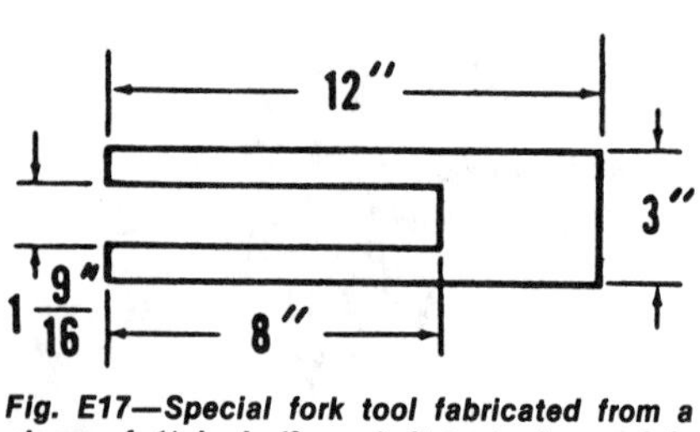

Fig. E17—Special fork tool fabricated from a piece of ⅛-inch (3 mm) flat stock, used in disassembly and reassembly of hydrostatic drive unit used on 1217 model. Refer to text.

HYDROSTATIC TRANSMISSION

MAINTENANCE

Model 1217

Check fluid level and clean all grass clippings, leaves, grease and dirt from cooling fins of hydrostatic drive unit before each use. Maintain fluid level at lower edge of check plug (CP–Fig. E18). Change transmission oil filter at 100 hour intervals and change fluid and filter at 500 hour intervals. Recommended fluid is a premium grade hydraulic fluid. Approximate hydrostatic drive unit and transaxle fluid capacity is 3.5 quarts (3.3 L).

R&R HYDROSTATIC TRANSMISSION

Model 1217

Remove rear fender/body assembly from tractor. Drain fluid from transaxle housing. Disconnect driveshaft coupler at transmission input shaft and all control linkage at transmission. Disconnect hydraulic lines and cap openings. Remove hydrostatic drive unit retaining cap screws and carefully separate drive unit from transaxle housing.

Reinstall by reversing removal procedure. Note retaining cap screw in upper right position is ¼-inch (6.35 mm) shorter than the other three. Fill with recommended fluid and adjust control linkage as outlined in appropriate CLUTCH section in this manual.

OVERHAUL HYDROSTATIC TRANSMISSION

Model 1217

Remove hydrostatic drive unit as previously outlined and drain unit.

Place transmission in holding fixture with input shaft pointing up. Remove dust shield (1–Fig. E16) and snap ring (3). Remove cap screws from charge pump body (7). One cap screw is ½-inch longer than the others and must be installed in original position. Remove charge pump body (7) with ball bearing (4). Ball bearing and oil seal (6) can be removed after removing retaining ring (2). Remove snap rings (5 and 8) and charge pump rotor assembly. Remove "O" rings (10) and pump plate (11). Turn hydrostatic unit over in fixture and remove output gear. Unscrew the two cap screws until two threads are engaged. Raise body (42) until it contacts cap screw heads. Insert a special fork tool (Fig. E17) between motor rotor (39–Fig. E16) and pintle (28). Remove cap screws, lift off body and motor assembly with fork tool and place assembly on a bench or in a holding fixture with output shaft pointing down. Remove fork and place a wide rubber band around motor rotor to hold ball pistons (38) in their bores. Carefully remove motor rotor assembly and lay aside for later disassembly. Remove motor race (41) and output shaft (40). Remove retainer (45), bearing (44) and

Illustrations courtesy Engineering Products Co.

oil seal (43). With housing assembly (12) resting in holding fixture, remove pintle assembly (28).

CAUTION: Do not allow pump to raise with pintle as ball pistons (22) may fall out of rotor (21). Hold pump in position by inserting a finger through hole in pintle.

Remove plug (37), spring (36) and charge relief ball (35). To remove directional check valves, drill through pintle with a drill bit that will pass freely through roll pins. Redrill holes from opposite side with a 1/4-inch drill bit. Press roll pin from pintle. Newer units are drilled at factory. Using a 5/16-18 tap, thread inside of valve bodies (34), then remove valve bodies using a draw bolt or slide hammer puller. Remove check valve balls (33) and retaining ring (32). To remove acceleration valves, remove retaining pin, insert a 3/16-inch (5 mm) rod 8 inches (203 mm) long through passage in pintle and carefully drive out spring (29), body (30) and ball (31). To remove dampening pistons (26), carefully tap outside edge of pintle on work bench to jar pistons free.

NOTE: If pintle journal is damaged, pintle must be renewed.

Remove pump cam ring (24) and pump race (23). Place a wide rubber band around pump rotor to prevent ball pistons (22) from falling out. Carefully remove pump assembly and input shaft (15).

To remove control shaft (19), drill a 11/32-inch hole through aluminum housing (12) directly in line with center line of dowel pin. Press dowel pin from control shaft, then withdraw control shaft. Remove oil seal (18). Thread drilled hole in housing with a 1/8-inch pipe tap. Apply a light coat of "Loctite" grade 635 to a 1/8-inch pipe plug, install plug and tighten until snug. Do not overtighten.

Number piston bores (1 through 5) on pump rotor and on motor rotor. Use a plastic ice cube tray or equivalent and mark cavities 1P through 5P for pump ball pistons and 1M through 5M for motor ball pistons. Remove ball pistons (22) one at a time, from pump rotor and place each ball in the correct cavity in tray. Remove ball pistons (38) and springs from motor rotor in the same manner.

Clean and inspect all parts and renew any showing excessive wear or other damage. Renew all gaskets, seals and "O" rings. Ball pistons are a select fit to 0.0002-0.0006 inch (0.0051-0.0152 mm) clearance in rotor bores and must be reinstalled in their original bores. If rotor bushing to pintle journal clearance is 0.002 inch (0.051 mm) or more, bushing wear or scoring is excessive and pump rotor or motor rotor must be renewed.

Check clearance between input shaft (15) and housing bushing. Normal clearance is 0.0013-0.0033 inch (0.033-0.0838 mm). If clearance is excessive, renew input shaft and/or housing assembly.

Install ball pistons (22) in pump rotor (21) and ball pistons (38) and springs in motor rotor (39), then use wide rubber bands to hold pistons in their bores.

Install charge relief valve ball (35) and spring (36) in pintle. Screw plug (37) into pintle until just below outer surface of pintle. Install acceleration valve springs (29) and bodies (30) making sure valves move freely. Tap balls (31) into pintle until roll pins will go into place. Install snap rings (32), check valve balls (33) and valve bodies (34) in pintle and secure with new roll pins.

NOTE: When installing oil seals (6, 18 or 43), apply a light coat of "Loctite" grade 635 to seal outer diameter.

Renew oil seal (18) and install control shaft (19) in housing. Install special washer (17), then press dowel pin through control shaft until 1¼ inches (32 mm) of pin extends from control shaft. Renew oil seal (43) and reinstall output shaft (40), bearing (44), retainer (45), output gear and snap ring.

Insert input shaft (15) in housing (12). Install snap ring (8) in its groove on input shaft. Place "O" ring (10), pump plate (11) and "O" ring in housing, then install charge pump drive key (14), charge pump rotor (9) and snap ring (8). Apply light grease or vaseline to pump

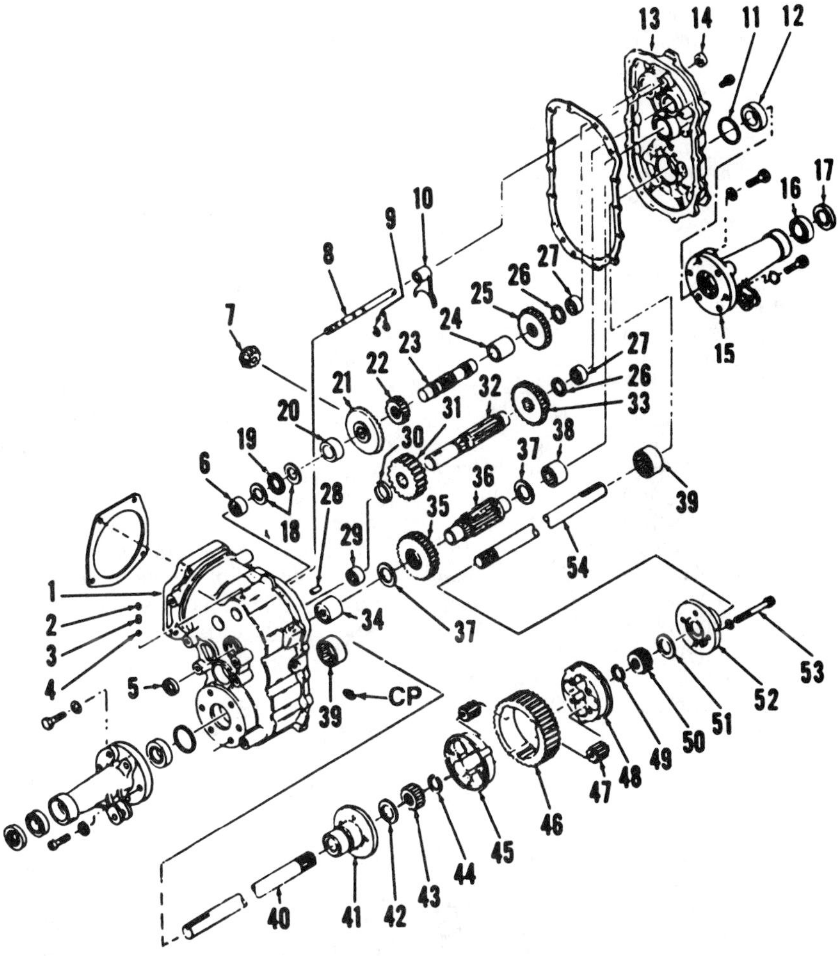

Fig. E18—Exploded view of transaxle and differential assembly used on 1217 model.

CP. Check plug
1. Case
2. Setscrew
3. Spring
4. Ball
5. Seal
6. Needle bearing
7. Transmission output gear
8. Shift rail
9. Snap rings
10. Shift fork
11. Quad ring
12. Tapered roller bearing
13. Cover
14. Seal
15. Axle housing
16. Ball bearing
17. Seal
18. Thrust washers
19. Thrust bearing
20. Spacer
21. Bevel gear
22. Gear (low range)
23. Shaft
24. Spacer
25. Gear (high range)
26. Thrust washer
27. Needle bearing
28. Dowel pin
29. Needle bearing
30. Spacer
31. Gear
32. Brake shaft
33. Sliding gears
34. Needle bearing
35. Gear
36. Output shaft
37. Thrust washers
38. Needle bearing
39. Needle bearing
40. Axle shaft L.H.
41. Differential carrier L.H.
42. Thrust washer
43. Axle gear
44. Snap ring
45. Body core
46. Ring gear
47. Pinion gears (8)
48. Body core
49. Snap ring
50. Axle gear
51. Thrust washer
52. Differential carrier R.H.
53. Cap screw
54. Axle shaft R.H.

Illustrations courtesy Engineering Products Co.

rollers and place rollers in rotor slots. Install oil seal (6) and pump race in charge pump body (7), then install body assembly. Secure with the five cap screws, making certain long cap screw is installed in its original location (in heavy section of pump body). Tighten cap screws to 28-30 ft.-lbs. (38-41 N·m). Install snap ring (5), bearing (4), retaining ring (2), snap ring (3) and dust shield (1).

Place charge pump and housing assembly in a holding fixture with input shaft pointing downward. Install pump race (23) and insert (25) in cam ring (24), then install cam ring assembly over cam pivot pin (13) and control shaft dowel pin. Turn control shaft (19) back and forth and check movement of cam ring. Cam ring must move freely from stop to stop. If not, check installation of insert (25) in cam ring.

Install pump rotor assembly and remove rubber band used to retain pistons. Install pintle assembly (28) over cam pivot pin (13) and into pump rotor. Place "O" ring (20) in position on housing.

Place body assembly (42) in a holding fixture with output gear down. Install motor race (41) in body, then install motor rotor assembly and remove rubber band used to retain pistons in rotor.

Using special fork tool (Fig. E17) to retain motor assembly in body, carefully install body and motor assembly over pintle journal. Remove fork tool, align bolt holes and install the two cap screws. Tighten cap screws to 15 ft.-lbs. (20 N·m).

Place hydrostatic unit on holding fixture with reservoir adapter opening and venting plug opening facing upward. Fill unit with recommended fluid until fluid flows from fitting hole in body. Plug all openings to prevent dirt or other foreign material from entering hydrostatic unit.

Reinstall unit on reduction and differential housing, using a new gasket, and tighten cap screws to 20 ft.-lbs. (27 N·m). Run engine and check fluid level in differential housing. Check adjustment of transmission linkage.

GEAR TRANSMISSIONS

MAINTENANCE

Model 1217

Fluid reservoir for hydrostatic transmission and the two speed transaxle is the transaxle housing. Refer to MAINTENANCE paragraph in HYDROSTATIC TRANSMISSION section in this manual for maintenance information.

Models 1618-2418-Models 1617-2417 Prior To S/N 62750

Three speed transmission fluid level should be checked daily. Maintain fluid at lower edge of check plug (CP–Fig. E20) opening. Recommended fluid is SAE 90 EP gear lubricant. Approximate fluid capacity is 0.5 pints (0.2 L).

Models 1617-2417 After S/N 62750

Check four speed transmission fluid level before each use. Maintain fluid level at round punch mark on folding dipstick (2–Fig. E21).

NOTE: Dipstick must be unfolded to check fluid level and folded during tractor operation.

Recommended fluid is Power King Special Blend Fluid. Approximate fluid capacity is 1⅜ pints (0.65 L).

R&R TRANSAXLE/ TRANSMISSION

Model 1217

Remove hydrostatic transmission as previously outlined. Disconnect brake rod at brake lever and transaxle shift lever at transaxle. Raise and support rear of tractor. Support transaxle housing on a jack and remove tow bar and bracket. Remove bolts securing transaxle, lower transaxle assembly to the ground and roll it away from tractor.

Reinstall by reversing removal procedure. Fill with recommended fluid and adjust control linkage as outlined in appropriate CLUTCH section in this manual. Adjust brake.

Models 1618-2418 – Models 1617-2417 Prior To S/N 62750

Remove engine as previously outlined. Drain fluid from transmission. Remove clutch cover assembly mounted on front of transmission case and the four cap screws retaining transmission to drive tube at the rear of transmission housing. Pull transmission forward and lift from tractor.

Reinstall by reversing removal procedure.

Models 1617-2417 After S/N 62750

Remove engine as previously outlined. Drain fluid from transmission and remove the four cap screws securing drive tube to rear of transmission. Slide transmission forward, out of drive tube and remove transmission.

NOTE: Use care not to lose spacer and washer between transmission and spline coupler on driveshaft.

Reinstall by reversing removal procedure.

OVERHAUL TRANSAXLE/ TRANSMISSION

Model 1217

To disassemble transaxle, separate hydrostatic drive and transaxle assembly. Remove brake caliper assembly and brake disc. Remove tires and wheels and remove hub assemblies. Thoroughly clean exterior of unit.

Remove axle housings (15–Fig. E18). Position unit with cover up, then remove cover (13). Lift out differential assembly and axles (40 through 54). Remove output shaft (36), gear (35) and thrust washers (37). Unscrew setscrew (2) and remove spring (3) and ball (4). Remove brake shaft (32), sliding gear (33), shift fork (10) and rod (8). Remove input shaft and gear components (18 through 25).

To disassemble differential assembly, remove cap screws (53), differential carriers (41 and 52) and axles from ring gear assembly. Remove snap ring to separate axle gear, carrier and axle. Remove pinions (47) and separate body cores (45 and 48) from ring gear (46).

Inspect parts for damage or excessive wear. To reassemble unit, reverse disassembly procedure. Check movement of shift rod when tightening setscrew (2). Install gears (22 and 25) so bevels of gears face together as shown in Fig. E19. Install carrier cap screws (53–Fig. E18) so head of cap screw is on side of shorter carrier (52). Do not rotate axle housings after housing has been pressed tight against seal (11) as seal may be cut.

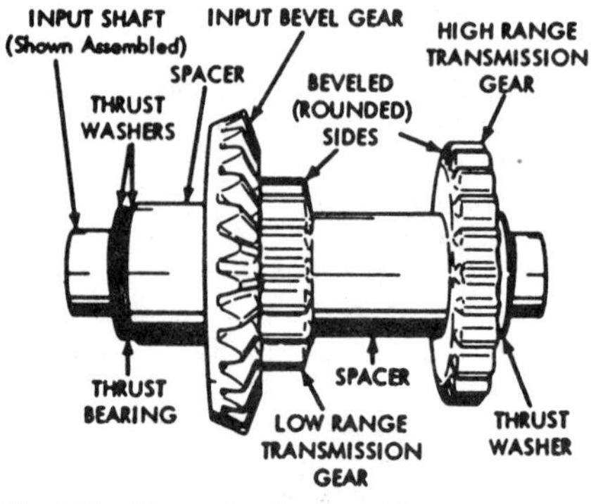

Fig. E19—View of reduction drive input shaft and gears. Note position of bevels on gears and location of thrust washers and thrust bearing.

Illustrations courtesy Engineering Products Co.

Models 1618-2418-Models 1617-2417 Prior To S/N 62750

Early models may have an optional tandem transmission mounted behind regular transmission and used as a range transmission. This transmission is identical to the front transmission and service procedures are the same.

To overhaul either transmission, remove transmission top cover and shift control assembly. Remove cap screws from input shaft bearing retainer (11–Fig. E20) and withdraw input shaft (17) and bearing assembly.

NOTE: Watch for center bearing (18), consisting of 13 rollers, as they can fall into lower part of transmission case as input shaft is removed.

Remove snap ring retaining main shaft bearing (23) in case and bump shaft out of case. Sliding gears (20 and 21) will be removed as main shaft (19) is withdrawn. Unbolt and remove lock plate (31) retaining idler shaft (35) and countershaft (30) in transmission case. Remove idler shaft and reverse idler gear (33) and countershaft and cluster gear (28). Drive pins from shift forks and remove expansion plugs from top cover. Remove second and third shift rail (10) and fork (9); then, first and reverse shift rail (6) and fork (7). A poppet spring and ball are located outside each rail and an interlock plunger (36) is positioned between rails. Further disassembly procedure is evident after examination of unit.

Reassemble by reversing disassembly procedure. Fill to proper level with recommended fluid.

Models 1617-2417 After S/N 62750

Remove the three cap screws securing shift lever assembly and remove complete assembly. Separate housing (51–Fig. E21) from housing (48). Remove the thrust bearings (38 and 46). Remove shaft (27) from housings and remove gears and bearings as required. Slide two-gear cluster (36) off countershaft (41). Remove pinion shaft (45), 22 tooth gear (44) and thrust bearing (43). Note locations of shift assemblies and remove forks and shift rails. Remove reverse idler shaft (32), spacer (31) and 12 tooth gear (30). Remove three-gear cluster (34), countershaft (41) and thrust washer (40). Remove snap ring and six-pin coupler plate from input shaft and gently drive input shaft through bearing (15), spacer (16) and gear (17). Remove snap ring and front bearing (15).

Reassemble by reversing disassembly procedure.

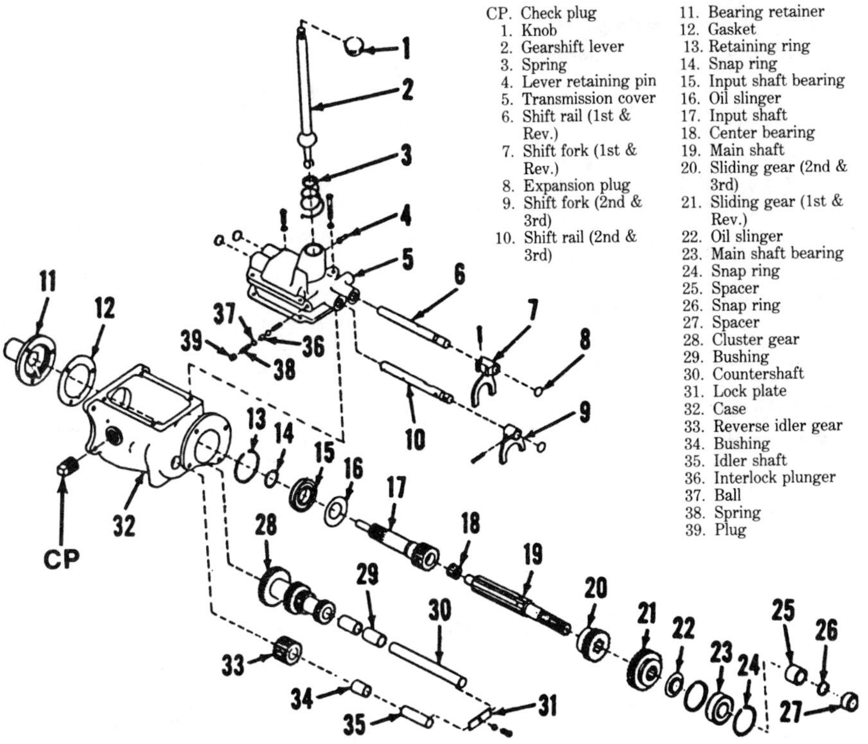

CP. Check plug
1. Knob
2. Gearshift lever
3. Spring
4. Lever retaining pin
5. Transmission cover
6. Shift rail (1st & Rev.)
7. Shift fork (1st & Rev.)
8. Expansion plug
9. Shift fork (2nd & 3rd)
10. Shift rail (2nd & 3rd)
11. Bearing retainer
12. Gasket
13. Retaining ring
14. Snap ring
15. Input shaft bearing
16. Oil slinger
17. Input shaft
18. Center bearing
19. Main shaft
20. Sliding gear (2nd & 3rd)
21. Sliding gear (1st & Rev.)
22. Oil slinger
23. Main shaft bearing
24. Snap ring
25. Spacer
26. Snap ring
27. Spacer
28. Cluster gear
29. Bushing
30. Countershaft
31. Lock plate
32. Case
33. Reverse idler gear
34. Bushing
35. Idler shaft
36. Interlock plunger
37. Ball
38. Spring
39. Plug

Fig. E20—Exploded view of transmission used on 1618, 2418 models and 1617, 2417 models prior to S/N 62750. Tandem transmission, when installed for extra reduction, is identical except for shape of shift lever.

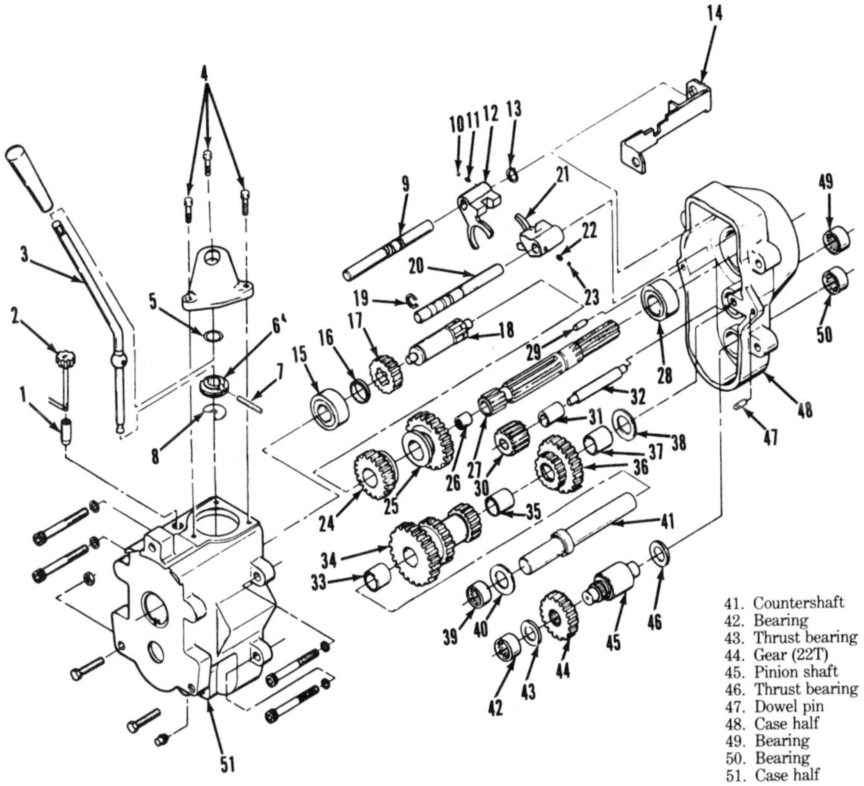

Fig. E21—Exploded view of four speed transmission used on 1617 and 2417 models after S/N 62750.

1. Dipstick tube
2. Folding dipstick
3. Shift lever
4. Cap screws
5. "O" ring
6. Retainer
7. Pin
8. Snap ring
9. Shift rod (high)
10. Detent ball
11. Detent spring
12. Shift fork
13. Snap ring
14. Shifter stop
15. Bearing
16. Spacer
17. Gear (16T)
18. Input shaft
19. Snap ring
20. Shift rod (low)
21. Shift fork
22. Detent spring
23. Detent ball
24. Gear (20T)
25. Gear (26T)
26. Bearing
27. Output shaft
28. Bearing
29. Dowel pin
30. Gear (Rev. idler)
31. Spacer
32. Reverse idler shaft
33. Bearing
34. Three gear cluster
35. Bearing
36. Two gear cluster
37. Bearing
38. Thrust bearing
39. Bearing
40. Thrust bearing
41. Countershaft
42. Bearing
43. Thrust bearing
44. Gear (22T)
45. Pinion shaft
46. Thrust bearing
47. Dowel pin
48. Case half
49. Bearing
50. Bearing
51. Case half

Illustrations courtesy Engineering Products Co.

NOTE: Some bushings must be finish reamed to fit shafts if renewed.

Fill transmission to proper level with recommended fluid.

FINAL DRIVES

MAINTENANCE

Models 1617-1618-2417-2418

Check fluid level in final drives at 50 hour intervals. Maintain fluid level at lower edge of check plug opening (Fig. E22). Change fluid at 100 hour intervals. Recommended fluid is SAE 80-90 EP gear lubricant and approximate capacity is 1.5 pints (0.7 L) for each final drive.

R&R AND OVERHAUL

Models 1618-2418- Models 1617-2417 Prior To S/N 62750

Raise and support rear of tractor, then remove rear wheel and tire assemblies. Remove level check plug and drain plug (Fig. E22) and drain fluid from final drive housing. Remove fender, brake bands (3–Fig. E23), brake drum (5) and key. Remove cap screws securing cover (19) and remove cover, axle (18), gear (17) and wheel hub (23) as an assembly. Remove pinion shaft (16) and thrust washer (14). Remove snap ring (22) and press hub (23) from axle (18). Remove key and pull axle (18) and gear (17) assembly out of cover. Renew seal (21) as necessary. Press axle (18) out of gear (17). Remove thrust washer (8) and snap ring.

Reinstall by reversing removal procedure. Fill to correct level with recommended fluid and adjust brakes.

Models 1617-2417 After S/N 62750

Remove rear body section and raise and support rear of tractor. Remove rear tire and wheel assemblies. Remove level check plug and drain plug (Fig. E22) and drain fluid from each final drive housing. Remove cap screws and nuts securing cover (19–Fig. E23) to housing (9). Remove the two rear nuts on the opposite side gear case to allow fender body support to be moved out of

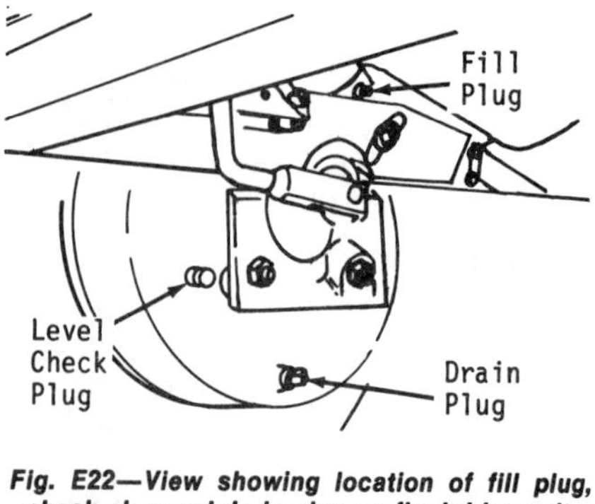

Fig. E22—View showing location of fill plug, check plug and drain plug on final drive unit.

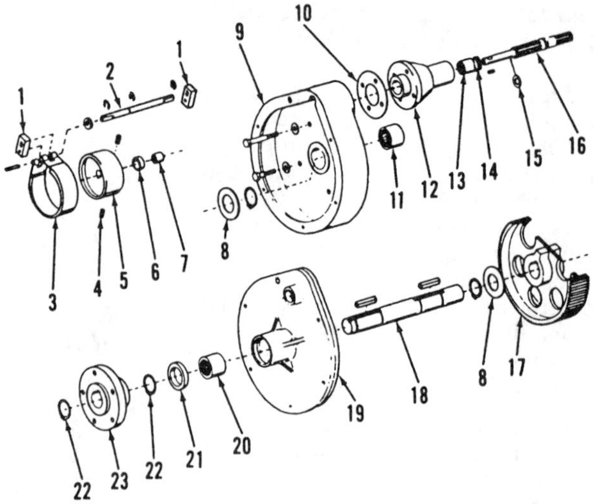

Fig. E23—Exploded view of final drive assembly.

1. Brake cams
2. Brake shaft
3. Brake band
4. Setscrew
5. Brake drum
6. Seal
7. Bearing
8. Thrust washers
9. Final drive housing
10. Gasket
11. Axle inner bearing
12. Bearing housing
13. Bearing
14. Seal
15. Thrust washer
16. Pinion shaft
17. Bull gear
18. Axle
19. Cover
20. Axle outer bearing
21. Seal
22. Snap rings
23. Wheel hub

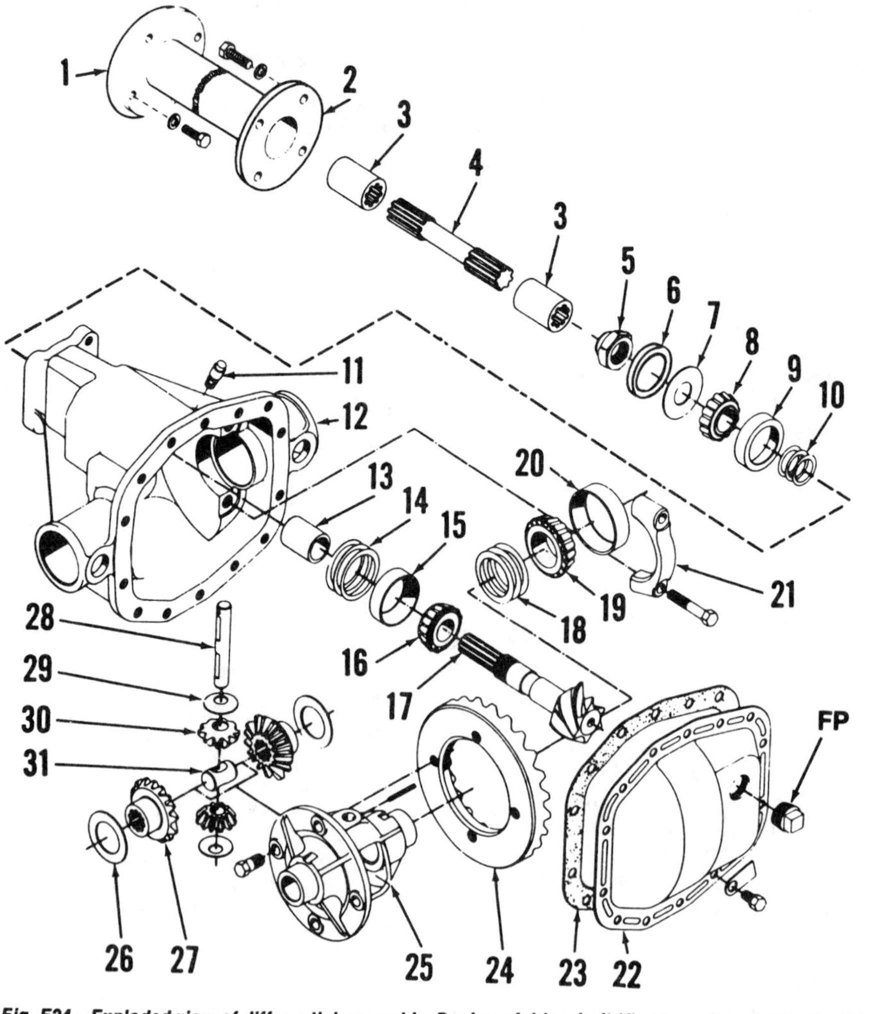

Fig. E24—Exploded view of differential assembly. Design of driveshaft (4) may vary and oil baffle (6) is not used on late production models.

1. Gasket
2. Shaft tube
3. Coupling
4. Driveshaft
5. Pinion nut
6. Oil baffle
7. Oil slinger
8. Bearing cone
9. Bearing cup
10. Shim pack
11. Breather
12. Housing
13. Spacer
14. Shim pack
15. Bearing cup
16. Bearing cone
17. Pinion
18. Shim pack
19. Bearing cone
20. Bearing cup
21. Bearing cap
22. Rear cover
23. Gasket
24. Ring gear
25. Differential case
26. Thrust washer
27. Side gear
28. Pinion shaft
29. Thrust washer
30. Differential pinion
31. Thrust block

Illustrations courtesy Engineering Products Co.

the way. Remove cover (19), axle (18), gear (17), and hub (23) as an assembly. Remove snap ring (22) and press hub (23) from axle (18). Remove key and pull axle (18) and gear (17) assembly out of cover. Renew seal (21) as necessary. Press axle (18) out of gear (17). Remove thrust washer (8) and snap ring.

Reinstall by reversing removal procedure. Fill to correct level with recommended fluid and adjust the brakes.

DIFFERENTIAL

MAINTENANCE

Model 1217

Model 1217 differential assembly is an integral part of the transaxle. Refer to appropriate MAINTENANCE paragraph in TRANSAXLE/TRANSMISSION section in this manual.

Models 1617-1618-2417-2418

Differential fluid level should be checked at 50 hour intervals. Maintain fluid level at lower edge of filler plug (FP–Fig. E24). Change differential fluid at 100 hour intervals. Recommended fluid is SAE 80-90 EP gear lubricant and approximate capacity is 2 pints (0.9 L).

R&R DIFFERENTIAL ASSEMBLY

Model 1217

Model 1217 differential assembly is an integral part of the transaxle. Refer to appropriate TRANSAXLE/TRANSMISSION section in this manual.

Models 1618-2418- Models 1617-2417 Prior To S/N 62750

Remove seat bracket and seat, drain lubricant and remove rear wheels, brakes and final drives. Unbolt differential housing from driveshaft tube flange and rearward movement of differential will disengage driveshaft splines at coupling (3–Fig. E24). Unit may now be lifted from tractor.

Reinstall by reversing removal procedure. Fill to correct level with recommended fluid.

Models 1617-2417 After S/N 62750

Remove transmission as previously outlined. Raise and support rear of tractor, then remove rear tires and wheels. Remove both final drive assemblies and lift differential assembly from tractor.

Reinstall by reversing removal procedure. Fill to correct level with recommended fluid.

OVERHAUL

Models 1617-1618-2417-2418

Remove rear cover (22—Fig. E24), then remove cap screws from bearing retainer clamps (21) and lift out ring gear (24) and differential case assembly (25) along with side bearings (19 and 20).

NOTE: Observe number and location of side adjusting shims (18) which will be needed in reassembly. The shims are placed between bearing cone (19) and differential case (25) in current production, but older models may have them inserted behind bearing cup (20) against carrier housing (12).

Remove lock pin and differential pinion shaft (28), then thrust block (31). Rotate differential side gears; then remove pinion gears (30), side gears (27) and thrust washers (26 and 29). After removing bevel pinion shaft nut (5) from pinion (17), bump pinion rearward and out of carrier housing. Oil baffle (6), slinger (7), bearing cone (8), shims (10) and spacer (13) will be removed along with bearing cone (16) and bevel pinion. Bevel pinion bearing cups (9 and 15) and shim pack (14) can now be removed from carrier housing (12).

Shims to be used for adjustment of differential are available in a variety of thicknesses.

When reassembling, adjust bevel pinion bearings by adding or removing shims (10) until pinion shaft will rotate freely with zero end play. Install ring gear and differential case assembly with shim packs (18) made up to eliminate differential case side play. Check backlash between ring gear and pinion and adjust by shifting shims (18) from one side of differential case to the other until backlash between gears is limited to approximately 0.005 inch (0.1270 mm). To check mesh of gear set, apply a thin coat of red lead or Prussian blue to ring gear teeth. Rotate gears and check pattern of coloring rubbed off ring gear teeth. A correct pattern is one which is well centered and shows ample tooth contact. If observed pattern is too high, move shims (14) as necessary from behind forward pinion bearing cup (9) to rear bearing shim pack. If pattern appears too low on ring gear teeth, move shims from rear pinion bearing to forward end of pinion shaft. After obtaining correct pattern, recheck backlash between ring gear and pinion. Adjust as necessary. Reassemble by reversing disassembly procedure. Fill to correct level with recommended fluid.

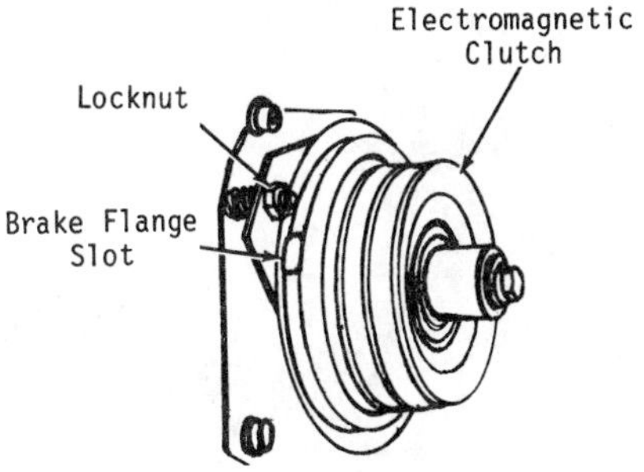

Fig. E25—View showing location of slots in brake flange of pto electric clutch. Refer to text for adjustment procedure.

POWER TAKE-OFF

ADJUSTMENT

All Models

Electric pto clutch is adjusted by placing 0.015 inch (0.381 mm) thick shims in each of the three slots in brake flange (Fig. E25). Engage pto (engine NOT running), loosen locknuts holding flange. Push flange until it bottoms and retighten locknuts using caution not to over torque them. Remove shims.

HYDRAULIC SYSTEM

MAINTENANCE

Models 1617-1618-2417-2418 So Equipped

Hydraulic fluid level should be checked daily. Maintain fluid level 5 inches (635 mm) below reservoir top. Recommended fluid is a good quality hydraulic fluid or Dexron Automatic Transmission Fluid. Approximate system fluid capacity is 5 quarts (4.7 L) for 1618 and 2418 models and 1617 and 2417 models prior to S/N 62750 and 2 gallons (7.6 L) for 1617 and 2417 models after S/N 62750.

Periodically adjust nut on belt tension rod so spring pressure on pump bracket allows ¼-inch (6.35 mm) belt deflection at a midpoint between pulleys is obtained.

HYDRAULIC PUMP AND CONTROL VALVE

Models 1617-1618-2417-2418 So Equipped

Service parts are not available for the hydraulic pump or control valve. If service is required, renew entire units.

A relief valve located in control valve body limits system pressure to 1000 psi (6895 kPa).

FORD

CONDENSED SPECIFICATIONS

	MODELS		
	YT 16	LGT 17	LGT 17H
Engine Make	B&S	Kohler	Kohler
Model	402707	KT-17	KT-17
Bore	3.44 in. (87.4 mm)	3.12 in. (79.3 mm)	3.12 in. (79.3 mm)
Stroke	2.16 in. (54.9 mm)	2.75 in. (69.9 mm)	2.75 in. (69.9 mm)
Displacement	40 cu. in. (659 cc)	42.18 cu. in. (690.5 cc)	42.18 cu. in (690.5 cc)
Power Rating	16 hp. (11.9 kW)	17 hp. (12.7 kW)	17 hp. (12.7 kW)
Slow Idle	1350 rpm	1200 rpm	1200 rpm
High Speed (No-Load)	3600 rpm	3600 rpm	3600 rpm
Capacities–			
Crankcase	3 pts. (1.4 L)	3 pts. (1.4 L)	3 pts. (1.4 L)
Hydraulic System	.....	.....	6 qts. (5.7 L)
Transaxle or Transmission	36 oz. (1065 mL)	64 oz. (1893 mL)	See Hyd.
Bevel Gear Box	.....	4 oz. (118 mL)	.....

FRONT AXLE AND STEERING SYSTEM

MAINTENANCE

All Models

It is recommended that steering spindles, tie rod ends, axle pivot, wheel bearings/bushings and steering gear be lubricated at 25 hour intervals. Use multi-purpose, lithium base grease and clean all fittings before and after lubrication. Clean all pivot points and linkages and lubricate with SAE 30 oil. Check for any looseness or wear and repair as necessary.

R&R AXLE MAIN MEMBER

Model YT-16

Raise and support front of tractor. Disconnect drag link end (8–Fig. F1) from steering arm (10). Disconnect any implement which may be attached to pivot bracket (7). Support axle and remove bolts connecting pivot bracket (7) to main member (5). Remove pivot bolt (4) and remove main member assembly, then pivot bracket.

Reinstall by reversing removal procedure.

Models LGT-17 And LGT-17H

Raise and support front of tractor. Disconnect drag link at left spindle (15–Fig. F2). Disconnect any implement which may be attached to pivot bracket (10). Support axle and remove bolts connecting pivot bracket (10) to main member (13). Remove pivot bolt (11) and remove axle main member (13), then pivot bracket (10).

Reinstall by reversing removal procedure.

TIE ROD AND TOE-IN

Model YT-16

Model YT-16 tie bar (1–Fig. F1) is made to correct specification by manufacturer and in nonadjustable. If steering becomes loose or tractor wanders or darts, check condition of tie bar, bolts and spacers, spindles, bushings, wheel bearings and drag link. Repair as necessary.

Models LGT-17 And LGT-17H

Tie rod ends (7–Fig. F2) are available for service. Adjust tie rod ends equally to provide 1/16 to 1/8-inch (1.6 to 3.2 mm) toe-in. Make certain jam nuts (6) are retightened after adjustment. If steering becomes loose or tractor wanders or darts, check condition of tie rod ends, spindles, bushings, wheel bearings and drag link. Repair as necessary.

R&R STEERING SPINDLES

Model YT-16

Raise and support front of tractor. Remove dust cap (18–Fig. F1) and cotter key. Remove washer (17) and slide wheel and hub assembly off spindle. Remove tie bar (1). Remove steering arm (10) from left spindle (13) or cotter key from right spindle (3). Slide spindle down out of axle main member.

Renew bushings (11), spacer (12) or spindle (3 or 13) as necessary.

Reinstall by reversing removal procedure.

Models LGT-17 And LGT-17H

Raise and support front of tractor. Remove dust cap (20–Fig. F2) and cotter key. Remove washer (19) and slide tire and hub assembly from spindle. Disconnect tie rod (14) from side to be serviced. If left side is to be removed, disconnect drag link. Remove spindle bolt (3) and spindle.

Renew bushings (12), spindles (2 or 15) or spindle bolts (3) as necessary.

Reinstall by reversing removal procedure.

FRONT WHEEL BEARINGS

All Models

Front wheel bushings (YT-16 model) or wheel bearings (LGT-17 and LGT-17H models) are pressed into wheel hubs. Renew bushings/bearings or hubs as necessary to correct excessive looseness. Lubricate using multipurpose, lithium base grease.

STEERING GEAR

Model YT-16

R&R AND OVERHAUL STEERING GEAR. Disconnect drag link end (17–Fig. F3) at steering sector (18). Remove steering wheel (4) and adapter (5). Remove bellows cover (7) and sleeve (8). Remove bearing bolts (13 and 19). Remove steering shaft (11). Remove "E" ring (20) and steering sector (18).

Reinstall by reversing removal procedure. Push steering shaft and flange bearing at lower end of steering shaft toward sector gear to provide full gear mesh before tightening bolts (13). Steering shaft should rotate smoothly with minimum backlash. If binding is encountered, loosen bolts (13) and move shaft away from sector gear (18) a small

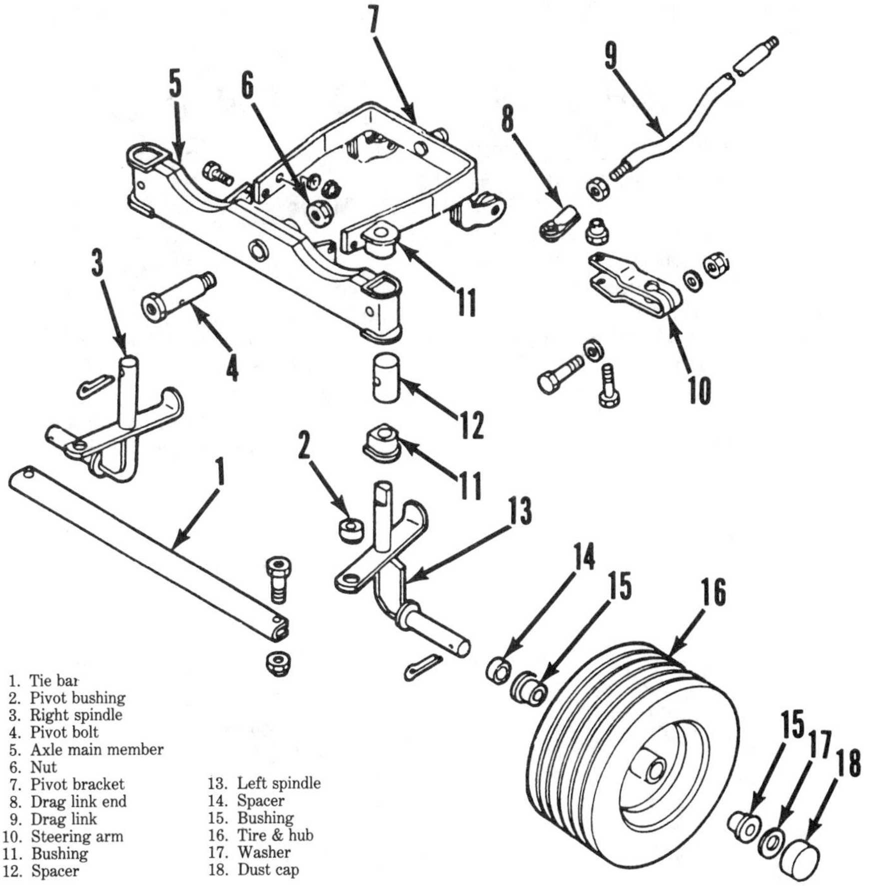

1. Tie bar
2. Pivot bushing
3. Right spindle
4. Pivot bolt
5. Axle main member
6. Nut
7. Pivot bracket
8. Drag link end
9. Drag link
10. Steering arm
11. Bushing
12. Spacer
13. Left spindle
14. Spacer
15. Bushing
16. Tire & hub
17. Washer
18. Dust cap

Fig. F1—Exploded view of front axle used on YT-16 models.

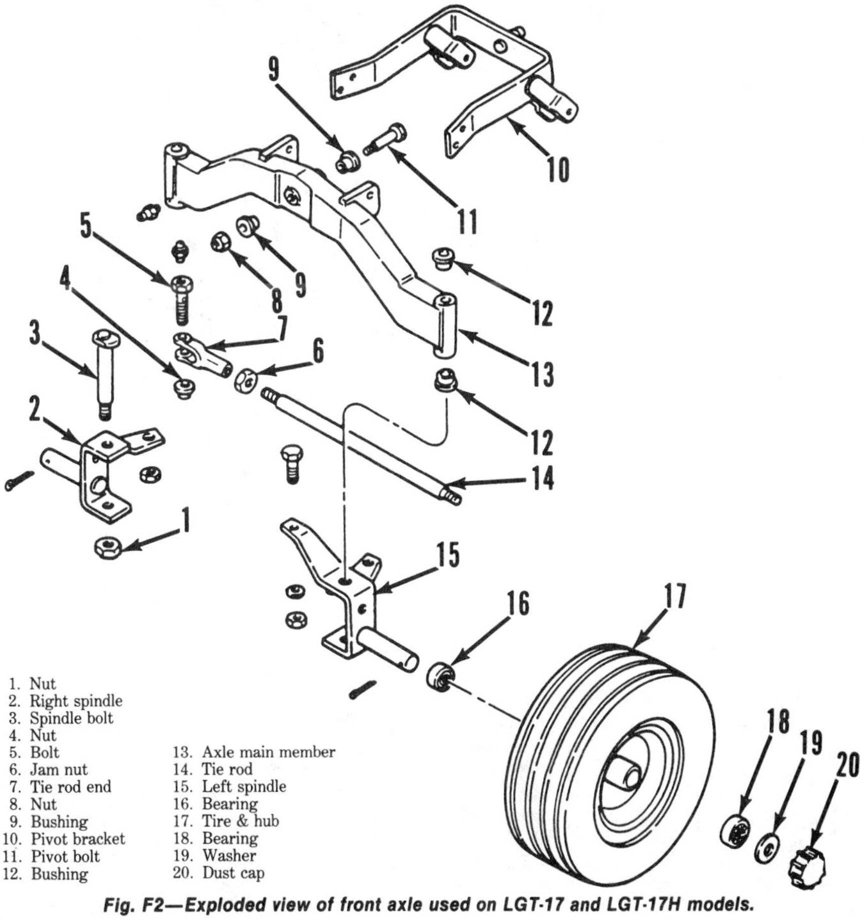

1. Nut
2. Right spindle
3. Spindle bolt
4. Nut
5. Bolt
6. Jam nut
7. Tie rod end
8. Nut
9. Bushing
10. Pivot bracket
11. Pivot bolt
12. Bushing
13. Axle main member
14. Tie rod
15. Left spindle
16. Bearing
17. Tire & hub
18. Bearing
19. Washer
20. Dust cap

Fig. F2—Exploded view of front axle used on LGT-17 and LGT-17H models.

Illustrations courtesy Ford.

amount. Retighten bolts and recheck steering.

Models LGT-17 And LGT-17H

R&R AND OVERHAUL STEERING GEAR. Disconnect drag link at quadrant gear (19–Fig. F4). Remove cover (13). Remove nut from the end of shaft (14) and withdraw shaft. Remove steering wheel (4), adapter (5), bellows (9) and sleeve (7). Drive out roll pin (23). Remove bolts (10) and remove steering shaft (12). Catch gear (22) and bushings (24 and 21) as shaft is removed.

Renew bearings, bushings and gears as necessary.

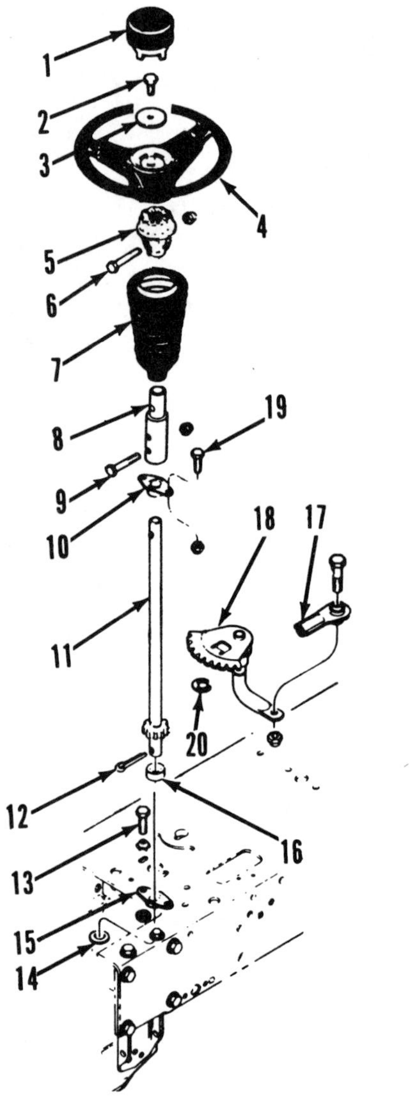

Fig. F3—Exploded view of front axle steering mechanism used on YT-16 models.

1. Dust cover
2. Bolt
3. Washer
4. Steering wheel
5. Adapter
6. Bolt
7. Bellows
8. Sleeve
9. Bolt
10. Flange bearing
11. Steering shaft
12. Cotter key
13. Bolt
14. Washer
15. Flange bearing
16. Spacer
17. Drag link end
18. Steering sector
19. Bolt
20. "E" ring

Reinstall by reversing removal procedure.

ENGINE

MAINTENANCE

All Models

Regular engine maintenance is required to maintain peak performance and long engine life.

Check oil level and clean air intake screen at 5 hour intervals under normal operating conditions.

Clean engine air filter at 25 hour intervals. Clean crankcase breather and engine cooling fins and lubricate governor linkage at 200 hour intervals.

Change engine oil and filter, perform tune-up, valve adjustment and clean carbon from cylinder heads as recommended by engine manufacturer.

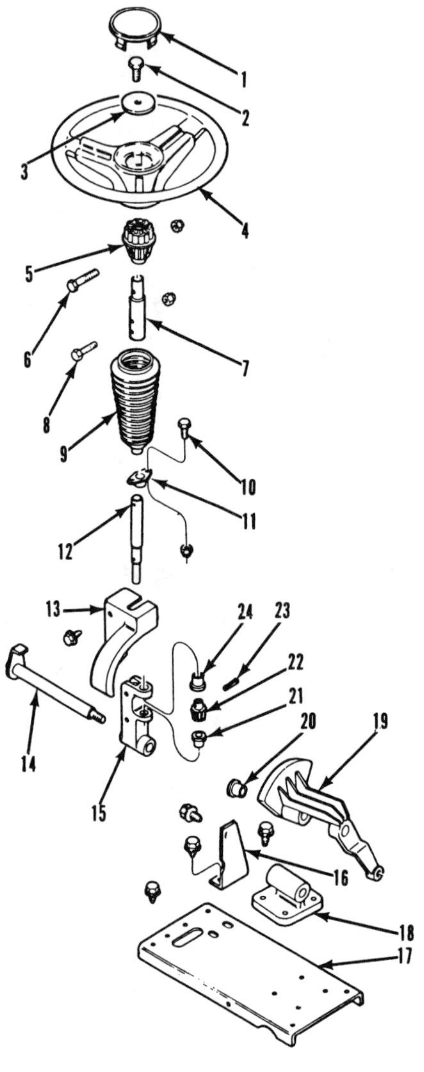

Fig. F4—Exploded view of steering gear used on LGT-17 and LGT-17H models.

1. Dust cover
2. Bolt
3. Washer
4. Steering wheel
5. Adapter
6. Bolt
7. Sleeve
8. Bolt
9. Bellows
10. Bolt
11. Flange bearing
12. Steering shaft
13. Sector & pinion cover
14. Shaft
15. Support
16. Brace
17. Support channel
18. Support
19. Sector gear
20. Bushing
21. Bushing
22. Gear
23. Roll pin
24. Bushing

REMOVE AND REINSTALL

All Models

Disconnect negative battery cable. Remove hood and side panels. Disconnect pto belts and drive belt or drive shaft, as equipped. Remove exhaust system and deflectors as necessary. Disconnect fuel line and all electrical connections. Remove engine mounting bolts and remove engine.

Reinstall by reversing removal procedure.

OVERHAUL

All Models

Engine make and model are listed at the beginning of this section. To overhaul engine or accessories, refer to appropriate engine section in this manual.

ELECTRICAL SYSTEM

MAINTENANCE AND SERVICE

All Models

Battery electrolyte level should be checked at 25 hour intervals of normal operation. If necessary, add distilled water until level is just below base of vent wells. **DO NOT** overfill. Keep battery posts clean and cable ends tight.

For alternator or starter service, refer to appropriate engine section in this manual.

Refer to Fig. F9 for YT-16 model wiring diagram and to Fig. F10 for LGT-17 and LGT-17H models wiring diagram.

BRAKE

ADJUSTMENT

All Models

Brake assembly on all models is located at left side of transaxle. Adjust brake by removing cotter key (Fig. F11) or loosening jam nut on YT-16 models and tightening nut until light drag is felt as tractor is pushed forward by hand. Loosen nut slightly and reinstall cotter key or tighten jam nut against adjustment nut on YT-16 models.

Illustrations courtesy Ford.

NOTE: On hydrostatic drive model, brakes must not engage before transmission is in neutral. Check neutral adjustment on transmission linkage.

R&R BRAKE PADS

Model YT-16

Remove jam nut (10–Fig. F12) and adjustment nut (9). Remove brake lever (7). Remove bolts (12 and 8) and spacer (11). Remove brake caliper (5), outer brake pad (3), metal backing plate (4), brake disc (2) and inner brake pad (1).

Make certain the two push pins (6) and metal backing plate (4) remain in place in brake pad holder and metal spacer (11) is in place on rear mounting bolt during reassembly.

Models LGT-17 And LGT-17H

Remove brake rod and brake actuating lever (10–Fig. F13). Remove the two cap screws (8) securing disc brake assembly and hold assembly together as it is removed. Note location of spacers (3 and 4) and remove. Remove worn outer brake pad (5), metal backing plate (6) and the two actuating pins (11) from caliper (7). Remove brake disc (2) and inner brake pad (1).

Make certain metal backing plate and outer brake pad are firmly seated in caliper and inner brake pad is seated in its recess. Reassemble by reversing disassembly procedure. Make certain spacers (3 and 4) are in their correct locations.

CLUTCH

MAINTENANCE AND ADJUSTMENT

Models YT-16 And LGT-17

Brake/clutch pedal is connected to a spring loaded, pivoting idler pulley. Belt tension is maintained by spring pressure. Periodically check condition of pulleys and location of belt guides and fingers. Refer to R&R DRIVE BELT section.

Model LGT-17H

Brake/clutch pedal is connected to hydrostatic transmission control valve. Refer to TRANSMISSION section for correct neutral position adjustment.

R&R DRIVE BELT

Model YT-16

Disconnect and remove any attachments from tractor. Set parking brake. Remove flat idler pulley, V-idler pulley and wire belt fingers from pivot bracket. See Fig. F14. Disconnect shift rod from gear select lever on transaxle. Remove belt.

Reinstall by reversing removal procedure. Block tractor wheels and release parking brake and adjust all belt fingers and guides as shown in Fig. F14. Fingers should be close to, but not touch belt or pulley.

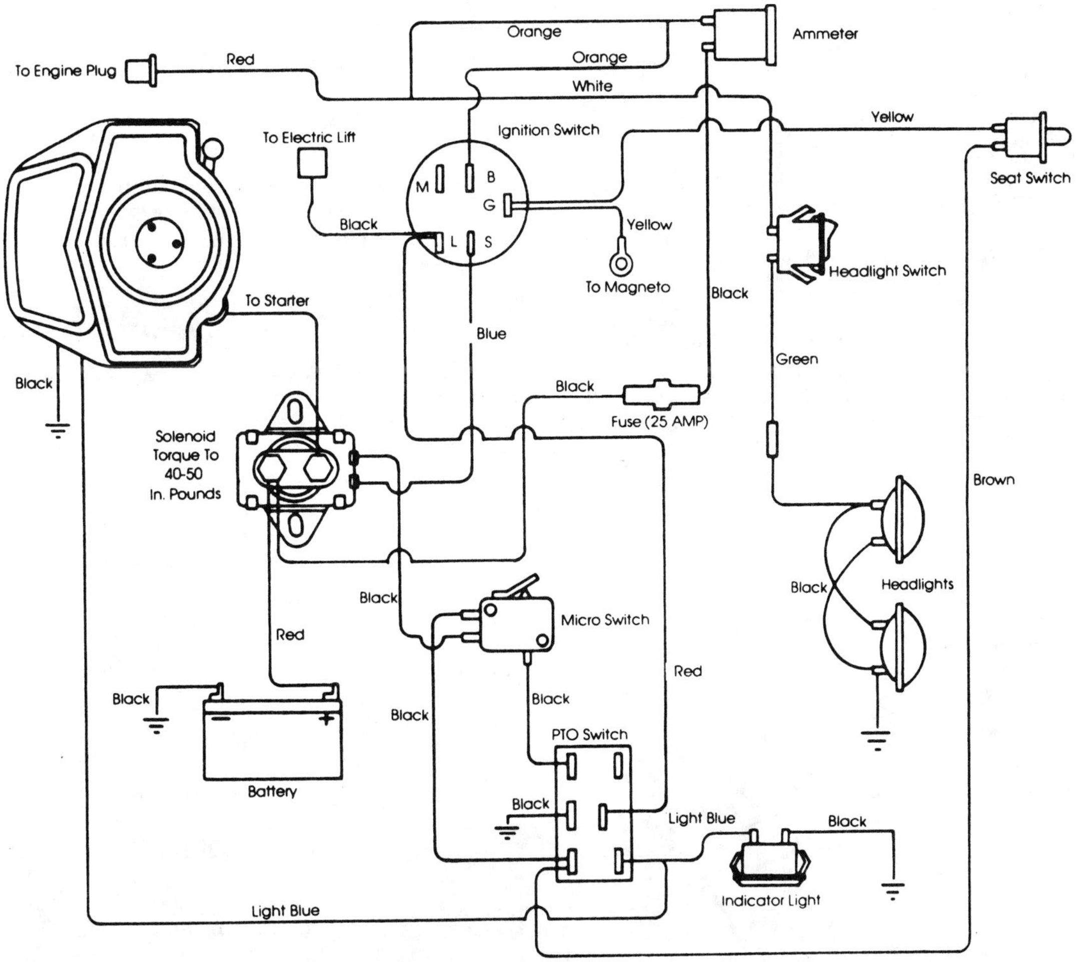

Fig. F9—Wiring diagram for YT-16 models.

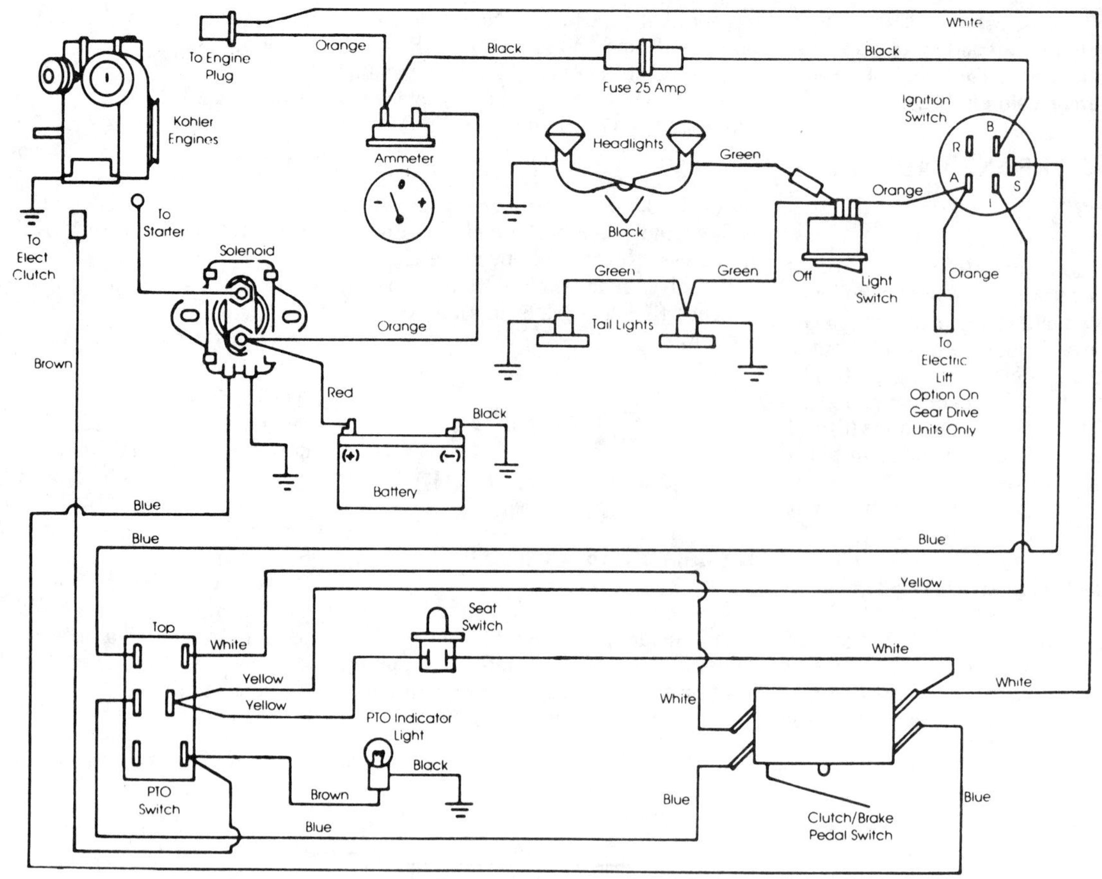

Fig. F10—Wiring diagram for LGT-17 and LGT-17H models.

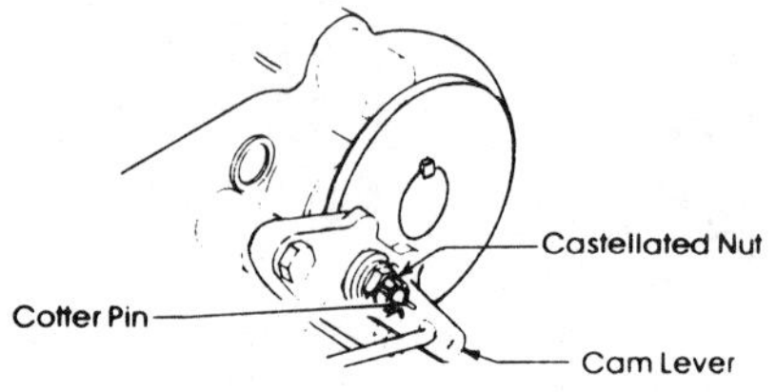

Fig. F11—View of typical disc brake system similar to unit used on all models. Refer to text for adjustment procedure.

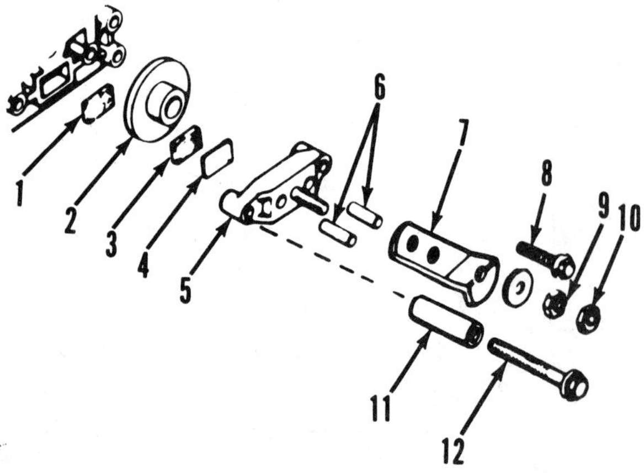

Fig. F12—Exploded view of disc brake unit used on YT-16 model. Refer to text.

1. Inner brake pad
2. Brake disc
3. Outer brake pad
4. Metal backing plate
5. Brake caliper
6. Pins
7. Brake lever
8. Bolt
9. Adjusting nut
10. Jam nut
11. Spacer
12. Bolt

Model LGT-17

Block wheels so tractor cannot move. Make certain parking brake is disengaged and remove flange nut (1–Fig. F15). Engage parking brake and lock in position. Slide V-idler pulley (3) outward until belt is clear of guide (4). Remove flat idler pulley (9) and belt finger (13) noting sequence in which they are installed. Remove belt.

Reinstall by reversing removal procedure. With parking brake disengaged, adjust all belt fingers and guides so they just have clearance between belt and/or pulleys.

TRANSMISSION

MAINTENANCE

Model YT-16

Transaxle is packed with 36 ounces (1065 mL) of EP lithium base grease during manufacture and should provide lifetime lubrication.

Model LGT-17

Transaxle fluid level should be checked at 100 hour intervals. Maintain fluid level at level check plug (Fig. F16). Fluid capacity is 64 ounces (1893 mL) and recommended fluid is SAE 140 EP gear lubricant.

Model LGT-17H

Hydrostatic transmission, hydraulic system and differential assembly fluid level should be checked at 100 hour intervals and fluid and filter should be changed at 250 hour intervals. Maintain fluid level at "FULL" mark on dipstick (Fig. F17). Fluid capacity is 6 quarts (5.7 L) and recommended fluid is SAE 20 detergent oil.

ADJUSTMENTS

Model LGT-17H (Hand Operated Transmission Control)

If tractor "creeps" forward or backward after brake/clutch pedal is fully applied and then released, drive lever neutral position should be adjusted.

Place blocks ahead and behind both front wheels and ahead and behind right rear wheel. Raise left rear wheel so tire clears the ground. Make certain tow control valve is in "DRIVE" position and range transmission selector lever is in "HI" position. Disconnect brake cable at brake lever. Loosen bolts holding neutral finger spring and neutral finger bracket at left side of tractor frame (Fig. F18). Depress brake/clutch pedal fully and engage parking brake latch. Locate turnbuckle under right side of frame just ahead of rear wheel and loosen jam nuts. Start engine and run at full speed.

CAUTION: Do not make adjustments while engine is running.

Wheel should not rotate. If wheel is turning, stop engine and adjust turnbuckle as necessary to position hydrostatic control linkage in "Neutral." Start engine and recheck adjustment. When adjustment is correct, position neutral finger bracket so neutral finger spring is centered in control arm notch. Tighten bracket and spring bolts in this position. Reconnect brake cable.

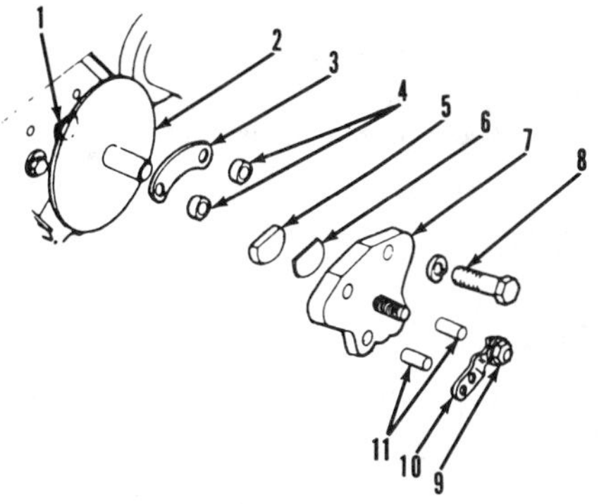

Fig. F13—Exploded view of disc brake unit used on LGT-17 and LGT-17H models. Refer to text.

1. Inner brake pad
2. Brake disc
3. Spacer
4. Spacers
5. Outer brake pad
6. Metal backing plate
7. Brake caliper
8. Bolt
9. Adjusting nut
10. Brake lever
11. Pins

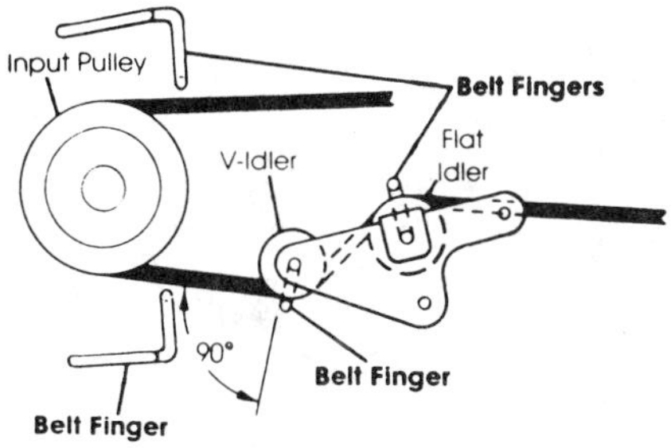

Fig. F14—View showing correct belt finger adjustment for YT-16 model drive belt.

Model LGT-17H (Foot Operated Transmission Control)

If tractor "creeps" forward or backward after foot control pedal is returned to neutral position, neutral position should be adjusted.

Place blocks ahead and behind both front wheels and ahead and behind right rear wheel. Raise left rear wheel so tire clears the ground. Make certain tow control valve is in "DRIVE" position and range transmission selector lever is in "HI" position. Disconnect brake cable at brake lever. Apply parking brake and set latch in "UP" position. Locate pivot bracket (2–Fig. F19) and loosen cap screw (4) slightly. Start engine and run at full speed.

CAUTION: Do not make adjustments while engine is running.

With foot pedal in "Neutral" position, wheel should not rotate. If wheel is rotating, stop engine and insert blade of screwdriver into adjustment notch (1). Twist screwdriver as necessary to position control linkage in "Neutral." Tighten cap screw (4), start engine and recheck adjustment. Reconnect brake cable.

R&R TRANSMISSION

Model YT-16

Raise and support rear of tractor. Depress brake/clutch pedal and lock in position. Disconnect brake rod and gear selector lever. Remove drive belt. Support transaxle and remove torque bracket (support strap) at front of left side, then remove transaxle retaining

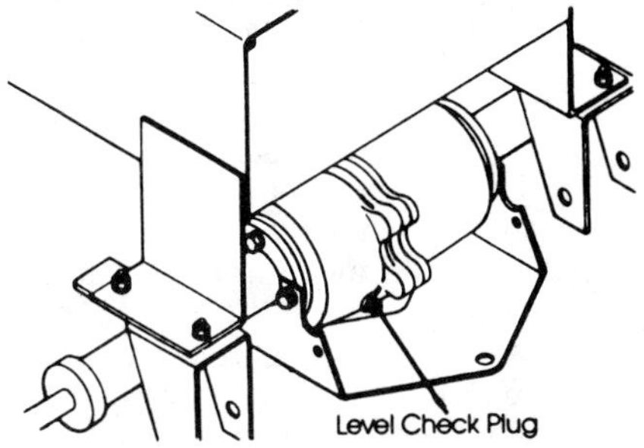

Fig. F16—View showing location of fluid level check plug for transaxle used on LGT-17 models.

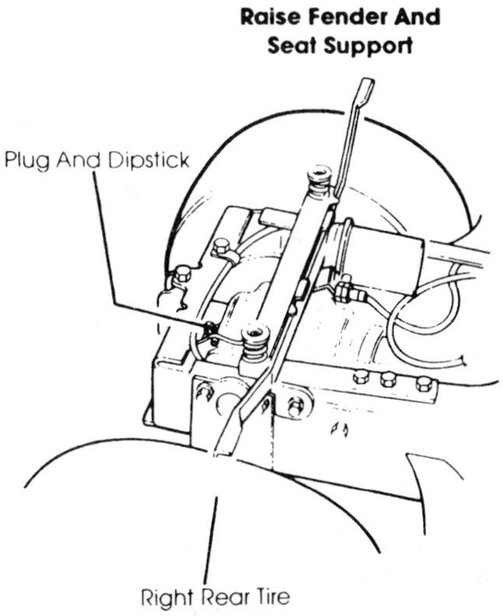

Fig. F17—View showing location of dipstick used to check fluid level of transmission and hydraulic system on LGT-17H models.

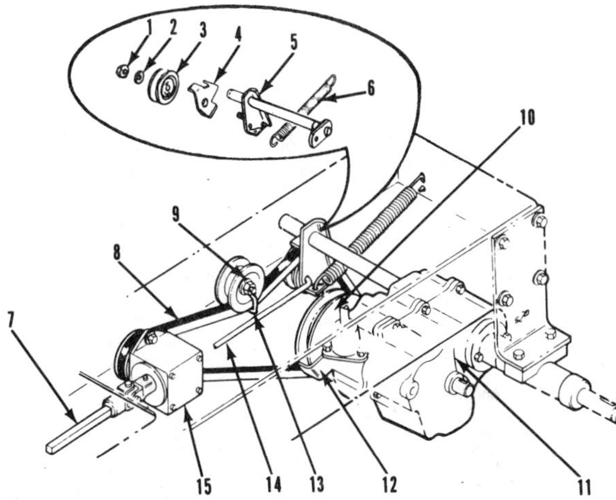

Fig. F15—View showing component parts of LGT-17 drive belt and pulley arrangement.

1. Flange nut
2. Washer
3. V-idler pulley
4. V-idler belt guide
5. Pivot shaft
6. Spring
7. Drive shaft
8. Drive belt
9. Flat idler pulley
10. Input pulley
11. Transaxle
12. Belt guide
13. Belt finger
14. Clutch rod
15. Bevel gear box

Illustrations courtesy Ford.

bolts at rear axle housings. Lower transaxle and roll out from under tractor.

Reverse removal procedure for reinstallation and refer to R&R DRIVE BELT section for drive belt installation.

Model LGT-17

Raise and support rear of tractor. Disconnect brake rod and remove drive belt. Remove the three bolts retaining shift lever and remove lever assembly. Support transaxle (transmission) and remove bolts retaining transaxle and brake support plate on left side of transaxle. Remove retaining bolts at rear axle housings, then lower and remove transaxle assembly.

Reinstall by reversing removal procedure. Refer to R&R DRIVE BELT section for drive belt installation.

Model LGT-17H

Remove seat/fender assembly and shift plate. Disconnect brake return spring and brake rod. Disconnect clevis from hydraulic control lever assembly at hydrostatic unit. Loosen setscrews and slide fan hub assembly toward transmission. Disconnect drive shaft at transmission end. Disconnect all hydraulic hoses and cap openings. Support transmission and loosen the two large side mount bolts. Remove cone locknuts from the four axle bolts and pull bolts out of axle mount assemblies. Remove the two large side mount bolts. Lift frame assembly off and away from transmission assembly.

Reinstall by reversing removal procedure. Fill to proper level with recommended fluid. Position hydrostatic control lever in neutral position and start and run engine at idle speed. Operate transmission in forward and reverse direction for short distances. Stop engine and check fluid level. Repeat procedure until all air has been bled from system. Adjust hydrostatic transmission control lever as outlined in ADJUSTMENT section. Adjust brake.

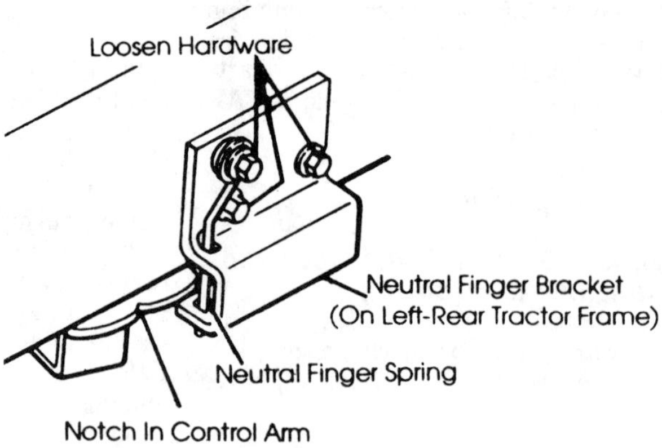

Fig. F18—View of neutral finger bracket and spring. Refer to text for adjustment procedure.

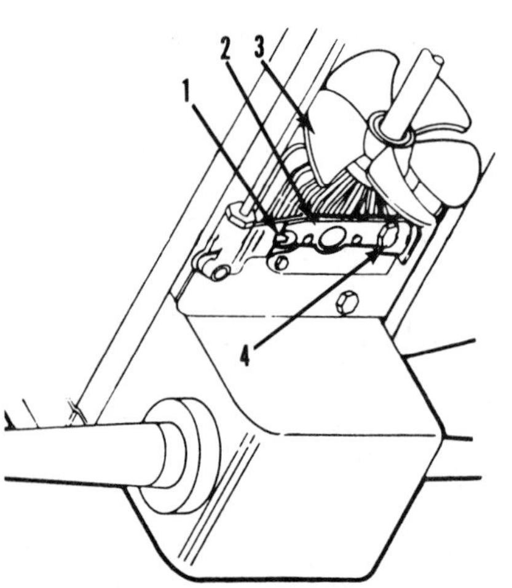

Fig. F19—View of pivot bracket and adjustment notch for LGT-17H models with foot controlled transmission. Refer to text.

1. Adjustment notch
2. Pivot bracket
3. Fan
4. Retaining bolt

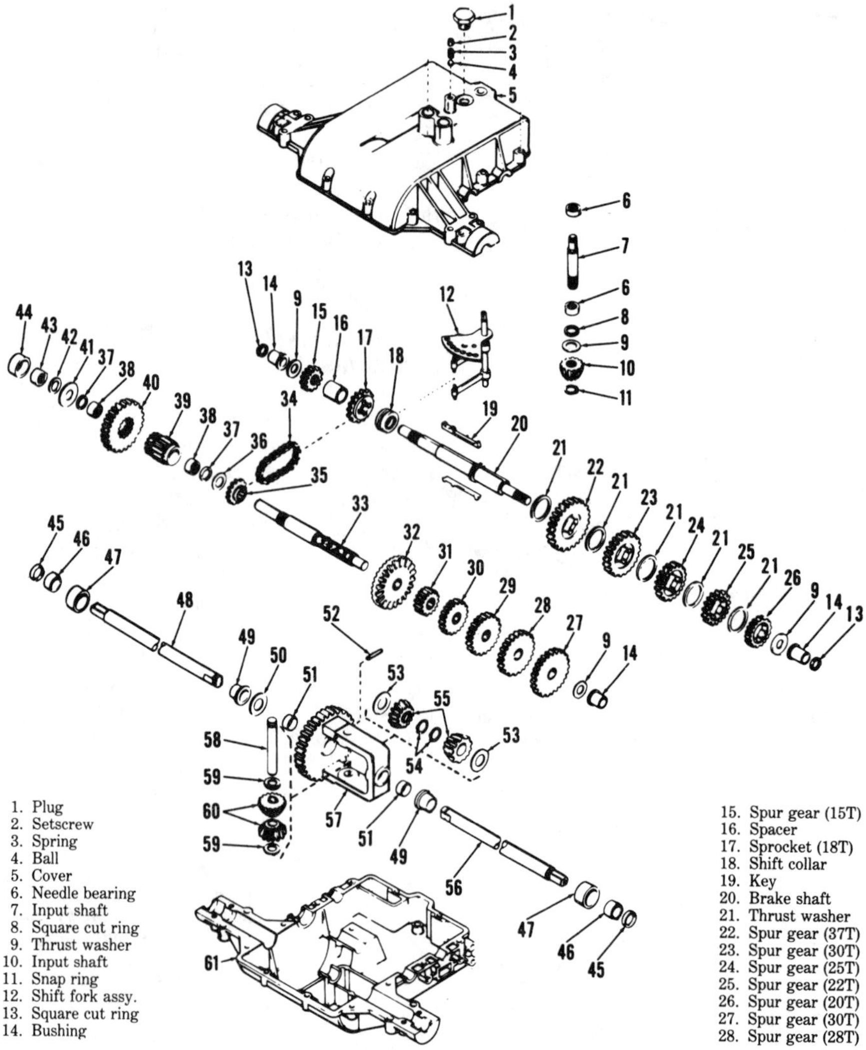

Fig. F20—Exploded view of transaxle assembly used on YT-16 models.

1. Plug
2. Setscrew
3. Spring
4. Ball
5. Cover
6. Needle bearing
7. Input shaft
8. Square cut ring
9. Thrust washer
10. Input shaft
11. Snap ring
12. Shift fork assy.
13. Square cut ring
14. Bushing
15. Spur gear (15T)
16. Spacer
17. Sprocket (18T)
18. Shift collar
19. Key
20. Brake shaft
21. Thrust washer
22. Spur gear (37T)
23. Spur gear (30T)
24. Spur gear (25T)
25. Spur gear (22T)
26. Spur gear (20T)
27. Spur gear (30T)
28. Spur gear (28T)
29. Spur gear (25T)
30. Spur gear (20T)
31. Spur gear (12T)
32. Bevel gear (42T)
33. Countershaft
34. Roller chain
35. Sprocket (9T)
36. Flat washer
37. Square cut seal
38. Needle bearing
39. Output pinion
40. Output gear
41. Flat washer
42. Square cut seal
43. Needle bearing
44. Spacer
45. Oil seal
46. Needle bearing
47. Spacer
48. Axle shaft
49. Bushing
50. Washer
51. Bushing
52. Pin
53. Thrust washer
54. Snap ring
55. Bevel gear
56. Axle shaft
57. Differential gear
58. Drive pin
59. Thrust washer
60. Bevel pinion
61. Case

Illustrations courtesy Ford.

OVERHAUL

Model YT-16

Remove transaxle assembly as outlined. Place shift lever in neutral and unbolt and remove shift lever. Remove setscrew (2 – Fig. F20), spring (3) and index ball (4). Unbolt cover (5) and push shift fork assembly (12) in while removing cover. Before removing gear and shaft assemblies, shifter fork (12) should be removed. Note position of parts before removal. Remove gear and shaft assemblies (Fig. F21) from case taking care not to disturb drive chain (34 – Fig. F20). Remove needle bearings (38 and 43), flat washer (41), square cut seals (37 and 42), output gear (40) and output pinion (39) from countershaft. Angle the two shafts together, then mark position of chain on sprocket collars and remove chain. Remove sprocket (35), bevel gear (32), spur gears (27, 28, 29, 30 and 31), thrust washer (9) and flange bushing (14). All gears are splined to the countershaft. Disassembly of shifter-brake shaft is apparent after inspection. Remove snap ring (11), input bevel gear and pull input shaft (7) through cover.

Disassemble differential by driving roll pin securing drive pin (58) out. Remove pinion gears (60) by rotating gears in opposite directions. Remove snap rings (54), side gears (55) and thrust washers (53), then slide axles out. Note axle shafts (48 and 56) are different lengths.

Clean and inspect all parts and renew any showing excessive wear or other damage. When installing new inner input shaft needle bearings, press bearing in to a depth of 0.135-0.150 inch (3.429-3.81 mm) below flush. When in-

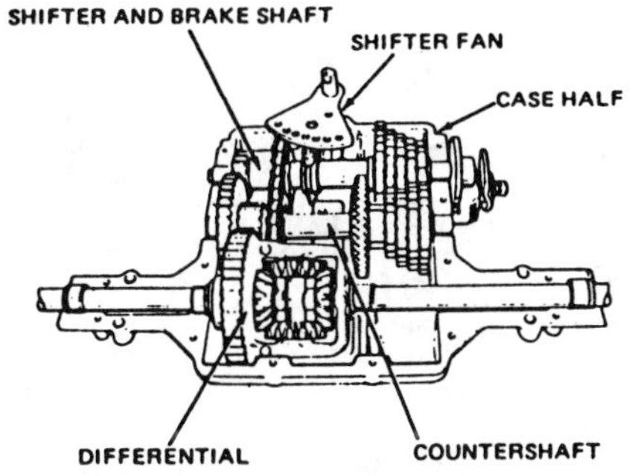

Fig. F21 — When disassembling transaxle used on YT-16 model, remove gear and shaft assemblies from case by lifting both shafts at once. Take care not to disturb chain.

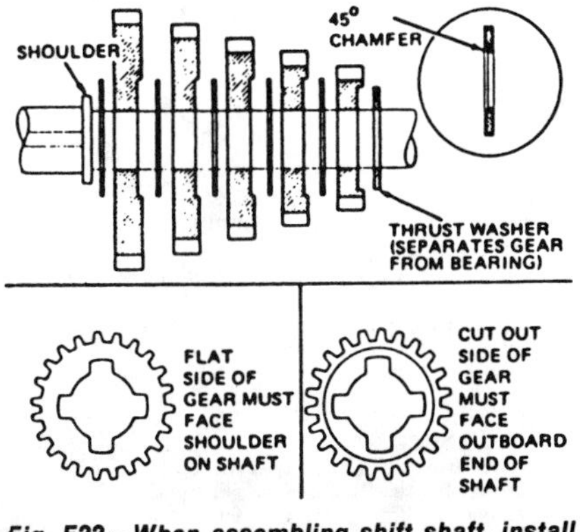

Fig. F22 — When assembling shift shaft, install thrust washers with 45° chamfer toward shoulder on output shaft. Flat side of gears also must face shoulder on output shaft.

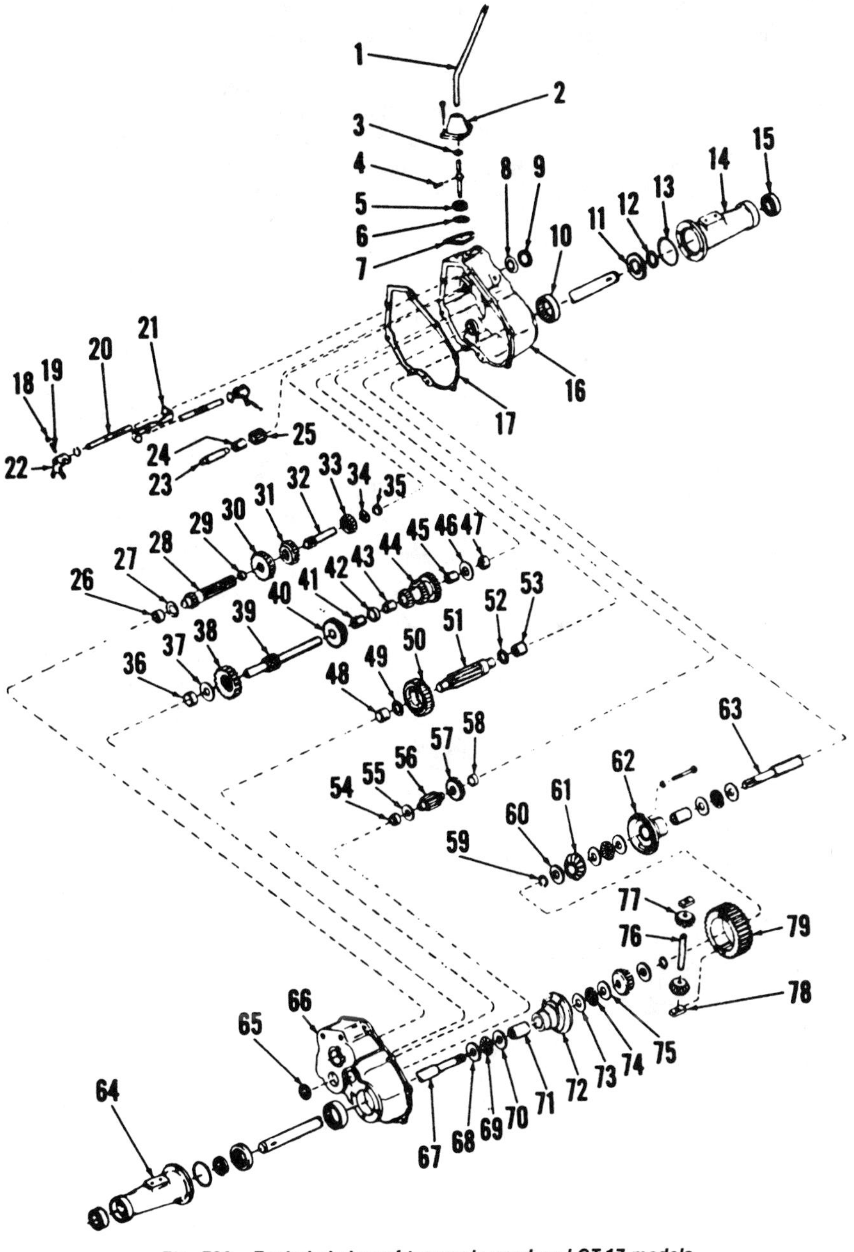

Fig. F23 — Exploded view of transaxle used on LGT-17 models.

1. Shift lever
2. Shift lever housing
3. Seal ring
4. Roll pin
5. Retainer
6. Snap ring
7. Gasket
8. Ball bearing
9. Oil seal
10. Carrier bearing
11. Seal retainer
12. Oil seal
13. "O" ring
14. Axle housing
15. Axle outer bearing
16. Transaxle case
17. Gasket
18. Detent ball
19. Spring
20. Shifter rod
21. Shifter stop
22. Shifter fork
23. Reverse idler shaft
24. Spacer
25. Reverse idler gear
26. Needle bearing
27. Thrust washer
28. Shifter shaft
29. Needle bearing
30. First, second & reverse gear
31. Third & fourth gear
32. Input shaft
33. Input gear
34. Thrust washer
35. Needle bearing
36. Needle bearing
37. Thrust washer
38. Idler gear
39. Brake & cluster shaft
40. Two-cluster gear
41. Bushing
42. Spacer
43. Bushing
44. Three-cluster gear
45. Bushing
46. Thrust washer
47. Needle bearing
48. Needle bearing
49. Thrust washer
50. Output gear
51. Output shaft
52. Thrust washer
53. Needle bearing
54. Needle bearing
55. Thrust washer
56. Low reduction shaft
57. Low reduction gear
58. Needle bearing
59. Snap ring
60. Thrust washer
61. Axle gear
62. Axle carriage (plain holes)
63. Axle shaft
64. Axle housing
65. Oil seal
66. Transaxle cover
67. Axle shaft
68. Thrust washer
69. Thrust bearing
70. Thrust washer
71. Bushing
72. Axle carriage (tapped holes)
73. Thrust washer
74. Thrust bearing
75. Thrust washer
76. Drive pin
77. Bevel pinion gear
78. Drive block
79. Ring gear

Illustrations courtesy Ford.

stalling thrust washers and shifting gears on brake shaft, the 45° chamfer on inside diameter of thrust washers must face shoulder on brake shaft. See Fig. F22. Flat side of gears must face shoulder on shaft. Complete reassembly and pack case with approximately 36 ounces (1065 mL) EP lithium base grease. Tighten case to cover cap screws to 90-100 in.-lbs. (10-11 N·m) torque.

Model LGT-17

Remove transaxle assembly as outlined. Remove drain plug and drain lubricant. Remove rear wheel and hub assemblies. Remove brake caliper and brake disc. Place shift lever (1–Fig. F23) in neutral position, then unbolt and remove shift lever assembly. Remove axle housings (14 and 64), then remove seal retainers (11) with oil seals (12) and "O" rings (13) by pulling each axle shaft out of case and cover as far as possible. Place transaxle unit on the edge of a bench with left axle shaft pointing downward. Remove cap screws securing case (16) to cover (66) and drive aligning dowel pins out of case. Lift case (16) up 1½ to 2 inches (40 to 50 mm), tilt case about 45°, rotate case clockwise and remove it from the assembly. Input shaft (32) and input gear (33) will be removed with case. Withdraw differential and axle shaft assembly and lay aside for later disassembly. Remove the three-cluster gear (44) with its thrust washer (46) and spacer (42). Lift out reverse idler gear (25), spacer (24) and shaft (23). Hold upper ends of shifter rods together and lift out shifter rods, forks, shifter stop (21), sliding gears (30 and 31) and shaft (28) as an assembly. Remove low reduction gear (57), reduction shaft (56) and thrust washer (55), then remove the two-cluster gear (40) from brake shaft. Lift out the output gear (50), shaft (51) and thrust washers (49 and 52). To remove brake shaft (39) and gear (38) from cover (66), block up under gear (38) and press shaft out of gear.

CAUTION: Do not allow cover or low reduction gear bearing boss to support any part of pressure required to press brake shaft from gear.

Remove input shaft (32) with input gear (33) and thrust washer (34) from case (16).

To disassemble differential, remove four cap screws and separate axle shaft and carriage assemblies from ring gear (79). Drive blocks (78), bevel pinion gears (77) and drive pin (76) can now be removed from ring gear. Remove snap rings (59) and withdraw axle shafts (63 and 67) from axle gears (61) and carriages (62 and 72).

Clean and inspect all parts and renew any showing excessive wear or other damage. When installing new needle bearings, press bearing (29) in spline shaft (28) to a depth of 0.010 inch (0.254 mm) below end of shaft and low reduction shaft bearings (54 and 58) 0.010 inch (0.254 mm) below thrust surfaces of bearing bases. Carrier bearings (10) should be pressed in from inside of case and cover until bearings are 0.290 inch (7.366 mm) below face of axle housing mounting surface. All other needle bearings are to be pressed in from inside of case and cover to a depth of 0.015-0.020 inch (0.381-0.508 mm) below thrust surfaces.

Renew all seals and gaskets. Reassemble by reversing disassembly procedure. When installing brake shaft (39) and idler gear (38), beveled edge of gear teeth must be up away from cover. Install reverse idler shaft (23), spacer (24) and reverse idler gear (25) with rounded end of gear teeth facing spacer. Install input gear (33) and shaft (32) so chamfered side of input gear is facing case (16).

Tighten transaxle cap screws to the following torque:

Differential cap screws . . . . 7 ft.-lbs. (9 N·m)
Case to cover cap screws . . . . . 10 ft.-lbs. (14 N·m)
Axle housing cap screws . . . . . 13 ft.-lbs. (18 N·m)
Shift lever housing cap screws . . . . 10 ft.-lbs. (14 N·m)

Fill transaxle, after unit is installed on tractor, to level plug opening with SAE 140 EP gear lubricant. Capacity is approximately 64 ounces (1893 mL).

Model LGT-17H

HYDROSTATIC DRIVE UNIT. Remove transmission as outlined and separate hydrostatic drive unit from range transmission. Thoroughly clean exterior of hydrostatic unit.

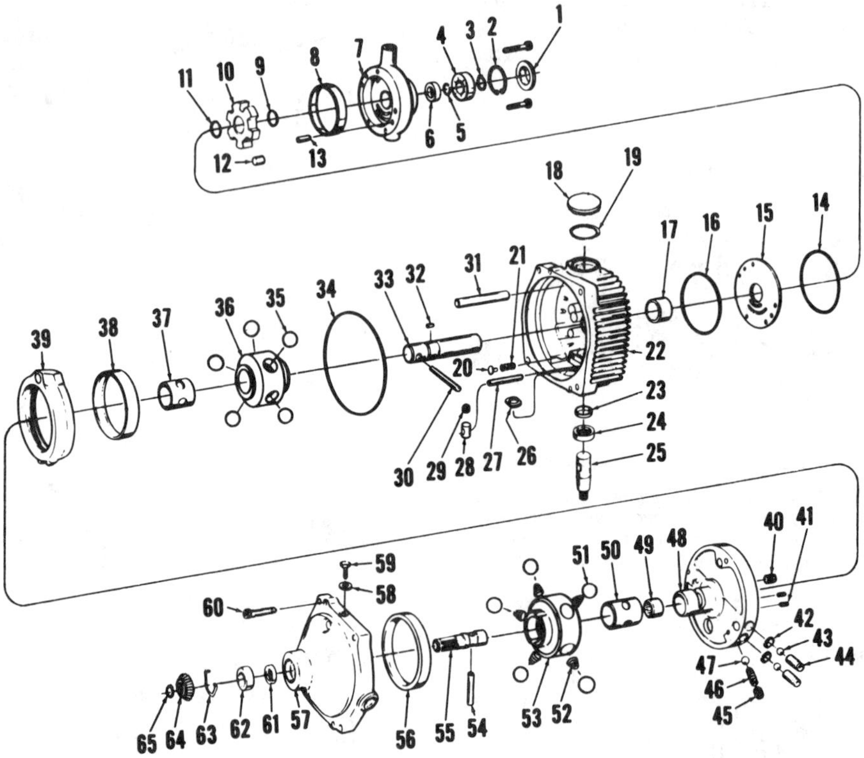

Fig. F24—Exploded view of hydrostatic transmission used on LGT-17H model.

1. Dust shield
2. Retaining ring
3. Snap ring
4. Ball bearing
5. Snap ring
6. Oil seal
7. Charge pump body
8. Charge pump race
9. Snap ring
10. Charge pump rotor
11. Snap ring
12. Pump roller (6)
13. Dowel pin
14. "O" ring
15. Pump plate
16. "O" ring
17. Bushing
18. Plug
19. Gasket
20. Neutral spring cap
21. Neutral spring (2)
22. Housing
23. Bushing
24. Oil seal
25. Control shaft
26. Washer
27. Dowel pin
28. Insert
29. Insert cap
30. Drive pin
31. Cam pivot pin
32. Charge pump drive key
33. Input shaft
34. "O" ring
35. Pump ball pistons
36. Pump rotor
37. Rotor bushing
38. Pump race
39. Pump cam ring
40. Plug (2)
41. Roll pins
42. Snap ring (2)
43. Check valve ball (2)
44. Directional check valve body (2)
45. Plug
46. Relief spring
47. Charge relief ball
48. Pintle
49. Needle bearing
50. Rotor bushing
51. Motor ball pistons
52. Springs
53. Motor rotor
54. Drive pin
55. Output shaft
56. Motor race
57. Body
58. Gasket
59. Venting plug
60. Cap screw
61. Oil seal
62. Ball bearing
63. Retainer
64. Output gear
65. Snap ring

Illustrations courtesy Ford.

Remove venting plug (59–Fig. F24) and plug (18), invert assembly and drain fluid. Place unit in a holding fixture so input shaft is pointing upward. Remove dust shield (1) and snap ring (3). Remove the five cap screws from charge pump body (7). One cap screw is ½-inch (12.7 mm) longer than the others and must be installed in original location (heavy section of pump body). Remove charge pump body (7) with ball bearing (4). Ball bearing and oil seal (6) can be removed from body (7) after first removing retaining ring (2). Remove the six pump rollers (12), snap rings (5, 9 and 11) and charge pump rotor (10). Remove "O" rings (14 and 16) and pump plate (15). Invert drive unit in holding fixture so output shaft is pointing upward. Remove snap ring (65) and output gear (64). Unscrew the two cap screws (60), then turn them in until two threads are engaged. Raise body (57) until it contacts the heads of cap screws (60). Insert a fork tool between motor rotor (53) and pintle (48) until the tool extends beyond opposite side.

NOTE: The special fork tool can be fabricated from a piece of ⅛-inch flat stock approximately 3 inches wide and 12 inches long. Cut a slot 1-9/16 inches wide and 8 inches long. Taper ends of prongs. Refer to Fig. F25.

Remove cap screws (60–Fig. F24) and by raising ends of forked tool, lift off body and motor assembly. Place body and motor assembly on a bench or in a holding fixture with output shaft pointing downward. Remove special fork tool and place a wide rubber band around motor rotor to hold ball pistons (51) and springs (52) in their bores. Carefully remove motor rotor assembly and lay aside for later disassembly. Remove motor race (56) and output shaft (55). Remove retainer (63), bearing (62) and oil seal (61).

With housing assembly (22) resting in holding fixture (input shaft pointing downward), remove pintle assembly (48).

CAUTION: Do not allow pump to raise with the pintle as the ball pistons (35) may fall out of rotor (36).

Hold pump in position by inserting a finger through hole in pintle. Remove plug (45), spring (46) and charge pump relief ball (47). To remove directional check valves from pintle (48), drill through pintle with a drill bit that will pass freely through roll pins (41). Redrill the holes from opposite side with a ¼-inch drill bit. Drive or press roll pins from pintle. Using a 5/16-18 tap, thread the inside of valve bodies (44). then remove valve bodies using a draw bolt or a slide hammer puller. Remove check valve balls (43) and snap rings (42). Do not remove plugs (40).

Remove pump cam ring (39) and pump race (38). Place a wide rubber band around pump rotor to prevent ball pistons (35) from falling out. Carefully remove pump assembly and input shaft (33).

To remove control shaft (25), drill a 11/32-inch hole through aluminum housing (22) directly in line with center line of dowel pin (27). Press dowel pin from control shaft, then withdraw control shaft. Remove oil seal (24). Thread the drilled hole in housing with a ⅛-inch pipe tap. Apply a light coat of "Loctite" grade 635 to ⅛-inch pipe plug, install plug and tighten it until snug.

Number piston bores (1 through 5) on pump rotor and on motor rotor. Use a plastic ice cube tray or equivalent and mark the cavities 1P through 5P for pump ball pistons and 1M through 5M for motor ball pistons. Remove ball pistons (35) one at a time, from pump rotor and place each ball in the correct cavity in tray. Remove ball pistons (51) and springs (52) from motor rotor in the same manner.

Clean and inspect all parts and renew any showing excessive wear or other damage. Ball pistons are a select fit to 0.0002-0.0006 inch (0.0051-0.0152 mm) clearance in rotor bores and must be reinstalled in their original bores. If rotor bushings (37 or 50) are scored or badly worn, 0.002 inch (0.051 mm) or more clearance on pintle journals, renew pump rotor or motor rotor assemblies. Check clearance between input shaft (33) and housing bushing (17). Normal clearance is 0.0013-0.0033 inch (0.0330-0.0838 mm). If clearance is excessive, renew input shaft and/or housing assembly.

Install ball pistons (35) in pump rotor (36) and ball pistons (51) and springs (52) in motor rotor (53), then use wide rubber bands to hold pistons in their bores. Install snap rings (42), check valve balls (43) and valve bodies (44) in pintle (48) and secure with new roll pins (41). Install charge pump relief valve ball (47), spring (46) and plug (45). When installing oil seals (6, 24 or 61), apply a light coat of "Loctite" grade 635 to the seal outer diameter. Renew oil seal (24) and install control shaft (25) in housing. Install special washer (26), then press dowel pin (27) into control shaft until end of dowel pin is 1¼ inches (31.75 mm) from control shaft. Renew oil seal (61), then reinstall output shaft (55) with drive pin (54), bearing (62), retainer (63), output gear (64) and snap ring (65) in body (57).

Insert input shaft (33) with drive pin (30) through bushing (17) in housing. Install snap ring (11) in its groove on input shaft. Place "O" ring (16), pump plate (15) and "O" ring (14) in housing, then install charge pump drive key (32), charge pump rotor (10) and snap ring (9). Apply light grease or vaseline to pump rollers (12) and place rollers in rotor slots. Install oil seal (6) and pump race (8) in charge pump body (7), then install body assembly. Secure with the five cap screws, making certain long cap screw is installed in its original location (in heavy section of pump body). Tighten cap screws to 28-30 ft.-lbs. (38-40 N·m). Install snap ring (5), bearing (4), retaining ring (2), snap ring (3) and dust shield (1).

Place charge pump and housing assembly in a holding fixture with input shaft pointing downward. Install pump race (38), insert cap (29) and insert (28) in cam ring (39), then install cam ring assembly over cam pivot pin (31) and control shaft dowel pin (27). Turn control shaft (25) back and forth and check movement of cam ring. Cam ring must move freely from stop to stop. If not, check installation of insert (28) and insert cap (29) in cam ring.

Install pump rotor assembly and remove rubber band used to retain pistons. Install pintle assembly (48) over cam pivot pin (31) and into pump rotor. Place "O" ring (34) in position on housing.

Place body assembly (57) in a holding fixture with gear (64) downward. Install motor race (56) in body, then install motor rotor assembly and remove rubber band used to retain pistons in rotor.

Using special fork tool (Fig. F25) to retain motor assembly in body, carefully install body and motor assembly over pintle journal. Remove fork tool, align bolt holes and install the two cap screws (60–Fig. F24). Tighten cap screws to 15-ft.-lbs. (20 N·m).

Place hydrostatic unit in holding fixture with plug (18) opening and venting plug opening facing upward. Fill unit with SAE 20 detergent oil until fluid flows from fitting hole in body. Install venting plug (59) with gasket (58), then install plug (18). Plug all openings to prevent dirt or other foreign material from entering hydrostatic unit. Reinstall on range transmission and differential assembly.

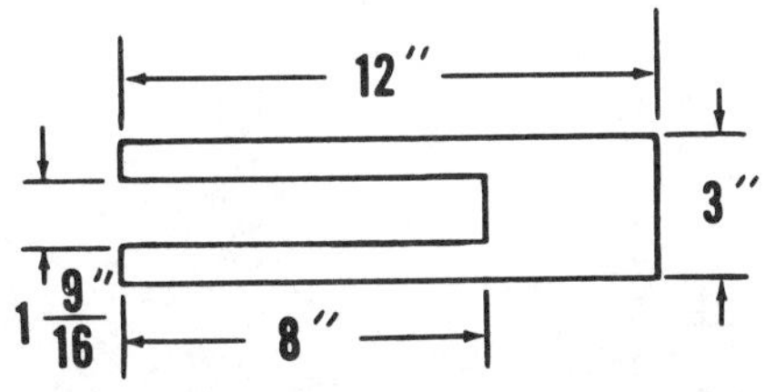

Fig. F25—View of special tool to be fabricated locally, used in disassembly and reassembly of hydrostatic transmission. Refer to text.

Illustrations courtesy Ford.

RANGE TRANSMISSION AND DIFFERENTIAL. Remove transmission as outlined and separate hydrostatic drive unit from range transmission and differential assembly.

Remove brake caliper assembly and brake disc. Remove tires and wheels and remove hub assemblies. Thoroughly clean exterior of unit.

Remove axle housings (15 – Fig. F26). Position unit with cover up, then remove cover (13). Lift out differential assembly and axles (40 through 54). Remove output shaft (36), gear (35) and thrust washers (37). Unscrew setscrew (2) and remove spring (3) and ball (4). Remove brake shaft (32), sliding gear (33), shift fork (10) and rod (8). Remove input shaft and gear components (18 through 25).

To disassemble differential assembly, remove cap screws (53), then remove differential carriers (41 and 52) and axles from ring gear assembly. Remove snap ring to separate axle gear, carrier and axle. Remove pinions (47) and separate body cores (45 and 48) from ring gear (46).

Inspect components for damage or excessive wear. To reassemble unit, reverse disassembly procedure. Check movement of shift rod when tightening setscrew (2). Install gears (22 and 25) so bevels of gears face together as shown in Fig. F27. Install carrier cap screws (53 – Fig. F26) so head of cap screw is on side of shorter carrier (52). Do not rotate axle housings after housing has been pressed tight against seal (11) as seal may be cut.

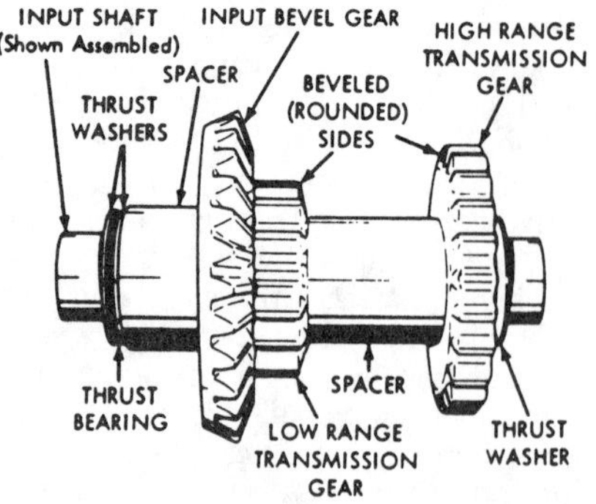

Fig. F27 — View of reduction drive input shaft and gears. Note position of bevels on gears.

Reinstall hydrostatic drive unit, brake disc and brake caliper.

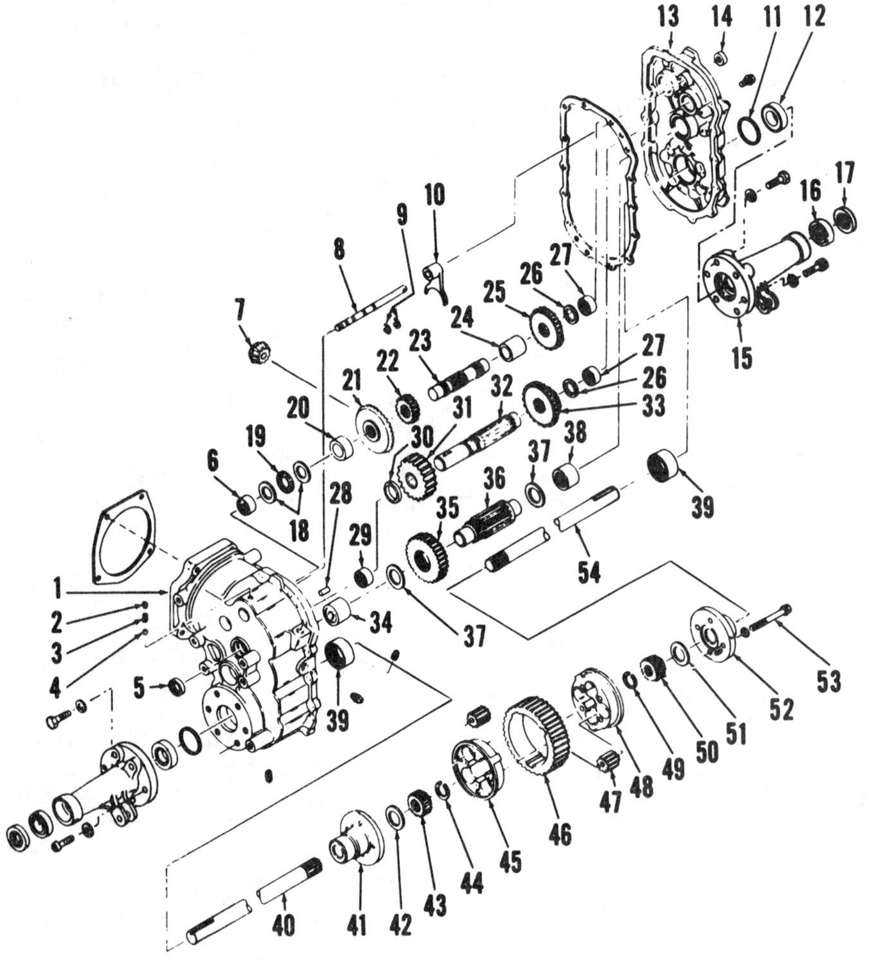

Fig. F26 — Exploded view of range transmission and differential assembly used on LGT-17H models.

1. Case
2. Setscrew
3. Spring
4. Ball
5. Seal
6. Needle bearing
7. Transmission output gear
8. Shift rail
9. Snap rings
10. Shift fork
11. Quad ring
12. Tapered roller bearing
13. Cover
14. Seal
15. Axle housing
16. Ball bearing
17. Seal
18. Thrust washers
19. Thrust bearing
20. Spacer
21. Bevel gear
22. Gear (low range)
23. Shaft
24. Spacer
25. Gear (high range)
26. Thrust washer
27. Needle bearing
28. Dowel pin
29. Needle bearing
30. Spacer
31. Gear
32. Brake shaft
33. Sliding gears
34. Needle bearing
35. Gear
36. Output shaft
37. Thrust washer
38. Needle bearing
39. Needle bearing
40. Axle shaft
41. Differential carrier
42. Thrust washer
43. Axle gear
44. Snap ring
45. Body core
46. Ring gear
47. Pinion gears (8)
48. Body core
49. Snap ring
50. Axle gear
51. Thrust washer
52. Differential carrier
53. Cap screw
54. Axle shaft

BEVEL GEAR BOX

R&R AND OVERHAUL

Model LGT-17

To remove right angle drive unit (bevel gear box), raise seat assembly and unbolt drive shaft rear coupling. Remove traction drive belt from pulley, then loosen set screws and remove pulley assembly. Unbolt and remove angle drive from tractor.

To disassemble unit, remove cover (1 – Fig. F28) and gasket (2). Unbolt and remove retainer (13) with oil seal (14) and gasket (12). Withdraw output shaft (10) and bearing (11), then remove driven gear (9) through cover opening. Remove oil seal (17) and snap ring (16).

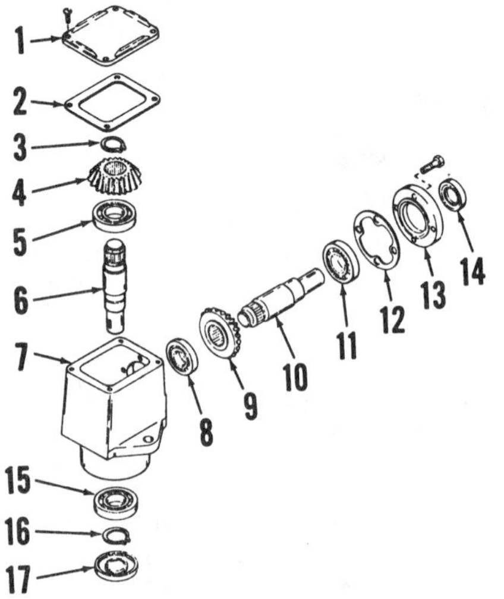

Fig. F28 — Exploded view of right angle drive unit used on LGT-17 models.

1. Cover
2. Gasket
3. Snap ring
4. Drive gear
5. Bearing
6. Input shaft
7. Housing
8. Bearing
9. Driven gear
10. Output shaft
11. Bearing
12. Gasket
13. Retainer
14. Oil seal
15. Bearing
16. Snap ring
17. Oil seal

Illustrations courtesy Ford.

Push input shaft (6), bearing (5) and drive gear (4) out through cover opening. Gear and bearing can be removed from input shaft after first removing snap ring (3). To remove bearing (8), either tap outside of housing (7) behind bearing with a mallet or apply heat to housing.

Clean and inspect all parts and renew any showing excessive wear or other damage.

Reassemble by reversing disassembly procedure. Use new oil seals and gaskets and pack 4 ounces (118 mL) of multi-purpose, lithium base grease in housing before installing cover (1).

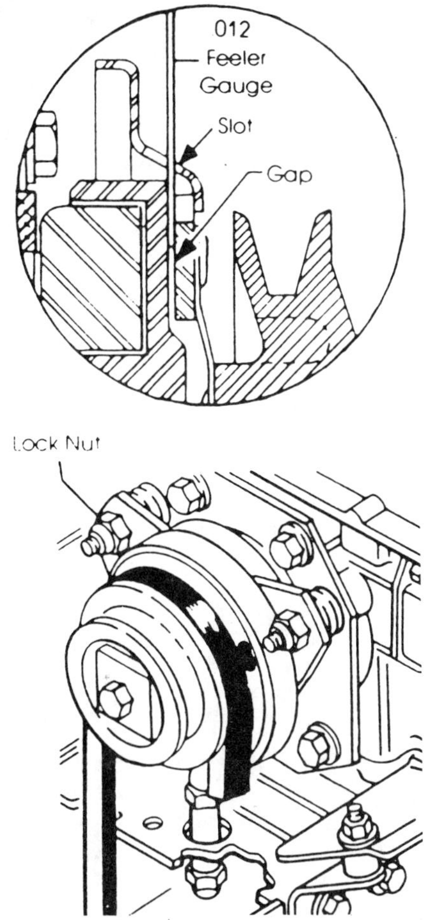

Fig. F29—View showing pto clutch on LGT-17 and LGT-17H models. YT-16 clutch is similar, but located at lower side of engine. Upper inset shows feeler gage inserted in slot. Refer to text.

POWER TAKE-OFF

ADJUSTMENT

All Models

The electric pto clutch is attached to the crankshaft at the front of the engine on LGT-17 and LGT-17H models and to the crankshaft at the lower side of the engine on YT-16 models.

Check clutch adjustment by inserting a 0.012 inch (0.305 mm) feeler gage into each of the three "SLOTS" (Fig. F29) in the clutch. Feeler gage should be a snug fit at all locations. If not, adjust locknuts equally (three on LGT-17 and LGT-17H models and four on YT-16 model) until correct clearance is obtained. Check clutch performance and note clutch must release with pto switch in "OFF" position.

R&R AND OVERHAUL

All Models

Remove belt and disconnect electrical connection. Remove cap screw and washers from end of crankshaft and remove clutch-armature assembly. Remove key. Unbolt and remove clutch field assembly.

Disassemble clutch-armature assembly by pressing armature from ball bearing, remove front retaining ring and press bearing from clutch.

Clean and inspect all parts. Normal resistance reading for clutch field winding is 2.75 to 3.60 ohms. If reading is higher or lower, field must be renewed.

Reassemble by reversing disassembly procedure. Tighten field retaining screws to 30 in.-lbs. (3.4 N•m). Tighten clutch-armature retaining cap screw in crankshaft to 20 ft.-lbs. (27 N•m).

HYDRAULIC SYSTEM

PRESSURE CHECK AND ADJUST

Model LGT-17H

Hydraulic system utilizes pressurized oil from hydrostatic drive unit. If hydraulic lift fails to function or lifts slowly, install a 0-1000 psi (7000 kPa) test gage and shut-off valve in line between hydrostatic transmission and tractor hydraulic control valve. With shut-off valve open, start and run engine at full rpm. Slowly begin to close shut-off valve until 800 psi (5516 kPa) is reached. If 800 psi (5516 kPa) cannot be obtained even with shut-off valve completely closed, pump must be renewed or repaired.

To check and adjust control valve pressure relief valve setting, install a

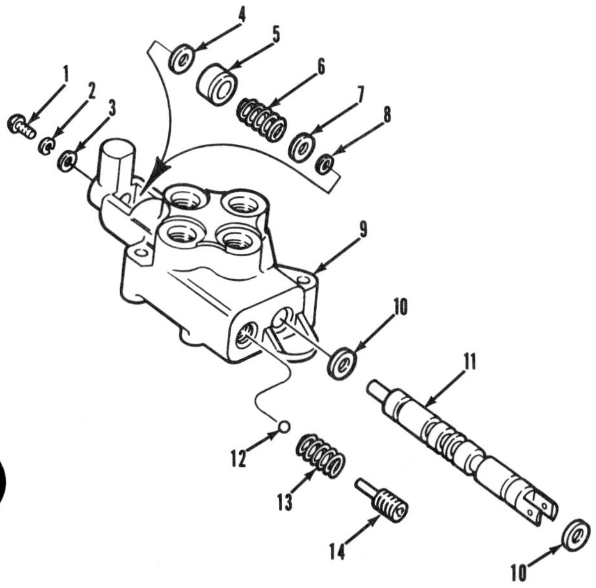

Fig. F30—Exploded view of hydraulic control valve used on LGT-17H models.

1. Screw
2. Washer (spring lock)
3. Washer
4. Washer
5. Spacer
6. Spring
7. Washer
8. Washer
9. Valve body
10. Seal
11. Spool
12. Ball
13. Relief valve spring
14. Relief valve adjustment plug

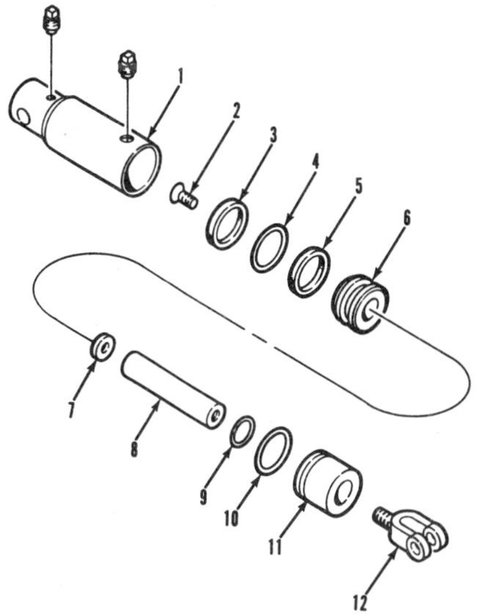

Fig. F31—Exploded view of hydraulic cylinder used on LGT-17H models.

1. Cylinder body
2. Screw
3. Back-up washer
4. "O" ring
5. Back-up washer
6. Piston
7. Rubber disc
8. Shaft
9. "O" ring
10. "O" ring
11. End cap & guide
12. Clevis

0-1000 psi (7000 kPa) test gage and shut-off valve in line running to lift side of cylinder. With shut-off valve open, start and run engine at full rpm. Close shut-off valve and hold lift lever in "UP" position. Adjust relief valve plug (14–Fig. F30) to obtain 800 psi (5516 kPa). Remove gage and shut-off valve.

CONTROL VALVE AND HYDRAULIC CYLINDER

Model LGT-17H

To disassemble and service control valve, refer to Fig. F30. Renew all "O" rings and sealing washers during reassembly.

To disassemble and service hydraulic cylinder, refer to Fig. F31. Renew all "O" rings and back-up washers during reassembly.

GILSON BROTHERS CO.

CONDENSED SPECIFICATIONS

	MODELS				
	52084	52084A	53042	53043	53044
Engine Make	B&S	B&S	B&S	B&S	B&S
Model	402707	402707	401417	401417	401417
Bore	3.44 in (87.4 mm)	3.44 in. (87.4 mm)	3.44 in. (87.4 mm)	3.44 in. (87.4 mm)	3.44 in. (87.4 mm)
Stroke	2.16 in. (54.9 mm)	2.16 in. (54.9 mm)	2.16 in. (54.9 mm)	2.16 in. (54.9 mm)	2.16 in. (54.9 mm)
Displacement	40 cu. in. (659 cc)	40 cu. in. (659 cc)	40 cu. in. (659 cc)	40 cu. in. (659 cc)	40 cu. in. (659 cc)
Power Rating	16 hp. (11.9 kW)	16 hp. (11.9 kW)	16 hp. (11.9 kW)	16 hp. (11.9 kW)	16 hp. (11.9 kW)
Slow Idle	1350 rpm	1350 rpm	1350 rpm	1350 rpm	1350 rpm
High Speed (No-Load)	3600 rpm	3600 rpm	3600 rpm	3600 rpm	3600 rpm
Capacities –					
Crankcase	3 pts. (1.4 L)	3 pts. (1.4 L)	3 pts. (1.4 L)	3 pts. (1.4 L)	3 pts. (1.4 L)
Hydraulic System	.....	.....	.....	.....	6 qts. (5.7 L)
Transaxle or Transmission	36 oz. (1065 mL)	36 oz. (1065 mL)	64 oz. (1893 mL)	64 oz. (1893 mL)	See Hyd.
Bevel Gear Box	.....	.....	4 oz. (118 mL)	4 oz. (118 mL)	.....

	MODELS				
	53045	53054	53055	53073A	53074A
Engine Make	B&S	B&S	B&S	B&S	B&S
Model	401417	421417	421417	422707	422707
Bore	3.44 in. (87.4 mm)	3.44 in. (87.4 mm)	3.44 in. (87.4 mm	3.44 in. (87.4mm)	3.44in. (87.4 mm)
Stroke	2.16 in. (54.9 mm)	2.28 in. (57.9 mm)	2.28 in. (57.9 mm)	2.28 in. (57.9 mm)	2.28 in. (57.9 mm)
Displacement	40 cu. in. (659 cc)	42.3 cu. in. (694 cc)	42.3 cu. in. (694 cc)	42.3 cu. in. (694 cc)	42.3 cu. in. (694 cc)
Power Rating	16 hp. (11.9 kW)	18 hp. (13.4 kW)	18 hp. (13.4 kW)	18 hp. (13.4 kW)	18 hp. (13.4 kW)
Slow Idle	1350 rpm	1400 rpm	1400 rpm	1400 rpm	1400 rpm
High Speed (No-Load)	3600 rpm	3600 rpm	3600 rpm	3600 rpm	3600 rpm
Capacities –					
Crankcase	3 pts. (1.4 L)	3 pts. (1.4 L)	3 pts. (1.4 L)	3 pts. (1.4 L)	3 pts. (1.4 L)
Hydraulic System	6 qts. (5.7 L)	6 qts. (5.7 L)	6 qts. (5.7 L	....	6 qts. (5.7 L)
Transaxle or Transmission	See Hyd.	See Hyd.	See Hyd.	64 oz. (1893 mL)	See Hyd.
Bevel Gear Box	.....	.....	.....	4 oz. (118 mL)	.....

FRONT AXLE AND STEERING SYSTEM

MAINTENANCE

All Models

It is recommended that steering spindles, tie rod ends, axle pivot, wheel bearings/bushings and steering gear be lubricated at 25 hour intervals. Use multi-purpose, lithium base grease. Clean all fittings before and after lubrication. Check assemblies for looseness or wear and repair as necessary.

R&R AXLE MAIN MEMBER

Models 52084-52084A-53073A-53074A

Raise and support front of tractor. Disconnect drag link end (8–Fig. G1) from steering arm (10). Disconnect any implement which may be attached to pivot bracket (7). Support axle and remove bolts connecting pivot bracket (7) to main member (5). Remove pivot bolt (4) and remove main member assembly, then pivot bracket.

Reinstall by reversing removal procedure.

Models 53042-53043-53044-53045-53054-53055

Raise and support front of tractor. Disconnect drag link attached to left spindle. Loosen set screw (14–Fig. G2), support axle and remove pivot shaft (6). Lower axle and roll it out from under tractor.

Reinstall by reversing removal procedure.

TIE ROD AND TOE-IN

Models 52084-52084A-53073A-53074A

Tie bar (1–Fig. G1) is made to correct specification by manufacturer and is non-adjustable. If steering becomes loose or tractor wanders or darts, check condition of tie bar, bolts and spacers, spindles, bushings, wheel bearings and drag link. Repair as necessary.

Models 53042-53043-53044-53045-53054-53055

Tie rod ends (10 and 13–Fig. G2) should be adjusted equally to provide 1/8 to 1/4-inch (3.175 to 6.35 mm) toe-in. Retighten jam nuts. If steering becomes loose or tractor wanders or darts, check condition of tie rod ends, spindles,

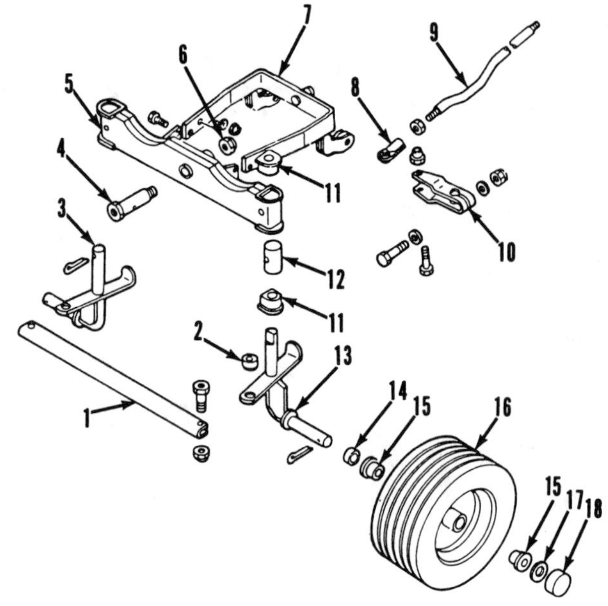

Fig. G1–Exploded view of front axle used on 52084, 52084A, 53073A and 53074A models.

1. Tie bar
2. Pivot bushing
3. Right spindle
4. Pivot bolt
5. Axle main member
6. Nut
7. Pivot bracket
8. Drag link end
9. Drag link
10. Steering arm
11. Bushing
12. Spacer
13. Left spindle
14. Spacer
15. Bushing
16. Tire & hub
17. Washer
18. Dust cap

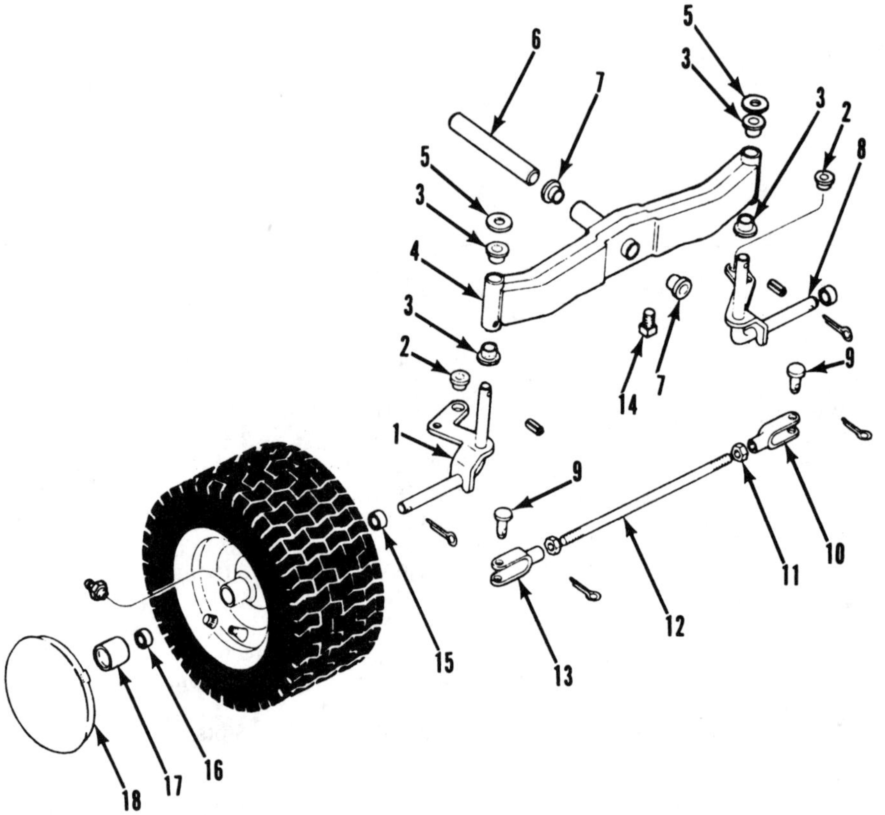

Fig. G2–Exploded view of front axle used on 53042, 53043, 53044, 53045, 53054 and 53055 models.

1. Spindle
2. Bushings
3. Bushings
4. Axle main member
5. Washer
6. Pivot pin
7. Bushings
8. Spindle
9. Pin
10. Tie rod end
11. Jam nut
12. Tie rod
13. Tie rod end
14. Setscrew
15. Bearing
16. Bearing
17. Dust cap
18. Hub cap

bushings, wheel bearings and drag link. Repair as necessary.

R&R STEERING SPINDLES

Models 52084-52084A-53073A-53074A

Raise and support front of tractor. Remove dust cap (18-Fig. G1) and cotter key. Remove washer (17), then slide wheel and hub assembly off spindle. Remove tie bar (1). Remove steering arm (10) from left spindle (13) or cotter key from right spindle (3). Slide spindle down out of axle main member.

Renew bushings (11), spacer (12) or spindle (3 or 13) as necessary.

Reinstall by reversing removal procedure.

Models 53042-53043-53044-53045-53054-53055

Raise and support front of tractor. Remove hub cap (18-Fig. G2). Remove cotter key, then remove wheel and hub assembly. Disconnect tie rod and if removing left spindle, disconnect drag link. Remove roll pin and slide spindle down out of axle main member.

Renew bushings (3), washers (5) or spindles as necessary.

Reinstall by reversing removal procedure.

FRONT WHEEL BEARINGS

All Models

According to model, front wheels may be equipped with either bushings or wheel bearings. Bushings or bearings are pressed into wheel hubs and should be renewed or hubs repaired to correct excessuive looseness. On models equipped with wheel bearings, note sealed bearing goes to the outside location (16–Fig. G2) and unsealed bearing goes to the inner location (15).

STEERING GEAR

Models 52084-52084A-53073-53074A

R&R AND OVERHAUL STEERING GEAR. Disconnect drag link end (17–Fig. G3) at steering sector (18). Remove steering wheel (4) and adapter (5). Remove bellows cover (7) and sleeve (8). Remove bearing bolts (13 and 19). Remove steering shaft (11). Remove "E" ring (20) and steering sector (18).

Reinstall by reversing removal procedure. Push steering shaft and flange bearing at lower end of steering shaft toward sector gear to provide full gear mesh before tightening bolts (13). Steering shaft should rotate smoothly with minimum backlash. If binding is encountered, loosen bolts (13) and move shaft away from sector gear (18) a small amount. Retighten bolts and recheck steering.

Models 53042-53043-53044-53045-53054-53055

R&R AND OVERHAUL STEERING GEAR. Disconnect drag link end (20–Fig. G4) from quadrant gear (11). Remove nuts (18 and 19) and unbolt supports (14) from frame. Lower assembly and withdraw shaft (13). Spacer (17), quadrant gear (11) and bushing (12) can be removed as shaft is withdrawn. Drive out roll pin (10), then remove pinion gear (9) and bracket (16) with bushings (8). Remove cap screw (1) and steering wheel (2), then unbolt bearing (4) and remove shaft (7) and bearing assembly.

Reassemble by reversing disassembly procedure.

To adjust steering gear play, loosen setscrew (15) and locknut (19). Turn adjusting nut (18) until steering wheel turns freely with minimum of play. Retighten locknut (19) and setscrew (15).

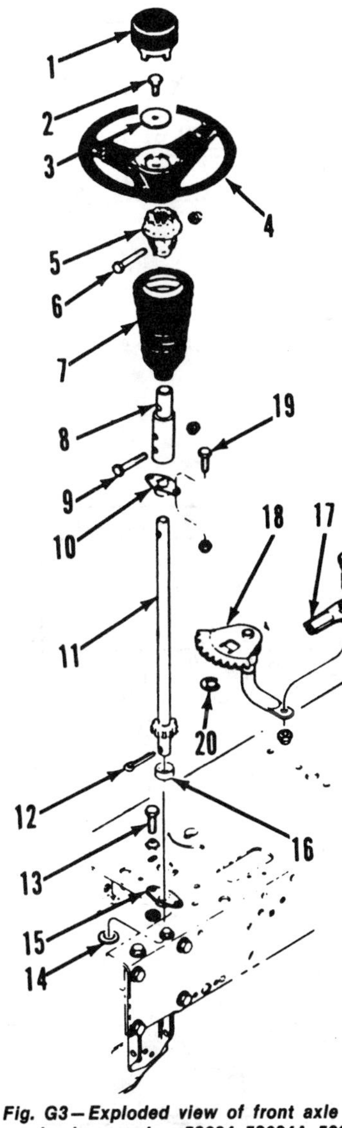

Fig. G3—Exploded view of front axle steering mechanism used on 52084, 52084A, 53073A and 53074A models.

1. Dust cover
2. Bolt
3. Washer
4. Steering wheel
5. Adapter
6. Bolt
7. Bellows
8. Sleeve
9. Bolt
10. Flange bearing
11. Steering shaft
12. Cotter key
13. Bolt
14. Washer
15. Flange bearing
16. Spacer
17. Drag link end
18. Steering sector
19. Bolt
20. "E" ring

ENGINE

MAINTENANCE

All Models

Regular engine maintenance is required to maintain peak performance and long engine life.

Check oil level and clean air intake screen at 5 hour intervals under normal operating conditions.

Clean engine air filter at 25 hour intervals and clean crankcase breather and engine cooling fins and lubricate governor linkage at 200 hour intervals.

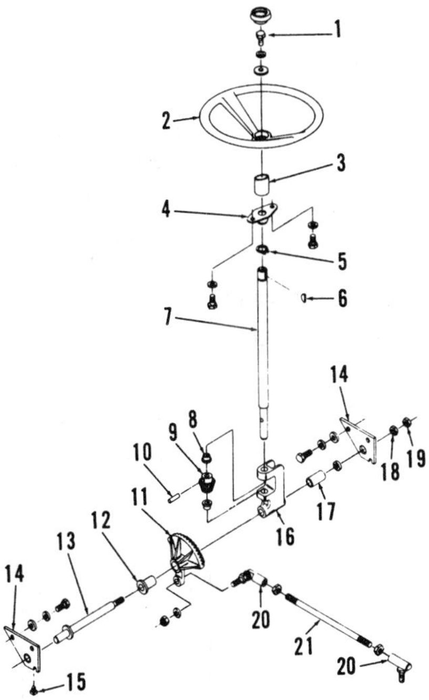

Fig. G4—Exploded view of steering gear used on 53042, 53043, 53044, 53045, 53054 and 53055 models.

1. Cap screw
2. Steering wheel
3. Sleeve
4. Flange bearing
5. Snap ring
6. Key
7. Steering shaft
8. Bushing
9. Pinion gear
10. Roll pin
11. Quadrant gear
12. Bushing
13. Shaft
14. Support
15. Setscrew
16. Bracket
17. Spacer
18. Adjusting nut
19. Locknut
20. Ball joint ends
21. Drag link

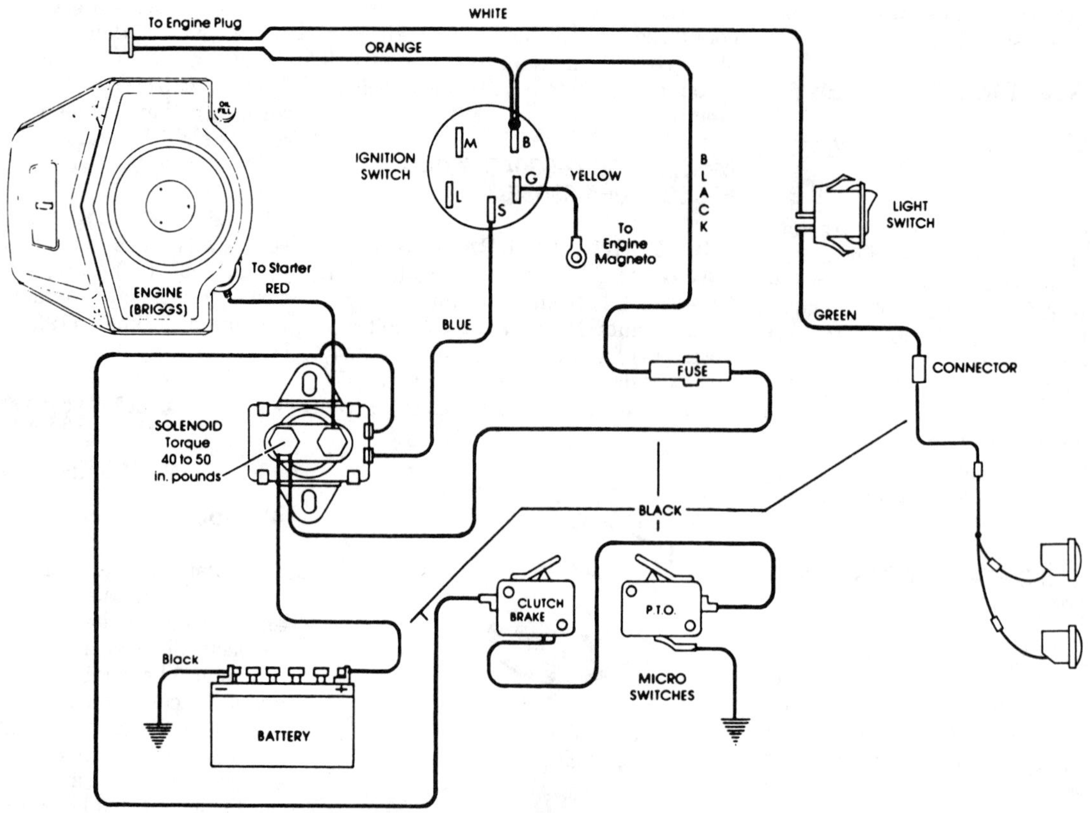

Fig. G9 – Wiring diagram for 52084 models.

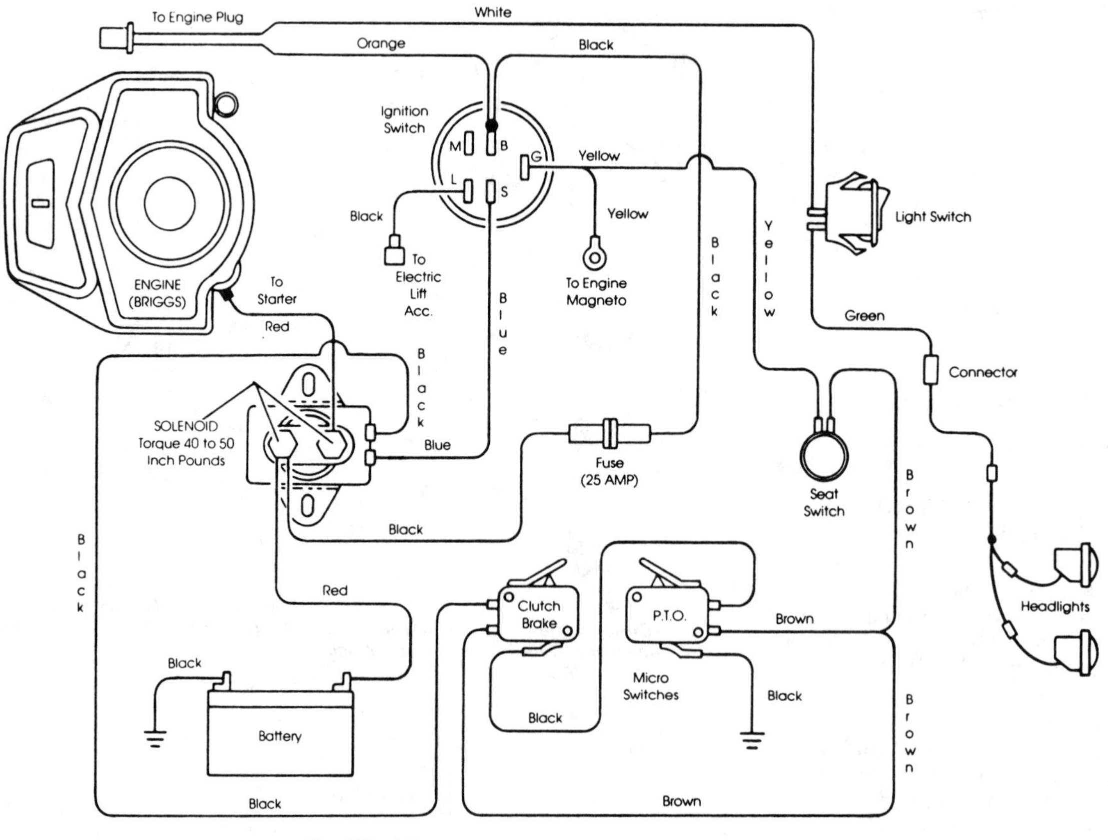

Fig. G10 – Wiring diagram for all models except 52084.

Change engine oil and filter, perform tune-up, valve adjustment and clean carbon from cylinder heads as recommended by engine manufacturer.

REMOVE AND REINSTALL

All Models

Disconnect negative battery cable. Remove hood and side panels. Disconnect pto belts and drive belt or drive shaft, as equipped. Remove exhaust system and deflectors as necessary. Disconnect fuel line and all electrical connections. Remove engine mounting bolts and remove engine.

Reinstall by reversing removal procedure.

OVERHAUL

All Models

Engine make and model are listed at the beginning of this section. To overhaul engine or accessories, refer to appropriate engine section of this manual.

ELECTRICAL SYSTEM

MAINTENANCE AND SERVICE

All Models

Battery electrolyte level should be checked at 25 hour intervals of normal operation. If necessary, add distilled water until level is just below base of vent wells. **DO NOT** overfill. Keep battery posts clean and cable ends tight.

For alternator or starter service, refer to appropriate engine section in this manual.

Refer to Fig. G9 for 52084 models wiring diagram and to Fig. G10 for wiring diagram for all remaining models.

BRAKE

ADJUSTMENT

All Models

Brake assembly on all models is located at left side of transaxle. Adjust brake by loosening jam nut or removing cotter key (as equipped) and tightening adjusting nut, or castelated nut (Fig. G11) until a light drag is felt as tractor is pushed forward by hand. Loosen nut slightly and reinstall cotter key or tighten jam nut against adjustment nut, as equipped.

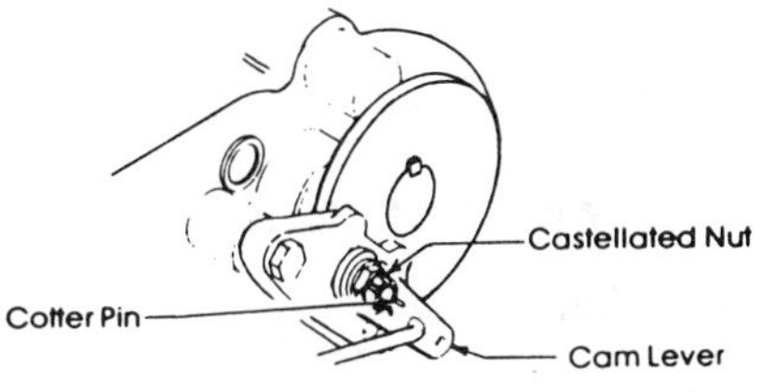

Fig. G11—View of typical disc brake system similar to unit on all models. Refer to text for adjustment procedure.

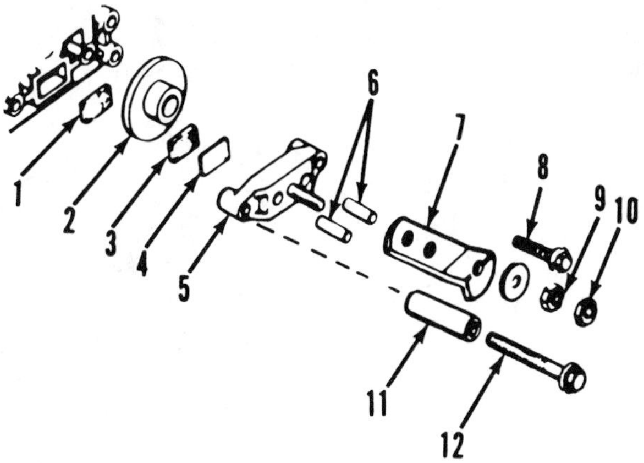

Fig. G12—Exploded view of disc brake unit used on 52084 and 52084A models.

1. Inner brake pad
2. Brake disc
3. Outer brake pad
4. Metal backing plate
5. Brake caliper
6. Pins
7. Brake lever
8. Bolt
9. Adjusting nut
10. Jam nut
11. Spacer
12. Bolt

NOTE: On hydrostatic drive models, brakes must not engage before transmission is in neutral. Check neutral adjustment on transmission linkage.

R&R BRAKE PADS

Models 52084-52084A

Remove jam nut (10–Fig. G12) and adjustment nut (9). Remove brake lever (7). Remove bolts (12 and 8) and spacer (11). Remove brake caliper (5), outer brake pad (3), metal backing plate (4), brake disc (2) and inner brake pad (1).

Make certain the two push pins (6) and metal backing plate (4) are in place in brake caliper (5) and metal spacer (11) is in place on rear mounting bolt during reassembly.

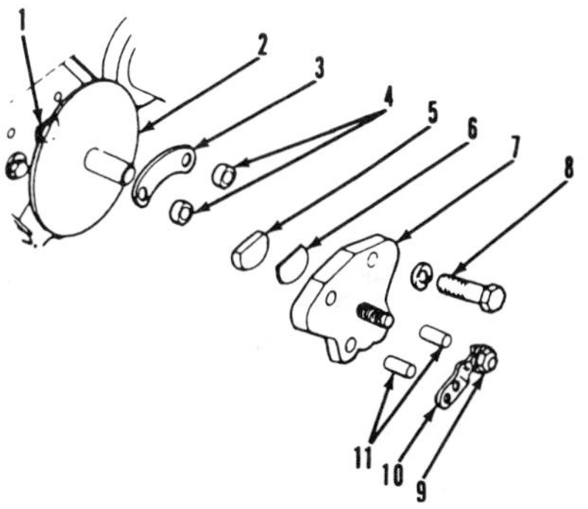

Fig. G13—Exploded view of disc brake unit used on all models except 52084 and 52084A. Refer to Fig. G12.

1. Inner brake pad
2. Brake disc
3. Spacer
4. Spacers
5. Outer brake pad
6. Metal backing plate
7. Brake caliper
8. Bolt
9. Adjusting nut
10. Brake lever
11. Pins

All Other Models

Remove brake rod and brake actuating lever (10–Fig. G13). Remove the two cap screws (8) securing disc brake assembly and hold assembly together as it is removed. Note location of spacers (3 and 4) and remove. Remove worn outer brake pad (5), metal backing plate (6) and the two actuating pins (11) from caliper (7). Remove brake disc (2) and inner brake pad (1).

Make certain metal backing plate and outer brake pad are firmly seated in caliper and inner brake pad is seated in its recess. Reassemble by reversing disassembly procedure. Make certain spacers (3 and 4) are in their correct locations.

CLUTCH

MAINTENANCE AND ADJUSTMENT

Models 53044-53045-53054-53055-53074A

Brake/clutch pedal is connected to hydrostatic transmission control valve. Refer to TRANSMISSION section for correct neutral position adjustment.

All Other Models

Brake/clutch pedal is connected to a

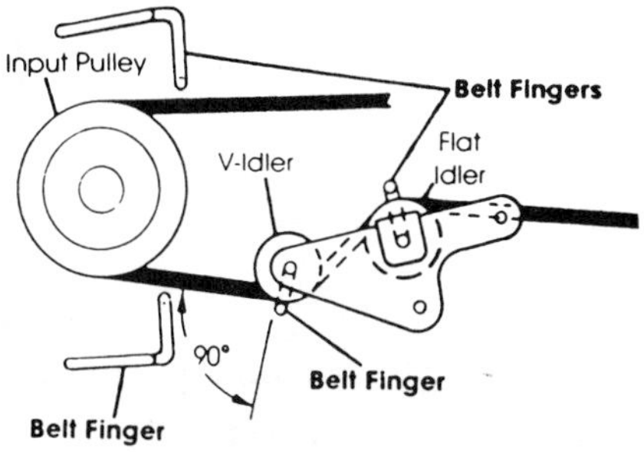

Fig. G14—View showing correct belt finger adjustment for 52084 and 52084A models drive belt.

spring loaded, pivoting idler pulley. Belt tension is maintained by spring pressure. Periodically check condition of pulleys and location of belt guides and fingers. Refer to R&R DRIVE BELT section.

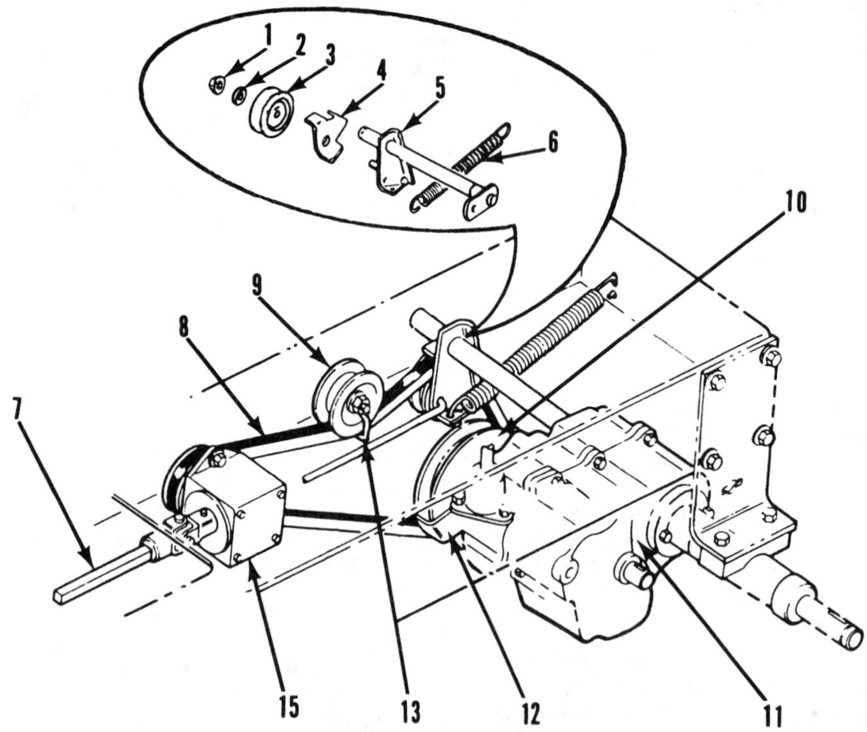

Fig. G15—View showing drive belt and pulley arrangement and component parts for all models except 52084, 52084A and 53074A models. Refer to Fig. G14 and G15A.

1. Flange nut
2. Washer
3. V-idler pulley
4. V-idler belt guide
5. Pivot shaft
6. Spring
7. Drive shaft
8. Drive belt
9. Flat idler pulley
10. Input pulley
11. Transaxle
12. Belt guide
13. Belt finger
15. Bevel gear box

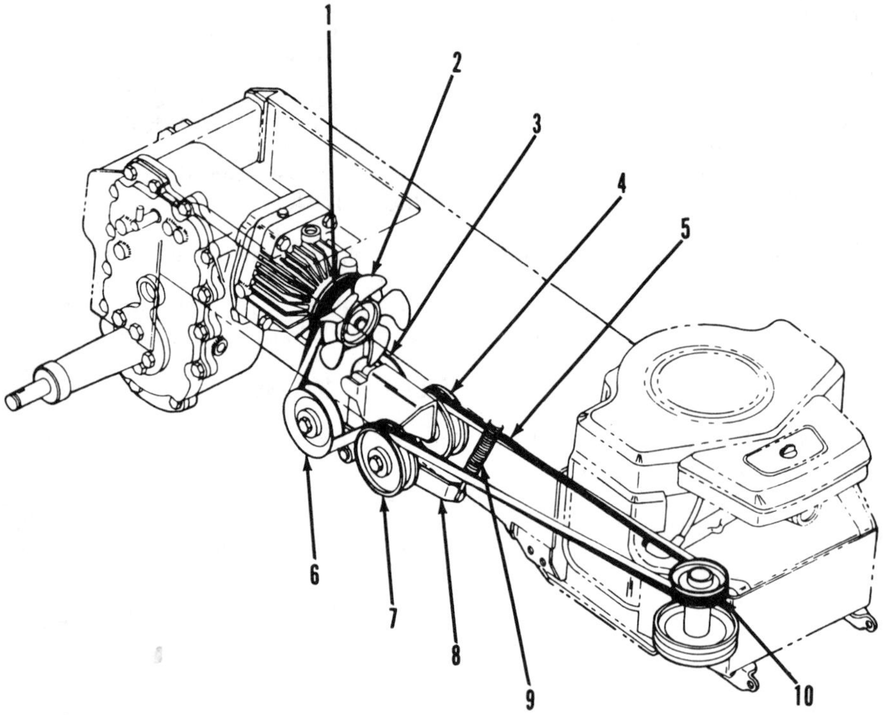

Fig. G15A—View showing hydrostatic drive belt and idler pulley arrangement on 53074A model.

1. Hydrostatic unit drive pulley
2. Cooling fan
3. Fixed "V" idler
4. Fixed flat idler
5. Drive belt
6. Spring loaded "V" idler
7. Spring loaded flat idler
8. Idler arm assy.
9. Spring
10. Engine pulley

R&R DRIVE BELT

Model 52084-52084A

Disconnect and remove any attachments from tractor. Set parking brake. Remove flat idler pulley, V-idler pulley and wire belt fingers from pivot bracket. See Fig. G14. Disconnect shift rod from gear select lever on transaxle. Remove belt.

Reinstall by reversing removal procedure. Block tractor wheels and release parking brake and adjust all belt fingers and guides as shown in Fig. G14. Fingers should be close to, but not touch belt or pulley.

Models 53042-53043-53073A

Block wheels so tractor cannot move. Make certain parking brake is disengaged and remove flange nut (1–Fig. G15). Engage parking brake and lock in position. Slide V-idler pulley (3) outward until belt is clear of guide (4). Remove flat idler pulley (9) and belt finger (13) noting sequence in which they are installed. Remove belt.

Reinstall by reversing removal procedure. With parking brake disengaged, adjust all belt fingers and guides so they just have clearance between belt and/or pulleys.

Model 53074A

Model 53074A is equipped with a belt driven hydrostatic drive transmission. A spring loaded idler pulley assembly maintains correct belt tension. To renew belt, remove any attachment which may be mounted to tractor that would make belt renewal difficult. Remove engine pulley belt cover. Disconnect spring (9–Fig. G15A) from idler lever assembly (8). Remove belt from V-idlers and flat idlers. Loosen belt around transmission input pulley and carefully work belt over the flexible fan blades (2) and remove belt.

Reinstall by reversing removal procedure.

All Other Models

All other models are equipped with hydrostatic drive transmissions which are driven by a driveshaft connected to engine crankshaft.

TRANSMISSION

MAINTENANCE

Models 52084-52084A

Transaxle type transmission is packed with 36 ounces (1065 mL) of EP lithium base grease during manufacture and, unless the case is damaged or leaking, this should provide lifetime lubrication.

Models 53042-53043-53073A

Transaxle type transmission fluid level should be checked at 100 hour intervals. Maintain fluid level at level check plug (Fig. G16). Fluid capacity is 64 ounces (1893 mL) and recommended fluid is SAE 140 EP gear lubricant.

Models 53044-53045-53054-53055-53074A

Hydrostatic transmission and hydraulic system fluid is drawn from the two speed range transmissions gear case. Fluid level should be checked at 100 hour intervals and fluid and filter should be changed at 250 hour intervals. Maintain fluid level at "FULL" mark on dipstick (Fig. G17). Fluid capacity is 6 quarts (5.7 L) and recommended fluid is SAE 20 detergent oil.

ADJUSTMENTS

Models 53044-53045-53054-53055-53074A

NEUTRAL POSITION ADJUSTMENTS. If tractor "creeps" forward or backward after brake/clutch pedal is depressed fully and then released, transmission control lever neutral position should be adjusted.

To adjust control lever neutral position, loosen jam nuts on turnbuckle (Fig. G18). If tractor creeps backward, turn turnbuckle 1/6-turn clockwise. If tractor creeps forward, turn turnbuckle 1/6-turn counter-clockwise. Retighten

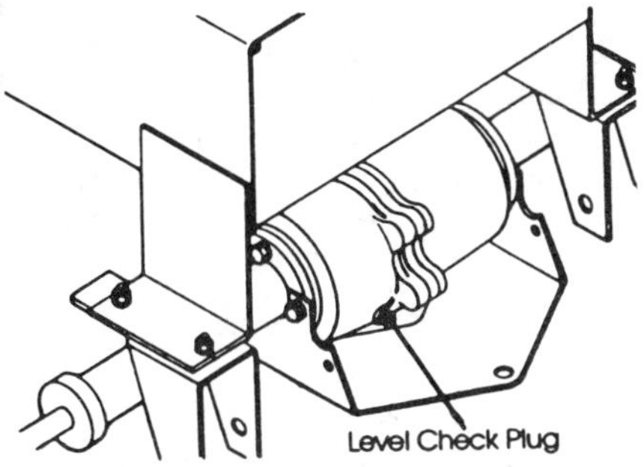
Level Check Plug

Fig. G16—View showing location of fluid level check plug for transaxle used on 53042, 53043 and 53073A models.

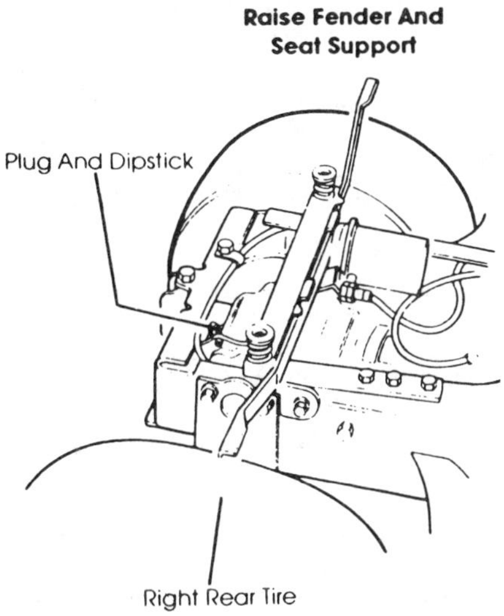
Raise Fender And Seat Support
Plug And Dipstick
Right Rear Tire

Fig. G17—View showing location of dipstick used to check fluid level of transmission and hydraulic system on 53044, 53045, 53054, 53055 and 53074A models.

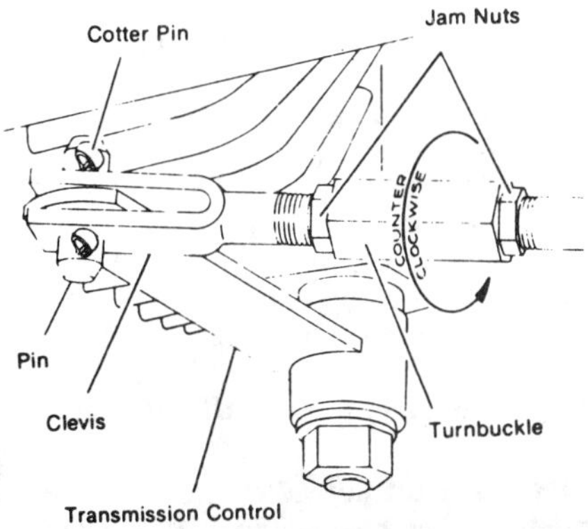
Cotter Pin
Jam Nuts
COUNTER CLOCKWISE
Pin
Clevis
Turnbuckle
Transmission Control

Fig. G18—View of neutral finger bracket and spring used on 53044, 53045, 53054, 53055 and 53074A models. Refer to text.

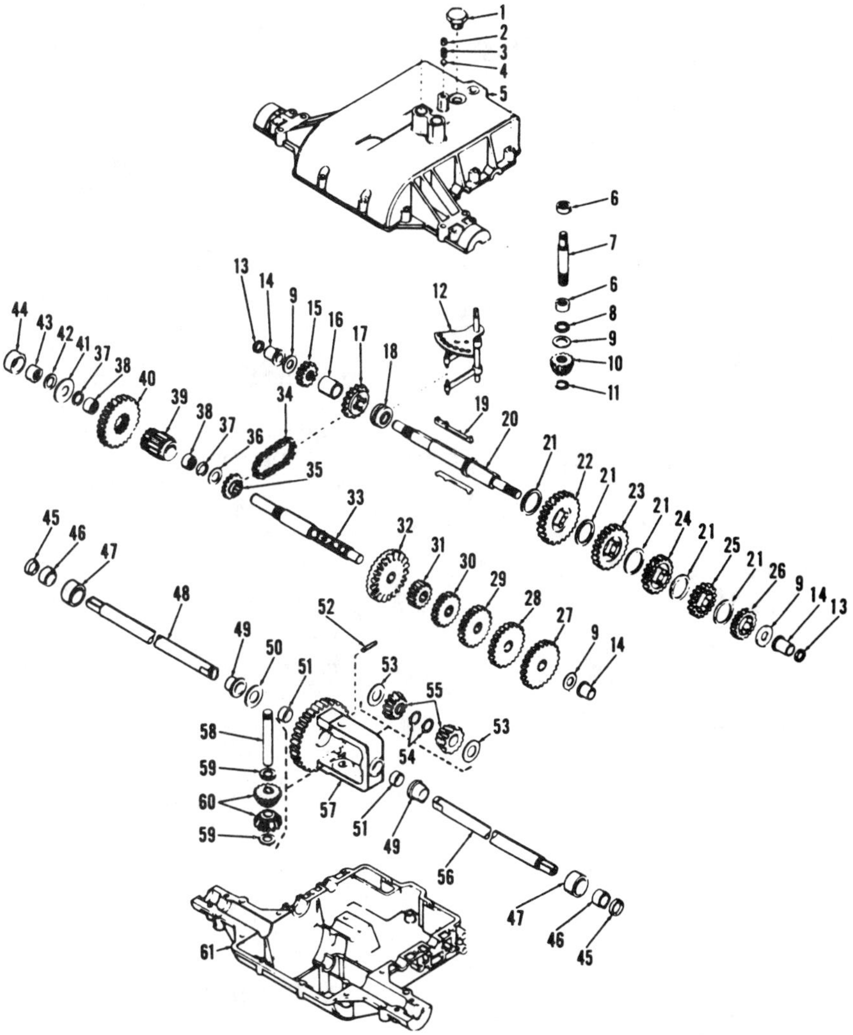

Fig. G20—Exploded view of transaxle assembly used on 52084 and 52084A models.

1. Plug
2. Setscrew
3. Spring
4. Ball
5. Cover
6. Needle bearing
7. Input shaft
8. Square cut ring
9. Thrust washer
10. Input gear
11. Snap ring
12. Shift fork assy.
13. Square cut ring
14. Bushing
15. Spur gear (15T)
16. Spacer
17. Sprocket (18T)
18. Shift collar
19. Key
20. Brake shaft
21. Thrust washer
22. Spur gear (37T)
23. Spur gear (30T)
24. Spur gear (25T)
25. Spur gear (22T)
26. Spur gear (20T)
27. Spur gear (30T)
28. Spur gear (28T)
29. Spur gear (29T)
30. Spur gear (20T)
31. Spur gear (12T)
32. Bevel gear (42T)
33. Countershaft
34. Roller chain
35. Sprocket (9T)
36. Flat washer
37. Square cut seal
38. Needle bearing
39. Output pinion
40. Output gear
41. Flat washer
42. Square cut seal
43. Needle bearing
44. Spacer
45. Oil seal
46. Needle bearing
47. Spacer
48. Axle shaft
49. Bushing
50. Washer
51. Bushing
52. Pin
53. Thrust washer
54. Snap ring
55. Bevel gear
56. Axle shaft
57. Differential gear
58. Drive pin
59. Thrust washer
60. Bevel pinion
61. Case

jam nuts against turnbuckle. Start engine and check for creep. Readjust as necessary until all creep is removed.

CAUTION: Do not make adjustments with engine running.

R&R TRANSMISSION

Models 52084-52084A

Raise and support rear of tractor. Depress brake/clutch pedal and lock in position. Disconnect brake rod and gear selector lever. Remove drive belt. Support transaxle and remove torque bracket (support strap) at front of left side and remove transaxle retaining bolts at rear axle housings. Lower transaxle and roll out from under tractor.

Reinstall by reversing removal procedure. Refer to R&R DRIVE BELT section for drive belt installation.

Models 53042-53043-53073A

Raise and support rear of tractor. Disconnect brake rod and remove drive belt. Remove the three bolts retaining shift lever and remove lever assembly. Support transaxle (transmission), then remove bolts retaining transaxle and brake support plate on left side of transaxle. Remove retaining bolts at rear axle housings and lower and remove transaxle assembly.

Reinstall by reversing removal procedure. Refer to R&R DRIVE BELT section for drive belt installation.

Models 53044-53045-53054-53055-53074A

Remove seat/fender assembly and shift plate. Disconnect clevis from hydraulic control lever assembly at hydrostatic unit. On 53074A model, remove drive belt. On all other models, loosen setscrews and slide fan hub assembly toward transmission. On all models, disconnect all hydraulic hoses and cap openings. Support transmission and loosen the two large side mount bolts. Remove cone locknuts from the four axle bolts and pull bolts out of axle mount assemblies. Remove the two large side mount bolts. Lift frame assembly off and away from transmission assembly.

Reinstall by reversing removal procedure. Fill to proper level with recommended fluid. Position hydrostatic control lever in neutral position, then start and run engine at idle speed. Operate transmission in forward and reverse direction for short distances. Stop engine and check fluid level. Repeat procedure until all air has been bled from system. Adjust hydrostatic transmission control lever as outlined in ADJUSTMENT section. Adjust brake.

OVERHAUL

Models 52084-52084A

Remove transaxle assembly as outlined. Place shift lever in neutral, then unbolt and remove shift lever. Remove setscrew (2 – Fig. G20), spring (3) and index ball (4). Unbolt cover (5) and push shift fork assembly (12) in while removing cover. Before removing gear and shaft assemblies, shifter for (12) should be removed. Note position of parts before removal. Remove gear and shaft assemblies (Fig. G21) from case taking care not to disturb drive chain (34 – Fig. G20). Remove needle bearings (38 and 43), flat washer (41), square cut seals (37 and 42), output gear (40) and output pinion (39) from countershaft. Angle the two shafts together, then mark position of chain on sprocket collars and remove chain. Remove sprocket (35), bevel gear (32), spur gears (27, 28, 29, 30 and 31), thrust washer (9) and flange bushing (14). All gears are splined to the countershaft. Disassembly of shifter-brake shaft is apparent after inspection. Remove snap ring (11), input bevel gear and pull input shaft (7) through cover.

Disassemble differential by driving roll pin securing drive pin (58) out. Remove pinion gears (60) by rotating gears in opposite directions. Remove snap rings (54), side gears (55) and thrust washers (53), then slide axles out. Note axle shafts (48 and 56) are different lengths.

Clean and inspect all parts and renew any showing excessive wear or other damage. When installing new inner input shaft needle bearings, press bearing in to a depth of 0.135-0.150 inch (3.429-3.81 mm) below flush. When installing thrust washers and shifting gears on brake shaft, the 45 degree chamfer on inside of diameter thrust washers must face shoulder on brake shaft. See Fig. G22. Flat side of gears must face shoulder on shaft. Complete reassembly and pack case with approximately 36 ounces (1065 mL) of EP lithium base grease. Tighten case to cover cap screws to 90-100 in.-lbs. (10-11 N·m).

Models 53042-53043-53073A

Remove transaxle assembly as outlined. Remove drain plug and drain lubricant. Remove rear wheel and hub assemblies. Remove brake caliper and brake disc. Place shift lever (1 – Fig. G23) in neutral position, then unbolt and remove shift lever assembly. Remove axle housings (14 and 64) and remove seal retainers (11) with oil seals (12) and "O" rings (13) by pulling each axle shaft out of case and cover as far as possible. Place transaxle unit on the edge of a bench with left axle shaft pointing downward. Remove cap screws securing case (16) to cover (66) and drive aligning dowel pins out of case. Lift case (16) up 1½ to 2 inches (40 to 50 mm), tilt case about 45 degrees, rotate case clockwise and remove it from the assembly. Input shaft (32) and input gear (33) will be removed with case. Withdraw differential and axle shaft assembly and lay aside for later disassembly. Remove the three-cluster gear (44) with its thrust washer (46) and spacer (42). Lift out reverse idler gear (25), spacer (24) and shaft (23). Hold upper ends of shifter rods together and lift out shifter rods, forks, shifter stop (21), sliding gears (30 and 31) and shaft (28) as an assembly. Remove low reduction gear (57), reduction shaft (56) and thrust washer (55), then remove the two-cluster gear (40) from brake shaft. Lift out the output gear (50), shaft (51) and thrust washers (49 and 52). To remove brake shaft (39) and gear (38) from cover (66), block up under gear (38) and press shaft out of gear.

CAUTION: Do not allow cover or low reduction gear bearing boss to support any part of pressure required to press brake shaft from gear.

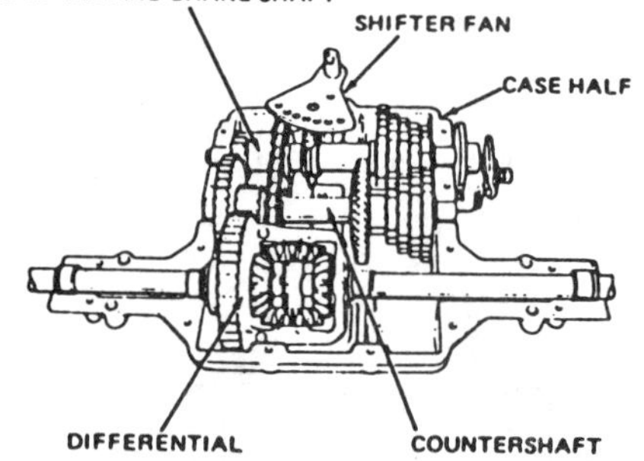

Fig. G21 – When disassembling transaxle used on 52084 and 52084A models, remove gear and shaft assemblies from case by lifting both shafts at once. Take care not to disturb chain.

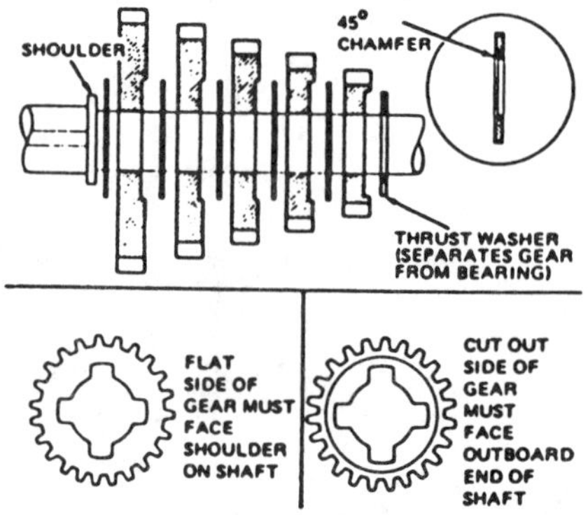

Fig. G22—When assembling shift shaft, install thrust washers with 45 degree chamfer toward shoulder on output shaft. Flat side of gears also must face shoulder on output shaft.

Remove input shaft (32) with input gear (33) and thrust washer (34) from case (16).

To disassemble differential, remove four cap screws and separate axle shaft and carriage assemblies from ring gear (79). Drive blocks (78), bevel pinion gears (77) and drive pin (76) can now be removed from ring gear. Remove snap rings (59) and withdraw axle shafts (63 and 67) from axle gears (61) and carriages (62 and 72).

Clean and inspect all parts and renew any showing excessive wear or other damage. When installing new needle bearings, press bearing (29) in spline shaft (28) to a depth of 0.010 inch (0.254 mm) below end of shaft and low reduction shaft bearings (54 and 58) 0.010 inch (0.254 mm) below thrust surfaces of bearing bases. Carrier bearings (10) should be pressed in from inside of case and cover until bearings are 0.290 inch (7.366 mm) below face of axle housing mounting surface. All other needle bearings are to be pressed in from inside of case and cover to a depth of 0.015-0.020 inch (0.381-0.508 mm) below thrust surfaces.

Renew all seals and gaskets. Reassemble by reversing disassembly procedure. When installing brake shaft (39) and idler gear (38), beveled edge of gear teeth must be up away from cover. Install reverse idler shaft (23), spacer (24) and reverse idler gear (25) with rounded end of gear teeth facing spacer. Install input gear (33) and shaft (32) so chamfered side of input gear is facing case (16).

Tighten transaxle cap screws to the following torque:

Differential cap screws .... 7 ft.-lbs. (9 N·m)
Case to cover cap screws ..... 10 ft.-lbs. (14 N·m)
Axle housing cap screws ..... 13 ft.-lbs. (18 N·m)
Shift lever housing cap screws .... 10 ft.-lbs. (14 N·m)

Fill transaxle, after unit is installed on tractor, to level plug opening with SAE 140 EP gear lubricant. Capacity is approximately 64 ounces (1893 mL).

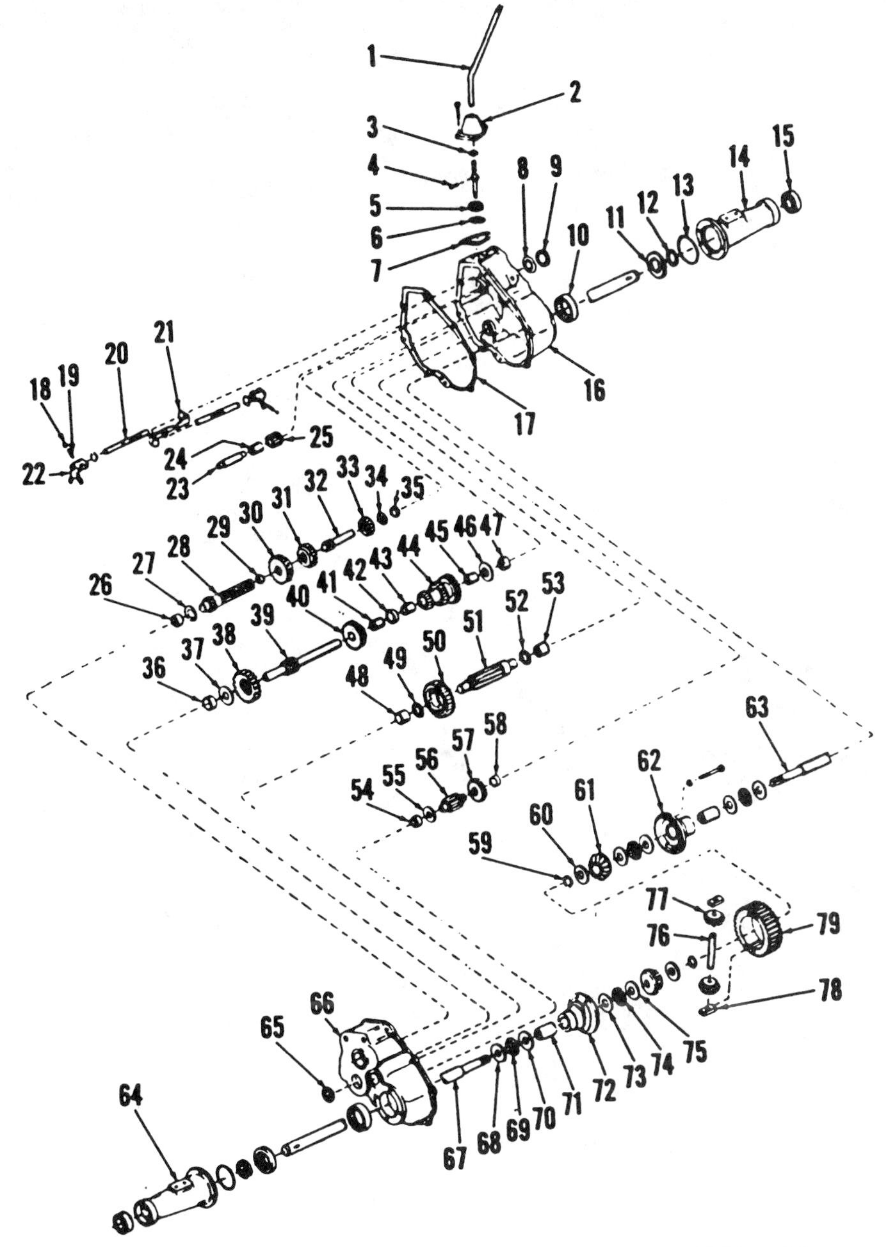

Fig. G23 — Exploded view of transaxle used on 53042, 53043 and 53073A models.

1. Shift lever
2. Shift lever housing
3. Seal ring
4. Roll pin
5. Retainer
6. Snap ring
7. Gasket
8. Ball bearing
9. Oil seal
10. Carrier bearing
11. Seal retainer
12. Oil seal
13. "O" ring
14. Axle housing
15. Axle outer bearing
16. Transaxle case
17. Gasket
18. Detent ball
19. Spring
20. Shifter rod
21. Shifter stop
22. Shifter fork
23. Reverse idler shaft
24. Spacer
25. Reverse idler gear
26. Needle bearing
27. Thrust washer
28. Shifter shaft
29. Needle bearing
30. First, second & reverse gear
31. Third & fourth gear
32. Input shaft
33. Input gear
34. Thrust washer
35. Needle bearing
36. Needle bearing
37. Thrust washer
38. Idler gear
39. Brake & cluster shaft
40. Two-cluster gear
41. Bushing
42. Spacer
43. Bushing
44. Three-cluster gear
45. Bushing
46. Thrust washer
47. Needle bearing
48. Needle bearing
49. Thrust washer
50. Output gear
51. Output shaft
52. Thrust washer
53. Needle bearing
54. Needle bearing
55. Thrust washer
56. Low reduction shaft
57. Low reduction gear
58. Needle bearing
59. Snap ring
60. Thrust washer
61. Axle gear
62. Axle carriage (plain holes)
63. Axle shaft
64. Axle housing
65. Oil seal
66. Transaxle cover
67. Axle shaft
68. Thrust washer
69. Thrust bearing
70. Thrust washer
71. Bushing
72. Axle carriage (tapped holes)
73. Thrust washer
74. Thrust bearing
75. Thrust washer
76. Drive pin
77. Bevel pinion gear
78. Drive block
79. Ring gear

Models 53044-53045-53054-53055-53074A

HYDROSTATIC DRIVE UNIT. Remove transmission as outlined and separate hydrostatic drive unit from range transmission. Thoroughly clean exterior of hydrostatic unit.

Remove venting plug (59 – Fig. G24) and plug (18), invert assembly and drain fluid. Place unit in a holding fixture so input shaft is pointing upward. Remove dust shield (1) and snap ring (3). Remove the five cap screws from charge pump body (7). One cap screw is ½-inch (12.7 mm) longer than the others and must be installed in original location (heavy section of pump body). Remove charge pump body (7) with ball bearing (4). Ball bearing and oil seal (6) can be removed from body (7) after first removing retaining ring (2). Remove the six pump rollers (12), snap rings (5, 9 and 11) and charge pump rotor (10). Remove "O" rings (14 and 16) and pump plate (15). Invert drive unit in holding fixture so output shaft is pointing upward. Remove snap ring (65) and output gear (64). Unscrew the two cap screws (60), then turn them in until two threads are engaged. Raise body (57) until it contacts the heads of cap screws (60). Insert a fork tool between motor rotor (53) and pintle (48) until the tool extends beyond opposite side.

NOTE: The special fork tool can be fabricated from a piece of 1/8-inch flat stock approximately 3 inches wide and 12 inches long. Cut a slot 1-9/16 inches wide and 8 inches long. Taper ends of prongs. Refer to Fig. G25.

Remove cap screws (60–Fig. G24) and by raising ends of forked tool, lift off body and motor assembly. Place body and motor assembly on a bench or in a holding fixture with output shaft pointing downward. Remove special fork tool and place a wide rubber band around motor rotor to hold ball pistons (51) and springs (52) in their bores. Carefully remove motor rotor assembly and lay aside for later disassembly. Remove motor race (56) and output shaft (55). Remove retainer (63), bearing (62) and oil seal (61).

With housing assembly (22) resting in holding fixture (input shaft pointing downward), remove pintle assembly (48).

CAUTION: Do not allow pump to raise with the pintle as the ball pistons (35) may fall out of rotor (36).

Hold pump in position by inserting a finger through hole in pintle. Remove plug (45), spring (46) and charge pump relief ball (47). To remove directional check valves from pintle (48), drill through pintle with a drill bit that will pass freely through roll pins (41). Redrill the holes from opposite side with a 1/4-inch drill bit. Drive or press roll pins from pintle. Using a 5/16-18 tap, thread the inside of valve bodies (44), then remove valve bodies using a draw bolt or a slide hammer puller. Remove check valve balls (43) and snap rings (42). Do not remove plugs (40).

Remove pump cam ring (39) and pump race (38). Place a wide rubber band around pump rotor to prevent ball pistons (35) from falling out. Carefully remove pump assembly and input shaft (33).

To remove control shaft (25), drill a 11/32-inch hole through aluminum housing (22) directly in line with center line of dowel pin (27). Press dowel pin from control shaft, then withdraw control shaft. Remove oil seal (24). Thread the drilled hole in housing with a 1/8-inch pipe tap. Apply a light coat of "Loctite" grade 635 to 1/8-inch pipe plug, install plug and tighten it until snug.

Number piston bores (1 through 5) on pump rotor and on motor rotor. Use a plastic ice cube tray or equivalent and mark the cavities 1P through 5P for pump ball pistons and 1M through 5M for motor ball pistons. Remove ball pistons (35) one at a time, from pump rotor and place each ball in the correct cavity in tray. Remove ball pistons (51) and springs (52) from motor rotor in the same manner.

Clean and inspect all parts and renew any showing excessive wear or other damage. Ball pistons are a select fit to 0.0002-0.0006 inch (0.0051-0.0152 mm) clearance in rotor bores and must be reinstalled in their original bores. If rotor bushings (37 or 50) are scored or badly worn, 0.002 inch (0.051 mm) or more clearance on pintle journals, renew pump rotor or motor rotor assemblies. Check clearance between input shaft (33) and housing bushing (17). Normal clearance is 0.0013-0.0033 inch (0.0330-0.0838 mm). If clearance is excessive, renew input shaft and/or housing assembly.

Install ball pistons (35) in pump rotor (36) and ball pistons (51) and springs (52) in motor rotor (53), then use wide rubber

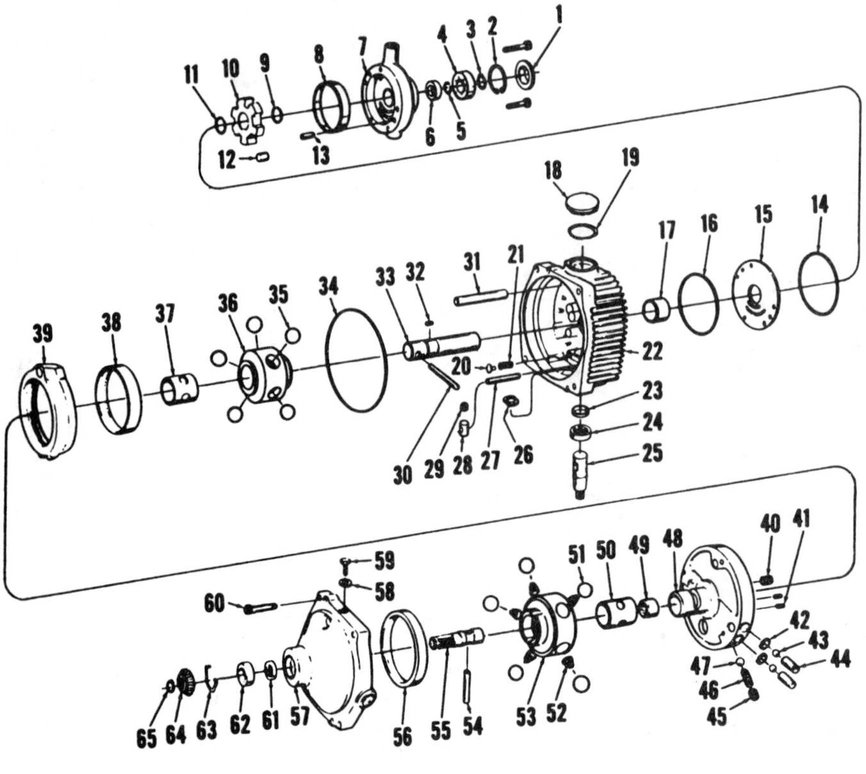

Fig. G24—Exploded view of hydrostatic transmission.

1. Dust shield
2. Retaining ring
3. Snap ring
4. Ball bearing
5. Snap ring
6. Oil seal
7. Charge pump body
8. Charge pump race
9. Snap ring
10. Charge pump rotor
11. Snap ring
12. Pump roller (6)
13. Dowel pin
14. "O" ring
15. Pump plate
16. "O" ring
17. Bushing
18. Plug
19. Gasket
20. Neutral spring cap
21. Neutral spring (2)
22. Housing
23. Bushing
24. Oil seal
25. Control shaft
26. Washer
27. Dowel pin
28. Insert
29. Insert cap
30. Drive pin
31. Cam pivot pin
32. Charge pump drive key
33. Input shaft
34. "O" ring
35. Pump ball pistons
36. Pump rotor
37. Rotor bushing
38. Pump race
39. Pump cam ring
40. Plug (2)
41. Roll pins
42. Snap ring (2)
43. Check valve ball (2)
44. Directional check valve body (2)
45. Plug
46. Relief spring
47. Charge relief ball
48. Pintle
49. Needle bearing
50. Rotor bushing
51. Motor ball pistons
52. Springs
53. Motor rotor
54. Drive pin
55. Output shaft
56. Motor race
57. Body
58. Gasket
59. Venting plug
60. Cap screw
61. Oil seal
62. Ball bearing
63. Retainer
64. Output gear
65. Snap ring

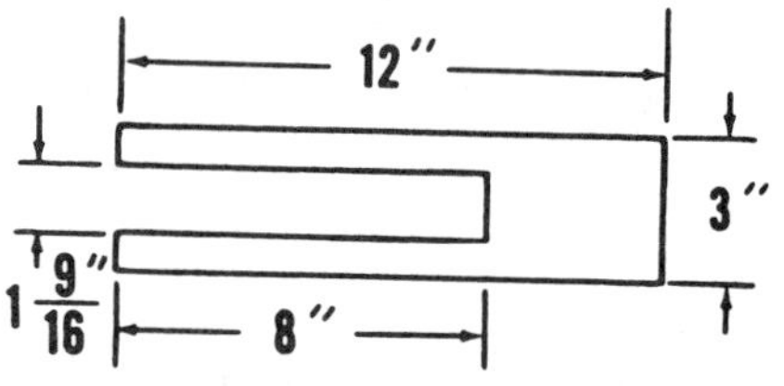

Fig. G25—View of special tool to be fabricated locally used in disassembly and reassembly of hydrostatic transmission. Use dimensions shown and taper ends of fork slightly.

bands to hold pistons in their bores. Install snap rings (42), check valve balls (43) and valve bodies (44) in pintle (48) and secure with new roll pins (41). Install charge pump relief valve ball (47), spring (46) and plug (45). When installing oil seals (6, 24 or 61), apply a light coat of "Loctite" grade 635 to the seal outer diameter. Renew oil seal (24) and install control shaft (25) in housing. Install special washer (26), then press dowel pin (27) into control shaft until end of dowel pin is 1¼ inches (31.75 mm) from control shaft. Renew oil seal (61), then reinstall output shaft (55) with drive pin (54), bearing (62), retainer (63), output gear (64) and snap ring (65) in body (57).

Insert input shaft (33) with drive pin (30) through bushing (17) in housing. Install snap ring (11) in its groove on input shaft. Place "O" ring (16), pump plate (15) and "O" ring (14) in housing, then install charge pump drive key (32), charge pump rotor (10) and snap ring (9). Apply light grease or vaseline to pump rollers (12) and place rollers in rotor slots. Install oil seal (6) and pump race (8) in charge pump body (7), then install body assembly. Secure with the five cap screws, making certain long cap screw is installed in its original location (in heavy section of pump body). Tighten cap screws to 28-30 ft.-lbs. (38-40 N·m). Install snap ring (5), bearing (4), retaining ring (2), snap ring (3) and dust shield (1).

Place charge pump and housing assembly in a holding fixture with input shaft pointing downward. Install pump race (38), insert cap (29) and insert (28) in cam ring (39), then install cam ring assembly over cam pivot pin (31) and control shaft dowel pin (27). Turn control shaft (25) back and forth and check movement of cam ring. Cam ring must move freely from stop to stop. If not, check installation of insert (28) and insert cap (29) in cam ring.

Install pump rotor assembly and remove rubber band used to retain pistons. Install pintle assembly (48) over cam pivot pin (31) and into pump rotor. Place "O" ring (34) in position on housing.

Place body assembly (57) in a holding fixture with gear (64) downward. Install motor race (56) in body, then install motor rotor assembly and remove rubber band used to retain pistons in rotor.

Using special fork tool (Fig. G25) to retain motor assembly in body, carefully install body and motor assembly over pintle journal. Remove fork tool, align bolt holes and install the two cap screws (60 – Fig. G24). Tighten cap screws to 15 ft.-lbs. (20 N·m).

Place hydrostatic unit on holding fixture with plug (18) opening and venting plug opening facing upward. Fill unit with SAE 20 detergent oil until fluid flows from fitting hole in body. Install venting plug (59) with gasket (58), then install plug (18). Plug all openings to prevent dirt or other foreign material from entering hydrostatic unit. Reinstall on range transmission and differential assembly.

RANGE TRANSMISSION AND DIFFERENTIAL. Remove transmission as outlined and separate hydrostatic drive unit from range transmission and differential assembly.

Remove brake caliper assembly and brake disc. Remove tires and wheels, then remove hub assemblies. Thoroughly clean exterior of unit.

Remove axle housings (15 – Fig. G26). Position unit with cover up, then remove cover (13). Lift out differential assembly and axles (40 through 54). Remove output shaft (36), gear (35) and thrust washers (37). Unscrew set screw (2) and remove spring (3) and ball (4). Remove brake shaft (32), sliding gear (33), shift fork (10) and rod (8). Remove input shaft and gear components (18 through 25).

To disassemble differential assembly, remove cap screws (53) and remove differential carriers (41 and 52) and axles from ring gear assembly. Remove snap ring to separate axle gear, carrier and axle. Remove pinions (47) and separate body cores (45 and 48) from ring gear (46).

Inspect components for damage or excessive wear. To reassemble unit, reverse disassembly procedure. Check movement of shift rod when tightening

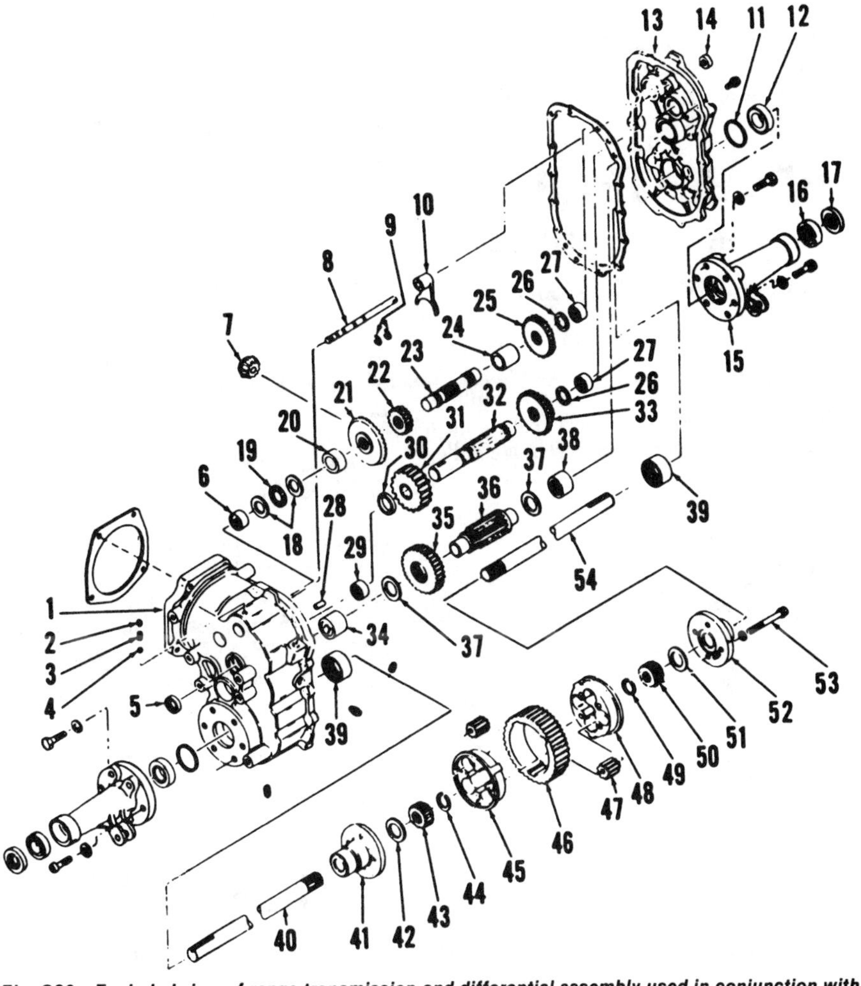

Fig. G26 – Exploded view of range transmission and differential assembly used in conjunction with hydrostatic drive transmission.

1. Case
2. Setscrew
3. Spring
4. Ball
5. Seal
6. Needle bearing
7. Transmission output gear
8. Shift rail
9. Snap rings
10. Shift fork
11. Quad ring
12. Tapered roller bearing
13. Cover
14. Seal
15. Axle housing
16. Ball bearing
17. Seal
18. Thrust washers
19. Thrust bearing
20. Spacer
21. Bevel gear
22. Gear (low range)
23. Shaft
24. Spacer
25. Gear (high range)
26. Thrust washer
27. Needle bearing
28. Dowel pin
29. Needle bearing
30. Spacer
31. Gear
32. Brake shaft
33. Sliding gears
34. Needle bearing
35. Gear
36. Output shaft
37. Thrust washer
38. Needle bearing
39. Needle bearing
40. Axle shaft
41. Differential carrier
42. Thrust washer
43. Axle gear
44. Snap ring
45. Body core
46. Ring gear
47. Pinion gears (8)
48. Body core
49. Snap ring
50. Axle gear
51. Thrust washer
52. Differential carrier
53. Cap screw
54. Axle shaft

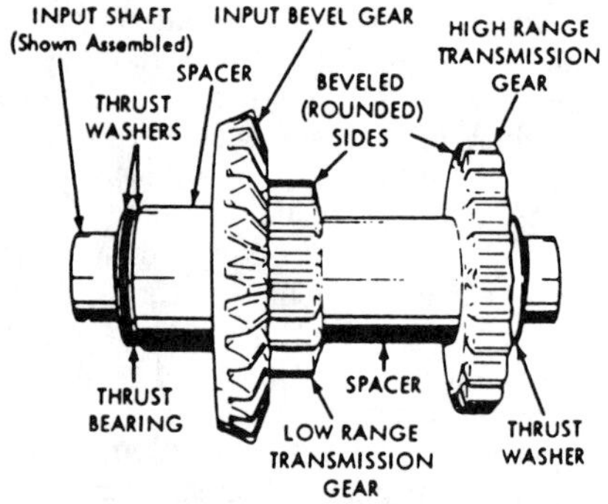

Fig. G27—View of reduction drive input shaft and gears. Note position of bevels on gears.

setscrew (2). Install gears (22 and 25) so bevels of gears face together as shown in Fig. G27. Install carrier cap screws (53–Fig. G26) so head of cap screw is on side of shorter carrier (52). Do not rotate axle housings after housing has been pressed tight against seal (11) as seal may be cut.

Reinstall hydrostatic drive unit, brake disc and brake caliper.

BEVEL GEAR BOX

R&R AND OVERHAUL

Models 53042-53043-53073A

To remove right angle drive unit (bevel gear box), raise seat assembly and unbolt drive shaft rear coupling. Remove traction drive belt from pulley, then loosen setscrews and remove pulley

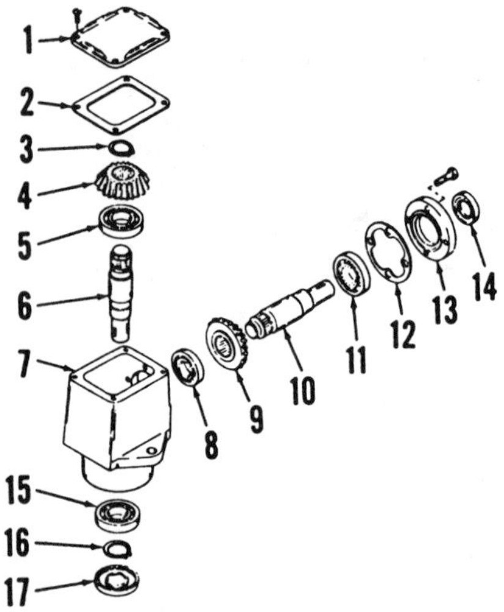

Fig. G28—Exploded view of right angle drive unit used on 53042, 53043 and 53073A models.

1. Cover
2. Gasket
3. Snap ring
4. Drive gear
5. Bearing
6. Input shaft
7. Housing
8. Bearing
9. Driven gear
10. Output shaft
11. Bearing
12. Gasket
13. Retainer
14. Oil seal
15. Bearing
16. Snap ring
17. Oil seal

assembly. Unbolt and remove right angle drive from tractor.

To disassemble unit, remove cover (1–Fig. G28) and gasket (2). Unbolt and remove retainer (13) with oil seal (14) and gasket (12). Withdraw output shaft (10) and bearing (11), then remove driven gear (9) through cover opening. Remove oil seal (17) and snap ring (16). Push input shaft (6), bearing (5) and drive gear (4) out through cover opening. Gear and bearing can be removed from input shaft after first removing snap ring (3). To remove bearing (8), either tap outside of housing (7) behind bearing with a mallet or apply heat to housing.

Clean and inspect all parts and renew any showing excessive wear or other damage. Reassemble by reversing disassembly procedure. Use new oil seals and gaskets and pack 4 ounces (118 mL) of multi-purpose, lithium base grease in housing before installing cover (1).

POWER TAKE-OFF ADJUSTMENT

Models 52084-52084A-53073A-53074A

These models incorporate a pivoting idler pulley and lever arrangement to loosen or tighten pto belt. Periodically check condition of belt and adjust belt guides and fingers to obtain 1/16 to 1/8-inch (1.588 to 3.175 mm) clearance between belt and guide or finger.

Models 53042-53043-53044-53045-53054-53055

An electric pto clutch is mounted to the engine at the front of tractor. Check clutch adjustment by inserting a 0.012 inch (0.305 mm) feeler gage into each of the three "SLOTS" (Fig. G29) in the clutch. Feeler gage should be a snug fit at all locations. If not, adjust locknuts equally until correct clearance is obtained. Check clutch performance and note clutch must release with pto switch in "OFF" position.

R&R AND OVERHAUL ELECTRIC CLUTCH

All Models So Equipped

Remove belt and disconnect electrical connection. Remove cap screw and washers from end of crankshaft, then remove clutch-armature assembly. Remove key. Unbolt and remove clutch field assembly.

Disassemble clutch-armature assembly by pressing armature from ball bearing, remove front retaining ring and press bearing from clutch.

Clean and inspect all parts. Normal resistance reading for clutch field winding is 2.75 to 3.60 ohms. If reading is higher or lower, field must be renewed.

Reassemble by reversing disassembly procedure. Tighten field retaining screws to 30 in.-lbs. (3.4 N•m).

HYDRAULIC SYSTEM

PRESSURE CHECK AND ADJUST

Models With Hydrostatic Drive

Hydraulic system utilizes pressurized oil from hydrostatic drive unit. If hydraulic lift fails to function or lifts slowly, install a 0-1000 psi (7000 kPa) test gage and shut-off valve in line between hydrostatic transmission and tractor hydraulic control valve. With shut-off valve open, start and run engine at full rpm. Slowly begin to close shut-off valve until 800 psi (5516 kPa) is reached. If 800 psi (5516 kPa) cannot be obtained even with shut-off valve com-

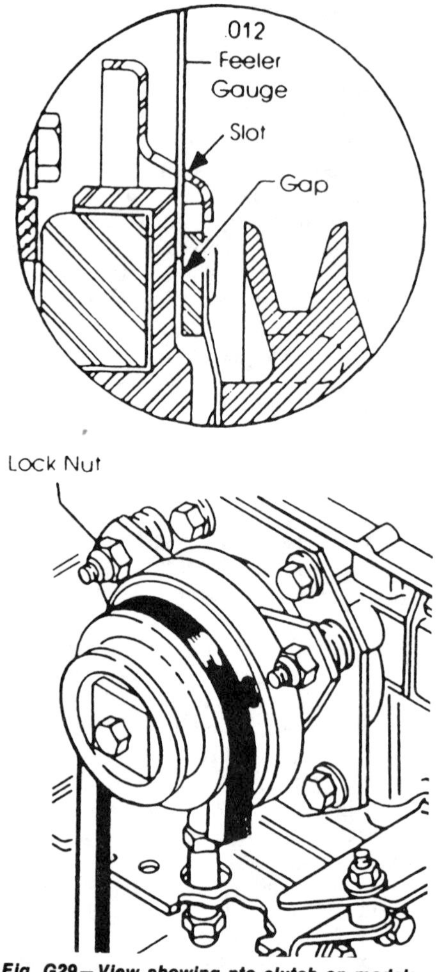

Fig. G29—View showing pto clutch on models equipped with electric pto. Upper inset shows feeler gage inserted in slot. Refer to text.

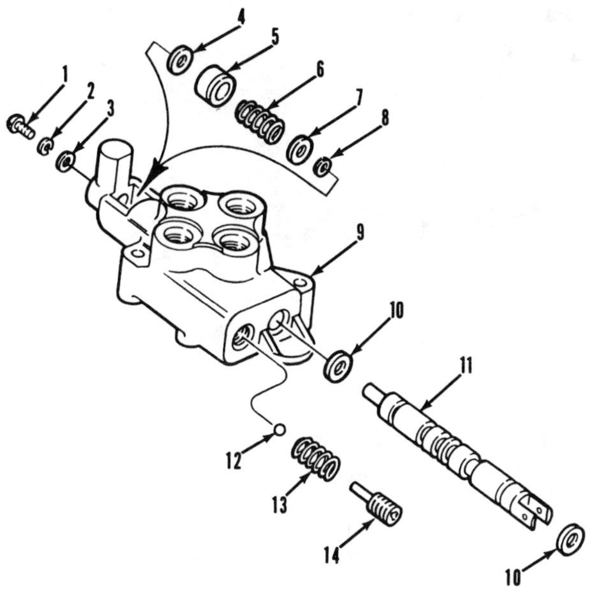

Fig. G30—Exploded view of hydraulic control valve.

1. Screw
2. Washer (spring lock)
3. Washer
4. Washer
5. Spacer
6. Spring
7. Washer
8. Washer
9. Valve body
10. Seal
11. Spool
12. Ball
13. Relief valve spring
14. Relief valve adjustment plug

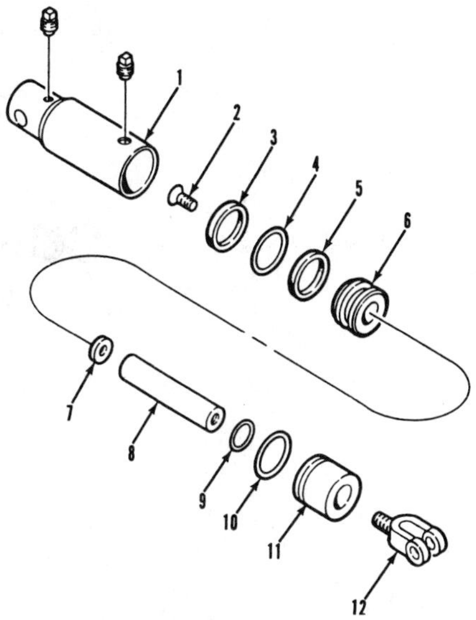

Fig. G31—Exploded view of hydraulic cylinder used on most models equipped for hydraulic lift.

1. Cylinder body
2. Screw
3. Back-up washer
4. "O" ring
5. Back-up washer
6. Piston
7. Rubber disc
8. Shaft
9. "O" ring
10. "O" ring
11. End cap & guide
12. Clevis

pletely closed, pump must be renewed or repaired.

To check and adjust control valve pressure relief valve setting, install a 0-1000 (0-7000 kPa) test gage and shut-off valve in line running to lift side of cylinder. With shut-off valve open, start and run engine at full rpm. Close shut-off valve and hold lift lever in "UP" position. Adjust relief valve plug (14–Fig. G30) to obtain 800 psi (5516 kPa). Remove gage and shut-off valve.

CONTROL VALVE AND HYDRAULIC CYLINDER

All Models So Equipped

To disassemble and service control valve, refer to Fig. G30. Renew all "O" rings and sealing washers during reassembly.

To disassemble and service hydraulic cylinder, refer to Fig. G31. Renew all "O" rings and back-up washers during reassembly.

GRAVELY

CONDENSED SPECIFICATIONS

	MODELS			
	450	816	8162-T	8163-T
Engine Make	Onan	Onan	Onan	Onan
Model	CCKA	CCKA	CCKA	CCKA
Bore	3.25 in. (82.55 mm)	3.25 in. (82.55 mm)	3.25 in. (82.55 mm)	3.25 in. (82.55 mm)
Stroke	3 in. (76.2 mm)	3 in. (76.2 mm)	3 in. (76.2 mm)	3 in. (76.2 mm)
Displacement	49.8 cu. in. (815.7 cc)	49.8 cu. in. (815.7 cc)	49.8 cu. in. (815.7 cc)	49.8 cu. in. (815.7 cc)
Power Rating	16.5 hp. (12.3 kW)	16.5 hp. (12.3 kW)	16.5 hp. (12.3 kW)	16.5 hp. (12.3 kW)
Slow Idle	1000 rpm	1000 rpm	1000 rpm	1000 rpm
High Speed (No-Load)	3600 rpm	3600 rpm	3600 rpm	3600 rpm
Capacities –				
Crankcase	4 pts. (1.9 L)	4 pts. (1.9 L)	4 pts. (1.9 L)	4 pts. (1.9 L)
Hydraulic System	.....	1.5 qts. (1.4 L)	.....	1.5 qts. (1.4 L)
Transaxle	5 qts. (4.7 L)	6 qts. (5.7 L)	6 qts. (5.7 L)	6 qts. (5.7 L)
Fuel Tank	5.3 gal. (20 L)	5.3 gal. (20 L)	5.3 gal. (20 L)	5.3 gal. (20 L)

	MODELS			
	8171-T	8173-KT	8179-T	8179-KT
Engine Make	Onan	Kohler	Onan	Kohler
Model	CCKA	KT-17	CCKA	KT-17
Bore	3.25 in. (82.55 mm)	3.12 in. (79.3 mm)	3.25 in. (82.55 mm)	3.12 in. (79.3 mm)
Stroke	3 in. (76.2 mm)	2.75 in. (69.9 mm)	3 in. (76.2 mm)	2.75 in. (69.9 mm)
Displacement	49.8 cu. in. (815.7 cc)	42.18 cu. in. (690.5 cc)	49.8 cu.in. (815.7 cc)	42.18 cu. in. (690.5 cc)
Power Rating	16.5 hp. (12.3 kW)	17 hp. (12.7 kW)	16.5 hp. (12.3 kW)	17 hp. (12.7 kW)
Slow Idle	1000 rpm	1200 rpm	1000 rpm	1200 rpm
High Speed (No-Load)	3600 rpm	3600 rpm	3600 rpm	3600 rpm
Capacities –				
Crankcase	4 pts. (1.9 L)	3 pts. (1.4 L)	4 pts. (1.9 L)	3 pts. (1.4 L)
Hydraulic System	1.5 qts. (1.4 L)	1.5 qts. (1.4 L)	1.5 qts. (1.4 L)	1.5 qts. (1.4 L)
Transaxle	5 qts. (4.7 L)	6 qts. (5.7 L)	6 qts. (5.7 L)	6 qts. (5.7 L)
Fuel Tank	5.3 gal. (20 L)	5.3 gal. (20 L)	5.3 gal. (20 L)	5.3 gal. (20 L)

	MODELS			
	8183-T	8193-KT	8199-T	8199-KT
Engine Make	Onan	Kohler	Onan	Kohler
Model	B48M	KT-19	B48G	KT-19
Bore	3.25 in. (82.55 mm)	3.12 in. (79.3 mm)	3.25 in. (82.55 mm)	3.12 in. (79.3 mm)
Stroke	2.875 in. (73.03 mm)	2.75 in. (69.9 mm)	2.875 in. (73.03 mm)	2.75 in. (69.9 mm)
Displacement	47.7 cu. in. (781.7 cc)	46.98 cu. in. (770.5 cc)	47.7 cu. in. (781.7 cc)	46.98 cu. in. (770.5 cc)
Power Rating	18 hp. (13.4 kW)	19 hp. (14.2 kW)	20 hp. (14.9 kW)	19 hp. (14.2 kW)
Slow Idle	1200 rpm	1200 rpm	1200 rpm	1200 rpm
High Speed (No-Load)	3600 rpm	3600 rpm	3600 rpm	3600 rpm
Capacities–				
Crankcase	4 pts. (1.86 L)	3 pts. (1.4 L)	4 pts. (1.86 L)	3 pts. (1.4 L)
Hydraulic System	1.5 qts. (1.4 L)	1.5 qts. (1.4 L)	1.5 qts. (1.4 L)	1.5 qts. (1.4 L)
Transaxle	6 qts. (5.7 L)	6 qts. (5.7 L)	6 qts. (5.7 L)	6 qts. (5.7 L)
Fuel Tank	5.3 gal. (20 L)	5.3 gal. (20 L)	5.3 gal. (20 L)	5.3 gal. (20 L)

FRONT AXLE AND STEERING SYSTEM

MAINTENANCE

Models 450-816

Lubricate left and right spindles and axle pivot at eight hour intervals. Clean and repack front wheel bearings at 200 hour intervals. Use multi-purpose, lithium base grease and clean all grease fittings before and after lubrication.

All Other Models

Lubricate left and right spindles, axle pivot, steering gear and pitman arm at 25 hour intervals. Clean and repack front wheel bearings at 200 hour intervals. Use multi-purpose, lithium base grease and clean all grease fittings before and after lubrication.

R&R AXLE MAIN MEMBER

Model 450

Disconnect tie rods (2–Fig. GR1) at steering arms (3). Support front of tractor and remove pivot pin (5). Raise front of tractor and roll axle assembly forward from tractor.

Reinstall by reversing removal procedure.

Model 816

Disconnect tie rods (2–Fig. GR2) at steering arms (3). Support front of tractor and remove pivot pin (5). Raise front of tractor and roll axle assembly forward from tractor.

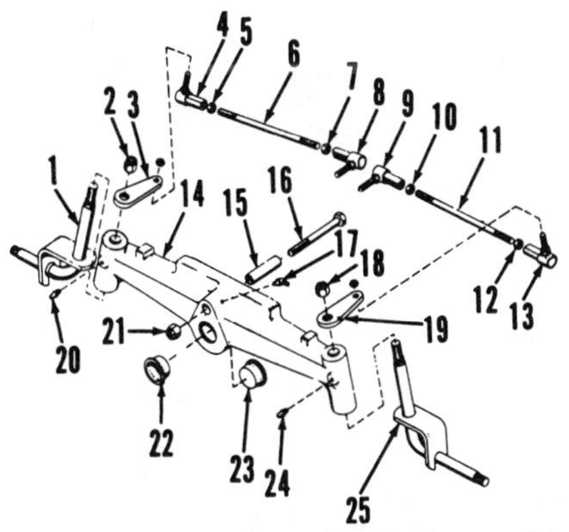

Fig. GR3 – Exploded view of front axle assembly used on all models except 450 and 816 models.

1. Spindle
2. Nut
3. Steering arm
4. Tie rod end
5. Jam nut
6. Tie rod
7. Jam nut
8. Tie rod end
9. Tie rod end
10. Jam nut
11. Tie rod
12. Jam nut
13. Tie rod end
14. Axle main member
15. Bushing
16. Pivot bolt
17. Grease fitting
18. Nut
19. Steering arm
20. Grease fitting
21. Nut
22. Plug
23. Plug
24. Grease fitting
25. Spindle

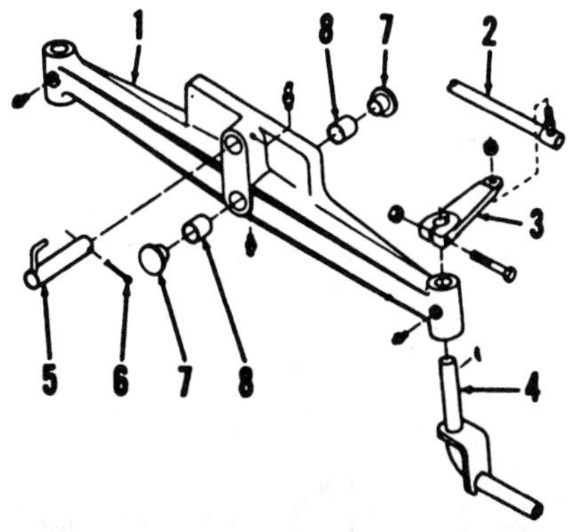

Fig. GR1 – Exploded view of front axle assembly used on 450 model.

1. Axle main member
2. Tie rod
3. Steering arm
4. Spindle
5. Pivot pin
6. Cotter pin
7. Plug
8. Bushings

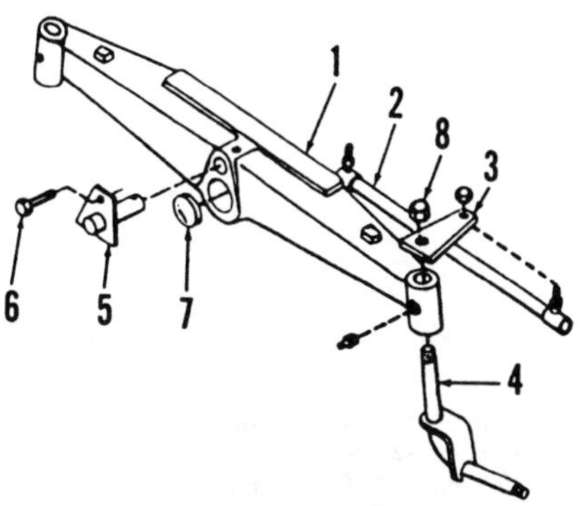

Fig. GR2 – Exploded view of front axle assembly used on 816 model.

1. Axle main member
2. Tie rod
3. Steering arm
4. Spindle
5. Pivot pin
6. Bolt
7. Plug
8. Nut

Illustrations courtesy Gravely.

Reinstall by reversing removal procedure.

All Other Models

Disconnect tie rod ends (4 and 13–Fig. GR3) at steering arms (3 and 19). Support front of tractor and remove pivot bolt (16). Raise front of tractor and roll axle assembly forward from tractor.

Reinstall by reversing removal procedure.

TIE ROD AND TOE-IN

Models 450-816

Tie rods and toe-in are nonadjustable on 450 and 816 models. Toe-in should be ¼ to ¾-inch (6.35 to 19 mm) and if incorrect indicates faulty tie rods or spindle bushings. Renew or repair as required.

All Other Models

Renewable tie rod ends (4,8,9 and 13–Fig. GR3) are threaded onto tie rods (6 and 11). Correct toe-in is ¼ to ¾-inch (6.35 to 19 mm). Adjust by loosening jam nuts (5,7,10 and 12) and lengthening or shortening tie rods (6 and 11) equally as required. Tighten jam nuts.

R&R STEERING SPINDLES

Model 450

Raise and support front of tractor. Remove front wheel and tire assemblies. Disconnect tie rods (2–Fig. GR1) at steering arms (3). Remove clamp bolts and steering arms. Remove keys from spindles and push spindles (4) out of axle main member (1).

Reinstall by reversing removal procedure.

All Other Models

Refer to Fig. GR2 for 816 model and to Fig. GR3 for all other models. Raise and support front of tractor. Remove front tire and wheel assemblies. Disconnect tie rod ends at steering arms. Remove nuts from spindles, mark steering arm position in relation to spindles and remove steering arms. Push spindles out of axle main member.

Reinstall by reversing removal procedure.

R&R AND OVERHAUL STEERING GEAR

Model 450 And Early Production 816 Models

Remove hood, fuel tank, battery and steering wheel. Disconnect tie rod ends from pitman arm (13–Fig. GR4) and remove bolts securing gearbox to frame. Work gearbox down through frame and out of tractor.

Loosen setscrew and remove pitman arm, then unbolt and remove steering plate cover and mounting plate. Clean and inspect all parts and renew any showing excessive wear or other damage. When reassembling, pack housing with multi-purpose grease.

All Other Models

Remove steering wheel (10–Fig. GR5), bushing (14), tube (15), bushing (16) and retaining ring (17). Remove steering shaft nut (24) and withdraw shaft. Disconnect tie rod ends from steering arms. Unbolt and remove steering pitman arms (5), rack (8) and bearing support (23).

Reassemble by reversing disassembly procedure. Make certain front wheels

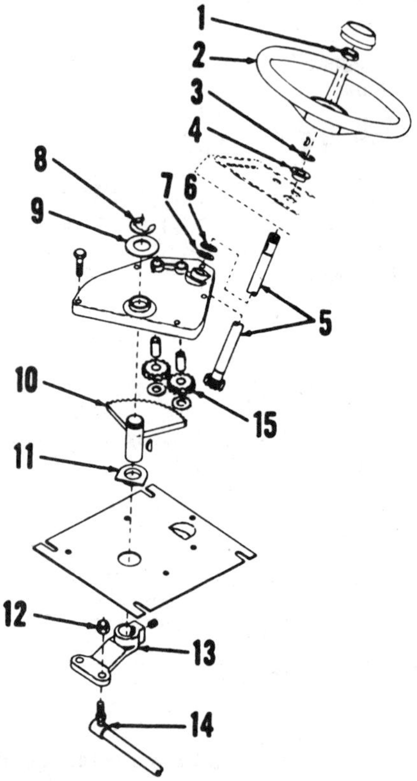

Fig. GR4—Exploded view of steering gear assembly used on all 450 models and early production 816 models.

1. Nut
2. Steering wheel
3. "E" ring
4. Bushing
5. Steering shaft & pinion
6. "E" ring
7. Seal
8. "E" ring
9. Washer
10. Quadrant gear
11. Bushing
12. Nut
13. Pitman arm
14. Tie rod
15. Spur gears

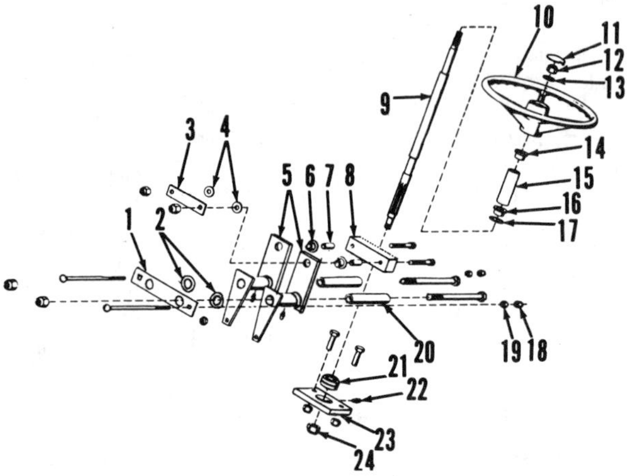

Fig. GR5—Exploded view of steering gear assembly used on all models except 450 model and early production 816 models.

1. Adjustment plate
2. Thrust bearing races
3. Strap
4. Washers
5. Steering pitman arms
6. Bearing
7. Bushings
8. Gear (rack)
9. Steering shaft
10. Steering wheel
11. Cap
12. Nut
13. Washer
14. Bushing
15. Tube
16. Bushing
17. Retaining ring
18. Jam nut
19. Adjustment nuts
20. Sleeves
21. Bearing
22. Grease fitting
23. Support
24. Nut

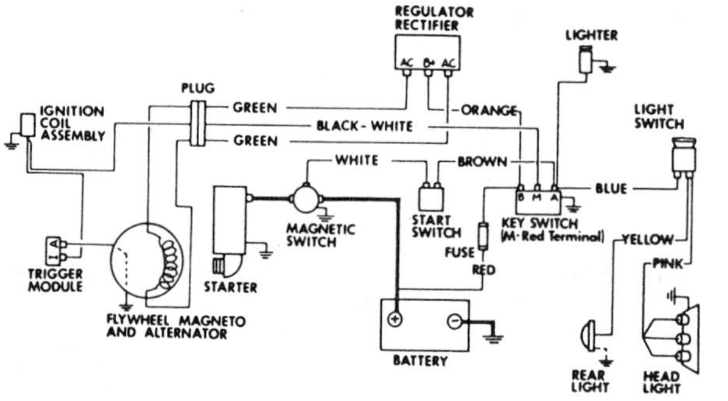

Fig. GR6—Typical wiring diagram for 450 model. Circuit for electric lift is not shown.

Illustrations courtesy Gravely.

are straight ahead when installing shaft (9). Tighten adjusting nuts (19) to obtain desired gear mesh, then tighten jam nuts (18) to maintain adjustment. Lubricate with multi-purpose grease.

ENGINE

MAINTENANCE

All Models

Regular engine maintenance is required to maintain peak performance and long engine life.

Check oil level at five hour intervals under normal operating conditions. Refer to appropriate engine section for maintenance schedule and procedures.

REMOVE AND REINSTALL

All Models

Remove hood and disconnect negative battery cable. Disconnect fuel line, choke cable, throttle cable and all necessary electrical connections at engine. Remove rear deck and fenders. Drain transmission and place a jack under engine for support. Remove the four bolts holding the adapter plate to the frame and slide engine rearward out of transmission. Refer to appropriate TRANSMISSION section in this manual for procedure to remove input gear, thrust bearing and engine adapter.

Reinstall by reversing removal procedure.

OVERHAUL

All Models

Engine make and model are listed at the beginning of this section. For tune-up specifications, engine overhaul procedures and engine maintenance, refer to appropriate engine section in this manual.

ELECTRICAL SYSTEM

MAINTENANCE AND SERVICE

All Models

Battery electrolyte level should be checked at 50 hour intervals of normal operation. If necessary, add distilled water until level is just below base of vent well. **DO NOT** overfill. Keep battery posts clean and cable ends tight.

For alternator or starter service, refer to appropriate engine section in this manual.

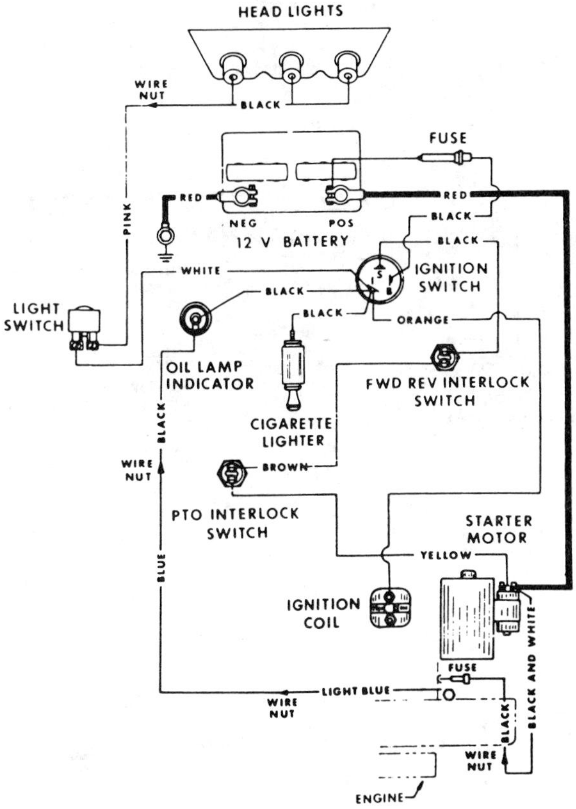

Fig. GR7—Wiring diagram for 816 model.

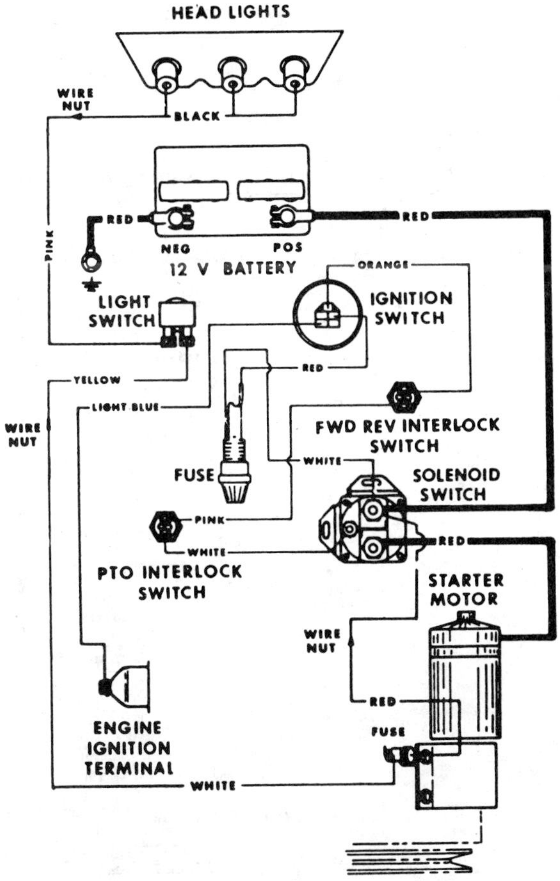

Fig. GR8—Wiring diagram for 8162-T and 8163-T models.

Illustrations courtesy Gravely.

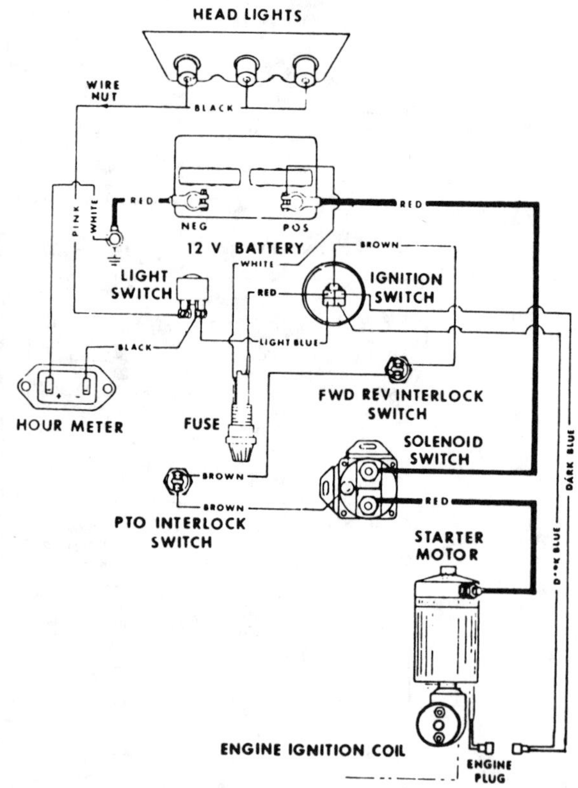

Fig. GR9 — Wiring diagram for 8173-KT, 8183-T and 8193-KT models.

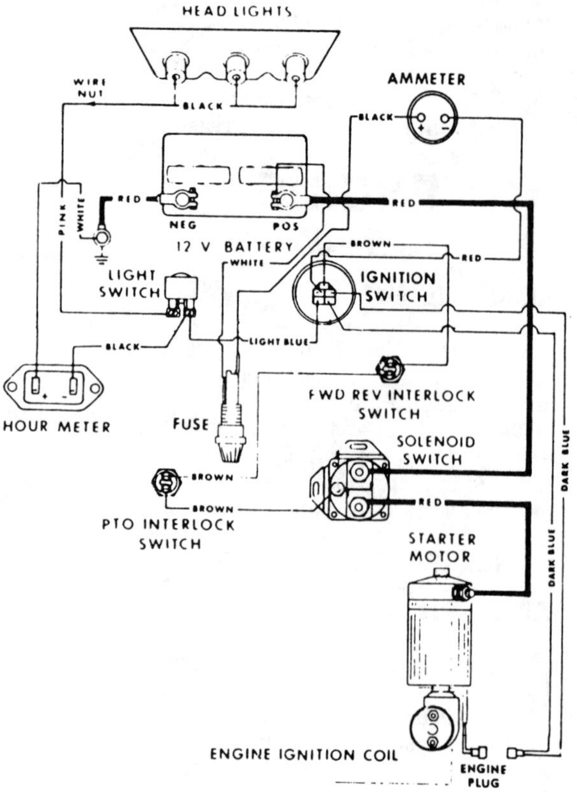

Fig. GR10 — Wiring diagram for 8171-T, 8179-T and 8179-KT models.

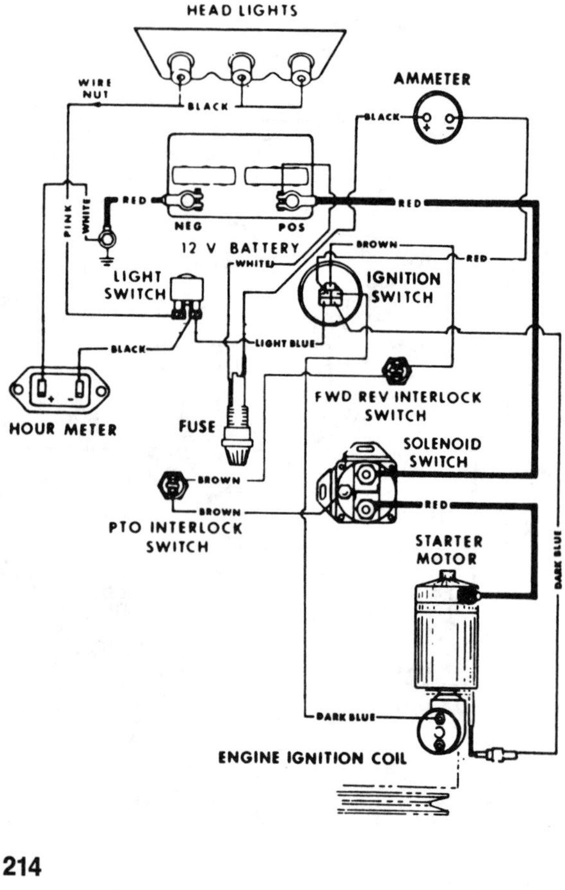

Fig. GR11 — Wiring diagram for 8199-T and 8199-KT models.

Refer to Figs. GR6 through GR11 for appropriate wiring diagram.

BRAKES

ADJUSTMENT

Transaxle Type Brake

To check adjustment, push direction control lever all the way forward. Watch

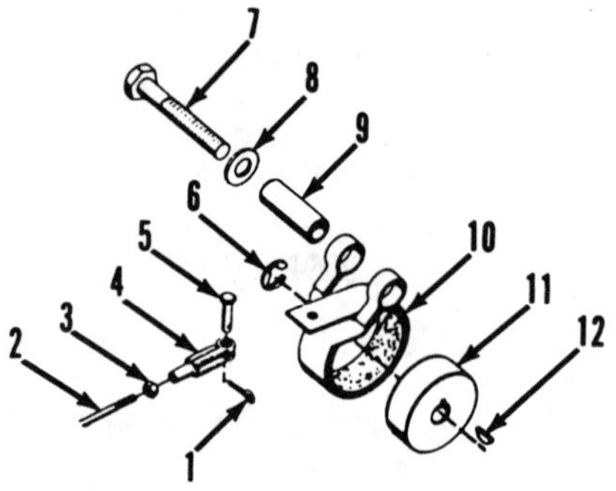

Fig. GR12 — Exploded view of transaxle type brake used on most models.

1. Cotter pin
2. Rod
3. Jam nut
4. Clevis
5. Pin
6. "E" ring (nut used on some models)
7. Bolt
8. Washer
9. Spacer
10. Brake band
11. Drum
12. Key

Illustrations courtesy Gravely.

motion of brake band as brake pedal is depressed. Brake is correctly adjusted when band becomes tight on drum as direction control lever moves to neutral position. If adjustment is incorrect, loosen jam nut (3–Fig. GR12) and remove pin (5) from clevis (4). Turn clevis clockwise to tighten brake or counter-clockwise to loosen brake as required. Reconnect clevis to brake band with pin and recheck adjustment. Brake pedal should stop approximately 3 inches (76.2 mm) from footrest.

Individual Wheel Brakes (Optional)

To adjust optional individual wheel brakes, turn nut (N–Fig. GR13) on 450 models or clevis (2–Fig. GR14) for all other models as necessary to provide good brake action.

R&R BRAKE BANDS

Transaxle Type Brake

Remove pin (5—Fig. GR12) and disconnect clevis (4). Remove bolt (7), washer (8) and bushing (9). Remove band (10). Reinstall by reversing removal procedure. Adjust brake.

Individual Wheel Brakes (Optional)

On 450 models, remove adjustment nut (N—Fig. GR13) and remove cotter key from pin in opposite end of brake band. Remove pin and band.

On all other models, remove cotter pin from actuating arm (9), remove the bolt retaining opposite end of brake band and remove band.

Reinstall by reversing removal procedure. Adjust brakes.

Fig. GR13—View showing location of brake band adjustment nut (N) on 450 models. Refer to text.

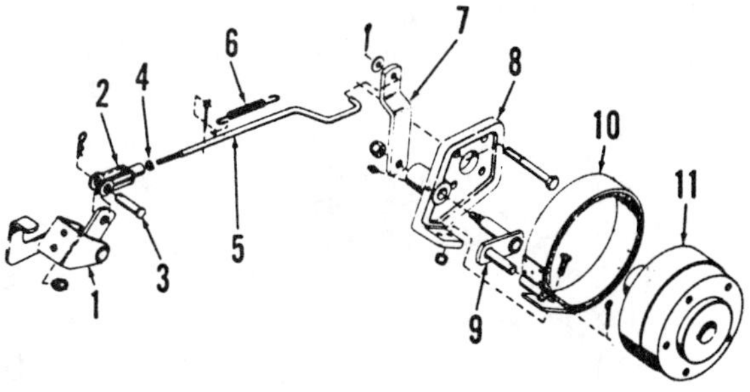

Fig. GR14—Exploded view of individual wheel brake system used on all models except 450 model. Model 450 is similar.

1. Brake arm
2. Clevis
3. Pin
4. Jam nut
5. Brake rod
6. Spring
7. Brake lever
8. Mounting plate
9. Actuating arm
10. Brake band
11. Drum

Fig. GR15—View of right side of transaxle used on 816 model showing forward clutch and transaxle type brake.

1. Adjusting nuts
2. Clutch discs
3. Clutch cam
4. Clutch rod
5. Yoke
6. Tranaxle brake
7. Brake arm

CLUTCH

SERVICE AND ADJUSTMENT

Model 450

Clutch is an integral part of transaxle. Refer to appropriate TRANSAXLE section in this manual for service procedure.

To adjust clutch, turn adjusting nuts on forward-reverse control rods so spring has 0.010 inch (0.254 mm) gap between spring coils with control lever locked into position.

Model 816

Clutches for forward and reverse speeds are mounted at sides of transaxle; forward clutch on right side, reverse clutch on left side. Refer to appropriate TRANSAXLE section in this manual for service procedure.

To adjust clutches, place direction control lever in neutral position. Disconnect clutch rods (4–Fig. GR15). Loosen or tighten nuts (1) until there is 0.020-0.030 inch (0.508-0.762 mm) gap between clutch discs (2). Make certain direction control lever is vertical. Adjust length of clutch rods by turning rods in yoke (5) until rod can be connected to clutch cam (3) without disturbing position of cam or direction control lever.

All Other Models

Clutches for forward and reverse

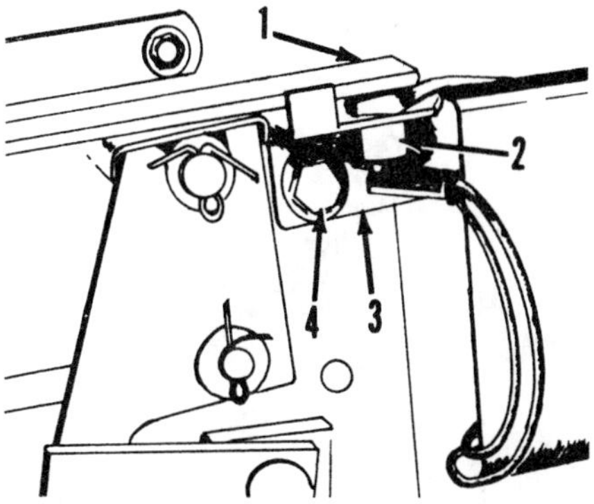

Fig. GR16—View showing location of safety-neutral switch. Refer to text for adjustment procedure.

1. Spring
2. Neutral switch
3. Switch bracket
4. Bolt

Fig. GR17—Edge of actuating lever must be parallel with gasket seam. Refer to text.

1. Gasket seam
2. Actuating lever edge
3. Actuating lever
4. Clevis

Illustrations courtesy Gravely.

speeds are mounted at sides of transaxle; forward clutch on right side, reverse clutch on left side. Refer to appropriate TRANSAXLE section in this manual for service procedure.

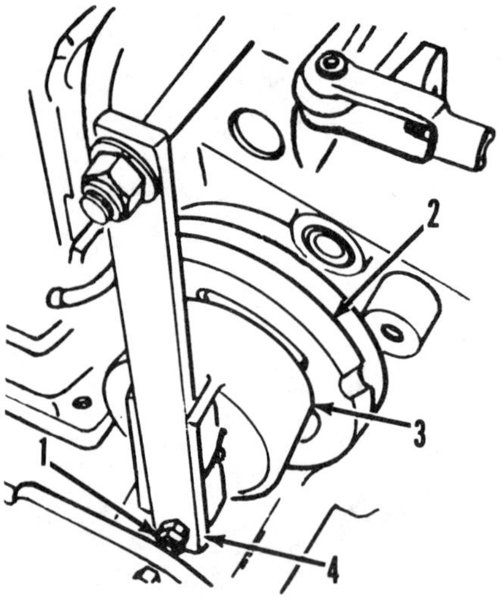

Fig. GR18—View of forward-reverse clutch adjustment points. Refer to text.

1. Bolt
2. Clutch disc gap
3. Housing
4. Clutch spring

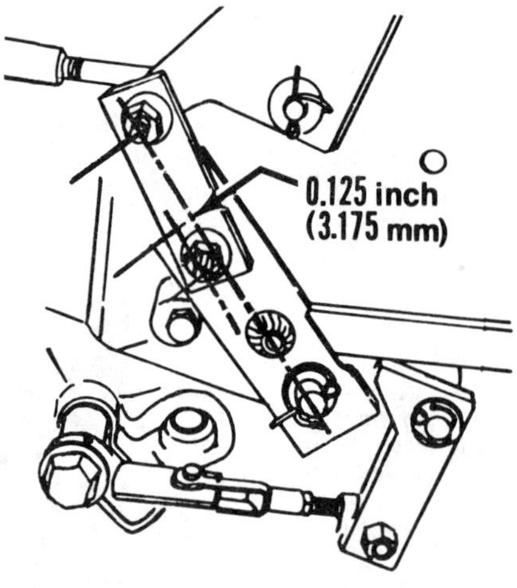

Fig. GR19—Linkage should go over center 0.125 inch (3.175 mm). Refer to text.

NEUTRAL SWITCH ADJUSTMENT. Place direction control lever in neutral position and engage brake lock. Loosen switch bracket mounting bolts (4–Fig. GR16) slightly and tap bracket (3) up until clutch detent spring lifts off actuating arm (keep switch bracket horizontal). Tap bracket down until leaf spring (1) rests on actuating arm. Tighten bolts (4).

ACTUATING LEVER ADJUSTMENT. Place direction control lever in neutral position. Adjust clevis (4–Fig. G17) so edge (2) of slide rod actuating lever (3) is parallel with transaxle case gasket seam (1).

CLUTCH SPRING ADJUSTMENT. Place direction control lever in neutral position. Push clutch spring (4–Fig. GR18) in until there is no clutch disc gap (2). Adjust bolt (1) until clutch spring is vertical. Repeat procedure for remaining clutch.

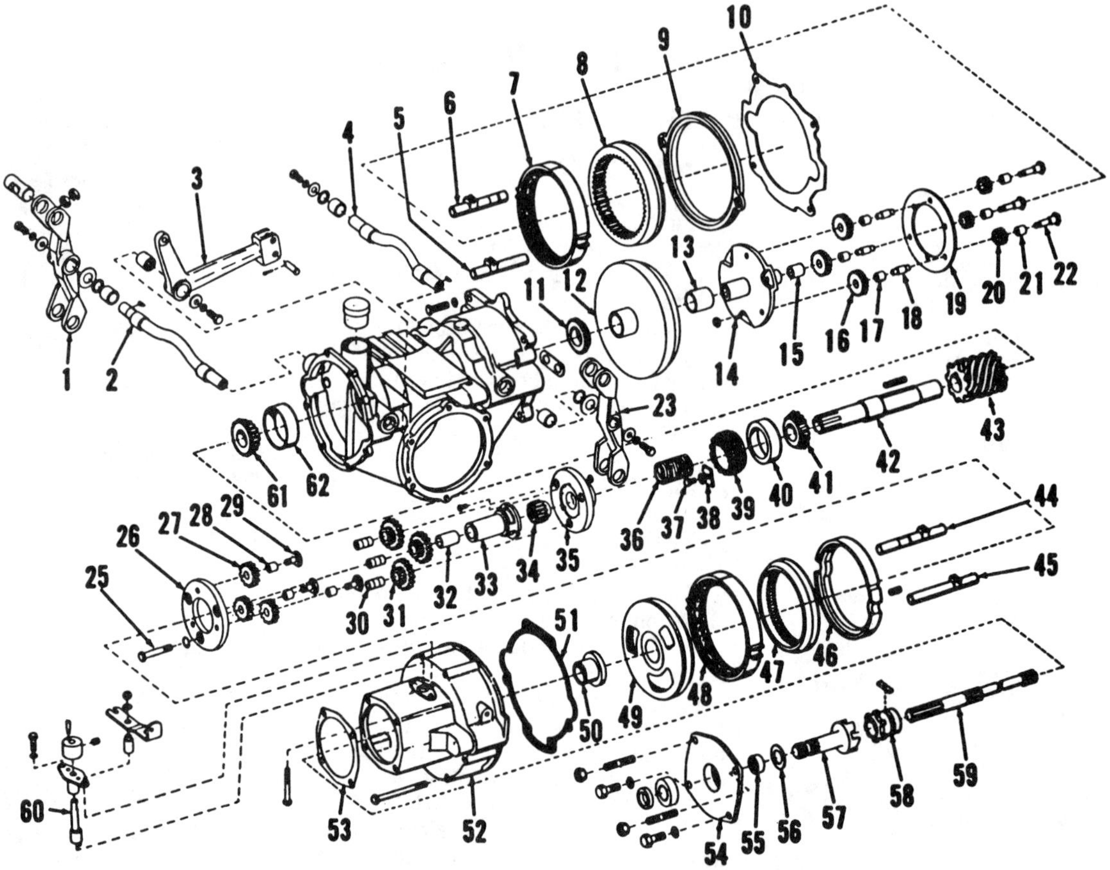

Fig. GR20—Exploded view of transaxle used on 450 model.

1. Fwd.-Rev. lever
2. Fwd.-Rev. actuating shaft
3. Shift link
4. High-Low actuating shaft
5. Clutch slide rod
6. Clutch slide rod
7. Clutch cup
8. Ring gear
9. Gear cup
10. Gasket
11. Thrust plate
12. Clutch plate
13. Bushing
14. Gear carrier
15. Bushing
16. Orbit gear
17. Bushing
18. Pin
19. Pin spacer
20. Orbit gear
21. Bushing
22. Pin
23. High-Low lever
25. Bolt (L.H. thread)
26. Pin plate
27. Reverse idler gear
28. Bushing
29. Bolt
30. Pin
31. Orbit gear
32. Bearing
33. Quill
34. Sun gear
35. Pin plate
36. Spring
37. Lock screw
38. Lock plate
39. Bearing adjustment nut
40. Bearing cup
41. Bearing cone
42. Shaft
43. Worm gear
44. Clutch slide rod
45. Clutch slide rod
46. Gear cup
47. Internal gear
48. Clutch cup
49. Reverse gear
50. Thrust bearing
51. Gasket
52. Advance housing
53. Gasket
54. Cover
55. Bearing
56. Thrust washer
57. Pto output shaft
58. Pto clutch dog
59. Pto shaft
60. Pto actuating shaft
61. Bearing cone
62. Bearing cup

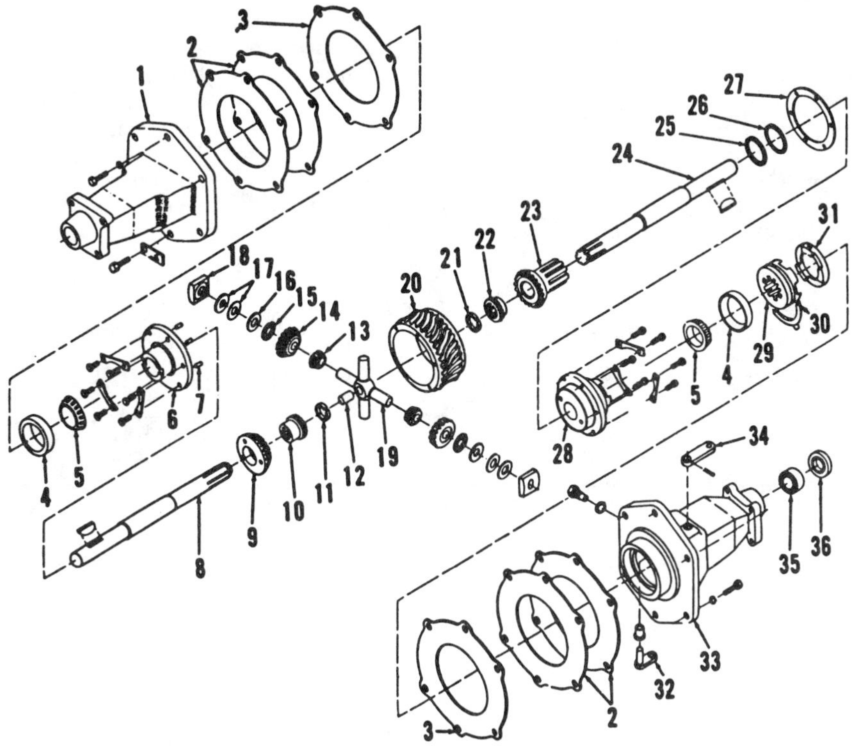

1. Axle housing L.H.
2. Shims
3. Gasket
4. Bearing cup
5. Bearing cone
6. Differential carrier
7. Pin
8. Axle shaft
9. Shifting gear
10. Side gear
11. Snap ring
12. Thrust pin
13. Spider gear
14. Shifting pinion
15. Thrust bearing
16. Thrust washer
17. Shims
18. Thrust washer
19. Cross pin
20. Ring gear
21. Snap ring
22. Side gear
23. Shifting gear
24. Axle shaft
25. Thrust bearing
26. Thrust washer
27. Shim
28. Differential carrier
29. Sliding clutch dog
30. Clutch yoke
31. Stationary clutch
32. Shift lever
33. Axle housing R.H.
34. Shift arm
35. Bearing
36. Oil seal

Fig. GR21—Exploded view of Swiftamatic differential used on 450 models.

CLUTCH GAP ADJUSTMENT. Place direction control lever in neutral position. Adjust nuts on slide rods so forward clutch gap (2–Fig. GR18) is 0.070 inch (1.8 mm) and reverse clutch gap is 0.125 inch (3.2 mm).

CAM BUSHING ADJUSTMENT. Place direction control level in neutral position. Loosen retaining bolt and rotate bushing until linkage goes over center 0.125 inch (3.175 mm). See Fig. GR19. Retighten bolt.

TRANSAXLE

MAINTENANCE

Model 450

Check transaxle fluid level at 100 hour intervals. Maintain level at "FULL" mark on transaxle dipstick. Change fluid at 200 hour intervals. Recommended fluid is SAE 90 EP gear lubricant and approximate capacity is 5 quarts (4.7 L).

All Other Models

Check transaxle fluid level at 100 hour intervals. Maintain fluid level at lower edge of check plug opening. Change fluid at 200 hour intervals. Recommended fluid is SAE 10W-30 motor oil with service classification SC, SD or SE. Approximate fluid capacity is 6 quarts (5.7 L).

R&R TRANSAXLE

Model 450

Remove seat and rear fender assembly. Remove engine as previously outlined. Disconnect brake control rods at brakes, lift attachment control rod and transaxle control rods. Disconnect attachment control rod at front end of transaxle. Support rear of tractor frame and remove transaxle retaining bolts. Roll transaxle away from tractor.

Reinstall by reversing removal procedure.

All Other Models

Remove engine as previously outlined.

Fig. GR22—View of spider gears used on 450 models.

Fig. GR23—View of differential used on 450 models.

Illustrations courtesy Gravely.

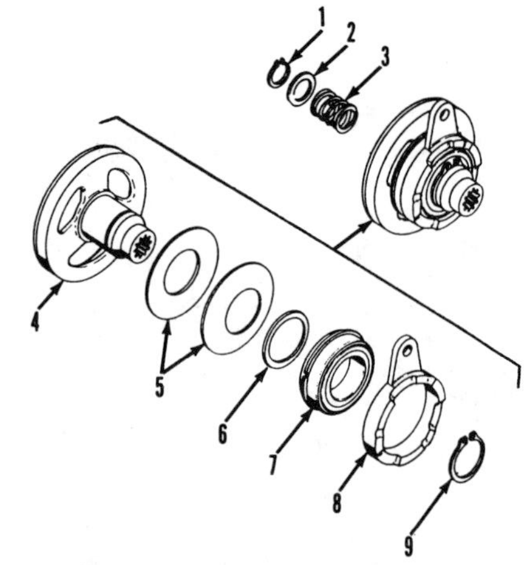

Fig. GR25—Exploded view of forward-reverse clutch used on 816 model.

1. Snap ring
2. Washer
3. Spring
4. Hub
5. Belleville washers
6. Shim
7. Bearing
8. Cam
9. Snap ring

Drain hydraulic reservoir and disconnect hydraulic lines. Cap all openings. Disconnect forward and reverse clutch rods from clutch cams (3–Fig. GR15) and the two speed rod from shifter arm. Remove shift rods from the 1-3 and 2-4 shift arms. Remove pto rod from pto lever and lift rod from cross shaft. Disconnect brake rod from brake band and remove bolts securing cross shaft to transmission.

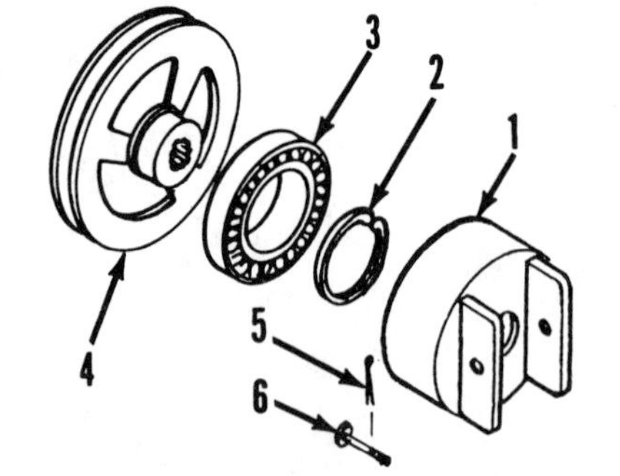

Fig. GR26—Exploded view of forward-reverse clutch used on all models except 450 and 816 models.

1. Housing
2. Snap ring
3. Bearing
4. Hub
5. Cotter pin
6. Pin

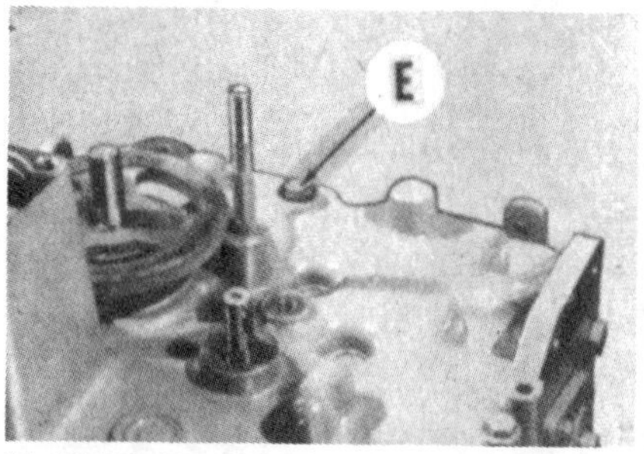

Fig. GR27—Retaining ring "E" must be removed before transaxle can be disassembled. Refer to text.

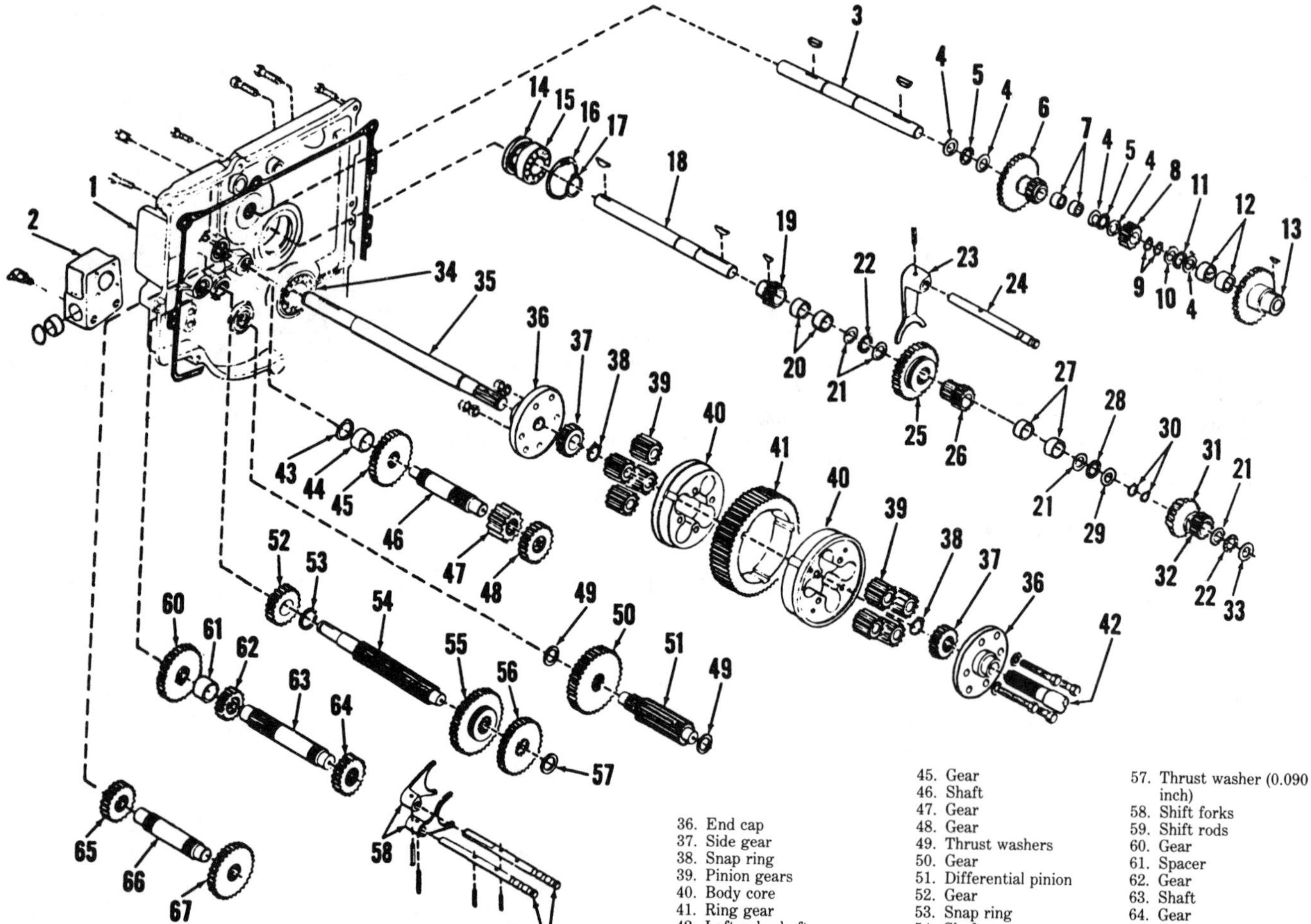

Fig. GR28—Exploded view of transaxle cover, shafts and gears used on all models except 450 model.

1. Cover
2. Right axle bearing retainer
3. Shaft
4. Bearing races (0.030 inch)
5. Thrust bearing
6. Idler gear
7. Needle bearings
8. Gear
9. Snap rings
10. Bearing race (0.090 inch)
11. Thrust bearing
12. Needle bearings
13. Reverse gear
14. Seal
15. Ball bearing
16. Snap ring
17. Snap ring
18. Shaft
19. Gear & bearing assy.
20. Needle bearings
21. Bearing races (0.030 inch)
22. Thrust bearing
23. High-Low shift fork
24. Shift rod
25. High-Low gear
26. Gear & bearing assy.
27. Needle bearings
28. Thrust bearing
29. Bearing race
30. Snap rings
31. Bevel gear
32. Gear
33. Bearing race (0.090 inch)
34. Ball bearing
35. Right axle shaft
36. End cap
37. Side gear
38. Snap ring
39. Pinion gears
40. Body core
41. Ring gear
42. Left axle shaft
43. Thrust washer
44. Spacer
45. Gear
46. Shaft
47. Gear
48. Gear
49. Thrust washers
50. Gear
51. Differential pinion
52. Gear
53. Snap ring
54. Shaft
55. Sliding gear
56. Sliding gear
57. Thrust washer (0.090 inch)
58. Shift forks
59. Shift rods
60. Gear
61. Spacer
62. Gear
63. Shaft
64. Gear
65. Gear
66. Shaft
67. Gear

Illustrations courtesy Gravely.

Block rear wheels and support front of transmission. Remove transmission to frame bolts and roll frame and forward section of tractor away from transmission.

Reinstall by reversing removal procedure.

OVERHAUL

Model 450

FORWARD-REVERSE SYSTEM. Forward-reverse clutch and gears are housed in the front part of the transaxle. Remove advance housing (52–Fig. GR20) and rotate forward-reverse actuating shaft (2) so planetary system is released. Remove clutch and planetary components (45 through 50 and 25 through 36). Note spacer plate bolts (25) have left hand threads. Inspect components for damage or excessive wear. Inspect clutch cup (48) friction surfaces and renew cup if excessively worn. Inspect surfaces of reverse gear (49) and internal gear (47) and renew as required.

Reassemble by reversing disassembly procedure. Orbit gears must be timed when installed. When orbit gears are meshed correctly with sun gear (34), small punch marks on orbit gears (31) will form an equilateral triangle.

HIGH-LOW SYSTEM. A high-low planetary gear set and clutch are located in rear of transaxle. To disassemble unit, rotate high-low actuating shaft (4–Fig. GR20) to release planetary system and remove components (5 through 22).

Inspect parts for damage or excessive wear. Inspect clutch cup (7) friction surface, clutch mating surface on clutch plate (12) and ring gear (8). Renew parts as necessary.

Reassemble by reversing disassembly procedure.

SWIFTAMATIC DIFFERENTIAL. Swiftamatic uses a two-speed axle with a differential assembly. To disassemble, remove retaining cap screws and remove right axle housing (33–Fig. GR21). Check to see that stationary clutch (31) does not move in axle housing and that sliding clutch (29) will slide freely on splines of shifter gear (23). To disassemble differential, remove left axle housing (1) and complete differential assembly components (5 through 28). Inspect ring gear (20) and worm gear (43–Fig. GR20). If renewal of worm gear is required, refer to FORWARD-REVERSE section and remove forward-reverse assembly. Remove bearing nut (39) and withdraw worm gear and shaft assembly.

Reassemble by reversing disassembly procedure. Tighten bearing adjustment nut (39) so end play is removed from worm gear shaft. Lock nut in place with screw and lock plate (38).

Note difference between early and late differential spider gear assemblies shown in Figs. GR21 and GR22.

NOTE: On models with ring gear assembly shown in Fig. GR21, some models have only two pinion assemblies (13 through 18) and should be converted by adding two additional pinion assemblies to the pins of cross pin (19) as shown in Fig. GR22.

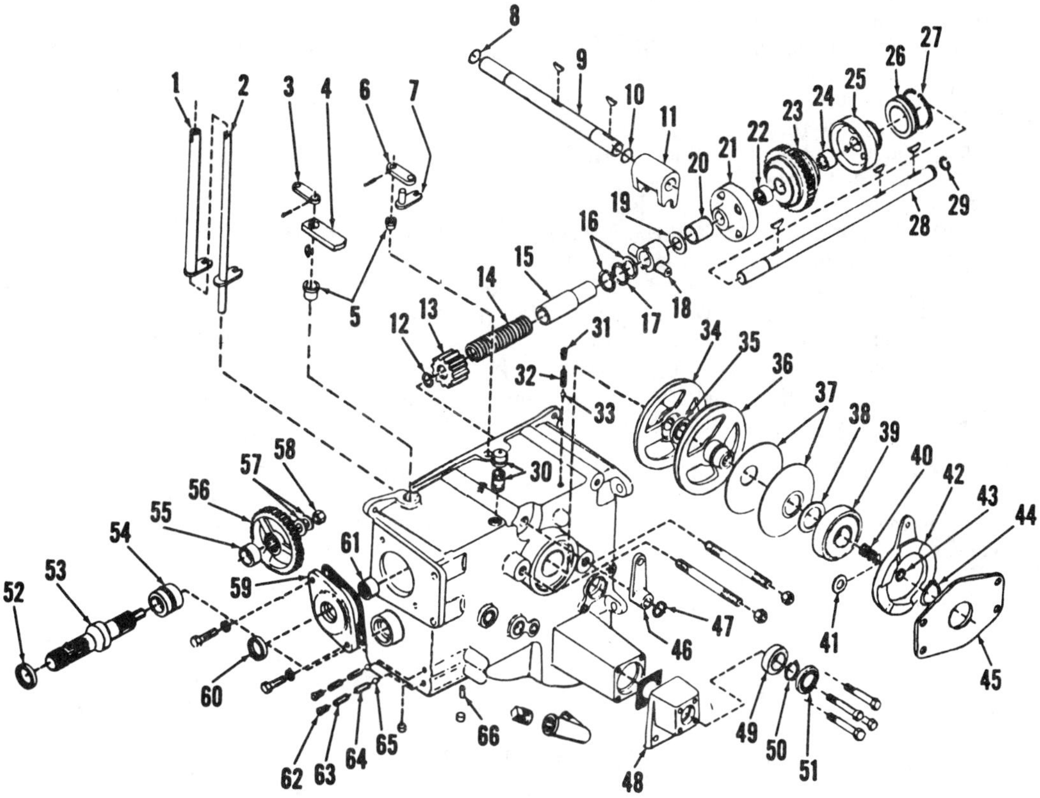

Fig. GR29—Exploded view of transaxle used on all models except 450 models.

1. Outer shift tube
2. Inner shift shaft
3. Shaft shifter arm
4. Tube shifter arm
5. Bushings
6. High-Low shift lever
7. High-Low shift arm
8. "O" ring
9. Shaft
10. "O" ring
11. Pto yoke
12. Snap ring
13. Pto gear
14. Spring
15. Pto throw-out sleeve
16. Bearing races
17. Thrust bearing
18. Pto shift collar
19. Thrust washer
20. Spacer
21. Clutch cup
22. Needle bearing
23. Clutch cone
24. Needle bearing
25. Clutch cup
26. Ball bearing
27. Snap ring
28. Pto shaft
29. Snap ring
30. Fill pipe & cap
31. Plug
32. Spring
33. High-Low detent ball
34. Clutch hub
35. Oil seal
36. Clutch disc
37. Belleville springs
38. Shim
39. Ball bearing
40. Spring
41. Washer
42. Clutch cam
43. Snap ring
44. Snap ring
45. Clutch plate
46. Pto arm
47. Snap ring
48. Left axle bearing retainer
49. Ball bearing
50. Snap ring
51. Seal
52. Seal
53. Pto shaft
54. Bearing
55. Spacer
56. Pto gear
57. Washers
58. Locknut
59. Bearing cap
60. Seal
61. Bearing
62. Plug
63. Plug
64. Spring
65. Detent ball
66. Pin

Illustrations courtesy Gravely.

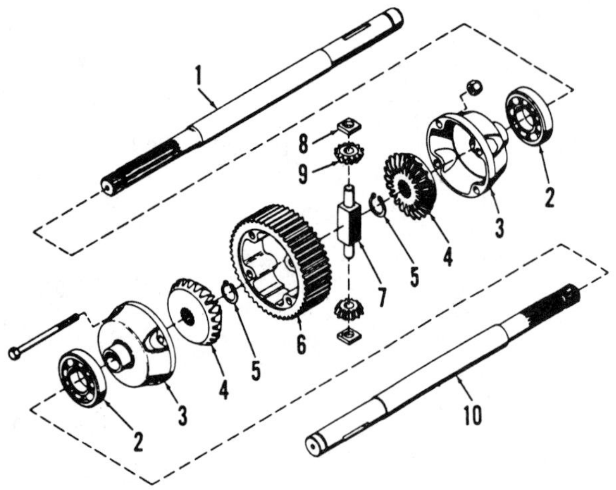

Fig. GR30—Exploded view of differential assembly used on 816 models after S/N 18930 and all other models except 450 model.

1. Axle shaft
2. Ball bearing
3. End cap
4. Bevel gear
5. Snap ring
6. Ring gear
7. Pinion shaft
8. Drive block
9. Spider gear
10. Axle shaft

On early type differential assemblies adjust gear mesh by installing pinion gear assembly components (13 through 18–Fig. GR21) in ring gear (20). Install gears (9 and 10) and carrier (6) on ring gear. Adjust thickness of shims (17) and until gears have solid contact without wobble.

On all models, tighten differential carrier cap screws to 20 ft.-lbs. (27 N•m). To determine correct thickness of shims (2) to be installed, install reassembled differential assembly and place equal thickness of shims (2) with axle housings (1 and 33) and install axle housings. Secure axle housings with 3 cap screws and measure axle end play. End play should be 0.020 inch (0.508 mm). Change thickness of shims (2) as required. Shim pack should be equal on both axle housings. Tighten axle housing cap screws to 45 ft.-lbs. (61 N•m). Fill transaxle housing to correct level with specified fluid.

All Other Models

FORWARD-REVERSE CLUTCHES. Remove transaxle as previously outlined. Remove forward and reverse clutches.

Refer to Fig. GR25 for 816 models or to Fig. GR26 for all other models.

To disassemble clutch shown in Fig. GR25, remove snap ring (1), washer (2) and spring (3). Compress Belleville springs (5), remove snap ring (9) and disassemble clutch.

Reassemble by reversing disassembly procedure. Note Belleville springs (5) are installed with cupped sides together at outer edges.

To disassemble clutch shown in Fig. GR26, press housing (1) off clutch hub (4) using spacers through the three slots in clutch hub. Remove snap ring (2) and press bearing (3) from hub.

Reassemble by reversing disassembly procedure.

TRANSAXLE. Remove right wheel and hub. Remove brake assembly from right side and "E" ring (E–Fig. GR27) from pto shaft. Clean dirt or paint from right axle and retainer and remove right axle bearing retainer (2–Fig. GR28).

Remove cover screws and lift transaxle cover (1) off case. Gear and bearing (19) should remain in bearing (15). Remove gear (6) and bearings. Remove plug (31–Fig. GR29), spring (32) and ball (33) and remove range shifter fork (23–Fig. GR28) along with high-low gear and shaft assembly (20 through 33). Remove gear (50) and pinion shaft (51) and slide gear (52) and shaft (54) out of sliding gears. Remove gears and shaft (60 through 64) assembly. Remove gear (65), then withdraw gears and shaft (43 through 48). Remove shaft (66) and gear (67). Remove plugs (62–Fig. GR29), remove shift detent balls and springs and remove shift forks (58–Fig. GR28) and gears (55 and 56) noting position of forks and gears. Note interlock pin (66–Fig. GR29) may be dislodged. Remove reverse shaft (3–Fig. GR28) and gears (8 and 13) and bearings.

Slide pto yoke shaft (9) through case and remove pto yoke (11). Remove bearing cap (59), snap ring (29) and pto shaft assembly (12 through 28). Disassemble pto assembly using exploded view in Fig. GR29 as a guide. Remove differential assembly from transaxle case. Disassemble differential as shown in Fig. GR28 or GR30. Bevel gear differential shown in Fig. GR30 is used on 816 models after S/N 18930.

To reassemble transaxle, reverse disassembly procedure. Tighten differential bolts to 25-30 ft.-lbs. (34-40 N•m). Note location of 0.090 inch thick bearing races (10, 33 and 57–Fig. GR28) and 0.030 inch thick bearing races (4 and 21).

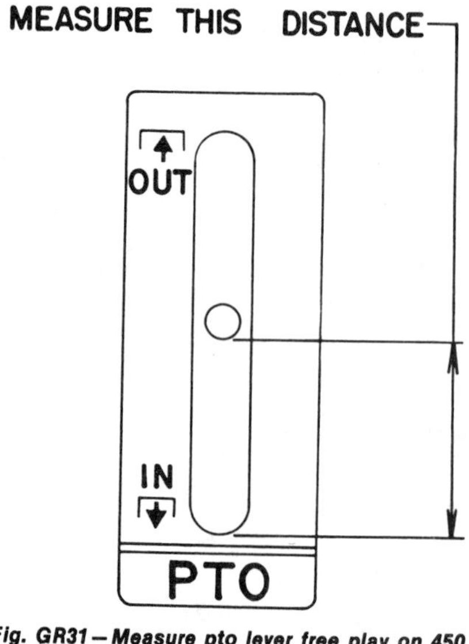

Fig. GR31—Measure pto lever free play on 450 and 816 models as indicated. Refer to text.

POWER TAKE-OFF

ADJUSTMENT

Models 450-816

Adjust length of pto rod to obtain 1³/₈ to 1¹/₂ inches (34.925 to 38.100 mm) pto lever free play with pto engaged (Fig. GR31).

All Other Models

Adjust length of pto rod to obtain ¹/₂ to 1 inch (12.7 to 25.4 mm) pto lever free play with pto engaged (Fig. GR32).

R&R AND OVERHAUL

All Models

Refer to appropriate TRANSAXLE section of this manual for pto service procedures.

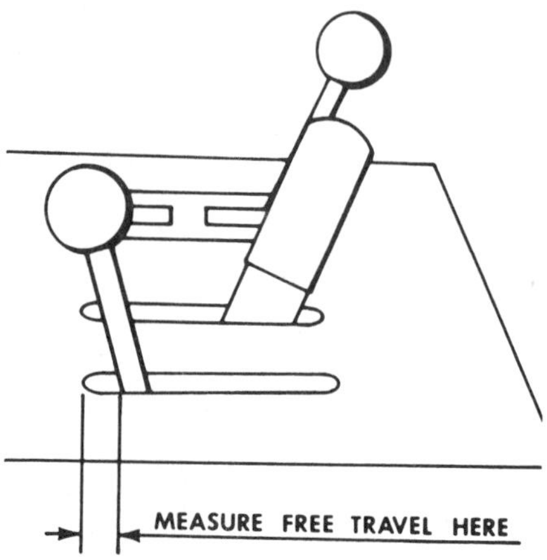

Fig. GR32—Measure pto lever free play on all models except 450 and 816 models as indicated. Refer to text.

Illustrations courtesy Gravely.

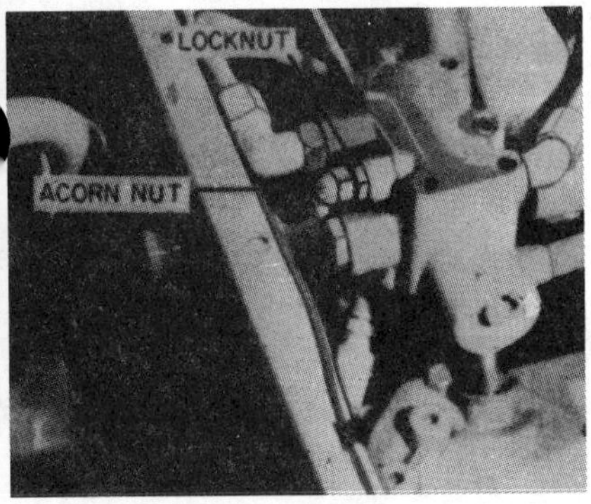

Fig. GR33—Remove acorn nut, loosen locknut and adjust screw to obtain 950-1000 psi (6550-6895 kPa). Pressure must not exceed 1000 psi (6895 kPa).

HYDRAULIC SYSTEM

MAINTENANCE

All Models So Equipped

Hydraulic reservoir fluid level should be checked at 200 hour intervals. Maintain fluid level at "FULL" mark on dipstick with dipstick just inserted in its opening. Do not screw dipstick in and then out to check level. Recommended hydraulic fluid is Automatic Transmission Fluid, Type A, Suffix A. Approximate capacity is 1.5 quarts (1.4 L).

PRESSURE CHECK

All Models Equipped With Hydraulic Lift

Install a 90° "T" fitting to "out" port of hydraulic pump. Connect hydraulic line at one side of "T" and a 1500 psi (10342 kPa) test gage in remaining side of "T". Start engine and run at full throttle. Operate lift control lever until cylinder reaches end of travel. Correct pressure is 950-1000 psi (6550-6895 kPa) when lift cylinder reaches end of travel. If pressure is low, remove acorn nut (Fig. GR33), loosen locknut and turn screw to obtain recommended pressure. Do not exceed 1000 psi (6895 kPa). If correct pressure cannot be obtained, renew pump and/or control valve as necessary.

PUMP AND CONTROL VALVE

All Models Equipped With Hydraulic Lift

Refer to Figs. GR34 and GR35 for view of hydraulic system pump, control valve, lift cylinder and reservoir. Pump and control valve are serviced as assemblies only.

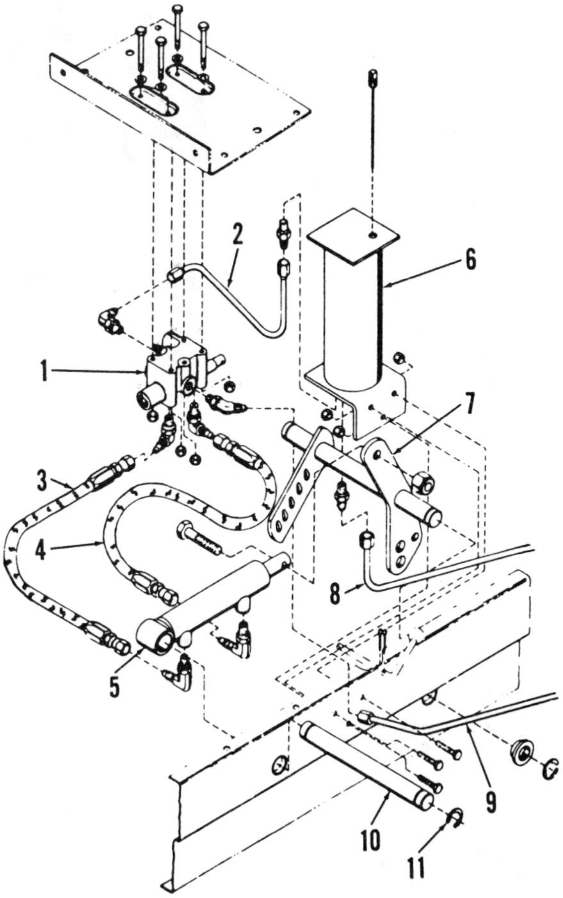

Fig. GR34—Exploded view of typical hydraulic system.

1. Control valve
2. Valve-to-reservoir line
3. Hydraulic hose
4. Hydraulic hose
5. Cylinder
6. Reservoir
7. Lift lever
8. Pump intake line
9. Pump exhaust line
10. Shaft
11. "E" ring

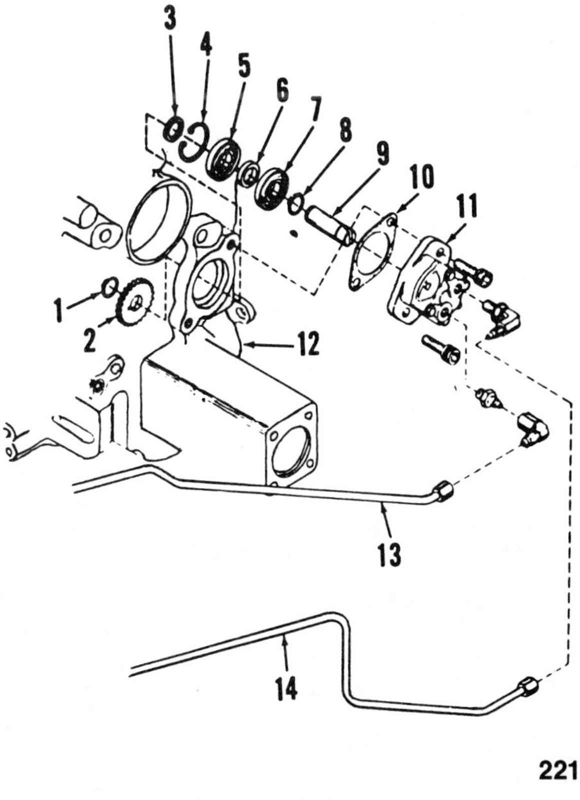

Fig. GR35—Exploded view of hydraulic pump drive.

1. Snap ring
2. Pump gear
3. Spacer
4. Snap ring
5. Ball bearing
6. Spacer
7. Ball bearing
8. Snap ring
9. Shaft
10. Gasket
11. Hydraulic pump
12. Transaxle case
13. Intake line
14. Exhaust line

INTERNATIONAL HARVESTER

CONDENSED SPECIFICATIONS

	MODELS			
	582	582 Special	682,782	982
Engine Make	B&S	B&S	Kohler	Onan
Model	401417	401417	KT-17	B48G
Bore	3.44 in. (87.3 mm)	3.44 in. (87.3 mm)	3.12 in. (79.3 mm)	3.25 in. (82.55 mm)
Stroke	2.16 in. (54.8 mm)	2.16 in. (54.8 mm)	2.75 in. (69.9 mm)	2.88 in. (76.03 mm)
Displacement	40 cu. in. (656 cc)	40 cu. in. (656 cc)	42.18 cu. in. (690.5 cc)	47.7 cu. in. (781.7 cc)
Power Rating	16 hp. (11.9 kW)	16 hp. (11.9 kW)	17 hp. (12.7 kW)	20 hp. (14.9 kW)
Idle Speed	1200 rpm	1200 rpm	1200 rpm	1200 rpm
High Speed (No-Load)	3600 rpm	3600 rpm	3600 rpm	3600 rpm
Capacities–				
Crankcase	3 pts.* (1.4 L)	3 pts.* (1.4 L)	3 pts. (1.4 L)	3.5 pts. (1.6 L)
Transmission Or Transaxle	8 pts. (3.8 L)	4 pts. (1.9 L)	7 qts. (6.6 L)	7 qts. (6.6 L)
Creeper Drive	½-pt. (0.24 L)	.....	.....	.....
Right Angle Gear Box	.....	4 oz. (118 mL)	.....	.....
Fuel Tank	4 gal. (15.1 L)	4 gal. (15.1 L)	4 gal. (15.1 L)	4 gal. (15.1 L)

*Early production engine crankcase capacity is 3.5 pints (1.6 L).

FRONT AXLE AND STEERING SYSTEM

MAINTENANCE

All Models

Lubricate steering spindles and steering arm at 10 hour intervals. Lubricate axle pivot bolt at 30 hour intervals and steering gear box at 100 hour intervals.

NOTE: Do not over lubricate steering gear box. Four strokes of a hand grease gun is sufficient. Do not use a high volume or high pressure grease gun.

Use multi-purpose, lithium base grease and clean all fittings before and after lubrication.

R&R AXLE MAIN MEMBER

Models 582 – 582 Special – 682 – 782

To remove front axle main member, disconnect drag link from left hand steering spindle. Support front of tractor and remove axle pivot bolt (3 – Fig. IH1). Raise tractor and roll axle out.

Reinstall by reversing removal procedure.

Model 982

To remove front axle main member (2 – Fig. IH2), disconnect drag link (7 – Fig. IH3) from steering lever (5). Support front of tractor and remove axle pivot bolt (4 – Fig. IH2). Raise front of tractor and roll axle out.

Reinstall by reversing removal procedure.

TIE ROD, STEERING LEVER AND TOE-IN

All Models

A single tie rod with renewable ends and a single drag link with renewable ends are used on 582, 582 Special, 682 and 782 models.

Two tie rods with renewable ends, a

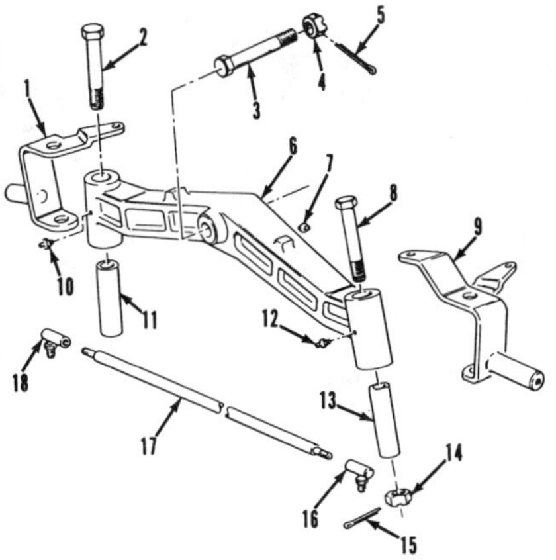

Fig. IH1—Exploded view of front axle assembly used on 582, 682 and 782 models.

1. Right steering spindle
2. Spindle bolt
3. Axle pivot bolt
4. Nut
5. Cotter pin
6. Axle main member
7. Grease fitting
8. Spindle bolt
9. Left steering spindle
10. Grease fitting
11. Bushing
12. Grease fitting
13. Bushing
14. Nut
15. Cotter pin
16. Tie rod end
17. Tie rod
18. Tie rod end

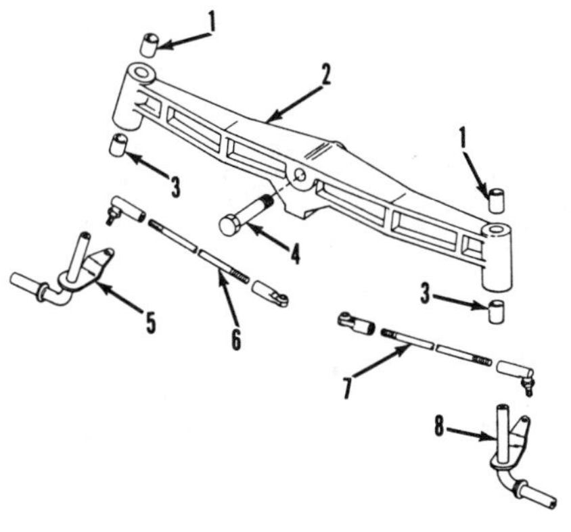

Fig. IH2—Exploded view of front axle assembly used on 982 models.

1. Upper bushing
2. Axle main member
3. Lower bushing
4. Axle pivot bolt
5. Right wheel spindle
6. Tie rod
7. Tie rod
8. Left wheel spindle

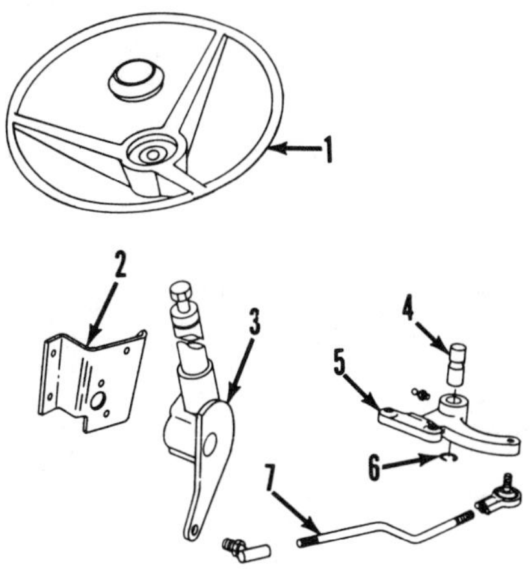

Fig. IH3—View of steering column assembly and steering lever.

1. Steering wheel
2. Bracket
3. Steering gear & column assy.
4. Steering lever pivot pin
5. Steering lever
6. Snap ring
7. Drag link

single drag link with renewable ends and a steering lever mounted on axle main member are used on 982 models.

Steering lever pivot pin (4–Fig. IH3) for 982 model is retained by a snap ring (6). Pin is threaded for slide hammer to aid removal.

Toe-in for all models should be 1/32 to 1/8-inch (0.794 to 3.175 mm) and is adjusted on 582, 582 Special, 682 and 782 models by lengthening or shortening tie rod (17–Fig. IH1) as necessary. Toe-in for 982 model is adjusted by lengthening and shortening each of the two tie rods (6 and 7–Fig. IH2) as necessary. Tie rods should be kept as nearly equal in length as possible.

R&R STEERING SPINDLES

Models 582—582 Special—682—782

Raise and support front of tractor. Remove front tires, wheels and hubs, disconnect tie rod (17–Fig. IH1) at steering arms (1 and 9) and drag link at steering arm (9). Remove bolts (2 and 8) and pull spindles (1 and 9) from axle main member (6).

Reinstall by reversing removal procedure. Renew bushings (11 and 13) and/or bolts (2 and 8) as necessary. Tighten nuts (14) on bolts to 80 ft.-lbs. (108 N·m).

Model 982

Raise and support front of tractor. Remove front tires, wheels and hubs. Disconnect tie rods (6 and 7–Fig. IH2) at spindles (5 and 8). Remove cap screws and flat washers from upper ends of spindles (5 and 8) and pull spindles from axle main member (2).

Reinstall by reversing removal procedure. Renew bushings (1 and 3) and/or spindles (5 and 8) as necessary.

FRONT WHEEL BEARINGS

All Models

Front wheel bearings are a press fit in wheel hub and a slip fit on spindle. Bearings should be removed and cleaned at 100 hour intervals. Repack bearings with multi-purpose, lithium base grease.

To remove bearings, raise and support front of tractor. Remove wheel and tire assembly. Bearings can be driven from wheel hub using a long drift punch and hammer. Bearings should be driven from inside wheel hub toward outside of hub. If bearings are a loose fit in wheel hub, renew or repair hub.

Reinstall by reversing removal procedure.

R&R STEERING GEAR

Models 582—582 Special—682—782

Remove steering wheel cover and steering wheel retaining nut. Use suitable puller and remove steering wheel.

On 782 model, disconnect hydraulic lines from transmission at control valve. Disconnect linkage at control valve and remove valve assembly. Mark location of control valve mounting plate on steering column and remove plate.

On all models, disconnect drag link from steering lever. Remove mounting bolts and pull steering assembly down through control panel and remove.

Reinstall by reversing removal procedure.

Model 982

Remove steering wheel cover and

Fig. IH4—Exploded view of 582, 682 and 782 models steering gear and column assembly.

1. Steering wheel
2. Bearing
5. Steering gear & shaft assy.
6. Steering pin
7. Jam nut
8. Drag link
9. Ball joint end
10. Ball joint end
11. Steering lever plate
12. Seal retainer
13. Seal
14. Housing
15. Washer
16. Adjusting nut
17. Jam nut
18. Snap ring
19. Bearing retainer
20. Bearing balls
21. Ball retainers
22. Adjusting plug
23. Cotter pin

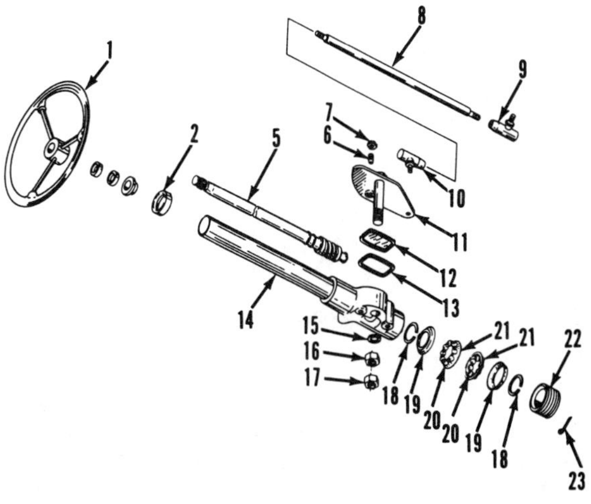

steering wheel retaining nut. Use suitable puller and remove steering wheel. Shut off fuel at tank, disconnect fuel line. Remove fire wall and fuel tank as an assembly. Disconnect drag link at steering lever. Disconnect hydraulic lines from transmission at control valve and remove valve assembly. Mark location of control valve mounting plate on steering column and remove plate. Remove the three steering column retaining bolts and pull steering assembly down through control panel and remove.

Reinstall by reversing removal procedure.

STEERING GEAR OVERHAUL

Models 582 – 582 Special – 682 – 782

Remove steering column and gear assembly as previously outlined. Remove steering lever jam nut (17 – Fig. IH4), adjusting nut (16) and washer (15). Remove steering lever (11), seal retainer (12) and seal (13). Remove cotter pin (23) and adjusting plug (22). Pull steering gear and bearing assembly from housing. Remove bearing race retainer snap rings (18). Remove bearing retainers (19), bearing balls (20) and retainers (21). Remove jam nut (7) and steering pin (6) from steering lever (11).

Inspect all parts for wear, cracks or damage and renew as necessary.

To reassemble, coat steering gear with multi-purpose, lithium base grease. Install retainers (21), bearing balls (20) and retainers (19) on steering gear (5). Install snap rings (18). Install steering gear and bearing assembly in housing (14). Make certain bearing retainers (19) enter bore of housing squarely and do not bind. Install adjusting plug (22) and tighten until all end play is removed from shaft and gear. Shaft and gear assembly should turn freely. Install cotter pin (23). Pack housing with multi-purpose, lithium base grease. Install seal (13), retainer (12) and steering lever plate (11). Place a 3/32-inch (2.4 mm) thick shim stock between steering lever plate (11) and seal retainer (12). Install washer (15) and adjusting nut (16). Tighten adjusting nut (16) snug and remove the 3/32-inch (2.4 mm) shim stock. Install jam nut (17) and tighten to 40 ft.-lbs. (54 N•m) while holding adjusting nut in place. Install steering pin (6) in steering lever plate until it just engages steering gear. Center steering gear by rotating shaft to position half-way between full left and full right turn. Adjust steering pin (6) inward until all backlash is removed but steering gear turns through full left/right cycle with no binding. Install jam nut (7) and tighten to 40 ft.-lbs. (54 N•m) while holding steering pin in proper position.

Model 982

To disassemble steering gear, remove jam nuts (5 – Fig. IH5). Remove steering lever plate (14). Remove cotter pin (11), adjustment plug (10) and slide steering shaft assembly (8) out of housing. Remove Belleville washer (9), retainer cups (6) and bearings (7).

Inspect parts for wear, scoring or damage and renew as necessary.

To reassemble, apply multi-purpose lithium base grease to bearings (7) and retainer cups (6) and install on steering shaft (8). Apply grease to steering gear and install assembly into housing (3). Install Belleville washer (9) and plug (10). Tighten plug to 4-6 ft.-lbs. (5-8 N•m) and install cotter pin (11). Shaft should turn freely in housing. Pack housing with grease and install new seal (16) and retainer (15) on housing. Remove steering pin (13) from steering lever plate (14) and install steering lever plate on housing. Place a 5/32-inch (4 mm) shim between steering lever plate and seal retainer, install washer (4) and tighten one jam nut (5) until just snug. Install second jam nut and tighten against first nut to 22-25 ft.-lbs. (30-34 N•m) while holding first nut in position on stud. Remove shim stock. Install steering pin in steering lever plate until it just engages steering gear. Center steering gear by rotating shaft to position half-way between full left and full right turn. Adjust steering pin (13) inward until all backlash is removed but steering gear turns through full left/right cycle with no binding. Tighten jam nut (12) to 35-45 ft.-lbs. (48-61 N•m).

ENGINE

MAINTENANCE

All Models

Regular engine maintenance is required to maintain peak performance and long engine life.

Check oil level and clean air intake screen at five hour intervals under normal operating conditions.

Clean engine air filter and cooling fins at 25 hour intervals and check for any loose nuts, bolts or linkage. Repair as necessary.

Change engine oil and filter, perform tune-up, valve adjustment and clean carbon from cylinder heads as recommended in appropriate ENGINE section of this manual.

REMOVE AND REINSTALL

Models 582 – 582 Special

To remove engine, disconnect battery ground cable. Raise hood and remove engine side panels which are secured with wing nuts and a spring. Disconnect headlight wiring and remove hood and grille as an assembly. Disconnect alternator-regulator wire, starter cables and pto clutch wire. Remove air cleaner assembly and disconnect choke and throttle cables. Disconnect engine shut-off wire. Shut off fuel and disconnect fuel line at carburetor. Remove engine mounting bolts, slide engine forward and remove.

Reinstall by reversing removal procedure.

Models 682 – 782

To remove engine, disconnect battery ground cable. Raise hood and remove engine side panels which are secured

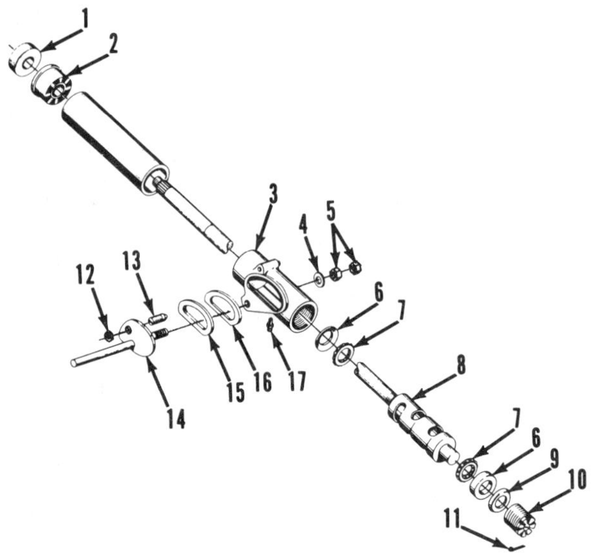

Fig. IH5 – Exploded view of 982 models steering gear and column assembly.

1. Dust cover
2. Upper bearing
3. Housing
4. Washer
5. Nuts
6. Cup
7. Bearing
8. Steering gear
9. Belleville washer
10. Adjustment plug
11. Cotter pin
12. Nut
13. Steering pin
14. Steering lever plate
15. Seal retainer
16. Seal
17. Grease fitting

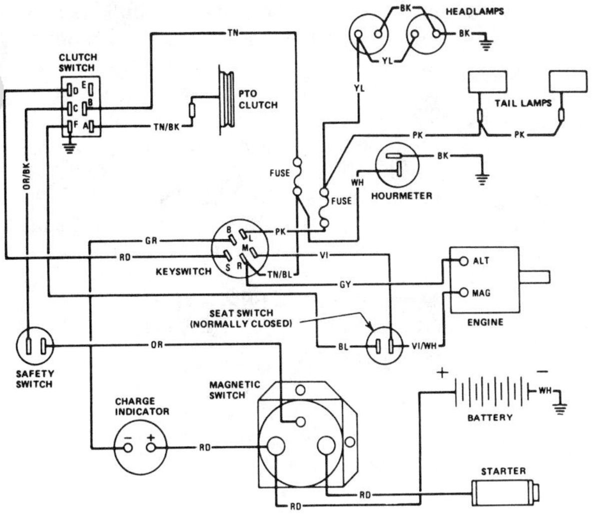

Fig. IH5A – Wiring diagram for 582 and 582 Special models.

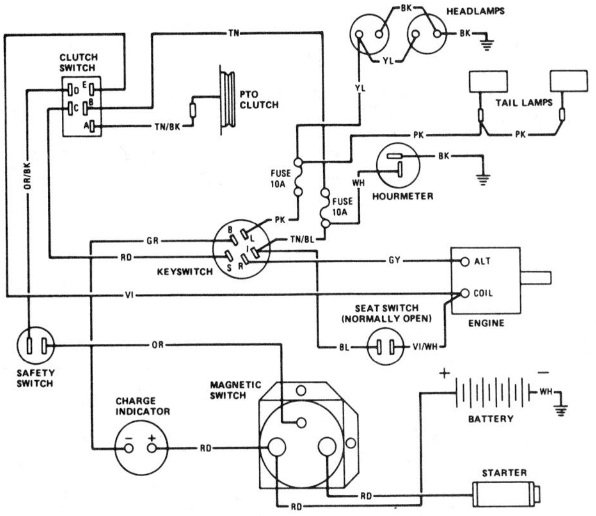

Fig. IH5B – Wiring diagram for 682 and 782 models.

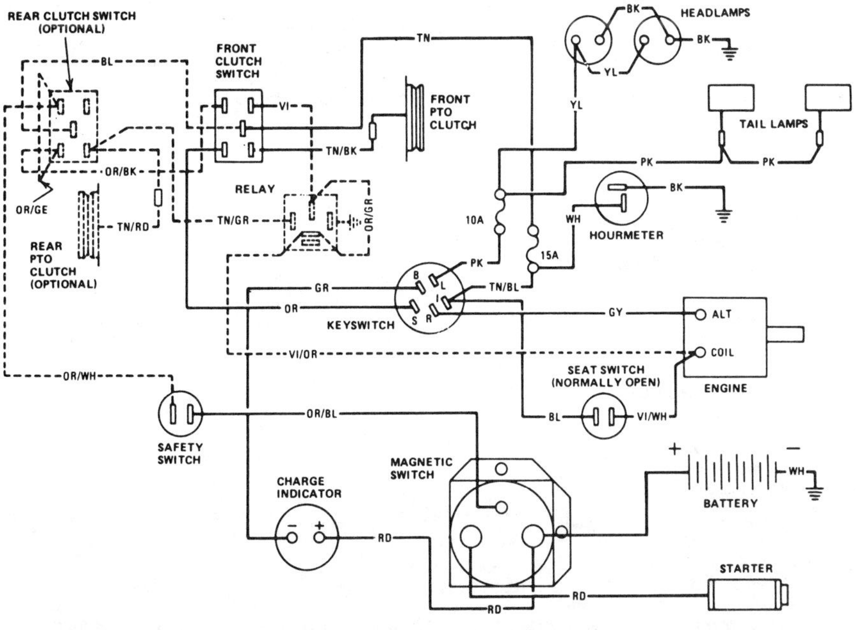

Fig. IH5C – Wiring diagram for 982 model.

with wing nuts and a spring. Disconnect headlight wiring and remove hood and grille as an assembly. Remove air cleaner assembly and disconnect choke cable, throttle cable and wiring harness. Disconnect pto clutch wire and starter cable. Shut off fuel and disconnect fuel line at tank. Remove front flex coupler to flywheel flange retaining nuts. Remove engine mounting bolts and remove engine.

Reinstall by reversing removal procedure.

Model 982

To remove engine, disconnect battery ground cable. Raise hood and remove engine side panels which are secured with wing nuts and a spring. Disconnect pto clutch wire and starter cable. Remove air cleaner assembly and disconnect choke and throttle cables. Shut off fuel and disconnect fuel line from fuel pump. Disconnect positive terminal coil wire and wire at rectifier. Remove engine mounting bolts and remove engine.

Reinstall by reversing removal procedure.

OVERHAUL

All Models

Engine make and model are listed at the beginning of this section. To overhaul engine or accessories, refer to appropriate engine section of this manual.

ELECTRICAL SYSTEM

MAINTENANCE AND SERVICE

All Models

Battery electrolyte level should be checked at 50 hour intervals of normal operation. If necessary, add distilled water until level is just below base of vent well. **DO NOT** overfill. Keep battery posts clean and cable ends tight.

For alternator or starter service, refer to appropriate engine accessory section in this manual.

Refer to Fig. 5A through 5C for appropriate wiring diagram for model being serviced.

BRAKES

ADJUSTMENT

Model 582 (Prior To S/N 720000)

Internal type brake on 582 model (prior to S/N 720000) should engage

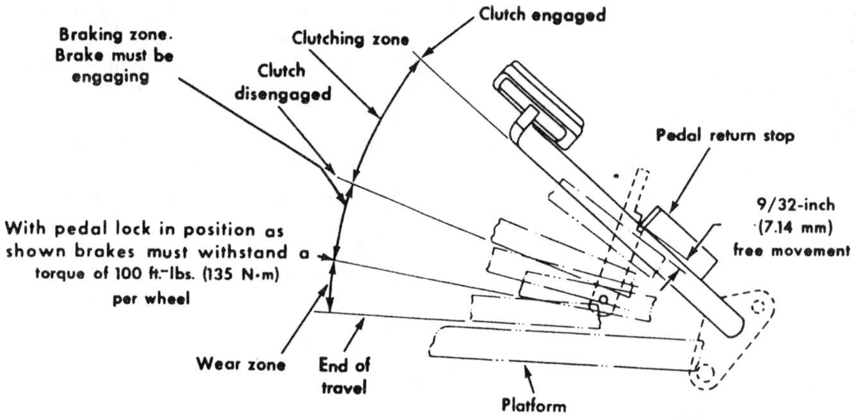

Fig. IH6—Brake should engage when pedal is pushed down to within a maximum of 1-5/16 inches (33 mm) and a minimum of 3/4-inch (19 mm) above top of platform.

when pedal arm is pressed down to within a maximum of 1-5/16 inches (33 mm) and a minimum of ¾-inch (19 mm) distance above top of platform (Fig. IH6).

To adjust, loosen locknut (9–Fig. IH7) and turn adjusting bolt (10) in or out as required. Brake pedal is spring loaded and pedal may be pushed down against platform even if correctly adjusted.

Model 582 Special

To adjust disc type brake on 582 Special model, place brake pedal in up position (Fig. IH8). Remove all slack from brake linkage by moving brake arm up and pulling brake rod down. Adjust nut to allow 1/4 to 5/16-inch (6.4 to 7.9 mm) clearance between spacer and brake arm. Check brake operation.

Models 682–782 (Prior To S/N 720000)

To adjust internal type brakes on 682 and 782 models (prior to S/N 720000), raise and support rear wheels off the ground. With brake pedal in full up position (Fig. IH9), loosen jam nut (14–Fig. IH10) and tighten adjustment bolt (13) to 8-10 in.-lbs. (0.9-1.1 N·m). Operate brake pedal through several full strokes and readjust. If brake drags with pedal in full up position after adjustment, loosen adjustment screw slightly. Tighten jam nut while holding adjustment screw in correct position.

With brake properly adjusted, it should require 100 ft.-lbs. (135 N·m) torque to turn wheel with brake pedal locked in down position (Fig. IH9).

Models 582–682–782 (After S/N 720000) And All 982 Models

Raise and support rear wheels off the ground. Adjust left brake first. Disconnect brake rod (13–Fig. IH11) and at clevis and shorten or lengthen rod by turning clevis (14) until gap between inner brake pad and brake disc is 0.030-0.035 inch (0.762-0.889 mm).

Repeat procedure for right side.

R&R BRAKE PADS/DISCS

Model 582 (Prior To S/N 720000)

Drain transmission lubricant. Remove brake adjustment bolt (10–Fig. IH7)

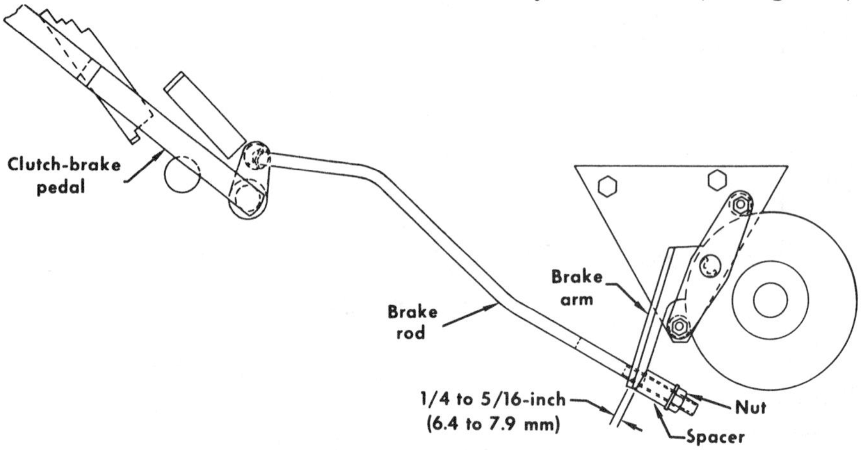

Fig. IH8—View of disc brake system used on 582 Special model. Refer to text for adjustment procedure.

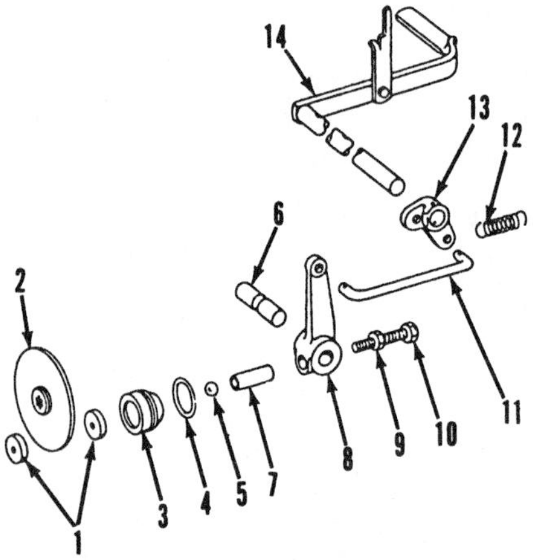

Fig. IH7—Exploded view of brake system used on 582 model prior to S/N 720000.

1. Brake pads
2. Brake disc
3. Retainer
4. "O" ring
5. Ball
6. Pivot pin
7. Push rod
8. Brake lever
9. Jam nut
10. Adjustment bolt
11. Brake rod
12. Return spring
13. Arm
14. Clutch & brake pedal

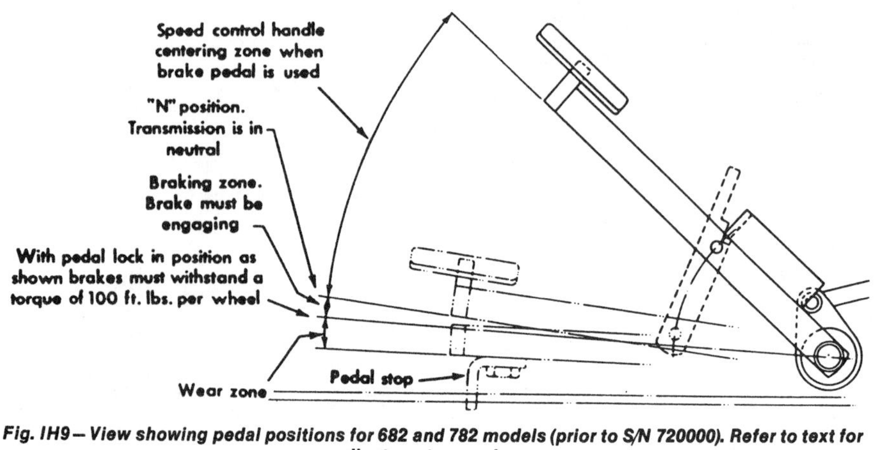

Fig. IH9—View showing pedal positions for 682 and 782 models (prior to S/N 720000). Refer to text for adjustment procedure.

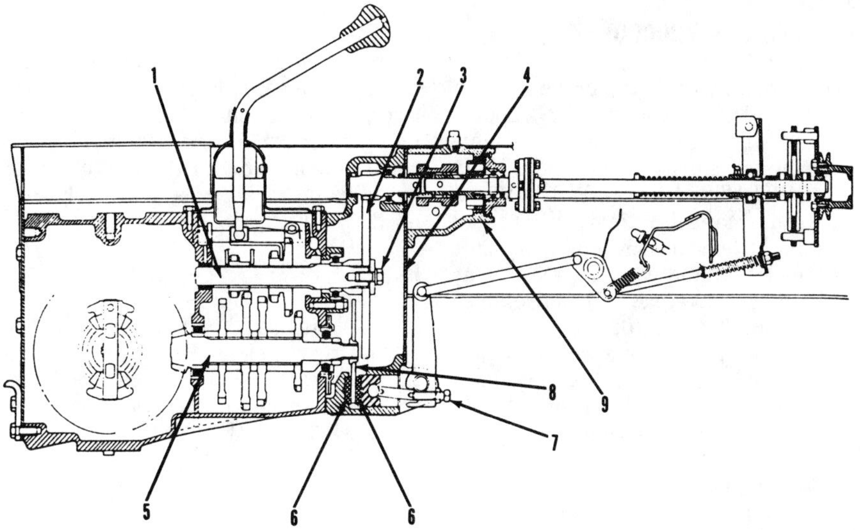

Fig. IH12—Sectional view of 582 model standard transmission.

1. Mainshaft
2. Reduction gear
3. Bolt
4. Cover
5. Countershaft
6. Brake pads
7. Adjustment bolt
8. Brake disc
9. Creeper drive unit

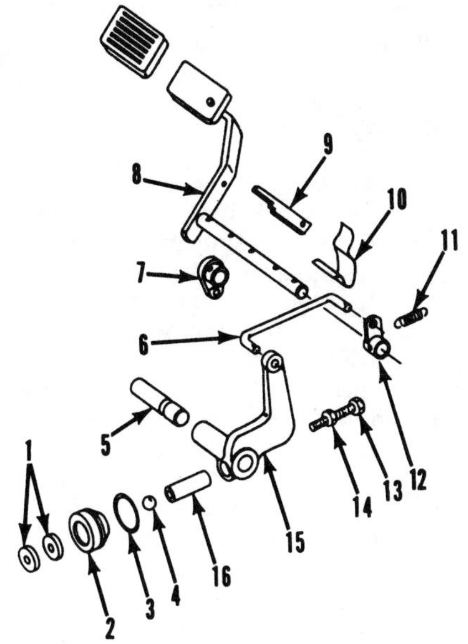

Fig. IH10—View showing component parts of 682 and 782 models (prior to S/N 720000) brake system. Refer to text.

1. Brake pads
2. Disc retainer
3. "O" ring
4. Ball
5. Pivot shaft
6. Brake rod
7. Neutral return lever
8. Pedal assy.
9. Brake lock
10. Spring
11. Brake return spring
12. Arm
13. Adjustment bolt
14. Jam nut
15. Brake lever
16. Push rod

and jam nut (9). Remove brake lever (8), pivot pin (6) and push rod (7). Remove creeper drive unit, if so equipped. Remove reduction housing front cover plate (4–Fig. IH12). Remove bolt (3) which secures reduction gear (2) to mainshaft (1). Remove reduction gear. Move brake disc (8) along countershaft (5) to remove brake pads (6). Both pads and disc may be removed without removing front retainer (3–Fig. IH7), however, "O" ring (4) should be inspected and renewed as necessary.

Reinstall by reversing removal procedure. Bolt (3–Fig. IH12) should be tightened to 55 ft.-lbs. (75 N·m). Make certain ball (5–Fig. IH7) is in place before installing push rod (7). Fill transmission to proper level with Hy-Tran, or equivalent. Adjust brake as previously outlined.

Model 582 Special

Remove nut and spacer from brake rod (Fig. IH8) and disconnect brake rod from actuator cam (5–Fig. IH13). Unbolt brake assembly from frame. Remove retaining ring from brake pad and disassemble brake.

Reinstall by reversing removal procedure. Adjust brake as previously outlined.

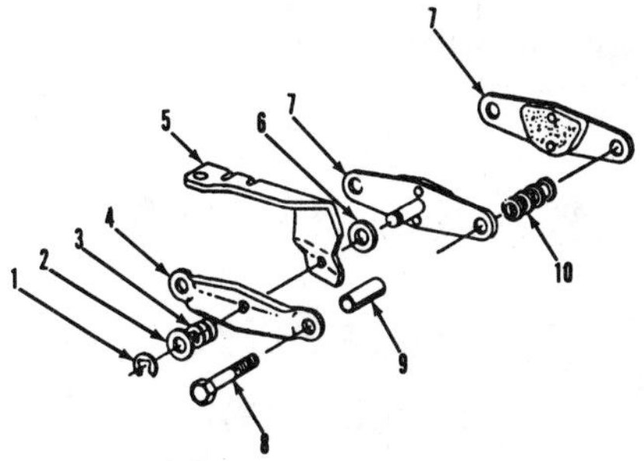

Fig. IH13—View of component parts of 582 Special model brake system.

1. Retaining ring
2. Thrust washer
3. Spring
4. Actuator plate
5. Cam
6. Hardened washer
7. Brake pads
8. Bolt
9. Spacer
10. Spring

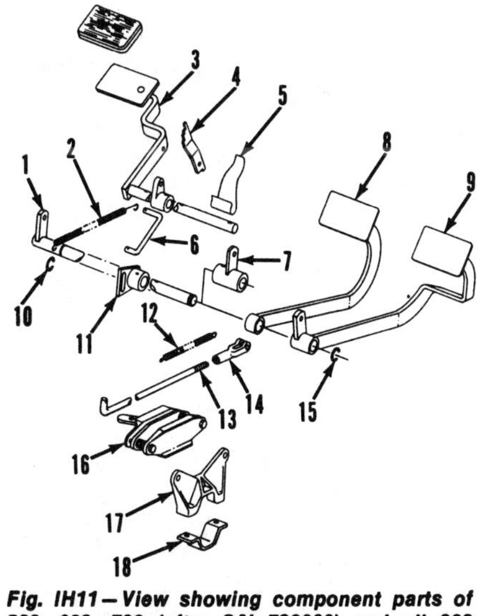

Fig. IH11—View showing component parts of 582, 682, 782 (after S/N 720000) and all 982 models. Single pedal system is standard, dual pedal system is optional.

1. Brake shaft
2. Return spring
3. Pedal assy.
4. Brake lock
5. Spring
6. Center brake rod
7. Right brake lever
8. Left brake pedal
9. Right brake pedal
10. Retaining ring
11. Arm assy.
12. Return spring
13. Brake rod
14. Clevis
15. Retaining ring
16. Brake assy.
17. Bracket
18. Flange

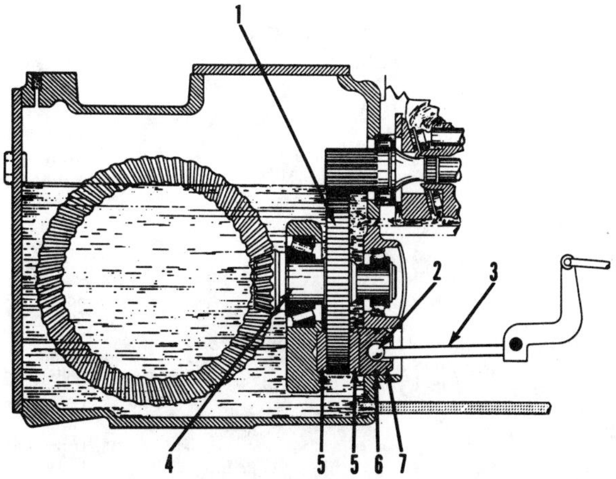

Fig. IH14—Sectional view of 682 and 782 models (prior to S/N 720000) reduction drive showing component parts of internal type brake system.

1. Constant mesh gear
2. Ball
3. Push rod
4. Pinion shaft
5. Brake pads
6. "O" ring
7. Retainer

Models 682 – 782 (Prior To S/N 720000)

To renew internal type brake pads on 682 and 782 models prior to S/N 720000, first drain transmission lubricant. Remove transmission as outlined in HYDROSTATIC DRIVE SYSTEM section of this manual. Remove differential assembly. Remove bevel pinion shaft (4 – Fig. IH14) and constant mesh gear (1). Remove brake pads (5) and retainer (7). Remove "O" ring (6).

Reinstall by reversing removal procedure. Fill transmission to proper level with Hy-Tran, or equivalent. Adjust brake as previously outlined.

Models 582 – 682 – 782 (After S/N 720000) And All 982 Models

Remove cotter pin and washer and disconnect brake rod from brake lever (1 – Fig. IH15). Remove cap screws from mounting flange and remove brake assembly from axle carrier. Remove cap screws (7) and disassemble caliper assembly.

Reassemble by reversing disassembly procedure. Make certain spacers (6) and springs (5) are correctly installed.

CLUTCH

ADJUSTMENT

Model 582

A clearance of 0.050 inch (1.27 mm) must be maintained between clutch release lever (10 – Fig. IH16) and clutch release bearing (9). Maintain 9/32-inch (7 mm) pedal free movement which is measured at pedal arm point of contact with front edge of pedal return stop. See Fig. IH6.

To adjust clearance, turn adjusting nut on clutch release rod (14 – Fig. IH16) as necessary.

Model 582 Special

Transaxle used on 582 Special models is belt driven. Clutch linkage which releases clutch belt tension is in conjunction with brake linkage. If belt is in good condition, proper adjustment is obtained by adjusting brakes as previously outlined. See Fig. IH8.

Models 682 – 782 – 982

Models 682, 782 and 982 have hydrostatic transmissions. Brake pedal, when fully depressed, should move speed control lever to neutral position. Refer to HYDROSTATIC DRIVE SYSTEM section of this manual for adjustment procedure.

R&R CLUTCH

Model 582

To remove clutch assembly, remove engine side panels and frame cover. Disconnect battery ground cable. Remove pivot pin and hanger assembly (11 – Fig. IH16). Remove bolts from flex coupling, drive roll pins out of driveshaft coupling and coupling arm. Slide couplings forward on clutch shaft and remove clutch shaft assembly including pressure plates, drive plate and clutch release lever. Remove drive plate, pressure plate and release lever from shaft. Clamp shaft in vise with loading spring (6) slightly compressed and remove roll pin. Carefully open vise jaws slowly. Remove spring and remaining parts.

Inspect all parts for wear or damage. Clutch loading spring (6 – Fig. IH16) free length should be 6.7 inches (170 mm) and should require 235-240 pounds (1053-1067 N) pressure to compress to 5.2 inches (132 mm). Teaser spring (17) free length should be 0.442 inch (11 mm) and should require 50 pounds (222 N) pressure to compress to 0.37 inch (9 mm).

Reinstall by reversing removal procedure. Adjust clutch as previously outlined.

Model 582 Special

Model 582 Special is equipped with a transaxle which is belt driven. Combination brake/clutch pedal, when depressed, releases drive belt tension. To renew drive belt, disconnect battery ground cable. Remove drawbar assembly and center frame cover. Depress clutch/brake pedal and lock in lowest position. Loosen drive belt guides and remove idler pulley. Remove the two mounting bolts securing right angle drive gear box to cross support and rotate gear box to bring pulley down. Remove drive belt.

Install new drive belt on drive pulley and transaxle pulley. Rotate gear box back into position and reinstall bolts. Install idler pulley. Release clutch/brake pedal and secure drive belt guides so there is 1/8 to 3/16-inch (3 to 5 mm) clearance between belt and guides. Install center frame cover, drawbar assembly and reconnect battery.

HYDROSTATIC DRIVE SYSTEM

MAINTENANCE

Models 682 – 782 – 982

Check fluid level and clean hydrostatic drive unit cooling fins at 30 hour intervals. Maintain fluid level at "FULL" mark on dipstick which is located just in front of tractor seat. Change transmission filter at 100 hour intervals and change fluid and filter at 500 hour intervals. Recommended fluid is International Harvester Hy-Tran, or equivalent. Approximate fluid capacity is 7 qts. (6.6 L).

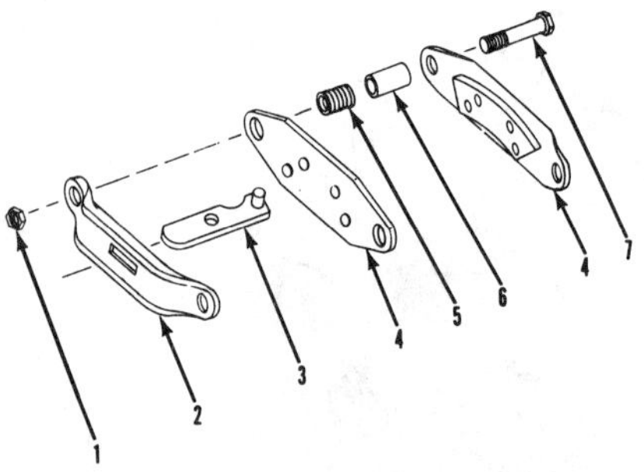

Fig. IH15 – Exploded view of brake caliper and component parts used on 582, 682, 782 models (after S/N 720000) and all 982 models.

1. Nut
2. Cam bracket
3. Cam
4. Pads
5. Spring
6. Spacer
7. Bolt

Fig. IH16 – Exploded view of clutch system used on 582 models with standard transmission. Bushing (7) and clutch release bearing (9) are a single unit on some models.

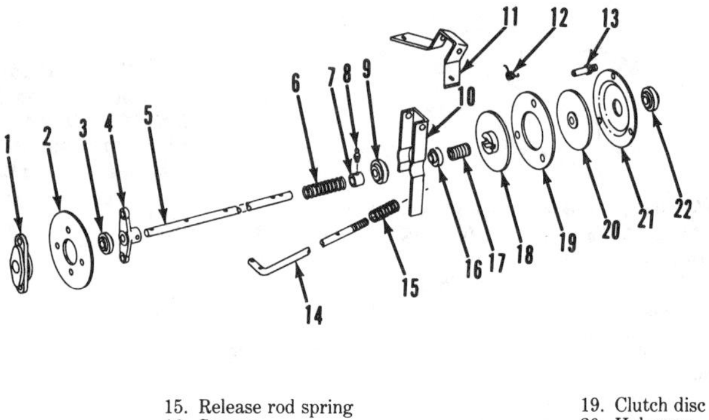

1. Coupler
2. Flex coupler disc
3. Bushing
4. Coupling arm
5. Shaft
6. Spring
7. Bushing
8. Grease fitting
9. Clutch release bearing
10. Release lever
11. Hanger
12. Spring
13. Drive stud
14. Release rod
15. Release rod spring
16. Spacer
17. Teaser spring
18. Hub assy.
19. Clutch disc
20. Hub assy.
21. Disc
22. Bushing

ADJUSTMENT

Models 682 – 782 – 982

SPEED CONTROL LEVER ADJUSTMENT. Speed control lever should require a pulling force of 7 to 8 pounds (3.2 to 3.6 kg) to maintain speed control lever position. Adjust by removing left side engine panel and tighten or loosen friction control nut as necessary.

CAM BRACKET ADJUSTMENT. If tractor creeps when speed control lever is in neutral position or if linkage has been removed and is being reinstalled, adjust brake pedal and speed control lever, then raise and support tractor so rear tires are off the ground. Remove frame cover and lubricate "T" slot (T – Fig. IH17). Move speed control lever to fast forward position. Loosen cam bracket mounting bolts and move cam bracket to its highest position in slotted holes and tighten bolts slightly to retain in this position. Start engine and use punch and hammer to adjust cam bracket downward until wheels stop turning.

CAUTION: Use care when working near rotating tires or parts.

Move speed and directional control lever to forward position. Depress brake pedal and lock in this position. If there is excessive transmission noise or vibration with brake pedal depressed, adjust cam bracket to position where noise or vibration stops. Release brake and shut off engine. Move speed control lever to fast forward position and tighten cam bracket retaining bolts. Start engine, move speed control lever to fast forward position, depress brake pedal fully and release. Wheels should stop turning and speed control lever should return to neutral position. If speed control lever is not in neutral position, loosen jam nut on neutral return rod and adjust until lever is in neutral position. Tighten jam nut. Speed control rod should not touch end of slot when brake pedal is fully depressed. If rod touches, disconnect clevis from brake cross shaft, loosen jam nut and lengthen rod until clearance is obtained. Tighten jam nut and connect clevis to brake cross shaft.

R&R HYDROSTATIC DRIVE UNIT

Models 682 – 782

Remove frame cover and an engine side panel. Remove hydraulic lines from center section to lift control valve, as equipped. On all models, disconnect flex coupling. Remove driveshaft coupling roll pin and slide coupling forward on shaft. Remove coupling arm roll pin and slide coupling arm rearward on transmission input shaft. Disconnect front flex coupling at flywheel flange and remove driveshaft. Remove retaining ring securing control cam assembly to damper spring plate. Remove locknut connecting stud ball joint and linkage rod to control cam. Remove cam bracket mounting bolts and move bracket and linkage up out of the way. Disconnect brake rod from brake lever and remove brake adjusting screw. Disconnect suction line and cap openings. Remove hydrostatic unit mounting bolts from rear frame housing, bring unit forward, tilting top of unit downward and bring it up and out.

Reinstall by reversing removal procedure. Tighten all hydrostatic unit to frame mounting bolts except those which hold cam bracket, to 30 ft.-lbs. (41 N·m). Adjust brakes and cam bracket assembly. Fill unit to proper level with Hy-Tran, or equivalent.

Model 982

Raise hood and remove engine side panels. Remove frame cover. Disconnect headlight wiring and remove hood and grille as an assembly. Remove hydraulic lines from center section to lift control valve. Remove engine oil filter. Remove coupling arm roll pin. Remove engine mounting bolts and slide engine forward until coupling arm is free of input shaft. Remove cam bracket mounting bolts and move bracket and linkage up out of the way. Remove brake rod and return spring. Disconnect suction line from hydrostatic unit. Remove hydrostatic mounting bolts from rear frame housing and bring unit forward, up and out.

Reinstall by reversing removal procedure. Tighten all hydrostatic unit to frame mounting bolts except those which hold cam bracket, to 30 ft.-lbs. (41 N·m). Fill unit to proper level with Hy-Tran, or equivalent.

OVERHAUL HYDROSTATIC DRIVE UNIT

Models 682 – 782 – 982

Thoroughly clean exterior of unit. A holding fixture, similar to one shown in Fig. IH18, makes servicing unit more convenient. Mark charge pump housing and center section before disassembly as it is possible to install pump incorrectly.

Remove charge pump housing (3 – Fig. IH20). Remove rotor assembly (4) and rotor drive pin, then remove port plate (6) and gasket (7).

NOTE: On later models, charge pump port plate and gasket are not used; see Fig. IH21.

Pry oil seal (1 – Fig. IH20) from charge pump housing and press needle bearing (2) out the front of housing. Remove both check valves (13), back-up washers (14) and "O" ring (11), shim (10), spring (9) and relief valve cone (8) from relief bore in center housing. Unbolt and remove center housing (57).

CAUTION: Valve plates (18 and 50) may stick to center housing. Be careful not to drop or damage them.

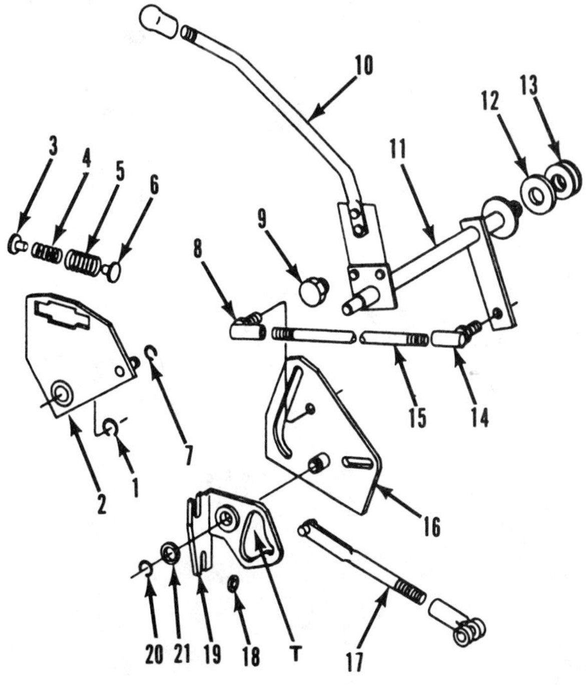

Fig. IH17 – Exploded view of hydrostatic transmission speed and direction control linkage.

1. Retaining ring
2. Damper spring plate
3. Damper spring guide pin
4. Damper spring (light)
5. Damper spring (heavy)
6. Damper spring guide pin
7. Retaining ring
8. Ball joint
9. Cross-shaft bearing
10. Speed control lever
11. Cross-shaft
12. Friction disc
13. Belleville washers
14. Ball joint
15. Linkage shaft
16. Cam bracket assy.
17. Rod
18. Washer
19. Bracket
20. Retaining ring
21. Washer
T. "T" slot

Remove and identify valve plates. Pump valve plate has two relief notches. Tilt housing (35) on its side and remove cylinder block and piston assemblies from pump and motor shafts. Lay cylinder block assemblies aside for later disassembly.

CAUTION: Some swashplates (26) have spring pin holes drilled through both walls. When removing stub shaft (33) and trunnion shaft (37), do not drive spring pins through the shaft and into holes in bottom of swashplate, as removal is then very difficult.

Mark or tape a punch exactly 15/32-inch (12 mm) from end. Carefully drive roll pins (53) into swashplate until mark on punch is even with top surface of swashplate; at this point spring pins should be centered in trunnion shaft. Remove swashplate and thrust plate (25), then withdraw pump shaft and bearing (27 and 28). Remove the two socket head screws securing motor swashplate and motor shaft (42). Bearings (36 and 40), bushing (34) and oil seals (29, 32 and 39) can now be removed from housing (35). Remove needle bearings (17 and 52) from center housing (57).

Carefully remove slipper retainer (23) with nine pistons (24) from pump cylinder block (22), then remove slipper retainer (45) with nine pistons (44) from motor cylinder block. Place each cylinder block on wood block in a press, compress spring (21 or 48) and remove retaining ring. Release press and remove spring and washers. Springs (21 and 48) should have a free length measurement of 1-3/64 to 1-1/16 inches (26.6 to 27 mm) and should test 63-75 pounds (280.24-333.62 N) when compressed to a length of 19/32-inch (15 mm). Check cylinder blocks (22 and 46) on valve face end and renew if deep scratches or other damage is evident. Inspect pistons and bores in cylinder blocks for excessive wear or scoring. Piston slippers can be lapped to remove light scratches. Minimum slipper thickness is 0.121 inch (3.073 mm) for both pump and motor. Slipper thickness must not vary more than 0.002 inch (0.051 mm) for all nine pistons in each block.

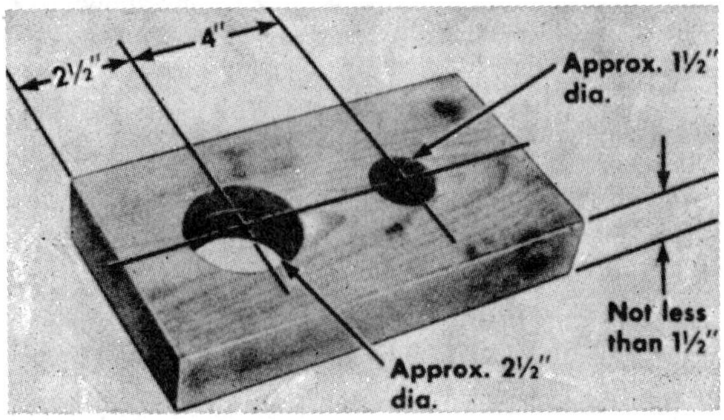

Fig. IH18—View showing dimensions of wooden fixture which may be used to hold hydrostatic transmission while servicing unit.

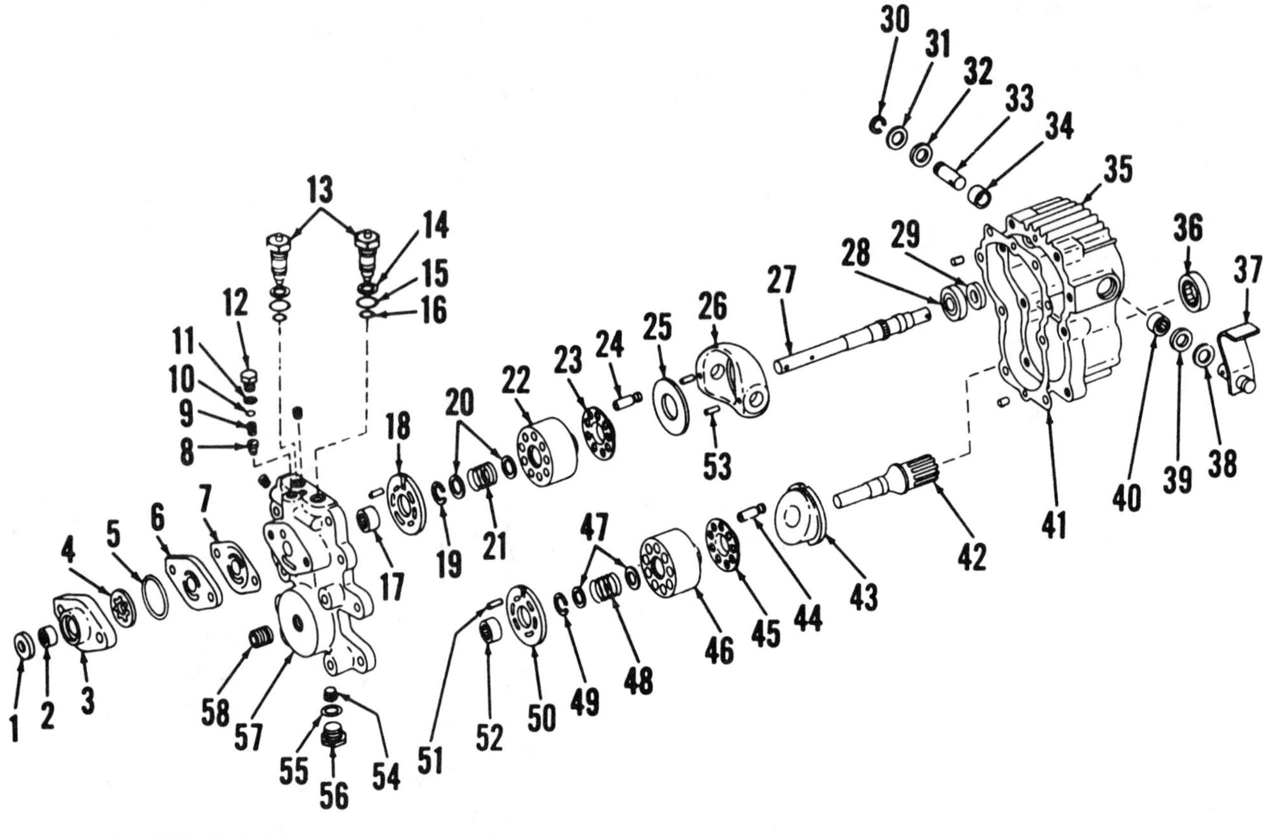

Fig. IH20—Exploded view of hydrostatic drive unit. Later models do not use port plate (6) and gasket (7). See Fig. IH21.

1. Oil seal
2. Needle bearing
3. Charge pump housing
4. Rotor assy.
5. "O" ring
6. Port plate
7. Gasket
8. Relief valve cone
9. Spring
10. Shim
11. "O" ring
12. Plug
13. Check valves
14. Back-up washer
15. "O" ring
16. "O" ring
17. Needle bearing
18. Valve plate
19. Retaining ring
20. Washers
21. Spring
22. Cylinder block
23. Slipper retainer
24. Piston (9 used)
25. Thrust plate
26. Swashplate
27. Pump shaft
28. Ball bearing
29. Oil seal
30. Snap ring
31. Washer
32. Oil seal
33. Stub shaft
34. Bushing
35. Housing
36. Roller bearing
37. Trunnion shaft
38. Washer
39. Oil seal
40. Needle bearing
41. Gasket
42. Motor shaft
43. Swashplate (motor)
44. Piston (9 used)
45. Slipper retainer
46. Cylinder block
47. Washers
48. Spring
49. Retaining ring
50. Valve plate
51. Pin
52. Needle bearing
53. Roll pin
54. Plug
55. "O" ring
56. Plug
57. Center housing
58. Oil filter fitting

Check pump valve plate and motor valve plate (Fig. IH22) for excessive wear and renew as necessary. Inspect motor swashplate (43 – Fig. IH20) and pump swashplate thrust plate (25) for wear, embedded material or scoring and renew as indicated.

Check charge pump housing (3), rotor assembly (4) and port plate (6), on models so equipped, for wear or scoring and renew as necessary. Relief valve cone (8) should be free of nicks and scratches. Relief valve spring (9) should have a free length of 1.057 inches (26.848 mm) and should test 7.0-7.6 pounds (31.14-33.81 N) when compressed to a length of 0.525 inch (13.33 mm).

Renew all "O" rings, gaskets and seals and reassemble by reversing removal procedure. Thoroughly lubricate all parts with Hy-Tran, or equivalent.

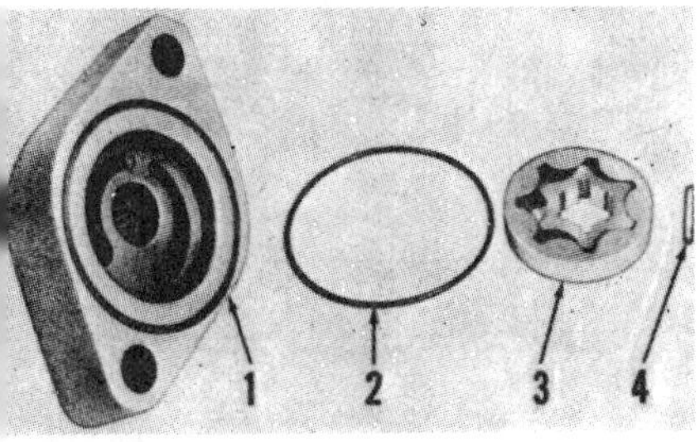
1
2
3
4

Fig. IH21 – Exploded view of charge pump assembly used on late models.

1. Charge pump housing
2. "O" ring
3. Rotor assy.
4. Rotor pin

When installing new bearings (17), press bearings in until 0.100 inch (2.54 mm) above machined surface of center housing. Check valves (13 – Fig. IH20) are interchangeable and are serviced only as an assembly. Pump swashplate (26) must be installed with thin pad towards top of transmission. When installing charge pump port plate gasket, on models so equipped, circular groove in gasket must be towards top of unit. Flat end of gasket and port plate on early models and charge pump housing (3) on all models, must be installed towards right side of unit. Motor swashplate socket head screws should be tightened to 67 ft.-lbs. (91 N·m), charge pump cap screws to 52 ft.-lbs. (70 N·m) and center housing to main housing cap screws to 35 ft.-lbs. (47 N·m).

Fill transmission to proper operating level with recommended fluid and adjust linkage as necessary.

GEAR TRANSMISSION

MAINTENANCE

Model 582

Check transmission fluid level at 100 hour intervals. Maintain fluid level at lower edge of fill plug opening (Fig. IH23). Recommended fluid is Hy-Tran, or equivalent and approximate capacity is 8 pints (3.8 L).

Model 582 Special

Check transmission fluid level at 100 hour intervals. Maintain fluid level at lower edge of fill plug opening (Fig. IH24). Recommended fluid is IH-135H EP Gear Lubricant, or equivalent. Approximate fluid capacity is 4 pints (1.9 L).

R&R GEAR TRANSMISSION

Model 582

Remove battery and disconnect electrical leads from solenoid, tail lights and seat safety switch. Remove rear fender to frame bolts and battery ground wire. Remove foot platform mounting screws and remove fender assembly. Remove frame cover. Disconnect brake rod and rear flex coupling. Remove the three point hitch lift lever, if so equipped. On all models, support frame, remove frame mounting bolts and roll differential and transmission assembly out of frame.

Reinstall by reversing removal procedure.

Model 582 Special

Disconnect and remove battery. Remove the four bolts in battery box which secure fender assembly to rear frame. Remove drawbar assembly. Remove foot platform mounting screws

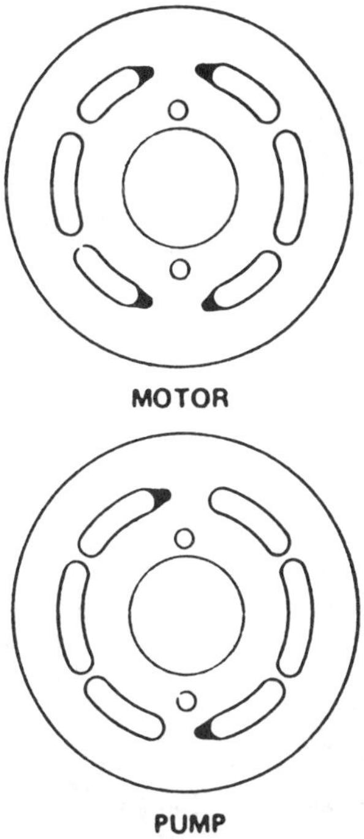
MOTOR
PUMP

Fig. IH22 – Check pump and motor valve plates for wear or other damage. Note motor valve plate has four notches (dark areas) and pump valve plate has two notches.

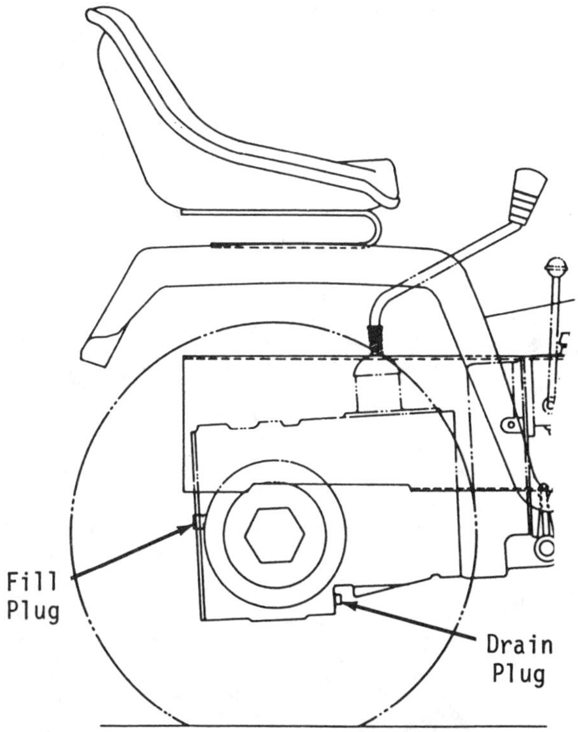
Fill Plug
Drain Plug

Fig. IH23 – View showing location of fill plug and drain plug on 582 models.

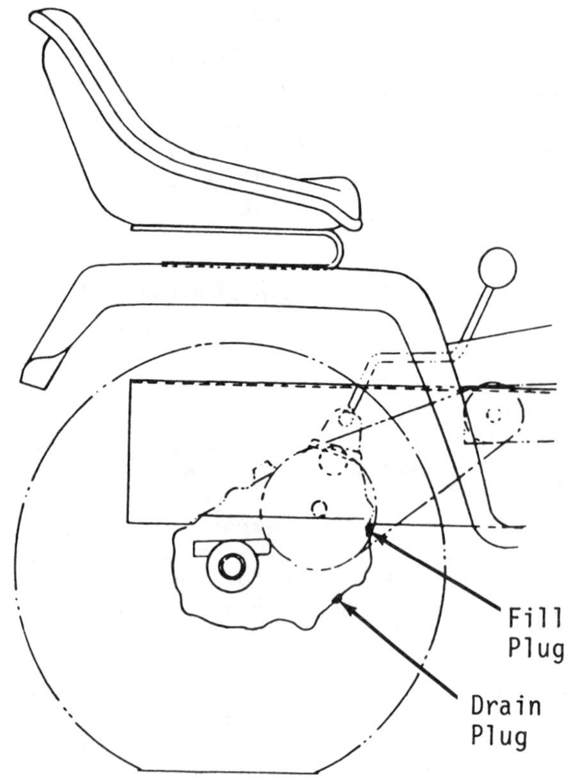

Fig. IH24—View showing location of fill plug and drain plug on 582 Special models.

and remove fender assembly. Remove center frame cover. Place wedges between front axle and frame at each side to prevent tractor from tipping. Depress clutch pedal and lock in lowest position. Loosen the two belt guides and work belt off transaxle input pulley. Remove bolts securing transaxle to frame and roll assembly out of frame.

Reinstall by reversing removal procedure. Adjust belt guides so there is 1/8 to 3/16-inch (3 to 5 mm) clearance between belt and guides.

OVERHAUL GEAR TRANSMISSION

Model 582

Drain lubricant and remove transmission and differential assembly from tractor. Disassemble reduction gear unit and if prior to S/N 720000, remove internal brake disc. On all models, remove reduction housing mounting bolts and remove housing. Remove differential assembly. Remove gear shift lever and cover assembly. Shift transmission into two speeds to lock transmission and remove nut (26–Fig. IH25) from countershaft (13). Remove shifter fork setscrews. Cover gearshift detent ball and spring holes and drive shift rods forward out of transmission. Remove detent balls and springs. Remove bolts from mainshaft front bearing retainer and pull mainshaft forward out of transmission as gears are removed. Push countershaft rearward out of transmission noting location of gears and spacers as they are removed. Pull mainshaft needle bearings from housing. Remove reverse idler shaft and gear. Remove countershaft front bearing, retainer and shims.

Reassemble by reversing disassembly procedure. Note beveled edge of spacer (24) is toward bearing (28). Rounded edge of gear teeth on first gear (19) are toward bearing (15), rounded edge of gear teeth on second gear (21) are toward bearing (28) and rounded edge of gear teeth on third gear (23) are toward bearing (15). Nut (26) should be tightened to 85 ft.-lbs. (115 N·m). Make certain pinion gear depth and backlash are properly adjusted.

Model 582 Special

Drain lubricant and remove transaxle assembly from tractor. Remove tires and wheels. Remove snap ring securing wheel hub to axle and remove hub and Woodruff key. Remove brake disc from brake shaft. Loosen setscrew and remove input pulley. Place shift lever in neutral position and remove shift lever housing and gasket.

NOTE: If disassembly of shift lever is necessary, scribe match marks to make certain shift lever is not reassembled 180° out of line.

Scribe match marks on axle housings and case and cover and remove axle housings. Remove "O" rings (54–Fig. IH26), oil seals (53 and 71) and retainers (52 and 72). Axle support bearings (56) are removed by driving them from the inside of axle housing outward. Support transaxle assembly on bench with left axle downward. Drive alignment dowels into cover, remove the eight cap screws and separate case from cover. Remove differential assembly. Remove thrust washer (33), three gear cluster (32) from brake shaft (28). Remove reverse idler gear (8), spacer (9) and shaft (10). Hold upper ends of shifter rods together and lift out shifter rods, forks, shifter stop (5), sliding gears (15 and 16) and shaft (18) as an assembly. Remove low reduction gear (48), reduction shaft (49) and thrust washer (50), then remove the two-cluster gear (29) from brake shaft. Lift out output gear (46), shaft (45) and

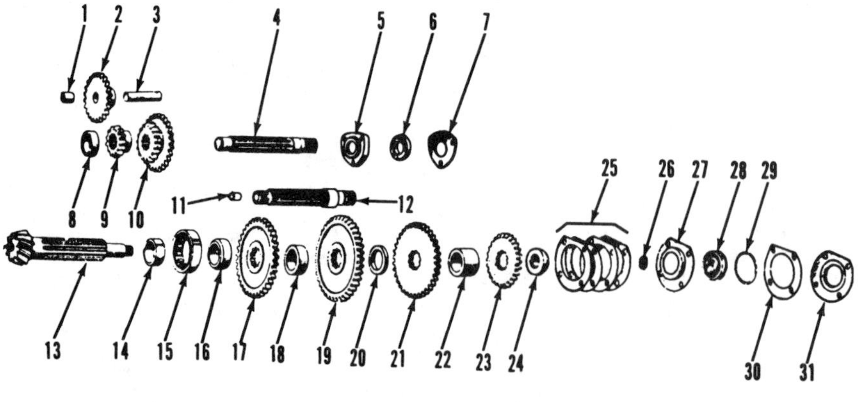

Fig. IH25—Exploded view of transmission used on 582 models (S/N 719999 and before). Models after S/N 720000 are similar but do not have internal brake mechanism.

1. Bushing
2. Reverse gear
3. Reverse idler shaft
4. Mainshaft
5. Bearing cage
6. Bearing
7. Bearing retainer
8. Bearing
9. Sliding gear (1st & Rev.)
10. Sliding gear (2nd & 3rd)
11. Bushing
12. Input shaft
13. Countershaft
14. Bearing race
15. Bearing
16. Spacer
17. Reverse gear
18. Spacer
19. First gear
20. Spacer
21. Second gear
22. Spacer
23. Third gear
24. Spacer
25. Shims
26. Nut
27. Retainer
28. Bearing
29. Snap ring
30. Gasket
31. Retainer

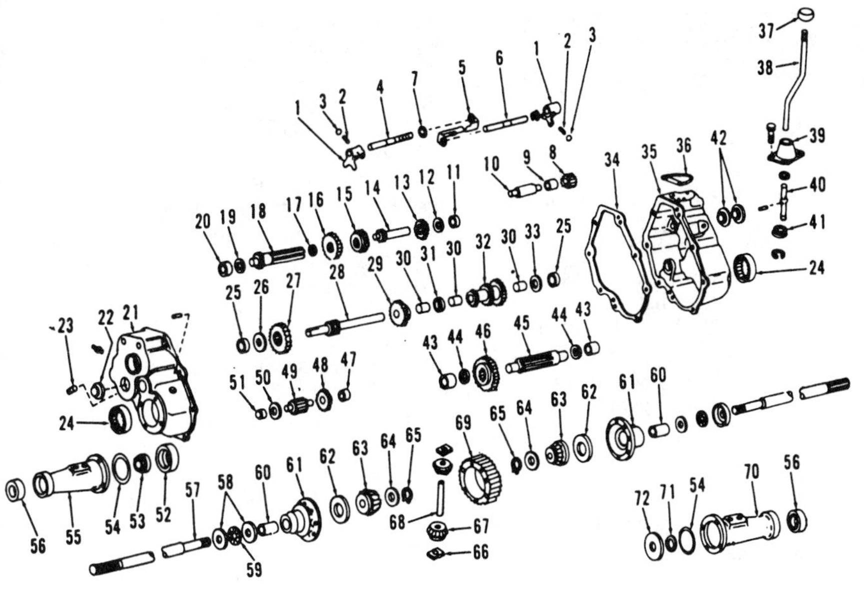

Fig. IH26—Exploded view of transaxle assembly used on 582 Special models.

1. Shift fork
2. Spring
3. Ball
4. Shift rod
5. Shifter stop
6. Shift rod
7. Snap ring
8. Reverse idler
9. Spacer
10. Reverse idler shaft
11. Bearing
12. Thrust washer
13. Input gear
14. Input shaft
15. Third & fourth gear
16. First, second & reverse gear
17. Bearing
18. Shifter shaft
19. Thrust washer
20. Needle bearing
21. Cover
22. Oil seal
23. Drain plug
24. Needle bearing
25. Bearing
26. Thrust washer
27. Idler gear
28. Brake shaft
29. Two-cluster gear
30. Bushing
31. Spacer
32. Cluster gear assy.
33. Thrust washer
34. Gasket
35. Case
36. Gasket
37. Knob
38. Shift lever
39. Shift lever housing
40. Shifter
41. Keeper
42. Bearing & seal
43. Needle bearing
44. Thrust washer
45. Shaft
46. Output gear
47. Bearing
48. Low reduction gear
49. Low reduction shaft
50. Washer
51. Bearing
52. Seal retainer
53. Oil seal
54. "O" ring
55. Axle housing
56. Ball bearing
57. Axle
58. Thrust race
59. Thrust bearing
60. Bushing
61. Carrier assy.
62. Washer
63. Axle gear
64. Thrust washer
65. Snap ring
66. Drive block
67. Bevel pinion gear
68. Drive pin
69. Ring gear
70. Axle housing
71. Oil seal
72. Seal retainer

thrust washers (44). To remove brake shaft (28) and gear (27) from cover (21), block up under gear (27) and press shaft out of gear.

CAUTION: Do not allow cover or low reduction gear bearing boss to support any part of pressure required to press brake shaft from gear.

Remove input shaft (14) with input gear (13) and thrust washer (12) from case (35).

To disassemble differential, remove four cap screws and separate axle shaft and carrier assemblies from ring gear (69). Drive blocks (66), bevel pinion gears (67) and drive pin (68) can now be removed from ring gear. Remove snap rings (65) and withdraw axle shaft (57) from axle gears (63) and carriers (61).

Clean and inspect all parts and renew any showing excessive wear or other damage. When installing new needle bearings, press bearing (17) in spline shaft (18) to a depth of 0.010 inch (0.254 mm) below end of shaft and low reduction shaft bearings (51 and 47) 0.010 inch (0.254 mm) below thrust surfaces of bearing bosses. Carrier bearings (24) should be pressed in from inside of case and cover until bearings are 0.290 inch (7.366 mm) below face of axle housing mounting surface. All other needle bearings are to be pressed in from inside of case and cover to a depth of 0.015-0.020 inch (0.381-0.508 mm) below thrust surfaces.

Renew all seals and gaskets and reassemble by reversing disassembly procedure. Note when installing brake shaft (28) and idler gear (27), beveled edge of gear teeth must be up away from cover. Install reverse idler shaft (10), spacer (9) and reverse idler gear (8) with rounded end of teeth facing spacer. Install input gear (13) and shaft (14) with chamfered side of input gear facing case (35).

Tighten transaxle cap screws to the following torque:

Differential cap screws	7-10 ft.-lbs. (9-14 N·m)
Axle housing cap screws	8-10 ft.-lbs. (11-14 N·m)
Shift lever housing cap screws	8-10 ft.-lbs. (11-14 N·m)

Reinstall transaxle and fill to proper level with specified fluid. Adjust brake and drive belt as necessary.

REDUCTION GEAR

R&R REDUCTION GEAR

Model 582

Reduction gear assembly is removed with the gear type transmission. Refer to appropriate GEAR TRANSMISSION section for removal procedure.

Models 682–782–982

Reduction gear assembly is removed with the hydrostatic drive, reduction gear and differential assembly as a unit. Refer to the HYDROSTATIC DRIVE SYSTEM section for removal procedure.

REDUCTION GEAR OVERHAUL

Model 582

Drain lubricant and remove creeper drive unit, as equipped.

On models having internal type transmission brakes (S/N 719999 or below) remove brake adjusting screw, pivot pin, brake lever and push rod.

On all models, remove reduction hous-

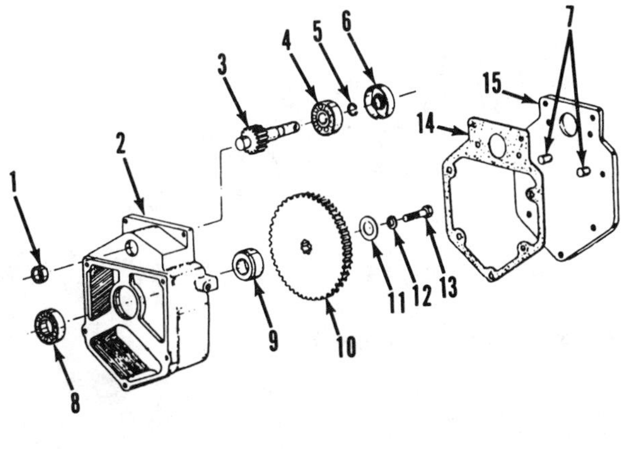

Fig. IH27—Exploded view of reduction gear assembly used on 582 models with S/N 720000 or after. Refer to Fig. IH12 for 582 models below S/N 720000 with internal type brake.

1. Bearing
2. Housing
3. Drive gear & shaft
4. Bearing
5. Snap ring
6. Seal
7. Dowels
8. Bearing
9. Spacer
10. Gear
11. Flat washer
12. Lock washer
13. Bolt
14. Gasket
15. Cover

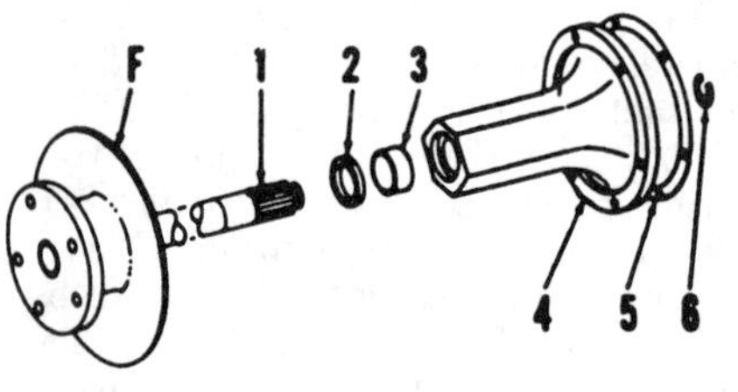

Fig. IH29 — Exploded view of rear axle assembly.

F. Brake disc
1. Axle
2. Oil seal
3. Needle bearing
4. Axle carrier
5. Gasket
6. "C" ring retainer

ing cover (15 – Fig. IH27). Remove bolt (13) and washers (12 and 11). Remove gear (10). Use suitable puller to remove reduction gear, shaft and bearing assembly from housing.

On models with creeper drive, press spring pin out of gear and shaft assembly and remove splined coupling.

On all models, press bearings off the gear and shaft assembly. Press needle bearing from rear of housing.

To reassemble, press needle bearing in until flush with housing. Continue to reverse removal procedure noting copper sealing washers are installed on the two lower mounting bolts. Bolts should be tightened to 80 ft.-lbs. (108 N·m). Bolt (13) is tightened to 55 ft.-lbs. (75 N·m). Fill to proper level with recommended fluid and adjust brakes as necessary.

Models 682 – 782 – 982

Reduction gear is serviced with the differential assembly. Refer to DIFFERENTIAL section. Note late production models (S/N 720000 or after) do not have an internal brake mechanism.

DIFFERENTIAL

R&R AND OVERHAUL

Model 582

Remove rear cover and remove "C" ring retainers (6 – Fig. IH29) from inner ends of axle shafts (1). Remove bolts securing axle carriers (4) to differential housing and remove axle and carrier assemblies. Remove differential carrier bearing cages (2 – Fig. IH30) noting location and thickness of shims (B). Angle ring gear carrier assembly (8) and remove from housing. Remove pin (12) and shaft (11). Rotate side gears (90° and remove spider gears (9) and side gears (10).

NOTE: Models with S/N 720000 or after use a ring gear and carrier and spider gears as shown in Fig. IH31.

Remove bearing cones (5 – Fig. IH30) as necessary. Ring gear is riveted to carrier and is available separately for some models. Always check part availability before removing. To remove pinion gear and shaft assembly, refer to GEAR TRANSMISSION section.

Check all bearings for looseness, or damage and check all gears for wear or damage.

To reassemble, install ring gear and carrier assembly in differential housing and install bearing cages (2 – Fig. IH30) and shims (B). Adjust bearing preload by adding or removing shims (B) until a preload of 1 to 8 pounds (0.5 to 3.6 kg) pull on a spring scale measured as shown in Fig. IH32 is attained. Remove ring gear and carrier assembly noting thicknesses of shim packs (B – Fig. IH30). Install pinion gear and shaft assembly as outlined in GEAR TRANSMISSION section and tighten nut on pinion shaft to 85 ft.-lbs. (115 N·m). Install ring gear and carrier assembly and adjust gear backlash by moving shims (B) from side to side as necesary to obtain 0.003-0.005 inch (0.076-0.127 mm) backlash. Paint ring gear teeth with Prussian Blue or red lead and rotate ring gear. Observe contact pattern on tooth surfaces. Refer to Fig. IH33. If pattern is too low (pinion depth too great) as "B", remove shims (25 – Fig. IH25) as necessary to correct pattern. If pattern is too high (pinion depth too shallow) as "C", add shims (25) to correct pattern. Make certain ring gear backlash is correct after each shim change. Adjust as necessary. Carrier bearing cage retaining bolts should be tightened to 20 ft.-lbs. (27 N·m) during final assembly. Right side axle carrier mounting bolt at 9 o'clock position should be coated with thread sealer to prevent fluid loss. Continue to reverse disassembly procedure and refill to correct level with recommended fluid.

Model 582 Special

Differential assembly is an integral part of the transaxle assembly. Refer to appropriate GEAR TRANSMISSION section for service procedure.

Models 682 – 782 – 982

Remove transmission and differential assembly as outlined in HYDROSTATIC DRIVE SYSTEM section.

Drain fluid from differential and remove the axles and their carriers. Remove left and right bearing cages (29 – Fig. IH34) and note thickness of

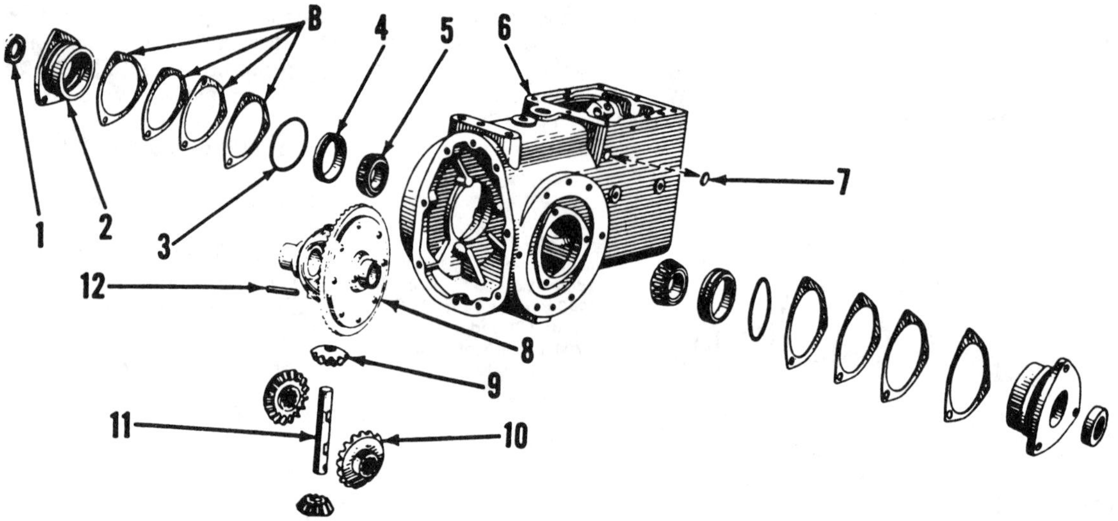

Fig. IH30 — Exploded view of differential assembly used for all models with S/N 719999 or below except 582 Special.

B. Shims
1. Oil seal
2. Bearing retainer
3. "O" ring
4. Bearing cup
5. Bearing cones
6. Case
7. Expansion plug
8. Ring gear & carrier
9. Spider gear
10. Side gear
11. Cross shaft
12. Pin

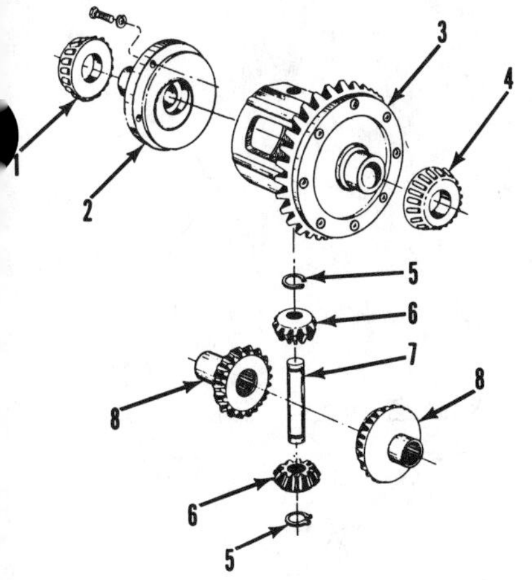

Fig. IH31—Exploded view of differential and spider gear assembly used on all models after S/N 720000 except 582 Special model.

1. Bearing
2. Flange
3. Ring gear & carrier
4. Bearing
5. Snap rings
6. Spider gears
7. Cross-shaft
8. Axle gears

each shim pack (31). Turn differential assembly in housing until it can be removed.

NOTE: It may be necessary on some models to remove one of the bearing cones (4 or 11) before differential can be removed.

Remove differential flange (5), snap rings (7), shaft (9) and spider gears (8). Remove axle gears (10).

Fig. IH32—Differential bearing preload is correct when a steady pull of 1 to 8 pounds (0.5 to 3.6 kg) on a spring scale is necessary to rotate differential as shown.

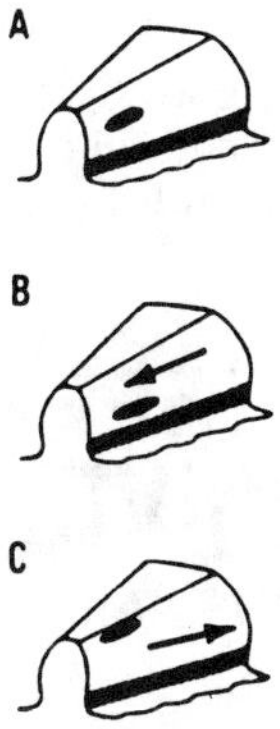

Fig. IH33—Bevel pinion tooth contact pattern is correct at "A", too low at "B" and too high at "C".

NOTE: On models with S/N 719999 or before, differential and spider gear assembly is similar to ones shown in Fig. IH30. Shaft (11—Fig. IH30) is retained by pin (12) in carrier and ring gear assembly (8).

Ring gear and case assembly (6—Fig. IH34) are serviced as an assembly only.

Remove hydrostatic drive unit and on models with S/N 719999 or before, remove expansion plug (28) and snap ring (27). Press bevel pinion shaft out of front bearing and constant mesh reduction gear. Remove rear bearing cup (14) and shim pack (15).

On models with S/N 720000 and after, remove expansion plug (28) and snap ring (27). Remove and cover (24), snap ring (22), spacer (21), gear (20) and snap ring (19). Remove bevel pinion shaft (12) from housing (18). Remove bearing cup (14) and shim pack (15).

To reassemble, install differential assembly in housing and adjust preload on bearings (4 and 11) by varying thickness of shim packs (31) until a steady pull of 1 to 8 pounds (0.5 to 3.6 kg) is required to rotate differential assembly. See Fig. IH32. Remove differential assembly keeping shim packs (31) with their respective bearing cage.

If a new bevel pinion shaft, transmission case or rear bearing and cup are to be installed, take the number stamped on the top of case just ahead of rear cover (2) and the number stamped on the end of the bevel pinion shaft (12), add these two figures together and add 0.015 inch to their total. Place shim pack of a thickness equal to this total under bearing cup (15).

On models with S/N 719999 or before, install pinion shaft and bearing assembly in housing (18) making certain it goes through reduction gear (20). Install front bearing (26).

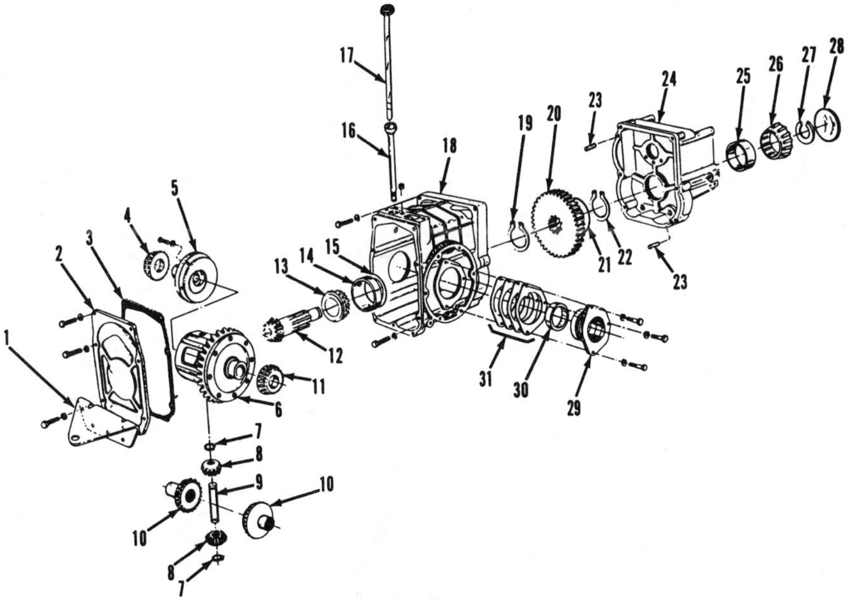

Fig. IH34—Exploded view of reduction gear and differential assembly used on 682, 782 and 982 models after S/N 720000.

1. Drawbar
2. Cover
3. Gasket
4. Bearing
5. Differential flange
6. Ring gear & carrier
7. Snap rings
8. Spacer gears
9. Cross-shaft
10. Axle gears
11. Bearing
12. Pinion shaft
13. Bearing
14. Bearing cup
15. Shims
16. Dipstick tube
17. Dipstick
18. Case
19. Snap ring
20. Reduction gear
21. Spacer
22. Snap ring
23. Dowel pin
24. Cover
25. Bearing cup
26. Bearing
27. Snap ring
28. Expansion plug
29. Cage
30. Bearing cup
31. Shim pack

On models with S/N 720000 or after, install pinion shaft and bearing assembly in housing. Install snap ring (19), gear (20), spacer (21) and snap ring (22). Install cover (24), bearing cup (25) and bearing (26).

On all models, install a snap ring (27) which will allow 0.003 inch (0.0762 mm) pinion shaft end play. Snap rings are available in a variety of thicknesses. Install a new expansion plug (28).

Install differential assembly and install a dial indicator to measure ring gear backlash. Move shims (31) from side to side until 0.003 to 0.005 inch (0.0762 to 0.1270 mm) backlash is obtained.

RIGHT ANGLE DRIVE GEAR BOX

MAINTENANCE

Model 582 Special

Right angle drive gear box is a sealed unit, packed by the manufacturer with 4 ounces (118 mL) of IH-251 EP, or equivalent, grease.

R&R RIGHT ANGLE DRIVE GEAR BOX

Model 582 Special

Disconnect negative battery cable and remove center frame cover. Lock clutch/brake pedal in depressed position. Drive roll pin from rear half of flex coupler. Remove the two mounting bolts retaining right angle drive gear box to cross support and remove gear box.

Reinstall by reversing removal procedure.

OVERHAUL RIGHT ANGLE GEAR BOX

Model 582 Special

Remove cover (2–Fig. IH35) and clean lubricant from inside housing. Remove snap ring retaining output pulley, loosen pulley setscrew and remove pulley and key. Remove cap (17). Remove seal (18) from cap. Remove bearing (15) and use wood dowel or brass rod to drive gear (13) off shaft (14). Remove shaft, gear and bearing (12). Remove seal (11) and snap ring (10). Press shaft (7) out of bearing (9) and housing (8). Remove snap ring (4), gear (5) and bearing (6).

Lubricate parts with IH-251 EP, or equivalent grease and reassemble by reversing disassembly procedure. Before installing cover (2), fill housing (8) with 4 ounces (118 mL) of IH-251 EP, or equivalent grease. Seal (11) should seat in housing and seal (18) should be 0.045 inch (1.1 mm) below edge of cap (17).

CREEPER DRIVE

MAINTENANCE

Model 582

Model 582 may be equipped with an optional creeper drive unit. Check fluid level at 100 hour intervals. Maintain fluid level at lower edge of check plug opening (P–Fig. IH36). Recommended fluid is Hy-Tran, or equivalent. Approximate capacity is ½-pint (0.24 L).

R&R CREEPER DRIVE

Model 582

Remove frame cover and drive roll pins from flex couplings and slide couplers forward on driveshaft. Place drain pan under tractor and remove unit mounting bolts. Bump unit to free from dowels and remove.

Reinstall by reversing removal procedure. Fill unit to proper level through breather opening (2–Fig. IH36) until level with lower edge of plug opening (P) in side of housing. Hy-Tran, or equivalent, is the recommended fluid. Reinstall breather (2) and plug (P).

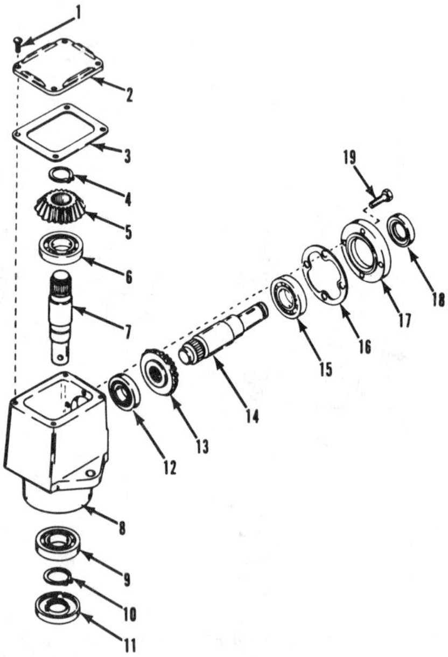

Fig. IH35—Exploded view of right angle drive gear box used on 582 Special model.

1. Screw
2. Cover
3. Gasket
4. Snap ring
5. Gear
6. Bearings
7. Shaft
8. Housing
9. Bearing
10. Snap ring
11. Seal
12. Bearing
13. Gear
14. Shaft
15. Bearing
16. Gasket
17. Cap
18. Seal
19. Bolt

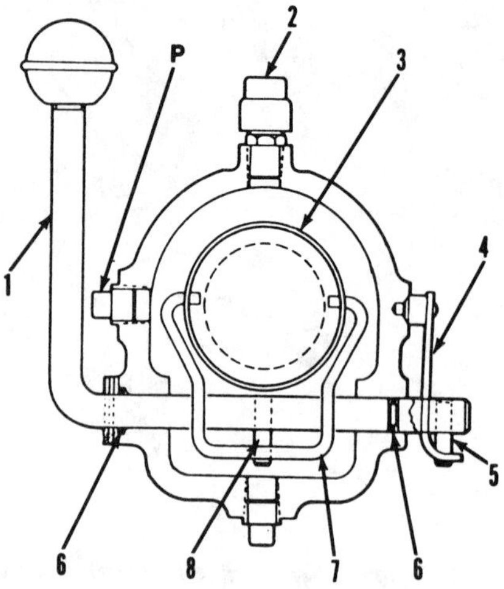

Fig. IH36—Sectional view of creeper drive unit.

1. Shift lever
2. Breather
3. Collar
4. Detent
5. Pin
6. "O" rings
7. Yoke
8. Pin
P. Plug

CREEPER DRIVE OVERHAUL

Model 582

Remove snap ring (11–Fig. IH37) and pull input shaft (8), bearing (6), planetary assembly and direct drive coupler (16) from housing (17).

NOTE: It may be necessary to rotate input shaft to align spline grooves of direct drive coupling with shift collar for removal.

Remove spring pin from drive coupling and remove coupling from shaft. Slide planet carrier (14) off input shaft

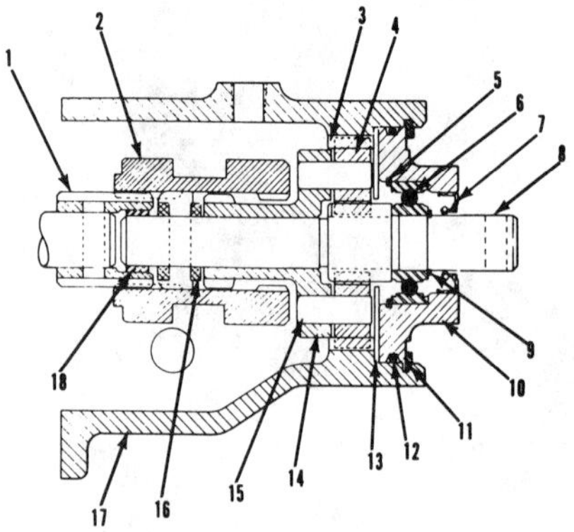

Fig. IH37—Sectional view of creeper drive unit.

1. Driven coupling
2. Shift collar
3. Ring gear
4. Planet gear
5. Snap ring
6. Bearing
7. Seal
8. Input shaft
9. Snap ring
10. Bearing housing
11. Snap ring
12. "O" ring
13. Thrust washer
14. Planet carrier
15. Carrier pin
16. Direct drive coupler
17. Housing
18. Pilot bushing

(8) and planet gears (4) off carrier pins (15). Remove thrust washer (13). Remove snap ring (5) and press bearing and shaft assembly out of bearing housing (10). Remove oil seal (7) from bearing housing (10). Remove snap ring (9) and press input shaft (8) out of bearing (6). Drive pin (5–Fig. IH36) from lever (1) and remove shift detent (4). Shift lever and shift collar toward rear of housing while lifting shift collar up to disengage shift collar (3) from yoke (7). Remove pin (8) and slide lever (1) out of housing and yoke. Remove yoke. Remove "O" rings (6).

Clean all parts and inspect for wear or damage. Ring gear (3–Fig. IH37) is an integral part of housing (17) and if damaged, complete assembly must be renewed.

Reassemble by reversing disassembly procedure. Fill to proper level with recommended fluid.

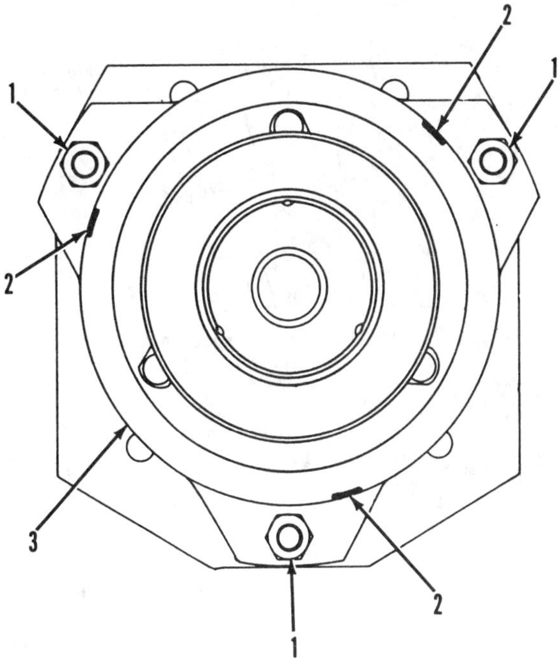

Fig. IH38—Front view of electric pto clutch.
1. Locknuts
2. Slots
3. Drum

POWER TAKE-OFF

ADJUSTMENT

All Models

Insert a 0.012 inch (0.30 mm) feeler gage between armature assembly and rotor through access holes (2—Fig. IH38) and adjust locknuts (1) until feeler gage can just be removed. Repeat procedure at all three access hole (2) locations.

R&R ELECTRIC PTO CLUTCH

All Models

Location and construction of electric pto clutch may vary from model to model, however, removal procedure is similar.

Remove flange (6–Fig. IH39), springs (4) and retaining bolt, washer and armature assembly (5). Remove rotor (3). Remove field assembly (2).

Reinstall by reversing removal procedure. Adjust as previously outlined.

HYDRAULIC SYSTEM

MAINTENANCE

Models 682–782–982

Pressurized oil is utilized from the charge pump in hydrostatic drive unit to provide hydraulic lift system and supply remote outlets. Refer to HYDROSTATIC DRIVE SYSTEM section for maintenance information.

PRESSURE CHECK AND ADJUST

Models 682–782–982

Install a 0-1000 psi (0-7000 kPa) pressure gage in test port (3–Fig. IH40). Start engine and allow transmission fluid to reach operating temperature. Run engine at full speed. With hydraulic control valve (as equipped) in neutral position, gage will indicate charge pump pressure as follows:

Model 682	90-165 psi (620-1138 kPa)
Models 782 & 982	90-200 psi (620-1379 kPa)

With control valve in raise position and the lift cylinder at end of stroke,

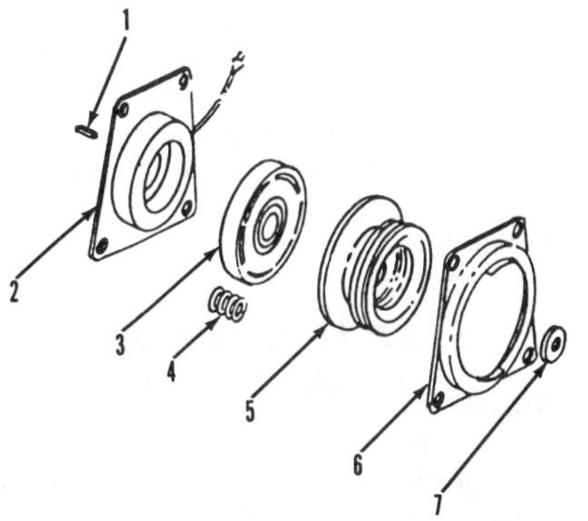

Fig. IH39—Exploded view of electric pto clutch.
1. Key
2. Field assy.
3. Rotor
4. Spring
5. Armature
6. Drum
7. Washer

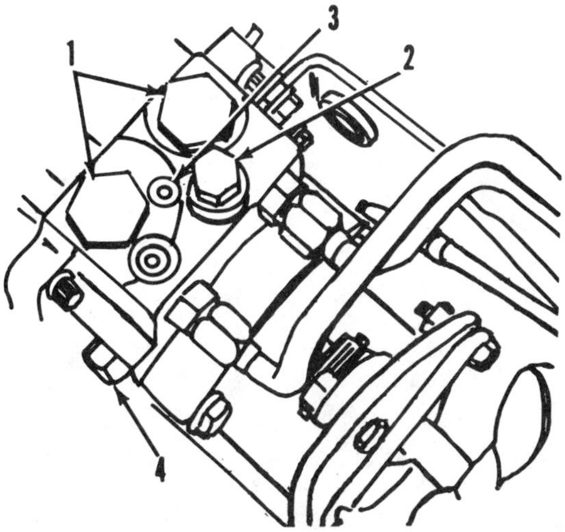

Fig. IH40—View of hydrostatic drive unit showing locations of check valves (1), implement relief valve (2), test port (3) and charge pump relief valve (4).

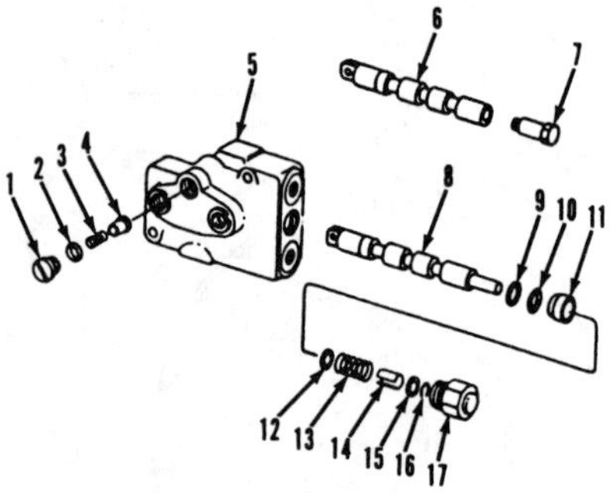

Fig. IH41 — Exploded view of double spool valve. Single spool valve is similar.

1. Plug
2. "O" ring
3. Spring
4. Plunger
5. Body
6. Spool (single)
7. Bolt
8. Spool (double)
9. "O" ring
10. "O" ring
11. Spacer
12. Washer
13. Spring
14. Spacer
15. Washer
16. Retainer ring
17. Cap

gage will indicate lift system pressure as follows:

Model 782 . . . . 500-625 psi (3447-4309 kPa)
Model 982 . . . . 700-900 psi (4826-6205 kPa)

If charge pressure is not as specified, remove charge pump relief valve (4) and add or remove shims as necessary.

If lift system pressure is not as specified, remove implement relief valve (2) and add or remove shims as necessary.

CONTROL VALVE

Models 782 – 982

Model 782 and 982 may be equipped with a single or double spool hydraulic control valve. Service procedures are similar for each valve.

To remove control valve, raise hood and remove an engine side panel. Mark locations of hydraulic lines at control valve and disconnect all lines and cap openings. Remove connecting links at valve spools, remove valve mounting bolts and remove valve.

To disassemble control valve, remove cap (17 – Fig. IH41) and pull spool (8 or 6) from body (5). Disassemble spool only if necessary to renew springs.

Renew all internal "O" rings and seals. Lubricate all parts during assembly and reassemble by reversing removal procedure.

LIFT CYLINDER

Models 782 – 982

Two different lift cylinders which are similar in appearance have been used. Make certain cylinder is correctly identified before ordering service parts.

To remove cylinder, remove frame cover, mark locations of hydraulic hoses and disconnect hoses and cap openings. Remove pin securing cylinder to lift bracket through access hole in right-hand side of frame. Remove cylinder mounting bolt and remove cylinder.

To disassemble cylinder, note location of 45° and 90° fittings and remove fittings. Place cylinder in a holding fixture and use a suitable spanner wrench to turn cylinder head (6 – Fig. IH42) until end of wire retainer (5) appears in access hole in cylinder body. Pry end of wire retainer out of hole and rotate cylinder head to remove wire retainer. Pull cylinder head out of cylinder body (1).

NOTE: Cylinder head "O" ring (4) may catch in wire retainer groove in cylinder body. If so, cut "O" ring through access hole and use wire retainer removal procedure to remove "O" ring.

Turn piston and rod assembly as it is pulled from cylinder body.

Lubricate all parts and renew all seals and "O" rings during reassembly. Reassemble by reversing disassembly procedure.

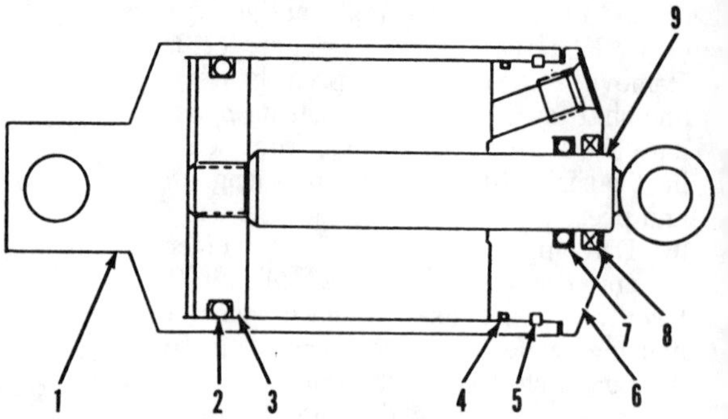

Fig. IH42 — Cut-a-way view of hydraulic lift cylinder. Refer to text.

1. Cylinder body
2. "O" ring
3. Piston
4. "O" ring
5. Wire retainer
6. Head
7. "O" ring
8. Seal
9. Rod

KUBOTA

CONDENSED SPECIFICATIONS

MODEL	G1800, G1800-S	G1900, G1900-S	G2000, G2000-S
Engine Make	Kubota	Kubota	Kubota
Model	D662	D722	WG750-G
Bore	64 mm (2.52 in.)	67 mm (2.64 in.)	68 mm (2.68 in.)
Stroke	68 mm (2.68 in.)	68 mm (2.68 in.)	68 mm (2.68 in.)
Displacement	655 cc (39.96 cu. in.)	719 cc (43.86 cu. in.)	740 cc (45.14 cu. in.)
Power Rating	11.9 kW (16 hp)	13.4 kW (18 hp)	15.7 kW (21 hp)
Slow Idle	800 rpm	800 rpm	1200 rpm
High Speed (No-Load)	3400 rpm	3400 rpm	3400 rpm
Capacities—			
Crankcase	3 L (3.2 qts.)	3 L (3.2 qts.)	3 L (3.2 qts.)
Hydrostatic Transmission	5.7 L (6 qts.)	5.7 L (6 qts.)	5.7 L (6 qts.)
Cooling System	3.3 L (3.5 qts.)	3.3 L (3.5 qts.)	3.3 L (3.5 qts.)
Fuel Tank	19 L (5 gals.)	19 L (5 gals.)	19 L (5 gals.)

FRONT AXLE AND STEERING SYSTEM

MAINTENANCE

All Models

The front wheel spindles and front axle should be lubricated after every 50 hours of operation. Jack up front axle so wheels are suspended and apply grease through grease fittings. Use a good quality, lithium base grease. Check for looseness and binding in front axle components.

R&R AXLE MAIN MEMBER

All Models

Detach pto springs, remove belt guides and remove pto belts. Remove pto driven pulley. Detach drag link from steering arm. Support front of tractor. Remove lower bolt (17—Fig. K1) and screws retaining axle pivot (1). Withdraw pivot (1), then lower and roll axle assembly away from tractor.

Maximum allowable clearance between pivot (1) and axle main member is 0.30 mm (0.012 in.). Axle inside diameter should be 40.00-40.04 mm (1.575-1.576 in.). Pivot outside diameter should be 39.950-39.975 mm (1.573-1.574 in.).

Axle end play should be 0.1-0.4 mm (0.004-0.016 in.). Maximum allowable end play is 1.0 mm (0.040 in.). Adjust end play using shims (6).

Tighten pivot mounting screws to 39-49 N·m (29-36 ft.-lbs.). Tighten lower bolt (17) to 7.8 N·m (69 in.-lbs.).

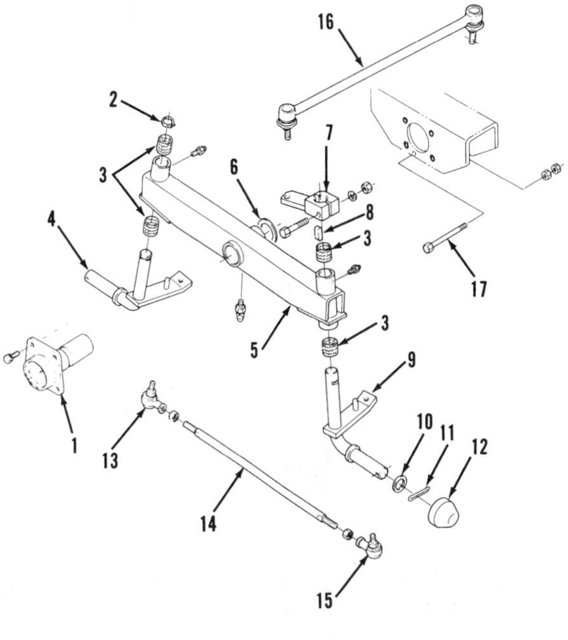

Fig. K1—Exploded view of front axle assembly.

1. Pivot
2. Snap ring
3. Bushing
4. Spindle (R.H.)
5. Axle main member
6. Shim
7. Steering arm
8. Key
9. Spindle (L.H.)
10. Washer
11. Cotter pin
12. Hub cap
13. Tie rod end (R.H.)
14. Tie rod
15. Tie rod end (L.H.)
16. Drag link
17. Bolt

TIE ROD AND TOE-IN

All Models

The tie rod is equipped with renewable ends that are not interchangeable.

Front wheel toe-in should be 0-5 mm

Illustrations courtesy Kubota.

(0 to 3/16 in.). Adjust toe-in by altering tie rod length. Tighten tie rod mounting nuts to 39-54 N·m (29-39 ft.-lbs.).

STEERING SPINDLES

All Models

Raise and support side to be serviced. Remove wheel and tire. Disconnect tie rod end, and if needed, drag link end. If removing right spindle (4—Fig. K1), detach snap ring (2) securing top of spindle and withdraw spindle from axle. If removing left spindle (9), remove steering arm clamp bolt. Remove steering arm (7) and withdraw spindle from axle.

Inspect components for excessive wear and damage. Renewable bushings are located in axle. Tighten steering arm clamp bolt to 39-54 N·m (29-39 ft.-lbs.). Tighten tie rod mounting nuts to 39-54 N·m (29-39 ft.-lbs.). Tighten drag link mounting nut to 74-103 N·m (54-75 ft.-lbs.).

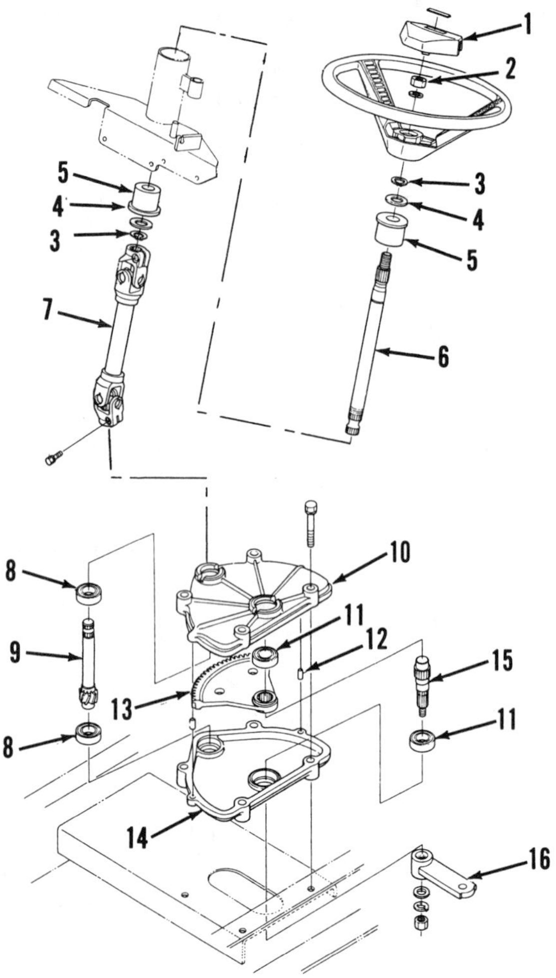

Fig. K2—Exploded view of steering shaft and gearbox. Pitman arm (16) on rear steer models is equipped with two holes for front and rear drag links.

1. Cap
2. Nut
3. Snap ring
4. Washer
5. Bushing
6. Shaft
7. Joint shaft
8. Bearing
9. Pinion
10. Gearbox half
11. Bearing
12. Dowel pin
13. Sector gear
14. Gearbox half
15. Pitman shaft
16. Pitman arm

FRONT WHEEL BEARINGS

All Models

Front wheel bearings are sealed and should not require lubrication. Renew bearings if excessive play is noted or bearing is damaged.

STEERING GEAR

All Models

To remove steering gear, remove steering wheel cap (1—Fig. K2) and unscrew steering wheel retaining nut (2). Pull steering wheel off shaft. Remove rear dash panel and remove clamp screws at both ends of joint shaft (7). Detach snap rings (3) and withdraw shaft (6), then remove joint shaft (7). Disconnect wiring harnesses under dash and remove dash panel. Disconnect interfering wires and cables and remove dash frame. Remove seat and fenders. Detach speed control pedal from shaft and remove tunnel cover. Detach drag link(s) from pitman arm (16), then remove pitman arm from shaft. Remove steering gearbox.

Separate gearbox halves (10 and 14) and remove shafts (9 and 15), sector gear (13) and bearings as necessary.

When assembling gearbox, align marks (M—Fig. K3) on sector gear and shaft. Lubricate gears with grease and use a liquid gasket to seal mating surfaces of gearbox halves. Align marks (S—Fig. K4) on pitman shaft (15) and pitman arm (16). Tighten nut retaining pitman arm to 74-103 N·m (54-75 ft.-lbs.). Tighten gearbox mounting screws to 39-54 N·m (29-39 ft.-lbs.). Tighten drag link nut to 88 N·m (65 ft.-lbs.). Tighten steering joint shaft (7—Fig. K2) clamp screws to 20-29 N·m (15-21 ft.-lbs.).

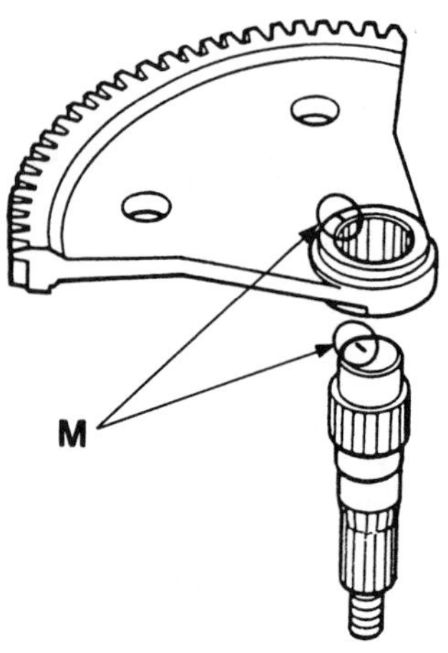

Fig. K3—Marks (M) must be aligned when installing sector gear on pitman shaft.

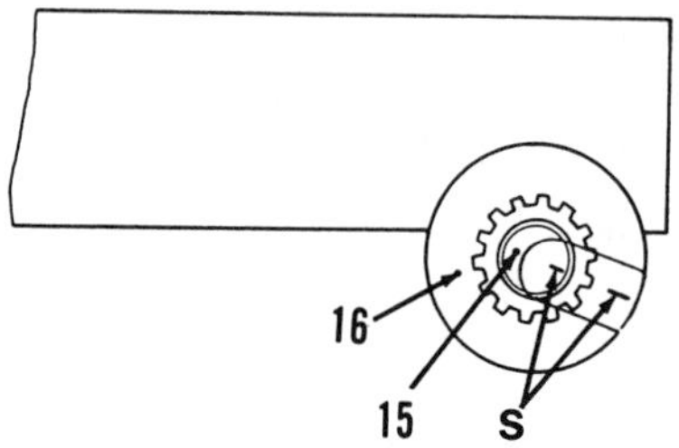

Fig. K4—Install pitman arm (16) on pitman shaft (15) so marks (S) are aligned.

ENGINE

Models G1800, G1800-S, G1900 and G1900-S are equipped with a three-cylinder, liquid-cooled diesel engine. Models G2000 and G2000-S are equipped with a three-cylinder, liquid-cooled spark ignition engine. Crankshaft rotation on all engines is counterclockwise as viewed from flywheel end. Firing order is 1-2-3.

Illustrations courtesy Kubota.

REMOVE AND REINSTALL

All Models

To remove engine, proceed as follows: Drain coolant and engine oil. Remove air cleaner and intake pipes, battery and hood. Remove muffler and disconnect all interfering wiring and cables. Detach pto belts and disconnect fuel line. On diesel engines, disconnect the fuel return line. Detach drive shaft from crankshaft pulley. Remove radiator support bolts. Unscrew engine retaining nuts and lift engine and radiator as an assembly out of tractor.

Reverse removal procedure when installing engine. Tighten drive shaft flange to crankshaft pulley screws to 20-29 N·m (15-21 ft.-lbs.).

MAINTENANCE

All Models

LUBRICATION. Check oil level before operation.

Manufacturer recommends oil with API classification CC or CD for diesel engines or API classification SF for gasoline engines. Use SAE 30 or 10W-30 oil for temperatures above 77° F (25° C); use SAE 20 or 10W-30 oil for temperatures between 32° F (0° C) and 77° F (25° C); use SAE 10W or 10W-30 for temperatures below 32° F (0° C).

Change oil and filter after first 50 hours of operation and then after every 100 hours of operation, or more frequently if operation is severe. Crankcase capacity is 3 L (3.2 qts.) for all models.

AIR FILTER. The air filter should be cleaned after every 50 hours of operation or more frequently if operating conditions are severe. Paper element type filter may be cleaned with low pressure compressed air blown from the inside toward the outside; air pressure must not exceed 205 kPa (30 psi).

The filter element can be washed in a solution of Kubota Filter Cleaner or Donaldson ND-1400 dissolved in water. Allow the filter to soak in the solution for 15 minutes, then rinse by agitating in clear water. Allow filter to dry before reusing.

Install a new filter each year or after six cleanings.

COOLING SYSTEM. Drain, flush and refill cooling system annually. Cooling system capacity is 3.3 L (3.5 qts.).

A thermostat is used to maintain cooling system temperature. The thermostat should begin to open at 82° C (180° F) and open fully at 95° C (203° F).

SPARK PLUG. Gasoline Engine. Recommended spark plug is NGK BCP6ES11. Spark plug gap should be 1.0-1.1 mm (0.039-0.043 in.). Tighten spark plug to 20-24 N·m (15-18 ft.-lbs.).

IGNITION SYSTEM. Gasoline Engine. A breakerless, solid-state ignition system is used. Ignition timing should be 20 degrees before top dead center. Ignition mark "IG" on flywheel should align with timing mark (TM—Fig. K5) on engine end plate. If not, loosen distributor retaining screw and rotate distributor as necessary. Specified air gap (G—Fig. K6) between distributor rotor (R) and trigger coil (C) is 0.2-0.4 mm (0.008-0.016 in.).

If distributor is removed from engine, align match mark on distributor shaft with match mark on end of distributor drive shaft when reinstalling distributor. The distributor is available only as an assembly.

Specified ignition coil primary resistance is 1.3-1.6 ohms. To check resistance, connect ohmmeter leads to the positive and negative terminals of coil. Specified ignition coil secondary resistance is 10.7k-14.5k ohms. To check resistance, connect ohmmeter leads to coil positive terminal and to high tension terminal.

CARBURETOR. Gasoline Engine. Gasoline engines are equipped with the float type carburetor shown in Fig. K7.

To adjust carburetor, engine must be at normal operating temperature. Turn idle mixture screw (9) so it is 1-1½ turns out from lightly seated. Run engine with speed control lever set to low idle speed. Turn idle mixture screw to obtain fastest engine speed. Adjust idle speed screw (7) so engine idles at 1200 rpm. High speed mixture is controlled by fixed main jet (16); no optional jet sizes are offered.

To disassemble carburetor, remove drain screw (21—Fig. K7) and drain fuel from float bowl. Remove fuel shut-off solenoid (28). Remove screw (26) and remove float bowl (20). Remove set screw (19), then withdraw pin (18) and remove float (17) and fuel inlet valve (11). Unscrew main jet (16), nozzle (15), idle mixture screw (9) and pilot jet (5). Remove screws from choke and throttle shafts (1 and 2) and remove shafts and plates (6 and 13).

Clean parts using suitable carburetor cleaner. Blow out passages with com-

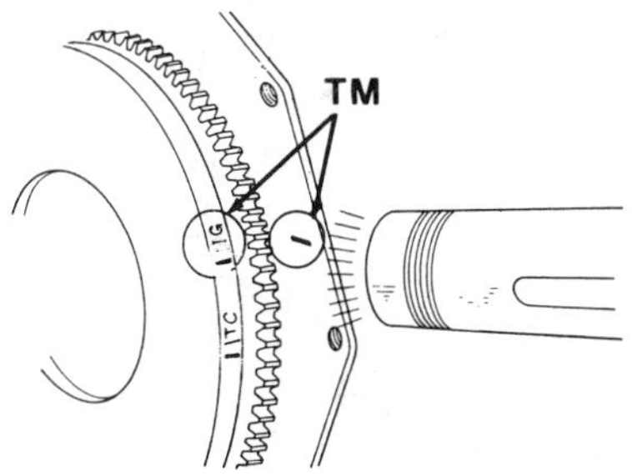

Fig. K5—Use a timing light to check for correct ignition timing on gasoline engine. Ignition should occur when "IG" mark on flywheel aligns with mark on end plate.

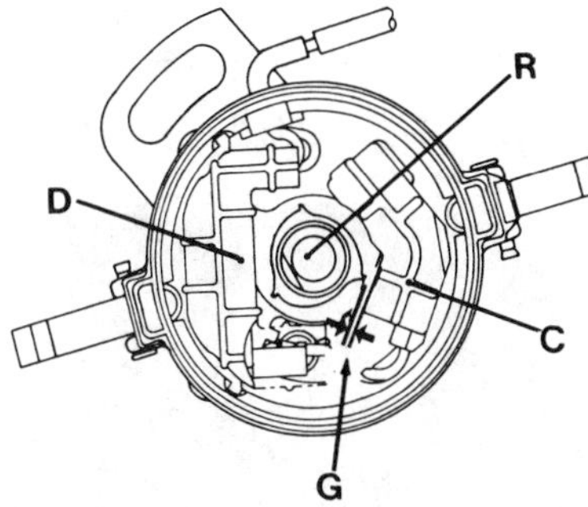

Fig. K6—Air gap (G) between distributor rotor (R) and trigger coil (C) should be 0.2-0.4 mm (0.008-0.016 in.).

Fig. K7—Exploded view of carburetor used on gasoline engine.

1. Choke shaft
2. Throttle shaft
3. Seal
4. Spring
5. Pilot jet
6. Choke plate
7. Idle speed screw
9. Idle mixture screw
11. Fuel inlet valve
12. Clip
13. Throttle plate
14. Gasket
15. Nozzle
16. Main jet
17. Float
18. Pin
19. Set screw
20. Fuel bowl
21. Drain screw
22. "O" ring
23. Gasket
24. Clip
25. Bracket
26. Screw
27. "O" ring
28. Fuel shut-off solenoid

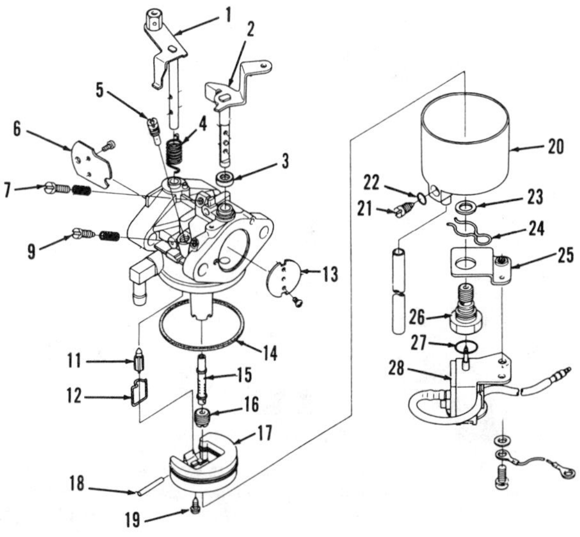

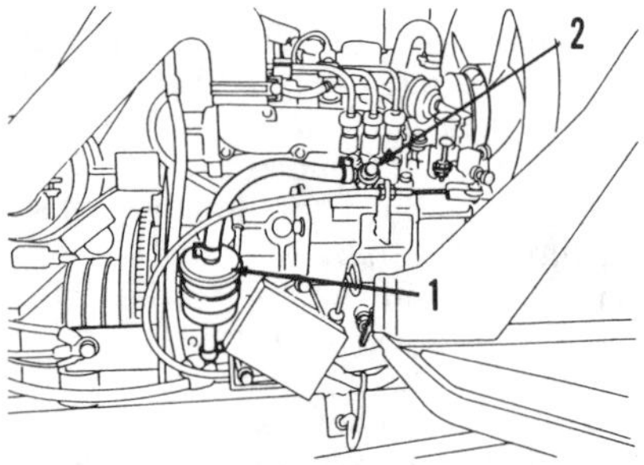

Fig. K8—View of diesel engine fuel filter (1) and bleed screw (2).

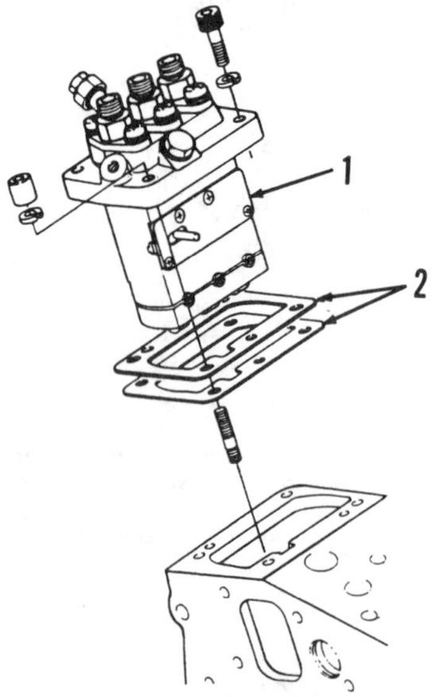

Fig. K9—Injection timing on diesel engines is adjusted by varying thickness of shims (2).

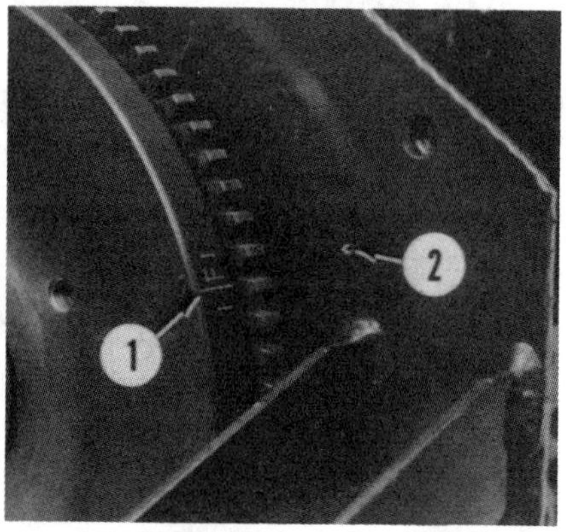

Fig. K10—Injection should occur when FI mark (1) on flywheel is aligned with timing mark (2) on end plate.

Fig. K11—View of diesel engine showing location of low speed adjusting screw (1) and high speed stop screw (3). Refer to text.

pressed air. Do not use drills or wires to clean jets or passages, as calibration of carburetor will be affected if openings are enlarged.

To reassemble, reverse the disassembly procedure.

FUEL FILTER. The fuel filter should be changed after every 500 hours of operation, or sooner if loss of engine power is evident. Clean fuel on diesel engines is particularly important due to problems that may occur if contaminated fuel is used.

On diesel engine models, bleed air from fuel system as outlined in following paragraph.

BLEED FUEL SYSTEM. Diesel Engine. Air must be bled from fuel system if fuel lines are disconnected or if tractor runs out of fuel. Do not bleed system when engine is hot.

To bleed fuel system, loosen injection pump bleed screw (2—Fig. K8). Turn key switch to "ON" position and allow fuel to flow out screw hole until bubbles are absent. Tighten bleed screw.

If engine fails to start, loosen high pressure fuel line fittings at the injectors. Crank engine with starter until fuel is discharged at the loosened fittings, then tighten fittings and start engine.

INJECTION PUMP TIMING. Diesel Engine. Injection pump timing is adjusted by varying thickness of shims (2—Fig. K9) between fuel injection pump and crankcase. To check injection pump timing, remove the fuel injection lines. With speed control lever at maximum speed position, turn flywheel in normal direction of rotation (counterclockwise) until fuel fills the open injection pump delivery valve holders. Continue turning flywheel slowly while observing number 1 delivery valve holder. When fuel just begins to flow over open delivery valve holder, stop turning engine and check flywheel timing mark. The "FI" mark (1—Fig. K10) on flywheel for that cylinder should then be aligned with punch mark (2) on engine end plate. If not, remove fuel injection pump and change shim thickness; decreasing shim thickness will advance timing and increasing shim thickness will retard timing. Changing shim thickness by 0.15 mm (0.006 in.) will change timing about 1.5 to 2 degrees.

Refer to preceding paragraph and bleed fuel system.

GOVERNOR. Diesel Engine. A flyball type governor is mounted on the fuel injection pump drive (timing) gear. Governor linkage is contained in the timing gear housing.

Slow idle speed for all models is 800 rpm. To adjust idle speed, remove cap nut (1—Fig. K11) from adjusting screw located at rear side of fuel injection pump (2). With engine at normal operating temperature, turn adjusting screw to obtain desired idle speed.

High idle speed stop screw (3) is sealed and should be adjusted by qualified personnel only. High idle (no-load) speed should be 3400 rpm. To adjust, move speed control lever to maximum speed position and turn high idle speed screw to obtain desired engine speed.

Gasoline Engine. A flyball type governor is mounted at rear side of the distributor drive (timing) gear. Refer to Fig. K12 for view of governor external linkage.

To adjust governor, loosen governor lever clamp nut (2). Move governor lever (3) and turn stop screw (4) as necessary so carburetor throttle plate is fully open. Use a screwdriver to rotate governor shaft (1) fully clockwise, then tighten clamp nut (2) while holding shaft and lever in this position.

Start engine and move speed control lever (5) to maximum speed position. Turn high speed adjusting screw (6) so maximum no-load engine speed is 3400 rpm.

VALVE CLEARANCE. Valve clearance can be checked after removing rocker arm cover. On all models, valve clearance with engine cold should be 0.15-0.18 mm (0.006-0.007 in.) for both intake and exhaust valves. Turn crankshaft to position number 1 piston at top dead center (1/TC mark on flywheel aligned with punch mark on engine end plate) of compression stroke, then check clearance with feeler gage between rocker arm and tip of valve stem. Loosen adjusting screw locknut, turn adjusting screw to obtain correct clearance, tighten nut and recheck clearance. Readjust if necessary.

Repeat procedure for remaining cylinders. Turning crankshaft ⅔ turn (240 degrees) will place next piston in firing order (1-2-3) at TDC on compression stroke.

REPAIRS

TIGHTENING TORQUES. Refer to the following table for special tightening torques:

Connecting rod	27-30 N·m (20-22 ft.-lbs.)
Crankshaft pulley cap screw	98-108 N·m (72-79 ft.-lbs.)
Crankshaft seal carrier	10-11 N·m (7-8 ft.-lbs.)

Illustrations courtesy Kubota.

Cylinder head	38-42 N·m (28-31 ft.-lbs.)
Exhaust manifold	12-14 N·m (9-10 ft.-lbs.)
Flywheel	54-59 N·m (40-43 ft.-lbs.)
Injection pump	10-11 N·m (7-8 ft.-lbs.)
Injector nozzle holder	34-39 N·m (25-29 ft.-lbs.)
Injector nozzle nut	49-68 N·m (36-50 ft.-lbs.)
Main bearing retainer screw	27-30 N·m (20-22 ft.-lbs.)
Main bearing caps	13-16 N·m (10-12 ft.-lbs.)
Rocker arm brackets	10-11 N·m (7-8 ft.-lbs.)
Timing gear cover	10-11 N·m (7-8 ft.-lbs.)

WATER PUMP. To remove water pump, drain coolant and, if necessary, relocate radiator for access to water pump. Remove fan belt, fan and pulley, then unbolt and remove pump from timing gear cover.

Using suitable puller, remove pulley flange (1—Fig. K14) from shaft. Press shaft and bearing assembly (2) out toward rear of housing. Remove the impeller (6) and seal (5) from shaft.

Reassemble using new shaft and bearing assembly and new seal. Press shaft and bearing assembly into housing, using a suitable installing tool that contacts only outer rim of bearing. Position new seal over rear end of shaft and press into place using hollow driver that contacts only outer rim of seal. Support front end of shaft and press impeller onto shaft. Support rear end of shaft and press flange onto shaft. Reinstall pump with new gasket.

COMPRESSION PRESSURE. Compression pressure should be checked with all injectors or spark plugs removed, battery in fully charged condition and with valve clearance properly adjusted. Move stop control lever to stop position on diesel engine. Fully open throttle on gasoline engine. Crank engine with starter and check compression pressure of each cylinder. Starter should turn engine at 200-300 rpm. Compression of engine in good condition should be 2840-3240 kPa (412-470 psi) for diesel engine and 1270 kPa (184 psi) for gasoline engine. Variation between cylinders should not exceed 10 percent. Low limit for compression pressure is 2260 kPa (327 psi) on diesel engine and 880 kPa (128 psi) on gasoline engine.

CYLINDER HEAD. To remove cylinder head, proceed as follows: Drain cooling system and remove upper radiator hose. Remove air intake tube. Unbolt and remove thermostat housing, intake manifold and exhaust manifold. On diesel engines, remove fuel injection lines, injector fuel return lines, injectors and glow plugs. Remove rocker arm cover, rocker arm shaft assembly and the push rods. Take care not to lose the lash caps from upper end of valve stems. Unbolt and remove cylinder head assembly from block. Take care not to damage the hollow dowel pin extending from corner of cylinder block. Remove the cylinder head gasket.

Before reinstalling cylinder head, completely clean all traces of old cylinder head gasket and carbon from both head and block surfaces. Be careful not to damage hollow dowel pin protruding from corner of cylinder block. Clean carbon from top of pistons, taking care to clean all loose carbon from cylinders. Closely check cylinder head for cracks. Use straightedge and feeler gage to check cylinder head for any warped or twisted condition. Do not place straightedge over combustion chamber. Maximum allowable deviation from flatness is 0.05 mm (0.002 in.) per 100 mm (3.94 in.). Surface grind head if not within limits. Refer to VALVE SYSTEM paragraphs for valve head recess.

Piston to cylinder head clearance should be 0.50-0.65 mm (0.020-0.026 in.) for diesel engine and 0.85-1.30 mm (0.033-0.051 in.) for gasoline engine. Clearance may be checked after cylinder head is installed. Turn crankshaft so that piston is at bottom of compression stroke and insert a soft lead wire into cylinder through injector or spark plug opening. Turn the crankshaft by hand so that piston goes past top dead center, then withdraw the lead wire and measure flattened thickness. This thickness is equal to piston to cylinder head clearance.

To reinstall cylinder head, reverse removal procedure and note the following: Do not apply any sealing compound to head gasket surfaces. Place a new "O" ring around hollow dowel pin. Oil the threads of the cylinder head fasteners and install all finger tight. Refer to tightening sequence in Fig. K15 and tighten the head bolts and nuts in several steps until final torque of 38-42 N·m (28-31 ft.-lbs.) is reached. Loosen locknuts and back rocker arm adjusting screws out. Be sure that lash caps (13—Fig. K16) are on valve stems. Install push rods and rocker arms and tighten rocker arm shaft support nuts to a torque of 10-11 N·m (7-8 ft.-lbs.). Adjust valve clearance as previously described in MAINTENANCE section. Complete reassembly by reversing disassembly procedure. Start engine and run until normal operating temperature is reached, then retorque cylinder head bolts to specified torque.

VALVE SYSTEM. Intake and exhaust valves seat directly in cylinder head. Lash caps are fitted to upper end of valve stems. Valve guides are renewable.

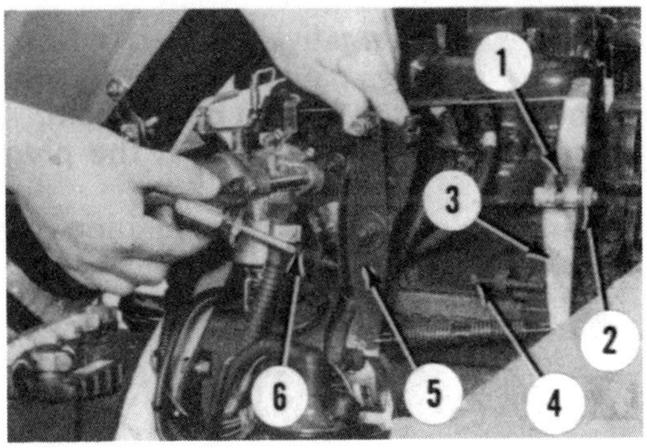

Fig. K12—View of gasoline engine showing governor linkage adjustment points. Refer to text.

1. Governor shaft
2. Clamp bolt
3. Governor lever
4. High speed stop screw
5. Speed control lever
6. High speed adjusting screw

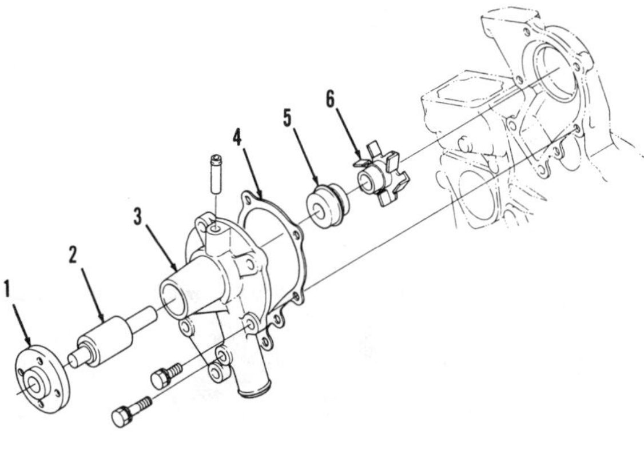

Fig. K14—Exploded view of water pump.

1. Flange
2. Bearing & shaft
3. Housing
4. Gasket
5. Mechanical seal
6. Impeller

Illustrations courtesy Kubota.

Intake and exhaust valve springs are alike. Rocker arm bushings are renewable.

Valve stem diameter is 5.968-5.980 mm (0.2350-0.2354 in.) and guide inside diameter is 6.010-6.025 mm (0.2366-0.2372 in.), providing a clearance of 0.030-0.057 mm (0.0012-0.0022 in.). Maximum allowable stem to guide clearance is 0.1 mm (0.004 in.). If necessary to renew guides, use suitable piloted valve guide driver and drive old guides out top of head. Install new guides from top side of head so that upper ends are flush with valve cover gasket surface of head. Note that exhaust valve guides have a counterbore in lower end of guide, but intake valve guides are not counterbored. Ream guides after installation for desired stem to guide clearance. Grind valve seats after installing new guides.

Valve face angle is 45 degrees and seat angle is 45 degrees for intake and exhaust valves. Desired valve seat width is 2.12 mm (0.083 in.). Seat width may be narrowed by reducing outside diameter of seat using a 15 degree stone.

After grinding valves and seats, install valves in head and check distance valve heads recess below cylinder head surface or protrude above head. On gasoline engine, specified distance is 0.75-0.95 mm (0.030-0.037 in.) below head surface with a maximum allowable recession of 1.2 mm (0.047 in.). On diesel engine, valve may protrude up to 0.10 mm (0.004 in.) or recess up to 0.10 mm (0.004 in.) with a maximum allowable recession of 0.30 mm (0.012 in.). If distance is excessive with new valve, manufacturer recommends replacement of cylinder head.

Valve spring free length should be 31.6 mm (1.244 in.). Renew springs if free length is 28.4 mm (1.118 in.) or less. Spring pressure at compressed height of 27.0 mm (1.063 in.) should be 64.7 N (14.6 lbs.). Minimum allowable spring pressure at 27.0 mm (1.063 in.) is 54.9 N (12.3 lbs.). Renew spring if distorted or otherwise damaged.

Rocker arm bushing ID should be 10.50-10.52 mm (0.413-0.414 in.). Rocker shaft diameter should be 10.47-10.48 mm (0.412-0.413 in.). Desired clearance between rocker arm shaft and bushings is 0.016-0.045 mm (0.0006-0.0018 in.); maximum allowable clearance is 0.15 mm (0.006 in.). When renewing rocker arm bushings, be sure oil hole in bushing is aligned with oil hole in rocker arm.

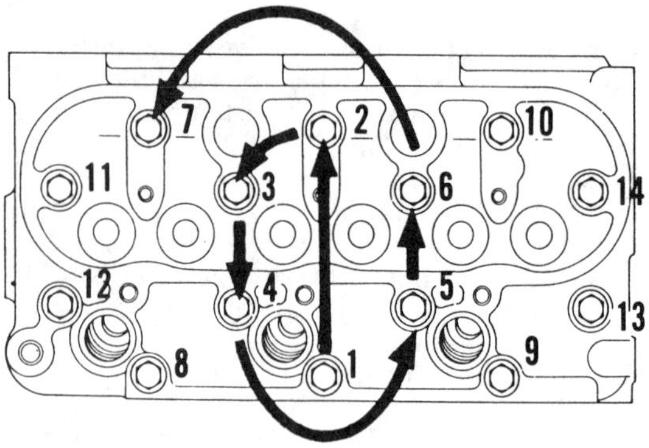

Fig. K15—Follow sequence shown above when tightening cylinder head screws.

INJECTORS. Diesel Engine. To remove injectors, first thoroughly clean injectors, fuel lines and surrounding area using suitable solvent and compressed air. Remove fuel leak-off line and high pressure lines from injectors. Cap or plug all openings immediately to prevent entry of dirt. Unscrew and remove injectors from cylinder head.

Prior to reinstalling injectors, remove old nozzle sealing washers from head (if not removed with injector) and be sure cylinder head bores are clean. Install new nozzle washers in head with corrugated side up. Install injector and tighten to a torque of 49-68 N·m (36-50 ft.-lbs.). When reinstalling fuel lines, leave fittings for high pressure lines loose at injectors and crank engine until fuel emerges from loose fittings, then tighten injector line fittings.

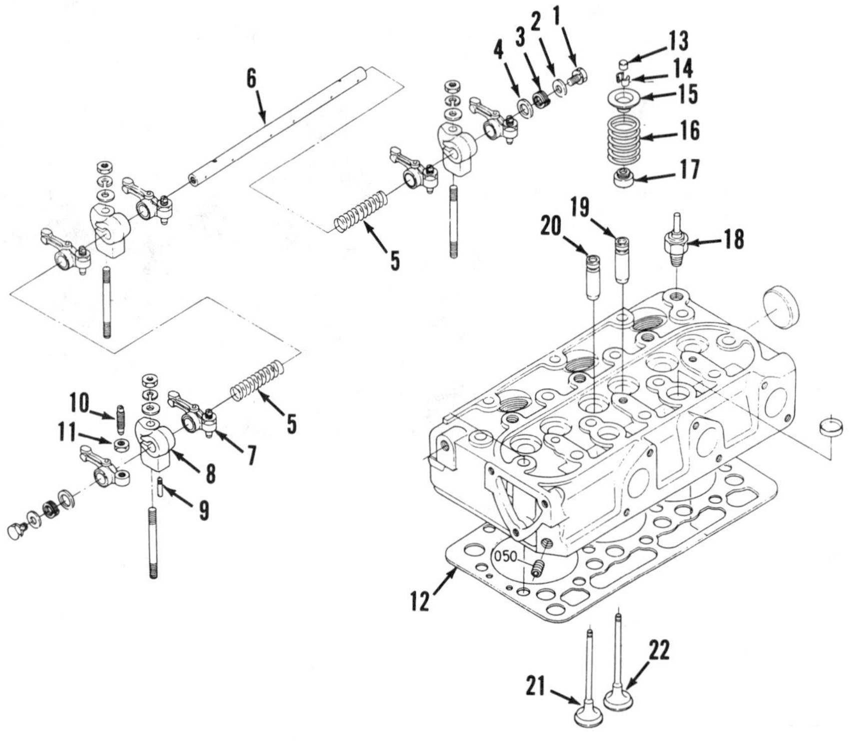

Fig. K16—Exploded view of rocker shaft and valve assemblies.

1. Screw
2. Washer
3. Spring
4. Washer
5. Spring
6. Rocker shaft
7. Rocker arm
8. Rocker shaft stand
9. Dowel pin
10. Adjusting screw
11. Locknut
12. Head gasket
13. Valve cap
14. Split lock
15. Spring retainer
16. Valve spring
17. Valve stem seal
18. Temperature sensor
19. Intake valve guide
20. Exhaust valve guide
21. Exhaust valve
22. Intake valve

TESTING. A faulty injector may be located on the engine by loosening high pressure fuel line fitting on each injector, in turn, with engine running. This will allow fuel to escape from the loosened fitting rather than enter the cylinder. The injector that least affects engine operation when its line is loosened is the faulty injector. If a faulty injector is found and considerable time has elapsed since injectors have been serviced, it is recommended that all injectors be removed and serviced, or that new or reconditioned units be installed.

A complete job of testing and adjusting fuel injectors requires use of special test equipment. Only clean, approved testing oil should be used in tester tank. Injection nozzle should be tested for opening pressure, spray pattern and seat leakage. Check injection nozzle as outlined in the following paragraphs.

Illustrations courtesy Kubota.

WARNING: Fuel emerges from injector with sufficient force to penetrate the skin. When testing injector, keep yourself clear of nozzle spray.

OPENING PRESSURE. Before conducting test, operate tester lever until fuel flows from tester line, then attach injection nozzle to tester. Pump tester lever a few quick strokes to purge air and to be sure nozzle valve is not stuck and spray hole is open.

Operate tester lever slowly and observe pressure gage reading when nozzle opens. Opening pressure should be 13730-14710 kPa (1990-2133 psi).

Opening pressure is adjusted by changing thickness of shim (4—Fig. K17). Shims are available in different thicknesses of 0.90-1.95 mm (0.035-0.077 in.) in steps of 0.025 mm (0.001 in.). Installing a thicker shim will increase opening pressure, a thinner shim will decrease pressure.

SPRAY PATTERN. The injector nozzle tip used on all models has one orifice. To check spray pattern, operate tester handle and observe spray. Spray pattern must be uniform, well atomized and slightly conical. If irregular pattern is observed, check for partially clogged or damaged spray hole or improperly seating nozzle valve.

SEAT LEAKAGE. Wipe nozzle tip dry, then operate tester lever slowly to maintain pressure of 12750 kPa (1850 psi) for ten seconds. There should be no drops or noticeable accumulation of oil on tip of injector. If drop forms, disassemble and overhaul injector. Slight wetting of tip is acceptable on used injector.

OVERHAUL. First, thoroughly clean outside of injector. Clamp nozzle nut (9—Fig. K17) in a soft-jawed vise. Remove nozzle holder (3) and separate components. Place all parts in clean calibrating oil or diesel fuel as they are removed, keeping parts from each injector together and separate from other units. Nozzle valve and body (8) are individually fitted and these parts must not be interchanged.

Clean exterior surfaces with a brass wire brush. Soak parts in an approved carbon solvent if necessary to loosen hard carbon deposits. Rinse parts in clean diesel fuel or calibrating oil to neutralize solvent and prevent etching of polished surfaces.

Clean nozzle spray hole from inside using a pointed hardwood stick or wood splinter as shown in Fig. K18. Scrape carbon from pressure chamber using hooked scraper as shown in Fig. K19. Clean valve seat using brass scraper as shown in Fig. K20.

Examine lapped pressure surfaces of nozzle and valve (8—Fig. K17), spacer (7) and nozzle holder (3) for obvious damage. Check to see that nozzle valve will slide smoothly downward in valve body by its own weight. If valve binds or slides too fast, renew nozzle assembly. Check spring (5) for distortion, corrosion or other damage. Check for worn pressure pin (6) and renew if damage is noted. Be sure nozzle spray orifice is clean.

Back flush nozzle using reverse flush adapter on injector tester. Thoroughly rinse all parts in clean diesel fuel. Make sure all sealing surfaces are absolutely clean. Reassemble injector by reversing disassembly procedure. Tighten nozzle holder (3) to a torque of 34-39 N·m (25-29 ft.-lbs.). Retest assembled injector as previously outlined.

GLOW PLUGS. Diesel Engine. Glow plugs are parallel connected with each glow plug grounding through its mounting threads. Indicator light should glow about 5 seconds after turning glow plug switch to "PREHEAT" if units are operating satisfactorily.

If indicator light fails to glow, check for burned out indicator lamp, an open circuit at switch, indicator lamp or glow plug connections, or defective combination box. Resistance between glow plug terminal and body should be about 0.9 ohms when glow plug is cold. If resistance is zero, glow plug is shorted. If resistance is infinite, glow plug element is burned out.

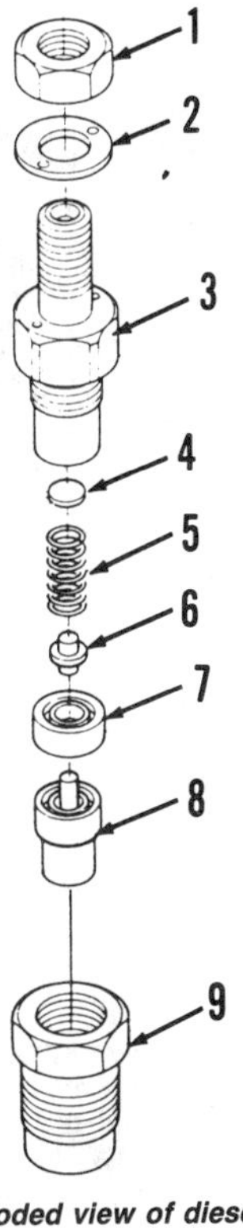

Fig. K17—Exploded view of diesel fuel injector.

1. Nut
2. Washer
3. Nozzle holder
4. Shim
5. Spring
6. Pressure pin
7. Spacer
8. Nozzle & valve
9. Nozzle nut

INJECTION PUMP. Diesel Engine. To remove injection pump, first thoroughly clean pump, injection lines and surrounding area. Close fuel supply valve. Disconnect and remove fuel injection lines and fuel supply line from pump. Immediately cap or plug all openings to prevent entry of dirt. Disconnect stop and throttle control linkage from levers on plate in front of pump. Unbolt pump from cylinder block and lift pump upward, taking care not to lose or damage shim between pump flange and cylinder block. Be sure to place pump in clean area.

Refer to Fig. K21 for an exploded view of fuel injection pump. The injection pump should be tested and overhauled

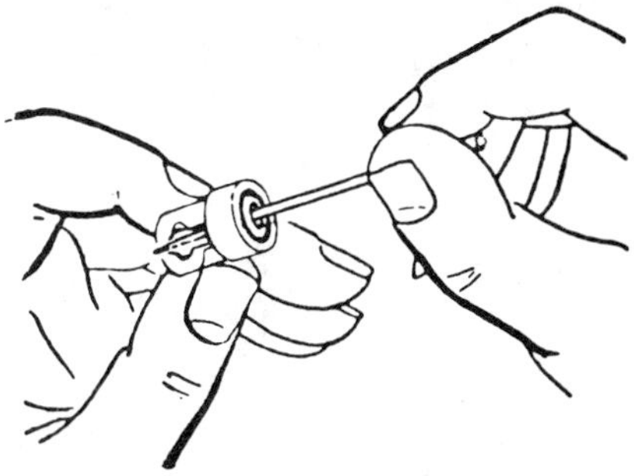

Fig. K18—Use a pointed hardwood stick to clean spray hole as shown.

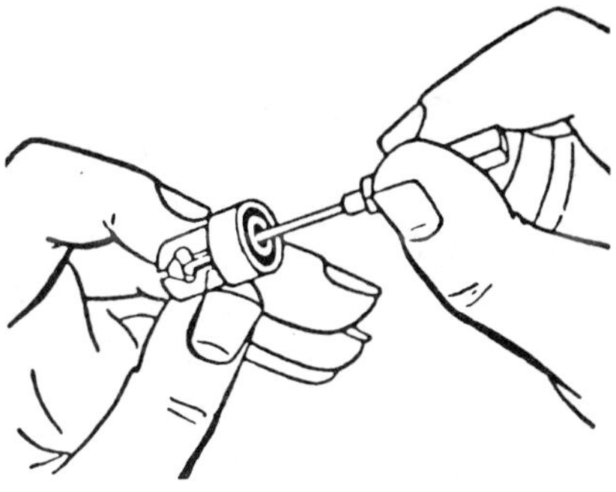

Fig. K19—Use a hooked scraper to clean carbon from pressure chamber.

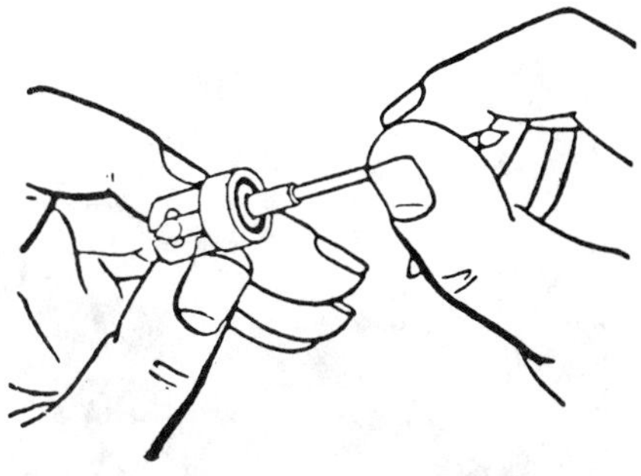

Fig. K20—Clean valve seat using brass scraper as shown.

Illustrations courtesy Kubota.

by a shop qualified in diesel injection equipment repair.

To reinstall pump, use same thickness shim as when removed unless change of timing is indicated. Move pump rack pin (P—Fig. K21A) forward so that rear end of rack is flush with pump housing. Move control levers so that slot (S) of governor lever (G) is aligned with notch (N) in crankcase. Insert pump in crankcase opening so that rack pin engages slot in governor fork lever. Release control levers and be sure pump is in place. Install retaining nuts and tighten to a torque of 10-11 N·m (7-8 ft.-lbs.). Install fuel lines and governor linkage, leaving injection line fittings at injectors loose to bleed the fuel system as outlined in MAINTENANCE section. Check injection pump timing as outlined in MAINTENANCE section.

TIMING GEAR COVER. Diesel Engine. To remove timing gear cover, proceed as follows: Drain cooling system and remove engine and radiator as previously outlined. Disconnect radiator hoses and remove radiator. Remove fan belt, fan and alternator. Disconnect linkage from throttle lever and stop control lever on control plate (11—Fig. K22) at front of fuel injection pump. Unbolt the control plate and carefully lift the plate and levers assembly, then disconnect the governor spring (27) from throttle lever arm (15) on bottom of plate. Disconnect the start spring (29) from pin (18) in front of timing cover. Lock engine from turning, flatten the lock tab washer and remove the crankshaft pulley cap screw. Remove crankshaft pulley. Remove cap screw (S) located inside throttle/stop lever plate opening. Remove the remaining timing gear cover bolts and remove cover from engine.

With cover removed, clean all traces of old gasket from cover and front of cylinder block. Inspect seal sleeve on crankshaft and renew if necessary. Install new crankshaft oil seal (21) in timing gear cover with lip to inside.

To install cover, use new gasket and "O" rings. Apply grease to inside of crankshaft seal and carefully position cover over crankshaft to avoid damage to crankshaft seal. Install the retaining bolts and tighten to a torque of 10-11 N·m (7-8 ft.-lbs.). Install crankshaft pulley with punch marks (M—Fig. K24) on pulley and crankshaft aligned. Oil pulley retaining cap screw threads and install cap screw with new lock tab washer. Tighten cap screw to a torque of 98-108 N·m (72-79 ft.-lbs.) and bend washer against one flat of cap screw head. Reconnect start spring (29—Fig. K22). Place new gasket on throttle/stop lever plate (11), hold plate in position and reconnect the governor spring (27). Install cover retaining cap screws with copper washers. Complete remainder of assembly by reversing disassembly procedure.

Gasoline Engine. To remove timing gear cover, proceed as follows: Drain cooling system and remove engine and radiator as previously outlined. Disconnect radiator hoses and remove radiator. Remove fan belt, fan and alternator. Disconnect governor linkage and remove governor shaft plate (9—Fig. K23). Lock engine to prevent crankshaft rotation, flatten the lock tab washer and remove the crankshaft pulley cap screw. Remove crankshaft pulley. Remove cap screw (S) located inside governor plate opening. Remove the remaining timing gear cover bolts and remove cover from engine.

With cover removed, clean all traces of old gasket from cover and front of cylinder block. Inspect seal sleeve on crankshaft and renew if necessary. Install new crankshaft oil seal (17) in timing gear cover with lip to inside.

To install cover, use new gasket and "O" rings. Apply grease to inside of crankshaft seal and carefully position cover over crankshaft to avoid damage to crankshaft seal. Install the retaining bolts and tighten to a torque of 10-11 N·m (7-8 ft.-lbs.). Install crankshaft pulley with punch marks (M—Fig. K24) on pulley and crankshaft aligned. Oil pulley retaining cap screw threads and install cap screw with new lock tab washer. Tighten cap screw to a torque of 98-108 N·m (72-79 ft.-lbs.) and bend washer against one flat of cap screw head. Before installing governor plate (9—Fig. K24), rotate governor shaft against spring tension so arm (15) is towards top as shown in Fig. K25. Install governor shaft and plate so pin on arm engages slot in governor lever (23—Fig. K23). Complete remainder of assembly by reversing disassembly procedure.

TIMING GEARS. The timing gears are accessible for inspection after removing timing gear cover. Remove the rocker arm shaft assembly, and, on

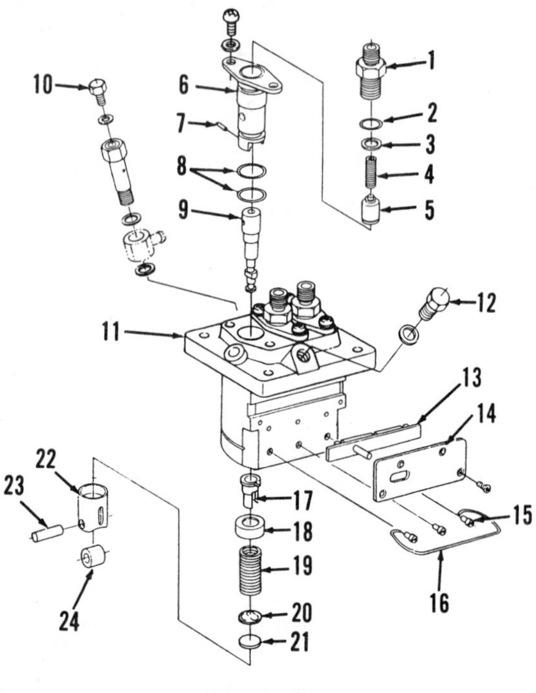

Fig. K21—Exploded view of fuel injection pump used on diesel engine.

1. Delivery valve holder
2. "O" ring
3. Shim
4. Spring
5. Delivery valve
6. Sleeve
7. Pin
8. "O" rings
9. Plunger
10. Air bleed screw
11. Pump body
12. Plug
13. Control rack
14. Plate
15. Tappet guide pin
16. Retainer
17. Control sleeve
18. Spring cap
19. Spring
20. Washer
21. Shim
22. Tappet
23. Pin
24. Roller

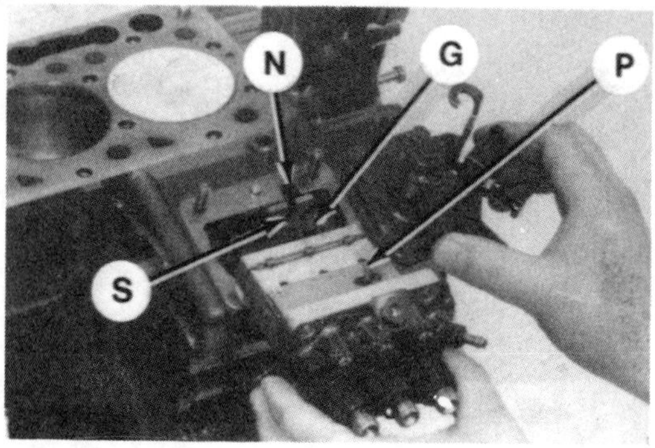

Fig. K21A—When installing fuel injection pump, be sure that pump rack pin (P) engages slot (S) of governor lever (G).

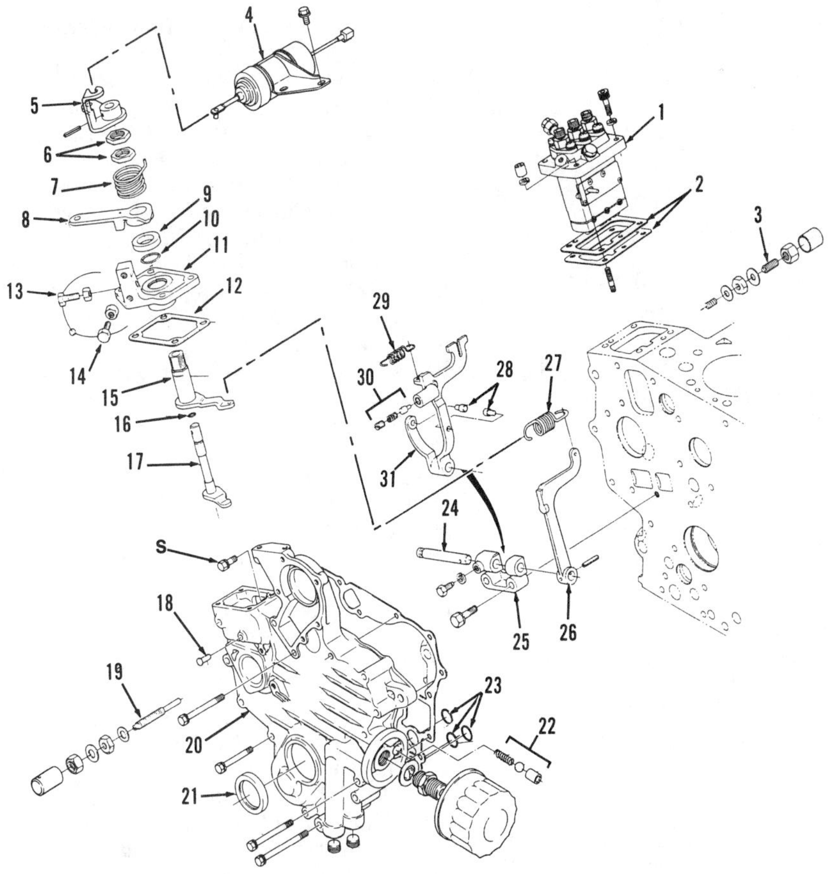

Fig. K22—Exploded view of governor mechanism used on diesel engine.

1. Injection pump assy.
2. Shims
3. Low idle speed adjusting screw
4. Fuel shut-off solenoid
5. Shut-off lever
6. Nuts
7. Return spring
8. Speed control lever
9. Collar
10. "O" ring
11. Plate
12. Gasket
13. High speed adjusting screw
14. Shut-off stop screw
15. Speed control arm
16. "O" ring
17. Shut-off arm
18. Start spring pin
19. Maximum fuel limiting screw
20. Timing gear housing
21. Oil seal
22. Oil pressure regulator valve
23. "O" rings
24. Pivot shaft
25. Bracket
26. Governor arm
27. Governor spring
28. Pins
29. Start spring
30. Torque overload screw
31. Governor lever

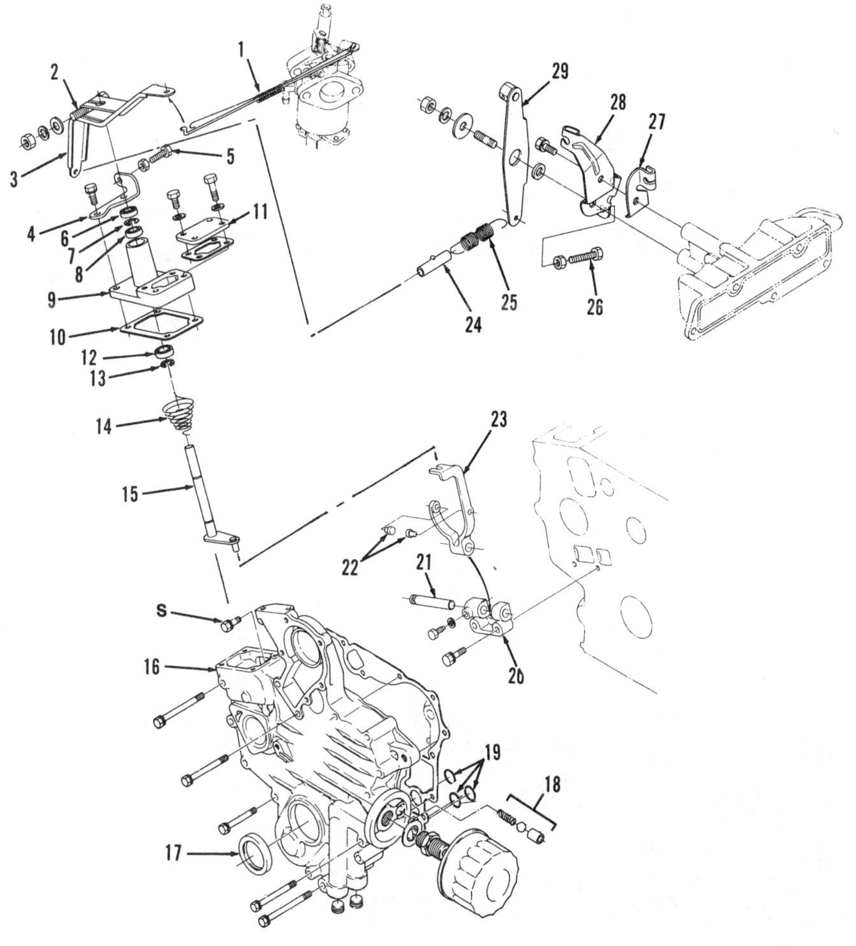

Fig. K23—Exploded view of governor mechanism used on gasoline engine.

1. Throttle rod
2. Clamp bolt
3. Governor lever
4. Bracket
5. Maximum speed screw
6. Seal
7. Snap ring
8. Bearing
9. Support
10. Gasket
11. Plate
12. Bearing
13. Snap ring
14. Spring
15. Governor shaft
16. Timing gear housing
17. Oil seal
18. Oil pressure regulating valve
19. "O" rings
20. Bracket
21. Pivot shaft
22. Pins
23. Governor lever
24. Damper
25. Governor spring
26. High speed screw
27. Holder
28. Bracket
29. Speed control lever

Illustrations courtesy Kubota.

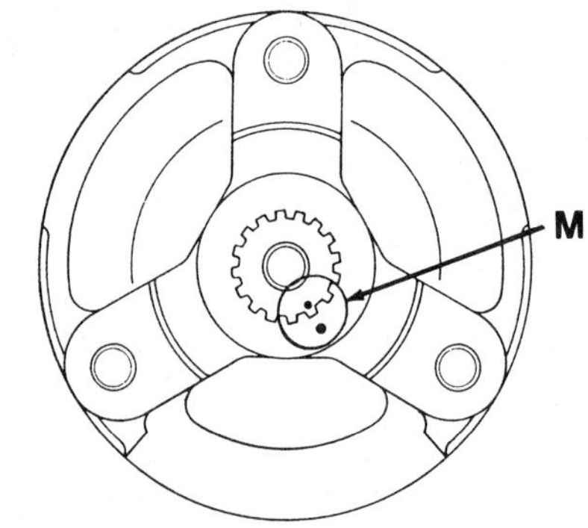

Fig. K24—Align marks (M) on fan drive pulley and crankshaft when installing pulley.

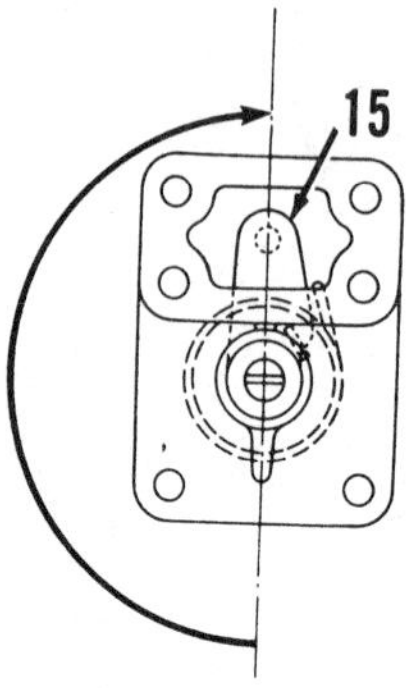

Fig. K25—Before installing governor plate, rotate governor shaft against spring tension so arm (15) is towards top.

diesel engine, remove the fuel injection pump to relieve pressure on timing gears. Check timing gear backlash with dial indicator. Desired timing gear backlash is 0.04-0.12 mm (0.0016-0.0047 in.); maximum allowable backlash between any two gears is 0.15 mm (0.006 in.). Desired end play of idler gear on shaft is 0.2-0.5 mm (0.008-0.020 in.); maximum allowable end play is 0.6 mm (0.024 in.). Oil clearance of idler gear on shaft should be 0.020-0.084 mm (0.0008-0.0033 in.) with maximum oil clearance being 0.10 mm (0.004 in.). Idler gear bushing, thrust collar and shaft are renewable.

To remove idler gear (15—Fig. K26), remove snap ring (12) and pull gear and thrust collar (13) from shaft. If renewing bushing (14), press new bushing in until flush to 0.2 mm (0.008 in.) below outside surfaces of gear. Unbolt and remove idler shaft (16) from front of crankcase if necessary to renew shaft. Reinstall gear with timing marks aligned as shown in Fig. K27.

Crankshaft timing gear (4—Fig. K28), washer (5), "O" ring (6) and crankshaft oil seal sleeve (7) can be removed using suitable pullers engaging back side of gear. When installing gear, install idler gear with timing marks aligned as shown in Fig. K27. Then install washer, "O" ring and oil seal sleeve with chamfered inside diameter of sleeve toward "O" ring.

To remove and install camshaft gear and fuel injection pump gear, refer to CAMSHAFT and INJECTION PUMP CAMSHAFT paragraphs.

OIL PUMP. The oil pump (1—Fig. K28) is located on front of engine cylinder block inside the timing gear cover. Oil pressure relief valve is located under the oil filter and is retained by the filter adapter nut and washer as shown in Figs. K22 and K23. Engine oil pressure should be 196-441 kPa (28-64 psi) at rated engine speed with minimum allowable pressure being 69 kPa (10 psi). To check pressure, remove the oil pressure switch and install a suitable oil pressure gage.

To remove oil pump, first remove timing cover as previously outlined, then unbolt and remove pump and gear assembly from front of crankcase. Remove screw retaining port plate to back of oil pump and remove the plate. Measure clearance between outer rotor and pump housing using a feeler gage; desired clearance is 0.07-0.15 mm (0.0028-0.0059 in.) with maximum allowable clearance of 0.3 mm (0.012 in.). Measure clearance between inner and outer rotors with feeler gage; maximum allowable clearance is 0.25 mm (0.010 in.). Rotor end clearance should be 0.08-0.13 mm (0.003-0.005 in.); maximum allowable clearance is 0.2 mm (0.008 in.). To measure rotor end clearance, place strip of Plastigage between rotors and port plate and install pump on crankcase. Remove the pump and check clearance with Plastigage indicator strip.

If any clearance is not within specification limits, renew the complete pump assembly.

CAMSHAFT. To remove engine camshaft (21—Fig. K26), first remove cylin-

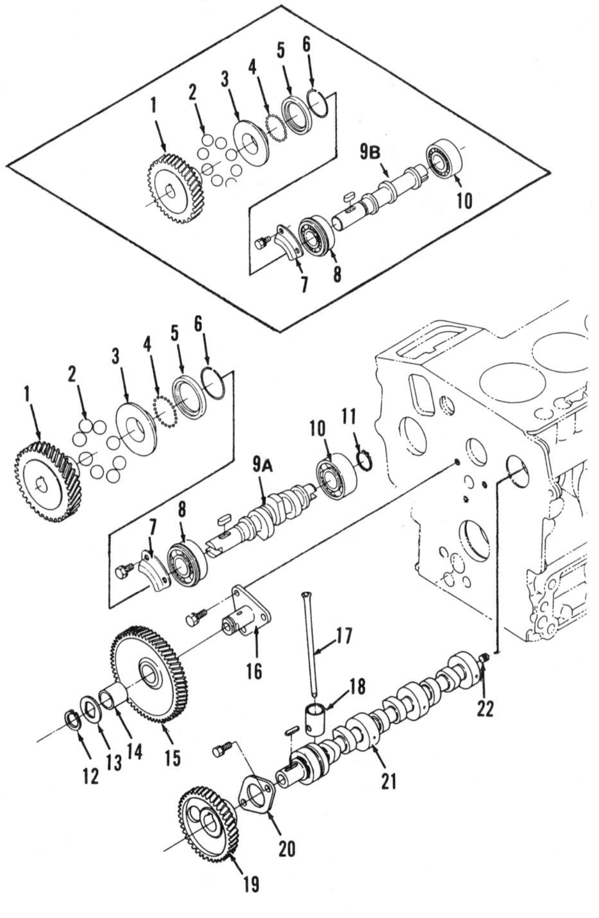

Fig. K26—Exploded view of engine timing gears and camshaft assemblies. Camshaft (9A) is used on diesel engines to actuate fuel injection pump. On gasoline engine, shaft (9B shown in inset) drives the ignition distributor.

1. Gear
2. Governor flyballs
3. Governor sleeve
4. Ball bearings
5. Bearing case
6. Snap ring
7. Bearing retainer
8. Bearing

9A. Fuel injection camshaft
9B. Distributor drive shaft

10. Bearing
11. Snap ring
12. Snap ring
13. Washer
14. Bushing
15. Idler gear
16. Stub shaft
17. Push rod
18. Tappet
19. Gear
20. Thrust plate
21. Engine camshaft
22. Set screw

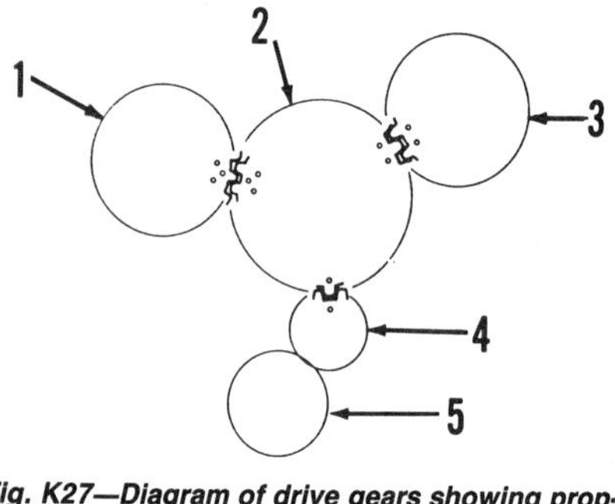

Fig. K27—Diagram of drive gears showing proper alignment of timing marks. Note three marks on injection pump gear (1) and idler gear (2); two marks on camshaft gear (3) and idler gear; single marks on crankshaft gear (4) and idler gear. No marks are used on oil pump gear (5).

Illustrations courtesy Kubota.

der head and timing gear cover as previously outlined. Remove cam followers (tappets) from their bores and identify each by position from which they were removed. Remove cap screws from camshaft thrust plate (20) through hole in camshaft gear. Withdraw camshaft, plate and camshaft gear from front of cylinder block. If necessary, press camshaft out of gear and retainer plate.

Camshaft lobe height should be 26.88 mm (1.0583 in.); renew camshaft if lobe height is less than 26.83 mm (1.0563 in.). Camshaft bearing journal diameter should be 32.934-32.950 mm (1.2966-1.2972 in.). Desired oil clearance of camshaft journals in bearing bore is 0.05-0.09 mm (0.002-0.0035 in.) with maximum allowable clearance being 0.15 mm (0.006 in.). With camshaft supported in V-blocks at outer bearing journals, camshaft runout at center journals should be 0.01 mm (0.0004 in.); maximum allowable runout is 0.08 mm (0.003 in.).

Check bottom of cam followers (tappets) for abnormal wear and if noted, carefully inspect camshaft lobes for scoring or abnormal wear pattern. Renew cam followers if renewing camshaft.

When installing camshaft gear on camshaft, first install retainer plate. Then heat gear to approximately 80° C (175° F) and push gear onto shaft, being sure that timing marks on gear are forward and that keyway aligns with key on camshaft. Install camshaft, retainer plate and gear with timing marks of gear and idler gear aligned as shown in Fig. K27. Complete reassembly of engine by reversing disassembly procedure.

INJECTION PUMP CAMSHAFT. Diesel Engine. To remove injection pump camshaft (9A—Fig. K26), first remove fuel injection pump and timing gear cover as previously outlined. Remove cap screws retaining governor fork lever bracket to front of cylinder block. Remove cap screws from bearing retainer plate (7), then withdraw injection pump camshaft, gear and governor fork assembly from front of cylinder block.

Inspect camshaft lobes, gear and bearings for excessive wear or other damage and renew parts as necessary.

When installing, first hook the governor spring to fork. Leave governor fork lever bracket bolts loose until fuel injection pump camshaft and injection pump are installed. Be sure injection pump camshaft gear and idler gear timing marks are aligned as shown in Fig. K27. Be sure that pins on governor lever fork are centered on governor sleeve (3—Fig. K26). Tighten governor fork bracket cap screws and complete reassembly of engine by reversing disassembly procedure.

DISTRIBUTOR DRIVE SHAFT. Gasoline Engine. To remove distributor drive shaft (9B—Fig. K26), remove distributor and timing gear cover as previously outlined. Remove cap screws from retainer plate (7) and governor fork lever bracket, then withdraw distributor drive shaft, gear and governor fork assembly from front of crankcase.

When installing, first hook the governor spring to fork. Leave governor fork lever bracket bolts loose until distributor drive shaft is installed. Be sure distributor drive shaft gear and idler gear timing marks are aligned as shown in Fig. K27. Be sure that pins on governor lever fork are centered on governor sleeve (3—Fig. K26). Tighten governor fork lever bracket screws and complete reassembly of engine by reversing disassembly procedure. When installing distributor, align match mark on end of distributor shaft with match mark on distributor drive shaft. Check and adjust ignition timing as outlined in MAINTENANCE section.

GOVERNOR. All models are equipped with a flyball type governor at back side of fuel injection pump gear or distributor drive shaft as shown in Fig. K26. Refer to FUEL INJECTION PUMP CAMSHAFT or DISTRIBUTOR DRIVE SHAFT paragraph for removal and installation procedure. Governor balls (2) are carried in bores of rear face of gear (1). Ball movement with variation in engine speed against governor sleeve (3) actuates the governor lever, which is connected to fuel injection pump control rack or carburetor linkage. Force of flyball movement through governor sleeve is balanced by the governor spring.

Governor components are accessible after removing fuel injection pump camshaft or distributor drive shaft as previously outlined. Inspect governor parts and renew any that are excessively worn or damaged. Be careful not to lose any of the thrust bearing balls (4) or governor flyballs (2) when disassembling fuel injection pump camshaft and governor unit.

PISTON AND ROD UNITS. Piston and connecting rods are removed from above as a unit after removing cylinder head, oil pan and oil pickup. Be sure that carbon and ring ridge (if present) are removed from top of cylinder before removing pistons. Identify piston and rod units so they can be reinstalled in original positions if found to be serviceable.

When installing piston and connecting rod on diesel engine, be sure that mark (N—Fig. K30) on piston crown faces camshaft side of cylinder block and identification mark (M) on connecting rod and cap faces away from camshaft side of cylinder block.

When installing piston and connecting rod on gasoline engine, be sure that slotted area of piston (S—Fig. K30) and identification mark (M) on connecting

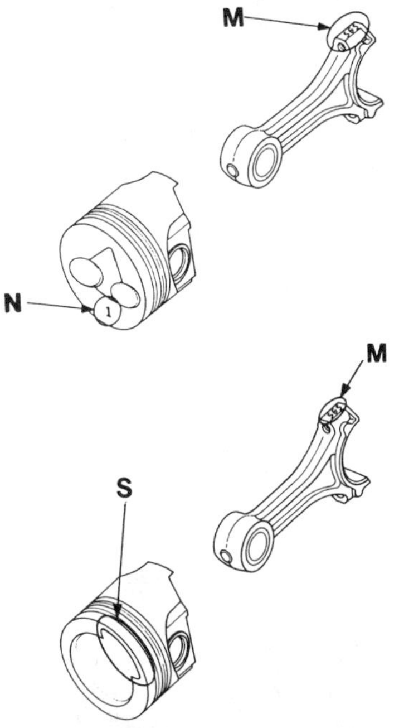

Fig. K30—On diesel engine, number (N) on piston crown and stamped number (M) on rod must be on opposite sides. On gasoline engine, notch (S) and stamped number (M) on rod must be on same side.

Fig. K28—Engine oil pump is mounted on front of cylinder block.

1. Oil pump
2. Drive gear
3. Nut
4. Crankshaft gear
5. Oil slinger
6. "O" ring
7. Collar

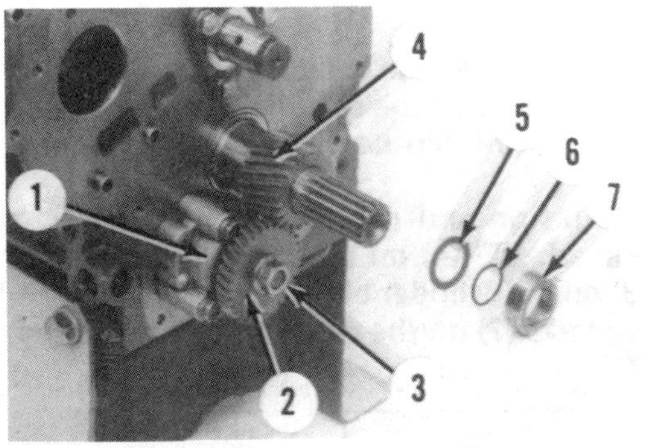

Illustrations courtesy Kubota.

rod and cap are on same side and face away from camshaft side of cylinder block.

Note that numbers stamped on rod and cap must be on same side when cap is installed on connecting rod. Lubricate threads on connecting rod cap screws and tighten to a torque of 15-18 N·m (11-14 ft.-lbs.).

PISTON, PIN AND RINGS. All models are equipped with cam ground aluminum pistons fitted with two compression rings and one oil control ring. Pistons and rings are available in oversize of 0.50 mm (0.020 in.) as well as standard size. Piston pins are available only in standard size of 20.002-20.011 mm (0.7875-0.7878 in.) for diesel engine and 18.000-18.005 mm (0.7087-0.7089 in.) for gasoline engine. On diesel engine, piston pin bore is 20.000-20.013 mm (0.7874-0.7879 in.) with wear limit of 20.05 mm (0.7894 in.). On gasoline engine, piston pin bore is 18.000-18.011 mm (0.7087-0.7091 in.) with wear limit of 18.05 mm (0.7106 in.). When inserting pin, first heat piston to a temperature of approximately 80° C (175° F). Install retaining snap rings in piston pin bore grooves with sharp edge of snap rings out.

On diesel engine, piston ring side clearance in groove should be 0.085-0.112 mm (0.003-0.004 in.) for second compression ring and 0.02-0.06 mm (0.0008-0.0024 in.) for the oil control ring. Top compression ring is keystone type. Maximum allowable side clearance in groove is 0.15 mm (0.006 in.) for all rings. Renew piston if ring side clearance is excessive. Ring end gap should be 0.15-0.30 mm (0.006-0.012 in.) for top compression ring and 0.30-0.45 mm (0.012-0.018 in.) for second compression ring with maximum allowable end gap being 1.2 mm (0.047 in.). Install rings with marked side up. Place oil ring spring expander in groove with ends butted together, then install oil ring on top of expander with ends 180 degrees away from expander ends. Position ring end gaps around piston as shown in Fig. K31.

On gasoline engine, piston ring side clearance in groove should be 0.02-0.06 mm (0.0008-0.0024 in.) for all rings. Maximum allowable side clearance in groove is 0.15 mm (0.006 in.) for all rings. Renew piston if ring side clearance is excessive. Ring end gap should be 0.15-0.35 mm (0.006-0.014 in.) for compression rings with maximum allowable end gap being 1.2 mm (0.047 in.). Install rings with marked side up. Position ring end gaps around piston as shown in Fig. K32.

CYLINDERS. Standard cylinder inside diameter is 64.000-64.019 mm (2.5197-2.5204 in.) for Models G1800 and G1800-S; 67.000-67.019 mm (2.6378-2.6385 in.) for Models G1900 and G1900-S; and 68.000-68.019 mm (2.6772-2.6779 in.) for Models G2000 and G2000-S. Rebore cylinders to 0.50 mm (0.020 in.) oversize if scored or wear exceeds 0.15 mm (0.006 in.) more than standard diameter.

CONNECTING RODS AND BEARINGS. Connecting rods are fitted with a renewable bushing (5—Fig. K33) at pis-

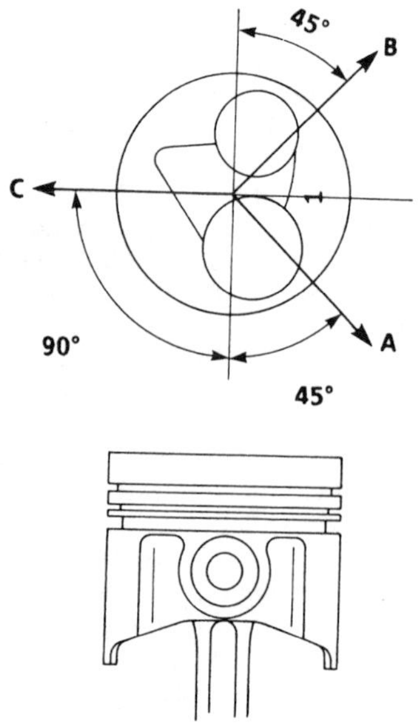

Fig. K31—On diesel engine, stagger piston ring end gaps around piston as shown.

A. Top compression ring
B. Second compression ring
C. Oil ring

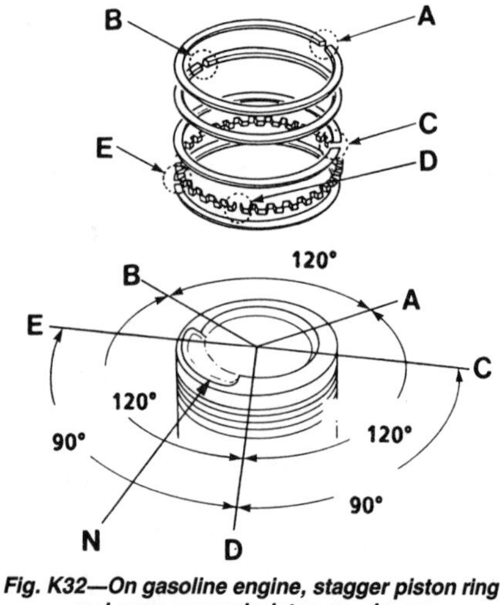

Fig. K32—On gasoline engine, stagger piston ring end gaps around piston as shown.

A. Top compression ring
B. Second compression ring
C. Side rail gap
D. Spacer gap
E. Side rail gap
N. Piston notch

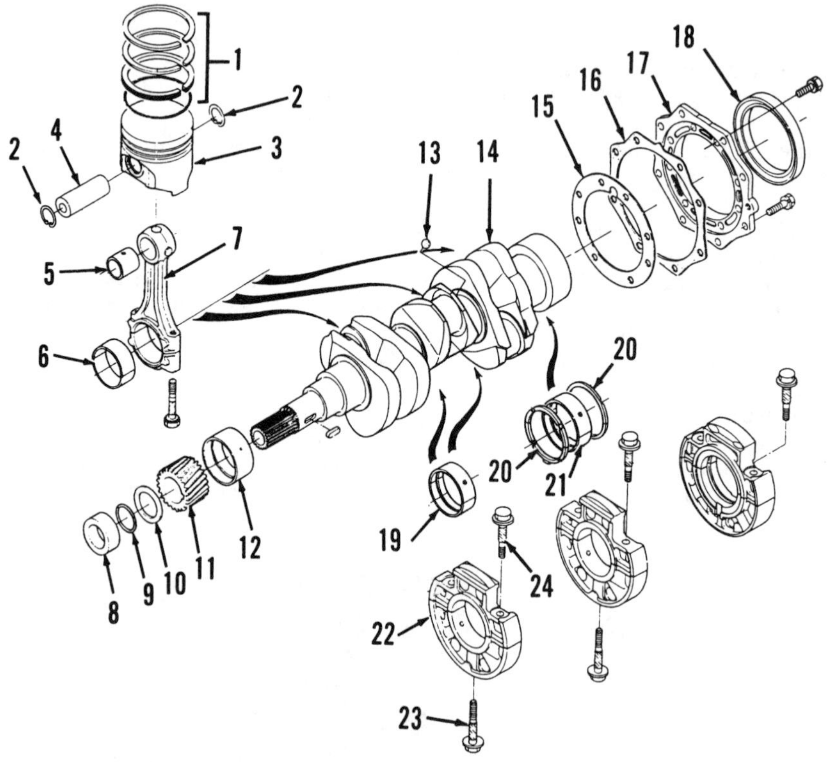

Fig. K33—Exploded view of piston, rod and crankshaft assemblies.

1. Piston rings
2. Snap ring
3. Piston
4. Piston pin
5. Bushing
6. Bearing insert
7. Connecting rod
8. Collar
9. "O" ring
10. Oil slinger
11. Gear
12. Bearing
13. Ball
14. Crankshaft
15. Gasket
16. Gasket
17. Seal carrier
18. Oil seal
19. Bearing
20. Thrust washers
21. Bearing
22. Bearing carrier
23. Locating screw
24. Clamp screws

ton pin end and a precision shell type insert bearing (6) at crankpin end.

Bushing inside diameter in small end of rod should be 18.025-18.040 mm (0.7096-0.7102 in.) for gasoline engine. Clearance between bushing and piston pin should be 0.020-0.040 mm (0.0006-0.0015 in.) with a maximum allowable clearance of 0.10 mm (0.004 in.).

Bushing inside diameter in small end of rod should be 20.002-20.011 mm (0.7875-0.7878 in.) for diesel engine. Clearance between bushing and piston pin should be 0.014-0.038 mm (0.0006-0.0015 in.) with a maximum allowable clearance of 0.10 mm (0.004 in.).

On gasoline and diesel engines, rod bearing inside diameter should be 33.994-34.040 mm (1.3383-1.3402 in.). Clearance between crankpin bearing and crankpin should be 0.019-0.081 mm (0.00075-0.0032 in.) with maximum allowable clearance of 0.15 mm (0.006 in.).

When checking connecting rod for twist or misalignment, maximum allowable twist or bend is 0.05 mm (0.002 in.) per 100 mm (3.94 in.).

If installing new piston pin bushing, be sure oil hole in bushing is aligned with hole in rod, then hone bushing to obtain desired bushing to piston pin clearance.

CRANKSHAFT AND BEARINGS. The crankshaft is carried in a sleeve type bearing (12—Fig. K33) pressed into front bearing bore of crankcase and in insert type bearings (19 and 21) in bearing carriers for intermediate (center) and rear main journals. Crankshaft end play is controlled by thrust washers (20) at front and rear of rear main bearing carrier. Before removing crankshaft, check crankshaft end play.

To remove crankshaft, remove timing gear cover, crankshaft timing gear, piston and rod units and flywheel. Remove the rear oil seal retainer (17). Note that two threaded holes (C—Fig. K35) for jackscrews are provided in seal retainer to facilitate removal of retainer. Remove the cap screw (23—Fig. K33) from bottom of each center bearing web, then carefully remove crankshaft and bearing carriers from rear end of crankcase taking care not to damage front main bearing remaining in crankcase. Note that the main bearing carriers are a tight fit in cylinder block to prevent oil loss between oil passages in block and carriers.

Carefully identify main bearing carriers as to their position on crankshaft to ensure correct reassembly. Bearing carrier halves are not interchangeable. Unbolt bearing carriers and remove the carriers and bearings from crankshaft. Note that crankshaft bearing thrust washers are located at rear main bearing carrier.

Crankshaft front and center main journal standard diameter is 39.934-39.950 mm (1.5722-1.5728 in.); rear main journal standard diameter is 43.934-43.950 mm (1.7297-1.7303 in.). Desired bearing oil clearance for all main journals is 0.034-0.092 mm (0.0013-0.0036 in.). Maximum allowable oil clearance for all main bearings is 0.2 mm (0.008 in.). Crankshaft end play should be 0.15-0.31 mm (0.006-0.012 in.) with maximum allowable end play being 0.5 mm (0.020 in.). Crankpin standard diameter is 33.959-33.975 mm (1.3370-1.3376 in.) with desired bearing oil clearance of 0.019-0.081 mm (0.00075-0.003 in.). Check crankshaft main journal by supporting shaft in V-blocks at front and rear journals and checking runout at a center journal with a dial indicator. Maximum allowable runout is 0.08 mm (0.003 in.).

Main bearings and crankpin bearings are available in undersizes of 0.2 and 0.4 mm (0.008 and 0.016 in.) and thrust bearings are available in oversize thickness of 0.2 and 0.4 mm (0.008 and 0.016 in.). Crankshaft journals may be reground undersize and with overwidth thrust surfaces at rear main journal to accommodate undersize bearings and oversize thrust washers.

To renew front main bearing, use suitable driver to remove old bearing from crankcase and install new bearing with oil hole in bearing aligned with oil passage in cylinder block. Bearing should be 0-0.3 mm (0-0.012 in.) below flush with front face of crankcase and positioned with seam toward exhaust side of engine.

Install bearing inserts in bearing carriers and stick thrust washers (20—Fig. K33) to rear bearing carrier with light grease. Match identification marks (M—Fig. K34) of carrier halves. Lubricate crankshaft and bearings, then install center carriers in correct position on crankshaft with face of carriers having Japanese character identification (C) toward flywheel end of crankshaft. Install rear carrier so side having bolt holes (H) is toward flywheel end of crankshaft. Tighten the bearing carrier screws to a torque of 13-16 N·m (10-12 ft.-lbs.). Lubricate front main bearing and insert crankshaft and bearing supports into crankcase from rear taking care not to damage front main bearing.

Install new rear oil seal with lip to inside in rear seal retainer plate. Lubricate seal and carefully install retainer plate over rear of crankshaft with arrow mark up as shown in Fig. K35. Be sure to remove the two jackscrews from retainer. Tighten retainer plate cap screws to a torque of 10-11 N·m (7-8 ft.-lbs.). Tighten the main bearing carrier retaining screw (23—Fig. K33) for center bearings to a torque of 27-30 N·m (20-22 ft.-lbs.). Complete reassembly of engine by reversing disassembly procedure.

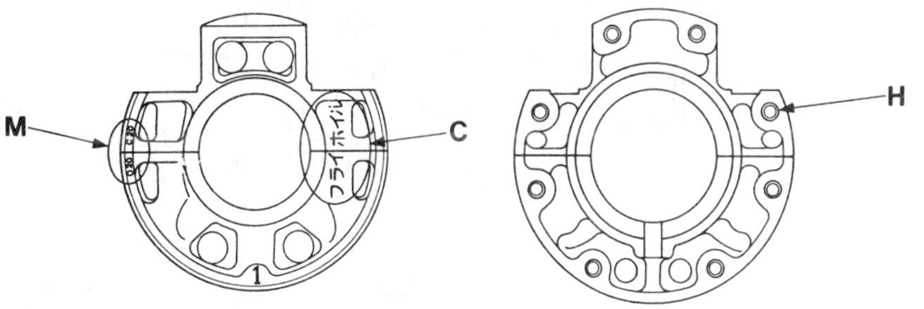

Fig. K34—Match alignment marks (M) of carrier halves. Install carriers with face of carriers having Japanese character identification (C) or side having bolt holes (H) toward flywheel end of crankshaft.

Fig. K35—Install jackscrews in threaded holes (C) to aid in removing seal carrier (B) from locating dowels. Install seal carrier so arrow (A) is up.

Illustrations courtesy Kubota.

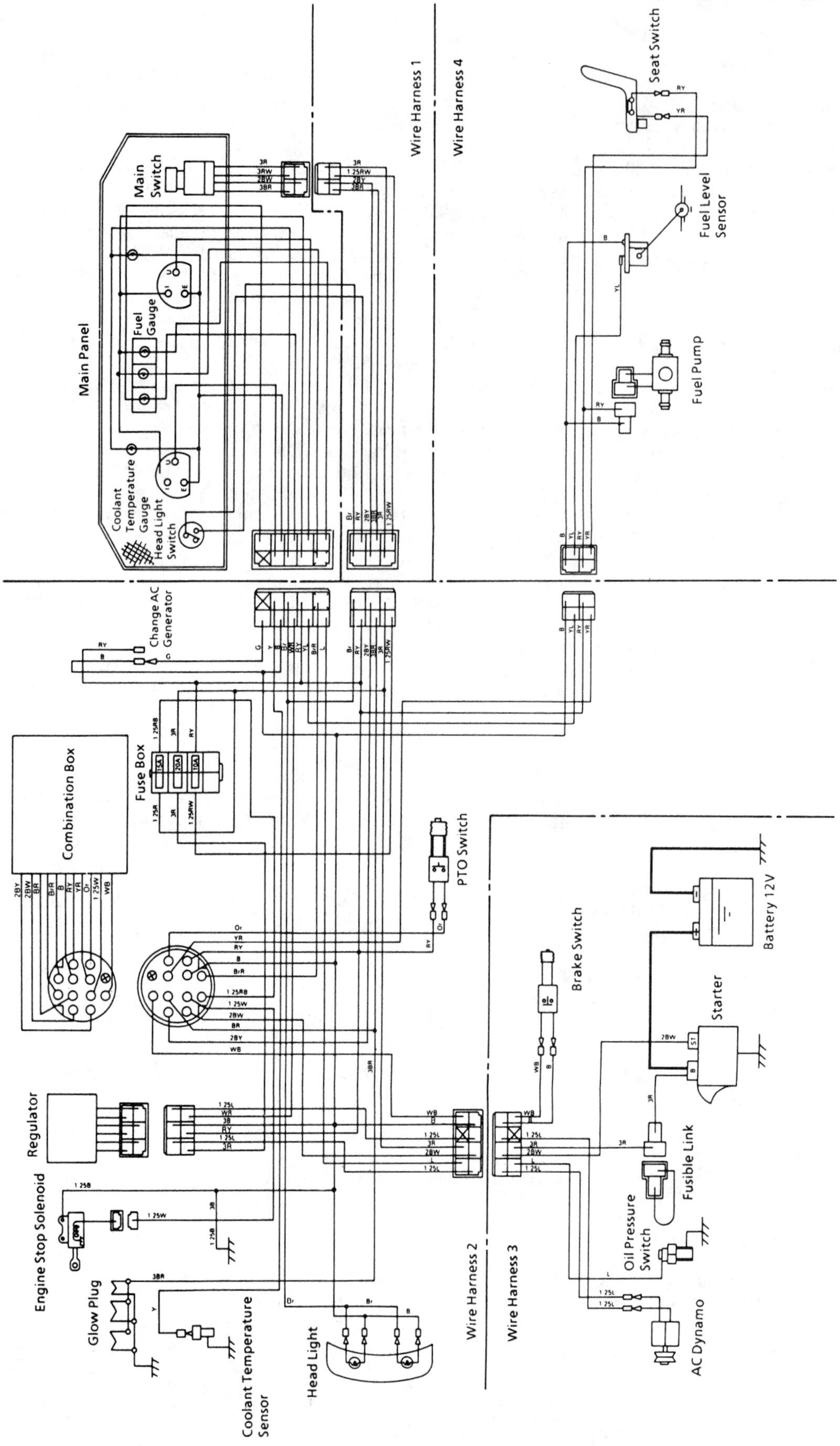

Fig. K36A—Typical wiring diagram of Models G1800, G1800-S, G1900 and G1900-S.

Or. Orange
R. Red
W. White
Y. Yellow

BR. Black/red
BW. Black/white
BY. Black/yellow
BrR. Brown/red

GR. Green/red
RB. Red/black
RW. Red/white
RY. Red/yellow

WB. White/black
WR. White/red
YL. Yellow/blue
YR. Yellow/red

Illustrations courtesy Kubota.

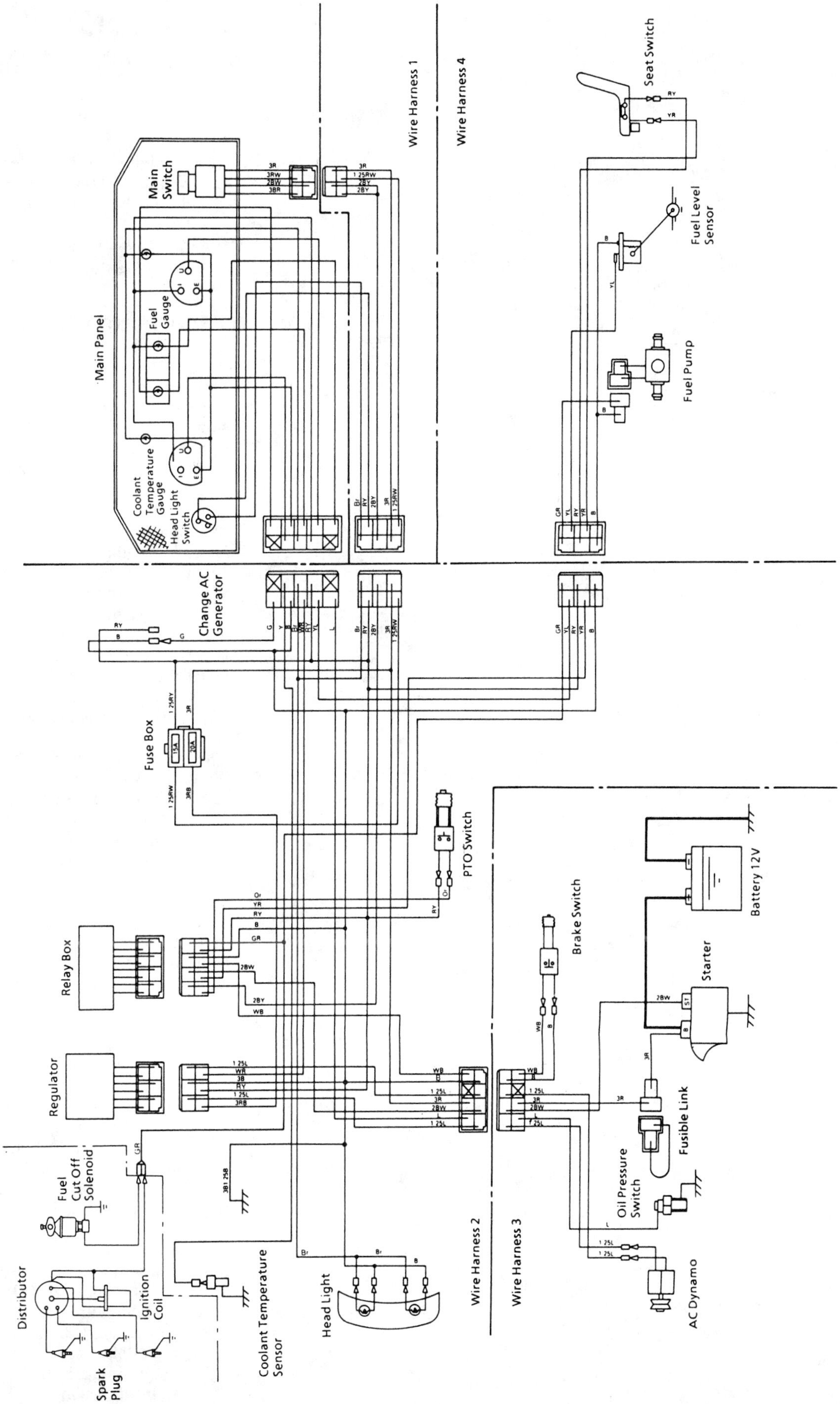

Fig. K36B—Typical wiring diagram of Models G2000 and G2000-S. Refer to Fig. K36A for wiring color code.

Illustrations courtesy Kubota.

ELECTRICAL SYSTEM

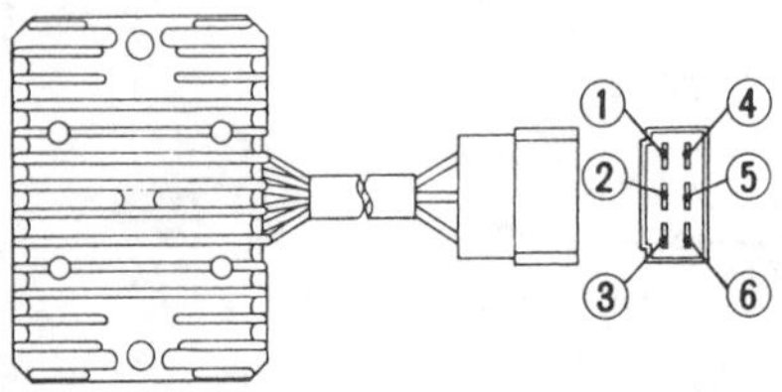

Fig. K37A—Drawing of voltage regulator connector plug.

1. Blue wire
2. Black wire
3. Blue wire
4. Red wire
5. Yellow wire
6. Green wire

Ohmmeter (−) probe \ Ohmmeter (+) probe		Wiring colors					
		Blue	Blue	Red	Black	Yellow	Green
Wiring colors	Blue		OFF	ON	OFF	OFF	OFF
	Blue	OFF		ON	OFF	OFF	OFF
	Red	OFF	OFF		OFF	OFF	OFF
	Black	OFF	OFF	OFF		OFF	OFF
	Yellow	ON	ON	ON	ON		OFF
	Green	OFF	OFF	OFF	OFF	OFF	

Fig. K37B—To test voltage regulator, check for continuity (ON) and infinity (OFF) at wiring leads as indicated in chart.

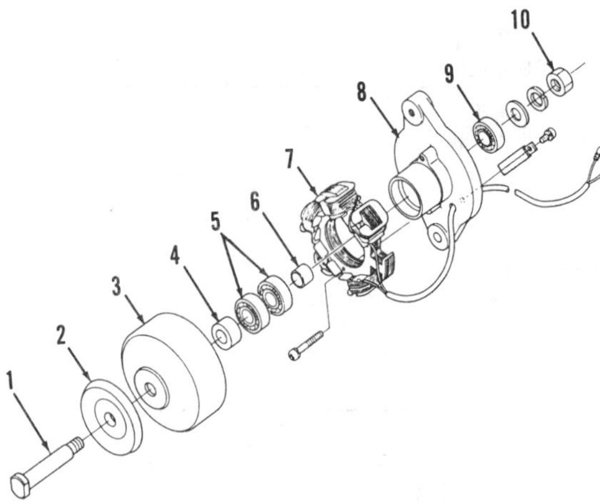

Fig. K38—Exploded view of alternator.

1. Shaft
2. Pulley half
3. Rotor
4. Collar
5. Bearings
6. Collar
7. Stator
8. Plate
9. Bearing
10. Nut

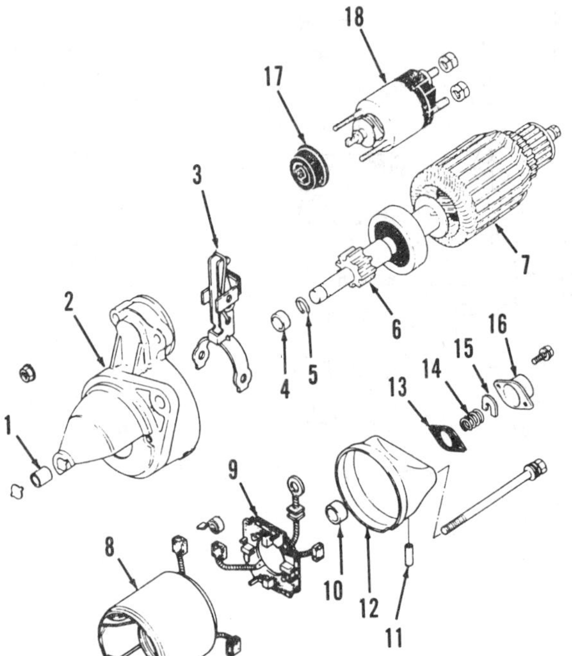

Fig. K39—Exploded view of starter used on diesel engine.

1. Bushing
2. Drive housing
3. Yoke
4. Collar
5. Snap ring
6. Clutch
7. Armature
8. Field coil frame
9. Brush assy.
10. Bushing
11. Vent tube
12. End frame
13. Gasket
14. Spring
15. Retainer
16. Cap
17. Boot
18. Solenoid

On Models G1800, G1800-S, G1900 and G1900-S, a 12 volt battery rated at 45 amp-hours is used. A 12 volt battery rated at 35 amp-hours is used on Models G2000 and G2000-S.

Refer to Fig. K36A or K36B for wiring diagram.

ALTERNATOR AND REGULATOR. To check alternator no-load output, disconnect the lead wires from the alternator. Place a voltmeter between the two alternator leads; the alternator should develop an AC voltage of 20 volts or more at 5200 rpm (generator speed). If AC voltage is less than 20 volts, repair or renew the alternator.

With the charging circuit fully connected and with a fully charged 12-volt battery, regulated voltage should be 14.2 to 14.8 volts. If not within specified voltage, disconnect wiring connector (Fig. K37A) from regulator and use ohmmeter to check for continuity or infinity at connector leads as indicated in chart shown in Fig. K37B. Renew the regulator if continuity or infinity readings are not obtained as indicated in chart.

To disassemble alternator, clamp pulley in a vise and remove nut (10—Fig. K38). Tap end of shaft (1) with a soft hammer to separate flywheel (3) from stator housing (8). Remove clamp from stator leads, remove stator mounting screws and remove stator (7) from housing.

Renew flywheel if magnets are loose or damaged. Check stator for indications of burned wiring or other damage and renew as necessary. Check bearings (5 and 9) for roughness or damage and renew as necessary.

To reassemble, reverse the disassembly procedure.

STARTING MOTOR. Diesel Engine. To disassemble starter, disconnect brush lead from solenoid. Unbolt and remove solenoid (18—Fig. K39). Remove cap (16), brake shoe (15) and spring (14) from rear of starter. Remove through-bolts and end cover (12). Disengage brushes from brush holder (9) and remove brush holder assembly. Separate field coil housing (8), armature (7), clutch and drive pinion, and shift lever (3) from drive end frame (2).

Inspect all parts for wear or damage. New brush length is 16 mm (0.630 in.) and minimum brush length is 10.5 mm (0.413 in.). New commutator diameter

Illustrations courtesy Kubota.

is 28 mm (1.102 in.) and wear limit is 27 mm (1.063 in.). Specified clearance between armature shaft and commutator end bushing is 0.03-0.10 mm (0.001-0.004 in.) and between shaft and drive end bushing is 0.05-0.10 mm (0.002-0.005 in.).

To reassemble, reverse the disassembly procedure.

STARTING MOTOR. Gasoline Engine. To disassemble starter, disconnect brush lead from solenoid. Unbolt and remove solenoid (14—Fig. K40). Remove through-bolts and screws from rear cover (13), then remove cover. Disengage brushes from brush holder (11) and remove brush holder from coil housing. Withdraw coil housing (10), armature (8), clutch and drive pinion (6) and shift lever (7) from drive end frame (2).

Inspect all parts for wear or damage. New brush length is 17 mm (0.67 in.) and minimum brush length is 11.5 mm (0.453 in.). Renew commutator if diameter is 31 mm (1.220 in.) or less.

To reassemble starter, reverse the disassembly procedure. Pinion end clearance with starter pinion in engaged position is 0.5-2.0 mm (0.020-0.079 in.). To check and adjust pinion clearance, connect battery positive terminal to "S" terminal of starter solenoid and connect battery negative terminal to starter frame to engage the pinion. Push pinion inward to remove free play and measure pinion end clearance. Adjust by adding or removing washers between solenoid switch and drive end frame.

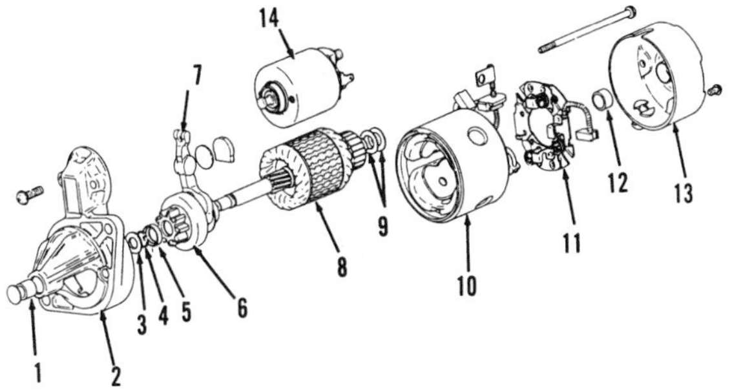

Fig. K40—Exploded view of starter used on gasoline engine.

1. Bushing
2. Drive housing
3. Washer
4. Clip
5. Collar
6. Clutch
7. Shift lever
8. Armature
9. Washers
10. Field coil frame
11. Brush plate
12. Bushing
13. End frame

PTO

ADJUSTMENT

All Models

With pto control lever in engaged position, length of spring (S—Fig. K44) should be 88-90 mm (3.46-3.54 in.). Adjust spring length (L) by turning adjusting nuts (N). When pto belts are new, spring length should be 91-92 mm (3.58-3.62 in.); after 50 hours of operation readjust spring length to 88-90 mm (3.46-3.54 in.).

To adjust pto brake, measure distance from end of adjusting screw (1—Fig. K45) to arm (2) with pto disengaged. Distance (A) should be 3-4 mm (0.12-0.16 in.). Turn screw (1) to obtain desired distance. With distance correct, length of spring (3) should be 138-142 mm (5.43-5.59 in.). Adjust spring length (B) by turning nuts (4).

After adjustment, check pto operation. Brake should release when pto is engaged.

Adjust height (H—Fig. K46) of pto switch so tip is 29-31 mm (1.14-1.22 in.) from bracket (3).

PTO DRIVE BELTS

All Models

The pto drive belts (3—Fig. K47) may be removed after removing belt guards and retaining bolt (5). Push belt idler (1) away from belts and slip belts off pto pulley.

When installing guard bolt (5—Fig. K47), be sure threads do not extend past inner surface of bracket (4).

PTO SHAFT ASSEMBLY

All Models

R&R AND OVERHAUL. Detach pto drive belts and remove drive pulley. Detach snap ring (3—Fig. K48) and pull shaft (2) out toward front. Unscrew mounting screws to remove carrier (6). Inspect components and renew any that are damaged or worn excessively.

PTO BRAKE AND BELT IDLER

All Models

To disassemble, first disconnect clutch spring, return spring and brake spring. Remove snap ring (10—Fig. K49) and withdraw pto brake assembly (8). Remove snap ring (5) and pull idler arm (3) and pulley (2) off stub shaft.

Minimum allowable thickness of brake pad is 4.0 mm (0.157 in.). Renew brake pad if soaked with oil or otherwise damaged. Inside diameter of brake shaft bushing (7—Fig. K49) when new is 10.00-10.07 mm (0.394-0.396 in.). Maximum allowable clearance between brake pivot shaft and bushing is 0.35 mm (0.014 in.). Inside diameter of idler arm bushings (6) when new is 18.00-18.07 mm (0.709-0.711 in.). Maximum allowable clearance between bushings and idler stub shaft is 0.35 mm (0.014 in.).

To reassemble, reverse the disassembly procedure. Adjust belt tension and pto brake as previously outlined.

BRAKE

ADJUSTMENT

All Models

To adjust brake pedal travel, apply a force of approximately 147 N (33 lbs.)

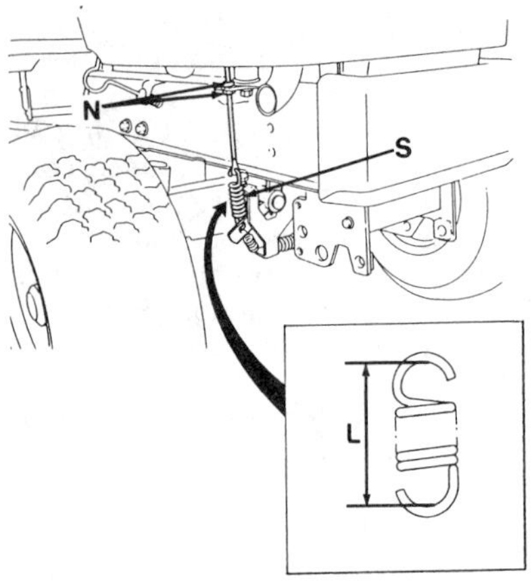

Fig. K44—With pto control lever in engaged position, length of spring (S) should be 88-90 mm (3.46-3.54 in.). Adjust spring length by turning adjusting nuts (N).

against pedal. Spring (2—Fig. K50) should compress to a length of 80-130 mm (3.15-5.12 in.). Loosen outer nut and turn adjusting nut (3) to obtain desired spring length.

To adjust brake pedal free play, depress pedal with a force of approximately 29 N (7 lbs.). Free play should be 20-40 mm (0.8-1.6 in.). Adjust free play by loosening stop mounting screw (6—Fig. K51) and repositioning stop (5).

After adjusting pedal travel and free play, check operation of cruise control lever. The cruise control lever should disengage when the brake pedal is depressed. If not, readjust pedal travel and free play.

With brake pedal depressed fully, distance (D—Fig. K52) from tip of brake switch (7) to bracket (9) should be approximately 32 mm (1.26 in.). Turn nuts (8) as needed to obtain desired distance.

R&R AND OVERHAUL

All Models

An internally expanding, drum type brake is mounted on the front of the differential case. To disassemble brake, detach acutating rod, remove brake cover retaining screws and remove brake assembly (Fig. K53). Detach snap ring (10) and pull drum (11) off shaft.

Minimum allowable brake pad thickness is 7.5 mm (0.3 in.). Maximum allowable inside diameter of brake drum is 82 mm (3.23 in.).

When installing brake assembly, turn brake lever (3) clockwise to seat brake shoes while tightening cover screws.

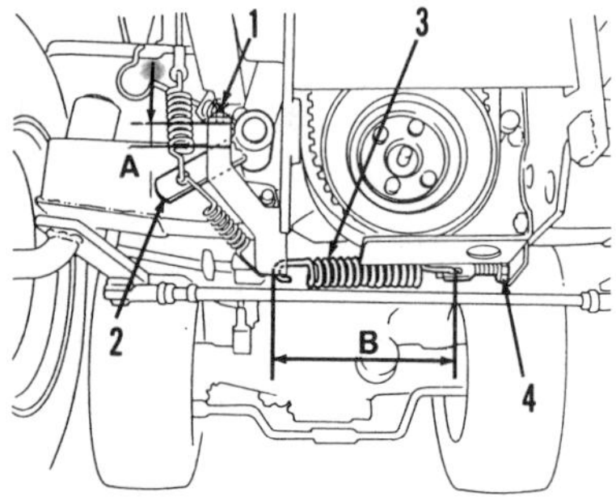

Fig. K45—Adjust pto brake as outlined in text.

1. Adjusting screw
2. Tension arm
3. Brake spring
4. Nut

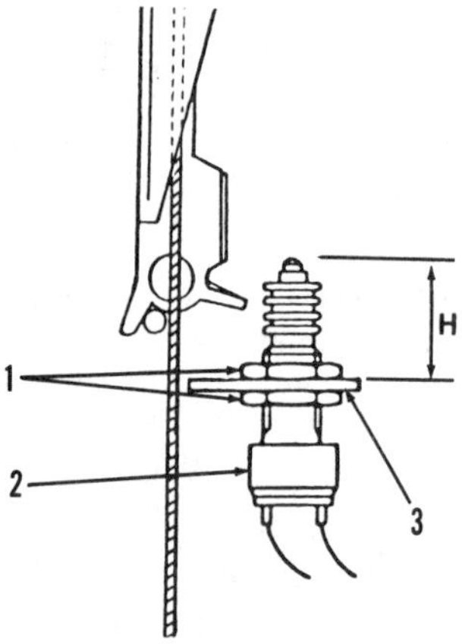

Fig. K46—Adjust height (H) of pto switch (2) so tip is 29-31 mm (1.14-1.22 in.) from bracket (3).

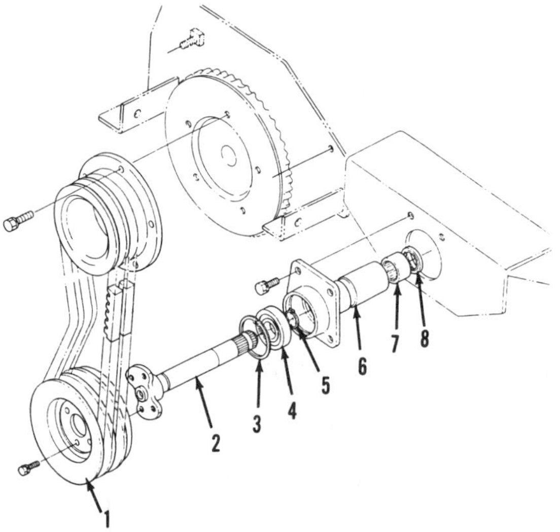

Fig. K48—Exploded view of pto shaft assembly.

1. Pulley
2. Pto shaft
3. Snap ring
4. Bearing
5. Snap ring
6. Axle pivot
7. Bearing
8. Seal

Fig. K47—Threads of guard screw (5) must not extend past inner surface of front axle bracket (4).

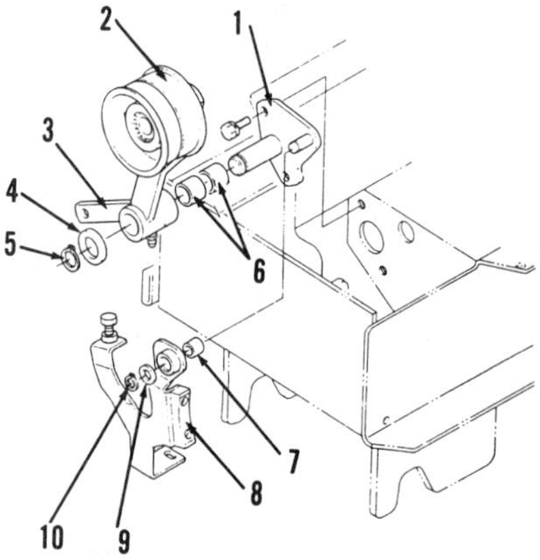

Fig. K49—Exploded view of pto brake and belt tension arm.

1. Stub shaft
2. Idler pulley
3. Tension arm
4. Washer
5. Snap ring
6. Bushings
7. Bushing
8. Pto brake
9. Washer
10. Snap ring

Illustrations courtesy Kubota.

REAR STEERING MECHANISM

Models G1800-S, G1900-S and G2000-S are equipped with a four-wheel steering system (see Fig. K54). A drag link (8) connected to the steering gearbox pitman arm (6) transmits steering motion to the right rear steering arm (9). A tie rod (10) connects the rear steering arms. When the steering wheel is turned, steering motion is passed to all wheels resulting in a tighter turning radius than a two-wheel steering system.

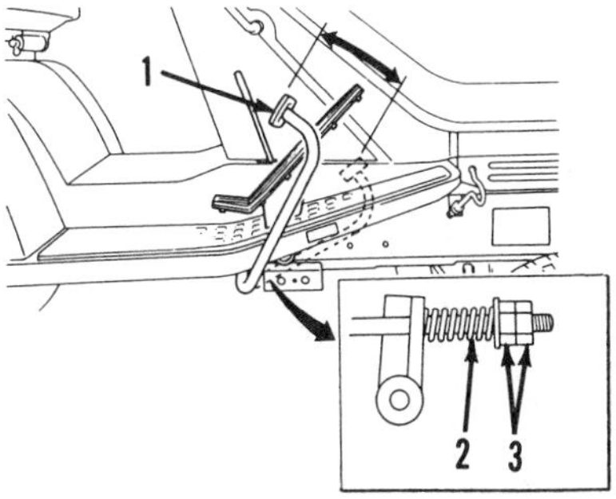

Fig. K50—Spring (2) should compress to a length of 80-130 mm (3.15-5.12 in.) when pedal (1) is depressed as outlined in text. Loosen outer nut and turn adjusting nut (3) to obtain desired spring length.

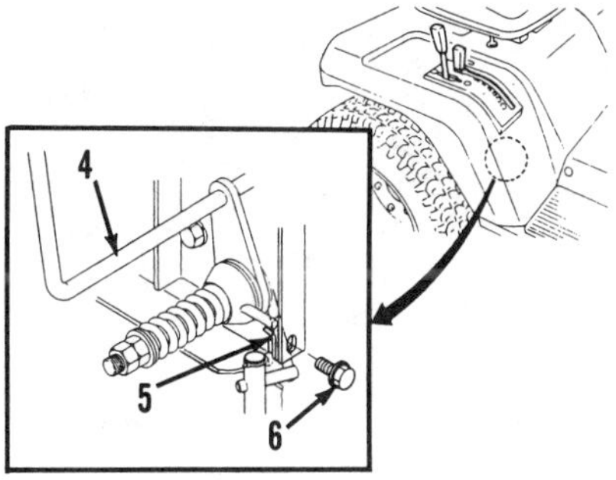

Fig. K51—Adjust pedal free play by loosening stop mounting screw (6) and repositioning stop (5).

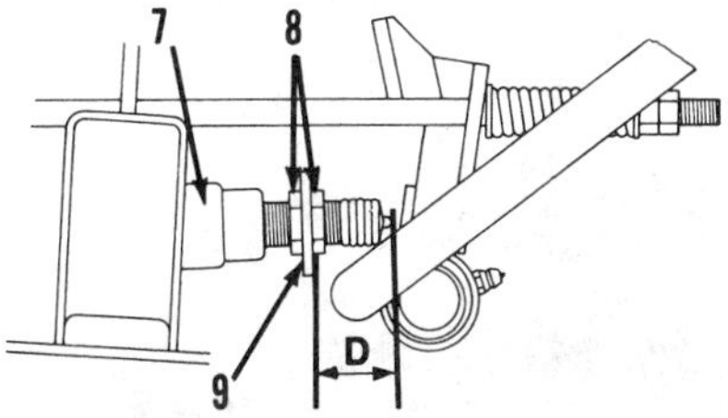

Fig. K52—With brake pedal depressed fully, distance (D) from tip of brake switch (7) to bracket (9) should be approximately 32 mm (1.26 in.). Turn nuts (8) as needed to obtain desired distance.

ADJUSTMENT

Models G1800-S, G1900-S and G2000-S

There should be zero toe-in at rear wheels. Support rear of tractor and remove rear wheels. Measure from face of axle flange to tractor frame as shown in Fig. K55. Adjust length of tie rod (10—Fig. K54) so distances are equal on each side.

With front wheels pointing straight ahead, adjust length of rear drag link (8) so face of axle flange is parallel to tractor frame as shown in Fig. K56. Install rear wheels.

Support front of tractor and turn steering so front wheels are at full travel to left. Lift front left wheel so axle contacts frame. Adjust stop screw (2—Fig. K57) on axle support so it contacts steering arm. Turn steering wheel to full right and repeat procedure for right side.

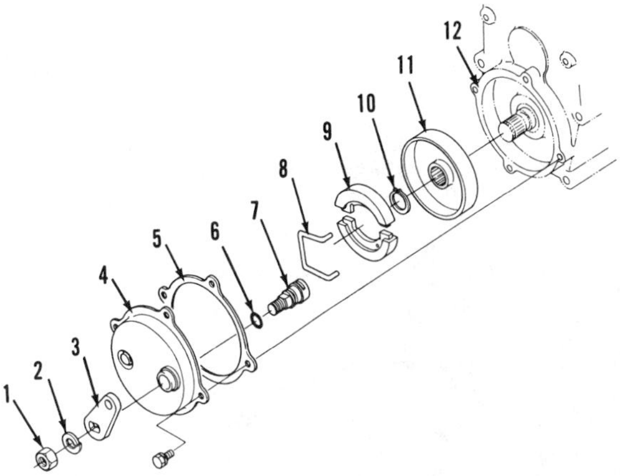

Fig. K53—Exploded view of brake assembly.

1. Nut
2. Lockwasher
3. Brake lever
4. Cover
5. Gasket
6. "O" ring
7. Cam
8. Return spring
9. Brake shoes
10. Snap ring
11. Brake drum
12. Differential housing

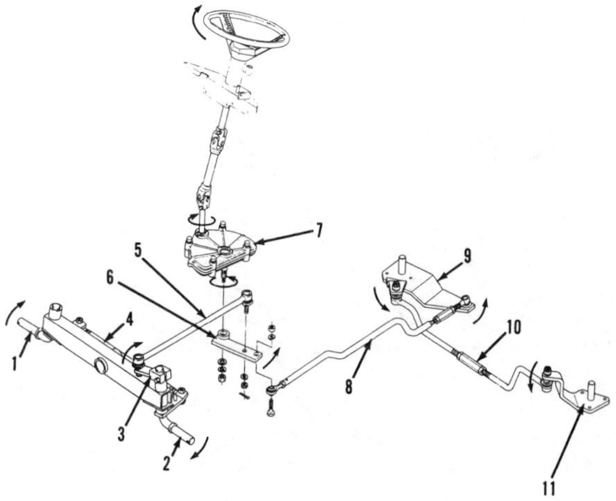

Fig. K54—Drawing of steering system on four-wheel steering models.

1. Spindle (R.H.)
2. Spindle (L.H.)
3. Steering arm
4. Tie rod
5. Front drag link
6. Pitman arm
7. Steering gearbox
8. Rear drag link
9. Steering arm (R.H.)
10. Tie rod
11. Steering arm (L.H.)

TRANSMISSION

MAINTENANCE

All Models

The transmission and differential utilize a common oil sump. Level should be checked daily or before operation. The transmission oil filter and oil should be changed after first 50 hours of operation and after every 300 hours of operation thereafter. Recommended oil is Exxon Torque Fluid 56, Shell Donax TD or TM, or an equivalent oil. Capacity is 5.7 liters (6 qts.).

ADJUSTMENT

All Models

To perform neutral adjustment, support rear of tractor and remove rear wheels. Loosen screw (S—Fig. K58). Run engine at rated speed and turn holder shaft (H) so axles just start to rotate and note position of holder shaft, then turn shaft in opposite direction so axles start to rotate in opposite direction and note position of shaft. Turn shaft to midpoint of these two positions to obtain neutral and tighten screw.

Engage parking brake and note if notch on parking brake lever (2—Fig. K59) engages pin on speed pedal shaft (3). If not, remove tunnel cover and adjust length of speed control rod (4—Fig. K60).

To check maximum forward speed, install rear wheels and support rear of tractor. Run engine at full speed with transmission in forward. Determine rotation rate of rear wheels. Wheels should rotate at 115-125 rpm. If not, remove seat and fender and adjust maximum speed screw (7—Fig. K61). Setting length (L) of adjusting screw should be approximately 22 mm (0.866 in.).

A force of 20-39 N (4-9 lbs.) should be required to move the cruise control lever (8—Fig. K62). To adjust control lever friction, turn adjusting nuts (10) as necessary. After adjustment, make sure that control lever releases automatically when brake pedal is fully depressed.

REMOVE AND REINSTALL

The transmission and rear axle assembly are removed as a unit. Drain oil from differential case. Remove seat and fender. Disconnect fuel level sensor wire and remove seat support. Remove fuel tank. Disconnect hydraulic cylinder hose from control valve. Detach drive shaft, transmission control rod and brake rod. Detach rear drag link. Support rear of frame, remove transmission mounting screws and roll rear drive unit away from frame.

To separate transmission from differential case, detach hydraulic tubes and hoses and remove control valve assembly. Remove plates on side and top of transmission. Unscrew transmission mounting screws and separate transmission from differential case.

Reinstall transmission and rear drive unit by reversing removal procedure. Apply Loctite to transmission mounting screws and tighten to 20-29 N·m (15-21 ft.-lbs.). Tighten axle housing to frame screws to 69-88 N·m (51-65 ft.-lbs.).

OVERHAUL

All Models

To disassemble transmission, remove output gear from motor shaft. Remove fan and fan flange from pump shaft. Remove speed control arm. Unbolt and remove auxiliary pump housing (4—Fig. K64) and spacer plate (11). Remove charge relief valve (9), snap ring (1) and oil seal (2) from pump housing. Remove charge pump rotors (13) and drive pin (49) from pump shaft. Remove screws securing center section (26) and lift center section from housing (41). Remove bypass valve (18) and check valves (25) from center section, being careful not to interchange parts or lose shims.

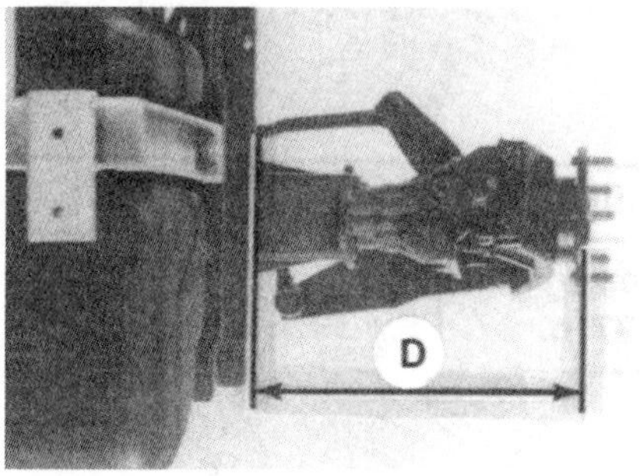

Fig. K55—Distance (D) from axle flange to frame should be equal on both sides. Adjust length of tie rod as needed.

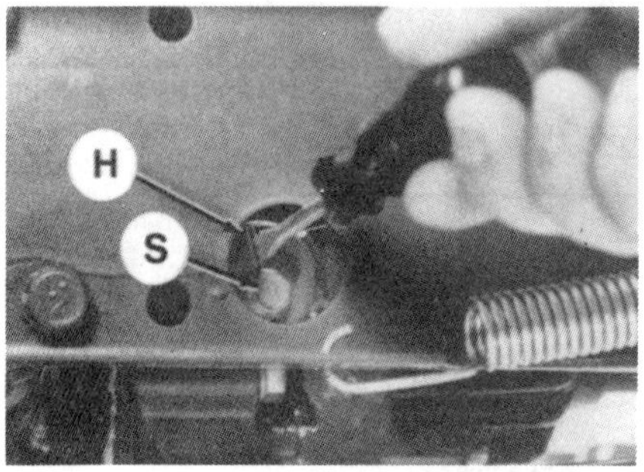

Fig. K58—Refer to text and perform neutral adjustment as outlined.

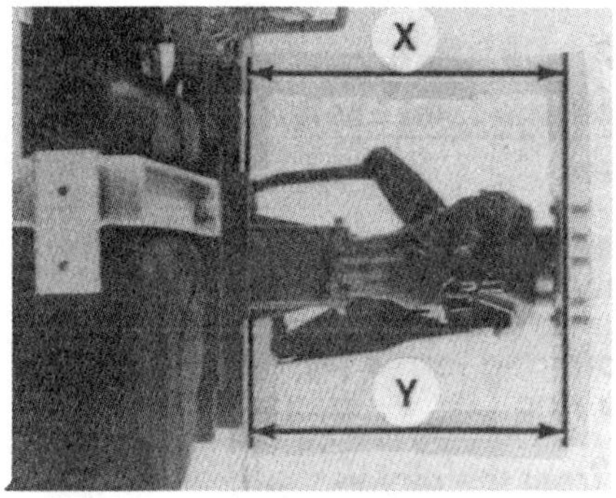

Fig. K56—Axle flange should be parallel to tractor frame (distance X equals distance Y). Adjust length of rear drag link as needed.

Fig. K59—Groove on parking brake lock (2) should engage pin of hydrostatic pedal shaft (3) when in neutral. If not, adjust transmission speed control rod length shown in Fig. K60.

Fig. K61—At maximum speed, wheels should rotate at 115-125 rpm. If not, remove fender and adjust maximum speed screw (7).

Fig. K57—Refer to text and adjust stop screws (2) on each axle support.

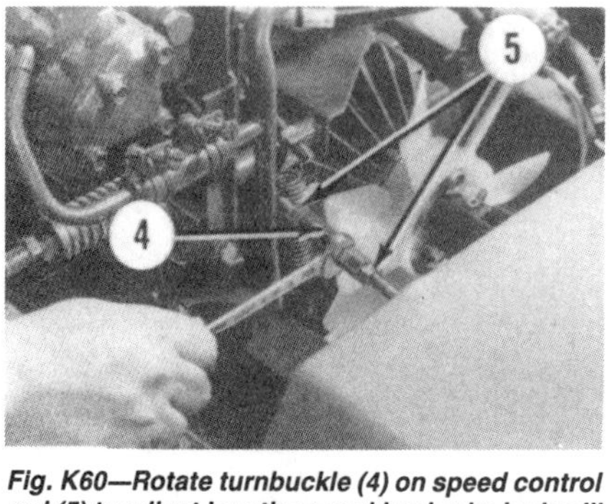

Fig. K60—Rotate turnbuckle (4) on speed control rod (5) to adjust length so parking brake lock will engage pin shown in Fig. K59.

Fig. K62—Use a spring scale (9) to measure force required to move cruise control lever (8). Turn adjusting nut (10) to adjust. Refer to text.

Illustrations courtesy Kubota.

Position housing (41) on its side. Remove pump cylinder block (28) and piston assembly. Remove motor cylinder block (43) and piston assembly and output shaft (44). Lay cylinder block assemblies aside for later disassembly.

Remove thrust washer (35), thrust bearing (36), thrust washer (37), swashplate (39) and bearings (40). Remove snap ring (56) and extract input shaft (50) assembly. Remove control arm (57).

Pump and motor pistons should be identified prior to removal so they can be reinstalled in original cylinder bore. Do not lose ball (31) in each piston.

Inspect components for damage and excessive wear. Inspect pistons and cylinder bores for scoring and scratches. Check polished faces of pump cylinder block, motor cylinder block and center section for scratches or excessive wear. Renew center section and pump and motor cylinder block assemblies if sealing surfaces are worn or damaged. Inspect all valves and their seats for nicks and scratches and renew as necessary. Inspect charge pump rotors and faces of center section (26) and auxiliary pump housing (4) for scoring and scratches. Renew all seals and "O" rings.

Coat parts with transmission oil during assembly. Be sure pistons are returned to original bore in cylinder. Note that thrust washer (37) is thinner than thrust washer (35). Tighten screws securing center section (26) to housing (41) to 23-24 N·m (17-18 ft.-lbs.). Install spacer (11) so flat edge (F—Fig. K65) of spacer is positioned opposite control arm (57). Install auxiliary pump housing (4—Fig. K64) and tighten retaining screws to 25-29 N·m (18-22 ft.-lbs.).

DIFFERENTIAL

MAINTENANCE

All Models

The transmission and differential utilize a common oil sump. Check oil level using dipstick on rear of differential case. Oil level should be checked daily or before operation. Oil and oil filter should be changed after every 300 hours of operation. Recommended oil is Exxon Torque Fluid 56, Shell Donax TD or TM, or an equivalent oil. Capacity is 5.7 liters (6 qts.).

R&R AND OVERHAUL

All Models

Separate rear drive unit and remove transmission as previously outlined. Remove rear wheels, and on Models G1800-S, G1900-S and G2000-S, remove tie rod. Remove front cover (1—Fig. K66) and gear assembly. Remove pinion gear (11). Unbolt and remove rear axle housings (15 and 30) while noting that differential will be loose in case. Remove differential assembly from transmission case. Keep shims (17) with their respective axle housing so they can be reinstalled in original positions.

Use a suitable puller to remove ball bearings (18). Remove cap screws retaining differential carrier (29) to ring gear (19). Remove dowel pins (20 and 25).

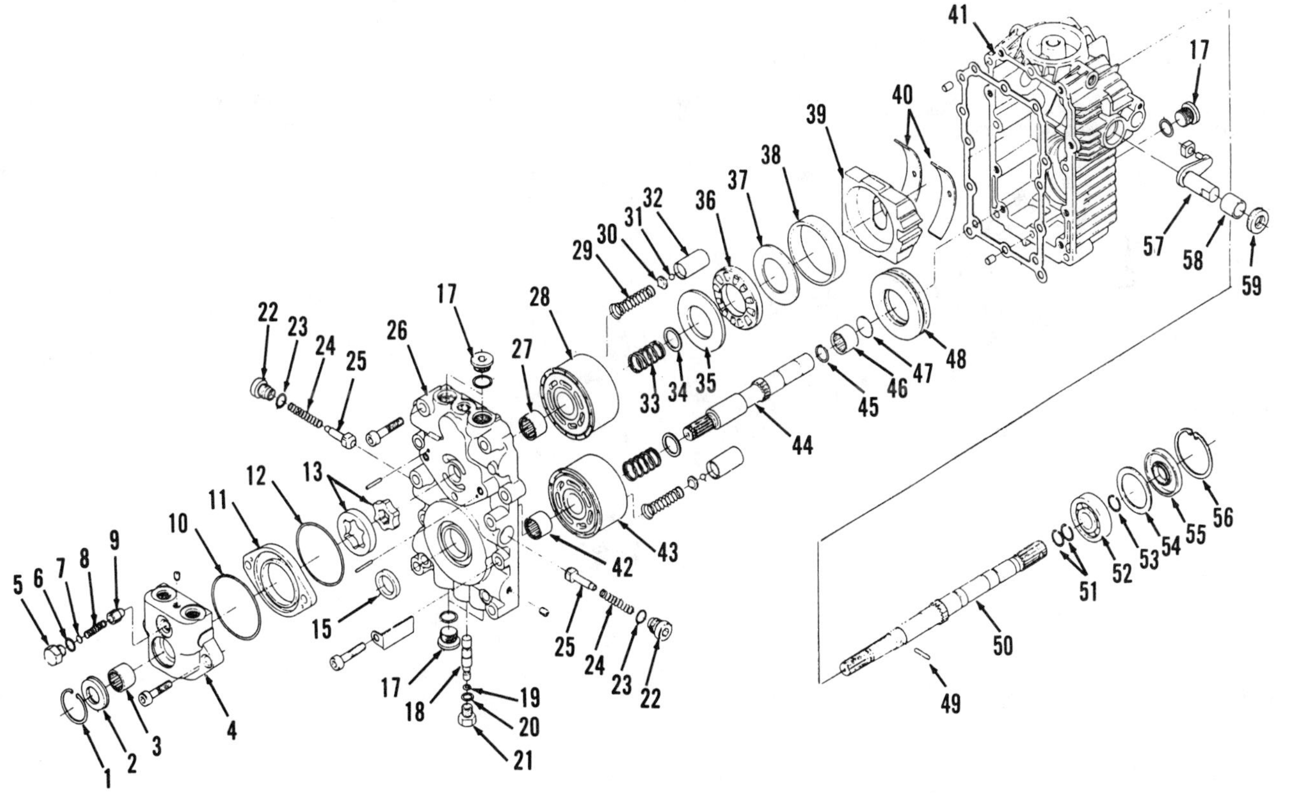

Fig. K64—Exploded view of hydrostatic transmission.

1. Snap ring
2. Seal
3. Bearing
4. Auxiliary pump housing
Plug
"O" ring
7. Shim
8. Spring
9. Charge pump relief valve
10. "O" ring
11. Spacer
12. "O" ring
13. Pump rotors
15. Seal
17. Plug
18. Bypass valve
19. "O" ring
20. "O" ring
21. Plug
22. Plug
23. "O" ring
24. Spring
25. Check valve
26. Cover
27. Bearing
28. Pump cylinder block
29. Spring
30. Spring seat
31. Ball
32. Piston
33. Spring
34. Washer
35. Thrust washer (thick)
36. Thrust bearing
37. Thrust washer (thin)
38. Sleeve
39. Swashplate
40. Bearing
41. Transmission housing
42. Bearing
43. Motor cylinder block
44. Motor output shaft
45. Snap ring
46. Bearing
47. Thrust washer
48. Bearing
49. Pin
50. Pump shaft
51. Snap rings
52. Bearing
53. Snap ring
54. Oil seal
56. Snap ring
57. Control arm
58. Bushing
59. Seal

Illustrations courtesy Kubota.

Push pinion shaft (28) out of differential carrier and separate side gears (26), thrust shaft (24), pinions (23) and thrust washers (22) from carrier.

Specified clearance between side gears (26) and differential carrier (29) or ring gear (19) bore is 0.025-0.066 mm (0.0010-0.0026 in.) with a maximum allowable clearance of 0.30 mm (0.012 in.). Specified inside diameter of differential carrier (29) and ring gear (19) is 32.000-32.025 mm (1.2598-1.2608 in.). Specified side gear (26) outside diameter is 31.959-31.975 mm (1.2582-1.2588 in.).

Specified clearance between pinion shaft (28) and pinions (23) is 0.016-0.052 mm (0.0006-0.0020 in.) with a maximum allowable clearance of 0.30 mm (0.012 in.). Specified pinion shaft diameter is 15.966-15.984 mm (0.6286-0.6292 in.).

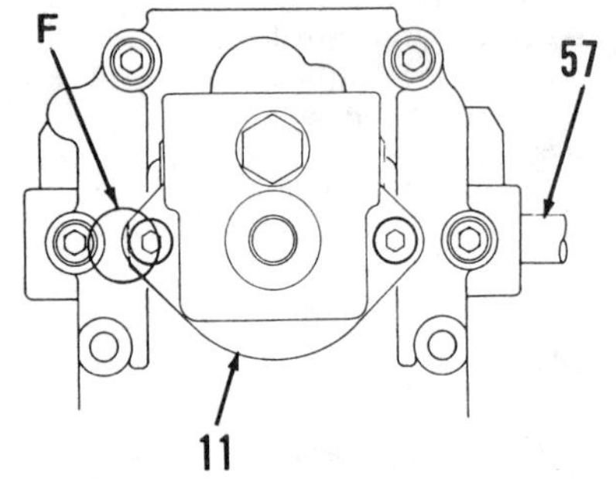

Fig. K65—Install auxiliary pump spacer (11) so flat edge (F) is opposite control arm (57).

Specified pinion inside diameter is 16.000-16.018 mm (0.6299-0.6306 in.).

Backlash between pinions and side gears should be 0.1-0.3 mm (0.004-0.012 in.) with a maximum allowable backlash of 0.4 mm (0.016 in.). Backlash is adjusted by changing thickness of side gear shims (27).

Install differential using original shims (17). Differential carrier bearings should have zero backlash and zero preload; adjust by changing thickness of shims (17) as required. Backlash between pinion gear (11) and ring gear (19) should be 0.02-0.20 mm (0.001-0.008 in.) with a maximum allowable backlash of 0.4 mm (0.016 in.). To adjust backlash, move shims (17) from one side of ring gear to the other side.

Tighten rear axle housing screws to 38-50 N·m (28-37 ft.-lbs.). Tighten front cover screws to 38-50 N·m (28-37 ft.-lbs.). Tighten tie rod attaching nuts to 63-72 N·m (46-53 ft.-lbs.).

REAR AXLES AND STEERING

STEERING KNUCKLE AND KING PINS

Models G1800-S, G1900-S and G2000-S

R&R AND OVERHAUL. Support rear of tractor and remove rear wheel. Detach tie rod, and if necessary, drag link. Unscrew and remove upper kingpin (5—Fig. K67) and lower king pin (6). Remove steering knuckle (4) and universal joint (10). Remove snap ring (1) and withdraw universal joint from wheel hub (2). Remove snap ring (9) and tap wheel hub out of bearings and steering knuckle. Remove snap rings (3 and 8) and remove bearings (7).

Inspect components for excessive wear and damage. Reverse disassembly procedure for reassembly.

AXLE SUPPORT AND UNIVERSAL JOINT

Models G1800-S, G1900-S and G2000-S

R&R AND OVERHAUL. Remove steering knuckle as previously outlined. Remove universal joint (10—Fig. K67). Unscrew and remove axle support (18) from axle housing.

Kingpin bushings (16) are renewable. Remove seal (11) and snap ring (12) for access to bearing (13).

Inspect components for excessive wear and damage. Reverse disassembly procedure for reassembly.

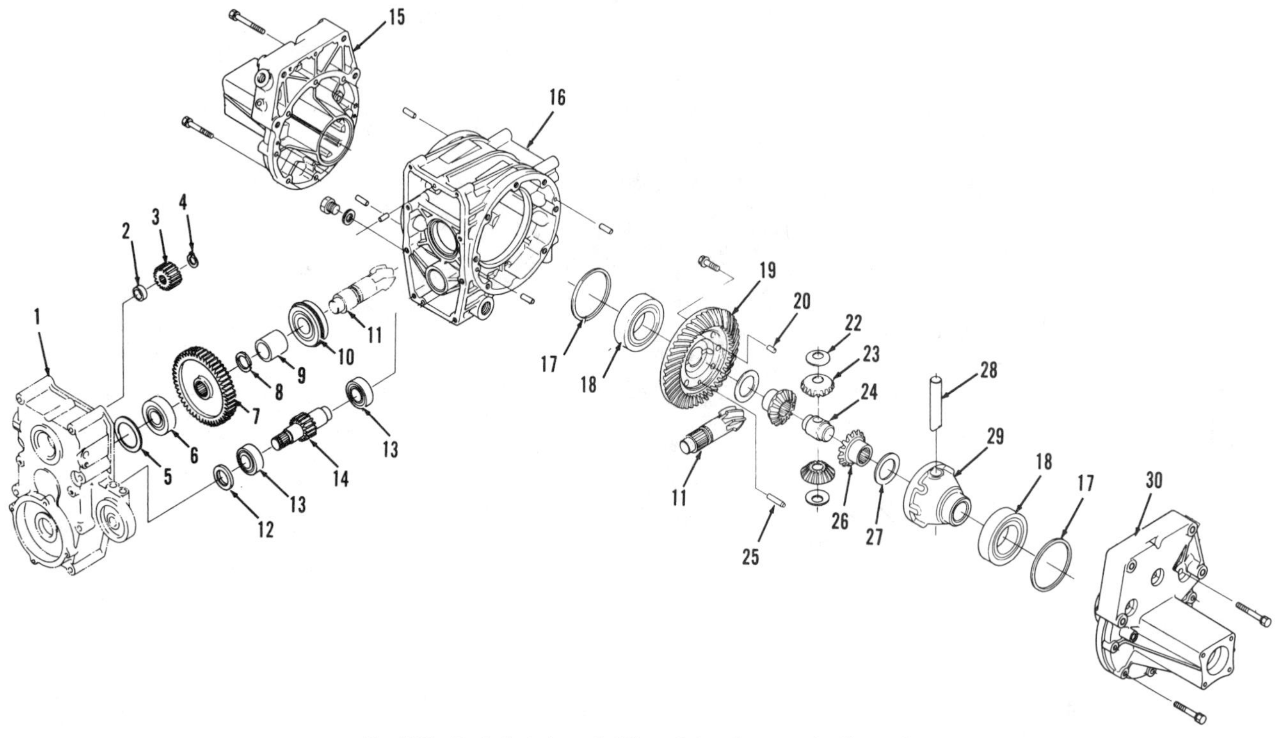

Fig. K66—Exploded view of differential and gear reduction unit.

1. Front cover
2. Spacer
3. Drive gear
4. Snap ring
5. Shim
6. Bearing
7. Driven gear
8. Stop collar
9. Spacer
10. Bearing
11. Bevel pinion
12. Oil seal
13. Bearing
14. Brake shaft
15. Axle housing (R.H.)
16. Case
17. Shims
18. Bearings
19. Bevel ring gear
20. Pin
22. Thrust washer
23. Pinion
24. Spacer
25. Pin
26. Side gear
27. Shim
28. Pinion shaft
29. Differential carrier
30. Axle housing (L.H.)

AXLE SHAFT

All Models

To remove axle shaft (20—Fig. K67), remove rear drive unit as outlined in TRANSMISSION section and remove axle housing (21). On Models G1800-S, G1900-S and G2000-S, remove axle support (18) as previously outlined. On all models, tap axle inward and remove from axle housing. Remove snap ring (24), spacer (23) and bearing (22) from axle shaft.

Inspect components for excessive wear and damage. Reverse disassembly procedure for reassembly.

HYDRAULIC SYSTEM

MAINTENANCE

All Models

The hydrostatic transmission provides pressurized oil for hydraulic system operation. Refer to TRANSMISSION section for maintenance and service information on hydrostatic transmission.

PRESSURE TESTS

All Models

Two relief valves are used in the hydraulic circuit (9 and 32—Fig. K68). To check relief operation, proceed as follows: Remove seat and fender. Disconnect lift cylinder hose from control valve adapter (33) and connect a suitable gage to adapter port. Run engine at rated speed and move hydraulic lever to "LIFT" position. Gage should read 2940-3430 kPa (427-497 psi). Install or remove shims (5) to obtain correct pressure. Reconnect lift cylinder hose.

Disconnect hydrostatic transmission hydraulic line that attaches to fitting (35) on adapter (33). Remove fitting (35) and attach a suitable gage to adapter port. Run engine at rated speed. Gage should read 1960-2450 kPa (284-355 psi). Install or remove shims (30) to obtain correct pressure. Install fitting and hydraulic line.

CONTROL VALVE

All Models

Refer to Fig. K68 for an exploded view of control valve assembly. Mark all hydraulic lines before removal so lines can be reconnected correctly. Cap all openings.

Unbolt and remove valve cover (24), control lever (26) and valve spool (15) from valve body. Note number and location of shims (5 and 30) when disassembling relief valves.

Carefully inspect components for damage and wear, particularly spool, valve bore, relief valve balls and seats for scoring, nicks and scratches. Install new "O" rings and thoroughly lubricate components with transmission oil.

LIFT CYLINDER

All Models

Hydraulic cylinder is a welded assembly and no service parts are available. Renew cylinder if internal or external leakage is apparent.

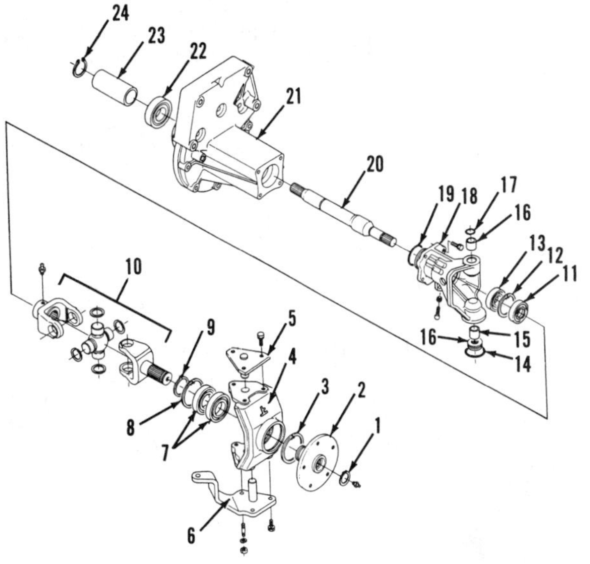

Fig. K67—Exploded view of rear axle assembly on four-wheel steering models. Components (20, 21, 22, 23 and 24) are also used on two-wheel steering models.

1. Snap ring
2. Wheel hub
3. Snap ring
4. Knuckle
5. Upper kingpin
6. Steering arm & lower kingpin
7. Bearings
8. Snap ring
9. Snap ring
10. Universal joint
11. Seal
12. Snap ring
13. Bearing
14. "O" ring
15. Thrust bearing
16. Bushing
17. "O" ring
18. Axle support
19. "O" ring
20. Axle shaft
21. Axle housing
22. Bearing
23. Spacer
24. Snap ring

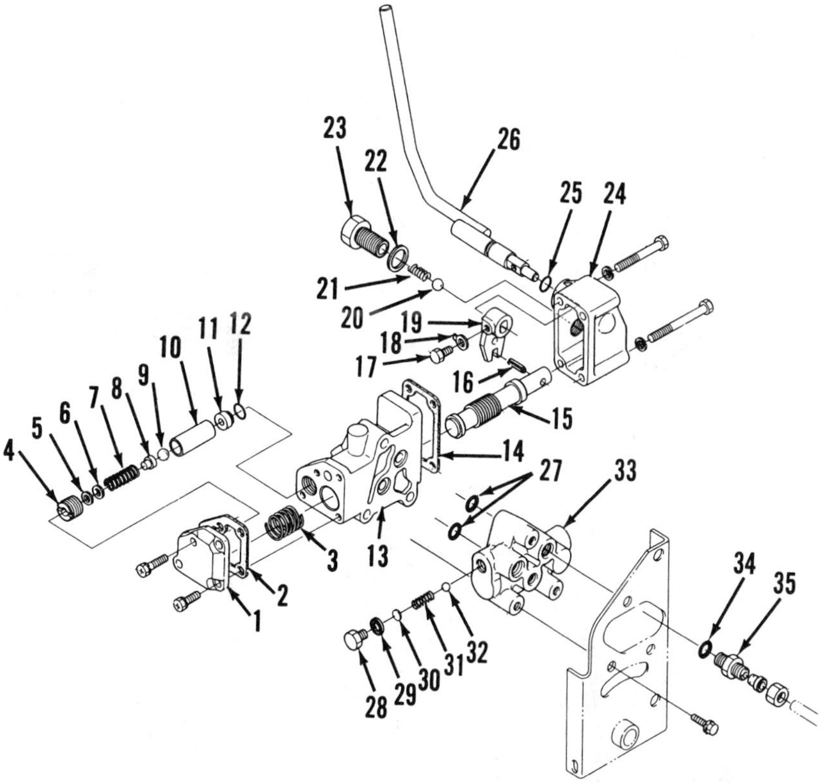

Fig. K68—Exploded view of hydraulic control valve assembly.

1. Cover
2. Gasket
3. Spring
4. Plug
5. Shim
6. Washer
7. Spring
8. Spring seat
9. Relief valve
10. Collar
11. Valve seat
12. "O" ring
13. Housing
14. Gasket
15. Spool
16. Pin
17. Screw
18. Tab washer
19. Arm
20. Detent ball
21. Spring
22. Gasket
23. Screw
24. Cover
25. "O" ring
26. Control lever
27. "O" rings
28. Plug
29. Washer
30. Shim
31. Spring
32. Relief valve
33. Adapter
34. "O" ring
35. Fitting

MTD

CONDENSED SPECIFICATIONS

Due to the numerous MTD models and the wide variety of engines installed, an accurate cross-reference table is not available. Determine the manufacturer and model number of the engine being serviced and refer to the appropriate engine section in the rear of this manual for service information.

FRONT AXLE AND STEERING SYSTEM

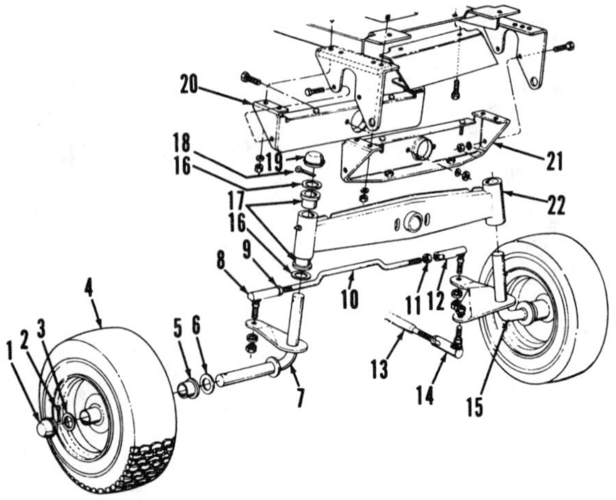

Fig. MTD1—Exploded view of front axle assembly used on Model 995A.

1. Dust cap
2. Cotter pin
3. Washer
4. Tire & hub assy.
5. Bushing
6. Washer
7. Spindle
8. Tie rod end
9. Jam nut
10. Tie rod
11. Jam nut
12. Tie rod end
13. Drag link
14. Drag link end
15. Spindle
16. Washers
17. Bushings
18. Cotter pin
19. Cap
20. Rear pivot bracket
21. Front pivot bracket
22. Axle main member

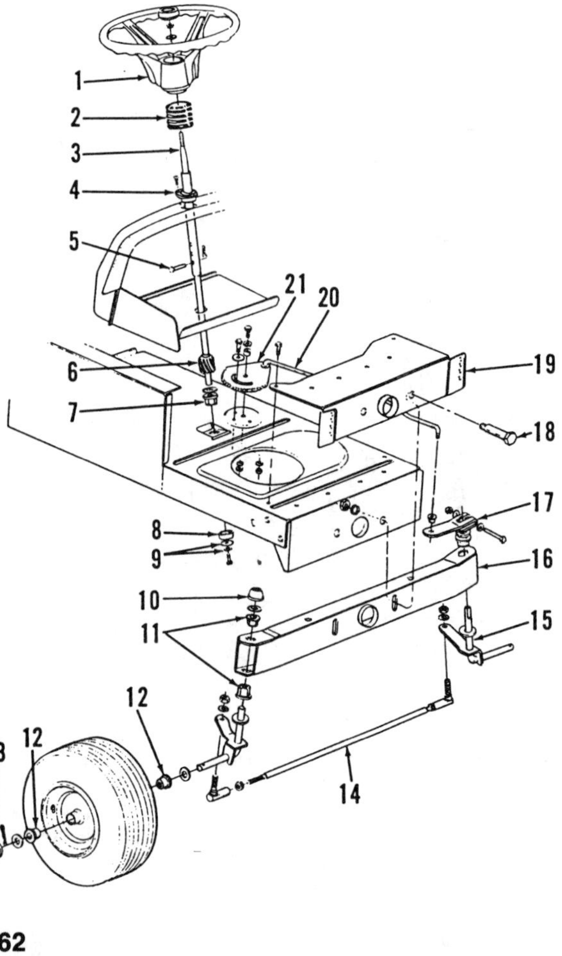

Fig. MTD2—Exploded view of front axle and steering system used on all 500-600 Series and Models 700-701-702-703-704-705-706-714-715.

1. Steering wheel
2. Bellow
3. Steering shaft extension
4. Bushing
5. Clevis pin
6. Steering shaft & gear
7. Bushing
8. Spacer
9. Washers
10. Push cap
11. Spindle bushings
12. Wheel bushings
13. Hub cap
14. Tie rod
15. Spindle
16. Axle
17. Steering arm
18. Shoulder bolt
19. Pivot bracket
20. Drag link
21. Gear segment

MAINTENANCE

All Models

Lubricate steering spindles and steering gear periodically with a lithium base, multipurpose grease. Clean away excess grease.

R&R AXLE MAIN MEMBER

Model 995A

To remove axle main member (22—Fig. MTD1), raise and support front of tractor frame and remove front wheels. Disconnect drag link end (14) at left spindle (15). Disconnect tie rod (10) from both spindles. Remove cap (19) and cotter pin (18) and withdraw spindles from axle. Support axle main member (22), remove front (21) and rear (20) pivot brackets and remove axle from tractor.

Reinstall by reversing removal procedure.

All 500-600 Series And Models 700-701-702-703-704-705-706-714-715

To remove axle main member (16—Fig. MTD2), raise and support front of tractor and remove front wheels. Disconnect tie rod (14) from spindles (15). Unbolt steering arm (17) from left spindle and remove push cap (10) from right spindle, then withdraw spindles and bushings (11) from axle ends. Unbolt and separate axle from frame.

Reassemble by reversing removal procedure.

All 700-800 Series Except Models 700-701-702-703-704-705-706-714-715-840 Through 849

To remove axle main member (22—Fig. MTD3), raise and support front of tractor. Remove front wheels. Disconnect drag link (14) at right axle spindle (7). Disconnect tie rod (9) from both spindles. Remove cap (19) and cotter pin (18) and withdraw spindles from axle.

Illustrations courtesy MTD.

Support axle main member (22) and remove pivot bolt (20). Lower axle assembly out of frame (21).

Reinstall by reversing removal procedure.

800 Series Models 840 Through 849

To remove axle main member (25—Fig. MTD3A), raise and support front of tractor frame and remove front wheels. Disconnect drag link end (12) at steering arm (16). Disconnect tie rod (9) from both spindles. Detach steering arm (16) from right spindle (6) and withdraw spindle. Remove cap (22) and cotter pin (23) and withdraw left spindle (26). Support axle main member (25), remove rear (13) and front (21) pivot brackets and remove axle from tractor.

Reinstall by reversing removal procedure.

TIE ROD

All Models

All models are equipped with adjustable tie rod. Inspect ball joints for excessive wear and looseness and renew if needed.

Adjust length of tie rod to obtain front wheel toe-in of 1/8 inch (3.2 mm). Secure tie rod ends with jam nuts.

STEERING SPINDLES

All Models

Raise and support front of tractor. Remove front wheels. Detach tie rod and drag link from steering spindle or steering arm. If so equipped, detach steering arm from upper end of spindle, otherwise, remove dust cap and withdraw cotter pin to free steering spindles. Pull spindle down out of axle main member. Inspect spindle bushings and renew as necessary.

Reinstall by reversing removal procedure. Apply lithium base, multipurpose grease to bushings before installing spindles.

STEERING GEAR

Model 995A

To remove steering gear, remove nut (8—Fig. MTD4), washer (7) and pull steering shaft (1) up and out of pinion gear (6). Remove cap screw (4) and nut (14). Drive sector shaft (9) out of bracket (2) and sector gear (10). Remove spacer (13) and gear (10).

Reinstall by reversing removal procedure. Make certain sector shaft (9) goes through steering lever (11).

All 500-600 Series And Models 700-701-702-703-704-705-706-714-715

To remove steering gears, remove steering wheel cap, retaining nut and washer, then lift off steering wheel (1—Fig. MTD2) and bellows (2). Disconnect battery cables. Remove clevis pin (5) and bushing (4) retaining screw, then withdraw shaft extension (3) and bushing. Disconnect drag link (20), then unbolt and remove gear segment (21). Remove retaining bolt from bottom of steering shaft (6) and withdraw washers (9), spacer (8), flange bushing (7) and steering shaft (6).

Inspect all parts for excessive wear and damage. Reassemble by reversing removal procedure. Lubricate steering gears with lithium base, multipurpose grease and bushings with SAE 30 oil.

All 700-800 Series Except Models 700-701-702-703-704-705-706-714-715-840 Through 849

To remove steering gears, remove nut (11—Fig. MTD5) and pull steering shaft (1) up and out of pinion gear (9). Disconnect drag link from steering arm (4). Remove cap screw (12) and washer (13). Remove starter solenoid from bracket (2), remove bolts securing bracket to frame and remove bracket and sector gear assembly.

Reinstall by reversing removal procedure. Adjust cap screw (3) so there is a minimum of steering wheel free play. Steering wheel must turn freely through entire left to right cycle.

800 Series Models 840 Through 849

To remove steering gear, first remove battery and battery tray. Remove any sheet metal panels that prevent access to steering components. Remove screw

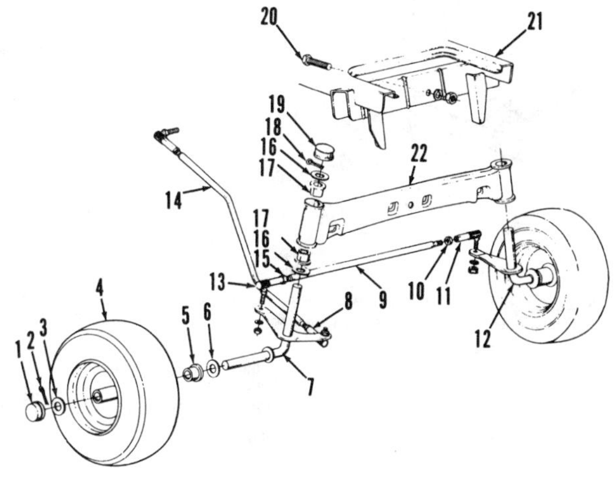

Fig. MTD3—Exploded view of front axle assembly used on all 700-800 Series except Models 700-701-702-703-704-705-706-714-715-840 through 849.

1. Dust cap
2. Cotter pin
3. Washer
4. Tire & hub assy.
5. Bushing
6. Washer
7. Spindle
8. Drag link end
9. Tie rod
10. Jam nut
11. Tie rod end
12. Spindle
13. Tie rod end
14. Drag link
15. Jam nut
16. Washer
17. Bushing
18. Cotter pin
19. Cap
20. Pivot bolt
21. Frame
22. Axle main member

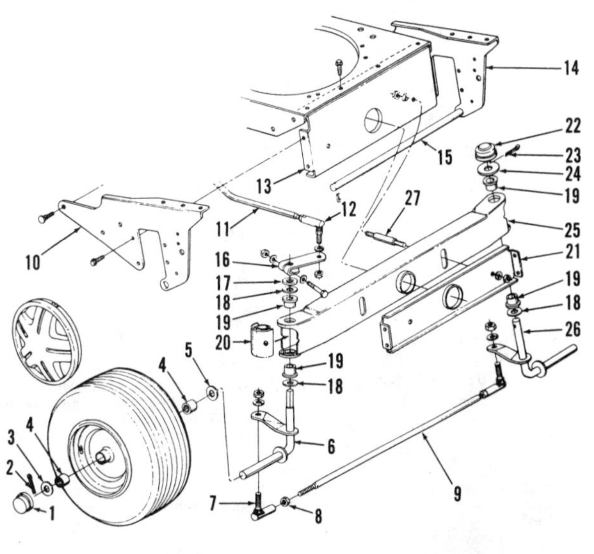

Fig. MTD3A—Exploded view of front axle assembly used on 800 Series Models 840 through 849.

1. Dust cap
2. Cotter pin
3. Washer
4. Bushing
5. Washer
6. Spindle
7. Tie rod end
8. Jam nut
9. Tie rod
10. Right side plate
11. Drag link
12. Drag link end
13. Rear pivot bracket
14. Left side plate
15. Connector rod
16. Steering arm
17. Spacer
18. Washer
19. Bushing
20. End cap
21. Front pivot bracket
22. Cap
23. Cotter pin
24. Washer
25. Axle main member
26. Spindle

retaining steering shaft bearing (6—Fig. MTD5A) to dash. Remove steering shaft coupling bolt (8) and withdraw steering wheel and upper steering shaft (7). Detach drag link (23) from sector gear (23). Remove cap screw (16) from bottom of sector shaft (21) and remove sector gear (23) and shaft. Unscrew retaining screw (15) at bottom of steering shaft (9). Remove bolts retaining shaft support (12). Move lower steering shaft (9) up out of support (12), then withdraw steering shaft down out of tractor. Inspect and renew bearings (11 and 18) as necessary.

Reassemble by reversing disassembly procedure.

ENGINE

MAINTENANCE

All Models

Regular engine maintenance is required to maintain peak operating performance and long engine life.

Check oil level at 8-hour intervals under normal operating conditions. Refer to appropriate engine section in this manual for maintenance schedule and procedures.

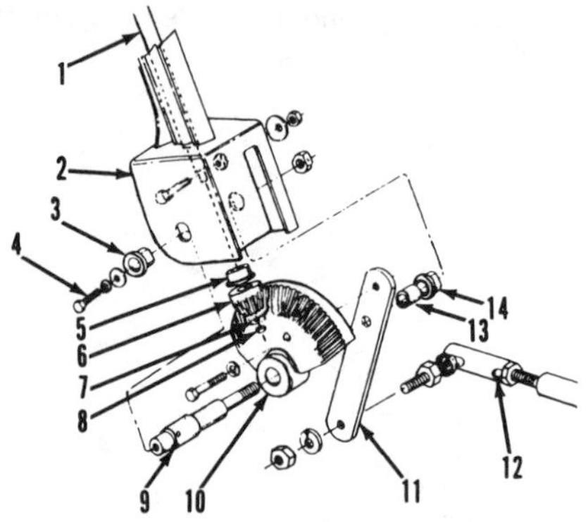

Fig. MTD4—Exploded view of steering gear assembly used on Model 995A.

1. Steering shaft
2. Bracket
3. Bushing
4. Cap screw
5. Bushing
6. Pinion gear
7. Washer
8. Nut
9. Shaft
10. Segment gear
11. Steering arm
12. Drag link end
13. Spacer
14. Nut

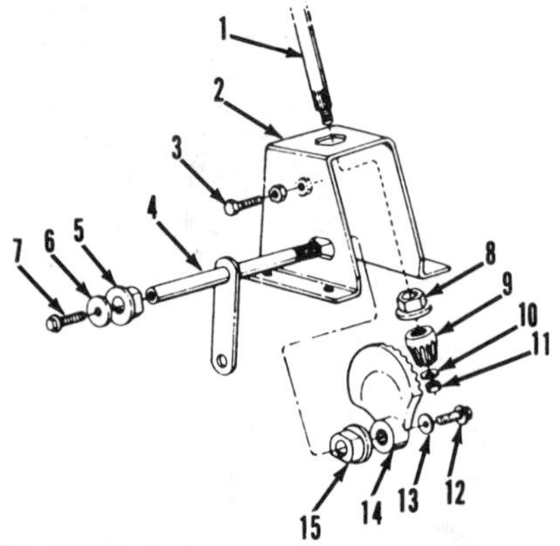

Fig. MTD5—Exploded view of steering gear assembly used on all 700-800 Series except Models 700-701-702-703-704-705-706-714-715.

1. Steering shaft
2. Bracket
3. Cap screw
4. Steering arm
5. Bushing
6. Washer
7. Cap screw
8. Bushing
9. Pinion gear
10. Washer
11. Nut
12. Cap screw
13. Washer
14. Segment gear
15. Bushing

REMOVE AND REINSTALL

All Models

Remove hood, grille brace, or side panels, and grille. Disconnect battery cables, starter wires and all necessary electrical connections at engine. Disconnect fuel line and, on early models, remove fuel tank. Disconnect choke and throttle control cables at engine. Remove drive belts and, on Model 995A, disconnect drive shaft. Remove engine mounting bolts and remove engine.

Reinstall by reversing removal procedure.

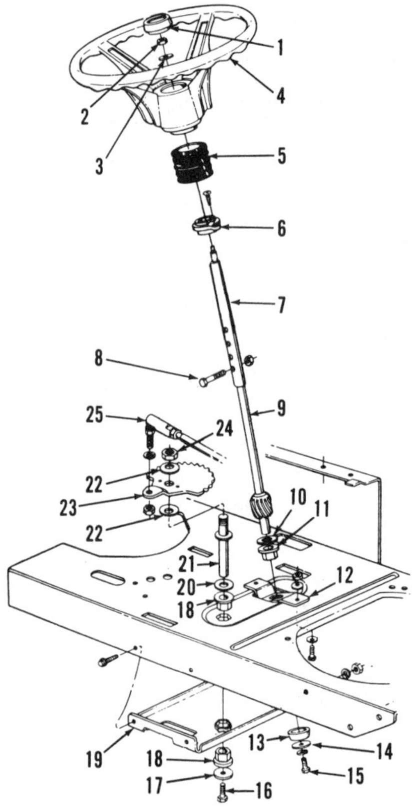

Fig. MTD5A—Exploded view of steering mechanism used on 800 Series Models 840 through 849.

1. Cap
2. Nut
3. Belleville washer
4. Steering wheel
5. Bellows
6. Bearing
7. Upper steering shaft
8. Bolt
9. Lower steering shaft
10. Washer
11. Hex bushing
12. Support
13. Spacer
14. Washer
15. Screw
16. Screw
17. Washer
18. Bearing
19. Idler support bracket
20. Washer
21. Segment gear shaft
22. Belleville washers
23. Segment gear
24. Nut
25. Drag link

OVERHAUL

All Models

Refer to appropriate engine section in this manual for tune-up specifications, engine overhaul procedures and engine maintenance.

ELECTRICAL SYSTEM

MAINTENANCE AND SERVICE

All Models

Battery electrolyte level should be checked after every 50 hours of operation. If necessary, add distilled water until level is just below base of vent well. Do not overfill. Keep battery posts clean and cable ends tight.

Refer to appropriate engine section for alternator or starter service.

Refer to Fig. MTD6 for wiring diagram typical of most models.

BRAKE

ADJUSTMENT

800 Series With Internal Disc Brake

Later Series 800 models may be equipped with a transaxle that uses a disc brake contained within the transaxle case. When brake is disengaged, brake arm on side of transaxle should be pulled by release spring against stop. When brake pedal is depressed (brake arm moves forward), there should be $^1/_4$-$^3/_8$ in. (6.4-9.5 mm) gap between end of brake arm and stop. To adjust brake, remove cotter pin that secures castle nut on transaxle brake arm and rotate castle nut to obtain desired gap.

Illustrations courtesy MTD.

Models Equipped With External Disc Brake And Castle Nut

This adjustment applies to disc brakes equipped with a castle nut (N—Fig. MTD7). To adjust disc brake, detach cotter pin and turn castle nut (N) so brake engagement begins when brake lever (L) is $^{1}/_{4}$-$^{5}/_{16}$ in. (6.4-7.9 mm) away from axle housing. Reinstall cotter pin.

Models 822A Through 836A

To adjust disc brake, loosen jam nut (6—Fig. MTD8) and tighten adjustment bolt (5) until pads are pushed against brake disc (12). Loosen adjustment bolt (5) one full turn. Check brake operation and when satisfactory, tighten jam nut (6).

Model 995A

Brake is located on the left side of transaxle and the adjustment access hole is above the left rear axle mounting bracket. Loosen jam nut (10—Fig. MTD9) and use a $^{7}/_{16}$-in. socket and extension to tighten adjustment bolt (11) until pads (3 and 4) are pushed against brake disc (2). Loosen adjustment bolt ½ turn and check brake operation. When satisfactory, tighten jam nut (10).

All Other Models

To adjust disc brake, located on right side of transaxle, loosen jam nut (11—Fig. MTD10) and turn adjusting nut (10) until disc is locked. Back off adjusting nut one complete turn and tighten jam nut.

R&R BRAKE PADS

800 Series With Internal Disc Brake

Later Series 800 models may be equipped with a transaxle that uses a disc brake contained within the transaxle case. To service brake, transaxle must be disassembled. Refer to REPAIRS section.

Models 822A Through 836A

Disconnect brake rod from brake lever (7—Fig. MTD8). Remove bolts (1) and separate caliper halves (2 and 3). Remove pad (10), metal backing plate (9) and actuating pin (8) from caliper half (2). Remove pad (10) from caliper half (3).

Reinstall by reversing removal procedure. Adjust brake.

Model 995A

Disconnect brake rod springs (12 and 13—Fig. MTD9) from brake lever (8). Remove bolts securing caliper (7) to transaxle. Remove pad (4), metal backing plate (5) and actuating pin (6) from caliper (7). Remove disc (2) and inner pad (3).

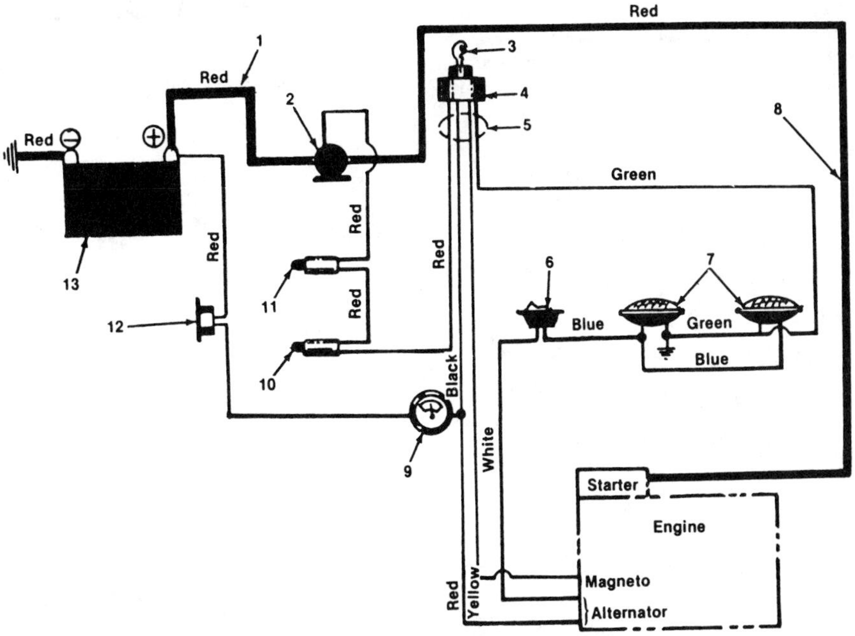

Fig. MTD6—Wiring diagram typical of most models.

1. Positive battery cable
2. Solenoid
3. Ignition key
4. Ignition switch
5. Wire harness
6. Light switch
7. Headlights
8. Starter cable
9. Ammeter
10. Safety switch
11. Safety switch
12. Circuit breaker
13. Battery

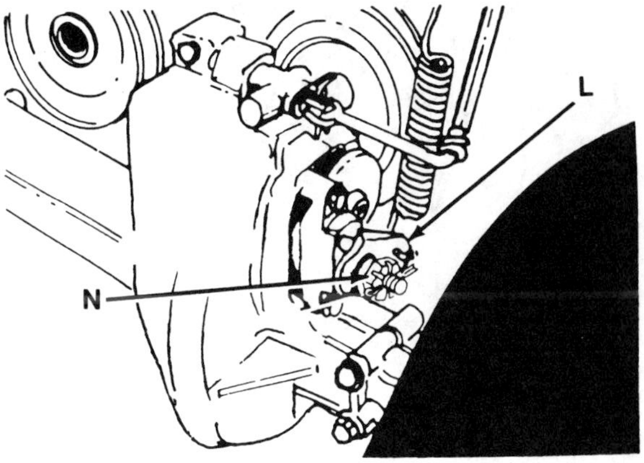

Fig. MTD7—Adjust brake on models equipped with castle nut (N) by turning castle nut. Brake should engage when brake lever (L) is $^{1}/_{4}$-$^{5}/_{16}$ in. (6.4-7.9 mm) from axle housing.

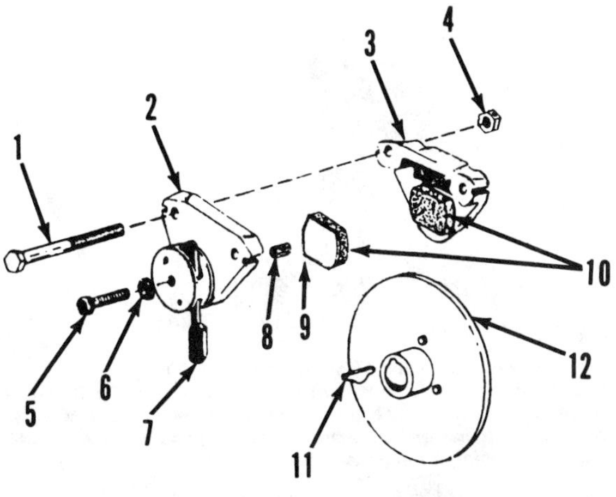

Fig. MTD8—Exploded view of brake assembly used on Models 822A through 836A.

1. Bolt
2. Caliper half
3. Caliper half
4. Nut
5. Adjustment bolt
6. Jam nut
7. Brake lever
8. Actuating pin
9. Metal backing plate
10. Brake pads
11. Key
12. Disc

Illustrations courtesy MTD.

Reinstall by reversing removal procedure. Adjust brake.

All Other Models

Disconnect brake linkage from cam lever. Unbolt and remove brake pad holder (5—Fig. MTD10). Slide brake disc (2) off transaxle shaft and remove inner brake pad (1) from holder slot in transaxle housing.

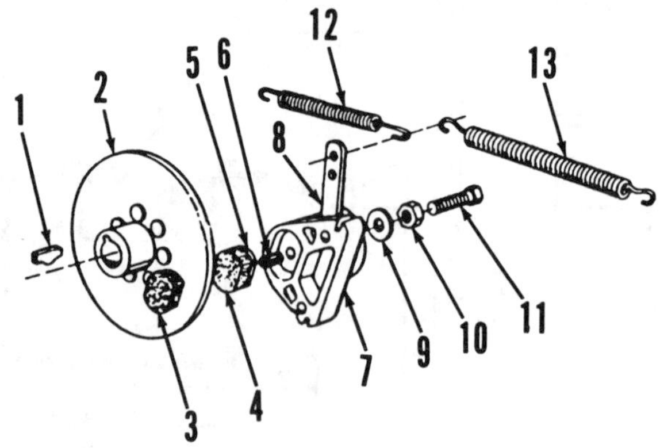

Fig. MTD9—Exploded view of brake system used on Model 995A.

1. Key
2. Disc
3. Brake pad
4. Brake pad
5. Metal backing plate
6. Pin
7. Caliper
8. Brake lever
9. Washer
10. Jam nut
11. Adjustment bolt
12. Spring
13. Spring

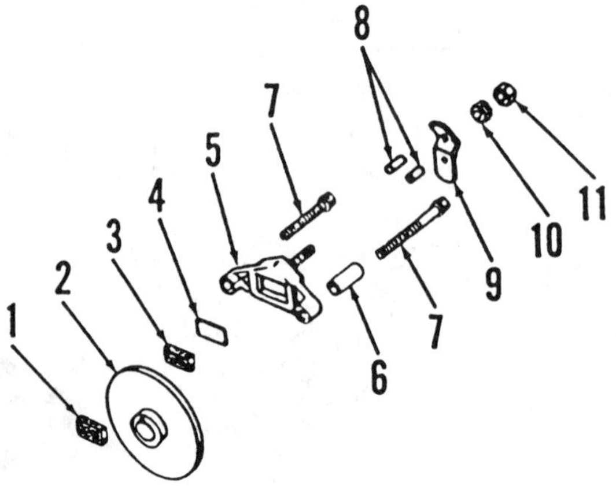

Fig. MTD10—Exploded view of brake assembly used on all other models. Note that a castle nut and cotter pin are used in place of adjusting nut (10) and jam nut (11) on some models.

1. Brake pad (inner)
2. Brake disc
3. Brake pad (outer)
4. Back-up plate
5. Carrier
6. Spacer
7. Cap screws
8. Dowel pins
9. Cam lever
10. Adjusting nut
11. Jam nut

Renew worn or damaged parts as needed. Reassemble by reversing disassembly procedure. Adjust as previously outlined.

DRIVE BELTS

R&R FRONT DRIVE BELT

500-600-700-800 Series With Variable Speed Transmission

Remove rear drive belt as outlined in following section. Position lift lever in disengaged position. Raise and support front of tractor. Unscrew belt guide bolts around engine drive pulley. If so equipped, detach mower deck belt from engine pulley. Remove belt guard for engine pulley. Move clutch/brake pedal to park position, push toward front on variable speed pulley (Fig. MTD11) and detach belt from engine pulley. If necessary, remove belt guide pins. Move clutch/brake pedal as needed to disengage belt from variable speed pulley.

Reverse removal procedure to install belt.

700-800 Series With Five-Speed Transaxle

Remove battery, and if so equipped, remove mower deck. Depress clutch pedal and engage parking brake. Raise and support front of tractor. Detach transaxle pulley belt guard and remove belt from transaxle pulley. Remove lower idler (Fig. MTD12), detach clutch shaft (Fig. MTD13) and remove upper idler. Detach and remove transaxle shift rod. Remove drive belt.

Install belt by reversing removal procedure.

Series 700 With Sundstrand Hydrostatic Transmission

To remove belt, remove mower deck, if so equipped. Depress clutch/brake pedal and set parking brake. Unscrew then move forward the belt guard around engine pulley. Unscrew hex bolt serving as belt guide near front idler pulley. Loosen rear idler belt pulley and move belt off idler pulleys. Detach pto clutch lead. Move belt from engine pul-

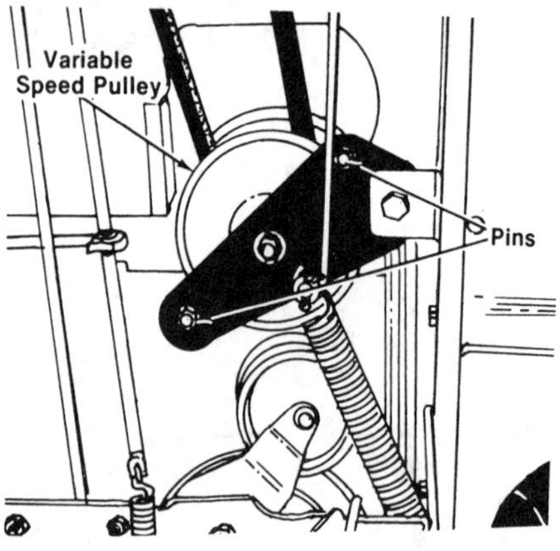

Fig. MTD11—On Series 500, 600, 700 and 800 tractors with variable speed drive, push forward on variable speed pulley and remove belt from engine pulley. It may be necessary to remove guide pins for access to belt.

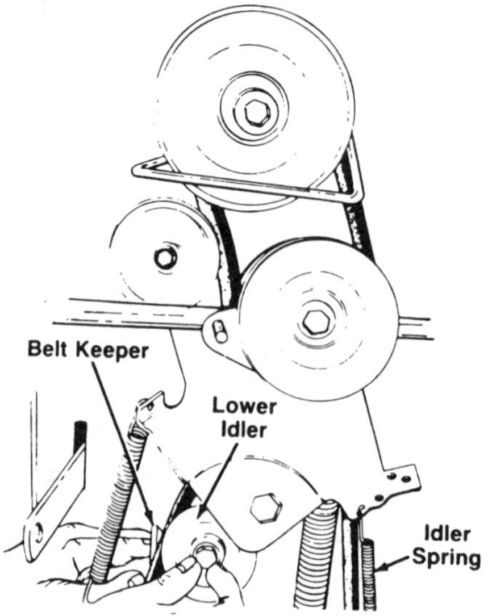

Fig. MTD12—To remove drive belt on Series 700 and 800 tractors with five-speed transaxle, remove lower belt idler pulley and refer to Fig. MTD13.

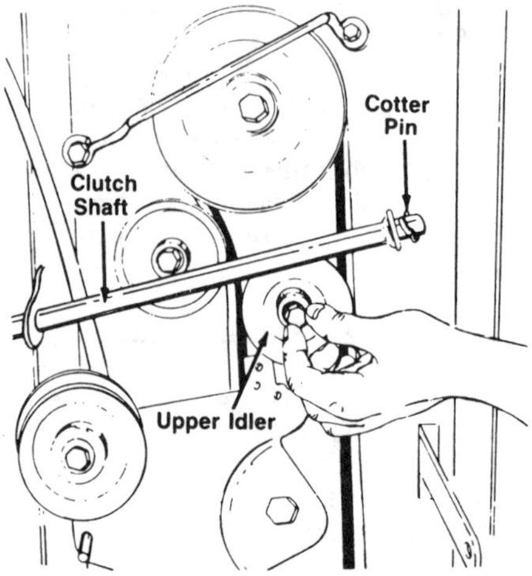

Fig. MTD13—After removing lower idler pulley (Fig. MTD12), remove clutch shaft and upper idler pulley, then remove shift rod and belt.

Illustrations courtesy MTD.

ley onto hub. Lift belt off transmission pulley and over fan, then remove belt from engine pulley. Be sure tractor cannot move, then disengage parking brake. Detach brake rod from pedal and pass belt through opening.

Reverse removal procedure to reinstall belt.

All 800 Series And 700 Series With Eaton Hydrostatic Transmission

Remove mower deck if so equipped. Raise and support front of tractor. Remove rear belt as outlined in following section. Loosen bolt securing engine pulley belt guard and swing guard forward toward front of tractor. Note direction of twist in old belt or refer to Fig. MTD14.

NOTE: If belt is installed incorrectly, tractor will run backward when in forward gears.

Pry jackshaft assembly (10—Fig. MTD14) toward front of tractor and remove belt (11).

Install new belt on engine pulley (12), twist it to the left and install on jackshaft pulley (10).

R&R REAR CLUTCH BELT

500-600-700-800 Series With Variable Speed Transmission

Remove front drive belt. Remove panel encircling transmission shift lever, or transmission panel on later 800 models. Move belt idler pulley away from belt and move belt over idler pulley. Detach belt from variable speed pulley. On later 800 models, belt can be removed. On other models, detach shift lever bracket from tractor frame and remove belt.

Reinstall belt by reversing removal procedure.

800 Series With Four-Speed Transaxle

Remove mower deck if so equipped. Raise and support front of tractor. Depress clutch/brake pedal and engage parking brake. Remove belt guard pins (8—Fig. MTD14), idler pulley (3) and transaxle pulley (1). Remove belt.

Reinstall by reversing removal procedure. Make certain belt clip (4) is correctly installed on idler pulley (3).

DRIVE SHAFT

R&R DRIVE SHAFT

Model 995A

The drive shaft connecting engine with hydrostatic drive unit is removed by loosening setscrew (7—Fig. MTD15) and sliding drive shaft assembly (6) forward until rear universal joint yoke slips off hydrostatic drive unit. Lift drive shaft and pull front universal joint and yoke (8) from crankshaft.

Reinstall by reversing removal procedure

HYDROSTATIC TRANSMISSION

MAINTENANCE

Series 700 And 800 So Equipped

The hydrostatic transmission is equipped with an integral reservoir. The reservoir is factory-filled with oil. Oil level should be at lower or "COLD" mark on reservoir with unit cold. Recommended oil is engine oil with an SF rating. On Series 700 and 800 models with an Eaton transmission (see Figs. MTD20 or MTD22), fill with SAE 20 oil. On Series 700 models with a Sundstrand transmission (see Fig. MTD21), fill with SAE 10W-30 oil.

If contamination is observed in reservoir tank or color of fluid has changed (black or milky), fluid should be drained and renewed.

To drain fluid from Eaton transmission, remove transmission from reduction gear housing, remove reservoir and vent plug from body. Turn unit upside down and allow to drain.

NOTE: Reservoir tank has left-hand threads.

When refilling Eaton transmission, add oil until fluid overflows vent hole in body. Then reinstall vent plug and fill reservoir tank to COLD mark on tank.

To drain fluid from Sundstrand transmission, unscrew hex drain plug on transmission underside.

NOTE: Reservoir cap has left-hand threads.

Fill transmission until oil reaches lower mark on reservoir.

Model 995A

Hydrostatic drive unit and two-speed transaxle share a common fluid reservoir located in the transaxle housing. Check fluid level with dipstick located just below seat and deck assembly at rear of tractor (1—Fig. MTD16). Check fluid level when fluid is cold, at 8-hour intervals. Maintain fluid level at "FULL" mark on dipstick. Change fluid and filter at 100-hour intervals. Approximate fluid capacity is 6 quarts (5.7 L). Recommended fluid is SAE 20W engine oil with service classification SF.

LINKAGE ADJUSTMENT

Series 700 With Eaton Hydrostatic Transmission

Raise and securely support rear of tractor so wheels are clear of ground. With transmission in neutral, run en-

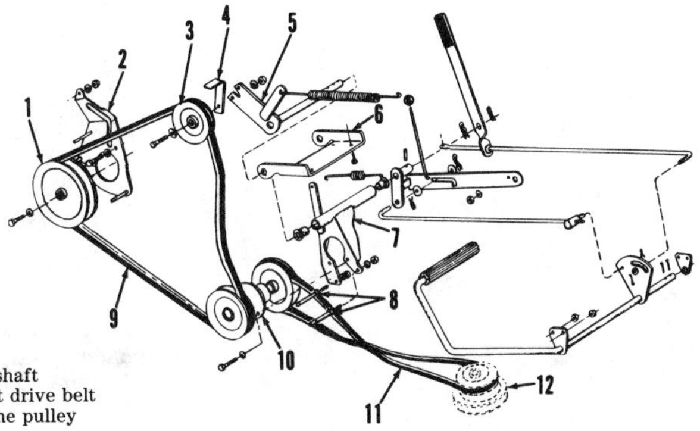

Fig. MTD14—Exploded view of drive belts and pulleys used on Series 800 tractors with four-speed transaxle. Series 700 and 800 tractors equipped with a hydrostatic transmission are similar.

1. Transaxle pulley
2. Belt guard
3. Idler pulley
4. Belt clip
5. Bracket
6. Idler mounting bracket
7. Jackshaft mounting bracket
8. Belt guard pin
9. Clutch belt
10. Jackshaft
11. Front drive belt
12. Engine pulley

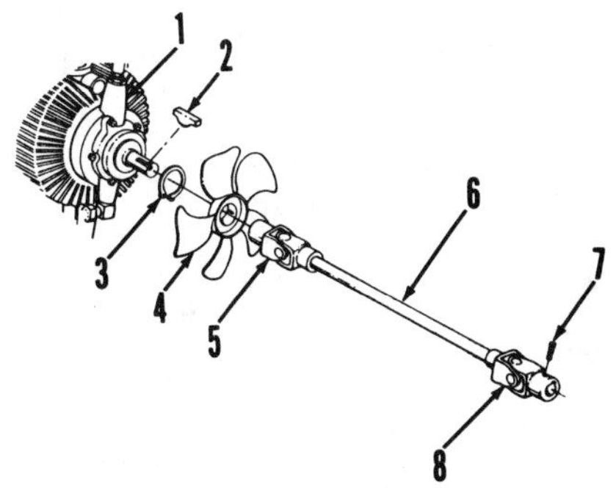

Fig. MTD15—Exploded view of drive shaft assembly used on Model 995A.

1. Hydrostatic transmission
2. Key
3. Snap ring
4. Cooling fan
5. Rear universal joint & yoke
6. Drive shaft
7. Setscrew
8. Front universal joint & yoke

Illustrations courtesy MTD.

gine. Wheels should not rotate. If wheels rotate, loosen jam nuts on transmission control rod (R—Fig. MTD17) and turn rod until wheels do not rotate with transmission in neutral.

CAUTION: Exercise caution when working around spinning wheels and drive components.

Tighten jam nuts and recheck operation.

Series 700 With Sundstrand Hydrostatic Transmission

Raise and securely support rear of tractor so wheels are clear of ground. With engine running and hydrostatic control lever in neutral, wheels should not rotate.

If wheels rotate, shut off engine and remove transmission panel. Loosen bolt attaching hydrostatic control lever bracket to ball joint bracket. Loosen bolt (B—Fig. MTD17A) attaching pintle arm extension (E) to transmission pintle arm (P). Start engine, fully depress clutch/brake pedal and engage parking brake lock.

CAUTION: Exercise caution when working around spinning wheels and drive components.

Move pintle arm until wheels stop turning, then stop engine and tighten pintle arm extension bolt. Place hydrostatic control lever in neutral position and tighten control lever bracket bolt.

Remove blocks supporting rear of tractor and recheck operation.

Series 800 So Equipped

Detach control rod (11—Fig. MTD18) from speed control lever (12). Loosen nuts (7) and push control rod (11) so spacer (5) contacts stop screw (6). Adjust stop screw so shift lever moves easily in neutral slot. Place shift lever in forward position and transmission pintle plate (9) in full forward position. While holding pintle plate, turn pintle nut (8) clockwise as far as possible and tighten nuts (7). Depress brake/speed pedal (3) full forward and adjust length of control rod (11) so spacer (5) contacts stop screw (6) and rod end fits into hole of speed control lever (12).

To perform neutral adjustment, remove shift lever knobs and remove panel over transmission. Raise and securely support rear of tractor so wheels are clear of ground. With transmission in neutral, run engine. Wheels should not rotate. If wheels rotate, detach brake spring (2) and reinstall cotter pin and washer. Lock pedal in full forward position, loosen nuts (7) and start engine. Rotate pintle nut (8) so wheels do not rotate.

CAUTION: Exercise caution when working around spinning wheels and drive components.

Tighten nuts (7), reconnect brake spring and check operation.

Model 995A

Raise and support rear of tractor so tires are off the ground. Disconnect rod end (13—Fig. MTD16) from control rod (16). Loosen jam nut (12) and the two cap screws (10). Move slide mounting plate (6) until alignment hole (5) is aligned with cam plate (7) alignment hole. Insert a 5/16 in. bolt through alignment holes. Start engine and run at idle speed.

CAUTION: Exercise caution when working around spinning wheels and drive components.

Rotate cam plate (7) until all wheel rotation stops. Tighten cap screws (10). Shut off engine and engage parking brake. Adjust length of rod (9) until rod

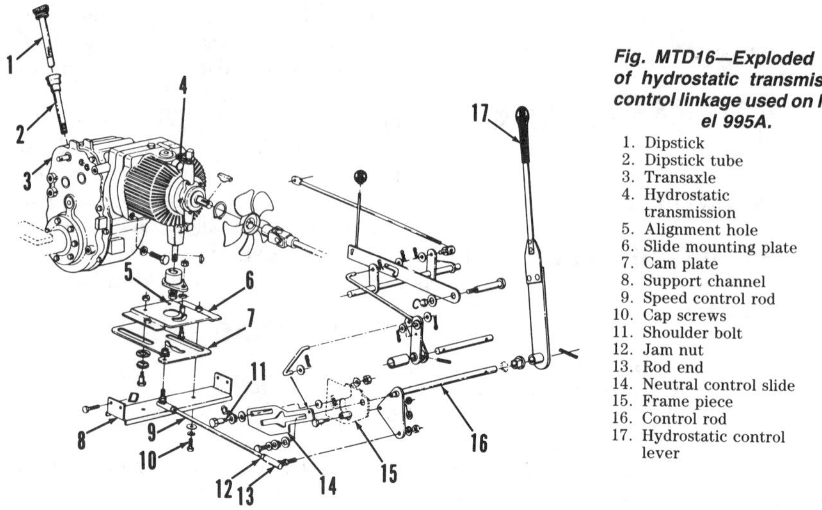

Fig. MTD16—Exploded view of hydrostatic transmission control linkage used on Model 995A.

1. Dipstick
2. Dipstick tube
3. Transaxle
4. Hydrostatic transmission
5. Alignment hole
6. Slide mounting plate
7. Cam plate
8. Support channel
9. Speed control rod
10. Cap screws
11. Shoulder bolt
12. Jam nut
13. Rod end
14. Neutral control slide
15. Frame piece
16. Control rod
17. Hydrostatic control lever

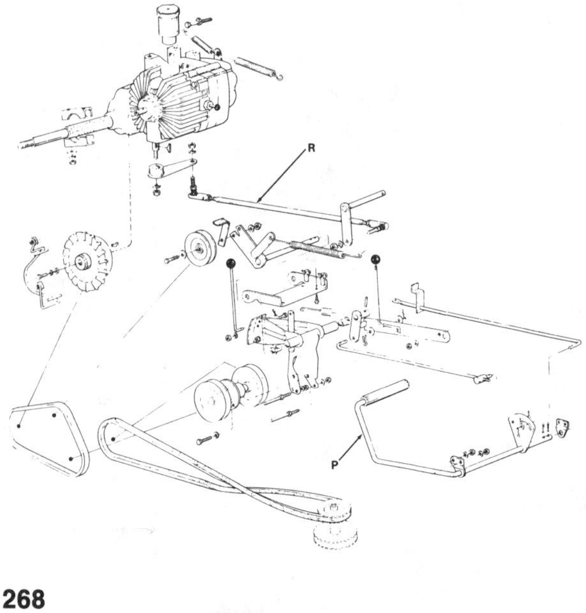

Fig. MTD17—Adjust length of control rod (R) on Series 700 tractors with Eaton hydrostatic transmission as outlined in text.

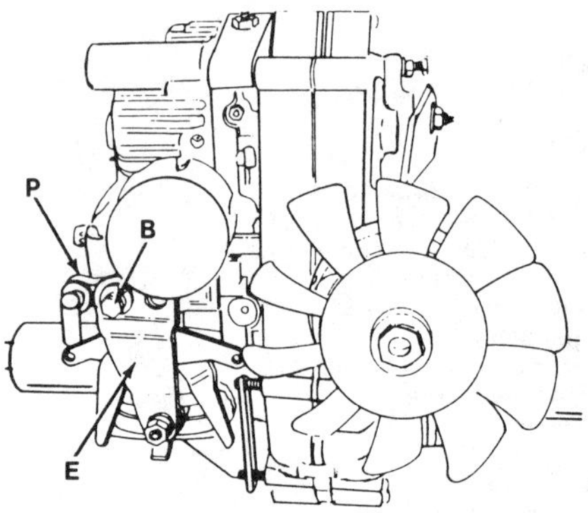

Fig. MTD17A—View of hydrostatic transmission control linkage on Series 700 tractors equipped with Sundstrand transmission. Refer to text for adjustment procedure.

Illustrations courtesy MTD.

end (13) can be inserted into control rod plate. Secure with lock washer and nut. Tighten jam nut (12) and remove the 5/16-in. bolt from alignment holes.

REMOVE AND REINSTALL

700 And 800 Series So Equipped

To remove hydrostatic transmission, remove seat and fenders. Remove final drive belt. Disconnect transmission speed control linkage at transmission. Disconnect brake actuating and return springs at brake lever and remove brake caliper. Remove reduction gear housing mounting bolts and axle housing clamp bolts, then remove transmission and reduction gear housing from tractor. Unbolt and separate hydrostatic transmission from gear housing.

To reinstall hydrostatic unit, reverse the removal procedure. Refill unit with recommended fluid and adjust control linkage as previously outlined.

Model 995A

To remove hydrostatic transmission, remove seat, fenders and drive shaft. Disconnect hydraulic hoses and plug or cap openings to prevent dirt from entering system. Disconnect hydrostatic drive control linkage. Remove bolts retaining hydrostatic unit to transaxle and remove unit.

To reinstall hydrostatic unit, reverse the removal procedure. Fill with recommended fluid and adjust control linkage as previously outlined.

OVERHAUL HYDROSTATIC TRANSMISSION

700 Series With Eaton Hydrostatic Transmission

These models are equipped with an Eaton Model 7 hydrostatic transmission. To overhaul transmission proceed as follows.

Place transmission in a holding fixture (see Fig. MTD19) with input shaft pointing up. Remove dust shield (1—Fig. MTD20) and snap ring (3). Remove cap screws from charge pump body (7). Note that one cap screw is 1/2 in. (13 mm) longer than the others and must be installed in original position. Remove charge pump body (7) with ball bearing (4). Ball bearing and oil seal (6) can be removed after removing retaining ring (2). Remove snap rings (5 and 8) and charge pump rotor assembly (9). Remove "O" rings (10) and pump plate (11).

Turn hydrostatic unit over in fixture and remove output gear. Unscrew the two cap screws until two threads remain engaged. Raise body (42) until it contacts cap screw heads. Insert a special fork tool (Fig. MTD20A) between motor rotor (39—Fig. MTD20) and pintle (28). Remove cap screws, lift off body and motor assembly with fork tool and place assembly on a bench or in a holding fixture with output shaft pointing down. Remove fork and place a wide rubber band around motor rotor to hold ball pistons (38) in their bores. Carefully remove motor rotor assembly and lay aside for later disassembly. Remove motor race (41) and output shaft (40). Remove retainer (45), bearing (44) and oil seal (43). With housing assembly (12) resting in holding fixture, remove pintle assembly (28).

CAUTION: Do not allow pump to raise with pintle as ball pistons (22) may fall out of rotor (21).

Hold pump in position by inserting a finger through hole in pintle. Remove plug (37), spring (36) and charge relief ball (35). To remove directional check valves, it may be necessary to drill through pintle with a drill bit that will pass freely through roll pins. Redrill holes from opposite side with a 1/4-in. drill bit. Newer units are drilled at factory. Press roll pin from pintle. Using a 5/16-18 tap, thread inside of valve bodies (34) then remove valve bodies using a draw bolt or slide hammer puller. Remove check valve balls (33) and retaining ring (32). To remove acceleration valves, remove retaining pin, insert a 3/16-in. (5 mm) rod 8 in. (203 mm) long through passage in pintle and carefully drive out spring (29), body (30) and ball (31). To remove dampening pistons (26), carefully tap outside edge of pintle on work bench to jar pistons free.

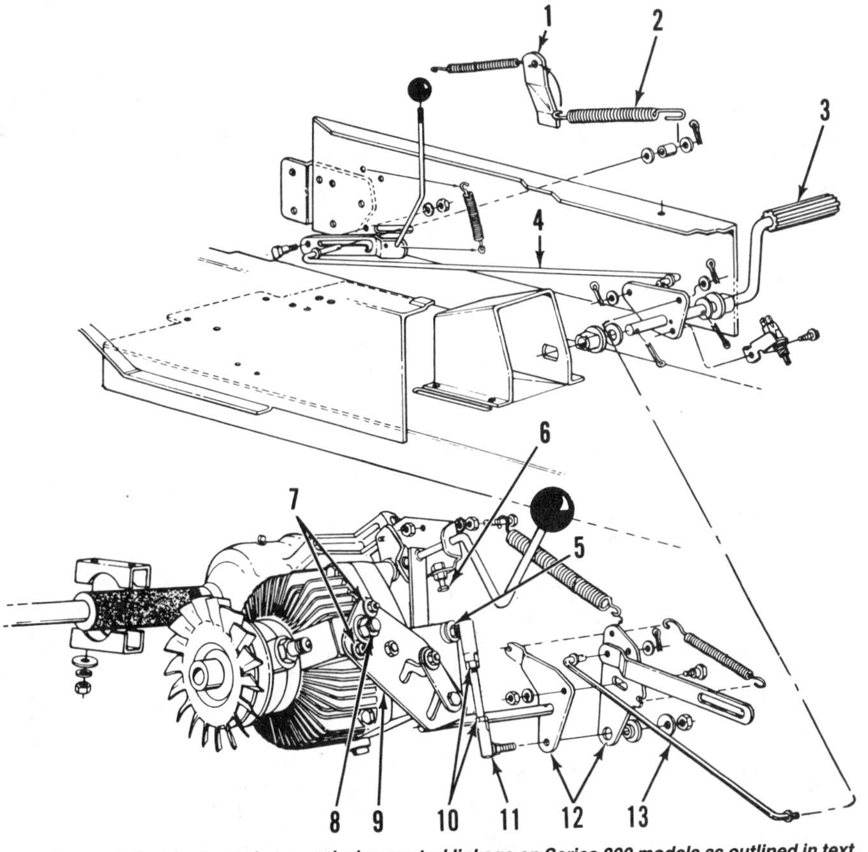

Fig. MTD18—Adjust hydrostatic transmission control linkage on Series 800 models as outlined in text.

1. Brake lever
2. Brake spring
3. Brake/speed control pedal
4. Brake rod
5. Spacer
6. Stop screw
7. Nuts
8. Pintle nut
9. Pintle plate
10. Jam nuts
11. Control rod
12. Speed control levers
13. Speed control rod

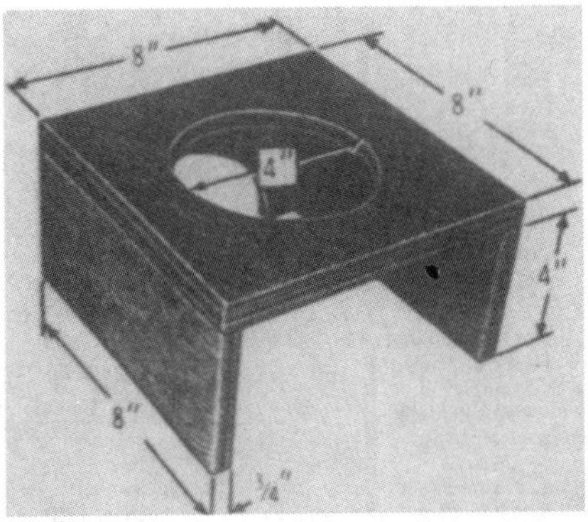

Fig. MTD19—The wooden stand shown above is helpful when repairing an Eaton Model 7 or 11 hydrostatic transmission.

Illustrations courtesy MTD.

NOTE: If pintle journal is damaged, pintle must be renewed.

Remove pump cam ring (24) and pump race (23). Place a wide rubber band around pump rotor to prevent ball pistons (22) from falling out. Carefully remove pump assembly and input shaft (15).

To remove control shaft (19), drill a $^{11}/_{32}$-in. hole through aluminum housing (12) directly in line with center line of dowel pin. Press dowel pin from control shaft, then withdraw control shaft. Remove oil seal (18). Thread drilled hole in housing with a $^{1}/_{8}$-in. pipe tap. Apply a light coat of Loctite RC/680 to a $^{1}/_{8}$-in. pipe plug, install plug and tighten until snug. Do not overtighten.

Number piston bores (1 through 5) on pump rotor and on motor rotor. Use a plastic ice cube tray or equivalent and mark cavities 1P through 5P for pump ball pistons and 1M through 5M for motor ball pistons. Remove ball pistons (22), one at a time, from pump rotor and place each ball in the correct cavity in tray. Remove ball piston (38) and springs from motor rotor in the same manner.

Clean and inspect all parts and renew any showing excessive wear or other damage. Renew all gaskets, seals and "O" rings. Ball pistons are a select fit to 0.0002-0.0006 in. (0.005-0.015 mm) clearance in rotor bores and must be reinstalled in their original bores. If rotor bushing to pintle journal clearance is 0.002 in. (0.05 mm) or more, pump rotor or motor rotor and/or pintle must be renewed. Check clearance between input shaft (15) and housing bushing. Normal clearance is 0.0013-0.0033 in. (0.033-0.084 mm). If clearance is excessive, renew input shaft and/or housing assembly.

Install ball pistons (22) in pump rotor (21) and ball pistons (38) and springs in motor rotor (39), then use wide rubber bands to hold pistons in their bores.

Install charge relief valve ball (35) and spring (36) in pintle. Screw plug (37) into pintle until just below outer surface of pintle. Install acceleration valve springs (29) and bodies (30) making sure valves move freely. Tap balls (31) into pintle until roll pins will go into place. Install snap rings (32), check valve balls (33) and valve bodies (34) in pintle and secure with new roll pins.

NOTE: When installing oil seals (6, 18 or 43), apply a light coat of Loctite RC/680 to seal outer diameter.

Renew oil seal (18) and install control shaft (19) in housing. Install special washer (17), then press dowel pin through control shaft until $1^{1}/_{4}$ in. (32 mm) of pin extend from control shaft. Renew oil seal (43) and reinstall output shaft (40), bearing (44), retainer (45), output gear and snap ring.

Insert input shaft (15) in housing (12). Install snap ring (8) in its groove on input shaft. Place "O" ring (10), pump plate (11) and "O" ring in housing, then install charge pump drive key (14). Apply light grease or Vaseline to pump rollers and place rollers in rotor slots (9). Install oil seal (6) and pump race in charge pump body (7), then install body assembly. Secure with the five cap screws, making certain long cap screw is installed in its original location (in heavy section of pump body). Tighten cap screws to 28-30 ft.-lbs. (38-41 N·m). Install snap ring (5), bearing (4), retaining ring (2), snap ring (3) and dust shield (1).

Place charge pump and housing assembly in a holding fixture with input shaft pointing downward. Install pump race (23) and insert (25) in cam ring (24), then install cam ring assembly over cam pivot pin (13) and control shaft dowel pin. Turn control shaft (19) back and forth and check movement of cam ring. Cam ring must move freely from stop to stop. If not, check installation of insert (25) in cam ring.

Install pump rotor assembly and remove rubber band used to retain pistons. Install pintle assembly (28) over cam pivot pin (13) and into pump rotor. Place "O" ring (20) in position on housing.

Place body assembly (42) in a holding fixture with output gear down. Install motor race (41) in body, then install motor rotor assembly and remove rubber band used to retain pistons in rotor.

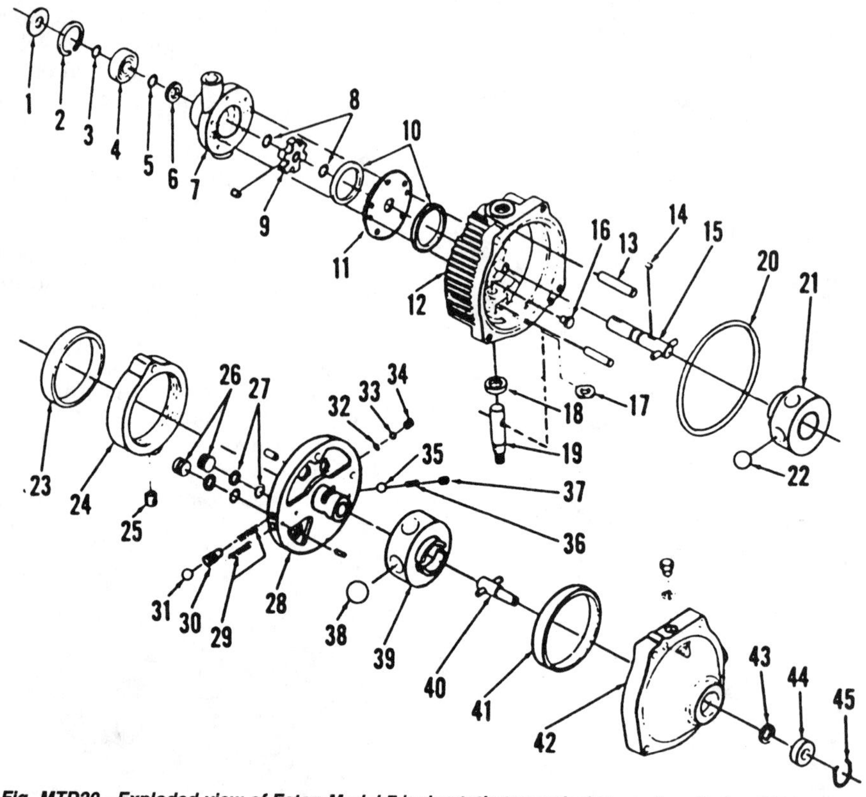

Fig. MTD20—Exploded view of Eaton Model 7 hydrostatic transmission used on Series 700 tractors so equipped.

1. Dust shield
2. Retaining ring
3. Snap ring
4. Ball bearing
5. Snap ring
6. Oil seal
7. Charge pump body
8. Snap ring
9. Charge pump rotor assy.
10. Square cut seals
11. Pump plate
12. Housing
13. Cam pivot pin
14. Key
15. Input shaft
16. Neutral spring cap
17. Washer
18. Oil seal
19. Control shaft
20. "O" ring
21. Pump rotor
22. Pump ball pistons
23. Pump race
24. Pump cam ring
25. Cam ring insert
26. Dampening pistons
27. "O" ring
28. Pintle
29. Spring
30. Acceleration valve body
31. Acceleration valve ball
32. Retaining ring
33. Check valve ball
34. Check valve body
35. Charge relief ball
36. Spring
37. Relief valve plug
38. Motor ball piston
39. Motor rotor
40. Output shaft
41. Motor race
42. Body
43. Oil seal
44. Ball bearing
45. Retainer

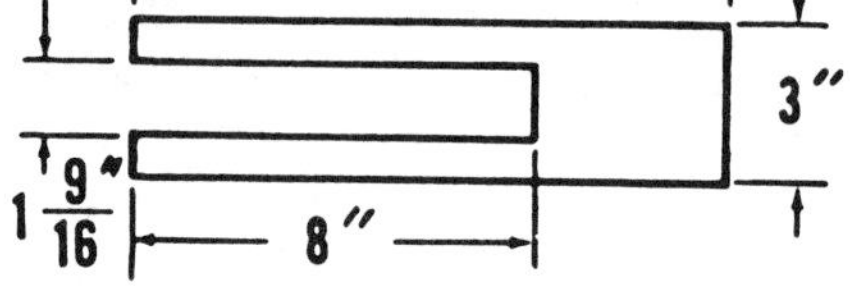

Fig. MTD20A—A special fork tool for Eaton hydrostatic transmission overhaul may be constructed from ⅛-in. flat stock using the dimensions shown above.

Illustrations courtesy MTD.

Using special fork tool (Fig. MTD20A) to retain motor assembly in body, carefully install body and motor assembly over pintle journal. Remove fork tool, align bolt holes and install two cap screws. Tighten cap screws to 15 ft.-lbs. (20 N·m).

Place hydrostatic unit on holding fixture with reservoir adapter opening and venting plug opening facing upward. Fill unit with recommended fluid until fluid flows from fitting hole in body. Plug all openings to prevent dirt or other foreign material from entering hydrostatic unit.

Series 700 With Sundstrand Hydrostatic Transmission

To disassemble transmission, remove transmission control arm. Unscrew (left-hand threads) oil reservoir (16—Fig. MTD21) and drain oil from transmission housing. Unbolt and separate end cover (32) from housing (17), being careful not to damage machined surfaces of end cover and pump and motor cylinder blocks. Remove check valves (26) and bypass valve (29) from end cover. Pry out oil seals (30) and (31).

Position housing (17) on its side and slide out pump cylinder block (34) and pistons and motor cylinder block (27) and pistons. The motor shaft (23) should slide out with the motor cylinder block. Remove motor thrust plates (20) and bearing (21). Remove spring (38), washer (39) and pump swashplate (42) with thrust plate (40) and bearing (41).

Remove slot guide (12) and the cradle bearings (43). Remove filter (11), washer (10) and spring (9). Remove retaining ring (1) and pry out oil seal (2) taking care not to damage pump shaft or housing. Tap inner end of pump shaft (6) with a soft hammer to remove shaft and bearing (4) from housing. Withdraw trunnion arm (15) and pry oil seal (13) from housing.

Inspect pistons and cylinder bores for excessive wear, scoring or other damage. Pump and motor cylinder blocks are serviced as complete assemblies. Check polished surfaces of end cover and cylinder blocks for wear, scoring or scratches. End cover and cylinder blocks must be renewed if wear exceeds 0.0004 inch (0.01 mm). Inspect check valve balls and their seats for damage or wear and renew as necessary.

Coat parts with 10W-30 engine oil. To reassemble, reverse the disassembly procedure. Be sure that pistons are returned to original bores in cylinder blocks and that they slide freely in cylinder block. Renew all "O" rings and oil seals.

Series 800 And 900

These models are equipped with an Eaton Model 11 hydrostatic transmission. To overhaul transmission proceed as follows:

If so equipped, remove venting plug (59—Fig. MTD22) and reservoir adapter (18 through 21), invert assembly and drain fluid. Place unit in a holding fixture similar to one shown in Fig. MTD19 so input shaft is pointing up.

Remove dust shield (1—Fig. MTD22) and snap ring (3). Place an identifying mark across joining edges of each section to aid in correct assembly. Remove five cap screws from charge pump body (7). Note that one cap screw is ½ in. (13 mm) longer than the others and must be installed in original position (heavy section of pump body). Remove charge pump body (7) with ball bearing (4). Ball bearing and oil seal (6) can be removed after removing retaining ring (2). Remove six charge pump rollers (12), snap rings (5, 9 and 11) and charge pump rotor (10). Remove "O" rings (14 and 16) and pump plate (15).

Turn hydrostatic unit over in fixture and remove snap ring (65) and output gear (64). Unscrew two cap screws (60) until two threads remain engaged. Raise body (57) until it contacts cap screw heads. Insert a special fork tool (Fig. MTD20A) between motor rotor (53—Fig. MTD22) and pintle (48). Remove cap screws, lift off body and motor assembly with fork tool and place assembly on a bench or in a holding fixture with output shaft pointing down. Remove fork and place a wide rubber band around motor rotor to hold ball pistons (51) in their bores. Carefully remove motor rotor assembly and lay aside for later disassembly. Remove motor race (56) and output shaft (55). Remove retainer (63), bearing (62) and oil seal (61). With housing assembly (22) resting in holding fixture, remove pintle assembly (48).

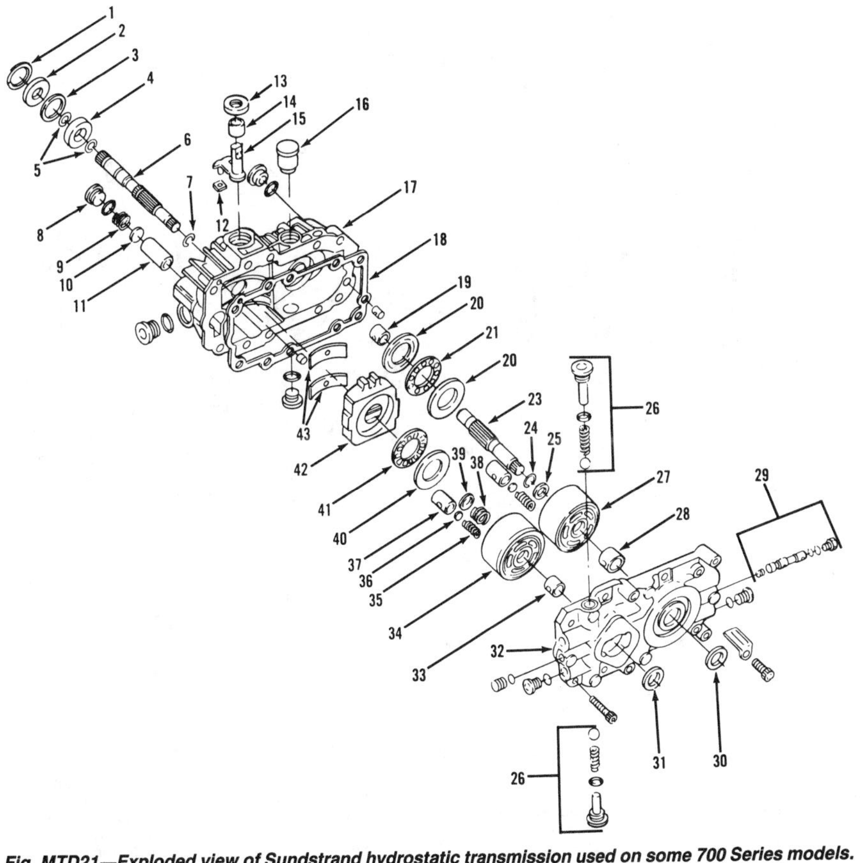

Fig. MTD21—Exploded view of Sundstrand hydrostatic transmission used on some 700 Series models.

1. Retaining ring
2. Oil seal
3. Spacer
4. Ball bearing
5. Retaining rings
6. Pump shaft
7. Retaining ring
8. Plug
9. Spring
10. Washer
11. Filter
12. Slot guide
13. Oil seal
14. Bearing
15. Trunnion arm
16. Reservoir
17. Transmission housing
18. Gasket
19. Bearing
20. Thrust washers
21. Thrust bearing
23. Motor shaft
24. Retaining ring
25. Washer
26. Check valves
27. Motor cylinder block
28. Bearing
29. Bypass valve assy.
30. Oil seal
31. Oil seal
32. End cover
33. Bearing
34. Pump cylinder block
35. Spring
36. Seat
37. Piston
38. Spring
39. Washer
40. Thrust washer
41. Thrust bearing
42. Swashplate
43. Bearings

Illustrations courtesy MTD.

CAUTION: Do not allow pump to raise with pintle as ball pistons (35) may fall out of rotor (36).

Hold pump in position by inserting a finger through hole in pintle. Remove plug (45), spring (46) and charge relief ball (47). To remove directional check valves, it may be necessary to drill through pintle with a drill bit that will pass freely through roll pins (41). Redrill holes from opposite side with a 1/4-in. drill bit. Newer units are drilled at factory. Press roll pin from pintle. Using a 5/16-18 tap, thread inside of valve bodies (44) then remove valve bodies using a draw bolt or slide hammer puller. Remove check valve balls (43) and retaining ring (42). Do not remove plugs (40).

Remove pump cam ring (39) and pump race (38). Place a wide rubber band around pump rotor to prevent ball pistons (35) from falling out. Carefully remove pump assembly and input shaft (33).

To remove control shaft (25), drill a 11/32 in. hole through aluminum housing (22) directly in line with center line of dowel pin (27). Press dowel pin from control shaft, then withdraw control shaft. Remove oil seal (24). Thread drilled hole in housing with a 1/8-in. pipe tap. Apply a light coat of Loctite RC/680 to a 1/8-in. pipe plug, install plug and tighten until snug. Do not overtighten.

Number piston bores (1 through 5) on pump rotor and on motor rotor. Use a plastic ice cube tray or equivalent and mark cavities 1P through 5P for pump ball pistons and 1M through 5M for motor ball pistons. Remove ball pistons (35) one at a time, from pump rotor and place each ball in the correct cavity in tray. Remove ball piston (51) and springs (52) from motor rotor in the same manner.

Clean and inspect all parts and renew any showing excessive wear or other damage. Renew all gaskets, seals and "O" rings. Ball pistons are a select fit to 0.0002-0.0006 in. (0.005-0.015 mm) clearance in rotor bores and must be reinstalled in their original bores. If rotor bushing to pintle journal clearance is 0.002 in. (0.051 mm) or more, pump rotor or motor rotor must be renewed. Check clearance between input shaft (33) and housing bushing. Normal clearance is 0.0013-0.0033 in. (0.033-0.084 mm). If clearance is excessive, renew input shaft and/or housing assembly.

Install ball pistons (35) in pump rotor (36) and ball pistons (51) and springs (52) in motor rotor (53), then use wide rubber bands to hold pistons in their bores.

Install charge relief valve ball (47) and spring (46) in pintle. Screw plug (45) into pintle until just below outer surface of pintle. Install snap rings (42), check valve balls (43) and valve bodies (44) in pintle (48) and secure with new roll pins (41).

NOTE: When installing oil seals (6, 24 or 61), apply a light coat of Loctite RC/680 to seal outer diameter.

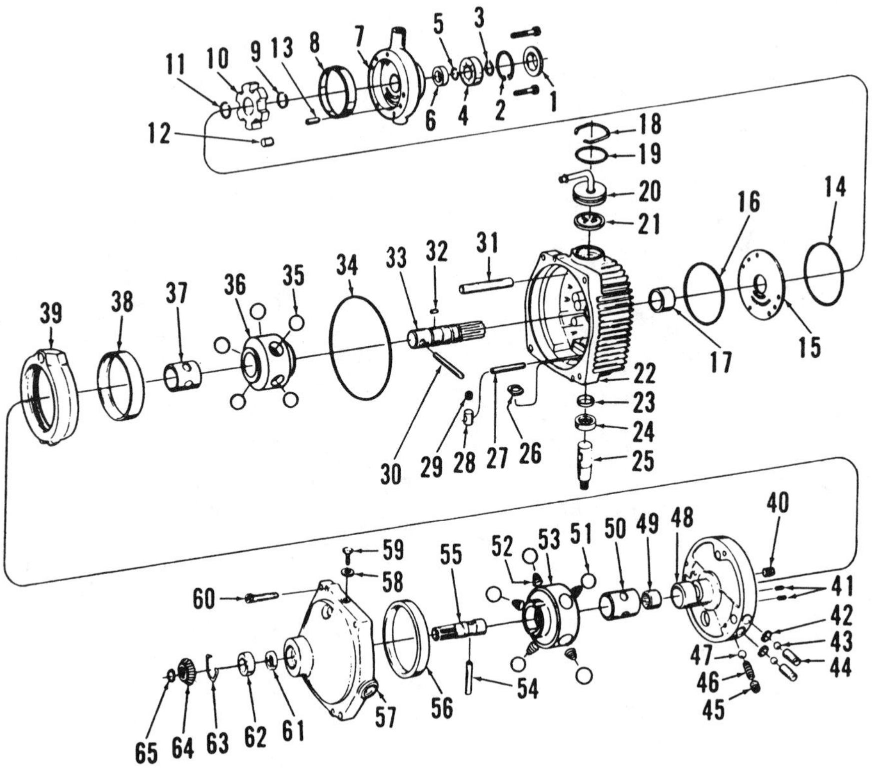

Fig. MTD22—Exploded view of Eaton Model 11 hydrostatic transmission.

1. Dust shield
2. Retaining ring
3. Snap ring
4. Ball bearing
5. Snap ring
6. Oil seal
7. Charge pump body
8. Charge pump race
9. Snap ring
10. Charge pump rotor
11. Snap ring
12. Pump roller (6 used)
13. Dowel pin
14. "O" ring
15. Pump plate
16. "O" ring
17. Bushing
18. Retainer
19. "O" ring
20. Reservoir adapter
21. Screen
22. Housing
23. Bushing
24. Oil seal
25. Control shaft
26. Washer
27. Dowel pin
28. Insert
29. Insert cap
30. Drive pin
31. Cam pivot pin
32. Charge pump drive key
33. Input shaft
34. "O" ring
35. Pump ball pistons
36. Pump rotor
37. Rotor bushing
38. Pump race
39. Pump cam ring
40. Plug (2 used)
41. Roll pins
42. Snap ring (2 used)
43. Check valve ball (2 used)
44. Directional check valve body (2 used)
45. Plug
46. Relief spring
47. Charge relief ball
48. Pintle
49. Needle bearing
50. Rotor bushing
51. Motor ball pistons
52. Springs
53. Motor rotor
54. Drive pin
55. Output shaft
56. Motor race
57. Body
58. Gasket
59. Venting plug
60. Cap screw
61. Oil seal
62. Ball bearing
63. Retainer
64. Output gear
65. Snap ring

Renew oil seal (24) and install control shaft (25) in housing. Install special washer (26), then press dowel pin through control shaft until 1 1/4-in. (32 mm) of pin extend from control shaft. Renew oil seal (61) and reinstall output shaft (55), bearing (62), retainer (63), output gear (64) and snap ring (65).

Insert input shaft (33) with drive pin (30) through bushing (17) in housing. Install snap ring (11) in its groove on input shaft. Place "O" ring (16), pump plate (15) and "O" ring (14) in housing, then install charge pump drive key (32). Apply light grease or Vaseline to pump rollers (12) and place rollers in rotor slots. Install oil seal (6) and pump race (8) in charge pump body (7), then install body assembly. Secure with the five cap screws, making certain longer cap screw is installed in its original location (in heavy section of pump body). Tighten cap screws to 28-30 ft.-lbs. (38-41 N·m). Install snap ring (5), bearing (4), retaining ring (2), snap ring (3) and dust shield (1).

Place charge pump and housing assembly in a holding fixture with input shaft pointing downward. Install pump race (38), insert cap (29) and insert (28) in cam ring (39), then install cam ring assembly over cam pivot pin (31) and control shaft dowel pin (27). Turn control shaft (25) back and forth and check movement of cam ring. Cam ring must move freely from stop to stop. If not, check installation of insert (28) in cam ring.

Install pump rotor assembly and remove rubber band used to retain

Illustrations courtesy MTD.

pistons. Install pintle assembly (48) over cam pivot pin (31) and into pump rotor. Place "O" ring (34) in position on housing.

Place body assembly (57) in a holding fixture with output gear down. Install motor race (56) in body, then install motor rotor assembly and remove rubber band used to retain pistons in rotor.

Using special fork tool (Fig. MTD20A) to retain motor assembly in body, carefully install body and motor assembly over pintle journal. Remove fork tool, align bolt holes and install two cap screws. Tighten cap screws to 15 ft.-lbs. (20 N·m).

Place hydrostatic unit on holding fixture with reservoir adapter opening and venting plug opening facing upward. Fill unit with recommended fluid until fluid flows from fitting hole in body. Install venting plug (59—Fig. MTD22) with gasket (58), then install reservoir adapter (20), screen (21), "O" ring (19) and retainer (18). Plug all openings to prevent dirt or other foreign material from entering hydrostatic unit.

REDUCTION GEARS AND DIFFERENTIAL

LUBRICATION

Series 700-800 With Eaton Hydrostatic Transmission

Oil level should be maintained even with oil level plug opening. Recommended oil is SAE 90 EP gear oil. Oil capacity is $2^3/_4$ pints (1.3 L).

Series 700 With Sundstrand Hydrostatic Transmission

The reduction gear and differential unit used on 700 series tractors with Sundstrand hydrostatic transmission is lubricated with grease at factory and should not require additional lubrication. Capacity is 16 ounces (473 mL) of Shell Darina #0 grease.

Model 995A

Hydrostatic transmission and reduction gear and differential unit share a common oil reservoir in gear housing. Check fluid level with dipstick located just below seat and deck assembly at rear of tractor (1—Fig. MTD16). Check fluid level when fluid is cold, at 8-hour intervals. Maintain fluid level at "FULL" mark on dipstick. Change fluid and filter at 100-hour intervals. Approximate fluid capacity is 6 quarts (5.7 L). Recommended fluid is SAE 20W engine oil with service classification SF.

R&R AND OVERHAUL

Series 700-800 With Eaton Hydrostatic Transmission

To remove reduction gear and differential unit, refer to hydrostatic transmission section and separate transmission from differential. Remove wheel and hub assemblies from axles. Remove park brake disc from brake shaft. Clean axle shaft and remove any burrs from shafts. Unscrew cap screws

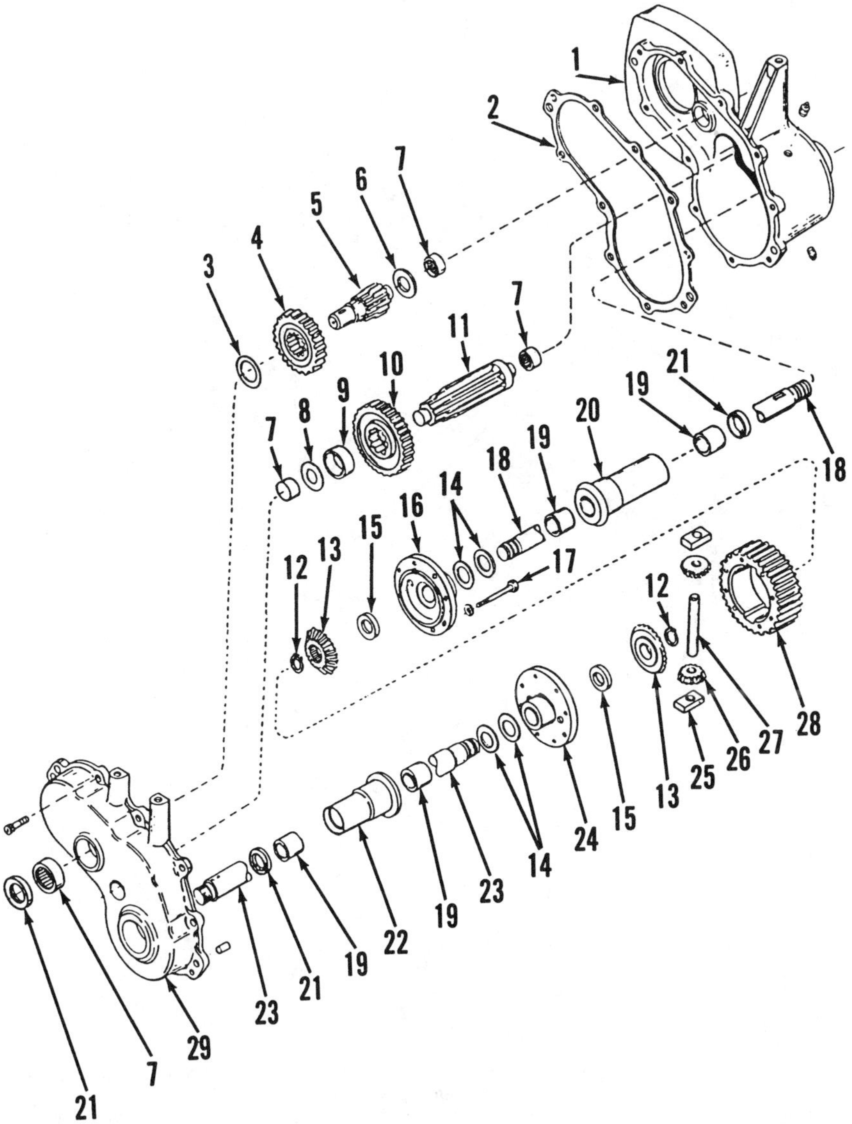

Fig. MTD23—Exploded view of Peerless reduction gear and differential unit used on Series 700 and 800 tractors with Eaton hydrostatic transmission.

1. Case
2. Gasket
3. Washer
4. Idler gear
5. Brake shaft
6. Washer
7. Bearing
8. Washer
9. Spacer
10. Output gear
11. Output shaft
12. Snap ring
13. Side gears
14. Thrust washers
15. Thrust washer
16. Differential carrier
17. Bolt
18. Axle shaft (R.H.)
19. Bushing
20. Axle housing
21. Oil seal
22. Axle housing
23. Axle shaft (L.H.)
24. Differential carrier
25. Drive block
26. Pinion gear
27. Pinion shaft
28. Ring gear
29. Cover

Illustrations courtesy MTD.

and drive out dowel pins in cover (29—Fig. MTD23). Lift cover off case and axle shaft. Withdraw brake shaft (5), idler gear (4) and thrust washers (3 and 6) from case. Remove output shaft (11), output gear (10), spacer (9), thrust washer (8) and differential assembly from case. Axle shaft housings (20 and 22) must be pressed from case and cover.

To disassemble differential, unscrew cap screws (17) and separate axle shaft and carrier assemblies from ring gear (28). Drive blocks (25), bevel pinion gears (26) and pinion shaft (27) can now be removed from ring gear. Remove snap rings (12) and slide axle shaft (18 and 23) from axle gears (13) and carriers (16 and 24).

Clean and inspect all parts and renew any parts damaged or excessively worn. When installing needle bearings, press bearings in from inside of case or cover until bearings are 0.015-0.020 inch (0.38-0.51 mm) below thrust surfaces. Be sure heads of differential cap screws (17) and right axle shaft (18) are installed in right carrier housing (16). Right axle shaft is installed through case (1). Tighten differential cap screws to 7 ft.-lbs. (9 N·m) and cover cap screws to 10 ft.-lbs. (14 N·m). Differential assembly and output shaft (11) must be installed in case at same time. Remainder of assembly is reverse of disassembly procedure. Refill differential unit with 2¾ pints (1.3 L) of SAE 90 EP gear oil.

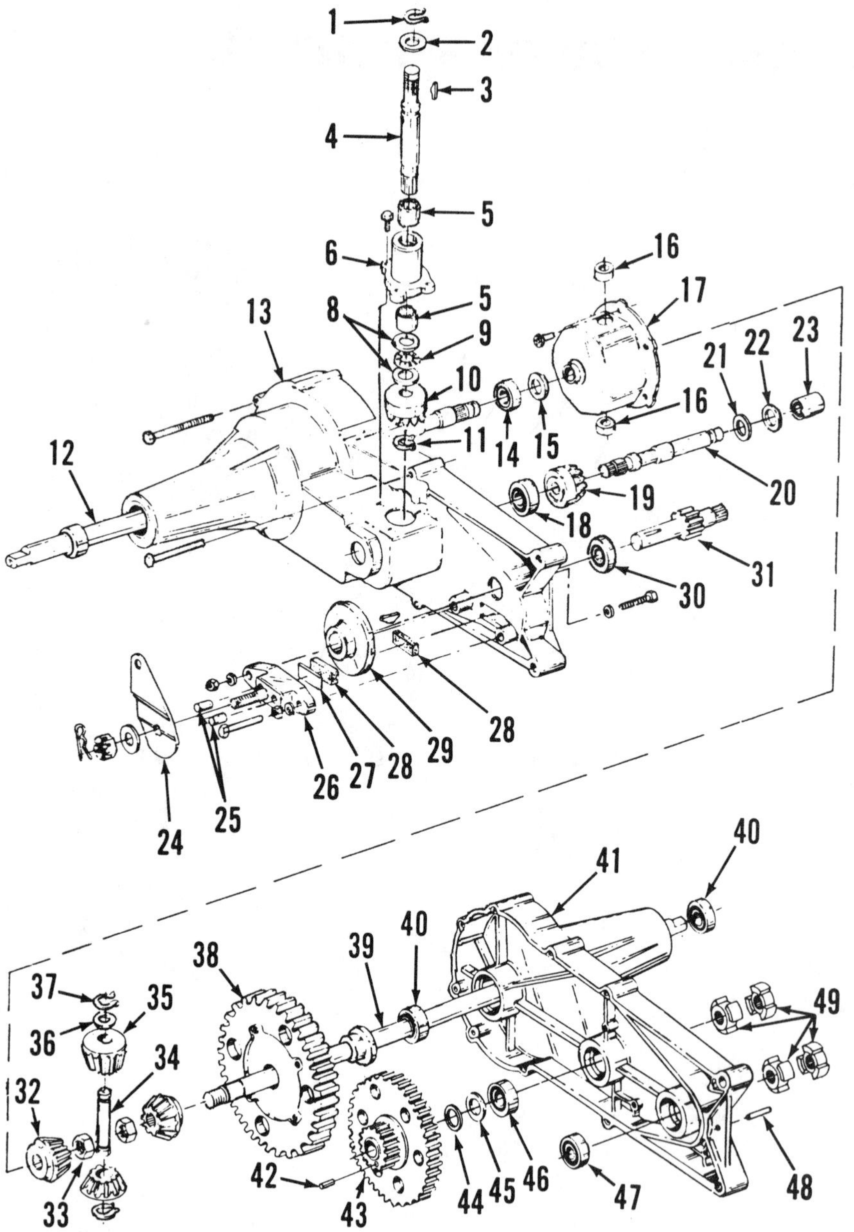

Fig. MTD23A—Exploded view of reduction gear and differential used on Series 700 Models with Sundstrand hydrostatic transmission.

1. Snap ring
2. Washer
3. Key
4. Input shaft
5. Bearings
6. Housing
8. Thrust washers
9. Thrust bearing
10. Bevel gear
11. Snap ring
12. Axle shaft
13. Case half
14. Bearing
15. Washer
16. Bushing
17. Differential housing
18. Bearing
19. Bevel gear
20. Drive shaft
21. Washer
22. Washer
23. Bearing
24. Brake arm
25. Actuating pins
26. Brake carrier
27. Backup plate
28. Brake pads
29. Brake disc
30. Bearing
31. Brake shaft
32. Axle gear
33. Nut
34. Shaft
35. Pinion gear
36. Thrust washer
37. Snap ring
38. Reduction gear
39. Axle shaft
40. Bearing
41. Case half
42. Dowel pin
43. Reduction gear
44. Washer
45. Washer
46. Bearing
47. Bearing
48. Dowel pin
49. Couplers

Series 700 With Sundstrand Hydrostatic Transmission

To remove reduction gear and differential unit, refer to hydrostatic transmission section and separate transmission from differential. Remove wheel and hub assemblies from axles. Remove park brake assembly (24 through 28—Fig. MTD23A) from brake shaft (31). Clean axle shafts (12 and 39) and remove any burrs from shafts. Remove input pulley and fan from input shaft (4). Remove screws securing input shaft housing (6) and remove input shaft assembly. Remove case screws while noting length and location of screws. Separate case halves; it may be necessary to tap against end of drive shaft (20).

Clean grease from internal components and remove reduction gears, differential and axle shafts, drive shaft, brake shaft and bearings from housing halves. Remove snap rings (1 and 11) and pull pinion gear (10), thrust bearing (8 and 9) and input shaft (4) from housing (6). Unbolt and remove differential housing (17) from reduction gear (38) and remove side gears (32), pinion gears (35) and pinion shaft (34).

When assembling unit, install bronze washer (22 and 45) on dowel pin (42) and face of reduction gear (43) so slot on washer faces outward. Fill unit with 16 ounces (473 mL) of Shell Darina #0 grease. Tighten screws securing case halves together to 80-120 in.-lbs. (9.0-13.6 N·m).

Model 995A

Remove hydrostatic transmission. Raise and support rear of tractor. Disconnect brake rod at brake lever and two-speed shift linkage at differential. Support differential and remove differential to frame retaining bolts. Lower differential and roll assembly away from tractor. Drain fluid from unit and remove brake assembly.

Remove axle housings (15—Fig. MTD24). Position unit with cover up, then remove cover (13). Lift out differential assembly and axles (40 through 54). Remove output shaft (36), gear (35) and thrust washers (37). Unscrew setscrew (2) and remove spring (3) and ball (4). Remove brake shaft (32), sliding gear (33) and shift fork (10) and rod (8). Remove input shaft and gear components (18 through 25).

To disassemble differential assembly, remove cap screws (53), differential carriers (41 and 52) and axles from ring gear assembly. Remove snap ring to separate axle gear, carrier and axle. Remove pinions (47) and separate body cores (45 and 48) from ring gear (46).

Inspect components for damage or excessive wear. To reassemble unit, reverse disassembly procedure. Check movement of shift rod when tightening setscrew (2). Install gears (22 and 25) so bevels of gears face together as shown in Fig. MTD25. Install carrier cap screws (53—Fig. MTD24) so head of cap screw is on side of shorter carrier (52). Do not rotate axle housings after housing has been pressed tight against seal (11) as seal may be cut.

TRANSAXLE

LUBRICATION

All Models So Equipped

Transaxle used on 500, 600 and 700 series tractors with variable speed transmission is lubricated with grease at factory and should not require additional lubrication. Capacity is 10 ounces (296 mL) of lithium base grease.

Transaxle used on 800 series tractors with variable speed transmission is lubricated with grease at factory and should not require additional lubrication. Capacity for single-speed transaxle is 32 ounces (947 mL) of Shell Darina grease. Capacity for two-speed transaxle is 16 ounces (473 mL) of Benalene 372-0 grease.

Five-speed transaxle used on 700 and 800 series tractors is lubricated with grease at factory and should not require additional lubrication. Capacity is 36 ounces (1.1 L) of Bentonite grease.

On 800 series tractors with four-speed transaxle, check fluid level at 100-hour intervals and maintain fluid level at lower edge of check plug opening. Recommended fluid is SAE 90 EP gear oil. Approximate oil capacity is 64 ounces (1.9 L).

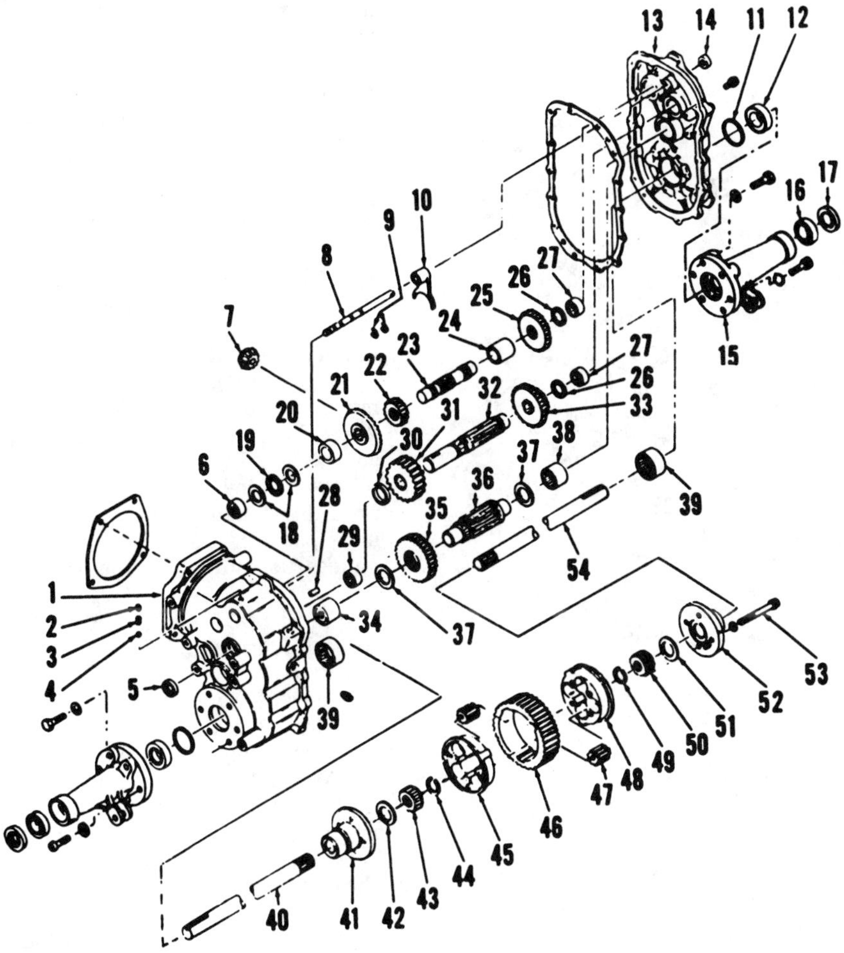

Fig. MTD24—Exploded view of reduction gear and differential unit used on Model 995A.

1. Case
2. Setscrew
3. Spring
4. Ball
5. Seal
6. Needle bearing
7. Transmission output gear
8. Shift rail
9. Snap rings
10. Shift fork
11. Quad ring
12. Tapered roller bearing
13. Cover
14. Seal
15. Axle housing
16. Ball bearing
17. Seal
18. Thrust washers
19. Thrust bearing
20. Spacer
21. Bevel gear
22. Gear (low range)
23. Shaft
24. Spacer
25. Gear (high range)
26. Thrust washer
27. Needle bearing
28. Dowel pin
29. Needle bearing
30. Spacer
31. Gear
32. Brake shaft
33. Sliding gear
34. Needle bearing
35. Gear
36. Output shaft
37. Thrust washer
38. Needle bearing
39. Needle bearing
40. Axle shaft (L.H.)
41. Differential carrier (L.H.)
42. Thrust washer
43. Axle gear
44. Snap ring
45. Body core
46. Ring gear
47. Pinion gears (8)
48. Body core
49. Snap ring
50. Axle gear
51. Thrust washer
52. Differential carrier (R.H.)
53. Cap screw
54. Axle shaft (R.H.)

REMOVE AND REINSTALL

500-600-700 Series

To remove transaxle assembly, first remove drive belt from transaxle pulley as previously outlined. Disconnect shift linkage and brake linkage at transaxle. Support rear of tractor and remove all fasteners securing transaxle to frame. Raise rear of tractor and roll transaxle assembly out from under tractor.

Reinstall by reversing removal procedure. Adjust brake as previously outlined.

800 Series With Four-Speed Transaxle

Remove gear selector lever knob and disconnect brake rod at brake lever. Remove belt from transaxle pulley as previously outlined. Support rear of tractor, remove transaxle support plate

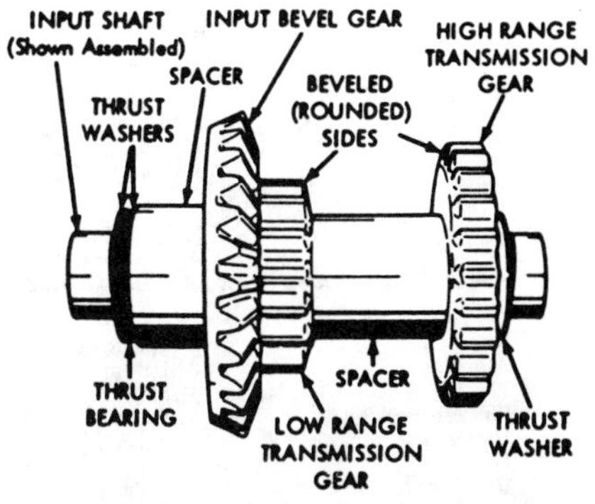

Fig. MTD25—View of reduction drive input shaft and gears. Note position of bevels on gears and location of thrust washers and thrust bearing.

Illustrations courtesy MTD.

at left side of transaxle and retaining bolts at each axle housing. Raise rear of tractor and roll transaxle out from under tractor.

Reinstall by reversing removal procedure. Adjust brake as previously outlined.

800 Series With Five-Speed Transaxle

To remove transaxle assembly, first remove drive belt from transaxle pulley as previously outlined. Disconnect shift linkage and brake linkage at transaxle. Support rear of tractor and remove all fasteners securing transaxle to frame. Raise rear of tractor and roll transaxle assembly out from under tractor.

Reinstall by reversing removal procedure. Adjust brake as previously outlined.

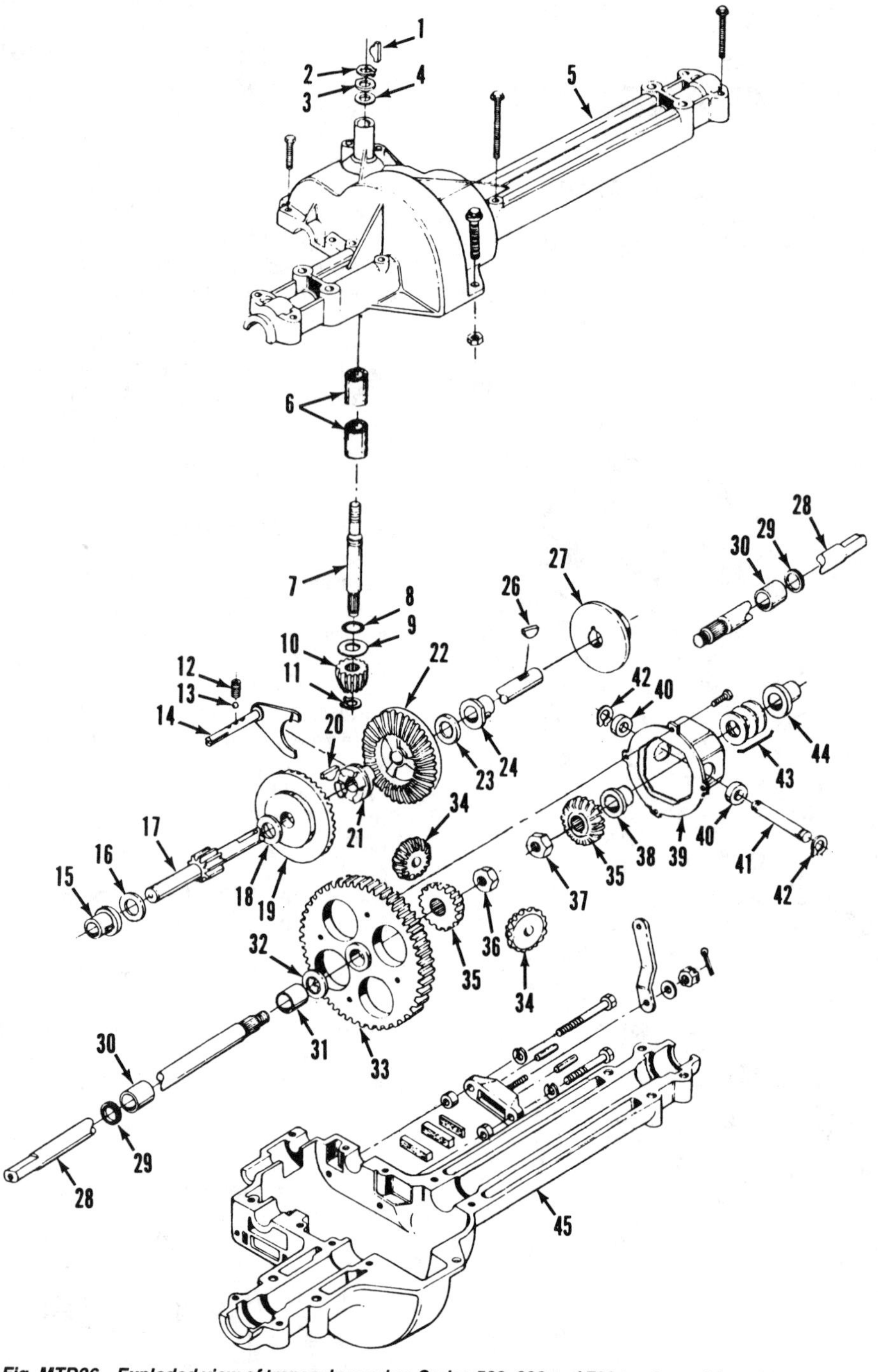

Fig. MTD26—Exploded view of transaxle used on Series 500, 600 and 700 tractors with variable speed transmission.

1. Key	12. Spring detent	23. Washer	35. Axle gear
2. Snap ring	13. Detent ball	24. Flange bearing	36. Nut
3. Washer (0.040 in.)	14. Shifter	26. Key	37. Nut
4. Washer (0.030 in.)	15. Flange bearing	27. Brake disc	38. Flange bearing
5. Upper case half	16. Washer	28. Axle shaft	39. Housing
6. Bearings	17. Drive/brake shaft	29. Seal	40. Thrust bearing
7. Input shaft	18. Washer	30. Sleeve bearing	41. Cross shaft
8. Square seal	19. Bevel gear	31. Sleeve bearing	42. Snap ring
9. Thrust washer	20. Key	32. Washer	43. Washers
10. Gear	21. Clutch collar	33. Differential gear	44. Flange bearing
11. Snap ring	22. Bevel gear	34. Spider gear	45. Lower case half

OVERHAUL

500-600-700 Series With Variable Speed Transmission

The transaxle used on these models provides forward and reverse motion and is coupled through a belt drive to the variable speed pulley.

Thoroughly clean exterior of transaxle case. Remove input pulley and key (1—Fig. MTD26). Remove brake caliper assembly, brake disc (27) and key (26). Remove all cap screws retaining upper case half (5) to lower case half (45). Carefully separate upper case half from lower case half. Remove snap rings (2 and 11), washers (3 and 4), gear (10), thrust washer (9) and square seal (8). Remove input shaft (7) and bearings (6) as needed. Remove shifter assembly (14). Lift drive /brake shaft (17) and gear assembly out as an assembly and disassemble as needed. Lift axle shafts (28) and differential assembly out as an assembly. Remove the cap screws retaining differential housing (39) to differential gear (33). Remove snap rings (42), cross shaft (41) and spider gears (34). Remove nuts (36 and 37) and remove axle gears (35). Pull axles out of housing and gear assembly.

Reassemble by reversing disassembly procedure. Before installing upper case half, pack transaxle housing with 10 ounces (296 mL) of lithium base grease.

700 And 800 Series With Five-Speed Transaxle

Place shift lever in neutral, then unbolt and remove shift lever. Remove brake assembly. Remove setscrew (2—Fig. MTD27), spring (3) and index ball (4). Unbolt cover (5) and push shift fork assembly (12) in while removing cover. Before removing gear and shaft assemblies, shifter fork (12) should be removed. Note position of parts before removal. Remove gear and shaft assemblies (Fig. MTD28) from case, taking care not to disturb drive chain (34—Fig. MTD27). Remove needle bearings (38 and 43), flat washer (41), square cut seals (37 and 42), output gear (40) and output pinion (39) from countershaft. Angle the two shafts together, then mark position of chain on sprocket collars and remove chain. Remove sprocket (35), bevel gear (32), spur gears (27, 28, 29, 30 and 31), thrust washer (9) and flange bushing (14). All gears are splined to the countershaft. Disassembly of shifter-brake shaft is apparent after inspection. Remove snap ring (11) and input bevel gear (10), then pull input shaft (7) through cover.

Disassemble differential by driving roll pin (52) out of pinion shaft (58). Remove pinion gears (60) by rotating gears in opposite directions. Remove snap rings (54), side gears (55) and thrust washers (53), then slide axles out. Note that axle shafts (48 and 56) are different lengths.

Clean and inspect all parts and renew any showing excessive wear or other damage. When installing new inner input shaft needle bearings, press lower bearing in to a depth of 0.135-0.150 inch (3.43-3.81 mm) below flush (Fig. MTD29) and upper bearing flush with housing. Pack all needle bearings with grease during assembly.

Install shift collar so thick segment is toward shoulder as shown in Fig. MTD30. Note that replacement shafts have four keyways and keys. When installing thrust washers (21—Fig. MTD27) and shifting gears on brake shaft of early units, the 45 degree chamfer on inside diameter of thrust washers must face shoulder on brake shaft. On later units and replacement parts, washers are curved and rounded side must face shaft shoulder. See Fig. MTD31. Flat side of gears must face shoulder on shaft. Install drive chain (34—Fig. MTD27) on sprockets, aligning match marks made prior to disassembly. Install gears and shaft assemblies as a unit in case. Note that "V" protrusion on bushings must index in "V" notch in case.

Complete reassembly and pack case with approximately 36 ounces (1065 mL) of multipurpose, lithium base grease. Tighten case to cover cap screws to 90-100 in.-lbs. (10-11 N·m).

800 Series With Four-Speed Transaxle

Remove transaxle assembly as previously outlined. Remove drain plug and drain lubricant. Remove rear wheel and hub assemblies. Remove brake caliper and brake disc. Place shift lever (1—Fig. MTD32) in neutral position, then unbolt and remove shift lever assembly. Remove axle housings (14 and 64) and remove seal retainers (11) with oil seals (12) and "O" rings (13) by pulling each axle shaft out of case and cover as far as possible. Place transaxle unit on the edge of a bench with left axle shaft pointing downward. Remove cap screws securing case (16) to cover (66) and drive aligning dowel pins out of case. Lift case (16) up 1 to 2 inches (40 to 50 mm), tilt case about 45 degrees, rotate case clockwise and remove it from the assembly. Input shaft (32) and input gear (33) will be removed with case. Withdraw differential and axle shaft assembly and lay aside for later disassembly. Remove the three-cluster gear (44) with its thrust washer (46) and spacer (42). Lift out reverse idler gear (25), spacer (24) and shaft (23). Hold upper ends of shifter rods together, then lift out shifter rods (20), forks (22), shifter stop (21), sliding gears (30 and 31) and shaft (28) as an assembly. Remove low reduction gear (57), reduction shaft (56) and thrust washer (55), then remove the two-cluster gear (40) from brake shaft. Lift out the output gear (50), shaft (51)

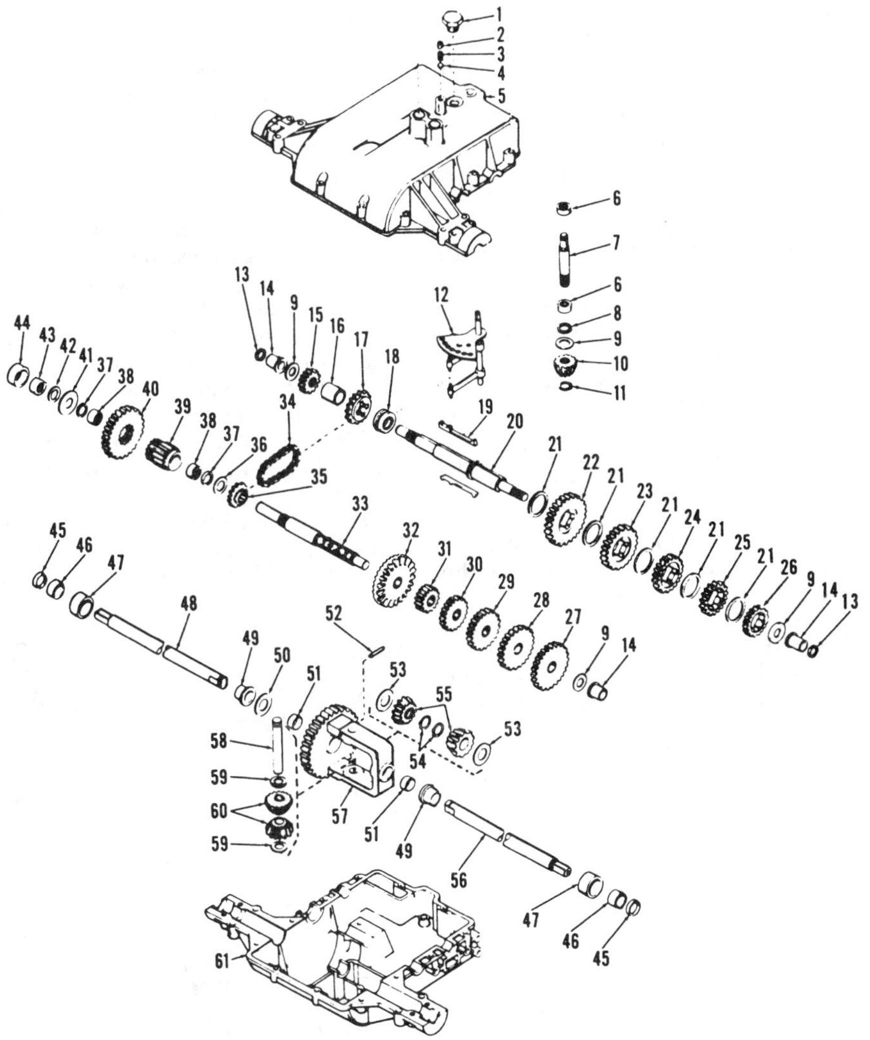

Fig. MTD27—Exploded view of five-speed transaxle used on Series 700 and 800 tractors so equipped.

1. Plug
2. Setscrew
3. Spring
4. Ball
5. Cover
6. Needle bearing
7. Input shaft
8. Square cut ring
9. Thrust washer
10. Input pinion
11. Snap ring
12. Shift fork assy.
13. Square cut ring
14. Bushing
15. Spur gear (15 T)
16. Spacer
17. Sprocket (18 T)
18. Shift collar
19. Key
20. Brake shaft
21. Thrust washer
22. Spur gear (37 T)
23. Spur gear (30 T)
24. Spur gear (25 T)
25. Spur gear (22 T)
26. Spur gear (20 T)
27. Spur gear (30 T)
28. Spur gear (25 T)
29. Spur gear (25 T)
30. Spur gear (20 T)
31. Spur gear (12 T)
32. Bevel gear (42 T)
33. Countershaft
34. Roller chain
35. Sprocket (9 T)
36. Flat washer
37. Square cut seal
38. Needle bearing
39. Output pinion
40. Output gear
41. Flat washer
42. Square cut seal
43. Needle bearing
44. Spacer
45. Oil seal
46. Needle bearing
47. Spacer
48. Axle shaft
49. Bushing
50. Washer
51. Bushing
52. Pin
53. Thrust washer
54. Snap rings
55. Pinion gears
56. Axle shaft
57. Differential gear
58. Pinion shaft
59. Thrust washer
60. Bevel pinion
61. Case

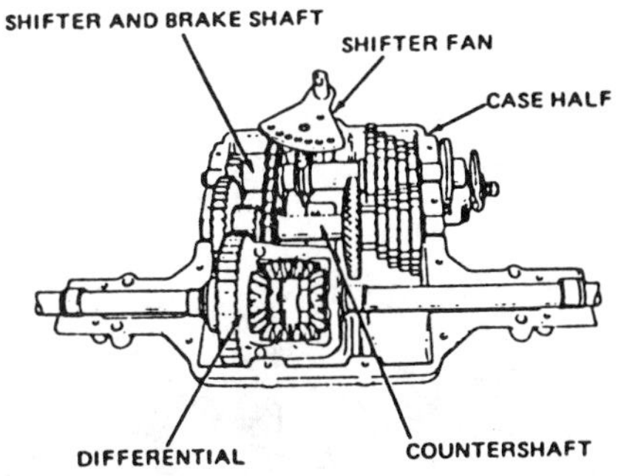

Fig. MTD28—When disassembling transaxle, remove gear and shaft assemblies from case by lifting both shafts at once. Be careful not to disturb chain.

Illustrations courtesy MTD.

and thrust washers (49 and 52). To remove brake shaft (39) and gear (38) from cover (66), block up under gear (38) and press shaft out of gear.

CAUTION: Do not allow cover or low reduction gear bearing boss to support any part of pressure required to press brake shaft from gear.

Remove input shaft (32) with input gear (33) and thrust washer (34) from case (16).

To disassemble differential, remove four cap screws and separate axle shaft and carriage assemblies from ring gear (79). Drive blocks (78), pinion gears (77) and pinion shaft (76) can now be removed from ring gear. Remove snap rings (59) and withdraw axle shafts (63 and 67) from axle gears (61) and carriages (62 and 72).

Clean and inspect all parts and renew any showing excessive wear or other damage. When installing new needle bearings, press bearing (29) into spline shaft (28) to a depth of 0.010 inch (0.25 mm) below end of shaft and low reduction shaft bearings (54 and 58) 0.010 inch (0.25 mm) below thrust surfaces of bearing bosses. Carrier bearings (10) should be pressed in from inside of case and cover until bearings are 0.290 inch (7.37 mm) below face of axle housing mounting surface. All other needle bearings are to be pressed in from inside of case and cover to a depth of 0.015-0.020 inch (0.38-0.51 mm) below thrust surfaces.

Renew all seals and gaskets. Reassemble by reversing disassembly procedures. When installing brake shaft (39) and idler gear (38), beveled edge of gear teeth must be up away from cover. Install reverse idler shaft (23), spacer (24) and reverse idler gear (25) with rounded end of gear teeth facing spacer. Install input gear (33) and shaft (32) so chamfered side of input gear is facing case (16).

Tighten transaxle cap screws to the following torque:

Differential cap screws . . . . . 7 ft.-lbs. (9 N·m)
Case to cover cap screws . . . 10 ft.-lbs. (14 N·m)
Axle housing cap screws . . . 13 ft.-lbs. (18 N·m)
Shift lever housing cap screws . . . . 10 ft.-lbs. (14 N·m)

Reinstall transaxle and fill with recommended fluid.

800 Series With Two-Speed Transaxle

This two-speed transaxle provides forward and reverse motion and is coupled through a belt drive to the variable speed pulley.

Refer to Fig. MTD33 for an exploded view of transaxle. Disassembly should be performed with unit inverted (input shaft pointing down). Remove plate (59) and extract detent springs (58). Unbolt and remove lower housing half (55). Locate and remove four detent balls (60). Lift out shifter forks (56 and 63). During disassembly, be sure to identify shims and washers so they can be returned to their original location.

Roll drive shaft (45) assembly rearward and remove shim washers, bearings (37) and shift collar (38) from ends of drive shaft. Rotate drive shaft down into housing and rotate output shaft (34) assembly up. Remove shim washers and bearings (27) from each end of shaft. Remove disc brake assembly from output shaft. Remove drive gear (29), spacer (30) and shim washer (31), then pull output shaft out of sprocket (32). Lift the drive shaft assembly from housing. Remove retaining rings (39 and 52) and withdraw gears (40 and 53), clutch collar (50), bevel gear (42) and sprocket (44) from drive shaft. Lift out differential and axles, remove cap screws holding differential case (17) to differential gear (25) and separate differential components (16 through 24). Remove snap rings (2 and 12) and pull pinion gear (11), thrust bearing (9 and 10) and input shaft (8) from housing.

Inspect components for excessive wear and damage. Pins (49) must move freely in clutch collar (50). Renew pins if broken or excessively rounded.

When assembling unit, note the following: Outer axle bearings (14) must be installed with sealed end toward outer end of axle. Hub side of sprocket (32) must be toward bevel gear (42). Drive shaft (45) end play should be 0.004-0.007 inch (0.10-0.18 mm). Install shims as needed to obtain desired end play.

Prior to fitting case half, fill with 16 ounces (473 mL) of Benalene 372-0 grease.

800 Series Single-Speed Transaxle

This single-speed transaxle provides forward and reverse motion and is coupled through a belt drive to the variable speed pulley.

Refer to Fig. MTD34 for an exploded view of transaxle. To disassemble, unbolt and remove input shaft (33) and housing (4) assembly. Remove snap rings (2 and 38) and withdraw bevel gear (37), thrust washer (35 and 36) and shaft from housing. Unbolt and remove axle carrier (9). Remove cap screws retaining housing cover (10) and separate cover from housing. Withdraw axle shafts (39 and 51) and differential assembly from housing. Unbolt and separate differential case (49) from differential gear (42). Remove nuts (45) and side gears (43) from axles. Remove brake disc holder (13), then remove drive shaft (16)

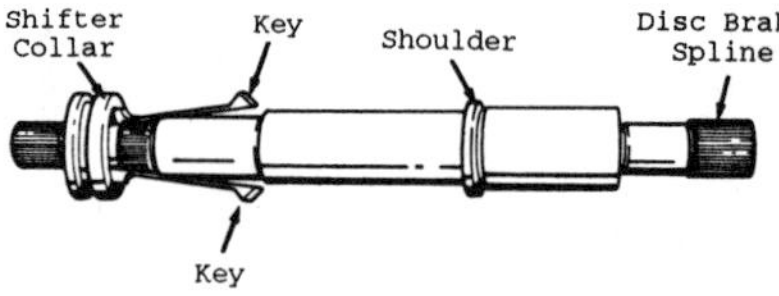

Fig. MTD30—Install shift collar so thick segment is nearer shaft shoulder.

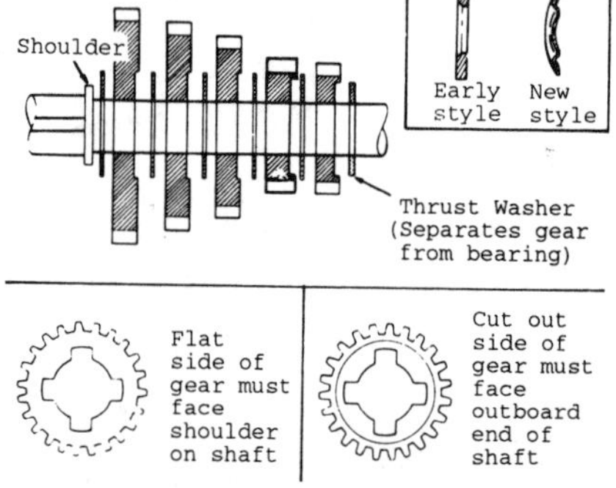

Fig. MTD31—When assembling shift shaft, install thrust washers with 45 degree chamfer toward shoulder on output shaft. Install curved washers so rounded side is toward shaft shoulder. Flat side of gears also must face shoulder on output shaft.

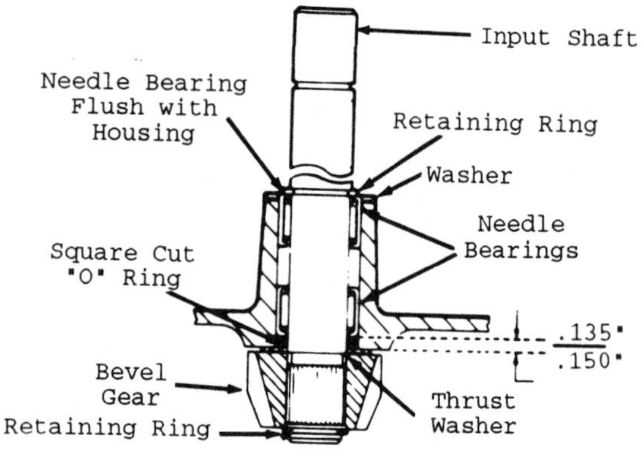

Fig. MTD29—When renewing transaxle input shaft bearings, lower bearing should be installed 0.135-0.150 in. (3.43-3.81 mm) below flush with housing. Upper bearing should be flush with housing.

Illustrations courtesy MTD.

and bevel gear (22) assembly. Unscrew set screw (27) in case to release detent spring and ball (25) before removing shift fork (24).

Inspect components for excessive wear and damage. Fill with 32 ounces (947 mL) of Shell Darina grease.

POWER TAKE-OFF

Model 995A is equipped with a front pto shaft that is belt driven off the engine crankshaft pulley. The clutch is actuated by a pivoting idler system.

All other models use a pivoting idler system under the tractor.

MAINTENANCE AND ADJUSTMENT

All Models

Lubricate pto engagement lever at 25-hour intervals. Use multipurpose, lithium base grease.

On Model 995A, adjust pto cable at either end until idler depresses safety switch plunger within 1/8-in. (3.2 mm) of bottoming out in switch with pto in "OFF" position. Refer to Fig. MTD35.

On all other models, pto adjustment procedures vary according to type of equipment installed.

HYDRAULIC SYSTEM

MAINTENANCE AND ADJUSTMENT

Model 995A

Pressurized oil from the hydrostatic drive unit is utilized for the hydraulic system. Refer to HYDROSTATIC TRANSMISSION section for maintenance information.

To check and adjust hydraulic system pressure, install a 0-1000 psi (0-7000 kPa) test gage in line with the lift cylinder (6—Fig. MTD36). Start and run tractor engine at 3600 rpm. Work control valve lever (3) in direction that provides a reading on test gage. Hold control lever in this direction to fully extend or retract lift cylinder. When cylinder reaches end of stroke, gage should read approximately 700 psi (4827 kPa). Pressure may be adjusted by turning adjustment screw (4) on control valve in or out as necessary.

CAUTION: Do not exceed 700 psi (4827 kPa) pressure as damage to hydrostatic drive unit may occur.

Hydraulic system control valve (5) and cylinder (6) are serviced as complete assemblies only.

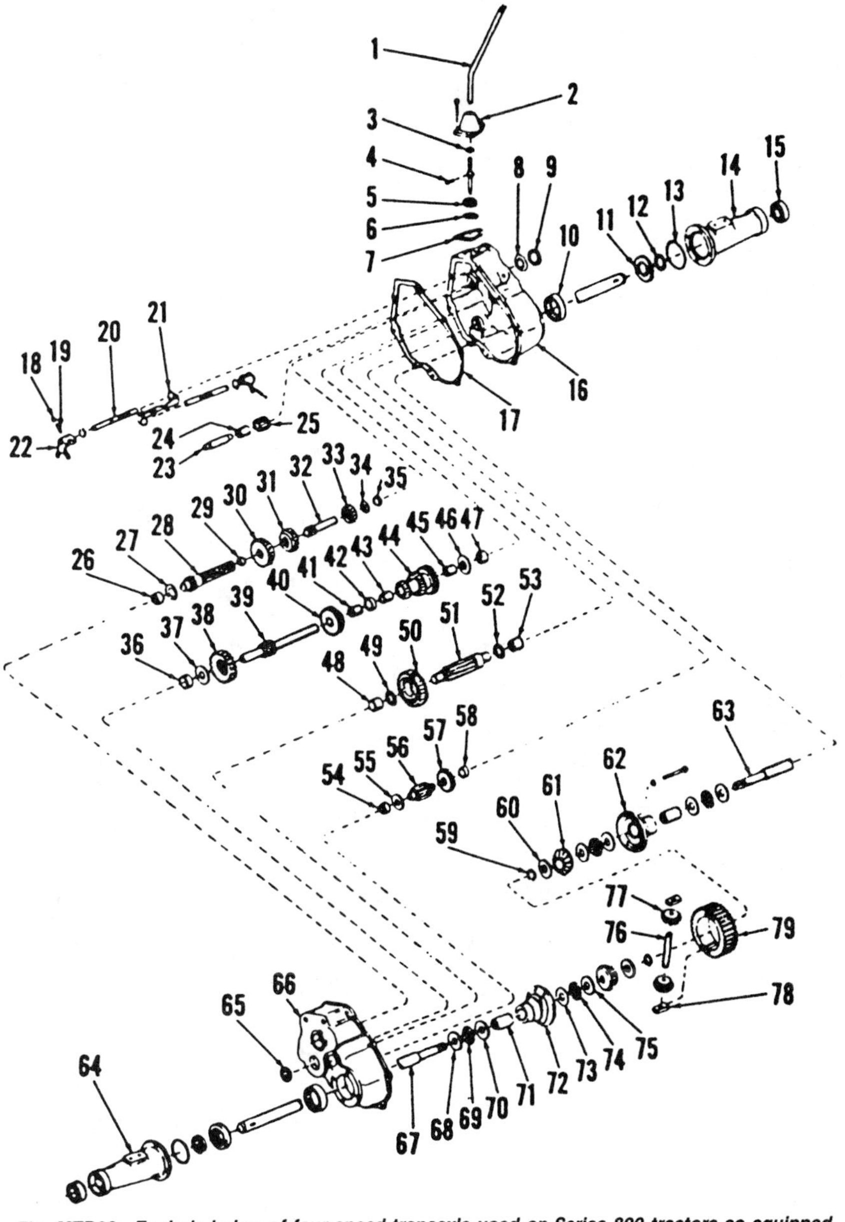

Fig. MTD32—Exploded view of four-speed transaxle used on Series 800 tractors so equipped.

1. Shift lever
2. Shift lever housing
3. Seal ring
4. Roll pin
5. Retainer
6. Snap ring
7. Gasket
8. Ball bearing
9. Oil seal
10. Carrier bearing
11. Seal retainer
12. Oil seal
13. "O" ring
14. Axle housing
15. Axle outer bearing
16. Transaxle case
17. Gasket
18. Detent ball
19. Spring
20. Shifter rod
21. Shifter stop
22. Shifter fork
23. Reverse idler shaft
24. Spacer
25. Reverse idler gear
26. Needle bearing
27. Thrust washer
28. Shifter shaft
29. Needle bearing
30. First, second & reverse gear
31. Third & fourth gear
32. Input shaft
33. Input gear
34. Thrust washer
35. Needle bearing
36. Needle bearing
37. Thrust washer
38. Idler gear
39. Brake & cluster shaft
40. Two-gear cluster
41. Bushing
42. Spacer
43. Bushing
44. Three-gear cluster
45. Bushing
46. Thrust washer
47. Needle bearing
48. Needle bearing
49. Thrust washer
50. Output gear
51. Output shaft
52. Thrust washer
53. Needle bearing
54. Needle bearing
55. Thrust washer
56. Low reduction shaft
57. Low reduction gear
58. Needle bearing
59. Snap ring
60. Thrust washer
61. Axle gear
62. Axle carriage (plain holes)
63. Axle shaft
64. Axle housing
65. Oil seal
66. Transaxle cover
67. Axle shaft
68. Thrust washer
69. Thrust bearing
70. Thrust washer
71. Bushing
72. Axle carriage (tapped holes)
73. Thrust washer
74. Thrust bearing
75. Thrust washer
76. Pinion shaft
77. Pinion gear
78. Drive block
79. Ring gear

Illustrations courtesy MTD.

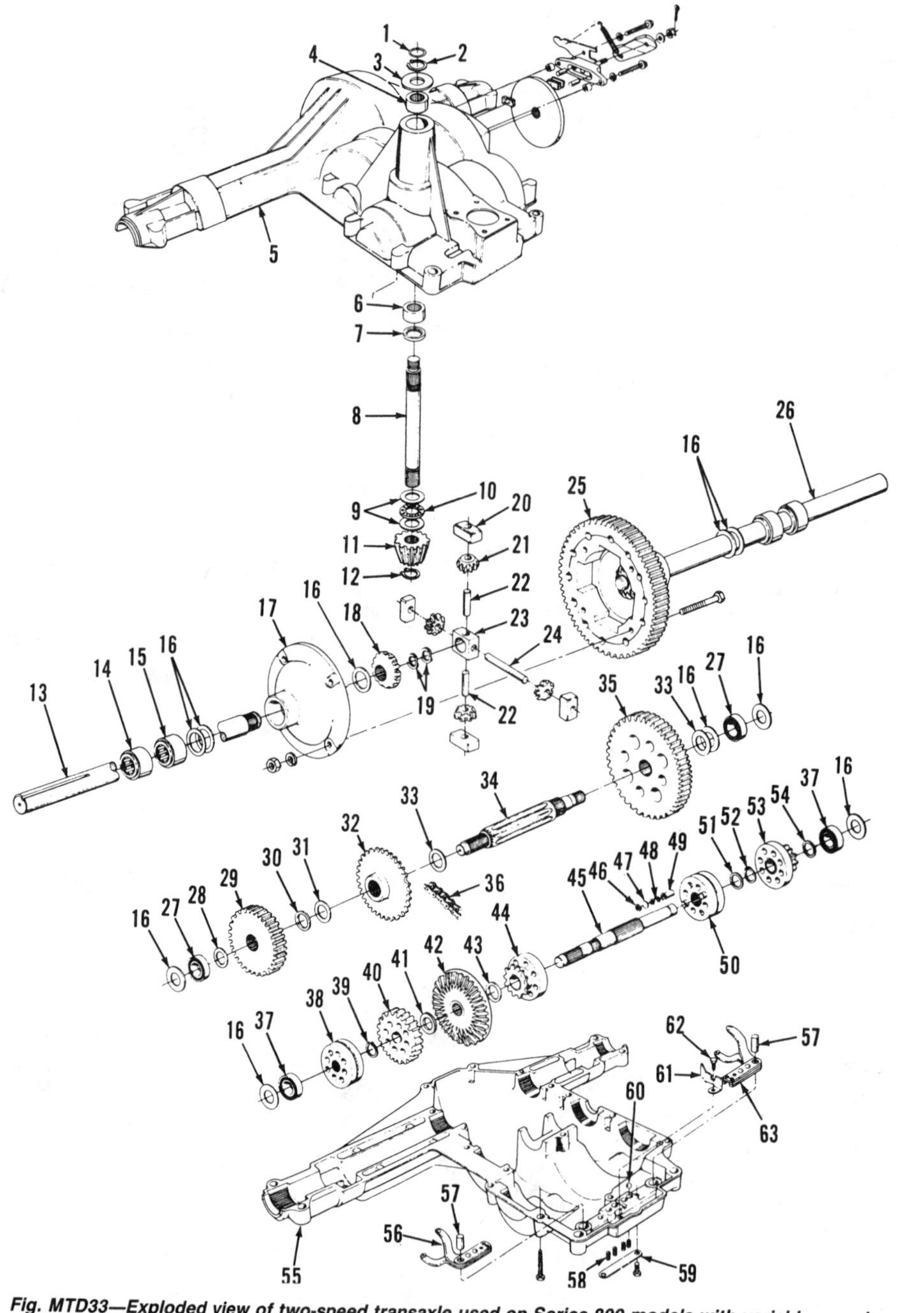

Fig. MTD33—Exploded view of two-speed transaxle used on Series 800 models with variable speed drive.

1. Wire ring
2. Snap ring
3. Thrust washer
4. Bearing
5. Upper case half
6. Bearing
7. Oil seal
8. Input shaft
9. Thrust washer
10. Thrust bearing
11. Gear
12. Snap ring
13. Axle shaft
14. Bearing
15. Bearing
16. Washer
17. Differential cover
18. Side gear
19. Retainers
20. Thrust block
21. Pinion gear
22. Shafts
23. Pinion block
24. Shaft
25. Ring gear
26. Axle shaft
27. Bearing
28. Washer
29. Gear (31 T)
30. Washer
31. Spacer
32. Sprocket
33. Washer
34. Output shaft
35. Gear (45 T)
36. Chain
37. Bearing
38. Forward shift collar
39. Retainer
40. Gear
41. Washer
42. Bevel gear
43. Washer
44. Sprocket
45. Drive shaft
46. Retainer
47. Plug
48. Spring
49. Pin
50. Creeper shift collar
51. Washer
52. Retainer
53. Plate assy.
54. Shim
55. Lower case half
56. Shift arm (right)
57. Pin
58. Detent spring
59. Plate
60. Detent ball
61. Shift guide
62. Screw
63. Shift arm (left)

1. Wire ring
2. Snap ring
3. Washer
4. Input shaft housing
5. Case
6. Dowel pin
7. Oil seal
8. Bushing
9. Axle housing
10. Cover
11. Plate
12. Brake pads
13. Brake holder
14. Bearing
15. Brake disc
16. Drive shaft
17. Snap ring
18. Washer
19. Spring retaining plate
20. Spring
21. Pin
22. Bevel gear
23. Shift collar
24. Shift fork
25. Detent spring & ball
27. Set screw
28. Bevel gear
29. Washer
30. Washer
31. Bushing
32. Bearing
33. Input shaft
34. Seal
35. Thrust washer
36. Thrust bearing
37. Gear
38. Snap ring
39. Axle shaft
40. Bushing
41. Washer
42. Gear
43. Side gear
44. Pinion
45. Nut
46. Bushing
47. Snap ring
48. Spacer
49. Differential case
50. Shaft
51. Axle

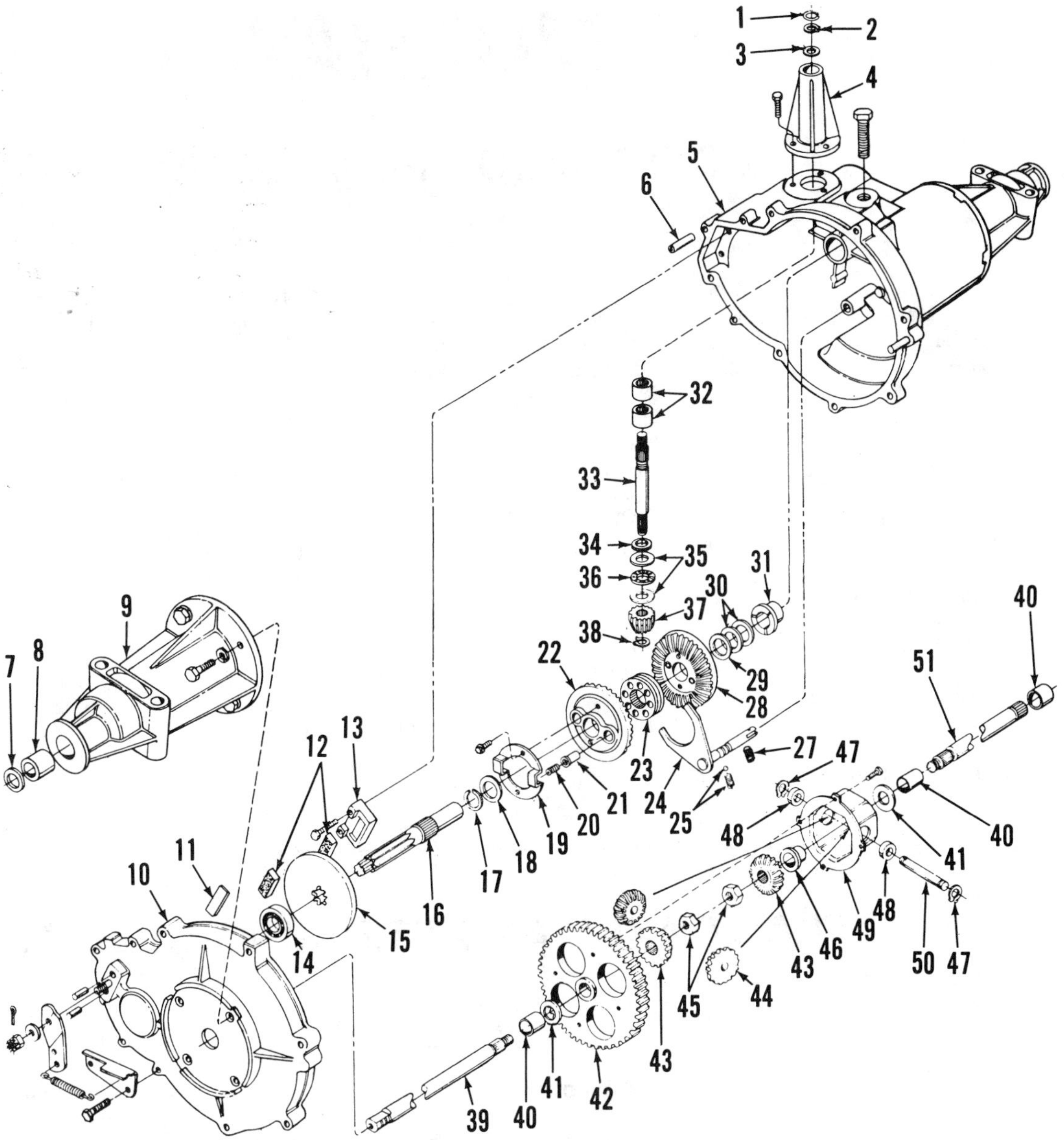

Fig. MTD34—Exploded view of single-speed transaxle used on Series 800 models with variable speed drive.

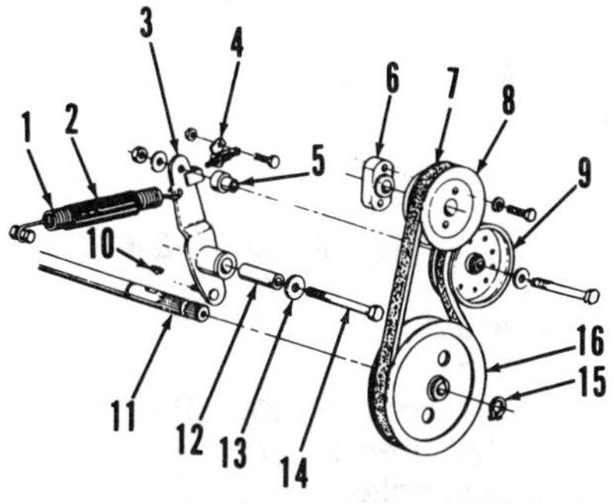

Fig. MTD35—Exploded view of pto system used on Model 995A.

1. Spring
2. Cover
3. Idler bracket
4. Safety switch
5. Idler adapter
6. Spacer
7. Belt
8. Engine pulley
9. Idler pulley
10. Key
11. Pto shaft
12. Spacer
13. Washer
14. Bolt
15. Snap ring
16. Pin drive pulley

Fig. MTD36—View showing locations of the various components of the hydraulic lift system used on Model 995A.

1. Hydrostatic drive unit
2. Filter
3. Control valve lever
4. Pressure adjustment screw
5. Control valve
6. Cylinder

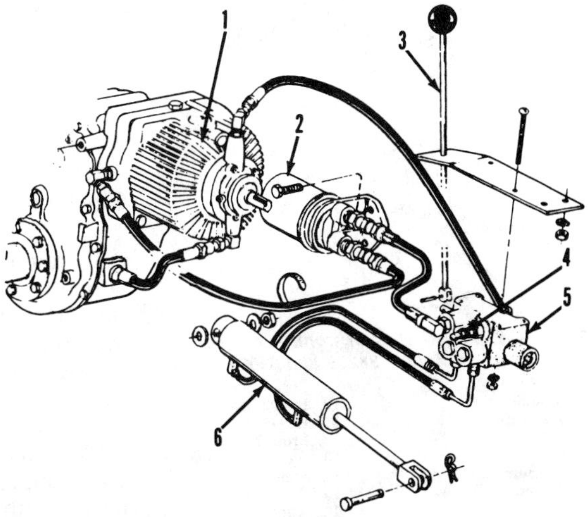

Illustrations courtesy MTD.

MURRAY

CONDENSED SPECIFICATIONS

	MODELS				
	38206	39004	46100A	46300A	46300B
Engine Make	B&S	B&S	B&S	B&S	B&S
Model	402707	422707	422707	422707	422707
Bore	3.44 in. (87.3 mm)	3.44 in. (87.3 mm)	3.44 in. (87.3 mm)	3.44 in. (87.3 mm)	3.44 in. (87.3 mm)
Stroke	2.16 in. (54.8 mm)	2.28 in. (57.9 mm)	2.28 in. (57.9 mm)	2.28 in. (57.9 mm)	2.28 in. (57.9 mm)
Displacement	40 cu.in. (656 cc)	42.3 cu.in. (694 cc)	42.3 cu.in. (694 cc)	42.3 cu.in. (694 cc)	42.3 cu.in. (694 cc)
Power Rating	16 hp (11.9 kW)	18 hp (13.4 kW)	18 hp (13.4 kW)	18 hp (13.4 kW)	18 hp (13.4 kW)
Slow Idle	1400 rpm	1400 rpm	1400 rpm	1400 rpm	1400 rpm
High Speed (No-Load)	3600 rpm	3600 rpm	3600 rpm	3600 rpm	3600 rpm
Capacities—					
Crankcase	3 pts.* (1.4 L)	3 pts.* (1.4 L)	3 pts.* (1.4 L)	3 pts.* (1.4 L)	3 pts.* (1.4 L)
Transaxle	36 oz. (1065 mL)	64 oz. (1892 mL)	36 oz. (1065 mL)	36 oz. (1065 mL)	36 oz. (1065 mL)

*Early model crankcase capacity is 3.5 pints (1.7 L).

FRONT AXLE AND STEERING SYSTEM

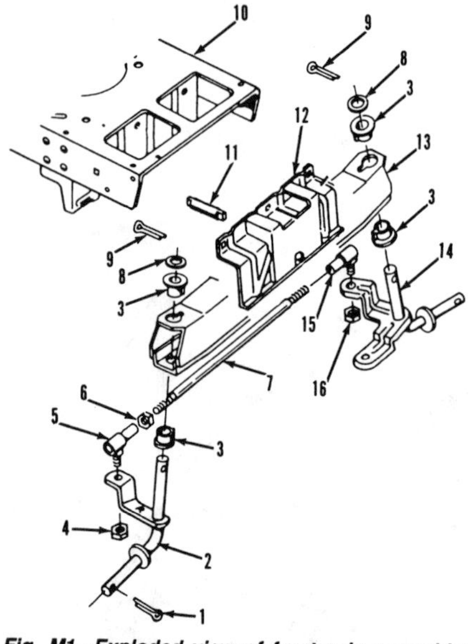

Fig. M1—Exploded view of front axle assembly used on Models 38206 and 39004.

1. Cotter pin
2. Spindle
3. Bushings
4. Nut
5. Tie rod end
6. Jam nut
7. Tie rod
8. Washers
9. Cotter pin
10. Frame
11. Pivot pin
12. Front hanger
13. Axle main member
14. Spindle
15. Tie rod end
16. Nut

MAINTENANCE

All Models

Steering system should be lubricated at 8-hour intervals. Lubricate spindles and axle pivot points with good quality motor oil. If tractor is operated in sandy areas, a dry graphite spray is recommended. Periodically apply multipurpose grease to steering gear. Do not lubricate nylon wheel bushings.

R&R AXLE MAIN MEMBER

Models 38206-39004

To remove axle assembly, raise and support front of tractor. Disconnect drag link at left steering spindle (14—Fig. M1). Support axle main member (13) and remove front hanger (12). Remove pivot pin (11), lower axle assembly and roll away from tractor.

Check axle and pivot pin for excessive wear and renew as necessary. Reinstall by reversing removal procedure.

Models 46100-46300A-46300B

To remove axle assembly, raise and support front of tractor. Disconnect drag link from steering arm (13—Fig. M2). Support axle main member (16) and remove front hanger (17). Remove pivot pin (18), lower axle assembly and roll away from tractor.

Check axle main member and pivot pin for excessive wear and renew as necessary. Reinstall by reversing the removal procedure.

TIE ROD AND TOE-IN

All Models

Remove tie rod (7—Fig. M1 or 20—Fig. M2) by disconnecting tie rod ends from steering spindles.

On Models 38206 and 39004, adjust front wheel toe-in by loosening left and right jam nuts (6—Fig. M1) and turning tie rod as required to obtain 1/8-in. (3.2 mm) toe-in. Tighten jam nuts.

Toe-in is not adjustable on 46100A, 46300A and 46300B models.

Illustrations courtesy Murray Ohio Mfg. Co.

R&R STEERING SPINDLES

Models 38206-39004

Raise and support front of tractor. Remove front tires and wheels. Detach tie rod (7—Fig. M1) from steering spindles (2 and 14). Disconnect drag link from left steering spindle. Remove cotter pin (9) and lower spindle from axle main member.

Renew bushings (3) and/or spindles (2 and 14) as required. Reinstall by reversing the removal procedure.

Models 46100A-46300A-46300B

Raise and support front of tractor. Remove front tires and wheels. Disconnect drag link (11—Fig. M2) from steering arm (13). Disconnect tie rod (20) from steering spindles (19 and 21). Remove "E" ring (12) from left spindle and remove cotter pin (23) from right spindle and lower spindles from axle main member.

Renew bushings (15) and/or spindles (19 and 21) as required. Reinstall by reversing the removal procedure.

R&R STEERING GEAR

Models 38206-39004

Remove screws securing steering column bearing to control panel, remove bolt (4—Fig. M3) and pull steering

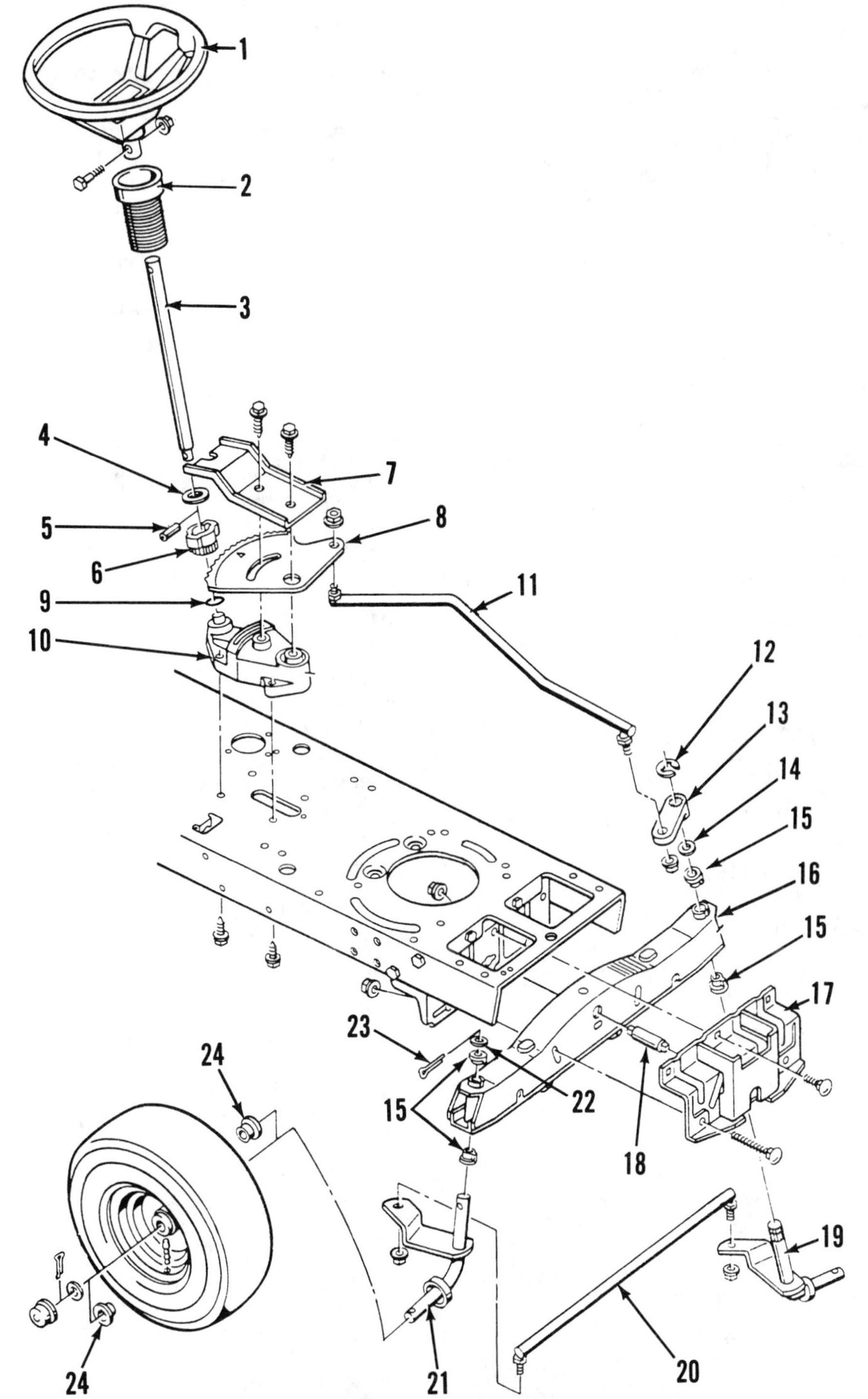

Fig. M2—Exploded view of front axle assembly and steering gear used on Models 46100A, 46300A and 46300B.

1. Steering wheel
2. Bellows
3. Steering shaft
4. Washer
5. Pin
6. Pinion gear
7. Retainer
8. Sector gear
9. "O" ring
10. Base
11. Drag link
12. "E" ring
13. Steering arm
14. Washer
15. Bushing
16. Axle main member
17. Front hanger
18. Pivot pin
19. Spindle
20. Tie rod
21. Spindle
22. Washer
23. Cotter pin
24. Wheel bearings

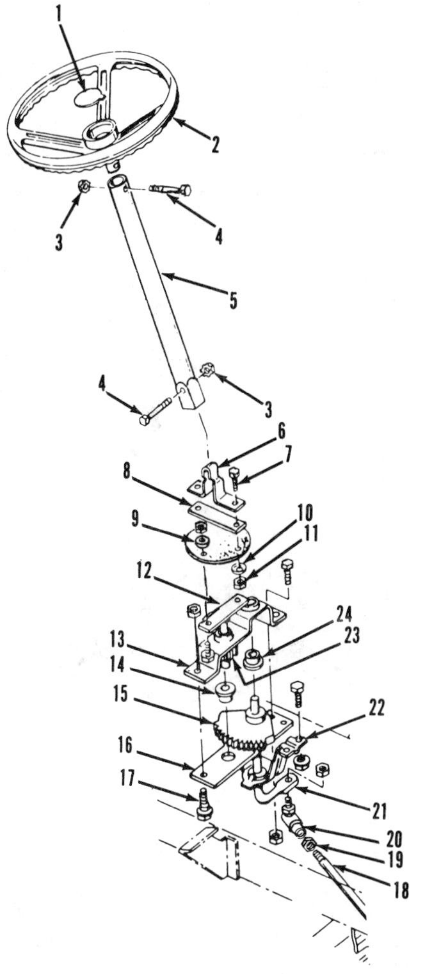

Fig. M3—Exploded view of steering column and gear assembly used on Models 38206 and 39004.

1. Cap
2. Steering wheel
3. Nut
4. Bolt
5. Steering column
6. Upper coupling
7. Bolt
8. Plate
9. Flex disc
10. Washer
11. Nut
12. Shaft
13. Bracket
14. Bushing
15. Sector gear
16. Plate
17. Bolt
18. Drag link
19. Jam nut
20. Drag link end
21. Steering lever
22. Support
23. Pinion gear
24. Bushing

Illustrations courtesy Murray Ohio Mfg. Co.

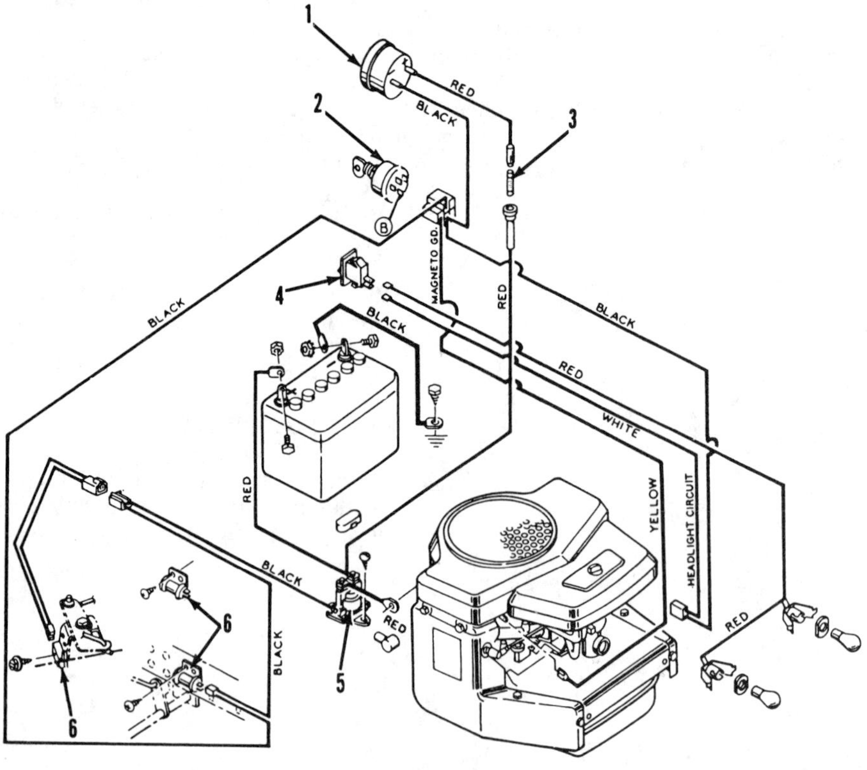

Fig. M5—Typical wiring diagram for Models 38206 and 39004.

1. Ammeter
2. Ignition switch
3. Fuse (15 amp)
4. Light switch
5. Solenoid
6. Safety switches

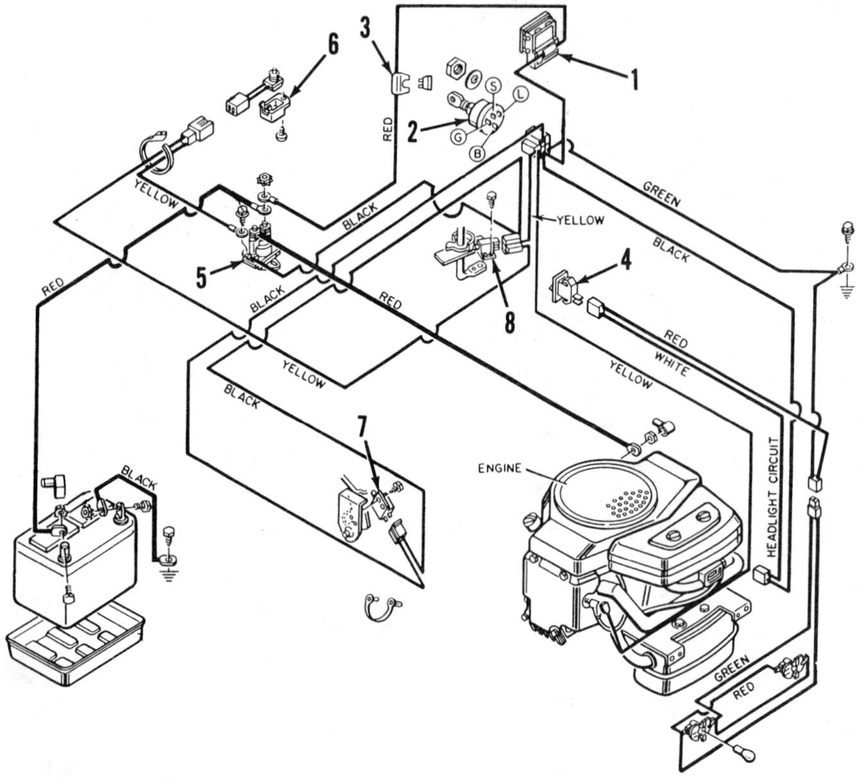

Fig. M6—Wiring Diagram for Models 46100A, 46300A and 46300B.

1. Ammeter
2. Ignition switch
3. Fuse (15 amp)
4. Light switch
5. Solenoid
6. Seat switch
7. Clutch switch
8. Pto switch

wheel and column assembly from control panel. Remove coupling (6), plate (8), flex disc (9) and bracket (13). Remove snap ring and pull shaft (12) out of bracket (13) and gear (23). Disconnect drag link end (20) from steering lever (21). Remove steering lever (21) and gear (15).

Reassemble by reversing disassembly procedure.

Models 46100A-46300A-46300B

Remove steering wheel (1—Fig. M2) and bellows (2). Detach steering column bearing on control panel. Detach side access panel, then drive pin (5) out of pinion gear (6). Withdraw steering shaft (3). Detach drag link (11) from sector gear (8). Remove screws securing base (10) and remove base assembly.

When assembling steering components, pinion gear (6) and sector gear (8) must be at mid-travel when meshed together.

ENGINE

MAINTENANCE

All Models

Regular engine maintenance is required to maintain peak performance and long engine life.

Check oil level after every 5 hours of operation under normal operating conditions.

Refer to engine section for specific maintenance recommendations.

REMOVE AND REINSTALL

All Models

Remove hood assembly. Disconnect negative battery cable. Detach all interfering wires and cables. Disconnect fuel line. If so equipped, remove heat shields. Disengage drive belts from engine pulleys and remove pulleys. Remove exhaust pipe. Unscrew engine mounting bolts and lift engine out of tractor.

Reinstall by reversing removal procedure.

OVERHAUL

All Models

Engine make and model are listed at the beginning of this section. For tuneup specifications, engine overhaul procedures and engine maintenance, refer to appropriate engine section in this manual.

Illustrations courtesy Murray Ohio Mfg. Co.

ELECTRICAL SYSTEM

MAINTENANCE AND SERVICE

All Models

Battery electrolyte level should be checked at 50-hour intervals of normal operation. If necessary, add distilled water until level is just below base of vent well. DO NOT overfill. Keep battery posts clean and cable ends tight.

Refer to appropriate engine service section for alternator or starter service.

Refer to Fig. M5 or M6 for wiring diagram.

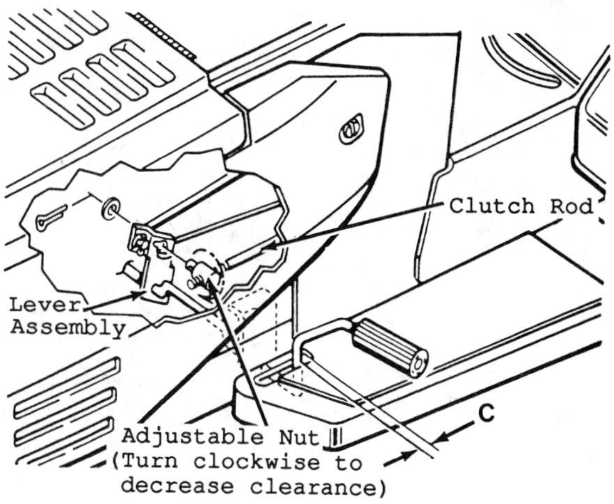

Fig. M7—View showing clearance (C) between clutch/brake pedal rod and end of slot in step plate. Refer to text.

BRAKES

ADJUSTMENT

Model 38206

Disconnect clutch rod (Fig. M7) from lever assembly, then turn adjustable nut to obtain $^3/_4$ inch (19 mm) clearance (C) between clutch/brake pedal rod and end of slot in step plate (Fig. M7) when clutch rod is connected. Push clutch/brake pedal completely forward and engage parking brake lock. Place shift lever in neutral position. Tighten hex nut (Fig. M8) until rear wheels are locked. Release brake and make certain wheels turn freely. Remove cotter pin and washer, then turn adjustable nut until center of brake rod is aligned with "V" notch as shown in Fig. M9 when rod is connected.

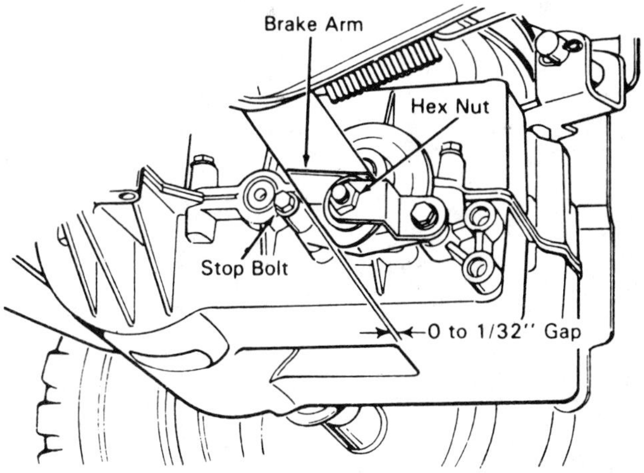

Fig. M8—View showing location of brake adjustment on Model 38206. Clearance between brake arm and stop bolt must be 0 to $^1/_{32}$ in. (0 to 0.80 mm).

Model 39004

Disconnect clutch rod (Fig. M7) from lever assembly, then turn adjustable nut to obtain $^3/_4$ inch (19 mm) clearance (C) between clutch/brake pedal rod and end of slot in step plate (Fig. M7) when clutch rod is connected. Push clutch/brake pedal completely forward and engage parking brake lock. Place shift lever in neutral position. Push link plate forward (Fig. M10). If adjustable nut is not aligned with "V" notch as shown in Fig. M10, remove cotter pin and washer from adjustable nut and turn adjustable nut until it is aligned with "V" notch. Reinstall washer and cotter pin. Make certain that wheels will not turn when brake is applied and that wheels will turn freely when brake is released.

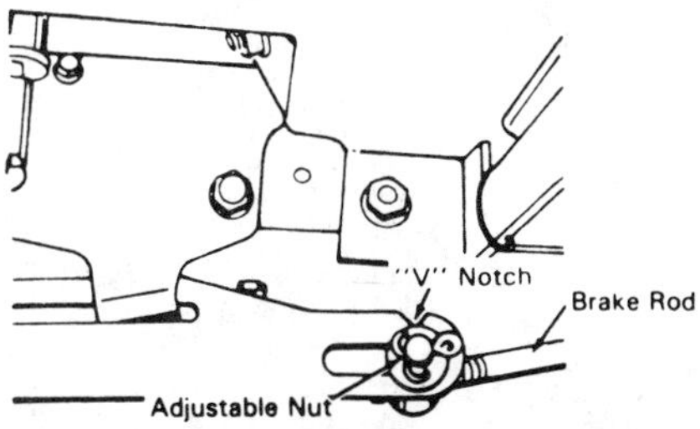

Fig. M9—View showing brake rod adjustment locations on Model 38206. Refer to text.

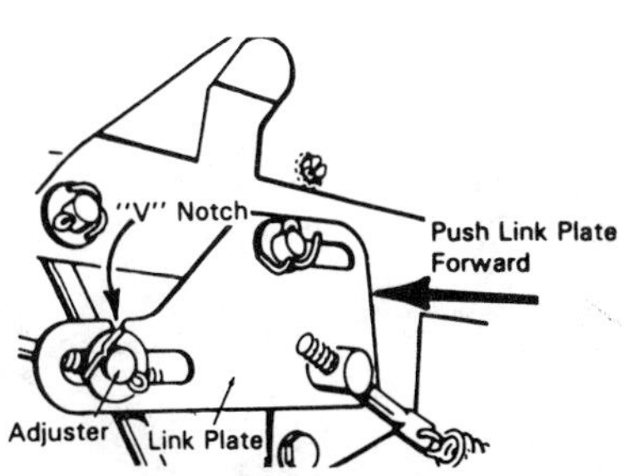

Fig. M10—View showing brake rod adjustment location on Model 39004. Refer to text.

Illustrations courtesy Murray Ohio Mfg. Co.

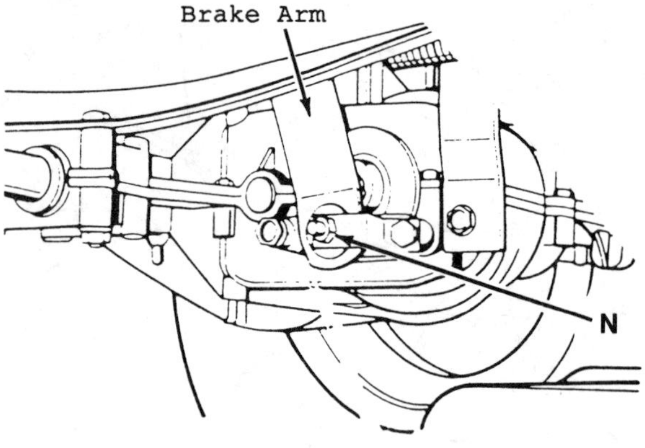

Fig. M11—Adjust brake on Models 46100A, 46300A and 46300B by turning nut (N) as outlined in text.

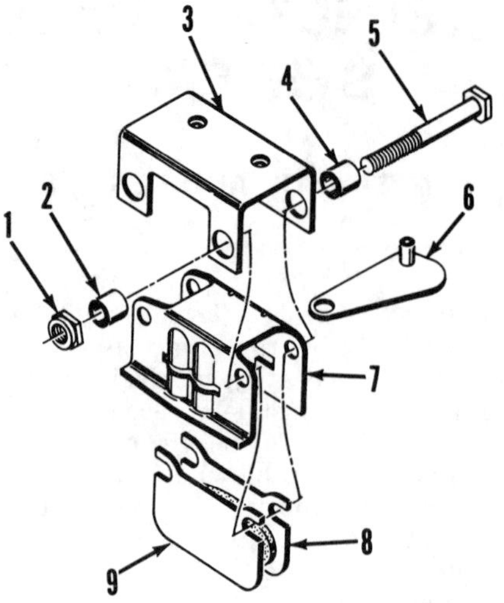

Fig. M12—Exploded view of brake assembly used on Model 39004.

1. Locknut
2. Spacer
3. Bracket
4. Spacer
5. Bolt
6. Lever
7. Bracket
8. Brake pad
9. Brake pad

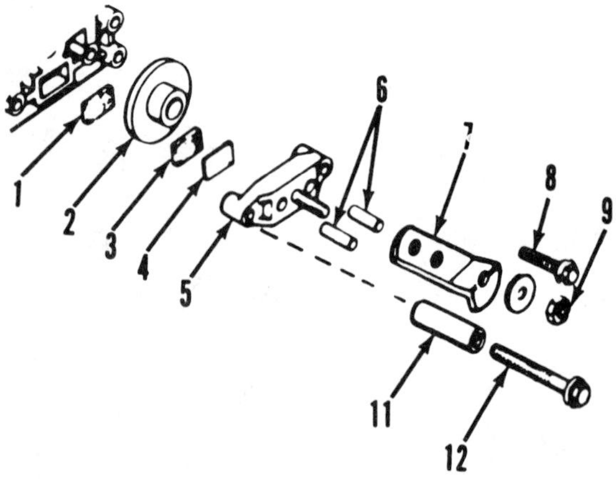

Fig. M13—Exploded view of brake assembly used on Model 38206.

1. Inner brake pad
2. Brake disc
3. Outer brake pad
4. Metal backing plate
5. Brake caliper
6. Pins
7. Brake lever
8. Bolt
9. Adjusting nut
11. Spacer
12. Bolt

All Other Models

To adjust brake, first check clearance (C—Fig. M7) between clutch/brake pedal rod and end of slot in step plate. If necessary, turn clutch rod adjustable nut to obtain clearance of $^{3}/_{16}$-$^{1}/_{4}$ in. (5-6 mm). Push clutch/brake pedal fully forward and set parking brake. Position gearshift lever in neutral. Turn nut (N—Fig. M11) so rear wheels will not turn when tractor is pushed. With parking brake off, rear wheels should turn when tractor is pushed. Turn nut (N) as needed.

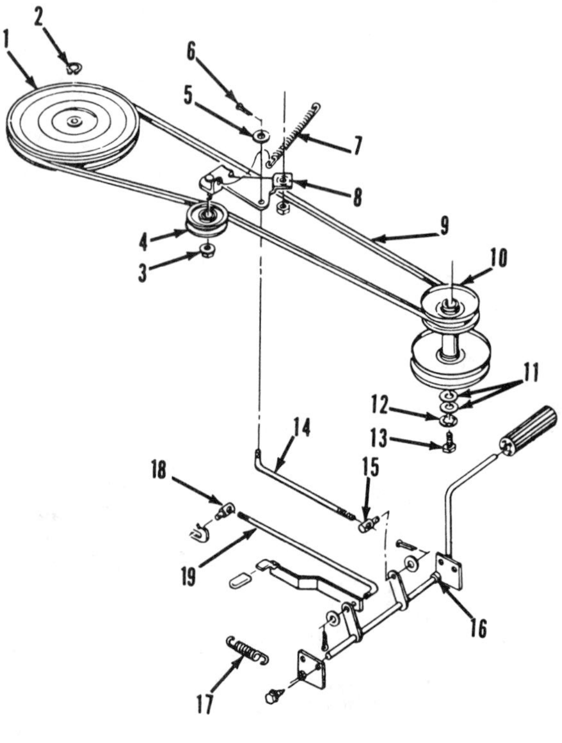

Fig. M14—View of drive belt and spring-loaded, pivoting idler arrangement that provides clutch on Model 38206. Refer to text.

1. Pulley
2. Snap ring
3. Nut
4. Idler pulley
5. Washer
6. Cotter pin
7. Spring
8. Pivot plate
9. Drive belt
10. Drive pulley
11. Flat washers
12. Lockwasher
13. Bolt
14. Clutch rod
15. Adjustable nut
16. Pedal assy.
17. Spring
18. Adjustable nut
19. Brake rod

R&R BRAKE PADS

Model 39004

Disconnect brake rod at lever (6—Fig. M12). Remove brake assembly from transaxle bracket. Remove bolts (5) and spacers (2). Separate all parts.

Reinstall by reversing removal procedure. Adjust brakes as previously outlined.

All Other Models

Remove adjustment nut (9—Fig. M13) and brake lever (7). Remove bolts (12 and 8) and spacer (11). Remove brake caliper (5), outer brake pad (3), metal backing plate (4), brake disc (2) and inner brake pad (1).

Make certain the two push pins (6) and metal backing plate (4) remain in place in brake pad holder and metal spacer (11) is in place on rear mounting bolt during reassembly.

CLUTCH

MAINTENANCE

All Models

Clutch/brake pedal is connected to a pivoting idler pulley which is spring loaded to maintain correct belt tension. Adjust clutch/brake pedal as outlined in ADJUSTMENT paragraph in BRAKE section.

R&R DRIVE BELT

All Models Except 39004

Remove mower deck. Set parking brake and remove idler pulley (4—Fig. M14 and M15). Loosen belt guides around drive pulley (10). Remove belt from transmission drive pulley and engine pulley, then push belt through shift lever hole and over shift lever as shown in Fig. M16. Pull belt from under tractor.

Reinstall by reversing removal procedure. Adjust clearance between belt guides and belt to $1/16$ to $1/8$ inch (1.6 to 3.2 mm). Adjust clutch/brake pedal as outlined in ADJUSTMENT paragraph in BRAKE section.

Model 39004

Remove clutch spring (3—Fig. M17) from frame. Disconnect adjustable nut and clutch rod (Fig. M9). Remove snap ring (8—Fig. M17) and pulley (7). Remove idler pulleys (12 and 15) and belt (11).

Reinstall belt in top groove of drive pulley (6). Route belt as shown in Fig. M17. Make certain belt is above front running board support and pedal lever. Put belt over rear running board support, place belt around pulley (7) and install pulley (7) and snap ring (8). Install idler pulley (15), then idler pulley (12). Connect spring (3) to frame and reassemble adjustable nut and pedal lever. Adjust clutch/brake pedal as outlined in ADJUSTMENT paragraph in BRAKE section.

TRANSAXLE

MAINTENANCE

Model 39004

Check fluid level after every 100 hours of operation and maintain fluid

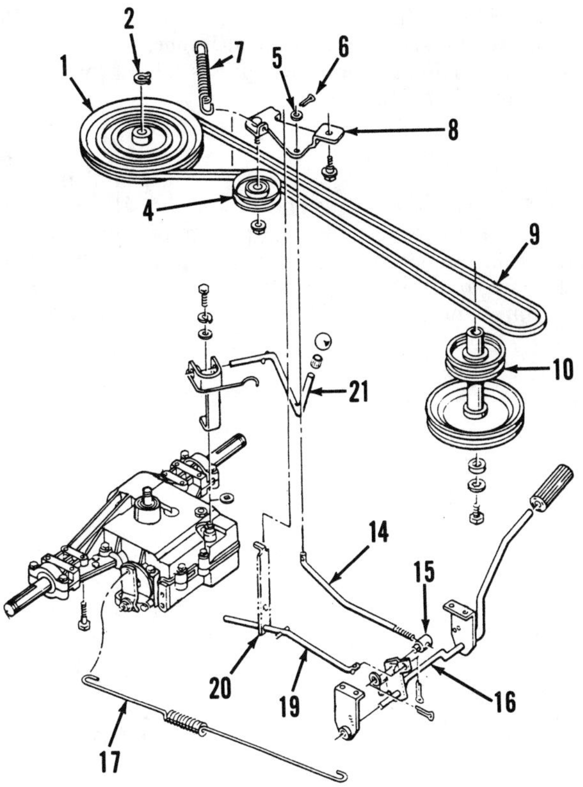

Fig. M15—View of spring-loaded idler type drive clutch system used on Models 46100A, 46300A and 46300B.

1. Pulley
2. Snap ring
4. Idler pulley
5. Washer
6. Cotter pin
7. Spring
8. Pivot plate
9. Drive belt
10. Drive pulley
14. Clutch rod
15. Adjustable nut
16. Pedal assy.
17. Spring
19. Brake rod
20. Park brake lever
21. Shift lever

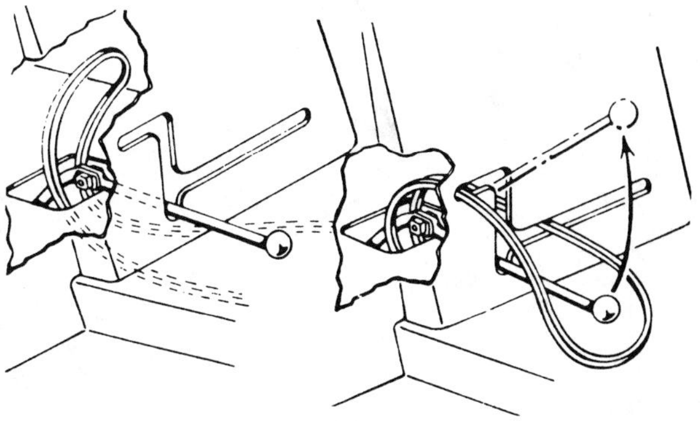

Fig. M16—Drive belt on Models 38206, 46100A, 46300A and 46300B must be pushed through gear shift lever opening and slipped over gear shift lever to remove belt. Refer to text.

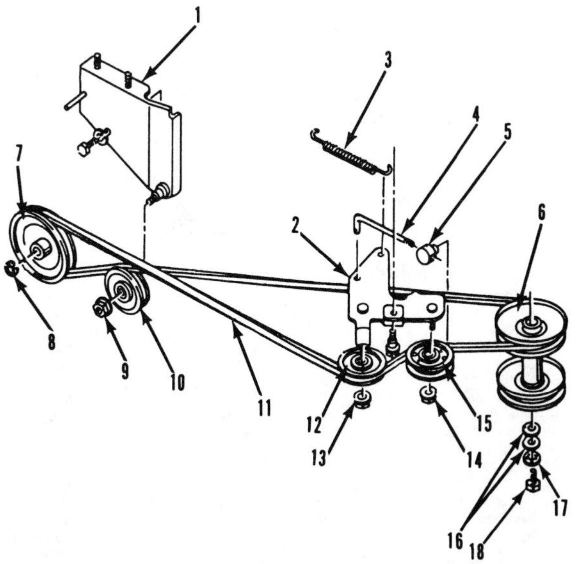

Fig. M17—View of drive belt and spring-loaded pivoting idler arrangement that provides clutch on Model 39004. Refer to text.

1. Bracket
2. Pivot plate
3. Spring
4. Clutch rod
5. Adjustable nut
6. Drive pulley
7. Pulley
8. Snap ring
9. Nut
10. Fixed idler pulley
11. Drive belt
12. Idler pulley
13. Nut
14. Nut
15. Idler pulley
16. Flat washers
17. Lockwashers
18. Bolt

level at lower edge of check plug opening. Recommend fluid is SAE 90 EP gear oil. Approximate fluid capacity is 64 ounces (1892 mL).

All Other Models

The five-speed transaxle is packed with 36 ounces (1065 mL) of multipurpose, lithium base grease during manufacture and does not require periodic maintenance.

R&R TRANSAXLE

Model 39004

Remove mower deck. Raise and support rear of tractor. Disconnect brake rod and remove drive belt. Remove the three bolts retaining shift lever and remove lever assembly. Support transaxle, then remove bolts retaining transaxle and brake support plate on left side of transaxle. Remove retaining bolts at rear axle housings, then lower and remove transaxle assembly.

Reinstall by reversing removal procedure.

All Other Models

Remove mower deck. Raise and support rear of tractor. Disconnect brake rod and gear selector lever. Remove drive belt. Support transaxle and unscrew transaxle retaining bolts. Lower transaxle and roll away from tractor.

Reinstall by reversing removal procedure.

OVERHAUL TRANSAXLE

All Models Except Model 39004

Remove transaxle assembly as previously outlined. Place shift lever in neutral, then unbolt and remove shift lever. Remove brake assembly. Remove setscrew (2—Fig. M18), spring (3) and index ball (4). Unbolt cover (5) and push shift fork assembly (12) in while removing cover. Before removing gear and shaft assemblies, shifter fork (12) should be removed. Note position of parts before removal. Remove gear and shaft assemblies (Fig. M19) by lifting both shafts at once from case taking care not to disturb drive chain (34—Fig. M18). Remove needle bearings (38 and 43), flat washer (41), square cut seals (37 and 42), output gear (40) and output pinion (39) from countershaft. Angle the two shafts together, then mark position of chain on sprocket collars and remove chain. Remove sprocket (35), bevel gear (32), spur gears (27, 28, 29, 30 and 31), thrust washer (9) and flange bushing (14). All gears are splined to the countershaft. Disassembly of shifter-brake shaft is apparent after inspection. Remove snap ring (11) and input bevel gear (10), then pull input shaft (7) through cover.

Disassemble differential by driving out roll pin (52) securing pinion shaft (58). Remove pinion gears (60) by rotating gears in opposite directions. Remove

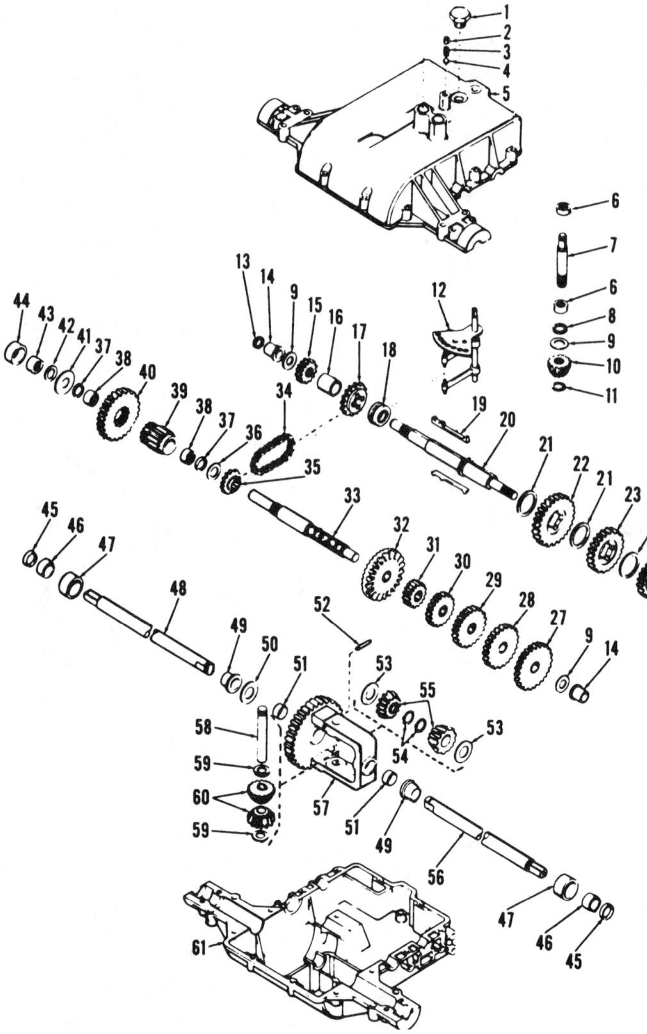

Fig. M18—Exploded view of transaxle used on all models except 39004.

1. Plug
2. Setscrew
3. Spring
4. Ball
5. Cover
6. Needle bearing
7. Input shaft
8. Square cut ring
9. Thrust washer
10. Input pinion
11. Snap ring
12. Shift fork assy.
13. Square cut ring
14. Bushing
15. Spur gear (15 T)
16. Spacer
17. Sprocket (18 T)
18. Shift collar
19. Key
20. Brake shaft
21. Thrust washer
22. Spur gear (37 T)
23. Spur gear (30 T)
24. Spur gear (25 T)
25. Spur gear (22 T)
26. Spur gear (20 T)
27. Spur gear (30 T)
28. Spur gear (25 T)
29. Spur gear (25 T)
30. Spur gear (20 T)
31. Spur gear (12 T)
32. Bevel gear (42 T)
33. Countershaft
34. Roller chain
35. Sprocket (9 T)
36. Flat washer
37. Square cut seal
38. Needle bearing
39. Output pinion
40. Output gear
41. Flat washer
42. Square cut seal
43. Needle bearing
44. Spacer
45. Oil seal
46. Needle bearing
47. Spacer
48. Axle shaft
49. Bushing
50. Washer
51. Bushing
52. Pin
53. Thrust washer
54. Snap rings
55. Bevel gear
56. Axle shaft
57. Differential gear & case assy.
58. Pinion shaft
59. Thrust washer
60. Bevel pinion
61. Case

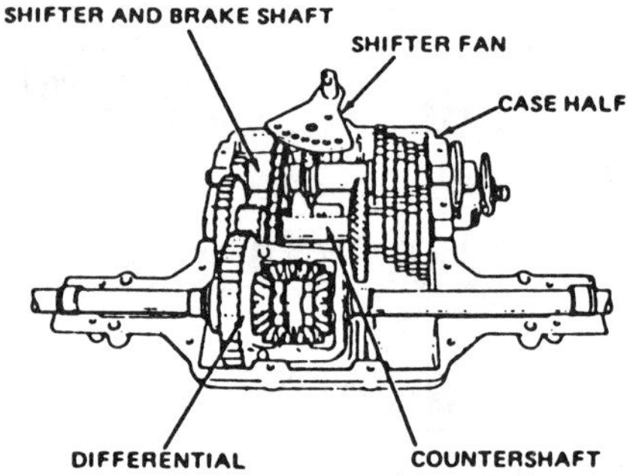

Fig. M19—When disassembling transaxle, remove gear and shaft assemblies from case by lifting both shafts at once. Be careful not to disturb chain.

Illustrations courtesy Murray Ohio Mfg. Co.

snap rings (54) and slide axles out of side gears (55) and thrust washers (53). Note that axle shafts (48 and 56) are different lengths.

Clean and inspect all parts and renew any showing excessive wear or other damage. When installing new input shaft needle bearings, press lower bearing in to a depth of 0.135-0.150 inch (3.43-3.81 mm) below flush (Fig. M20) and upper bearing flush with housing. Pack all needle bearings with grease during assembly.

Install shift collar (18—Fig. M18) so thick segment is toward shoulder as shown in Fig. M21. Note that replacement shafts have four keyways and keys. When installing thrust washers (21—Fig. M18) and shifting gears on brake shaft of early units, the 45 degree chamfer on inside diameter of thrust washers must face shoulder on brake shaft. On later units and replacement parts, washers (21) are curved and rounded side must face shaft shoulder. See Fig. M22. Flat side of gears must face shoulder on shaft. Install drive chain (34—Fig. M18) on sprockets, aligning match marks made prior to disassembly. Install gears and shaft assemblies as a unit in case. Note that "V" protrusion on flanged bushings must index in "V" notch in case.

Complete reassembly and pack case with approximately 36 ounces (1065 mL) of multipurpose, lithium base grease. Tighten case to cover cap screws to 90-100 in.-lbs. (10-11 N·m).

Model 39004

Remove transaxle assembly as previously outlined. Remove drain plug and drain lubricant. Remove rear wheel and hub assemblies. Remove brake caliper and brake disc. Place shift lever (1—Fig. M23) in neutral position, then unbolt and remove shift lever assembly. Remove axle housings (14 and 64) and remove seal retainers (11) with oil seals (12) and "O" rings (13) by pulling each axle shaft out of case and cover as far as possible. Place transaxle unit on the edge of a bench with left axle shaft pointing downward. Remove cap screws securing case (16) to cover (66) and drive aligning dowel pins out of case. Lift case (16) up 1 to 2 inches (40 to 50 mm), tilt case about 45 degrees, rotate case clockwise and remove it from the assembly. Input shaft (32) and input gear (33) will be removed with case. Withdraw differential and axle shaft assembly and lay aside for later disassembly. Remove the three-gear cluster (44) with its thrust washer (46) and spacer (42). Lift out reverse idler gear (25), spacer (24) and shaft (23). Hold upper ends of shifter rods together, then lift out shifter rods (20), forks (22), shifter stop (21), sliding gears (30 and 31) and shaft (28) as an assembly. Remove low reduction gear (57), reduction shaft (56) and thrust washer (55), then remove the two-gear cluster (40) from brake shaft. Lift out the output gear (50), shaft (51) and thrust washers (49 and 52). To remove brake shaft (39) and gear (38) from cover (66), block up under gear (38) and press shaft out of gear.

CAUTION: Do not allow cover or low reduction gear bearing boss to support any part of pressure required to press brake shaft from gear.

Remove input shaft (32) with input gear (33) and thrust washer (34) from case (16).

To disassemble differential, remove four cap screws and separate axle shaft and carriage assemblies from ring gear (79). Drive blocks (78), bevel pinion gears (77) and drive pin (76) can now be removed from ring gear. Remove snap rings (59) and withdraw axle shafts (63 and 67) from axle gears (61) and carriages (62 and 72).

Clean and inspect all parts and renew any showing excessive wear or other damage. When installing new needle bearings, press bearing (29) in spline shaft (28) to a depth of 0.010 inch (0.25 mm) below end of shaft and low reduction shaft bearings (54 and 58) 0.010 inch (0.25 mm) below thrust surfaces of bearing bosses. Carrier bearings (10) should be pressed in from inside of case and cover until bearings are 0.290 inch (7.37 mm) below face of axle housing mounting surface. All other needle bearings are to be pressed in from inside of case and cover to a depth of 0.015-0.020 inch (0.38-0.51 mm) below thrust surfaces.

Renew all seals and gaskets. Reassemble by reversing disassembly procedures. When installing brake shaft (39) and idler gear (38), beveled edge of gear teeth must be up away from cover. Install reverse idler shaft (23), spacer (24) and reverse idler gear (25) with rounded end of gear teeth facing spacer. Install input gear (33) and shaft (32) so chamfered side of input gear is facing case (16).

Tighten transaxle cap screws to the following torque:

Differential cap screws . . . . . 7 ft.-lbs. (9 N·m)
Case to cover cap screws . . . 10 ft.-lbs. (14 N·m)
Axle housing cap screws . . . 13 ft.-lbs. (18 N·m)
Shift lever housing cap screws . . . . 10 ft.-lbs. (14 N·m)

Reinstall transaxle and fill with recommended fluid.

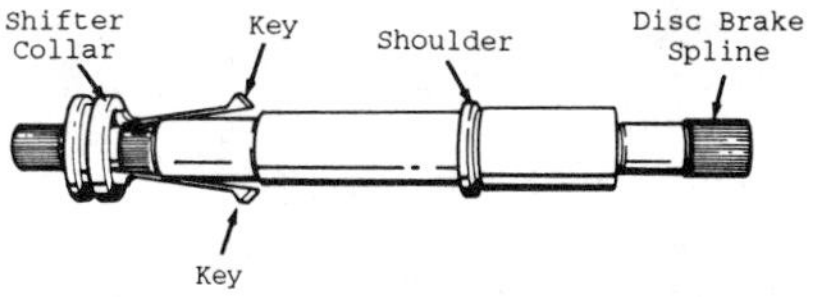

Fig. M21—Install shift collar on brake shaft so thick segment is nearer shaft shoulder.

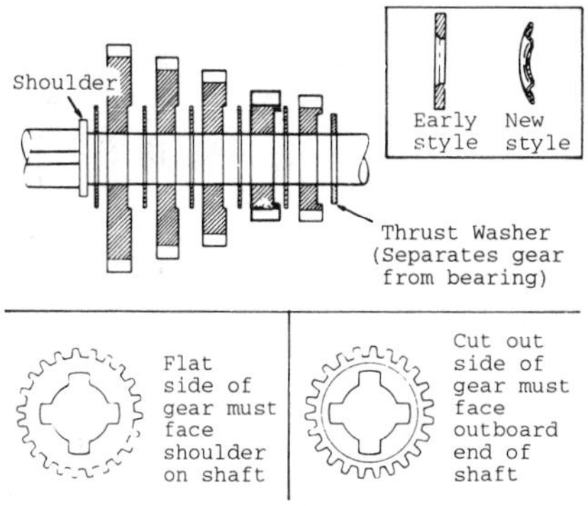

Fig. M22—When assembling brake shaft, install thrust washers with 45 degree chamfer toward shoulder on output shaft. Install curved washers so rounded side is toward shaft shoulder. Flat side of gears also must face shoulder on output shaft.

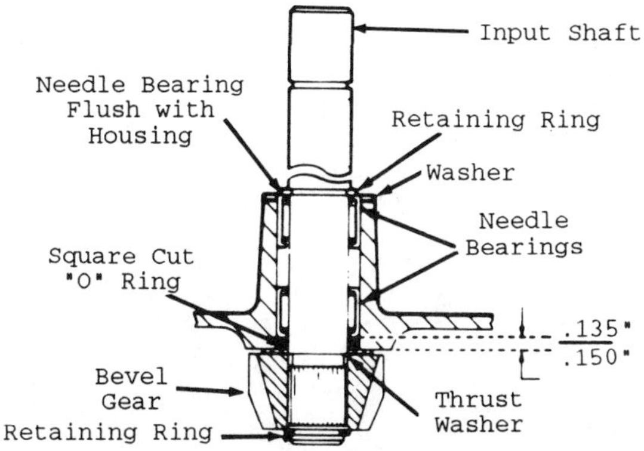

Fig. M20—When renewing transaxle input shaft bearings, lower bearing should be installed 0.135-0.150 in. (3.43-3.81 mm) below flush with housing. Upper bearing should be flush with housing.

Illustrations courtesy Murray Ohio Mfg. Co.

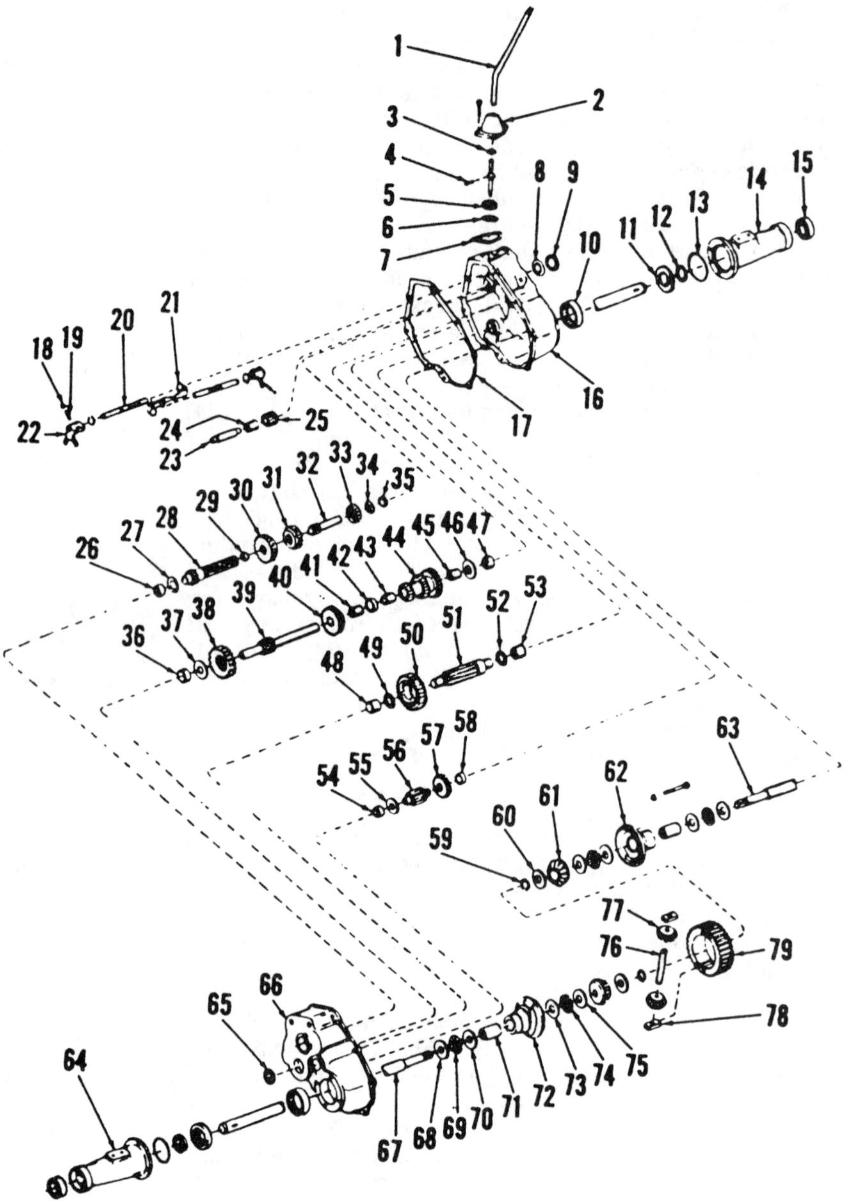

Fig. M23—Exploded view of transaxle used on Model 39004.

1. Shift lever
2. Shift lever housing
3. Seal ring
4. Roll pin
5. Retainer
6. Snap ring
7. Gasket
8. Ball bearing
9. Oil seal
10. Carrier bearing
11. Seal retainer
12. Oil seal
13. "O" ring
14. Axle housing
15. Axle outer bearing
16. Transaxle case
17. Gasket
18. Detent ball
19. Spring
20. Shifter rod
21. Shifter stop
22. Shifter fork
23. Reverse idler shaft
24. Spacer
25. Reverse idler gear
26. Needle bearing
27. Thrust washer
28. Shifter shaft
29. Needle bearing
30. First, second & reverse gear
31. Third & fourth gear
32. Input shaft
33. Input gear
34. Thrust washer
35. Needle bearing
36. Needle bearing
37. Thrust washer
38. Idler gear
39. Brake & cluster shaft
40. Two-gear cluster
41. Bushing
42. Spacer
43. Bushing
44. Three-gear cluster
45. Bushing
46. Thrust washer
47. Needle bearing
48. Needle bearing
49. Thrust washer
50. Output gear
51. Output shaft
52. Thrust washer
53. Needle bearing
54. Needle bearing
55. Thrust washer
56. Low reduction shaft
57. Low reduction gear
58. Needle bearing
59. Snap ring
60. Thrust washer
61. Axle gear
62. Axle carriage (plain holes)
63. Axle shaft
64. Axle housing
65. Oil seal
66. Transaxle cover
67. Axle shaft
68. Thrust washer
69. Thrust bearing
70. Thrust washer
71. Bushing
72. Axle carriage (tapped holes)
73. Thrust washer
74. Thrust bearing
75. Thrust washer
76. Pinion shaft
77. Bevel pinion gear
78. Drive block
79. Ring gear

Illustrations courtesy Murray Ohio Mfg. Co.

SIMPLICITY

CONDENSED SPECIFICATIONS

	MODELS		
	5216	**6216**	**6218**
Engine Make	Onan	B&S	B&S
Model	CCKA	402707	422707
Bore	3.25 in. (82.5 mm)	3.44 in. (87.4 mm)	3.44 in. (87.4 mm)
Stroke	3 in. (76.2 mm)	2.16 in. (54.9 mm)	2.28 in. (57.9 mm)
Displacement	49.8 cu. in. (816 cc)	40 cu. in. (659 cc)	42.3 cu. in. (695 cc)
Power Rating	16.5 hp. (12.3 kW)	16 hp. (11.9 kW)	18 hp. (13.4kW)
Slow Idle	1200 rpm	1350 rpm	1350 rpm
High Speed (No-Load)	3850 rpm	3600 rpm	3600 rpm
Capacities –			
Crankcase	3.5 qts. (3.3 L)	3 pts. (1.4 L)	3 pts. (1.4 L)
Hydraulic System	5.5 qts. (5.2 L)	.....	.....
Transmission	See Hyd.	3.5 pts. (1.6 L)	3.5 pts. (1.6 L)
Final Drive	0.5 qts. (0.47 L)	......	......
Fuel Tank	3.8 gal. (14.3 L)	2.2 gal. (8.3 L)	2.2 gal. (8.3 L)

	MODELS		
	7018	**7117**	**7119**
Engine Make	Kohler	Kohler	Kohler
Model	K361QS	KT17	KT19
Bore	3.75 in. (95.3 mm)	3.125 in. (79.4 mm)	3.125 in. (79.4 mm)
Stroke	3.25 in. (82.5 mm)	2.75 in. (69.8 mm)	3.063 in. (78 mm)
Displacement	35.9 cu. in. (589 cc)	42.18 cu. in. (691.4 cc)	47 cu. in. (770 cc)
Power Rating	18 hp. (13.4 kW)	17 hp. (12.7 kW)	19 hp. (14.2kW)
Slow Idle	1800 rpm	1200 rpm	1200 rpm
High Speed (No-Load)	3800 rpm	3600 rpm	3600 rpm
Capacities –			
Crankcase	2 qts. (1.9 L)	3 pts. (1.4 L)	3 pts. (1.4 L)
Hydraulic System	3 qts. (2.8 L)	3 qts. (2.8 L)	3 qts. (2.8 L)
Transmission	See Hyd.	See Hyd.	See Hyd.
Bevel Gear Box	*	*	*
Fuel Tank	3 gal. (11.4 L)	3 gal. (11.4L)	3 gal. (11.4L)

*Refer to **MAINTENANCE** paragraph in **BEVEL GEAR BOX** section.

FRONT AXLE AND STEERING SYSTEM

MAINTENANCE

All Models

It is recommended steering spindles, axle pivot and steering gear be lubricated at 25 hour intervals of normal operation and front wheel bearings should be cleaned and repacked at 100 hour intervals. Use multi-purpose, lithium base grease. Clean all fittings before and after lubrication. Check for any excessive looseness of parts, bearings, gears or tie rods and repair as necessary.

AXLE MAIN MEMBER

Model 5216

To remove axle main member (6–Fig. S1) assembly from tractor, raise and support front of tractor and disconnect drag link end (21) from pitman arm at steering gear box. Support axle main member and remove pivot bolts (1 and 10). Lower axle assembly and roll out from under tractor. Renew bushings (3, 5 and 16) and bolt (1) as necessary.

Reinstall by reversing removal procedure.

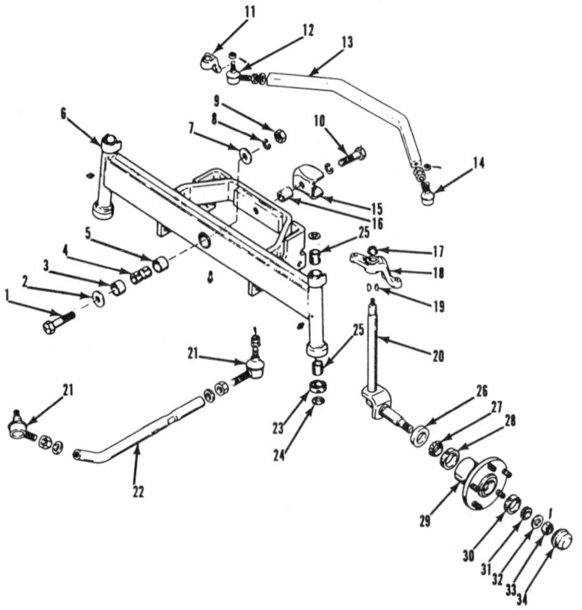

Fig. S1—Exploded view of front axle used on 5216 models.

1. Front pivot bolt
2. Washer
3. Pivot bushing
4. Spacer
5. Pivot bushing
6. Main member
7. Washer
8. Lockwasher
9. Nut
10. Rear pivot bolt
11. Right steering arm
12. Tie rod end
13. Tie rod
14. Tie rod end
15. Clevis
16. Spacer
17. Nut
18. Left steering arm
19. Keys
20. Spindle
21. Drag link ends
22. Drag link
23. Thrust bearing
24. Spacer
25. Bushings
26. Seal
27. Bearing
28. Bearing cup
29. Hub
30. Bearing cup
31. Bearing
32. Washer
33. Nut
34. Dust cap

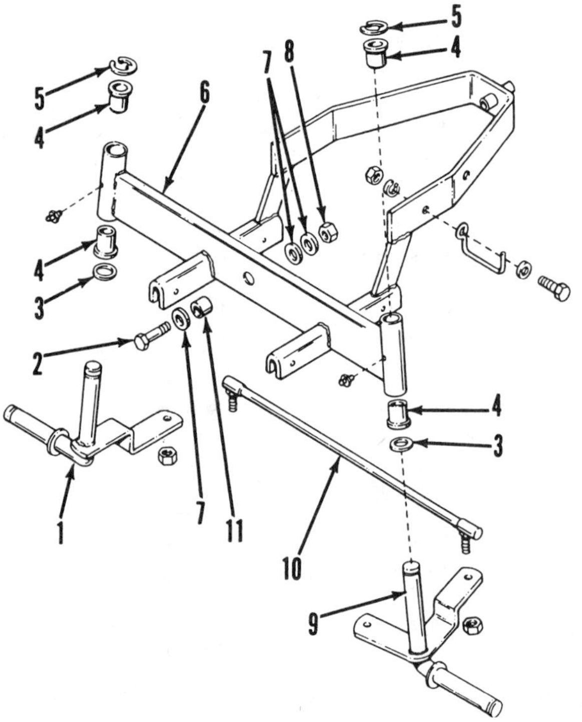

Fig. S2—Exploded view of front axle used on 6216 and 6218 models.

1. Spindle
2. Front pivot bolt
3. Washer
4. Bushings
5. Retaining ring
6. Main member
7. Washers
8. Nut
9. Spindle
10. Tie rod
11. Spacer

Models 6216-6218

To remove axle main member (6–Fig. S2) assembly from tractor, support front of tractor and disconnect drag link at left spindle arm. Remove pivot bolt (2), pull axle assembly forward out of center pivot and remove axle. Renew bushing (11) and bolt (2) as necessary.

Reinstall by reversing removal procedure.

Models 7018-7117-7119

To remove axle main member (1–Fig. S3) assembly from tractor, support front of tractor and disconnect drag link (12) at left spindle arm (14). Remove pivot bolt (4), pull axle assembly forward out of center pivot and remove axle. Renew bushing (2) and bolt (4) as necessary.

Reinstall by reversing removal procedure.

TIE ROD AND TOE-IN

All Models

On models equipped with renewable tie rod ends, toe-in should be adjusted by lengthening or shortening tie rod until 1/16 to 1/8-inch (1.588 to 3.175 mm) toe-in is obtained.

STEERING SPINDLES

Model 5216

To remove spindles (20–Fig. S1), raise and support front of tractor. Remove dust cap (34), cotter pin and nut (33). Remove wheel and hub assembly and bearings. Remove nut (17), steering arms (11 or 18) and keys (19). Remove spindle (20). Renew bushings (25) and spindle (20) as necessary.

Reinstall by reversing removal procedure.

Models 6216-6218

To remove spindle (1 or 9–Fig. S2), raise and support front of tractor. Remove front wheel and tire. Discon-

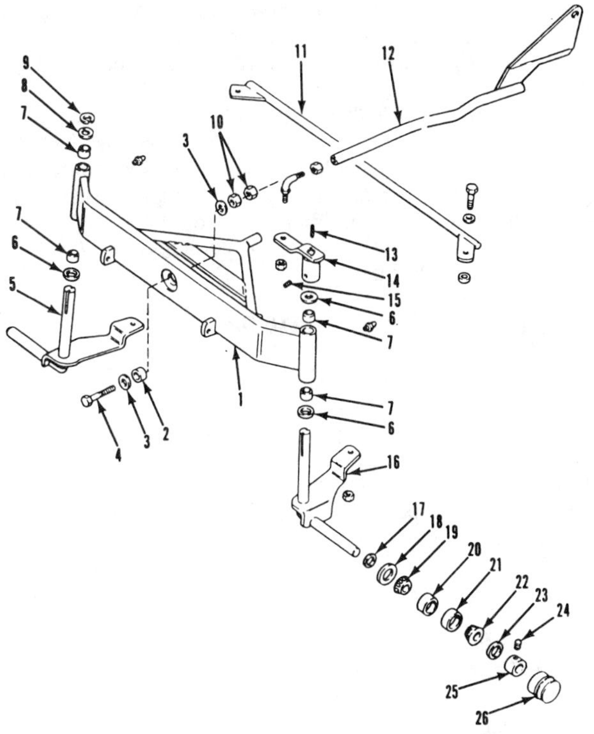

Fig. S3—Exploded view of front axle used on 7018, 7117 and 7119 models.
1. Main member
2. Spacer
3. Washer
4. Pivot bolt
5. Spindle
6. Washer
7. Bushings
8. Washer
9. Retaining ring
10. Nuts
11. Tie rod
12. Drag link
13. Key
14. Steering arm
15. Setscrew
16. Spindle
17. Spacer
18. Seal
19. Bearing
20. Bearing cup
21. Bearing cup
22. Bearing
23. Washer
24. Setscrew
25. Retaining collar
26. Dust cap

nect tie rod (10) from both sides and drag link from left side. Remove snap ring (5) and remove spindle. Renew bushings (4), washers (3) and spindles (1 or 9) as necessary.

Reinstall by reversing removal procedure.

Models 7018-7117-7119

To remove spindles (5 and 16–Fig. S3), raise and support front of tractor. Remove dust cap (26), loosen setscrew (24); then remove collar (25) and washer (23). Remove wheel, hub and bearings. Remove tie rod (11). On left side, disconnect drag link (12). Remove the two square head setscrews (15) and use a suitable puller to remove steering arm (14). Remove the two keys (13) and remove left spindle. On right side, remove snap ring (9) and remove right spindle. Renew washers (6 and 8), bushings (7) and spindles as necessary.

Reinstall by reversing removal procedure.

FRONT WHEEL BEARINGS

Model 5216

To remove front wheel bearings, raise and support side to be serviced. Remove dust cap (34–Fig. S1), cotter pin and nut (33). Remove washer (32), bearing (31) and hub (29). Remove seal (26) and inner bearing (27) from hub. Renew bearing cups (28 and 30) in hub as necessary.

Clean all parts thoroughly and inspect for wear or damage. Repack bearings using a good quality grease recommended for wheel bearings. Apply a coating of grease on spindle and in hub.

When reinstalling, seat bearing cups squarely in hub. Place inner bearing (27) in hub, then install seal. Install hub assembly on spindle. Install outer bearing (31) and washer (32). Install nut (33) and tighten until a slight preload is noted on bearing as wheel is being rotated. Install cotter pin to lock nut in this position. Install dust cap (34).

Models 6216-6218

To renew wheel bearing bushings, raise and support front of tractor. Remove wheel cover. Remove snap ring and washer from end of spindle. Remove wheel. Inner and outer bushings may be renewed as necessary. If bushings are loose in wheel, repair or renew wheel. Apply a coating of good quality multipurpose, lithium base grease to spindle and bushings.

Reinstall by reversing removal procedure.

Models 7018-7117-7119

To remove front wheel bearings, raise and support front of tractor. Remove dust cap (26–Fig. S3), loosen setscrew (24); then remove collar (25) and washer (23). Remove outer bearing (22) and wheel and hub assembly. Remove seal (18) and inner bearing (19) from wheel hub. Remove spacer (17) from spindle. Renew bearing cups (20 or 21) as necessary.

Clean all parts thoroughly and inspect for wear or damage. Repack bearings using a good quality grease recommended for wheel bearings. Apply a coating of grease to spindle and in hub.

When reinstalling, seat bearing cups squarely in hub. Place inner bearing (19) in cup in hub. Install seal (18). Install spacer (17) on spindle, then install wheel and hub and bearing assembly. Install outer bearing (22) and washer (23). Install collar. Push collar against washer until all end play is removed from bearings. Tighten setscrew (24). Install dust cap (26).

STEERING GEAR

Model 5216

R&R AND OVERHAUL. To remove the rotating ball nut and sector type steering gear box, disconnect battery, remove oil cooler, rear engine shroud and fuel tank. Disconnect clutch rod. Unbolt pto clutch and drive shaft assembly. Remove steering wheel. Disconnect choke and throttle cables.

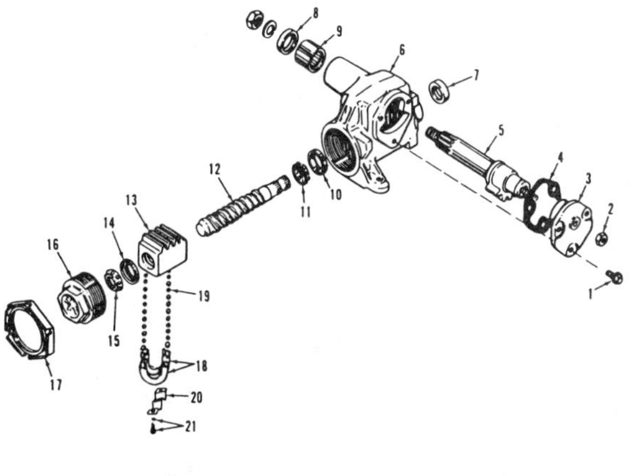

Fig. S4—Exploded view of steering gear assembly used on 5216 models.
1. Side cover bolts
2. Lash adjuster locknut
3. Side cover & needle bearing assy.
4. Side cover gasket
5. Pitman shaft & lash adjuster
6. Steering gear housing
7. Worm shaft seal
8. Pitman shaft seal
9. Pitman shaft needle bearing
10. Bearing race
11. Bearing (upper)
12. Worm shaft
13. Ball nut
14. Bearing race
15. Bearing (lower)
16. Adjuster plug
17. Adjuster plug locknut
18. Ball guides
19. Balls
20. Ball guide clamp
21. Clamp screw & washer

Remove instrument panel. Remove roll pin in steering shaft "U" joint and pull shaft up into support tube. Loosen clutch pivot arm and tilt it away from frame. Remove shims from pivot support assembly.

NOTE: When reinstalling pivot, shims must be placed in their original positions.

Remove nut from pitman shaft and pitman arm. Unbolt steering gear case from tractor frame. Remove "U" joint from steering gear. Lift assembly from frame.

To disassemble steering gear, clamp gear housing in a vise with worm shaft horizontal. Determine center of travel on worm shaft by rotating shaft from stop to stop. Place a container in position to catch lubricant from case as cover is removed. Remove side cover plate cap screws. Tap pitman shaft and cover out of gear case. Remove locknut (17 – Fig. S4) and adjuster plug (16) with lower bearing (15) and bearing retainer (14) from gear case. Pull worm shaft (12) with ball nut (13) from case through lower end. Remove upper bearing (11) from case. Remove ball guides (18) from nut and allow the 48 steel balls to drop into a container. Remove nut from worm shaft.

Clean all parts thoroughly and inspect all bearings, cups, worm shaft and ball nut for wear or damage. Bearing cup for bearing (15) and adjusting plug (16) are serviced as an assembly only. Renew pitman shaft needle bearing (9) and seal (8) as needed. Use seal protectors when installing shafts through seals and lubricate all bearings prior to installation.

To reassemble ball nut, place ball nut as near center of worm shaft as possible with nut holes aligned with races in worm shaft. Install ball guides (18) in ball nut. Drop 24 steel balls into each circuit of ball nut while oscillating the nut to ensure proper seating of balls in worm shaft races. Apply a heavy coating of multi-purpose, lithium base grease to worm shaft races and rotate ball nut assembly on worm shaft until steel balls are well lubricated. Install upper shaft seal (7) in case (6). Install worm shaft and ball nut assembly into case. Install bearing and retainer in adjusting plug and install assembly into case. Preload bearings until 3 to 6 in.-lbs. (0.339 to 0.678 N·m) force is required to rotate worm shaft. Install locknut and tighten while holding adjustment plug from turning. Pack 9 ounces (266 mL) of multi-purpose, lithium base grease into gear case. Place seal protector on splines of pitman shaft and install shaft in case so center tooth of sector gear enters center groove of ball nut. Install side cover (3) and gasket (4). Rotate lash adjustment screw counter-clockwise to allow cover to seat properly, then install side cover cap screws and tighten. Center ball nut on worm shaft and turn lash adjustment screw clockwise to remove all ball nut to sector lash without binding. Tighten adjuster screw locknut while holding screw in this location. Ball nut should travel through entire cycle with no binding or roughness.

Reinstall by reversing removal procedure.

Models 6216-6218

R&R AND OVERHAUL. To remove steering gear, remove steering wheel, upper dash, fuel tank and lower dash. Disconnect drag link (11 – Fig. S5) from steering rod (10). Unbolt and remove upper steering plate (6) and steering shaft (8). Remove lower plate (9) and steering rod (10) as an assembly. Remove the two setscrews (13) and use a suitable puller or press to remove steering gear (12) from steering rod (10).

Clean and inspect all parts. Renew bushings (5) as necessary.

To reassemble, install key (14) with rounded end toward bent end of steering rod (10) until top edge of gear is flush with end of key (14). Steering rod should extend approximatey ¼-inch (6 mm)

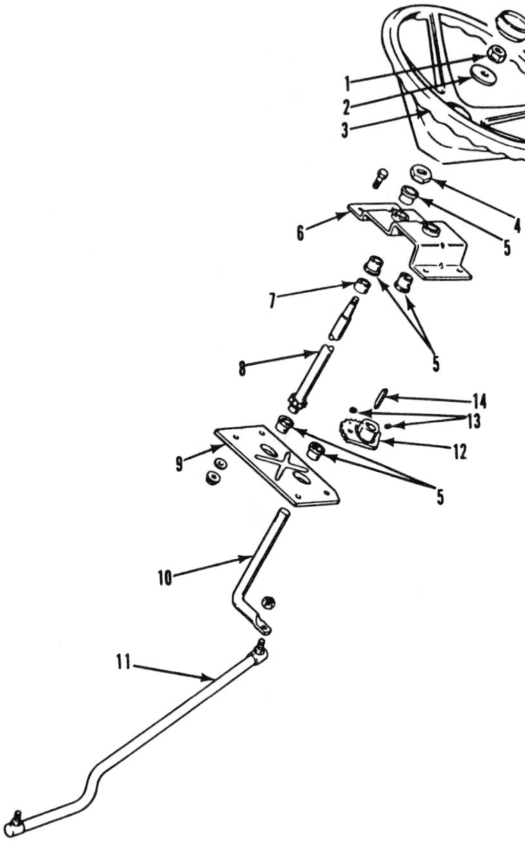

Fig. S5—Exploded view of steering shaft and mechanism used on 6216 and 6218 models.

1. Locknut
2. Washer
3. Steering wheel
4. Hex washer
5. Bushings
6. Upper steering plate
7. Spacer
8. Steering shaft
9. Lower steering plate
10. Steering rod
11. Drag link
12. Steering gear
13. Setscrews
14. Key

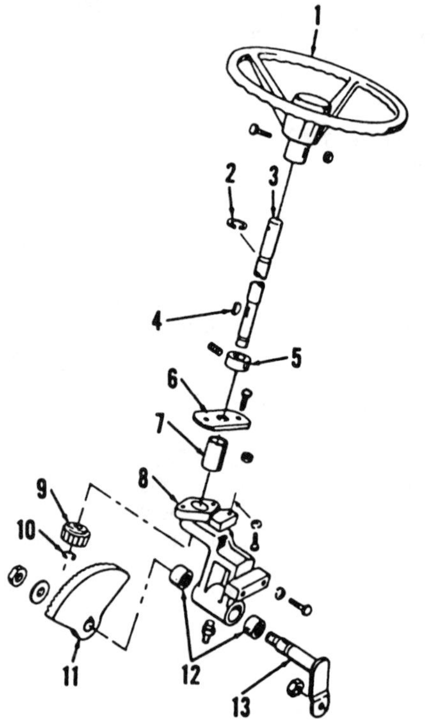

Fig. S6—Exploded view of steering shaft and mechanism used on 7018, 7117 and 7119 models.

1. Steering wheel
2. Retaining ring
3. Steering shaft
4. Key
5. Retaining collar
6. Steering plsc
7. Bushing
8. Casting
9. Pinion
10. "E" ring
11. Bevel gear
12. Needle bearing
13. Steering arm assy.

above steering gear. Install assembly on lower plate (9). Place spacer (7) on steering shaft (8) with bevel towards steering gear. Center steering gear on lower plate and install steering shaft so hole in shaft faces front and back.

Reassemble by reversing disassembly procedure.

Models 7018-7117-7119

R&R AND OVERHAUL. To remove steering gear, remove steering wheel and battery. Disconnect drag link from steering arm (13–Fig. S6) and turn steering arm to allow access to mounting bolts. Remove mounting bolts and move steering gear assembly forward until casting lug clears edge of frame opening and lower entire assembly.

To disassemble steering gear, clamp support casting in a vise and remove locknut and washer from steering arm (13). Use a plastic mallet to separate steering arm (13) from bevel gear (11) and pull steering arm out of casting (8). Position steering shaft (3) in a vise. Remove retaining ring (10) and use a suitable puller to remove pinion gear (9). Remove key from shaft and slide shaft out of bushing. Remove adjusting plate (6) and bushing (7). Use a bearing puller to remove needle bearings from casting.

Clean and inspect all parts for wear or damage. During reassembly, press needle bearings into each end of bore with end of bearing with manufacturers name facing outward. Bearing at gear end must be 1/8-inch (3 mm) below surface of casting and bearing at steering arm end must fit flush with casting. Install bushing (7) and adjusting plate assembly, then tighten adjusting plate until approximately 1/64-inch (0.4 mm) of bushing is above casting surface.

Reinstall by reversing removal procedure.

ENGINE

MAINTENANCE

All Models

Regular engine maintenance is required to maintain peak performance and long engine life.

Check oil level and clean air intake screen at 5 hour intervals under normal operating conditions.

Clean engine air filter at 25 hour intervals and clean crankcase breather, governor linkage and engine cooling fins at 200 hour intervals.

Change engine oil and filter, perform tune-up, valve adjustment and clean carbon from cylinder head or heads as recommended by engine manufacturer.

REMOVE AND REINSTALL

Model 5216

Remove hood and battery and disconnect throttle and choke cables. Disconnect electrical connections. Disconnect fuel lines at tank. Remove oil cooler and shrouds, as equipped. If equipped with pto, remove pto yoke from engine flywheel. Depress clutch pedal and remove clutch belts from engine pulley. Disconnect clutch rod and lower clutch assembly. Remove bottom cover attached to tractor frame and remove engine mounting bolts. Remove engine.

Reinstall by reversing removal procedure.

Models 6216-6218

Disconnect battery ground cable. Disconnect fuel line at carburetor. Disconnect throttle and choke cables and all necessary electrical wiring. Remove drive belts. Remove engine mounting bolts and remove engine.

Reinstall by reversing removal procedure.

Models 7018-7117-7119

Disconnect fuel line from fuel pump and cap opening. Disconnect battery ground cable, all necessary electrical wires and choke and throttle cables. Disconnect drive shaft from engine flywheel. Remove engine mounting bolts and remove engine.

Reinstall by reversing removal procedure.

OVERHAUL

All Models

Engine make and model are listed at the beginning of this section. To overhaul engine or accessories, refer to appropriate engine section of this manual.

ELECTRICAL SYSTEM

MAINTENANCE AND SERVICE

All Models

Battery electrolyte level should be checked at 25 hour intervals of normal operation. If necessary, add distilled water until level is just below base of vent well. **DO NOT** overfill. Keep battery posts clean and cable ends tight.

For alternator or starter service, refer to appropriate engine section of this manual.

Refer to Fig. S7, S9 or S10 for view of typical wiring diagram. Wiring may vary slightly according to optional equipment installed.

BRAKE

ADJUSTMENT

Model 5216

Unbolt and remove frame top cover. Loosen locknuts (6 and 8–Fig. S12) at each end of turnbuckles (7) and adjust

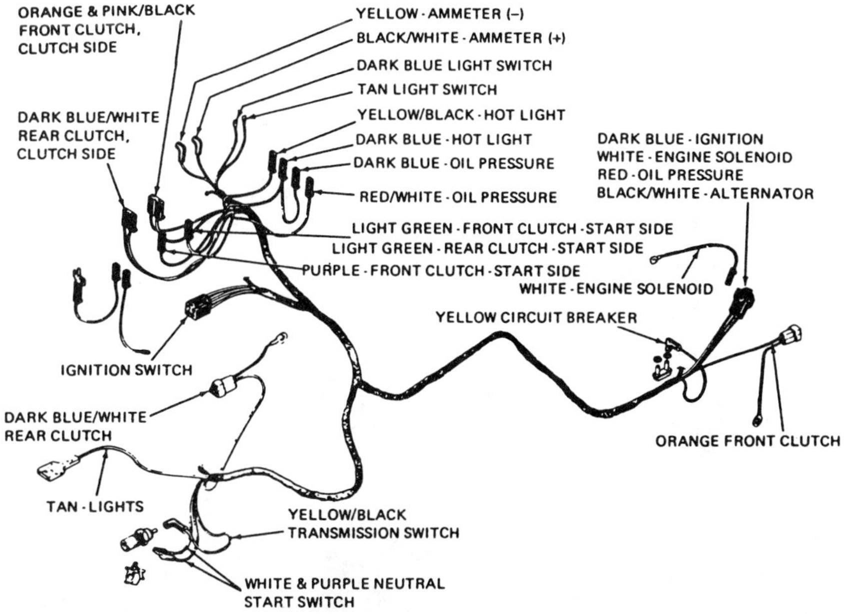

Fig. S7—View of typical 5216 models wiring diagram. Wiring may vary slightly when electric options are added.

turnbuckles until brake pedal free travel is 1¼ inches (31.75 mm) for each pedal. Brakes must be adjusted equally so both brakes are activated at the same time when pedals are locked together. Retighten locknuts and reinstall frame cover.

Models 6216-6218

Locate brake band which is to the inside of left rear tire. Pull forward on brake band to remove all slack. Make certain brake pedal is in full "up" position and measure gap (E – Fig. S13) between spacer (C) and brake band (B). Gap (E) should be ⅝ to ¾-inch (16 to 19 mm). Adjust by tightening or loosening brake rod nut (D) as necessary. A minimum of two full threads must extend beyond nut to properly secure nut on rod.

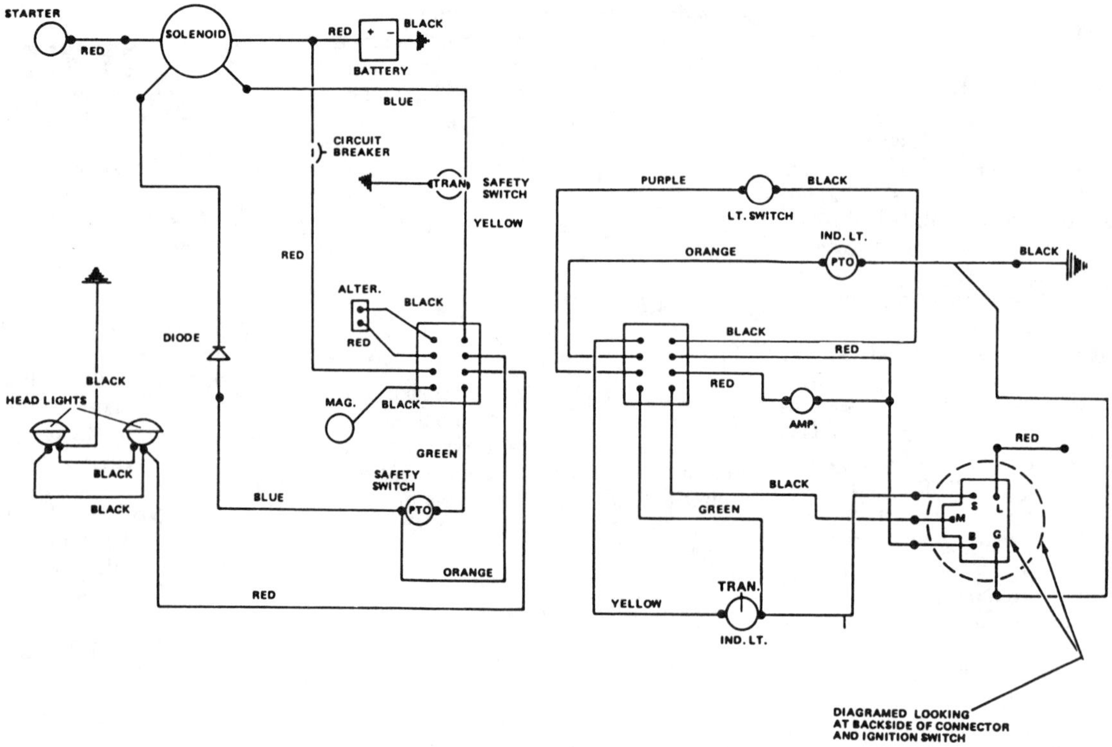

Fig. S9—View of typical 6216 and 6218 models wiring diagram. Wiring may vary slightly when electric options are added.

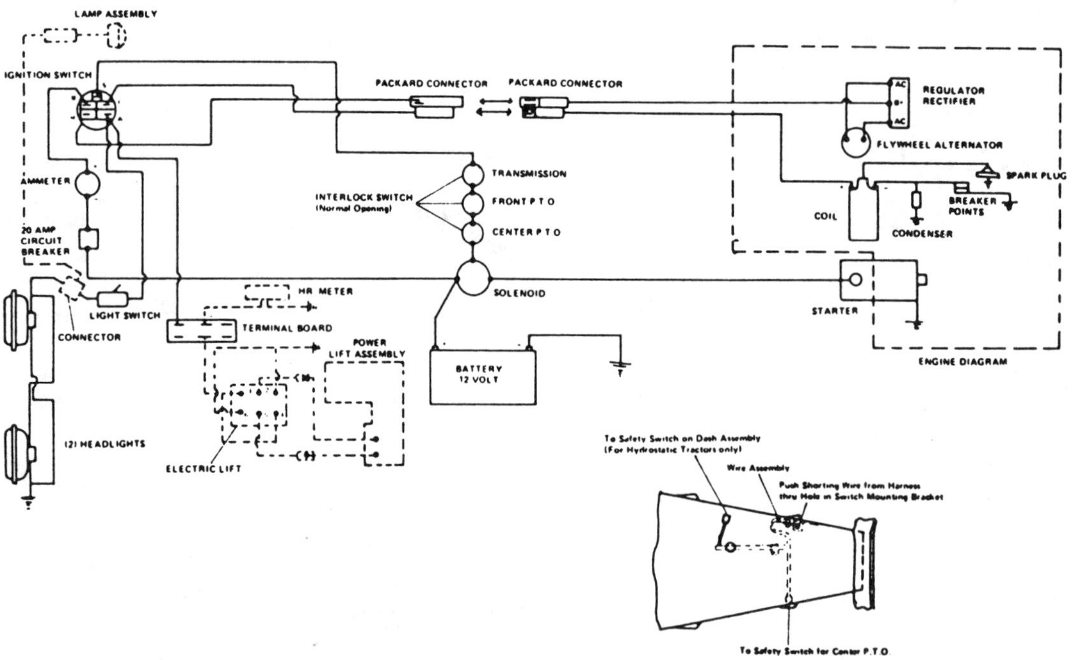

Fig. S10—View of typical 7018, 7117 and 7119 models wiring diagram. Wiring may vary slightly when electric options are added.

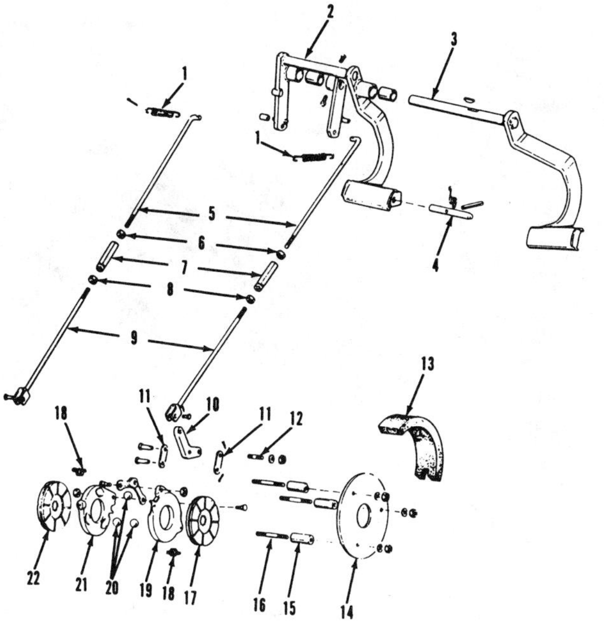

Fig. S12—Exploded view of brake mechanism used on 5216 models.

1. Return spring
2. Left pedal assy.
3. Right pedal assy.
4. Pedal lock rod
5. Front brake rods
6. Jam nuts
7. Turnbuckles
8. Jam nuts
9. Rear brake rods
10. Brake lever
11. Link plate
12. Stud
13. Brake pad
14. Disc assy.
15. Spacer
16. Stud
17. Brake disc
18. Spring
19. Actuating disc
20. Steel balls
21. Actuating disc
22. Brake disc

Models 7018-7117-7119

Loosen jam nut on front end of parking brake rod and turn handle and rod end until parking brake is tight when brake handle is against fender in brake lock position. With parking brake engaged and foot pedal released, adjust jam nuts on front of brake rod to obtain ¾-inch (19 mm) clearance between jam nut and rod guide (Fig. S14).

R&R BRAKES

Model 5216

Raise and support rear of tractor, then remove rear tires and wheels. Remove wheel guards. Remove cap screws retaining each final drive (drop) housing to axle extension, slide drop housing away from axle extension until pinion shaft clears coupling and remove housing. Disconnect hydraulic lift cylinder from left axle extension and lift shaft and move out of the way. Remove lift shaft assembly, drawbar assembly and transmission filter. Remove the four cap screws retaining axle extensions to transmission case and the hex nuts and lockwashers from stationary brake disc studs. Remove axle extension and coupling. Remove outer stationary disc (14–Fig. S12), outer rotating disc (17) and spacer (15). Remove pin attaching link plate (11) to yoke link assembly. Remove disc actuating assembly and inner disc (22). Remove springs (18) from actuating assembly. Remove oil seals from differential supports.

Clean and inspect all parts. Steel balls (20) should be smooth and round. Ramp area of actuating plates (19 and 21) should be smooth with no grooves, pits or worn spots. Renew discs (17 and 22) and actuating plates (19 and 21) if worn, cracked or scored.

Install new oil seals in differential supports and install supports in transmission case. Install inner disc (22). Install the two spacers (15) with flat edges toward yoke link assembly on studs (16). Install spacer (15) which is round, on remaining stud. Reassemble actuating plates and install assembly in brake housing. Connect yoke link assembly and actuating link to link plate using a pin and cotter pin. Install retaining ring and outer rotating disc on differential output shaft, lubricate splines and carefully install shaft. Do not damage oil seal. Install outer stationary disc (14) and axle extension.

NOTE: If axle extension does not have a drain hole in bottom area, drill a ⅜-inch hole to allow water and any oil leakage to drain.

Install coupling on differential shaft, spring on pinion shaft and install drop housing. Reinstall by reversing removal procedure. Adjust brakes as previously outlined.

Models 6216-6218-7018-7117-7119

Brake band and drum are located on left side of transmission and renewal procedure is obvious after examination. Adjust brake as previously outlined.

CLUTCH

MAINTENANCE

All Models

Clutch belts should be checked and adjusted at 100 hour intervals or whenever it is suspected drive belts may be slipping. Periodically check all belts, stops, idler pulley, drive pulley and linkage for wear, looseness or damage.

ADJUSTMENT

Model 5216

Unbolt hydraulic oil cooler and raise left end of cooler to expose clutch belt tension adjuster. Make certain clutch pedal is in engaged "up" position. Refer to Fig. S15 and adjust plug at top of spring until spring length is 8 inches (203 mm). Adjust belt stop on right side of tractor so there is a clearance of 1/16 to ⅛-inch (1.588 to 3.175 mm) between belt stop and belts when clutch is engaged. Unbolt and remove cover from bottom of frame. Adjust locknut on clutch rod until clutch pedal free travel is 1½ to 1¾ inches (38 to 45 mm). Fully depress clutch pedal and check clearance be-

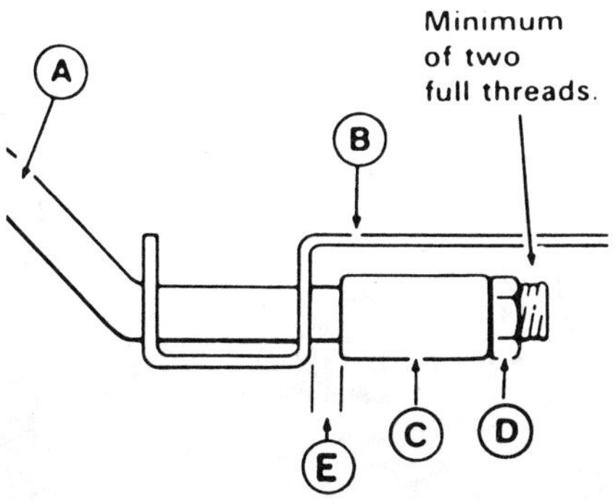

Fig. S13—Adjust brake on 6216 and 6218 models by loosening or tightening locknut (D) until gap (E) between spacer (C) and band (B) is @ to ¾-inch (16 to 19 mm). Refer to text.

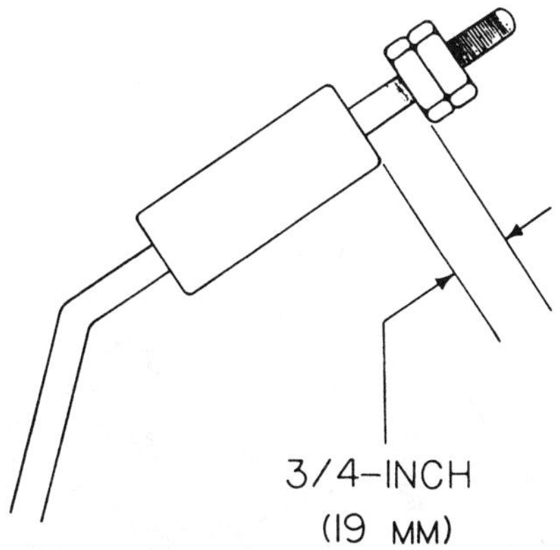

Fig. S14—When adjusting brake on 7018, 7117 and 7119 models, gap between brake rod guide and adjustment nuts should be ¾-inch (19 mm). Refer to text.

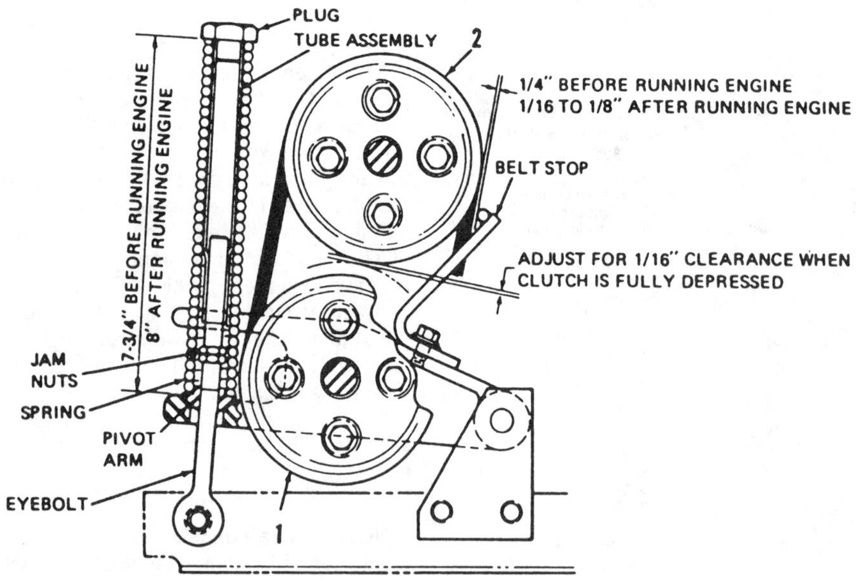

Fig. S15—View showing clutch (drive) belt tension adjustment and belt stop adjustment on 5216 models. With clutch pedal depressed, a clearance of 1/16-inch (1.6 mm) should exist between engine crankshaft pulley (2) and drive shaft pulley (1). Refer to text for adjustment procedure.

tween engine pulley (2 – Fig. S15) and drive shaft pulley (1). Clearance should be 1/16-inch (1.588 mm). If not, remove plug and belt tension spring and adjust jam nuts on eyebolt until correct clearance is obtained. Readjust belt tension and clutch pedal free travel. Bolt oil cooler in place and install bottom frame cover.

Model 6216-6218

Variable speed control and clutch are adjusted together. Place speed control lever up in full speed position, then loosen shoulder bolt (4 – Fig. S16). With transmission in neutral, start engine, depress clutch pedal, set parking brake and stop engine. Unlatch parking brake and allow pedal to come up slowly, then measure distance from pedal shaft to forward edge of foot rest. Distance should be 5½ inches (140 mm); if not, adjust nut (2 – Fig. S17) toward spring to increase measurement. Place speed control lever down in low speed position, then pull upper handle (1 – Fig. S16) only upward and hold to lock lever in position. Push bar (3) down and tighten shoulder bolt.

Models 7018-7117-7119

Jam nuts on clutch rod should be adjusted so spring length is ½-inch (12.7 mm) between washers when clutch brake pedal is in "up" position.

TRANSMISSION

MAINTENANCE

Model 5216

Hydrostatic transmission, tractor hydraulic system and three speed gear transmission all use oil from the three speed gear type transmission case. To check fluid level on early 5216 models, open check cock (Fig. S18) counterclockwise two turns or until fluid runs out. If no fluid runs out of check cock, remove filler plug and add Dexron automatic transmission fluid, or equivalent, until oil drips from check cock. Install filler plug and tighten. Close check cock. To check fluid level on all other models, a dipstick is located behind seat. Fluid should be maintained at the "FULL" mark on dipstick.

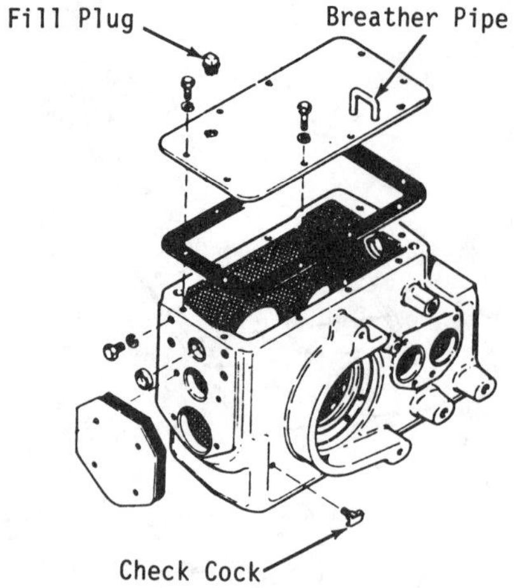

Fig. S18 – View of filler plug and check cock on early 5216 models. Refer to text.

On all models, transmission oil should be checked and oil cooler cleaned at 5 hour intervals. Change fluid and filter at 400 hour intervals. System oil capacity is 5.5 quarts (5.2L) when filter is changed.

Models 6216-6218

Transmission fluid level should be checked at 25 hour intervals or if excessive leakage becomes apparent. To check fluid level, remove oil level check cap screw (A – Fig. S20). Oil level should be even with lower edge of cap screw hole. Recommended fluid is SAE 90 EP gear lube. Fluid capacity is 3.5 pints (1.6L).

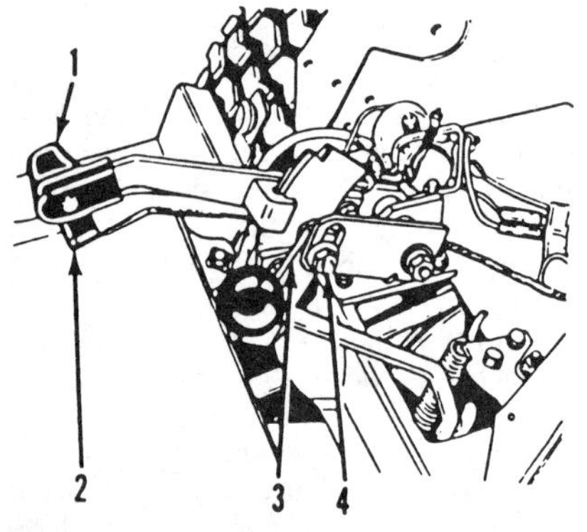

Fig. S16 – View of variable speed control linkage used on 6216 and 6218 models. Refer to text.

1. Control lever
2. Handle
3. Bar
4. Shoulder bolt

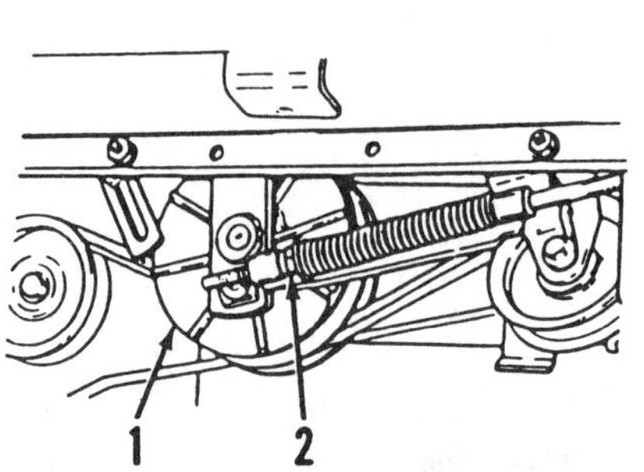

Fig. S17 – Variable speed pulley (1) and clutch rod used on 6216 and 6218 models. Adjusting nut (2) is used to adjust clutch. Refer to text.

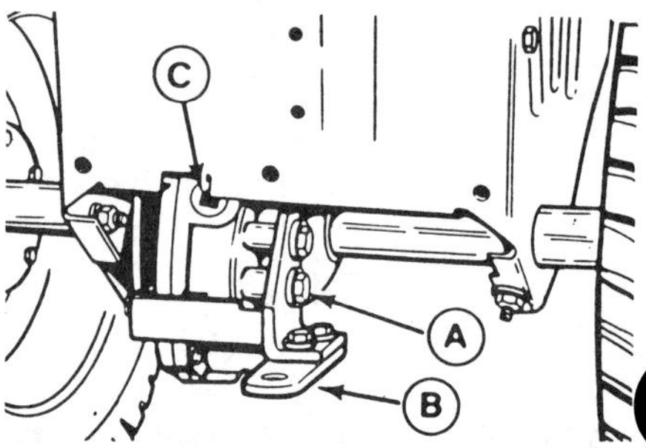

Fig. S20 – View showing location of filler plug (C), level check cap screw (A) and drain plug (B) on 6216 and 6218 models.

All belts, guides, stops and pulleys should be inspected for wear, looseness or damage at 25 hour intervals and repaired as necessary.

Models 7018-7117-7119

Hydrostatic transmission and tractor hydraulic system use fluid from gear transmission (reduction gear) case. Dexron automatic transmission fluid, or equivalent, is recommended and fluid level should be maintained 1/8-inch (3.175 mm) below top rear edge of fill tube (Fig. S21) when upper, forward edge of tube is 1/2-inch (12.7 mm) lower than rolled edge of pump support plate (Fig. S21). Fluid should be checked at 25 hour intervals. Fluid and filter should be changed at 400 hour intervals. Fluid capacity is 3 quarts (2.8 L) when filter is changed.

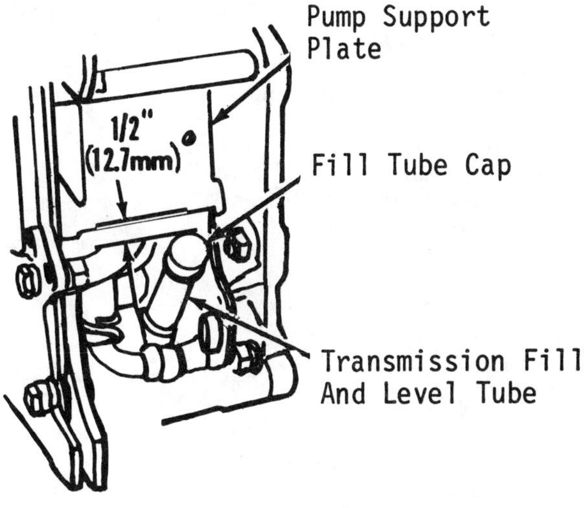
Pump Support Plate
1/2" (12.7mm)
Fill Tube Cap
Transmission Fill And Level Tube

Fig. S21 — View showing location of fill and level tube on 7018, 7117 and 7119 models. Refer to text.

ADJUSTMENTS

Model 5216

To adjust hydrostatic transmission for neutral position, remove top frame cover. Raise and support tractor so rear wheels are off the ground. Loosen locknuts (A – Fig. S22). Start engine and operate at high idle speed. Move hydrostatic control lever to forward position, then into neutral position. Rear wheels should not rotate when control lever is in neutral position. Turn turnbuckle (B) as required. Tighten locknuts and recheck adjustment.

Models 6216-6218

Refer to CLUTCH ADJUSTMENT section.

Models 7018-7117-7119

Raise and support tractor so rear wheels are off the ground. Start engine and run at high idle. Move hydrostatic control lever to forward position, then neutral position. If wheels rotate when lever is in neutral position, stop engine. Raise seat deck and check if pump control arm (E – Fig. S23) is exactly centered with centering mark (D). If not, loosen bolt (C) and move control cam (B) until centering mark (D) is centered on roller (E). Tighten bolt (C). If control arm is centered with centering mark, but rear wheels still rotate with lever in neutral position, loosen jam nut (H) on end of cam pivot shaft (G). If wheel rotation is in reverse, turn adjusting nut (I) 1/8 to 1/4-turn clockwise as viewed from right side of tractor. If rotation is forward, turn adjusting nut (I) 1/8 to 1/4-turn counter-clockwise. Lock jam nuts and recheck adjustment.

R&R TRANSMISSION

Model 5216

GEAR TYPE TRANSMISSION. Remove seat supports and frame top and bottom covers. Disconnect lower end of hydrostatic control linkage and remove wire from transmission oil temperature sending unit. Remove sending unit and drain transmission. Disconnect oil filter hose from charge pump inlet elbow and remove oil cooler return hose and hydraulic lift pressure line from top of pump housing. Plug or cap all openings. Disconnect drive shaft rear coupling and, if so equipped, unbolt rear pto shaft hub from electric clutch. Place a floor jack under transmission housing, support tractor under frame and remove rear wheels, drop housings, axle extensions and brakes. Remove lift cylinder, lift shaft and drawbar. Remove transmission retaining cap screws and remove transmission assembly.

HYDROSTATIC DRIVE UNIT. Remove frame top and bottom covers. Disconnect lower end of control linkage and remove control arm from hydrostatic unit. Remove wire from transmission oil temperature sending unit. Remove sending unit and drain

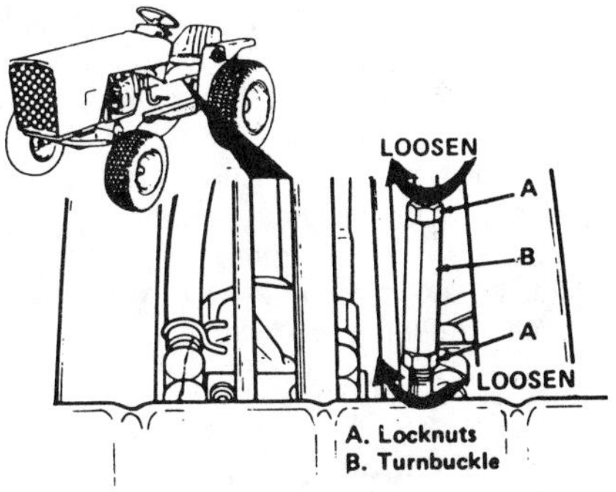
LOOSEN
A
B
A
LOOSEN
A. Locknuts
B. Turnbuckle

Fig. S22 — View showing adjustment location for hydrostatic neutral on 5216 models. Refer to text.

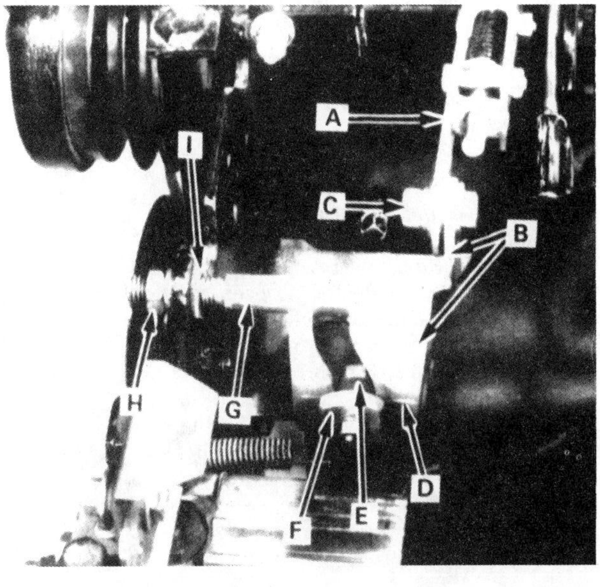
A
I
C
B
H
G
F
E
D

Fig. S23 — View showing location of hydrostatic transmission neutral adjustment points on 7018, 7117 and 7119 models. Refer to text.

transmission. Disconnect oil filter hose from charge pump elbow. Remove oil cooler return hose and hydraulic lift pressure line from top of pump housing. Plug or cap all openings. Unbolt drive shaft rear coupling. Remove flexible disc, slide yoke from drive shaft, then loosen clamp screws and remove yoke and key from pump input shaft. Remove access hole cover in left side of tractor frame. Remove the two flange nuts securing hydrostatic drive unit to the three speed transmission and remove unit through opening in bottom of frame.

Reinstall by reversing removal procedure. Fill to proper level with recommended fluid. Position hydrostatic control lever in neutral position and start and run engine at idle speed. Operate transmission in forward and reverse direction for short distances. Stop engine and check fluid level. Repeat procedure until all air has been bled from system.

Models 6216-6218

Remove seat and fender assembly. Disconnect brake rod from brake band and clutch rod from idler bracket. Remove drive belt and transmission pulley. Unbolt and remove shift lever assembly. Remove "U" bolt clamp from right axle and remove frame support and transmission support. Lower transmission and slide it clear of tractor.

Reinstall by reversing removal procedure. Adjust clutch and brake and check transmission fluid level.

Models 7018-7117-7119

GEAR CASE. Raise and support rear of tractor. Drain transmission fluid and remove hydrostatic drive unit from gear case. Remove brake linkage and brake drum. Disconnect oil suction line from bottom of gear case. Place jack under gear case and remove rear wheels, rear axles and differential assembly. Unbolt and remove gear case.

Reinstall by reversing removal procedure. Fill system to correct level.

HYDROSTATIC DRIVE UNIT. Remove seat and fender assembly. Support left rear side of tractor and remove rear wheel. Remove transmission fan and shroud and deflector. Loosen and remove drive belt, input pulley and fan. Disconnect pump control arm spring. Remove bolt, control arm roller, washer and nut. Remove oil filter and drain lubricant from reduction gear housing. Disconnect hydraulic hoses at oil filter assembly. Remove mounting bolts and slide transmission out of gear case.

Reinstall by reversing removal procedure. Lift pin in relief valve (Fig. S24)

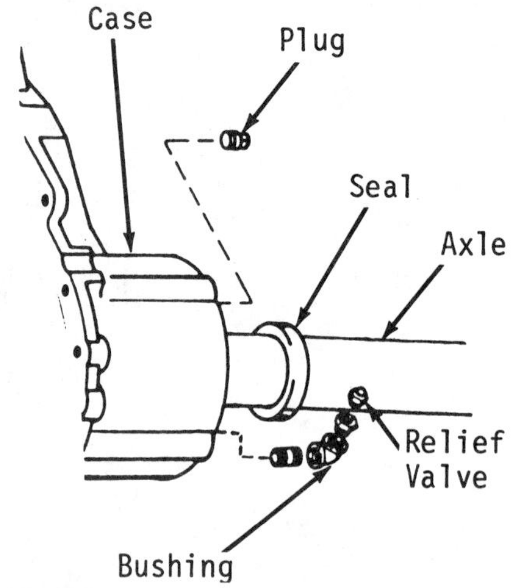

Fig. S24 — View showing location of relief valve on 7018, 7117 and 7119 models. Refer to text.

and fill system with recommended fluid until fluid is visible in filler tube (Fig. S21). Remove spark plug cables. Raise rear wheels off the ground. Set speed control lever halfway forward and crank engine for short intervals. When wheels start to rotate, connect spark plugs. Start and run tractor for 1-2 minutes. Stop engine and lift pin in relief valve. Add fluid until at proper level. Recheck fluid level after 5 hours of operation.

OVERHAUL

Model 5216

GEAR TYPE TRANSMISSION. Remove top cover. Remove center cap screw and washers, then remove hub from front of top pto shaft. Pry out oil seal, remove bearing retaining ring and move top pto shaft forward. Remove rear snap ring and gear from shaft and withdraw pto shaft from front of transmission housing. Remove cap screws securing differential supports (39 and 56–Fig. S25) to transmission housing. Hold up on differential assembly, remove differential support assemblies (39 through 42 and 56 through 59), then remove differential assembly from transmission housing.

Unbolt and remove shifter stem (2) and withdraw shift level and shaft (1). Remove oil seal (8). Unbolt and remove shift guide (3) with two spacers. Remove cap screws securing shifter shaft (9) to housing and withdraw shifter shaft with "O" rings (10 and 12) and gasket (11).

CAUTION: Cover holes in forks (4 and 5) to prevent detent balls and springs from flying out when shifter shaft is removed.

Remove shift forks. Unbolt and remove cap (18) with "O" ring (19) and bearing cup (20). Remove bearing cone (21) and keep with cap (18) and bearing cup (20). Remove sliding gear (28), then move shaft (29) to the left and remove bearing cone (35) and sliding gear (30). Lift out sliding gear shaft (29). Unbolt and remove cap (38) with shims (37) and bearing cup (36) and place with bearing cone (35). Remove cap screws securing cap (14) to housing and withdraw cap with shims (15) and "O" ring, if so equipped. Remove bearing cup (16) from

Fig. S25 — Exploded view of three speed gear type transmission and differential used on 5216 models.

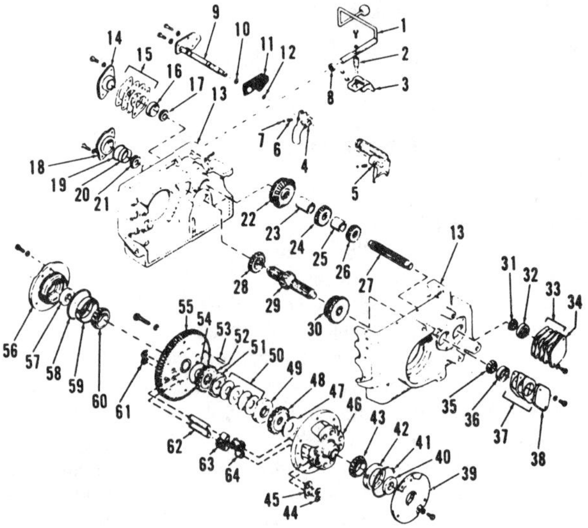

1. Shift lever & shaft
2. Shifter stem
3. Shift guide
4. Third shift fork
5. First & second shift fork
6. Detent spring
7. Detent ball
8. Oil seal
9. Shifter shaft
10. "O" ring
11. Gasket
12. "O" ring
13. Transmission housing
14. Bearing cap
15. Shims
16. Bearing cup
17. Bearing
18. Bearing cap
19. "O" ring
20. Bearing cup
21. Bearing
22. Bevel & third gear
23. Spacer
24. Second gear
25. Spacer
26. First gear
27. Shaft
28. Third sliding gear
29. Sliding gear shaft
30. First & second sliding gear
31. Bearing
32. Bearing cup
33. Shims
34. Bearing cap
35. Bearing
36. Bearing cup
37. Shims
38. Bearing cap
39. Differential support
40. Oil seal
41. "O" ring
42. Bearing cup
43. Bearing
44. Snap rings
45. Lock plate (4)
46. Differential case
47. Thrust washer
48. Differential side gear
49. Clutch plate
50. Belleville washers
51. Clutch plate
52. Differential side gear
53. Dowel pin
54. Thrust washer
55. Output gear
56. Differential support
57. Oil seal
58. "O" ring
59. Bearing cup
60. Bearing
61. Snap rings
62. Pinion shafts (8)
63. Pinion gears (8)
64. Spacers (8)

housing and place it with cap (14) and shims (15). Move shaft (27) to the left as far as possible, then remove bearing cone (31), gears (24 and 26) and spacers (23 and 25) from shaft. Lift out shaft (27) with bevel gear (22) and bearing cone (17). Unbolt and remove cap (34) with shims (33) and bearing cup (32) and place with bearing cone (31).

Clean and inspect all parts and renew any showing excessive wear or other damage.

NOTE: Do not remove the press fit bearing cone (17) from shaft (27) unless bearing is to be renewed.

If new bearing cup and cone (16 and 17) are to be installed, check and adjust position of bevel gear (22) as follows. Assemble new bearing assembly on cap (14) as shown in Fig. S26. Use a depth gage to measure distance from cap flange to top of cone inner race as shown. This dimension will be 1.25+ inches. A four digit number (3.11+) is stamped on the transmission case. Refer to Fig. S27 and use chart to determine correct thickness shim pack (15-Fig. S25) to be used with new bearing assembly (16 and 17). For example: If dimension "B" in Fig. S26 measures 1.265 inches and dimension "A" stamped on transmission case is 3.122 inches, correct shim pack thickness to be installed is 0.018 inch. Shims are available in a variety of thicknesses.

When reassembling, place bevel gear (22–Fig. S25) on shaft (27), then press bearing cone (17) on end of shaft. Place assembly in housing (13), move shaft to the left as far as possible, then install long spacer (23), second speed gear (24) with bevel edge of teeth toward right side, short spacer (25), first speed gear (26) with bevel edge of teeth toward left side and bearing cone (31). Install bearing cup (16), coat shim pack (26) with a

DIMENSION "A" TRANSMISSION NUMBER (On Case)

DIMENSION "B"	3.118	3.119	3.120	3.121	3.122	3.123	3.124	3.125	3.126	3.127	3.128
1.255	.012	.011	.010	.009	.008	.007	.006	.005	.004	.003	.002
1.257	.014	.013	.012	.011	.010	.009	.008	.007	.006	.005	.004
1.259	.016	.015	.014	.013	.012	.011	.010	.009	.008	.007	.006
1.261	.018	.017	.016	.015	.014	.013	.012	.011	.010	.009	.008
1.263	.020	.019	.018	.017	.016	.015	.014	.013	.012	.011	.010
1.265	.022	.021	.020	.019	.018	.017	.016	.015	.014	.013	.012
1.267	.024	.023	.022	.021	.020	.019	.018	.017	.016	.015	.014
1.269	.026	.025	.024	.023	.022	.021	.020	.019	.018	.017	.016
1.271	.028	.027	.026	.025	.024	.023	.022	.021	.020	.019	.018
1.272	.030	.029	.028	.027	.026	.025	.024	.023	.022	.021	.020

Fig. S27—Chart used in determining thickness of shim pack (15—Fig. S25) when installing new bearing assembly (16 and 17). Refer to text.

light coat of gasket sealer and install shims, cap (14) with "O" ring if so equipped. Apply a suitable sealer to cap screw threads, install and tighten cap screws to a torque of 20 ft.-lbs. (27 N·m). Install bearing cup (32), make up shim pack (33) to a thickness of 0.040 inch and install shim pack and bearing cap (34). Tighten cap screws evenly to 20 ft.-lbs. (27 N·m). Check input shaft (27) for end play. Add or remove shims as required until shaft will rotate freely with zero end play. Do

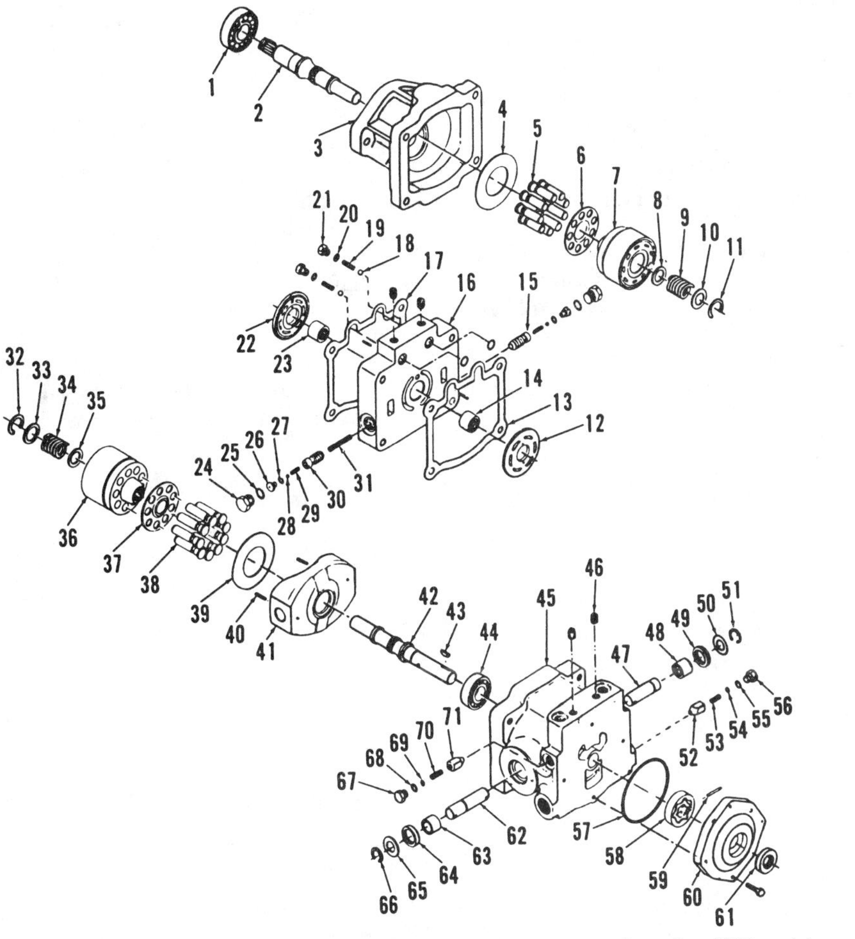

Fig. S29—Exploded view of typical Sundstrand hydrostatic drive unit used on 5216 models.

1. Ball bearing
2. Motor (output) shaft
3. Motor housing
4. Thrust plate
5. Pistons (9)
6. Shoe plate
7. Cylinder block
8. Washer
9. Spring
10. Washer
11. Snap ring
12. Pump valve plate
13. Gasket
14. Needle bearing
15. Reverse acceleration valve
16. Center section
17. Gasket
18. Check valve ball (2)
19. Spring (2)
20. "O" ring
21. Plug (2)
22. Motor valve plate
23. Needle bearing
24. Plug (2)
25. "O" ring (2)
26. Cap (2)
27. "O" ring (2)
28. Ball (2)
30. Forward acceleration valve
31. Acceleration spring
32. Snap ring
33. Washer
34. Spring
35. Washer
36. Cylinder block
37. Shoe plate
38. Pistons (9 used)
39. Thrust plate
40. Roll pin (2)
41. Swashplate
42. Pump (input) shaft
43. Key
44. Ball bearing
45. Pump housing
46. Pressure check port plug
47. Stub shaft
48. Needle bearing
49. Oil seal
50. Washer
51. Snap ring
52. Hydraulic lift relief valve
53. Spring
54. Shim
55. "O" ring
56. Plug
57. "O" ring
58. Charge pump assy.
59. Drive pin
60. Charge pump cover
61. Oil seal
62. Control shaft
63. Needle bearing
64. Oil seal
65. Washer
66. Snap ring
67. Plug
68. "O" ring
69. "O" ring
70. Spring
71. Charge pressure relief valve

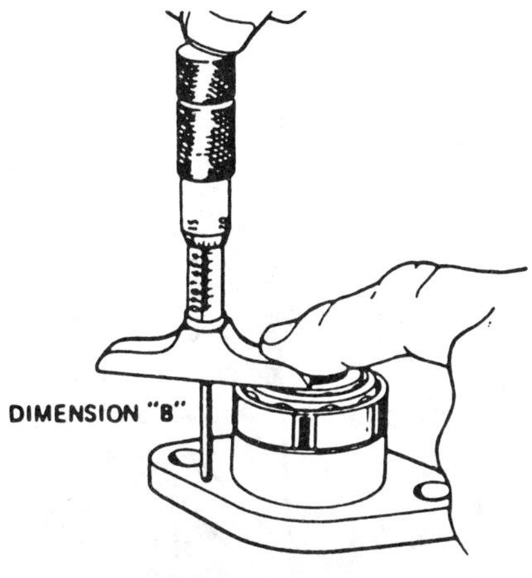

Fig. S26—Use a depth gage to measure dimension "B" when determining thickness of shim pack (15—Fig. S25) to be used with new bearing assembly (16 and 17).

not preload bearings. When adjustment is correct, remove cap and shims. Apply a light coat of sealer to shims and cap screw threads, reinstall cap and shims, then tighten cap screws to 20 ft.-lbs. (27 N•m). Place first and second sliding gear (30) on sliding gear shaft (29), then position assembly in transmission housing. Install third speed sliding gear (28) with bevel edge of teeth to right side and install bearing cone (21) on shaft. Install bearing cap (18) with "O" ring (19) and bearing cup (20). Use sealer on cap screw threads and tighten cap screws to 20 ft.-lbs. (27 N•m). Install bearing cone (35) and bearing cup (36). Make up a 0.040 inch thick shim pack (37) and install shim pack and bearing cap (38). Tighten cap screws evenly to 20 ft.-lbs. (27 N•m). Check sliding gear shaft for end play. Add or remove shims as required until the shaft will rotate freely with zero end play. Do not preload bearings. When adjustment is correct, remove bearing cap (38) and shim pack (37). Apply a light coat of sealer to shims and cap screw threads, reinstall shim pack and cap, then tighten cap screws to 20 ft.-lbs. (27 N•m).

Using new oil seal (8), gasket (11) and "O" rings (10 and 12) install shifter components (1 through 9). Reinstall differential assembly, top pto shaft and hydrostatic drive unit.

Reassemble by reversing disassembly procedure.

HYDROSTATIC DRIVE UNIT. Scribe mark across charge pump cover (60-Fig. S29) and pump housing (45), pump housing and center section (16) and center section and motor housing (3) for aid in reassembly. Remove all rust, paint or burrs from end of pump shaft (42). Unbolt and remove charge pump cover (60), then remove charge pump (58) and withdraw drive pin (59) from pump shaft. Remove oil seal (61) from cover. Remove hydraulic lift relief valve assembly (52 through 56) and charge pressure relief valve (67 through 71).

CAUTION: Keep valve assemblies separated. Do not mix component parts.

Remove the four through-bolts and separate pump assembly, center section and motor assembly. Valve plates (12 and 22) may stick to cylinder blocks (7 and 36). Be careful not to let them drop. Remove pump cylinder block and piston assembly (32 through 38) and lay aside for later disassembly. Remove thrust plate (39), drive out roll pins (40), then withdraw control shaft (62) with snap ring (66) and washer (65) and stub shaft (47) with snap ring (51) and washer (50). Lift out swashplate (41). Using a hot air gun or by dipping bearing end of housing in oil heated to approximately 250°F (121°C), heat housing and drive out input shaft (42) with bearing (44). Press bearing from input shaft. Oil seals (49 and 64) and needle bearings (48 and 63) can now be removed from housing (45).

If valve plates (12 and 22) stayed on center section (16), identify plates and lay them aside. Remove check valve assemblies (18 through 21), then remove forward acceleration valve assembly (24 through 30) and spring (31). Remove reverse acceleration valve from opposite side. Acceleration valves are not interchangeable. Reverse acceleration valve can be identified by small orifice hole in side of valve body (15). Needle bearings (14 and 23) can now be removed.

Remove motor cylinder block and piston assembly (5 through 11) and thrust plate (4) from motor housing. Heat bearing end of housing (3), using a hot air gun or by dipping in oil heated to approximately 250°F (121°C), then drive output shaft and bearing assembly from housing. Remove output bevel pinion and shims, then press bearing (1) from shaft (2).

Carefully remove shoe plate and pistons (5 and 6) and (37 and 38) from cylinder blocks (7 and 36). Place each cylinder block on wood blocks in a press, compress spring (9 or 34) and remove snap ring (11 or 32). Release press and remove spring and washers.

Inspect pistons and bores in cylinder blocks for excessive wear or scoring. Light scratches on piston shoes can be removed by lapping. Inspect valve plates (12 and 22) and valve plate contacting surfaces of cylinder blocks (7 and 36) for excessive wear or scoring and renew as necessary. Check thrust plates (4 and 39) for excessive wear or other damage.

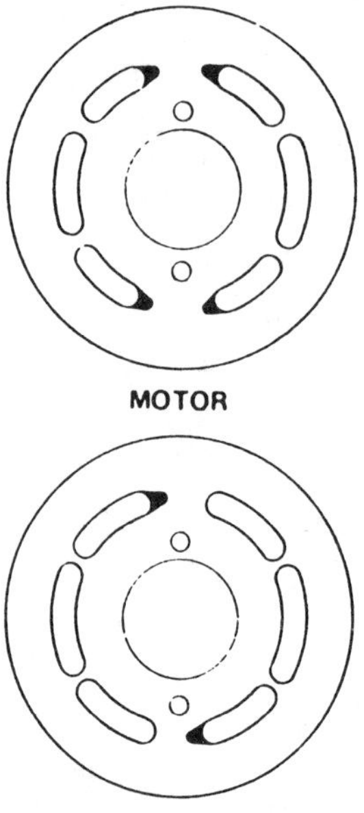
MOTOR
PUMP

Fig. S30—Motor plate has four notches (dark areas) and pump plate has two notches. Refer to text.

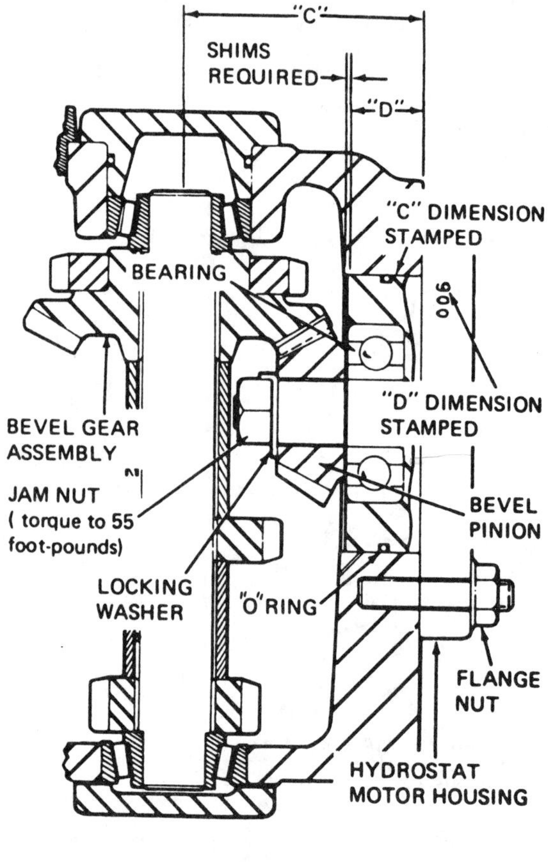
"C"
SHIMS REQUIRED
"D"
"C" DIMENSION STAMPED
906
BEARING
"D" DIMENSION STAMPED
BEVEL GEAR ASSEMBLY
JAM NUT (torque to 55 foot-pounds)
BEVEL PINION
LOCKING WASHER
"O" RING
FLANGE NUT
HYDROSTAT MOTOR HOUSING

Fig. S31—View showing location of "C" and "D" dimension code numbers stamped on three speed transmission case and hydrostatic motor housing. Stamped code numbers are used with chart in Fig. S31A to determine shim pack thickness to be installed between bearing and output bevel pinion. Refer to text.

Inspect charge pump (58) for wear, pitting or scoring.

Renew all oil seals, "O" rings and gaskets, lubricate all internal parts with new transmission fluid. Reassemble by reversing disassembly procedure. Heat bearing end of housings (3 and 45) before installing shaft and bearing assemblies. Install needle bearings (14 and 23) in center section (16) until they are 0.100 inch (2.54 mm) above machined surfaces. Install pump valve plate (12) and motor valve plate (22) in original positions with retaining pins in grooves. Pump valve plate (12) has two relief notches and motor valve plate (22) has four relief notches. See Fig. S30. Check valve assemblies are interchangeable and are installed at rear side of center section. When installing acceleration valves, reverse acceleration valve (with small orifice hole in the side) must be installed in left side of center section. Install charge pressure relief valve assembly (67 through 71 – Fig. S29) using original shim pack (69). Install lift relief valve assembly (52 through 56) with original shim pack (54). Tighten the four through-bolts to a torque of 35 ft.-lbs. (48 N•m). Install drive pin (59), charge pump (58) and cover (60). Tighten cover retaining cap screws to a torque of 20 ft.-lbs. (27 N•m).

If original motor housing (3), output shaft (2), bearing (1) and output bevel pinion are being reused, install output bevel gear with original shim pack and tighten retaining nut to 55 ft.-lbs. (75 N•m).

If new output bevel pinion, output shaft, bearing, motor housing or complete new hydrostatic assembly is being installed, correct shim pack must be installed between output shaft bearing (1) and output bevel pinion. To determine correct thickness shim pack, refer to Fig. S31 and note dimension (D) code number stamped on motor housing flange and dimension (C) code number stamped on three speed transmission case. Refer to chart in Fig. S31A and using dimension (D and C – Fig. S31) code numbers, determine correct thickness shim pack from chart. For example: If dimension (D) number on motor housing is 905 and dimension (C) number is 3012, correct thickness shim pack should be 0.045 inch. Install shim pack between output shaft bearing (1 – Fig.S29) and output bevel pinion and tighten nut to 55 ft.-lbs. (75 N•m).

Before installing hydrostatic drive unit, place unit on a bench with charge pump inlet port upward. Rotate pump shaft counter-clockwise while pouring new fluid into charge pump inlet port. When resistance is felt on rotation of input (pump) shaft, enough fluid has been installed to prime system and lubricate unit during initial operation. Plug all openings until unit is installed.

Models 6216-6218

Remove tires, wheels and hubs from both axles. Clean ends of axle shafts sufficiently so that axles can be withdrawn through bearings (61 and 64) and seals (62 and 63). Remove retaining ring (32 – Fig.S32), brake drum (30) and key (31). Remove cap screws and separate case halves (4 and 6). Lift out axles and differential assembly. Remove shaft and gear assembly (25). Remove nuts from shift rails (11 and 33) and reverse gear shaft (44). Remove shift rail (33), shift fork (37), shaft (44) and gear (46). Remove shift rail (11), shift fork (15), shaft (10) and gear (16). The balance of disassembly is evident after examination.

Clean and inspect all parts for wear or damage. Renew as necessary.

Reassemble by reversing disassembly procedure. Note differential bolts are tightened to 20 ft.-lbs. (27 N•m) torque. Nuts securing shift rails and reverse shaft are tightened to 50 ft.-lbs. (68 N•m). Refill transmission with SAE 90 EP gear lube.

Models 7018-7117-7119

GEAR CASE. Remove all paint, rust or burrs from brake drum shaft (1 – Fig.

DIMENSION "D"
HYDROSTATIC PUMP-MOTOR (On Motor Housing)

DIMENSION "C" TRANSMISSION NUMBER (On Case)	900	901	902	903	904	905	906	907	908	909	910	911	912	913	914	915
3000	.038	.037	.036	.035	.034	.033	.032	.031	.030	.029	.028	.027	.026	.025	.024	.023
3001	.039	.038	.037	.036	.035	.034	.033	.032	.031	.030	.029	.028	.027	.026	.025	.024
3002	.040	.039	.038	.037	.036	.035	.034	.033	.032	.031	.030	.029	.028	.027	.026	.025
3003	.041	.040	.039	.038	.037	.036	.035	.034	.033	.032	.031	.030	.029	.028	.027	.026
3004	.042	.041	.040	.039	.038	.037	.036	.035	.034	.033	.032	.031	.030	.029	.028	.027
3005	.043	.042	.041	.040	.039	.038	.037	.036	.035	.034	.033	.032	.031	.030	.029	.028
3006	.044	.043	.042	.041	.040	.039	.038	.037	.036	.035	.034	.033	.032	.031	.030	.029
3007	.045	.044	.043	.042	.041	.040	.039	.038	.037	.036	.035	.034	.033	.032	.031	.030
3008	.046	.045	.044	.043	.042	.041	.040	.039	.038	.037	.036	.035	.034	.033	.032	.031
3009	.047	.046	.045	.044	.043	.042	.041	.040	.039	.038	.037	.036	.035	.034	.033	.032
3010	.048	.047	.046	.045	.044	.043	.042	.041	.040	.039	.038	.037	.036	.035	.034	.033
3011	.049	.048	.047	.046	.045	.044	.043	.042	.041	.040	.039	.038	.037	.036	.035	.034
3012	.050	.049	.048	.047	.046	.045	.044	.043	.042	.041	.040	.039	.038	.037	.036	.035
3013	.051	.050	.049	.048	.047	.046	.045	.044	.043	.042	.041	.040	.039	.038	.037	.036
3014	.052	.051	.050	.049	.048	.047	.046	.045	.044	.043	.042	.041	.040	.039	.038	.037
3015	.053	.052	.051	.050	.049	.048	.047	.046	.045	.044	.043	.042	.041	.040	.039	.038
3016	.054	.053	.052	.051	.050	.049	.048	.047	.046	.045	.044	.043	.042	.041	.040	.039
3017	.055	.054	.053	.052	.051	.050	.049	.048	.047	.046	.045	.044	.043	.042	.041	.040
3018	.056	.055	.054	.053	.052	.051	.050	.049	.048	.047	.046	.045	.044	.043	.042	.041
3019	.057	.056	.055	.054	.053	.052	.051	.050	.049	.048	.047	.046	.045	.044	.043	.042
3020	.058	.057	.056	.055	.054	.053	.052	.051	.050	.049	.048	.047	.046	.045	.044	.043
3021	.059	.058	.057	.056	.055	.054	.053	.052	.051	.050	.049	.048	.047	.046	.045	.044
3022	.060	.059	.058	.057	.056	.055	.054	.053	.052	.051	.050	.049	.048	.047	.046	.045
3023	.061	.060	.059	.058	.057	.056	.055	.054	.053	.052	.051	.050	.049	.048	.047	.046
3024	.062	.061	.060	.059	.058	.057	.056	.055	.054	.053	.052	.051	.050	.049	.048	.047

Fig. S31A – Chart used in determining shim pack thickness to be installed between bearing and output bevel pinion. Refer to text for procedure.

S33) and axle tube (37) to prevent bearing or seal damage during removal or installation. Make certain all grease fittings and keys have been removed from axle tube and remove cover (33), intermediate shaft (26), gear (27) and washers (25 and 28). Remove drive gear (17), axle tube (37) and thrust washers. Remove all seals from case halves. Continue disassembly as necessary.

Reassemble by reversing disassembly procedure.

HYDROSTATIC DRIVE UNIT. Clean exterior of unit and remove hydraulic hoses, fan and control cam assembly from unit. Scribe a mark on control side of housing, center section and charge pump as an assembly aid. Remove charge pump housing (23 – Fig. S34). Remove gerotor assembly (22) and drive pin. Pry oil seal (25) from housing and press needle bearing (24) out front of housing. Remove both check valves (37), back-up washers and "O" rngs. Remove relief valves (20 and 28) from bores in center section (27). Unbolt and remove center section.

CAUTION: Valve plates (19 and 30) may stick to center section. Taking care not to drop them, remove and identify valve plates.

Lift out both cylinder block and piston assemblies. Drive roll pins (12) into swashplate (14) and remove stub shaft (2), control shaft (8), swashplate (14) and thrust plate (15). Withdraw pump shaft (13) and bearing (11). Remove cap screws securing motor swashplate (16) to housing, then lift out swashplate and motor shaft (17). Bearings and oil seals can now be removed from housing as required.

To disassemble cylinder block and piston assemblies, carefully withdraw slipper retainer (35) with pistons (36) from cylinder block (34). Place cylinder block on a wood block in a press; compress spring (33) and remove retaining ring (31). Spring (33) should have a free length of 1-3/64 to 1-1/16 inches (26.6 to 27.9 mm) and should test 63-75 pounds (280-334 N) force when compressed to a length of 19/32-inch (15 mm). Check cylinder blocks for scratches or other damage and renew as necessary. Inspect pistons and bores for excessive wear or scoring. Piston slippers can be lapped to remove light scratches. Minimum slipper thickness is 0.021 inch (0.533 mm) for both pump and motor. Slipper

1. Bushings
2. Shift lever
3. Pivot rod
4. Case
5. Gasket
6. Case
7. Seal
8. Bearing
9. Washer
10. Input shaft
11. Shift rail (2nd & 3rd)
12. Detent ball
13. Detent spring
14. Spring
15. Shift fork (2nd & 3rd)
16. Pinion gear assy.
17. Key
18. Retaining ring
19. Washer
20. Bearing
21. Key
22. Bearing
23. Washer
24. Retaining ring
25. Shaft & gear assy.
26. Washer
27. Bearing
28. Seal
29. Washer
30. Brake drum
31. Key
32. Retaining ring
33. Shift rail (1st & Rev.)
34. Spring
35. Detent spring
36. Detent ball
37. Shift fork (1st & Rev.)
38. Bearing
39. Washer
40. Gear assy. (1st)
41. Washer
42. Bearing
43. Locknut
44. Reverse gear shaft
45. Bushing
46. Gear (1st & Rev.)
47. "O" ring
48. Brake band
49. Brake lining
50. Bearing
51. Washer
52. Drive gear
53. Differential gears
54. Snap ring
55. Differential pinion gears
56. Differential studs
57. Differential plate
58. Washer
59. Bearing
60. Axle shaft
61. Bearing
62. Seal
63. Seal
64. Bearing
65. Axle tube
66. "O" ring
67. Axle shaft
68. Spring
69. Differential pinion
70. Spacer

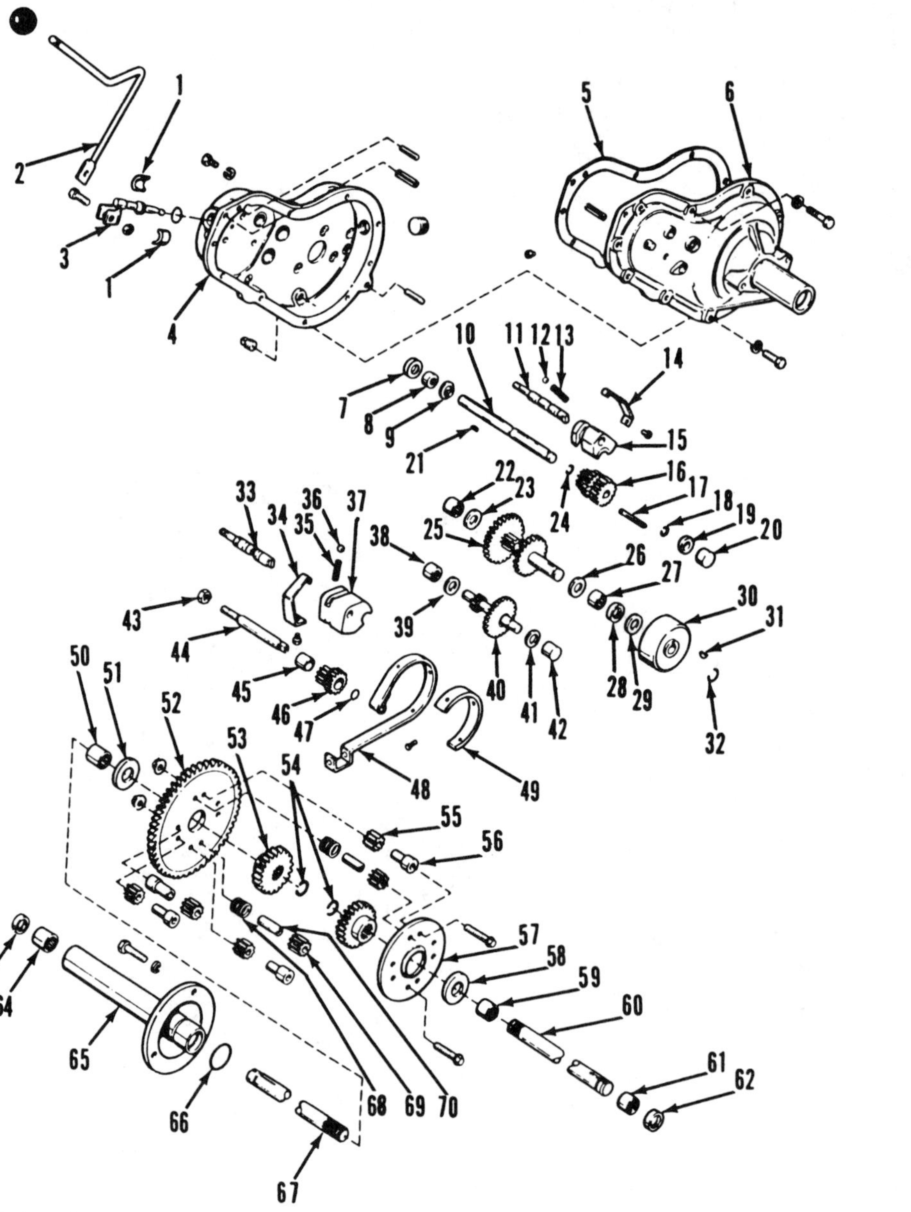

Fig. S32 – Exploded view of three speed transmission used for 6216 and 6218 models.

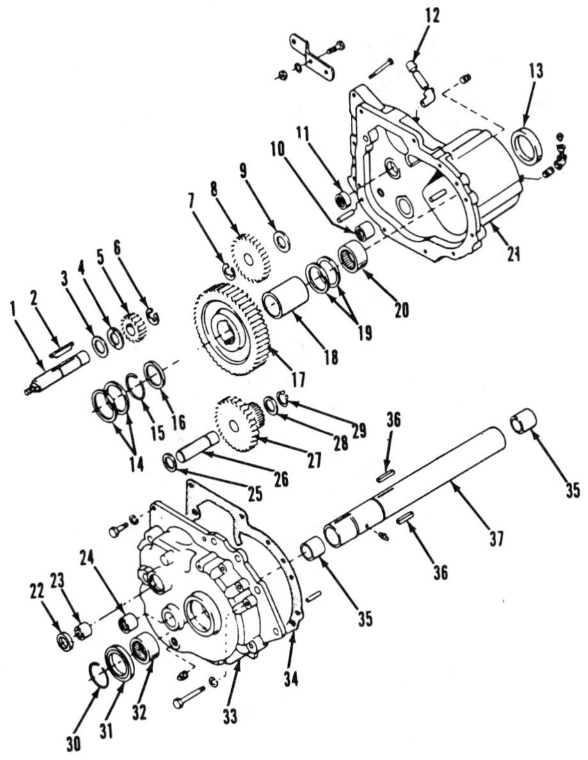

Fig. S33—Exploded view of gear case used for 7018, 7117 and 7119 models in conjunction with the hydrostatic transmission.

1. Brake shaft
2. Key
3. Washer
4. Washer
5. Gear
6. Retaining ring
7. Retaining ring
8. Gear (1st)
9. Washer
10. Bearing
11. Bearing
12. Cap
13. Seal
14. Washers
15. Retaining ring
16. Washer
17. Drive gear
18. Spacer
19. Washers
20. Bearing
21. Case
22. Seal
23. Bearing
24. Bearing
25. Washer
26. Intermediate shaft
27. Gear assy.
28. Washer
29. Retaining ring
30. Retaining ring
31. Seal
32. Bearing
33. Cover
34. Gasket
35. Bushings
36. Keys
37. Axle tube

thickness must not vary more than 0.002 inch (0.051mm) for all nine pistons in each block.

Check pump and motor valve plates for excessive wear and note pump valve plate has two notches and motor valve plate has four notches. See Fig. S30. Valve plates are not interchangeable. Inspect motor swashplate and pump swashplate thrust plate for wear, embedded material or scoring. Check charge pump housing (23–Fig. S34) and rotor assembly (22) for wear or scoring and renew as necessary.

Renew all "O" rings, gaskets and seals. Reassemble by reversing disassembly procedure. Lubricate all parts with clean oil when reassembling. When installing new bearings (29), press bearings into center section (27) until 1/16 to 1/8-inch (1.5 to 3 mm) of bearing protrudes; bearings are pilots for valve plates when unit is reassembled. Check valves (37) are interchangeable and are serviced only as an assembly. Pump swashplate (14) must be installed with thin pad towards top of transmission. Press pins (3 and 4–Fig. S35) into swashplate (7) to dimension (D), which should be 1/4-inch (6 mm) below swashplate surface. Two pins (3) are used in control shaft (1) side of swashplate. Install motor swashplate (16–Fig. S34) with notch at top and high point of cam angle towards bottom. Flat end of charge pump housing (23) must be installed towards right side of unit. Position screws which hold center section (27) and housing (6) together in their respective holes and tighten evenly to 20-30 ft.-lbs. (27 N·m). Tighten charge pump mounting bolts to 50-55 ft.-lbs. (68-75 N·m).

DIFFERENTIAL

MAINTENANCE

All Models

Differential assembly on all models except 7018, 7117 and 7119 models is an integral part of transmission assemblies and fluid level is maintained according to TRANSMISSION MAINTENANCE section.

Models 7018, 7117 and 7119 use a differential assembly which is a sealed unit located on right axle tube.

R&R DIFFERENTIAL

Models 5216-6216-6218

Differential is an integral part of the transmission. Refer to appropriate TRANSMISSION section for model being serviced.

Models 7018-7117-7119

Raise and support rear of tractor. Remove both rear tires and wheels. Loosen setscrew on hub retaining collar for right wheel hub, then remove collar and hub. Loosen left collar setscrew and slide axle to the right until collar is against left hub. Remove recessed washer, retaining ring and the two keys from right end of axle shaft. Slide differential assembly off and remove axle shaft.

Reinstall by reversing removal procedure.

OVERHAUL DIFFERENTIAL

Models 5216-6216-6218

Differential is an integral part of the transmission. Refer to appropriate TRANSMISSION section for model being serviced.

Models 7018-7117-7119

Remove the six bolts (12–Fig. S36) and nuts (3) from around outer rim of differential covers. Slowly and evenly loosen, then remove the eight locknuts (2). Remove inner differential cover (4) and all internal parts.

Clean and inspect all parts for wear or damage. Renew parts as necessary.

Generously coat all internal parts with multi-purpose, lithium base grease.

Reassemble by reversing disassembly procedure.

FINAL DRIVE

Model 5216 is equipped with final drive assemblies.

MAINTENANCE

Left and right final drive fluid levels should be checked at 50 hour intervals. Fluid level should be maintained at lower edge of fill plug (Fig. S37). Recommended fluid is SAE 90 EP gear lube and capacity of each final drive housing is 0.5 quarts (0.47 L).

R&R FINAL DRIVE

Raise and support rear of tractor, then remove wheel and tire from side to be serviced. Remove fender assembly, then unbolt and remove final drive assembly. Retaining ring and coupling (2 and 3—Fig. S38) will remain in axle housing.

OVERHAUL FINAL DRIVE

Drain fluid from final drive unit. Remove spring (4-Fig. S38), then unbolt and remove housing plate (1) with bearing (5), slinger (6), bull pinion (7) and thrust washer (13). Press bull pinion and bearing assembly from plate (1), then press bull pinion from bearing and remove slinger. Remove locknut (8) and

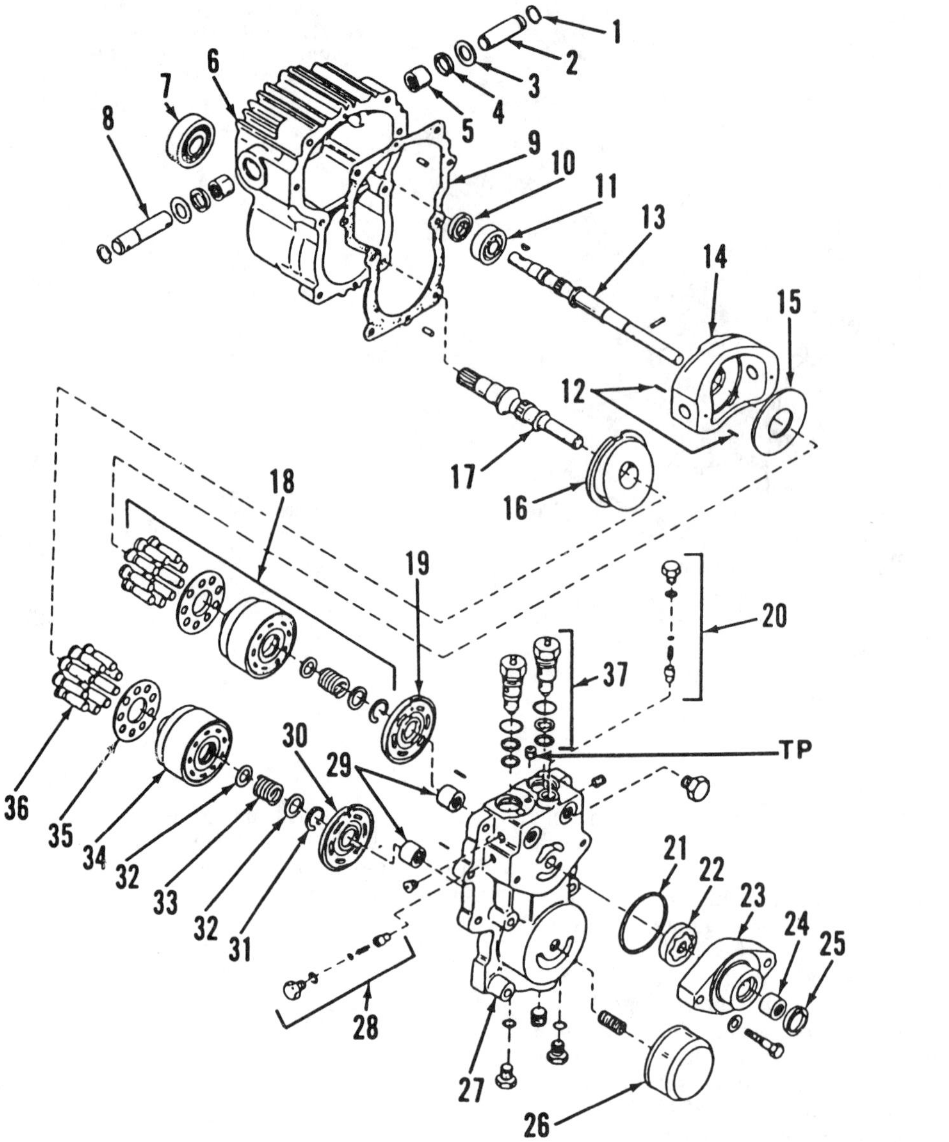

TP. Test port
1. Retaining ring
2. Stub shaft
3. Washer
4. Seal
5. Needle bearing
6. Housing
7. Ball bearing
8. Control shaft
9. Gasket
10. Seal
11. Ball bearing
12. Roll pins (3)
13. Pump shaft
14. Swashplate (pump)
15. Thrust plate
16. Swashplate (motor)
17. Motor shaft
18. Pump assy.
19. Valve plate (pump)
20. Relief valve (implement)
21. "O" ring
22. Gerotor assy.
23. Charge pump housing
24. Needle bearing
25. Seal
26. Oil filter
27. Center section
28. Relief valve (charge)
29. Needle bearings
30. Valve plate (motor)
31. Retaining ring
32. Washers
33. Spring
34. Cylinder block (motor)
35. Slipper retainer
36. Pistons (motor)
37. Check valves

Fig. S34 – Exploded view of typical Sundstrand hydrostatic transmission used on 7018, 7117 and 7119 models. Early models use a remote mounted oil filter in place of filter (26) shown. On tractors not equipped with hydraulic lift, relief valve (28) is not used and valve (20) is charge pressure relief valve.

washer (9) and lift out bull gear (10). Withdraw axle (20), then remove bearing cones (11 and 18) and oil seal (19) from housing. If necessary, press roller bearing (16) and bearing cups (12 and 17) from housing.

Clean and inspect all parts and renew any showing excessive wear or other damage.

To reassemble, press roller bearing (16) and bearing cups (12 and 17) in housing, if removed. Lubricate bearing cone (18) with SAE 90 EP gear lube and install in cup (17). Install oil seal (19)

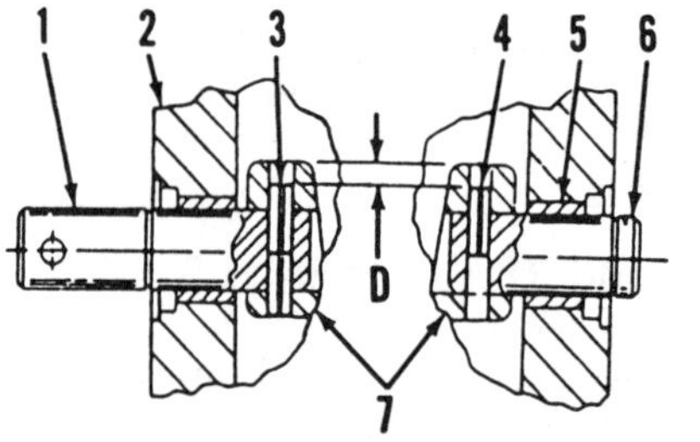

Fig. S35 – When reinstalling pump swashplate (7), drive roll pins in until 1/4-inch (6 mm) "D" below swashplate surface. Two pins are used on control shaft (1) side.

1. Control shaft
2. Housing
3. Roll pins (2)
4. Roll pin (1)
5. Bearing
6. Stub shaft
7. Swashplate

Fig. S36 – Exploded view of differential assembly used on 7018, 7117 and 7119 models.

1. Seal
2. Locknuts
3. Nuts
4. Differential cover
5. Spacer plate
6. Axle washer
7. Snap ring
8. Keys
9. Differential gear
10. Axle washers
11. Differential cover
12. Cap screws
13. Carrier
14. Cap screws
15. Pinion gears
16. Springs
17. Shafts

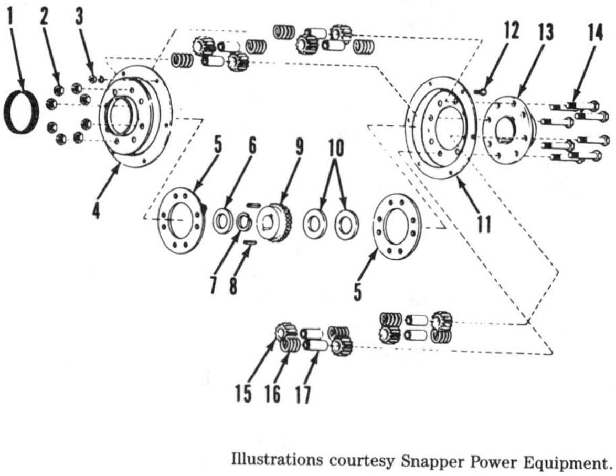

Illustrations courtesy Snapper Power Equipment.

with lip inward, in housing (15). Insert axle (20) through seal and bearing. Lubricate bearing cone (11) with SAE 90 EP gear lube and install it over axle until it is seated in bearing cup (12). Install bull gear (10), washer (9) and locknut (8). Tighten locknut until a force of 12 in.-lbs. (1.5 N) is required to rotate axle.

Reassemble by reversing disassembly procedure.

BEVEL GEAR BOX

On 7018, 7117 and 7119 models, main drive from engine is connected to a bevel gear box located under the operators seat.

MAINTENANCE

Fluid level in bevel gear box should be checked at 25 hour intervals. Fluid level should be maintained at a level which just touches end of roll pin in check plug when plug is just contacting gear box housing. **DO NOT** screw plug in to check level. Recommended fluid is SAE 90 EP gear lube. Fluid should be changed at 400 hour intervals.

R&R AND OVERHAUL

To remove bevel gear unit, first remove complete drive unit, pulleys and belts. Refer to appropriate TRANSMISSION section for removal procedure. Refer to POWER TAKE-OFF section and remove pto clutch assembly and idler pulley assembly. Remove pulley from right side of bevel gear unit. Disconnect drive shaft from bevel gear input shaft. Remove cap screws securing gear box to side plates. Pull side plates apart and withdraw bevel gear unit.

Reinstall by reversing removal procedure.

Drain lubricant and remove case cover (17–Fig. S39). Remove nut (1) and key from shaft (26). Support gear (23) and drive shaft (26) through gear and case to expose key (25). Remove key. Remove retaining ring (24) from groove and slide it toward gear. Remove shaft (26) from case and retaining ring and gear from inside case. Remove cap screw and clamp plate (12). Remove key from external portion of shaft (8) and drive shaft, gear and ball bearing assembly out of case. Remove cap screw (15) and press shaft from gear (13). Remove key and press shaft from bearing.

Reassemble by reversing disassembly procedure.

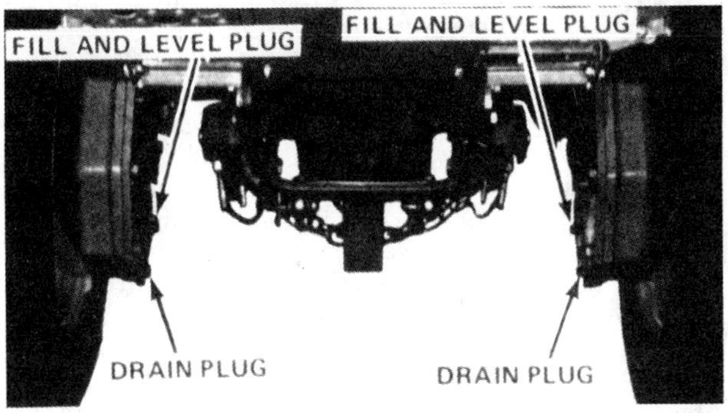

Fig. S37—View showing fill and drain plugs on final drive units of 5216 model. Refer to text.

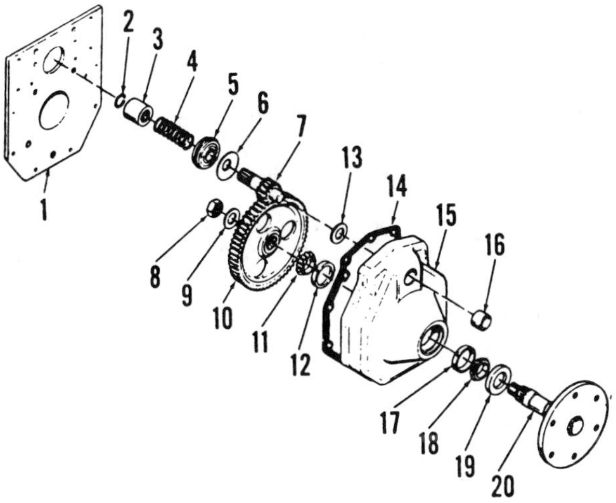

Fig. S38—Exploded view of final drive unit used on 5216 models.

1. Housing plate
2. Retaining ring
3. Coupling
4. Spring
5. Bearing
6. Oil slinger
7. Bull pinion
8. Locknut
9. Washer
10. Bull gear
11. Bearing
12. Bearing cup
13. Thrust washer
14. Gasket
15. Housing
16. Roller bearing
17. Bearing cup
18. Bearing
19. Oil seal
20. Axle

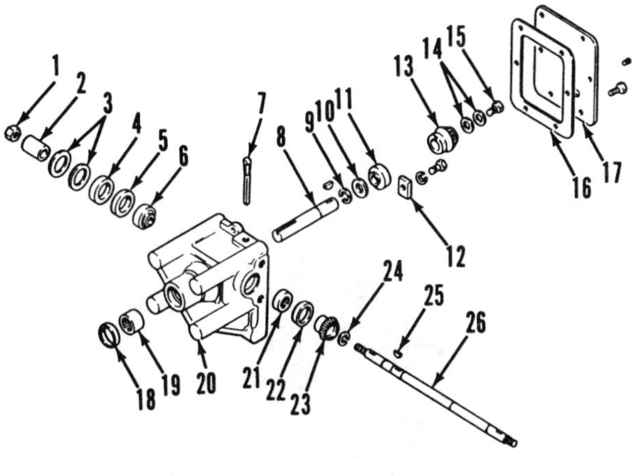

Fig. S39—Exploded view of bevel gear box used on 7018, 7117 and 7119 models.

1. Nut
2. Spacer
3. Shims
4. Seal
5. Spacer
6. Bearing
7. Dipstick
8. Input shaft
9. Retaining ring
10. Washer
11. Bearing
12. Clamp plate
13. Bevel gear (drive)
14. Washers
15. Cap screw
16. Gasket
17. Cover
18. Seal
19. Bearing
20. Housing
21. Bearing
22. Seal
23. Bevel gear (driven)
24. Retaining ring
25. Key
26. Output shaft

POWER TAKE-OFF

ADJUSTMENTS

Model 5216

FRONT CLUTCH ADJUSTMENT. To check front electric clutch adjustment, insert a 0.010 inch (0.254 mm) feeler gage (Fig. S40) through each of the openings approximately 1/16-inch (1.6 mm). Nuts (A) should each be turned 1/12-turn until feeler gage just does enter 1/16-inch (1.6 mm).

Models 6217-6218

BELT STOP ADJUSTMENT. Belt stops (A – Fig. S41) are adjusted by loosening belt stop mounting bolts and positioning them so there is 1/16-inch (1.6 mm) clearance between each stop and belt. Retighten mounting bolts.

Models 7018-7117-7119

POWER TAKE-OFF CLUTCH ADJUSTMENT. When pto clutch is properly adjusted, clutch pulley (Fig. S42) will move exactly ⅛-inch (3.2 mm) toward clutch cone when pto lever is moved from fully disengaged to fully engaged position. Adjust jam nuts (A) on end of pto clutch rod to obtain correct movement.

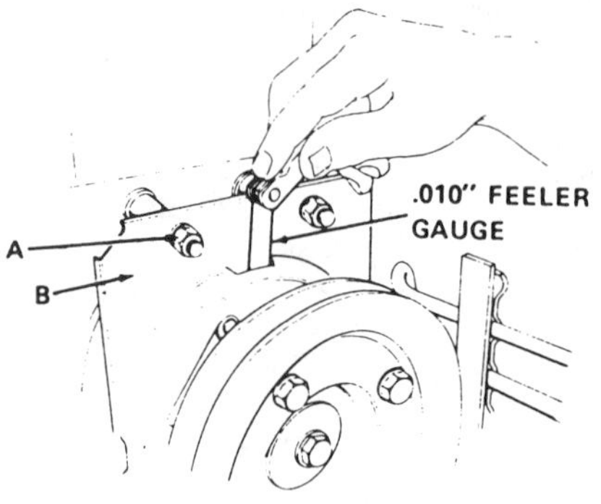

Fig. S40 – View showing front electric clutch adjustment on 5216 models. Refer to text.

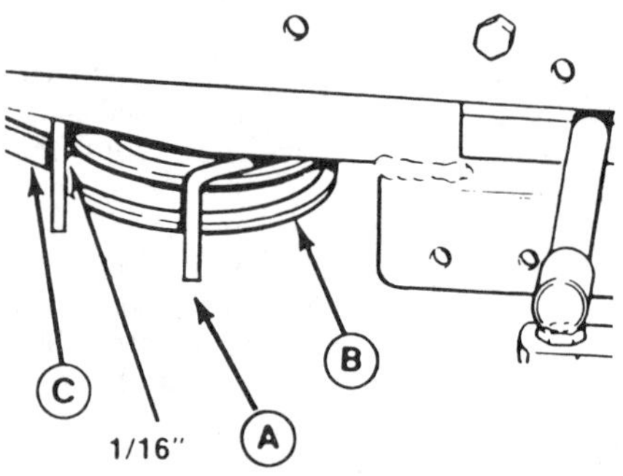

Fig. S41 – View showing pto belt stop adjustment on 6216 and 6218 models. Refer to text.

R&R AND OVERHAUL

Model 5216

FRONT PTO CLUTCH. Remove belt, cap screws and pulley from pto clutch (Fig. S43). Remove cap screw and washers from end of crankshaft, then withdraw clutch-armature assembly and remove key. Disconnect clutch wires at connector, then unbolt and remove clutch field unit from engine.

Disassemble clutch-armature assembly by pressing armature from ball bearing, then remove front retaining ring and press bearing from clutch.

Clean and inspect all parts. Normal resistance reading for clutch field winding is 2.75 to 3.60 ohms. If reading is higher or lower, field must be renewed.

Reassemble by reversing disassembly procedure. Tighten field retaining screws to 30 in.-lbs. (3.4 N•m). Tighten clutch-armature retaining cap screw in crankshaft to 20 ft.-lbs. (27 N•m) and pulley retaining cap screws to 10 ft.-lbs. (14 N•m).

REAR PTO AND CLUTCH. Remove seat support and top and bottom frame covers. Clean top of transmission and surrounding area. Unbolt and remove flexible coupling (1- Fig. S44), yoke (2), felt ring (17) and rubber washer (16). Disconnect electric wires and link (9). Unbolt electric clutch assembly (7) from hub (3). Lower front end of pto drive shaft (15), then remove clutch and shaft assembly through frame top opening. Remove cap screw (5) with lockwasher and collar (6). Remove clutch-armature assembly and Woodruff key. Press armature from bearing, remove bearing retaining ring and press clutch from bearing. Remove the four socket head screws securing field to bearing housing (12) and withdraw field and washer (10). Remove retaining ring (14) and washer (13) and remove bearing (11) and housing (12) from drive shaft (15).

Unbolt and remove transmission top cover and pto shield (34). Remove cap screw (4) and washers, then remove hub (3). Pry out oil seal (18), remove bearing retaining ring (20) and move shaft (22) forward. Remove snap ring (24) and gear (23) from shaft, then withdraw shaft from front of transmission housing. Spacer (19) and bearing (21) can now be removed from shaft (22), then needle bearing (25) can be removed from rear of transmission housing.

Remove the two studs, then withdraw idler shaft (33) with "O" ring (32) and spring (31). Remove idler gear (29), bearing (28), thrust washers (30 and 27) and spacer (26).

Spread retaining ring in collar (35), then remove collar, retainers (36 and 39), spring (38) and balls (37). Remove snap ring (40), then unbolt and remove bearing cap (42) with oil seal (41), "O" ring (43) and shims (44). Withdraw output shaft (47) with bearing cup (45) and cone (46). Remove output gear (48) and bearing cone (52) from above. If necessary, remove bearing cup (53) from housing. Remove snap ring (51), plug (50) and "O" ring (49) from output shaft (47).

Clean and inspect all parts and renew any showing excessive wear or other damage. To check clutch field, connect leads from an ohmmeter across field connector terminals. Normal resistance reading should be between 2.75 and 3.60 ohms. If reading is higher or lower, coil winding is defective and field must be renewed.

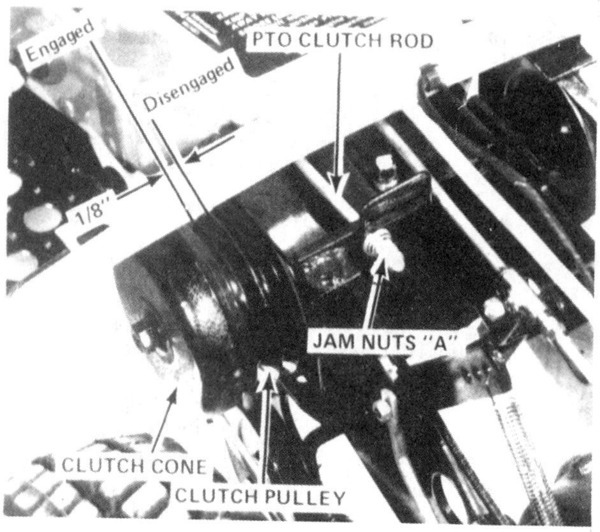

Fig. S42 – View showing clutch (pto) adjustment location on 7018, 7117 and 7119 models. refer to text.

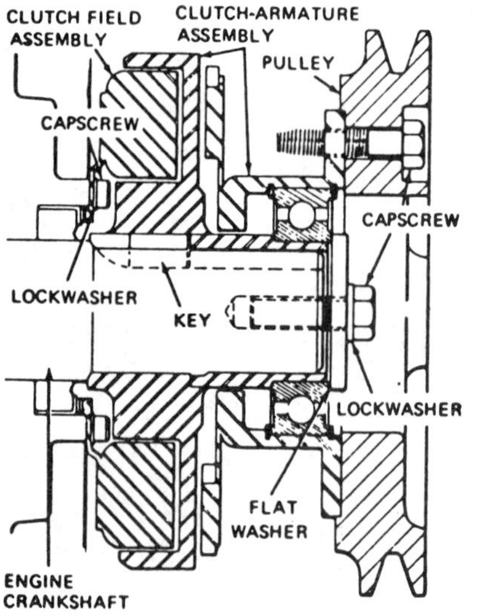

Fig. S43 – Cross-sectional view of electric pto clutch used on some models. Refer to text.

When reassembling, renew all oil seals, "O" rings and gaskets. Reassemble by reversing disassembly procedure. Output shaft (47) should have an end play of 0.002-0.005 inch (0.0508-0.1270 mm). Use a dial indicator to check end play and if necessary, add or remove shims (44) to correct end play. When installing cap screws (4 and 5), use "Loctite", or equivalent, on threads and tighten to 45 ft.-lbs. (61 N·m).

Models 6216-6218

A spring loaded idler is used to apply tension to pto belt. Correct belt tension is maintained by adjusting idler tension spring length. Tension of idler pulley against drive belt should be sufficient to drive implement without belt slippage. Excessive tension will cause premature failure of belts and pulley bearings. Service procedures are apparent after examination.

Models 7018-7117-7119

Remove drive belt from pto pulley. Remove retaining nut (15 – Fig. S45) and withdraw clutch plate (14). Protect threads on pto shaft and pry key out of shaft keyway.

CAUTION: Pto pulley is spring loaded. Install C-clamp to compress spring and hold pulley before attempting to remove pulley retaining ring.

Compress pulley internal spring with a C-clamp, then remove retaining ring. Remove pivot assembly retaining screws. Slowly release clamp and remove pto clutch assembly. Remove cotter pin retaining idler pivot to right side plate, then withdraw idler assembly.

Check pto shaft runout at outer retaining ring groove with a dial indicator. If runout exceeds 0.010 inch (0.254 mm), shaft should be renewed or straightened.

Inspect all parts for excessive wear or damage and renew as necessary.

Reassemble by reversing disassembly procedure. Make certain bearing (11) is installed with locking groove facing outward. If clutch plate retaining nut (15) is 3/4-inch, tighten to 70 ft.-lbs. (95 N·m) torque; if 1/2-inch nut is used, tighten to 50 ft.-lbs. (68 N·m). Adjust clutch as previously outlined.

HYDRAULIC SYSTEM

Hydraulic system utilizes pressurized oil from charge pump of hydrostatic drive unit. Refer to HYDROSTATIC

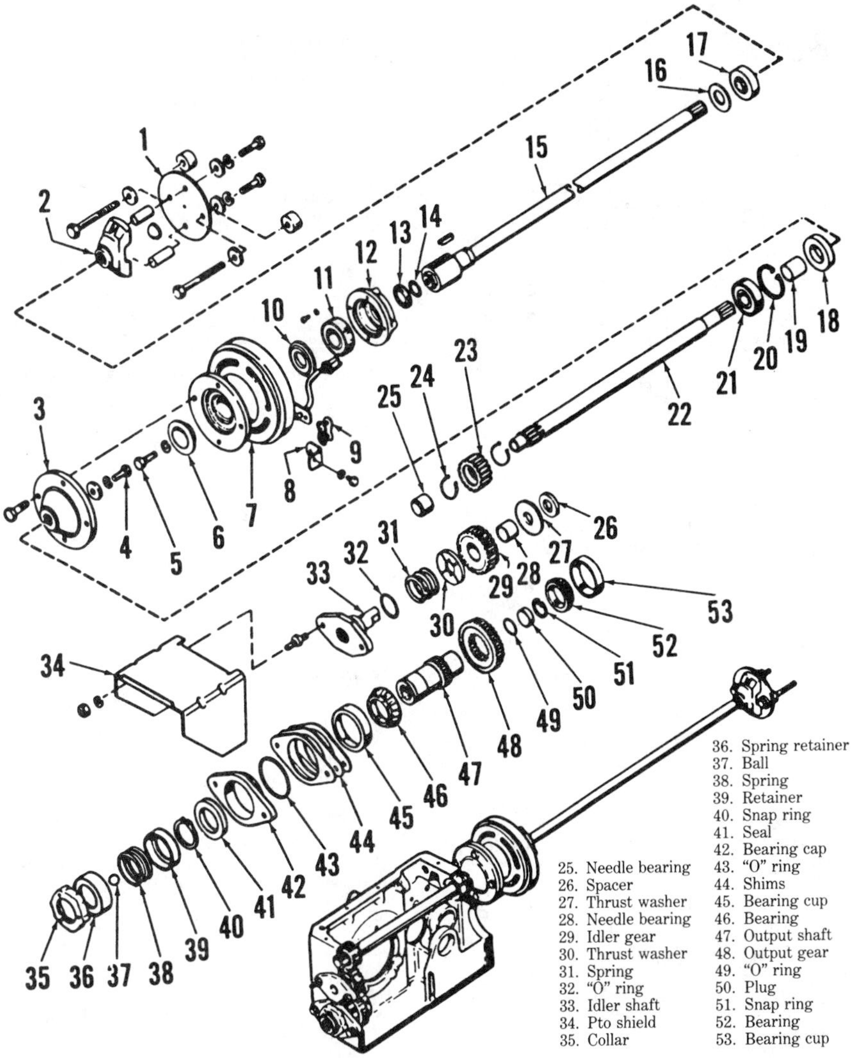

25. Needle bearing
26. Spacer
27. Thrust washer
28. Needle bearing
29. Idler gear
30. Thrust washer
31. Spring
32. "O" ring
33. Idler shaft
34. Pto shield
35. Collar
36. Spring retainer
37. Ball
38. Spring
39. Retainer
40. Snap ring
41. Seal
42. Bearing cap
43. "O" ring
44. Shims
45. Bearing cup
46. Bearing
47. Output shaft
48. Output gear
49. "O" ring
50. Plug
51. Snap ring
52. Bearing
53. Bearing cup

Fig. S44 – Exploded view of rear pto shafts and gears similar to ones used on 5216 models. Refer to text.

1. Flexible coupling
2. Yoke
3. Hub
4. Cap screw
5. Cap screw
6. Collar
7. Pto clutch assy.
8. Bracket
9. Connecting link
10. Washer
11. Bearing
12. Bearing housing
13. Washer
14. Retaining ring
15. Pto drive shaft
16. Rubber washer
17. Felt ring
18. Seal
19. Spacer
20. Retaining ring
21. Bearing
22. Top pto shaft
23. Gear
24. Snap ring

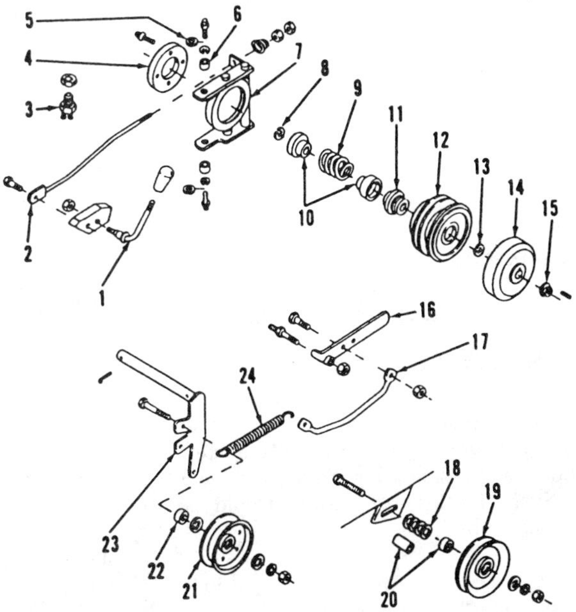

Fig. S45 – Exploded view of pto clutch and idler pulley assembly used on 7018, 7117 and 7119 models.

1. Clutch control handle
2. Control handle
3. Interlock switch
4. Brake disc
5. Spring washer (2)
6. Spacer (2)
7. Pivot assy.
8. Retaining ring
9. Spring
10. Spring guides
11. Bearing
12. Pulley
13. Retaining ring
14. Clutch plate
15. Retaining nut
16. Tensioning lever
17. Tension rod
18. Washers (4)
19. Rear idler pulley
20. Spacers
21. Front idler pulley
22. Spacer
23. Pivot arm
24. Tension spring

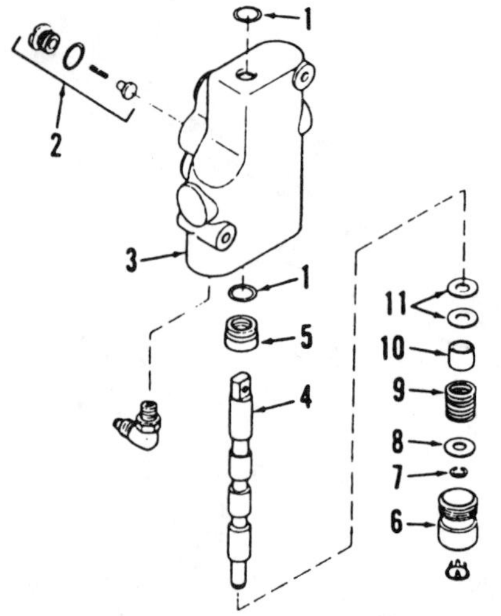

Fig. S46—Exploded view of CESSNA lift control valve used on some models.

1. "O" rings
2. Check valve assy.
3. Valve body
4. Valve spool
5. Bushing
6. End cap
7. Retaining ring
8. Washer
9. Spring
10. Spool
11. Washers

Fig. S48—Exploded view of single spool control valve with a spring and ball type detent mechanism used on late models.

1. End cap
2. Detent spring
3. Detent balls
4. Detent adapter
5. Washer
6. Spring
7. Washer
8. Washer
9. Bushing
10. "O" ring
11. Valve body
12. Plug
13. "O" ring
14. Spring
15. Plunger
16. Spool
17. "O" ring

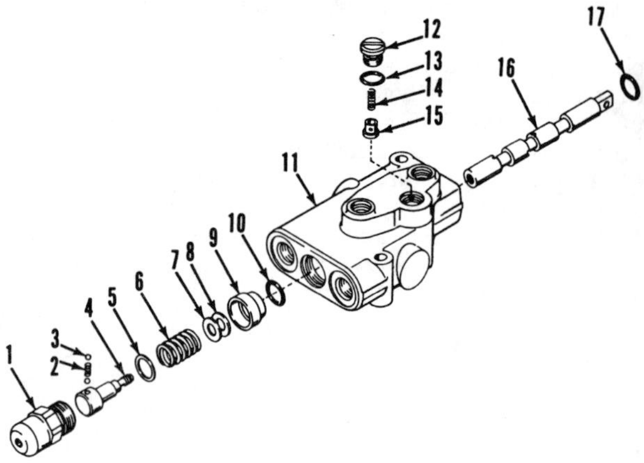

Fig. S49—Exploded view of two spool valve with a spring and ball type detent mechanism used on late models.

1. End cap
2. Detent spring
3. Detent balls
4. Detent adapter
5. Washer
6. Spring
7. Washer
8. Washer
9. Bushing
10. "O" ring
11. Valve body
12. Plug
13. "O" ring
14. Spring
15. Plunger
16. Spool
17. "O" rings
18. Spool

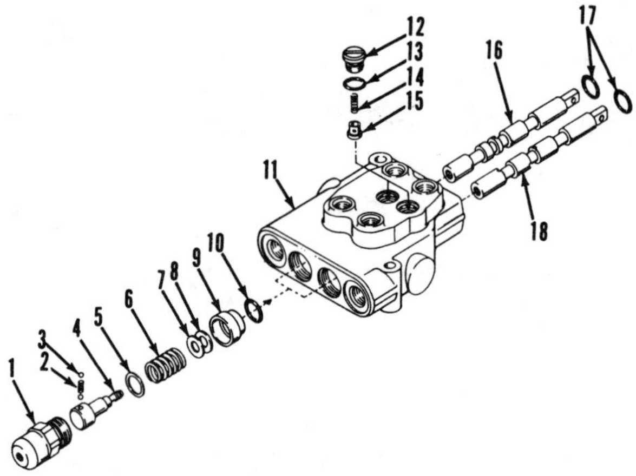

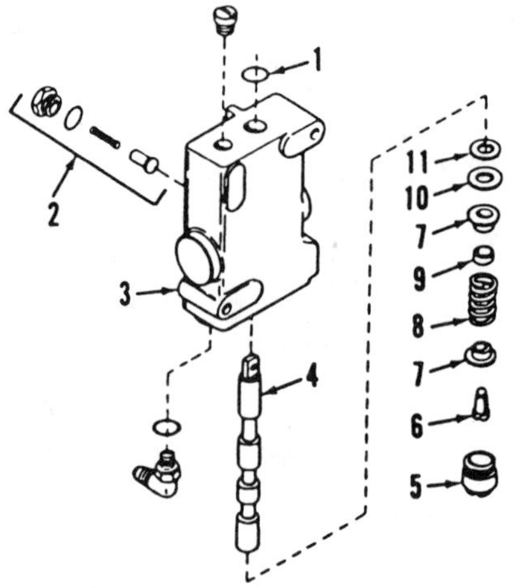

Fig. S47—Exploded view of AICO lift control valve used on some models.

1. "O" ring
2. Check valve assy.
3. Valve body
4. Valve spool
5. End cap
6. Spool stem
7. Spring seats
8. Spring
9. Spool stop
10. Washer
11. Quad ring

DRIVE paragraphs in TRANSMISSION section for maintenance and service procedure.

TROUBLESHOOTING

All Models So Equipped

Some problems which may occur during operation of hydraulic lift system and their possible causes are as follows:

1. System lifts load slowly. Could be caused by:
   a. Excessively worn charge pump.
   b. Faulty lift relief valve.
   c. Relief valve pressure adjusted low.
   d. Damaged lift cylinder or lines.
2. Load lowers with control valve in raise position. Could be caused by:
   a. Faulty lift check plunger or seat in control valve.
3. Load lowers with control valve in neutral position. Could be caused by:
   a. External leakage (lines or fittings).
   b. Internal leakage (worn spool or valve body).

PRESSURE CHECK AND ADJUST

Model 5216

Remove pressure check test port plug (46-Fig. S29) and install a 0-1000 psi (0-6895 kPa) test gage. Start engine and operate tractor until fluid reaches normal operating temperature. Test gage reading should be 70-150 psi (483-1034 kPa). Hold lift control valve with lift cylinder at end of stroke. Test gage should read 550-700 psi (3792-4826 kPa). If not, stop engine and add or remove shims (54) behind plug (56) to obtain correct life pressure relief setting. Remove test gage and reinstall plug.

NOTE: If lift relief pressure (550-700 psi (3792-4826 kPa) cannot be reached, remove and inspect charge pump as outlined in HYDROSTATIC DRIVE UNIT paragraph in TRANSMISSION section.

Models 7018-7117-7119

Remove test port plug (TP-Fig. S34) and install a 0-1000 psi (0-6895 kPa) test gage. Start engine and run at near full speed. Gage reading should be 70-150 psi (482-1034 kPa). Renew relief valve (28) or vary shims as necessary to obtain correct charge pump.

CONTROL VALVE

All Models So Equipped

A variety of hydraulic control valves have been used according to model or date of manufacturer. Overhaul of valves is apparent after inspection and reference to Fig. S46, S47, S48 or S49, according to model being serviced. Renew all "O" rings and lubricate all parts during reassembly. Tighten caps to 20 ft.-lbs. (27 N·m) torque.

SNAPPER

CONDENSED SPECIFICATIONS

	MODELS				
	LT12 (prior 1990)	YT12	LT16	YT16	YT18, HYT18
Engine Make	B&S	B&S	B&S	B&S	Kohler
Model	404437	404437	402437	402437	MV18
Bore	3.44 in. (87.3 mm)	3.44 in. (87.3 mm)	3.44 in. (87.3 mm)	3.44 in. (87.3 mm)	3.12 in. (79.2 mm)
Stroke	2.16 in. (54.8 mm)	2.16 in. (54.8 mm)	2.16 in. (54.8 mm)	2.16 in. (54.8 mm)	2.75 in. (69.85 mm)
Displacement	40 cu. in. (656 cc)	40 cu. in. (656 cc)	40 cu. in. (656 cc)	40 cu. in. (656 cc)	42.18 cu. in. (691.3 cc)
Power Rating	12.5 hp (9.3 kW)	12.5 hp (9.3 kW)	16 hp (11.9 kW)	16 hp (11.9 kW)	18 hp (13.4 kW)
Slow Idle Speed	1200 rpm	1200 rpm	1200 rpm	1200 rpm	1200 rpm
High Idle (No-Load) Speed	3600 rpm	3600 rpm	3600 rpm	3600 rpm	3750 rpm
Capacities—					
Crankcase	3 pts.* (1.4L)	3 pts.* (1.4 L)	3 pts.* (1.4 L)	3 pts.* (1.4 L)	3.5 pts. (1.6 L)
w/filter	....	....	....	....	4.5 pts. (2.1 L)
Transmission	See text	See text	See text	See text	See text

*Early production engine crankcase capacity is 3.5 pints (1.6 L).

FRONT AXLE AND STEERING SYSTEM

MAINTENANCE

All Models

The front wheel spindles and axle pivot pin should be lubricated after every 25 hours of operation. Jack up front axle so wheels are suspended and apply grease to spindles and axle through grease fittings. Use a good quality multipurpose grease. Check for looseness and binding in front axle components.

After 50 hours of operation, or twice each year, lubricate steering shaft bushings and sector gear bushings by injecting two shots of multipurpose grease through grease fittings.

R&R AXLE MAIN MEMBER

All Models

Raise and support front of tractor. Disconnect drag link (26—Fig. SN1) at steering arm (28). Unscrew pivot bolt (36) and roll axle assembly away from tractor.

Remove pivot sleeve (38) and inspect components for excessive wear and damage. Bushings (37) in axle are renewable.

Reinstall axle and lubricate through grease fitting in axle main member until grease appears at bearings.

TIE ROD AND TOE-IN

All Models

The tie rod (35—Fig. SN1) is equipped with renewable ends (31). Front wheel toe-in should be 3/16 inch (4.8 mm). Adjust toe-in by altering tie rod length.

STEERING SPINDLES

All Models

Raise and support front of tractor. Remove wheel and tire. Disconnect tie rod end, and if needed, drag link end. If removing right spindle (41—Fig. SN1), detach snap ring (40) securing top of spindle and withdraw spindle from axle. If removing left spindle (34), drive out pin (29), then drive spindle down and out of steering arm (28) and axle.

Inspect components for excessive wear and damage. Renewable bushings (30) are located in axle.

To reassemble, reverse the disassembly procedure. Lubricate bushings with multipurpose grease until grease is visible at the bushings.

DRAG LINK

All Models

A drag link (26—Fig. SN1) is located between steering arm and pitman arm. Inspect drag link ends for looseness and renew as necessary. Length of drag link is not adjustable.

FRONT WHEEL BEARINGS

All Models

To remove front wheel bushings, raise and support front of tractor. Remove dust cap from wheel hub and cotter pin from spindle end. Slide wheel hub assembly off spindle. Wheel bushings are a press fit in wheel hub. If bushings are loose in hub, renew or repair hub as necessary.

Lubricate bushings through grease fitting in wheel hub using a good quality multipurpose grease. Inject grease until visible around bushings.

STEERING GEAR

All Models

ADJUSTMENT. To adjust sector gear backlash, loosen nuts (25—Fig. SN1). Tighten inner nut until backlash is removed but gears rotate without binding. Tighten outer nut against inner nut.

R&R AND OVERHAUL. To disassemble steering gear assembly, drive out pin (2—Fig. SN1) and remove steering wheel. Remove collar (3), retainer (4) and spring (5). Remove snap ring (19) and drive roll pin (11) out of steering gear (10). Pull steering shaft up until disengaged from support (14), then move shaft down out of dash panel. Detach drag link (26) from steering arm (23). Unscrew support screws (15) and remove support assembly (14) from bottom of tractor. Remainder of disassembly is evident after inspection of unit. Note that bushings are renewable.

Inspect components for damage and excessive wear. Reassemble by reversing disassembly procedure. Steering lock bushing (8) must be centered in dash panel and move freely before installing spring (4) and tightening screws (15). Move support (14) in slots in side panels to properly position bushing (8), then tighten screws (15). Lubricate with grease after assembly. Adjust steering gear backlash as previously outlined.

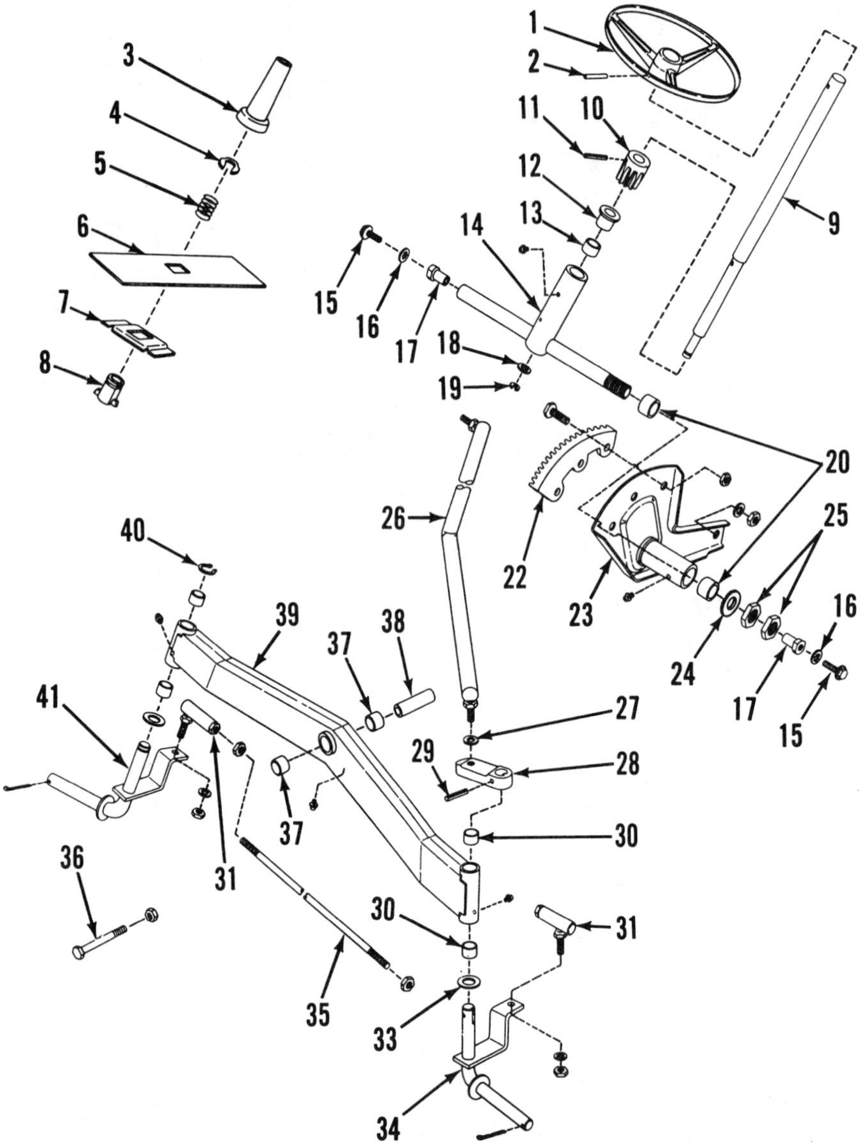

Fig. SN1—Exploded view of front axle and steering gear assembly used on all models.

1. Steering wheel
2. Roll pin
3. Collar
4. Retainer
5. Spring
6. Cover
7. Riser
8. Bushing
9. Steering shaft
10. Pinion gear
11. Roll pin
12. Bushing
13. Bushing
14. Support
15. Screw
16. Lockwasher
17. Pivot
18. Washer
19. Snap ring
20. Bushing
22. Sector gear
23. Steering arm
24. Washer
25. Nuts
26. Drag link
27. Lockwasher
28. Steering arm
29. Roll pin
30. Bushing
31. Tie rod end
33. Thrust washer
34. Spindle
35. Tie rod
36. Pivot bolt
37. Bushings
38. Sleeve
39. Axle main member
40. Snap ring
41. Spindle

ENGINE

MAINTENANCE

All Models

Regular engine maintenance is required to maintain peak performance and long engine life.

Check oil level prior to each operating interval and after 5 hours of operation. Inspect air cleaner element periodically depending on operating conditions.

Change engine oil, perform tune-up, adjust valves and clean carbon from cylinder heads as recommended by engine manufacturer. Refer to appropriate engine section in this manual for additional information.

REMOVE AND REINSTALL

All Models

To remove engine, remove hood and disconnect battery cables. Detach drive shaft from engine. Disconnect all interfering cables, wires and hoses. Detach belt from pto pulley. Unscrew engine mounting screws and lift engine out of tractor.

Install engine by reversing removal procedure. Note that flat washers must be against flex disc when reinstalling drive shaft. Tighten drive shaft coupling set screws to 15 ft.-lbs. (20 N·m).

ELECTRICAL SYSTEM

MAINTENANCE AND SERVICE

All Models

Battery electrolyte level should be checked once each month. If necessary, add distilled water until level is just below base of vent well. DO NOT overfill. Keep battery posts clean and cable ends tight.

For alternator or starter service, refer to appropriate engine service section in this manual.

Refer to Fig. SN3 for a typical wiring diagram.

BRAKES

ADJUSTMENT

Model HYT18

To adjust parking brake, first pull roll release rod (2—Fig. SN4) up to disengage hydrostatic transmission. (Roll release rod is located under the seat.) Fully depress neutral return/brake pedal and raise parking brake lever to lock pedal down. Push tractor forward and observe if rear wheels turn. If necessary, adjust brake control rod (6) until rear wheels do not turn when tractor is pushed by hand. Release the parking brake and make sure that wheels turn freely.

All Other Models

The brake/clutch pedal should engage the brake when the pedal is depressed into the lower portion of its travel. There should be approximately ¾ inch (19 mm) of free pedal travel between point where the clutch is engaged and the brake is engaged. To adjust brake engagement, move spacer ferrules (F—Fig. SN5) at end of brake cable. Placing ferrules between anchor (A) and lever (L) will cause earlier engagement.

R&R BRAKE

Model HYT18

The brake is located inside the transmission unit. Refer to HYDROSTATIC DRIVE SYSTEM section for brake service.

All Other Models

To disassemble brake, unbolt and remove pivot bracket (19—Fig. SN6). Detach brake cable (30) and remove pivot (21) and brake band (24) as an assembly. If necessary, unscrew retaining nut and remove brake drum (23).

Inspect components for excessive wear and damage. The brake band should be renewed if oil-soaked or worn to a thickness of $^{1}/_{16}$-in. (1.6 mm) or less.

Reassemble by reversing disassembly procedure. Check brake adjustment.

CLUTCH

All models except Model HYT18 are equipped with a clutch system that consists of a drive disc connected to the en-

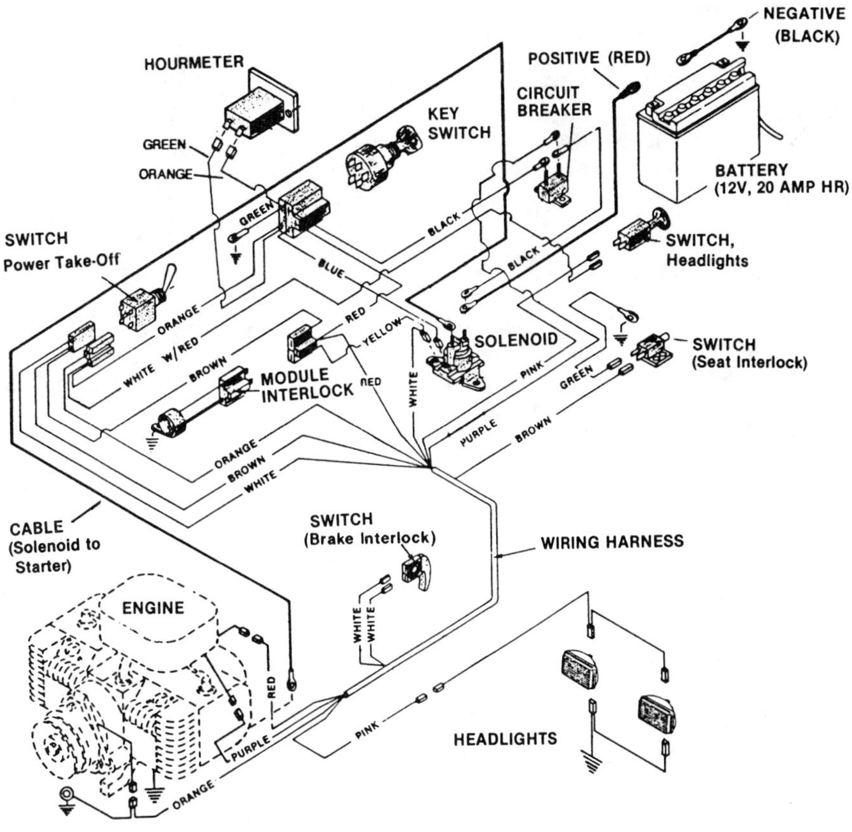

Fig. SN3—Typical schematic wiring diagram.

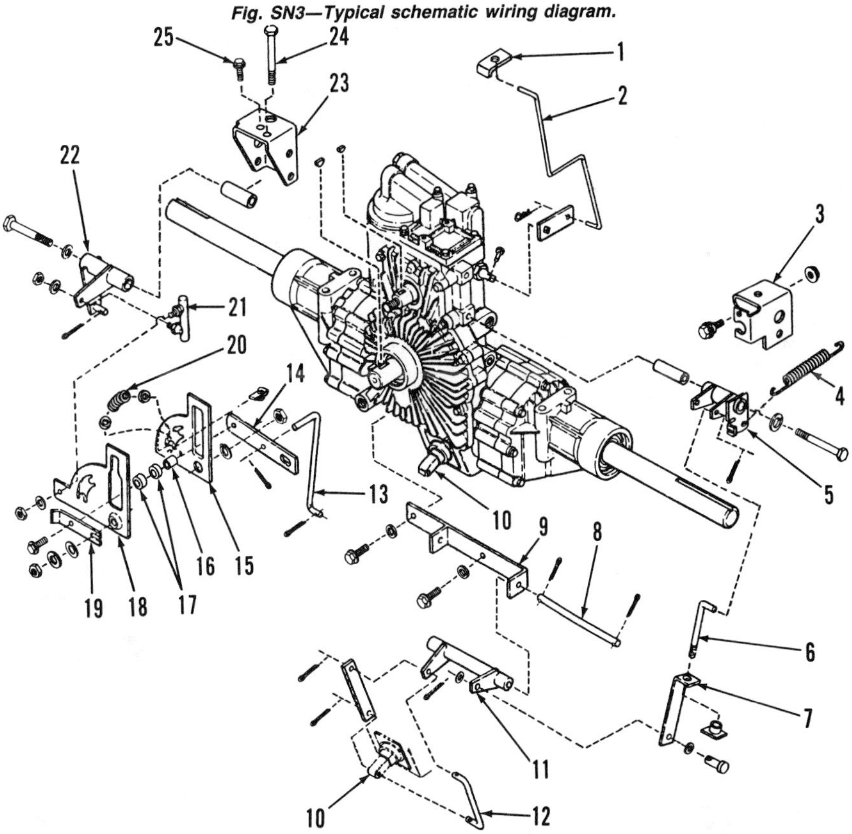

Fig. SN4—Exploded view of neutral return/brake linkage used on Model YT18H.

1. Latch
2. Roll release rod
3. Neutral return bracket
4. Return spring
5. Neutral return bellcrank
6. Brake adjusting link
7. Bracket
8. Pivot pin
9. Pivot bracket
10. Parking brake lever
11. Parking brake pivot
12. Brake link
13. Neutral return link
14. Neutral return arm
15. Roller guide
16. Spacer
17. Rollers
18. Neutral return cam
19. Retainer bracket
20. Spring
21. Shift link
22. Speed control bellcrank
23. Neutral adjusting bracket
24. Neutral adjusting bolt
25. Lock bolt

Illustrations courtesy Snapper Power Equipment.

gine drive shaft and a driven disc that transmits power to the transmission. The driven disc may be moved laterally against the drive disc to vary tractor speed. Depressing the brake/clutch pedal moves the driven disc away from the drive disc while applying the brake.

R&R AND OVERHAUL

All Models Except HYT18

The driven disc should be renewed if rubber is worn to a thickness of $^1/_{16}$-in. (1.6 mm) or less. To remove rubber disc, remove rear cover plate and seat support. Move shifter to neutral and unscrew nuts retaining disc to hub.

After installing new disc, check position of stop pin (P—Fig. SN8), which should be $^3/_{16}$-in. (4.8 mm) from edge of shift yoke (Y). The stop pin prevents metal-to-metal contact when rubber is excessively worn.

With driven disc in reverse position (right side of drive disc), center of driven disc should be $^1/_{16}$-in. (1.6 mm) below center of drive disc (see Fig. SN9). To obtain desired position, loosen screws securing bearing support (7—Fig. SN10) and relocate position of drive disc.

The shift handle should be located in neutral detent slot when driven disc is positioned in center of drive disc. If shift lever is not in neutral slot, loosen nut (N—Fig. SN8) and move shift arm to correctly position shift lever.

The drive disc should be renewed if warped, excessively worn or damaged. Maximum allowable runout is 0.020 inch (0.51 mm). If drive disc must be removed, proceed as follows: Remove drive shaft. Unscrew spindle (5—Fig. SN10) from disc (9) by turning coupler (1). Note that disc has left-hand threads. Unscrew coupler to remove front bearing. Coupler has left-hand threads.

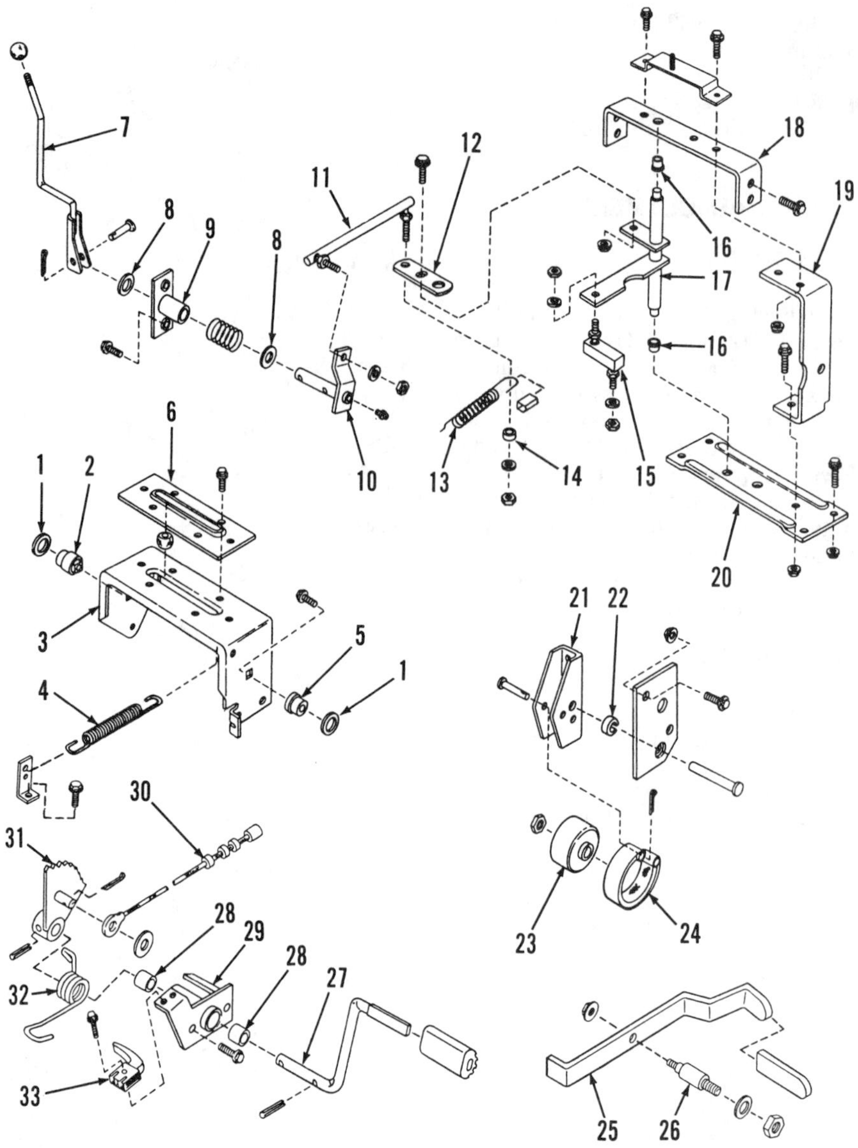

Fig. SN6—Exploded view of brake and clutch mechanism used on all models except YT18H.

1. Washer
2. Bushing
3. Clutch yoke
4. Spring
5. Bushing
6. Retainer plate
7. Shift shaft
8. Thrust washer
9. Shift pivot
10. Shift arm
11. Shift link
12. Shift arm
13. Spring
14. Roller
15. Shift link
16. Bushing
17. Shift arm
19. Brake pivot bracket
20. Bracket
21. Brake pivot lever
22. Ring
23. Brake drum
24. Brake band
25. Park brake lever
26. Stub shaft
27. Clutch/brake pedal
28. Bushing
29. Pivot bracket
30. Cable
31. Parking brake ratchet
32. Spring
33. Clutch interlock switch

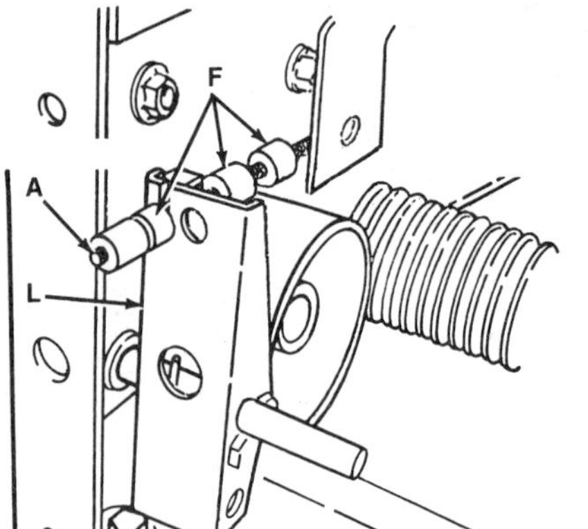

Fig. SN5—To adjust brake engagement on all models except YT18H, detach brake cable from brake lever (L) and relocate spacer ferrules (F) as needed between anchor ferrule (A) and lever. Placing more ferrules between anchor and lever will result in earlier engagement.

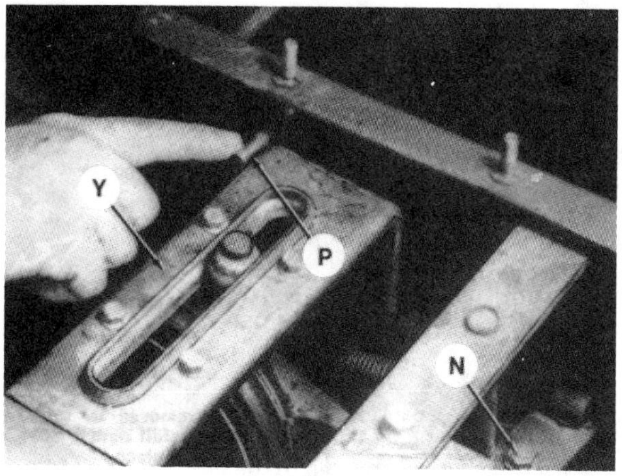

Fig. SN8—Position of stop pin (P) should be $^3/_{16}$-in. (4.8 mm) from edge of shift yoke (Y).

Illustrations courtesy Snapper Power Equipment.

When assembling drive disc, install shims (8) as needed so face of disc is $2^{3}/_{16}$ in. (55.6 mm) from rear of support (7) as shown in Fig. SN10.

PRIMARY CHAIN CASE

MAINTENANCE

All Models So Equipped

Check for presence of grease in chain case by removing plug (1—Fig. SN12). If lubricant is required, add 1 ounce of Snapper OO grease. Maximum capacity is 2 ounces. Excessive lubricant will result in leakage.

R&R AND OVERHAUL

All Models So Equipped

The primary chain case is accessible after removing transmission.

To disassemble unit, remove boots from each side. Unscrew locknut (17—Fig. SN12) and remove driven disc hub (12). Unscrew retaining fasteners and separate case halves (2 and 10). Remove input shaft (9), sprocket (6) and chain as a unit.

Inspect components for damage and excessive wear. If bearing renewal is necessary, be sure to support area around bearing so case half is not distorted. Press bearings (7) into case until snap ring contacts case. Press needle bearings (4) into case from inside of case; press against lettered side of bearing. Hex hub (6) must slide freely on hex shaft.

When assembling chain case, install nylon thrust washers (5) so cupped side is toward hub (6). Install Belleville washer (16) with cupped side toward hub (12). Fill case with 2 ounces of Snapper OO grease.

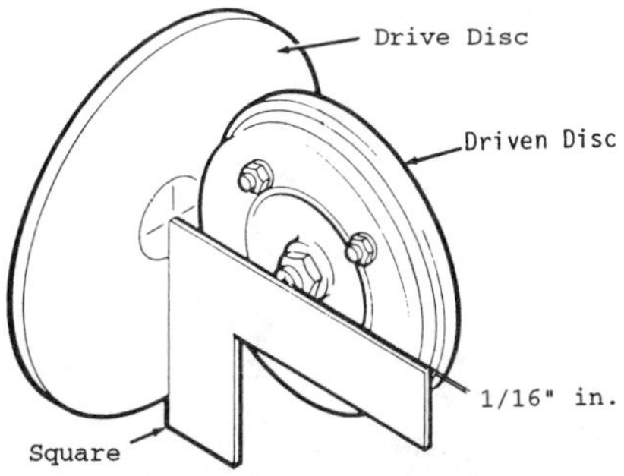

Fig. SN9—With driven disc in reverse position (right side of drive disc), center of driven disc should be $^{1}/_{16}$-in. (1.6 mm) below center of drive disc.

GEAR TYPE TRANSAXLE

MAINTENANCE

All Models So Equipped

The oil level should be checked every season or after 100 hours of operation, whichever occurs first. Oil should be even with lower edge of plug opening (Fig. SN13). Add oil as needed through fill plug opening. Recommended oil is Snapper OO.

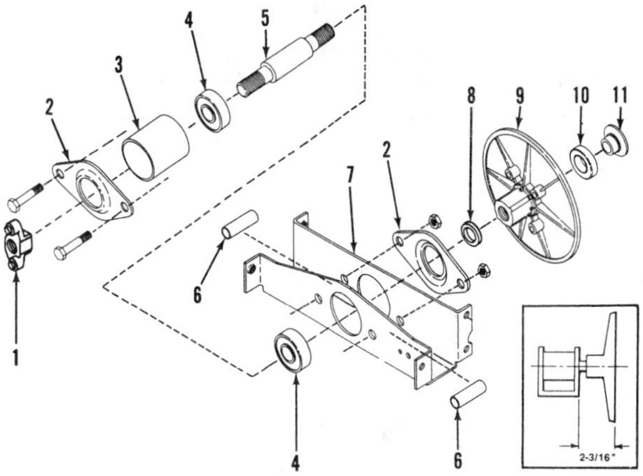

Fig. SN10—Exploded view of drive disc and shaft assembly used on all models except YT18H. Face of drive disc should be $2^{3}/_{16}$ in. (55.6 mm) from rear of support. Install shims (8) as needed.

1. Coupler
2. Retainers
3. Tube
4. Bearings
5. Shaft
6. Spacers
7. Bearing support
8. Shim
9. Drive disc
10. Bearing
11. Button

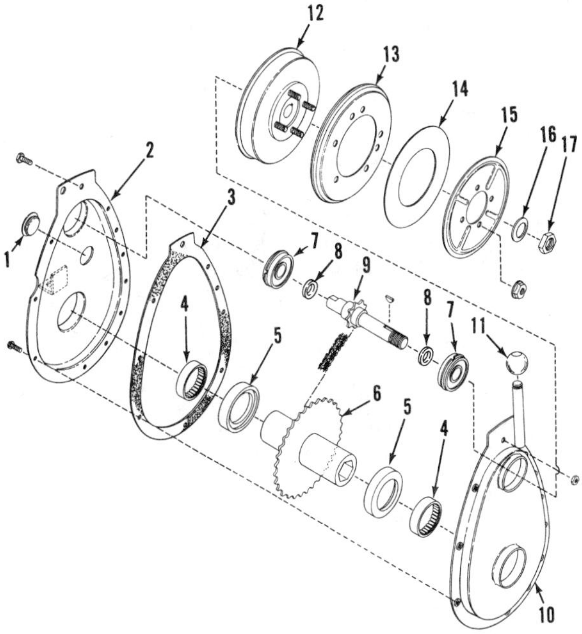

Fig. SN12—Exploded view of primary chain case used on all models except YT18H.

1. Plug
2. Case half
3. Gasket
4. Bearing
5. Thrust washer
6. Sprocket & hub
7. Bearings
8. "O" rings
9. Input shaft
10. Case half
11. Bushing
12. Driven disc hub
13. Driven disc
14. Thrust washer
15. Plate
16. Belleville washer
17. Locknut

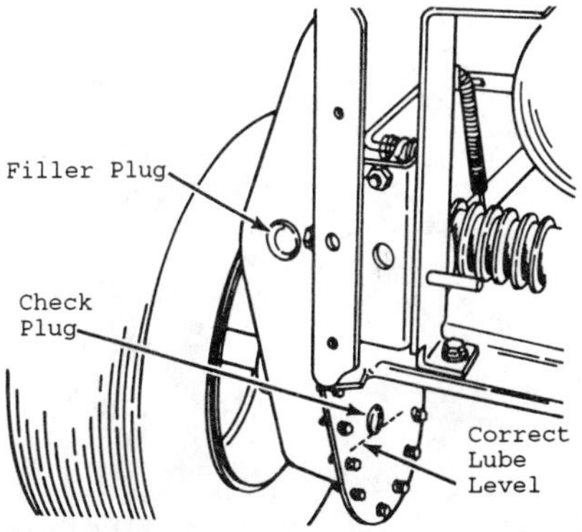

Fig. SN13—Add oil as needed through fill plug opening so oil level is even with lower edge of check plug opening.

R&R AND OVERHAUL

All Models So Equipped

To remove transaxle, remove rear console panel, fenders and brake drum. Support rear of tractor and remove rear wheels. Unscrew locknut then drive out taper pins (4—Fig. SN14) and remove wheel hubs (3) from axles. Remove set screws (2) and drive off lock collar (1) on right axle by rotating in direction opposite of axle forward rotation. Remove boot clamp (43—Fig. SN15), dust seal (42), oil seal (41) and lock collar (40) from left axle. Remove cap screw (44) from end of drive hex tube (29). Clean ends of axles. Loosen dust seal boot clamps on each end of hex tube. Detach lift hanger bracket on tractor underside, which may interfere with transaxle removal. Remove transaxle retaining screws from tractor frame and remove transaxle from tractor.

To disassemble transaxle, remove screws and nut securing cover (7) and remove cover. Disassembly is evident after inspection and referral to Fig. SN15 while noting the following: Note location of all thrust washers and shim washers as they are removed to ensure correct reassembly. Lift axle assembly (32), drive sprocket (22) and chain (31) out as an assembly. Shift fork (33) rides on guide pins (34) and must be pulled straight out. Lift hex shaft (29), drive chain and short hex tube (12) and sprocket (13) out as an assembly. Shaft (20) is retained by a screw at end of shaft. If shaft cannot be pulled out of bearing, reinstall screw and tap against it to dislodge shaft. If bearings must be removed, force bearings toward inside of case while supporting area around bearings so case will not be deformed.

Refer to Fig. SN16 for view of shift fork assembly. Ball (51) and spring (52) are located between roll pin (53) and detent pin (34). Renew shift fork if fingers are spread excessively.

Refer to Fig. SN17 for an exploded view of axle and differential assembly. Prior to disassembly note presence of alignment marks on sprocket (61) and differential plate (67). The eight cap screws retaining differential plate and pinions (65) must be discarded and new screws must be installed. Discard spacers (66) if wear is indicated after comparing with a new spacer. Renew spacer (63) if worn. When assembling

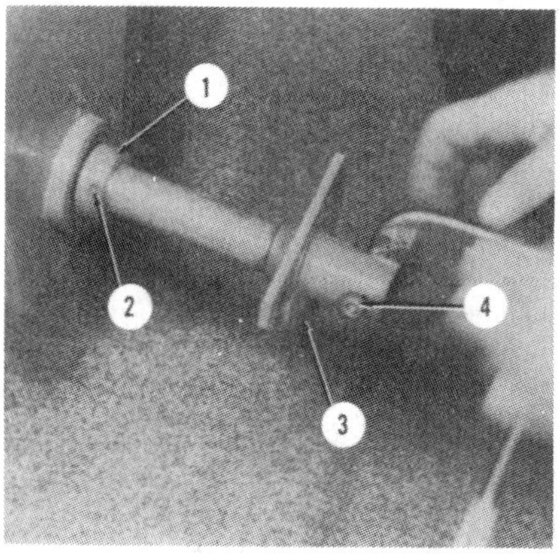

Fig. SN14—To remove wheel hub (3), unscrew locknut then drive out taper pin (4). Remove set screws (2) and drive off collar (1) by rotating in direction opposite of axle forward rotation.

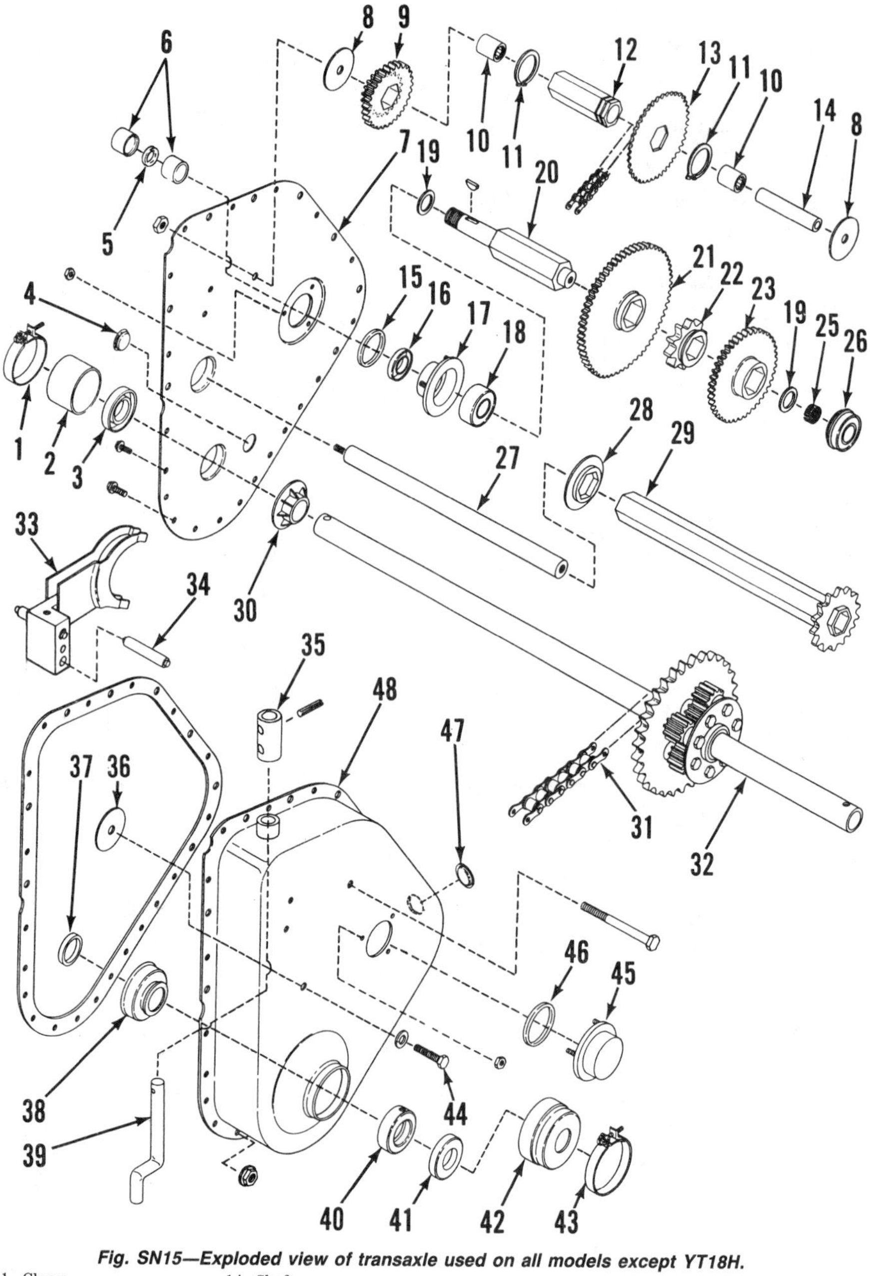

Fig. SN15—Exploded view of transaxle used on all models except YT18H.

1. Clamp
2. Dust seal cap
3. Oil seal
4. Oil level/fill plug
5. "O" ring
6. Spacers
7. Cover
8. Thrust washer
9. Cluster gear
10. Bearing
11. Snap ring
12. Hex tube
13. Sprocket
14. Shaft
15. "O" ring
16. Seal
17. Bearing retainer
18. Bearing
19. Shim washers
20. Hex shaft
21. Gear (67 T)
22. Sprocket
23. Gear (47 T)
25. Tolerance ring
26. Bearing
27. Support shaft
28. Hex washer
29. Hex shaft & sprocket
30. Washer
31. Drive chain
32. Differential & axle assy.
33. Shift fork
34. Guide pin
35. Coupling
36. Thrust washer
37. Spacer
38. Bearing
39. Shift rod
40. Lock collar
41. Seal
42. Dust seal cap
43. Clamp
44. Support shaft retaining bolt
45. Bearing retainer
46. "O" ring
47. Oil fill/check plug
48. Transmission case

Illustrations courtesy Snapper Power Equipment.

differential, align marks on sprocket and plate. Tighten cap screws to 18-25 ft.-lbs. (25-34 N·m).

Inspect components for excessive wear and damage. Press needle bearings (10—Fig. SN15) into hex tube (12) until flush with tube ends. Press against lettered side of bearing. Reassemble by reversing disassembly procedure.

After installation, fill with oil through fill plug opening (Fig. SN13). Oil should be even with lower edge of plug opening. Recommended oil is Snapper OO.

HYDROSTATIC DRIVE SYSTEM

MAINTENANCE

All Models So Equipped

Check level in oil reservoir at least twice each year (reservoir is located under center console panel). Do not allow any dirt or other foreign material to enter oil reservoir. Oil level should be 1-2 inches (38-51 mm) below top of filler neck. Recommended oil is SAE 20W motor oil.

ADJUSTMENT

All Models So Equipped

If tractor continues to creep when forward/reverse pedal is released, adjust hydrostatic control linkage as follows: Raise and support rear of tractor so rear wheels are free to turn. Remove the seat assembly and the access cover from top of fender. Temporarily connect a jumper wire to seat interlock switch to permit engine to be started. Start engine and completely depress speed control pedal to start wheel rotation in forward direction. Release speed control pedal and observe if wheels stop turning. If wheels continue to turn, loosen lock bolt (25—Fig. SN19) and turn neutral adjusting bolt (24) as necessary until rear wheels stop turning. Tighten the lock bolt to lock the neutral adjustment. Reinstall seat and check tractor operation.

TROUBLE-SHOOTING

All Models So Equipped

The following problems and possible causes should be used as an aid in locating and correcting transmission problems.

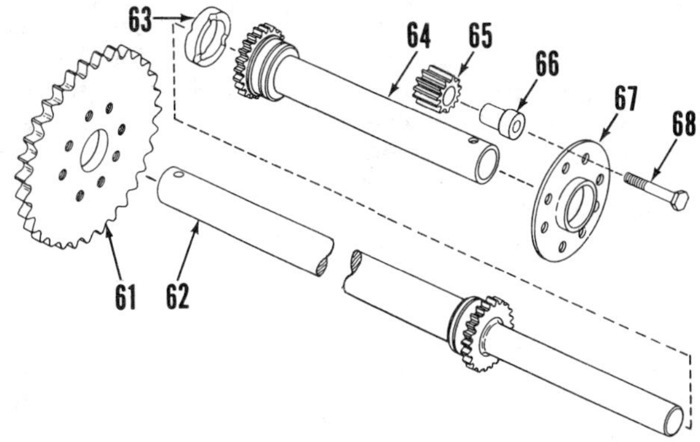

Fig. SN17—Exploded view of axle and differential assembly.

61. Sprocket
62. Long axle
63. Spacer
64. Short axle
65. Pinion (8)
66. Spacer (8)
67. Plate
68. Self-locking cap screw

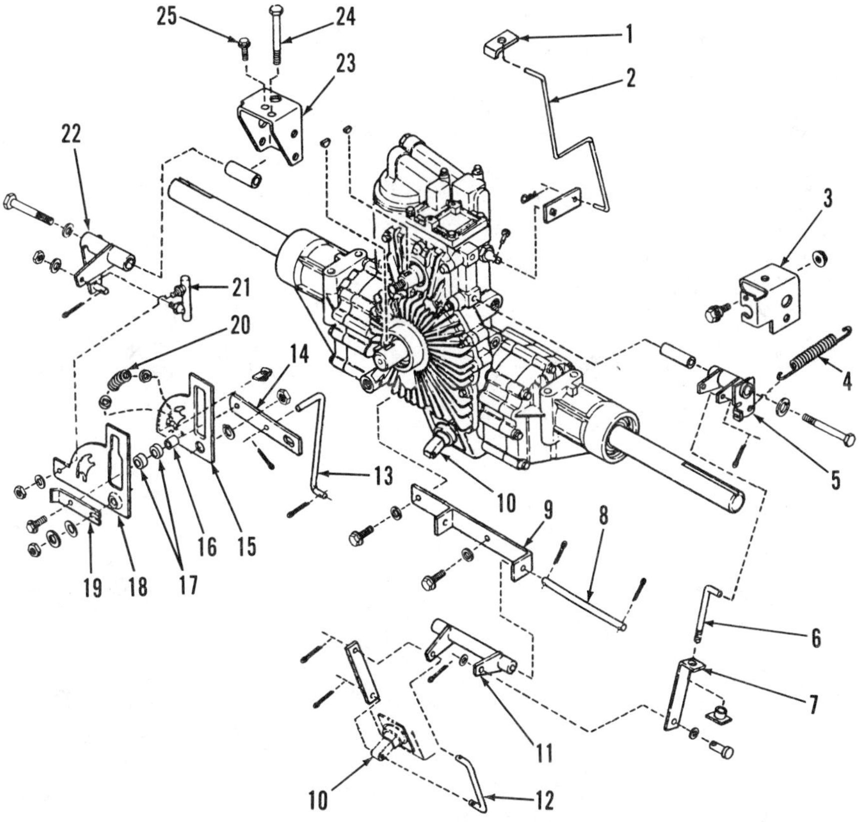

Fig. SN19—Exploded view of hydrostatic transmission control linkage and neutral return/brake linkage used on Model YT18H.

1. Latch
2. Roll release rod
3. Neutral return bracket
4. Return spring
5. Neutral return bellcrank
6. Brake adjusting link
7. Bracket
8. Pivot pin
9. Pivot bracket
10. Parking brake lever
11. Parking brake pivot
12. Brake link
13. Neutral return link
14. Neutral return arm
15. Roller guide
16. Spacer
17. Rollers
18. Neutral return cam
19. Retainer bracket
20. Spring
21. Shift link
22. Speed control bellcrank
23. Neutral adjusting bracket
24. Neutral adjusting bolt
25. Lock bolt

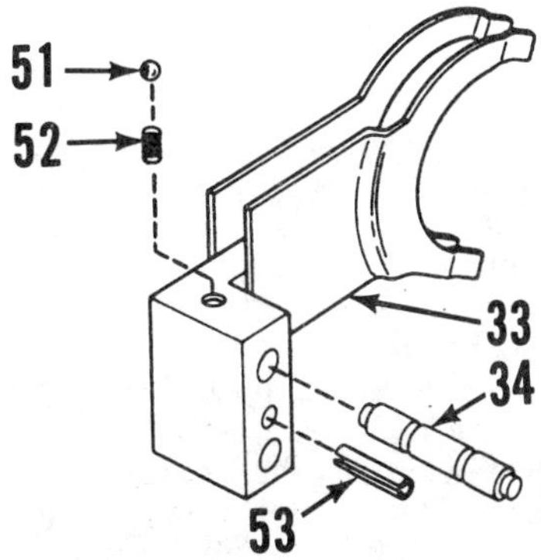

Fig. SN16—Exploded view of shift fork assembly.

33. Shift fork
34. Detent guide pin
51. Detent ball
52. Spring
53. Roll pin

Illustrations courtesy Snapper Power Equipment.

1. Loss of power or transmission will not operate in either direction. Could be caused by:
   a. Drive belt slipping or broken.
   b. Speed control linkage broken.
   c. Transmission oil level low.
   d. Wrong type of transmission oil or oil contaminated with water.
   e. Transmission oil temperature too hot.
   f. Transmission roll release valve in wrong position.
   g. Oil filter plugged.
   h. Charge pump or charge pressure relief valve faulty.
   i. Transmission pump and/or motor worn or damaged.
   j. Drive pulley slipping on transmission input shaft.
   k. Internal damage to reduction gear assembly.
2. Transmission operating too hot. Could be caused by:
   a. Transmission oil level low.
   b. Wrong type of oil in transmission.
   c. Cooling fan defective.
   d. Transmission cooling fins blocked.
   e. Roll release valve faulty.
   f. Tractor is being overloaded.
   g. Transmission pump and/or motor worn.
3. Transmission jerks when starting or operates in one direction only. Could be caused by:
   a. Faulty speed control linkage.
   b. Faulty charge check valve.
4. Tractor creeps when in neutral. Could be caused by:
   a. Worn or misadjusted control linkage.

R&R AND OVERHAUL

All Models So Equipped

To remove hydrostatic unit, remove center cover and fenders. Disconnect drive shaft. Drain oil from hydraulic reservoir. Disconnect hydraulic hoses and plug all openings to prevent entry of dirt. Disconnect neutral return/brake control linkage and speed control linkage. Raise rear of tractor. Remove cap screws retaining transaxle to tractor frame and roll unit away from tractor.

Prior to disassembly, remove fan, pulley, and all external brackets, levers and fittings. Thoroughly clean outside of unit. Drain oil from unit, remove oil filter and detach oil filter base. During disassembly mark components so they may be returned to their original position.

Service procedure is the same for either axle housing assembly. Remove cap screws securing axle housing (6—Fig. SN20) and withdraw axle housing and planetary assembly (components 1 through 21) from hydrostatic housing (29).

CAUTION: Be very careful not to dislodge hydrostatic motor assembly. The spring-loaded balls (22) are a selective fit in motor rotor (24) and must be installed in matching bore in rotor if removed.

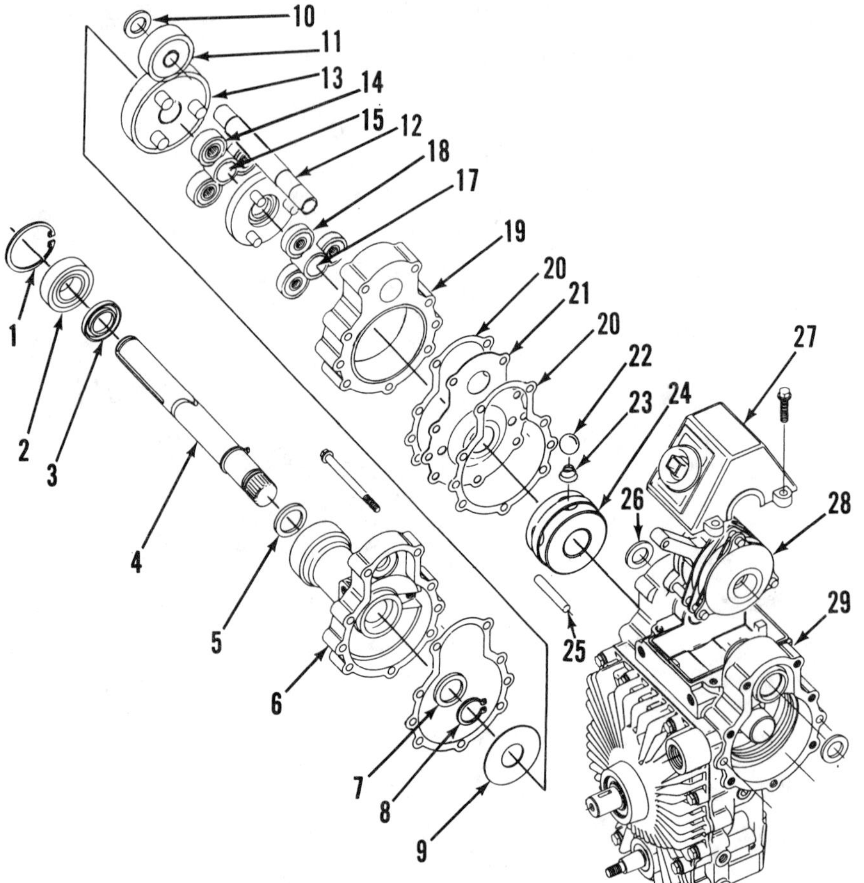

Fig. SN20—Exploded view of Eaton 850 hydrostatic transmission left axle housing assembly used on Model YT18H.

1. Snap ring
2. Bearing
3. Seal
4. Axle shaft
5. Thrust washer
6. Axle housing
7. Thrust washer
8. Snap ring
9. Thrust washer
10. Spacer
11. Brake gear
12. Brake shaft
13. Planetary carrier
14. Planetary gear (second)
15. Sun gear (second)
16. Planetary carrier
17. Sun gear (first)
18. Planetary gear (first)
19. Ring gear
20. Gaskets
21. Wear plate
22. Ball piston
23. Spring
24. Motor rotor
25. Dowel
26. Seal
27. Brake cover
28. Parking brake assy.
29. Transmission housing

To disassemble axle assembly, remove wear plate (21), first planetary assembly (16, 17 and 18), ring gear (19), brake gear (11) and second planetary assembly (13, 14 and 15). Remove large thrust washer (9), snap ring (8) and small thrust washer (7) from axle shaft. Remove outer snap ring (1), then remove axle shaft (4), oil seal (3) and bearing (2) from axle housing (6) by tapping on splined end of shaft with a soft hammer.

Detach brake housing (27) and remove parking brake assembly (28) being careful not to move brake lever. The brake is self-adjusting and moving lever will expand brake assembly making installation more difficult. Brake assembly is available only as a unit and should not be disassembled.

To remove hydrostatic motors, place a large rubber band or similar device around the motor rotor (24) to hold spring-loaded ball pistons (22) in their bores. Carefully withdraw motor assembly. Due to selective fit, if balls are removed they must be returned to original rotor bore.

To disassemble hydrostatic pump, unbolt and remove oil filter base from housing. Unscrew fasteners retaining pump cover (35—Fig. SN21). Note that there is a screw located in the drain hole. Separate cover from housing and remove snap ring (30). Tap or press input shaft (32) and bearing (31) out of cover by forcing shaft toward outside of cover. Shaft and bearing assembly is available only as a unit assembly. Drive input shaft seal (33) out of cover bore. Remove charge pressure relief valve spring (45) and ball (46). Remove but-

Illustrations courtesy Snapper Power Equipment.

tons (36), control shaft (37) and cam ring insert (38) from cam ring (40). Remove pump rotor drive ring (41). Place a wide rubber band or similar device around pump rotor (42) to hold the ball pistons (43) in their bores. The balls are a selective fit in rotor and must be installed in their matching bore if removed from rotor. Carefully lift pump rotor from housing.

Dump valve (51) can be removed from housing after removing nut (48). Check valves (Fig. SN22) should be flushed, but not removed, and are available only with housing. Removal of dampening pistons (39—Fig. SN21) is also not recommended.

Charge pump assembly (1—Fig. SN23) may be removed from housing for inspection after removing pump plate (2). Charge pump components are available only with the housing (47) as an assembly.

Inspect components for excessive wear and damage. Pump and motor rotors and balls are unit assemblies as balls are selectively fitted to rotor bores.

Assemble by referring to appropriate exploded view and reversing disassembly procedure while noting the following: Lubricate all assemblies with clean transmission oil. Install dump valve spring (52—Fig. SN21) so right angle bend points toward "O" ring (50). Tighten dump valve nut (48) to 150 in.-lbs. (17 N·m). Install charge pump cover so arrow (A—Fig. SN23) points toward "CW" on housing. Do not tighten cover screws until pump rotor is installed. Note that pins on bottom of pump rotor (42—Fig. SN21) must engage holes in charge pump inner rotor. Rotate pump rotor clockwise to center pump cover then tighten cover screws in a crossing pattern to 44 in.-lbs. (5.0 N·m). Install cam ring (40) so flush side is out. Install seals (33 and 34) so lip is to inside of cover. Be sure snap ring (30) is firmly seated in cover after installing input shaft assembly (32). Hold buttons (36) in place with petroleum jelly. Be sure charge relief valve ball (46) is not dislodged when installing cover assembly on housing. Splines on input shaft must engage splines of drive ring (41); do not force cover onto housing. Tighten cover screws to 105 in.-lbs. (11.9 N·m) using pattern shown in Fig. SN24. Carefully install motor rotors in housing and remove rubber band used to retain ball pistons in rotors.

Install brake assembly so lever is toward cover side of housing. Note that axle housings and sides of transaxle housing are identified with an "A" or "B" as shown in Fig. SN25. The "B" side of transaxle should be assembled first. Note that "B" side brake shaft (12—Fig. SN20) is longer than the "A"

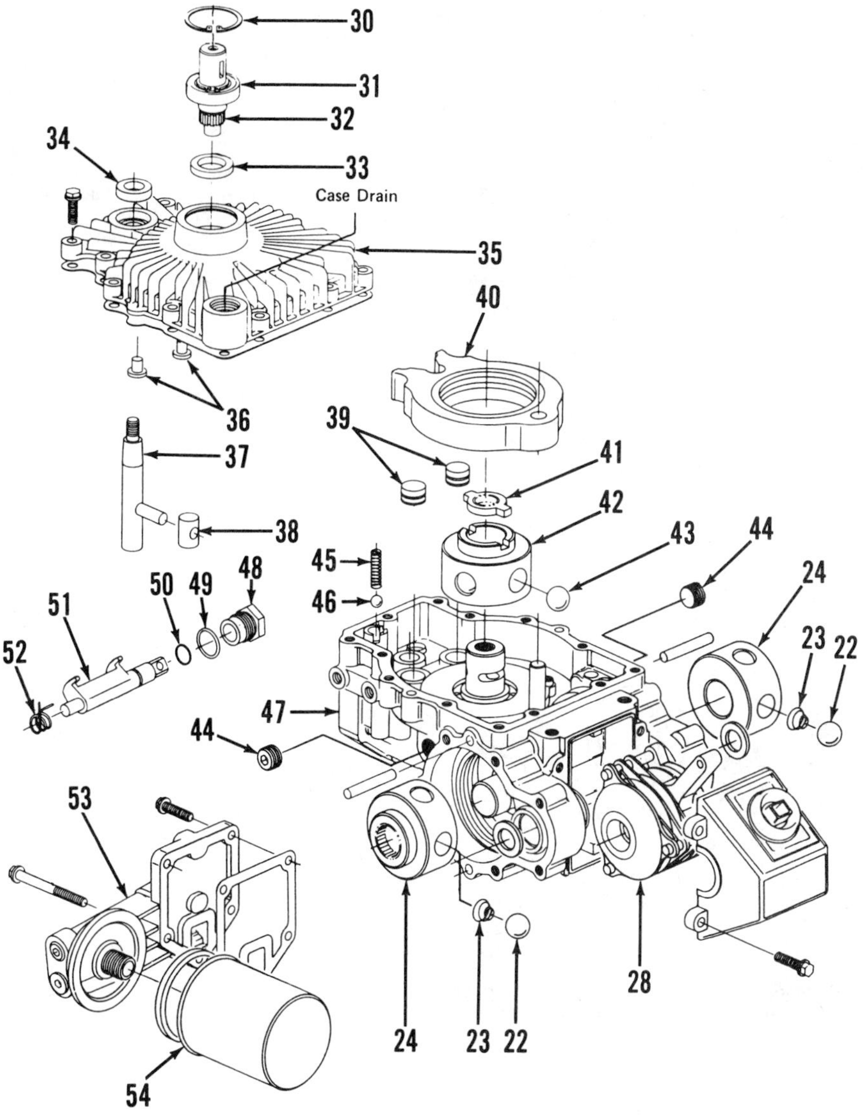

Fig. SN21—Exploded view of pump and housing components on Eaton 850 hydrostatic transmission.

22. Ball piston
23. Spring
24. Motor rotor
28. Brake assy.
30. Snap ring
31. Bearing
32. Input shaft assy.
33. Oil seal
34. Oil seal
35. Cover
36. Buttons
37. Control shaft
38. Insert
39. Dampening pistons
40. Cam ring
41. Drive ring
42. Pump rotor
43. Ball piston
44. Plugs
45. Charge relief valve spring
46. Charge relief valve ball
47. Housing
48. Plug
49. "O" ring
50. "O" ring
51. Dump valve bracket
52. Spring
53. Oil filter base
54. Oil filter

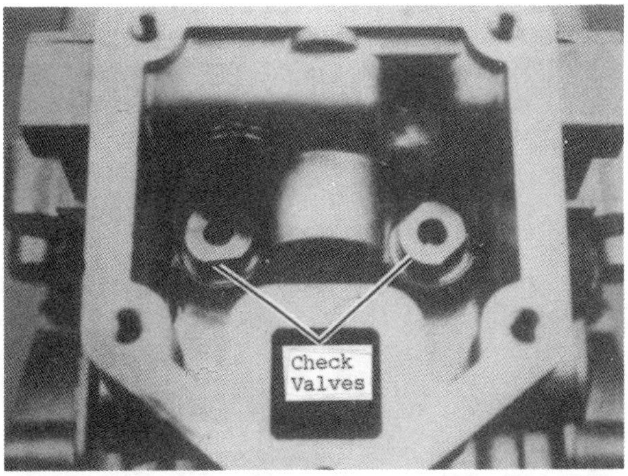

Fig. SN22—Check valves are available only with housing and should be cleaned by flushing with solvent, but not removed.

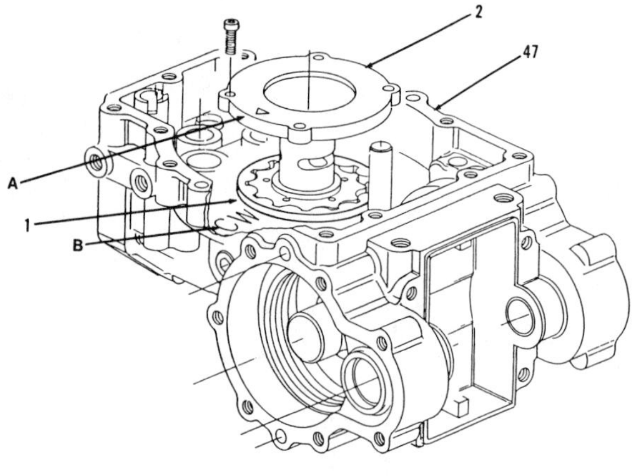

Fig. SN23—Install charge pump cover (2) so arrow (A) points towards "CW" on housing.

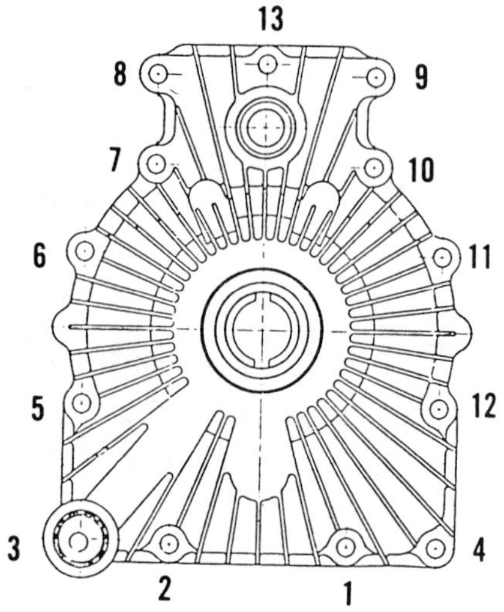

Fig. SN24—Tighten cover screws to 105 in.-lbs. (11.9 N·m) using sequence shown above.

side shaft. Insert the "B" side brake shaft into the brake assembly and rotate so the splines are engaged. Install "A" side brake shaft. When both shafts are properly installed, the shafts will rotate independently.

Install axle components while noting the following: Inner sun gear (17) is taller than outer sun gear (15). Install ring gear (19) so side with needle bearing mates with axle housing (6). Install outer sun gear (15) so convex (pointed) end is toward inner sun gear (17). Install axle housing seal (3) so lip points inward. Install wear plate (21) so bowed side is facing motor rotor. Tighten axle housing screws to 125 in.-lbs. (14 N·m). Tighten oil filter base screws to 125 in.-lbs. (14 N·m).

Fig. SN25—Note that axle housings and sides of transaxle housing are identified with an "A" or "B" as shown above. Refer to text.

Illustrations courtesy Snapper Power Equipment.

POWER TAKE-OFF

ADJUSTMENT

All Models So Equipped

Adjust electric pto clutch by turning each of the adjusting nuts (N—Fig. SN26) until clearance between rotor and armature is 0.010 inch (0.25 mm). Do not overtighten nuts. Operate clutch several times and recheck clearance.

OVERHAUL

Refer to Fig. SN27 for an exploded view of pto clutch. Clutch may be removed after unscrewing center screw. The bearing of the Ogura clutch may be removed by pressing out bearing after removing the snap ring. Field coil (4) resistance should be 3.4-5.5k ohms. If installing a new clutch or components, cycle clutch several times at half throttle and then several times at full throttle to burnish clutch surfaces.

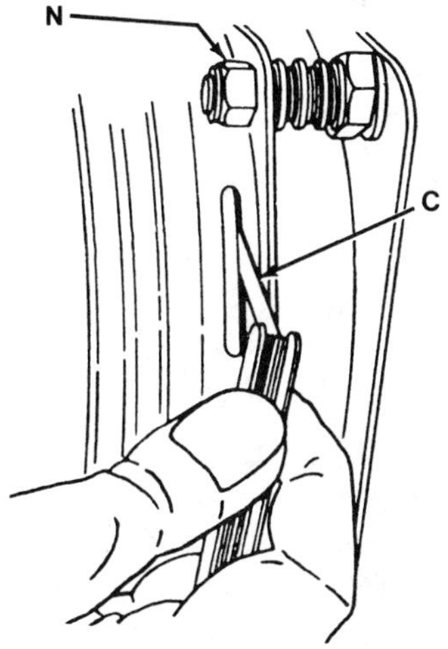

Fig. SN26—To adjust pto clutch, turn each adjusting nut (N) of pto clutch so clearance (C) between rotor and armature is 0.010 inch (0.025 mm).

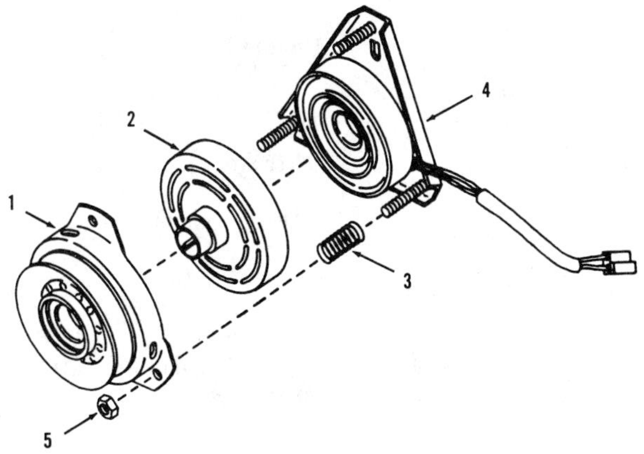

Fig. SN27—Exploded view of electric pto clutch.

1. Pulley assy.
2. Rotor
3. Spring
4. Field coil

SPEEDEX

CONDENSED SPECIFICATIONS

	MODEL 1832
Engine Make	B&S
Model	422437
Bore	3.44 in. (87.3 mm)
Stroke	2.28 in. (57.9 mm)
Displacement	42.3 cu. in. (694 cc)
Power Rating	18 hp. (13.4 kW)
Slow Idle	1400 rpm
High Speed (No-Load)	3600 rpm
Capacities–	
Crankcase	3 pts.* (1.4 L)
Transmission	1 qt. (0.9 L)
Final Drive	1 qt. (0.9 L)
Hyraulic System	1 gal. (3.8 L)
Fuel Tank	3 gal. (11.4 L)

*Early model crankcase capacity is 3.5 pints (1.7 L).

FRONT AXLE AND STEERING SYSTEM

MAINTENANCE

Lubricate left and right steering spindles (fittings 27 and 29–Fig. SX1), axle pivot (fitting 25), pitman shaft support (fittings 17 and 22) and steering column (fitting 16) at 20 hour intervals. Use multi-purpose, lithium base grease and clean fittings before and after lubrication.

Clean and repack front wheel bearings at 250 hour intervals. Use a good quality grease recommended for wheel bearing applications.

R&R AXLE MAIN MEMBER

Raise and support front of tractor. Disconnect drag link (9–Fig. SX1) at steering arm (5). Support axle main member (3) and pull axle main member forward out of pivot housing.

Reinstall by reversing removal procedure.

TIE ROD AND TOE-IN

Remove tie rod (8–Fig. SX1) by disconnecting tie rod ends (6) at right and left steering arms (5 and 26).

Reinstall by reversing removal procedure and adjust toe-in by loosening jam nuts (7) and turning tie rod (8) as required to obtain 1/8-inch (3.175 mm) toe-in. Tighten jam nuts.

R&R STEERING SPINDLES

Raise and support front of tractor. Remove front tires and wheels. Disconnect tie rod (8–Fig. SX1) at left and right steering arms (5 and 26). Disconnect drag link at right steering arm (5). Drive tapered pins (4) out of steering arm and spindle assemblies. Remove spindles.

Renew bushings (2) and/or spindles (1) as necessary. Reinstall by reversing removal procedure.

R&R STEERING GEAR

Disconnect drag link (9–Fig. SX1) at pitman shaft (12). Drive roll pin (21) from gear (20) and pitman shaft (12), pull pitman shaft out of support (14) and gear (20). Drive pin (15) out of steering wheel (13) and steering shaft (19). Remove steering wheel and pull steering shaft (19) out from lower side of tractor. Remove spacers (18).

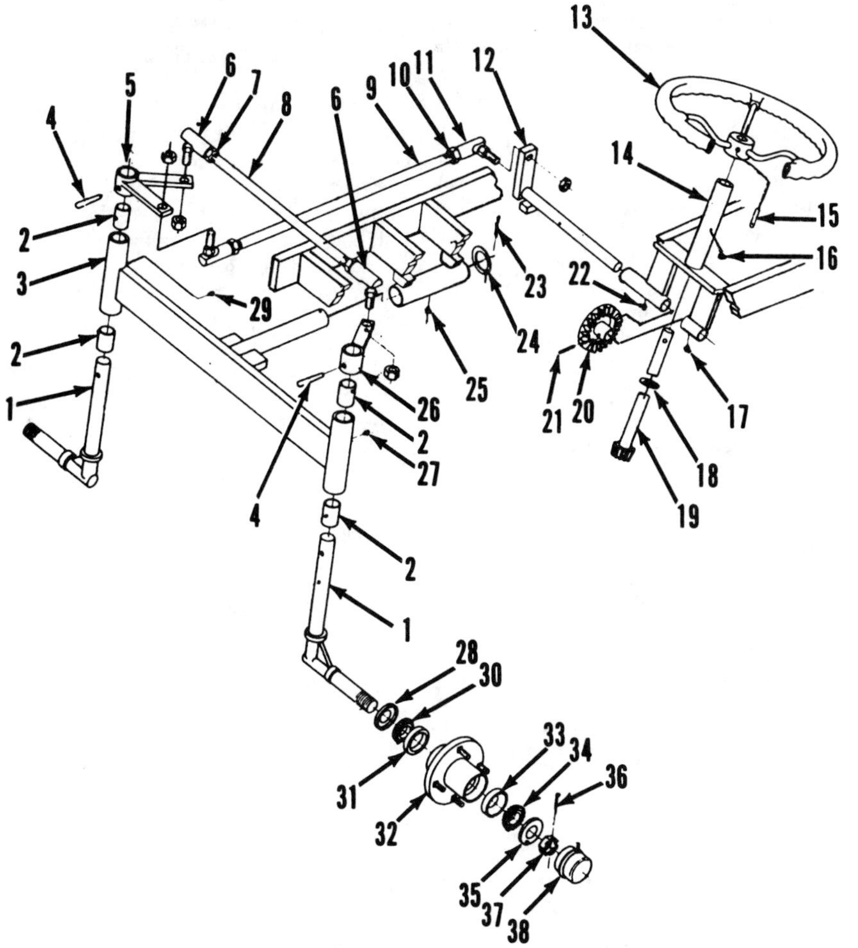

Fig. SX1—Exploded view of front axle assembly.

1. Spindle
2. Bushing
3. Axle main member
4. Taper pin
5. Steering arm (RH)
6. Tie rod end
7. Jam nut
8. Tie rod
9. Drag link
10. Jam nut
11. Drag link end
12. Pitman shaft
13. Steering wheel
14. Pitman shaft support
15. Pin
16. Grease fitting
17. Grease fitting
18. Spacer
19. Steering shaft & gear
20. Gear
21. Pin
22. Grease fitting
23. Cotter pin
24. Washer
25. Grease fitting
26. Steering arm (LH)
27. Grease fitting
28. Seal
29. Grease fitting
30. Bearing
31. Bearing cup
32. Hub
33. Bearing cup
34. Bearing
35. Washer
36. Cotter pin
37. Nut
38. Dust cover

Reinstall by reversing removal procedure. Install spacers (18) to provide minimum steering wheel backlash and smooth stop-to-stop turning cycle.

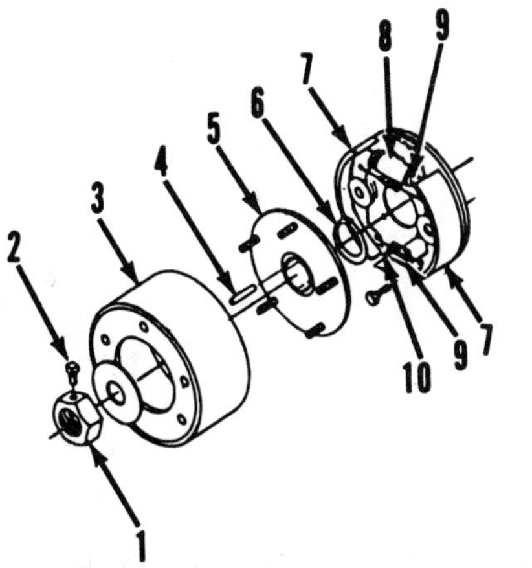

Fig. SX2—Exploded view of individual wheel brake assembly.

1. Nut
2. Setscrew
3. Brake drum
4. Key
5. Hub assy.
6. Snap ring
7. Brake shoes
8. Wheel cylinder
9. Springs
10. Adjuster assy.

ENGINE

MAINTENANCE

Regular engine maintenance is required to maintain peak performance and long engine life.

Check oil level at five hour intervals under normal operating conditions.

Clean and re-oil foam precleaner at 25 hour intervals and clean or renew dry filter element at 100 hour intervals.

REMOVE AND REINSTALL

Remove hood assembly and disconnect negative battery cable. Disconnect all necessary electrical leads and fuel line. Disconnect throttle and choke linkage at carburetor. Remove main drive belt. Remove hydraulic pump or disconnect lines as necessary. Remove engine mounting bolts and remove engine.

Reinstall by reversing removal procedure.

OVERHAUL

Engine make and model are listed at the beginning of this section. For tune-up specifications, engine overhaul procedures and engine maintenance, refer to appropriate engine section in this manual.

ELECTRICAL SYSTEM

MAINTENANCE AND SERVICE

Battery electrolyte level should be checked at 50 hour intervals of normal operation. If necessary, add distilled water until level is just below base of vent well. **DO NOT** overfill. Keep battery posts clean and cable ends tight.

For alternator or starter service, refer to appropriate engine section in this manual.

BRAKES

MAINTENANCE AND ADJUSTMENT

Individual Wheel Brakes

Check fluid level in each master cylinder at 20 hour intervals. Maintain fluid level ¼-inch (6.35 mm) below top edge of master cylinder. Recommended fluid is a good quality automotive type brake fluid.

To adjust wheel brakes, use a brake adjustment tool to turn adjuster assembly (10–Fig. SX2) until it cannot be tightened further (move tool handle upward to tighten). Back adjuster off seven notches. Repeat procedure for each side.

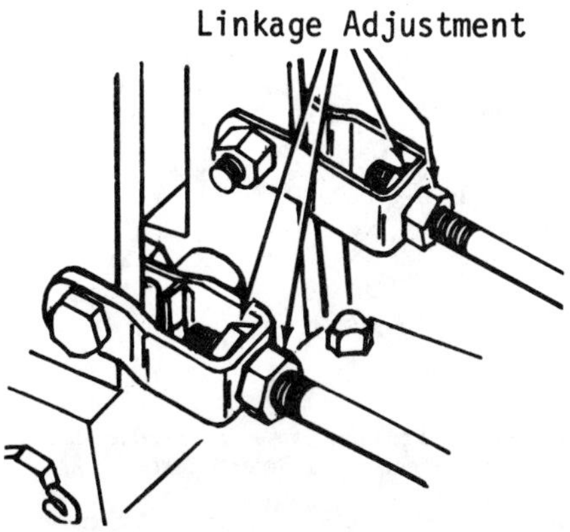

Fig. SX3—View of adjusting nuts used to adjust individual wheel brake pedal height.

Adjust individual pedals to equal height by turning adjustment nuts on master cylinder shafts as shown in Fig. SX3.

Service Brake

Service brakes and clutch rods must be adjusted so transmission drive belt is fully engaged when clutch/brake pedal is released. Rods must be adjusted so clutch is fully disengaged before brake engagement. Refer to Fig. SX4 for clutch rod adjustment points.

After clutch/brake pedal is correctly adjusted, brake band may be tightened by turning adjustment nuts (3–Fig. SX5).

R&R BRAKE SHOES/BAND

Individual Wheel Brake Shoes

Raise and support rear of tractor. Remove tires and wheels. Loosen setscrews (2–Fig. SX2) and remove nuts (1). Pull brake drum (3). Remove hold down springs, brake shoes (7), springs (9) and adjuster assembly (10).

Reinstall by reversing removal procedure. Adjust brakes as previously outlined.

Service Brake Band

To remove brake band, remove cotter pin (4–Fig. SX5), nuts (3) and adjusting knuckle (5). Remove keeper (7) and anchor (6). Remove brake band.

To remove drum (2), remove cap screw (8) and pull drum (2).

Reinstall by reversing removal procedure.

CLUTCH

ADJUSTMENT

Transmission drive belt runs on a pivoting idler pulley arrangement to provide transmission clutch action.

Adjust clutch/brake pedal linkage as outlined in ADJUSTMENT-SERVICE BRAKE paragraph in BRAKE section in this manual.

R&R TRANSMISSION DRIVE BELT

Loosen brake rod adjustment setscrew and front belt guard at frame. Depress clutch/brake pedal and allow engine to slide to rear ½-inch (12.7 mm). Remove belt from transmission pulley. Pull belt from engine pulley end through open end of belt guard.

Install new belt through front and around engine pulley. Loosen clutch adjusting setscrew (Fig. SX4) as necessary to install belt on transmission pulley. Adjust belt retaining rods to obtain ⅛-inch (3.175 mm) clearance between belt and rods. Adjust clutch so drive belt has ½-inch (12.7 mm) deflection with moderate pressure at a point midway between pulleys.

TRANSMISSION

MAINTENANCE

Transmission fluid level should be

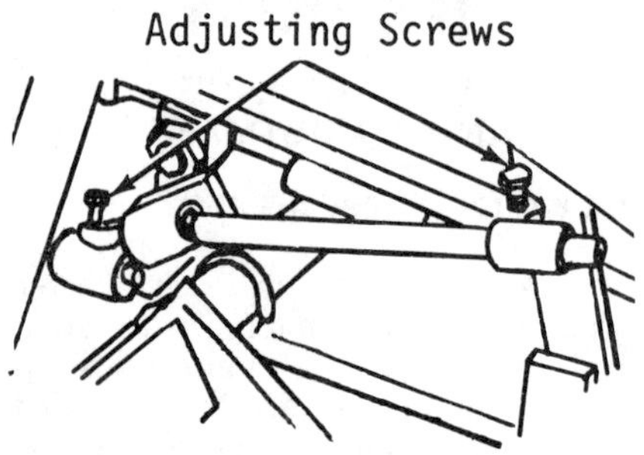

Fig. SX4—View showing adjusting screws used on clutch rod. Refer to text.

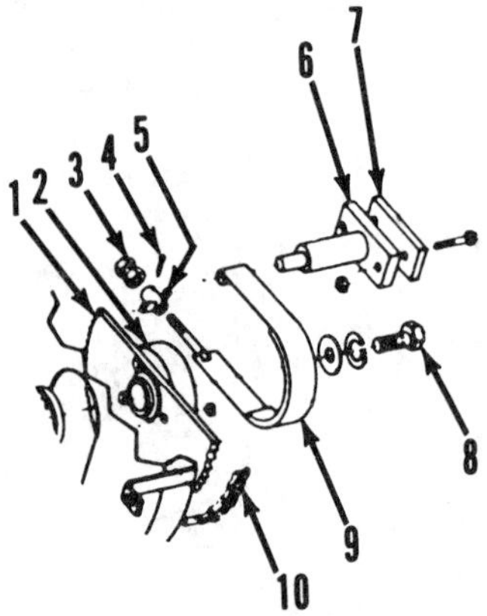

Fig. SX5—Exploded view of service brake mechanism. Refer to text.

1. Sprocket
2. Drum
3. Nut
4. Cotter pin
5. Adjusting knuckle
6. Anchor
7. Keeper
8. Cap screw
9. Brake band
10. Drive chain

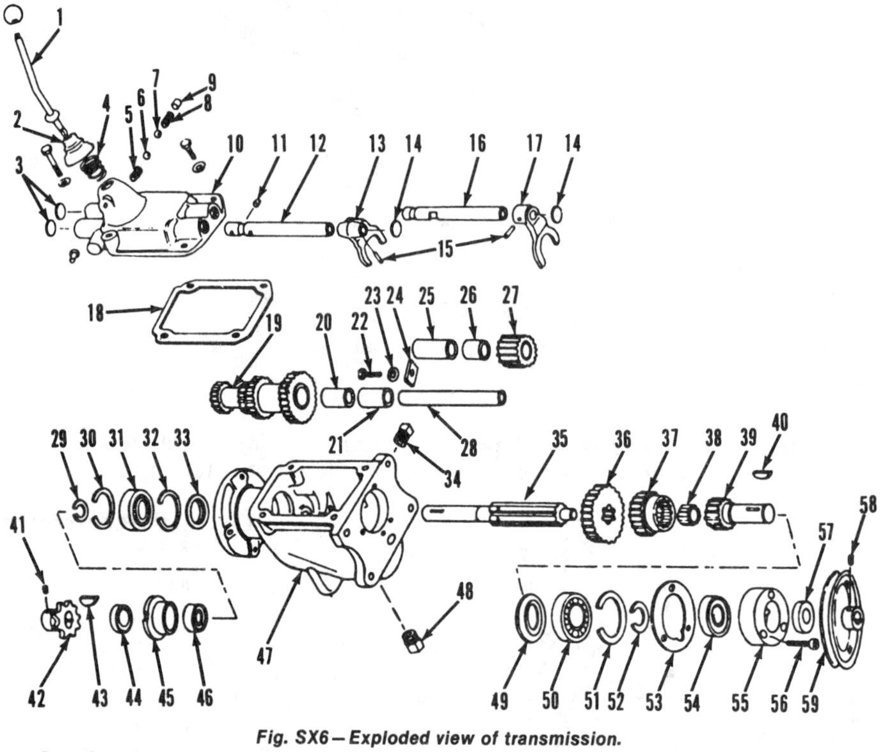

Fig. SX6—Exploded view of transmission.

1. Gear selector lever
2. Dust boot
3. Expansion plugs
4. Spring
5. Detent spring
6. Detent ball
7. Detent ball
8. Detent spring
9. Poppet
10. Cover
11. Interlock plunger
12. Shift rail (1st & Rev.)
13. Shift fork
14. Expansion plug
15. Pin
16. Shift rail (2nd & 3rd)
17. Shift fork
18. Gasket
19. Cluster gear
20. Bushing
21. Bushing
22. Cap screw
23. Washer
24. Lock plate
25. Reverse idler shaft
26. Bushing
27. Reverse idler gear
28. Countershaft
29. Snap ring
30. Snap ring
31. Bearing
32. Snap ring
33. Washer
34. Filler plug
35. Main shaft
36. Gear (1st & Rev.)
37. Gear (2nd)
38. Roller bearings (13)
39. Main drive gear
40. Key
41. Setscrew
42. Sprocket
43. Key
44. Seal
45. Bearing retainer
46. Bearing
47. Case
48. Drain plug
49. Washer
50. Bearing
51. Snap ring
52. Snap ring
53. Gasket
54. Bearing
55. Retainer
56. Cap screw
57. Seal
58. Setscrew
59. Pulley

1. Bushing
2. Pinion gear
3. Bearing
4. Key
5. Bearing cup
6. Back-up ring
7. Cover
8. Chain guard
9. Sprocket
10. Chain
11. Back-up ring
12. Bearing cup
13. Bearing
14. Slinger
15. Seal
16. Bearing
17. Support
18. Housing
19. Bearing
20. "O" ring
21. Plug
22. Axle shaft & gear
23. Spider gears
24. Bull gear
25. "O" ring
26. Axle shaft & gear
27. Thrust washer

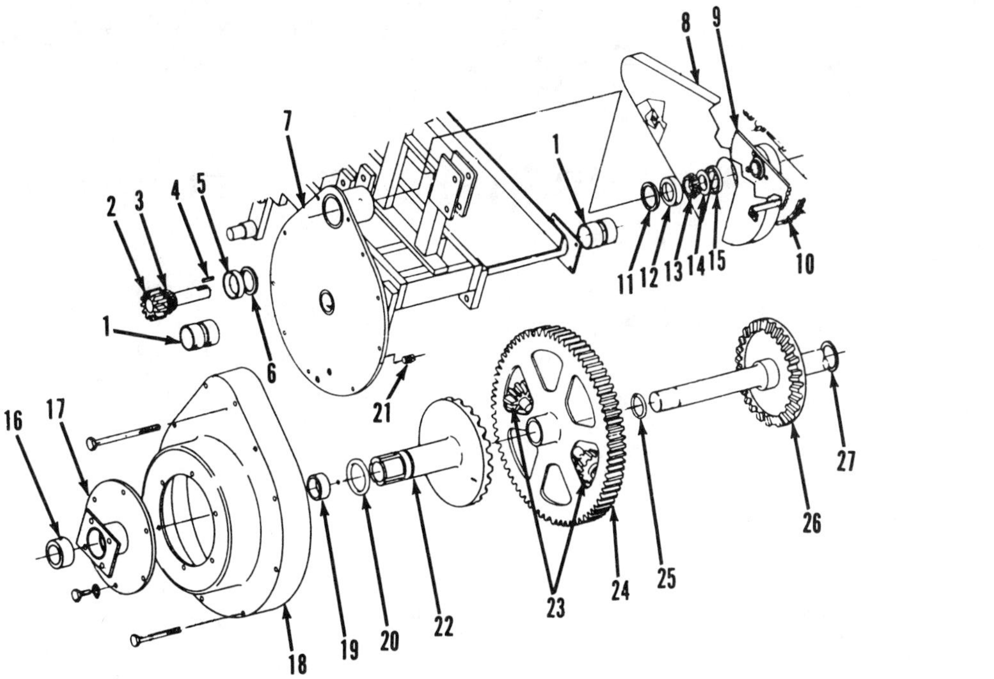

Fig. SX7—Exploded view of final drive assembly.

checked at 50 hour intervals. Maintain fluid level at filler plug (34–Fig. SX6) opening.

Change fluid at 500 hour intervals. Recommended fluid is SAE 90 EP gear lubricant. Approximate capacity is 1 quart (0.9 L).

R&R TRANSMISSION

Remove transmission drive belt as previously outlined. Disconnect drive chain at output sprocket (42–Fig. SX6). Remove transmission mounting bolts and remove transmission.

Reinstall by reversing removal procedure. Adjust drive belt and clutch/brake linkage.

OVERHAUL

Remove cap screws securing top cover (10–Fig. SX6) to case (47), then remove cover and shift fork assembly. Remove pins (15) and expansion plugs (3 and 14), then remove shift rails (12 and 16) and forks (13 and 17) as necessary. Remove detent balls (6 and 7) and springs (5 and 8). Loosen setscrews (58) and remove pulley (59) and key (40). Remove cap screws (56), bearing retainer (55) and snap rings (51 and 52). Remove bearing retainer (45) and snap rings (29 and 30). Remove input shaft (39) and output shaft (3) and separate gears as necessary. Remove cap screw (22), washer (23) and lock-plate (24). Push shaft (28) out and remove cluster gear (19). Remove reverse idler shaft (25) and gear (27).

Clean and inspect all parts for wear or damage. Renew parts and bearings as necessary. Renew all seals and gaskets. Reassemble by reversing disassembly procedure.

FINAL DRIVE AND DIFFERENTIAL ASSEMBLY

MAINTENANCE

Lubricate drive chain (10–Fig. SX7) and sprockets sparingly with appropriate roller chain lubricant and check final drive fluid level at 50 hour intervals. Maintain final drive fluid level at edge of plug (21) opening. Change final drive fluid at 500 hour intervals. Recommended fluid is SAE 90 EP gear lubricant. Approximate capacity is 1 quart (0.9 L).

OVERHAUL

Raise and support rear of tractor. Remove rear tires and wheels. Drain fluid from final drive. Remove drive chain (10–Fig. SX7), service brake drum and brake band. Remove wheel hub, brake drum and brake shoes. Disconnect brake line at wheel cylinder, then remove backing plate and wheel cylinder assembly. Remove cover (7) and housing (18). Remove gear and shaft (22). Remove key (4) and pinion (2). Renew bearings (3 and 13) and cups (5 and 12) as necessary. Remove bull gear (24), spider gears (23) and gear and shaft (26).

Renew all seals and "O" rings. Check all bearings and bushings for roughness or excessive wear. Reassemble by reversing disassembly procedure.

HYDRAULIC SYSTEM

MAINTENANCE AND SERVICE

Fluid level should be checked at 20 hour intervals. Maintain hydraulic reservoir 3/4 full of fluid.

Change fluid at 500 hour intervals. Recommended fluid is "Type A" automatic transmission fluid. Approximate fluid capacity of system is 1 gallon (11.4 L).

A gear type hydraulic pump is belt driven from a pulley on engine crankshaft. Pump drive belt tension is correct when belt does not slip when system is under a load.

Service parts are not available for hydraulic pump, control valve or cylinder. Renew complete assemblies as necessary.

WHEEL HORSE

CONDENSED SPECIFICATIONS

	MODELS				
	D-160	D-180	D-200	18 HP AUTOMATIC	417
Engine Make	Onan	Kohler	Kohler	Kohler	Kohler
Model	B43M	K-482S	K-532S	K-482S	KT-17
Bore	3.25 in. (82.55 mm)	3.25 in. (82.55 mm)	3.38 in. (85.73 mm)	3.25 in. (82.55 mm)	3.12 in. (79.3 mm)
Stroke	2.62 in. (66.55 mm)	2.875 in. (73.03 mm)	3.0 in. (76.2 mm)	2.875 in. (73.03 mm)	2.75 in. (69.9 mm)
Displacement	43.3 cu. in. (712.4 cc)	47.8 cu. in. (781.73 cc)	53.67 cu. in. (879.7 cc)	47.8 cu. in. (781.73 cc)	42.18 cu. in. (690.5 cc)
Power Rating	16 hp. (11.9 kW)	18 hp. (13.4 kW)	20 hp. (14.9 kW)	18 hp. (13.4 kW)	17 hp. (12.7 kW)
Slow Idle	1350 rpm	1350 rpm	1350 rpm	1350 rpm	1200 rpm
High Speed (No-Load)	3600 rpm	3600 rpm	3600 rpm	3600 rpm	3400 rpm
Capacities –					
Crankcase	2 qts. (1.9 L)	1¾ qts. (1.6 L)	3½ qts. (3.3 L)	1¾ qts. (1.6 L)	1¾ qts. (1.6 L)
Hydraulic System	6 qts. (5.7 L)	6 qts. (5.7 L)	6 qts. (5.7 L)	6 qts. (5.7 L)	.....
Transaxle or Transmission	See Hyd.	See Hyd.	See Hyd.	See Hyd.	2 qts. (1.9 L)
Fuel Tank	5¾ gal. (21.9 L)	8 gal. (30.4 L)	8 gal. (30.4 L)	8 gal. (30.4 L)	3 gal. (11.4 L)

	MODELS				
	417-A	GT-1642	GT-1848	LT-1637	LT-1642
Engine Make	Kohler	B&S	B&S	B&S	B&S
Model	KT-17	402417	422437	402707	402707
Bore	3.12 in. (79.3 mm)	3.44 in. (87.3 mm)	3.44 in. (87.3 mm)	3.44 in. (87.3 mm)	3.44 in. (87.3 mm)
Stroke	2.75 in. (69.9 mm)	2.16 in. (54.8 mm)	2.28 in. (57.9 mm)	2.16 in. (54.8 mm)	2.16 in. (54.8 mm)
Displacement	42.18 cu. in. (690.5 cc)	40 cu. in. (656 cc)	42.33 cu. in. (694 cc)	40 cu. in. (656 cc)	40 cu. in. (656 cc)
Power Rating	17 hp. (12.7 kW)	16 hp. (11.9 kW)	18 hp. (13.4 kW)	16 hp. (11.9 kW)	16 hp. (11.9 kW)
Slow Idle	1200 rpm	1450 rpm	1450 rpm	1400 rpm	1400 rpm
High Speed (No-Load)	3400 rpm	3300 rpm	3300 rpm	3300 rpm	3300 rpm
Capacities –					
Crankcase	1¾ qts. (1.6 L)	1½ qts.* (1.4 L)	1½ qts.* (1.4 L)	1½ qts.* (1.4 L)	1½ qts.* (1.4 L)
Hydraulic System	5 qts. (4.7 L)	***	***	.....	.....
Transaxle or Transmission	See Hyd.	2 qts.*** (1.9 L)	2 qts.*** (1.9 L)	¾-qts. (0.7 L)	36 oz. (1065 mL)
Fuel Tank	3 gal. (11.4 L)	3 gal. (11.4 L)	3 gal. (11.4 L)	1½ gal. (5.7 L)	1½ gal. (5.7 L)

*Early production engine crankcase capacity is 1¾ quarts (1.6 L).
***All "C" and "GT" series equipped with hydrostatic drive have a common fluid reservoir for hydrostatic drive unit and reduction gear unit. Fluid capacity is 5 quarts (4.7 L).

	MODELS				
	216-5	B-165	C-161	C-175	C-195
Engine Make	B&S	B&S	B&S	Kohler	Kohler
Model	402707	401707	401707	KT-17	KT-19
Bore	3.44 in. (87.3 mm)	3.44 in. (87.3 mm)	3.44 in. (87.3 mm)	3.12 in. (79.3 mm)	3.12 in. 79.3 mm)
Stroke	2.16 in. (54.8 mm)	2.16 in. (54.8 mm)	2.16 in. (54.8 mm)	2.75 in. (69.9 mm)	3.06 in. (78 mm)
Displacement	40 cu. in. (656 cc)	40 cu. in. (656 cc)	40 cu. in. (656 cc)	42.18 cu. in. (690.5 cc)	46.98 cu. in. (770.5 cc)
Power Rating	16 hp. (11.9 kW)	16 hp. (11.9 kW)	16 hp. (11.9 kW)	17 hp. (12.7 kW)	19 hp. (14.2 kW)
Slow Idle	1400 rpm	1400 rpm	1450 rpm	1200 rpm	1200 rpm
High Speed (No-Load)	3300 rpm	3300 rpm	3300 rpm	3400 rpm	3400 rpm
Capacities –					
Crankcase	1½ qts* (1.4 L)	1½ qts.* (1.4 L)	1½ qts.* (1.4 L)	1¾ qts. (1.6 L)	1¾ qts. (1.6 L)
Hydraulic System	.....	**	***	***	***
Transaxle or Transmission	36 oz. (1065 mL)	36 oz.** (1065 mL)	2 qts.*** (1.9 L)	2 qts.*** (1.9 L)	2 qts.*** (1.9 L)
Fuel Tank	1½ gal. (5.7 L)	1½ gal. (5.7 L)	3 gal. (11.4 L)	3 gal. (11.4 L)	3 gal. (11.4 L)

*Early production engine crankcase capacity is 1¾ quarts (1.6 L).
**Model B-165 fluid reservoir for hydrostatic drive unit and reduction gear unit are separate. Hydrostatic drive unit fluid capacity is ¾-quarts (0.7 L) and reduction gear unit capacity is 1⅜ quarts (1.3 L).
***All "C" and "GT" series equipped with hydrostatic drive have a common fluid reservoir for hydrostatic drive unit and reduction gear unit. Fluid capacity is 5 quarts (4.7 L).

FRONT AXLE AND STEERING SYSTEM

MAINTENANCE

All Models

It is recommended that steering gear, front wheel bearings, steering spindles and axle pivot be lubricated at 25 hour intervals. Use multi-purpose, lithium base grease. Clean all fittings before and after lubrication. Clean all other pivot points and linkages and lubricate with SAE 30 oil. Check for any looseness or wear and repair as necessary.

R&R AXLE MAIN MEMBER

Models B-165 – LT-1637 – LT-1642 – 216-5

Raise and support front of tractor. Disconnect tie rods. Support axle main member (2–Fig. WH1) and remove snap ring (21). Remove bolts securing front hanger and pivot pin assembly (19) and remove assembly. Lower axle to the ground and roll from under tractor.

Reinstall by reversing removal procedure.

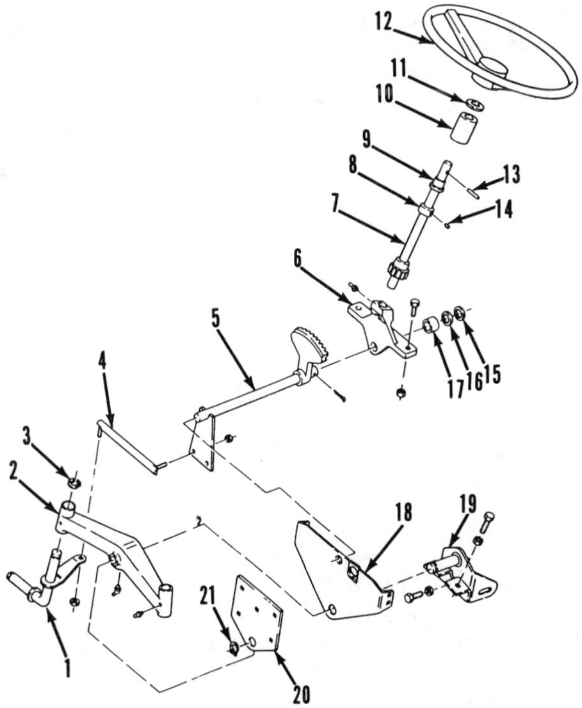

Fig. WH1 – Exploded view of front axle and steering assembly used on B-165, LT-1637, LT-1642 and 216-5 models.

1. Spindle
2. Axle main member
3. Snap ring
4. Tie rod
5. Lower steering shaft
6. Steering support
7. Upper steering shaft
8. Collar
9. Bushing
10. Collar
11. Spring washer
12. Steering wheel
13. Roll pin
14. Setscrew
15. Shim washer
16. Shim washer
17. Bearing
18. Support
19. Hanger & pivot pin assy.
20. Support
21. Snap ring

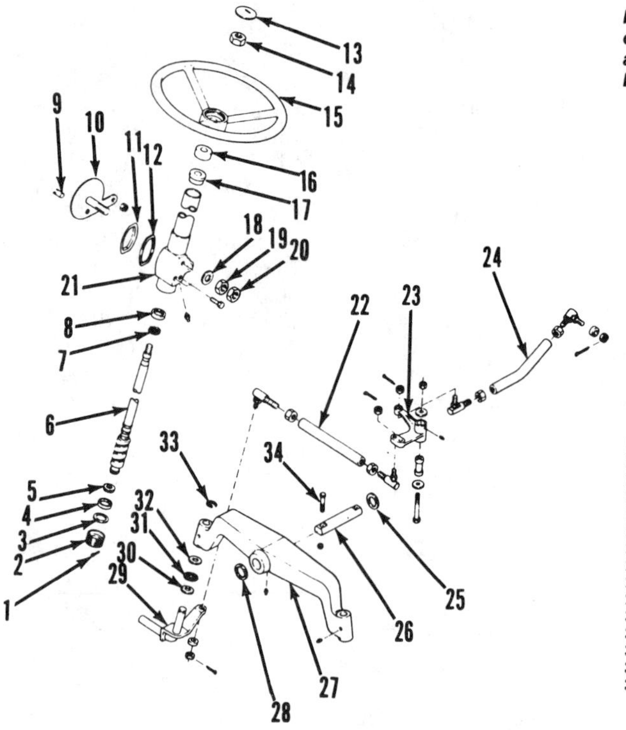

Fig. WH2— Exploded view of front axle and steering assembly used on D-160, D-180, D-200 and 18 HP Automatic models.

1. Cotter key
2. Adjustment plug
3. Bevel washer
4. Bearing cup
5. Bearing
6. Steering shaft & worm gear
7. Bearing
8. Bearing cup
9. Steering pin
10. Steering lever plate
11. Seal retainer
12. Seal
13. Dust cap
14. Nut
15. Steering wheel
16. Dust cover
17. Bearing
18. Washer
19. Adjustment nut
20. Jam nut
21. Housing assy.
22. Tie rod
23. Steering lever
24. Drag link
25. Washer
26. Pivot pin
27. Axle main member
28. Washer
29. Spindle
30. Washer
31. Thrust bearing
32. Washer
33. Snap ring
34. Retaining bolt

Models D-160 – D-180 – D-200 – 18 HP Automatic

Raise and support front of tractor. Disconnect tie rods. Support axle main member (27 – Fig. WH2) and remove retaining bolt (34). Remove pivot pin (26). Lower axle and roll assembly away from tractor.

Reinstall by reversing removal procedure.

Models 417 – 417A – C-161 – C-175 – C-195 – GT-1642 – G-1848

Raise and support front of tractor. Disconnect tie rods. Support axle main member (1 – Fig. WH3) and remove retaining bolt (18) and snap ring (20). Remove pivot pin (19). Pivot pin may be inserted from front side of axle on some models. Lower axle and roll assembly away from tractor.

Reinstall by reversing removal procedure.

TIE ROD AND TOE-IN

Models D-160 – D-180 – D-200 – 18 HP Automatic

Tie rods (22 – Fig. WH2) and drag link (24) have renewable ends. Front wheel toe-in should be 1/16 to 1/8-inch (1.588 to 3.175 mm) and is obtained by equally adjusting tie rod lengths. Make certain all jam nuts are tight after adjustments are correct.

All Other Models

Tie rods are fixed in length and toe-in is nonadjustable. Renew complete tie rod assemblies if excessive looseness is apparent.

R&R STEERING SPINDLES

All Models

Raise and support front of tractor, then remove wheel and hub assembly. Disconnect tie rod ends and remove snap ring located on upper end of steering spindle. Slide spindle out of axle main member. Note thrust washer and bearing (30, 31 and 32 – Fig. WH2) on D-160, D-180, D-200 and 18 HP Automatic models.

Reinstall by reversing removal procedure.

FRONT WHEEL BEARINGS

All Models

Front wheels may be equipped with bushings or ball bearings according to model. Bearings/bushings are a press fit in wheel hubs. Renew bearings/bushings and/or repair wheel hubs as necessary to correct excessive looseness.

STEERING GEAR

Models B-165 – LT-1637 – LT-1642 – 216-5

R&R AND OVERHAUL. Remove steering wheel retaining nut and use a suitable suitable puller to remove steering wheel. Drive out roll pin (13 – Fig. WH1). Loosen setscrew (14) and move

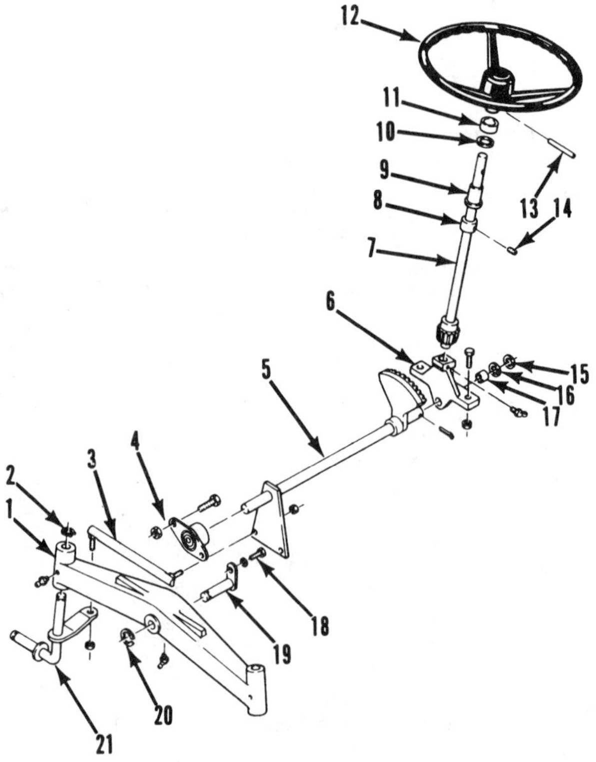

Fig. WH3 – Exploded view of front axle and steering assembly used on 417, 417A, C-161, C-175, C-195, GT-1642 and GT-1848 models.

1. Axle main member
2. Snap ring
3. Tie rod
4. Flange bearing
5. Lower steering shaft
6. Steering support
7. Upper steering shaft
8. Collar
9. Bushing
10. Washer
11. Collar
12. Steering wheel
13. Roll pin
14. Setscrew
15. Shim washer
16. Shim washer
17. Bearing
18. Retaining bolt
19. Pivot pin
20. Snap ring
21. Spindle

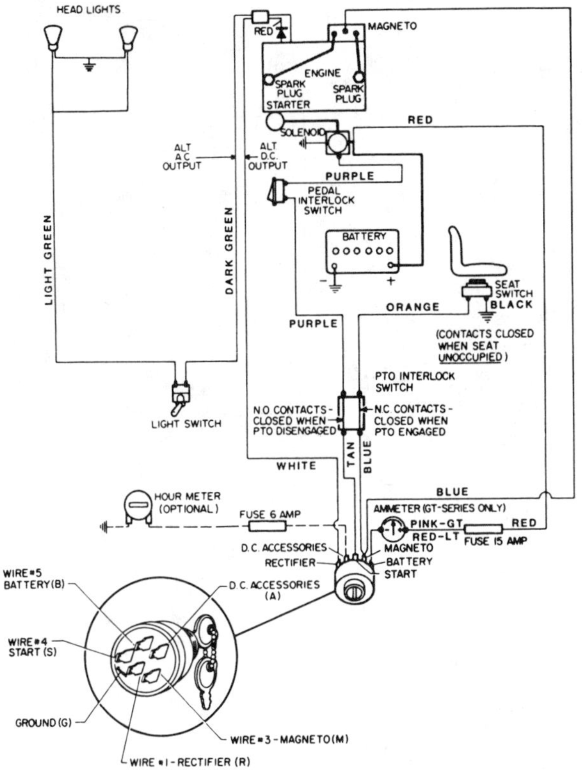

Fig. WH4—View showing wiring diagram for B series tractors.

upper steering shaft (7) up to clear support (6). Disconnect inner ends of tie rods from arm of lower steering shaft (5). Remove cotter pin and shims (15 and 16) at rear of lower shaft, then unbolt and remove support (6). Upper shaft (7) and lower shaft (5) can now be pulled out from underneath tractor.

Inspect parts for wear, scoring or other damage and renew as necesary.

Reinstall by reversing removal procedure. Before tightening setscrew (14), apply downward pressure on steering wheel, slide collar (8) up against bushing (9) and tighten setscrew (14). Adjust gear mesh by adding or removing shims (15 and 16) until 0 to 0.015 inch (0 to 0.4 mm) end play is present in lower shaft (5).

Models D-160 – D-180 – D-200 – 18 HP Automatic

R&R AND OVERHAUL. Remove steering wheel and foam dust cover. Disconnect drag link from steering lever plate (10–Fig. WH2). Remove through-bolts which secure gear housing to bracket and remove assembly from the bottom of tractor.

To disassemble steering gear, remove jam nut (20–Fig. WH2), adjustment nut (19) and washer (18). Remove steering lever plate (10). Remove cotter pin (1), adjustment plug (2) and slide steering shaft assembly (6) out of housing. Remove Belleville washer (3), bearing cups (4 and 8) and bearings (5 and 7).

Inspect parts for wear, scoring or other damage. Renew as necessary.

To reassemble, apply heavy grease to bearings (5 and 7) and bearing cups (4 and 8) and install on steering shaft (6). Apply heavy grease to steering gear and install assembly into housing (21). Install Belleville washer (3) and plug (2). Tighten plug to 4-6 ft.-lbs. (5-8 N·m) and install cotter pin (1). Shaft must turn freely in housing. Pack housing with multi-purpose, lithium base grease and install seal (12) and retainer (11) on housing. Remove steering pin (9) from steering lever plate (10) and install steering lever plate on housing. Place a 3/32-inch (2.381 mm) shim between steering lever plate and seal retainer, install washer (18) and adjustment nut (19). Tighten adjustment nut (19) until snug. Install jam nut (20) and tighten against adjustment nut to 22-25 ft.-lbs. (30-34 N·m) while holding adjustment nut in position on stud. Remove shim stock. Install steering pin in steering lever plate until it just does engage steering gear. Center steering gear by rotating shaft to position half-way between full left and full right turn. Adjust steering pin (9) inward until all backlash is removed but steering gear turns through full left/right cycle with no binding. Hold steering pin in this position and tighten jam nut to 35-45 ft.-lbs. (48-61 N·m).

Reinstall by reversing removal procedure.

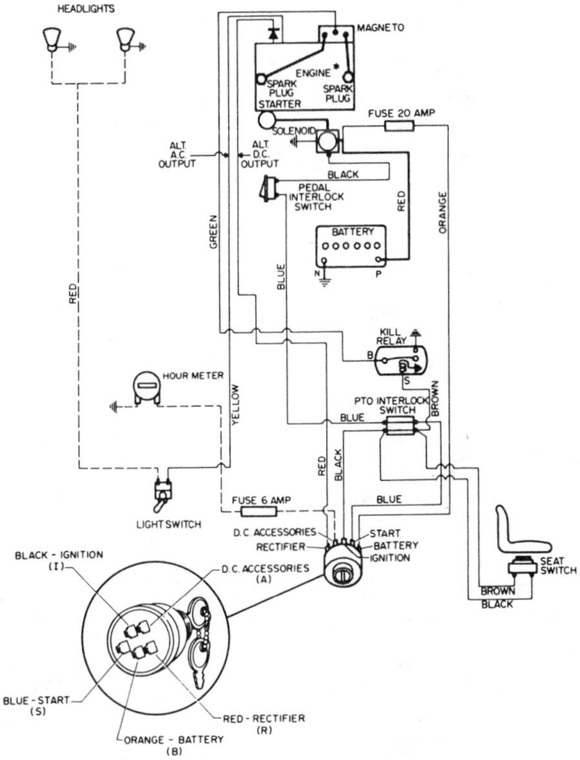

Fig. WH5—View showing wiring diagram for LT and GT series tractors.

Models 417—417A—C-161— C-175—C-195—GT-1642— GT-1848

R&R AND OVERHAUL. Remove steering wheel retaining nut and use a suitable puller to remove steering wheel. Drive out roll pin (13–Fig. WH3).

Loosen setscrew (14) and move upper steering shaft (7) up to clear support (6). Disconnect inner ends of tie rods from arm of lower steering shaft (5). Remove cotter pin and shims (15 and 16) at rear of lower shaft, then unbolt and remove support (6). Upper shaft (7) and lower shaft (5) can now be pulled out from underneath tractor. Unbolt and remove flange bearing (4).

Inspect parts for wear, scoring or other damage and renew as necessary.

Reinstall by reversing removal procedure. Before tightening setscrew (14), apply downward pressure on steering wheel, slide collar (8) up against bushing (9) and tighten setscrew (14). Adjust gear mesh by adding or removing shims (15 and 16) until 0 to 0.015 inch (0 to 0.4 mm) end play is present in lower steering shaft (5).

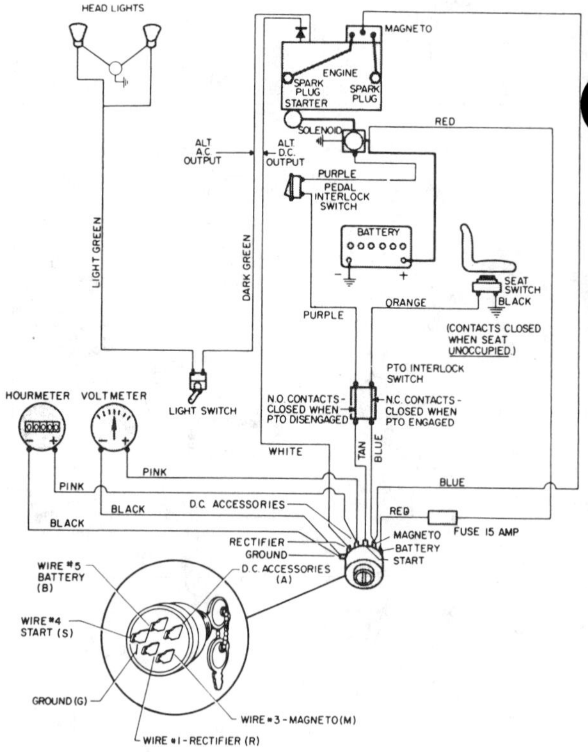

Fig. WH6—View showing wiring diagram for 216-5 model.

ENGINE

MAINTENANCE

All Models

Regular engine maintenance is required to maintain peak performance and long engine life.

Check oil level and clean air intake screen at 5 hour intervals under normal operating conditions.

Clean engine air filter and cooling fins at 25 hour intervals and check for any loose nuts, bolts or linkage. Repair as necessary.

Change engine oil and filter, perform tune-up, valve adjustment and clean carbon from cylinder heads as recommended in appropriate engine section of this manual.

REMOVE AND REINSTALL

Models B-165—LT-1637—LT-1642—216-5

Remove negative battery cable. Disconnect choke cable, throttle cable, fuel line and all electrical connections. Disconnect pto linkage at pto clutch turnbuckle. Remove pto brake adjustment screw and bracket. Remove pto clutch cone assembly. Remove crankshaft bolt and bearing race. Depress clutch pedal to relieve belt tension and remove remaining pto clutch housing from crankshaft. Remove engine mounting bolts and remove engine.

Reinstall by reversing removal procedure.

Models D-160 – D-180 – D-200 – 18 HP Automatic

Remove negative battery cable. Disconnect choke and throttle controls and disconnect fuel line at fuel pump. Disconnect all electrical connections. Remove mufflers and complete hood and grille shroud assembly. Disconnect pto rod at clutch bar. Remove engine retaining bolts, slide engine forward until flex coupling is free of pump drive shaft and remove engine.

Reinstall by reversing removal procedure.

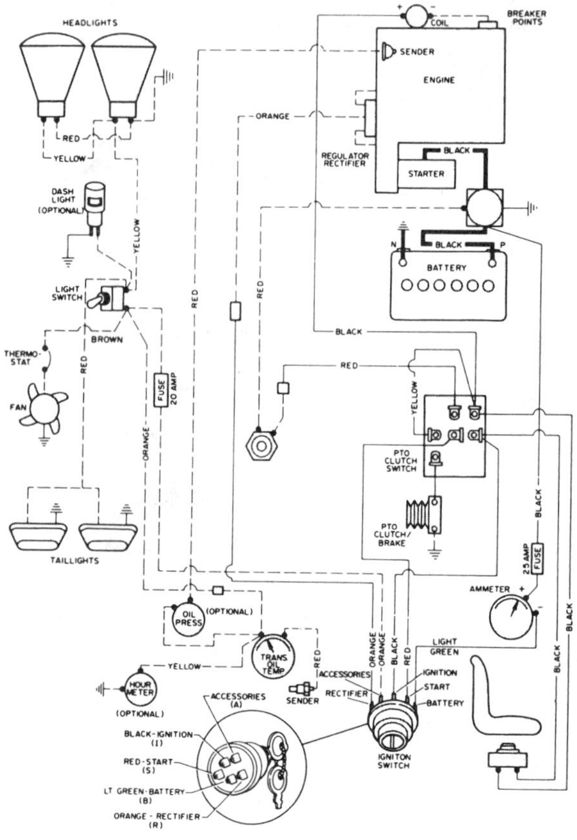

Fig. WH7—View showing wiring diagram for D-160 model.

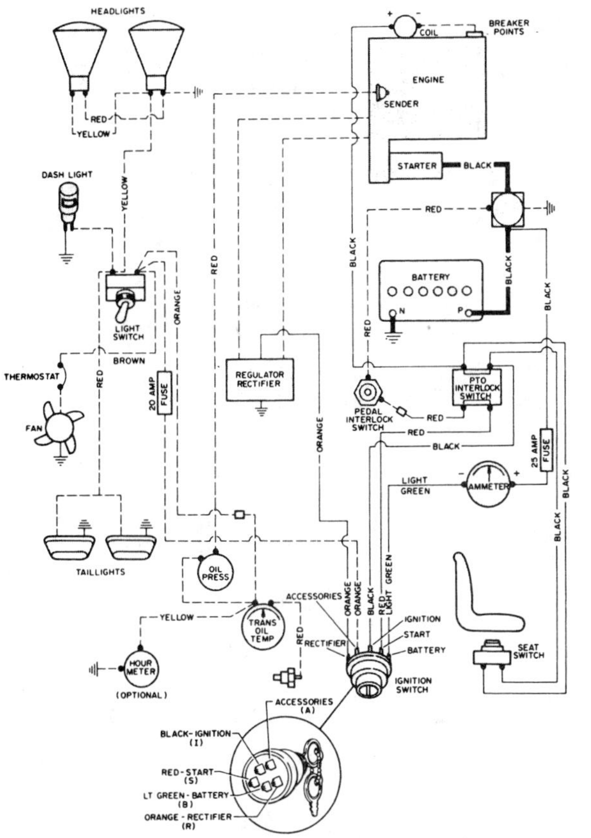

Fig. WH8—View showing wiring diagram for D-180 and D-200 models.

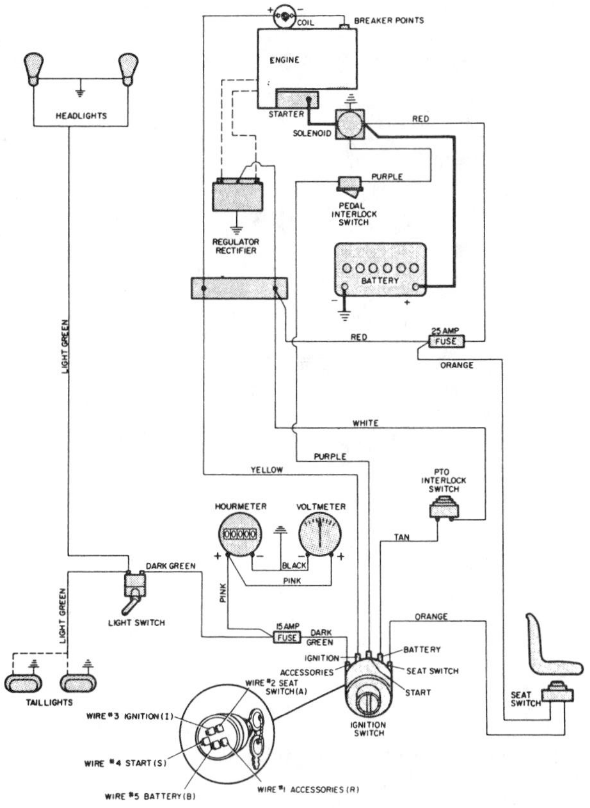

Fig. WH9—View showing wiring diagram for 417 and 417A models.

Models 417 – 417A – C-161 – C-175 – C-195 – GT-1642 – GT-1848

Remove negative battery cable. Disconnect choke and throttle cables, fuel lines and all electrical connections. Remove belt guard. Disconnect pto clutch rod and yoke as necessary. Disengage transmission drive clutch on hydrostatic drive models or depress clutch and remove drive belt from engine pulley on standard transmission models. On all models, remove engine retaining bolts and remove engine.

Reinstall by reversing removal procedure.

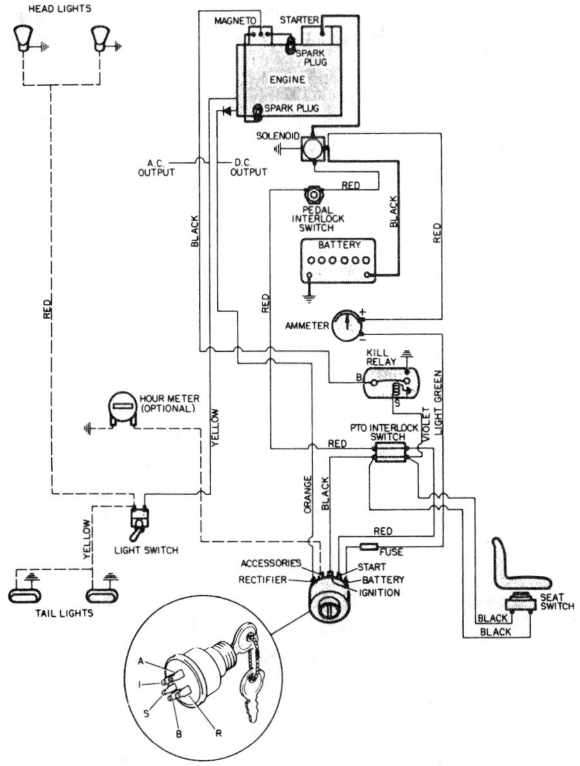

Fig. WH10—View showing wiring diagram for C-161 model.

OVERHAUL

All Models

Engine make and model are listed at the beginning of this section. To overhaul engine or accessories, refer to appropriate engine section in this manual.

ELECTRICAL SYSTEM

MAINTENANCE AND SERVICE

All Models

Battery electrolyte level should be checked at 50 hour intervals of normal operation. If necessary, add distilled water until level is just below be of vent well. **DO NOT** overfill. Keep battery posts clean and cable ends tight.

For alternator or starter service, refer to appropriate engine section in this manual.

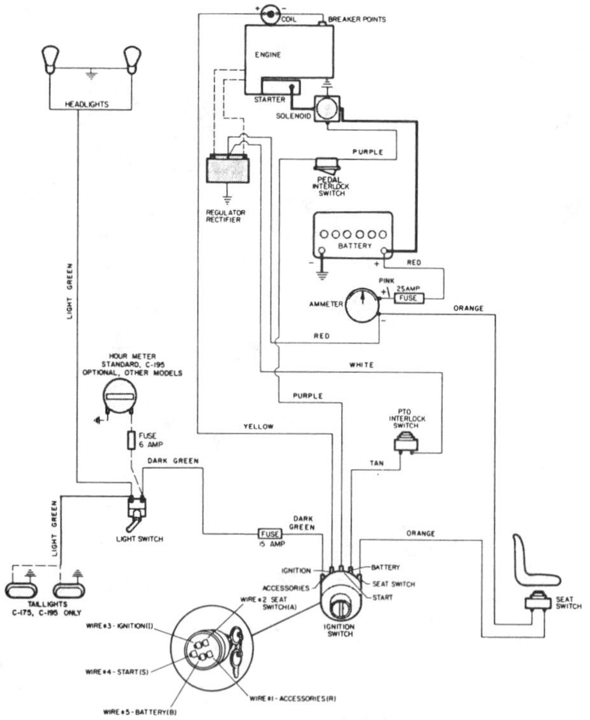

Fig. WH11—View showing wiring diagram for C-175 model.

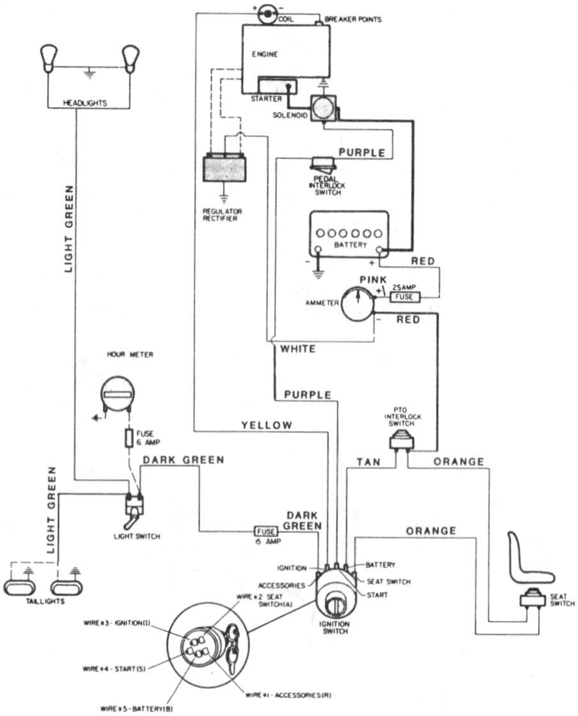

Fig. WH12—View showing wiring diagram for C-195 model.

Wiring diagrams are shown in Fig. WH4 through WH12. Refer to diagram for appropriate model.

BRAKE

ADJUSTMENT

Models B-165 – LT-1642 – 216-5

FIVE SPEED TRANSMISSION. Make certain transmission brake lever contacts back stop plate (or rod) when brake pedal is released (Fig. WH13). Loosen locknut and tighten adjustment nut until a light drag is felt as tractor is pushed forward by hand. Loosen adjustment nut slightly and tighten jam nut. After adjusting brake, turn adjustment nut on brake rod to obtain 2-11/16 to 2-3/4 inches (68 to 70 mm) distance between inside of nut and washer as shown in Fig. WH13.

HYDROSTATIC DRIVE TRANSMISSION. Loosen jam nut (5–Fig. WH14) and tighten adjustment nut (4) until a light drag is felt as tractor is pushed forward by hand. Loosen nut slightly and tighten jam nut (5). Depress brake pedal and note brake lever (3) should travel 3/8 to 3/4-inch (9.5 to 19 mm) before overtravel spring (1) begins to compress. Overtravel spring retaining nut (6) should be adjusted so spring side of nut (6) is 7/8-inch (22 mm) from end of brake rod.

Model LT-1637

Disc brake assembly is located at left side of three speed transaxle (Fig. WH15). Loosen locknut and tighten adjustment nut until a light drag is felt as tractor is pushed forward by hand. Tighten locknut.

Models D-160 – D-180 – D-200 – 18 HP Automatic

With foot and park brake released, tighten band adjustment nut (Fig. WH16) until brake band can no longer be moved side to side on the brake drum. Loosen adjustment nut until side to side movement is just restored. Adjust jam nuts (Fig. WH17) on brake rod until

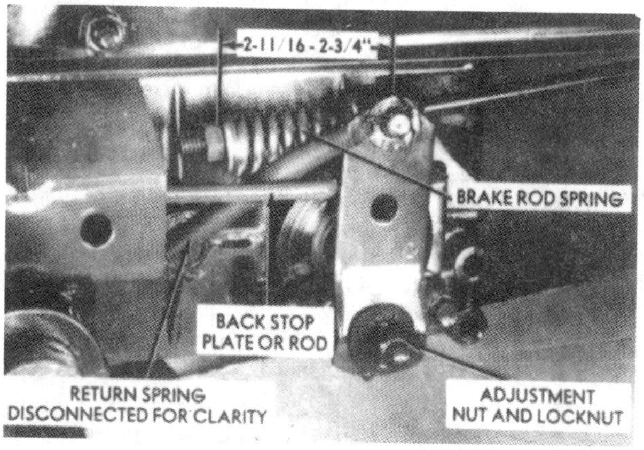

Fig. WH13—View showing brake adjustment locations on B-165 and 216-5 models with five speed transmission. Refer to text.

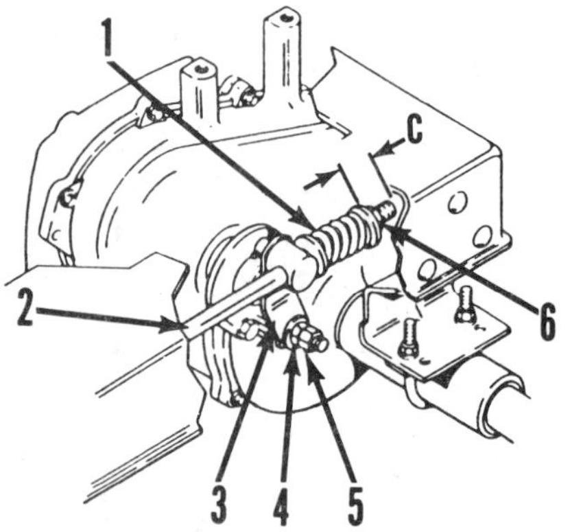

Fig. WH14—View showing brake adjustment locations on B-165 model with hydrostatic drive transmission. Refer to text.

1. Overtravel spring
2. Brake rod
3. Brake lever
4. Adjustment nut
5. Jam nut
6. Spring retaining nut

C. 7/8-inch (22 mm) distance

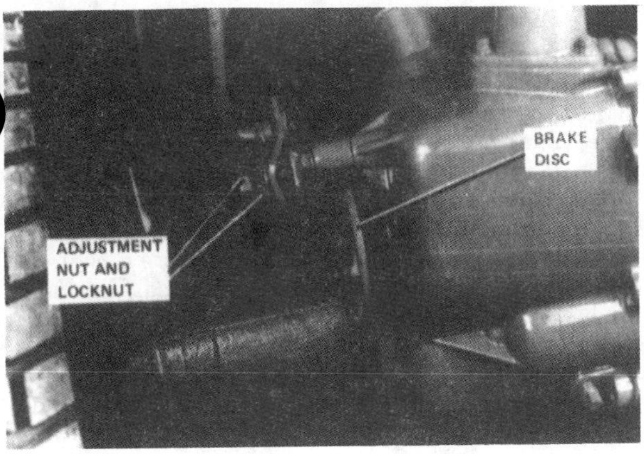

Fig. WH15—View showing brake adjustment locations on LT-1637 and LT-1642 models. Refer to text.

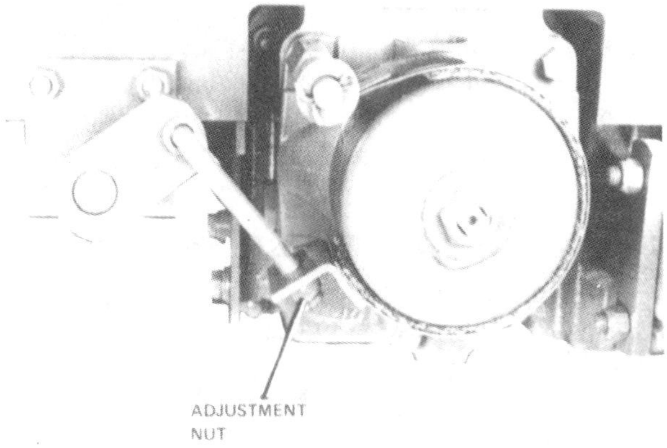

Fig. WH16—View showing brake band, drum and adjustment nut on D series tractors. Refer to text.

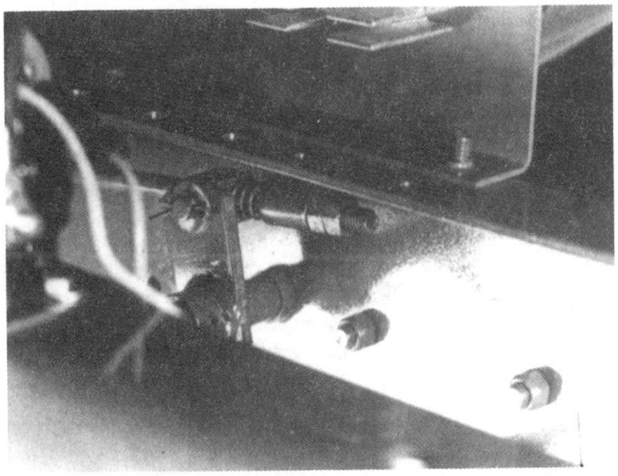

Fig. WH17—View showing location of front brake rod spring, spacer and jam nuts. Refer to text.

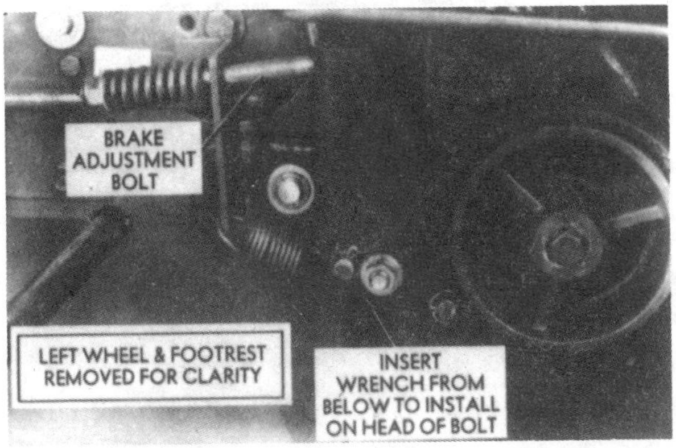

Fig. WH18—View showing brake mechanism and adjustment nut on 417, GT-1642, GT-1848, C-161, C-175 and C-195 models with eight speed transmission. Refer to text.

spacer contacts spring, but spring remains free to turn.

Models 417 – 417A – GT-1642 – GT-1848 – C-161 – C-175 – C-195

EIGHT SPEED TRANSMISSION. Push brake pedal down and engage parking brake. Tighten adjustment nut (Fig. WH18) until both rear wheels skid when tractor is pushed by hand. Release brake and tighten adjustment nut another ½-turn. After adjustment, parking brake lever should not travel to the rear end of lever slot when engaged and brake band should not "drag" on brake drum when brake is fully released.

HYDROSTATIC DRIVE TRANSMISSION. Set parking brake lever in the first engagement notch. Tighten the nut on brake linkage bolt until coils of the heavy spring are fully compressed, then back nut off ½-turn (Fig. WH19) Release parking brake and make certain brake band does not "drag" on brake drum when brake is released.

R&R BRAKE PADS OR BAND

All Models With Disc Brake

Disconnect brake rod from brake lever. Remove locknut, adjustment nut and brake lever. Remove caliper retaining bolts, note location of spacers, if used and remove caliper assembly. Remove brake disc and inner brake pad.

Reverse removal procedure for reassembly. Make certain brake actuating pins move freely in caliper. Adjust brakes as outlined for appropriate model.

All Models With Band Brake

Remove brake band actuating rod. Remove cotter key and washer from brake band stationary pin and remove band.

Reinstall by reversing removal procedure. Adjust brake as outlined for appropriate model.

CLUTCH

MAINTENANCE

All Models With Gear Type Transmission

Clutch pedal is connected to a spring loaded, pivoting idler pulley. Belt tension is maintained by spring pressure. Periodically check condition of pulleys and location of belt guides and fingers. Refer to R&R DRIVE BELT section.

All Hydrostatic Drive Models

Clutch pedal returns hydrostatic drive unit to neutral position when depressed. Refer to appropriate ADJUSTMENT paragraph located in TRANSMISSION section for neutral position adjustment.

R&R DRIVE BELT

Models B-165 – LT-1642 – 216-5

FIVE SPEED TRANSMISSION. Remove shift knob and shift cover plate from front fender. Loosen the two rear belt guides, depress clutch and remove belt from pto pulley. Loosen belt guides and remove belt from idler and engine pulleys. Remove belt through shift lever opening.

Reverse removal procedure for reinstallation. Adjust all belt guides and fingers to obtain 1/16 to 1/8-inch (1.588 to 3.175 mm) clearance between belt and guide or finger.

HYDROSTATIC DRIVE TRANSMISSION. Disengage pto and transmission clutch. Move speed control lever fully forward. Loosen pto adjustment bolt locknut and back bolt out of pto bracket. Remove pto pulley from engine and loosen nut retaining flat idler pulley. Loosen the two nuts on right hand clamp holding rear footrest cross rod. Work belt off transmission input pulley, clutch idler pulley and V-idler pulley. Loosen belt guide and remove belt from engine pulley.

Reverse removal procedure and position belt guides and fingers 1/16 to 1/8-inch (1.588 to 3.175 mm) away from belt with clutch engaged. Operate the tractor for 15 minutes and readjust flat idler pulley to prevent belt slippage.

Model LT-1637

Loosen all belt guides and fingers. Depress clutch pedal and work belt off idler pulley and transmission pulley. Work belt off engine pulley and remove belt.

Reinstall by reversing removal procedure. Position belt guides and fingers 1/16 to 1/8-inch (1.588 to 3.175 mm) away from belt with clutch engaged.

Models 417 – C-161 – C175 – C-195 – GT-1642 – GT-1848

Remove belt guard and right side footrest. On eight speed transmission models, loosen clutch idler pulley. On all models, depress clutch and work belt off rear drive pulley. Disconnect pto rod, remove yoke clevis and pull yoke aside. Remove pto brake. Remove belt.

Reinstall by reversing removal procedure. Position belt guides or fingers 5/32-inch (3.969 mm) away from belt with clutch engaged. Adjust pto brake.

TRANSMISSION

MAINTENANCE

Three Speed Transmission

Three speed transmission fluid level should be checked at 25 hour intervals of normal operation. Maintain fluid level at lower edge of filler plug located just below drawbar. Change fluid at 200 hour intervals. Recommended fluid is SAE 90 EP gear lubricant and transmission capacity is ¾-quart (0.7 L).

Five Speed Transmission

Five speed transmission is packed with 36 ounces (1065 mL) of EP lithium base grease during manufacture and should provide lifetime lubrication.

Eight Speed Transmission

Eight speed transmission fluid level should be checked at 25 hour intervals of normal operation. Maintain fluid level at "FULL" mark on dipstick which is located under operators seat. Change fluid at 200 hour intervals. Recommended fluid is SAE 140 EP gear lube and fluid capacity is approximately 2 quarts (1.9 L). **DO NOT** overfill.

Hydrostatic Drive Transmission

Hydrostatic drive transmission fluid level should be checked before each use. Check before engine is started. Maintain fluid level at "COLD FULL" line on filler tube of B-165 models or at "FULL" mark on dipstick of all other models. Fluid and filter (as equipped) should be changed at 100 hour intervals. Approximate fluid capacity is ¾-quart (0.7 L) for B-165 model, 5 quarts (4.7 L) for C-161, C-175, C-195, GT-1642 and GT-1848 models and 6 quarts (5.7 L) for 18 HP Automatic, D-160, D-180 and D-200 models. **DO NOT** overfill. Recommended fluid for B-165 model is SAE 20 motor oil and recommended fluid for all other models is SAE 10-30 motor oil. Refer to GEAR REDUCTION section.

ADJUSTMENT

Models D-160 – D-180 – D-200 – 18 HP Automatic

NEUTRAL ADJUSTMENT. Raise and support rear of tractor. Start and run engine at half throttle. Move motion control lever to full speed forward position.

CAUTION: Use care when working around rotating rear tires.

Depress clutch/brake pedal fully and release. Wheels and tires should stop when pedal is depressed and remain stopped even after pedal is released. If wheel rotation continues after pedal is released, turn eccentric screw (Fig. WH20) until wheel rotation stops. Turn eccentric screw until wheels just begin to rotate in forward direction, then turn screw back until wheels just begin to rotate in reverse direction. Center eccentric midway between these two points.

BRAKE
ADJUSTMENT

Fig. WH19—View showing brake mechanism and adjustment nut on C-161, C-175 and C-195 models with hydrostatic drive transmission. Models 417A, GT-1642 and GT-1848 models with hydrostatic drive transmission are similar. Refer to text.

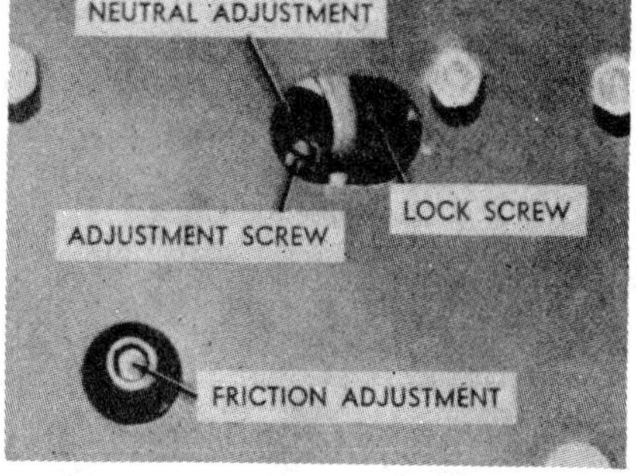
NEUTRAL ADJUSTMENT
LOCK SCREW
ADJUSTMENT SCREW
FRICTION ADJUSTMENT

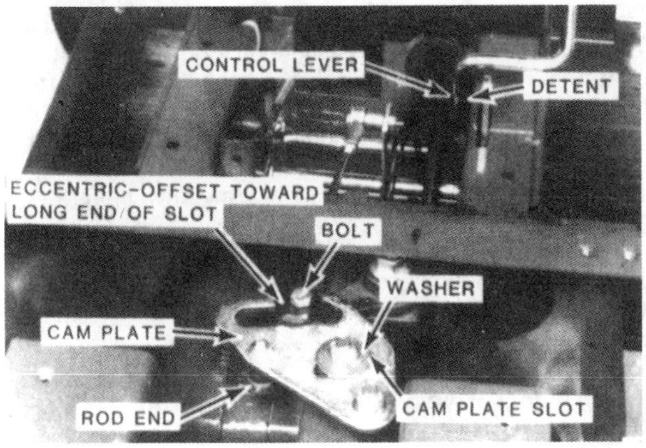

Fig. WH21—View of neutral position adjustment points on some models with hydrostatic drive transmission. Refer to text.

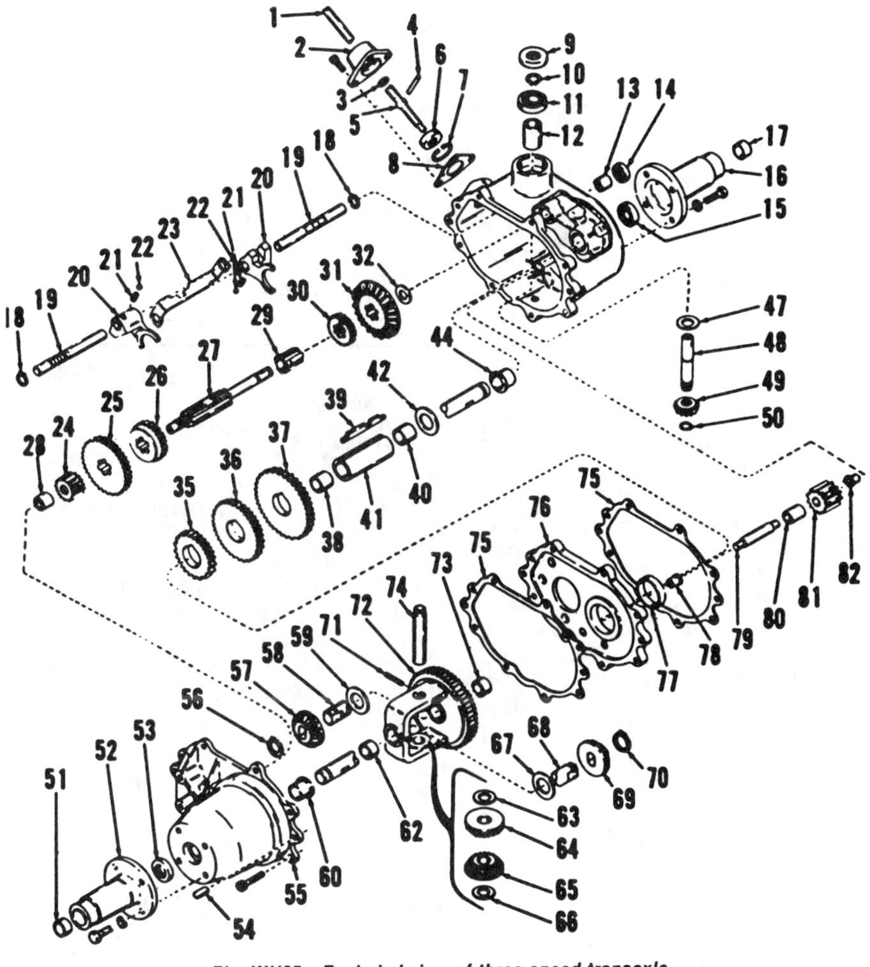

Fig. WH25—Exploded view of three speed transaxle.

1. Shift lever
2. Lever housing
3. Quad ring
4. Roll pin
5. Shift lever
6. Retainer
7. Snap ring
8. Gasket
9. Oil seal
10. Snap ring
11. Ball bearing
12. Bushing
13. Roller bearing
14. Oil seal
15. Oil seal
16. Axle housing
17. Bushing
18. Snap ring
19. Shift rod
20. Shift fork
21. Spring
22. Steel ball
23. Shifter stop
24. Spur gear
25. Sliding gear (1st & Rev.)
26. Sliding gear (2nd & 3rd)
27. Shift & brake shaft
28. Needle bearing
29. Idler gear
30. Gear
31. Bevel gear
32. Washer
35. Gear (25T)
36. Gear (34T)
37. Gear (39T)
38. Bushing
39. Key
40. Bushing
41. Countershaft sleeve
42. Thrust washer
44. Bushing
47. Washer
48. Shaft
49. Pinion gear
50. Snap ring
51. Bushing
52. Axle housing
53. Oil seal
54. Dowel pin
55. Cover
56. Snap ring
57. Slide gear
58. Axle shaft
59. Thrust washer
60. Bushing
62. Bushing
63. Thrust washer
64. Pinion gears
65. Pinion gears
66. Thrust washer
67. Thrust washer
68. Axle shaft
69. Side gear
70. Snap ring
71. Roll pin
72. Differential carrier & gear.
73. Bushing
74. Drive pin
75. Gasket
76. Center plate
77. Bushing
78. Bushing
79. Reverse idler gear
80. Spacer
81. Reverse idler
82. Bushing

CONTROL LEVER FRICTION ADJUSTMENT. If transmission speed control lever does not hold firm at selected speed position, increase tension on lever by tightening friction adjusting nut (Fig. WH20) a little at a time until lever will retain its set position.

All Other Models

NEUTRAL ADJUSTMENT. Raise and support rear of tractor so wheels are free to turn. Start engine and run at full throttle. Move motion control lever to forward position.

CAUTION: Use care when working near rotating tires, driveshaft or cooling fan.

After oil reaches operating temperature, fully depress clutch/brake pedal and release. Loosen bolt (Fig. WH21) and turn eccentric until tires stop rotating. Continue turning eccentric until wheels just begin to rotate in the opposite direction. Set eccentric midway between these two points. Tighten bolt and recheck neutral position in each direction.

R&R TRANSMISSION

All Three, Five And Eight Speed Transmissions

Unbolt and remove belt guard. Depress clutch pedal and remove drive belt at transmission pulley. Disconnect brake rod from clutch linkage on models which have a common brake/clutch pedal and on models with separate brake pedal, disconnect and remove brake linkage rod at transaxle. Do not overlook retracting springs used on some tractors. Remove lever knobs from gear shift and range selector (if so equipped) so cover plates can be removed or transaxle may be lowered without interference from console deck. Raise and support frame securely at side points just forward of transaxle, then remove transmission mounting cap screws and lower and remove entire unit from tractor.

To reinstall, reverse removal procedure. Refer to appropriate paragraph for brake adjustment procedure for model being serviced after transaxle is reinstalled.

Model B-165 With Hydrostatic Drive

Remove hydrostatic unit drive belt and disconnect transmission control rod. Unbolt and remove actuating pins from disc brake. Raise and support rear of tractor. Support transmission and remove transaxle mounting "U" bolts. Raise rear of frame and roll tractor away from transaxle. Separate hydrostatic drive unit from transaxle assembly.

Reinstall by reversing removal procedure. Refer to appropriate paragraph for brake adjustment and neutral position adjustment for model being serviced.

All Other Models With Hydrostatic Drive

Remove seat, rear fender and control cover plate. Support rear of tractor with stands under footrest rear cross rods. Drain oil from transaxle. Remove rear wheels. Shut off fuel and disconnect fuel line. Unbolt oil filter base from seat support. Remove two bolts, bracket and lift tube from top of transaxle. Remove two bolts securing seat support bracket and withdraw fuel tank and seat support as an assembly. Remove cooling fan and slip drive belt off input pulley. Disconnect speed control rod from cam plate. Disconnect all hydraulic hoses. Unbolt and remove transmission from transaxle housing.

Reinstall by reversing removal procedure. Tighten transmission mounting bolts to 30-35 ft.-lbs. (41-47 N·m). Reinstall cooling fan making sure side with manufacturer's name faces out. Refer to appropriate paragraph for brake adjustment and neutral position adjustment for model being serviced.

OVERHAUL

Three Speed Transmission

Remove transmission as previously outlined. Drain fluid and clean exterior of transmission thoroughly. Remove input pulley and brake disc assembly. Remove wheel and hub assemblies. Place shift lever in neutral position, unscrew retaining cap screws and remove shift lever assembly. Remove axle housings (16 and 52–Fig. WH25). Place unit in a vise so socket head cap screws are pointing up. Remove all rust, paint and burrs from axle shafts. Drive dowel pins out of case and cover. Unscrew socket head cap screws and lift cover from case. Screw two socket head screws into case to hold center plate (76) down while removing differential assembly. Pull differential assembly straight up out of case. It may be necessary to gently bump lower axle shaft to loosen differential assembly. Remove center plate. Hold shifter rods together and lift out shifter rods, forks, shifter stop, shaft (27), sliding gears (25 and 26) and spur gear (24) as an assembly. Be careful when removing shaft (27) as rollers in bearing (13) may be loose and fall out. Remove idler shaft (29) and gear (30). Remove reverse idler shaft (79), spacer (80), gear (81), cluster gears (35, 36 and 37) on countershaft sleeve (41) and thrust washer (42). Remove input shaft oil seal (9), snap ring (10), input shaft (48) and bevel gear (31), washer (32) and gear (49). Washer (32) is removed with input shaft and gear. Remove bearing (11) and bushing (12).

To disassemble differential assembly, drive roll pin (71) out of drive pin (74). Remove drive pin, thrust washers (63 and 66) and pinion gears (64 and 65). Remove snap rings (56 and 70) and withdraw axle shafts from side gears (57 and 69). Remove side gears.

To disassemble cluster gear assembly, press gears from shaft. Note beveled edge of gears (35 and 36) is on side closest to large gear (37). Remove gears from key (39). Note short raised portion of key (39) is between middle gear (36) and large gear (37). Long raised section of key (39) is between small gear (35) and middle gear (36).

Clean and inspect components for excessive wear or damage. Renew all seals and gaskets. Check for binding of shift forks on shift rods. Position shift forks in neutral position by aligning notches on shift forks with notch in shifter stop (23–Fig. WH25) as shown in Fig. WH26.

To assemble transaxle, install input shaft assembly by reversing disassembly procedure. Position case so open side is up. Install bearing (13–Fig. WH25) and seal (14) if removed during disassembly. Install idler shaft (29), gear (30), bevel gear (31) and washer (32). Reverse idler shaft (79) may be used to temporarily hold idler gear assembly in position. Position cluster gears (35, 36 and 37) on key (39) so bevel on gears (35 and 36) is toward larger gear (37) as shown in Fig. WH27. Press gears and key on countershaft sleeve (41–Fig. WH25). Install shifter assembly components (18 through 27) in case being sure shifter rods are properly seated. Install reverse idler shaft (79), gear (81) and spacer (80). Beveled edge of gear should be up. Install gasket, center plate (76) and gasket on case. Assemble differential assembly by reversing disassembly procedure. Install differential assembly in case with longer axle pointing down. Be sure shift shaft gear (24) mates correctly with ring gear (72). Install locating dowel pins and secure cover (55) to case. Install seals (15 and 53), axle housings (52 and 16) and shifter assembly (1 through 8). Fill to correct operating level with specified fluid.

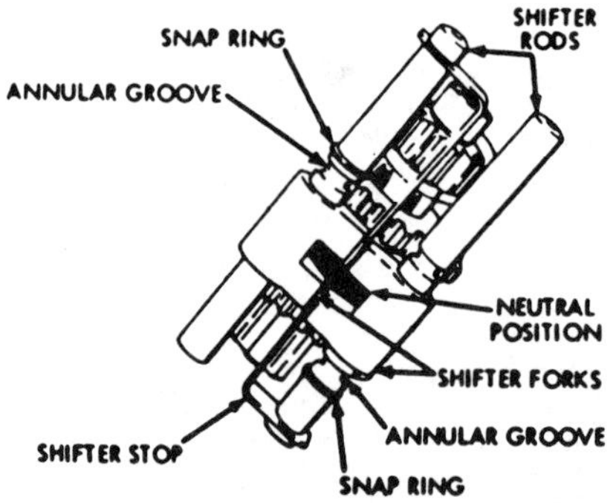

Fig. WH26–To position shifter assembly in neutral for reassembly, align notches in shifter forks with notch in shifter stop.

Five Speed Transmission

A variety of Peerless 800 series transaxles have been used and position of input shaft may vary according to model or date of manufacture. Input shaft described in this section is located to the left of bevel gear, however, procedure is similar for all Peerless 800 series transaxles.

Remove transaxle assembly as outlined. Place shift lever in neutral and unbolt and remove shift lever. Remove setscrew (2–Fig. WH28), spring (3) and index ball (4). Unbolt cover (5) and push shift fork assembly (12) in while removing cover. Before removing gear and shaft assemblies, shifter fork (12) should be removed. Note position of parts before removal. Remove gear and shaft assemblies (Fig. WH29) from case taking care not to disturb drive chain (34–Fig. WH28). Remove needle bearings (38 and 43), flat washer (41), square cut seals (37 and 42), output gear (40) and output pinion (39) from countershaft. Angle the two shafts together, then mark position of chain on sprocket collars and remove chain. Remove sprocket (35), bevel gear (32), spur gears (27, 28, 29, 30 and 31), thrust washer (9) and flange bushing (14). All gears are splined to the countershaft. Disassembly

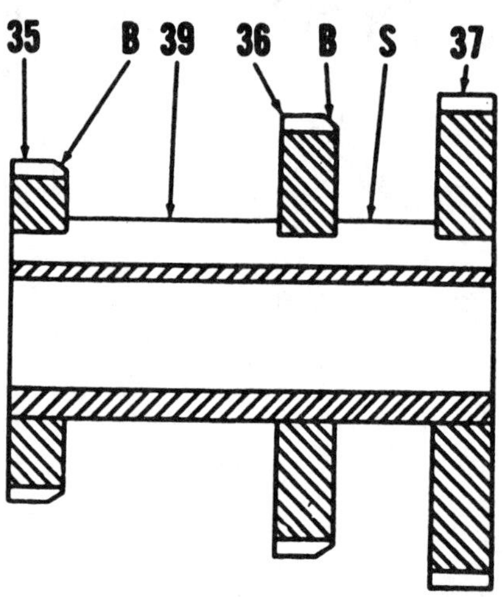

Fig. WH27–Note position of bevels (B) on cluster gears and short section (S) of key (39) between gears (36 and 37) on three speed transaxle. Refer to text and Fig. WH25.

of shifter-brake shaft is apparent after inspection. Remove snap ring (11), input bevel gear and pull input shaft (7) through cover.

Disassemble differential by driving roll pin securing drive pin (58) out. Remove pinion gears (70) by rotating gears in opposite directions. Remove snap ring (54), side gears (55) and thrust washers (53), then slide axles out. Note axle shafts (48 and 56) are different lengths.

Clean and inspect all parts and renew any showing excessive wear or other damage. When installing new inner input shaft needle bearings, press bearing in to a depth of 0.135-0.150 inch (3.429-3.81 mm) below flush. When installing thrust washers and shifting gears on brake shaft, the 45° chamfer on inside diameter of thrust washers must face shoulder on brake shaft. See Fig. WH30. Flat side of gears must face shoulder on shaft. Complete reassembly and pack case with approximately 36 ounces (1065 mL) EP lithium base grease. Tighten case to cover cap screws to 90-100 in.-lbs. (10-11 N•m).

Eight Speed Transmission

Remove transmission as previously outlined. Drain fluid and clean exterior of transmission. Remove wheel and hub assemblies. Clamp transaxle in holding jig or large shop vise so right axle is pointed down. Remove brake band assembly and brake drum. Remove input pulley and back out setscrew (41–Fig. WH31) so gear shift lever (43) can be removed. Clean exposed ends of axles and remove any burrs near keyway edges. Unbolt left transmission case and lift off over axle. Differential assembly and both axles can now be lifted out and set aside for later disassembly. Pull out spline shaft (32), low-reverse gear (31) and second-third gear (27). Lift out cluster/brake shaft (36) with cluster gear (38) and reduction gear (39). Remove reverse idler gear (34) and idler shaft (33). First and reverse gear shift rail (26) and its shift fork should be removed using caution so stop balls (4) are not dropped and lost when out of engagement with shift rail detents. Balls, as well as spring (1) and guide pin (2) may be caught by holding free hand over case

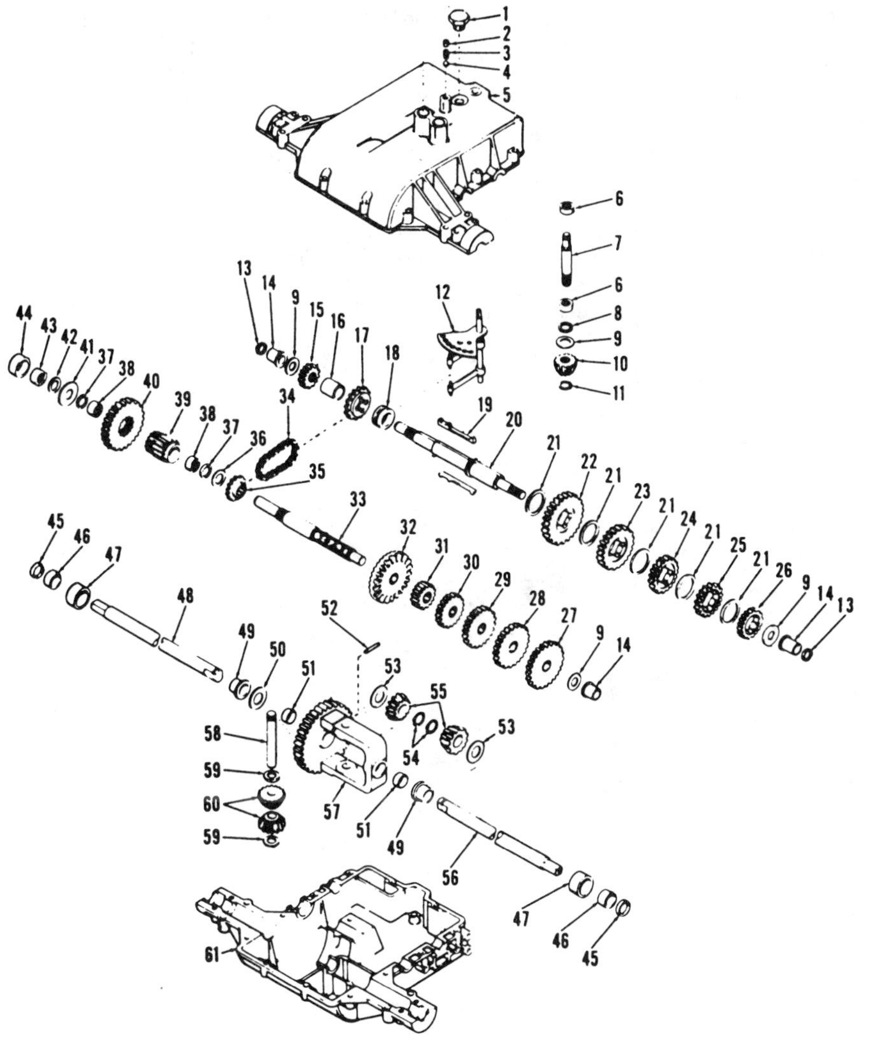

Fig. WH28—Exploded view of five-speed transaxle.

1. Plug
2. Setscrew
3. Spring
4. Ball
5. Cover
6. Needle bearing
7. Input shaft
8. Square cut ring
9. Thrust washer
10. Input pinion
11. Snap ring
12. Shift fork assy.
13. Square cut ring
14. Bushing
15. Spur gear
16. Spacer
17. Sprocket (18T)
18. Shift collar
19. Key
20. Brake shaft
21. Thrust washer
22. Spur gear (35T)
23. Spur gear (30T)
24. Spur gear (25T)
25. Spur gear (22T)
26. Spur gear (20T)
27. Gear (30T)
28. Gear (28T)
29. Gear (25T)
30. Gear (20T)
31. Spur gear
32. Bevel gear (42T)
33. Countershaft
34. Roller chain
35. Sprocket (9T)
36. Flat washer
37. Square cut ring
38. Needle bearing
39. Output pinion
40. Output gear
41. Flat washer
42. Square cut seal
43. Needle bearing
44. Spacer
45. Oil seal
46. Needle bearing
47. Spacer
48. Axle shaft (short)
49. Bushing
50. Washer
51. Bushing
52. Pin
53. Thrust washer
54. Snap rings
55. Bevel gears
56. Axle shaft (long)
57. Differential gear assy.
58. Drive pin
59. Thrust washer
60. Bevel pinions
61. Case

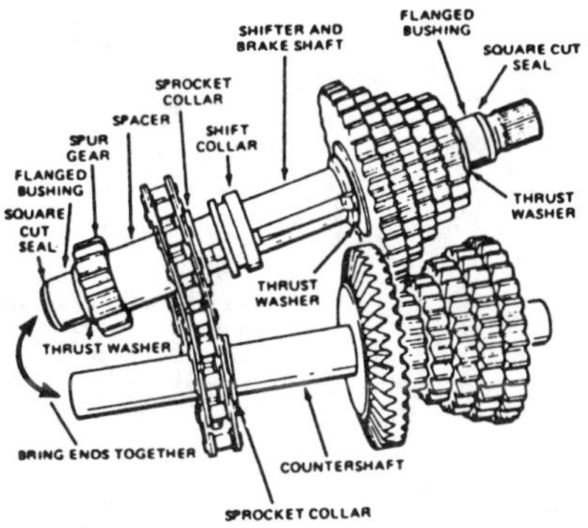

Fig. WH29—Mark position of chain on sprocket collars, angle shafts together and remove chain.

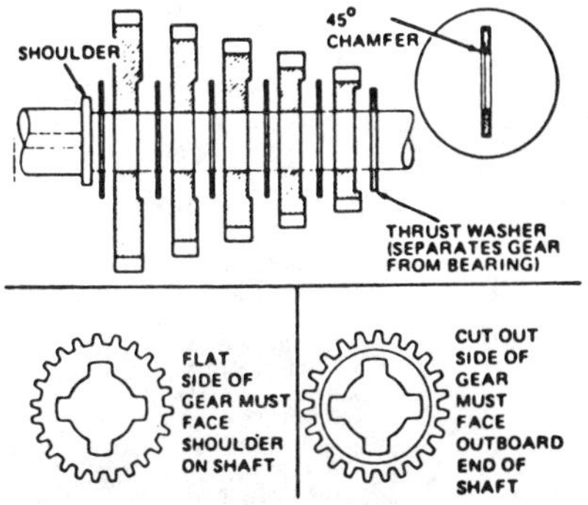

Fig. WH30—When installing thrust washers and gears on brake shaft, 45° chamfer on inside diameter of thrust washers must face shoulder on brake shaft.

opening. Remove second and third gear (rear) shift rail (25) and fork. Remove output gear drive set (15 and 14). Back out detent bolt (60) and withdraw input shaft (23) which carries input gear and spline (22) and range sliding gear (20). When moving these gears, watch for stop spring (59) and detent ball (58) when range shifter fork (61) is dislodged. Remove range cluster gear (16) and its shaft (18).

Disassemble differential assembly by unbolting through-bolts so end caps (51 and 54) can be separated from ring gear (53). Remove differential pinions (54).

NOTE: Models equipped with limited slip differential or spur gear differential will vary in construction. However, after removal of through-bolts, units may be disassembled and reassembly is obvious after initial inspection.

After snap rings (50) have been removed, axle shafts (49) can be pulled from internal splines of axle gears (52) and out of end caps.

Inspect all parts for damage or wear. Renew all gaskets and seals. Reassemble by reversing disassembly procedure. Fill transmission to correct operating level with recommended fluid.

Model B-165 Hydrostatic Drive Unit

Remove hydrostatic drive unit as previously outlined and thoroughly clean exterior of unit.

Remove venting plug (41 – Fig. WH32), invert assembly and drain fluid. Remove cap screws (53) and place unit in a holding fixture so output shaft is pointing downward. Remove by-pass plunger (11). Lift off finned aluminum housing (8) with control and input shaft assemblies, taking care pump and cam ring assemblies (24 through 27) are not removed with housing. If pump and rotor are raised, ball pistons are likely to fall. These ball pistons are a select fit to rotor bores with a clearance range of 0.0002-0.0006 inch (0.0051-0.0152 mm) and are not interchangeable. A wide rubber band may be used to hold ball pistons in position during removal. Remove cam ring (27) and pump race (26), then carefully remove pump assembly. Hold down on motor rotor (31) and remove pintle assembly (30). Remove dampening pistons (28). Remove free wheeling valve bracket (21) and springs (22). Remove plunger guide (19) and "O" ring (20). Place a wide rubber band around motor rotor (31) to retain ball pistons and springs, then remove motor assembly and motor race (40).

Remove snap ring (50), gear (49), spacer (48), retainer (47), snap ring (46) and key (39). Support body (43) and press output shaft (38) out of bearing (45) and oil seal (44). Ball bearing (45) and oil seal (44) can now be removed from body (43).

Remove retainer (3) and withdraw ball bearing (5) and input shaft (6). Remove snap ring (4) and press bearing (5) off of input shaft (6). Remove oil seal (7). Carefully drill an 11/32-inch hole through housing (8) directly in line with center dowel pin (14). Press dowel pin

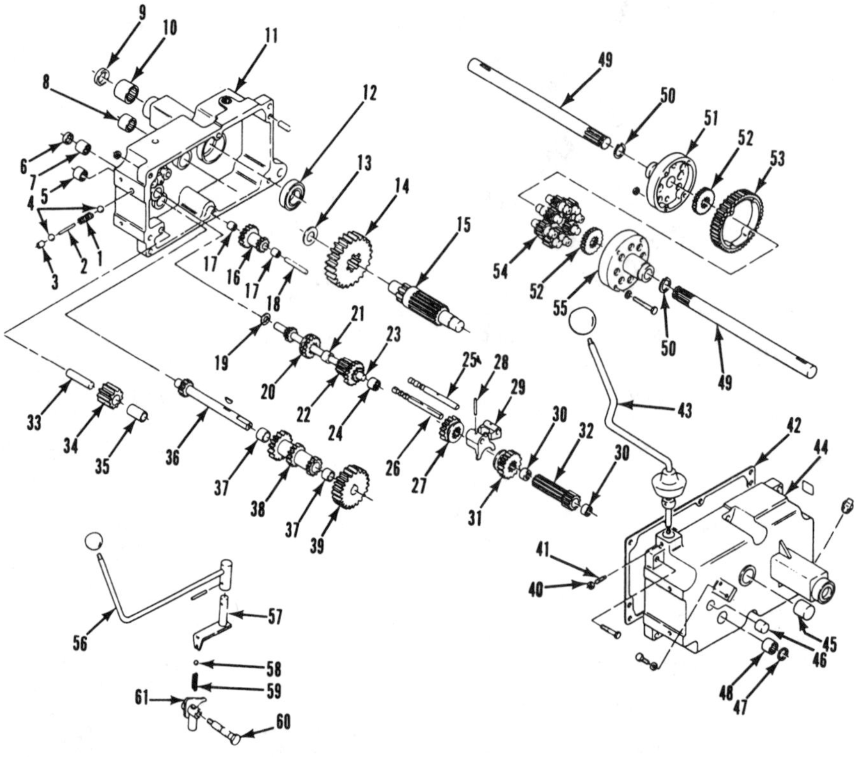

Fig. WH31 – Exploded view of eight-speed transmission.

1. Detent spring
2. Detent pin
3. Plug
4. Detent balls
5. Needle bearing
6. Seal
7. Needle bearing
8. Needle bearing
9. Axle seal
10. Needle bearing
11. Case
12. Ball bearing
13. Thrust washer
14. Gear (44T)
15. Spline/gear
16. Reduction cluster
17. Needle bearings
18. Reduction shaft
19. Thrust washer
20. Sliding gear
21. Needle bearing
22. Gear & spline
23. Input shaft
24. Needle bearing
25. Rear shift rail
26. Front shift rail
27. High – 2nd gear
28. Pin
29. Shift forks
30. Needle bearings
31. Low – reverse gear
32. Pinion & spline
33. Reverse idler shaft
34. Reverse idler gear
35. Bronze bushing
36. Brake/cluster shaft
37. Bearings
38. Cluster gear
39. Reduction gear
40. Jam nut
41. Setscrew
42. Gasket
43. Selector lever
44. Case
45. Bearing
46. Bearing
47. Seal (brake shaft)
48. Bearing
49. Axle shaft
50. Snap ring
51. Differential case half
52. Axle gear
53. Ring gear
54. Pinion gears
55. Differential case half
56. Range gear selector lever
57. Shift lever
58. Detent ball
59. Detent spring
60. Detent bolt
61. Shift fork

from control shaft, remove snap ring (16) and washer (17) and withdraw control shaft (13). Remove oil seal (12). Use an 1/8-inch pipe tap to thread 11/32-inch hole drilled in housing to be plugged during reassembly.

To remove directional check valves from pintle (30), drill through pintle using a drill bit which will pass freely through roll pins (37). From opposite side redrill holes using a 1/4-inch drill bit. Press roll pins (37) from pintle. Use a 5/16-inch, 18-thread tap to thread inside each check valve body (35) to accept a draw bolt to pull valve bodies. Remove check valve balls (34) and snap rings (33).

Number piston bores (1 through 5) on pump rotor and on motor rotor. Use a plastic ice cube tray or equivalent and mark cavities 1P through 5P for pump ball pistons and 1M through 5M for motor ball pistons. Remove ball pistons (25) one at a time, from pump rotor and place each ball in the correct cavity in tray. Remove ball pistons (25) and springs (32) from motor rotor (31) in the same manner.

Clean and inspect all parts and renew any showing excessive wear or other damage. Ball pistons are a select fit to 0.0002-0.0006 inch (0.0051-0.0152 mm) clearance in rotor bores and must be reinstalled in their original bores. If bushings inside pump rotor (24) or motor rotor (31) are scored or if bushing clearance exceeds 0.002 inch (0.051 mm), entire pump rotor or motor rotor assembly must be renewed.

To reassemble, install ball pistons (25) in pump rotor (24) and ball pistons (25) and springs (32) in motor rotor making certain each ball piston is installed in its original location. Use a wide rubber band to retain ball pistons in rotors. Install snap rings (33), check valve balls (34) and valve bodies (35) in pintle (30). Use new roll pins (37) to secure. Install dampening pistons (28) and new "O" rings (29). Renew oil seals (7 and 12).

NOTE: When installing oil seals (7, 12 and 44), apply a thin coat of "Loctite" grade 635 to outer rim of seal before installation.

Install control shaft (13) and input shaft (6) in housing (8) and press dowel pin (14) into control shaft (13) until end of pin extends $1\frac{1}{8}$ inch (28.5 mm) from control shaft. Apply "Loctite" grade 635 to threads of a 1/8-inch pipe plug and install plug in hole which was drilled for disassembly. Tighten plug until snug. Renew oil seal (44) and install output shaft (38), bearing (45), snap ring (46), retainer (47) and spacer (48). Install key (39), gear (49) and snap ring (50).

Place housing (8) in a holding fixture with input shaft pointing downward. Install pump cam ring (27) and race (26) on pivot pin (18) and dowel pin (14). Insert (36) must be fitted in cam ring (27) with hole facing out. Cam ring should move freely from stop to stop. Install pump rotor assembly and remove rubber band retainer. Install free-wheeling valve parts (10, 9, 19 and 20). Install free-wheeling valve bracket (21) and springs (22) against valve bodies (35), making sure each pin of bracket (21) engages a valve body (35). Install pintle assembly over cam pivot pin and into pump rotor. Renew "O" ring (23) at housing joint and place housing assembly on its side on a clean surface.

Place body assembly (43) in holding fixture with output shaft pointing downward. Install motor race (40), then install motor rotor assembly with rotor slot aligned on drive pin of output shaft (38). Remove rubber band retainer. Place body and motor assembly on its side on bench top so motor rotor is facing pintle in housing. Slide body and housing assemblies together with assembly bolt holes in alignment. Install cap screws and tighten to 15 ft.-lbs. (20 N·m). Rotate input and output shafts by hand to check for freedom of movement. If there is binding, disassemble unit and locate and correct any problem. Install free-wheeling valve plunger (11).

Reassemble by reversing disassembly procedure.

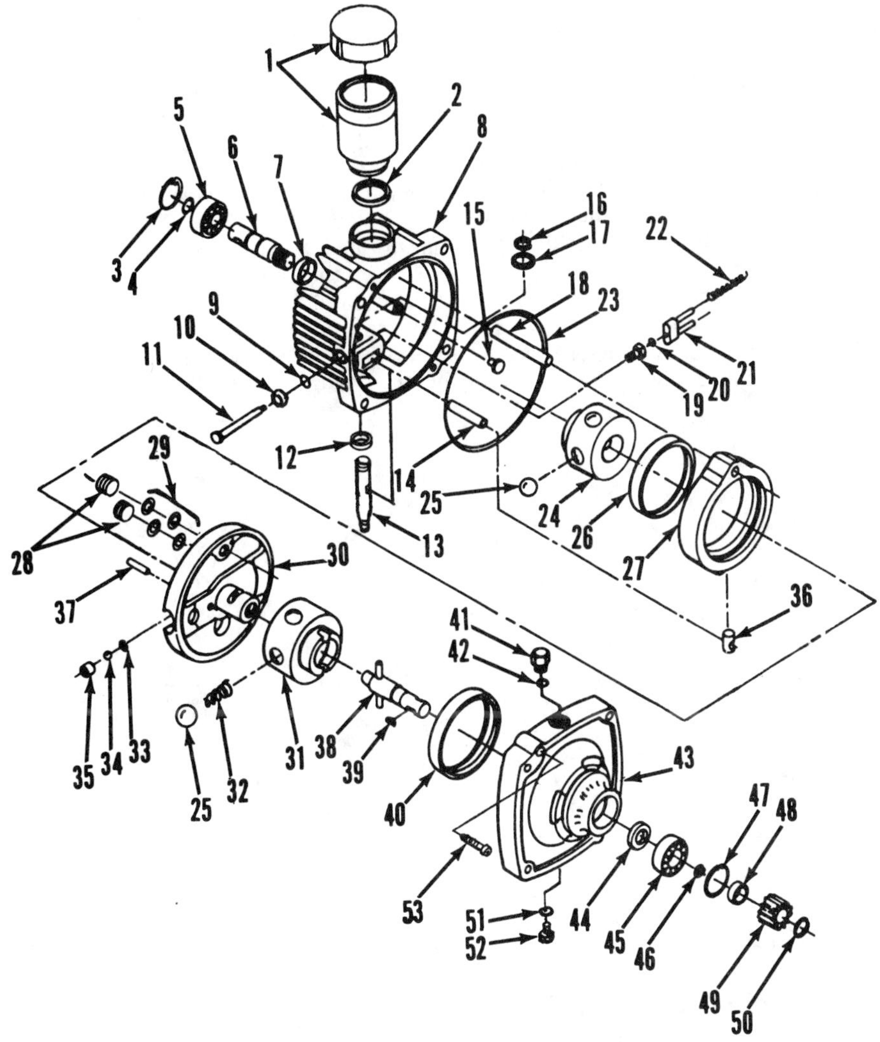

Fig. WH32—Exploded view of hydrostatic drive unit used on B-165 model.

1. Reservoir & cap
2. Gasket
3. Retaining ring
4. Retaining ring
5. Bearing
6. Input shaft
7. Seal
8. Cover
9. "O" ring
10. Nut
11. Shaft (dump valve)
12. Seal
13. Control shaft
14. Pin
15. Cap
16. Snap ring
17. Washer
18. Pivot pin
19. Plunger guide
20. "O" ring
21. Dump valve pin & bracket
22. Spring
23. "O" ring
24. Pump rotor
25. Pump ball piston
26. Pump race
27. Cam ring
28. Dampening pistons
29. "O" rings
30. Pintle
31. Motor rotor
32. Spring
33. Retaining ring
34. Check valve ball
35. Check valve body
36. Insert
37. Pin
38. Output shaft
39. Key
40. Motor race
41. Plug
42. "O" ring
43. Body
44. Seal
45. Bearing
46. Retaining ring
47. Retaining ring
48. Spacer
49. Output gear
50. Snap ring
51. "O" ring
52. Plug
53. Cap screw

Models C-161 – 18 HP Automatic – D-160 – D-180 – D-200

Remove hydrostatic drive unit as previously outlined and thoroughly clean exterior of unit. Remove cooling fan and input pulley from input shaft. Refer to Fig. WH33 and remove control arm assembly. Separate pump and motor by removing the four Allen head screws.

To disassemble pump, unbolt pump housing (1–Fig. WH34) from pump end cap (26). If necessary, tap sides of housing with a plastic hammer to break the gasket loose. Make sure as pump housing and end cap separate that cylinder block assembly (21) stays with input shaft. Carefully slide cylinder block assembly off shaft and place on a lint free towel to prevent damage.

NOTE: If any of the pistons (23) slip out, return them to their original bores.

To remove variable swashplate (12) and input shaft, remove thrust plate (13). Use a sharp awl and remove pump shaft seal and retainer. Remove snap ring from shaft and tap lightly on input end. Use a 3/16-inch punch to drive trunnion shaft roll pins out (Fig. WH35). One roll pin is used in short stub shaft and two pins are used in control shaft. Remove trunnion shaft retaining rings (7–Fig. WH34) and drive stub shaft (2) inward. Drive control shaft out from inside. Remove swashplate (12) and trunnion seals (5). Press needle bearing out of housing from inside.

Remove valve plate (17) from charge pump housing. Remove charge pump housing mounting bolts and carefully remove pump assembly. If needle bearing (16) in charge pump housing is damaged, remove gerotor assembly (14) and press bearing out. Install new bearing with identification number showing from valve plate side. Press bearing until 0.100 inch (2.54 mm) is left out of the bore.

To disassemble motor, remove cover plate (48) and snap ring. Mark motor housing (45) and end cap (35) to aid in reassembly. Remove cap screws and separate housing and end cap. Remove thrust plate and ball bearing (46). Remove centering ring (43) and place motor end cap on press with cylinder block assembly up. Press motor shaft until retaining spring clip (32) pops loose.

CAUTION: Operation requires moving shaft only a short distance. DO NOT press shaft through cylinder block.

Slide cylinder block assembly and retaining clip from motor shaft. Remove valve plate (33) and withdraw shaft. If motor end cap bearings are rough or damaged, press seal retainer and both needle bearings out the cylinder block side. When reinstalling bearings, press output side bearing in first with lettered end out. Press seal retainer insert (42) in with bearing until flush with end cap. Install second bearing from cylinder block side and press bearing (lettered end outward) until 0.100 inch (2.54 mm) is protruding from face of end cap.

Remove acceleration valves (36 through 40) from opposite sides of motor end cap (35), implement relief valve (24), charge relief valve (27) and free wheeling valve (30). Remove all port plugs to facilitate cleaning of all internal passages.

Clean and inspect all parts and renew any showing excessive wear or damage. Renew all seals, gaskets and "O" rings.

To reassemble, install relief valve assembly (24) in its bore and tighten plug to 22-26 ft.-lbs. (30-35 N·m). Install charge relief valve (27) and check valve assembly, then tighten plugs to 22-26 ft.-lbs. (30-35 (N·m). Insert free-wheeling valve. When installing acceleration valves, pay close attention to metering plug. Any blockage of meter-

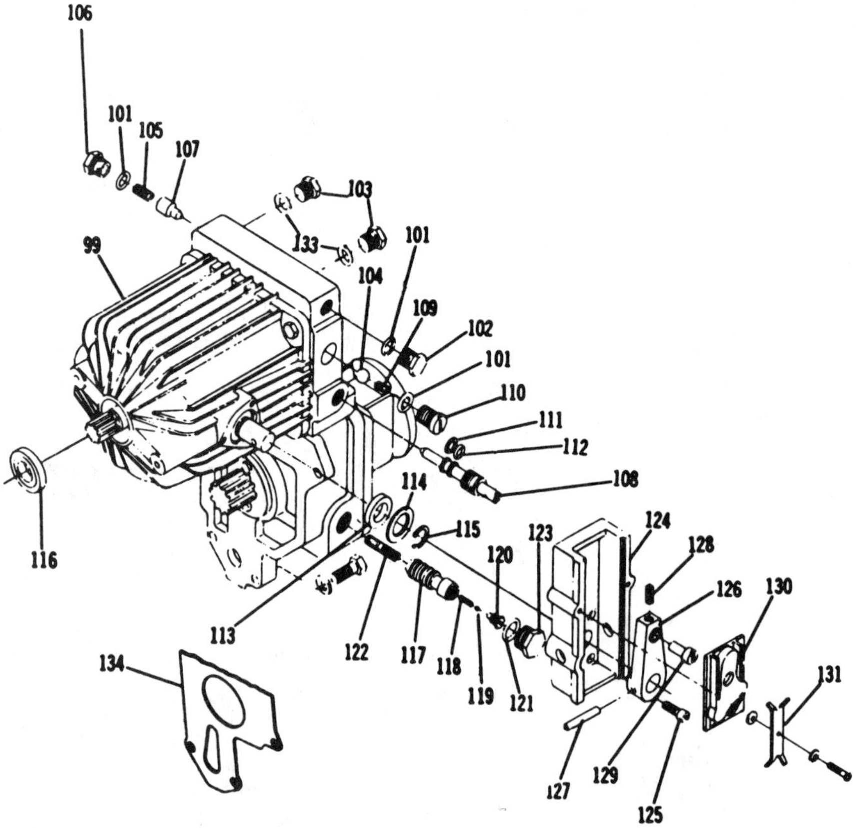

99. Pump housing
101. "O" ring
102. Plug
103. Plugs
104. Check valve ball
105. Charge relief spring
106. Charge relief plug
107. Charge relief valve
108. By-pass
109. Check valve spring
110. Check valve plug
111. "O" ring
112. Back-up ring
113. Control shaft seal
114. Trunnion washer
115. Retainer ring
116. Input shaft seal
117. Sleeve
118. Spring
119. Ball
120. Valve
121. "O" ring
122. Acceleration valve spring
123. Plug
124. Cam block support
125. Socket head screw (3)
126. Cam follower arm
127. Roll pin
128. Setscrew
129. Cam follower pin
130. Cam
131. Tension plate (2)
133. "O" ring
134. Gasket

Fig. WH33 – Exploded view of typical hydrostatic drive unit center section and control cam parts used on C-161 and 18 HP Automatic models and all D-series. On models equipped with hydraulic lift, charge relief valve (107) is on opposite side of center section and lift relief valve is located in its place.

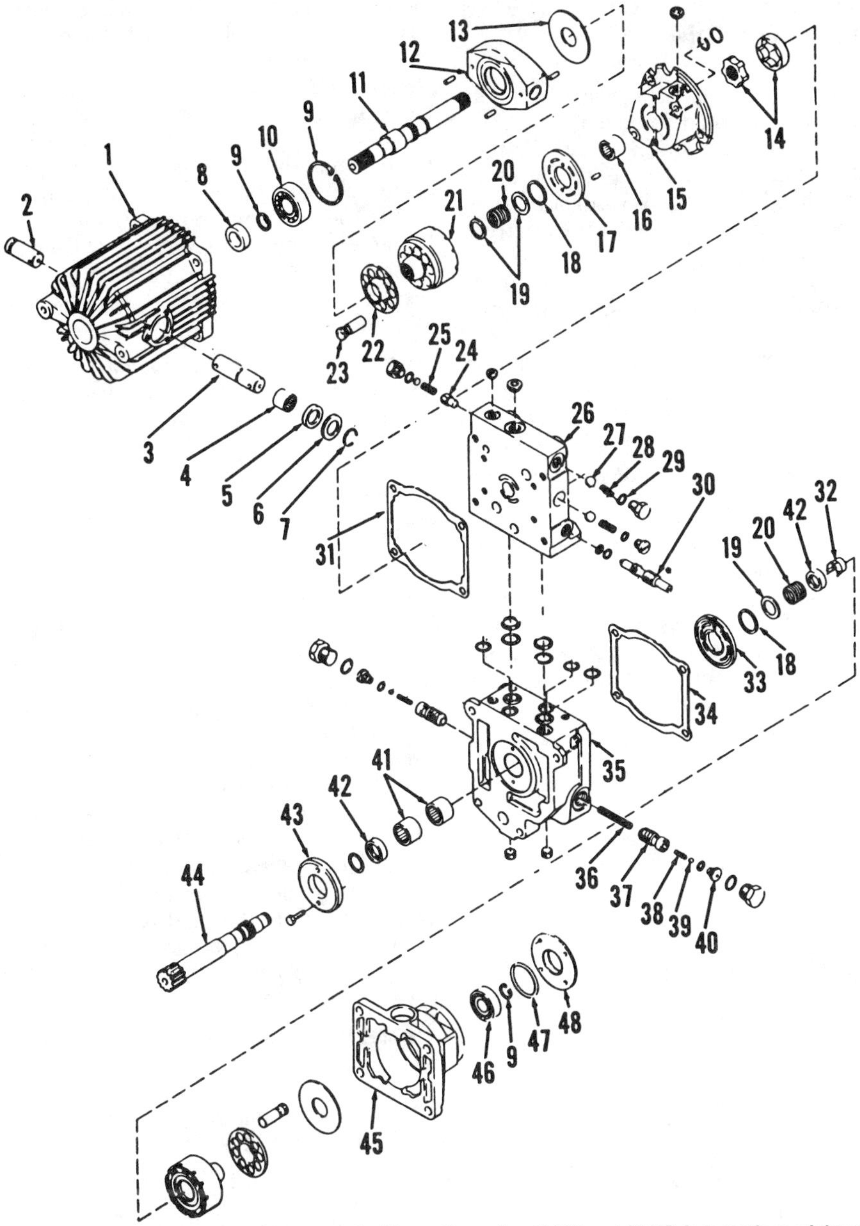

Fig. WH34 – Exploded view of hydrostatic drive unit used on C-161 and 18 HP Automatic models and all D-series.

1. Pump housing
2. Stub shaft
3. Control shaft
4. Bearing
5. Seal
6. Washer
7. Retainer ring
8. Pump shaft seal
9. Snap ring
10. Bearing
11. Pump shaft
12. Variable swashplate
13. Thrust plate
14. Gerotor assy.
15. Charge pump housing
16. Needle bearing
17. Valve plate
18. Retainer ring
19. Washers
20. Spring
21. Cylinder block
22. Slipper retainer
23. Piston
24. Implement relief valve
25. Spring
26. Pump end cap
27. Charge relief valve
28. Spring
29. "O" ring
30. Free-wheeling valve
31. Gasket
32. Retainer clip
33. Valve plate
34. Gasket
35. Motor end cap
36. Spring
37. Acceleration valve body
38. Spring
39. Ball
40. Metering plug
41. Needle bearings
42. Retainer
43. Centering ring
44. Motor shaft
45. Motor housing
46. Ball bearing
47. "O" ring
48. Cover plate

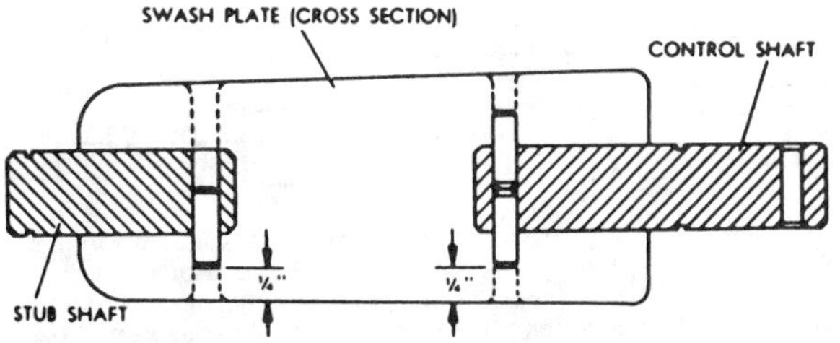
SWASH PLATE (CROSS SECTION)
CONTROL SHAFT
STUB SHAFT
¼"
¼"

Fig. WH35 – Cross-sectional view of variable swashplate, stub shaft and control shaft. Two roll pins are used to retain control shaft.

ing plug orifice formed by the small orifice ball will cause loss of power and slow response. Tighten metering plug to 65-70 ft.-lbs. (88-95 N·m) and install remaining plugs.

Inspect pump housing end cap (26) and charge pump gerotor set for excessive wear or deep scratches and renew as necessary. Superficial scratches on pump end cap may be removed by hand lapping. Check bronze side of valve plate (Fig. WH36) for scratches and wear. If scratches or wear can be felt with fingernail, renew plate. Pump and motor valve plates are not interchangeable. Bronze side of valve plates must face pump or motor cylinder block.

Inspect cylinder blocks (21 – Fig. WH34), pistons (23) and thrust plates (13) for wear or damage. Thrust plates must be checked for flatness, scoring and imbedded material and renew as required. Slippers are bronze bonnets crimped to the pistons. Piston assemblies should be renewed if slipper surface damage is deeper than 0.005 inch (0.127 mm). A lapping plate with 1500 grit compound or 4/0 grit emery paper should be used for lapping slippers. After lapping, slipper thickness should not vary over 0.002 inch (0.051 mm) for all pistons. See Fig. WH37. Slipper should be free on piston with a maximum end play of 0.004 inch (0.102 mm). Cylinder block should be renewed if bores show excessive wear or severe scratches. Block face (balance land end) is a lapped surface. Light scratches can

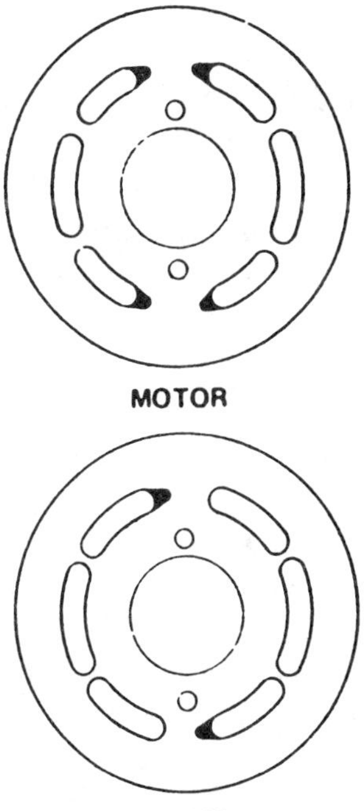
MOTOR
PUMP

Fig. WH36 – Motor valve plate has four notches (dark areas) and pump valve plate has two notches (dark areas). Refer to text.

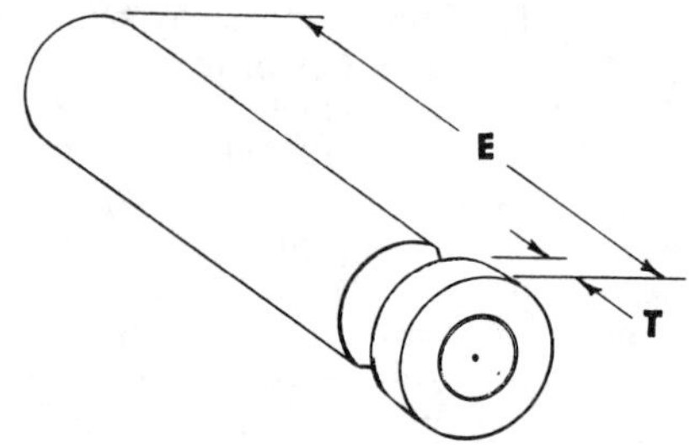

Fig. WH37—Slipper thickness (T) must not vary more than 0.002 inch (0.051 mm) for all nine pistons and maximum end play (E) is 0.004 inch (0.102 mm). Slipper must move freely on piston.

be removed by hand lapping. Renew blocks which require removal of 0.005 inch (0.127 mm) or more stock to remove scratches.

Apply a light coat of transmission fluid to seals and "O" rings and lubricate mating surfaces of valve plate, cylinder block and bearing surfaces. Reassemble unit by reversing disassembly procedure.

Models GT-1642—GT-1848 With Hydrostatic Drive

Remove hydrostatic drive unit as previously outlined and drain unit.

Place transmission in holding fixture with input shaft pointing up. Remove dust shield (1–Fig. WH40) and snap ring (3). Remove cap screws from charge pump body (7). One cap screw is 1/2-inch longer than the others and must be installed in original position. Remove charge pump body (7) with ball bearing (4). Ball bearing and oil seal (6) can be removed after removing retaining ring (2). Remove snap rings (5 and 8) and charge pump rotor assembly. Remove seal rings (10) and pump plate (11). Turn hydrostatic unit over in fixture and remove output gear. Unscrew the two cap screws until two threads are engaged. Raise body (42) until it contacts cap screw heads. Insert a special fork tool (Fig. WH41) between motor rotor (39–Fig. WH40) and pintle (28). Remove cap screws, lift off body and motor assembly with fork tool and place assembly on a bench or in a holding fixture with output shaft pointing down. Remove fork and place a wide rubber band around motor rotor to hold ball pistons (38) in their bores. Carefully remove motor rotor assembly and lay aside for later disassembly. Remove motor race (41) and output shaft (40). Remove retainer (45), bearing (44) and oil seal (43). With housing assembly (12) resting in holding fixture, remove pintle assembly (28).

CAUTION: Do not allow pump to raise with pintle as ball pistons (22) may fall out of rotor (21). Hold pump in position by inserting a finger through hole in pintle.

Remove plug (37), spring (36) and charge relief ball (35). To remove directional check valves, drill through pintle with a drill bit that will pass freely through roll pins. Redrill holes from opposite side with a 1/4-inch drill bit. Press roll pin from pintle. Newer units are drilled at factory. Using a 5/16-18 tap, thread inside of valve bodies (34), then remove valve bodies using a draw bolt or slide hammer puller. Remove check valve balls (33) and retaining ring (32). To remove acceleration valves, remove retaining pin, insert a 3/16-inch (5 mm) rod 8 inches (203 mm) long through passage in pintle and carefully drive out spring (29), body (30) and ball (31). To remove dampening pistons (26), carefully tap outside edge of pintle on work bench to jar pistons free.

NOTE: If pintle journal is damaged, pintle must be renewed.

Remove pump cam ring (24) and pump race (23). Place a wide rubber band around pump rotor to prevent ball pistons (22) from falling out. Carefully remove pump assembly and input shaft (15).

To remove control shaft (19), drill a 11/32-inch hole through aluminum housing (12) directly in line with center line of dowel pin. Press dowel pin from control shaft, then withdraw control shaft. Remove oil seal (18). Thread drilled hole in housing with a 1/8-inch pipe tap. Apply a light coat of "Loctite" grade 35 to a 1/8-inch pipe plug, install plug and

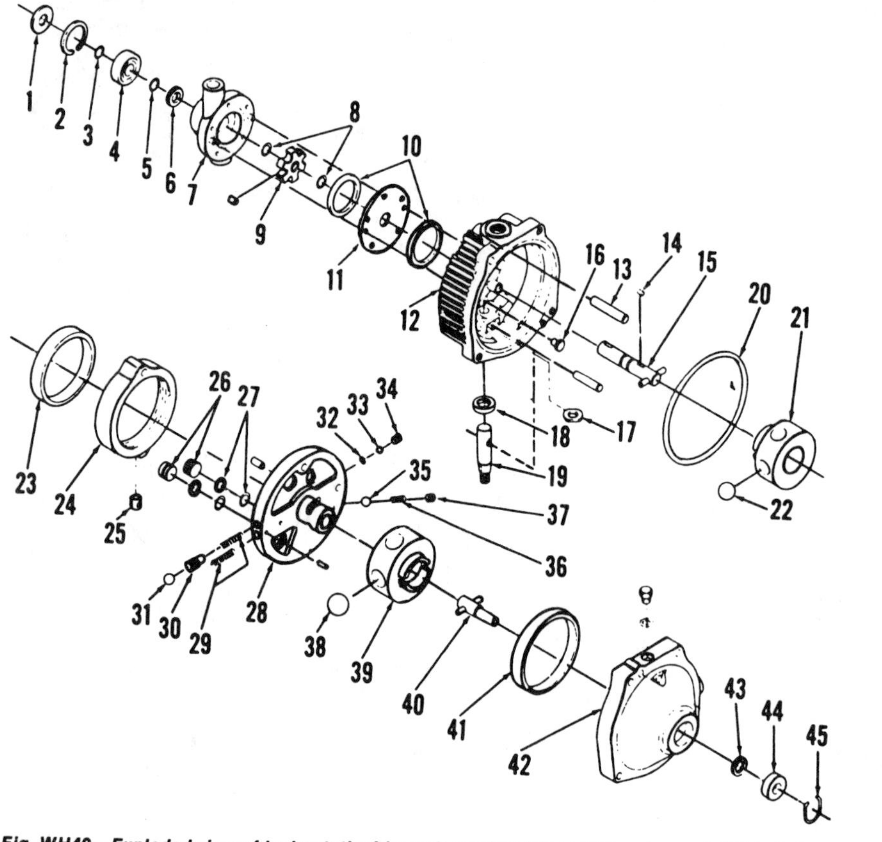

Fig. WH40—Exploded view of hydrostatic drive unit used on GT and LT series and 417-A, C-175 and C-195 models.

1. Dust shield
2. Retaining ring
3. Snap ring
4. Ball bearing
5. Snap ring
6. Oil seal
7. Charge pump body
8. Snap ring
9. Charge pump rotor assy.
10. Square cut seals
11. Pump plate
12. Housing
13. Cam pivot pin
14. Key
15. Input shaft
16. Neutral spring cap
17. Washer
18. Oil seal
19. Control shaft
20. "O" ring
21. Pump rotor
22. Pump ball pistons
23. Pump race
24. Pump cam ring
25. Cam ring insert
26. Dampening pistons
27. "O" ring
28. Pintle
29. Spring
30. Acceleration valve body
31. Acceleration valve ball
32. Retaining ring
33. Check valve ball
34. Check valve body
35. Charge relief ball
36. Spring
37. Relief valve plug
38. Motor ball piston
39. Motor rotor
40. Output shaft
41. Motor race
42. Body
43. Oil seal
44. Ball bearing
45. Retainer

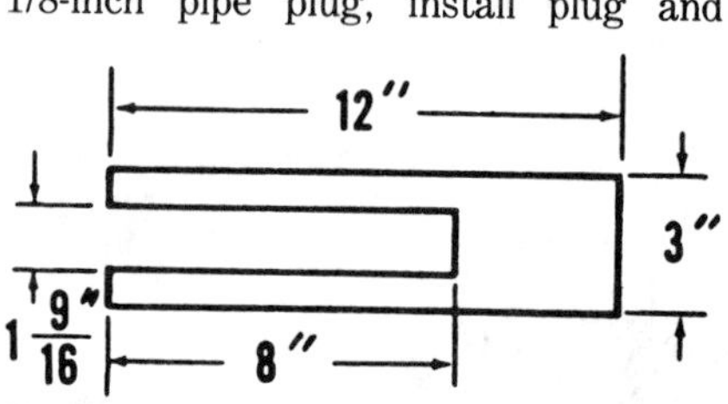

Fig. WH41—Special fork tool fabricated from a piece of 1/8-inch (3 mm) flat stock, used in disassembly and reassembly of hydrostatic drive unit used on GT and LT series and 417-A, C-175 and C-195 models.

tighten until snug. Do not overtighten.

Number piston bores (1 through 5) on pump rotor and on motor rotor. Use a plastic ice cube tray or equivalent and mark cavities 1P through 5P for pump ball pistons and 1M through 5M for motor ball pistons. Remove ball pistons (22) one at a time, from pump rotor and place each ball in the correct cavity in tray. Remove ball pistons (38) and springs from motor rotor in the same manner.

Clean and inspect all parts and renew any showing excessive wear or other damage. Renew all gaskets, seals and "O" rings. Ball pistons are a select fit to 0.0002-0.0006 inch (0.0051-0.0152 mm) clearance in rotor bores and must be reinstalled in their original bores. If rotor bushing to pintle journal clearance is 0.002 inch (0.051 mm) or more, bushing wear or scoring is excessive and pump rotor or motor rotor must be renewed. Check clearance between input shaft (15) and housing bushing. Normal clearance is 0.0013-0.0033 inch (0.033-0.0838 mm). If clearance is excessive, renew input shaft and/or housing assembly.

Install ball pistons (22) in pump rotor (21) and ball pistons (38) and springs in motor rotor (39), then use wide rubber bands to hold pistons in their bores.

Install charge relief valve ball (35) and spring (36) in pintle. Screw plug (37) into pintle until just below outer surface of pintle. Install acceleration valve springs (29) and bodies (30) making sure valves move freely. Tap balls (31) into pintle until roll pins will go into place. Install snap rings (32), check valve balls (33) and valve bodies (34) in pintle and secure with new roll pins.

NOTE: When installing oil seals (6, 18 or 43), apply a light coat of "Loctite" grade 635 to seal outer diameter.

Renew oil seal (18) and install control shaft (19) in housing. Install special washer (17), then press dowel pin through control shaft until 1-1/4 inches (32 mm) of pin extends from control shaft. Renew oil seal (43) and reinstall output shaft (40), bearing (44), retainer (45), output gear and snap ring.

Insert input shaft (15) in housing (12). Install snap ring (8) in its groove on input shaft. Place both seal rings (10) and pump plate (11) in housing, then install charge pump drive key (14), charge pump rotor (9) and snap ring (8). Apply light grease or Vaseline to pump rollers and place rollers in rotor slots. Install oil seal (6) and pump race in charge pump body (7), then install body assembly. Secure with the five cap screws, making certain long cap screw is installed in its original location (in heavy section of pump body). Tighten cap screws to 28-30 ft.-lbs. (38-41 N•m). Install snap ring (5), bearing (4), retaining ring (2), snap ring (3) and dust shield (1).

Place charge pump and housing assembly in a holding fixture with input shaft pointing downward. Install pump race (23) and insert (25) in cam ring (24), then install cam ring assembly over cam pivot pin (13) and control shaft dowel pin. Turn control shaft (19) back and forth and check movement of cam ring. Cam ring must move freely from stop to stop. If not, check installation of insert (25) in cam ring.

Install pump rotor assembly and remove rubber band used to retain pistons. Install pintle assembly (28) over cam pivot pin (13) and into pump rotor. Place "O" ring (20) in position of housing.

Place body assembly (42) in a holding fixture with output gear down. Install motor race (41) in body, then install motor rotor assembly and remove rubber band used to retain pistons in rotor.

Using special fork tool (Fig. WH41) to retain motor assembly in body, carefully install body and motor assembly over pintle journal. Remove fork tool, align bolt holes and install the two cap screws. Tighten cap screws to 15 ft.-lbs. (20N•m).

Place hydrostatic unit on holding fixture with reservoir adapter opening and venting plug opening facing upward. Fill unit with recommended fluid until fluid flows from fitting hole in body. Plug all openings to prevent dirt or other foreign material from entering hydrostatic unit.

Reinstall unit on reduction and differential housing, using a new gasket, and tighten cap screws to 20 ft.-lbs. (27 N•m). Run engine and check fluid level in differential housing. Check adjustment of transmission linkage.

REDUCTION DRIVE AND DIFFERENTIAL

MAINTENANCE

All Models

Differential assembly for three, five and eight speed transmissions are an integral part of transmission assembly. Refer to appropriate TRANSMISSION section for maintenance and service procedures.

Model B-165 with hydrostatic drive is equipped with a separate reduction gear and differential assembly. Reduction gear and differential assembly fluid level should be checked at 25 hour intervals and fluid level should be maintained at lower edge of oil fill plug at rear of unit (Fig. WH42). Fluid should be changed at 100 hour intervals and approximate capacity is 1-3/8 quart (1.3 L). Recommended fluid is SAE 90 EP gear lubricant.

On all other models, hydrostatic drive unit is connected to a two-speed reduction gear unit with an integral differential assembly. Refer to appropriate TRANSMISSION section for maintenance procedure.

Fig. WH42—View showing location of fill plug and drain plug for reduction gear and differential assembly on B-165 hydrostatic drive models. Refer to text.

R&R AND OVERHAUL

Model B-165 With Hydrostatic Drive

Remove seat and fender assembly. Remove brake adjusting nuts and slide brake actuating lever off disc brake caliper. Disconnect control linkage at transmission. Remove drive belt. Support reduction gear assembly, remove retaining bolts and raise and support rear of frame. Move reduction gear assembly away from tractor.

Remove wheels and hubs and drain lubricant from differential. Separate transmission from differential unit. Remove brake disc from brake shaft.

Clean axle shafts and remove any burrs on shafts. Unscrew cap screws and drive out dowel pins in cover (29–Fig. WH43). Lift cover off case and axle shaft. Withdraw brake shaft (5), input gear (4) and thrust washers (3 and 6) from case. Remove output shaft (11), output gear (10), space (9), thrust washer (8) and differential assembly from case. Axle shaft housings (20 and 22) must be pressed from case and cover.

To disassemble differential, unscrew four cap screws and separate axle shaft and carriage assemblies from ring gear (28). Drive blocks (25), bevel pinion gears (26) and drive pin (27) can now be removed from ring gear. Remove snap rings (12) and slide axle shafts (18 and 23) from axle gears (13) and carriers (16 and 24).

Clean and inspect all parts and renew any parts damaged or excessively worn. When installing needle bearings, press bearings in from inside of case or cover until bearings are 0.015-0.020 inch (0.381-0.508 mm) below thrust surfaces. Be sure heads of differential cap screws and right axle shaft (18) are installed in right carrier (16). Right axle shaft is installed through case (1). Tighten differential cap screws to 7 ft.-lbs. (9 N•m) and cover cap screws to 10 ft.-lbs. (14 N•m). Differential assembly and output shaft (11) must be installed in case at the same time.

Reassemble by reversing disassembly procedure.

Reinstall reduction gear assembly by reversing removal procedure. Adjust brakes and transmission linkage.

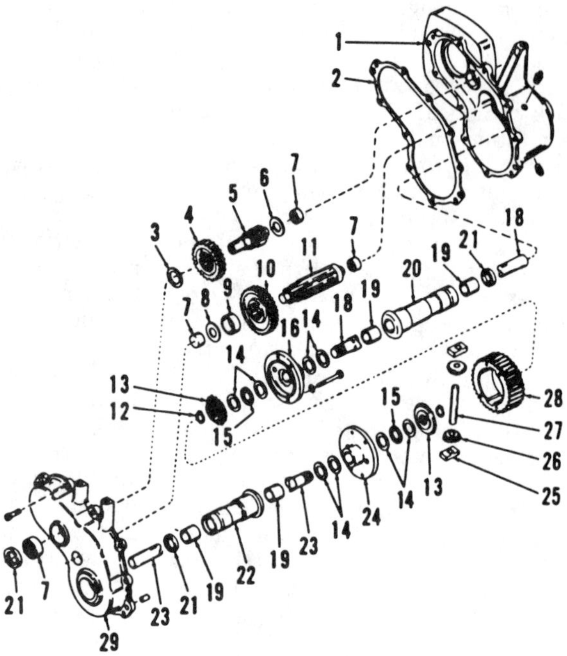

Fig. WH43—Exploded view of reduction gear and differential used on B-165 model.

1. Case
2. Gasket
3. Washer
4. Idler gear
5. Brake shaft
6. Washer
7. Bearing
8. Washer
9. Spacer
10. Output gear
11. Output shaft
12. Snap ring
13. Bevel gear
14. Thrust washers
15. Thrust bearing
16. Differential carrier
18. Axle R.H.
19. Bushing
20. Axle housing
21. Oil seal
22. Axle housing
23. Axle L.H.
24. Differential carrier
25. Drive block
26. Pinion
27. Pinion shaft
28. Ring gear
29. Cover

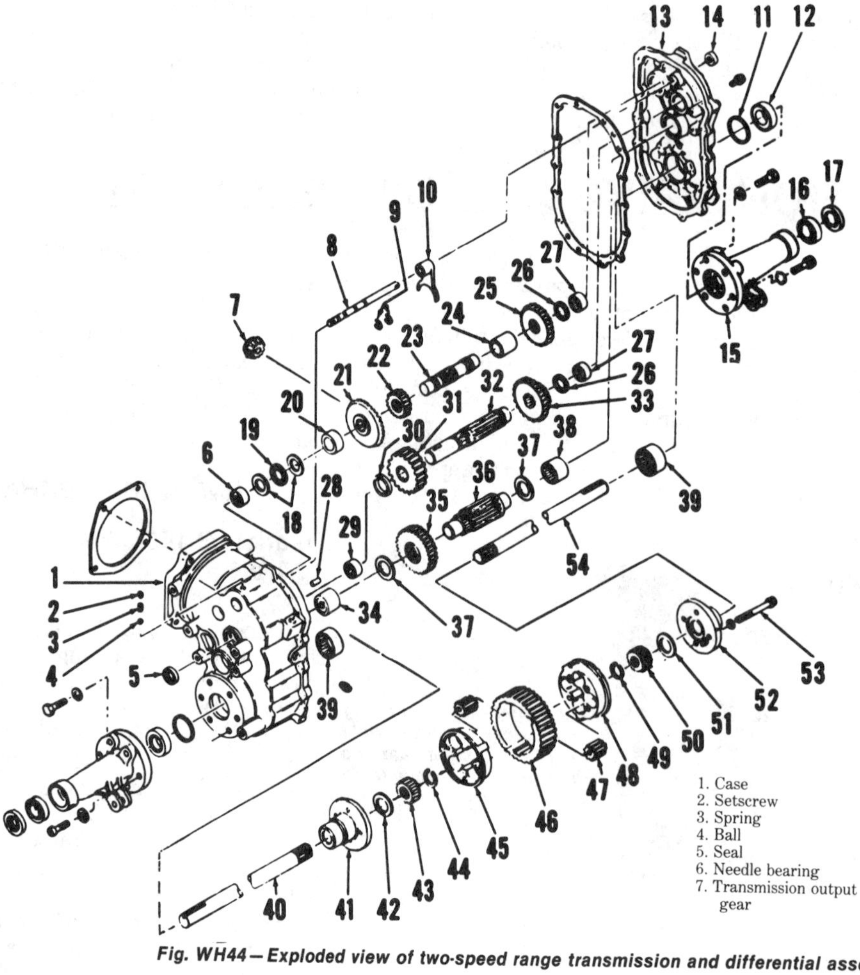

1. Case
2. Setscrew
3. Spring
4. Ball
5. Seal
6. Needle bearing
7. Transmission output gear
8. Shift rail
9. Snap rings
10. Shift fork
11. Quad ring
12. Tapered roller bearing
13. Cover
14. Seal
15. Axle housing
16. Ball bearing
17. Seal
18. Thrust washers
19. Thrust bearing
20. Spacer
21. Bevel gear
22. Gear (low range)
23. Shaft
24. Spacer
25. Gear (high range)
26. Thrust washer
27. Needle bearing
28. Dowel pin
29. Needle bearing
30. Spacer
31. Gear
32. Brake shaft
33. Sliding gears
34. Needle bearing
35. Gear
36. Output shaft
37. Thrust washer
38. Needle bearing
39. Needle bearing
40. Axle shaft L.H.
41. Differential carrier L.H.
42. Thrust washer
43. Axle gear
44. Snap ring
45. Body core
46. Ring gear
47. Pinion gears (8)
48. Body core
49. Snap ring
50. Axle gear
51. Thrust washer
52. Differential carrier R.H.
53. Cap screw
54. Axle shaft R.H.

Fig. WH44—Exploded view of two-speed range transmission and differential assembly.

Models C-161 – 18 HP Automatic – D-160 – D180 – D-200

Refer to R&R TRANSMISSION section for procedure used to remove hydrostatic drive and reduction gear and differential assembly.

Separate hydrostatic drive and two-speed range transmission assembly. Remove brake caliper assembly and brake disc. Remove tires and wheels and remove hub assemblies. Thoroughly clean exterior of unit.

Remove axle housings (15-Fig. WH44). Position unit with cover up, then remove cover (13). Lift out differential assembly and axles (40 through 54). Remove output shaft (36), gear (35) and thrust washers (37). Unscrew setscrew (2) and remove spring (3) and ball (4). Remove brake shaft (32), sliding gear (33), shift fork (10) and rod (8). Remove input shaft and gear components (18 through 25).

To disassemble differential assembly, remove cap screws (53), differential carriers (41 and 52) and axles from ring gear assembly. Remove snap ring to separate axle gear, carrier and axle. Remove pinions (47) and separate body cores (45 and 48) from ring gear (46).

Inspect components for damage or excessive wear. To reassemble unit, reverse disassembly procedure. Check movement of shift rod when tightening setscrew (2). Install gears (22 and 25) so bevels of gears face together as shown in Fig. WH45. Install carrier cap screws (53 – Fig. WH44) so head of cap screw is on side of shorter carrier (52). Do not rotate axle housings after housing has been pressed tight against seal (11) as seal may be cut.

Reinstall hydrostatic drive unit, brake disc and brake caliper.

Models GT-1642 – GT-1848 With Hydrostatic Drive

To disassemble transaxle, remove rear wheels and hubs. Remove burrs and corrosion from axle shafts. Remove brake drum and remove burrs and corrosion from brake shaft. Remove eight retaining bolts and lift off left case half. A soft mallet may be used to break gasket seal.

NOTE: Some early models used two gaskets. Be sure to use two new gaskets when reassembling these units.

Withdraw pinion shaft (17 – Fig. WH46) and reduction gear (16); brake shaft (6) and gears (4 and 5); and differential assembly from right case half (2). Remove axle seals (8) and brake shaft seal (10). New seals are not installed until after transaxle is reassembled.

Inspect bearings in case halves and renew as necessary. Axle needle bearings (9) can be driven out from inside case. All other bearings should be driven out from outside case. To remove upper bearing in right case, use a 1/4-inch punch through small opening behind bearing.

Axle needle bearings (9) should be pressed in from outside. All other bearings are pressed in from inside case. All bearings should be flush with machined surface on inside of case.

To disassemble differential, remove four bolts and separate case halves (19). Remove pinion gears (21) and ring gear (22). Remove snap rings (24), then separate axles (18 and 23) from end caps (19) and gears (20).

Clean and inspect all parts and renew as necessary. Reduction gear (16) is a press fit on pinion shaft (17).

To reassemble, reverse disassembly procedure. Be sure differential pinions (21) are installed in opposite directions (teeth up, teeth down, teeth up, etc.). Be sure beveled edge of axle gears face end cap. Use new high-tensile nuts and grade 8 bolts when reassembling differential, making sure hardened washers are under bolt heads and nuts are on same side as long axle. Tighten bolts to 30-35 ft.-lbs. (41-47 N•m). Tighten case retaining bolts to 30-35 ft.-lbs. (41-47 N•m). After assembling case, install new oil seals. Lubricate seals and protect from nicks. Use a suitable seal driver to install seals flush with outside of case.

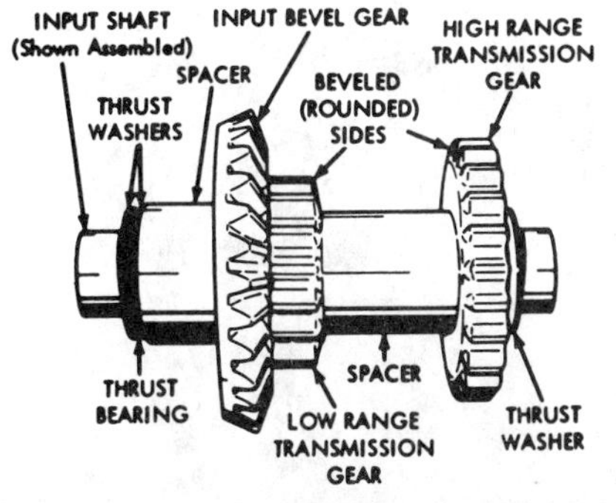
INPUT SHAFT
(Shown Assembled)
INPUT BEVEL GEAR
HIGH RANGE TRANSMISSION GEAR
SPACER
THRUST WASHERS
BEVELED (ROUNDED) SIDES
THRUST BEARING
SPACER
LOW RANGE TRANSMISSION GEAR
THRUST WASHER

Fig. WH45 – View of reduction drive input shaft and gears. Note position of bevels on gears and location of thrust washers and thrust bearing.

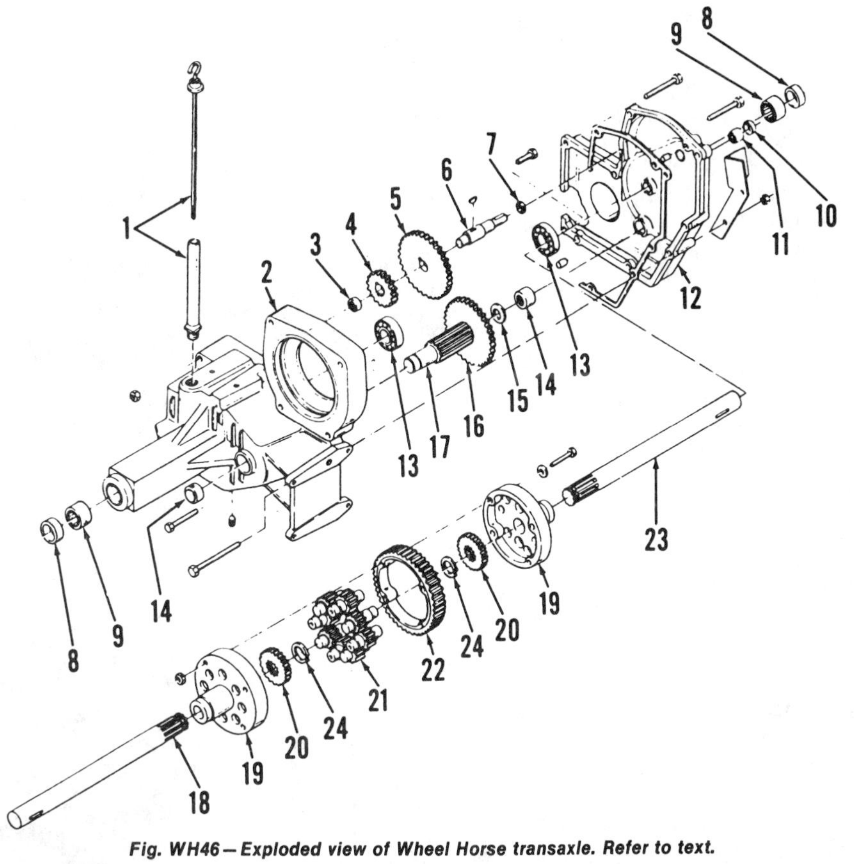

Fig. WH46 – Exploded view of Wheel Horse transaxle. Refer to text.

1. Dipstick & filler tube
2. Case half R.H.
3. Needle bearing
4. Gear
5. Gear
6. Brake shaft
7. Thrust washer
8. Axle seal
9. Needle bearing
10. Seal
11. Needle bearing
12. Case half L.H.
13. Ball bearing
14. Needle bearing
15. Thrust washer
16. Reduction gear
17. Pinion gear
18. Axle R.H.
19. End caps
20. Axle gears
21. Pinion gears
22. Ring gear
23. Axle L.H.
24. Snap rings

POWER TAKE-OFF

MAINTENANCE AND ADJUSTMENT

Models 216-5 – B-165 And All LT Series

Periodically check condition of pto drive belt and lubricate pivot points on linkage with SAE 30 oil. To adjust, engage pto clutch and turn adjustment bolt (7 – Fig. WH47) to obtain 0.010 inch (0.254 mm) clearance between brake pad (BP) and clutch pulley face. There should be a 3/8-inch (9.5 mm) gap between threaded spacer and clutch brake bracket as shown in Fig. WH48. If not, loosen locknut and turn threaded spacer (4 – Fig. WH47) until correct gap is obtained. Tighten locknut and recheck brake pad adjustment.

Models 417 – 417A And All GT And C Series Except C-195 Model

Periodically check condition of pto drive belt and lubricate pivot points on linkage with SAE 30 oil. If clutch slippage becomes apparent, turn trunnion (Fig. WH49) farther onto clutch rod (in one turn increments) until the slippage is eliminated. To adjust pto clutch brake, engage pto clutch and loosen the two bolts holding brake pad bracket to support bracket. Place a 0.012 inch (0.3 mm) feeler gage between brake pad and clutch pulley, hold brake pad against feeler gage and pulley and tighten the two brake bracket bolts. Remove feeler gage and check pto clutch performance.

Model C-195

Periodically check condition of pto drive belt and lubricate pivot points on linkage with SAE 30 oil. To adjust, engage pto clutch. Loosen jam nut on clutch rod (Fig. WH50) and turn threaded spacer until a 1/8-inch (3.2 mm) gap is obtained between hex head of spacer and trunnion. Disengage pto clutch and adjust nut on pto clutch rod end to obtain 1/4-inch (6.4 mm) gap between nut and brake bracket.

Model 18 HP Automatic And All D Series

FRONT MOUNTED PTO. Periodically check condition of pto drive belt and lubricate pivot points on linkage with SAE 30 oil. To adjust pto clutch, pull control lever (28 – Fig. WH52) back to its limit. Loosen locknuts on clutch rods (25 and 27), then turn threaded turnbuckle (26) to adjust rod length so pulley (15) clears rear clutch plate (1) and can be rotated easily by hand. Tighten locknuts against turnbuckle. With clutch engaged, washers (22) on clutch rod forward of trunnion should rotate freely. To adjust pto brake, loosen the two brake bracket bolts so bracket (17) can be shifted in its mounting holes. Adjust bracket position so there is 0.010 inch (0.254 mm) clearance between pad surface and pulley with clutch engaged.

REAR PTO. Model D-200 is equipped with a rear pto and shaft alignment, bearing condition and mounting brackets should be checked periodically. Lubricate bearing (9 – Fig. WH53) at regular intervals.

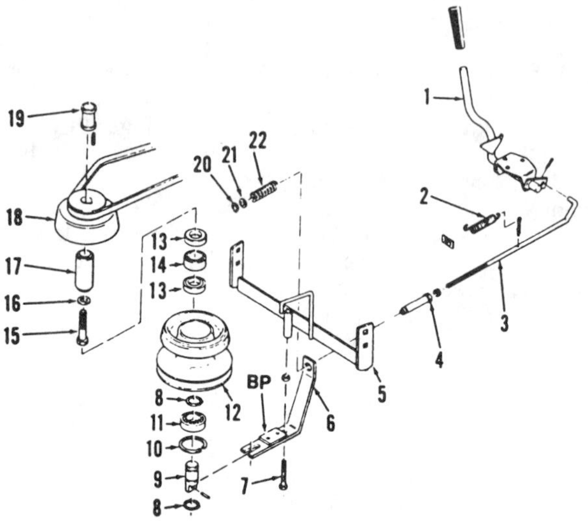

Fig. WH47 – Exploded view of manual pto clutch and operating linkage used on 216-5 and B-165 models and all LT-series.

1. Control lever
2. Spring
3. Clutch rod
4. Threaded spacer
5. Clutch mounting bracket
6. Brake bracket
7. Adjusting bolt
8. Snap ring
9. Clutch shaft
10. Snap ring
11. Ball bearing
12. Pto pulley & clutch cone
13. Seals
14. Needle bearing
15. Bolt
16. Special washer
17. Bearing race
18. Engine pulley & clutch
19. Spacer tube
20. Snap ring
21. Special washer
22. Spring

BP. Brake pad

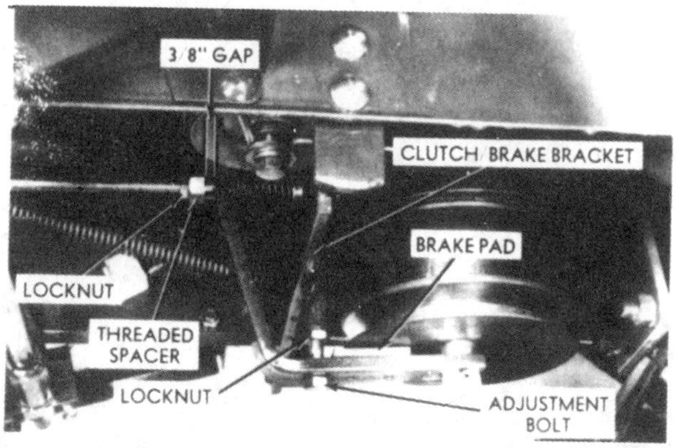

Fig. WH48 – View showing correct gap between threaded spacer and brake bracket on 216-5 and B-165 models and all LT-series. Refer to text.

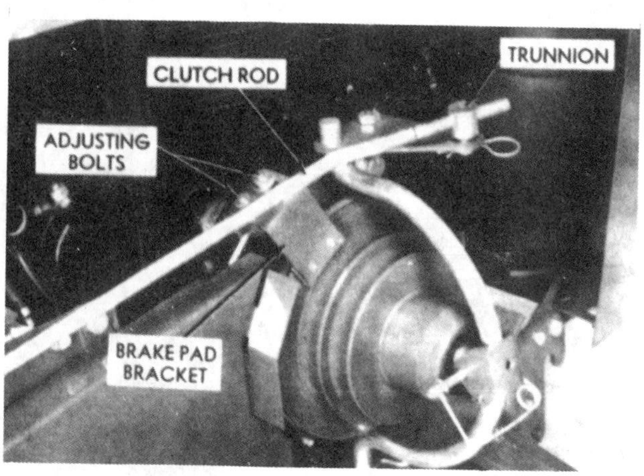

Fig. WH49 – View showing pto linkage adjustment points on 417 and 417-A models and all GT and C series. Refer to text.

R&R PTO ASSEMBLY

Models 216-5 – B-165 And All LT Series

Remove adjusting bolt (7 – Fig. WH47) and brake bracket (6). Withdraw pto pulley and clutch cone (12). Remove crankshaft bolt (15) and bearing race (17). Depress clutch pedal and slip drive belt off engine pulley (18). Remove pulley from crankshaft.

Reinstall by reversing removal procedure. Adjust as previously outlined.

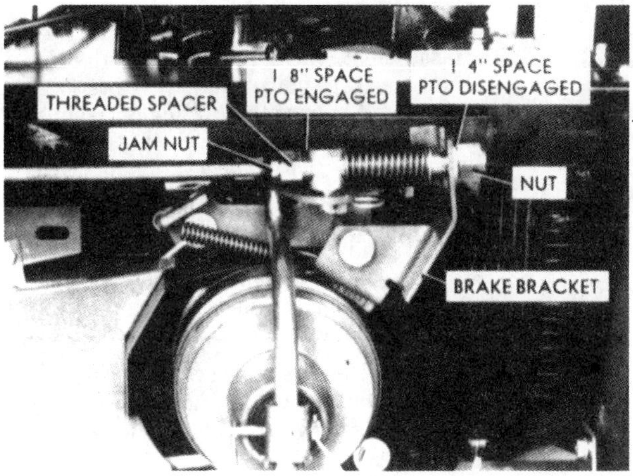

Fig. WH50 – View showing pto linkage adjustment points on C-195 model. Refer to text.

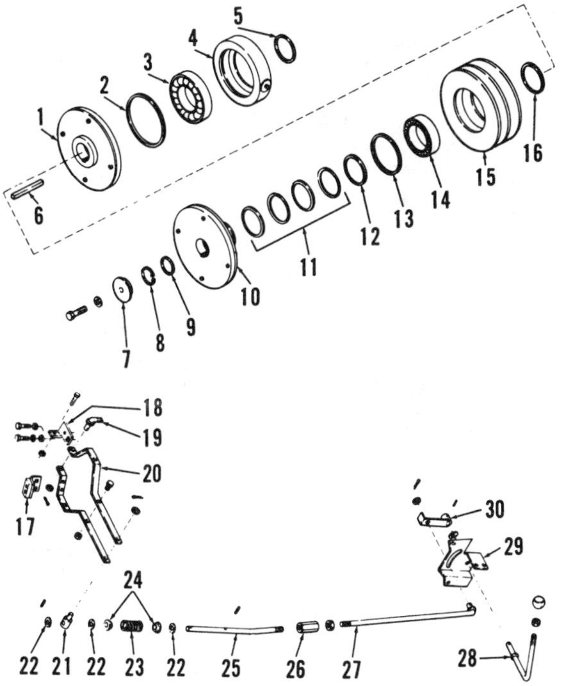

Fig. WH52 – Exploded view of pto clutch and control linkage used on 18 HP Automatic and all D-series.

1. Inner clutch plate
2. Spiroloc ring
3. Bearing
4. Collar
5. Spiroloc ring
6. Crankshaft key
7. End cap
8. Snap ring
9. Washer
10. Outer clutch plate
11. Spring washer (4)
12. Spacer
13. Spiroloc ring
14. Bearing
15. Clutch pulley
16. Spiroloc ring
17. Pulley brake
18. Bracket mount (D-160)
19. Hook bracket (D-180, D-200)
20. Clutch bar
21. Trunnion
22. Washers (3)
23. Tension spring
24. Spring seat bushing (2)
25. Clutch rod (front)
26. Turnbuckle
27. Clutch rod (rear)
28. Operating handle assy.
29. Mounting bracket
30. Clutch arm

Models 417 – 417A And All GT and C Series Except C-195 Model

Disconnect trunnion (9 – Fig. WH54) from plate (10) and pin (12) from clevis (13). Press inward on pulley assembly (19) and swing housing rod (14) rearward so drive belts can be removed. Remove snap rings (16 and 17). Note sequence of parts and continue clutch disassembly. Clutch plate (24) will separate from engine pulley when bearing race retainers (23) are unbolted.

Reinstall by reversing removal procedure.

Model C-195

Except for obvious differences in linkage and adjustment procedure, removal of C-195 model pto clutch is similar to procedure outlined for all other C series.

Models 18 HP Automatic And All D Series

FRONT MOUNTED PTO. Remove brake bracket (17 – Fig. WH52). Remove the 7/16-inch cap screw and special washer securing clutch hub and remove outer clutch plate (10) and pulley assembly (15). Place assembly on a solid surface, clutch plate (10) down and depress pulley (15) against internal springs for removal of Spiroloc ring (16). Pulley (15) may now be lifted from hub of outer clutch plate (10) and spacer (12) and set of four spring washers (11) can be removed.

NOTE: Spring washers must be reinstalled in their original positions.

Remainder of clutch disassembly is obvious after examination.

Reverse disassembly and removal procedure for reassembly and reinstallation.

REAR PTO. Refer to Fig. WH53 for exploded view of rear pto used on D-200 models. Removal and disassembly procedure is obvious after examination.

HYDRAULIC SYSTEM

MAINTENANCE

All Hydrostatic Drive Models

Hydraulic system utilizes pressurized oil from the hydrostatic drive unit. Refer to appropriate TRANSMISSION section for maintenance information.

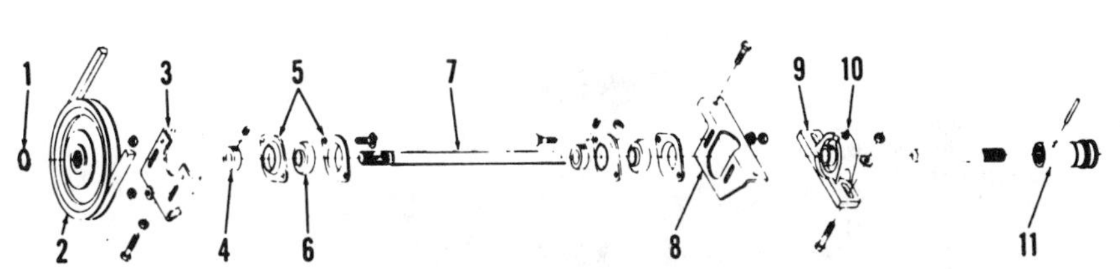

Fig. WH53 – Exploded view of rear pto assembly showing parts arrangement and sequence.

1. Snap ring
2. Pulley
3. Bearing bracket (front)
4. Collar (2)
5. Bearing flange (4)
6. Bearing (2)
7. Shaft
8. Bearing bracket (rear)
9. Pillow block
10. Grease fitting
11. Coupler assy.

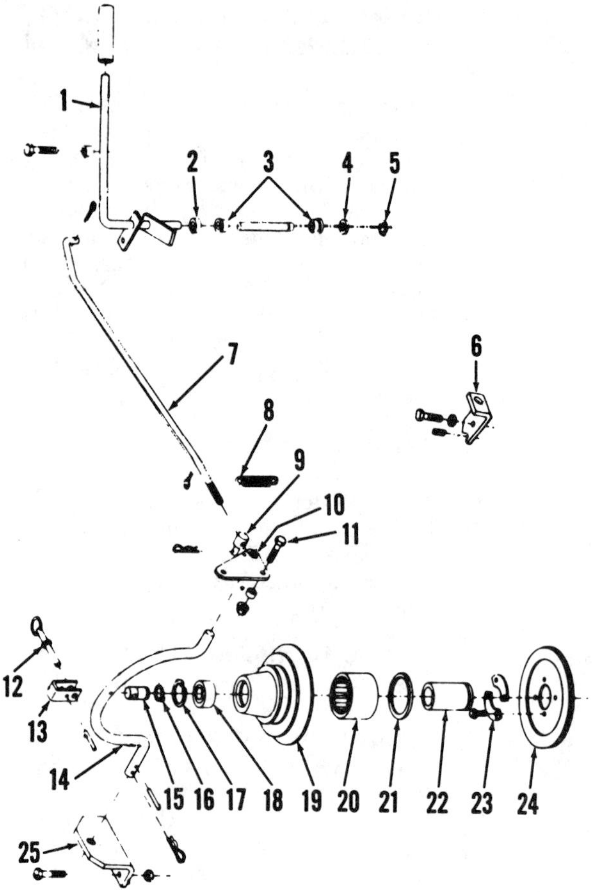
1
2
3
4
5
6
7
8
9
10
11
12
13
14
15 16 17 18 19 20 21 22 23 24
25

Fig. WH54—Exploded view of manual pto clutch and linkage used on 417 and 417-A models and all GT and C-series except C-195. Refer to text.

1. Lever
2. Spacer
3. Bushing (2)
4. Washer
5. "E" ring
6. Bracket
7. Clutch rod
8. Spring
9. Trunnion
10. Plate
11. Pivot bolt
12. Pin
13. Clevis
14. Housing rod
15. Clutch shaft
16. Snap ring
17. Snap ring
18. Ball bearing
19. Pulley & housing
20. Bearing
21. Seal
22. Bearing race
23. Bearing race retainer (2)
24. Clutch plate
25. Support

PRESSURE TEST

All Hydrostatic Drive Models

Connect a 0-1000 psi (0-7000 kPa) pressure gage to lift cylinder front hose as shown in Fig. WH55. With engine running at 3/4 throttle, hold lift control valve lever in "DOWN" position. Pressure gage should read 500 to 700 psi (3448 to 4827 kPa).

Implement life system pressure is controlled by relief valve (Fig. WH57) located in control valve on B-165 model and by relief valve (24–Fig. WH34) on all other models. Add or remove shims or turn adjustment screw (as equipped) to obtain correct pressure.

CAUTION: Serious transmission damage may result if pressure is above 700 psi (4827 kPa).

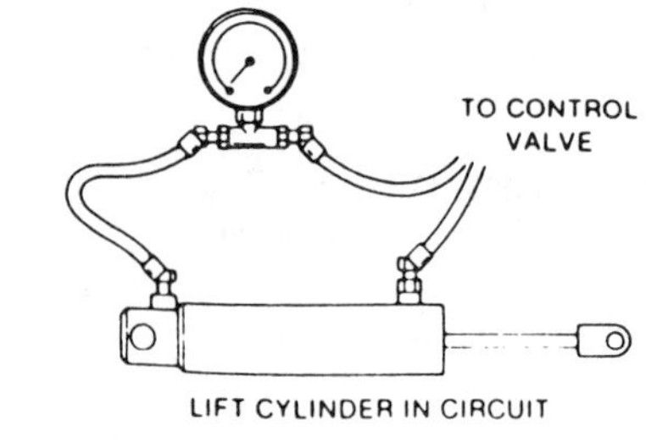
TO CONTROL VALVE
LIFT CYLINDER IN CIRCUIT
OR
TO CONTROL VALVE
PLUG
LIFT CYLINDER OUT OF CIRCUIT

Fig. WH55—Connect a 0-1000 psi (0-7000 kPa) pressure gage to lift cylinder front hose as shown to check implement lift system pressure.

CONTROL VALVE

All Hydrostatic Drive Models

Hydraulic lift system control valves have been supplied by three different manufacturers: OMCO, VICTOR and AICO. Identify manufacturer of control valve to be serviced and refer to appropriate paragraph for service procedures.

OMCO CONTROL VALVE. To disassemble valve, remove through-bolts and separate valve sections. Note location and quantity of mylar shims, located around bolt holes if used, for reassembly. Disconnect control valve handle and remove spool end cap (Fig. WH56). Remove screw from end of spool and withdraw spring and retainers, detent assembly (if so equipped) and spacer block. Push spool out cap end of valve body.

NOTE: Spool is select fit to valve body. When servicing two spool valves, identify parts so spool will be returned to original body.

If valve is equipped with relief valve (Fig. WH57), remove it for cleaning and inspection. Remove and discard all "O" rings. If spool or valve body bore is deeply scratched or scored, complete valve must be renewed.

Reassemble valve by reversing disassembly procedure. Lubricate all parts with clean oil prior to assembly. Tighten through-bolts evenly to 72-75 in.-lbs. (8-8.5 N•m). On valves equipped with float detent, adjust setscrew (Fig. WH56) so spool will lock into float position when engaged, but release easily when light pressure is applied to handle. Adjust relief valve setting (if so equipped) as outlined in PRESSURE TEST paragraph.

VICTOR CONTROL VALVE. To disassemble valve, remove snap ring (Fig. WH58) from valve body. Remove retainer screw, cap, spring and washer. Remove handle from spool and pull spool out of valve body. Remove and

Fig. WH56—View of OMCO control valve equipped with float detent.

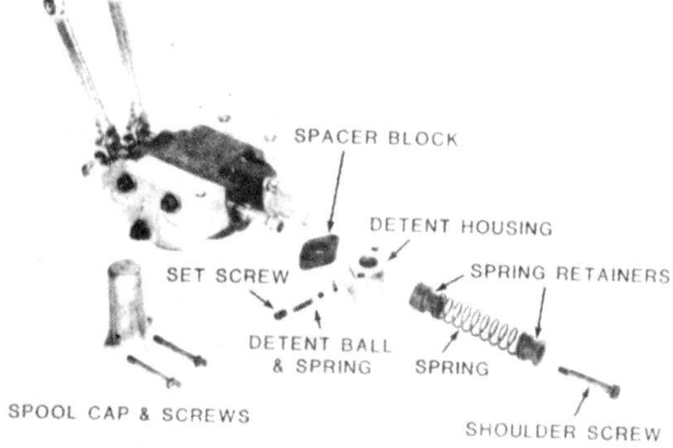
SPACER BLOCK
DETENT HOUSING
SET SCREW
SPRING RETAINERS
DETENT BALL & SPRING
SPRING
SPOOL CAP & SCREWS
SHOULDER SCREW

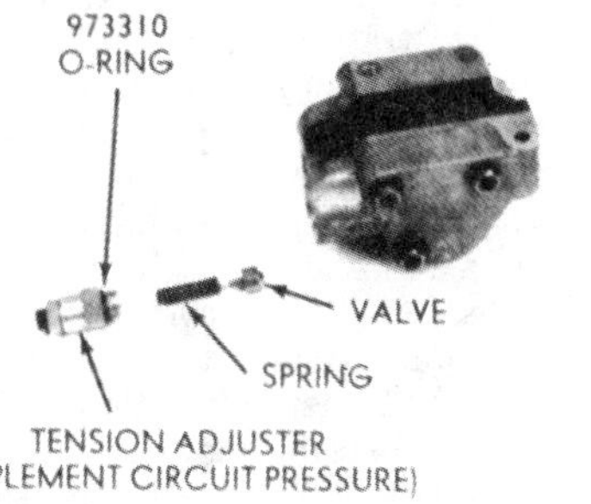
973310
O-RING
VALVE
SPRING
TENSION ADJUSTER
(IMPLEMENT CIRCUIT PRESSURE)

Fig. WH57—View of OMCO control valve equipped with implement relief valve.

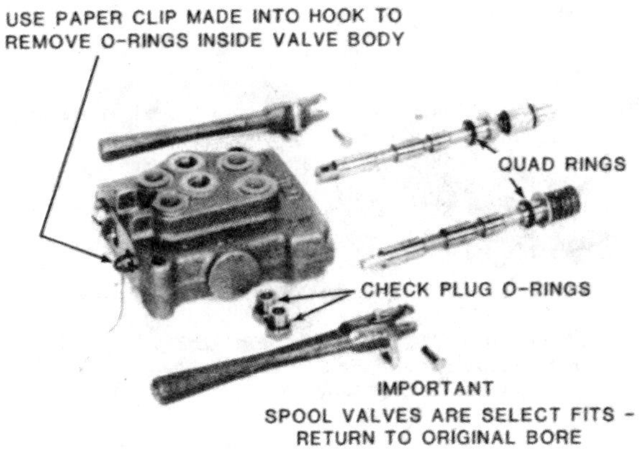
USE PAPER CLIP MADE INTO HOOK TO
REMOVE O-RINGS INSIDE VALVE BODY
QUAD RINGS
CHECK PLUG O-RINGS
IMPORTANT
SPOOL VALVES ARE SELECT FITS -
RETURN TO ORIGINAL BORE

Fig. WH60—View of AICO control valve. Refer to text.

RETAINER CAP
SCREW & WASHER
WASHER
SNAP RING
SPRING

Fig. WH58—View of partially disassembled VICTOR control valve.

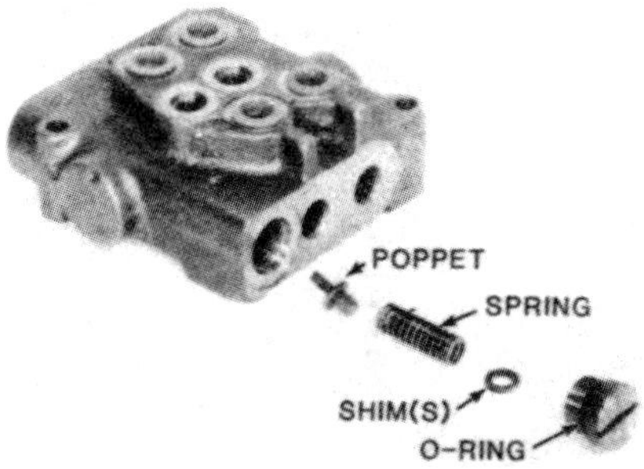
POPPET
SPRING
SHIM(S)
O-RING

Fig. WH61—Implement relief valve assembly used on AICO control valve is adjusted by adding or removing shims.

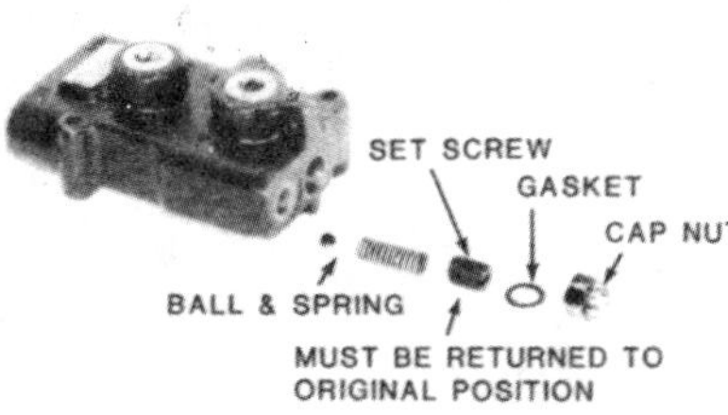
SET SCREW
GASKET
CAP NUT
BALL & SPRING
MUST BE RETURNED TO
ORIGINAL POSITION

Fig. WH59—View of implement relief valve assembly used on VICTOR control valve. Pressure setting is adjusted by position of setscrew. Refer to text.

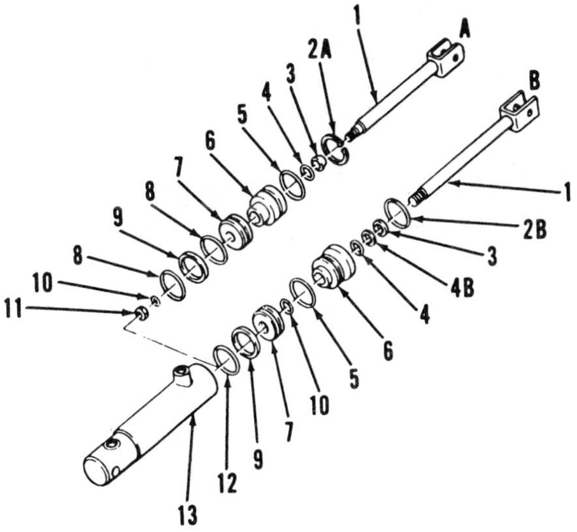
1
2A
A
3
4
5
B
6
7
8
9
1
2B
3
8
4B
10
11
4
6
5
10
7
9
12
13

Fig. WH62—Exploded view of two styles of lift cylinders used on tractors so equipped.

1. Piston rod.
2A. Snap ring
2B. Wire retaining ring
3. Wiper
4. "O" ring
4B. Back-up ring
5. "O" ring
6. Rod guide bushing
7. Piston
8. Back-up rings
9. Seal ring
10. "O" ring
11. Retaining nut
12. "O" ring
13. Cylinder tube

discard all "O" rings. Remove relief valve assembly (Fig. WH59) for cleaning and inspection. If valve spool or body bore is deeply scratched or scored, renew complete valve.

Reassemble valve by reversing disassembly procedure. Lubricate all parts with clean oil prior to assembly. Check and adjust relief valve pressure setting as outlined in PRESSURE TEST paragraph.

AICO CONTROL VALVE. To disassemble valve, first remove valve handle. Unscrew and remove spool end cap. On valves equipped with float detent, catch detent balls and spring as end cap is removed. Push spool out cap end of valve body. See Fig. WH60.

NOTE: Spool is select fit to body bore. On two spool valves, identify spools so they can be returned to original bores.

Unscrew spool stem and separate centering spring assembly from spool. Remove relief valve assembly (Fig. WH61) for cleaning and inspection. Remove and discard all "O" rings. If valve spool or body bore is deeply scratched or scored, renew complete valve.

Reassemble valve by reversing disassembly procedure. Lubricate all parts with clean oil prior to assembly. Check and adjust relief valve pressure setting as outlined in PRESSURE

(HOLES 1/4 in. DEEP)
PISTON IS LIGHTLY STAKED TO ROD - TO REMOVE, USE STRAP WRENCH OR DRILL TWO HOLES IN END, AS SHOWN, FOR PINS OR SPANNER WRENCH. RE-STAKE AFTER ASSEMBLY.

Fig. WH63—On "B" style cylinder (Fig. WH62), remove piston as shown. Be careful not to mar piston or rod surfaces.

TEST paragraph.

LIFT CYLINDER

All Models So Equipped

OVERHAUL. To disassemble cylinder, refer to Fig. WH62 and remove snap ring (2A) or wire retaining ring (2B) from cylinder tube. Withdraw cylinder tube from piston and rod assembly. On "A" style cylinder, unscrew retaining nut (11) and withdraw piston (7) and guide (6). On "B" style cylinders, piston is threaded to rod and lightly staked. To remove piston, use a strap wrench or drill two holes 1/4-inch (6 mm) deep in end of piston for spanner wrench or pins as shown in Fig. WH63. Be careful not to mar piston or piston rod. Remove and discard all "O" rings and seals.

Reassemble by reversing disassembly procedure. Lubricate all parts with clean oil prior to assembly. Be careful not to cut "O" rings during installation. Be sure rod guide bushing is seated against retaining ring.

BRIGGS & STRATTON

BRIGGS & STRATTON CORPORATION
Milwaukee, Wisconsin 53201

Model	No. Cyls.	Bore	Stroke	Displacement	Power Rating
400400	2	3.44 in. (87.3 mm)	2.16 in. (54.8 mm)	40 cu. in. (656 cc)	14 hp. (10.4 kW)
400700	2	3.44 in. (87.3 mm)	2.16 in. (54.8 mm)	40 cu. in. (656 cc)	14 hp. (10.4 kW)
401400	2	3.44 in. (87.3 mm)	2.16 in. (54.8 mm)	40 cu. in. (656 cc)	16 hp. (11.9 kW)
401700	2	3.44 in. (87.3 mm)	2.16 in. (54.8 mm)	40 cu. in. (656 cc)	16 hp. (11.9 kW)
402400	2	3.44 in. (87.3 mm)	2.16 in. (54.8 mm)	40 cu. in. (656 cc)	16 hp. (11.9 kW)
402700	2	3.44 in. (87.3 mm)	2.16 in. (54.8 mm)	40 cu. in. (656 cc)	16 hp. (11.9 kW)
404400	2	3.44 in. (87.3 mm)	2.16 in. (54.8 mm)	40 cu. in. (656 cc)	16 hp. (11.9 kW)
404700	2	3.44 in. (87.3 mm)	2.16 in. (54.8 mm)	40 cu. in. (656 cc)	12.5 hp. (11.9 kW)
421400	2	3.44 in. (87.3 mm)	2.28 in. (57.9 mm)	42.33 cu. in. (694 cc)	18 hp. (13.4 kW)
421700	2	3.44 in. (87.3 mm)	2.28 in. (57.9 mm)	42.33 cu. in. (694 cc)	18 hp. (13.4 kW)
422400	2	3.44 in. (87.3 mm)	2.28 in. (57.9 mm)	42.33 cu. in. (694 cc)	18 hp. (13.4 kW)
422700	2	3.44 in. (87.3 mm)	2.28 in. (57.9 mm)	42.33 cu. in. (694 cc)	18 hp. (13.4 kW)

Engines in this section are four-stroke, two-cylinder, air-cooled engines. The crankshaft is supported at each end either with ball bearings or plain bearings. The plain bearings are either integral with crankcase, cover or sump, or renewable DU type bearings. Cylinder block and crankcase are a single aluminum casting. Some models are equipped with integral cast iron cylinder liners.

Connecting rods for all models ride directly on crankpin journals. Models 402440, 402770, 422440 and 422770 are pressure lubricated by a rotor type oil pump. All other models are splash lubricated. Vertical crankshaft models that are splash lubricated are equipped with a gear-driven oil slinger while horizontal crankshaft models use an oil dipper attached to number one cylinder connecting rod cap.

Early models are equipped with a flywheel magneto ignition with breaker points, condenser and coil mounted externally on engine. Later models are equipped with a Magnetron breakerless ignition.

All models use a float type carburetor with an integral fuel pump.

Always provide engine model and serial number when ordering parts or special tools.

MAINTENANCE

SPARK PLUG. Recommended spark plug for all models is Champion RJ12. Electrode gap is 0.030 inch (0.76 mm).

CARBURETOR. A downdraft float type carburetor is used. The carburetor is equipped with an integral diaphragm type fuel pump. Refer to Fig. B70 for an exploded view of carburetor and fuel pump.

Initial setting of idle mixture screw (11—Fig. B70) for all models is 1½ turns out. Initial setting of the high speed mixture screw (10), if equipped, is 1½ turns out.

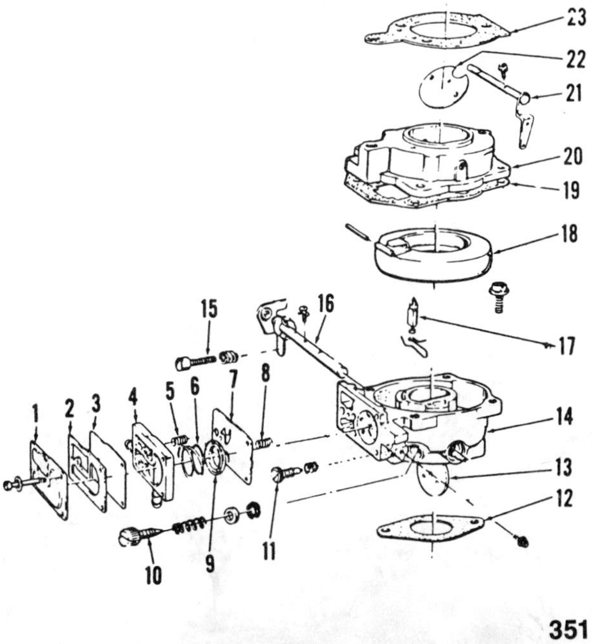

Fig. B70—Exploded view of downdraft Flo-Jet with integral fuel pump.

1. Diaphragm cover
2. Gasket
3. Damping diaphragm
4. Pump body
5. Spring
6. Pump spring
7. Diaphragm
8. Spring
9. Spring cap
10. Main fuel valve assy.
11. Idle valve assy.
12. Mounting gasket
13. Throttle plate
14. Lower carburetor body
15. Throttle adjustment screw
16. Throttle shaft
17. Fuel inlet valve
18. Float
19. Carburetor body gasket
20. Upper carburetor body
21. Choke shaft
22. Choke plate
23. Air cleaner gasket

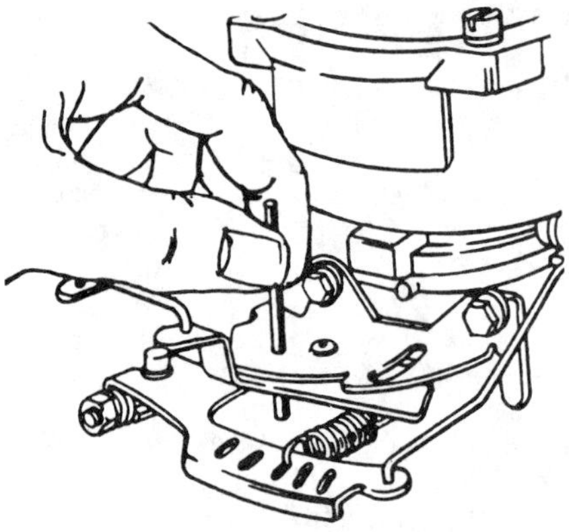

Fig. B71—Insert a 1/8-inch (3.18 mm) diameter pin through the two holes in governor control plate to correctly set governed idle position.

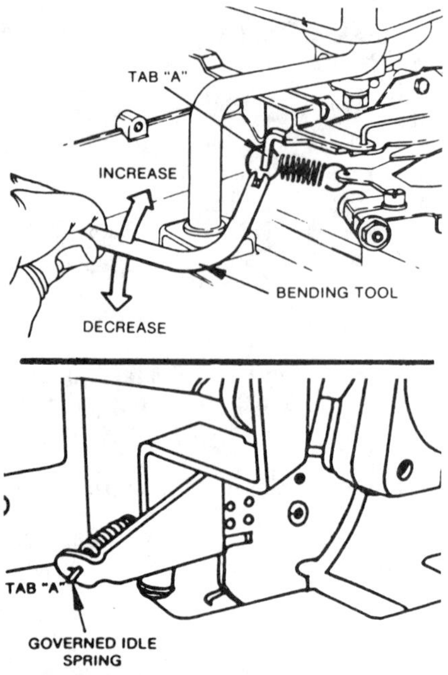

Fig. B72—With governor plate locked with a 1/8-inch (3.18 mm) pin (Fig. B71), bend tab "A" to obtain 1200 rpm. Upper view is for a horizontal crankshaft engine and lower view is for vertical crankshaft engines.

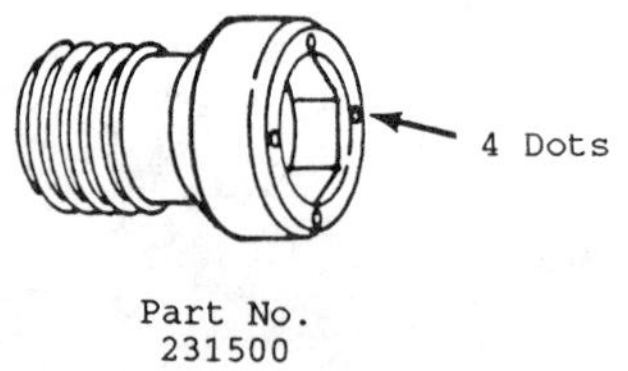

Fig. B73—Optional high speed jet to provide richer fuel-air mixture is identified by four dots on the jet face.

MODELS PRIOR TO TYPE 1100. These carburetors can be distinguished from later production carburetors by the absence of a Welch plug above the idle mixture screw.

Run engine until operating temperature is reached. To adjust idle mixture screw, place remote speed control in idle position. Hold carburetor throttle lever against idle speed adjusting screw and adjust idle speed to 1200 rpm. With throttle lever against idle speed adjusting screw, turn idle mixture screw clockwise until a reduction in engine speed is noted. Back out idle mixture screw 1/2 turn. With throttle lever against idle speed screw, readjust idle speed to 900 rpm. Release throttle lever. Move remote speed control to a position where a 1/8-inch diameter pin can be inserted through two holes in governor control plate (Fig. B71). With remote control in governed idle position, bend tab "A," Fig. B72, to obtain 1200 rpm. Remove pin.

If equipped with a high speed mixture screw, place remote speed control in fast position and adjust high speed mixture screw for leanest setting that will allow satisfactory acceleration and steady governor operation. Note: Some early models are equipped with a fixed main jet needle that appears the same as the adjustable mixture screw and is located in the same position (10—Fig. B70). The fixed main jet needle is identified by a small hole in the tip. The fixed main jet needle must be seated fully in the carburetor; backing out main jet needle will result in an overly rich mixture.

Some models are equipped with a fixed high speed jet. The jet is located behind a plug in the same location as the high-speed mixture adjusting screw (10) of other models. The fixed main jet is not adjustable. If throttle response is poor or if engine hesitates when it is accelerated or placed under load due to a lean fuel mixture, a richer jet may be required. The optional richer main jet is identified by four dots on the jet face (Fig. B73). Optional jet is also available for high altitude operation.

MODELS TYPE 1100 AND ABOVE. These carburetors can be distinguished from previous carburetors by the Welch plug located above the idle mixture screw.

Run engine until engine operating temperature is reached. To adjust idle mixture screw, place remote speed control in idle position. Hold carburetor throttle lever against idle speed adjusting screw and adjust idle speed to 1200 rpm. With throttle lever against idle speed adjusting screw, turn idle mixture screw clockwise until a reduction in engine speed is noted and mark screw position. Back out idle mixture screw until engine speed lessens again and mark screw position. Rotate screw so it is halfway between the two marked positions. With throttle lever against idle speed screw, readjust idle speed to 900 rpm. Release throttle lever. With remote control in governed idle position, bend tang (Fig. B74) to obtain 1200 rpm.

If equipped with a high speed mixture screw, place remote speed control in fast position and adjust high speed mixture screw for leanest setting that will allow satisfactory acceleration and steady governor operation. The fixed main jet is not adjustable. If throttle response is poor or if engine hesitates when it is accelerated or placed under load due to a lean fuel mixture, a richer jet may be required. The optional richer main jet is identified by four dots on the jet face (Fig. B73). Optional jet is also available for high altitude operation.

OVERHAUL

To disassemble carburetor, remove idle and main fuel mixture screws. Remove anti-afterfire solenoid from carburetor, if so equipped. Remove fuel pump body (4—Fig. B70) and upper carburetor body (20). Remove float assembly (18) and fuel inlet valve (17). Inlet valve seat is a press fit in upper carburetor body. Use a self-threading screw to remove seat. New seat should be pressed into upper body until flush with body. Remove retaining screws from throttle plate (13) and choke plate (22) and withdraw throttle shaft (16) and choke shaft (21).

If necessary to renew throttle shaft bushings, use a 1/4 × 20 tap to remove old bushings.

NOTE: If carburetor body has a plug in one side of throttle shaft bore, bushing on that side is not renewable and plug should not be removed.

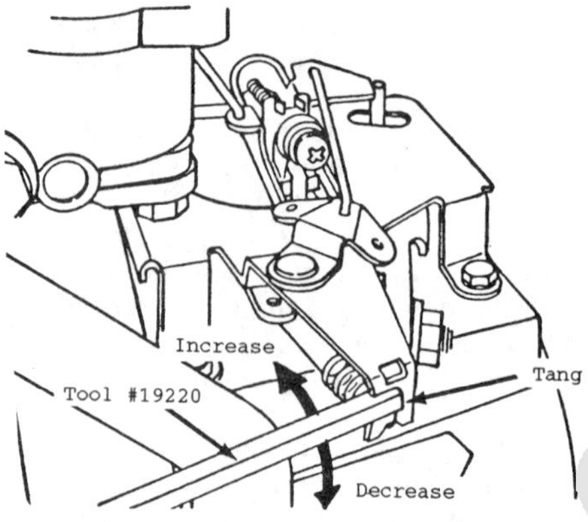

Fig. B74—With remote control in governed idle position, bend governor lever tang to obtain 1200 rpm.

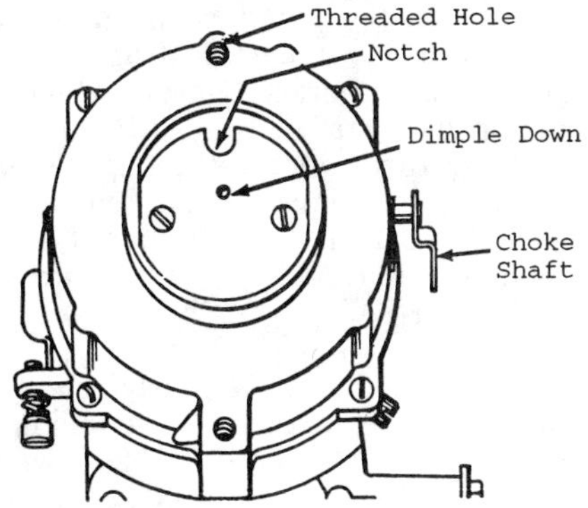

Fig. B75—Install the choke plate so the notch is toward the threaded hole.

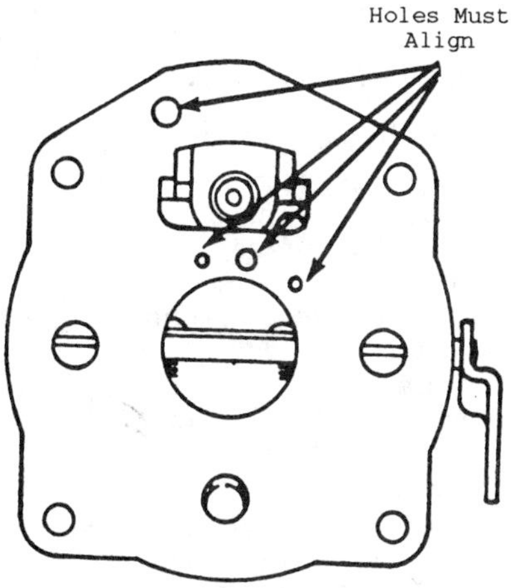

Fig. B76—Holes in body gasket of late models must be properly aligned with holes in upper carburetor body as shown.

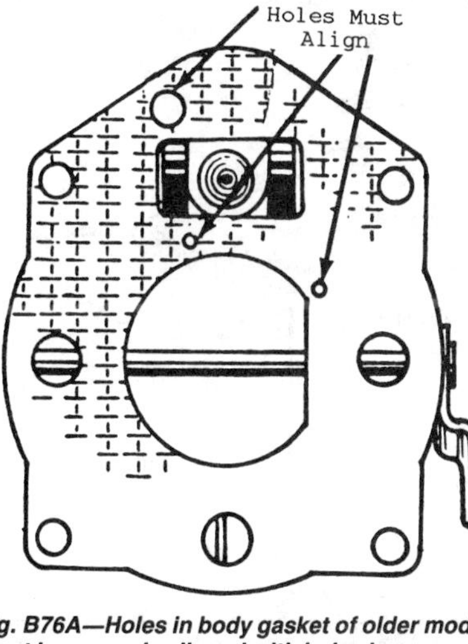

Fig. B76A—Holes in body gasket of older models must be properly aligned with holes in upper carburetor body as shown.

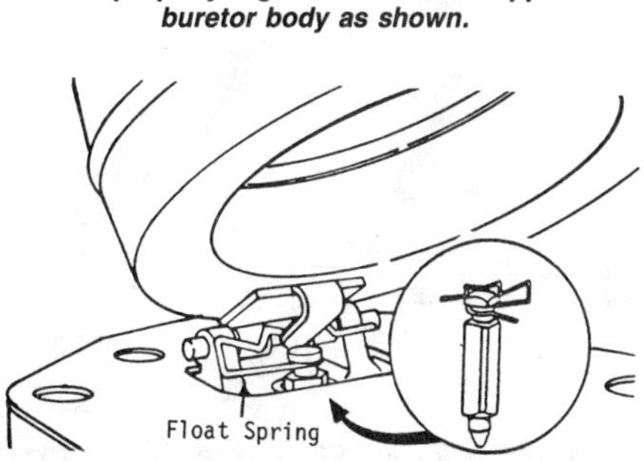

Fig. B77—Install fuel inlet valve clip as shown.

Press new bushings in using a vise, then ream with B&S tool 19056. Align-ream bushings with a 7/32-inch drill if throttle shaft binds.

When assembling the carburetor, note the following: Install the choke plate so the notch is toward the threaded hole in upper body as shown in Fig. B75. Be sure holes in body gasket are properly aligned with holes in upper carburetor body as shown in Fig. B76 or Fig. B76A. Note proper installation of fuel inlet valve clip in Fig. B77. To check float level, invert carburetor body and float assembly. Float should be parallel to body as shown in Fig. B78. Adjust float by bending float lever tang that contacts inlet valve. Install throttle plate so that, with throttle closed, dimples will be on opposite side of shaft from idle port, and convex sides of dimples are facing outward (see Fig. B79). When installing fixed jet plug or anti-afterfire solenoid, tighten plug or solenoid to 100 in.-lbs. (11.3 N·m). Be sure springs (5 and 8—Fig. B70) fit onto spring bosses shown in Fig. B80.

ANTI-AFTERFIRE SYSTEM. Some models are equipped with an anti-afterfire system that stops fuel flow through the carburetor when the ignition switch is in off position. Two different systems have been used. Early systems use a vacuum solenoid that blocks vacuum to the fuel bowl thereby stopping fuel flow through fuel passages. Later systems are found on carburetors with a fixed main fuel jet. A solenoid inserts a plunger into the jet to stop fuel flow.

A view of the vacuum type anti-afterfire system is shown in Fig. B81. The solenoid is controlled electrically by the ignition switch. The solenoid blocks the fuel bowl vent circuit and allows crankcase vacuum to enter the circuit when the ignition switch is off. Solenoid should "click" if operating properly. All hoses must be clamped tight, without kinks and undamaged. Vacuum block in air cleaner must fit properly.

Refer to Fig. B82 for drawing of anti-afterfire system used on carburetor with fixed jet. Solenoid may be removed and tested by connecting a 9-volt battery to solenoid. Solenoid plunger should retract when energized and should extend freely when battery is

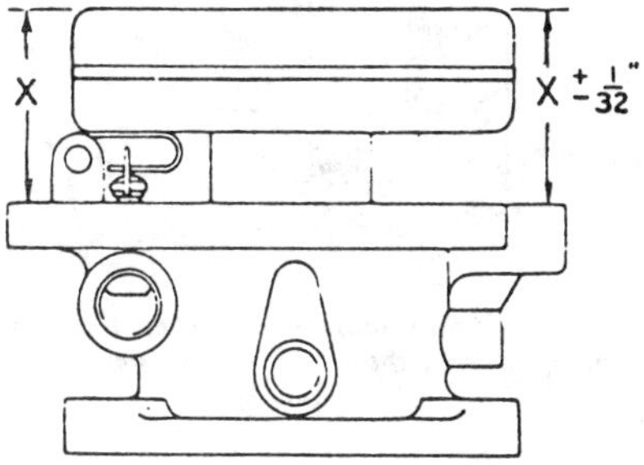

Fig. B78—Check carburetor float setting as shown. Bend tang, if necessary, to adjust float level.

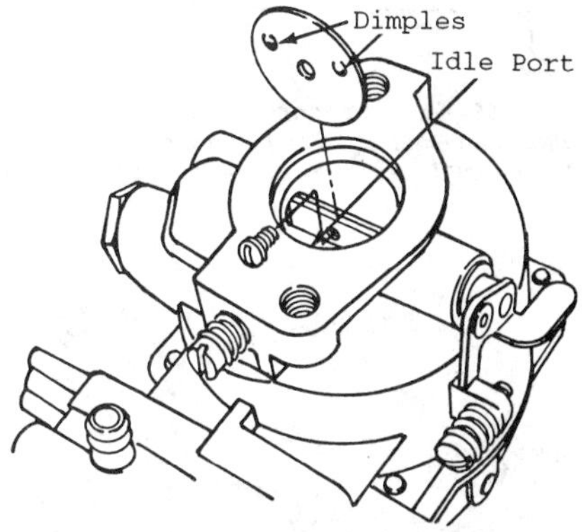

Fig. B79—Install throttle plate so that with throttle closed dimples will be on opposite side of shaft from idle port, and convex sides of dimples are toward carburetor base surface.

Fig. B80—Fuel pump check valve springs must fit onto spring bosses.

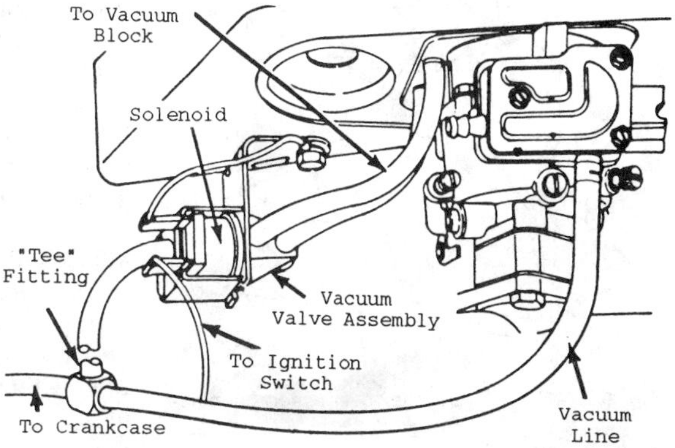

Fig. B81—A view of the vacuum type anti-afterfire system.

disconnected. A faulty solenoid will cause hard starting and low power output.

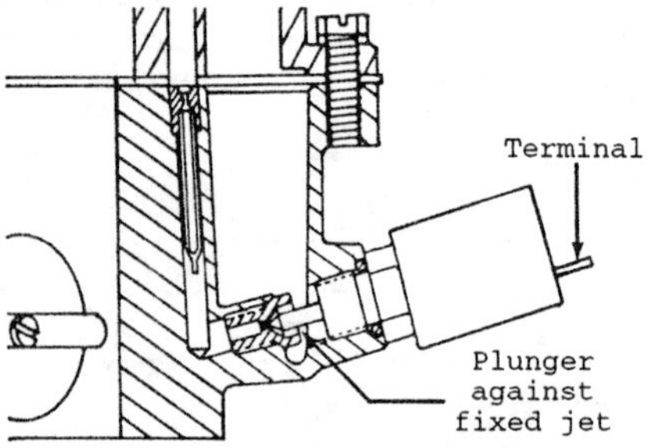
Terminal
Plunger against fixed jet

Fig. B82—Carburetors with fixed main jet may be equipped with the anti-afterfire system shown above.

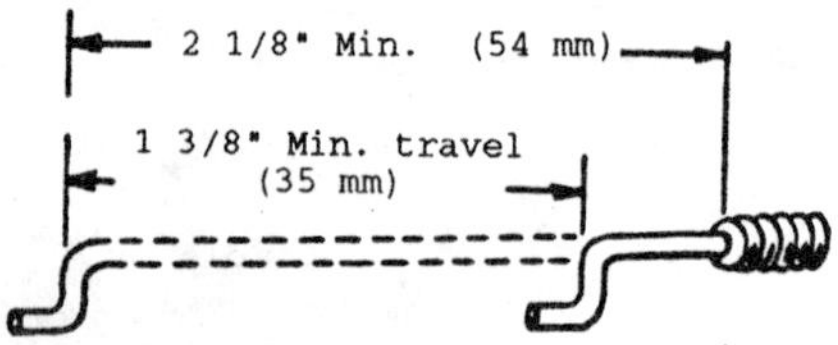
2 1/8" Min. (54 mm)
1 3/8" Min. travel (35 mm)

Fig. B83—For proper operation, remote control wire must extend to dimension shown above and have a minimum travel of $1^3/_8$ inches (34.9 mm).

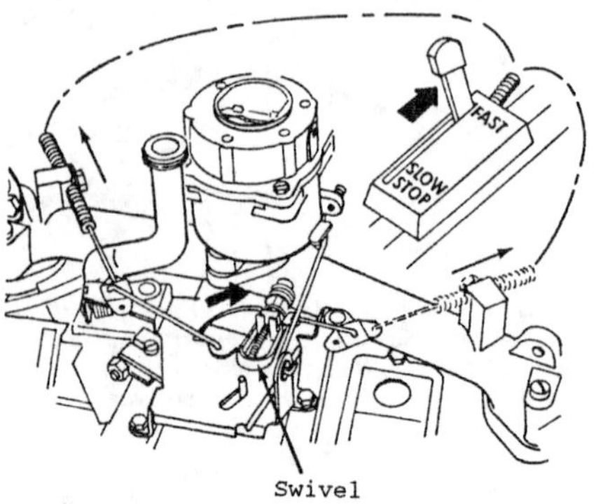
FAST
SLOW STOP
Swivel

Fig. B84—On engines with type number 1100 and above, swivel should be against side of quarter circle when remote control lever is in FAST position.

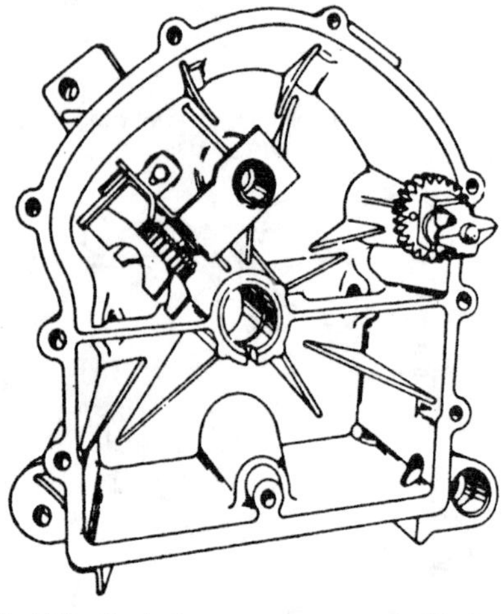

Fig. B85—Typical governor gear assembly position in crankcase.

FUEL PUMP. All parts of the vacuum-diaphragm type pump are serviced separately. When disassembling pump, care must be taken to prevent damage to pump body (plastic housing) and diaphragm. Inspect diaphragm for punctures, wrinkles or wear. All mounting surfaces must be free of nicks, burrs and debris.

To assemble pump, position diaphragm on carburetor. Place spring and cup on top of diaphragm. Install flapper valve springs. Carefully place pump body, remaining diaphragm, gasket and cover plate over carburetor casting and install mounting screws. Tighten screws in a staggered sequence to avoid distortion.

CARBURETOR CONTROL MECHANISM. To assure proper speed control, measure travel of remote control wire with remote control unit installed. Minimum wire travel is $1^3/_8$ inches as shown in Fig. B83.

To adjust speed control cable on engines with a type number below 1100, move remote control lever to idle position. Carburetor throttle lever should contact idle speed screw. If not, loosen cable housing clamp and reposition cable housing.

To adjust speed control cable on engines with type number 1100 and above, move remote control lever to FAST position. Swivel shown in Fig. B84 should be against side of quarter circle. If not, loosen cable housing clamp and reposition cable housing.

Remote choke control should completely close carburetor choke plate when remote control is in "CHOKE" position. If necessary, loosen cable clamp and reposition cable to synchronize carburetor choke and remote control.

GOVERNOR. All models are equipped with a gear-driven mechanical governor. Governor gear and weight assembly is enclosed within the engine and is driven by the camshaft gear. Refer to Fig. B85.

To adjust governor, loosen nut securing governor lever to governor shaft. Push governor lever counterclockwise until throttle is wide open. Hold lever in this position while rotating governor shaft counterclockwise as far as it will go. Tighten governor lever nut to 100 in.-lbs. (11.3 N·m).

To adjust top governed speed on engines with a type number below 1100, first adjust carburetor and governed low idle speed as outlined in CARBURETOR section. Install governor spring end loop in appropriate hole in governor lever for desired engine rpm as shown in Fig. B86. Check engine top governed speed using an accurate tachometer.

To adjust top governed speed on engines with type number 1100 and above, first adjust carburetor and governed low idle speed as outlined in CARBURETOR section. Turn top governed speed adjusting screw shown in Fig. B87 until desired engine rpm is obtained. Check engine rpm using an accurate tachometer.

IGNITION SYSTEM. Early production engines were equipped with a flywheel type magneto ignition with points, condenser and coil located externally on engine. Late production engines are equipped with a "Magnetron" breakerless ignition system.

Refer to appropriate following paragraph for model being serviced.

FLYWHEEL MAGNETO IGNITION. The flywheel magneto system consists of a permanent magnet cast into flywheel, armature and coil assembly, breaker points and condenser.

Breaker points and condenser are located under or behind intake manifold and are protected by a metal cover which must be sealed around edges and at wire entry location to prevent entry of dirt or moisture.

Breaker point gap should be 0.020 inch (0.50 mm) for all models.

Breaker points are actuated by a plunger (Fig. B88) which is installed with the smaller diameter end toward breaker points. Renew plunger if length

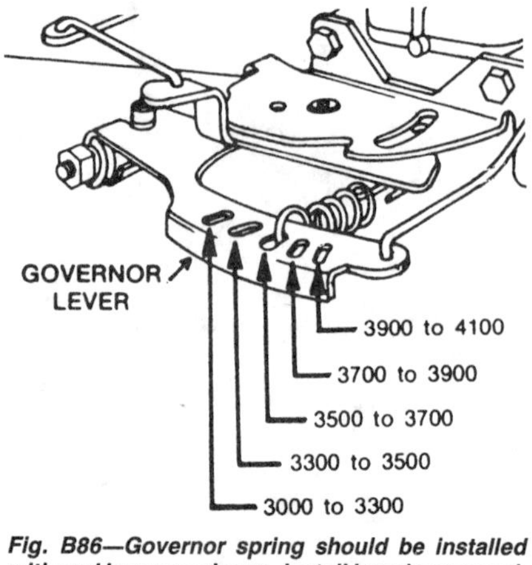
GOVERNOR LEVER
3900 to 4100
3700 to 3900
3500 to 3700
3300 to 3500
3000 to 3300

Fig. B86—Governor spring should be installed with end loops as shown. Install loop in appropriate governor lever hole for engine speed (rpm) desired.

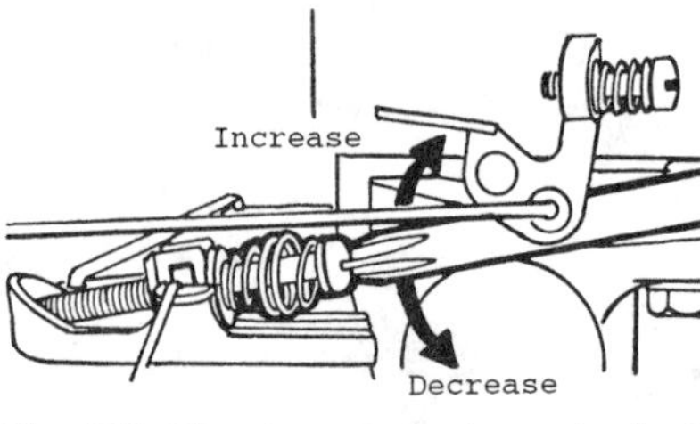
Increase
Decrease

Fig. B87—Turn top governed speed adjusting screw to obtain desired engine rpm.

is 1.115 inch (28.32 mm) or less. Renew plunger seal by installing seal on plunger (make certain it is securely attached) and installing seal and plunger assembly into plunger bore. Slide seal over plunger bore. Slide seal over plunger boss until seated against casting at base of boss.

Armature air gap should be 0.010-0.014 inch (0.25-0.36 mm) and is adjusted by loosening armature retaining bolts and moving armature as necessary on slotted holes. Tighten armature retaining bolts.

MAGNETRON IGNITION. "Magnetron" ignition consists of permanent magnets cast in flywheel and a self-contained transistor module mounted on the ignition armature.

To check ignition, attach B&S tester number 19051 to each spark plug lead and ground tester to engine. Spin flywheel at least 350 rpm. If spark jumps the 0.166 inch (4.2 mm) tester gap, ignition system is operating satisfactorily.

Armature air gap should be 0.008-0.012 inch (0.20-0.30 mm) and is adjusted by loosening armature retaining bolts and moving armature as necessary on slotted holes. Tighten armature bolts.

Flywheel does not need to be removed to service Magnetron ignition except to check condition of flywheel key or keyway.

To remove Magnetron module from armature, remove stop switch wire, module primary wire and armature primary wire from module by using a $^{3}/_{16}$-inch (5 mm) rod to release spring and retainer (Fig. B89). Remove spring and retainer clip. Unsolder wires. Remove module by pulling out on module retainer while pushing down on module until free of armature laminations.

During installation, use 60/40 rosin core solder and make certain all wires are held firmly against coil body with tape, Permatex No. 2 or equivalent gasket sealer.

LUBRICATION. Models 402440, 402770, 422440 and 422770 are equipped with a pressure lubrication system and an oil filter. Oil is routed from an oil pump driven by the camshaft to the crankshaft. All other models are splash lubricated using a gear-driven oil slinger (5—Fig. B94) on vertical crankshaft models or an oil dipper attached to number one connecting rod (27—Fig. B94A) on horizontal crankshaft models.

Manufacturer recommends using oil with an API service classification of SE or SF. SAE 10W-30 oil may be used for temperatures between 0° F (−18° C) and 100° F (38° C). SAE 30 oil may be used when temperatures will be above 40° F (4° C). When temperature will be below 20° F (−7° C), use petroleum based SAE 5W-20 or a suitable synthetic oil.

Check oil at regular intervals and maintain at "FULL" mark on dipstick. Dipstick should be pushed or screwed in completely for accurate measurement. DO NOT overfill.

Change oil after first 5 hours of operation. Thereafter, recommended oil change interval is every 25 hours on splash lubricated models, or every 50 hours on Models 402440, 402770, 422440 and 422770. On all models, oil change interval should be more frequent if usage is severe.

Crankcase oil capacity for early production engines is 3.5 pints (1.65 L). Oil capacity for late production engines is 3 pints (1.42 L). Check oil level with dipstick.

On Models 402440, 402770, 422440 and 422770 a low oil pressure switch may be located on the oil filter adapter. Switch should be closed at zero pressure and open at 5 psi (34.5 kPa). Switch may be connected to a warning device or into the ignition circuit.

Splash lubricated models may be equipped with a low oil level warning system called Oil Gard. A sensor is screwed into the base of the crankcase and monitors oil level. If the oil level falls below a certain level, the sensor activates a warning device or disables the ignition circuit, depending on the equipment. To check sensor, disconnect sensor lead, then connect ohmmeter leads to sensor lead and crankcase. Ohmmeter should indicate 50-80 ohms. If not, check condition of connections and sensor wire, and if found satisfactory, renew sensor.

CRANKCASE BREATHER. Crankcase breathers are built into engine valve covers. Horizontal crankshaft models have breather valve in each cover assembly. Vertical crankshaft models have only one breather in cover of number one cylinder.

Breathers maintain a partial vacuum in crankcase to prevent oil from being forced out past oil seals and gaskets or past breaker point plunger (early models) or piston rings.

Fiber disc of breather assembly must not be stuck or binding. A 0.045 inch (1.14 mm) wire gage SHOULD NOT enter space between fiber disc valve and body. Check with gage at 90 degree intervals around fiber disc.

When installing breathers make certain side of gasket with notches is toward crankshaft.

REPAIRS

CYLINDER HEADS. Combustion deposits should be removed from piston and cylinder head at 100 to 300 hour intervals, or whenever cylinder head is removed.

When removing cylinder heads, note locations from which different length bolts are removed as they must be reinstalled in their original positions. Also note that cylinder heads and cylinders are numbered and heads must be reinstalled on corresponding cylinders. See Fig. B90.

Always use a new gasket when reinstalling a cylinder head. Do not use sealer on gasket. Lubricate cylinder head

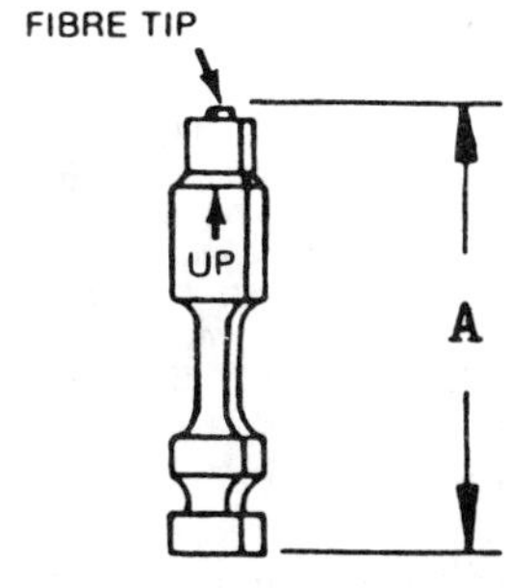

Fig. B88—Plunger must be renewed if plunger length (A) is worn to 1.115 inch (28.32 mm) or less.

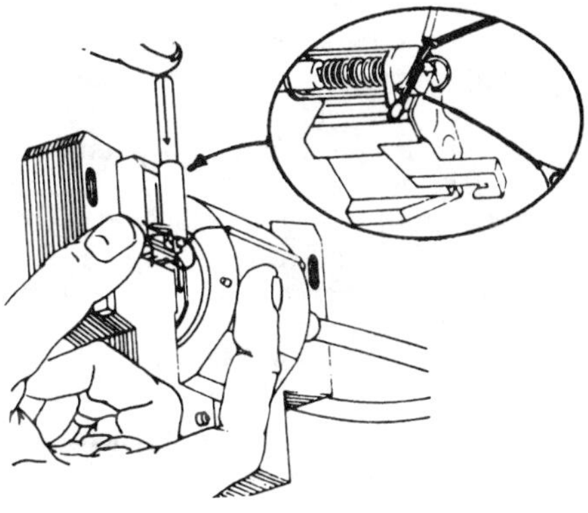

Fig. B89—Use a $^{3}/_{16}$-inch (5 mm) diameter rod to release wires from Magnetron module. Refer to text.

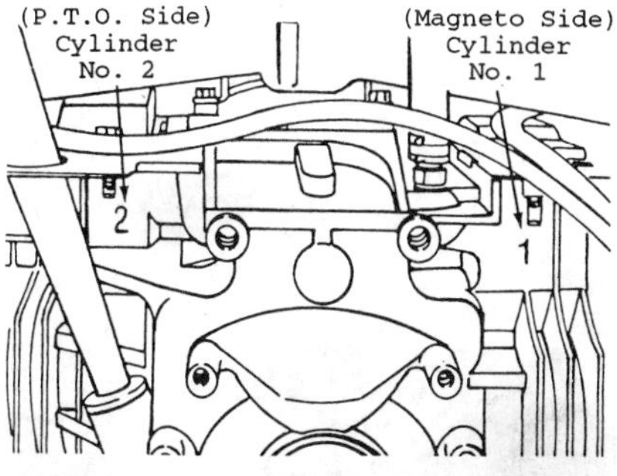

Fig. B90—Cylinder numbers are cast on the tappet chambers on flywheel side of engine. Cylinder heads are also identified (Fig. B91) and must be installed on the corresponding cylinders.

bolt threads with graphite grease. Install bolts in correct locations and tighten evenly in several steps following sequence shown in Fig. B91 to final torque of 160 in.-lbs. (18 N·m).

CONNECTING RODS. Connecting rods and pistons are removed from cylinder head end of block as an assembly after first removing cylinder head, crankcase cover and connecting rod caps. Be sure to remove carbon or ring ridge (if present) from top of cylinder prior to pushing connecting rod and piston out of cylinder. Identify pistons and connecting rods so they can be reinstalled in same cylinder from which they are removed.

Aluminum alloy connecting rods ride directly on crankpin journals. Connecting rod should be renewed if crankpin bearing bore measures 1.627 inches (41.33 mm) or more, if bearing surfaces are scored or damaged, or if pin bore measures 0.802 inch (20.37 mm) or more. A 0.005 inch (0.13 mm) oversize piston pin is available.

NOTE: Connecting rods of Models 402440, 402770, 422440 and 422770 (pressure lubricated) are not interchangeable with other models (splash lubricated).

Connecting rod should be installed on piston so oil hole in connecting rod is toward cam gear side of engine when notch on top of piston is toward flywheel with piston and rod assembly installed. See Fig. B92. Make certain match marks on connecting rod and cap are aligned and if so equipped, install oil dipper on number one rod. Install special washers and nuts and tighten to 190 in.-lbs. (22 N·m) for all models.

NOTE: Some horizontal crankshaft models that operate at slow speeds and under high load are equipped with an oil dipper attached to number two connecting rod as well as number one rod.

PISTONS, PINS AND RINGS. Pistons used in engines with cast iron cylinder liners (Series 400400, 400700, 402400, 402440, 402700, 402770, 422400, 422440, 422700 and 422770) have a shiny finish, and on early models, were marked with an "L" on piston crown. A chrome plated aluminum piston is used in aluminum bore (Kool-Bore) cylinders. Due to the two different cylinder bore materials, pistons and piston rings WILL NOT interchange.

Pistons for all models should be renewed if they are scored or damaged, or if a 0.009 inch (0.23 mm) feeler gage can be inserted between a new top ring and ring groove. Pistons and rings are available in several oversizes as well as standard bore size.

If piston pin bore measures 0.801 inch (20.35 mm) or more, piston should be renewed or pin bore reamed for 0.005 inch (0.13 mm) oversize pin. Standard piston pin diameter is 0.799 inch (20.30 mm) and should be renewed if worn or out-of-round more than 0.0005 inch (0.013 mm). A 0.005 inch (0.13 mm) oversize pin is available for all models. Piston pin is a hand push fit in piston pin bore. Heating piston in hot water or oil will make removal and installation of pin easier.

Refer to the following table for piston ring end gap rejection sizes:

Model Series	Compression Ring	Oil Ring
401400, 401700 421400, 421700	0.035 in. (0.889 mm)	0.045 in. (1.14 mm)
All Other Models	0.030 in. (0.762 mm)	0.035 in. (0.889 mm)

Install piston rings so end gaps are staggered around piston. Refer to Fig. B93 for correct installation of piston rings. Lubricate pistons and rings with oil and install pistons so notch in piston crown is toward flywheel side of engine as shown in Fig. B92.

CYLINDERS. Cylinder bores may be either aluminum or a cast iron liner that is an integral part of the cylinder block casting. Pistons and piston rings are not interchangeable between the two types. Series 400400, 400700, 402400, 402440, 402700, 402770, 422400, 422440, 422700 and 422770 have cast iron cylinder liners as an integral part of cylinder block casting. Series 401400, 401700, 421400 and 421700 have aluminum cylinder bores (Kool-Bore).

Standard cylinder bore diameter for all models is 3.4365-3.4375 inches (87.29-87.31 mm). Cylinder should be resized if more than 0.003 inch (0.076 mm) oversize or 0.0015 inch (0.038 mm) out-of-round for cast iron bore, or 0.003 inch (0.076 mm) oversize or 0.0025 inch (0.0635 mm) out-of-round for aluminum bore. Use a suitable hone such as B&S tool 19205 for aluminum bore or B&S tool 19211 for cast iron bore. Resize to nearest oversize for which piston and rings are available.

CRANKSHAFT. Crankshaft and camshaft used on pressure lubricated en-

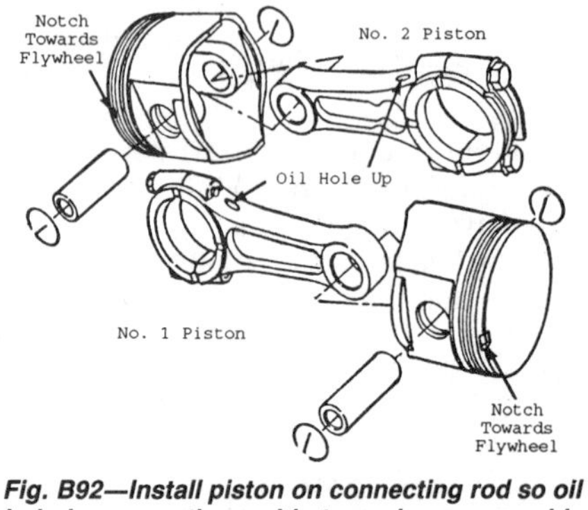

Fig. B92—Install piston on connecting rod so oil hole in connecting rod is toward cam gear side of engine when notch on top of piston is toward flywheel.

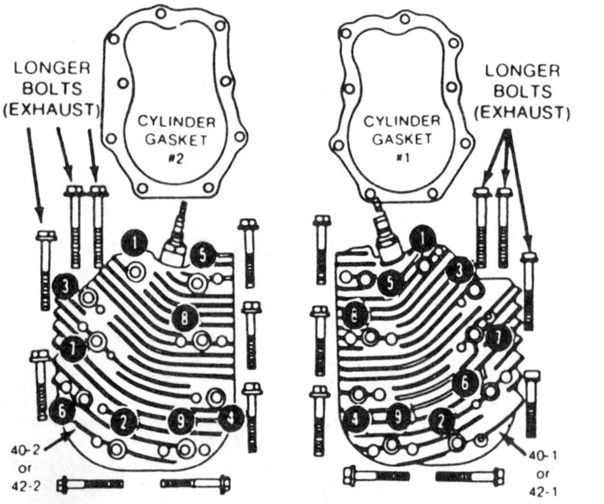

Fig. B91—Note locations of various length cylinder head bolts. Long bolts are used around exhaust valve area. Bolts should be tightened in sequence shown to 160 in.-lbs. (18 N·m) torque.

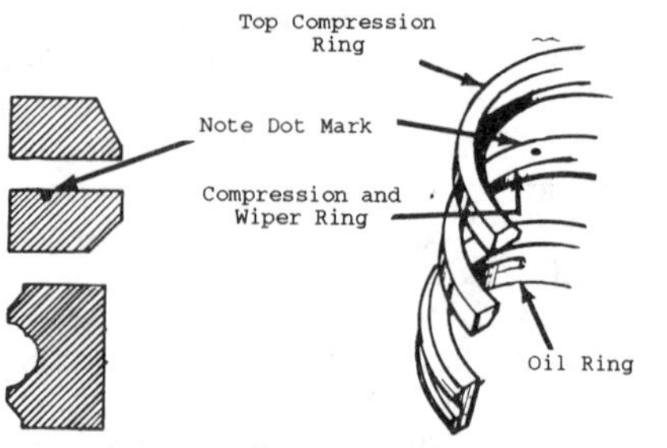

Fig. B93—Drawing of piston rings showing correct installation. Be sure that dot mark on second ring faces piston crown.

Illustrations courtesy Briggs & Stratton.

gines are dissimilar and not interchangeable with crankshaft and camshaft of splash lubricated engines.

To remove crankshaft from engines with integral type or DU type bearings, remove necessary air shrouds. Remove flywheel and front gear cover or sump. Remove connecting rod caps. Remove cam gear making certain valve tappets clear camshaft lobes. Remove crankshaft.

To reinstall crankshaft, reverse removal procedure making certain timing marks are aligned as shown in Fig. B95.

To remove crankshaft from engines with ball bearing main bearing, remove all necessary air shrouds and remove flywheel. Remove front gear cover or sump. Remove connecting rod caps. Compress exhaust and intake valve springs on number two cylinder to provide clearance for camshaft lobes. Remove crankshaft and camshaft together.

To reinstall crankshaft, reverse removal procedure making certain timing marks are aligned as shown in Fig. B96.

Renew crankshaft if main bearing journal at either end is 1.376 inches (34.95 mm) or less. Renew or regrind crankshaft if crankpin journal diameter is 1.622 inches (41.15 mm) or less. A 0.020 inch (0.508 mm) undersize connecting rod is available.

Crankshaft end play for all models should be 0.002-0.008 inch (0.05-0.20 mm). At least one 0.015 inch cover or sump gasket must be installed. Additional gaskets of 0.005 and 0.009 inch thickness are available if end play is less than 0.002 inch (0.05 mm). If end play is over 0.008 inch (0.20 mm), metal shims are available for installation between crankshaft gear and cover or sump. Do not use shims on engines with double ball bearings.

CRANKCASE AND MAIN BEARINGS. Crankshaft may be supported at each end in main bearings which are an integral part of crankcase, cover or sump. Crankshaft on some models may be supported by DU type bearings. Some models may use ball bearings in the crankcase, sump or cover to support the crankshaft.

Ball bearing main bearing is a press fit on crankshaft and must be removed by pressing crankshaft out of bearing. Renew ball bearing if worn or rough. Expand new bearing by heating in oil and install with shield side toward crankpin. If ball bearing is loose in crankcase, cover or sump bores, renew crankcase, cover or sump.

Integral type main bearings should be reamed out and service bushings installed if bearings are 0.0007 inch (0.018 mm) or more out-of-round or if diameter is 1.382 inches (35.10 mm) or more. Special reamers are available from Briggs & Stratton.

DU type main bearings should be renewed if bearings are 0.0007 inch (0.018 mm) or more out-of-round or if diameter is 1.382 inches (35.10 mm) or more.

Worn DU type bearings are pressed out of bores using B&S cylinder support 19227 and driver 19226. Note that pto main bearing of pressure lubricated models has an oil groove in inner diameter. Make certain oil holes in bearings align with oil holes in block and cover or sump. Press bearings in until they are 1/8 inch (3.2 mm) below thrust face, except on pto main bearing of vertical crankshaft models discussed in following paragraph. Stake bearings into place. See Fig. B97.

The pto DU bearing on vertical crankshaft models must be installed so oil hole is nearer thrust face of crankcase. See Fig. B98. Install bearing so oil hole is aligned and end of bearing is 1/32 inch (0.8 mm) from thrust face on pressure lubricated models, or 5/32 inch (4.0 mm) on splash lubricated models. Stake bearing in two places after installation.

Models 421400 and 422400 are equipped with an oil trough in the crankcase that encloses the oil dipper on number one connecting rod. Install the trough so the notch is toward the flywheel side of engine.

Tighten cover or sump screws to 225 in.-lbs. (25.4 N·m).

CAMSHAFT. Camshaft and camshaft gear are an integral part that ride on journals at each end of camshaft. Camshaft on pressure lubricated models has an oil passage that extends through the center of the camshaft. Note that camshafts cannot be interchanged between pressure lubricated and splash lubricated models.

To remove camshaft, refer to appropriate paragraph in CRANKSHAFT section for model being serviced.

Camshaft should be renewed if gear teeth or lobes are worn or damaged. Reject camshaft if either bearing journal diameter is 0.623 inch (15.82 mm) or less. Reject dimensions for cam lobes are 1.150 inch (29.21 mm) for intake and 1.120 inches (28.45 mm) for exhaust.

Be sure oil passage in camshaft on pressure lubricated models is open and

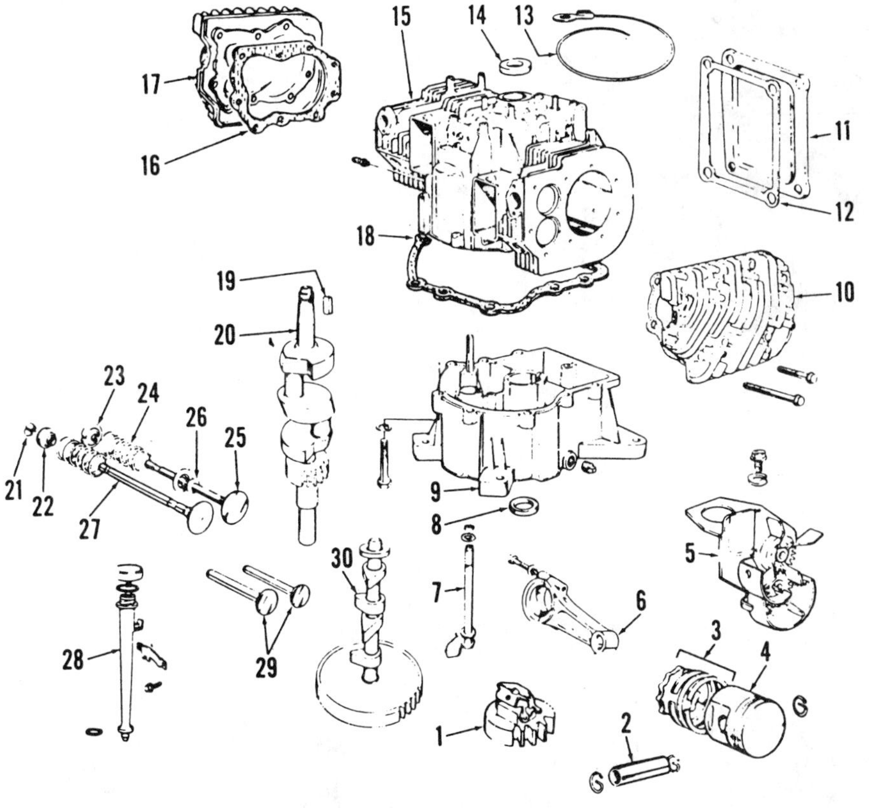

Fig. B94—Exploded view of vertical crankshaft engine assembly.

1. Governor gear
2. Piston pin & clips
3. Piston rings
4. Piston
5. Oil slinger assy.
6. Connecting rod
7. Governor shaft assy.
8. Oil seal
9. Crankcase assy.
10. Cylinder head
11. Crankcase cover plate
12. Crankcase gasket
13. Ground wire
14. Oil seal
15. Cylinder assy.
16. Head gasket
17. Cylinder head
18. Crankcase gasket
19. Key
20. Crankshaft
21. Split retainer
22. Rotocoil (exhaust valve)
23. Retainer (intake valve)
24. Valve spring (2)
25. Intake valve
26. Seal & retainer assy.
27. Exhaust valve
28. Oil dipstick assy.
29. Valve tappet (2)
30. Camshaft assy.

clean. When reinstalling camshaft, be sure that timing marks are aligned as shown in Fig. B95 or Fig B96.

GOVERNOR. To remove governor shaft, loosen nut and remove governor lever. Remove crankcase cover or sump. To obtain maximum clearance, rotate crankshaft until timing mark on crankshaft gear is at approximately 10 o'clock position. If equipped with ball bearings, rotate crankshaft to obtain clearance with crankshaft counterweight. On early models (see Fig. B99), remove "E" ring and thick washer on outer end of governor shaft and slide shaft out. On models equipped with two "E" rings (see Fig. B100), detach the two rings and slide out shaft. On models with type number 1100 and above (shown in Fig. B101), detach "E" ring and slide out shaft. On all models it may be necessary to rotate crankshaft to gain clearance for governor shaft removal. DO NOT force shaft against crankshaft when removing.

Later models (Figs. B100 and B101) use a renewable governor shaft bushing. Use suitable tools for removal and installation of bushing. Drive or pull bushing out toward outside of crankcase. Bushing should bottom against shoulder in bore during installation.

After installation, rotate governor shaft so governor arm is against cylinder rib.

Governor gear assembly is available only as an unit assembly. Be sure assembly rotates and operates freely without binding. Note that there is a thrust washer located between gear assembly and boss. Refer to MAINTENANCE section for governor adjustment.

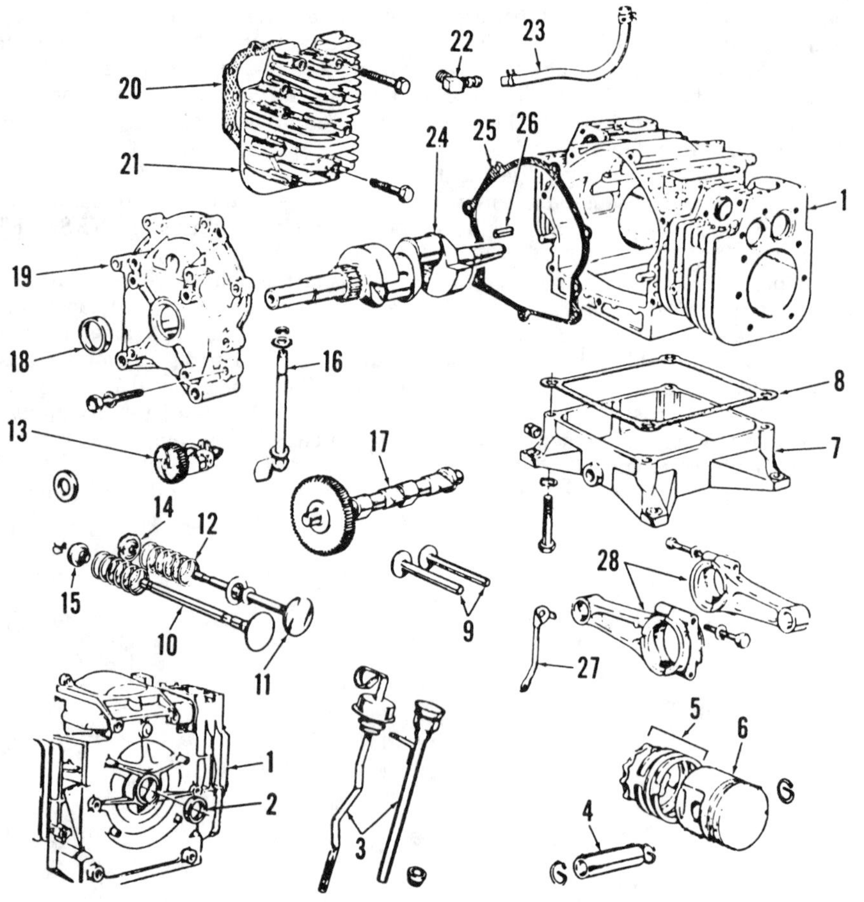

Fig. B94A—Exploded view of horizontal crankshaft engine assembly.

1. Cylinder assy.
2. Oil seal
3. Dipstick assy.
4. Piston pin & retainer clips
5. Piston rings
6. Piston
7. Crankcase
8. Crankcase gasket
9. Valve tappets
10. Exhaust valve
11. Intake valve
12. Valve spring
13. Governor gear assy.
14. Intake valve retainer
15. Rotocoil (exhaust valve)
16. Governor shaft
17. Camshaft
18. Oil seal
19. Crankcase cover
20. Head gasket
21. Cylinder head
22. Elbow connector
23. Fuel line
24. Crankshaft
25. Gasket
26. Key
27. Oil dipper
28. Connecting rods

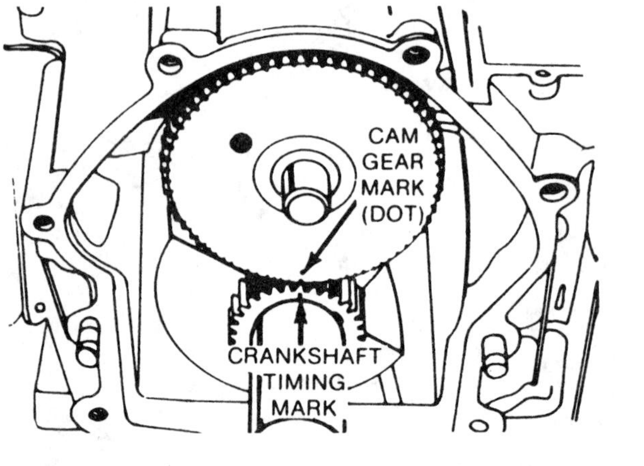

Fig. B95—Align timing marks as shown on models having main bearings as an integral part of crankcase, cover or sump. Refer to text.

VALVE SYSTEM. Valves seat in renewable inserts pressed into cylinder head surfaces of block. Valve seat width should be $^3/_{64}$ to $^1/_{16}$ inch (1.17 to 1.57 mm). Seat angle is 45 degrees for exhaust seat and 30 degrees for intake seat. If seats are loose, check clearance between seat and cylinder block counterbore using a 0.005 inch (0.13 mm) feeler gage. If feeler gage cannot be inserted between seat and cylinder block, peen seat as shown in Fig. B102. If feeler gage can be inserted between seat and cylinder block, renew seat. Use a suitable puller such as B&S tool No. 19138 to remove valve seat inert. Drive new seat into cylinder block until it bottoms, peen seat as shown in Fig. B102 and grind to correct angle.

Valves should be refaced at 45 degree angle for exhaust valves and at 30 degree angle for intake valves. Valves should be rejected if margin is less than $^1/_{64}$ inch (0.10 mm). See Fig. B103.

Valve guides should be checked for wear using valve guide gage (B&S tool 19151). If gage enters guide $^5/_{16}$ inch (7.9 mm) or more, guide should be reconditioned or renewed.

To recondition aluminum valve guides, use B&S tool kit 19232. Place reamer 19231 and guide 19234 in worn guide and center with valve seat. Mark reamer $^1/_{16}$ inch (1.57 mm) above top edge of service bushing (Fig. B104). Ream worn guide until mark on reamer is flush with top of guide bushing. DO NOT ream completely through guide. Place service bushing, part 231218, on driver 19204 so grooved end of bushing will enter guide bore first. Press bushing into guide until it bottoms. Finish ream completely through guide using reamer 19233 and guide 19234. Lubricate reamer with kerosene during reaming procedure.

To renew brass or sintered iron guides, use tap 19264 to thread worn guide bushing approximately ½ inch (12.7 mm) deep. DO NOT thread more than 1 inch (25.4 mm) deep. Install puller washer 19240 on puller screw 19238 and thread screw and washer assembly into worn guide. Center washer on valve seat and tighten puller nut 19239 against washer, continue to tighten while keeping threaded screw from turning (Fig. B104) until guide has been removed. Identify guide using Fig. B105 and find appropriate replacement. Place correct service bushing on driver 19204 so the two grooves on service bushings 231218 are down. Remaining bushing types can be installed either way. Press bushing in until it bottoms. Finish ream with reamer 19233 and reamer guide 19234. Lubricate reamer with kerosene and ream completely through new service bushing.

When reinstalling valves, note exhaust valve spring is shorter, has heavier diameter coils and is usually painted red. Intake valve has a stem seal, which should be renewed if valve has been removed.

To adjust valve clearance on all models, position piston of cylinder to be adjusted at top dead center on compression stroke. If valve springs are installed, stem end clearance should be 0.007-0.009 inch (0.18-0.23 mm) for exhaust valves and 0.004-0.006 inch (0.10-0.15 mm) for intake valves. Without valve springs, stem end clearance should be 0.009-0.011 inch (0.23-0.28 mm) for exhaust valves and 0.006-0.008 inch (0.15-0.23 mm) for intake valves. If clearance is less than specified, grind end of valve stem as necessary. If clearance is excessive, grind valve seat deeper or renew valve as necessary.

OIL PUMP. Pressure lubricated engines are equipped with rotor type oil pump. The pump is driven by the end of the camshaft on vertical crankshaft engines (see Fig. B106) or by a separate gear that is driven by the camshaft gear on horizontal engines (see Fig. B107).

The oil pump on vertical crankshaft models is accessible by removing pump cover on bottom of sump. On horizontal crankshaft models it is necessary to remove crankcase cover for access to oil pump. Inspect pump components and renew if damaged.

The pump relief valve is located in the oil filter adapter shown in Fig. B108. Remove adapter from crankcase while being careful not to drop spring and ball. Spring free length must be 1.091-1.159 inches (27.71-29.44 mm), otherwise reject spring.

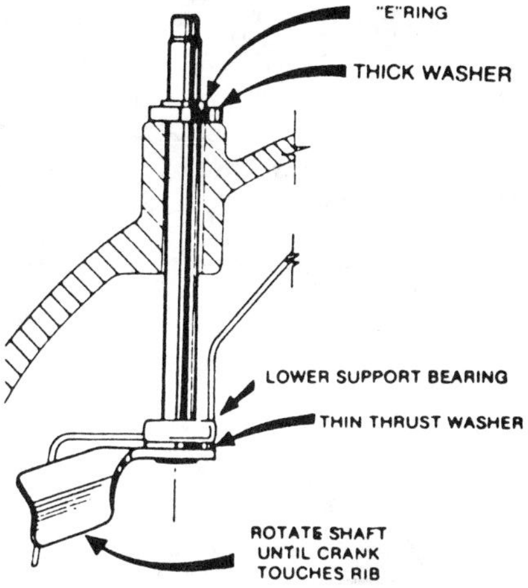

Fig. B99—Cross-sectional view of early style governor shaft assembly. To disassemble, remove "E" ring and washer. Carefully guide shaft down past crankshaft.

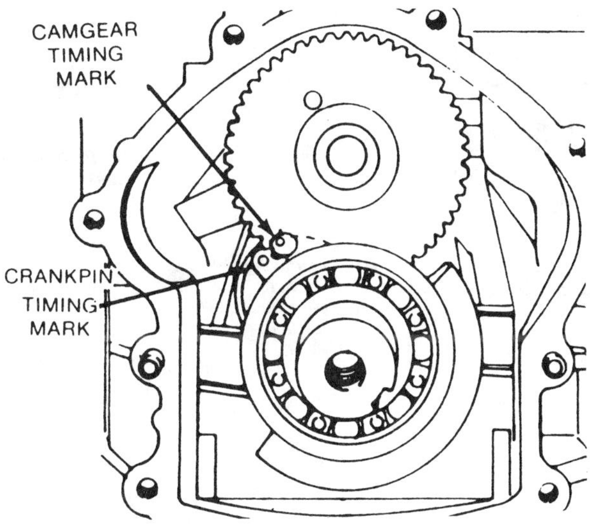

Fig. B96—Align timing marks as shown on models having ball bearing type main bearings. Refer to text.

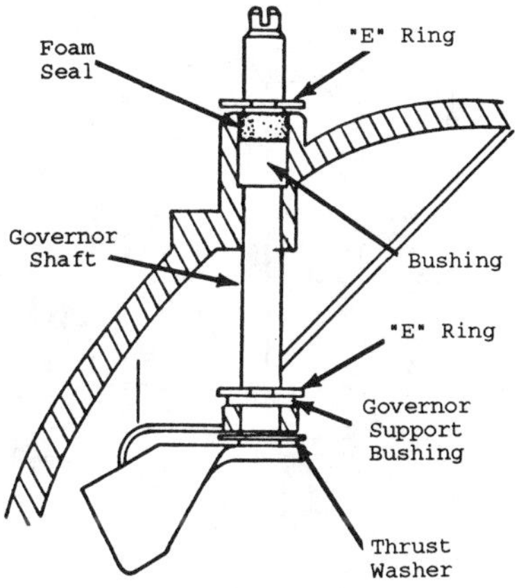

Fig. B100—Drawing of governor shaft assembly used on models below type number 1100 equipped with two "E" rings.

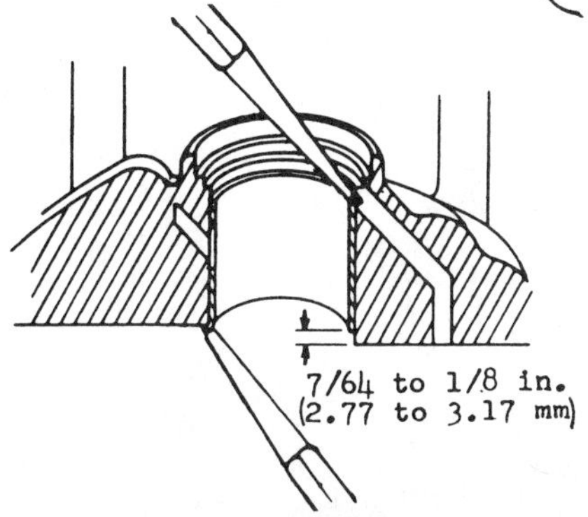

Fig. B97—Press DU type bearings in until ⅛ inch (3.2 mm) from thrust surface. Make certain oil holes are aligned. Stake bearing as shown. Refer to text.

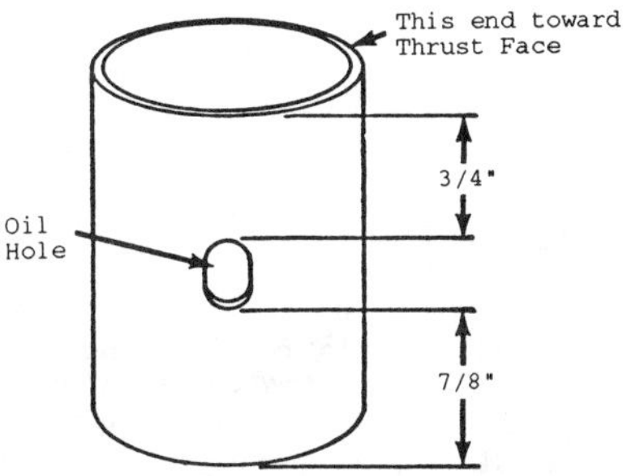

Fig. B98—The pto DU bearing on vertical crankshaft models must be installed so oil hole is nearer thrust face of crankcase.

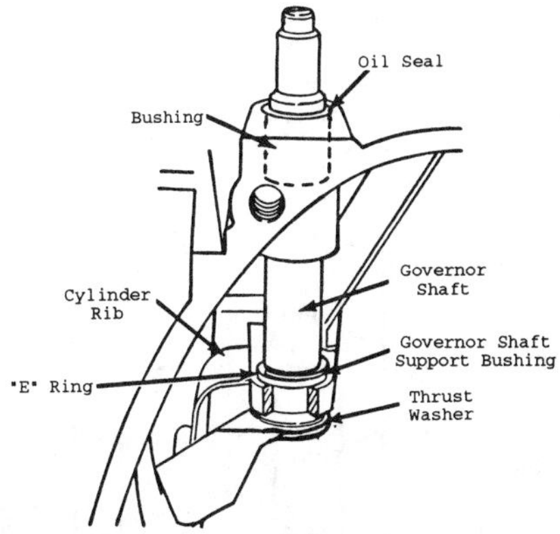

Fig. B101—Drawing of governor shaft assembly used on models with type number 1100 and above.

Illustrations courtesy Briggs & Stratton.

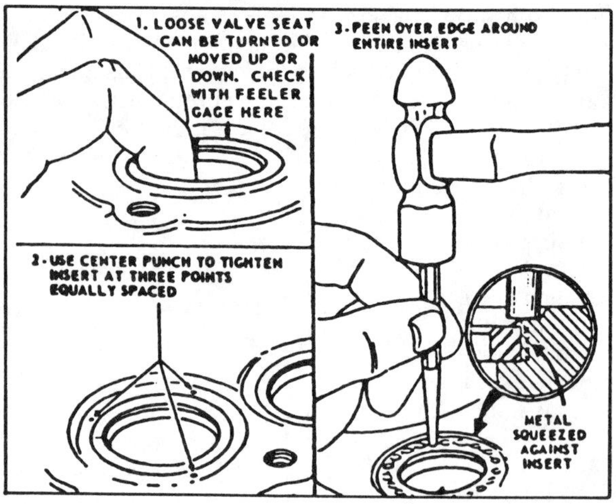

Fig. B102—Loose valve seat inserts can be tightened by peening as shown. If a 0.005 inch (0.13 mm) feeler gage can be inserted between seat and seat bore, install oversize insert or renew cylinder.

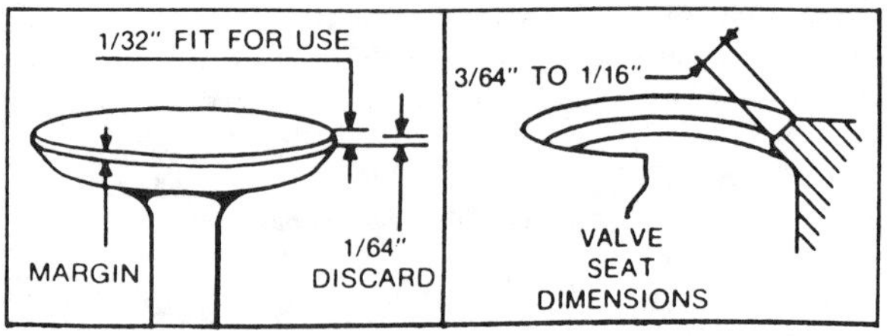

Fig. B103—Drawing showing correct valve face and seat dimensions. Refer to text.

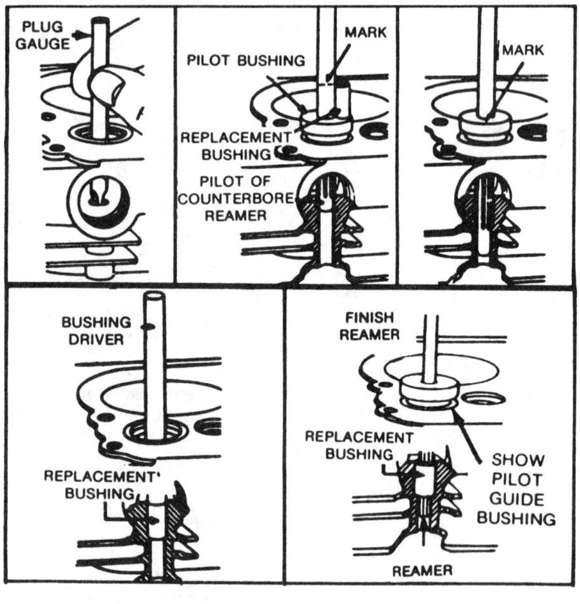

Fig. B104—Drawing showing correct procedure for reconditioning valve guides. Refer to text.

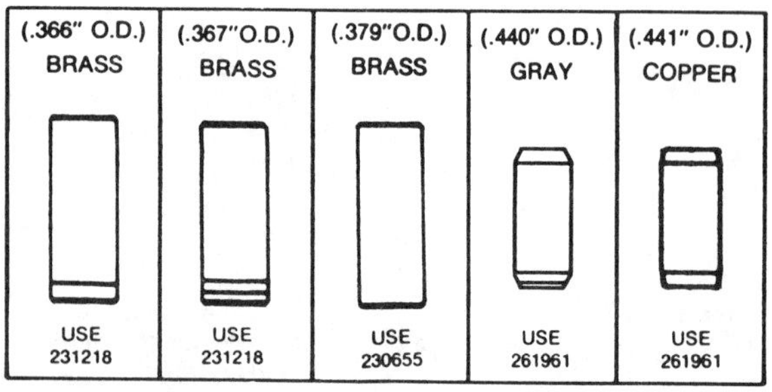

Fig. B105—Use correct bushing when renewing valve guide bushings.

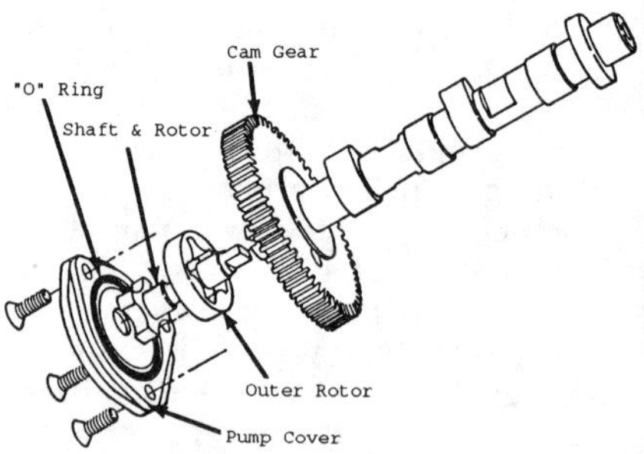

Fig. B106—View of oil pump used on vertical crankshaft engines.

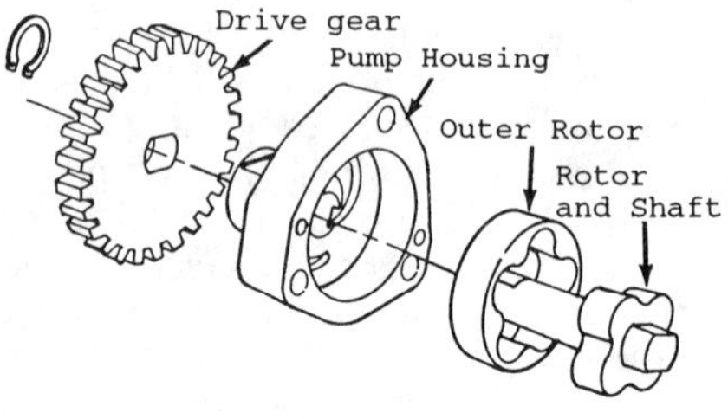

Fig. B107—View of oil pump used on horizontal crankshaft engines.

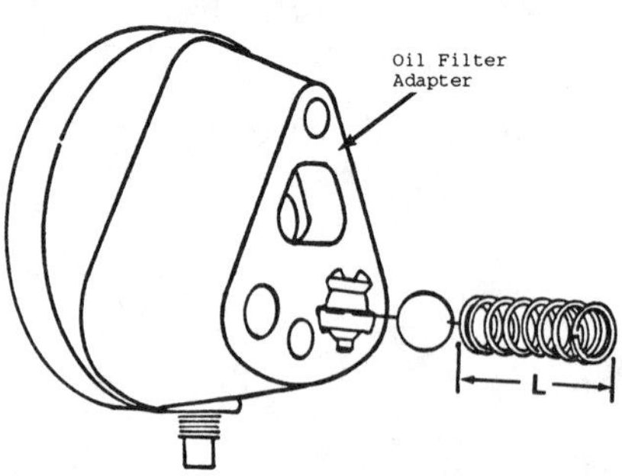

Fig. B108—Relief valve spring length (L) must be 1.091-1.159 inches (27.71-29.44 mm), otherwise reject spring.

Illustrations courtesy Briggs & Stratton.

BRIGGS & STRATTON

VANGUARD OHV ENGINES

Model	Bore	Stroke	Displacement	Power Rating
290400, 294400, 290700, 294700, 303400, 303700	68 mm (2.68 in.)	66 mm (2.60 in.)	480 cc (29.3 cu.in.)	11.9 kW (16 hp)

ENGINE INFORMATION

NOTE: Metric fasteners are used throughout engine except threaded hole in pto end of crankshaft, flange mounting holes and flywheel puller holes, which are U.S. threads.

The Vanguard models included in this section are air-cooled, four-stroke, twin-cylinder engines. Models 290400, 294400 and 303400 have a horizontal crankshaft while models 290700, 294700 and 303700 have a vertical crankshaft. All models are equipped with an overhead valve system. Number 1 cylinder is nearer flywheel. Cylinder number is marked on cylinder side nearest flywheel.

MAINTENANCE

SPARK PLUG. Recommended spark plug is either an Autolite 3924 or Champion RC12YC. Specified spark plug electrode gap is 0.76 mm (0.030 in.). Tighten spark plug to 19 N·m (165 in.-lbs.).

CAUTION: Briggs & Stratton does not recommend using abrasive blasting to clean spark plugs as this may introduce some abrasive material into the engine that could cause extensive damage.

FUEL FILTER. The fuel tank is equipped with a filter at the outlet and an inline filter may also be installed. Check filters periodically during operating season and clean or renew as necessary.

CARBURETOR. Initial adjustment of idle mixture screw (IM—Fig. B201 or Fig. B202) is $1^1/_4$ turns out from seated position. Remove air cleaner, carburetor cover and valley cover, if so equipped, for access to governor linkage. Place remote speed control in idle position. Bend throttle restrictor tang shown in Fig. B203 so throttle cannot open greater than $^1/_4$ open. Run engine until operating temperature is reached. Place remote speed control in idle position. Hold carburetor throttle lever against idle speed adjusting screw and adjust idle speed to 1400 rpm if governor idle spring (S—Fig. B204) is red, or to 1100 rpm if governor idle spring is white. With throttle lever against idle speed adjusting screw, turn idle mixture screw clockwise until there is a reduction in engine speed and note screw position. Back out idle mixture screw until engine speed lessens again and note screw position. Rotate screw so it is halfway between the lean and rich positions. With throttle lever against idle speed screw, readjust idle speed to 1200 rpm if governor idle spring is red, or to 900 rpm if governor idle spring is white. Release throttle lever. With remote control in idle position, bend tab (T—Fig. B204), to obtain 1400 rpm if governor idle spring is red, or to 1100 rpm if governor idle spring is white.

All models are equipped with a fixed high speed jet. An optional jet for high altitude operation is offered on vertical crankshaft models.

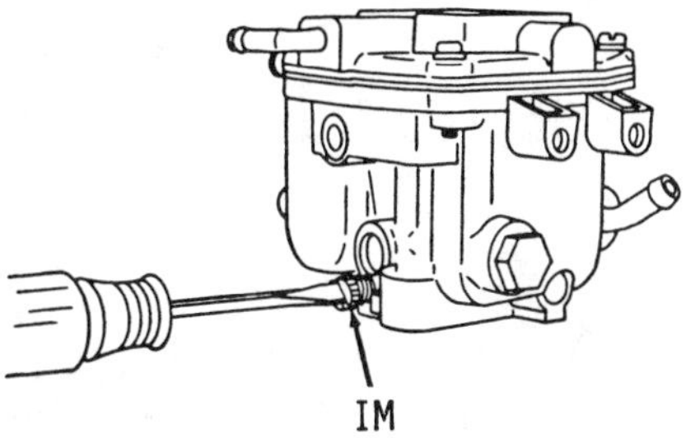

Fig. B201—View showing location of carburetor idle mixture screw (IM) on horizontal crankshaft models.

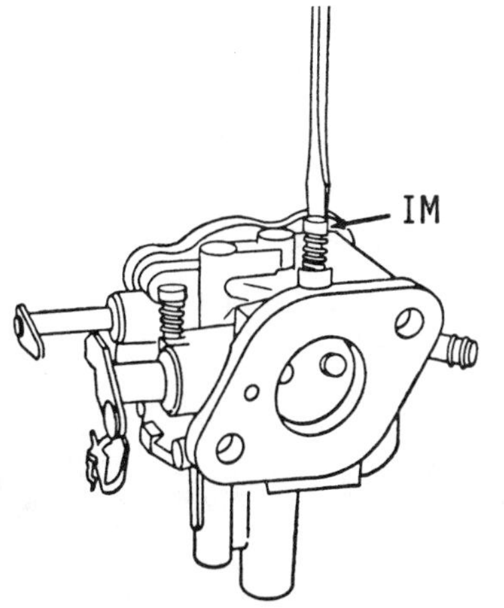

Fig. B202—View showing location of carburetor idle mixture screw (IM) on vertical crankshaft models.

OVERHAUL

HORIZONTAL CRANKSHAFT MODELS. Refer to Fig. B205 for an exploded view of carburetor. Clean parts in a suitable carburetor cleaner. Do not use wire or drill bits to clean fuel passages as carburetor calibration may be affected if passages are enlarged. Inspect parts for wear or damage and renew as necessary.

Note the following when assembling carburetor: Install choke plate so hole

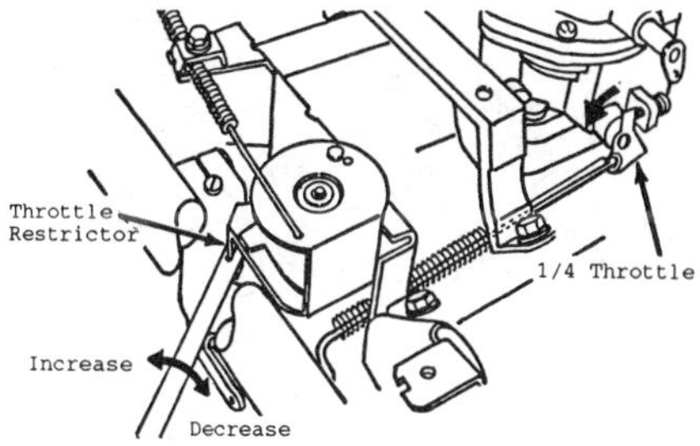

Fig. B203—When adjusting carburetor, bend throttle restrictor tang so throttle cannot open greater than $^1/_4$ open as outlined in text.

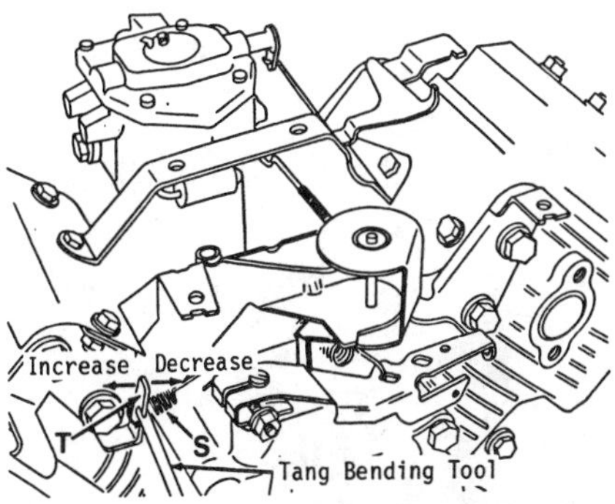

Fig. B204—Adjust governed idle speed as outlined in text.

in plate is toward vent tube (Fig. B206). Install throttle plate so flat portion (not sharp edge) of chamfers on plate fit against carburetor bore when plate and shaft are installed in carburetor. With fuel inlet valve and float installed in body, press lightly against fuel valve so it bottoms against seat. Top of float should be parallel with body gasket surface (see Fig. B207). Adjust float level by bending float tang. Note that spacer between carburetor and intake manifold has an indexing pin that must fit into intake manifold. Tighten carburetor mounting screws to 7 N·m (62 in.-lbs.).

VERTICAL CRANKSHAFT MODELS. Refer to Fig. B208 for an exploded view of carburetor. Clean parts in a suitable carburetor cleaner. Do not use wire or drill bits to clean fuel passages as carburetor calibration may be affected if passages are enlarged.

To check float level, invert carburetor body as shown in Fig. B209. Float should be parallel with bowl mating surface. Bend float tang to adjust float level. Tighten carburetor mounting screws to 7 N·m (62 in.-lbs.).

ANTI-AFTERFIRE SYSTEM. Some models are equipped with an anti-afterfire system that stops fuel flow through the carburetor when the ignition switch is in off position. A solenoid inserts a plunger into the main jet to stop fuel flow. Refer to Fig. B210 for drawing of anti-afterfire system. Solenoid may be removed and tested by connecting a 9-volt battery to solenoid. A faulty solenoid will affect engine performance.

FUEL PUMP. Some models are equipped with a vacuum operated, diaphragm type fuel pump. Individual components are not available; service pump as a unit assembly.

CARBURETOR CONTROL MECHANISM. To assure proper speed control, measure travel of remote control wire with remote control unit installed. Minimum wire travel is $1^3/_8$ inches as shown in Fig. B211.

To adjust speed control cable, move remote control lever to idle position. Carburetor throttle lever should contact idle speed screw. If not, loosen cable housing clamp and reposition cable housing.

Remote choke control should completely close carburetor choke plate when remote control is in "CHOKE" position. If necessary, loosen cable clamp and reposition cable to synchronize carburetor choke and remote control.

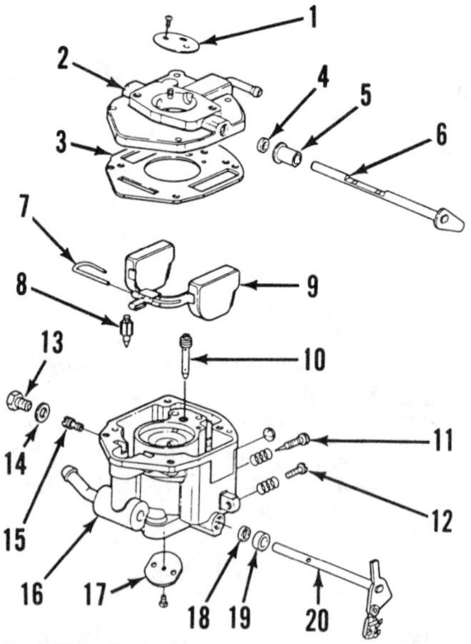

Fig. B205—Exploded view of carburetor used on horizontal crankshaft models.

1. Choke plate
2. Cover
3. Gasket
4. Seal
5. Collar
6. Choke shaft
7. Float pin
8. Fuel inlet valve
9. Float
10. Nozzle
11. Idle mixture screw
12. Idle speed screw
13. Plug
14. Gasket
15. Main jet
16. Body
17. Throttle plate
18. Seal
19. Collar
20. Throttle shaft

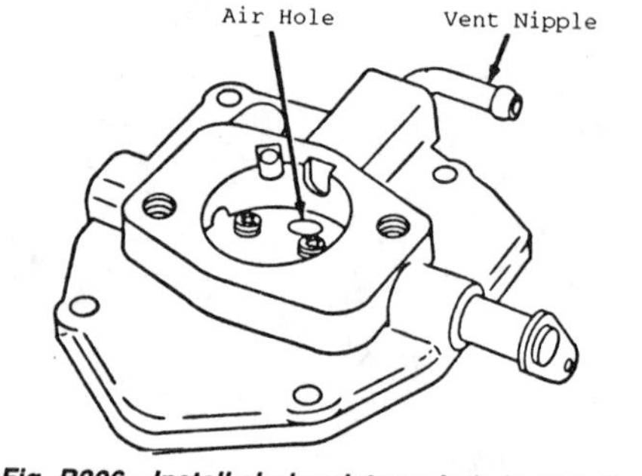

Fig. B206—Install choke plate so hole in plate is toward vent.

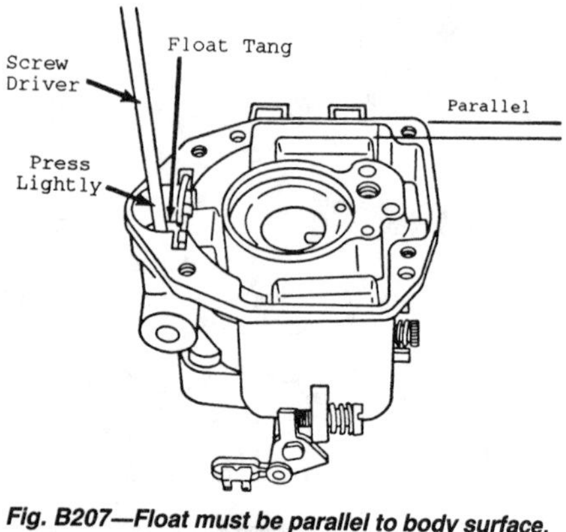

Fig. B207—Float must be parallel to body surface. Bend float tang to adjust.

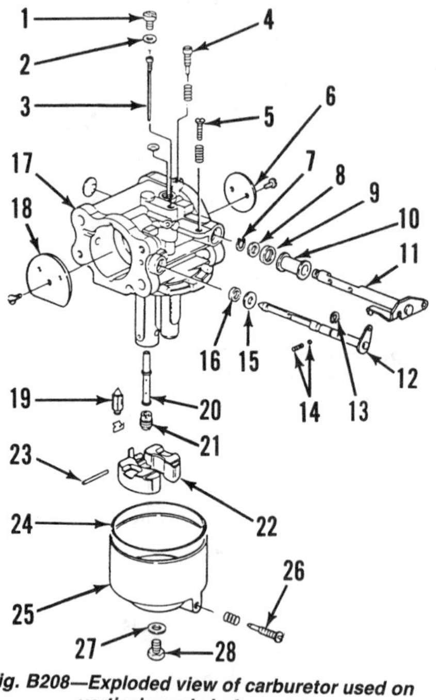

Fig. B208—Exploded view of carburetor used on vertical crankshaft models.

1. Plug
2. Washer
3. Idle jet
4. Idle mixture screw
5. Idle speed screw
6. Throttle plate
7. Clip
8. Washer
9. Seal
10. Collar
11. Throttle shaft
12. Choke shaft
13. Clip
14. Choke detent spring & ball
15. Washer
16. Seal
17. Body
18. Choke plate
19. Fuel inlet valve
20. Nozzle
21. Main jet
22. Float
23. Pin
24. Gasket
25. Fuel bowl
26. Drain screw
27. Gasket
28. Screw

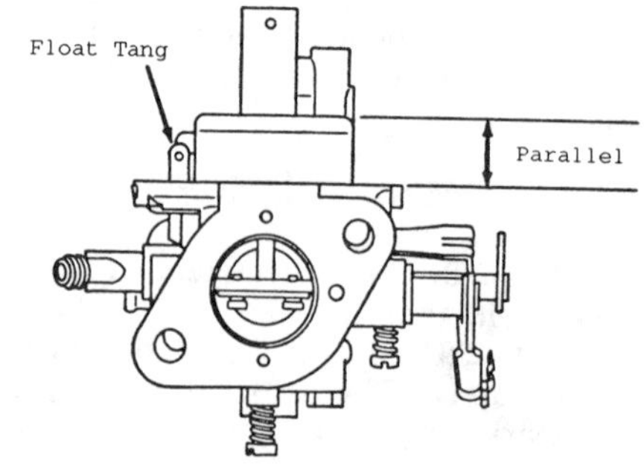

Fig. B209—Float must be level with body surface. Bend float tang to adjust.

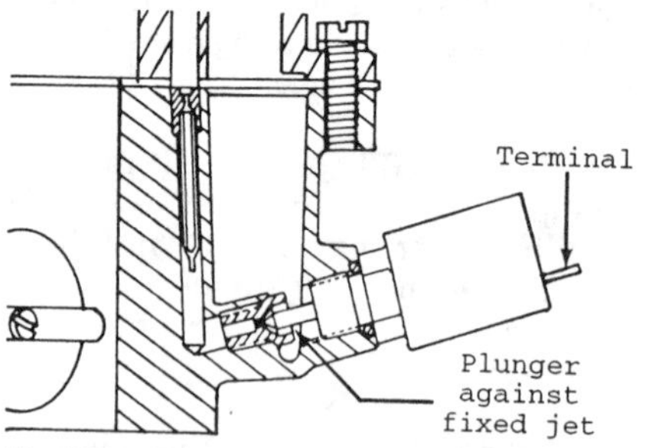

Fig. B210—Drawing of anti-afterfire solenoid used on some carburetors.

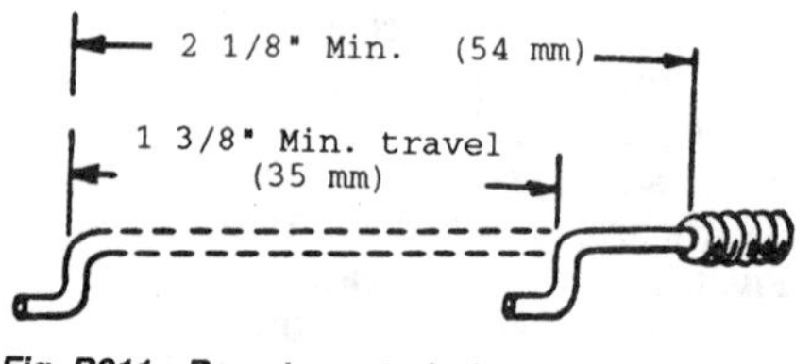

Fig. B211—Remote control wire must extend and travel to dimensions shown above.

GOVERNOR. The engine is equipped with a mechanical, flyweight type governor. To adjust governor linkage, proceed as follows: Remove air cleaner for access to governor linkage. Loosen governor lever clamp nut (N—Fig. B212), rotate governor lever (L) so throttle plate is fully open and hold lever in place. Turn governor shaft (S) counterclockwise as far as possible, then tighten nut (N) to 8 N·m (71 in.-lbs.).

No-load governed top speed is adjusted by bending tang at end of governor spring (G). Normal governed speed is 3600 rpm unless specified otherwise by equipment manufacturer.

The governor flyweight assembly is mounted on the end of the camshaft. Refer to CAMSHAFT and GOVERNOR SHAFT sections if internal service is required.

IGNITION SYSTEM. All models are equipped with a Magnetron ignition system.

To check spark, remove spark plug and connect spark plug cable to B&S tester 19051, then ground remaining tester lead to engine. Spin engine at 350 rpm or more. If spark jumps the 4.2 mm (0.166 in.) tester gap, system is functioning properly.

To remove armature and Magnetron module, remove flywheel shroud. Disconnect stop switch wire from module. Remove armature retaining screws and remove Magnetron unit.

Position armature so air gap between armature legs and flywheel surface is 0.20-0.30 mm (0.008-0.012 in.).

VALVE ADJUSTMENT. Adjust valve clearance with engine cold. Remove rocker arm covers. Remove spark plugs. Rotate crankshaft so piston is at top dead center on compression stroke. Using a suitable measuring device inserted through spark plug hole, continue to rotate crankshaft clockwise as viewed at flywheel end so piston is 6.35 mm (0.250 in.) past TDC. Clearance between rocker arm and valve stem should be 0.10-0.16 mm (0.004-0.006 in.) for intake and exhaust. Loosen lock nut (Fig. B213) and turn adjusting screw to obtain desired clearance. Tighten lock screw to 7 N·m (62 in.-lbs.). Tighten valve cover screws to 3 N·m (27 in.-lbs.).

LUBRICATION. The engine is lubricated with oil supplied by a rotor type oil pump attached to the crankcase cover.

Periodically check oil level; do not overfill. Oil dipstick should be screwed in until bottomed for correct oil level reading. Change oil after first eight hours of operation and every 50 hours thereafter under normal operating conditions. Recommended oil change interval is 25 hours if severe service is encountered.

The engine may be equipped with a spin-on type oil filter. If so equipped, manufacturer recommends changing oil filter after every 100 hours of operation. Filter should be changed more frequently if engine is operated in a severe environment.

Manufacturer recommends using oil with an API service classification of SE or SF. SAE 30 oil may be used when temperatures are above 40° F (4° C); SAE 10W-30 oil may be used for temperatures between 0° F (−18° C) and 100° F (38° C); below 0° F (−18° C) use petroleum based SAE 5W-20 or a suitable synthetic oil.

Crankcase capacity is 1.65 liters (3.5 pints) if equipped with an oil filter, 1.42 liters (3 pints) if not equipped with a filter.

A low oil pressure switch may be located on the oil filter housing, if so equipped. Switch should be closed at zero pressure and open at 31 kPa (4.5 psi).

CRANKCASE BREATHER. The engine is equipped with a crankcase breather that provides a vacuum for the crankcase. Vapor from the crankcase is evacuated to the air cleaner. A fiber disk acts as a one-way valve to maintain crankcase vacuum. The breather system must operate properly or excessive oil consumption may result.

The breather assembly is located in the valley between the cylinders. It should be possible to insert a 0.045 inch (1.14 mm) wire between fiber disc and breather body; do not use excessive force when measuring gap. Disc should not stick or bind during operation. Renew if distorted or damaged. Inspect breather tube for leakage.

CYLINDER HEAD. Manufacturer recommends that after every 500 hours of operation the cylinder heads should be removed and cleaned of deposits.

REPAIRS

TIGHTENING TORQUES. Recommended tightening torque specifications are as follows:

Carburetor mounting screws	7 N·m (62 in.-lbs.)
Connecting rod	13 N·m (115 in.-lbs.)
Crankcase cover	17 N·m (150 in.-lbs.)
Cylinder head	19 N·m (165 in.-lbs.)
Flywheel nut	175 N·m (129 ft.-lbs.)
Oil pump	7 N·m (62 in.-lbs.)
Rewind starter	7 N·m (62 in.-lbs.)
Rocker arm cover	3.0 N·m (27 in.-lbs.)
Rocker arm lock screw	7 N·m (62 in.-lbs.)
Rocker shaft stud	16 N·m (142 in.-lbs.)
Spark plug	22.5 N·m (200 in.-lbs.)

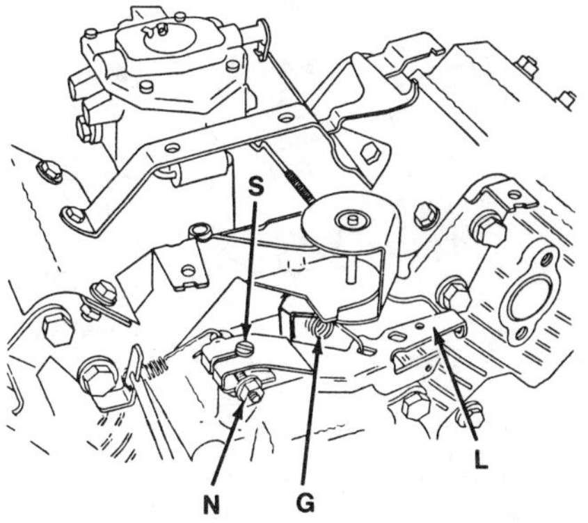

Fig. B212—To adjust governor, loosen governor lever clamp nut (N), rotate governor lever (L) so throttle plate is fully open and hold lever in place. Turn governor shaft (S) counterclockwise and retighten nut.

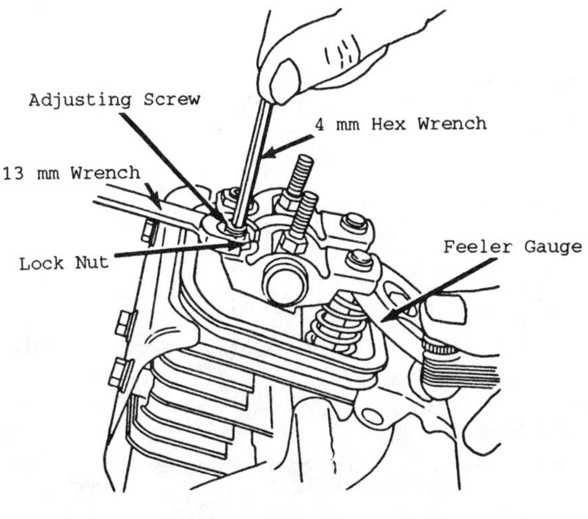

Fig. B213—After loosening lock nut, turn rocker arm adjusting screw so valve clearance is 0.10-0.16 mm (0.004-0.006 in.).

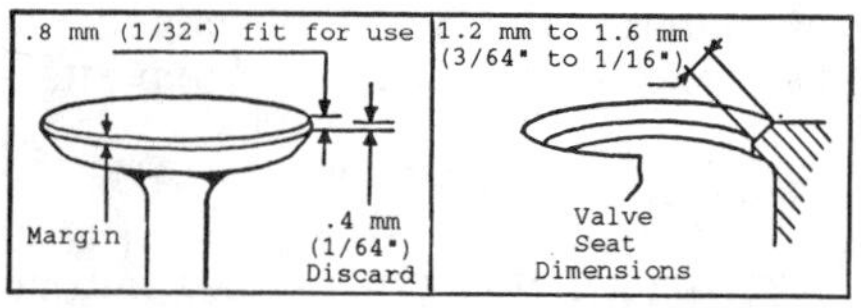

Fig. B214—Valve and valve seat dimensions.

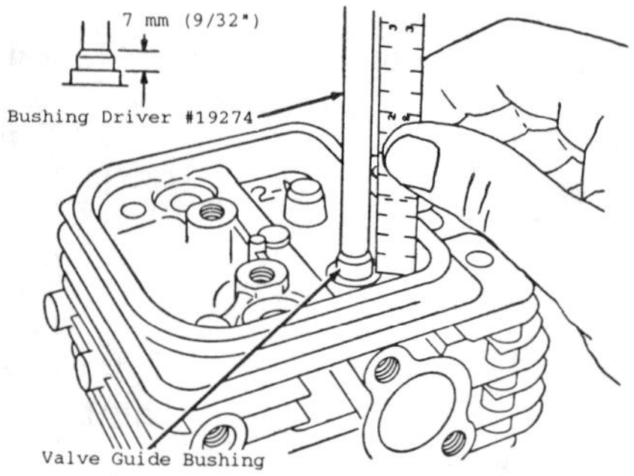

Fig. B215—Install valve guide so standout is 7 mm (0.276 in.) as shown above.

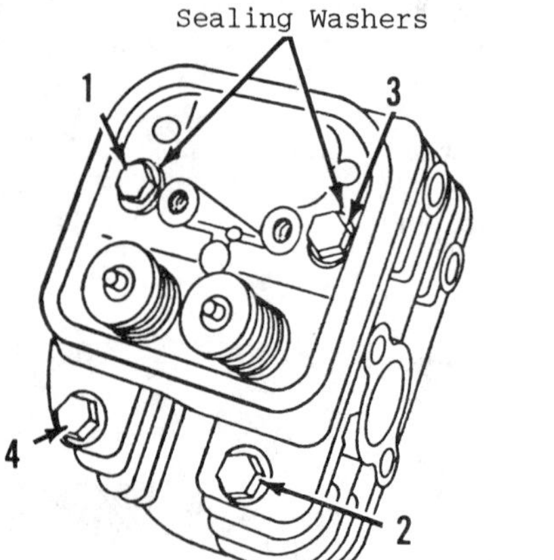

Fig. B216—Follow sequence shown above when tightening cylinder head bolts.

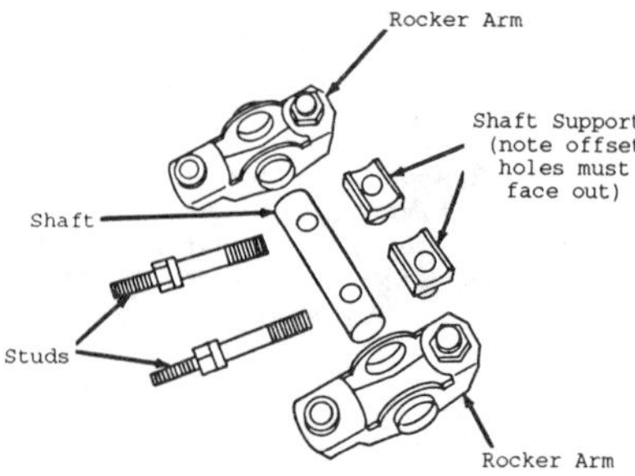

Fig. B217—Offset hole in rocker shaft supports must be nearer end of rocker shaft.

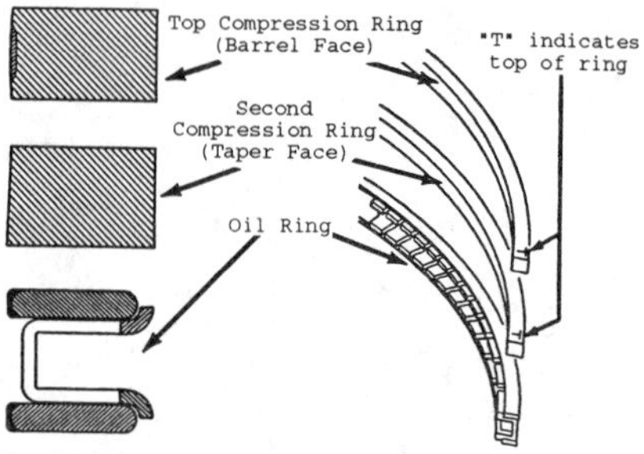

Fig. B218—Drawing showing correct installation of piston rings.

CYLINDER HEAD. Cylinder heads are not interchangeable. Cylinder number is cast in area adjacent to valve springs. To remove the cylinder head, first remove carburetor, intake manifold, muffler, exhaust manifold, air baffles and shields. Remove rocker arm cover, rocker arms and push rods. Unscrew cylinder head bolts and remove cylinder head.

Valve face and seat angles are 45 degrees. Specified seat width is 1.2-1.6 mm (0.047-0.063 in.). Minimum allowable valve margin is 0.4 mm (0.016 in.). See Fig. B214

The cylinder head is equipped with renewable valve guides for both valves. Reject guide if inside diameter is 6.057 mm (0.2385 in.) or more. Use B&S tool 19274 to remove and install guides. Guides may be installed either way. Top of guide should protrude 7 mm (0.276 in.) as shown in Fig. B215. Use B&S tools 19345 and 19346 to ream valve guide to correct size.

Note the following when reinstalling cylinder head: Do not apply sealer to cylinder head gasket. Sealing washers are used under the heads of the two cylinder head bolts adjacent to the rocker shaft. Tighten cylinder head bolts using sequence shown in Fig. B216 until final torque reading of 19 N·m (165 in.-lbs.) is obtained.

Rocker shaft studs are threaded into cylinder head. Note that rocker shaft stands have offset holes. Hole must be nearer end of shaft as shown in Fig. B217. Tighten studs to 16 N·m (142 in.-lbs.).

PISTON, PIN, RINGS AND CONNECTING ROD. To remove piston and rod assemblies, drain engine oil and remove engine from equipment. Remove cylinder heads as previously outlined. Remove carbon or ring ridge (if present) from top of cylinder bore. Clean pto end of crankshaft and remove any burrs or rust. Unscrew fasteners and remove crankcase cover. Rotate crankshaft so timing marks on crankshaft and camshaft gears are aligned (this will position valve tappets out of way). Remove camshaft. Mark rods and caps so they can be installed in original positions. Unscrew connecting rod screws and remove piston and rod.

Inspect pistons for scoring, scuffing or excessive wear and renew as necessary. Check piston ring grooves for wear by inserting new rings in grooves and measure clearance between ring and ring land using a feeler gage. Renew piston if ring side clearance exceeds 0.10 mm (0.004 in.) for compression rings and 0.20 mm (0.008 in.) for oil ring.

Piston pin is a slip fit in piston and rod. Reject size for piston pin diameter is 17.056 mm (0.6715 in.). Reject size for pin bore in piston is 17.109 mm (0.6736 in.).

The connecting rod rides directly on crankpin. Reject size for rod big end diameter is 37.109 mm (1.461 in.) and reject size for small end diameter is 17.081 mm (0.6725 in.). A connecting rod with 0.51 mm (0.020 in.) undersize big end diameter is available to accommodate a worn crankpin (machining instructions are included with new rod).

Install compression rings on piston with "T" side toward piston crown (Fig. B218). When assembling piston and rod, be sure notch in piston crown is toward flywheel and "OUT" side of rod is toward pto end of engine (Fig. B219). Install rod cap so marks on rod and cap align and tighten rod screws to 13 N·m (115 in.-lbs.).

Install camshaft while aligning timing marks (Fig. B220) on crankshaft and camshaft gears. Be sure governor arm is in proper position to contact governor slider on camshaft. Be sure "O" ring (Fig. B221) is in place. Install crankcase cover and tighten cover screws to 17 N·m (150 in.-lbs.) in sequence shown in Fig. B222. Do not force mating of cover with crankcase. Reassemble remainder of components.

CAMSHAFT. Camshaft and camshaft gear are an integral casting. The governor weight assembly and compression release mechanism are attached to the camshaft. Camshaft, governor and compression release are available only as a unit assembly.

To remove camshaft proceed as follows: Remove engine from equipment and drain crankcase oil. Clean pto end of crankshaft and remove any burrs or rust. Remove rocker arm push rods and mark them so they can be returned to original position. Unscrew fasteners and remove crankcase cover. Rotate crankshaft so timing marks on crankshaft and camshaft gears are aligned (this will position valve tappets out of way). Remove camshaft and tappets.

Reject size for camshaft bearing journal at flywheel end is 15.913 mm (0.6265 in.); reject size for bearing journal at pto end is 19.913 mm (0.7840 in.). Reject size for camshaft lobes is 30.25 mm (1.191 in.).

Be sure compression release and governor components operate freely without binding. Check for loose and excessively worn parts.

Reverse removal procedure to install components. Install camshaft while aligning timing marks (Fig. B220) on crankshaft and camshaft gears. Be sure governor arm is in proper position to contact governor slider on camshaft. Be sure "O" ring (Fig. B221) is in place. In-

Illustrations courtesy Briggs & Stratton.

stall crankcase cover and tighten cover screws to 17 N·m (150 in.-lbs.) in sequence shown in Fig. B222. Do not force mating of cover with crankcase. Reassemble remainder of components.

CRANKSHAFT. The crankshaft is supported at the flywheel end by a bushing in the cylinder block and by a ball bearing at the pto end. Reject crankshaft if worn to 34.943 mm (1.3757 in.) or less at pto bearing journal or if worn to 29.95 mm (1.179 in.) or less at flywheel bearing end. Renew main bearing at flywheel side if bore is 30.08 mm (1.184 in.) or less. Install new bearing bushing so notch is out and locating pin hole and oil hole are aligned. Renew cover if bearing bore is 35.06 mm (1.380 in.) or less. Regrind crankpin or renew crankshaft if crankpin diameter is worn to 36.957 mm (1.4550 in.) or less. A 0.51 mm (0.020 in.) undersize connecting rod is available for use on reground crankpin.

CYLINDER BLOCK. Standard cylinder bore diameter is 68.000-68.025 mm (2.6772-2.6781 inches). If cylinder bore wear exceeds 0.076 mm (0.003 in.) or if out-of-round of bore exceeds 0.038 mm (0.0015 in.), then cylinder should be bored to the next oversize.

OIL PUMP. The rotor type oil pump is located in the crankcase cover and driven by the camshaft. Remove crankcase cover for access to pump. Note that rotors are installed so dimples (D—Fig. B223) are on same side. Renew any components that are damaged or excessively worn. Tighten oil pump mounting screws to 7 N·m (62 in.-lbs.).

When installing crankcase cover, be sure governor arm is in proper position to contact governor slider on camshaft. Be sure "O" ring (Fig. B221) is in place. Install crankcase cover and tighten cover screws to 17 N·m (150 in.-lbs.) in sequence shown in Fig. B222. Do not force mating of cover with crankcase. Reassemble remainder of components.

GOVERNOR SHAFT. The governor shaft located in the crankcase cover transmits motion from the governor assembly on the camshaft to the governor linkage. The shaft rides in two bushings in the cover. Remove crankcase cover for access to shaft and bushings. Upper bushing is renewable, but lower bushing is not. If lower bushing is damaged or excessively worn, renew crankcase cover.

When reinstalling crankcase cover, be sure governor arm is in proper position to contact governor slider on camshaft. Be sure "O" ring (Fig. B221) is in place. Install crankcase cover and tighten cover screws to 17 N·m (150 in.-lbs.) in sequence shown in Fig. B222. Do not force mating of cover with crankcase. Reassemble remainder of components.

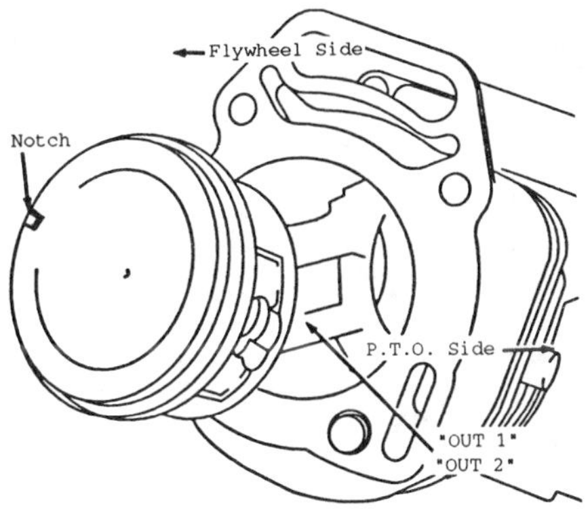

Fig. B219—Install piston and rod so notch in piston crown is toward flywheel and "OUT" mark on rod is toward pto end of crankshaft.

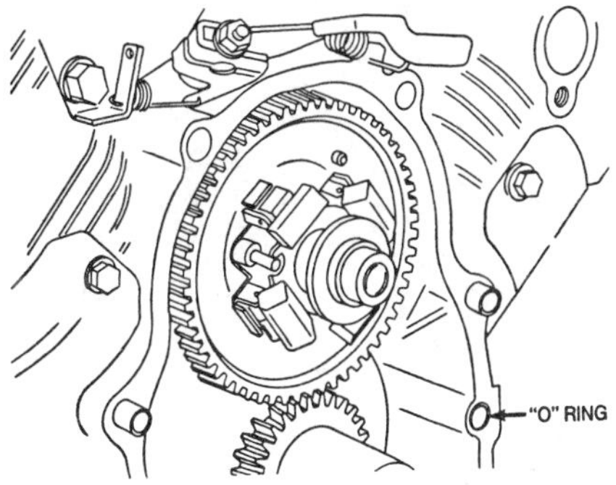

Fig. B221—Be sure "O" ring is in place before installing crankcase cover.

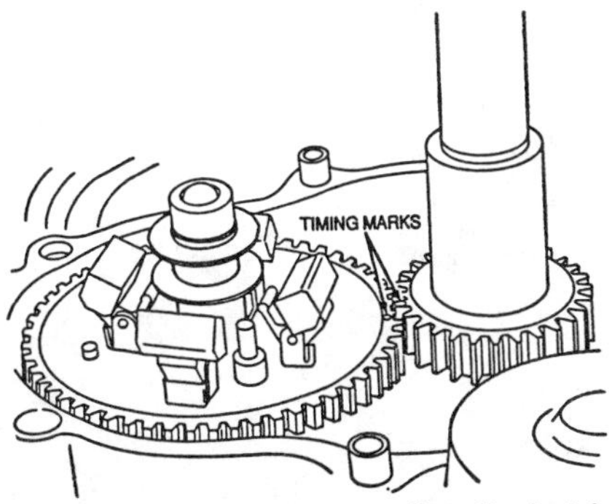

Fig. B220—Timing marks must align after installation of camshaft and crankshaft.

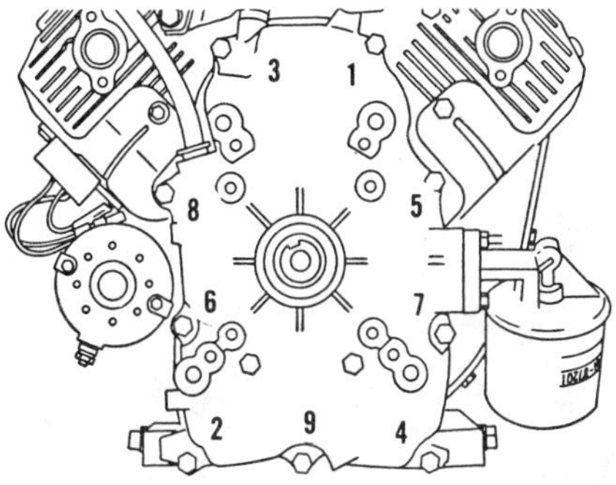

Fig. B222—Follow sequence shown above when tightening crankcase cover screws.

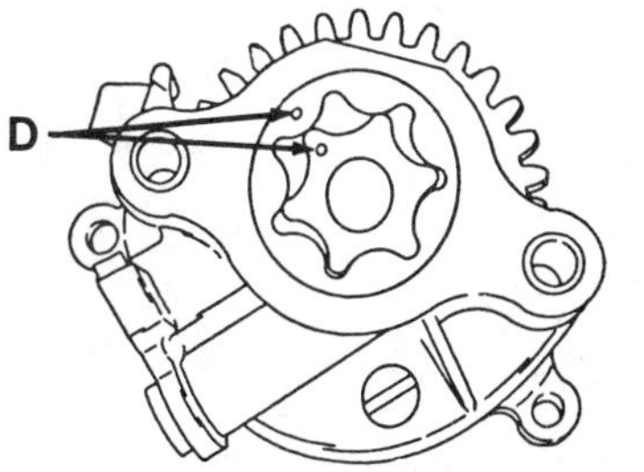

Fig. B223—Install oil pump rotors so dimples (D) are on same side.

Illustrations courtesy Briggs & Stratton.

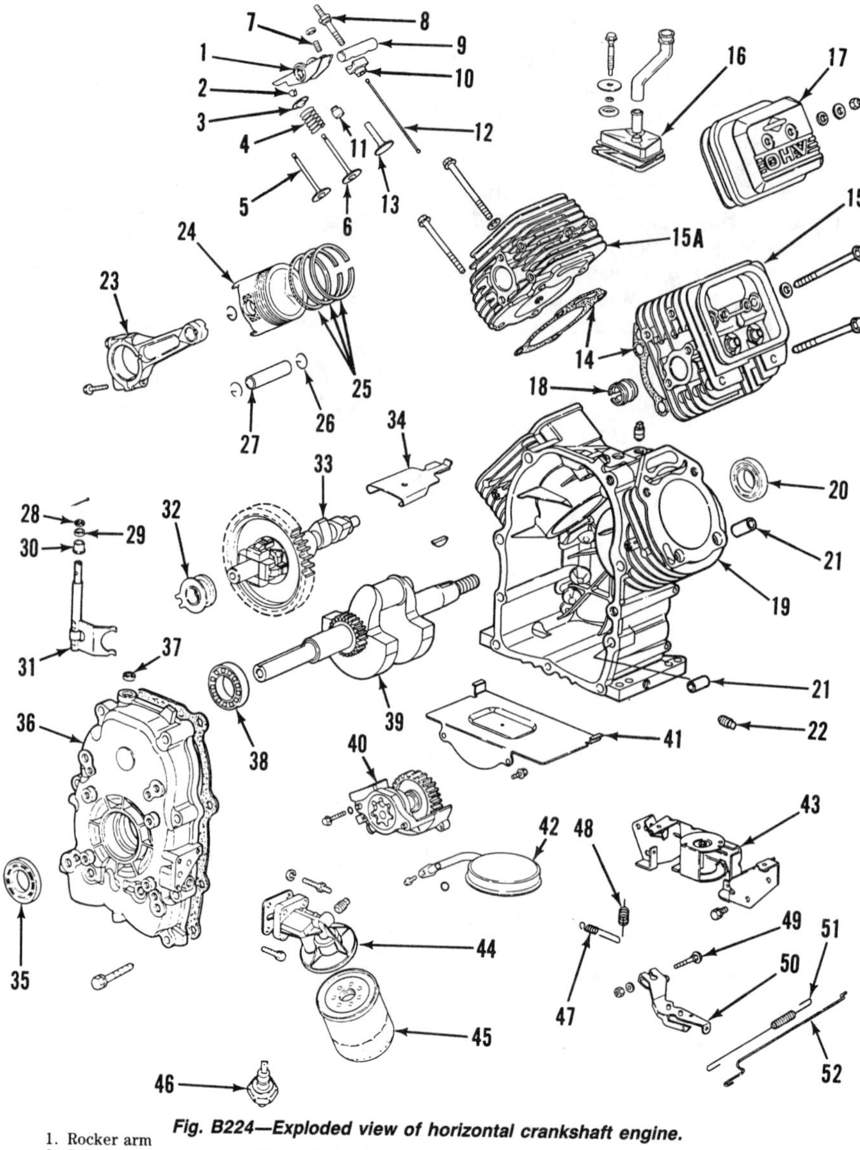

Fig. B224—Exploded view of horizontal crankshaft engine.

1. Rocker arm
2. Split retainer
3. Spring retainer
4. Spring
5. Intake valve
6. Exhaust valve
7. Adjusting screw
8. Rocker stud
9. Rocker shaft
10. Support
11. Seal
12. Push rod
13. Tappet
14. Head gasket

15A. Cylinder head, No. 1

15B. Cylinder head, No. 2

16. Crankcase breather
17. Valve cover
18. Exhaust port liner
19. Cylinder block
20. Oil seal
21. Dowels
22. Oil drain plug
23. Connecting rod
24. Piston
25. Piston rings
26. Retaining rings
27. Piston pin
28. Seal
29. Bushing
30. Bushing
31. Governor shaft
32. Governor slider
33. Camshaft & governor assy.
34. Baffle
35. Oil seal
36. Crankcase cover
37. Washer
38. Main bearing
39. Crankshaft
40. Oil pump
41. Windage plate
42. Oil pickup screen
43. Governor bracket
44. Oil filter adaptor
45. Oil filter
46. Oil pressure switch
47. Governor idle spring
48. Governor spring
49. Clamp bolt
50. Governor lever
51. Link spring
52. Governor link

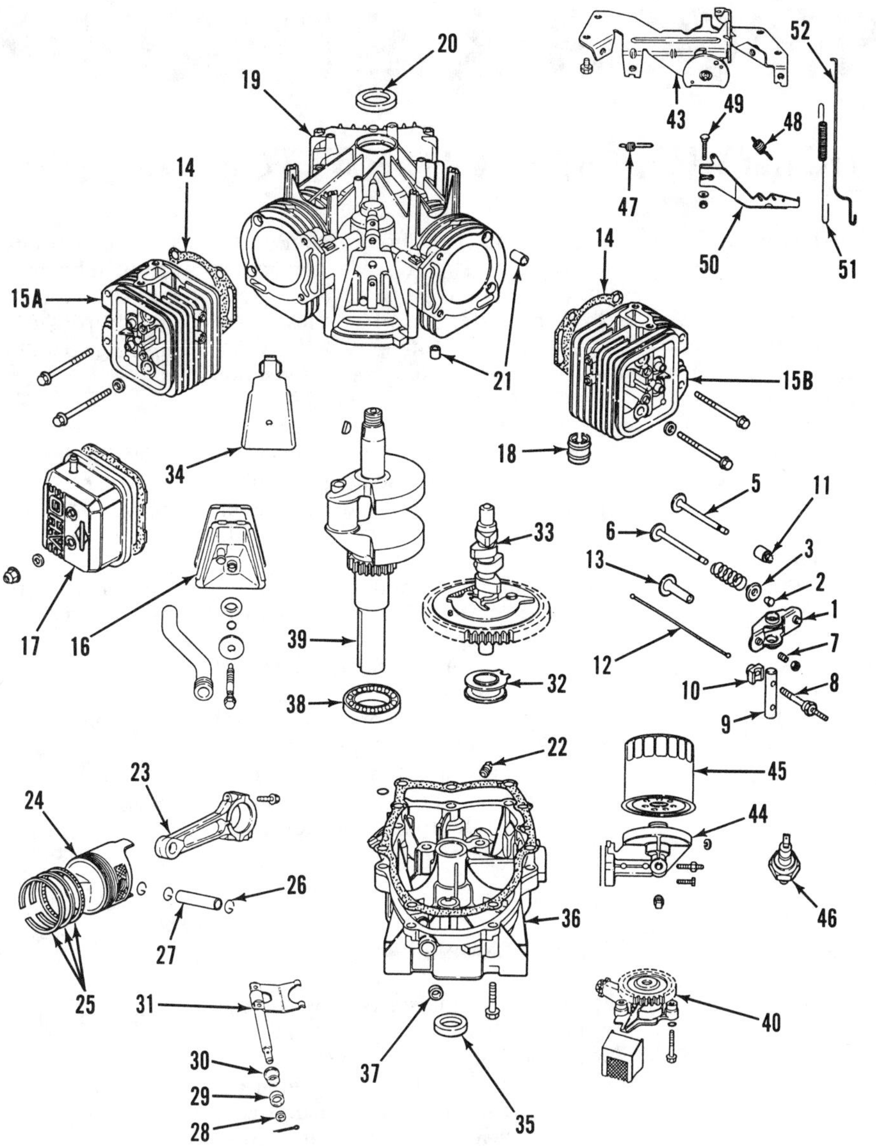

Fig. B225—Exploded view of vertical crankshaft engine. Refer to Fig. 224 for parts identification.

BRIGGS & STRATTON

SERVICING BRIGGS & STRATTON ACCESSORIES

REWIND STARTER

Vanguard Models

Refer to Fig. BA1 for exploded view of rewind starter. When installing the starter, position starter on blower housing, then pull out starter rope until dogs engage starter cup. Continue to place tension on rope and tighten starter mounting screws to 7 N·m (60 in.-lbs.).

To install a new rope, proceed as follows. Rope length should be 178 cm (70 in.). Remove starter and extract old rope from pulley. Allow pulley to unwind, then turn pulley counterclockwise until spring is tightly wound. Allow pulley to unwind until rope hole in pulley is aligned with rope outlet in housing, then pass rope through pulley hole and housing outlet and tie a temporary knot near handle end of rope. Release pulley and allow rope to wind onto pulley. Install rope handle, release temporary knot and allow rope to enter starter.

To disassemble starter, remove rope handle and allow pulley to totally unwind. Unscrew pulley retaining screw (14—Fig. BA1). Remove retainer (12), dogs (11), springs (10) and brake spring (9). Wear appropriate safety eyewear and gloves before disengaging pulley from starter as spring may uncoil uncontrolled. Place a shop towel around pulley and lift pulley out of housing; spring should remain with pulley. Do not attempt to separate spring (6) from cup (5) as they are a unit assembly.

Inspect components for damage and excessive wear. Reverse disassembly procedure to install components. Be sure inner end of rewind spring engages spring retainer adjacent to housing center post. Tighten screw (14) to 8 N·m (70 in.-lbs.). Install rope as previously outlined.

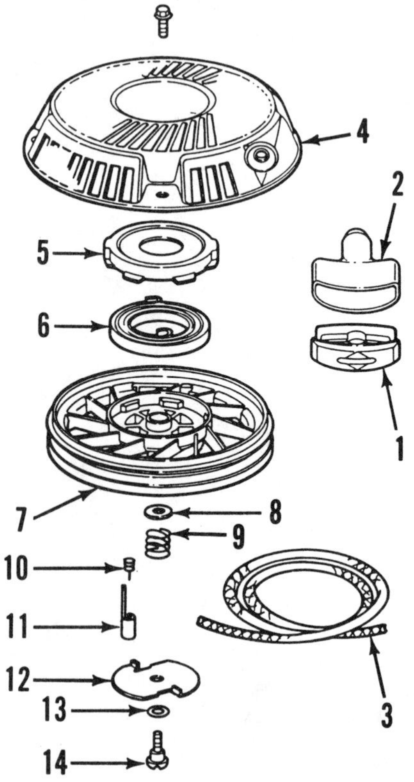

Fig. BA1—Exploded view of rewind starter used on Vanguard models.

1. Insert
2. Rope handle
3. Rope
4. Housing
5. Spring cup
6. Rewind spring
7. Pulley
8. Washer
9. Brake spring
10. Spring
11. Dog
12. Retainer
13. Washer
14. Screw

All Other Models

To remove rewind starter, remove the four nuts and lockwashers from studs in blower housing. Separate starter assembly from blower housing, then separate assembly from blower housing and starter clutch assembly. To disassemble, remove handle and pin and allow rope to rewind into housing. Grip end of rope in knot cavity and remove rope. Grip outer end of spring with pliers (Fig. BA2) and pull spring out of housing as far as possible. Turn spring $^1/_4$ turn and remove from pulley or bend one of the tangs with B&S tool 19229 and lift out starter pulley to disconnect spring.

Clean spring and housing and oil spring sparingly before reassembly. If pulley was removed, place a small amount of multipurpose grease on pulley, ratchet spring and ratchet spring adapter (Fig. BA3). Place ratchet spring, spring adapter and pulley into rewind housing and bend tang using B&S tool 19229 to adjust tang gap to $^1/_{16}$ inch (1.6 mm) minimum.

Fabricate a rewind tool (Fig. BA4) and wind pulley counterclockwise until spring is wound tightly. Unwind one turn or until hole in pulley for rope knot and eyelet in blower housing are aligned. Lock spring securely in smaller portion of tapered hole. Reinstall rope.

Sealed clutch (Fig. BA5) can be disassembled by prying cover (2) from housing (4). Clean housing and lubricate with one drop of clean engine oil. Tighten housing to 150 in.-lbs. (16 N·m).

ELECTRIC STARTER

Two styles of 12 volt starters have been used and may be identified by measuring housing and checking end cap material. Refer to Figs. BA6 and BA7.

To check starter performance, remove starter and connect to a 12 volt battery

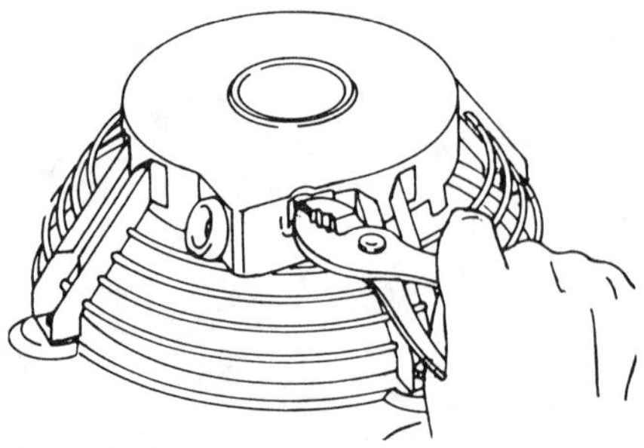

Fig. BA2—Grip outer end of spring with pliers and pull spring out of housing as far as possible. Refer to text.

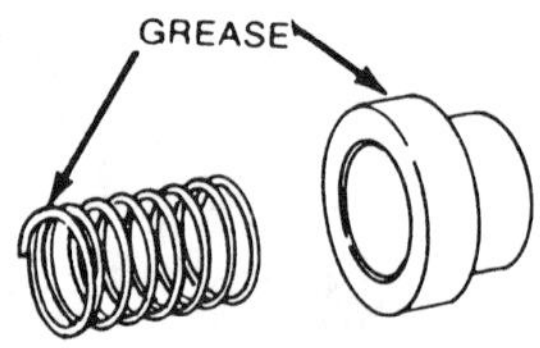

Fig. BA3—Lubricate pulley, spring and adapter with multipurpose grease. Refer to text.

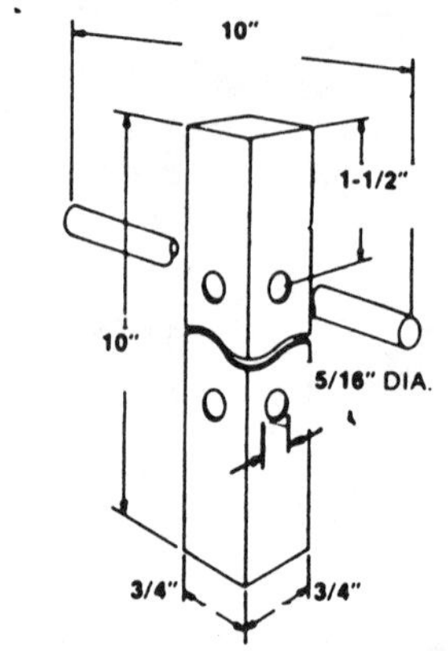

Fig. BA4—Use ¾-inch square stock to fabricate a spring rewind tool to the dimensions shown.

Illustrations courtesy Briggs & Stratton.

with a starter switch and ammeter in-line with positive cable. Activate starter motor and use a suitable tachometer to measure armature rpm. If starter housing length is 3 inches (76.2 mm) or $3^5/_8$ inches (92 mm), starter should draw 18 amps maximum when turning 6500 rpm minimum. If starter housing length is $4^1/_2$ inches (114.3 mm), starter should draw 35 amps maximum when turning 6500 rpm minimum. If starter fails to perform as specified, disassemble and

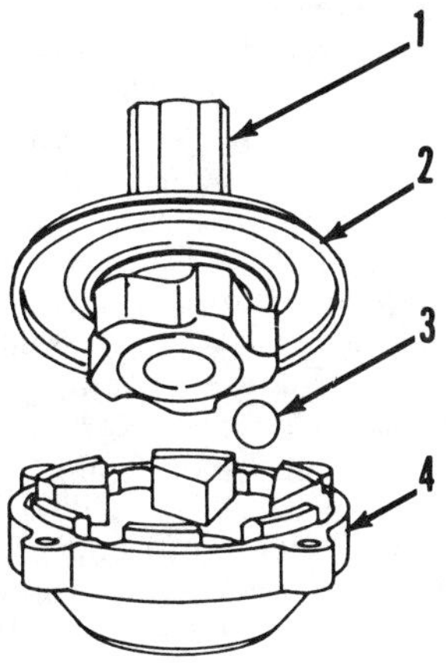

Fig. BA5—Exploded view of recoil starter clutch.
1. Ratchet
2. Cover
3. Ball
4. Housing

check for binding in armature bearings, worn or sticking brushes, faulty armature or weak magnets.

To remove starter drive, pry plastic cap off early style; then, on all models, drive roll pin out using a $^5/_{32}$ inch (4 mm) pin punch. Refer to Fig. BA8 or BA9 for exploded views and location of parts. DO NOT lubricate starter gear. When installing starter drive, be sure that beveled edges of pinion gear face away from starter motor. Roll pin must be centered in armature shaft.

To disassemble starter, remove through-bolts and separate all components.

Check armature for shorted, open or grounded winding and renew as necessary. Commutator may be machined and minimum diameter is 1.23 inches (31.24 mm). Commutator may be polished using fine sand paper. DO NOT use emery cloth to polish armature. Armature end play on current design starters is controlled by a spring washer and is not adjustable. End play on early design starter armature shaft should be 0.006-0.012 inch (0.15-0.30 mm) and is controlled by shim washers. Brushes should be renewed if worn to within $^1/_8$ inch (3 mm) or less from brush wire. Brush spring pressure should be 4 to 6 ounces (113.4 to 170.1 g).

Ring gear is secured to flywheel by rivets. To renew, drill out rivets using a $^3/_{16}$ inch drill. Attach the new ring gear with four screws and locknuts provided with gear.

ALTERNATOR

Type of alternator used can be identified by noting color of stator leads and wire connector. Refer to Figs. BA10, BA11, BA12, BA13, BA14 and BA15. Identify type to be serviced and refer to appropriate following section.

3 Amp DC Alternator

The 3 amp DC alternator (Fig. BA10) is regulated only by engine speed and provides 2 to 3 amp charging current to maintain battery state of charge.

To check output, connect an ammeter in series with red lead, start engine and

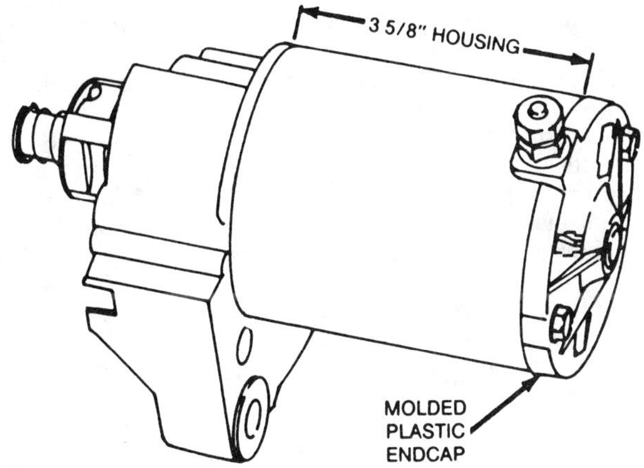

Fig. BA6—View of electric starter used on current production models. Note that endcap is made of molded plastic. Housing length may be $3^5/_8$ inches (92 mm) or $4^1/_2$ inches (114.3 mm).

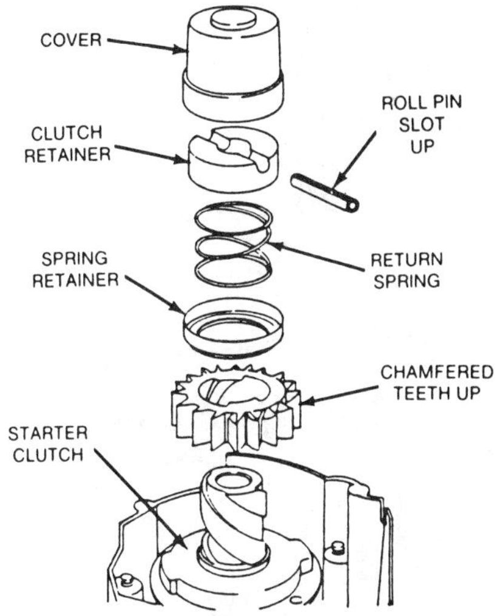

Fig. BA8—Exploded view of starter drive used on early production starters with 3 inch (76.2 mm) housing.

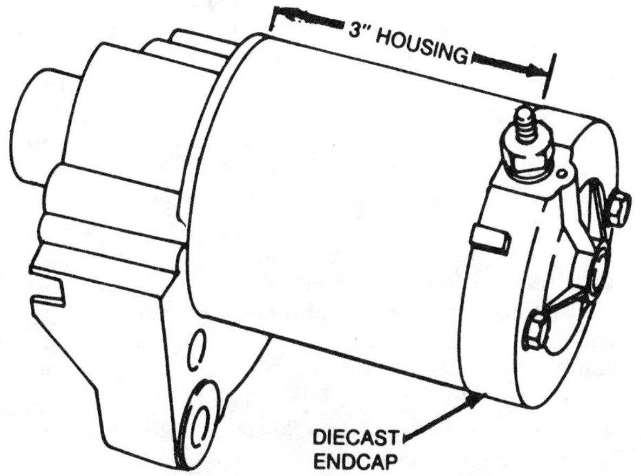

Fig. BA7—View of electric starter used on early production engines. Note that endcap is die cast metal. Housing length is 3 inches (76.2 mm).

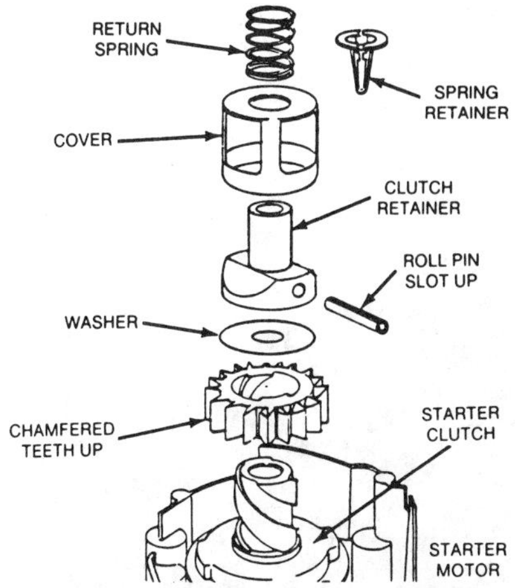

Fig. BA9—Exploded view of starter drive used on current production starters with housing length of $3^5/_8$ inches (92 mm) or $4^1/_2$ inches (114.3 mm).

run at 2400 rpm. Ammeter should show 2 amps charging current. Increase engine speed to 3600 rpm. Ammeter should show 3 amps charging current. If charging current is not as specified, stop engine and check for faulty diode in stator lead as follows: Insert a pin into stator lead wire just behind the diode. Connect ohmmeter test leads to connector output lead and to the pin and note meter reading. Then reverse the tester leads and again note meter reading. The meter should show continuity in one direction only. If there is no continuity, or if there is continuity in both directions, renew diode. If diode checks satisfactorily, renew the stator assembly.

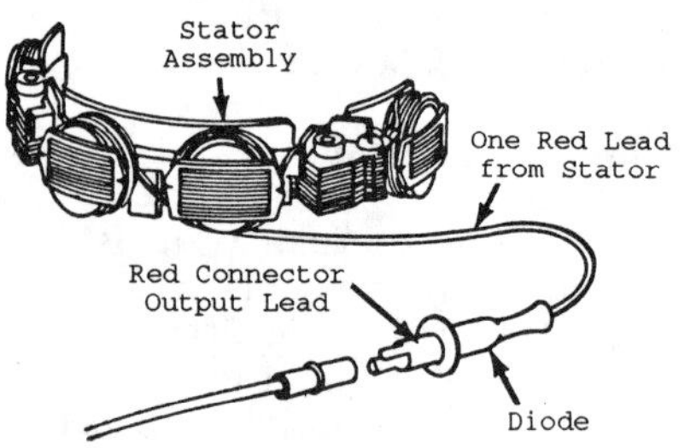

Fig. BA10—View of 3 amp direct current alternator stator used on some models.

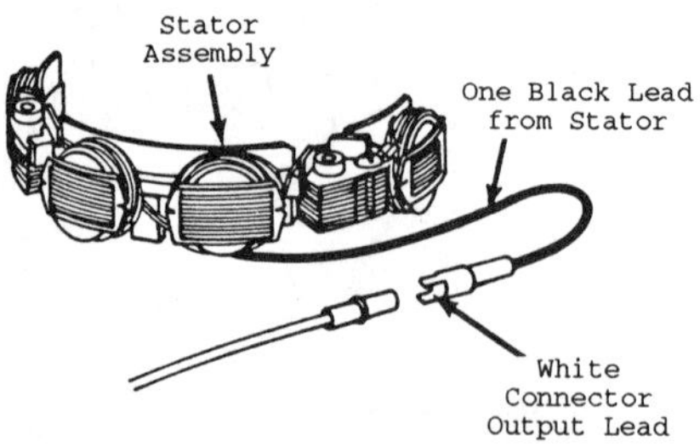

Fig. BA11—View of 5 amp alternating current alternator stator used on some models.

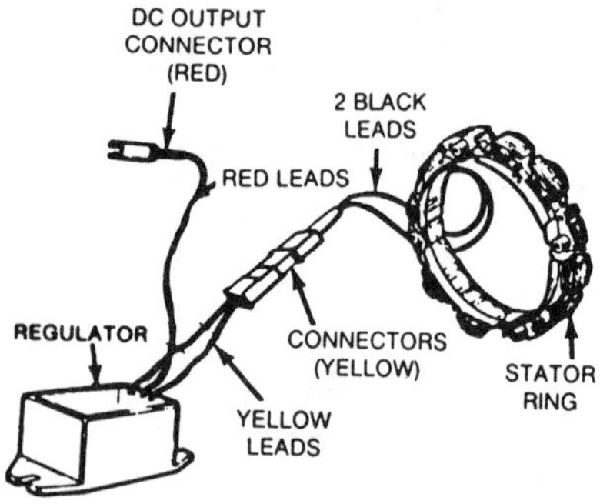

Fig. BA12—View of 10, 13 or 16 amp alternator stator and regulator used on some models.

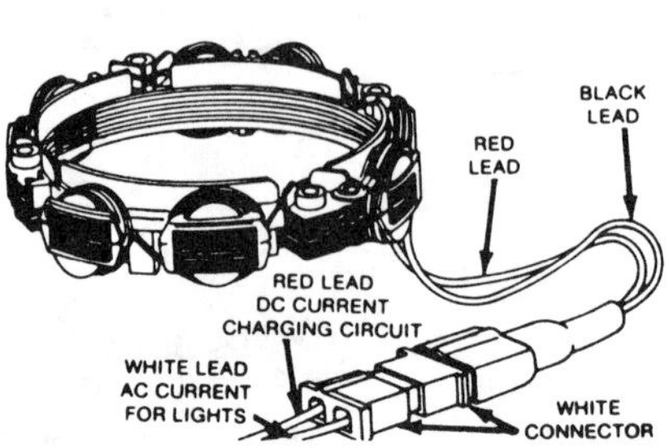

Fig. BA13—View of dual circuit alternator stator used on some models with lights.

5 Amp AC Alternator

The 5 amp alternator shown in Fig. BA11 provides alternating current that is regulated only by engine speed.

To check alternator, connect a voltmeter to black stator lead and check voltage reading with engine running at 3600 rpm. Voltage reading should be at least 14 volts ac. If not, renew stator.

10, 13 And 16 Amp Alternators

The 10, 13 and 16 amp alternators (Fig. BA12) provide regulated charging current. Charging rate is determined by state of charge in battery. Stator is located under the flywheel and output capacity is determined by the size of magnets cast into the flywheel.

Alternator output is determined by the size of the flywheel magnets. Magnets are $^{11}/_{16}$ inch $\times$ $^{7}/_{8}$ inch (18 mm $\times$ 22 mm) on 10 amp flywheels, $^{11}/_{16}$ inch $\times$ $1^{1}/_{16}$ inch (18 mm $\times$ 27 mm) on 13 amp flywheels, and $^{15}/_{16}$ inch $\times$ $1^{1}/_{16}$ inch (24 mm $\times$ 27 mm) on 16 amp flywheels.

To check stator output, disconnect yellow connector and connect voltmeter leads to pins for stator leads. With engine running at 3600 rpm, voltmeter should indicate at least 20 volts ac for 10 and 13 amp systems, and 30 volts ac for 16 amp systems. If not, renew stator.

To test regulator, the 12 volt battery must have a minimum of 5 volt charge. Connect an ammeter in series with charging circuit positive (red) lead and run engine at normal operating rpm. Ammeter should indicate a charge that will vary according to battery state of charge and capacity of alternator. If no charging current is indicated, renew regulator.

Dual Circuit Alternator

Dual circuit alternator (Fig. BA13) has one circuit to provide charging current to maintain battery state of charge and a separate circuit to provide alternating current for lights. The amount of current produced is regulated only by engine speed.

The charging circuit supplies alternating current through a solid state rectifier that converts the alternating current to direct current to maintain battery state of charge.

The lighting circuit provides alternating current to the lights.

The stator is located under the flywheel. A single ring of magnets cast into the flywheel creates the magnetic field for both circuits.

To test charging circuit output, connect an ammeter in series with the charging circuit lead (red wire). Start and run engine at 2400 rpm. Ammeter should indicate 2 amp charging current. Increase engine speed to 3600 rpm. Ammeter should indicate 3 amp charging current. If no charging current is indicated, check the diode in the connector. Attach ohmmeter lead to charging circuit connector pin in plug (a bump on the plug identifies the diode). Stick a pin through the red charging circuit wire just behind the plug and connect remaining ohmmeter lead to pin. Note reading, then reverse leads. Ohmmeter should indicate continuity in only one position. If not, renew plug assembly. If diode is good, but system still does not show a charge, renew stator.

To test lighting circuit, connect an ac voltmeter in series with stator lighting circuit lead (black wire) and ground. Run engine at 3600 rpm. Voltmeter reading should be at least 14 volts ac. Renew stator if there is insufficient voltage.

Tri-Circuit Alternator

The tri-circuit alternator (Fig. BA14) consists of a single ring of magnets cast into the flywheel that provides a magnetic field for the stator located under the flywheel. The stator produces alternating current and has a single output lead. Circuit separation is achieved by the use of a positive (+) diode and a negative (−) diode. The charging lead diode rectifies negative (−) 12 volts dc (5 amps at 3600 rpm) for lighting. This same charge lead contains a second diode that rectifies positive (+) 12 volts dc (5 amps at 3600 rpm) for battery charging and external loads. Some equipment manufacturers incorporate one or both diodes in wiring harness. Check wiring diagram for models being serviced for diode location.

To test alternator output, connect an ac voltmeter in series between stator output lead and ground. Start and run engine at 3600 rpm. Voltmeter should register 28 volts ac or more. Voltage will vary with engine rpm. If charge current is not indicated, renew stator.

To check diodes, disconnect charge lead from stator output lead. Connect ohmmeter lead to connector pin and connect remaining lead to the white (lighting circuit) wire. Reverse connections. Ohmmeter should indicate continuity in one position only. If not, renew diode. Repeat the procedure on red wire (charging circuit).

Illustrations courtesy Briggs & Stratton.

Quad Circuit Alternator

The quad circuit alternator (Fig. BA15) provides 8 amps positive (+) dc from the red regulator lead and 8 amps negative (−) dc from the black regulator lead. Note that the black regulator wire changes to a white wire at white connector. Charging rate is determined by state of charge in battery. Stator is located under the flywheel and output capacity is determined by the size of magnets cast into the flywheel.

To check stator output, disconnect yellow connector and connect voltmeter leads to pins for stator leads. With engine running at 3600 rpm, voltmeter should indicate at least 30 volts ac. If not, renew stator.

To test regulator output, the 12 volt battery must have a minimum of 5 volt charge. Connect an ammeter in series with charging circuit positive (red) lead and start and run engine at normal operating rpm. Ammeter should indicate a charge that will vary according to battery state of charge and capacity of alternator. If no charging current is indicated, renew regulator.

To check lighting circuit of regulator, obtain a 1 ohm, 20 watt resistor. Use a suitable jumper wire and connect to white connector of regulator as shown in Fig. BA16. Connect an ammeter between resistor and battery as shown. Run engine at 3600 rpm just long enough to produce test reading. Ammeter should indicate approximately 8 amps. If not, renew regulator.

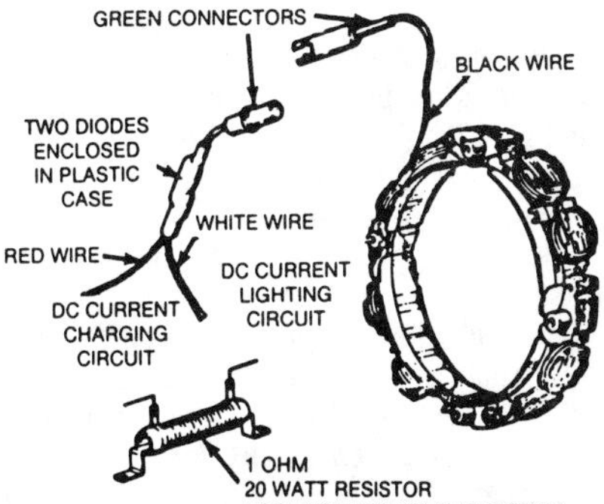

Fig. BA14—View of tri-circuit alternator stator used on some models with lights and electric pto clutches.

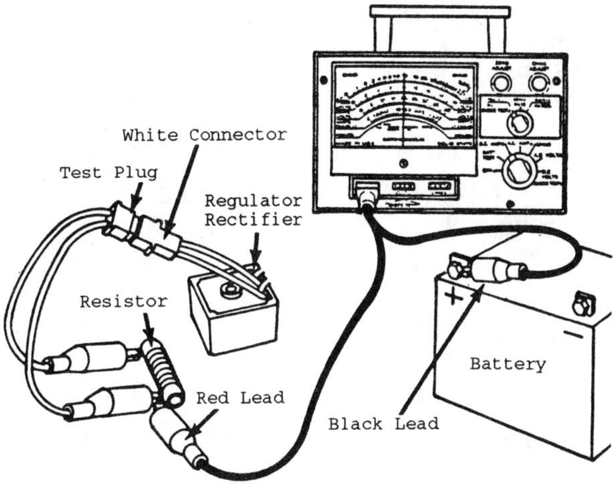

Fig. BA16—The regulator lighting circuit on quad circuit system can be checked using a 1 ohm, 20 watt resistor connected as shown above. Refer to text.

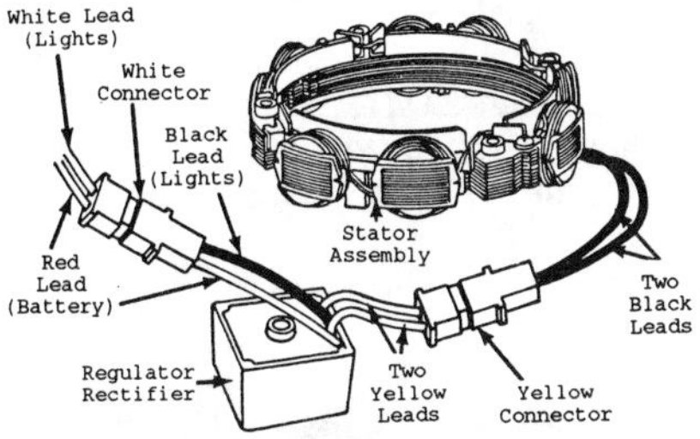

Fig. BA15—View of quad circuit alternator stator and regulator used on some models.

KOHLER

KOHLER COMPANY
Kohler, Wisconsin 503044

Model	No. Cyls.	Bore	Stroke	Displacement	Power Rating
K-361	1	3.75 in. (95.25 mm)	3.25 in. (82.55 mm)	35.89 cu. in. (588.2 cc)	18 hp. (13.4 kW)

Model K-361 is a one cylinder, four-cycle, horizontal crankshaft engine in which crankshaft is supported by ball bearing type main bearings.

Connecting rod rides directly on crankpin journal and is lubricated by oil splash system.

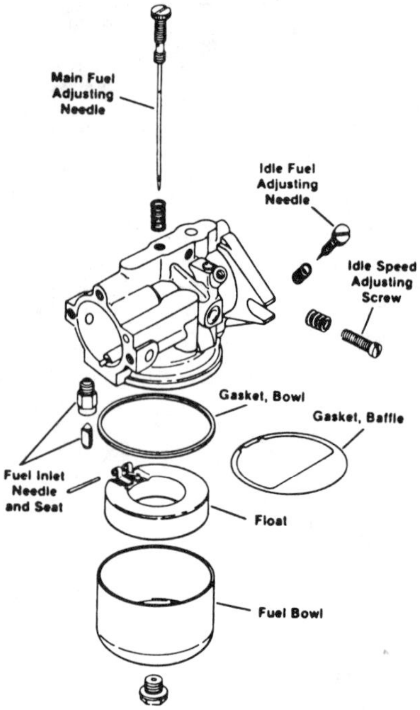

Main Fuel Adjusting Needle
Idle Fuel Adjusting Needle
Idle Speed Adjusting Screw
Gasket, Bowl
Gasket, Baffle
Fuel Inlet Needle and Seat
Float
Fuel Bowl

Fig. KO45—Exploded view of Kohler side draft carburetor used on K-361 engines.

A battery type ignition system with externally mounted breaker points is used.

Engines are equipped with Kohler side draft carburetors.

Special engine accessories or equipment is noted by a suffix letter on engine model number. Key to suffix letters is as follows:

A – Oil pan type
C – Clutch model
P – Pump model
Q – Quiet model
S – Electric start
T – Retractable start

MAINTENANCE

SPARK PLUG. Recommended spark plug is Champion J8 or equivalent and for radio noise reduction, Champion EH10 or equivalent. Electrode gap is 0.035 inch (0.889 mm).

CARBURETOR. Refer to Fig. KO45 for exploded view of Kohler side draft carburetor. For initial adjustment, open main fuel needle 2½ turns and open idle fuel needle ¾-1 turn. Make final adjustment with engine at operating temperature and running. Place engine under load and adjust main fuel needle to leanest mixture that will allow satisfactory acceleration and steady governor operation.

Adjust idle speed stop screw so engine idles at 1725-1875 rpm, then adjust idle fuel needle for smoothest idle operation. As each adjustment affects the other, adjustment procedure may have to be repeated.

To check float level, invert carburetor body and float assembly. There should be 11/64-inch (4.366 mm) clearance plus or minus 1/32-inch (0.794 mm) between machined surface of body casting and free end of float. Adjust float by bending float lever tang that contacts inlet valve. See Fig. KO46.

AUTOMATIC CHOKE. Some models are equipped with a Thermo-Electric automatic choke and fuel shutdown solenoid (Fig. KO47).

When installing or adjusting choke unit, position unit leaving mounting

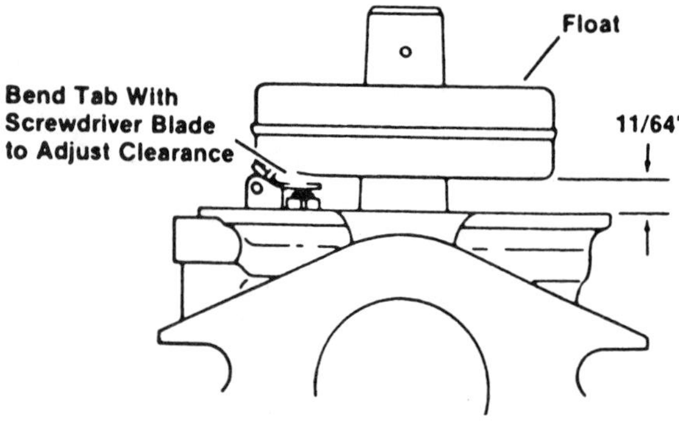

Float
Bend Tab With Screwdriver Blade to Adjust Clearance
11/64"

Fig. KO46—Check float setting by inverting carburetor throttle body and float assembly and measuring distance from free end of float to machined surface of casting. Correct clearance is 11/64-inch (4.366 mm).

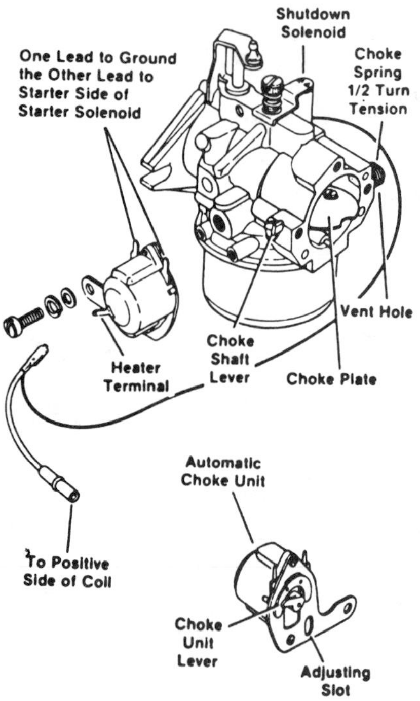

Shutdown Solenoid
One Lead to Ground the Other Lead to Starter Side of Starter Solenoid
Choke Spring 1/2 Turn Tension
Vent Hole
Choke Shaft Lever
Choke Plate
Heater Terminal
Automatic Choke Unit
To Positive Side of Coil
Choke Unit Lever
Adjusting Slot

Fig. KO47—Typical carburetor with Thermo-Electric automatic choke and shutdown control. Refer to text for details.

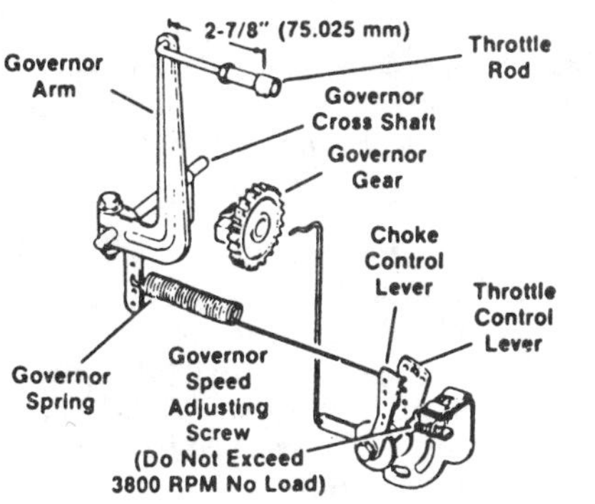

Fig. KO48 – View of typical governor linkage used on K-361 model.

screw slightly loose. Hold choke plate wide open and rotate choke unit clockwise with slight pressure until it can no longer be rotated. Hold choke unit and tighten mounting screws. Choke valve should close 5°-10° at 75° F (23.8° C).

With ignition on, shutdown solenoid plunger raises and opens float bowl vent to air horn. With ignition off, solenoid plunger closes vent and stops fuel flow. To check solenoid, remove solenoid and plunger from carburetor body. With lead wire connected, pull plunger approximately ¼-inch (6.35 mm) from solenoid and ground casing to engine surface. Turn ignition switch on. Renew solenoid unit if solenoid does not draw plunger in. After renewing solenoid, reset main fuel adjusting screw.

GOVERNOR. Model K-361 is equipped with a centrifugal flyweight mechanical governor. Governor flyweight mechanism is mounted within crankcase and driven by camshaft. Maximum no load speed is 3800 rpm.

Governor sensitivity can be adjusted by repositioning governor spring in governor arm and throttle control lever. If too sensitive, surging will occur with change of load. If a big drop in speed occurs when normal load is applied, sensitivity should be increased. Normal spring position is in third hold from bottom on governor arm and second hole from top on throttle control lever. To increase sensitivity, move spring end upward on governor arm. Refer to Fig. KO48.

Governor unit is accessible after removing crankshaft and camshaft. Gear and flyweight assembly turn on a stub shaft pressed into crankcase. To remove gear assembly, unscrew governor stop pin and slide gear off stub shaft. Remove governor cross-shaft by unscrewing governor bushing nut and removing shaft from inside. Renew gear assembly, cross-shaft or stub shaft if excessively worn or broken. If stub shaft must be renewed, press new shaft into block until it protrudes ⅜-inch (9.525 mm) above boss. Governor gear to stub shaft clearance is 0.0005-0.002 inch (0.0127-0.0508 mm).

IGNITION TIMING. A battery ignition system with externally mounted breaker points (Fig. KO49) is actuated by camshaft through a push rod. System is equipped with an automatic compression release located on camshaft and does not have an automatic timing advance. Ignition occurs at 20° BTDC at all engine speeds.

Initial breaker point gap is 0.020 inch (0.508 mm) but should be varied to obtain exact ignition timing as follows:

For static timing, disconnect spark plug lead from coil to prevent engine starting. Remove breaker point cover and rotate engine, by hand, in normal direction of rotation (clockwise, from flywheel side). Points should just begin to open when "S" mark appears in center of timing sight hole. Continue rotating engine until points are fully open. Check gap with a feeler gage. Gap setting can vary from 0.017-0.022 inch (0.4138-0.5588 mm), set to achieve smoothest running.

To set timing, using a power timing light, connect light to spark plug and with engine running at 1200-1800 rpm adjust points so "S" mark on flywheel is centered in timing sight hole.

Fig. KO49 – Ignition breaker points are located externally on engine crankcase on all models equipped with magneto or battery ignition systems.

LUBRICATION. A splash lubrication system is used. Oils approved by manufacturer must meet requirements of API service classification SE or SF.

For operation in temperatures above 32°F (0°C), use SAE 30 or SAE 10W-30 oil. For temperatures below 32° F (0°C) use SAE 5W-20 or 5W-30 oil.

It is recommended oil be changed at 25 hour intervals under normal operating conditions. Maintain oil level at "FULL" mark on dipstick but **DO NOT** overfill. Oil capacity of K-361 model is 2 quarts (1.9 L), K-361A model capacity is 1.5 quarts (1.4 L).

AIR CLEANER. Dry element type air cleaner should be cleaned every 50 hours of normal operation or more frequently if operating in dusty conditions. Remove dry element and tap element lightly on a flat surface to remove surface dirt. Do not wash element or attempt to clean element with compressed air. Renew element if extremely dirty or if it is bent, crushed or otherwise damaged. Make certain sealing surfaces of element seal effectively against back plate and cover.

REPAIRS

TIGHTENING TORQUES. Recommended tightening torques, with lightly lubricated threads, are as follows:

Spark plug . . . . 18-22 ft.-lbs. (24-30 N•m)
Connecting rod cap screws . . . 25 ft.-lbs. (34 N•m)
Cylinder head cap screws . . 30-35 ft.-lbs. (40-48 N•m)
Flywheel retainer nut . . . . . 60-70 ft.-lbs. (81-95 N•m)
Fuel pump screws (plastic pump) . . . . 5.8 ft.-lbs. (7.9 N•m)
Rocker arm housing screw . . . 15 ft.-lbs. (20 N•m)

CONNECTING ROD. Connecting rod and piston assembly is removed after removing oil pan and cylinder head. Aluminum alloy connecting rod rides directly on crankpin. Connecting rods are available in standard size as well as one for 0.010 inch (0.254 mm) undersize crankshafts. Desired running clearances are as follows:

Connecting rod to crankpin . . . . 0.001-0.002 in. (0.0254-0.0508 mm)
Connecting rod to piston pin . . . . 0.0003-0.0008 in. (0.0076-0.0203 mm)
Rod side play on crankpin . . . . 0.007-0.0175 in. (0.1778-0.4445 mm)

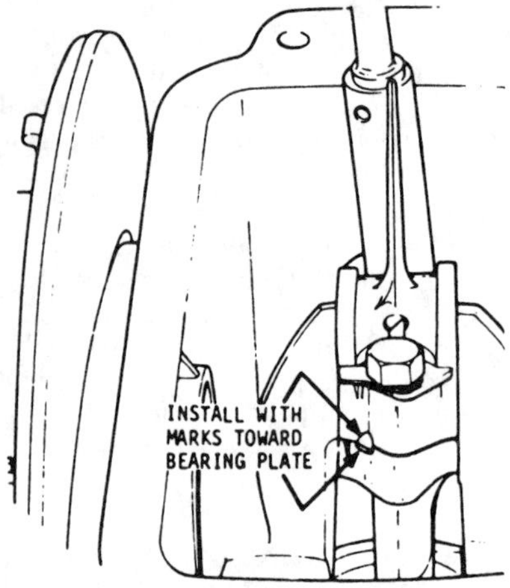

Fig. KO51 — When installing connecting rod and piston unit, be sure marks on rod and cap are aligned and are facing toward flywheel side of engine.

Standard crankpin diameter is 1.4495-1.500 inch (38.09-38.10 mm).

When reinstalling connecting rod and piston assembly, piston can be installed either way on rod, but make certain match marks (Fig. KO51) on rod and cap are aligned and are toward flywheel side of engine. Kohler recommends connecting rod cap screws to be torqued to 30 ft.-lbs. (40 N•m), loosened and retorqued to 25 ft.-lbs. (34 N•m). Cap screw threads should be lightly lubricated before installation.

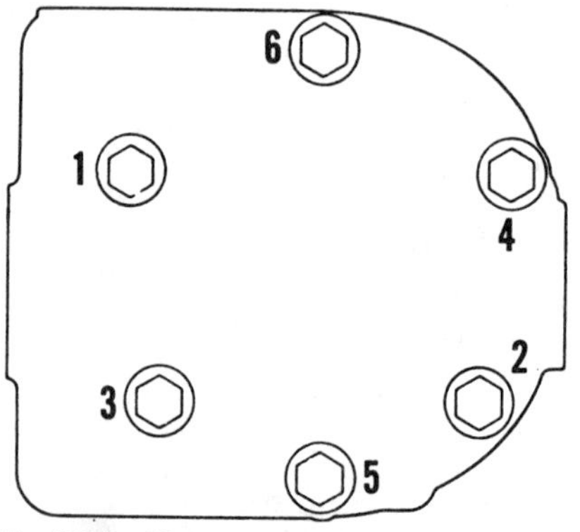

Fig. KO52 — Tighten cylinder head cap screws evenly, in sequence shown, to 30-35 ft.-lbs. (40-48 N•m).

PISTON, PIN AND RINGS. Aluminum alloy piston is fitted with two 0.093 inch (2.3622 mm) wide compression rings and one 0.187 inch (4.798 mm) wide oil control ring. Renew piston if scored or if side clearance of new ring in piston top groove exceeds 0.005 inch (0.127 mm). Pistons and rings are available in oversizes of 0.010, 0.020 and 0.030 inch (0.254, 0.508 and 0.762 mm) as well as standard size. Piston pin fit in piston bore should be from 0.0001 inch (0.0025 mm) interference to 0.0003 inch (0.0076 mm) loose. Standard piston pin diameter is 0.8753 inch (22.23 mm).

Recommended piston to cylinder bore clearance measured at thrust side at bottom of skirt is 0.003-0.0045 inch (0.0762-0.1143 mm). Recommended clearance when measured at thrust side just below oil ring is 0.007-0.0095 inch (0.1778-0.2413 mm).

Kohler recommends piston rings always be renewed if they are removed.

Piston ring specifications are as follows:

Ring end gap . . . . 0.010-0.020 in.
(0.254-0.508 mm)

Ring side clearance –

Compression ring . . . . . 0.002-0.004 in.
(0.0508-0.1016 mm)

Oil control ring . . . . 0.001-0.003 in.
(0.0254-0.0762 mm)

If compression ring has a groove or bevel on outside surface, install ring with groove or bevel down. If groove or bevel is on inside surface of compression ring, install ring with groove or bevel up. Oil control ring can be installed either side up.

CYLINDER BLOCK. Refer to Fig. KO59 for exploded view of engine. If cylinder wall is scored or bore is tapered more than 0.0015 inch (0.0381 mm) or out-of-round more than 0.005 inch (0.127 mm), cylinder should be bored to nearest suitable oversize of 0.010, 0.020, 0.030 inch (0.254, 0.508 or 0.762 mm). Standard cylinder bore is 3.750 inches (95.25 mm).

CYLINDER HEAD. Always use a new head gasket when installing cylinder head. Lightly lubricate cylinder head cap screws and tighten them evenly and in equal steps using sequence shown in Fig. KO52.

CRANKSHAFT. Crankshaft is supported by two ball bearings. Bearings have an interference fit with cylinder block of 0.0006-0.0022 inch (0.0152-0.0559 mm) and with bearing plate of 0.0012-0.0028 inch (0.0305-0.0711 mm). Ball bearing to crankshaft clearance is 0.0004 inch (0.0102 mm) interference to 0.0005 inch (0.0127 mm) loose.

Renew ball bearings if excessively loose or rough. Crankshaft end play should be 0.003-0.020 inch (0.0762-0.508 mm) and is controlled by varying thickness of bearing plate gaskets. Bearing plate gaskets are available in a variety of thicknesses.

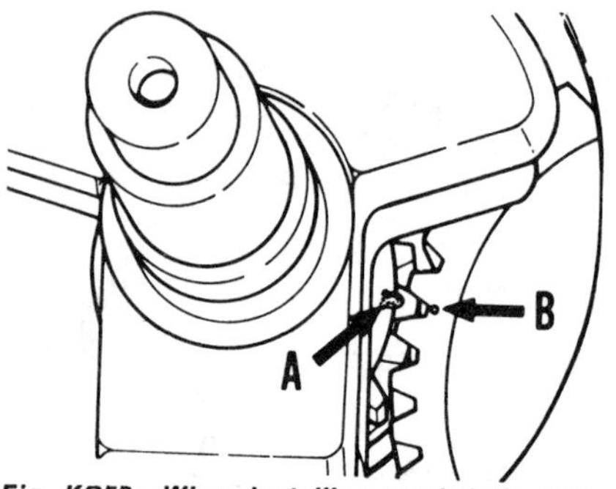

Fig. KO53 — When installing crankshaft, make certain timing mark (A) on crankshaft is aligned with timing mark (B) on camshaft gear.

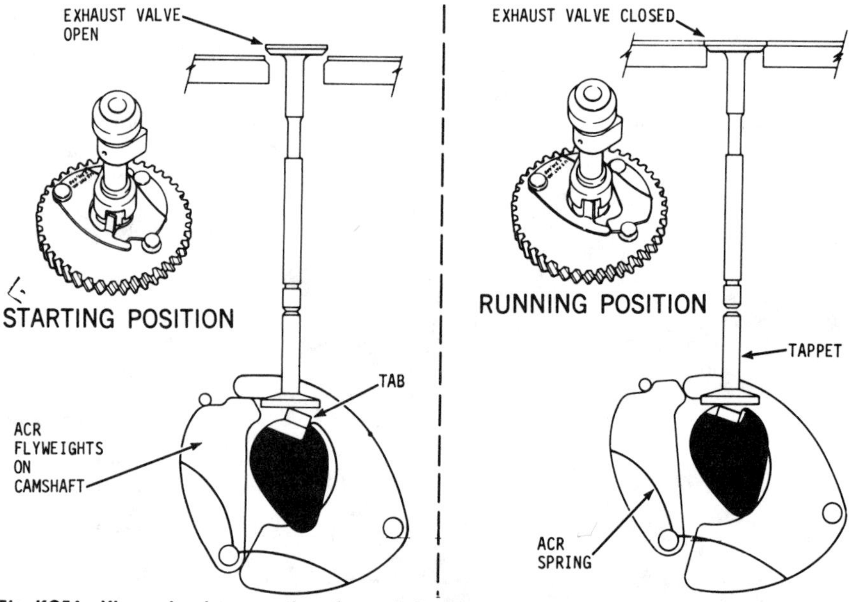

Fig. KO54 — Views showing operation of camshaft with automatic compression release. In starting position, spring has moved control lever which moves cam lever upward so tang is above exhaust cam lobe. This tang holds exhaust valve open slightly on a portion of the compression stroke to relieve compression while cranking engine. At engine speeds of 650 rpm or more, centrifugal force moves control lever outward allowing tang to move below lobe surface.

Standard crankpin journal diameter is 1.4995-1.500 inches (38.09-38.10 mm) and may be reground to 0.010 inch (0.254 mm) undersize.

When installing crankshaft, align timing marks on crankshaft and camshaft gears as shown in Fig. KO53. Refer to **DYNAMIC BALANCER** paragraph for installation and timing of balancer gears.

Kohler recommends crankshaft seals be installed in crankcase and bearing plate after crankshaft and bearing plate are installed. Carefully work oil seals over crankshaft and drive seals into place with hollow driver that contacts outer edge of seals.

CAMSHAFT. The hollow camshaft and integral camshaft gear turn on a pin that is a slip fit in flywheel side of crankcase and a drive fit in closed side of crankcase. Remove and install pin from open side (bearing plate side) of crankcase. Desired camshaft to pin running clearance is 0.001-0.0035 inch (0.0254-0.0889 mm). Desired camshaft end play of 0.005-0.010 inch (0.127-0.254 mm) is controlled by the use of 0.005 and 0.010 inch (0.127 and 0.254 mm) thick spacer washers between camshaft and cylinder block at bearing plate side of crankcase.

Camshaft is equipped with automatic compression release shown in Fig. KO54. Automatic compression release mechanism holds exhaust valve open during first part of compression stroke, reducing compression pressure and allowing easier cranking. Refer to Fig. KO54 for operational details. At speeds above 650 rpm, release mechanism is inactive. Release mechanism weights and weight pivot pins are not renewable separately from camshaft gear, but weight spring is available for service.

To check compression on engine equipped with automatic compression release, engine must be cranked at 650 rpm or higher to overcome release mechanism. A reading can also be obtained by rotating flywheel in reverse direction with throttle wide open. Compression reading for engine in good condition should be 110-120 psi (758-827 kPa). When reading falls below 100 psi (690 kPa), it indicates leaking rings or valves.

VALVE SYSTEM. Valve seats are hardened steel inserts cast into head. If seats become worn or damaged entire cylinder head must be renewed. Refer to Fig.KO56 for valve seat and guide details. Refer to Fig. KO60 for exploded view of valves and cylinder head.

Valve stem to guide operating clearance should be 0.0029-0.0056 inch

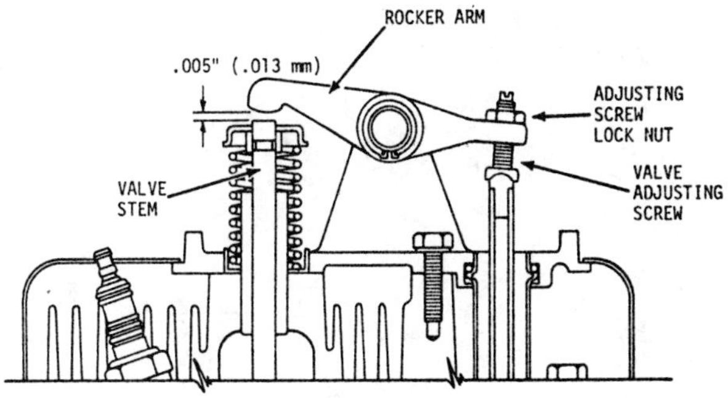

Fig. KO55—View of adjusting points on K-361 valve systems. Both intake and exhaust valve tappet clearance should be adjusted cold, to 0.005 inch (0.127 mm).

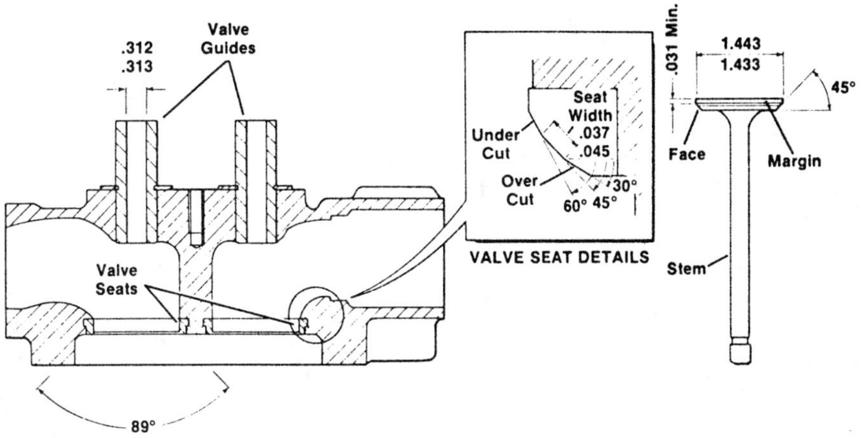

Fig. KO56—Principal valve service specifications illustrated. Refer to text for further details.

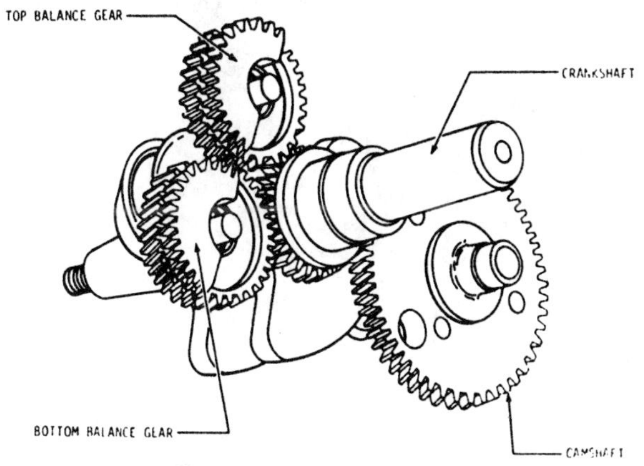

Fig. KO57—View showing components of dynamic balancer system.

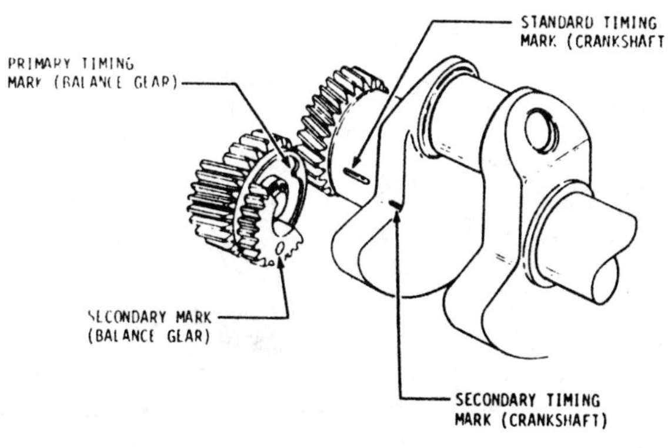

Fig. KO58—View showing timing marks for installing dynamic balance gears.

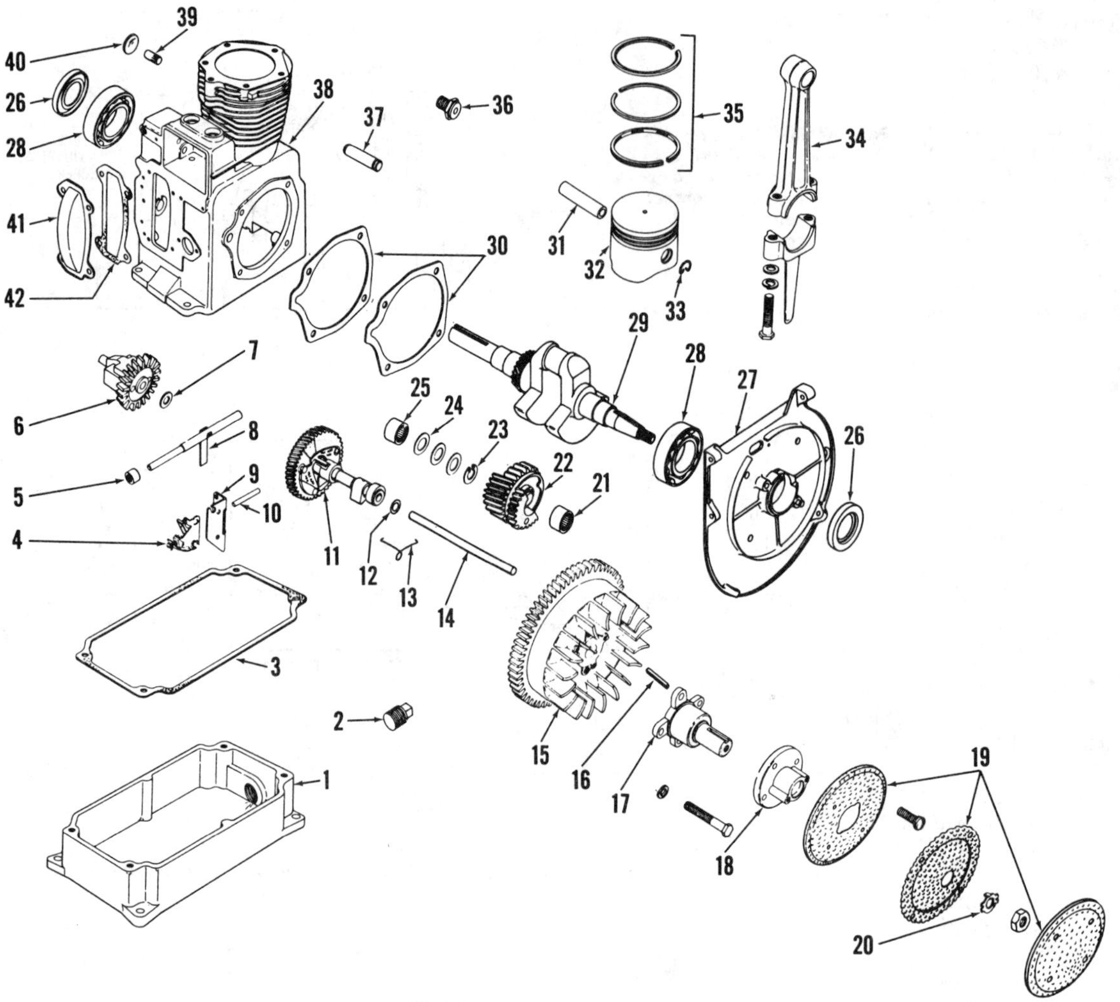

1. Oil pan
2. Oil drain plug
3. Oil pan gasket
4. Breaker points
5. Spacer
6. Governor gear
7. Thrust washer
8. Governor cross-shaft
9. Breaker plate
10. Breaker rod
11. Camshaft
12. Spacer
13. Actuating spring
14. Camshaft pin
15. Flywheel assy.
16. Key
17. Front drive shaft
18. Front drive adapter
19. Grass screen
20. Flywheel washer
21. Needle bearing
22. Balancer gear (2)
23. Retainer
24. Spacer
25. Needle bearing
26. Oil seal
27. Bearing plate
28. Main ball bearing
29. Crankshaft
30. Gasket
31. Piston pin
32. Piston
33. Retainer
34. Connecting rod assy.
35. Piston rings
36. Bushing
37. Shaft
38. Cylinder block assy.
39. Governor shaft
40. Expansion plug
41. Camshaft cover
42. Camshaft cover gasket

Fig. KO59 — Exploded view of K-361 basic engine assembly.

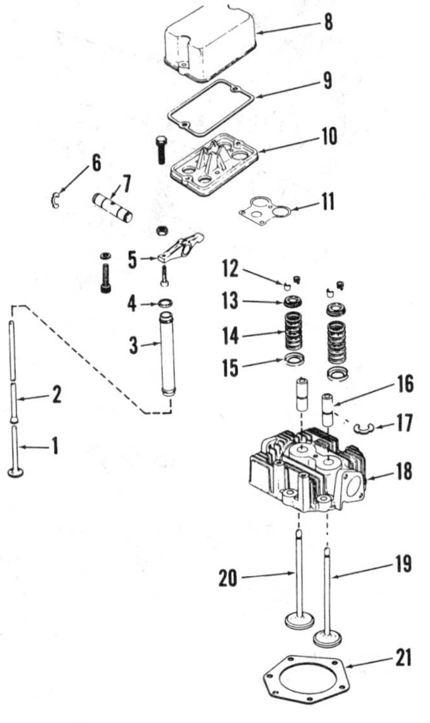

1. Tappet
2. Push rod
3. Push rod tube
4. "O" rings
5. Rocker arm
6. Retainer ring
7. Rocker arm shaft
8. Valve cover
9. Gasket
10. Rocker arm housing
11. Housing gasket
12. Valve keepers
13. Valve rotators
14. Valve springs
15. Spring retainers
16. Valve guides
17. Retainer ring
18. Cylinder head
19. Intake valve
20. Exhaust valve
21. Head gasket

Fig. KO60 — Exploded view of overhead valve system used on K-361 engines.

(0.0762-0.1422 mm) for exhaust valve and 0.001-0.0027 inch (0.0254-0.0686 mm) for intake valve.

Intake and exhaust valve clearance should be adjusted to 0.005 inch (0.127 mm) while cold. To adjust, disconnect spark plug wire and rotate engine by hand until both valves are seated and piston is at TDC. Loosen adjusting screw locknut (Fig. KO55) and turn adjusting screw in or out to obtain correct clearance. Tighten locknuts and recheck clearance.

DYNAMIC BALANCER. A dynamic balance system (Fig. KO57) is used on all K-361 engines. The two balance gears, equipped with needle bearings, rotate on two stub shafts which are pressed into bosses on pto side of crankcase. Snap rings secure gears on stub shafts and shim spacers are used to control gear end play. Balancer gears are driven by crankshaft in opposite direction of crankshaft rotation. Use following procedure to install and time dynamic balancer components.

To install new stub shafts, press shafts

Illustrations courtesy Kohler Co.

into special bosses in crankcase until they protrude 1.110 inches (28.194 mm) above thrust surface of bosses.

To install top balance gear-bearing assembly, first place one 3/8-inch spacer and one 0.010 inch shim spacer on stub shaft, then slide top gear assembly on shaft. Timing marks must face flywheel side of crankcase. Install one 0.005, one 0.010 and one 0.020 inch shim spacers in this order, then install snap ring. Using a feeler gage, check gear end play. Correct end play of balance gear is 0.002-0.010 inch (0.0508-0.254 mm). Add or remove 0.005 inch thick spacers as necessary to obtain correct end play. Always make certain a 0.020 inch thick spacer is next to snap ring.

Install crankshaft in crankcase and align primary timing mark on top balance gear with standard timing mark on crankshaft. See Fig. KO58. With primary timing marks aligned, engage crankshaft gear 1/16-inch (1.588 mm) into narrow section of top balancer gear. Rotate crankshaft to align timing marks on camshaft gear and crankshaft as shown in Fig. KO53. Press crankshaft into crankcase until it is seated firmly into ball bearing main.

Rotate crankshaft until crankpin is approximately 15° past bottom dead center. Install one 3/8-inch spacer and one 0.010 inch shim spacer on stub shaft. Align secondary timing mark on bottom balance gear with secondary timing mark on crankshaft. See Fig. KO58. Slide gear assembly into position on stub shaft. If properly timed, secondary timing mark on bottom balance gear will be aligned with standard timing mark on crankshaft after gear is fully on stub shaft. Install one 0.005 inch and one 0.020 inch shim spacer, then install snap ring. Check bottom balance gear end play and add or remove 0.005 inch thick spacers as required to obtain proper end play of 0.002-0.010 inch (0.0508-0.254 mm).

Always make certain a 0.020 inch thick spacer is next to snap ring.

KOHLER

KOHLER COMPANY
Kohler, Wisconsin 53044

Model	No. Cyls.	Bore	Stroke	Displacement	Power Rating
K-482	2	3.250 in. (82.55 mm)	2.875 in. (73.03 mm)	47.8 cu. in. (781.73 cc)	18 hp. (13.4 kW)
K-532	2	3.375 in. (85.73 mm)	3.0 in. (76.2 mm)	53.67 cu. in. (879.7 cc)	20 hp. (14.9 kW)
K-582	2	3.5 in. (88.9 mm)	3.0 in. (76.2 mm)	57.7 cu. in. (946 cc)	23 hp. (17.2 kW)
K-662	2	3.625 in. (92.08 mm)	3.250 in. (82.55 mm)	67.2 cu. in. (1099.4 cc)	24 hp. (17.9 kW)

All engines in this section are four cycle, two cylinder opposed, horizontal crankshaft engines. Crankshaft for K-482, K-532 and K-582 models is supported by a ball bearing at output end and a sleeve type bearing at output end and a sleeve type bearing at flywheel end of crankshaft. Crankshaft for K-662 model is supported at each end by tapered roller bearings.

Connecting rod for K-482, K-532 and K-582 models rides directly on crankpin journal and K-662 model connecting rod is equipped with a renewable type precision bearing insert. Pressure lubrication is provided by a gear type oil pump driven by crankshaft gear.

Either a battery type or magneto type ignition system is used according to model and application.

A side draft, float type carburetor is used for K-482, K-532 and K-582 models and a down draft, float type carburetor is used on K-662 model.

Engine model, serial and specification numbers must be given when ordering parts or service material.

MAINTENANCE

SPARK PLUG. Recommended spark plug is Champion H10 or equivalent for K-482, K-532 and K-582 models or Champion J8 or equivalent for K-662 model. Electrode gap is 0.025 inch (0.635 mm) for all models except K-582. Electrode gape for K-582 is 0.035 inch (0.880 mm).

CARBURETOR. Refer to Fig. KO61 for exploded view of Kohler side draft carburetor used on Models K-482, K-532 and K-582 engines and to Figs. KO62 or KO63 for Carter or Zenith downdraft carburetors used on Model K-662 engines. For initial adjustment on all models, open idle fuel needle 1¼ turns and open main fuel needle 2 turns. Make final adjustment with engine at operating temperature and running. Operate engine under load and adjust main fuel needle for leanest mixture that will allow satisfactory acceleration and steady governor operation.

Adjust idle speed stop screw to maintain a low idle speed of 1000 rpm. Then, adjust idle mixture needle for smoothest idle operation.

Since main fuel and idle fuel adjustments affect each other, recheck

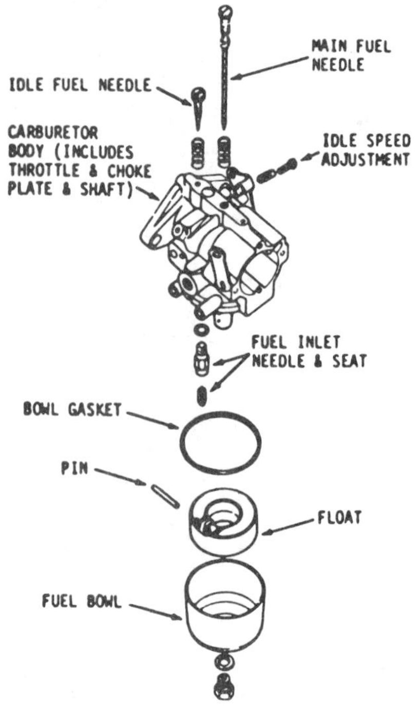

Fig. KO61 – Exploded view of Kohler carburetor used on K-482, K-532 and K-582 models.

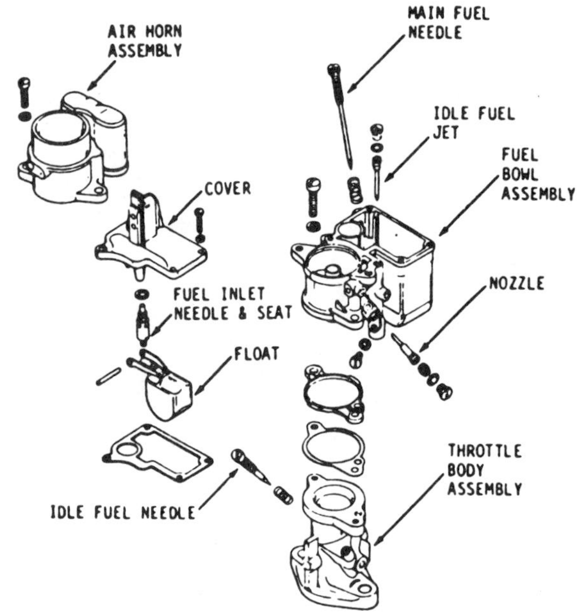

Fig. KO62 – Exploded view of Carter downdraft carburetor used on some K-662 engines.

Illustrations courtesy Kohler Co.

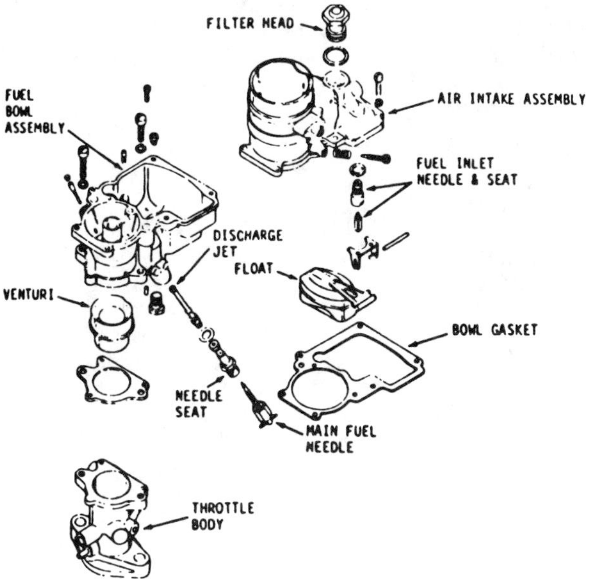

Fig. KO63—Exploded view of Zenith downdraft carburetor used on some K-662 engines.

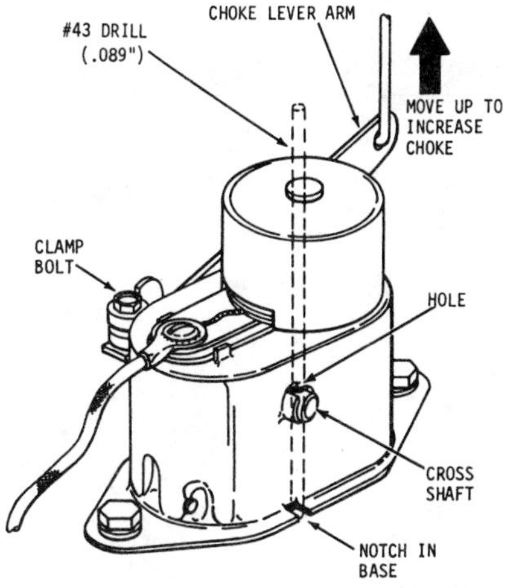

Fig. KO63B—To adjust choke, remove air cleaner and insert round rod (number 43 drill) through cross-shaft until it engages notch in base. Loosen clamp bolt on choke lever and adjust choke plate to desired setting. Tighten clamp bolt.

engine operation and readjust fuel needles as necessary.

To check float level, invert carburetor body and float assembly. There should be 11/64-inch (4.366 mm) clearance between machined surface of body casting and free end of float on Kohler carburetor used on Models K-482, K-532 and K-582. On Carter or Zenith carburetors used on Model K-662, there should be a distance of 1½ inches (38.1 mm) between machined surface of casting and top of float. Adjust as necessary by bending float lever tang that contacts inlet valve.

AUTOMATIC CHOKE. One of two different types automatic choke may be used according to model and application. One is an integral part of the downdraft carburetor while the other is mounted on exhaust manifold and choke plate is controlled through external linkage.

To adjust type which is integral with carburetor, loosen the three screws (Fig. KO63A) and rotate housing in clockwise direction to close, or counterclockwise to open.

To adjust exhaust manifold mounted type, remove air cleaner and move choke arm (Fig. KO63B) until hole in brass shaft aligns with slot in bearing and insert a small round rod as shown. Loosen clamp bolt on choke lever and adjust choke plate to desired position. Tighten clamp bolt and remove rod.

The electrical lead on this choke is connected so current flows to thermostatic element when ignition is switched on. Tension of thermostatic spring should be set to allow full choke at starting. Current through heating element controls tension of thermostatic spring, which gradually opens choke as engine warms up.

Choke on all models should be fully open when engine has reached operating temperature.

GOVERNOR. An externally mounted centrifugal flyweight type governor is used on all engines. Governor is driven by the camshaft gear and is lubricated by an external oil line connected to engine lubrication system.

GOVERNOR INSTALLATION AND TIMING (K-482, K-532 and K-582). Ignition breaker points are mounted on governor housing on Models K-482, K-532 and K-582 and are activated by a push rod which rides on a cam on governor drive shaft. Therefore, governor must be timed to engine.

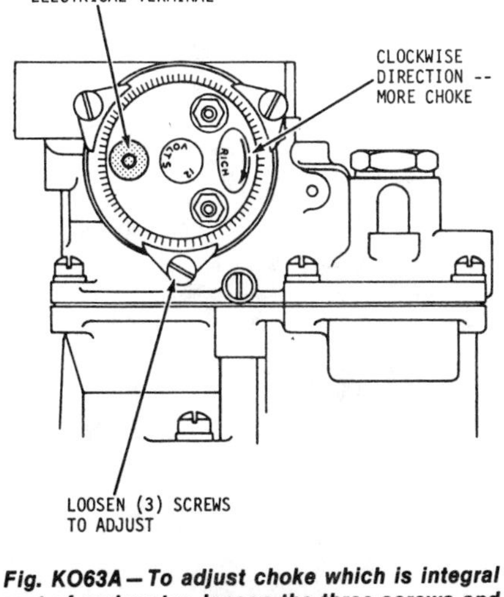

Fig. KO63A—To adjust choke which is integral part of carburetor, loosen the three screws and rotate housing clockwise to increase or counterclockwise to decrease choke action.

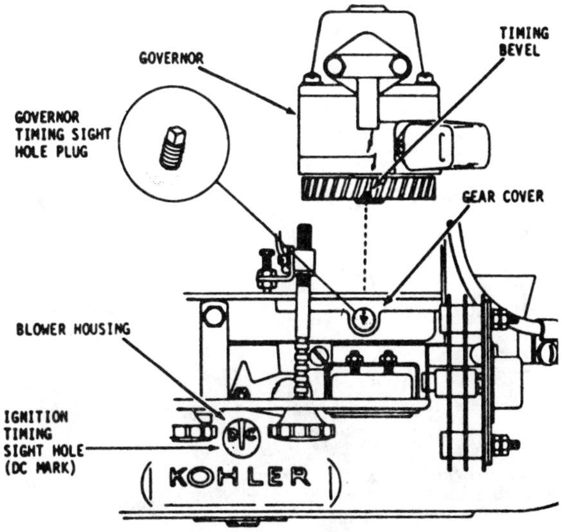

Fig. KO64—Top view showing location of sight holes and timing marks used when installing governor assembly on K-482, K-532 or K-582 models.

Illustrations courtesy Kohler Co.

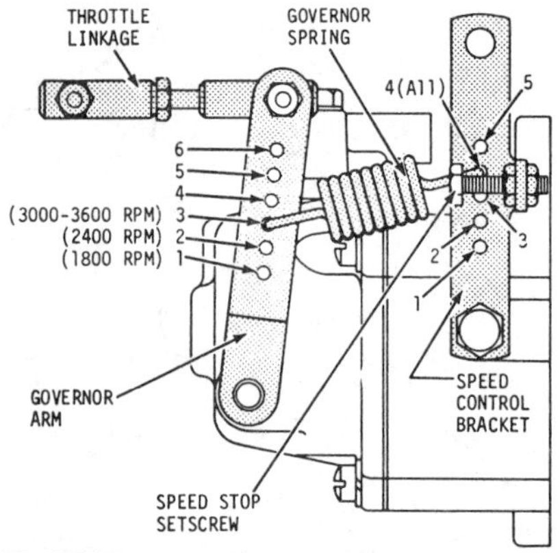

Fig. KO65 — View of governor linkage, maximum speed stop set screw and governor spring used on K-482, K-532 and K-582 models. Governor sensitivity is adjusted by moving governor spring to alternate holes.

To install governor assembly, first uncover ignition timing sight hole in top center of blower housing. See Fig. KO64. Rotate engine crankshaft until "DC" mark on flywheel is centered in sight hole. Remove governor timing sight hole plug from engine gear cover. Install governor assembly so tooth with special timing bevel on governor gear is centered in governor timing sight hole. Install governor retaining cap screws. Recheck to make certain both the "DC" mark on flywheel and beveled tooth on governor gear are centered in sight holes, then install sight hole plug and cover.

After governor is installed, readjust ignition timing as outlined in **IGNITION TIMING** paragraph.

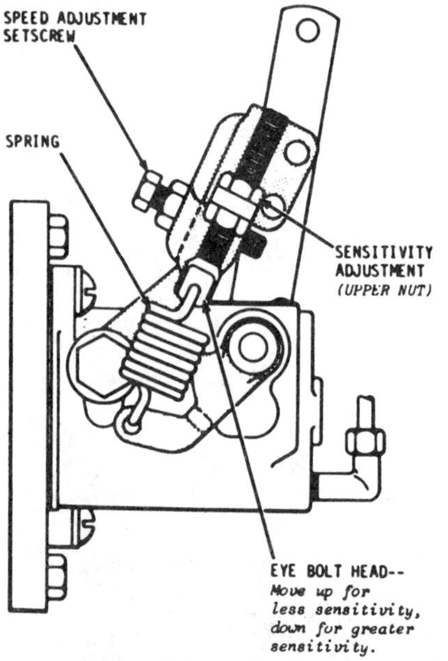

Fig. KO66 — View of governor spring, maximum speed adjustment set screw and sensitivity adjustment on K-662 model governor.

GOVERNOR INSTALLATION (K-662). Since breaker points are not mounted on or driven off governor on Model K-662, governor does not have to be timed to the engine. Installation of governor assembly is obvious.

GOVERNOR ADJUSTMENT. Model K-662 has a slightly different adjustment procedure due to the use of a different governor assembly. Make certain correct procedure is followed for model being serviced and set idle and high speed adjustments.

To adjust governor, first move governor arm forward to high idle position and check to see that throttle linkage moves carburetor throttle shaft to wide open position. If not, adjust length of throttle linkage as necessary. Make certain throttle linkage moves freely.

Start engine and move speed control lever to high idle position. Using a tachometer, check engine speed. Loosen locknuts and adjust high speed stop set screw on speed control bracket (Fig. KO65 or KO66) to obtain maximum no load speed (3800 rpm). Retighten locknuts.

Refer to Fig. KO65 for proper placement of governor spring in hole of governor arm and control bracket. If governor surging or hunting occurs, governor is too sensitive and governor spring should be moved to a set of holes closer together. If a big drop in speed occurs when normal load is applied, governor should be set for greater sensitivity. Increase spring tension by placing spring in holes spaced further apart. Move spring one hole at a time and recheck engine operation and high idle rpm after each adjustment.

On Model K-662, governor sensitivity is adjusted by loosening one nut and tightening opposite nut on spring eye bolt. See Fig. KO66. If governor hunting or surging occurs, governor is too sensitive. To correct, loosen bottom nut and tighten top nut. To increase sensitivity, loosen top nut and tighten bottom nut. After adjustment, recheck engine high idle rpm as changing sensitivity will affect high idle speed.

IGNITION TIMING (K-482, K-532 and K-582). Ignition points are externally mounted on governor housing and are activated by a push rod which rides on a cam on governor drive shaft. An automatic spark advance-retard mechanism is incorporated in the governor which allows engine to start at 8° BTDC and run (1200 rpm or above) at 27° BTDC.

To adjust ignition timing, first remove breaker point cover and adjust breaker contact gap to 0.020 inch (0.508 mm). Install breaker point cover, then remove cover from ignition timing sight hole in top center of blower housing. Attach timing light to number "1" spark plug (cylinder nearest to flywheel). Start engine and operate at 1200 rpm or above. Aim timing light at ignition timing sight hole. The "SP" mark on flywheel should be centered in sight hole when light flashes. See Fig. KO67. If "SP" mark is not centered in sight hole, loosen governor mounting cap screws and rotate governor assembly until the "SP" (27° BTDC) mark is centered as light flashes. Retighten governor mounting cap screws.

If slotted holes in governor flange will not allow enough rotation to center the timing mark, check governor timing as outlined in **GOVERNOR INSTALLATION and TIMING** paragraph.

IGNITION TIMING (K-662). Two types of external, self-contained magnetos have been used on Model M-662 engines. Standard magneto (Fig.

Fig. KO67 — Breaker points are externally mounted on K-482, K-532 and K-582 models governor housings. "SP" (27° BTDC) mark should be centered in sight hole at 1200 rpm or above. Refer to text.

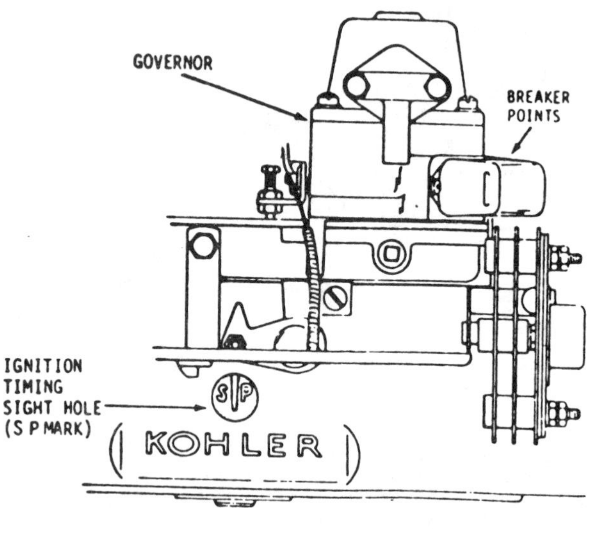

Illustrations courtesy Kohler Co.

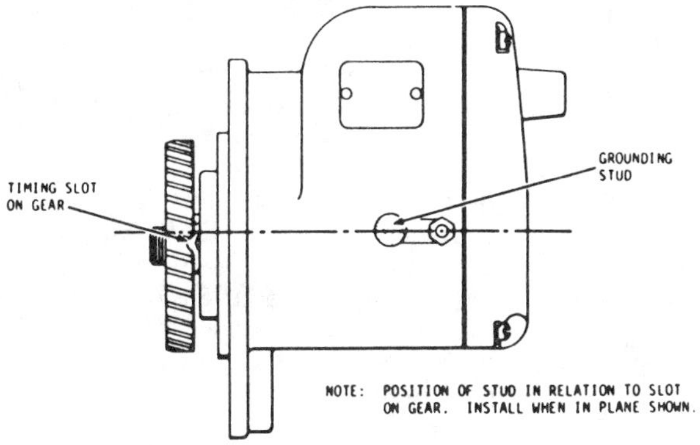

Fig. KO68—View of standard (simultaneous firing) magneto used on some K-662 engines.

KO68) is referred to as a simultaneous firing magneto. This type fires both spark plugs at the same time. Ignition occurs only in cylinder in which piston is on compression stroke. At this time, piston in opposite cylinder is on exhaust stroke and that spark is ineffective.

The other magneto (Fig. KO69) is classified as a distributor type magneto. This magneto is equipped with a distributor rotor with directs ignition voltage to the proper spark plug at appropriate time.

Breaker point gap on either type magneto is 0.015 inch (0.381 mm).

To install and time standard type magneto, first crank engine over slowly by hand until "DC" mark on flywheel is centered in timing sight hole (A-Fig. KO70). Rotate magneto drive gear until impulse coupling trips. At this time, "DC" mark on flywheel should be just past center of timing sight hole. To check and adjust running timing, attach a timing light, start engine and operate at 1200 rpm or above. Aim timing light at timing sight hole (A–Fig. KO70). The "SP" mark on flywheel should be centered in sight hole when timing light flashes. If not, loosen magneto mounting cap screws and rotate magneto assembly until the "SP" (22° BTDC) mark is centered in sight hole. Tighten mounting cap screws.

To install and time distributor type magneto (Fig. KO69), first remove spark plug from No. 1 cylinder (cylinder closest to flywheel). Slowly crank engine over until No. 1 piston is on compression stroke and "M" mark on engine flywheel is centered in ignition sight hole (A-Fig. KO70). Turn magneto drive gear counter-clockwise (facing gear) until white mark is centered in timing window on magneto. See Fig. KO69. Using a new flange gasket, carefully install magneto assembly. Tighten retaining cap screws securely. Slowly crank engine in normal direction of rotation until impulse coupling trips. At this time, "DC" mark on flywheel should be just center of timing sight hole. To check and adjust running timing, attach a timing light, start engine and operate at 1200 rpm or above. Aim timing light at timing sight hole (A-Fig. KO70). The "SP" (22° BTDC) mark should be centered in sight hole. Tighten mounting cap screws.

COMPRESSION TEST. Results of a compression test can be used to partially determine the condition of the engine. To check compression, remove spark plugs, make certain air cleaner is clean and set throttle and choke in wide open position. Crank engine with starting motor to a speed of about 1000 rpm. Insert gage in spark plug hole and take several readings on both cylinders. Consistent readings in the 110-120 psi (758-827 kPa) range indicate good compression. When compression reading falls below 100 psi (690 kPa) on either cylinder, valve leakage or excessive wear in cylinder, or ring wear are indicated.

If compression reading is higher than 120 psi (827 kPa), it indicates excessive carbon deposits have built up in combustion chamber. Remove cylinder heads and clean carbon from heads.

LUBRICATION. A gear type oil pump supplies oil through internal galleries to front main bearing, camshaft bearings, connecting rods and other wear areas. A full flow, spin-on type oil filter, located on crankcase in front of number "2" cylinder on Models K-482, K-532 and K-582, or just to the rear of number "1" cylinder on Model K-662, or a remote mounted canister type oil filter connected to engine via two pressure hoses located on left side of engine block, are used in the oil pressure system. Normal oil pressure for all models when operating at normal temperature should be 25 psi (172 kPa) at 1200 rpm, 30-50 psi (207-345 kPa) at 1800 rpm, 35-55 psi (241-379 kPa) at 2200 rpm and 45-65 psi (310-448 kPa) at 3200-3600 rpm. On Models K-482, K-532 and K-582, an oil pressure relief valve is located on crankcase just forward of number "1" cylinder (cylinder nearest to flywheel). To adjust oil pressure, loosen locknut and turn adjusting screw clockwise to increase pressure or counter-clockwise to decrease pressure. Retighten locknut.

High-quality detergent-type oil, having API classification "SE" or "SF," is recommended for use in Kohler engines. Use SAE 30 oil if temperature is above 32° F (0° C) and use SAE 5W-20 or 5W-30 if temperature is below 32° F (0° C).

Maintain crankcase oil level at full mark on dipstick, but do not overfill.

Recommended oil change interval is every 50 hours of normal operation. Oil capacity is 3.0 qts. (2.8 L). Add an extra 0.5 qt. (0.47 L) when filter on K-482, K-532 and K-582 is changed and an extra 1.0 qt. (0.95 L) to K-662 when filter is changed.

CRANKCASE BREATHER (K-482, K-532 and K-582). A one-way breather valve (2–Fig. KO71) located on top of governor housing is connected by a hose to air inlet side of carburetor. The breather system maintains a slight

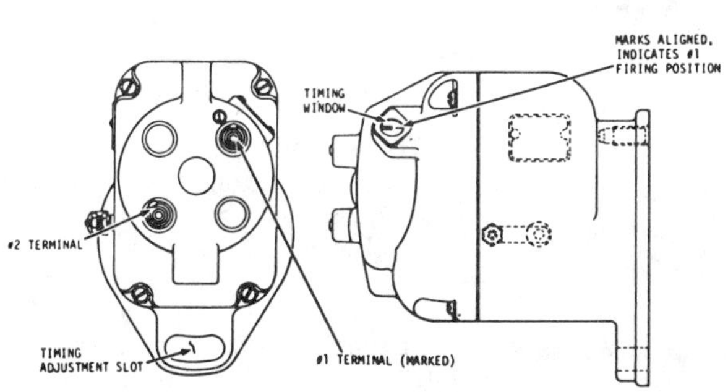

Fig. KO69—View of distributor type magneto used on some K-662 engines.

Fig. KO70—Location of ignition timing sight hole (A) on K-662 model engine.

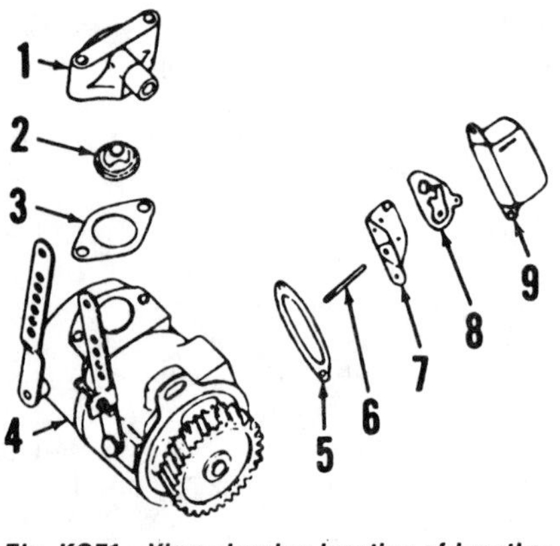

Fig. KO71 — View showing location of breather, valve assembly and breaker points on K-482, K-532 or K-582 models governor assembly. Breather valve (2) and valve housing (1) are available as an assembly.

1. Valve housing
2. Breather valve
3. Gasket
4. Governor assy.
5. Gasket
6. Breaker push rod
7. Bracket
8. Breaker points
9. Cover

vacuum within the crankcase. Vacuum should be approximately 12 inches on a water manometer with engine operating at 3600 rpm. If manometer shows crankcase pressure, breather valve is faulty, oil seals or gaskets are worn or damaged or engine has excessive blow-by due to worn rings.

Breather valve is pressed into valve housing (1) and is available only as an assembly with the housing.

CRANKCASE BREATHER (K-662). A reed type breather valve (Fig. KO72) located on top of gear cover maintains a partial vacuum within engine crankcase. Vacuum should be approximately 16 inches on a water manometer with engine operating at 3200 rpm. If manometer shows crankcase pressure, breather valve is faulty, oil seals or gaskets are worn or damaged, or engine has excessive blow-by due to worn rings.

Breather components must be correctly installed to function properly. See Fig. KO72. Reed must be free of rust, dents or cracks and must lay flat on breather ports in gear cover. Tabs on reed stop must be 0.020 inch (0.508 mm) above reed.

A closed system was also used on some K-662 models and all parts are the same except for a special housing which is connected to air inlet side of carburetor by a tube and provides a positive draw on breather whenever engine is in operation.

REPAIRS

TIGHTENING TORQUES. Recommended tightening torques with lightly lubricated threads are as follows:

Spark plug . . . . 22 ft.-lbs (30 N•m)
Camshaft nut –
K-482, K-532 & K-582 . . . . . . 40 ft.-lbs. (54 N•m)
K-662 . . . . 25 ft.-lbs. (34 N•m)
Connecting rod cap screws –
K-482, K-532 & K-582 . . . . . . 25 ft.-lbs. (34 N•m)
K-662 . . . . 35 ft.-lbs. (48 N•m)
Cylinder head cap screws –
K-482, K-532 & K-582 . . . . . . 35 ft.-lbs. (48 N•m)
K-662 . . . . 40 ft.-lbs. (54 N•m)
Closure plate cap screws –
K-482, K-532, K-582 . . . . 30 ft.-lbs. (40 N•m)
K-662 . . . . 50 ft.-lbs. (68 N•m)
Flywheel nut –
K-482, K-532 & K-582 . . . . . 115 ft.-lbs. (156 N•m)
K-662 . . . . 130 ft.-lbs. (176 N•m)
Oil pan cap screws –
K-482, K-532 & K-582 . . . . . . 30 ft.-lbs. (40 N•m)
K-662 . . . . 45 ft.-lbs. (61 N•m)

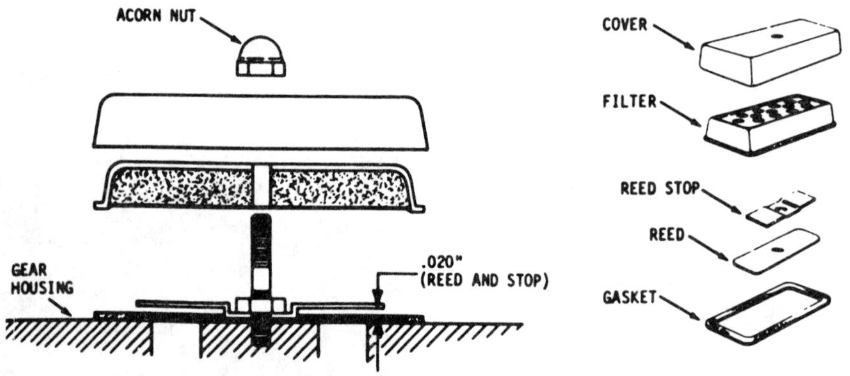

Fig. KO72 — Reed type crankcase breather used on K-662 models. Tabs on reed stop must be 0.020 inch (0.508 mm) above reed as shown.

CONNECTING RODS. Connecting rod and piston assemblies can be removed after removing oil pan (engine base) and cylinder heads. Identify each rod and piston assembly so they can be reinstalled in their original cylinders and do not intermix connecting rod caps.

On Models K-482, K-532 and K-582, connecting rods ride directly on the crankpins. Connecting rods 0.010 inch (0.254 mm) undersize are available for undersize reground crankshafts. Standard crankpin diameter is 1.6245-1.6250 inches (41.262-41.275 mm). Desired running clearances are as follows:

Connecting rods to crankpins . . . . 0.001-0.0035 in. (0.0254-0.0889 mm)
Connecting rods to piston pins . . . . 0.0003-0.0008 in. (0.0076-0.0203 mm)
Rod side play on crankpins . . . . 0.005-0.014 in. (0.1270-0.3556 mm)

On K-662 model connecting rods are equipped with renewable insert type bearings and a renewable piston pin bushing. Bearings are available for 0.002, 0.010 and 0.020 inch (0.0508, 0.254 and 0.508 mm) undersize crankshafts. Standard crankpin diameter is 1.8745-1.8750 inches (47.612-47.625 mm). Desired running clearances are as follows:

Connecting rods to crankpins . . . . 0.0003-0.0035 in. (0.0076-0.0889 mm)
Connecting rods to piston pins . . . . 0.0001-0.0006 in. (0.0025-0.0152 mm)
Rod side play on crankpins . . . . 0.007-0.011 in. (0.1778-0.2794 mm)

On all models, when installing connecting rod and piston units, make certain raised match marks on rods and rod caps are aligned and are toward flywheel side of engine. Kohler recommends connecting rod cap screws of K-482, K-532 and K-582 be tightened evenly to a torque of 30 ft.-lbs. (40 N•m), then loosened and retorqued to 25 ft.-lbs. (34 N•m). Kohler recommends connecting rod cap screws on K-662 engines be tightened to 40 ft.-lbs. (54 N•m), then loosened and retorqued to 35 ft.-lbs. (48 N•m).

On all models connecting rod cap screw threads should be lightly lubricated before installation.

PISTONS, PINS AND RINGS. Aluminum alloy pistons are fitted with two compression rings and one oil control ring. Renew pistons if scored or if side clearance of new ring in piston top ring groove exceeds 0.006 inch (0.1524 mm). Pistons are available in oversizes of 0.010, 0.020 and 0.030 inch (0.254,

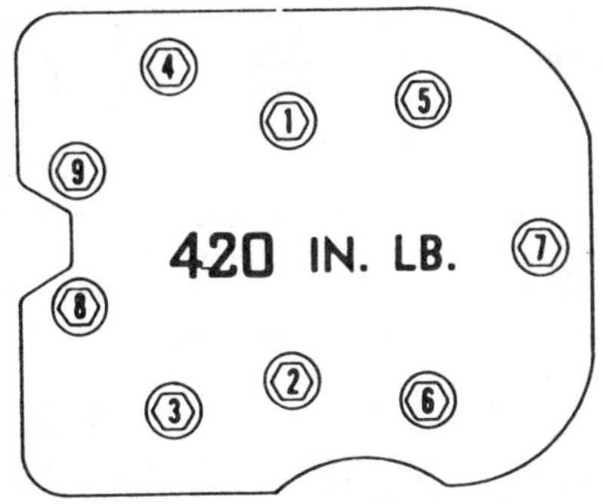

Fig. KO73—Tighten cylinder head cap screws on K-482, K-532 and K-582 models to 35 ft.-lbs. (48 N·m) using sequence shown.

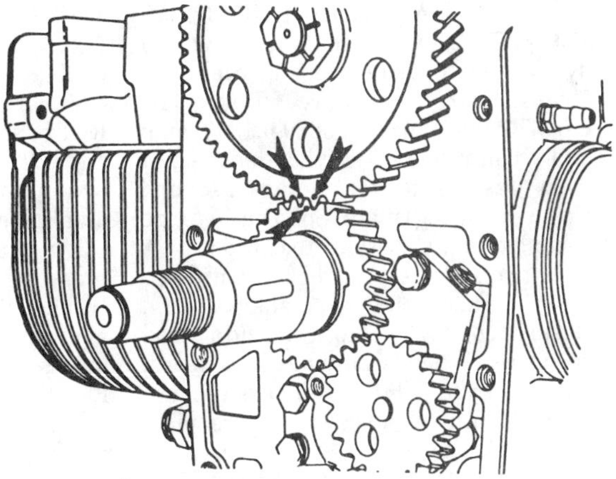

Fig. KO75—View showing timing marks on camshaft gear and crankshaft gear on K-482, K-532 and K-582 models. Model K-662 is similar.

0.508 and 0.762 mm) as well as standard size. Recommended piston to cylinder bore clearance when measured just below oil ring at right angle to pin is as follows:

K-482 . . . . 0.007-0.009 in. (0.1778-0.2286 mm)
K-532 . . . . 0.0065-0.0095 in. (0.1651-0.2413 mm)
K-582 . . . . 0.007-0.010 in. (0.178-0.254 mm)
K-662 . . . . 0.001-0.003 in. (0.0254-0.0762 mm)

Piston pin fit in piston boss on K-482, K-532 and K-582 should be from 0.000 (light interference) to 0.0005 inch (0.01270 mm) loose and for K-662 should be 0.0001-0.0003 inch (0.0025-0.0076 mm) loose.

Piston pin fit in rod on K-482, K-532 and K-582 should be 0.0003-0.0008 inch (0.0076-0.0203 mm) loose. Model K-662 connecting rod has a renewable bushing for piston pin and piston pin fit should be 0.0001-0.0006 inch (0.0025-0.0152 mm) loose. Always renew piston pin retaining rings.

Kohler recommends piston rings should always be renewed when they are removed.

Piston ring end gap for all models should be 0.010-0.020 inch (0.254-0.508 mm) for new bores and a maximum of 0.030 inch (0.762 mm) for used bores.

Piston ring specifications are as follows:

Ring width–
Compression rings . . . . 0.093 in. (2.36 mm)
Oil ring . . . . 0.187 in. (4.750 mm)

Ring side clearance (K-482, K-532 and K-482)–
Top ring . . . . 0.002-0.004 in. (0.0508-0.1016 mm)
2nd ring . . . . 0.0015-0.0035 in. (0.0381-0.0889 mm)
Oil ring . . . . 0.001-0.003 in. (0.0254-0.0762 mm)

Ring side clearance (K-662)–
Top ring . . . . 0.0025-0.0045 in. (0.0635-0.1143 mm)
2nd ring . . . . 0.0025-0.0045 in. (0.0635-0.1143 mm)
Oil ring . . . . 0.002-0.0035 in. (0.0508-0.0889 mm)

Piston rings are available in standard size and oversizes of 0.010, 0.020 and 0.030 inch (0.254, 0.508 and 0.762 mm).

CYLINDERS. If cylinder walls are scored or bores are tapered or out-of-round more than 0.005 inch (0.127 mm), cylinder should be bored to nearest suitable oversize of 0.010, 0.020 or 0.030 inch (0.254, 0.508 or 0.762 mm). Standard cylinder bore for K-482 is 3.250 inches (82.55 mm), for K-532 is 3.375 inches (85.73 mm), for K-582 is 3.5 inches (88.9 mm) and for K-662 is 3.625 inches (92.08 mm).

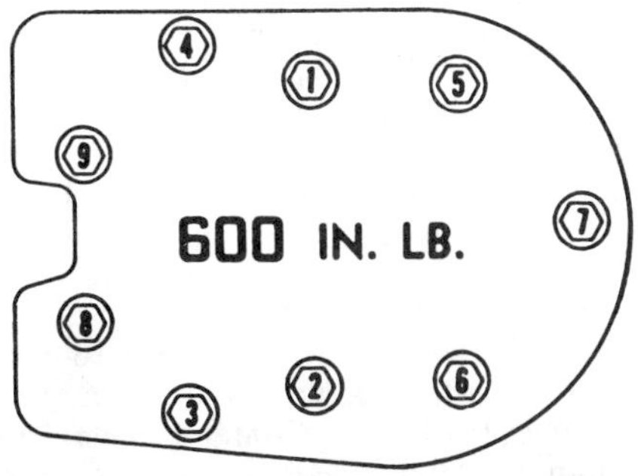

Fig. KO74—Tighten cylinder head cap screws on K-662 model to 50 ft.-lbs. (68 N·m) using sequence shown.

CYLINDER HEADS. Always use new head gaskets when installing cylinder heads. Tighten cylinder head cap screws evenly to a torque of 35 ft.-lbs. (48 N·m) on Models K-482, K-532 and K-582 using sequence shown in Fig. KO73 or 50 ft.-lbs. (68 N·m) on Model K-662 using sequence shown in Fig. KO74.

CRANKSHAFT AND MAIN BEARING (K-482, K-532 and K-582). Crankshaft is supported by a ball bearing at pto end of crankshaft and a sleeve type bearing at flywheel end of shaft.

The sleeve bearing is pressed into front of crankcase with an interference fit of 0.0015-0.0045 inch (0.038-0.114 mm). Standard front crankshaft journal diameter is 1.7490-1.750 inches (44.44-44.45 mm) and normal running clearance is 0.002-0.0035 inch (0.0508-0.0889 mm) in sleeve bearing.

The rear main (ball bearing) is secured to crankshaft and closure plate with retaining rings. If retaining rings, grooves and bearing are in good condition, crankshaft end play will be within recommended range of 0.004-0.010 inch (0.1016-0.254 mm).

Standard crankpin diameter is 1.6245-1.6250 inches (41.26-41.28 mm). If crankpin journals are scored or out-of-round more than 0.0005 inch (0.0127 mm) they may be reground 0.010 inch (0.254 mm) undersize for use with connecting rod for undersize shaft.

Renewable crankshaft gear is a light press fit on crankshaft. When installing crankshaft gear, align single punch mark on crankshaft gear with two punch marks on camshaft gear as shown in Fig. KO75.

Crankshaft oil seals can be installed after gear cover and closure plate are installed. Carefully work oil seals over crankshaft and drive seals into place with hollow driver that contacts outer edge of seals.

Fig. KO76—Front crankshaft seal on Model K-662 rotates with the crankshaft and seals against the gear cover.

CRANKSHAFT AND MAIN BEARING (K-662). Crankshaft is supported by tapered roller bearings at each end. Crankshaft end play should be 0.0035-0.0055 inch (0.0889-0.1397 mm) and is controlled by varying thickness of shim gaskets between crankcase and closure (bearing) plate. Shim gaskets are available in a variety of thicknesses.

Oil transfer sleeves which are pressed into flywheel end of crankcase and closure convey oil to connecting rod bearings. When installing oil sleeves, make certain oil holes in sleeves are aligned with oil holes in closure plate and crankcase.

Standard crankpin diameter is 1.8745-1.8750 inches (47.61-47.63 mm). Connecting rod bearings of 0.002 inch (0.0508 mm) oversize are available for use with moderately worn crankpin. Rod bearings of 0.010 inch (0.254 mm) and 0.020 inch (0.508 mm) oversize are available for use with reground crankpins.

Renewable crankshaft gear is a light press fit on crankshaft. When installing crankshaft gear, align single punch mark on camshaft gear with two punch marks on camshaft gear as shown in Fig. KO75.

To install rear (pto end) oil seal, carefully work seal, with lip inward, over end of shaft. Use a close fitting seal driver and tap seal into closure plate. Front oil seal (Fig. KO76) is a compression type seal which rotates with crankshaft and seals against gear cover.

CAMSHAFT. Camshaft is supported in two renewable bushings. To remove camshaft, first remove flywheel, gear cover, governor assembly and fuel pump. On Model K-662, remove magneto assembly. On all models, remove valve covers, valve spring keepers, spring retainers and valve springs. Wedge a piece of wood between camshaft gear and crankshaft gear and remove gear retaining nut and washer. Attach a suitable puller and remove camshaft gear and Woodruff key. Unbolt and remove gear cover plate from crankcase. Move tappets away from camshaft and carefully withdraw camshaft.

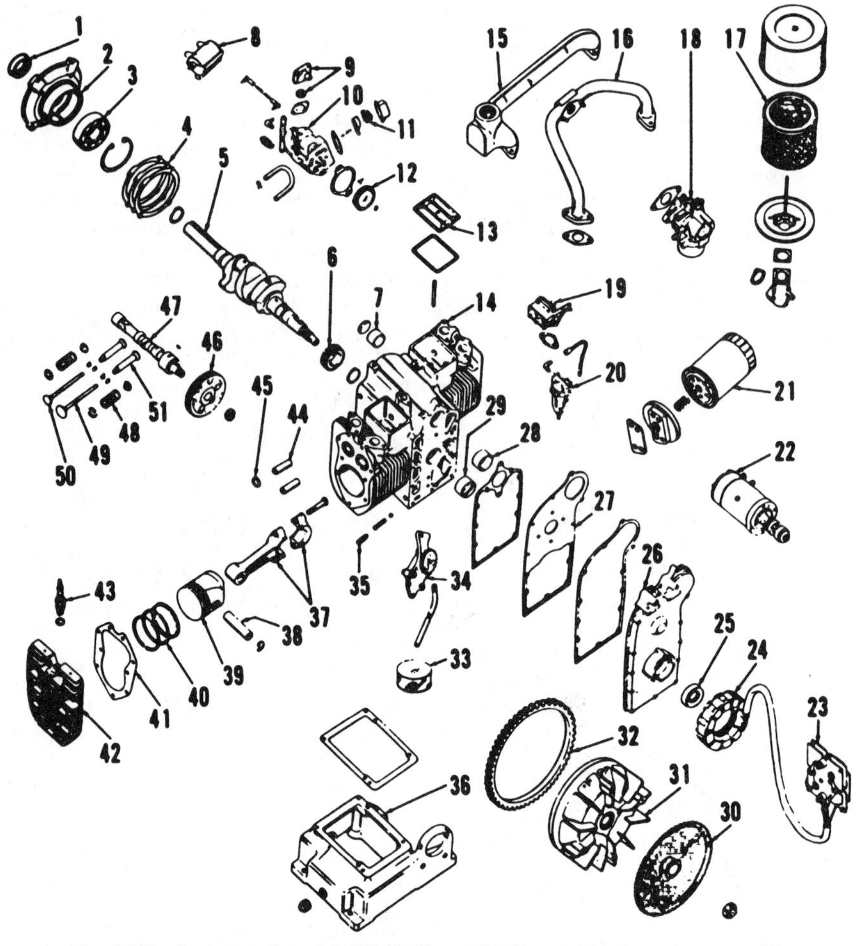

Fig. KO77—Exploded view of K-482, K-532 and K-582 model basic engine assembly.

1. Oil seal
2. Closure plate
3. Rear main (ball) bearing
4. Gaskets
5. Crankshaft
6. Crankshaft gear
7. Camshaft rear bushing
8. Ignition coil
9. Breather valve assy.
10. Governor assy.
11. Breaker points
12. Governor gear
13. Valve cover
14. Crankcase & cylinder block
15. Exhaust manifold
16. Intake manifold
17. Air cleaner element
18. Carburetor
19. Fuel pump
20. Fuel pump
20. Fuel filter
21. Oil filter
22. Starting motor
23. Rectifier-regulator
24. Alternator stator
25. Oil seal
26. Gear cover
27. Gear cover plate
28. Camshaft front bushing
29. Front main (sleeve)
30. Flywheel screen
31. Flywheel
32. Starter ring gear
33. Oil strainer
34. Oil pump
35. Oil pressure adjusting screw
36. Oil pan
37. Connecting rod
38. Piston pin
39. Piston
40. Piston rings
41. Head gasket
42. Cylinder head
43. Spark plug
44. Valve guides
45. Exhaust valve seat
46. Camshaft gear
47. Camshaft
48. Valve spring
49. Intake valve
50. Exhaust valve
51. Valve tappets.

NOTE: To prevent tappets from sliding into crankcase, tie a wire around adjusting nuts on each set of tappets.

Camshaft bushings can now be renewed. Camshaft bushings are presized and if carefully installed need no final sizing. Normal camshaft to bushing clearance is 0.0005-0.0035 inch (0.0127-0.0889 mm). Install new expansion plug behind rear camshaft bushing. Align two punch marked teeth on camshaft gear with single punch marked tooth on crankshaft gear (Fig. KO75) and reassemble engine by reversing disassembly procedure.

Camshaft end play for K-582 models is 0.004-0.010 inch (0.1016-0.254 mm) and end play for all other models is 0.017-0.038 inch (0.432-0.965 mm). End play is controlled by gear cover plate and gasket. If end play is excessive check plate for excessive wear and renew as necessary.

Tighten camshaft gear retaining nut and flywheel retaining nut to torque specified in **TIGHTENING TORQUES** paragraph for model being serviced.

Adjust valves as outlined in **VALVE SYSTEM** paragraph.

VALVE SYSTEM. Exhaust valve seats are renewable hardened valve seat inserts on all models. Intake valves in K-662 models seat on renewable inserts but intake valves in K-482, K-532 and K-582 models may seat on renewable inserts or seat may be machined directly into block casting. Seating surfaces should be 1/32-inch (0.794 mm) in width must be reconditioned if over 1/16-inch (1.588 mm) wide. Valves and seats are ground to 45° angle and Kohler recommends lapping valves to ensure proper valve to seat seal.

Maximum intake valve stem clearance

Illustrations courtesy Kohler Co.

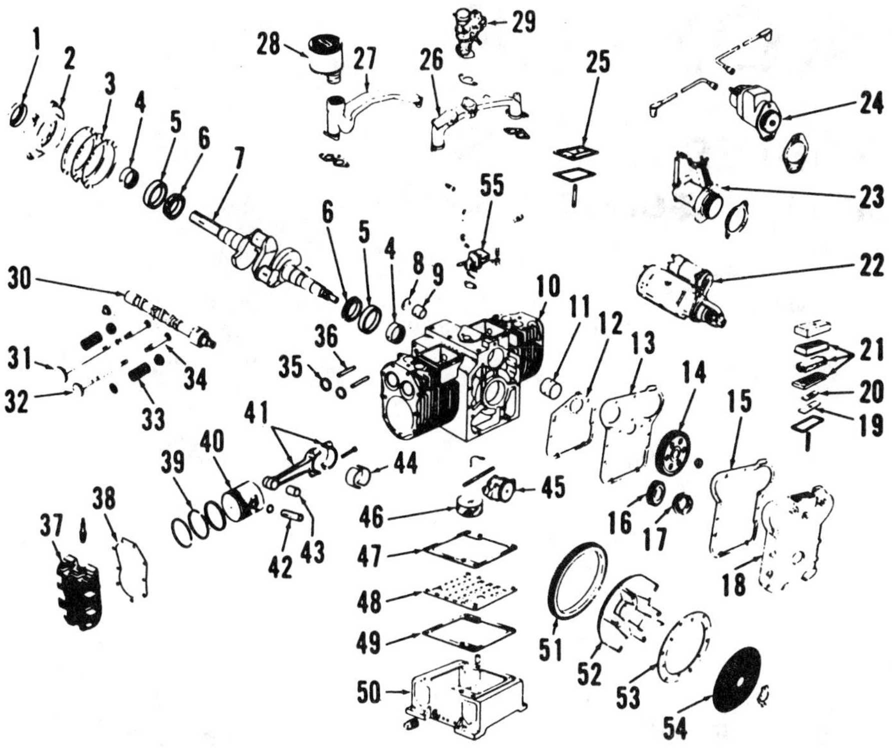

Fig. KO78—Exploded view of K-662 model basic engine assembly.

1. Rear oil seal
2. Closure plate
3. Shim gaskets
4. Oil transfer sleeve
5. Bearing cup
6. Bearing cone
7. Crankshaft
8. Expansion plug
9. Camshaft bushing (rear)
10. Crankcase & cylinder block
11. Camshaft bushing (front)
12. Gasket
13. Gear cover plate
14. Camshaft gear
15. Gasket
16. Crankshaft gear
17. Front oil seal
18. Gear cover
19. Breather reed
20. Reed stop
21. Breather filter
22. Starting motor
23. Governor assy.
24. Magneto
25. Valve cover
26. Intake manifold
27. Exhaust manifold
28. Muffler
29. Carburetor
30. Camshaft
31. Intake valve
32. Exhaust valve
33. Valve springs
34. Valve tappets
35. Valve seat inserts
36. Valve guides
37. Cylinder head
38. Head gasket
39. Piston rings
40. Piston
41. Connecting rod
42. Piston pin
43. Pin bushing
44. Connecting rod bearing
45. Oil pump
46. Oil strainer
47. Gasket
48. Baffle
49. Gasket
50. Oil pan
51. Starter ring gear
52. Flywheel
53. Flywheel plate
54. Flywheel screen
55. Fuel pump

in guide is 0.0045 inch (0.1143 mm) and maximum exhaust valve stem clearance in guide is 0.0065 inch (0.1651 mm) for all models. All models have renewable intake and exhaust valve guides which must be reamed to size after installation. K-662 models have a 0.010 inch (0.254 mm) oversize outside diameter guide available for crankcases with damaged guide bores.

Remove guides by driving them into valve stem chamber and breaking the end off until guide is completely removed. Use care not to damage engine block. Press new guides of K-482, K-532 and K-582 models in to a depth of 1-11/32 inch (34.131 mm) measured from cylinder head surface of block to end of guide. Press new guides of K-662 model in to a depth of 1-7/16 inch (35.513 mm) measured from cylinder head surface of block to end of guide.

Ream new guides installed in K-482, K-532 and K-582 models to 0.312-0.313 inch (7.8248-7.9502 mm) and ream new guides installed in K-662 model to 0.3430-0.3445 inch (8.7122-8.7503 mm).

Valve spring free length for K-482, K-532 and K-582 models should be 1.793 inches (45.54 mm) and for K-662 model should be 2.250 inches (57.15 mm). If rotator is used on exhaust valve, exhaust valve spring free length for K-482, K-532 and K-582 models should be 1.531 inches (38.887 mm) and for K-662 model should be 1.812 inches (46.03 mm).

Valve tappet gap (clearance) for K-482, K-532 and K-582 models should be adjusted so intake valves have 0.008-0.010 inch (0.203-0.254 mm) clearance and exhaust valves have 0.017-0.020 inch (0.432-0.508 mm) clearance when cold. Valve tappet gap (clearance) for K-662 model should be adjusted so intake valves have 0.006-0.008 inch (0.152-0.203 mm) clearance and exhaust valves have 0.015-0.017 inch (0.381-0.432 mm) clearance when cold.

OIL PUMP. A gear type oil pump supplies oil via internal galleries to front main bearing, camshaft bearings, connecting rods and other wear areas. Model K-662 uses transfer sleeves to direct oil to the tapered roller main bearings.

Oil pressure on Models K-482, K-532 and K-582 is controlled by an adjustable pressure valve. Refer to **LUBRICATION** paragraph for operating pressures.

If faulty, oil pump must be renewed as it is serviced as an assembly only.

FUEL PUMP. A mechanically operated diaphragm type fuel pump is used. Pump is actuated by a lever which rides on an eccentric on camshaft. A priming lever is provided on the pump and a repair kit is available for reconditioning pump.

KOHLER

KOHLER COMPANY
Kohler, Wisconsin 53044

Model	No. Cyls.	Bore	Stroke	Displ.	Power Rating
KT-17*	2	3.12 in. (79.3 mm)	2.75 in. (69.9 mm)	42.18 cu. in. (690.5 cc)	17 hp. (12.7 kW)
KT-19*	2	3.12 in. (79.3 mm)	3.06 in. (78 mm)	46.98 cu. in. (770.5 cc)	19 hp. (14.2 kW)
KT-21	2	3.31 in. (84.07 mm)	3.06 in. (78 mm)	52.76 cu. in. (866 cc)	21 hp. (15.7 kW)

*Series II engines included.

Engines in this section are four-stroke, twin-cylinder opposed, split crankcase type. Model KT-17 engines with specification number 24299 or below and KT-19 engines with specification number 49199 or below have a pressure spray lubrication system. KT-17 engines with specification number 24300 or above and KT-19 engines with specification number 49200 or above are designated Series II engines and have full pressure lubrication systems and improved crankshafts and rods. All KT-21 engines have full pressure lubrication systems.

Engine identification and serial number decals are located on top of engine shrouding usually around ignition coil. Significance of each number is explained in Fig. KO79.

Series II engines have **SERIES II** logo on the front shroud from the factory. When ordering parts or service material always give model number, specification number and serial number of engine being serviced.

MAINTENANCE

SPARK PLUG. Recommended spark plug is Champion RBL15YC or equivalent. Electrode gap is 0.025 inch (0.635 mm).

CARBURETOR. A single Kohler side draft carburetor is used on all models and carburetor adjustments should be made only after engine has reached normal operating temperature.

For initial adjustment stop engine and turn main fuel and idle fuel (Fig. KO80) mixture screws in until **LIGHTLY** seated. On KT-17 and KT-19 models turn main fuel adjusting screw out 2½ turns and on KT-21 models turn main fuel adjusting screw out 3 turns.

On all models turn idle mixture screw out 1 to 1¼ turns.

For final adjustment run engine at half throttle and when operating temperature has been reached turn main fuel mixture screw in until speed decreases and note position. Now turn

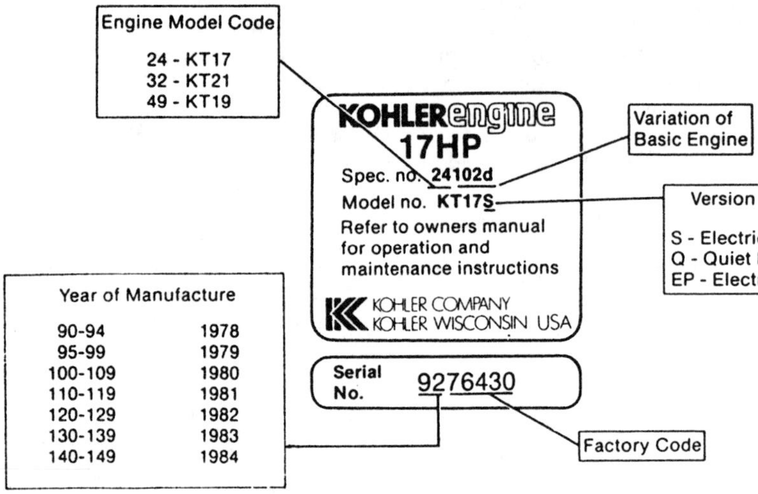

Fig. KO79—Significance of each digit of the specification, model and serial numbers are explained as shown.

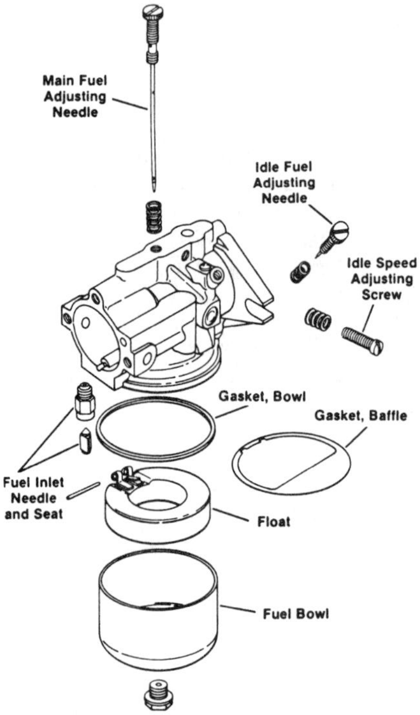

Fig. KO80—Exploded view of Kohler side draft carburetor used on KT series engines.

screw out past point where speed increases until speed again decreases. Note position of screw. Set mixture screw midway between the two points noted. Set throttle so engine is at idle and adjust idle mixture screw in the same manner.

Set idle speed screw so engine is idling at 1200 rpm.

If carburetor is to be cleaned, refer to Fig. KO80 and disassemble. Clean carburetor with suitable solvent making certain fiber and rubber seals are not damaged. Blow out all passages with compressed air and renew any worn or damaged parts.

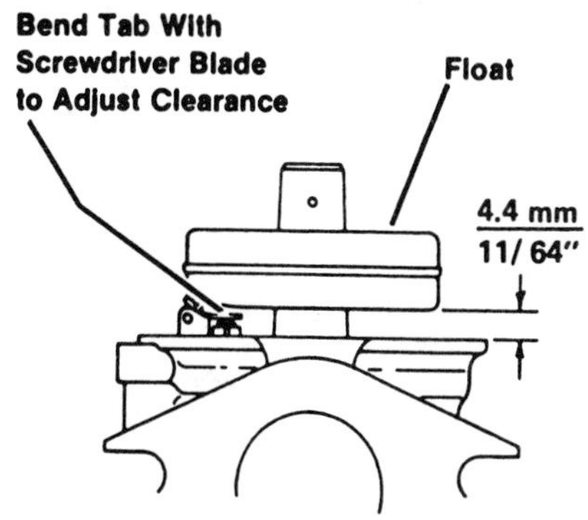

Fig. KO81—Check float setting by inverting carburetor throttle body and float assembly and measuring distance from free end of float to machined surface of casting. Correct clearance is 11/64-inch (4.4 mm).

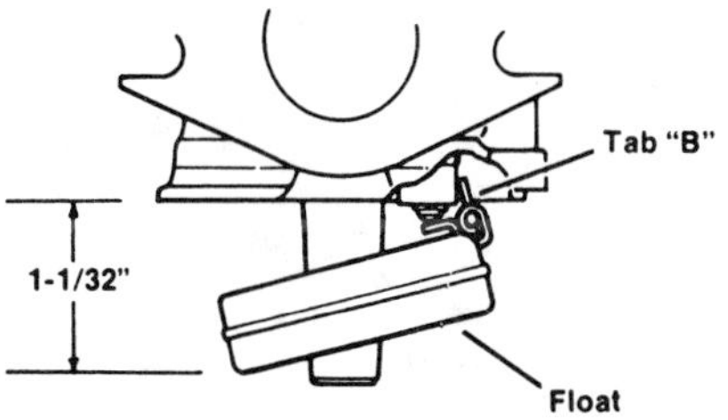

Fig. KO81A—Float drop must be adjusted by bending tab "B" to obtain 1-1/32 inch (26.194 mm) travel as shown.

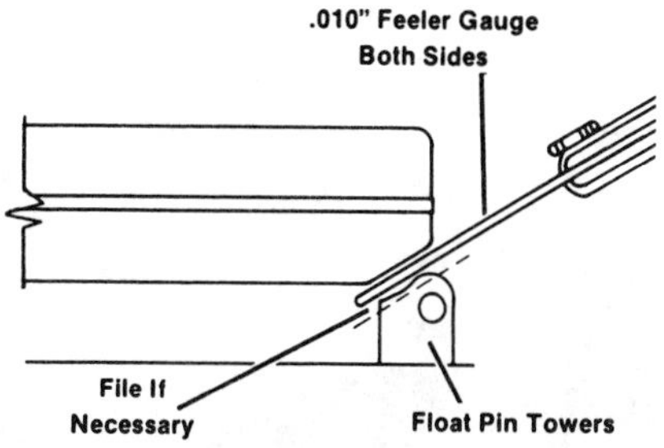

Fig. KO81B—Check clearance between float and float pin towers and file tower if necessary to obtain 0.010 inch (0.254 mm) clearance as shown.

For reassembly torque new fuel needle seat to 35 in.-lbs. (4 N·m). Install needle, float and pin. Set float to 11/64-inch (4.366 mm) as shown in Fig. KO81. Late production carburetors have a tab to limit float travel (B–Fig. KO81A) and float drop must be set to 1-1/32 inch (26.194 mm) by bending tab (B). Float clearance should be checked as shown in Fig. KO81B. File float pin tower as necessary to obtain 0.010 inch (0.254 mm) clearance. Reassemble carburetor, install and adjust.

AUTOMATIC CHOKE. Some models are equipped with a "Thermo-Electric" automatic choke and fuel shutdown solenoid (Fig. KO82).

When installing or adjusting choke unit, position unit leaving mounting screw slightly loose. Hold choke plate wide open and rotate choke unit clockwise with slight pressure until it can no longer be rotated. Hold choke unit and tighten mounting screws. Choke valve should close 5-10 degrees at 75° F (24° C). As temperature decreases, choke will close even more. To check choke, remove spark plug lead and crank engine. Choke valve should close a minimum of 45 degrees at 75° F (24° C).

With ignition on, shutdown solenoid plunger raises and opens float bowl vent to air horn. With ignition off, solenoid plunger closes vent and stops fuel flow. To check solenoid, remove solenoid and plunger from carburetor body. With lead wire connected, pull plunger approximately ¼-inch (6.35 mm) from solenoid and ground casing to engine surface. Turn ignition switch on. Renew solenoid if solenoid does not draw plunger in. After renewing solenoid, reset main fuel mixture adjusting screw.

GOVERNOR. KT-series engines are equipped with centrifugal flyweight mechanical governor with weight gear mounted within crankcase and driven by camshaft gear.

Governors are adjusted at factory and further adjustment should not be necessary unless governor arm or linkage works loose or becomes disconnected. Readjust linkage if engine speed surges or hunts with changing load or if speed drops considerably when normal load is applied.

Maximum no load speed for all models is 3600 rpm. If adjustment is necessary, loosen governor speed adjusting screw and pivot high speed stop tab until desired speed is reached, then tighten screw. Governor sensitivity can be adjusted by repositioning governor spring. Normally, governor spring is placed in the fifth hole from pivot of governor

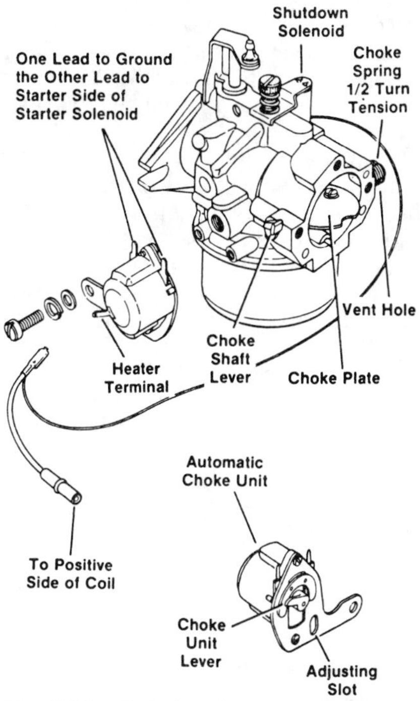

Fig. KO82—Typical carburetor with Thermo-Electric automatic choke and shutdown control. Refer to text for details.

arm and in the sixth hole from pivot on throttle control lever. To increase sensitivity, move spring end closer to center of governor arm. To allow a wider range of governor control with less sensitivity, move spring toward end of arm. Refer to Fig. KO83.

IGNITION TIMING. Breaker points are externally mounted on crankcase and are activated by a push rod which rides on camshaft. A timing sight hole is located in number "1" side of blower housing.

Two different timing location marks are stamped on flywheel. "T" mark locates TDC while "S" mark locates 23° BTDC, which is where timing is set.

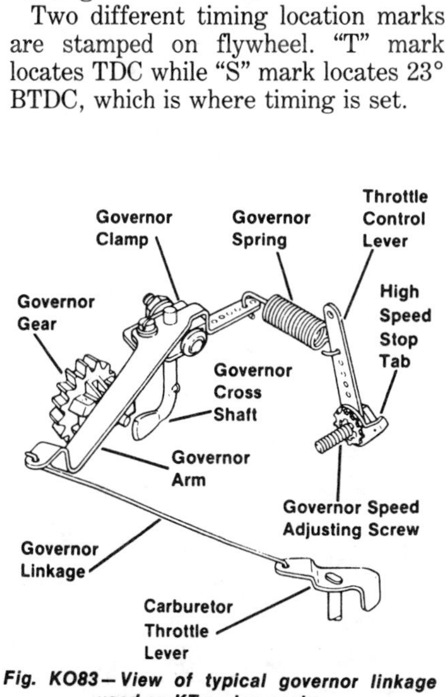

Fig. KO83—View of typical governor linkage used on KT series engines.

To set timing, remove point cover and reinstall cover screws to prevent oil loss. Attach timing light to number one spark plug, start engine and run at 1200 rpm. Vary point gap between 0.017 and 0.023 inch (0.432 and 0.584 mm) until "S" mark is in line with roll pin in cylinder barrel. If double image is seen through sight hole, time engine so roll pin is exactly between the two images. Apply thread sealant to screws when reinstalling point cover. See Fig. KO84.

COMPRESSION TEST. To check compression, remove spark plugs, make certain air cleaner is clean and set throttle and choke in wide open position. Crank engine with starting motor. Compression should test approximately 115-125 psi (793-862 kPa).

LUBRICATION. Oil pump (Fig. KO85) is driven by crankshaft and is located behind closure plate on pto side of engine on all models.

KT-17 engines with specification number 24299 and below and KT-19 engines with specification number 49199 and below use a pressure spray lubrication system. In this system oil is supplied to main bearings under pressure and travels through the hollow camshaft and sprays out two holes in camshaft to lubricate rods. Rods on these engines have oil hole drilling (Fig. KO91A) through which oil reaches bearing surface of rod.

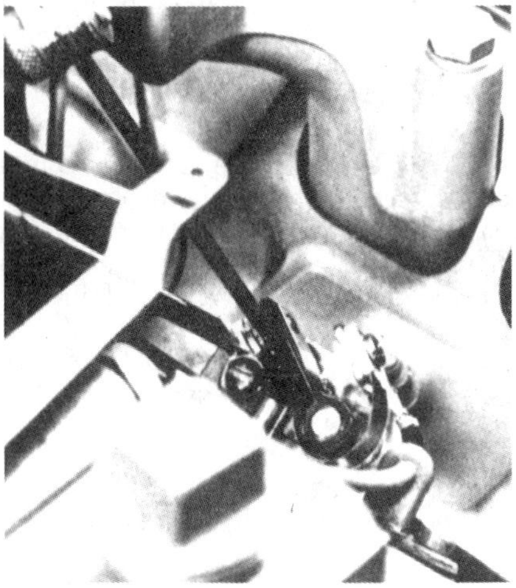

Fig. KO84 — Ignition breaker points are externally located on engine crankcase. Breaker point gap will vary from 0.017-0.023 inch (0.432-0.584 mm) depending on engine performance.

KT-17 Series II engines, specification number 24300 and above, KT-19 Series II engines, specification number 49200 and above and all KT-21 engines use a full pressure lubrication system with oil supplied to rods under pressure via drillings in crankshaft. In these engines oil pressure is regulated by a ball and spring type oil pressure relief valve located in crankcase behind closure plate (Fig. KO85A). Remove the 1/16-inch pipe plug (Fig. KO85B) in pto end of crankcase and install a 0-100 psi (0-690 kPa) pressure gage to check oil pressure. Pressure should be 20-30 psi (138-207 kPa).

Oil pump rotors and cover on all models and oil pressure relief valve and spring on models so equipped can be serviced without splitting crankcase.

Oil pump shaft to crankcase clearance should be 0.001-0.0025 inch (0.025-0.064 mm) and oil pump drive gear end play should be 0.010-0.029 inch (0.254-0.736 mm).

Free length of oil pressure relief valve spring should be 0.940 inch plus or minus 0.010 inch (23.87 mm plus or minus 0.254 mm).

High quality detergent type oil having API classification SC, CC, SD or SE should be used. Use SAE 30, 10W-30 or 10W-40 oil in temperatures of 32° F (0° C) and above and SAE 5W-30 oil in temperatures of 32° F (0° C) and below. Maintain crankcase oil level at full mark on dipstick, but do not overfill.

It is recommended oil be changed at 25 hour intervals under normal operating conditions. Crankcase capacity for all models is 3 pints (1.4 L).

REPAIRS

TIGHTENING TORQUES. Recommended tightening torques are as follows:

Spark plugs . . . . 10-15 ft.-lbs. (14-20 N·m)
Flywheel retaining screws . . . . 40 ft.-lbs. (54 N·m)
Manifold screws . . . . 13 ft.-lbs. (17 N·m)
Closure plate screws . . . . 13 ft.-lbs. (17 N·m)
Cylinder barrel nuts . . . . 22 ft.-lbs. (29 N·m)
Cylinder head bolts* . . . . . . 15-20 ft.-lbs. (20-27 N·m)
Connecting rod bolts** . . . . . . 17 ft.-lbs. (23 N·m)
Connecting rod nuts –
New . . . . 12 ft.-lbs. (16 N·m)
Used . . . . 8 ft.-lbs. (11 N·m)
Crankcase stud nuts . . . . 22 ft.-lbs. (29 N·m)
Crankcase slot head screws . . . . 3 ft.-lbs. (4 N·m)
Crankcase cap screws . . . . See Fig. KO89 for sequence & torque

*Lubricate with oil at assembly.
**Rod bolts should be overtorqued to 20 ft.-lbs. (27 N·m), loosened and retorqued to 17 ft.-lbs. (23 N·m) torque.

CYLINDER HEADS AND VALVE SYSTEM. If head is warped more than 0.003 inch (0.0762 mm) or has "hot spots", both head and head screws should be renewed. Tighten cylinder

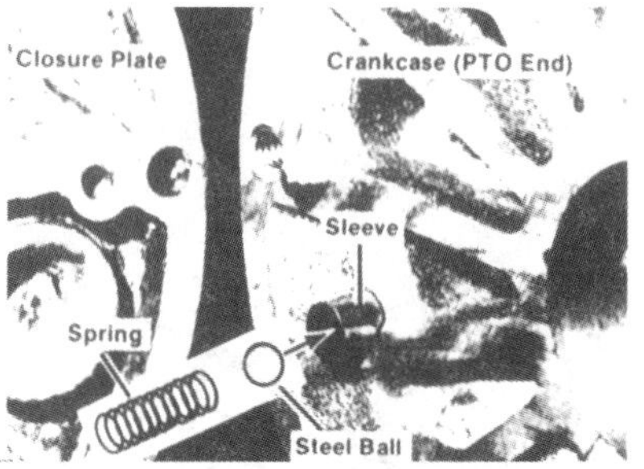

Fig. KO85A — View of relief valve spring, ball and sleeve on Series II engines.

Fig. KO85B — View of test port plug in oil gallery of Series II engines. Refer to text.

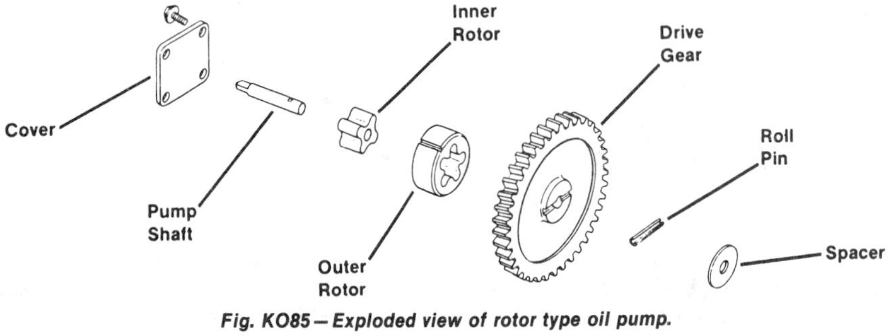

Fig. KO85 — Exploded view of rotor type oil pump.

Illustrations courtesy Kohler Co.

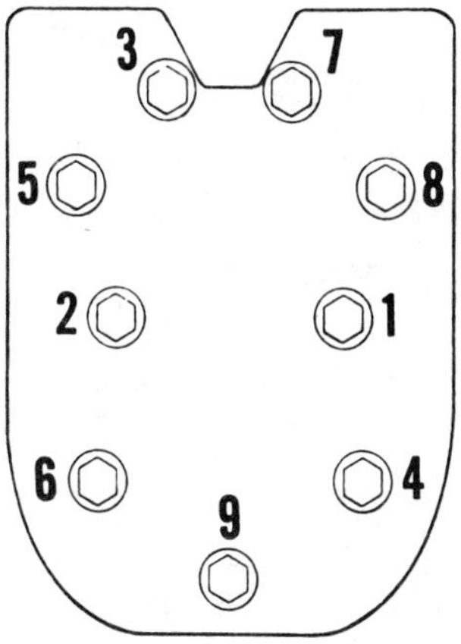

Fig. KO86 — Tighten cylinder head cap screws to 15-20 ft.-lbs. (20-27 N·m) torque using sequence shown. Lubricate cap screws during assembly.

head cap screws evenly to 15-20 ft.-lbs. (20-27 N·m) torque using sequence shown in Fig. KO86.

Intake valve seats are machined surfaces of cylinder casting on most KT series engines, however certain applications use a hard alloy renewable insert.

All models have renewable hardened alloy exhaust valve seats which should be measured for width before removal. Two different widths have been used due to changes in cylinder castings. Widths are 0.199 inch (0.711 mm) and 0.220 inch (0.558 mm) and are not interchangeable.

Valve face and seat angle is 45°. Desired seat width is 0.037-0.045 inch (0.9398-1.1430 mm). Valve tappet clearance (cold) is 0.003-0.006 inch (0.076-0.152 mm) for intake and 0.011-0.014 inch (0.2794-0.3556 mm) for exhaust. If adjustment is necessary, grind end of valve stem. Make certain end of valve stem is ground perfectly flat. See Fig. KO90.

Valve stem to guide clearance is 0.0025-0.0045 inch (0.0635-0.1143 mm) for intake and 0.0045-0.0065 inch (0.1143-0.1651 mm) for exhaust. To renew valve guides, press old guides into valve chamber and carefully break off protruding ends until guides are removed. New guides should have an interference fit of 0.0005-0.002 inch (0.0127-0.051 mm). Press new guides in and ream to 0.312-0.313 inch (7.9248-7.9502 mm). Depth from top of block to valve guide for KT-17, KT-17 Series II and KT-19 engines is 1.125 inch (28.58 mm) and is 1.390 inch (26.39 mm) for KT-19 Series II and KT-21 engines.

Valve spring free length (no rotators) should be 1.6876 inches (42.865 mm). If rotator is used on exhaust valve, spring free length should be 1.542 inches (39.1668 mm).

CYLINDERS, PISTONS, PINS AND RINGS. If cylinder walls are scored or bores are tapered 0.0015 inch (0.0381 mm) or out-of-round more than 0.002 inch (0.0508 mm), cylinders should be honed to nearest suitable oversize of 0.010, 0.020 or 0.030 inch (0.254, 0.508 or 0.762 mm). Standard cylinder bore diameter for KT-17 and KT-19 is 3.1245-3.1255 inch (79.3623-79.3877 mm), KT-21 standard bore is 3.312-3.315 inch (84.125-84.201 mm). During reassembly, use a new gasket and tighten cylinder barrel retaining nuts to a torque of 30 ft.-lbs. (41 N·m). Refer to Fig. KO87 for tightening sequence.

Cam ground aluminum alloy pistons are fitted with two compression rings and one oil control ring. Renew pistons if scored or if side clearance of new ring in top groove exceeds 0.004 inch (0.1016 mm). Pistons are available in oversizes

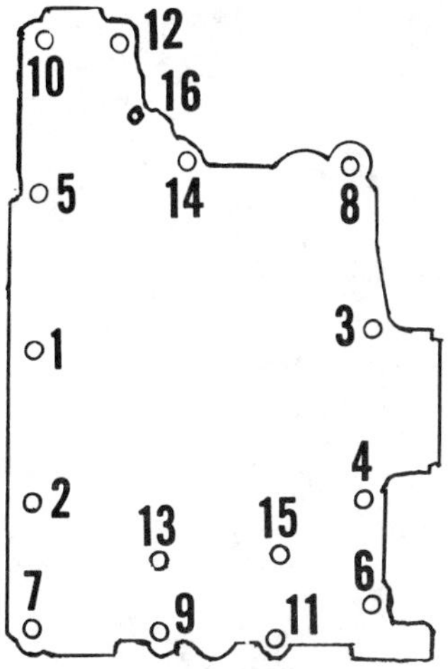

Fig. KO89 — Tighten crankcase bolts on KT-17 engines with serial numbers prior to 9755085 in the following sequence: 1 through 4, 12 and 14 to 260 in.-lbs. (29 N·m) torque. Tighten bolts 5 through 11, 13 and 15 to 150 in.-lbs. (17 N·m) torque. Tighten bolt 16 to 35 in.-lbs. (4 N·m) torque. Tighten bolts in sequence shown.

of 0.010, 0.020 and 0.030 inch (0.254, 0.508 and 0.762 mm) as well as standard size. Recommended piston to cylinder bore clearance measured just below oil ring at right angle to pin is 0.006-0.008 inch (0.1524-0.2032 mm).

Piston pin fit in piston should be from 0.000 inch (0.000 mm) (light interference) to 0.0003 inch (0.0076 mm) (loose). Piston pin fit in connecting rod should be 0.0006-0.0011 inch (0.0152-0.0279 mm). Piston pins of 0.005

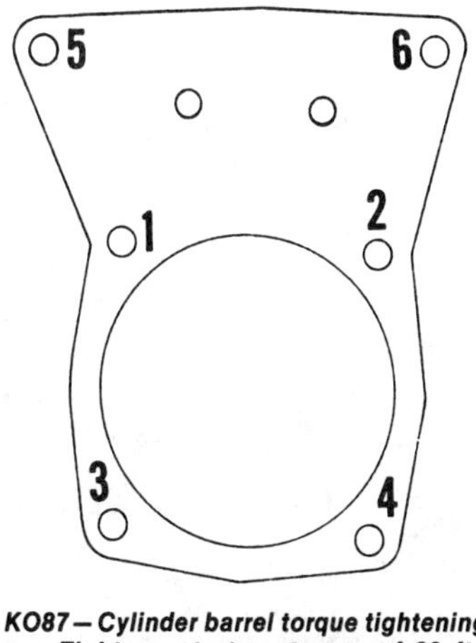

Fig. KO87 — Cylinder barrel torque tightening sequence. Tighten nuts to a torque of 22 ft.-lbs. (29 N·m).

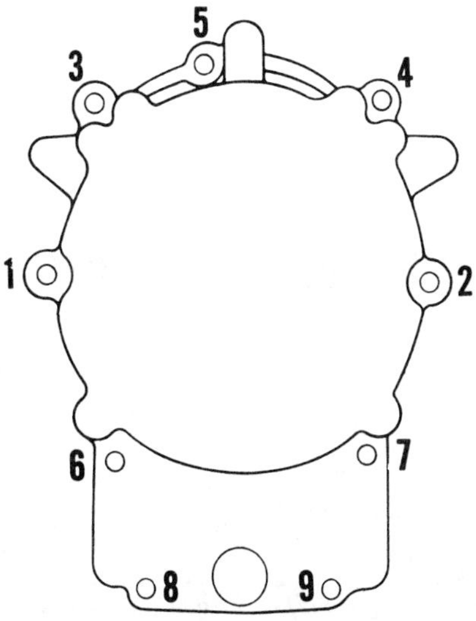

Fig. KO88 — Closure plate torque tightening sequence. Torque screws to 150 in.-lbs. (17 N·m).

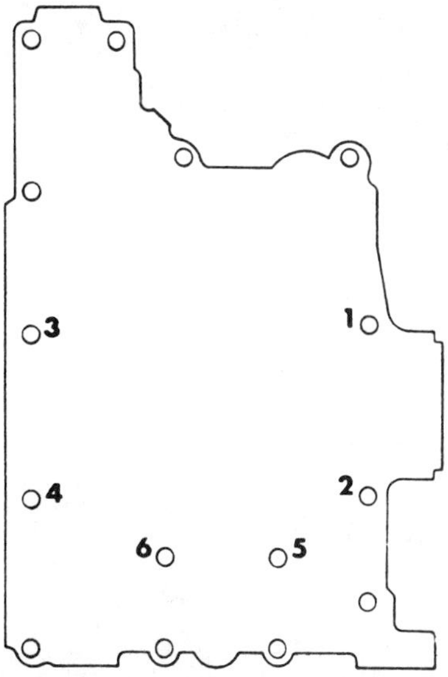

Fig. KO89A — Tighten crankcase bolts on KT-17 engines with serial number 9755085 and above, all KT-19 and KT-21 engines in the following sequence: Number 1 through 4 to 260 in.-lbs. (29 N·m) torque. Tighten 5 and 6 to 200 in.-lbs. (23 N·m) torque and tighten remaining bolts to 200 in.-lbs. (23 N·m) torque.

Illustrations courtesy Kohler Co.

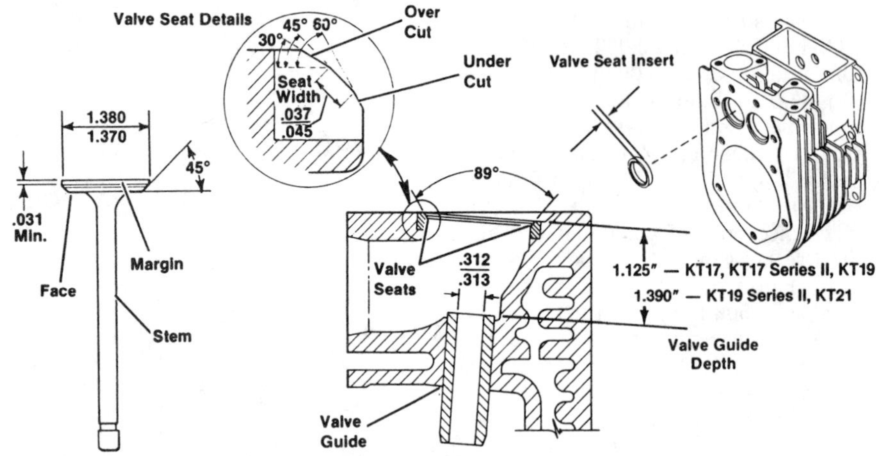

Fig. KO90 – Cross-sectional view of valve seat, guide and valve.

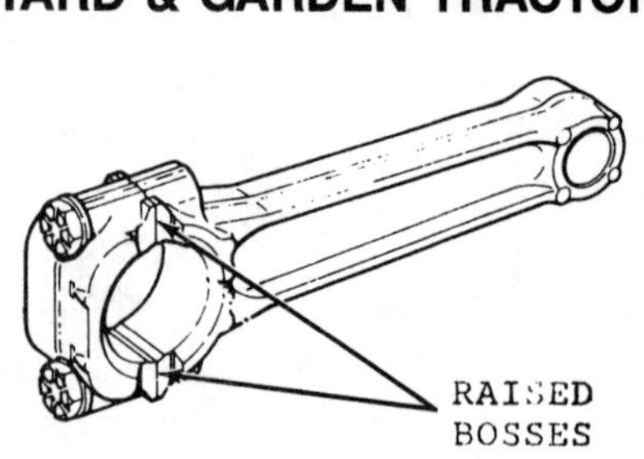

Fig. KO91B – Raised bosses on rods of Series II engines must be installed towards flywheel.

inch (0.127 mm) oversize are available. Always renew pin retaining rings.

Kohler recommends piston rings always be renewed when they are removed. Piston ring specifications are as follows:

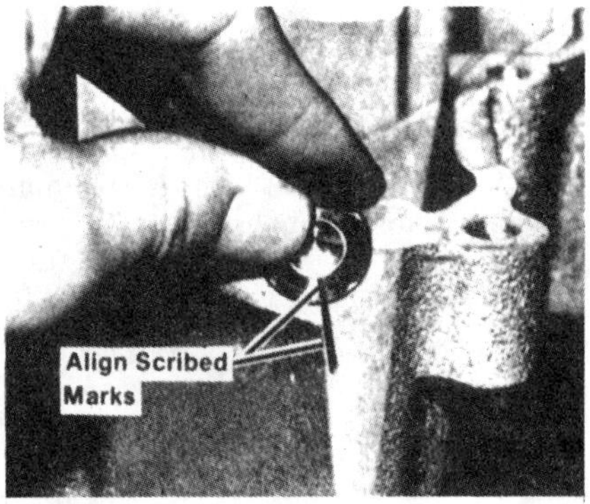

Fig. KO90A – Scribe mark across machined camshaft plug so it can be reinstalled in its original position.

Piston ringside clearance –
- Top ring . . . . 0.002-0.004 in. (0.051-0.102 mm)
- Middle ring . . . . 0.001-0.003 in. (0.025-0.076 mm)
- Oil ring . . . . 0.001-0.003 in. (0.025-0.076 mm)

Piston ring end gap, top and middle ring –
- Used cylinder . . . . 0.010-0.030 in. (0.254-0.762 mm)
- New cylinder bore . . . . 0.010-0.020 in. (0.254-0.508 mm)

Piston ring end gap, oil ring . . . . 0.060 in. (1.52 mm)

CRANKCASE, CRANKSHAFT, CONNECTING ROD. When making repairs on crankcase assembly, it will be necessary to identify one side of engine from the other. Number "1" side contains governor assembly, oil pump gear and dipstick assembly. To service internal parts, remove closure plate retaining screws and slide it off crankshaft. Lay crankcase down so number "2" side is up. Put tape around tappet stems to prevent them from falling into case when halves are split. On Series II engines with full pressure lubricating system, mark across machined camshaft plug and crankcase so plug may be reinstalled in its original position. See Fig. KO90A. Locate crankcase splitting notches. Place a large flat blade screwdriver in one notch and carefully pry halves apart (Fig. KO91).

Connecting rods must be removed with crankshaft. Identify each rod and piston unit so they can be reinstalled in correct cylinder and do not intermix connecting rod caps. Connecting rods ride directly on crankshaft. If crankpin is 0.0005 inch (0.0127 mm) out-of-round or is tapered 0.001 inch (0.0254 mm), regrind crankpin. Rods with 0.010 inch (0.254 mm) undersize crankpin bore are available for reground crankshaft. KT-17 and KT-19 standard crankpin diameter is 1.3735-1.3740 inches (34.8869-34.8996 mm). KT-21 standard

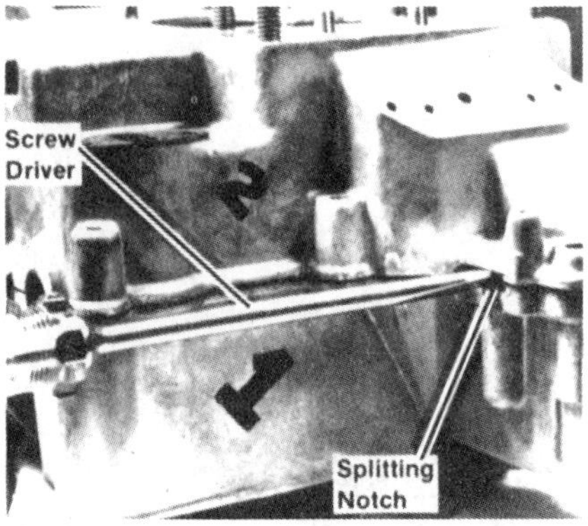

Fig. KO91 – Identify one side of engine from the other before making repairs. Number one side contains governor assembly, oil pump and dipstick. To split crankcase, lay on number one side, place large flat blade screwdriver in notch and carefully pry apart.

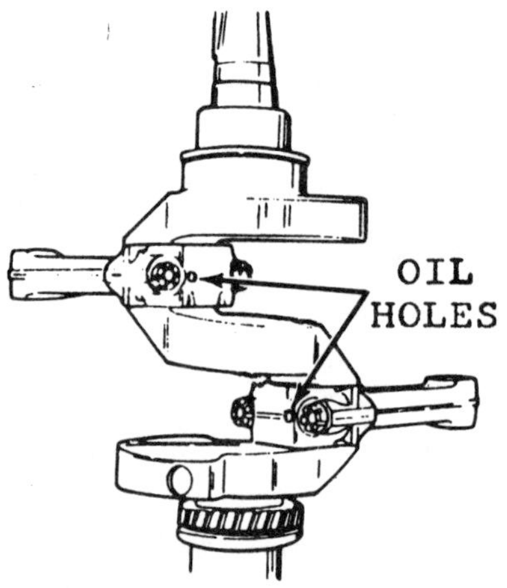

Fig. KO91A – Oil holes in rod caps of engines with pressure spray lubrication system must be facing up with crankcase in upright position when installed.

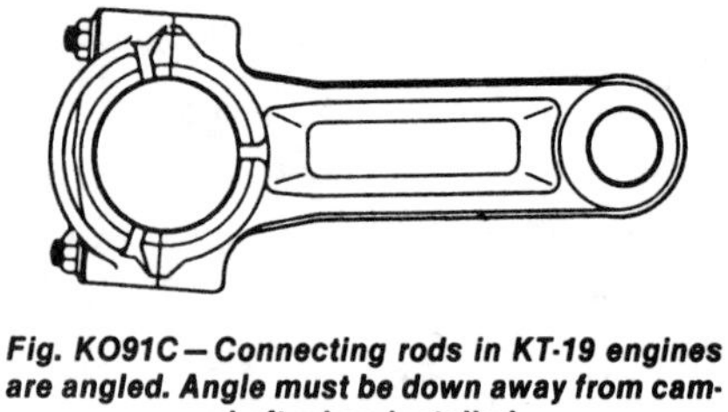

Fig. KO91C – Connecting rods in KT-19 engines are angled. Angle must be down away from camshaft when installed.

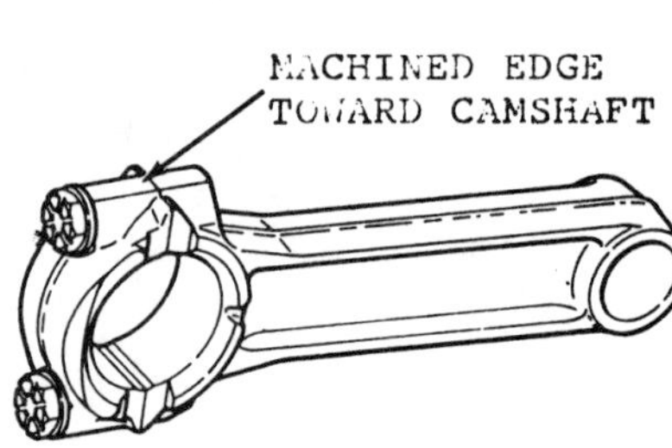

Fig. KO91D – Machined edge of KT-21 connecting rod must be toward camshaft when installed.

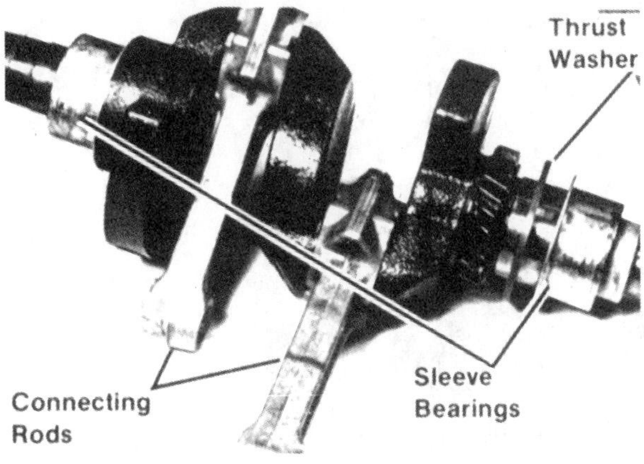

Fig. KO92—View showing sleeve type bearings and thrust washer used to control crankshaft end play.

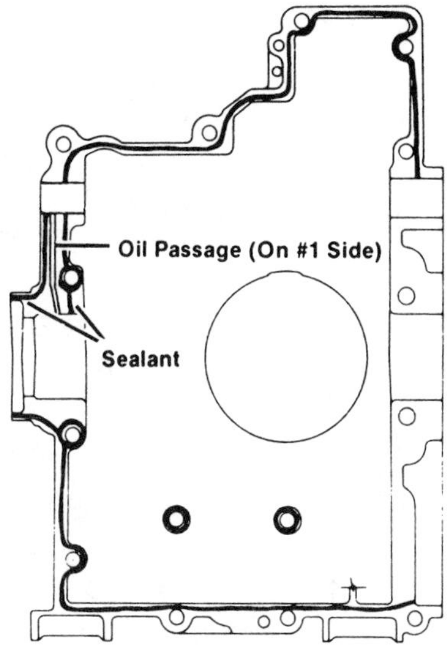

Fig. KO92C—Apply silicone base sealant around edge of crankcase, as indicated by the heavy dark line, making certain sealant does not enter oil passage on number one side of Series II engines.

crankpin bore is 1.4993-1.4998 inches (38.082-38.095 mm). Desired running clearances are as follows:

Connecting rods to
crankpins ....0.0012-0.0024 in.
(0.030-0.061 mm)

Connecting rod to
piston pin ....0.0006-0.0011 in.
(0.015-0.028 mm)

Rod side play on
crankpin ....0.005-0.016 in.
(0.127-0.406 mm)

Lubricate connecting rod journals, then assemble on crankshaft. Connecting rods and caps must be assembled so match marks on both are aligned.

Early production K-17 engines with pressure spray lubrication system use connecting rods which have small oiling holes in rod caps and must be installed facing up with crankcase in upright position. See Fig. KO91A.

Series II engines use connecting rods with raised bosses which must be installed towards flywheel. See Fig. KO91B.

Model KT-19 connecting rods are angled as shown in Fig. KO91C. Rods must be installed on piston so they will be angled down away from camshaft when installed on crankshaft.

Model KT-21 connecting rods have a machined edge (Fig. K091D) which must be toward camshaft when installed.

On all models, if connecting rod bolts are used to retain connecting rod cap, overtorque bolts to 20 ft.-lbs. (27 N•m), loosen and retorque to 17 ft.-lbs. (23 N•m). If "Posi-Lock" nuts are used to retain connecting rod caps, tighten nuts to 12 ft.-lbs. (16 N•m) for new rods and nuts and to 8 ft.-lbs. (11 N•m) for used rods and nuts.

Crankshaft may be supported by either ball bearings or sleeve type bearings. Crankshaft end play if supported by ball bearings is 0.002-0.023 inch (0.051-0.584 mm) and is adjusted by varying thickness of spacers on flywheel end of crankshaft. Spacers are available in a variety of thicknesses.

Sleeve type bearings have a running clearance of 0.0013-0.033 inch (0.0330-0.838 mm). On early production engines with pressure spray lubrication system and sleeve type bearings, crankshaft end play should be 0.003-0.013 inch (0.0762-0.3302 mm) and is adjusted by varying thickness of thrust washer (Fig. KO92) on pto end of crankshaft. Install washer with chamfer toward inside. Some models have a notched spacer between lip of bearing and crankcase on flywheel end. The bearing tab should fit into notch. Make certain oil hole in sleeve type bearings are aligned with hole in crankcase.

Some Series II engines use a roller thrust bearing and flat washers on flywheel end of crankshaft. Refer to Fig. KO92A for assembly sequence and note shaft locating washer is 0.1575 inch (4.0 mm) thick.

Early production engines with

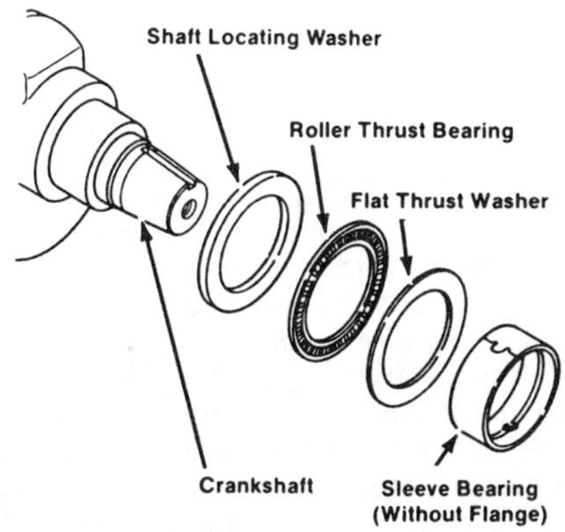

Fig. KO92A—View showing location of roller thrust bearing used on some Series II engines.

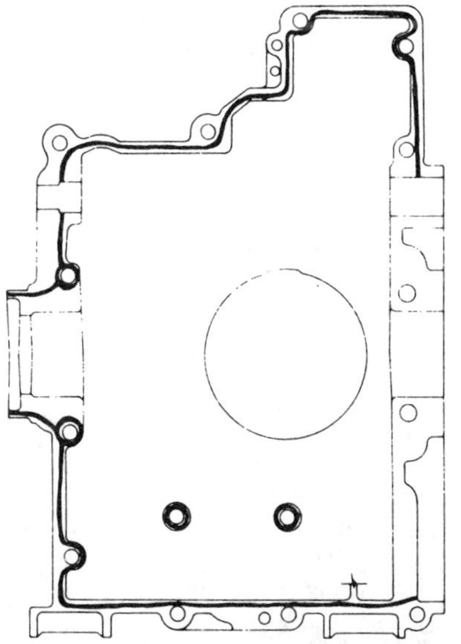

Fig. KO92B—Apply silicone base sealant around edge of crankcase, as indicated by the heavy dark line, on early production engines with pressure spray lubrication system.

Fig. KO93—Assemble crankshaft and camshaft in number one side of crankcase, making certain timing marks are aligned.

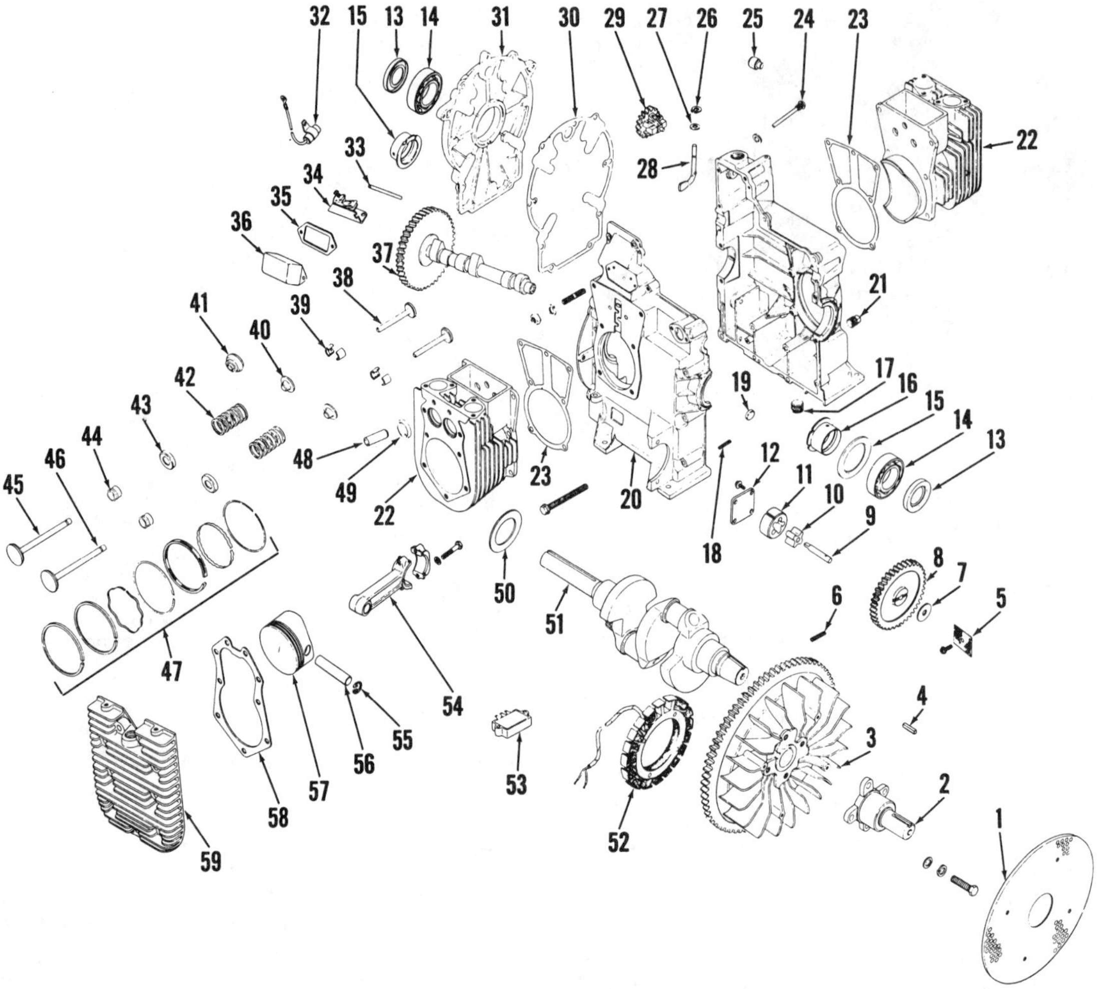

Fig. KO94—Exploded view of basic engine assembly for KT-17 series engines. Series KT-19 and KT-21 are similar.

1. Grass screen
2. Front drive shaft
3. Flywheel
4. Key
5. Screen
6. Roll pin
7. Spacer
8. Oil pump gear
9. Oil pump shaft
10. Inner rotor
11. Outer rotor
12. Pump cover
13. Oil seal
14. Ball bearing
15. Spacer (0.018-0.022 inch)
16. Sleeve bearing
17. Drain plug
18. Roll pin
19. Cap plug (3 used)
20. Crankcase halves
21. Drain plug
22. Cylinder barrel
23. Cylinder barrel gasket
24. Governor stop pin
25. Governor shaft
26. Retainer
27. Washer
28. Governor cross shaft
29. Governor gear
30. Closure plate gasket
31. Closure plate
32. Condenser
33. Breaker push rod
34. Breaker points
35. Gasket
36. Breaker cover
37. Camshaft
38. Tappet (4)
39. Retainer
40. Spring retainer
41. Valve rotator
42. Valve spring (4)
43. Spring retainer
44. Valve seal (4)
45. Exhaust valve
46. Intake valve
47. Piston rings
48. Valve guide (4)
49. Valve seat insert (4)
50. Thrust washer
51. Crankshaft
52. Stator
53. Rectifier-regulator
54. Connecting rod
55. Retainer
56. Piston pin
57. Piston
58. Head gasket
59. Cylinder head

pressure spray lubrication system use a cupped plug to seal camshaft opening at flywheel end. Bolt crankcase halves together applying sealant to edges as shown in Fig. KO92B. Tighten crankcase bolts as shown in Fig. KO89. Apply sealant to plug and drive in until flush with crankcase. Cupped end of plug must face outward.

All Series II and KT-21 engines use a machined steel plug and "O" ring to seal camshaft opening (Fig. KO90A). Apply silicone base sealant around edge of crankcase as indicated in Fig. KO92C, install "O" ring on machined plug and hold plug in opening while bolting halves together. Tighten bolts as shown in Fig. KO89A. Stake plug around edges of crankcase at original stake marks (Fig. KO90A).

Crankshaft oil seals can be installed after crankcase and closure plate are assembled. Carefully work oil seals over crankshaft and drive seals into place with hollow driver that contacts outer edge of seals.

CAMSHAFT. Camshaft is supported directly in machined bores in crankcase halves. Camshaft bearing running clearance is 0.0010-0.0025 inch (0.0254-0.0635 mm).

Renew camshaft if gear teeth or lobes are badly worn or chipped. Check camshaft end play with a feeler gage. End play should be 0.003-0.013 inch (0.0762-0.3302 mm). If end play exceeds 0.013 inch (0.3302 mm), renewal of camshaft or crankcase will be necessary. Assemble camshaft in number "1" side of crankcase, making sure timing mark on cam gear aligns with mark on crankshaft. See Fig. KO93.

FUEL PUMP. A mechanically operated diaphragm type fuel pump is used. Pump is actuated by a lever which rides on an eccentric on camshaft. Repair kit is available for reconditioning pump. Replacement pumps are nonmetallic in construction and only thin mounting gaskets can be used. Use flat washers under mounting screws to prevent damaging flange. Never install fuel pump on crankcase before case halves are assembled as pump lever may be below camshaft. Proper position is above the camshaft. Pump breakage and possible engine damage could result if lever is positioned below camshaft.

KOHLER

Model	No. Cyls.	Bore	Stroke	Displacement	Power Rating
MV16	2	3.12 in. (79.2 mm)	2.75 in. (69.85 mm)	42.18 cu. in. (691.3 cc)	16 hp (11.9 kW)
M18, MV18	2	3.12 in. (79.2 mm)	2.75 in. (69.85 mm)	42.18 cu. in. (691.3 cc)	18 hp (13.4 kW)
M20, MV20	2	3.12 in. (79.2 mm)	3.06 in. (78.0 mm)	46.98 cu. in. (769.8 cc)	20 hp (14.9 kW)

ENGINE IDENTIFICATION

All models are four-stroke, twin-cylinder, air-cooled engines. Models M18 and M20 are horizontal crankshaft engines. Models MV16, MV18 and MV20 are vertical crankshaft engines. All models are pressure lubricated and equipped with an oil filter.

Engine identification and serial numbers are located on a decal (or decals) located as shown in Fig. KO120 for Models M18 and M20, or as shown in Fig. KO121 for Models MV16, MV18 and MV20. Refer to Fig. KO122 to determine engine features and date of manufacture.

Always furnish engine model and serial number when ordering parts.

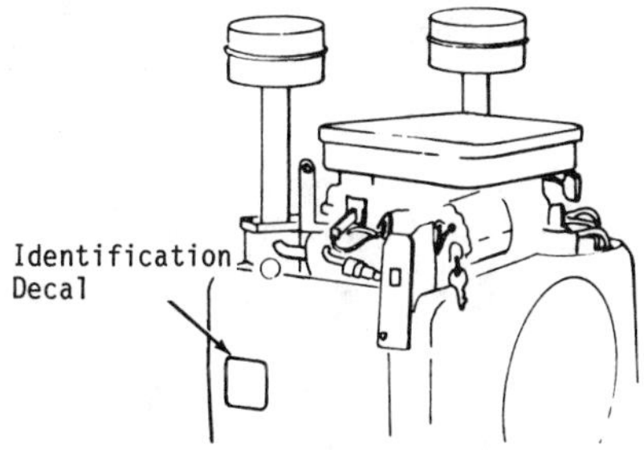
Identification
Decal

Fig. KO120—View showing location of decal for identification of engine model and serial numbers on Models M18 and M20.

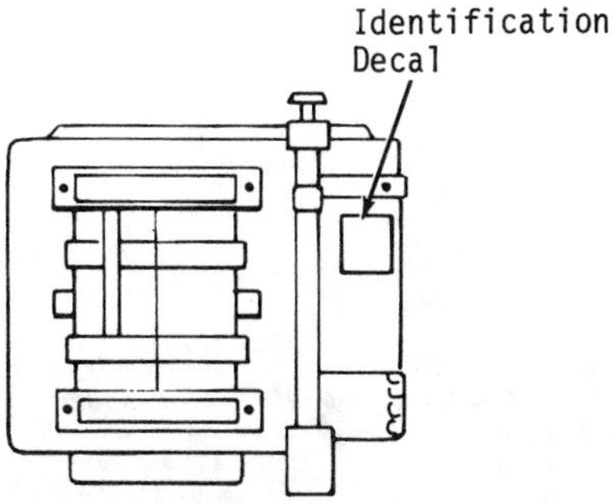
Identification
Decal

Fig. KO121—View showing location of decal for identification of engine model and serial numbers on Models MV16 and MV18.

MAINTENANCE

SPARK PLUG. Recommended spark plug for all models is Champion RV15YC or equivalent. Recommended spark plug electrode gap is 0.025 inch (0.635 mm).

CARBURETOR. Early Models M18 and M20 are equipped with a Kohler adjustable jet float type side draft carburetor. Model M18 after serial number 1629500986, Model M20 after serial number 1619504206 and all Models MV16, MV18 and MV20 are equipped with a float type side draft carburetor manufactured by Walbro to meet Kohler specifications. Refer to the appropriate service information for type being serviced.

Kohler Adjustable Jet Carburetor. Initial adjustment for main mixture nee-

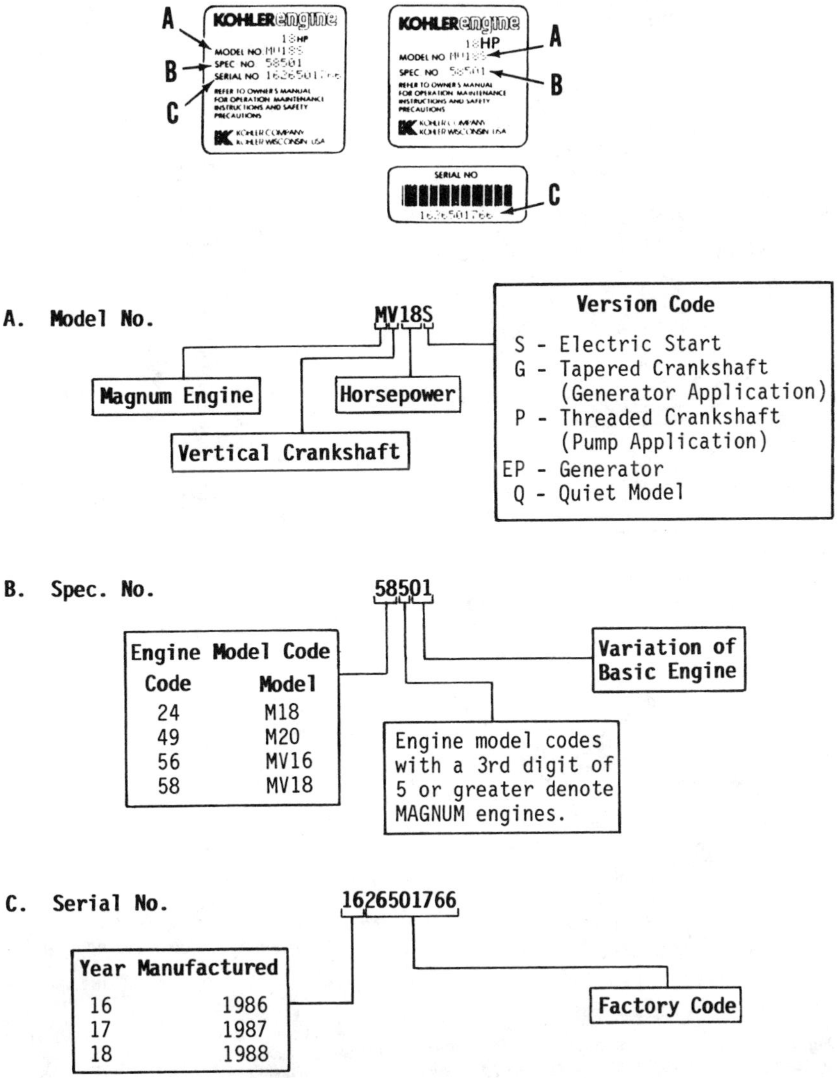
A
B
C
MODEL NO
SPEC NO
SERIAL NO
SERIAL NO
A. Model No.
MV18S
Magnum Engine
Horsepower
Vertical Crankshaft
Version Code
S - Electric Start
G - Tapered Crankshaft (Generator Application)
P - Threaded Crankshaft (Pump Application)
EP - Generator
Q - Quiet Model
B. Spec. No.
58501
Engine Model Code
Code Model
24 M18
49 M20
56 MV16
58 MV18
Variation of Basic Engine
Engine model codes with a 3rd digit of 5 or greater denote MAGNUM engines.
C. Serial No.
1626501766
Year Manufactured
16 1986
17 1987
18 1988
Factory Code

Fig. KO122—Engine model and serial number interpretation.

Illustrations courtesy Kohler Co.

dle (MN—Fig. KO123) should be $2^1/_2$ turns open from a lightly seated position. Initial adjustment for idle mixture needle (IN) should be 1 turn open from a lightly seated position.

Final adjustments should be made with engine at normal operating temperature and running. Place engine under load with throttle in wide-open position. Slowly turn main mixture needle (MN) out (counterclockwise) until engine rpm decreases. Note position of needle. Slowly turn main mixture needle in (clockwise). Engine speed will increase. However, as main mixture needle is turned further in, engine speed will again decrease. Note needle position. Adjust main mixture needle to a position midway between the two noted positions. Remove load from engine and place throttle in idle position. Adjust idle mixture needle (IN) using the same procedure used to adjust main mixture needle. Adjust throttle stop screw (TS) so engine idles at 1125-1275 rpm.

Walbro Carburetor. Initial adjustment of idle mixture needle (IN-Fig. KO124) should be $1^1/_4$ turns out from a lightly seated position for Models M18 and M20, and 1 turn out for Models MV16, MV18 and MV20.

On models so equipped, initial setting of high speed mixture screw is $1^1/_4$ turns out from a lightly seated position on Models MV16 and MV18, and 1 turn out on Model MV20. Some models are equipped with a fixed, nonadjustable main jet.

Final adjustments on all models must be performed with engine at normal operating temperature. Run engine and adjust idle speed by turning idle speed screw (S) to obtain an idle speed of 1200 rpm. Run engine at idle speed and turn idle mixture screw (I) counterclockwise until engine speed decreases and mark screw position. Turn screw clockwise until engine speed decreases again and mark screw position. Turn screw so it is halfway between the two marked positions. If necessary, readjust idle speed.

If equipped with adjustable high speed screw, proceed as follows: Run engine at full speed under load and turn high speed mixture screw counterclockwise until engine speed decreases and mark screw position. Turn screw clockwise until engine speed decreases again and mark screw position. Turn screw so it is halfway between the two marked positions. On models equipped with fixed main jet, a high altitude jet is available.

AIR FILTER. Three types of air filter assemblies have been used: Square type (Fig. KO125), Dome type (Fig. KO126) and an Anti-Icing type (Fig. KO127). For service information, refer to the ap-

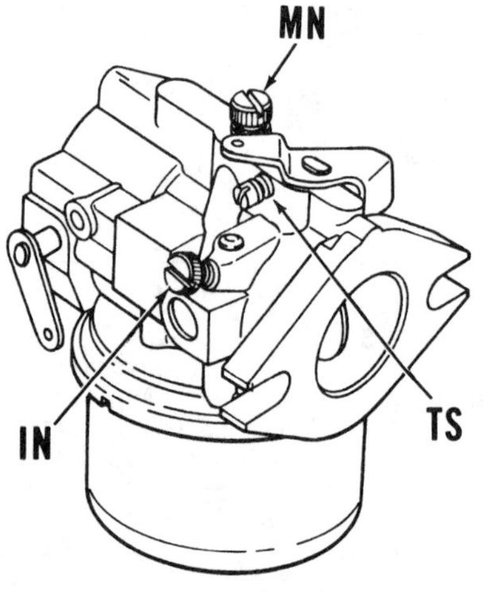

Fig. KO123—View showing location of main mixture needle (MN), idle mixture needle (IN) and throttle stop screw (TS) on Kohler adjustable carburetor.

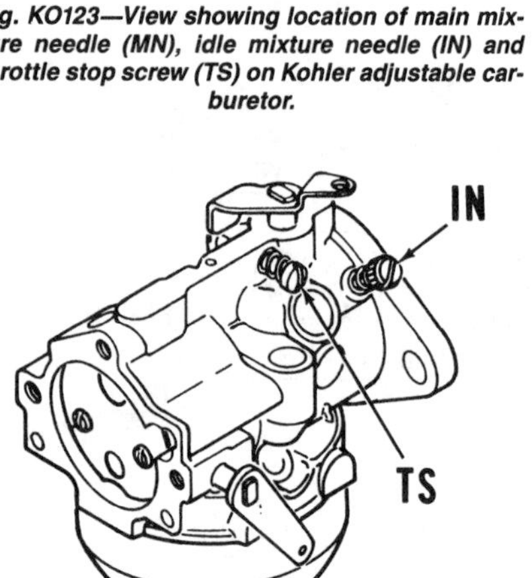

Fig. KO124—View showing location of idle mixture needle (IN) and throttle stop screw (TS) on Walbro fixed jet carburetor.

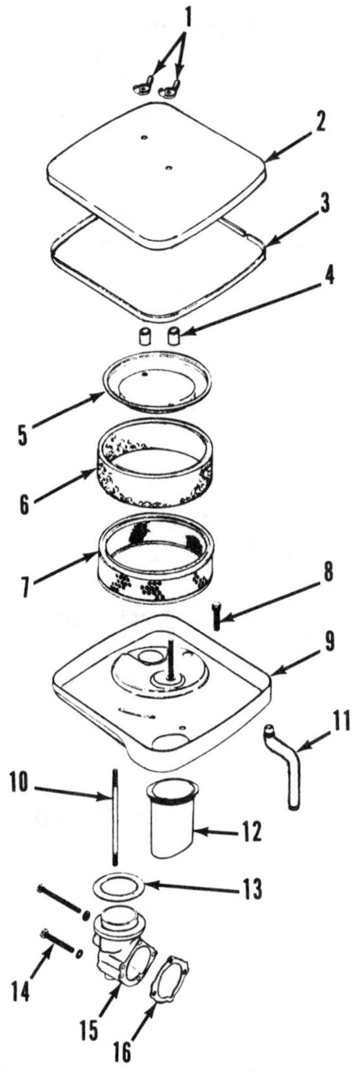

Fig. KO125—Exploded view of the square type air filter assembly used on some models.

1. Wing nuts
2. Cover
3. Gasket
4. Seals
5. Element cover
6. Foam precleaner
7. Paper element
8. Screw
9. Base
10. Stud
11. Breather hose
12. Intake tube
13. Gasket
14. Bolts
15. Intake elbow
16. Gasket

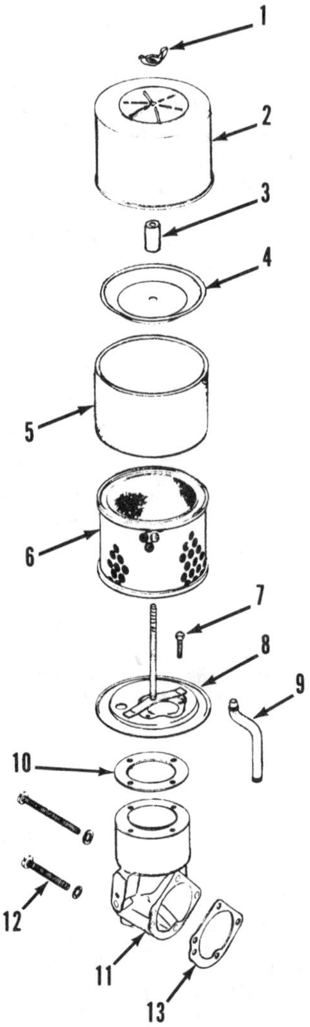

Fig. KO126—Exploded view of dome type air filter assembly used on some models.

1. Wing nut
2. Cover
3. Seal
4. Element cover
5. Foam precleaner
6. Paper element
7. Screw
8. Base
9. Breather hose
10. Gasket
11. Intake elbow
12. Bolts
13. Gasket

Illustrations courtesy Kohler Co.

propriate following service information for type being serviced.

Square Type Air Filter. Optional foam precleaner element (6—Fig. KO125) should be removed and cleaned after every 25 hours of normal use. Clean foam precleaner in a suitable solvent, rinse and air dry. Soak element in clean engine oil and squeeze out excess oil before reinstalling.

Paper element (7) should be removed and cleaned after every 100 hours of normal use. Paper element can be cleaned using a vacuum or very low pressure compressed air.

Dome Type Air Filter. Optional foam precleaner element (5—Fig. KO126) should be removed and cleaned after every 25 hours of normal use. Clean foam precleaner in a suitable solvent, rinse and air dry. Soak element in clean engine oil and squeeze out excess oil before reinstalling.

Paper element (6) should be removed and cleaned after every 100 hours of normal use. Paper element can be cleaned using a vacuum or very low pressure compressed air.

Anti-Icing Air Filter. Optional foam precleaner element (7—Fig. KO127) should be removed and cleaned after every 25 hours of normal use. Clean foam precleaner in a suitable solvent, rinse and air dry. Soak element in clean engine oil and squeeze out excess oil before reinstalling.

Paper element (8) should be removed and cleaned after every 100 hours of normal use. Paper element can be cleaned using a vacuum or very low pressure compressed air.

When installing air filter cover (2), refer to Fig. KO128 and install cover as shown in the left view when ambient temperature is below 45° F (10° C).

GOVERNOR. All models are equipped with a centrifugal flyweight mechanical governor. Governor gear and flyweight mechanism is located inside engine crankcase and is driven by the camshaft gear. For an exploded view of the governor assembly and external linkage, refer to Fig. KO129 for Models M18 and M20, and to Fig. KO130 for Models MV16, MV18 and MV20.

On Models M18 and M20, make initial adjustment of governor linkage whenever governor lever is loosened or removed from cross shaft. Make certain throttle linkage is connected to governor lever and throttle lever on carburetor. Loosen governor lever clamp bolt nut (1—Fig. KO129). Pull governor lever (10) away from carburetor as far as possible to fully open carburetor throttle plate. Rotate governor shaft (2) clockwise as far as possible and tighten clamp nut (1) to 4-5 ft.-lbs. (5-6 N·m). Maximum engine speed with no-load is 3600 rpm and is controlled by the high speed stop screw.

Governor sensitivity is controlled by changing the position of spring (4) in the holes in governor lever (10) and speed control lever (5). Standard spring position is the fifth hole from the cross shaft in governor lever and in the sixth hole from the hex head cap screw in the speed control lever. Always mark spring location prior to spring removal. To increase governor sensitivity, increase spring tension by moving spring in governor lever toward cross shaft. To decrease sensitivity and allow broader control, decrease spring tension by moving spring in governor lever away from cross shaft.

On Models MV16, MV18 and MV20, make initial adjustment of governor linkage whenever the governor lever is loosened or removed from cross shaft. Make certain governor linkage is con-

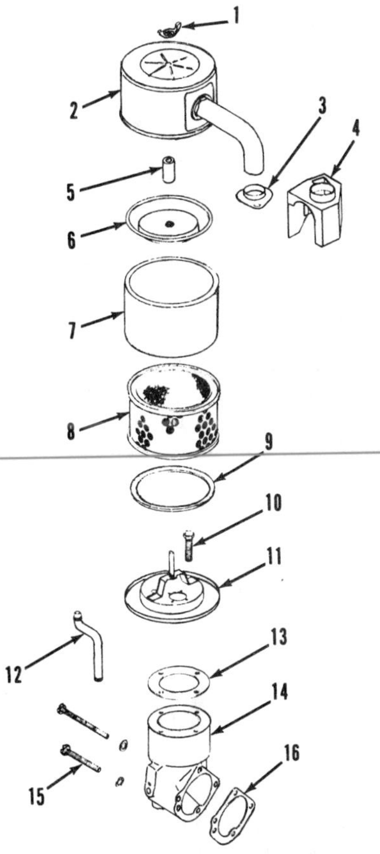

Fig. KO127—Exploded view of the anti-icing air filter assembly used on some models.

1. Wing nut
2. Cover
3. Heater plate
4. Heater cover
5. Seal
6. Element cover
7. Foam precleaner
8. Paper element
9. Base seal
10. Screw
11. Base
12. Breather hose
13. Gasket
14. Intake elbow
15. Bolts
16. Gasket

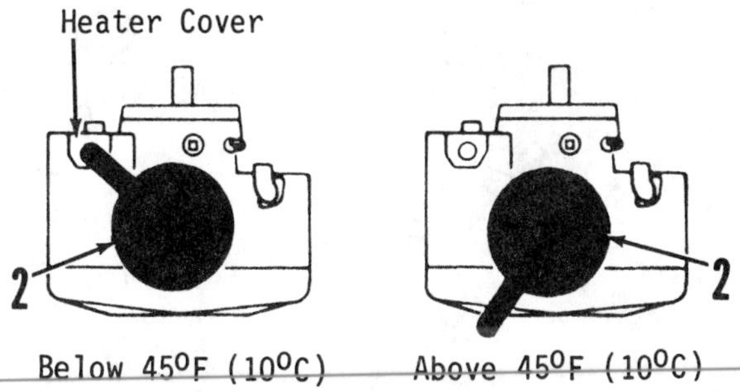

Fig. KO128—Cover (2) should be installed on air filter as shown according to ambient temperature.

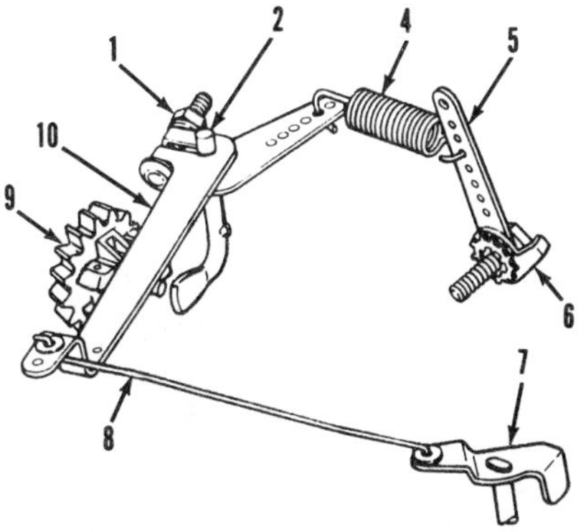

Fig. KO129—Exploded view of governor linkage used on Models M18 and M20.

1. Clamp nut
2. Governor shaft
4. Spring
5. Speed control lever
6. High speed stop
7. Carburetor throttle lever
8. Governor arm-to-carburetor rod
9. Governor gear & weight assy.
10. Governor arm

nected to governor lever and throttle lever on intake manifold. Throttle linkage is connected to throttle lever on the intake manifold and throttle lever on carburetor as shown in Fig. KO130. Loosen governor lever clamp bolt nut (7). Pull governor lever (5) down and away from the carburetor as far as possible to fully open carburetor throttle plate. Turn governor shaft (6) clockwise as far as possible and tighten clamp bolt nut (7) to 4-5 ft.-lbs. (5-6 N·m). Maximum engine speed with no-load is 3750 rpm and is controlled by high speed stop screw (10).

Governor sensitivity is controlled by changing the position of spring (8) in speed control lever (9) or governor lever (5). Standard spring position is the third hole from the cross shaft in governor lever and the second hole from the pivot point in the speed control lever. Always mark spring location prior to removal. To increase governor sensitivity, increase governor spring tension by moving spring in governor lever toward cross shaft. To decrease sensitivity and allow broader control, decrease spring tension by moving spring in governor lever away from cross shaft.

IGNITION SYSTEM. All models are equipped with an electronic magneto ignition system, which consists of a magnet assembly permanently affixed to the flywheel, an electronic magneto ignition module mounted on number one cylinder and a kill switch that stops the engine by grounding the ignition module.

Electronic magneto ignition module should be installed so a 0.008-0.012 inch (0.21-0.31 mm) air gap is between the module and flywheel.

To test ignition module, connect negative ohmmeter test lead to kill switch terminal (B—Fig. KO131) and positive ohmmeter test lead to coil laminations (A). Ohmmeter should register 5-1000 ohms. Reverse the positions of ohmmeter test leads to check primary resistance. Ohmmeter should register a minimum of 30,000 ohms. To check secondary resistance, connect one ohmmeter lead to terminal end of each spark plug cable. Ohmmeter should register 22,000-42,000 ohms. If ohmmeter readings are not as specified, renew ignition module.

VALVE ADJUSTMENT. Valve stem-to-tappet clearance should be checked after every 500 hours of operation. To check clearance, remove breather/valve cover assemblies from each cylinder. Rotate flywheel to position piston for No. 1 cylinder at top dead center of compression stroke. Use a feeler gage to measure clearance between valve stem ends and tappets. On all models, intake valve clearance should be 0.003-0.006 inch (0.08-0.15 mm). Required exhaust valve clearance is determined by engine serial number as follows: 0.016-0.019 inch (0.41-0.48 mm) before serial number 1816500656; 0.011-0.0.014 inch (0.30-0.35 mm) after serial number 1816500646 but before 1917809296 and 0.013-0.016 inch (0.33-0.41 mm) after serial number 1917809286. Rotate flywheel one complete turn (360 degrees) to position piston for No. 2 cylinder at top dead center and measure valve clearance for No. 2 cylinder.

If clearance is too small, remove valves and grind valve stem to increase clearance. If clearance is too large, grind valve seat or install a new valve to decrease clearance.

CYLINDER HEAD AND COMBUSTION CHAMBER. Cylinder heads should be removed and carbon and other deposits removed from cylinder heads and combustion chambers at 250 hour intervals if leaded gasoline is used or at 500 hour intervals if unleaded gasoline is used. Clean combustion deposits using a wooden or plastic scraper. Install new head gaskets and tighten cylinder head bolts to 15-20 ft.-lbs. (20-27 N·m) as outlined under CYLINDER HEAD in the REPAIRS section.

LUBRICATION. All models are equipped with a pressure lubrication system. Refer to OIL PUMP in the REPAIRS section for oil pump service information. Most engines are equipped with an oil filter.

Periodically check oil level; do not overfill. Oil dipstick should be screwed in until bottomed for correct oil level reading. Change oil after first 5 hours of operation. Thereafter change oil after every 50 hours if equipped with an oil filter or after every 25 hours on engines not equipped with a filter. Oil in all engines used in heavy duty operation should be changed after every 25 hours of operation.

Oil capacity for Models M18 and M20 is 3 pints (1.4 L) without filter or 3.5 pints (1.7 L) with filter. Oil capacity for Models MV16, MV18 and MV20 is 3.5

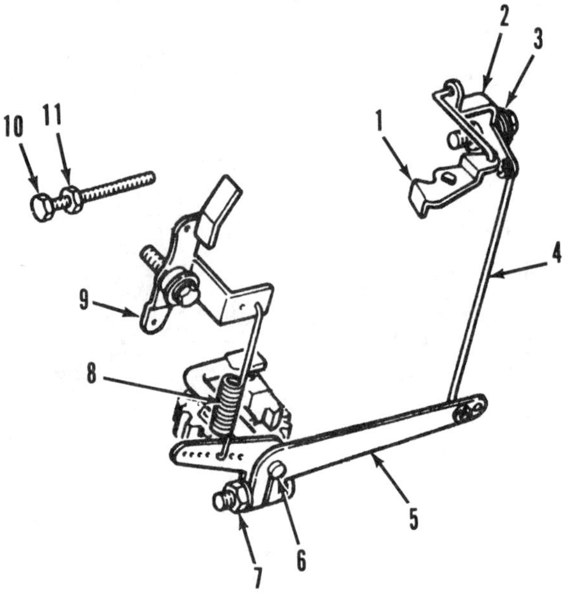

Fig. KO130—Exploded view of governor linkage used on Models MV16, MV18 and MV20.

1. Carburetor throttle lever
2. Throttle linkage
3. Torsion spring
4. Governor arm-to-throttle linkage rod
5. Governor arm
6. Governor shaft
7. Clamp nut
8. Spring
9. Speed control lever
10. High speed stop screw
11. Jam nut

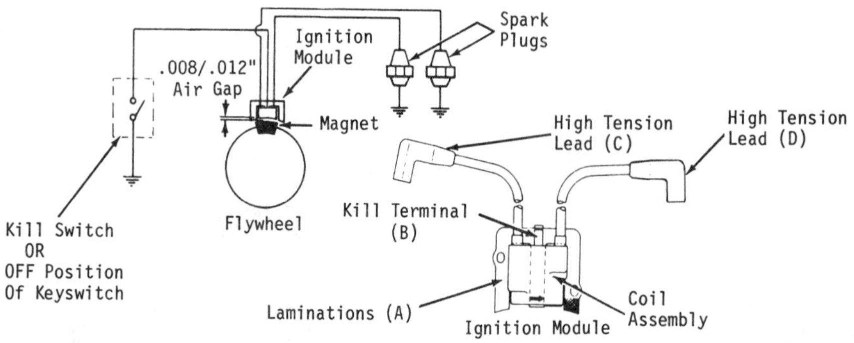

Fig. KO131—Schematic of the electronic magneto ignition system used on all models.

Illustrations courtesy Kohler Co.

pints (1.7 L) without filter or 4 pints (1.9 L) with filter.

Manufacturer recommends using oil with an API service classification of SF. Use SAE 30 oil for temperatures above 32° F (0° C); use SAE 5W-20, 5W-30, 10W-30 or 10W-40 oil for temperatures between 0° F (−18° C) and 32° F (0° C); below 0° F (−18° C) use only SAE 5W-20 or 5W-30.

REPAIRS

TIGHTENING TORQUES. Recommended tightening torque specifications are as follows:

Spark plug .... 10-15 ft.-lbs. (13-20 N·m)
Main fuel jet:
  Fixed type .... 12-16 in.-lbs. (1.4-1.8 N·m)
Carburetor fuel
  bowl nut. .... 45-55 in.-lbs. (5.1-6.2 N·m)
Cylinder head .... 15-20 ft.-lbs. (20-27 N·m)
Cylinder assembly .... 200 in.-lbs. (22.6 N·m)
Connecting rod:
  New nuts .... 140 in.-lbs. (15.8 N·m)
  Used nuts .... 100 in.-lbs. (11.3 N·m)
Flywheel .... 40 ft.-lbs. (54 N·m)
Oil filter bypass
  cover .... 125 in.-lbs. (14 N·m)
Closure plate .... 150 in.-lbs. (17 N·m)
Crankcase nuts .... 200 in.-lbs. (22.6 N·m)
Crankcase screws:
  5/16 in. .... 200 in.-lbs. (22.6 N·m)
  3/8 in. .... 260 in.-lbs. (29.4 N·m)

CARBURETOR. Kohler Adjustable Jet Carburetor. To disassemble carburetor, refer to Fig. KO132 and remove the float bowl retaining plug (16) and float bowl (15). Remove float pin (14), float assembly (13) and fuel inlet needle (10). Remove fuel inlet needle seat (9) and gasket (8). Remove idle mixture needle (5) and spring (6). Remove main fuel needle (1) and spring (2).

If throttle or choke shaft bores in carburetor are worn, a bushing kit is available for repair. Follow directions provided with kit carefully and use a suitable drill press to drill out shaft bores for bushing renewal.

Clean carburetor in carburetor cleaner only after all rubber and fiber washers and seals have been removed. Rinse and blow out all passages with compressed air. Do not use wire or a drill bit to clean fuel passages as calibration of carburetor could be affected if passages are enlarged.

To adjust float level, invert carburetor throttle body so the float tab rests on the fuel inlet needle. Float height (H—Fig. KO133) should be 11/64-in. (4.4 mm) from the carburetor mating surface to the float. Carefully bend float height tab to adjust.

After float level is adjusted, turn carburetor throttle body over and check the float drop. Float drop should be limited to 1 1/32 in. (26.2 mm) between the carburetor mating surface and the edge of float (Fig. KO134). Carefully bend the float drop limiting tab to adjust.

Invert carburetor throttle body and use a feeler gage to measure clearance between the float and the float hinge towers. Clearance should be at least 0.010 inch (0.25 mm) as shown in Fig. KO135. If necessary, file float tower to obtain the recommended clearance.

Walbro Carburetor. To disassemble carburetor, refer to Fig. KO136 and remove the bowl retaining plug (20) and float bowl (19). Remove float pin (17), float assembly (18) and fuel inlet needle (16). Do not attempt to remove the fuel inlet seat as it is not a serviceable item. Remove idle mixture needle (7), spring (8), throttle stop screw (10), spring (9) and the main fuel jet (14).

To clean the bowl vent channel and the off-idle ports, use a small punch to pierce and remove the Welch plugs (11 and 12). Use care not to damage carburetor body during this procedure. Throttle and choke shaft may be removed as required. Use care not to lose the choke detent ball and spring (5) as choke shaft (13) is removed.

Clean carburetor in a suitable carburetor cleaner only after all rubber and fiber washers and seals have been removed. Rinse and blow out all passages with compressed air. Do not use wire or a drill bit to clean passages as calibration of carburetor could be affected if passages are enlarged.

To reassemble, reverse the disassembly procedure. When installing Welch plugs, apply a suitable sealant to plugs and allow to dry. Install main fuel jet and tighten to 12-18 in.-lbs. (1.4-2.0 N·m).

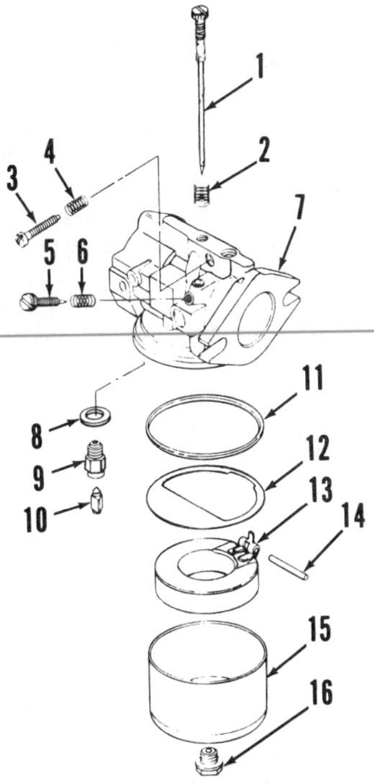

Fig. KO132—Exploded view of Kohler adjustable jet carburetor.

1. Main mixture needle
2. Spring
3. Throttle stop screw
4. Spring
5. Idle mixture needle
6. Spring
7. Carburetor body
8. Gasket
9. Inlet needle seat
10. Fuel inlet needle
11. Gasket
12. Baffle gasket
13. Float
14. Float pin
15. Float bowl
16. Plug & gasket

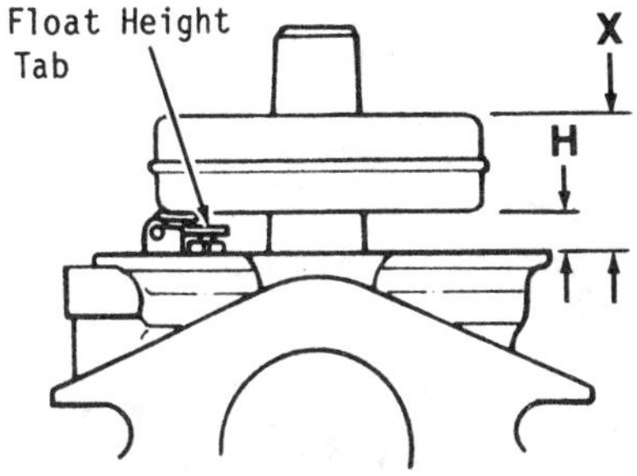

Fig. KO133—View of correct procedure to determine float height. Refer to text for specifications for model being serviced.

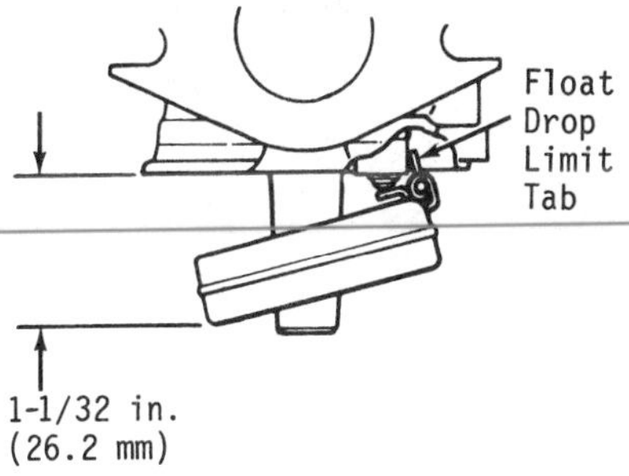

Fig. KO134—Float drop should be limited to 1 1/32 in. (26.2 mm).

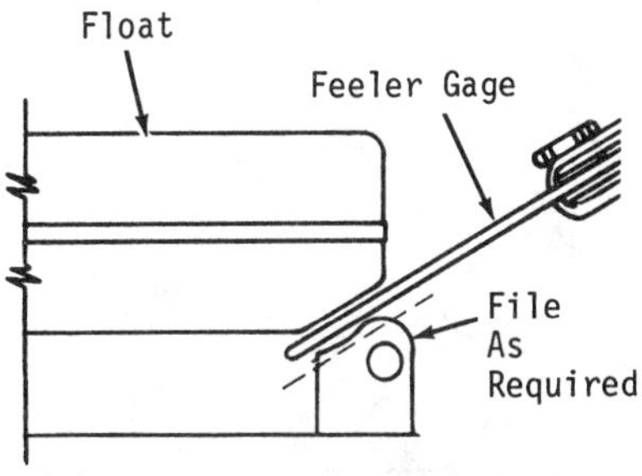

Fig. KO135—Check clearance between float tower and float as shown. Refer to text for specifications for model being serviced.

Illustrations courtesy Kohler Co.

To adjust float level, invert carburetor throttle body so that float tab rests on the fuel inlet needle. Float height (X—Fig. KO133) should be 0.690-0.720 inch (17.53-18.29 mm) from the carburetor mating surface to the float. Carefully bend float height tab to adjust.

CYLINDER HEAD. To remove cylinder heads, remove blower housing and cylinder head air baffles. Remove cylinder head bolts and remove cylinder heads and head gaskets. Note that cylinders are identified by a number 1 or number 2 mark stamped near the valve spring chamber. Mark cylinder heads with a corresponding number so they can be reinstalled in original location.

Clean carbon and other combustion deposits from cylinder heads, then check heads for distortion or other damage. If head is warped more than 0.003 inch (0.076 mm), both head and head bolts should be renewed.

To reinstall cylinder heads, reverse the removal procedure using new head gaskets. Tighten head bolts in two steps to specified torque of 15-20 ft.-lbs. (20-27 N·m) following sequence shown in Fig. KO137.

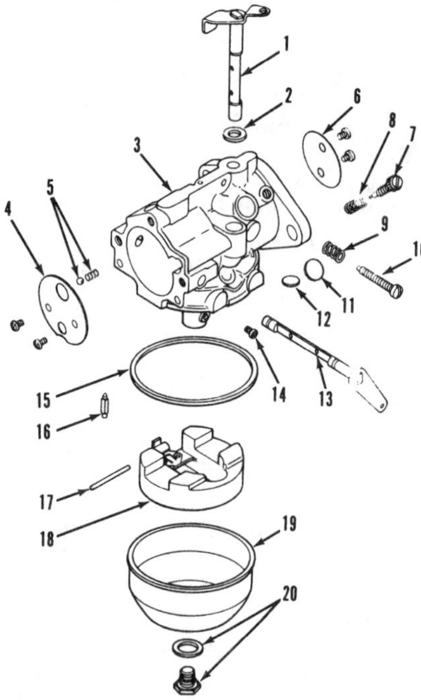

Fig. KO136—Exploded view of Walbro fixed jet carburetor.

1. Throttle shaft
2. Dust seal
3. Carburetor body
4. Choke plate
5. Choke detent ball & spring
6. Throttle plate
7. Idle mixture needle
8. Spring
9. Spring
10. Throttle stop screw
11. Welch plug
12. Welch plug
13. Choke shaft
14. Main jet
15. Gasket
16. Fuel inlet needle
17. Float pin
18. Float
19. Float bowl
20. Plug & gasket

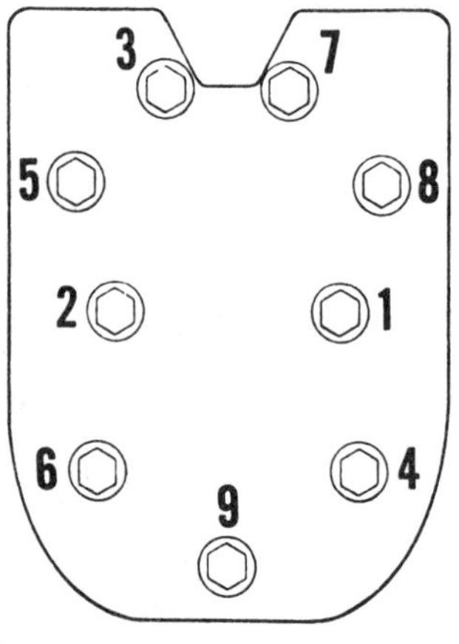

Fig. KO137—Cylinder head bolts should be loosened and tightened following the sequence shown.

CONNECTING ROD. To remove connecting rods, first mark the cylinder heads and crankcase halves as No. 1 side and No. 2 side as shown in Fig. KO138. Drain engine oil and remove oil filter. Remove mufflers, exhaust elbows, air cleaner, intake manifold and carburetor. Disconnect and remove all wiring, regulator, kill switch and solenoid. Remove blower housing and all cooling shrouds. Remove breather and valve tappet chamber covers. Remove fuel pump, ignition module and module bracket. Remove flywheel using a suitable puller. Remove starter motor. Remove cylinder heads, then remove cylinder retaining nuts and pull cylinder barrels off piston and connecting rod assemblies. Mark pistons and connecting rods as No. 1 and No. 2, then remove piston pin retaining rings, piston pins and pistons. Remove crankcase closure plate, oil pressure relief valve and oil filter adapter. Scribe a line across the camshaft bore plug and the No. 1 crankcase side. This line is used to align the plug during reassembly. Wrap a rubber band or tape around valve tappets stems in No. 2 crankcase side to prevent tappets from falling out when crankcase is split. Remove crankcase retaining nuts and screws. With No. 2 crankcase half facing upward, insert screwdriver into splitting notches provided in the crankcase halves and carefully pry apart crankcase halves. Remove camshaft and camshaft bore plug from No. 1 crankcase half. Mark tappet locations, then remove tappets. Remove crankshaft, connecting rods, front oil seal and bearings as an assembly. Mark connecting rods and rod caps as No. 1 and No. 2, then remove from crankshaft.

The aluminum alloy Posi-Lock connecting rods ride directly on the crankpin journal. Standard connecting rod-to-crankpin journal clearance is 0.0012-0.0024 inch (0.031-0.061 mm). If clearance exceeds 0.003 inch (0.075 mm), renew connecting rod and/or recondition crankshaft. A connecting rod with 0.010 inch (0.25 mm) undersize big end bore is available. The 0.010 inch (0.25 mm) undersize connecting rod is identified by a drilled hole in the connecting rod just above the big end bore (Fig. KO139). Standard connecting rod-to-piston pin clearance is 0.0006-0.0011 inch (0.015-0.028 mm) for all models. Standard connecting rod side play on crankpin is 0.005-0.016 inch (0.13-0.41 mm).

Reinstall connecting rods by reversing the removal procedure. Number one connecting rod should be installed on the crankshaft journal closest to the flywheel end of crankshaft, and number two connecting rod should be installed on crankshaft journal closest to the pto end of crankshaft. On Model M20, install the connecting rods with the shanks angled down and away from the camshaft. Make certain that match marks on connecting rods and caps are aligned (Fig. KO140). Tighten rod cap nuts to 140 in.-lbs. (15.8 N·m) for new connecting rods or 100 in.-lbs. (11.3 N·m) for used connecting rods. When installing crankshaft assembly, be sure that locating tab of sleeve bearing on pto end of shaft is positioned in notch in crankcase and that oil hole in bearing on flywheel end of shaft is aligned with oil passage in

Fig. KO138—Prior to disassembly, identify engine side to ensure correct reassembly. No. 1 side of engine has cylinder closest to flywheel end of crankshaft.

Illustrations courtesy Kohler Co.

crankcase. Make certain that camshaft gear timing mark is aligned with crankshaft timing mark.

Prior to joining crankcase halves or installing closure plate (horizontal crankshaft), oil pan (vertical crankshaft) or cylinders, apply silicone type sealant as shown in Figs. KO141, KO142, KO143 and KO144. Tighten crankcase fasteners (1 through 6—Fig. KO141) in sequence shown to 100 in.-lbs. (11.3 N·m), and tighten remaining fasteners (any sequence) to 100 in.-lbs. (11.3 N·m). Then tighten crankcase fasteners (1 through 4) to 260 in.-lbs. (29.4 N·m), and tighten remaining fasteners to 200 in.-lbs. (22.6 N·m). Tighten oil pan or closure plate fasteners to 150 in.-lbs. (17 N·m) following the sequence shown in Fig. KO142 or Fig. KO143. Tighten cylinder barrel nuts first to 100 in.-lbs. (11.3 N·m), then to 200 in.-lbs. (22.6 N·m) following sequence shown in Fig. KO144. Be sure that cam plug scribe mark is aligned with mark on crankcase half. Stake cam plug in place using the previous stake marks.

PISTON, PIN AND RINGS. The aluminum alloy piston is fitted with two compression rings and one oil control ring. To remove pistons, remove mufflers, exhaust elbows, air cleaner, intake manifold and carburetor. Disconnect and remove all wiring, regulator, kill switch and solenoid. Remove blower housing and all cooling shrouds. Remove breather and valve tappet chamber covers. Remove fuel pump, ignition module and module bracket and starter motor. Unbolt and remove cylinder heads and cylinder barrels. Remove piston pin retainers, piston pins and pistons from connecting rods.

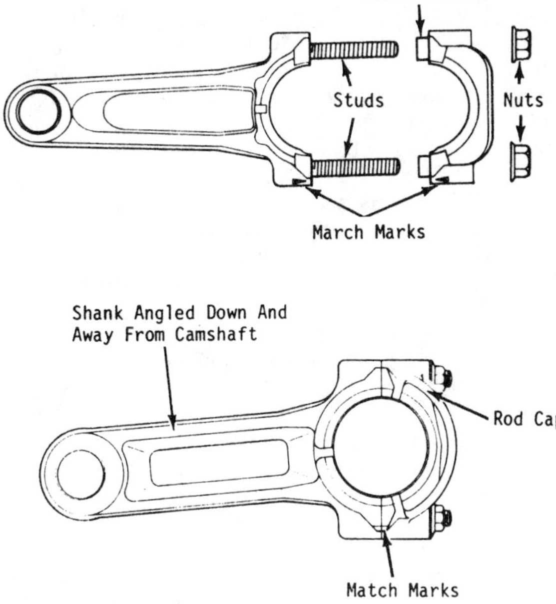

Fig. KO140—Align match marks on connecting rod and cap prior to installation. Connecting rods used on Model M20 (lower view) must be installed with shank angled away from camshaft (top of crankcase).

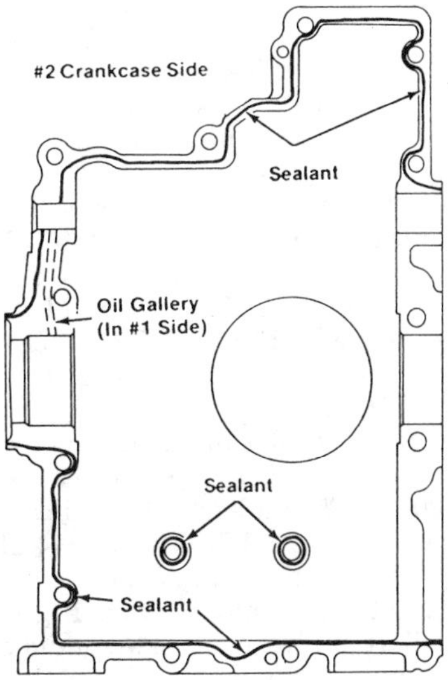

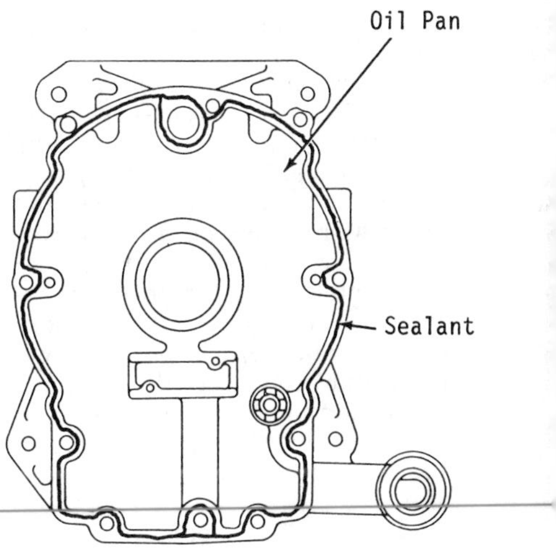

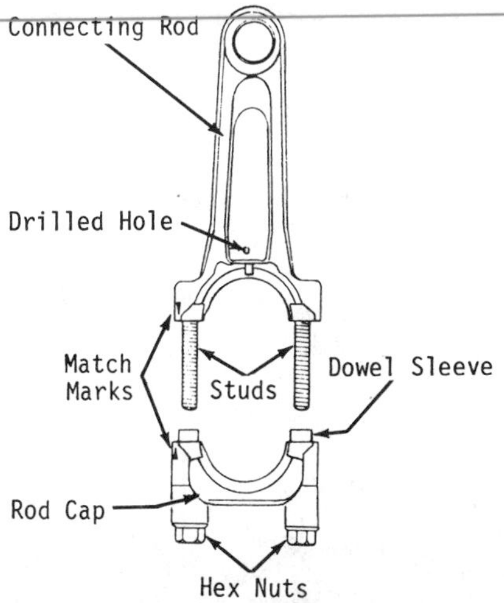

Fig. KO139—Connecting rods are available in 0.010 inch (0.25 mm) undersize for reconditioned crankshaft. Undersize connecting rod can be identified by the hole just above crankpin bearing bore.

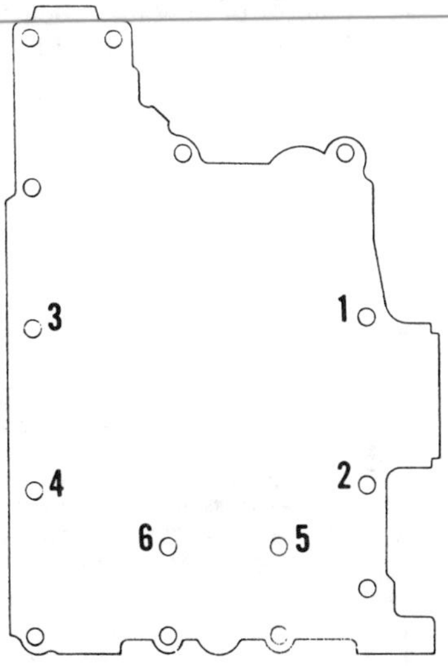

Fig. KO141—Prior to assembling crankcase, apply a silicone type sealant to No. 2 crankcase half as shown. Refer to text for fastener tightening sequence.

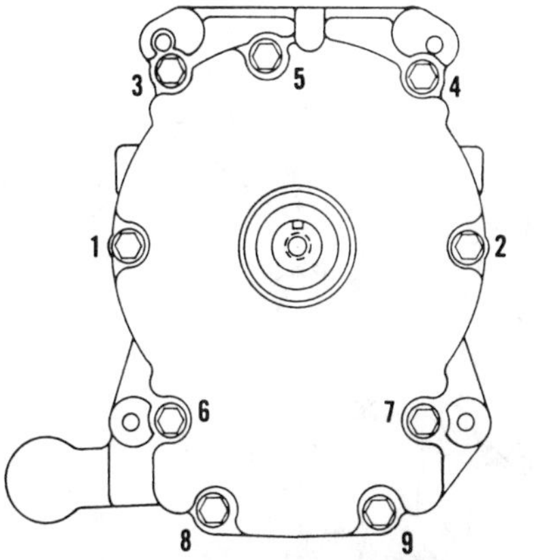

Fig. KO142—Prior to installing oil pan (vertical crankshaft), apply a silicone type sealant around all mating surfaces. Tighten oil pan fasteners to specified torque following sequence shown.

Remove piston rings from piston and clean ring grooves. Check piston for scoring, scuffing and excessive wear and renew as necessary.

Standard piston diameter is 3.1208-3.1215 inches (79.268-79.286 mm) for Model M20 and 3.1203-3.1210 inches (79.256-79.273 mm) for all other models. Piston diameter is measured ½ in. above lower edge of skirt perpendicular to piston pin. If skirt diameter is 3.1186 inches (79.212 mm) or less on Model M20, or 3.1181 inches (79.200 mm) on all other models, renew piston.

Piston-to-cylinder clearance should be 0.0030-0.0047 inch (0.076-0.117 mm) for Model M20 and 0.0035-0.0052 inch (0.089-0.132 mm) for all other models.

Insert each compression ring into cylinder bore and measure ring end gap using a feeler gage. Piston ring end gap should be 0.010-0.023 inch (0.25-0.58 mm) for all models.

Position new rings on piston and measure side clearance of piston ring in piston ring groove using a feeler gage. If side clearance exceeds 0.006 inch (0.15 mm), renew piston.

Standard piston pin diameter is 0.7499-0.7501 inch (19.048-19.053 mm) for Model M20 and 0.6247-0.6249 inch (15.867-15.872 mm) for all other models.

Piston rings should be installed on piston with "PIP" marked side of ring facing upward (Fig. KO145). Ring end gaps should be spaced equally around piston diameter. Lubricate piston, rings and cylinder with engine oil prior to reassembly.

Use a suitable ring compressor to install pistons in cylinder barrels. Be sure that arrow stamped on piston crown points toward flywheel side of engine. Install piston pin using new retaining rings. Complete installation by reversing the removal procedure.

CYLINDER. Cylinders can be removed from crankcase as separate assemblies. Number one cylinder is the cylinder closest to the flywheel and number two cylinder is the cylinder closest to the pto side of crankshaft.

Standard cylinder bore diameter is 3.1245-3.1255 inches (79.362-79.388 mm) for all models. If cylinder bore diameter is 3.128 inches (79.451 mm) or larger, cylinder must be bored to nearest oversize for which piston and rings are available, or renewed.

If cylinder is out-of-round 0.002 inch (0.05 mm) or more, cylinder should be rebored oversize, or renewed.

CRANKSHAFT, MAIN BEARINGS AND SEALS. The crankshaft is supported at each end in a sleeve type bearing. To remove crankshaft assembly, refer to CONNECTING ROD section.

Crankshaft end play is controlled by a thrust washer located at flywheel side of crankshaft on most models (Fig. KO146). Thrust washers are available in three sizes: 0.121, 0.130 and 0.139 inch. The 0.130 inch thick washer is considered standard. Crankshaft end play should be 0.002-0.014 inch (0.05-0.35 mm).

Some models may be equipped with a roller thrust bearing (Fig. KO146). Install the 0.039 inch thrust washer, the bearing, then the 0.156 inch shaft locating washer with the chamfered side of washer toward sleeve bearing as shown in Fig. KO146.

Standard crankshaft main bearing journal diameter is 1.7488-1.7510 inches (44.420-44.475 mm) for all models. If journal diameter is 1.7407 inches (44.214 mm) or less, renew crankshaft.

Crankshaft main bearing-to-crankshaft journal clearance should be 0.005 inch (0.125 mm) for all models. If clearance exceeds 0.006 inch (0.15 mm), renew bearing and/or crankshaft.

Inside diameter of new main bearing after installation should be 1.7439-1.7461 inches (44.295-44.351 mm) for all models.

Crankshaft crankpin diameter on Model M20 should be 1.4993-1.4998 inches (38.073-38.095 mm); recondition crankshaft if crankpin diameter is 1.4988 (38.069 mm) or less. Crankpin diameter for all other models should be 1.3733-1.3738 inches (34.882-34.895 mm); recondition crankshaft if crankpin

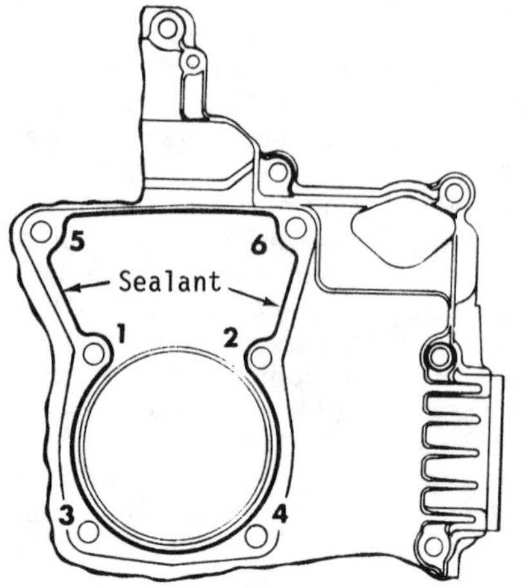

Fig. KO144—Prior to cylinder barrel installation, apply a silicone type sealant as shown. Tighten retaining nuts to specified torque following sequence shown.

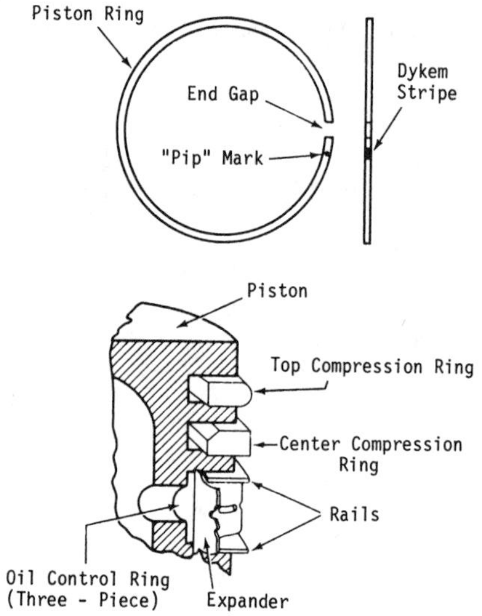

Fig. KO145—View showing correct piston ring installation.

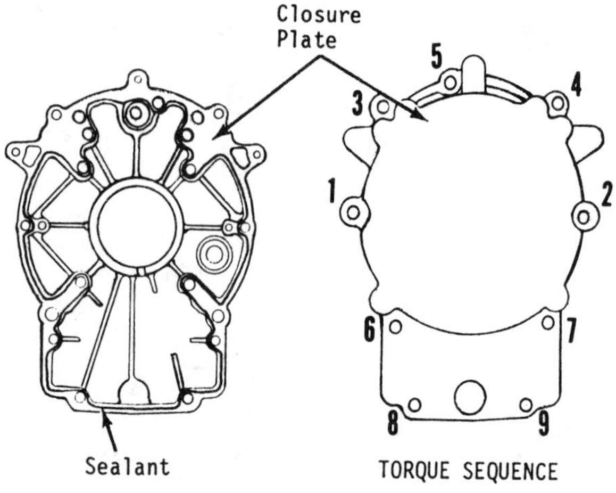

Fig. KO143—Prior to closure plate installation (horizontal crankshaft), apply a silicone type sealant as shown. Tighten retaining bolts and nuts to specified torque following sequence shown.

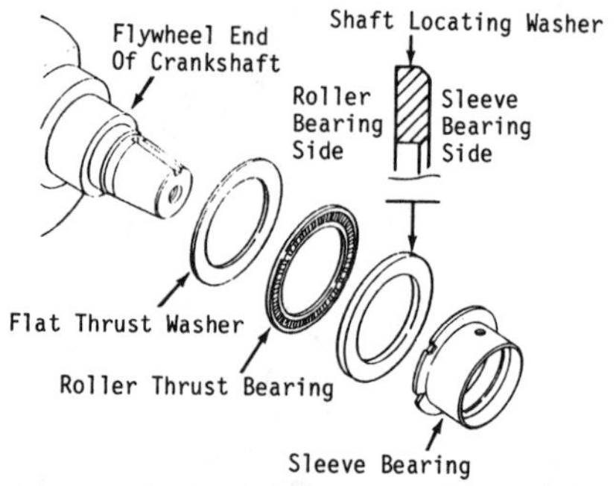

Fig. KO146—Some models are equipped with a crankshaft thrust bearing as shown. Refer to text.

Illustrations courtesy Kohler Co.

diameter is 1.3728 inches (34.869 mm) or less.

On all models, if crankpin journal is 0.0005 inch (0.013 mm) or more out-of-round, or if journal has 0.001 inch (0.025 mm) or more taper, recondition or renew crankshaft.

When installing crankshaft, make certain that timing marks on crankshaft and camshaft are aligned as shown in Fig. KO147. Be sure that locating tab of main bearing on pto end of shaft engages notch in crankcase and that oil hole in main bearing on flywheel end of shaft is aligned with oil passage in crankcase. Crankshaft oil seals should be installed so they are 1/32 in. (0.79 mm) below flush with seal bores. Refer to CONNECTING ROD section for remainder of reassembly procedure.

CAMSHAFT. The camshaft is supported at each end in bearings that are an integral part of the crankcase castings. To remove camshaft, split crankcase halves as outlined in CONNECTING ROD section. Be sure to identify tappets so they can be reinstalled in their original positions if reused.

Clearance between camshaft journals and bearing bores should be 0.0010-0.0025 inch (0.025-0.064 mm) for all models. Camshaft end play should be 0.003-0.013 inch (0.08-0.33 mm) and is controlled by shims installed between crankcase and camshaft at gear side of camshaft.

When installing camshaft, make certain that crankshaft timing mark and camshaft timing mark are aligned as shown in Fig. KO147.

Fig. KO147—When installing crankshaft and camshaft, align timing marks as shown.

VALVE SYSTEM. Refer to VALVE ADJUSTMENT under MAINTENANCE section for valve clearance and adjustment information. To remove valves, remove breather/valve cover assemblies and cylinder heads. Compress valve springs and remove valve spring retainers.

All models have renewable hardened alloy exhaust valve seats. Intake valve face and seat angles are 45 degrees for all models. Exhaust valve face and seat angles are 45 degrees for early models, and 30 degrees for later models and replacement exhaust valves and seats. Standard seat width for all valves is 0.037-0.045 inch (0.94-1.14 mm). See Fig. KO148. To renew valve seats, use suitable puller (NU-11726 or NU-11913) to remove valve seats from cylinder barrel. Be sure new seat is bottomed in cylinder barrel counterbore.

Valve stem-to-guide clearance should be 0.0025-0.0045 inch (0.064-0.114 mm) for the intake valve and 0.0045-0.0065 inch (0.114-0.165 mm) for the exhaust valve on all models. If intake valve stem diameter is 0.3103 inch (7.882 mm) or less, renew valve. If exhaust valve stem diameter is 0.3088 inch (7.843 mm) or less, renew valve.

Standard inside diameter for intake and exhaust valve guides is 0.3125 inch (7.94 mm). If intake valve guide inside diameter is worn 0.005 inch (0.13 mm) or more, or if exhaust valve guide inside diameter is worn 0.007 inch (0.18 mm) or more, renew guides.

Use a suitable puller to remove old valve guide, and install new valve guide using a suitable valve guide driver. Guide should be installed so end surface of guide is 1.125 inches (28.57 mm) below top edge of valve seat on Models M18, MV16 and MV18, or 1.390 inches (35.31 mm) on Models M20 and MV20. See Fig. KO148.

Install new intake valve seals on intake valve guides. Position valves in cylinder barrel and check valve clearance as outlined in MAINTENANCE section. Install valve springs with close coils facing away from spring retainers. Complete installation by reversing the removal procedure.

OIL PUMP. All models are equipped with a rotor type oil pump driven by the crankshaft (Fig. KO149). Oil pump assembly is located behind closure plate (horizontal crankshaft) or oil pan (vertical crankshaft) on pto side of engine. Oil pump rotors, cover and pressure relief valve on all models can be serviced without splitting the crankcase.

Oil pump shaft-to-crankcase clearance should be 0.0010-0.0025 inch (0.025-0.064 mm). Oil pump drive gear end play should be 0.010-0.029 inch (0.25-0.74 mm).

Free length of oil pressure relief valve spring should be 0.992 inch (25.20 mm).

Oil pump cover (Fig. KO149) outer side is marked "OUT" and is retained by self-tapping screws.

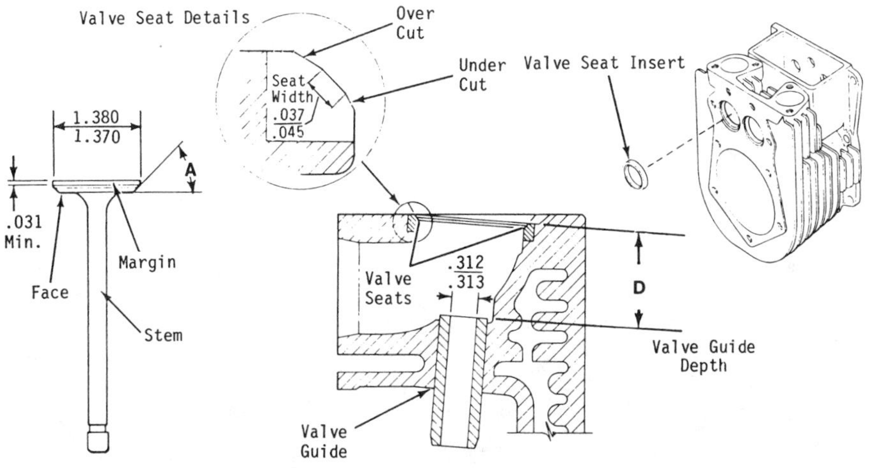

Fig. KO148—View showing valve specifications and information.

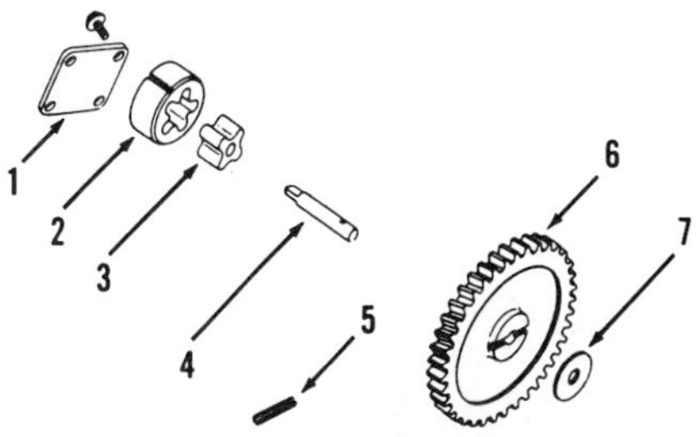

Fig. KO149—Exploded view of the rotor type oil pump used on all models.

1. Pump cover
2. Outer rotor
3. Inner rotor
4. Pump shaft
5. Roll pin
6. Drive gear
7. Spacer

Illustrations courtesy Kohler Co.

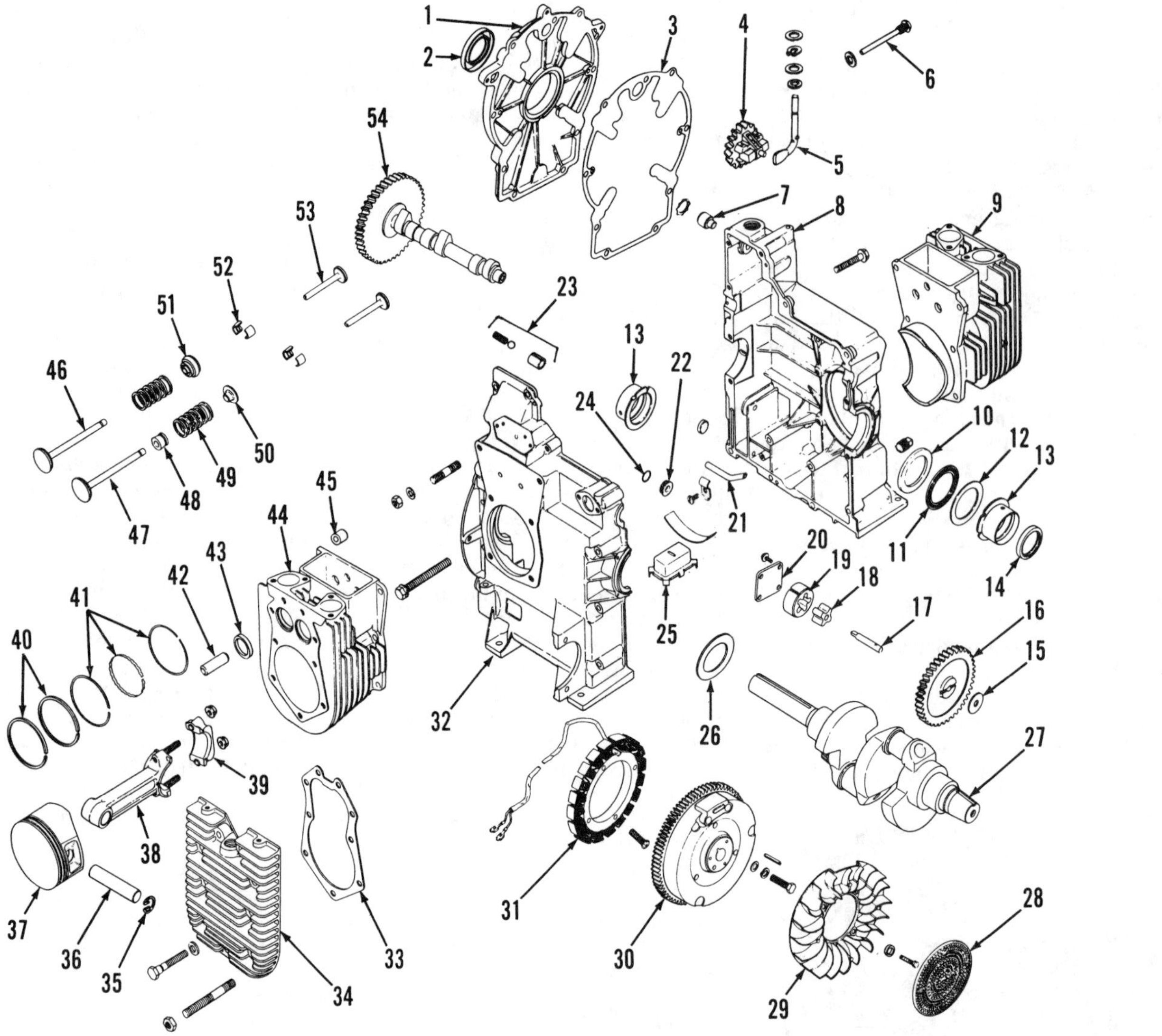

Fig. KO150—Exploded view of Kohler twin-cylinder, horizontal crankshaft engine.

1. Closure plate
2. Oil seal
3. Gasket
4. Governor assy.
5. Governor cross shaft
6. Governor stop pin
7. Governor stub shaft
8. Crankcase half
9. Cylinder barrel
10. Thrust washer
11. Thrust bearing
12. Thrust washer
13. Main bearing
14. Oil seal
15. Spacer
16. Oil pump gear
17. Oil pump shaft
18. Inner rotor
19. Outer rotor
20. Pump cover
21. Oil pickup tube
22. Camshaft bore plug
23. Oil pressure relief valve
24. "O" ring
25. Oil pickup screen
26. Thrust washer
27. Crankshaft
28. Screen
29. Fan
30. Flywheel
31. Stator
32. Crankcase half
33. Head gasket
34. Cylinder head
35. Retaining ring
36. Piston pin
37. Piston
38. Connecting rod
39. Rod cap
40. Compression rings
41. Oil ring
42. Valve guide
43. Valve seat insert
44. Cylinder barrel
45. Alignment bushing
46. Exhaust valve
47. Intake valve
48. Valve seal
49. Spring
50. Spring retainer
51. Valve rotator
52. Split retainer
53. Tappet
54. Camshaft

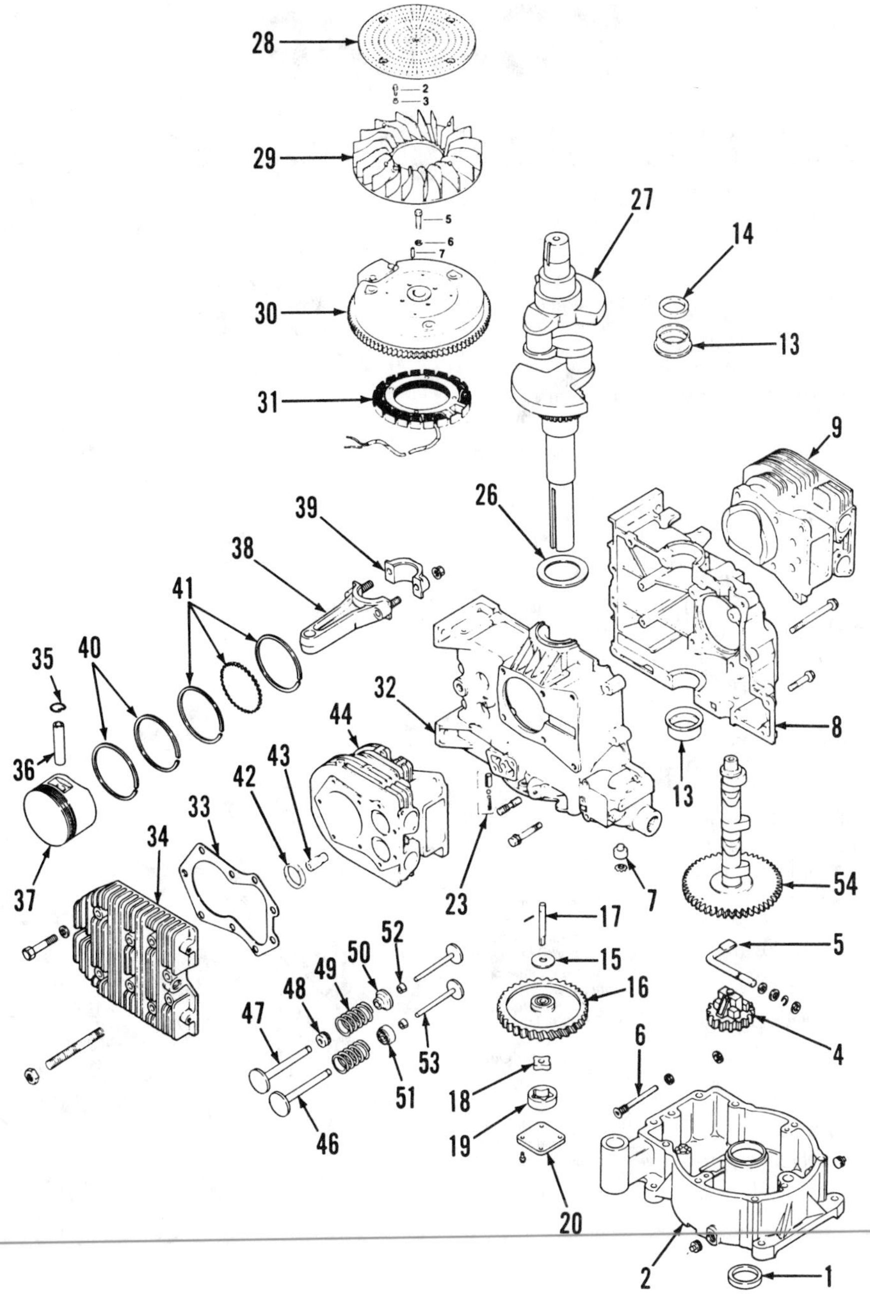

Fig. KO151—Exploded view of Kohler twin-cylinder, vertical crankshaft engine. Refer to Fig. KO150 for legend except for oil seal (1) and oil pan (2).

KOHLER

SERVICING KOHLER ACCESSORIES

12-VOLT GEAR DRIVE STARTERS

TWO BRUSH COMPACT TYPE. To disassemble starting motor, clamp mounting bracket in a vise. Remove through-bolts (H – Fig. KO105) and slide commutator end plate (J) and frame assembly (A) off armature. Clamp steel armature core in a vise and remove Bendix drive (E), drive end plate (F), thrust washer (D) and spacer (C) from armature (B).

Renew brushes if unevenly worn or worn to a length of 5/16-inch (7.938 mm) or less. To renew ground brush (K), drill out rivet, then rivet new brush lead to end plate. Field brush (P) is soldered to field coil lead.

Reassemble by reversing disassembly procedure. Lubricate bushings with a light coat of SAE 10 oil. Inspect Bendix drive pinion and splined sleeve for damage. If Bendix is in good condition, wipe clean and install completely dry. Tighten Bendix drive retaining nut to a torque of 130-150 in.-lbs. (15-18 N·m). Tighten through-bolts (H) to a torque of 40-55 in.-lbs. (4-7 N·m).

PERMANENT MAGNET TYPE. To disassemble starting motor, clamp mounting bracket in a vise and remove through-bolts (19 – Fig. KO106). Carefully slide end cap (10) and frame (11) off armature. Clamp steel armature core in a vise and remove nut (18), spacer (17), anti-drift spring (16), drive assembly (15), end plate (14) and thrust washer (13) from armature (12).

The two input brushes are part of terminal stud (6). Remaining two brushes (9) are secured with cap screws. When reassembling, lubricate bushings with American Bosch lubricant #LU3001 or equivalent. Do not lubricate starter drive. Use rubber band to hold brushes in position until started in commutator, then cut and remove rubber band. Tighten through-bolts to a torque of 80-95 in.-lbs. (8-10 N·m) and nut (18) to a torque of 90-110 in.-lbs. (11-12 N·m).

FOUR BRUSH BENDIX DRIVE TYPE. To disassemble starting motor, remove screws securing drive end plate (K – Fig. KO107) to frame (I). Carefully withdraw armature and drive assembly from frame assembly. Clamp steel armature core in a vise and remove Bendix drive retaining nut, then remove drive

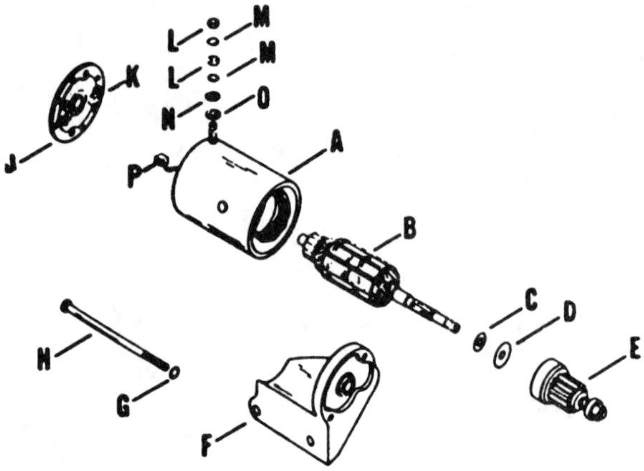

Fig. KO105 – Exploded view of two brush compact gear drive starting motor.

A. Frame & field coil assy.
B. Armature
C. Spacer
D. Thrust washer
E. Bendix drive assy.
F. Drive end plate & mounting bracket
G. Lockwasher
H. Through-bolt
J. Commutator end plate
K. Ground brush
L. Terminal nuts
M. Lockwashers
N. Flat washer
O. Insulating washer
P. Field brush

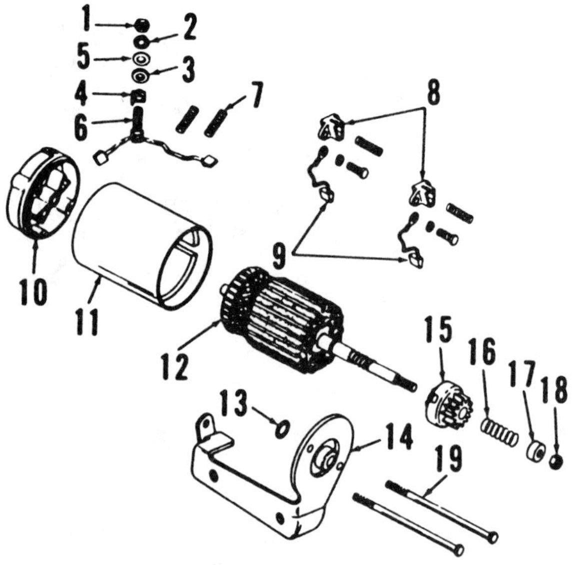

Fig. KO106 – Exploded view of permanent magnet type starting motor.

1. Terminal nut
2. Lockwasher
3. Insulating washer
4. Terminal insulator
5. Flat washer
6. Terminal stud & input brushes
7. Brush springs (4)
8. Brush holders
9. Brushes
10. Commutator end cap
11. Frame & permanent magnets
12. Armature
13. Thrust washer
14. Drive end plate & mounting bracket
15. Drive assy.
16. Anti-drift spring
17. Spacer
18. Nut
19. Through-bolts

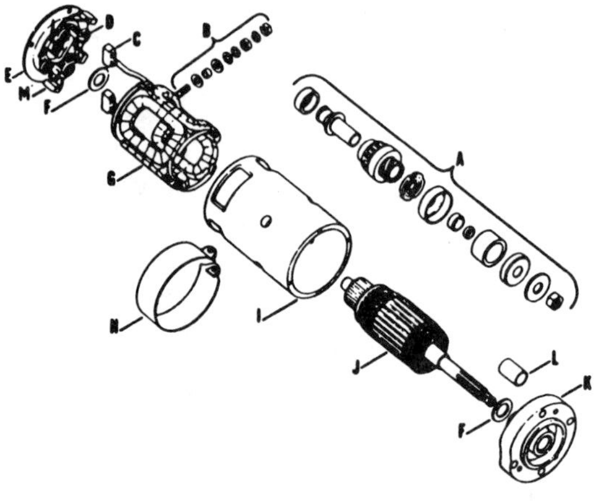

Fig. KO107 – Exploded view of conventional four brush starting motor with Bendix drive.

A. Bendix drive assy.
B. Terminal stud set
C. Field brushes
D. Brush springs
E. Commutator end plate
F. Thrust washers
G. Field coils
H. Cover
I. Frame
J. Armature
K. Drive end plate
L. Bushing
M. Ground brushes

Illustrations courtesy Kohler Co.

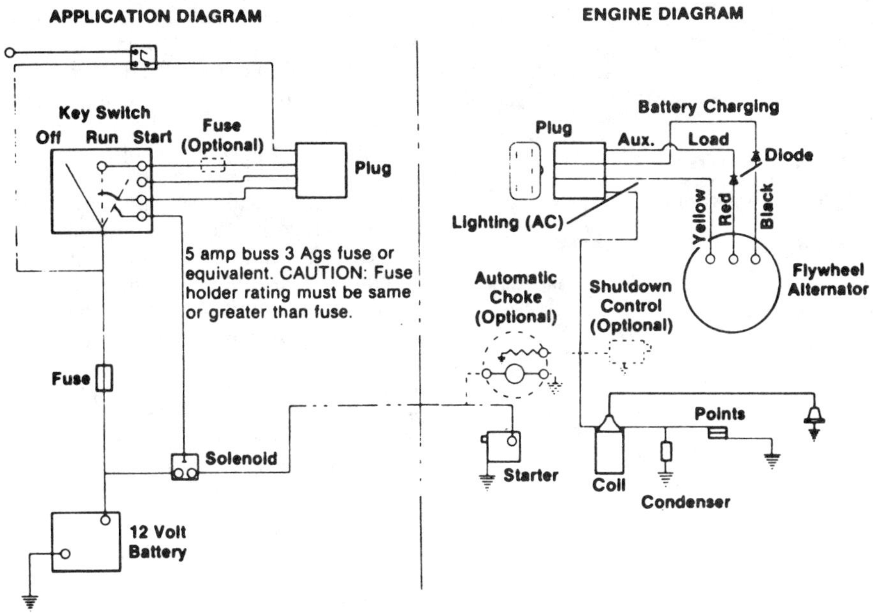

Fig. KO109—Typical electrical wiring diagram for engines equipped with 3 amp alternator.

unit (A), end plate (K) and thrust washer from armature (J). Remove cover (H) and screws securing end plate (E) to frame. Pull field brushes (C) from brush holders and remove end plate assembly.

The two ground brush leads are secured to end plate (E) and the two field brush leads are soldered to field coils. Renew brush set if excessively worn.

Inspect bushing (L) in end plate (K) and renew bushing if necessary. When reassembling, lubricate bushings with light coat of SAE 10 oil. Do not lubricate Bendix drive assembly.

Note starter may be reinstalled with Bendix in engaged or disengaged position. Do not attempt to disengage Bendix if it is in the engaged position.

FLYWHEEL ALTERNATORS

3 AMP ALTERNATOR. The 3 amp alternator consists of a permanent magnet ring with five or six magnets on flywheel rim, a stator assembly attached to crankcase and a diode in charging output lead. See Fig. KO109.

To avoid possible damage to charging system, the following precautions must be observed:

1. Negative post of battery must be connected to engine ground and correct battery polarity must be observed at all times.
2. Prevent alternator leads (AC) from touching or shorting.
3. Remove battery or disconnect battery cables when recharging battery with battery charger.
4. Do not operate engine for any length of time without a battery in system.
5. Disconnect plug before electric welding is done on equipment powered by and in common ground with engine.

TROUBLESHOOTING. Defective conditions and possible causes are as follows:

1. No output. Could be caused by:
   - A. Faulty windings in stator.
   - B. Defective diode.
   - C. Broken lead wire.
2. No lighting. Could be caused by:
   - A. Shorted stator wiring.
   - B. Broken lead.

If "no output" condition is the trouble, run following tests:

1. Connect ammeter in series with charging lead. Start engine and run at 2400 rpm. Ammeter should register 2 amp charge. Run engine at 3600 rpm. Ammeter should register 3 amp charge.
2. Disconnect battery charge lead from battery, measure resistance of lead to ground with an ohmmeter. Reverse ohmmeter leads and take another reading. One reading should be about mid-scale with meter set at R x 1. If both readings are high, diode or stator is open.
3. Expose diode connections on battery charge lead. Check resistance on stator side of ground. Reading should be 1 ohm. If 0 ohms, winding is shorted. If infinity ohms, stator winding is open or lead wire is broken.

If "no lighting" condition is the trouble, use an AC voltmeter and measure open circuit voltage from lighting lead to ground with engine running at 3000 rpm. If 15 volts, wiring may be shorted.

Check resistance of lighting lead to ground. If 0.5 ohms, stator is good, 0 ohms indicates shorted stator and a reading of infinity indicates stator is open or lead is broken.

3/6 AMP ALTERNATOR. The 3/6 amp alternator consists of a permanent magnet ring with six magnets on flywheel rim, a stator assembly attached to crankcase and two diodes located in battery charging lead and auxiliary load lead. See Fig. KO110.

To avoid possible damage to charging system, the following precautions must be observed.

1. Negative post of battery must be connected to engine ground and correct battery polarity must be observed at all times.

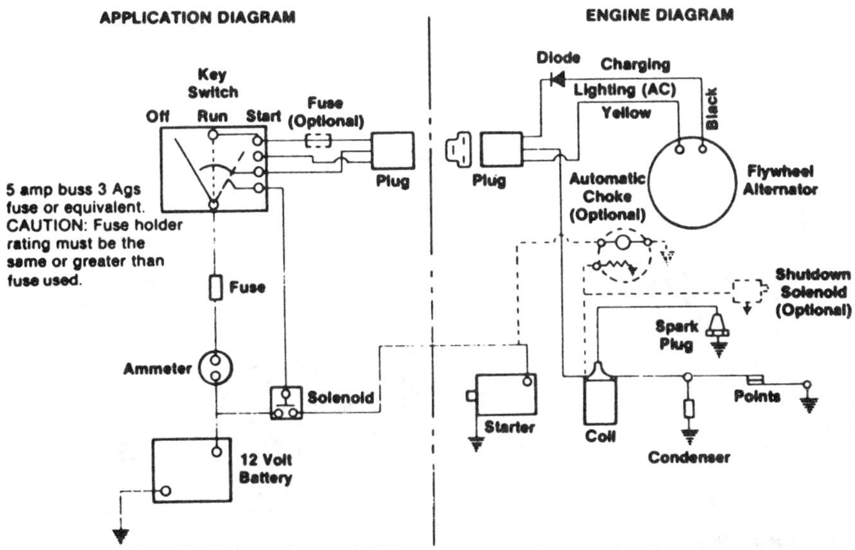

Fig. KO110—Typical electrical wiring diagram for engines equipped with 3/6 amp alternator.

Illustrations courtesy Kohler Co.

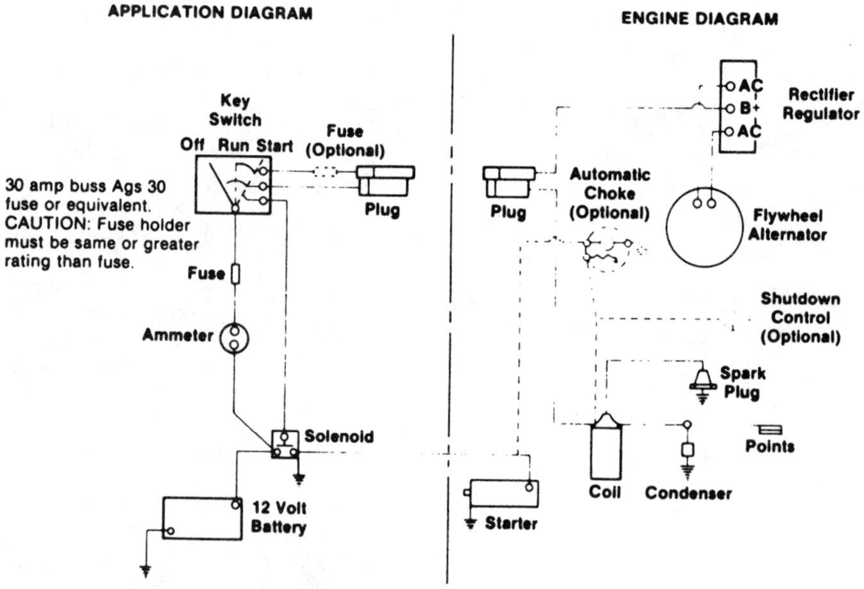

Fig. KO111—Typical electrical wiring diagram for engines equipped with 15 amp alternator and breaker point ignition. The 10 amp alternator is similar.

2. Prevent alternator leads (AC) from touching or shorting.
3. Do not operate for any length of time without a battery in system.
4. Remove battery or disconnect battery cables when recharging battery with battery charger.
5. Disconnect plug before electric welding is done on equipment powered by and in common ground with engine.

TROUBLESHOOTING. Defective conditions and possible causes are as follows:

1. No output. Could be caused by:
   A. Faulty windings in stator.
   B. Defective diode.
   C. Broken lead
2. No lighting. Could be caused by:
   A. Shorted stator wiring.
   B. Broken lead.

If "no output" condition is the trouble, run the following tests:

1. Disconnect auxiliary load lead and measure voltage from lead to ground with engine running 3000 rpm. If 17 volts or more, stator is good.
2. Disconnect battery charging lead from battery. Measure voltage from charging lead to ground with engine running at 3000 rpm. If 17 volts or more, stator is good.
3. Disconnect battery charge lead from battery and auxiliary load lead from switch. Measure resistance of both leads to ground. Reverse ohmmeter leads and take readings again. One reading should be infinity and the other reading should be about mid-scale with meter set at R x 1. If both readings are low, diode is shorted. If both readings are high, diode or stator is open.
4. Expose diode connections on battery charging lead and auxiliary load lead. Check resistance on stator side of diodes to ground. Readings should be 0.5 ohms. If reading is 0 ohms, winding is shorted. If infinity ohms, stator winding is open or lead wire is broken.

If "no lighting" condition is the trouble, disconnect lighting lead and measure open circuit voltage with AC voltmeter from lighting lead to ground with engine running at 3000 rpm. If 22 volts or more, stator is good. If less than 22 volts, wiring may be shorted.

Check resistance of lighting lead to ground. If 0.5 ohms, stator is good, 0 ohms reading indicates shorted stator and an infinity reading indicates an open stator winding or broken lead wire.

10 AND 15 AMP ALTERNATOR. Either a 10 or 15 amp alternator is used on some engines. Alternator output is controlled by a solid state rectifier-regulator. See Fig. KO113.

To avoid possible damage to charging system, the following precautions must be observed:

1. Negative post of battery must be connected to engine ground and correct battery polarity must be observed at all times.

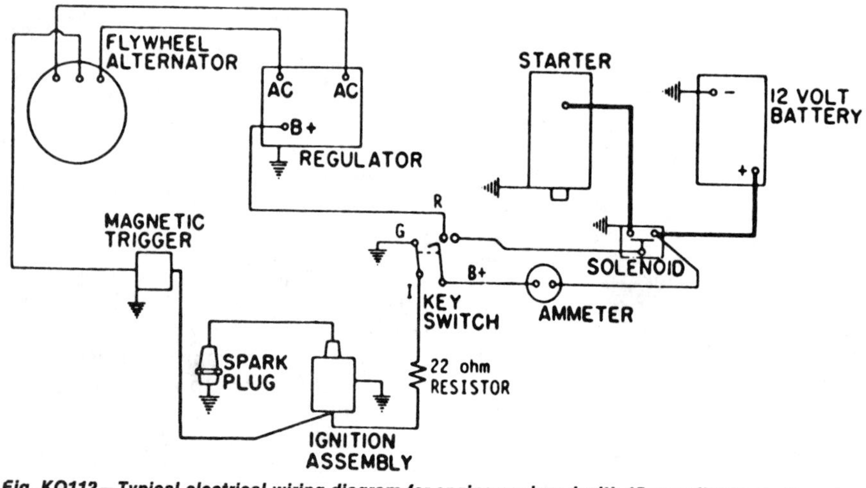

Fig. KO112—Typical electrical wiring diagram for engine equipped with 15 amp flywheel alternator and breakerless ignition system. The 10 amp alternator is similar.

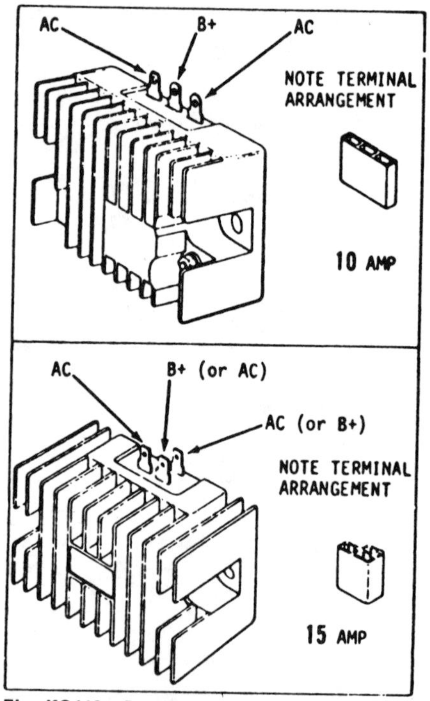

Fig. KO113—Rectifier-regulators used with 10 amp and 15 amp alternators. Although similar in appearance, units must not be interchanged.

Illustrations courtesy Kohler Co.

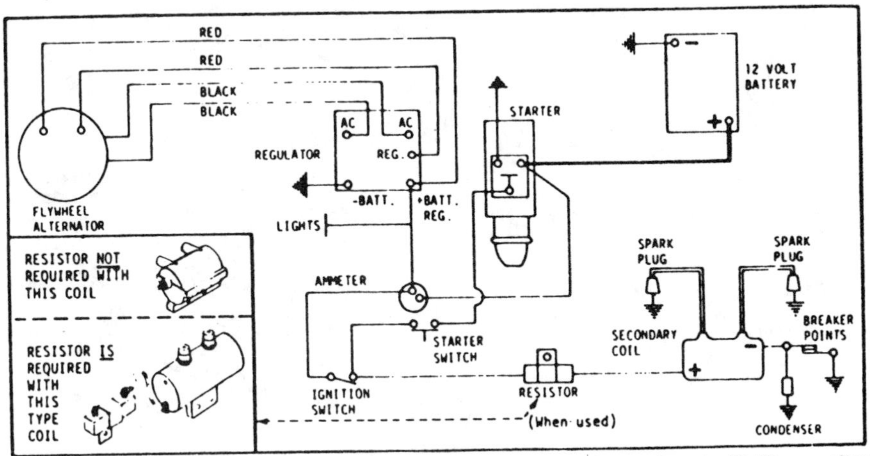

Fig. KO114—Typical electrical wiring diagram of two cylinder engine equipped with 30 amp alternator charging system. The 30 amp alternator on single cylinder engines is similar.

2. Rectifier-regulator must be connected in common ground with engine and battery.

3. Disconnect leads at rectifier-regulator if electric welding is to be done on equipment in common ground with engine.

4. Remove battery or disconnect battery cables when recharging battery with battery charger.

5. Do not operate engine with battery disconnected.

6. Make certain AC leads are prevented from being grounded at all times.

OPERATION. Alternating current (AC) produced by alternator is changed to direct current (DC) in rectifier-regulator. See Fig. KO112. Current regulation is provided by electronic devices which "sense" counter-voltage created by battery to control or limit charging rate. No adjustments are possible on alternator charging system. Faulty components must be renewed. Refer to the following troubleshooting paragraph to help locate possible defective parts.

TROUBLESHOOTING. Defective conditions and possible causes are as follows:

1. No output. Could be caused by:
   A. Faulty windings in stator.
   B. Defective diode(s) in rectifier.
   C. Rectifier-regulator not properly grounded.
2. Full charge-no regulation. Could be caused by:
   A. Defective rectifier-regulator.
   B. Defective battery.

If "no output" condition is the trouble, disconnect B+ cable from rectifier-regulator. Connect a DC voltmeter between B+ terminal on rectifier-regulator and engine ground. Start engine and operate at 3600 rpm. DC voltage should be above 14 volts. If reading is above 0 volts but less than 14 volts, check for defective rectifier-regulator. If reading is 0 volts, check for defective rectifier-regulator or defective stator by disconnecting AC leads from rectifier-regulator and connecting an AC voltmeter to the two AC leads. Check AC voltage with engine running at 3600 rpm. If reading is less than 20 volts (10 amp alternator) or 28 volts (15 amp alternator), stator is defective. If reading is more than 20 volts (10 amp alternator) or 28 volts (15 amp alternator), rectifier-regulator is defective.

If "full charge-no regulation" is the condition, use a DC voltmeter and check B+ to ground with engine operating at 3600 rpm. If reading is over 14.7 volts, rectifier-regulator is defective. If reading is under 14.7 volts but over 14.0 volts, alternator and rectifier-regulator are satisfactory and battery is probably defective (unable to hold a charge).

30 AMP ALTERNATOR. A 30 amp flywheel alternator consisting of a permanent field magnet ring (on flywheel) and an alternator stator (on bearing plate on single cylinder engines or gear cover on two cylinder engines) is used on some models. Alternator output is controlled by a solid state rectifier-regulator.

To avoid possible damage to charging system, the following precautions must be observed:

1. Negative post of battery must be connected to engine ground and correct battery polarity must be observed at all times.

2. Rectifier-regulator must be connected in common ground with engine and battery.

3. Disconnect wire from rectifier-regulator terminal marked "BATT. NEG." if electric welding is to be done on equipment in common ground with engine.

4. Remove battery or disconnect battery cables when recharging battery with battery charger.

5. Do not operate engine with battery disconnected.

6. Make certain AC leads are prevented from being grounded at all times.

OPERATION. Alternating current (AC) produced by alternator is carried by two black wires to full wave bridge rectifier where it is changed to direct current (DC). Two red stator wires serve to complete a circuit from regulator to secondary winding in stator. A zener diode is used to sense battery voltage and it controls a Silicon Controlled Rectifier (SCR). SCR functions as a switch to allow current to flow in secondary winding in stator when battery voltage gets above a specific level.

An increase in battery voltage increases current flow in secondary wind-

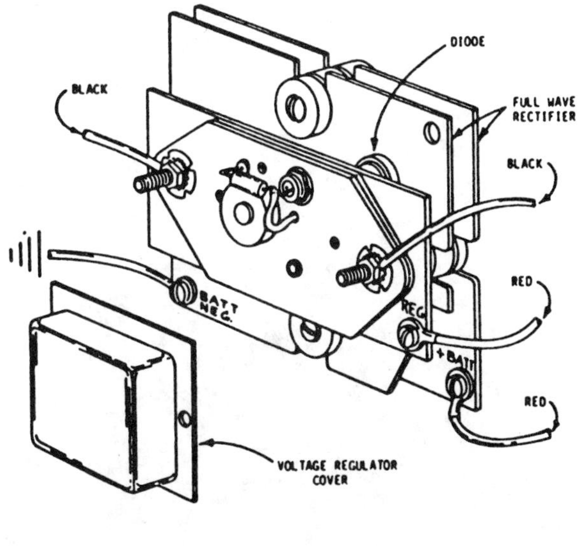

Fig. KO115—Rectifier-regulator used with 30 amp flywheel alternator, showing stator wire connections. Refer also to Fig. KO114.

Illustrations courtesy Kohler Co.

ing in stator. This increased current flow in secondary winding brings about a corresponding decrease in AC current in primary winding, thus controlling output.

When battery voltage decreases, zener diode shuts off SCR and no current flows to secondary winding. At this time, maximum AC current is produced by primary winding.

TROUBLESHOOTING. Defective conditions and possible causes are as follows:

1. No output. Could be caused by:
   A. Faulty windings in stator.
   B. Defective diode(s) in rectifier.
2. No charge (when normal load is applied to battery). Could be caused by:
   A. Faulty secondary winding in stator.
3. Full charge-no regulation. Could be caused by:
   A. Faulty secondary winding in stator.
   B. Defective regulator.

If "no output" condition is the trouble, check stator windings by disconnecting all four stator wires from rectifier-regulator. Check resistance on R x 1 scale of ohmmeter. Connect ohmmeter leads to the two red stator wires. About 2.0 ohms should be noted. Connect ohmmeter leads to the two black stator wires. Approximately 0.1 ohm should be noted. If readings are not at test values, renew stator. If ohmmeter readings are correct, stator is good and trouble is in rectifier-regulator. Renew rectifier-regulator.

If "no charge when normal load is applied to battery" is the trouble, check stator secondary winding by disconnecting red wire from "REG" terminal on rectifier-regulator. Operate engine at 3600 rpm. Alternator should now charge at full output. If full output of at least 30 amps is not attained, renew stator.

If "full charge-no regulation" is the trouble, check stator secondary winding by removing both red wires from rectifier-regulator and connecting ends of these two wires together. Operate engine at 3600 rpm. A maximum 4 amp charge should be noted. If not, stator secondary winding is faulty. Renew stator. If maximum 4 amp charge is noted, stator is good and trouble is in rectifier-regulator. Renew rectifier-regulator.

Refer to Fig. KO114 and KO115 for correct rectifier-regulator wiring connections.

CLUTCH Some models are equipped with either a dry disc clutch (Fig. KO116) or a wet type clutch (Fig. KO118). Both type clutches are lever operated. Refer to the following

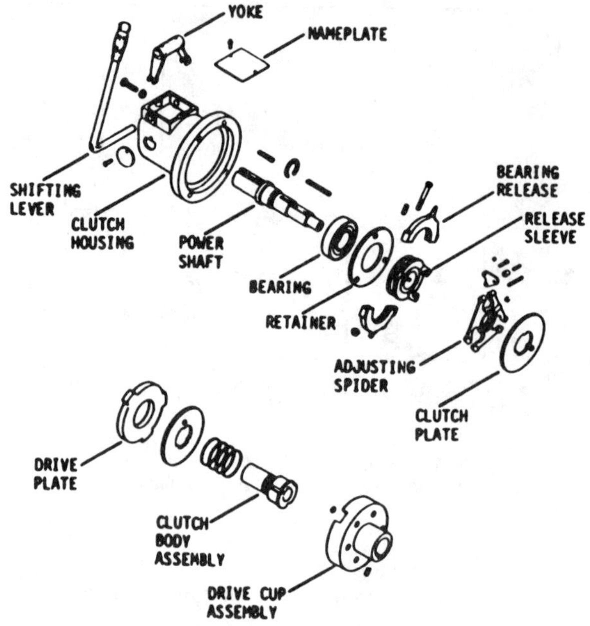

Fig. KO116—Exploded view of dry disc type clutch used on some models.

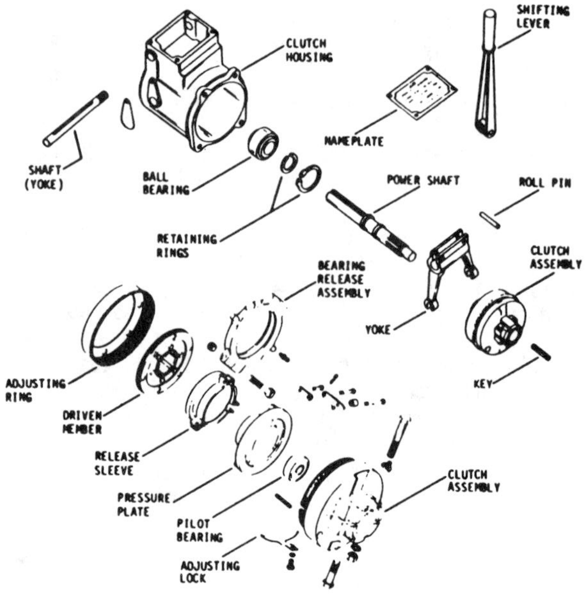

Fig. KO118—Exploded view of wet type clutch used on some models.

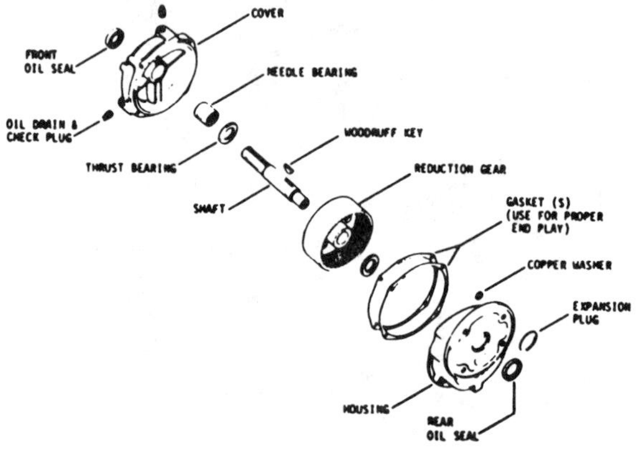

Fig. KO119—Exploded view of gear reduction drive used on some models.

Illustrations courtesy Kohler Co.

Fig. KO120—Combination clutch and chain type reduction drive used on some models.

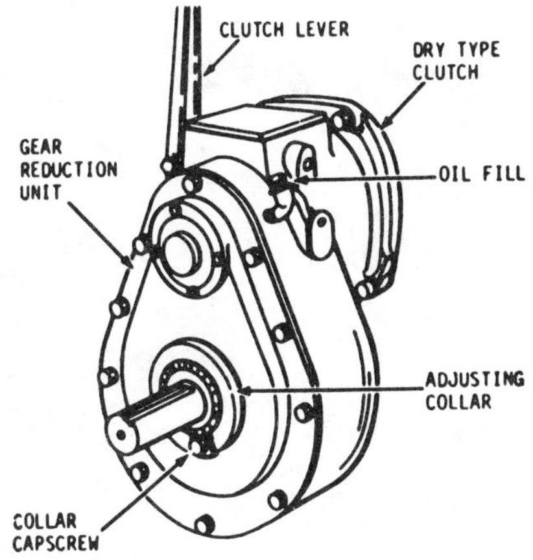

Fig. KO121—Output shaft end play on combination clutch and reduction drive must be adjusted to 0.0015-0.003 inch (0.0381-0.0762 mm). To adjust end play, loosen cap screw and rotate adjusting collar.

paragraphs for adjustment procedure.

DRY DISC TYPE. A firm pressure should be required to engage overcenter linkage. If clutch is slipping, remove nameplate (Fig. KO116) and locate adjustment lock by turning flywheel. Release clutch, back out adjusting lock screw, then turn adjusting spider clockwise until approximately 20 pounds (9 kg) pull is required to snap clutch overcenter. Tighten adjusting lock screw. Every 50 hours, lubricate clutch bearing collar through inspection cover opening.

WET TYPE CLUTCH. To adjust wet type clutch, remove nameplate and use a screwdriver to turn adjusting ring (Fig. KO118) in clockwise direction until a pull of 40-50 pounds (18-23 kg) at hand grip lever is required to snap clutch overcenter.

NOTE: Do not pry adjusting lock away from adjusting ring as spring type lock prevents adjusting ring from backing off during operation.

Change oil after each 100 hours of normal operation. Fill housing to level plug opening with non-detergent oil. Use SAE 30 oil in temperatures above 50° F (10° C), SAE 20 oil in temperatures 50° F (10° C) to freezing and SAE 10 oil in temperatures below freezing.

REDUCTION DRIVE (GEAR TYPE). The 6:1 ratio reduction gear unit (Fig. KO119) is used on some models. To remove unit, first drain lubricating oil, then unbolt cover from housing. Remove cover and reduction gear. Unbolt and remove gear housing from engine. Separate reduction gear, shaft and thrust washer from cover. Renew oil seals and needle bearings (bronze bushings on early units) as necessary.

When reassembling, wrap tape around gear on crankshaft to protect oil seal and install gear housing. Use new copper washers on two cap screws on inside of housing. Wrap tape on shaft to prevent keyway from damaging cover oil seal and install thrust washer, shaft and reduction gear in cover. Install cover and gear assembly using new gaskets as required to provide a shaft end play of 0.001-0.006 inch (0.0254-0.1524 mm). Gaskets are available in a variety of thicknesses. Fill unit to oil check plug opening with same grade oil as used in engine.

CLUTCH AND REDUCTION DRIVE (CHAIN TYPE). Some models are equipped with a combination clutch and reduction drive unit. Clutch is a dry type and method of adjustment is the same as for clutch shown in Fig. KO116. Clutch release collar should be lubricated each 50 hours of normal operation. Remove clutch cover for access to lubrication fitting. Reduction drive unit is a chain and sprocket type. See Fig. KO120. Fill reduction housing to level hole with same grade oil as used in engine. Capacity is 3 pints (1.4 L) and should be changed each 50 hours of normal operation. The tapered roller bearings on output shaft should be adjusted to provide 0.0015-0.003 inch (0.0381-0.0762 mm) shaft end play. Adjustment is by means of a collar which is locked in position by a 5/16-inch cap screw. See Fig. KO121.

ONAN

A DIVISION OF ONAN CORPORATION
1400 73rd Avenue, N.E.
Minneapolis, Minnesota 55432

Model	No. Cyls.	Bore	Stroke	Displacement	Power Rating
BF	2	3.125 in. (79.38 mm)	2.625 in. (66.68 mm)	40.3 cu. in. (660 cc)	16 hp. (11.9 kW)
BG	2	3.250 in. (82.55 mm)	3 in. (76.2 mm)	49.8 cu. in. (815.7 cc)	18 hp. (13.4 kW)
B43M, B43E	2	3.250 in. (82.55 mm)	2.260 in. (66.55 mm)	43.3 cu. in. (712.4 cc)	16 hp. (11.9 kW)
B43G	2	3.250 in. (82.55 mm)	2.620 in. (66.55 mm)	43.3 cu. in. (712.4 cc)	18 hp. (13.4 kW)
B48G	2	3.250 in. (82.55 mm)	2.875 in. (73.03 mm)	47.7 cu. in. (781.7 cc)	20 hp. (14.9 kW)
B48M	2	3.250 in. (82.55 mm)	2.875 in. (73.03 mm)	47.7 cu. in. (781.7 cc)	18 hp. (13.4 kW)

Engines in this section are four-stroke, twin-cylinder opposed, horizontal crankshaft type in which crankshaft is supported at each end in precision sleeve type bearings.

Connecting rods ride directly on crankshaft journals and all except early BF engines are pressure lubricated by a gear type oil pump driven by crankshaft gear. Early BF engines were splash lubricated.

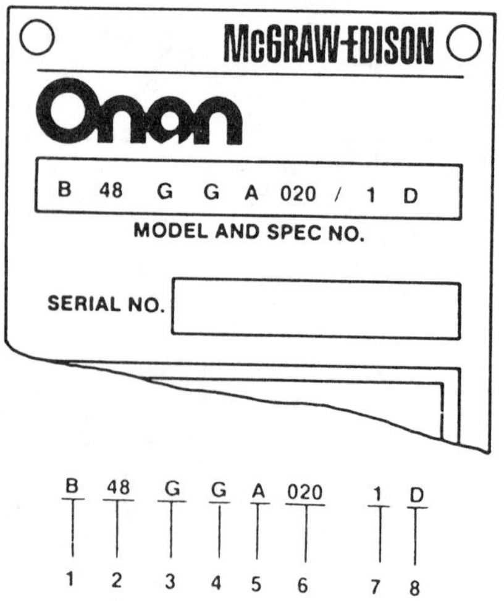

Fig. O19—Typical model, specification and serial number plate on Onan engine showing digit interpretation.

1. Identification of basic engine series.
2. Displacement (in cubic inches).
3. Engine duty cycle.
4. Fuel required.
5. Cooling system designation.
6. Power rating (in BHP).
7. Designated optional equipment.
8. Production modifications.

All models use a battery ignition system consisting of points, condenser, coil, battery and spark plug. Timing is adjustable only by varying point gap.

All models use a single down draft float type carburetor and may have a vacuum operated fuel pump attached to carburetor or mounted separately according to model and application.

Refer to model and specificaton number (Fig. O19) for engine model identification and interpretation.

Always give model, specification and serial numbers when ordering parts or service information.

MAINTENANCE

SPARK PLUG. Recommended spark plug is Champion H8 or equivalent. Electrode gap for all models is 0.025 inch (0.0635 mm).

CARBURETOR. According to model and application all engines use a single down draft float type carburetor manufactured by Marvel Schebler, Walbro or Nikki. Exploded view of each carburetor is shown in either Fig. O20, Fig. O20A or Fig. O20B. Refer to appropriate Figure for model being serviced.

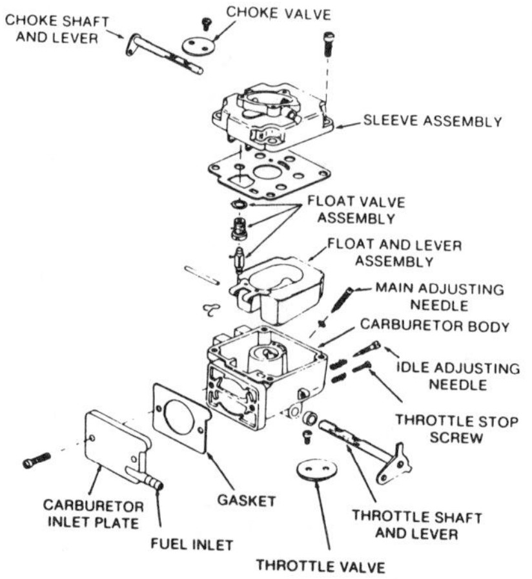

Fig. O20—Exploded view of Marvel Schebler carburetor showing component parts and their relative positions.

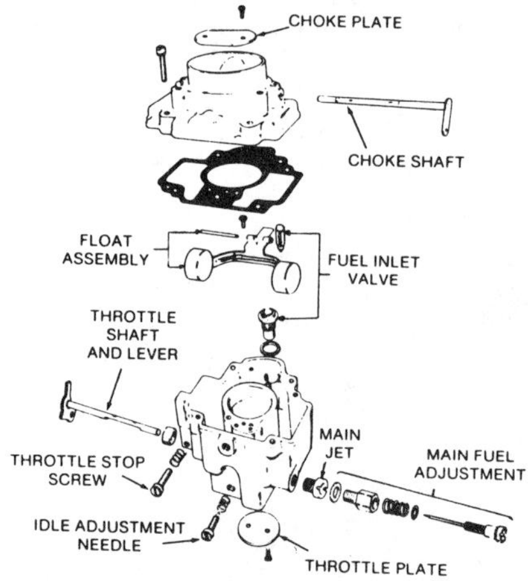

Fig. O20A—Exploded view of Walbro carburetor showing component parts and their relative positions.

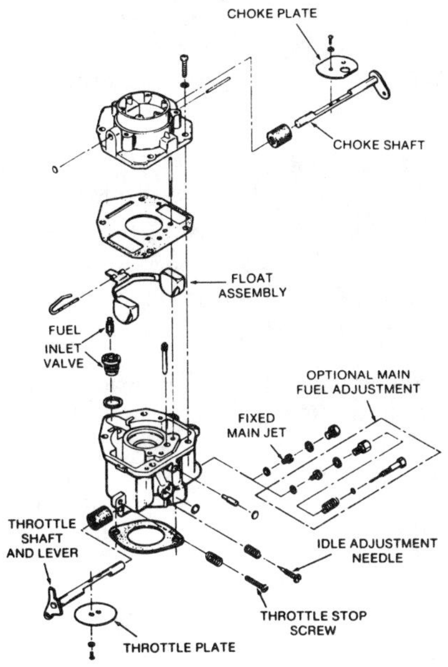

Fig. O20B—Exploded view of Nikki carburetor showing component parts and their relative positions.

For initial adjustment of Marvel Schebler carburetor refer to Fig. O20. Open idle mixture screw 1 turn and main fuel mixture screw 1¼ turns.

For initial adjustment of Walbro carburetor refer to Fig. O20A. Open idle mixture screw 1⅛ turns. If equipped with optional main fuel adjustment open main fuel mixture screw 1½ turns.

For initial adjustment of Nikki carburetor refer to Fig. O20B. Open idle mixture screw ¾-turn. If equipped with optional main fuel adjustment open main fuel mixture screw 1½ turns.

Make final adjustment to all models with engine at operating temperature and running. Place engine under load and adjust main fuel mixture screw to leanest mixture that will allow satisfactory acceleration and steady governor operation.

Run engine at idle speed, no load and adjust idle mixture screw for smoothest idle operation. As each adjustment affects the other, adjustment procedure may have to be repeated.

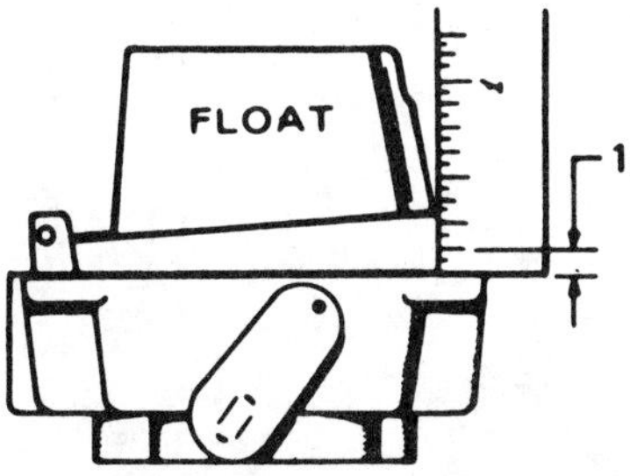

Fig. O21—Float valve setting for Marvel Schebler carburetor. Measure between inverted float and surface of gasket.

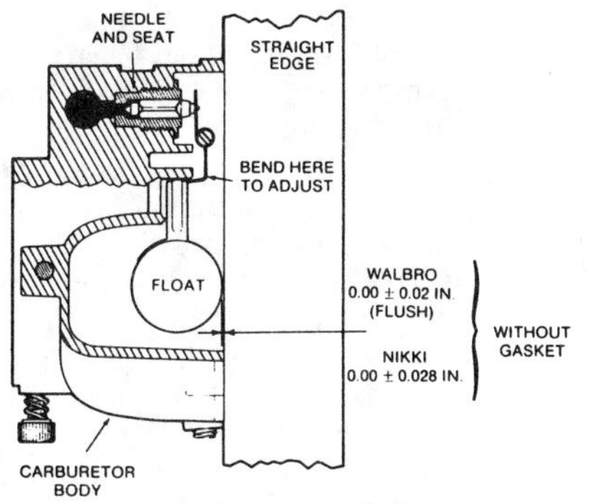

Fig. O22—To adjust float level on Walbro and Nikki carburetors, position carburetor as shown. Clearance should be measured from machined surface without gasket to float edge.

To check float level of Marvel Schebler Carburetor, refer to Fig. O21. Invert carburetor throttle body and float assembly. There should be ⅛-inch (3.175 mm) clearance between gasket and float as shown. Adjust float by bending float lever tang that contacts inlet valve.

To check float level of Walbro or Nikki carburetor, refer to Fig. O22. Position carburetor as shown. Walbro carburetor should have 0.00 inch plus or minus 0.02 inch (0.00 mm, plus or minus 0.508 mm) clearance and Nikki carburetor should have 0.00 inch, plus or minus 0.028 inch (0.00 mm, plus or minus 0.7112 mm) clearance between machined surface without gasket, to float edge. Adjust float by bending float lever tang that contacts inlet valve.

All models use a pulsating diaphragm fuel pump. Refer to SERVICING ONAN ACCESSORIES section for service information.

GOVERNOR: All models use a flyball weight governor located under timing gear cover on camshaft gear.

Front pull (Fig. O25) and side pull (Fig. O26) linkage arrangements are used and service procedures are similar.

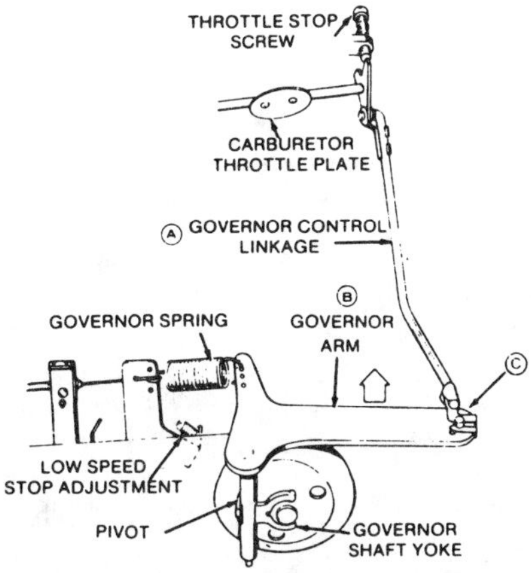

Fig. O25—View of front pull variable speed governor linkage.

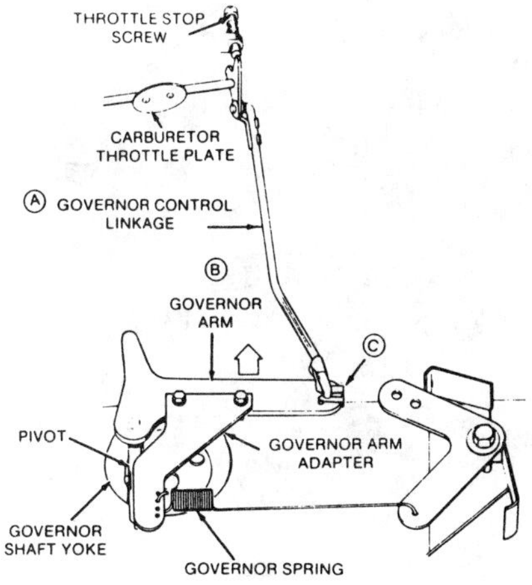

Fig. O26—View of side pull variable speed governor linkage.

For correct governor link installation on either front or side pull system, disconnect linkage (A) from hole (C) and push linkage and governor arm (B) toward carburetor as far as they will go. While held in this position insert end of linkage into nearest hole in governor arm. If between two holes insert in next hole out.

Normal factory setting is in third hole from pivot for side pull linkage and second hole from pivot for front pull linkage.

Sensitivity is increased by connecting governor spring closer to pivot and decreased by connecting governor spring further from pivot.

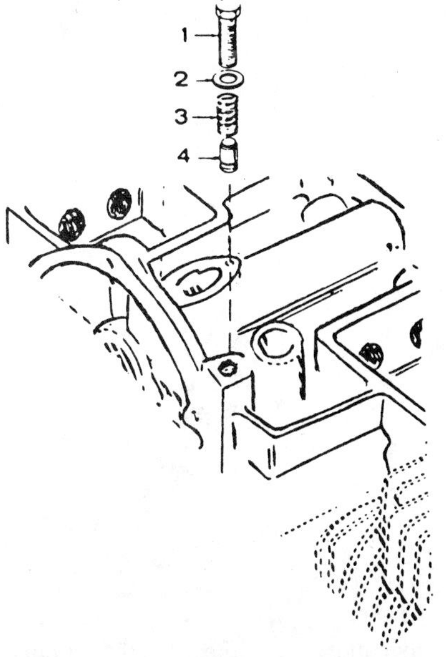

Fig. O27—View showing location of oil pressure relief valve and spring.

1. Cap screw
2. Sealing washer
3. Spring
4. Valve

Illustrations courtesy Onan.

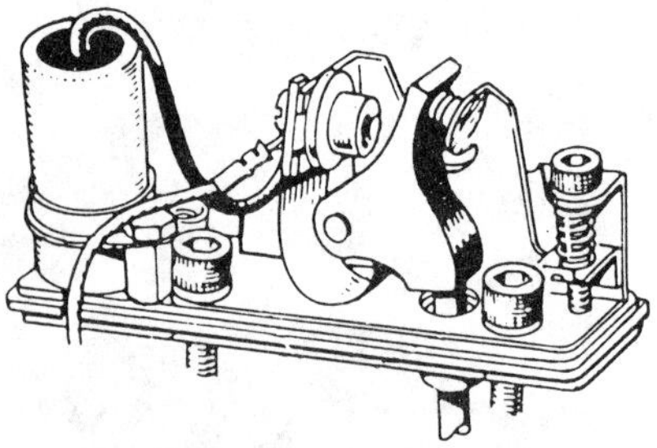

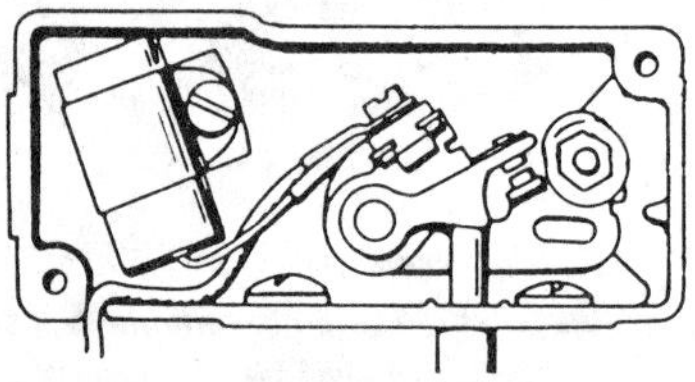

Fig. O28—Two different type point boxes are used according to model and application. Upper illustration shows top adjust point models while lower illustration shows side adjust point model.

Engine speed should be set to manufacturers specifications and speed should be checked after linkage adjustments.

IGNITION. A battery type ignition system is used on all models. Breaker point box is nonmovable and only means of changing timing is to vary point gap slightly. See Fig. O28. Static timing should be checked if point gap is changed or new points are installed. Use a continuity test light across breaker points to check timing and remove air intake hose from blower housing for access to timing marks. Refer to ignition specifications as follows:

Model	Initial Point Gap	Static Timing BTDC
BF	0.025 in. (0.635 mm)	21°
BF*	0.025 in. (0.635 mm)	21°
BF**	0.025 in. (0.635 mm)	26°
BF, B43M, B48M	0.021 in. 0.533 mm)	21°
B48G	0.020 in. (0.508 mm)	20°
B48G*	0.016 in. (0.406 mm)	16°

*Specification letter "C" and after.
**"Power Drawer" models.

LUBRICATION. Early model BF is splash lubricated and all other models are pressure lubricated by a gear type pump driven by crankshaft gear. Internal pump parts are not serviced separately so entire pump must be renewed if worn or damaged. Clearance between pump gear and crankshaft gear should be 0.002-0.005 inch (0.0508-0.127 mm) and normal operating pressure should be 30 psi (207 kPa). Oil pressure is regulated by an oil pressure relief valve (Fig. O27) located in engine block near timing gear cover. Spring (3) free length should be 1.00 inch (25.4 mm) and valve (4) diameter should be 0.3365-0.3380 inch (8.55-8.59 mm).

Oils approved by manufacturer must meet requirements of API service classification "SE" or "SE/CC".

Use SAE 5W-30 oil in below freezing temperatures and SAE 20W-40 oil in above freezing temperatures.

Check oil level with engine stopped and maintain at "FULL" mark on dipstick. **DO NOT** overfill.

Recommended oil change intervals for models without oil filter is every 25 hours of normal operation. If equipped with a filter, change oil at 50 hour intervals and filter every 100 hours.

Crankcase capacity without filter is 3.5 pints (1.66 L) for BG model, 2 quarts (1.86 L) for BF model, 1.5 quarts (1.4 L) for B43E, B43G and B48G models. If filter is changed add an additional 0.5 pint (0.24 L) for BF and BG models and an additional 1 pint (0.47 L) for B43E, B43G and B48G models. Oil filter is screw on type and gasket should be lightly lubricated before installation. Screw filter on until gasket contacts, then turn an additional ½-turn. Do not overtighten.

REPAIRS

TIGHTENING TORQUES. Recommended tightening torques are as follows:

Cylinder heads–
BF,
BG .... 14-16 ft.-lbs. (19-22 N·m)
B43E,*
B43G,*
B48G,*
B43M,*
B48M* .... 16-18 ft.-lbs. (22-24 N·m)

*If graphoil gasket is used, tighten to 14-16 ft.-lbs. (19-22 N·m).

Rear bearing plate .... 25-27 ft.-lbs. (34-37 N·m)

Connecting rod–
BG,
BF .... 14-16 ft.-lbs. (19-22 N·m)
All others .... 12-14 ft.-lbs. (16-19 N·m)

Flywheel .... 35-40 ft.-lbs. (48-54 N·m)

Oil base .... 18-23 ft.-lbs. (24-31 N·m)

Timing cover .... 8-10 ft.-lbs. (11-13 N·m)

Oil pump .... 7-9 ft.-lbs. (10-12 N·m)

CYLINDER HEADS. It is recommended cylinder heads be removed and carbon cleaned at 200 hour intervals.

CAUTION: Cylinder heads should not be unbolted from block when hot due to danger of head warpage.

When installing cylinder heads torque bolts in sequence shown in Fig. O29 in gradual, even steps until correct torque for model being serviced is obtained. Note spark plug location on some models varies but torque sequence is the same.

CONNECTING RODS. Connecting rod and piston are removed from cylinder head surface end of block after removing cylinder head and oil base. Connecting rods ride directly on crankshaft journal on all models and standard journal diameter should be 1.6252-1.6260 inches (41.28-41.30 mm). Connecting rods are available for 0.005, 0.010, 0.020, 0.030 and 0.040 inch (0.127, 0.254, 0.508, 0.762 and 1.016 mm) undersize crankshafts as well as standard.

Connecting rod to crankshaft journal running clearance should be 0.002-0.0033 inch (0.0508-0.0838 mm) and side play should be 0.002-0.016 inch (0.0508-0.406 mm).

When reinstalling connecting rods make certain rods are installed with rod bolts offset toward outside of block and tighten to specified torque.

PISTON, PIN AND RINGS. Aluminum pistons are fitted with two compression rings and one oil control ring. Pistons in BF models should be renewed if top compression ring has 0.008 inch (0.2032 mm) or more side clearance and pistons in all other models should be renewed if 0.004 inch (0.1016 mm) or more side clearance is present.

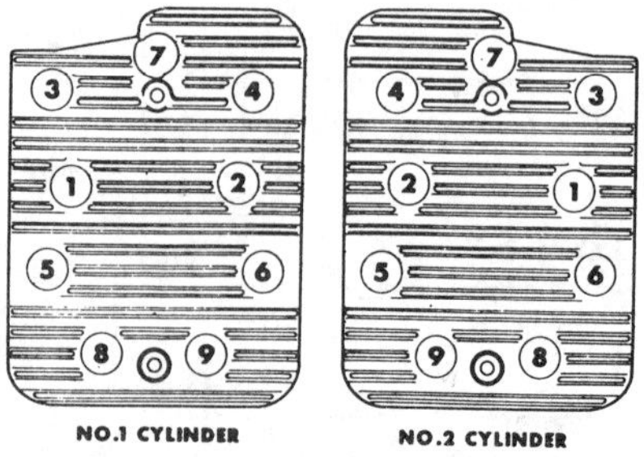

Fig. O29—Torque sequence shown is for all models even though spark plug may be in different location in head.

Illustrations courtesy Onan.

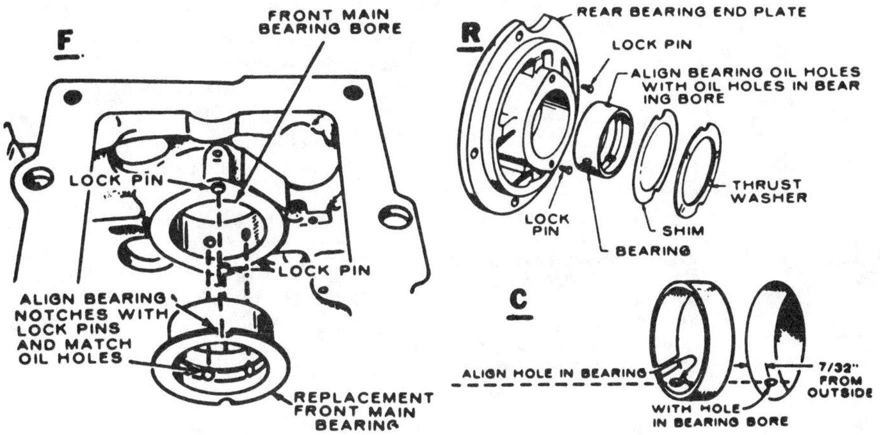

Fig. O31 — Alignment of precision main and camshaft bearing at rear (View R). Note shim use for end play adjustment. View C shows placement of camshaft bearings.

Pistons and rings are available in 0.005, 0.010, 0.020, 0.030 and 0.040 inch (0.127, 0.254, 0.508, 0.762 and 1.016 mm) oversize as well as standard.

Recommended piston to cylinder wall clearance for BF engines when measured 0.10 inch (2.54 mm) below oil control ring 90° from pin should be 0.001-0.003 inch (0.0254-0.0762 mm).

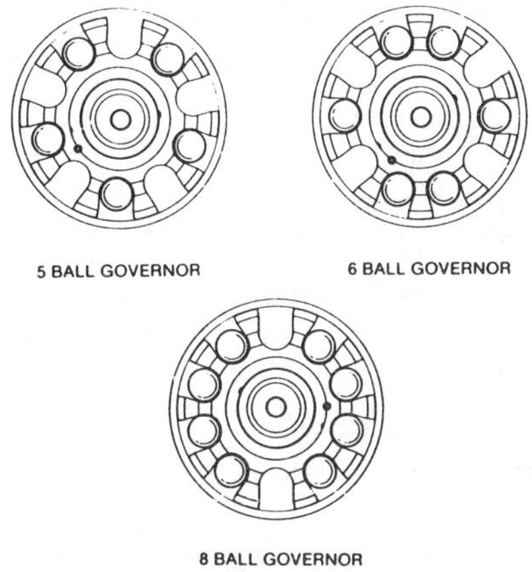

Fig. O32 — Flyballs must be arranged in pockets as shown, according to total number used to obtain governor sensitivity desired. Refer to text.

Recommended piston to cylinder wall clearance for B43M and B48M engines when measured 0.35 inch (8.89 mm) below oil control ring 90° from pin should be 0.0033-0.0053 inch (0.0838-0.1346 mm).

Recommended piston to cylinder wall clearance for B43E, B43G and B48G engines when measured 0.35 inch (8.89 mm) below oil control ring 90° from pin should be 0.0044-0.0064 inch (0.1118-0.1626 mm).

Ring end gap should be 0.010-0.020 inch (0.254-0.508 mm) clearance in standard bore.

Piston pin to piston bore clearance should be 0.002-0.004 inch (0.0508-0.1016 mm) and pin to rod clearance should be 0.0002-0.0007 inch (0.005-0.0178 mm) for all models.

Standard cylinder bore of BF engine is 3.1245-3.1255 inches (79.36-79.39 mm) and standard cylinder bore for all other models is 3.249-3.250 inches (82.53-82.55 mm).

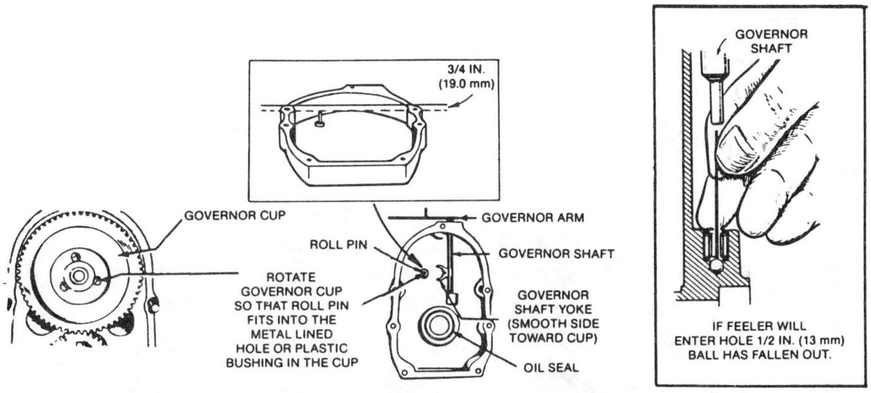

Fig. O33 — View of governor shaft, timing cover and governor cup showing assembly details. Refer to text.

Engines should be rebored if taper exceeds 0.005 inch (0.127 mm) or if 0.003 inch (0.0762 mm) out-of-round.

CRANKSHAFT, BEARINGS AND SEALS. Crankshaft is supported at each end by precision type sleeve bearings located in cylinder block housing and rear bearing plate. Main bearing journal standard diameter should be 1.9992-2.000 inches (50.7797-50.800 mm) and bearing operating clearance should be 0.0025-0.0038 inch (0.0635-0.0965 mm). Main bearings are available for a variety of crankshaft undersizes as well as standard size.

Oil holes in bearing and bore **MUST** be aligned when bearings are installed and bearings should be pressed into bores so inside edge of bearing is 1/16 to 3/32-inch (1.588 to 2.381 mm) back from inside end of bore to allow clearance for radius of crankshaft. Oil grooves in thrust washers must face crankshaft. The two alignment notches on thrust washers must be in good condition or excessive crankshaft end play will result.

NOTE: Replacement front bearing and thrust washer is a one-piece assembly, do not install a separate thrust washer.

Recommended crankshaft end play should be 0.006-0.012 inch (0.1524-0.3048 mm) and is adjusted by varying thrust washer thickness or placing shim between thrust washer and rear bearing plate. Shims are available in a variety of thicknesses. See Fig. O31.

It is recommended rear bearing plate and front timing gear cover be removed for seal renewal. Rear seal is pressed in until flush with seal bore. Timing cover seal should be driven in until it is 1-1/32 inch (26.19 mm) from mounting face of cover.

CAMSHAFT, BEARINGS AND GOVERNOR. Camshaft is supported at each end in precision type sleeve bearings. Make certain oil holes in bearings are aligned with oil holes in block, press front bearing in flush with outer surface of block and press rear bearing in until flush with bottom of counterbore. Camshaft to bearing clearance should be 0.0015-0.0030 inch (0.0381-0.0762 mm).

Camshaft end play should be 0.003 inch (0.0762 mm) and is adjusted by varying thickness of shim located between camshaft timing gear and engine block.

Camshaft timing gear is a press fit on end of camshaft and is designed to accommodate 5, 8 or 10 flyballs arranged as shown in Fig. O32. Number of flyballs is varied to alter governor sensitivity. Fewer flyballs are used on engines with variable speed applications and greater number of flyballs are used for con-

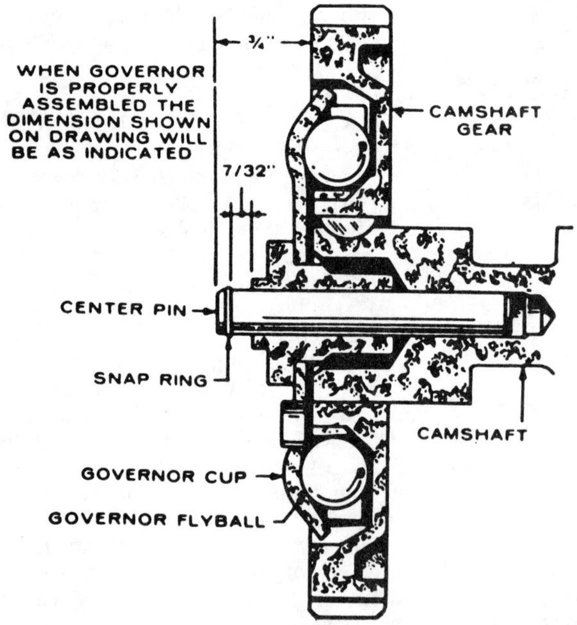

Fig. O34—Cross-sectional view of camshaft gear and governor assembly showing correct dimensions for center pin extension from camshaft. Further detail in text.

tinuous speed application. Flyballs must be arranged in "pockets" as shown according to total number used. Make certain timing mark on cam gear aligns with timing mark on crankshaft gear when reinstalling.

Center pin (Fig. O34) should extend ¾-inch (19 mm) from end of camshaft to allow 7/32-inch (5.6 mm) in-and-out movement of governor cup. If distance is incorrect remove pin and press new pin in to correct depth.

Always make certain governor shaft pivot ball is in timing cover by measuring as shown in Fig. O33 inset and check length of roll pin which engages bushed hole in governor cup. This pin must extend 25/32-inch (19.844 mm) from timing cover mating surface.

VALVE SYSTEM. Valve seats are renewable insert type and are available in a variety of oversizes as well as standard. Seats are ground at 45° angle and seat width should be 1/32 to 3/64-inch (0.794 to 1.191 mm).

Valves should be ground at 44° angle to provide an interference angle of 1°. Stellite valves and seats are available and rotocaps are available for exhaust valves.

Recommended valve stem to guide clearance is 0.001-0.0025 inch (0.0254-0.0635 mm) for intake valves and 0.0025-0.004 inch (0.0635-0.1016 mm) for exhaust valve. Renewable valve guides are shouldered and are pushed out from above. Early models use a gasket between guide and cylinder block and some models may be equipped with valve stem seal on intake valve which must be renewed if valve is removed.

Recommended valve tappet gap (cold) for B43E model is 0.003 inch (0.0762 mm) for intake valves and 0.010 inch (0.254 mm) for exhaust valves.

Recommended valve tappet gap (cold) for all remaining models is 0.008 inch (0.2032 mm) for intake valves and 0.013 inch (0.3302 mm) for exhaust valves.

Adjustment is made by turning tappet adjusting screw as required and valves of each cylinder must be adjusted with cylinder at "top dead center" on compression stroke. At this position both valves will be fully closed.

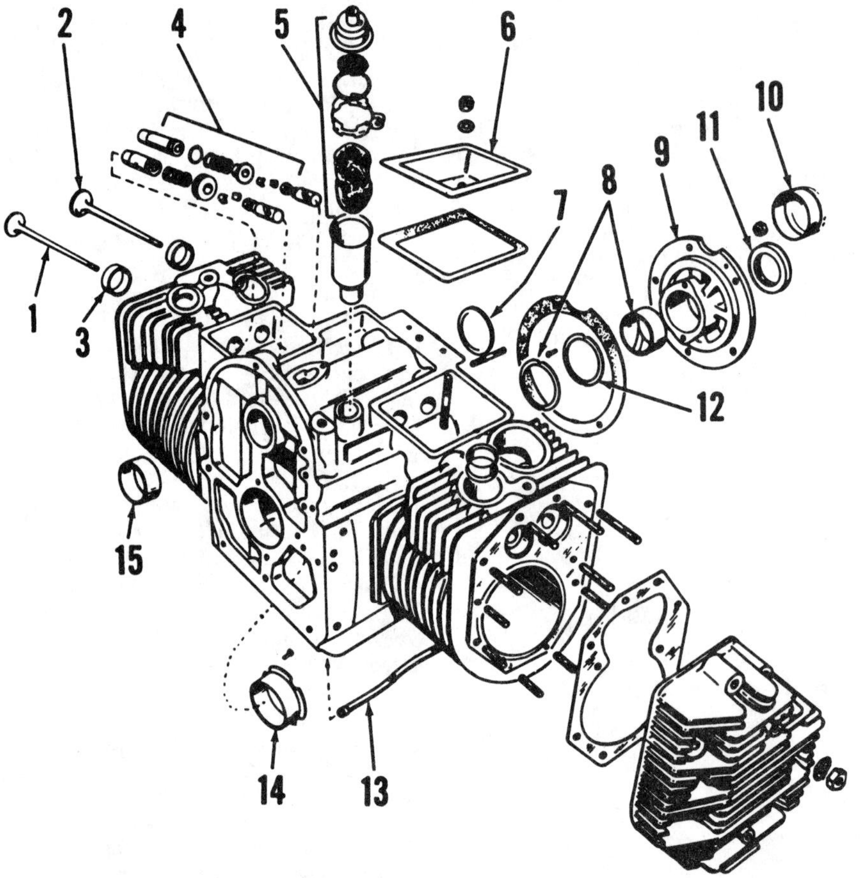

1. Exhaust valve
2. Intake valve
3. Seat insert
4. Guide, spring, tappet group
5. Crankcase breather group
6. Valve compartment cover
*7. Camshaft expansion plug
8. Two-piece main bearing
9. Rear main bearing plate
*10. Timing control cover
11. Crankshaft seal
12. Crankshaft thrust shim
13. Oil tube
14. One-piece main bearing
15. Camshaft bearing (2)

*Plug is installed on engine without timing control.

Fig. O35—Exploded view of cylinder and crankcase assembly typical of all models.

Illustrations courtesy Onan.

ONAN

A DIVISION OF ONAN CORPORATION
1400 73rd Avenue, N.E.
Minneapolis, Minnesota 55432

Model	No. Cyls.	Bore	Stroke	Displacement	Power Rating
CCK	2	3.25 in. (82.55 mm)	3 in. (76.2 mm)	49.8 cu. in. (815.7 cc)	12.9 hp. (9.6 kW)
CCKA	2	3.25 in. (82.55 mm)	3 in. (76.2 mm)	49.8 cu. in. (815.7 cc)	16.5 hp. (12.3 kW)
CCKB	2	3.25 in. (82.55 mm)	3 in. (76.2 mm)	49.8 cu. in. (815.7 cc)	20 hp. (14.9 kW)

Engines in this section are four-stroke, twin cylinder opposed, horizontal crankshaft engines. Crankshaft is supported at each end in precision sleeve type bearings.

CCK and CCKA engines prior to specification letter "D" have aluminum connecting rods which ride directly on crankshaft journal.

CCK and CCKA engines with specification letter "D" or after and all other models have forged steel rods equipped with renewable bearing inserts.

All models are pressure lubricated by a gear type oil pump. A "spin-on" oil filter is available.

Most models are equipped with a battery ignition system, however a magneto system is available for manual start models.

Engines are equipped with a side draft or a downdraft carburetor. Mechanical, pulsating diaphragm or electric fuel pumps are available.

Refer to Fig. O59 for interpretation of engine specification and model numbers.

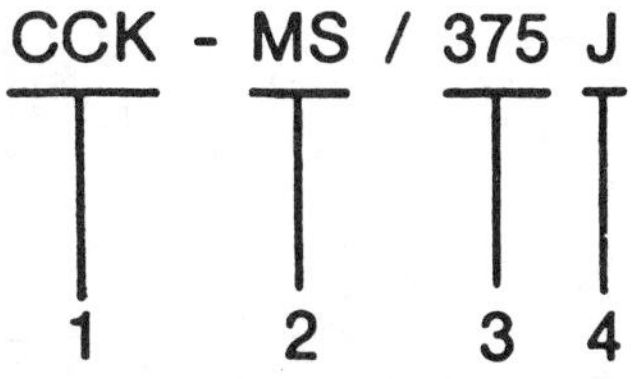
CCK - MS / 375 J
1
2
3
4

Fig. O59—Interpretation of engine model and specification number as an aid to identification of various engines.

1. General engine model identification
2. Starting type
   S. Manual
   MS. Electric
3. Optional equipment identification
4. Specification letter which advances with factory production modifications.

CCKA engines may be visually distinguished by the cup-shaped advance mechanism cover, rather than the flat-shaped expansion plug, in rear camshaft opening just below ignition point breaker box.

High compression heads are identified by a 3/32-inch radius boss located on corner of head nearest spark plug.

Always give specification, model and serial numbers when ordering parts or service material.

MAINTENANCE

SPARK PLUG. Recommended spark plug is Onan part number 167-0241 for non-resistor plug and 167-0237 for a resistor type plug or their equivalent. Electrode gap for all models is 0.025 inch (0.635 mm).

CARBURETOR. According to model and application, engine may be equipped with either a side draft or a downdraft carburetor. Refer to appropriate paragraph for model being serviced.

SIDE DRAFT CARBURETOR. Side draft carburetor is used on some CCKB models. Refer to Fig. O60 for exploded view of carburetor and location of mixture adjustment screws.

For initial carburetor adjustment,

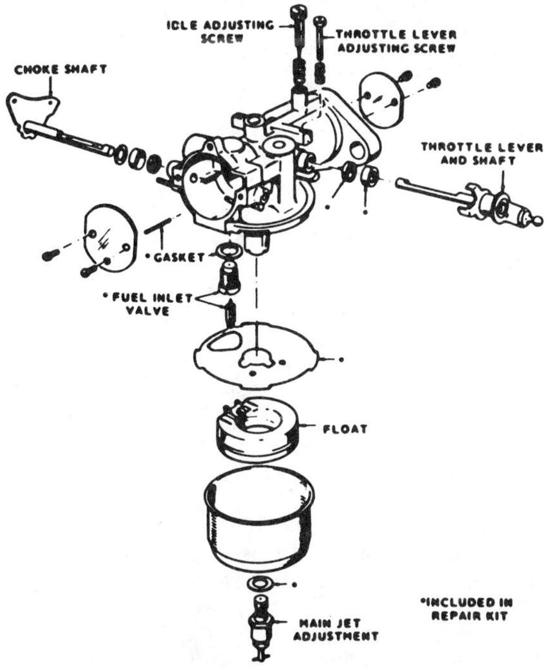
IDLE ADJUSTING SCREW
THROTTLE LEVER ADJUSTING SCREW
CHOKE SHAFT
THROTTLE LEVER AND SHAFT
•GASKET
•FUEL INLET VALVE
FLOAT
MAIN JET ADJUSTMENT
•INCLUDED IN REPAIR KIT

Fig. O60—Exploded view of side draft carburetor used on some models.

Illustrations courtesy Onan.

open idle mixture screw and main fuel mixture screw 1 to 1½ turns. Make final adjustments with engine at normal operating temperature and running. Place engine under load and adjust main fuel mixture screw for leanest setting that will allow satisfactory acceleration and steady governor operation. Set engine at idle speed, no load and adjust idle mixture screw for smoothest idle operation.

As each adjustment affects the other, adjustment procedure may have to be repeated.

To check float level, refer to Fig. O61. Invert carburetor throttle body and float assembly. Float clearance should be 11/64-inch (4.366 mm) for models prior to specification letter "H" and 1/8 to 3/16-inch (3.2 to 4.8 mm) for models with specification letter "H" or after.

Adjust float by bending float lever tang that contacts inlet valve.

DOWNDRAFT CARBURETOR. A downdraft carburetor is used on most CCK and CCKA models. Refer to Fig. O63 for exploded view of carburetor and location of mixture adjustment screws.

For initial carburetor adjustment, open idle mixture screw 1 turn. If equipped with main fuel adjustment open main fuel mixture screw about 2 turns.

Make final adjustments with engine at normal operating temperature and running. Place engine under load and adjust main fuel mixture screw (if equipped) for leanest setting that will allow satisfactory acceleration and steady governor operation. Onan special tool number 420-0169 (Fig. O64) aids main fuel mixture screw adjustment. Set engine at idle speed, no load and adjust idle mixture screw for smoothest idle operation.

As each adjustment affects the other, adjustment procedure may have to be repeated.

To check float level, refer to Fig. O65. Invert carburetor throttle body and float assembly. Float clearance should be ¼-inch (6.35 mm) for metal float or 5/16-inch (7.94 mm) for foam float when measured from gasket to closest edge of float as shown.

Adjust float by bending float lever tang that contacts inlet valve.

CHOKE. One of three types of automatic choke may be used according to model and application. Refer to appropriate paragraph for model being serviced.

THERMAL MAGNETIC CHOKE. This choke (Fig. O67) uses a strip heating element and a heat-reactive bimetallic spring to control choke valve position. When engine is cranked, solenoid shown at lower portion of choke

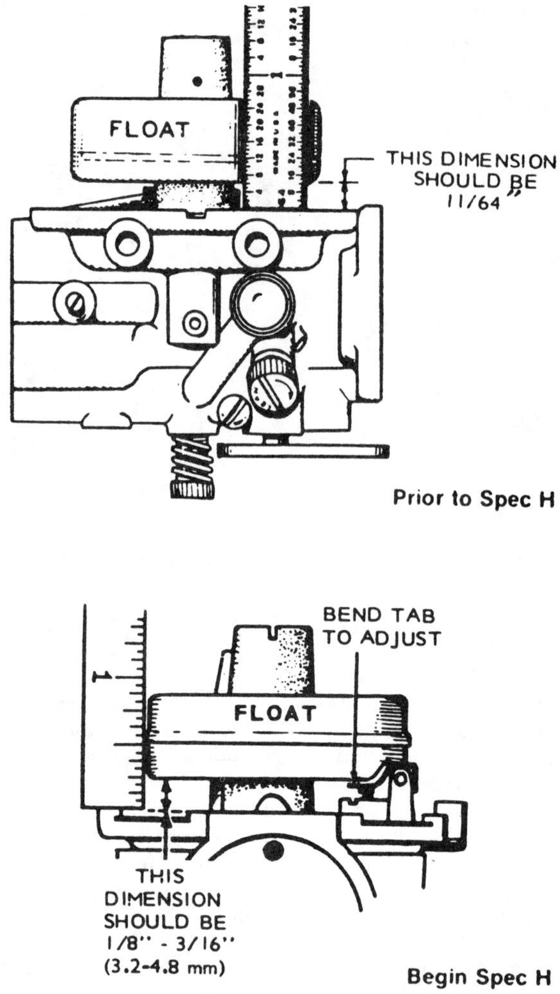
FLOAT
THIS DIMENSION SHOULD BE 11/64"
Prior to Spec H
BEND TAB TO ADJUST
FLOAT
THIS DIMENSION SHOULD BE 1/8" - 3/16" (3.2-4.8 mm)
Begin Spec H

Fig. O61—Measure and set float level as shown noting upper view is for engines with specification letter prior to "H" and lower view is for engines with specification letter "H" and after.

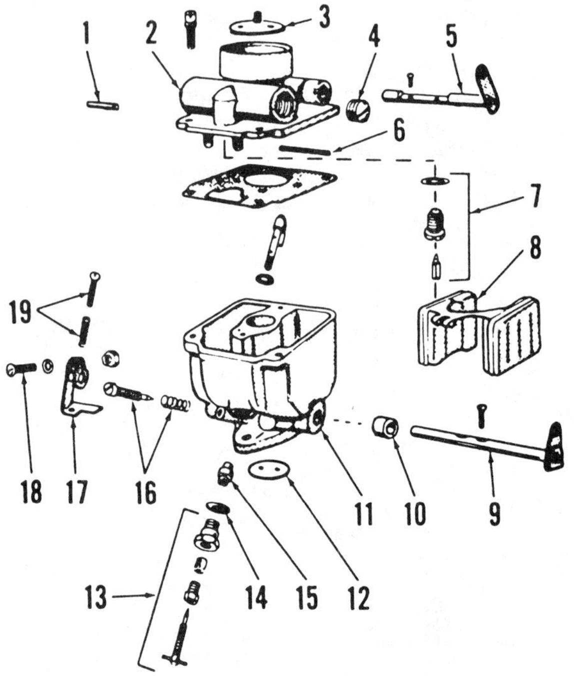
1
2
3
4
5
6
7
8
19
18
17
16
11
10
9
13
14
15
12

Fig. O63—Exploded view of downdraft carburetor used on CCK and CCKA models.

1. Choke stop pin
2. Cover assy.
3. Choke valve
4. Gas inlet plug
5. Choke shaft
6. Float pin
7. Inlet valve assy.
8. Float
9. Throttle shaft
10. Shaft bushing
11. Body assy.
12. Throttle plate
*13. Main fuel valve assy.
14. Gasket
*15. Main jet
16. Idle mixture adjustment needle & spring
17. Idle stop lever
18. Clamp screw
19. Stop screw & spring

*Either 13 or 15 is used.

Illustrations courtesy Onan.

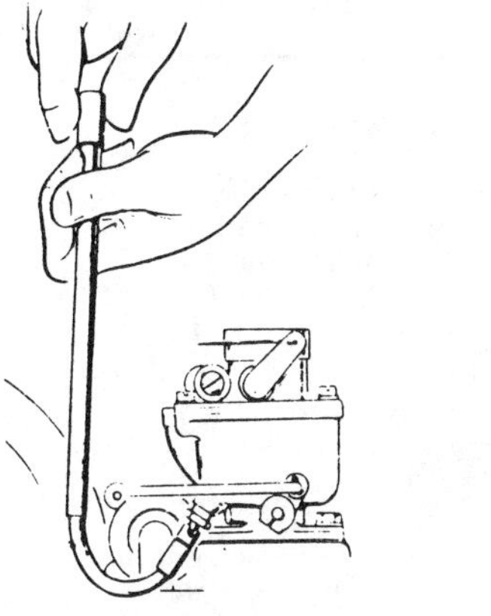

Fig. O64—View showing use of Onan main fuel adjusting tool, part number 420-0169, used to adjust main fuel mixture needle. Refer to text.

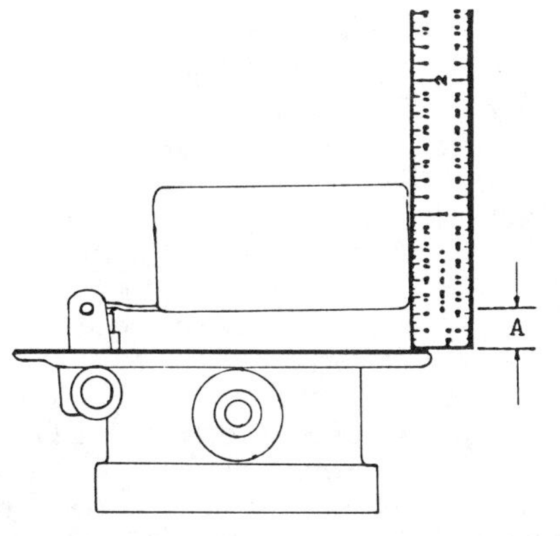

Fig. O65—View of typical float assembly of downdraft type carburetor. Clearance at "A" should be ¼-inch (6.35 mm) if float is made of metal and 5/16-inch (7.94 mm) if float is made of styrofoam. Clearance is measured with gasket in place.

body is activated and choke valve is closed, fully or partially, according to temperature conditions. When engine starts and runs, solenoid is released and bi-metallic spring controls choke opening. Refer to Fig. O70 for table of choke settings based on ambient temperature and adjustment procedure. Adjustment is made by rotating entire cover as shown.

BI-METALLIC CHOKE. This choke (Fig. O68) uses an electric heating element located inside its cover to activate a bi-metallic spring which opens or closes choke valve according to temperature. Electric current is supplied to heating element by leads from generator exciter. Refer to Fig. O70 for choke settings according to ambient temperature and adjustment procedure. Adjustment is made by rotating cover to open or close choke valve.

SISSON CHOKE. This choke (Fig. O69) uses a bi-metallic strip which reacts to manifold temperatures and a magnetic solenoid which is wired in series with starter switch circuit.

To adjust, pull choke lever up and insert a 1/16-inch diameter rod through shaft hole to engage notch in mounting flang and lock shaft against rotation. Loosen choke lever clamp screw and adjust choke lever so choke plate is completely closed, or not more than 5/16-inch (7.94 mm) open. Tighten clamp screw and remove 1/16-inch rod.

FUEL PUMP. Various mechanical, electric and pulse type fuel pumps are used according to model and application. Refer to SERVICING ONAN ACCESSORIES.

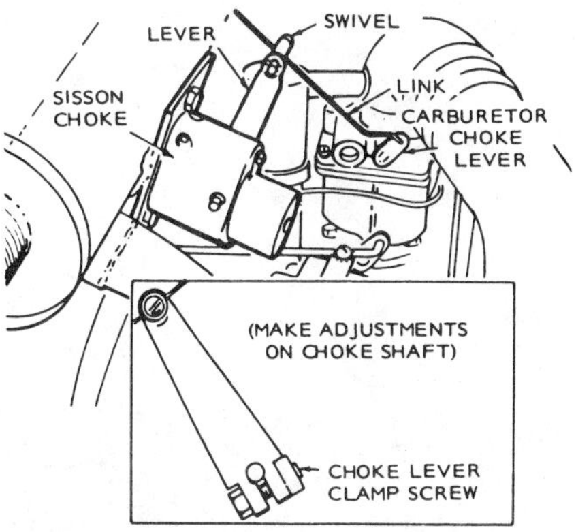

Fig. O69—View showing various locations of component parts for Sisson choke. Refer to text for adjustment procedure.

GOVERNOR. All models use a flyball weight governor located under timing gear cover on camshaft gear.

AMBIENT TEMP.	CHOKE SETTING
60° F (16° C)	1/8-in. (3.2 mm)
65° F (18° C)	9/64-in. (3.6 mm)
70° F (21° C)	5/32-in. (4 mm)
75° F (24° C)	11/64-in. (4.4 mm)
80° F (27° C)	3/16-in. (4.8 mm)
85° F (29° C)	13/64-in. (5.2 mm)
90° F (32° C)	7/32-in. (5.6 mm)
95° F (35° C)	15/64-in. (6 mm)
100° F (38° C)	1/4-in. (6.4 mm)

Fig. O70—Let engine cool at least one hour before adjusting choke setting. Drill bits may be used as gages to set choke opening according to ambient temperature.

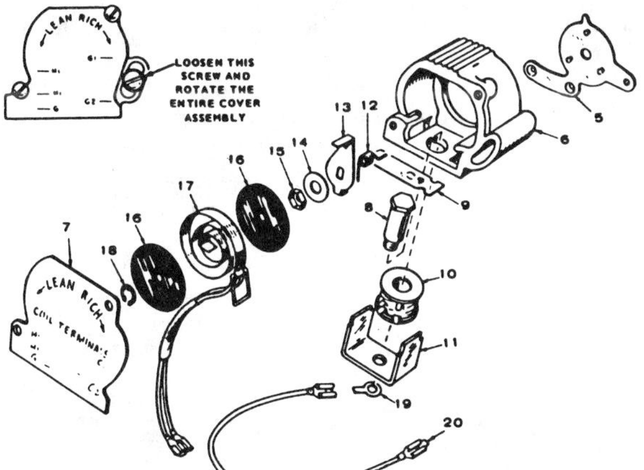

Fig. O67—Exploded view of thermal magnetic choke used on some engines.

5. Mounting plate
6. Body housing
7. Cover
8. Solenoid core
9. Armature
10. Solenoid coil
11. Frame
12. Lever spring
13. Lever
14. Washer
15. Pal nut
16. Insulator disc (2)
17. Heater & spring
18. Retainer (to shaft)
19. Ground terminal
20. Choke lead wire

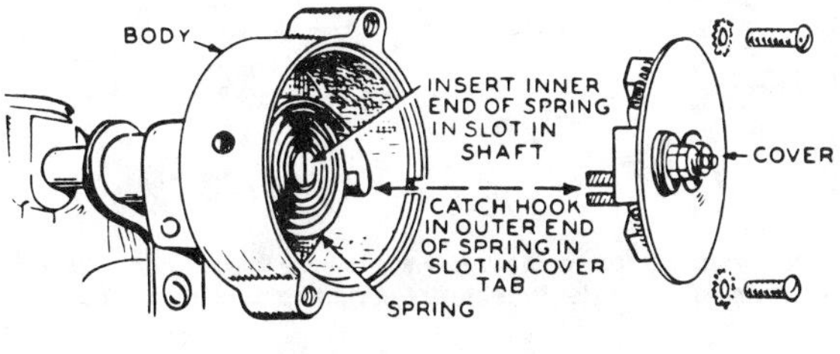

Fig. O68—Bi-metallic choke used on some models. Refer to text.

Illustrations courtesy Onan.

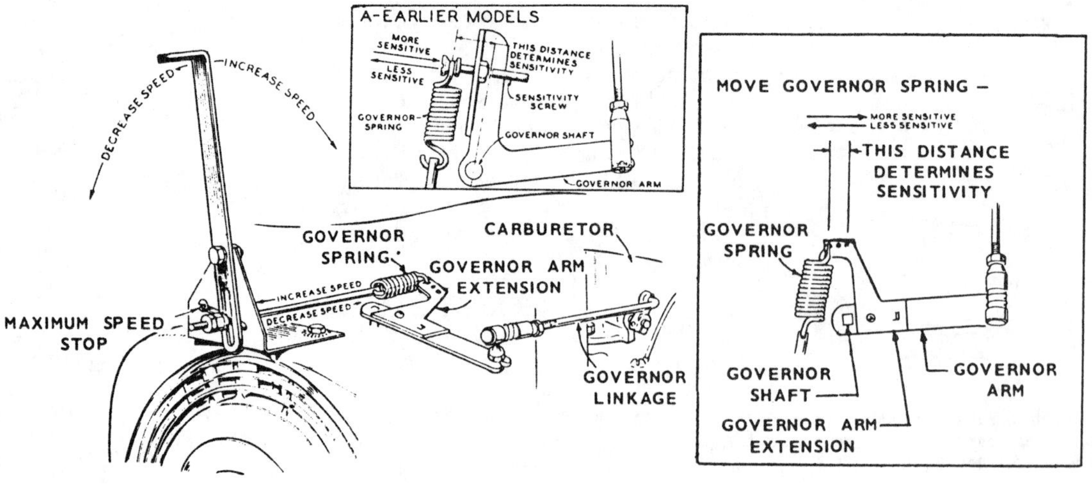

Fig. O71 — View of typical governor linkage arrangement.

Before any governor linkage adjustment is made, make certain all worn or binding linkage is repaired and carburetor is properly adjusted. Clean plastic ball joints, but do not lubricate. Beginning with specification letter "J", metal ball joints are used and these should be cleaned and lubricated with graphite.

To adjust governor linkage, adjust throttle stop screw on carburetor so engine idles at 1450 rpm, then adjust governor spring tension so engine idles at 1500 rpm when manual control lever is at minimum speed position.

Adjust sensitivity with engine running at minimum speed to obtain smoothest engine operation by moving governor spring outward on extension arm to decrease sensitivity or inward to increase sensitivity. Refer to Fig. O71.

Maximum full-load speed should not exceed 3000 rpm for continuous operation. To adjust, apply full load to engine and move control lever to maximum speed position. Adjust set screw in bracket slot to stop lever travel at desired speed.

IGNITION SYSTEM. Most models are equipped with battery ignition system, however a magneto system is available for manual start models. Ignition timing specifications are stamped on crankcase adjacent to breaker box.

Refer to appropriate paragraph for model being serviced.

BATTERY IGNITION. Ignition points and condenser are located in a breaker box at center, rear of engine. Refer to Fig. O72 for view of ignition wiring layout and Fig. O72A for view of points box and timing information.

To check timing, remove air intake hose from blower housing on pressure cooled engines or remove sheet metal plug in air shroud over right cylinder of "Vacu-Flo" engines to gain access to timing marks. Make certain points are set at 0.020 inch (0.51 mm) for all models.

Static timing is checked by connecting a test light across ignition points and rotating engine in direction of normal rotation (clockwise) until light comes on, then just goes out. This should be at 19° BTDC for all CCK models, 20° BTDC for CCKA and CCKB models without automatic spark advance and 1° ATDC for CCKA model with automatic spark advance (Fig. O73). Adjust by moving breaker box as required (Fig. O72A).

To check running timing, connect timing light to either spark plug and start and run engine. Operate engine at 1500 rpm or over. Light should flash when flywheel mark is aligned with 19° BTDC mark for all CCK models, 20° BTDC mark for CCKA models without automatic spark advance, 24° BTDC for CCKA model with automatic spark advance and 24° BTDC mark for CCKB model. Adjust by moving breaker box as required (Fig. O72A).

MAGNETO IGNITION. If engine is equipped with automatic spark advance mechanism (Fig. O73), magneto coil, mounted on gear cover (B – Fig. O72), contains both primary and secondary windings.

If engine is not equipped with automatic spark advance, stator contains primary (low voltage) windings only and a separate coil is used to develop secondary voltage (A – Fig. O72).

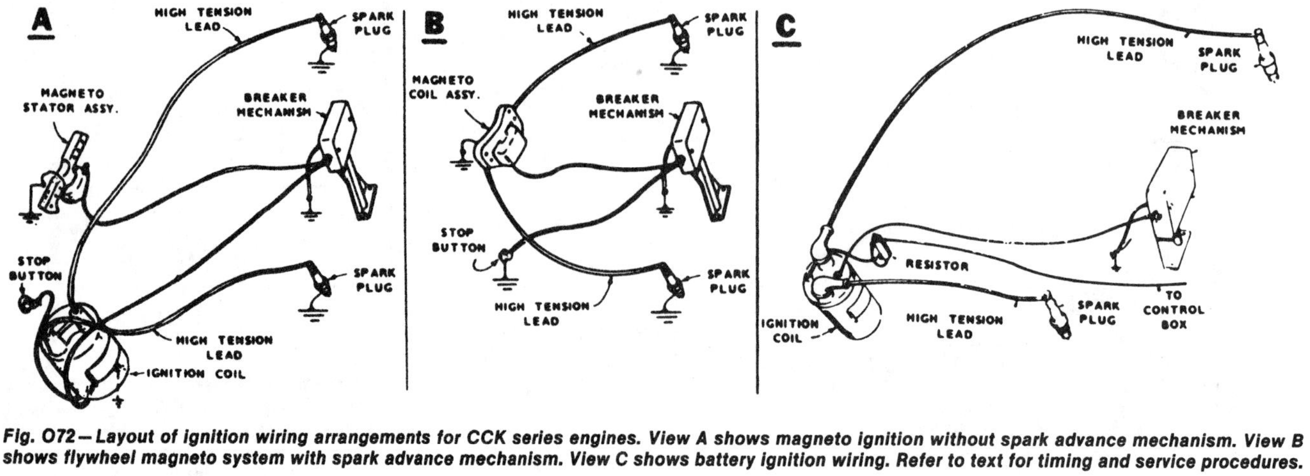

Fig. O72 — Layout of ignition wiring arrangements for CCK series engines. View A shows magneto ignition without spark advance mechanism. View B shows flywheel magneto system with spark advance mechanism. View C shows battery ignition wiring. Refer to text for timing and service procedures.

Illustrations courtesy Onan.

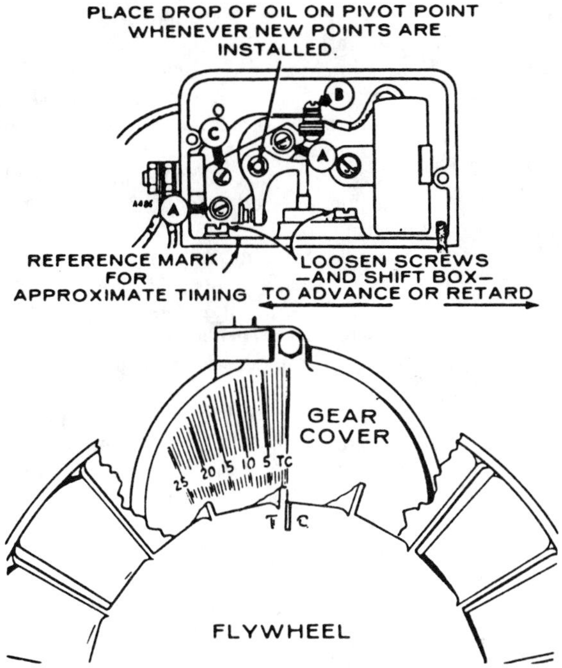
PLACE DROP OF OIL ON PIVOT POINT WHENEVER NEW POINTS ARE INSTALLED.
B
C
A
A
REFERENCE MARK FOR APPROXIMATE TIMING
LOOSEN SCREWS —AND SHIFT BOX— TO ADVANCE OR RETARD
GEAR COVER
25 20 15 10 5 TC
T C
FLYWHEEL

Fig. O72A — View showing location of points in breaker box of all models and location of timing marks on pressure cooled engines.

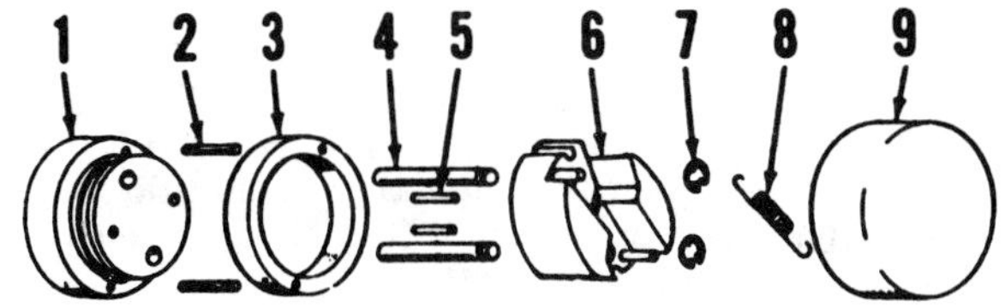
1
2
3
4
5
6
7
8
9

Fig. O73 — Exploded view of spark advance (timing control) used on some series engines.

1. Camshaft
2. Cam roll pin
3. Control cam
4. Groove pin
5. Roll pin
6. Weights (2)
7. Retainers (2)
8. Control spring
9. Cover

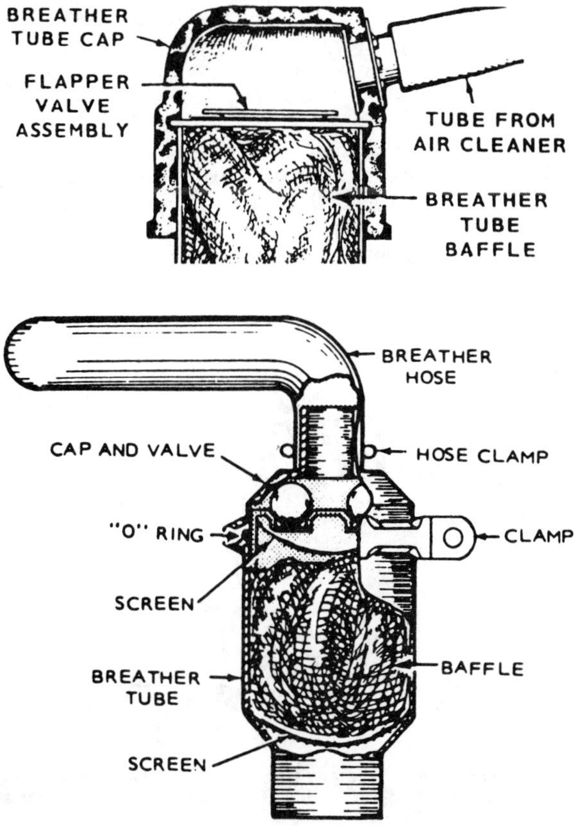
BREATHER TUBE CAP
FLAPPER VALVE ASSEMBLY
TUBE FROM AIR CLEANER
BREATHER TUBE BAFFLE
BREATHER HOSE
CAP AND VALVE
HOSE CLAMP
"O" RING
CLAMP
SCREEN
BREATHER TUBE
BAFFLE
SCREEN

Fig. O73A — Crankcase breather assembly should be cleaned at 200 hour intervals of normal engine operation. Upper breather is used on engines before specification letter "H" and lower breather is used on engines after specification letter "H".

Ignition points and condenser are located in a breaker box at center, rear of engine. Refer to Fig. O72 for view of ignition wiring layout and Fig. O72A for view of points box and timing information.

To check timing, remove air intake hose from blower housing on pressure cooled engines or remove sheet metal plug in air shroud over right cylinder of "Vacu-Flo" engines to gain access to timing marks. Make certain points are set at 0.020 inch (0.51 mm) for all models.

Static timing is checked by connecting a test light across ignition points and rotating engine in direction of normal rotation (clockwise) until light comes on, then just goes out. This should be at 19° BTDC for CCK model, 1° ATDC for CCKA model with automatic spark advance, 20° BTDC for CCKA model without automatic spark advance, 5° BTDC for CCKB model with electric start and 5° BTDC to 1° ATDC for CCKB model with manual start. Adjust by moving breaker box as required (Fig. O72A).

To check running timing, connect timing light to either spark plug and start and run engine. Operate engine at 1500 rpm or over. Light should flash when flywheel mark is aligned with 19° BTDC mark for all CCK models, 20° BTDC mark for CCKA model without automatic spark advance, 24° BTDC for CCKA model with automatic spark advance and 24° BTDC mark for CCKB models. Adjust by moving breaker box as required (Fig. O72A).

LUBRICATION. All models are pressure lubricated by a gear type pump driven by crankshaft gear. Internal pump parts are not serviced separately so entire pump must be renewed if worn or damaged. Clearance between pump gear and crankshaft gear should be 0.002-0.005 inch (0.0508-0.127 mm) and normal operating pressure should be 30 psi (207 kPa). Oil pressure is regulated by an oil pressure relief valve located on top of engine block near timing gear cover. Relief valve spring free length should be 2-5/16 inch (58.74 mm) and valve diameter should be 0.3365-0.3380 inch (8.55-8.59 mm).

Oils approved by manufacturer must meet requirements of API service classification "SE" or "SE/CC".

Use SAE 5W-30 oil in below freezing temperatures and SAE 20W-40 oil in above freezing temperatures.

Check oil level with engine stopped and maintain at "FULL" mark on dipstick. **DO NOT** overfill.

Recommended oil change interval for all models is every 100 hours of normal operation and change filter at 200 hour intervals. Oil capacity for all models with electric start is 3.5 quarts (3.3 L) without filter change and 4 quarts (3.8 L) with filter change. Oil capacity for all manual start models is 3 quarts (2.8 L) without filter change and 3.5 quarts (3.3 L) with filter change.

Oil filter is a screw on type and gasket should be lightly lubricated before installation. Screw filter on until gasket contacts, then turn an additional ½-turn. Do not overtighten.

CRANKCASE BREATHER. Engines are equipped with a crankcase breather (Fig. O73A) which helps maintain a partial vacuum in crankcase during engine operation to help control oil loss and to ventilate crankcase.

Clean at 200 hour intervals of normal engine operation. Wash valve in suitable solvent and pull baffle out of breather tube to clean. Reinstall valve with perforated disc toward engine.

REPAIRS

TIGHTENING TORQUES. Recommended tightening torques are as follows:

Cylinder head	29-31 ft.-lbs. (39-42 N•m)
Connecting rod – Aluminum	24-26 ft.-lbs. (33-35 N•m)
Forged steel	27-29 ft.-lbs. (37-39 N•m)
Rear bearing plate	20-25 ft.-lbs. (27-34 N•m)
Flywheel	35-40 ft.-lbs. (48-54 N•m)
Oil base	43-48 ft.-lbs. (58-65 N•m)
Blower housing screws	8-10 ft.-lbs. (11-14 N•m)
Exhaust manifold	15-20 ft.-lbs. (20-27 N•m)
Intake manifold	15-20 ft.-lbs. (20-27 N•m)
Timing gear cover	10-13 ft.-lbs. (14-18 N•m)
Valve cover nut	4-8 ft.-lbs. (6-11 N•m)
Starter bolts	25-28 ft.-lbs. (34-38 N•m)
Magneto stator screws	15-20 ft.-lbs. (20-27 N•m)
Spark plug	25-30 ft.-lbs. (34-41 N•m)

CYLINDER HEAD. Cylinder heads should be removed and carbon and lead deposits cleaned at 200 hour intervals (400 hours if using unleaded fuel) or as a reduction in engine power or excessive pre-ignition occurs.

CAUTION: Cylinder heads should not be unbolted from block when hot due to danger of head warpage.

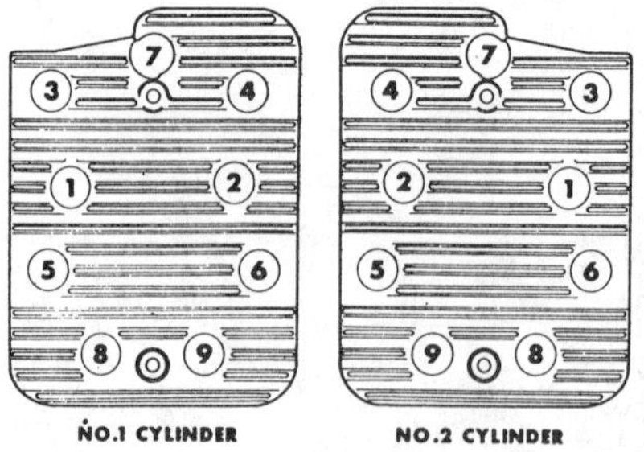

Fig. O74 – Tightening sequence for CCK series cylinder head cap screws. Procedure is outlined in text.

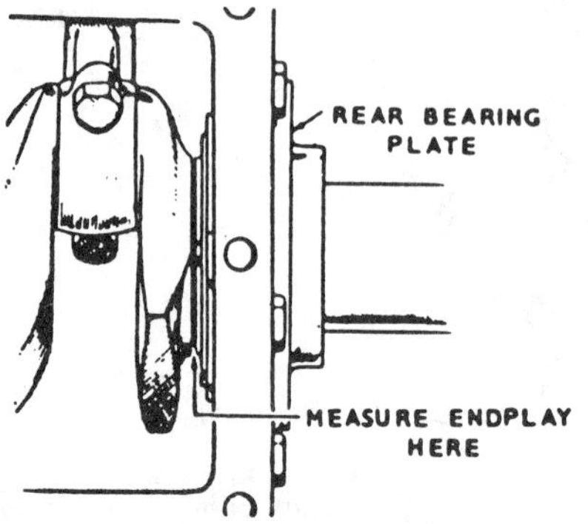

Fig. O75 – Procedure for crankshaft end play measurement. Refer to text.

Always install new head gaskets and torque retaining cap screws in 5 ft.-lbs. (7 N•m) steps in sequence shown (Fig. O74) until 29-31 ft.-lbs. (39-42 N•m) torque is obtained. Operate engine at normal operating temperature and light load for a short period, allow to cool and re-torque head bolts.

CONNECTING ROD. Connecting rod and piston are removed from cylinder head surface end of block after removing cylinder head and oil base.

Connecting rods for CCK and CCKA engines with specification letter "C" or before are aluminum rods which ride directly on crankpin journal and piston pin operates in unbushed bore of connecting rod.

Connecting rods for CCK and CCKA engines with specification letter "D" or after and all other models are equipped with precision type insert rod bearings and piston pin operates in renewable bushing pressed into forged steel connecting rod.

Standard crankpin journal diameter is 1.6252-1.6260 inches (41.28-41.30 mm) for all models and connecting rod or bearing inserts are available in a variety of sizes for undersize crankshafts as well as standard.

Connecting rod to crankshaft journal running clearance for aluminum rod should be 0.002-0.0033 inch (0.0508-0.0838 mm) and running clearance for bearings in forged steel rod should be 0.0005-0.0023 inch (0.0127-0.0584 mm).

Connecting rod caps should be reinstalled on original rod with raised lines (witness marks) aligned and connecting rod caps must be facing oil base after installation. Tighten to specified torque.

PISTON, PIN AND RINGS. Engines may be equipped with one of three different type three ring pistons. Do not intermix different type pistons in engines during service.

STRUT TYPE PISTON. Strut type piston may be visually identified by struts cast in underside of piston which run parallel to pin bosses. Piston to cylinder clearance when measured just below oil ring, 90° from pin, should be 0.0025-0.0045 inch (0.0635-0.1143 mm).

CONFORMATIC PISTON. Conformatic piston may be visually identified by smooth contours in underside of piston and no struts are visible. Top ring groove is 5/32-inch (3.969 mm) from top of piston and slots on opposite sides of

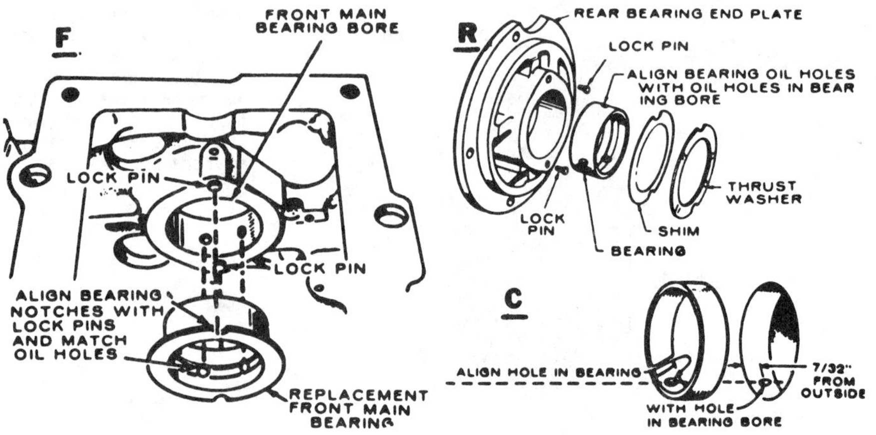

Fig. O76 – Alignment of precision main and camshaft bearings. One-piece bearing is used at front (view F), two-piece bearing at rear (view R). Note shim use for end play adjustment. View C shows placement of camshaft bearing.

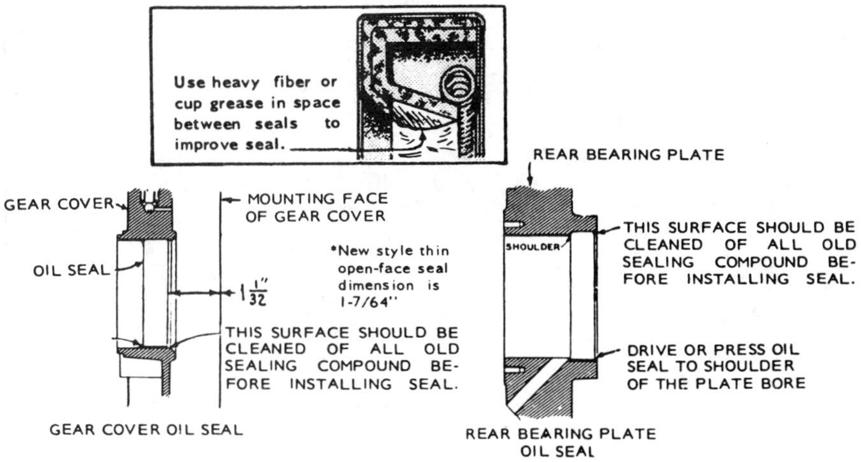
Use heavy fiber or cup grease in space between seals to improve seal.
REAR BEARING PLATE
GEAR COVER
MOUNTING FACE OF GEAR COVER
OIL SEAL
*New style thin open-face seal dimension is 1-7/64"
1 1/32"
THIS SURFACE SHOULD BE CLEANED OF ALL OLD SEALING COMPOUND BEFORE INSTALLING SEAL.
SHOULDER
THIS SURFACE SHOULD BE CLEANED OF ALL OLD SEALING COMPOUND BEFORE INSTALLING SEAL.
DRIVE OR PRESS OIL SEAL TO SHOULDER OF THE PLATE BORE
GEAR COVER OIL SEAL
REAR BEARING PLATE OIL SEAL

Fig. O77—View showing correct seal installation procedure. Fill cavity betwen lips of seal with heavy grease to improve sealing efficiency.

piston, behind oil control ring allow oil return and expansion. Piston to cylinder clearance when measured just below oil ring, 90° from pin, should be 0.0015-0.0035 inch (0.0381-0.0889 mm).

VANASIL PISTON. Vanasil type piston may be visually identified by smooth contours in underside of piston and no struts are visible. Top ring groove is 9/32-inch (7.144 mm) from top of piston and round holes behind oil control ring allow oil return and expansion. Piston to cylinder clearance when measured just below oil ring, 90° from pin, should be 0.006-0.008 inch (0.1524-0.2032 mm).

Side clearance of top ring in piston groove for all models should be 0.002-0.008 inch (0.051-0.203 mm).

Ring end gap for all models should be 0.010-0.023 inch (0.254-0.584 mm) and end gaps should be staggered around piston circumference.

Piston pin to piston pin bore clearance is 0.0001-0.0005 inch (0.0025-0.0127 mm) for all models and pin to rod clearance is 0.0002-0.0008 inch (0.005-0.020 mm) for aluminum rod and 0.00005-0.00055 inch (0.001-0.014 mm) for bushing in forged steel rod.

Standard piston pin diameter is 0.7500-0.7502 inch (19.05-19.06 mm) for all models and pins are available in 0.002 inch (0.0508 mm) oversize for aluminum rod.

Standard cylinder bore diameter for all models is 3.249-3.250 inches (82.53-82.55 mm) and cylinders should be bored to nearest oversize for which piston and rings are available if cylinder is scored or out-of-round more than 0.003 inch (0.0762 mm) or taper exceeds 0.005 inch (0.127 mm). Pistons and rings are available in a variety of oversizes as well as standard.

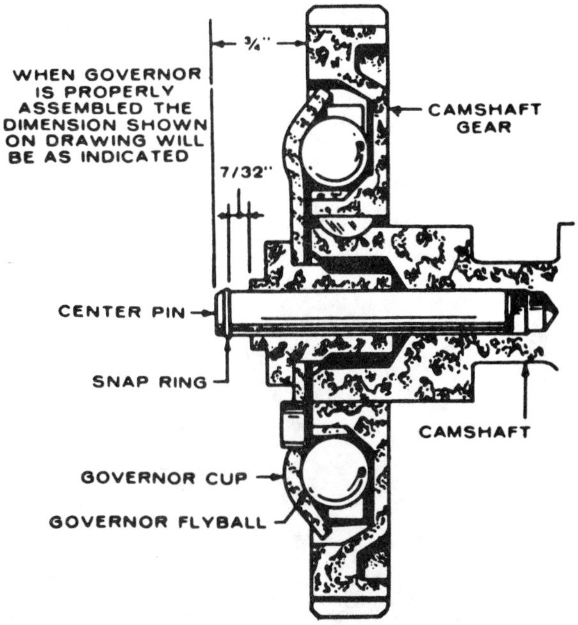
3/4"
WHEN GOVERNOR IS PROPERLY ASSEMBLED THE DIMENSION SHOWN ON DRAWING WILL BE AS INDICATED
7/32"
CAMSHAFT GEAR
CENTER PIN
SNAP RING
CAMSHAFT
GOVERNOR CUP
GOVERNOR FLYBALL

Fig. O78—Cross-sectional view of camshaft gear and governor assembly showing correct dimensions for center pin extension from camshaft. Note shim placement for adjusting end play in camshaft. Refer to text.

CRANKSHAFT, BEARINGS AND SEALS. Crankshaft is supported at each end by precision type sleeve bearings located in cylinder block housing and rear bearing plate. Main bearing journal standard diameter should be 1.9992-2.000 inches (50.7797-50.800 mm) and bearing operating (running) clearance for early model flanged type bearing (before specification letter "F") should be 0.002-0.003 inch (0.0508-0.0762 mm) and for all remaining models clearance should be 0.0024-0.0042 inch (0.061-0.107 mm). Main bearings are available for a variety of sizes for undersize crankshafts as well as standard.

Oil holes in bearing and bore **MUST** be aligned when bearings are installed and rear bearing should be pressed into bearing plate until flush, or recessed into bearing plate 1/64-inch (0.40 mm). Make certain bearing notches are aligned with lock pins (Fig. O76) during installation.

Apply "Loctite Bearing Mount" to front bearing and press bearing in until flush with block. Make certain bearing notches are aligned with lock pins (Fig. O76) during installation.

NOTE: Replacement front bearing and thrust washer is a one-piece assembly, do not install a separate thrust washer.

Rear thrust washer is installed with oil grooves toward crankshaft. Measure crankshaft end play as shown in Fig. O75 and add or remove shims or renew thrust washer (Fig. O76) as necessary to obtain 0.006-0.012 inch (0.15-0.30 mm) end play.

It is recommended rear bearing plate and front timing gear cover be removed for seal installation. Rear seal is pressed in until flush with seal bore. Timing cover seal should be driven in until it is 1-1/32 inch (26.19 mm) from mounting

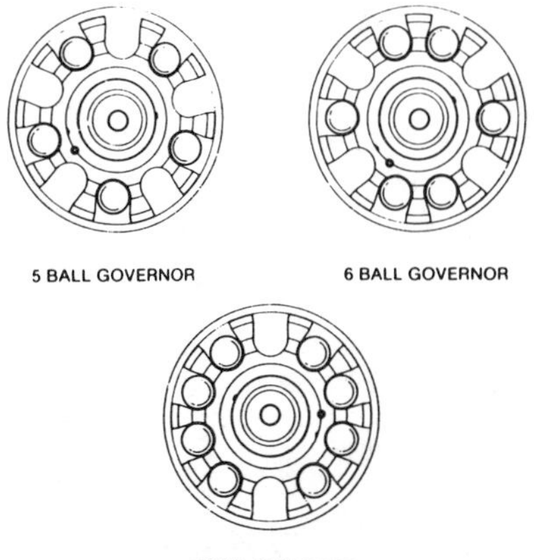
5 BALL GOVERNOR
6 BALL GOVERNOR
8 BALL GOVERNOR

Fig. O79—View showing correct spacing of governor flyballs according to total number of balls used.

Illustrations courtesy Onan.

face of cover for old style seal and 1-7/64 inch (28.18 mm) from mounting face of cover for new style, thin, open faced seal. See Fig. O77.

CAMSHAFT, BEARINGS AND GOVERNOR. Camshaft is supported at each end in precision type sleeve bearings. Make certain oil holes in bearings are aligned with oil holes in block (Fig. O76), press front bearing in flush with outer surface of block and press rear bearing in until flush with bottom of counterbore. Camshaft to bearing clearance should be 0.0015-0.0030 inch (0.0381-0.0762 mm).

Camshaft end play should be 0.003-0.012 inch (0.0762-0.305 mm) and is adjusted by varying thickness of shim located between camshaft timing gear and engine block. See Fig. O78.

Camshaft timing gear is a press fit on end of camshaft and is designed to accommodate 5, 8 or 10 flyballs arranged as shown in Fig. O79. Number of flyballs is varied to alter governor sensitivity. Fewer flyballs are used on engines with variable speed applications and greater number of flyballs are used for continuous (fixed) speed application. Flyballs must be arranged in "pockets" as shown according to total number used.

Center pin (Fig. O78) should extend ¾-inch (19 mm) from end of camshaft to allow 7/32-inch (5.6 mm) in-and-out movement of governor cup. If distance is incorrect, remove pin and press new pin in to correct depth.

Always make certain governor shaft pivot ball is in timing cover by measuring as shown in Fig. O80 inset and check length of roll pin which engages bushed hole in governor cup. This pin must extend to within ¾-inch (19 mm) of timing gear cover mating surface. See Fig. O80.

Make certain timing mark on camshaft gear is aligned with timing mark on crankshaft gear during installation.

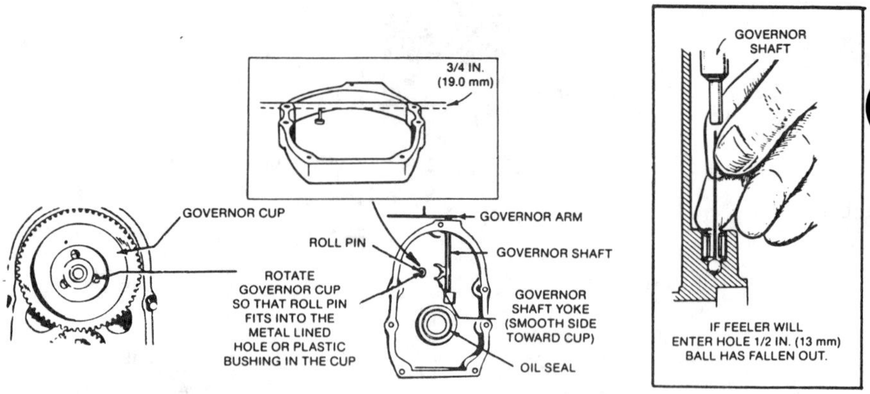

Fig. O80—View of governor shaft, timing gear cover and governor cup showing assembly details. Refer to text.

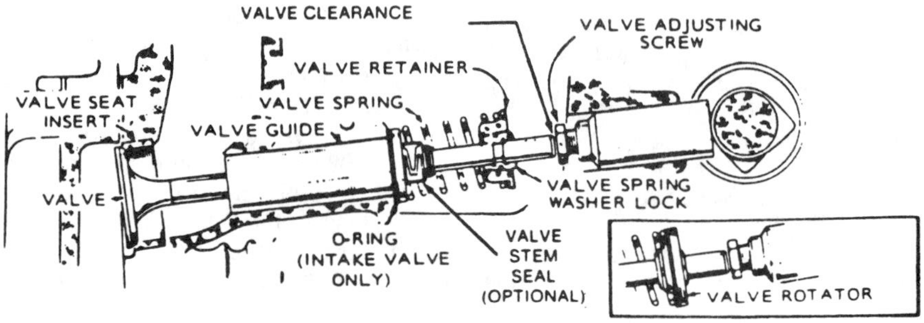

Fig. O81—View of typical valve train. Refer to text.

VALVE SYSTEM. Valve seats are renewable insert type and are available in a variety of oversizes as well as standard. Seats are ground at 45° angle and seat width should be 1/32 to 3/64-inch (0.794 to 1.191 mm).

Valves should be ground at 44° angle to provide an interference angle of 1°.

Recommended valve stem to guide clearance is 0.001-0.0025 inch (0.0254-0.0635 mm) for intake valves and 0.0025-0.004 inch (0.0635-0.1016 mm) for exhaust valves. Renewable shouldered valve guides are pushed out from head surface end of block. An "O" ring is installed on intake valve guide of some models and a valve stem seal is also available for intake valve. See Fig. O81.

Recommended valve tappet gap (cold) is 0.010-0.012 inch (0.254-0.305 mm) for both intake and exhaust valves on early models. Current models should have valve tappet gap (cold) of 0.006-0.008 inch (0.152-0.203 mm) for intake valves and 0.015-0.017 inch (0.381-0.432 mm) for exhaust valves.

Adjustment is made by turning tappet adjusting screw as required and valves of each cylinder must be adjusted with piston at "top dead center" on compression stroke. At this position both valves will be fully closed.

Illustrations courtesy Onan.

ONAN

SERVICING ONAN ACCESSORIES

STARTERS

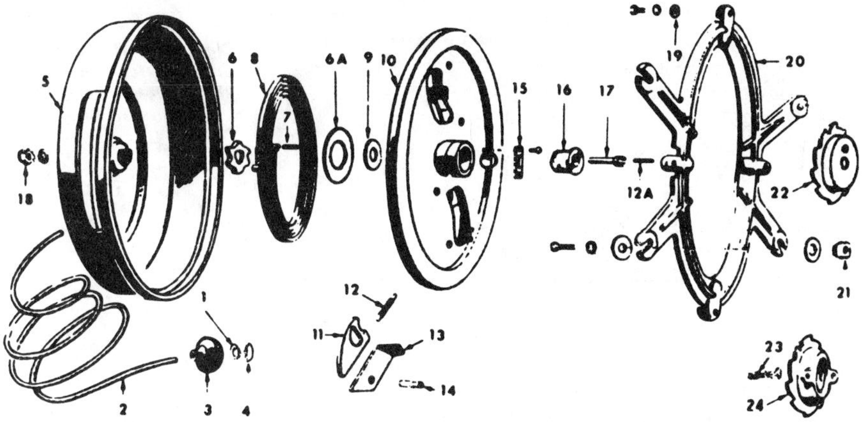

Fig. O140—Exploded view of "Ready-Pull" manual starter which is installed on some engines.

1. Rope retainer
2. Starter rope
3. Starter grip
4. Grip plug
5. Cover
6. Anti-backlash cog
6A. Spring washer
7. Roll pin
8. Recoil spring
9. Thrust washer
10. Rope sheave
11. Pawl (2)
12. Pawl spring
12A. Spiral pin
13. Ratchet arm (2)
14. Pivot roll pin (2)
15. Rope clamp
16. Hub bearing
17. Recessed screw
18. Flexlock nut
19. Washer
20. Mounting ring
21. Speed grip nut
22. Ratchet wheel (late)
23. Special cap screw
24. Ratchet wheel (early)

MANUAL STARTER. Engines may be equipped with manual "Readi-Pull" starter shown in exploded view in Fig. O140. For convenience, direction of starter rope pull may be adjusted to suit special cases by loosening clamps while holding starter cover (5) in position on its mounting ring (20), then turning cover (5) so rope (2) exits in desired direction.

Mounting ring (20) must be firmly attached to engine blower housing which must be as rigid as possible. If blower housing is damaged or if mounting holes for starter are misshaped or worn it may be necessary to renew entire blower housing. See Fig. O141 for cross-section detail of starter mounting ring attachment to blower housing.

To attach starter to earlier production engines, refer to Fig. O142, and use a pair of 10-penny (3 inch) common nails passed through holes in cover to insert into recesses in heads of special screws which retain ratchet wheel to engine flywheel. In later production (after specification D), spirol pin (12A–Fig. O140) is centered upon and engages drilled head screws (17) which secure ratchet wheel (22) and rope sheave hub bearing (16) for alignment during assembly.

Common repairs to manual starter can be made with minimum disassembly. Mounting ring (20) is left in place and only cover assembly (5) need be removed after its four clamps (19) are released. To renew starter rope (2), remove cover (5) from mounting ring (20), release clamp (15) to remove old rope from sheave (10). Then rotate sheave (10) in normal direction of crankshaft rotation so as to tighten spring (8) all the way. Align rope hole in sheave with rope slot in cover, secure new rope by its clamp (15), and when sheave (10) is released, spring (8) will recoil and wind rope on sheave. If renewal of recoil spring (8) is required, sheave (10) must be lifted out of cover (5). Remove starter rope (2) from sheave. Starting at outer end, wrap new spring (8) into a coil small enough to fit into recess in cover with loop at inner end of spring engaging roll pin (7) in cover. It may be necessary to secure wound-up spring with a temporary restraint such as a wire while doing so. Then reinstall rope sheave (10) so tab on sheave fits into loop at outer end of recoil spring. Install thrust washer (9) and spring washer (6A) if starter is so equipped.

Whenever starter is disassembled for any service, it is advisable to add a small amount of grease to factory-packed sheave hub bearing (16) and to clean and lubricate pawls (11) and ratchet arms (13) at pivot and contact points. If ratchet arms (13) require renewal due to wear, pawls (11) must first be removed. If kept clean, securely mounted and lightly lubricated, this starter will give long term reliable service.

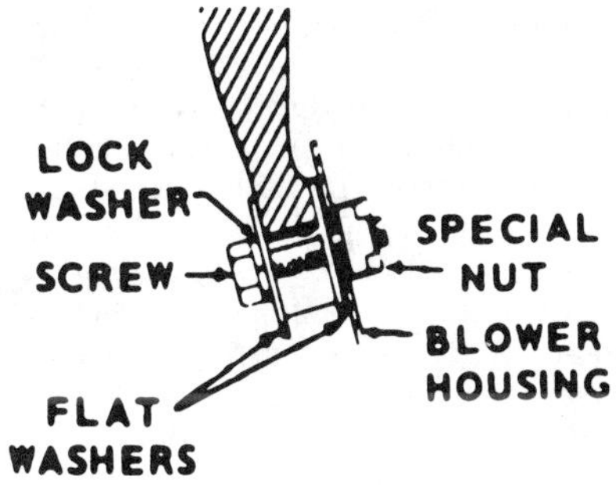

Fig. O141—Cross-section of mounting ring bolt to show housing attachment detail.

AUTOMOTIVE TYPE ELECTRIC STARTERS. Two styles of battery-driven electric starter motors are used on ONAN engines. Bendix-drive starter shown at A–Fig. O143 is designed to engage teeth of flywheel ring gear when

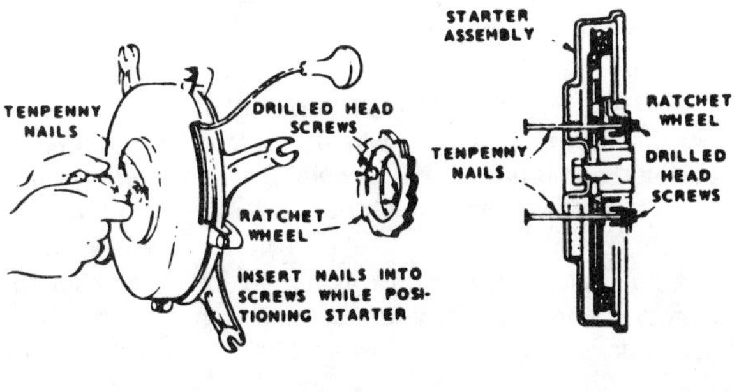

Fig. O142—Technique for mounting older style starter to ratchet wheel on engine flywheel. Refer to text.

Illustrations courtesy Onan.

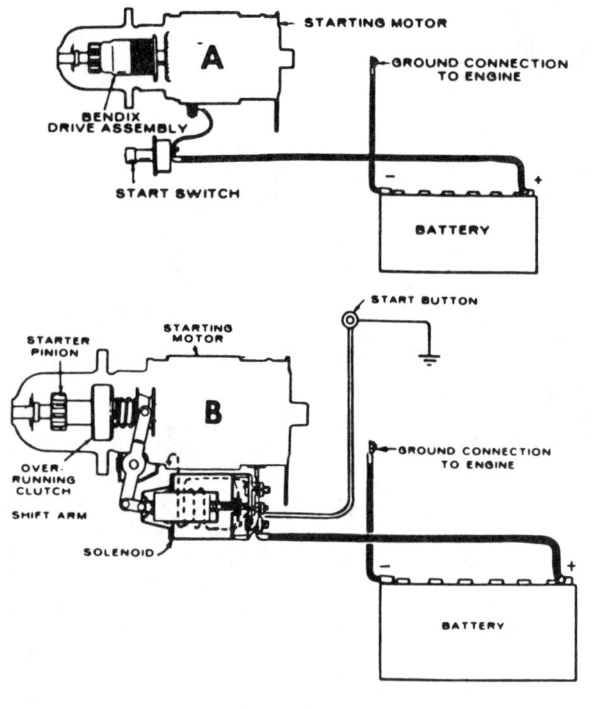

Fig. O143 — Views of Bendix-drive starter and solenoid-shift starter in basic electric circuits. Refer to text.

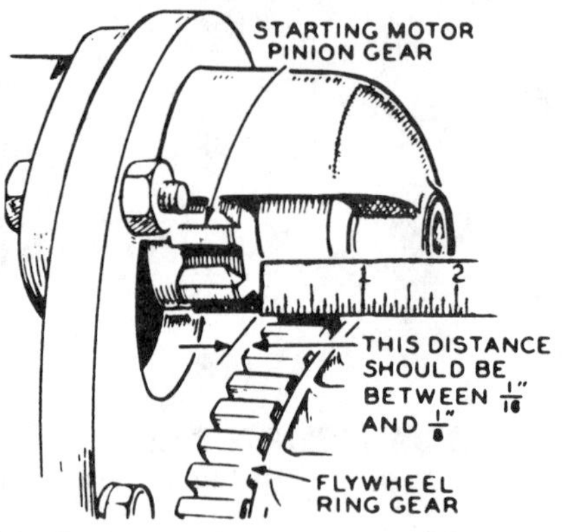

Fig. O144 — Check starter pinion to ring gear clearance as shown. Adjustment details in text.

starter switch is depressed to close circuit causing starter motor to turn. Engagement of starter pinion with ring gear by means of spiral shaft screw within Bendix pinion is cushioned by action of its coiled drive spring so starting motor can absorb sudden loading shock of engagement. Engine manufacturer recommends complete Bendix drive unit be renewed in case of failure, however, if a decision is made to overhaul starter drive by obtaining parts from manufacturer of starter (Prestolite), be sure correct drive spring is used. Length of spring is critical to mesh and engagement of starter pinion to flywheel ring gear. There are no procedures for adjustment of this starter.

Service is generally limited to cleaning and careful lubrication, renewal of starting motor brushes (4 used), brush tension springs and starter motor and drive housing bearings. See STARTER MOTOR TESTS for electrical check-out procedures for starter armature and field windings.

Solenoid-shift style starter (B – Fig. O143) uses a coil solenoid to shift starter pinion into mesh with flywheel ring gear and an over-running (one-way) clutch to ease disengagement of pinion from flywheel as engine starts and runs.

NOTE: Starter clutch will burn out if held in contact with flywheel for long periods and starter switch must be released quickly as engine starts.

All parts of starter are available for service. Solenoid unit and starter clutch are renewed as complete assemblies.

Refer to Figs. O144 and O145 for adjustment check points to measure flywheel ring gear to starter pinion clearance and gap between pinion and pinion stop on starter shaft. Starter pinion to ring gear clearance (Fig. O144) is adjusted during starter assembly by proper selection of spacer washers fitted to armature shaft as installed in housing. To adjust pinion clearance against pinion stop (Fig. O145), remove mounting screws which attach solenoid magnetic coil to front bracket and pinion housing assembly and select a proper thickness of fiber packing gaskets to set required clearance. Be sure plunger is pressed inward as shown when measuring.

Starting motor brushes require renewal when worn away by 0.3 inch (7.62 mm). Original brush length is 0.55 inches (13.97 mm).

Commutator must be clean and free from oil. Use No. 00 sandpaper to clean and lightly polish segments of commutator; never use emery cloth or any abrasive which may have a metallic content. Starter motor commutators do not need to have mica separators between segments undercut. Mica may be flush with surface.

STARTER MOTOR TESTS

Armature Short Circuit. Place armature in growler as shown in Fig. O146 and hold a hack saw blade or similar piece of thin steel stock above and parallel to core. Turn growler "ON". A short circuit is indicated by vibration of blade and attraction to core. If this condition appears, renew armature.

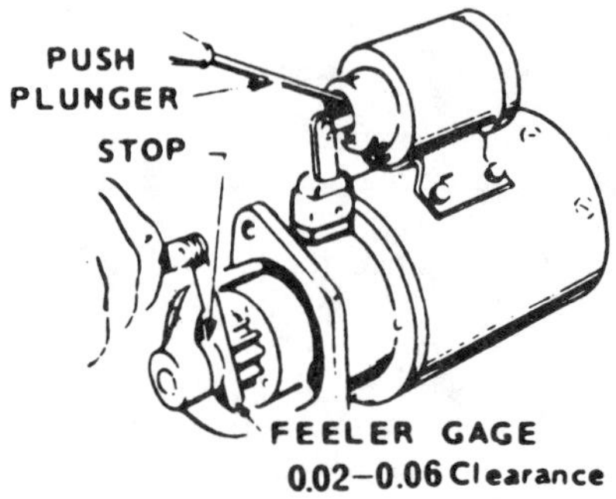

Fig. O145 — Measure clearance between starter pinion and pinion stop with feeler gage as shown.

Grounded Armature. Check each segment of commutator for grounding to shaft (or core) using ohmmeter setup as shown in Fig. O147. A low (Rx1 scale)

Fig. O146 — Use of growler to check armature for short circuit. Follow procedure in text.

Illustrations courtesy Onan.

Fig. O147—Test for grounded armature commutator by placing ohmmeter test probes as shown. Probe shown touching shaft may also be held against core. See text.

continuity reading indicates armature is grounded and renewal is necessary.

It is good procedure to mount armature on a test bench or between lathe centers to check for runout of commutator shaft. If shaft is worn badly, renewal is recommended. If commutator runout exceeds 0.004 inch (0.1016 mm), reface by turning.

Grounded Field Coils. Refer to Fig. O148 and touch one ohmmeter probe to a clean, unpainted spot on frame and the other to connector as shown, after unsoldering field coil shunt wire. A low range reading (Rx1 scale) indicates grounded coil winding. Be sure to check for possible grounding at connector lead which can be corrected, while grounded field coil cannot be repaired and calls for renewal.

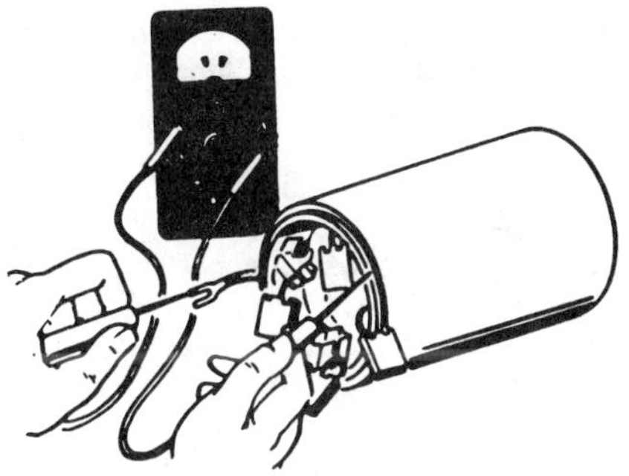

Fig. O148—Use ohmmeter as shown to check field coil for suspected internal grounding.

Fig. O149—Ohmmeter used to check for breaks or opens in field coil windings. Be sure to check lead wires.

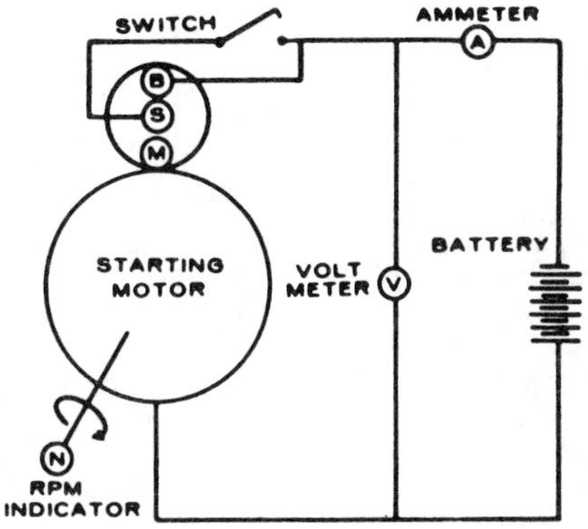

Fig. O150—Schematic for bench test of starting motor. Note ammeter is connected in series with load and voltmeter is connected "across" the load in parallel. Refer to text for test values.

Open in Field Coils. Use procedure shown in Fig. O149 and check all four brush holders for continuity. If there is no continuity or if a high resistance reading appears, renewal is necessary.

No Load Test. When starter is considered ready to return to service, connect motor on bench top as shown in Fig. O150. Acceptable test readings are:

Minimum speed	3700 rpm
Voltmeter reading	11.5 volts
Maximum current draw	60 amps

If starter motor does not check out as satisfactory on this test, make further checks for:

Weak brush springs.

Brushes not squarely seated.

Dirty commutator.

Poor electrical connections. May be caused by "cold" or corroded solder joints.

Tight armature. Not sufficient end play. End play should be 0.004-0.020 inch (0.1016-0.508 mm).

Open or ground in field coil.

Short circuit, open or ground in armature.

EXCITER CRANKING. Exciter cranking, with cranking torque furnished by switching battery current through a separate series winding of generator field coils and DC brushes, using DC portion of generator armature is wired as shown in Fig. O151.

This starting procedure may be standard for some CCK series. In cases where exciter cranking is inoperative, due to battery failure or other cause, unit may be started by use of a manual rope starter. In some cases, recoil type "Readi-Pull" starter, as covered in preceding section, may be furnished for standby use.

In case exciter cranking system will not operate, isolate starter solenoid switch and battery from DC field windings and perform a routine continuity check of all components by use of a voltohmmeter. See Fig. O151 for possible test points. If problem does not become apparent as caused by battery (low voltage), defective starter solenoid, short or open circuit in lead wires or DC brushes, it will be necessary to check out generator in detail and may involve factory service.

BATTERY CHARGING

FLYWHEEL ALTERNATOR. This battery charging system is simple and basically trouble-free. Flywheel-mounted permanent magnet rotor provides a rotating magnetic field to induce AC voltage in fixed stator coils. Current is then routed through a two-step mechanical regulator to a full-wave rectifier which converts this regulated alternating current to direct current for battery charging. Later models are equipped with a fuse between negative (−) side of rectifier and ground to protect rectifier from accidental reversal of battery polarity. See schematic Figs. O152 and O153. Maintenance services are limited to keeping components clean and ensuring that wire connections are secure.

TESTING. Check alternator output by connecting an ammeter in series between positive (+), red terminal of rectifier and ignition switch. Refer to Fig.

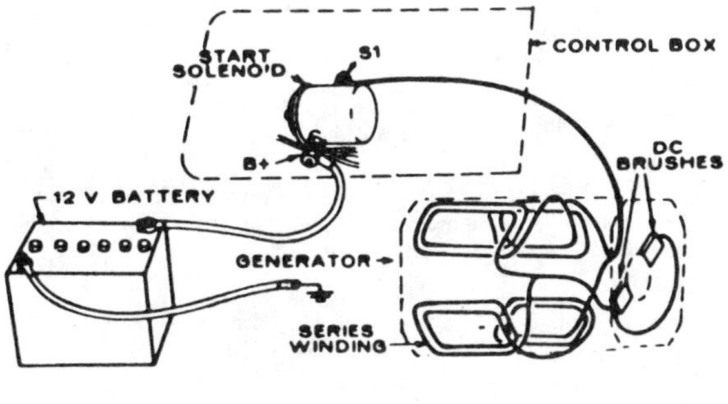

Fig. O151—Schematic of DC circuit portion of AC generator used for exciter cranking. Refer to text.

Illustrations courtesy Onan.

O153. At 1800 engine rpm, a discharged battery should cause about 8 amps to register on a meter so connected. As battery charge builds up, current should decrease. Regulator will switch from high charge to low charge at about 14½ volts with low charge current of about 2 amps. Switch from low charge to high charge occurs at about 13 volts. If output is inadequate, test as follows:

Check rotor magnetism with a piece of steel. Attraction should be strong.

Check stator for grounds after disconnecting by grounding each of the three leads through a 12-V test lamp. If grounding is indicated by lighted test lamp, renew stator assembly.

To check stator for shorts or open circuits, use an ohmmeter of proper scale connected across open leads to check for correct resistance values. Identify leads by reference to schematic.

From lead 7 to lead 8 . . . . 0.25 ohms
From lead 8 to lead 9 . . . . 0.95 ohms
From lead 9 to lead 7 . . . . 1.10 ohms

Variance by over 25% from these values calls for renewal of stator.

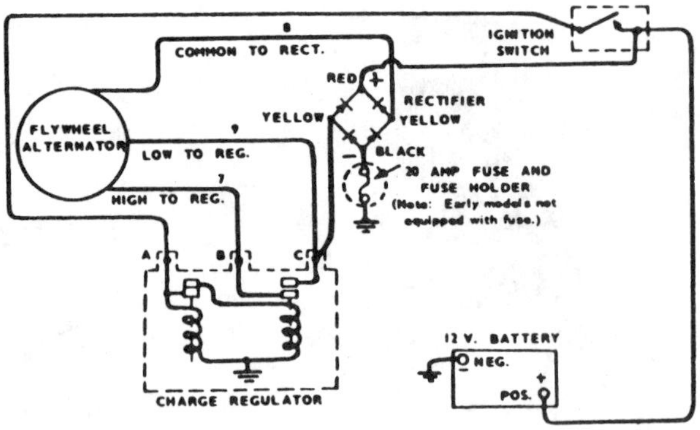

Fig. O153—Schematic of flywheel alternator circuits for location of test and check points. Refer to text for procedures.

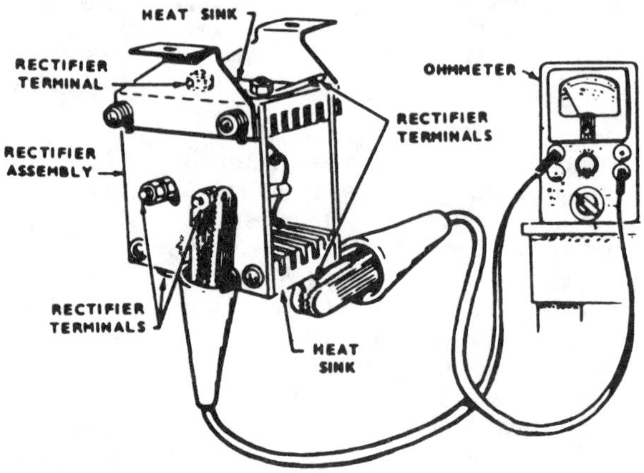

Fig. O154—Test each of four diodes in rectifier using Volt-Ohmmeter hookup as shown. See text for procedure.

RECTIFIER TESTS. Use an ohmmeter connected across a pair of terminals as shown in Fig. O154. All rectifier leads should be disconnected when testing. Check directional resistance through each of the four diodes by comparing resistance reading when test leads are reversed. One reading should be much higher than the other.

NOTE: Forward-backward ratio of a diode is on the order of several hundred ohms.

If a 12-V test lamp is used instead of an ohmmeter, bulb should light, but dimly. Full bright or no light at all indicates diode being tested is defective.

Voltage regulator may be checked for high charge rate by installing a jumper lead across regulator terminals (B and C–Fig. O153). With engine running, battery charge rate should be about 8 amperes. If charge rate is low, alternator or its wiring is defective.

If charge rate is correct (near 8 amps), defective regulator or its power circuit is indicated. To check, use a 12-V test lamp to check input at regulator terminal (A). If lamp lights, showing adequate input, regulator is defective and should be renewed.

NOTE: Regulator, being mechanical, is sensitive to vibration. Be sure to mount it on bulkhead or firewall separate from engine for protection from shock and pulsating motion.

Engine should not be run with battery disconnected, however, this alternator system will not be damaged if battery terminal should be accidentally separated from binding post.

FUEL SYSTEMS

ELECTRIC FUEL PUMP. Some engines may be furnished with Bendix Electric Fuel Pump, code R-8 or after.

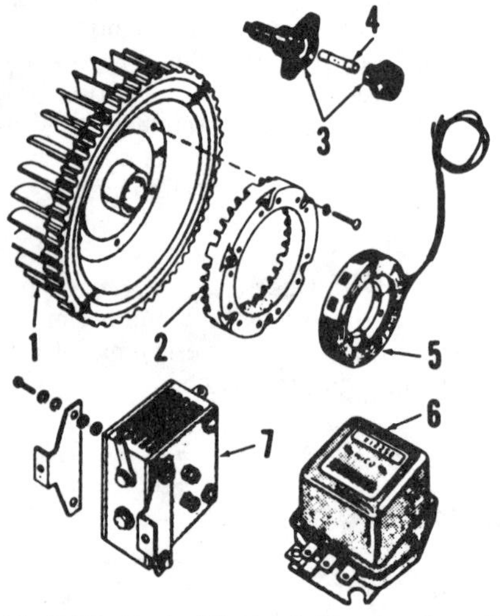

Fig. O152—Typical flywheel alternator shown in exploded view. In some models, regulator (6) and rectifier (7) are combined in a single unit.

1. Flywheel
2. Rotor
3. Fuse holder
4. Fuse
5. Stator & leads
6. Regulator
7. Rectifier assy

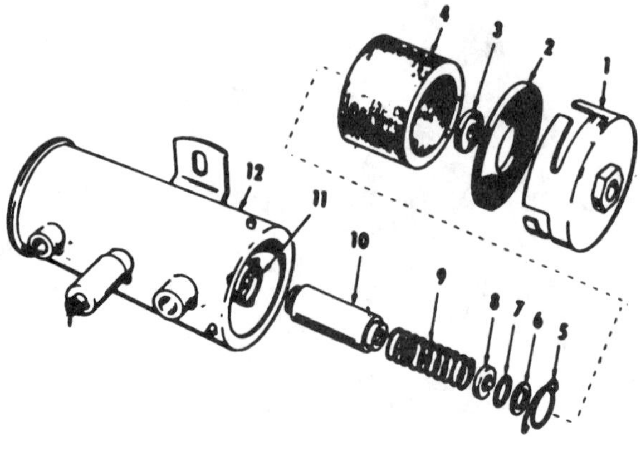

Fig. O155—Exploded view of electric fuel pump used on some engine models.

1. Cover
2. Cover gasket
3. Magnet
4. Filter element
5. Retainer spring
6. Washer
7. "O" ring
8. Cup valve
9. Plunger spring
10. Plunger
11. Plunger tube
12. Pump housing

Illustrations courtesy Onan.

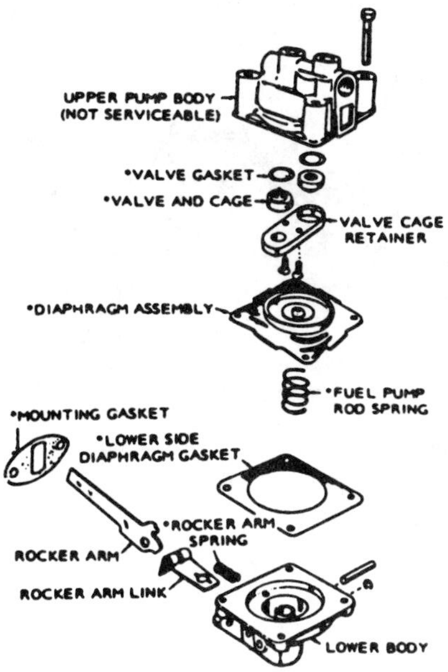
UPPER PUMP BODY (NOT SERVICEABLE)
•VALVE GASKET
•VALVE AND CAGE
VALVE CAGE RETAINER
•DIAPHRAGM ASSEMBLY
•FUEL PUMP ROD SPRING
•MOUNTING GASKET
•LOWER SIDE DIAPHRAGM GASKET
•ROCKER ARM SPRING
ROCKER ARM
ROCKER ARM LINK
LOWER BODY

• - PARTS INCLUDED IN REPAIR KIT.

Fig. O155A — Exploded view of mechanical type fuel pump used on some models. Refer to text for assembly information.

Maintenance service to these electric pumps is generally limited to simple disassembly and cleaning of removable components, not electrical overhaul. Stored gasoline is prone to deterioration and gum residues formed can foul internal precision units of a fuel system so as to cause sticking and sluggish operation of functional parts.

Refer to Fig. O155 for sequence of disassembly, beginning with cover (1) by turning 5/8-inch hex to release cover from bayonet lugs after fuel lines have been disconnected. Wash parts in solvent and blow dry using air pressure. Renew damaged or deteriorated parts, especially gasket (2) or filter element (4). Use needlenose pliers to remove retainer (5) and withdraw remainder of parts (6 through 10) from plunger tube (11). Clean and dry each item and inspect carefully for wear or damage. Plunger (10) calls for special attention. Clean rough spots very gently using crocus cloth if necessary. Clean bore of plunger tube (11) thoroughly and blow dry. For best results, use a swab on a stick to remove stubborn deposits from inside tube.

During reassembly, check fit of plunger (10) in tube (11). Full, free, in-and-out motion with no binding or sticking is required. If movement of plunger does not produce an audible click, interrupter assembly within housing (12) is defective. Renew entire pump.

If all parts appear serviceable, reassemble pump parts in order shown.

Pump output pressure can be raised or lowered by selection of a different plunger return spring (9). Consult authorized parts counter for special purpose spring or other renewable electric fuel pump parts.

NOTE: Seal at center of pump case mounting bracket retains a dry gas in pump electrical system. Be sure this seal is not damaged during disassembly for servicing.

MECHANICAL FUEL PUMP. A mechanical fuel pump (Fig. O155A) is used on some models. Pump operation may be checked by disconnecting fuel line at carburetor and slowly cranking engine by hand. Fuel should discharge from line.

CAUTION: Make certain engine is cool and there is nothing present to ignite discharged fuel.

To recondition pump, scribe a locating mark across upper and lower pump bodies and remove retaining screws. Noting location for reassembly, remove upper pump body, valve plate screw and washer, valve retainer, valves, valve springs and valve gasket. To remove lower diaphragm, hold mounting bracket and press down on diaphragm to compress spring. Turn bracket 90° to unhook diaphragm. Clean all parts thoroughly.

To reassemble, hold pump cover with diaphragm surface up. Place valve gasket and assembled valve springs and valves in cavity. Assemble valve retainer and lock in position by inserting and tightening valve retainer screw. To reassemble lower diaphragm section hold mounting bracket and press down on diaphragm to compress spring. Turn bracket 90° to hook diaphragm. Assemble pump upper and lower bodies, but do not tighten screws. Push pump lever to its limit of travel, hold in this position and tighten the four screws. This prevents stretching diaphragm. Reinstall pump.

PULSATING DIAPHRAGM FUEL PUMP. A pulsating diaphragm type fuel pump (Fig. O155B) is used on some models. Pump may be mounted directly to side of carburetor or at a remote mounting location.

Pump relies on a combination of crankcase pressure and spring pressure for correct operation.

Refer to Fig. O155B for disassembly and reassembly noting air bleed hole (10) must be open for correct pump operation.

CLUTCH. When optional Rockford clutches are furnished with these engines, an adaptor flange is fitted to engine output shaft for mounting clutch unit and a variety of housings are used dependent upon application and model of engine or clutch used. Refer to Fig. O172 for guidance in adjustment and proceed as follows:

Remove plate from top of housing and rotate engine manually until lock screw (1 – Fig. O172) is at top of ring (2) as shown. Loosen lock screw and turn adjusting ring clockwise (as facing through clutch toward engine) until toggles cannot be locked over center. Then, turn ring in reverse direction until toggles can just be locked over center by a very firm pull on operating lever. If a new clutch plate has been installed, slip under load to knock off "fuzz" and readjust. Lubricate according to instructions on unit plate.

REDUCTION GEAR ASSEMBLIES. Typical reduction gear unit is shown in Fig. O173. Ratio of 1:4 is common in industrial applications. Lubrication calls for use of SAE 50 motor oil or SAE 90 gear oil. Refer to instructions printed on gear case for guidance. In

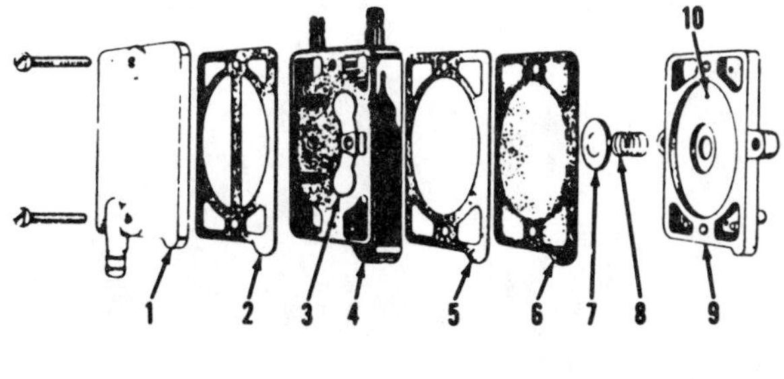
10
1
2
3
4
5
6
7
8
9

Fig. O155B — Exploded view of pulsating diaphragm fuel pump used on some models. Hole (10) must be open before reassembly.

1. Pump cover
2. Gasket
3. Reed valve
4. Valve body
5. Gasket
6. Diaphragm
7. Pump plate
8. Spring
9. Pump base
10. Air bleed hole

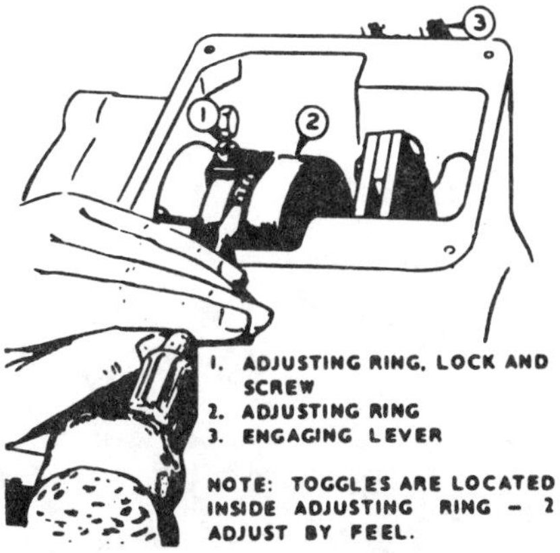
1. ADJUSTING RING, LOCK AND SCREW
2. ADJUSTING RING
3. ENGAGING LEVER
NOTE: TOGGLES ARE LOCATED INSIDE ADJUSTING RING – 2 ADJUST BY FEEL.

Fig. O172 — Procedure for adjustment of Rockford clutch.

1. Ring lock & screw
2. Adjusting ring
3. Clutch lever

Illustrations courtesy Onan.

most cases, a total of six plugs are fitted into case for lubricant fill or level check. Plug openings to be used are determined by positioning of gear box in relation to horizontal or vertical. It is recommended that square plug heads be cut off those plugs not to be used to fill, check or drain so as to eliminate chance of error by overfill or underfill. All parts shown are available for renewal if needed in overhaul.

NOTE: In some installations, no shaft seal is fitted between engine crankcase and reduction gear housing. In these cases, with a common oil supply, engine oil lubricates gears and bearings of reduction gear unit and gear oil is not used. Be sure to check nameplate or operator's manual.

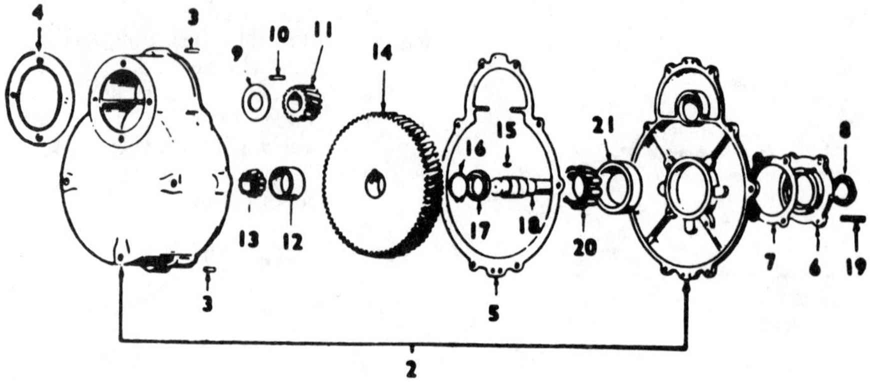

Fig. O173 — Exploded view of typical reduction gear set. See text for service details.

2. Housing & cover
3. Dowel pins (2)
4. Gasket (engine)
5. Cover gasket
6. Bearing retainer
7. Shims
8. Oil seal
9. Pinion washer
10. Pinion key
11. Pinion gear
12. Bearing cup
13. Bearing cone
14. Driven gear
15. Gear key
16. Snap ring
17. Bearing spacer
18. Shaft
19. Key
20. Bearing cone
21. Bearing cup

YANMAR

Model	No. Cyls.	Bore	Stroke	Displacement	Power Rating
3TG66UJ	3	66 mm (2.6 in.)	64.2 mm (2.53 in.)	659 cc (40.2 cu. in.)	13.4 kW (18 hp)

ENGINE IDENTIFICATION

The Yanmar 3TG66UJ is a four-stroke, three-cylinder engine. The engine is liquid cooled and equipped with overhead valves. The engine identification plate is located on the rocker cover.

MAINTENANCE

LUBRICATION. The engine is equipped with an oil filter and an oil pump that provides pressurized oil to the engine.

Check oil level before operation and after every 10 hours of operation. Oil level must be between marks on dipstick; do not operate engine if oil level is at or below lower mark.

Manufacturer recommends oil with API classification SE or SF. Use SAE 30 oil for temperatures above 0° C (32° F); use SAE 10W-30 oil for temperatures between −20° C (−4° F) and 30° C (86° F); SAE 5W-30 may be used between 10° C (50° F) and −35° C (−31° F).

Change oil and filter after first 50 hours of operation and then after every 200 hours of operation, or more frequently if operation is severe. Crankcase capacity is 3 L (3.1 qts.) with filter or 2.8 L (2.9 qts.) without filter.

AIR FILTER. The air filter should be cleaned after every 200 hours of operation. Filter is paper; do not wash. Tap filter lightly to remove debris, or direct low-pressure air, approximately 210 kPa (30 psi), from inside to outside to blow away debris.

FAN BELT. Adjust fan belt tension so a 107 N (24 lb.) force applied midway between fan and alternator pulleys will deflect belt 13 mm (½ in.). Reposition alternator to adjust fan belt tension.

COOLING SYSTEM. Drain, flush and refill cooling system annually. Manufacturer recommends low silicate antifreeze that does not contain stop leak additive. Cooling system capacity is 2.8 L (2.9 qts.).

A thermostat is used to maintain cooling system temperature. The thermostat should begin to open at 71° C (160° F) and open fully at 84° C (184° F).

SPARK PLUG. Recommended spark plug is NGK BPR4BS or Champion RN11YC. Spark plug electrode gap should be 0.8 mm (0.032 in.). Tighten spark plug to 20 N·m (180 in.-lbs.).

CARBURETOR. A fixed, nonadjustable main jet controls high speed mixture on standard equipment carburetors. A variable (adjustable) main jet kit is available for installation on tractors operating at high altitudes.

Initial adjustment of idle mixture screw (M—Fig. Y101) is 1½ turns out from a lightly seated position. Final adjustment of idle mixture screw must be performed with engine at normal operating temperature. Run engine at idle speed and turn idle mixture screw (M) counterclockwise until engine speed decreases. Turn screw clockwise until engine speed decreases again, then turn screw to obtain fastest engine speed. Turn screw out an extra ⅛ turn.

Shut off engine, loosen locknut (1—Fig. Y102) and turn speed adjusting screw (2) so point on throttle shaft flange (3) aligns with projection (4) on side of carburetor. Disconnect throttle cable, start engine and hold throttle shaft flange against stop screw (5). If necessary, adjust idle speed by turning stop screw (5) to obtain recommended speed of 1300 rpm.

Refer to GOVERNOR paragraph for high idle (no-load) speed adjustment.

GOVERNOR. The engine is equipped with a flyweight type governor (4—Fig. Y103) located on the backside of the timing gear case. Refer to REPAIRS section for service.

To adjust governor linkage, proceed as follows: Run engine until normal operating temperature is achieved. Disconnect throttle cable. Move governor lever against high speed stop screw (2—Fig. Y103) and check engine speed using an accurate tachometer. Specified engine speed is 3600—3800 rpm. If necessary, loosen nut (1) and turn high speed stop screw (2) to obtain desired speed. Tighten nut and recheck engine speed. Move throttle control lever to high speed position, loosen throttle cable clamp, pull

Fig. Y101—Initial adjustment of idle mixture screw (M) is 1½ turns out from a lightly seated position. Refer to text for adjustment procedure.

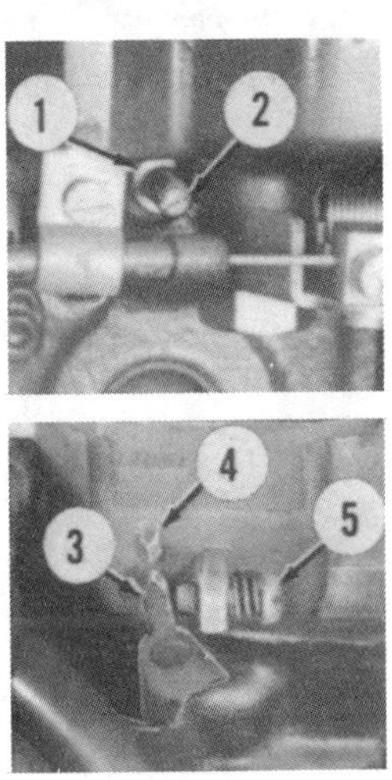

Fig. Y102—Loosen jam nut (1) and turn adjusting screw (2) so pointer on throttle shaft flange (3) is aligned with projection (4) on carburetor. Refer to text for adjustment of low idle stop screw (5).

Fig. Y103—Adjust high idle speed by turning high speed stop screw (2) as outlined in text.

1. Jam nut
2. High speed stop screw
3. Spring
4. Governor linkage & housing

cable to remove slack, reconnect throttle cable and tighten cable clamp.

IGNITION SYSTEM. The ignition system is an electronic type that uses three pulsars, an ignition module and three ignition coils. The pulsars are mounted equidistantly around the flywheel and trigger the ignition for each cylinder. Each cylinder is served by an ignition coil.

Ignition timing is not adjustable, but a viewing port is provided in flywheel housing. Ignition timing is determined by ignition module. However, timing should be the same for all cylinders.

If an ignition malfunction is suspected, refer to REPAIRS section for service procedure.

Fig. Y104—View showing index mark (M) on flywheel housing and TDC mark (F) for No. 1 cylinder on flywheel.

VALVE ADJUSTMENT. To adjust valve clearance, remove plug from flywheel viewing port on side of flywheel housing. Rotate flywheel so mark (F—Fig. Y104) indicating number 1 cylinder is aligned with mark (M) on flywheel housing and valves for number one cylinder are both closed (rocker arms will be loose). Use a feeler gage to measure clearance between rocker arm and valve stem end of both valves for number one cylinder. Specified clearance (engine cold) for both intake and exhaust valves is 0.2 mm (0.008 in.). If necessary, loosen locknut and turn rocker arm adjusting screw to obtain desired clearance. Rotate flywheel so mark on flywheel indicating number 2 cylinder is aligned with mark (M) on flywheel housing and valves for number 2 cylinder are closed. Adjust valve clearance for number 2 cylinder, then repeat procedure for number 3 cylinder.

COMPRESSION PRESSURE. At a cranking speed of 300 rpm, minimum compression pressure is 779 kPa (113 psi). Maximum variation between cylinders is 97 kPa (14 psi).

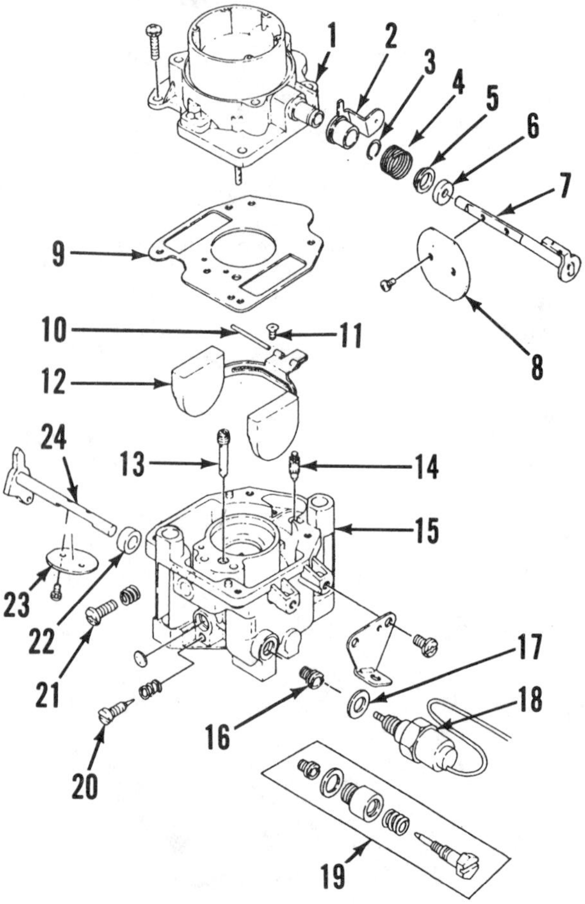

Fig. Y105—Exploded view of carburetor.

1. Cover
2. Lever
3. Snap ring
4. Spring
5. Bushing
6. Seal
7. Choke shaft
8. Choke plate
9. Gasket
10. Pin
11. Screw
12. Float
13. Low speed jet
14. Fuel inlet valve
15. Body
16. Main jet
17. Gasket
18. Fuel shut-off solenoid
19. High altitude main jet kit
20. Idle mixture screw
21. Idle speed adjustment screw
22. Seal
23. Throttle plate
24. Throttle shaft

REPAIRS

TIGHTENING TORQUES. Recommended tightening torque specifications are as follows:

Camshaft retainer	10-12 N·m (89-106 in.-lbs.)
Connecting rod	23 N·m (17 ft.-lbs.)
Cylinder head	34 N·m (25 ft.-lbs.)
Exhaust manifold	11 N·m (97 in.-lbs.)
Flywheel	83 N·m (61 ft.-lbs.)
Flywheel housing	
8 mm	26 N·m (19 ft.-lbs.)
10 mm	49 N·m (36 ft.-lbs.)
Intake manifold	11 N·m (97 in.-lbs.)
Main bearing	54 N·m (40 ft.-lbs.)
Oil pump	11 N·m (97 in.-lbs.)
Rear main seal	11 N·m (97 in.-lbs.)
Rocker arm shaft	25 N·m (18 ft.-lbs.)
Spark plug	20 N·m (15 ft.-lbs.)
Starter	49 N·m (36 ft.-lbs.)
Stub shaft	59 N·m (43 ft.-lbs.)
Timing cover	9 N·m (80 in.-lbs.)
Timing gear housing	
Aluminum	9 N·m (80 in.-lbs.)
Iron	11 N·m (97 in.-lbs.)
Water pump	26 N·m (19 ft.-lbs.)

CARBURETOR. Refer to exploded view of carburetor in Fig. Y105 when overhauling carburetor. Do not clean passages in carburetor with drill bits or wire as any enlargement of passages may change the calibration and performance of the carburetor. Do not polish or attempt to reshape idle jet (13 and 20), main jet (16) or fuel inlet valve (14). Renew jets if they are worn or damaged. Use low pressure air to dry out passages. Do not use cloth or paper towels to dry parts.

To check float level, invert carburetor. End of float (F—Fig. Y106) should be flush with body surface. Bend float tab (T) to adjust float setting. With carburetor upright, float drop should be at least 5 mm (0.20 in.) as measured from highest point on float to carburetor body surface as shown in Fig. Y107.

Bend float arm (A) as needed, then recheck float level.

Before tightening throttle plate and choke plate screws, rotate shaft so plate is in closed position, then tighten screws. This will center plate on shaft.

IGNITION SYSTEM. Refer to Fig. Y108 for wiring diagram. To test ignition components, first turn key switch ON and place hydrostatic lever in STOP position and pto switch in OFF position. Check for voltage at double pink/black wires at engine harness connector.

If there is no voltage at double pink/black wires, turn key switch OFF. Connect a jumper wire across seat safety switch terminals. Check for continuity between black wire at TDC (Time Delay Control Module) two-pin connector and ground. Turn key switch ON and check for voltage at pink wire at TDC two-pin connector. Check for voltage at pink wire, yellow/black wire, and purple/black wire at TDC eight-pin connector. If continuity and voltage checks are normal, pink wire from TDC eight-pin connector to engine connector is open. If there is no continuity at black wire, ground wire is faulty. If there is no voltage at pink wire at two-pin connector, check for faulty wire or connections. If

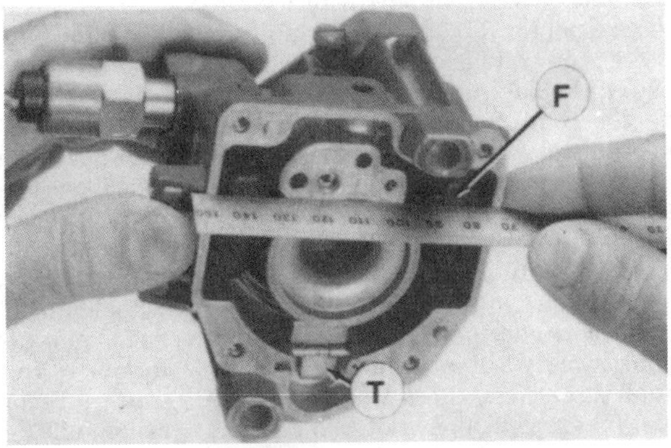

Fig. Y106—Refer to text and measure float level as shown. Bend float tab (T) to adjust float level.

Fig. Y107—Measure float drop from tip of float to body mating surface as shown. Bend float arm (A) to adjust.

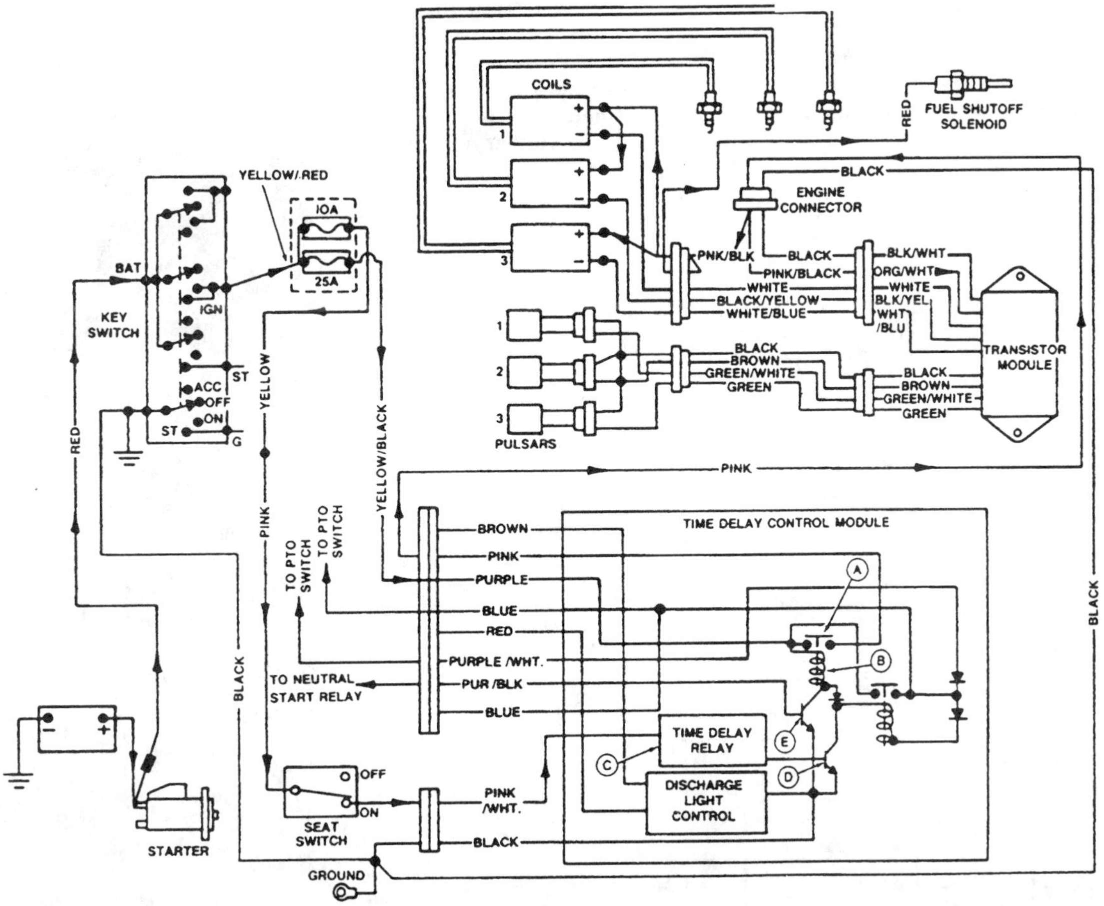

Fig. Y108—Wiring diagram of ignition circuit.

there is no voltage at pink wire at eight-pin connector, TDC module is faulty. If there is no voltage at yellow/black wire or purple/black wire at eight-pin connector, check for faulty wires or connections.

If there is voltage at double pink/black wires, perform an operational test of ignition system as follows: Connect a test light to negative terminal of each of the ignition coils in turn. Start engine and run at low idle speed. Test light should blink rapidly. If test light fails to blink at all three coils, check transistor module. If test light does not blink at one coil, test pulsar for that coil. If test light does blink but there is no spark, test coil.

To test pulsars, disconnect four-pin pulsar wiring connector. Use ohmmeter to check resistance between black ground wire and each of the remaining pulsar wires in the connector. Specified resistance is 15.5-23.3 ohms. If resistance is not as specified or there is no resistance, renew pulsar or wiring as necessary.

To test ignition coils, turn key switch OFF and disconnect wires from coil. Measure primary winding resistance across coil terminals. Specified resistance is 3.8-5.2 ohms. Disconnect spark plug wire and measure resistance between coil positive terminal and spark plug boot end of high tension lead. Specified resistance is 10.8K-16.2K ohms. Check for continuity between negative terminal of coil and core. There should be infinite resistance. Renew coil as necessary.

To check transistor module, disconnect transistor module wiring connectors and check resistance across terminals as indicated in chart shown in Fig. Y109. Note that readings are approximate and may vary due to condition and accuracy of ohmmeter.

Ignition timing is not adjustable but may be checked at the flywheel inspection port using a timing light. The flywheel has three BTDC marks and a TDC mark for each cylinder. If ignition timing is correct, the center BTDC mark on flywheel should align with mark on flywheel housing for all three cylinders. If BTDC mark is not in the same position for one cylinder, check input voltage to transistor module as follows: Run engine at low idle speed and measure voltage across black wire in module four-pin connector and each of the three remaining wires. Voltage at the three wires must be the same. If not, check for faulty wire or pulsar.

CYLINDER HEAD. The cylinder head is removable with engine in tractor. To remove head, drain engine coolant and disconnect battery cables. remove exhaust manifold, water pump, carburetor, ignition coils and spark plugs. Remove rocker arm cover, rocker arms and push rods. Mark rocker arms and push rods so they can be returned to original position. Remove cylinder head screws and lift head off engine. Remove intake manifold.

Inspect cylinder head. Head surface must be flat within 0.10 mm (0.004 in.), otherwise machine surface. Maximum thickness that may be removed is 0.2 mm (0.008 in.). Check valve recession if head is machined.

When installing cylinder head refer to Fig. Y110 for screw tightening sequence. Tighten cylinder head screws in several steps to final torque of 34 N·m (25 ft.-lbs.).

VALVE SYSTEM. The valves are located in the cylinder head. When disassembling valve components, mark parts as they are removed so they can be installed in original positions if reused. Valve seats (17 and 18—Fig. Y111) and guides (14) are renewable.

Intake valve face and seat angles should be 30 degrees. Exhaust valve face and seat angles should be 45 degrees. Intake valve seat width should be 1.14 mm (0.045 in.). Exhaust valve seat width should be 1.37 mm (0.054 in.). Minimum valve margin is 0.51 mm (0.020 in.). Minimum valve stem diameter is 5.40 mm (0.213 in.).

Maximum allowable valve guide inside diameter is 5.57 mm (0.219 in.). Guide may be knurled if valve stem-to-guide clearance exceeds 0.15 mm (0.006 in.) but is less than 0.20 mm (0.008 in.). Renew valve or guide if clearance is 0.20 mm (0.008 in.) or greater. Use suitable driver to remove and install guides. Install guide so grooved end is up. Valve guide standout above valve spring seating surface is 7 mm (0.276 in.). Ream guide to obtain desired clearance after installation.

Valve spring pressure should be 125 N (28 lbs.) at a length of 17 mm (0.669 in.) for both intake and exhaust. Minimum valve spring length is 27.59 mm (1.083 in.). Install spring so end with paint mark (less distance between coils at end) is next to head.

BLACK LEAD / RED LEAD	BLACK	BROWN	GREEN WHITE	GREEN	BLACK YELLOW	WHITE	WHITE BLUE	ORANGE WHITE	BLACK WHITE
BLACK		O (2.2K)	O (2.2K)	O (2.2K)	O (2.4K)	O (2.4K)	O (2.4K)	X	O (0Ω)
BROWN	O (2.2K)		O (4.4K)	O (4.4K)	O (5.1K)	O (5.1K)	O (5.1K)	X	O (2.2K)
GREEN WHITE	O (2.2K)	O (4.4K)		O (4.4K)	O (5.1K)	O (5.1K)	O (5.1K)	X	O (2.2K)
GREEN	O (2.2K)	O (4.4K)	O (4.4K)		O (5.1K)	O (5.1K)	O (5.1K)	X	O (2.2K)
BLACK YELLOW	X	X	X	X		X	X	X	X
WHITE	X	X	X	X	X		X	X	X
WHITE BLUE	X	X	X	X	X	X		X	X
ORANGE WHITE	O (15K)	O (20K)	O (20K)	O (20K)	O (50K)	O (50K)	O (50K)		O (15K)
BLACK WHITE	O (0Ω)	O (2.2K)	O (2.2K)	O (2.2K)	O (2.4K)	O (2.4K)	O (2.4K)	X	

Fig. Y109—The chart shown above may be used to trouble-shoot the ignition module. "O" denotes continuity and "X" denotes infinite resistance.

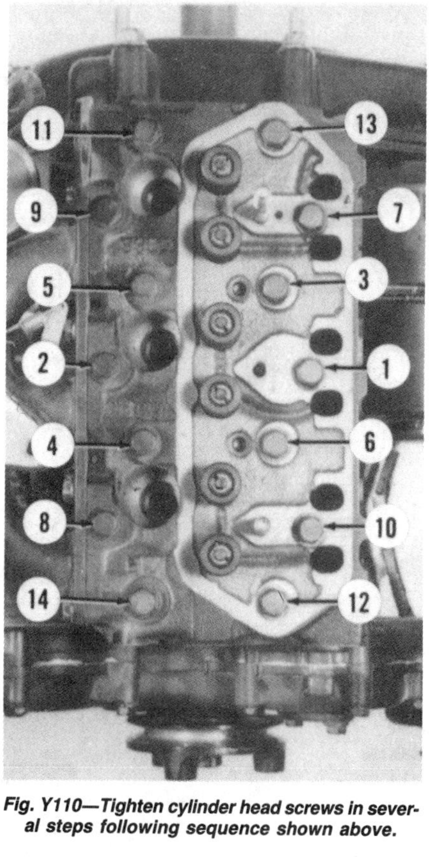

Fig. Y110—Tighten cylinder head screws in several steps following sequence shown above.

Maximum push rod runout measured with push rod supported at ends is 0.30 mm (0.012 in.). Minimum rocker shaft diameter is 9.9 mm (0.390 in.). Maximum rocker arm inner diameter is 10.1 mm (0.398 in.). Maximum rocker shaft support inner diameter is 10.1 mm (0.398 in.).

When reassembling, install new valve stem seals (13). Note that center rocker shaft support (8—Fig. Y111) is located on rocker shaft by a setscrew (7).

FLYWHEEL AND STUB SHAFT. Remove engine for access to flywheel. Remove pto clutch assembly. Remove flywheel housing (1—Fig. Y112). Remove stub shaft (3) and flywheel (4).

Stub shaft runout should not exceed 0.2 mm (0.008 in.). Mounting surfaces of flywheel and stub shaft should be flat within 0.05 mm (0.002 in.).

Flywheel retaining cap screws should not be reused. Install new flywheel cap screws and tighten to 83 N·m (61 ft.-lbs.). Tighten stub shaft screws to 59 N·m (43 ft.-lbs.). Tighten flywheel housing 10 mm screws to 49 N·m (36 ft.-lbs.) and 8 mm screws to 26 N·m (230 in.-lbs.).

PISTON, PIN, RINGS AND CONNECTING ROD. To remove piston and connecting rod units, remove engine from tractor. Remove pto clutch, flywheel, starter, pulsars, rear mounting plate, cylinder head and oil pan. Be sure to remove ring ridge (if present) from top of cylinder before removing piston. Mark piston, connecting rod and rod cap so they can be returned to original position, remove rod cap and push piston and rod unit out of cylinder block.

Inspect pistons for cracks, scoring or scuffing and renew as necessary. To check piston ring grooves for wear, insert new rings in piston grooves and use a feeler gage to measure ring side clearance. Renew piston if clearance exceeds 0.25 mm (0.010 in.) for all rings. Minimum allowable piston diameter measured at bottom of skirt and perpendicular to piston pin is 65.85 mm (2.593 in.). Maximum allowable standard cylinder bore diameter is 66.12 mm (2.603 in.). Maximum allowable piston-to-bore clearance is 0.15 mm (0.006 in.). Pistons and rings are available in standard size and 0.25 mm oversize.

Maximum allowable ring end gap for compression rings is 1.30 mm (0.051 in.). Maximum end gap for oil ring is 1.80 mm (0.071 in.). Maximum allowable piston ring side clearance is 0.25 mm (0.010 in.) for all rings.

Maximum allowable pin bore in piston is 20.1 mm (0.791 in.). Minimum allowable piston pin diameter is 19.9 mm (0.783 in.). Maximum allowable clearance between pin and piston bore is 0.10 mm (0.004 in.).

Maximum allowable small end diameter of connecting rod is 20.1 mm (0.791 in.). Maximum allowable clearance between pin and connecting rod bushing is 0.15 mm (0.006 in.). Small end bushing is renewable.

Maximum allowable big end bearing inside diameter is 36.07 mm (1.420 in.). Minimum allowable crankpin diameter is 35.93 mm (1.415 in.). Maximum allowable crankpin bearing clearance is 0.12 mm (0.005 in.). Crankpin may be reground to accept 0.25 mm undersize rod bearing. Maximum allowable rod side clearance on crankshaft is 0.8 mm (0.031 in.). Manufacturer recommends discarding used connecting rod screws.

Install oil ring expander in piston ring groove so ends are located above either end of piston pin. Install oil ring so ends are on opposite side from expander ends. Install second compression ring so inside chamfer is toward piston crown and end gap is positioned 120 degrees away from oil ring end gap. Install top piston ring so marked side is toward piston crown and end gap is 120 degrees away from second ring end gap.

Lubricate piston, rings, cylinder and crankpin bearing with engine oil, then install piston and rod assembly using a suitable piston ring compressor. Lubricate crankpin bearing insert with engine oil, then install rod cap with tangs on cap and connecting rod on same side. Tighten rod cap screws to 23 N·m (203 in.-lbs.).

TIMING GEARS, GEAR COVER AND HOUSING. Remove engine from

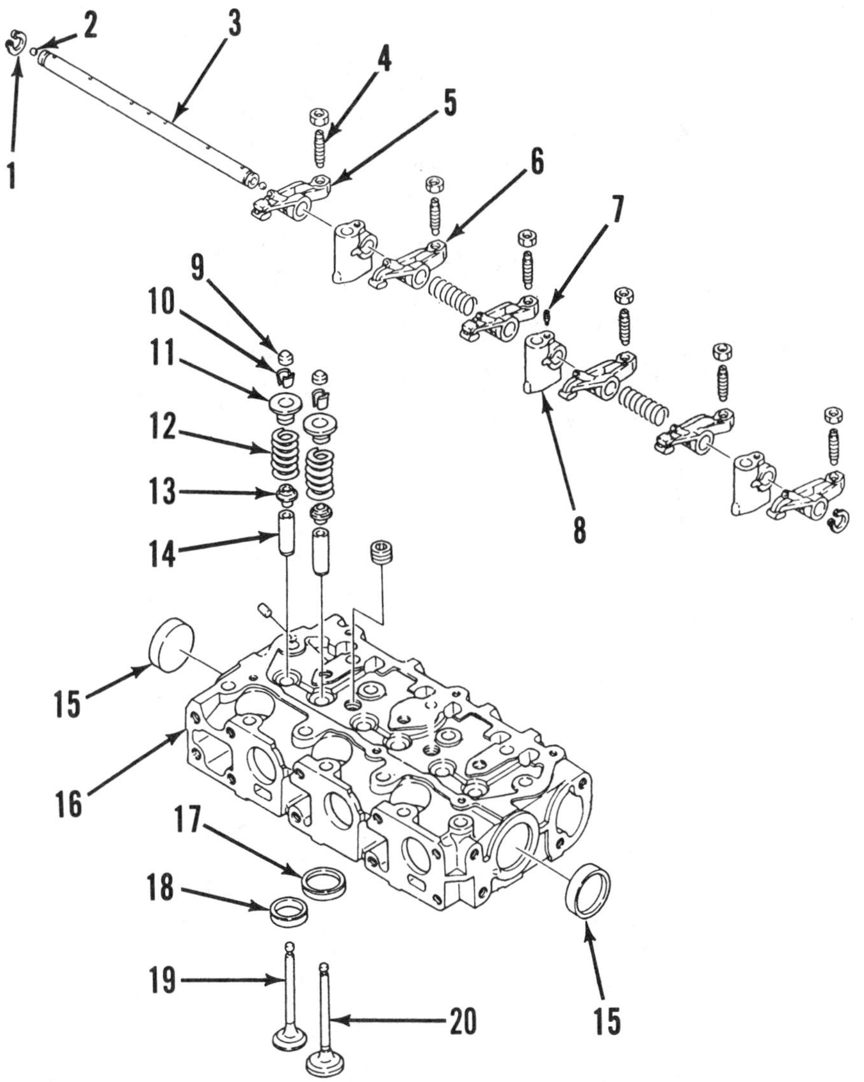

Fig. Y111—Exploded view of cylinder head and rocker arms. Center shaft support (8) is located in center of rocker shaft (3) by setscrew (7).

1. Snap ring
2. Ball
3. Rocker shaft
4. Adjusting screw
5. Rocker arm, exhaust
6. Rocker arm, intake
7. Setscrew
8. Shaft support
9. Valve stem cap
10. Split retainer
11. Spring retainer
12. Valve spring
13. Valve seal
14. Valve guide
15. Plugs
16. Cylinder head
17. Seat insert
18. Seat insert
19. Exhaust valve
20. Intake valve

tractor for access to gear housing. Remove fan, alternator belt and alternator. Unscrew crankshaft retaining screw and using a suitable puller remove crankshaft pulley. Detach timing gear cover.

Refer to Fig. Y114 for gear identification. Camshaft and crankshaft gears are pressed onto their respective shafts. Note timing marks (M) on crankshaft, idler and camshaft gears that must align during installation.

Specified gear backlash is 0.2 mm (0.008 in.) for camshaft, crankshaft and idler gears; 0.3 mm (0.012 in.) for oil pump gear; 0.38 mm (0.015 in.) for governor gear. Renew gear if backlash is greater than specified.

The idler gear rides on a stub shaft attached to the cylinder block. Maximum allowable inside diameter of idler gear bushing is 20.08 mm (0.790 in.). Bushing is renewable. Minimum allowable idler shaft diameter is 19.9 mm (0.783 in.). Maximum allowable clearance is 0.1 mm (0.004 in.).

To remove gear housing, remove all interfering geared components. Note that camshaft must be removed as gear is pressed on shaft. When installing gear housing, tighten mounting screws to 9 N·m (80 in.-lbs.) if housing is aluminum, or to 11 N·m (97 in.-lbs.) if housing is iron.

Install timing cover crankshaft seal so lip is toward inside of cover. Tighten cover retaining screws to 9 N·m (80 in.-lbs.).

CAMSHAFT AND LIFTERS. To remove camshaft, remove engine from tractor. Remove rocker arm cover. Remove push rods and mark so they can be returned to original position. Remove cylinder head if lifters (tappets) are to be removed. Remove timing gear cover.

Prior to removing camshaft, measure camshaft end play. If end play is greater than 0.5 mm (0.020 in.), renew thrust plate. Remove screws securing camshaft retainer. If lifters were not removed, use suitable magnetic tools to hold lifters away from camshaft or tilt engine so lifters move away from camshaft. Withdraw camshaft from engine.

Minimum allowable camshaft bearing journal diameter is 35.84 mm (1.411 in.) for end journals and 35.81 mm (1.410 in.) for the two inner journals. Minimum allowable lobe height is 29.7 mm (1.169 in.). Maximum bushing diameter in block is 36.115 mm (1.4218 in.) for front bushing (nearest timing gears) and 36.075 mm (1.4203 in.) for inner and rear bushings. Maximum allowable camshaft bearing clearance is 0.18 mm (0.007 in.). Camshaft bushings are renewable and must be installed so oil hole aligns with oil hole in block.

Minimum allowable lifter diameter is 17.85 mm (0.703 in.). Maximum lifter bore in block is 18.1 mm (0.713 in.). Maximum allowable lifter clearance is 0.1 mm (0.004 in.). Lifters should be renewed if a new camshaft is installed.

If necessary to remove camshaft gear, use a suitable press. Heat gear to 150° C (300° F) before pressing gear on camshaft.

When installing camshaft be sure to align timing marks (M—Fig. Y114). Tighten camshaft retainer plate screws to 10-12 N·m (89-106 in.-lbs.).

CRANKSHAFT AND MAIN BEARINGS. To service crankshaft, remove engine from tractor. Remove cylinder head, oil pan and pistons and connecting rods. Remove timing gears, timing gear housing, flywheel, starter, ignition pulsars and engine mounting plate. Remove rear oil seal case.

Measure crankshaft end play before disassembly. Maximum allowable end play is 0.30 mm (0.012 in.). Mark bearing caps prior to removal so they can be reinstalled in original position. Notice that arrow (A—Fig. Y115) on cap points toward flywheel end of crankshaft.

Minimum allowable main bearing journal diameter is 39.93 mm (1.572 in.). Maximum bearing inside diameter is 40.93 mm (1.611 in.). Maximum allowable bearing clearance is 0.12 mm (0.005 in.). Main bearings are available in standard size and 0.25 mm (0.010 in.) undersize.

Crankshaft thrust is controlled by renewable thrust rings located at front main bearing web. If crankshaft end play is excessive, install new thrust rings. Install thrust rings so oil grooves face out and tabs on bottom rings index in notches on bearing cap as shown in Fig. Y116.

If necessary to remove crankshaft gear, use a suitable press. Heat gear to 150° C (300° F) before pressing gear on crankshaft.

Lubricate main bearing inserts with engine oil. Be sure that main bearing caps are installed in their original locations (caps and cylinder block are numbered). Lubricate main bearing cap screws with engine oil, then tighten screws to 54 N·m (40 ft.-lbs.). Make certain crankshaft turns freely as main bearing cap screws are tightened.

Install new oil seal in seal retainer so lip of seal is toward engine. Seal should be installed flush with outer surface oil seal retainer. However, if crankshaft flange is grooved at original oil seal contact point, new seal may be pressed 3 mm (1/8 in.) farther into retainer so that seal lip will contact unworn portion of crankshaft. Tighten oil seal retainer cap screws to 11 N·m (97 in.-lbs.). When installing timing gears be sure to align timing marks (M—Fig. Y114).

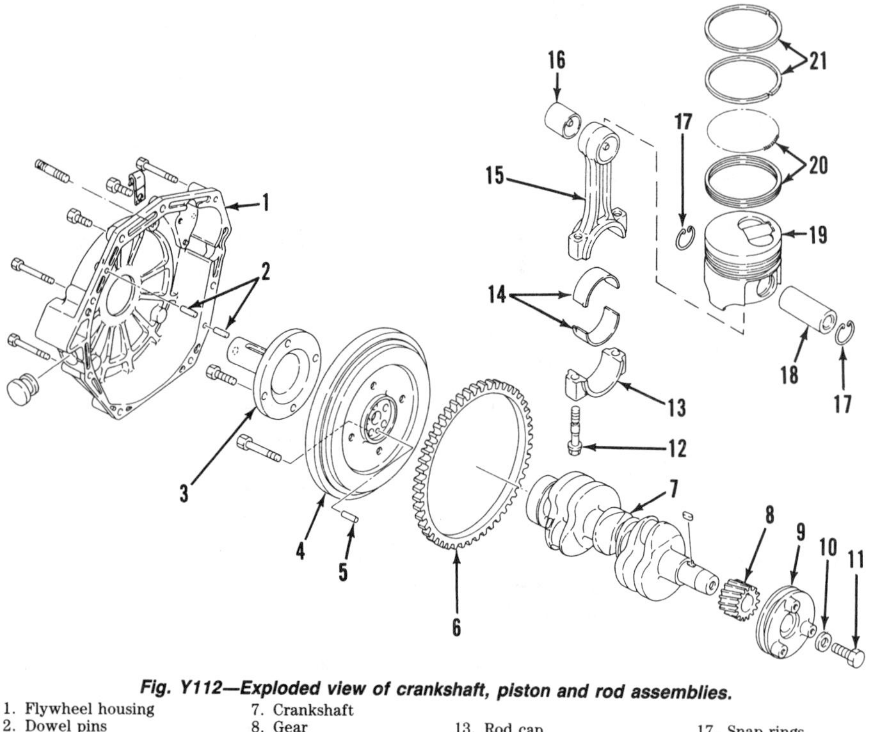

Fig. Y112—Exploded view of crankshaft, piston and rod assemblies.

1. Flywheel housing
2. Dowel pins
3. Stub shaft
4. Flywheel
5. Dowel pin
6. Ring gear
7. Crankshaft
8. Gear
9. Drive pulley
10. Washer
11. Screw
12. Rod bolt
13. Rod cap
14. Crankpin bearing insert
15. Connecting rod
16. Bushing
17. Snap rings
18. Piston pin
19. Piston
20. Oil ring
21. Compression rings

GOVERNOR. The governor assembly is driven by governor gear (G—Fig. Y114). The flyweight assembly is mounted on the back of the timing gear housing. To remove governor unit, remove engine from tractor. Remove timing gear cover and unscrew special bolt (27—Fig. Y117) retaining governor gear (25). Detach governor linkage, unscrew governor housing retaining screws and withdraw governor housing (16) and linkage assembly. Remove screws retaining bearing flange (23) and remove governor weight assembly from timing gear housing. Remove two screws from governor lever yoke (17), then remove governor lever (9), spacer (13) and ball (14) from governor housing. Drive oil seal (12) from housing if necessary.

Minimum allowable diameter of governor shaft (19) is 7.90 mm (0.311 in.). Maximum allowable inside diameter of flyweight carrier (22) is 8.15 mm (0.321 in.). Maximum allowable clearance between shaft and bore is 0.18 mm (0.007 in.). Inspect governor weights (20) and pivot pins (21) for excessive wear and renew as necessary. Weights must pivot freely on pins.

Bearings (24) are pressed on carrier (22). Specified backlash for governor gear is 0.38 mm (0.015 in.). Renew gear if backlash is greater than specified.

To reassemble, reverse the disassembly procedure. Adjust governor linkage as outlined in MAINTENANCE section.

OIL PUMP. A gerotor type oil pump is located on the front of the engine and driven by the crankshaft gear.

To check oil pressure, remove oil pressure switch in side of cylinder block and connect an oil pressure gage. With engine at operating temperature and running at half throttle setting, oil pressure should be 294-440 kPa (43-64 psi).

The oil pressure relief valve is located in the oil filter nipple (1—Fig. Y118). To remove nipple, remove oil filter, loosen nut (N) and unscrew nipple from cylinder block. Unscrew plug (5) for access to valve components. Relief valve pressure setting may be adjusted by adding or deleting shims (4). Spring free length should be 21.9-24.5 mm (0.862-0.965 in.). Compressed spring length should be 14.7 mm (0.579 in.) at 12 N (2.7 lbs.).

Remove timing gear cover for access to oil pump. Specified oil pump drive gear backlash is 0.3 mm (0.012 in.). Renew gear if backlash is greater than specified. Pump rotors must be no more than 0.25 mm (0.010 in.) below face of pump (see Fig. Y119). Maximum allowable clearance between outer pump rotor and pump body is 0.25 mm (0.010 in.). Maximum allowable clearance between inner rotor tip and outer rotor lobe is 0.25 mm (0.010 in.) as shown in Fig. Y120.

Pump drive gear is pressed onto pump shaft. Install outer rotor so mark (M—Fig. Y121) is toward inside of pump. Tighten oil pump retaining screws to 11 N·m (97 in.-lbs.).

WATER PUMP AND THERMOSTAT. Thermostat (11—Fig. Y123) is located in the water pump coolant outlet. To remove thermostat, drain coolant and remove thermostat cover (9). Thermostat should start to open at 71° C (160° F) and be fully open at 85° C (184° F).

To remove water pump, detach drive belt and remove fan. Unbolt and remove water pump from engine.

Refer to Fig. Y123 for an exploded view of water pump. To disassemble pump, use a suitable puller and remove pulley flange (8). Detach rear cover (1), support housing and press shaft (7), seal (5) and impeller (4) as an assembly out impeller end of housing. Press impeller off shaft.

When assembling pump, press shaft and bearing assembly (7) into housing so outer face of bearing is flush with front of housing. To prevent possible damage to bearing, be sure to use a suitable bearing driver tool that contacts only the outer race of the bearing. Press seal (5) into housing until bottomed. Support rear of shaft and press pulley flange (8) onto front of shaft until outer face of flange is flush with end of shaft. Install ceramic seal into impeller, making certain that lapped sealing surface of insert is clean and dry. Press impeller onto shaft so outer face of impeller and end of shaft are flush. Tighten cover screws to 9 N·m (80 in.-lbs.). Tighten thermostat housing screws to 9 N·m (80 in.-lbs.). Tighten water pump mounting screws to 26 N·m (230 in.-lbs.). Tighten fan screws to 11 N·m (97 in.-lbs.).

Fig. Y114—View showing location of timing gears. Note timing marks (M) on crankshaft, idler and camshaft gears that must align for correct valve timing.

C. Crankshaft gear
G. Governor gear
B. Idler gear
P. Oil pump gear
T. Camshaft gear

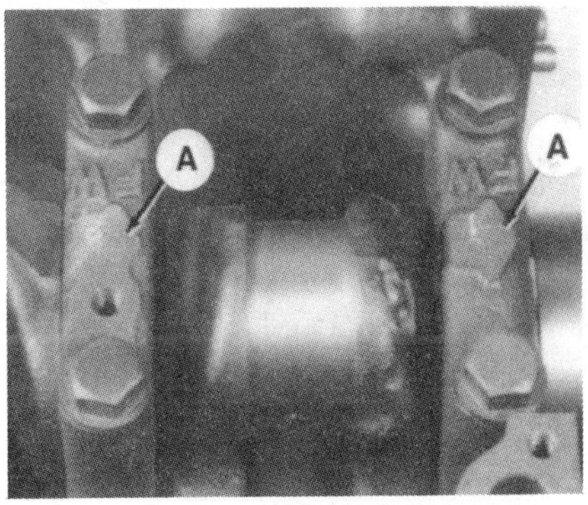

Fig. Y115—Arrow (A) on main bearing cap must point toward flywheel end of crankshaft.

Fig. Y116—Install thrust ring (R) on main bearing cap so tab indexes in notch of cap and oil grooves face outward.

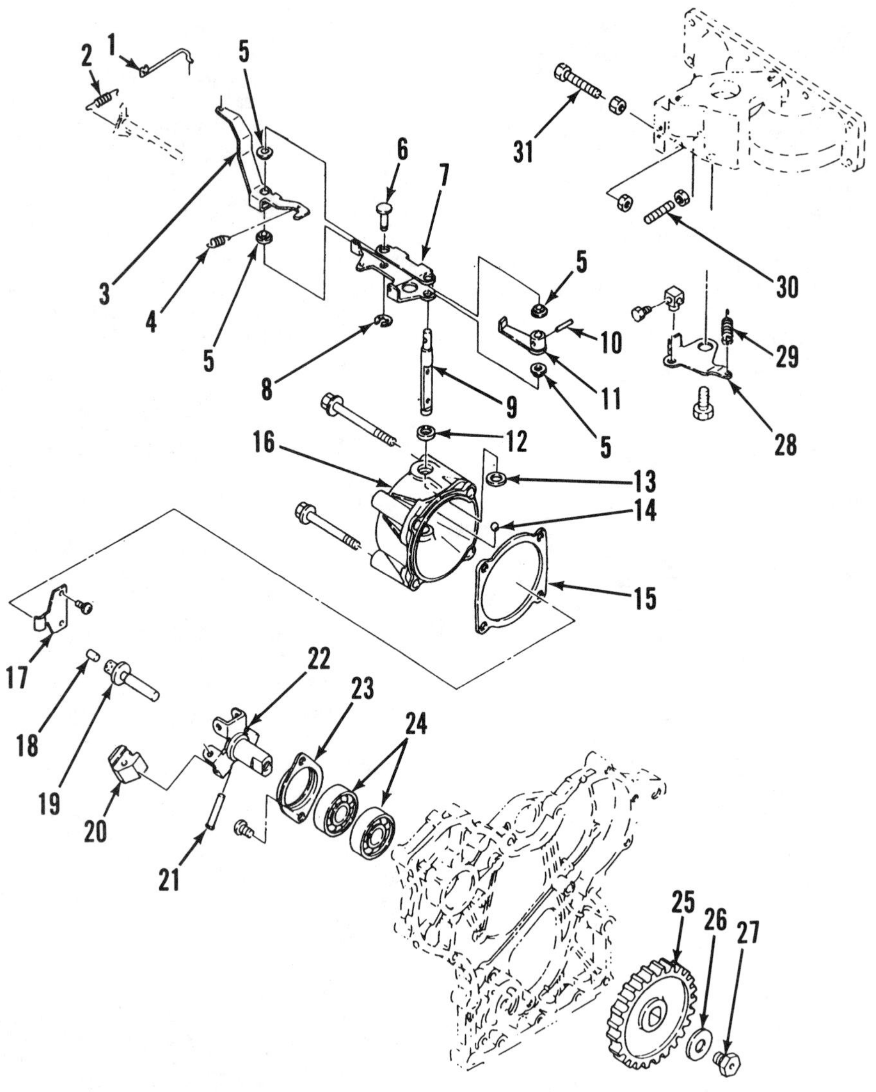

Fig. Y117—Exploded view of governor linkage and flyweights.

1. Carburetor link
2. Spring
3. Governor lever
4. Spring
5. Washers
6. Pin
7. Bracket
8. "E" ring
9. Governor lever
10. Pin
11. Arm
12. Oil seal
13. Spacer
14. Ball
15. Gasket
16. Governor housing
17. Yoke
18. Pin
19. Governor shaft
20. Flyweight
21. Pin
22. Flyweight carrier
23. Flange
24. Bearings
25. Governor gear
26. Washer
27. Special bolt

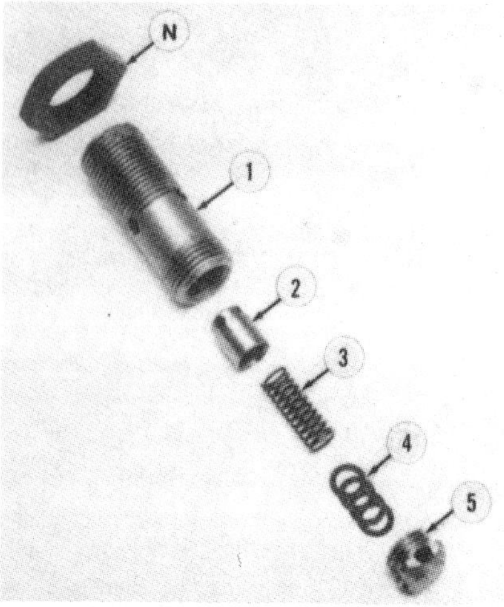

Fig. Y118—Exploded view of relief valve assembly contained in oil filter nipple.

N. Nut
1. Nipple
2. Relief valve
3. Spring
4. Shims
5. Plug

ELECTRIC STARTER. Starter no-load current draw should be 60 amps or less at starter speed of 7000 rpm. Starter current draw under load should be 230 amps at starting rpm.

Refer to Fig. Y124 for an exploded view of electric starter. Place alignment marks on pinion housing, frame and end cover before disassembly so they can be reinstalled in original position.

To disassemble starter, disconnect solenoid wire. Remove through-bolts and retaining screws from end cover. Pry plastic cap (18) from end cover (14), remove "E" ring (17) shims (16) and end cover. Pry springs away from negative brushes, pull brushes up and release springs to hold brushes in this position.

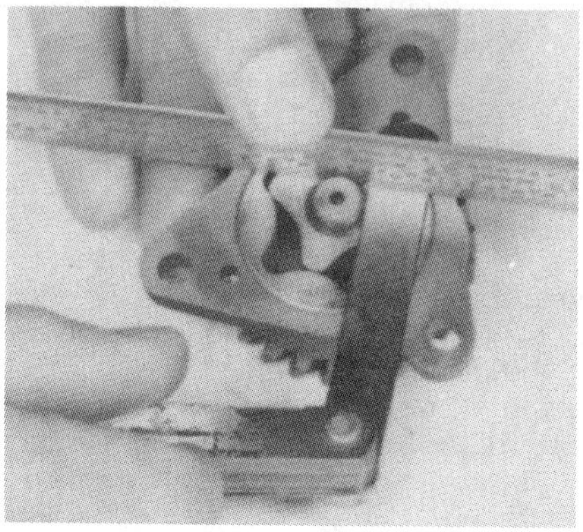

Fig. Y119—Pump rotors must be no more than 0.25 mm (0.010 in.) below face of pump body.

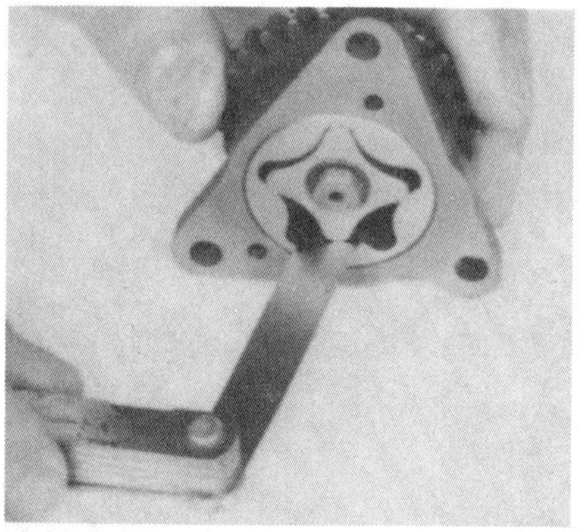

Fig. Y120—Maximum allowable clearance between inner rotor tip and outer rotor lobe is 0.25 mm (0.010 in.).

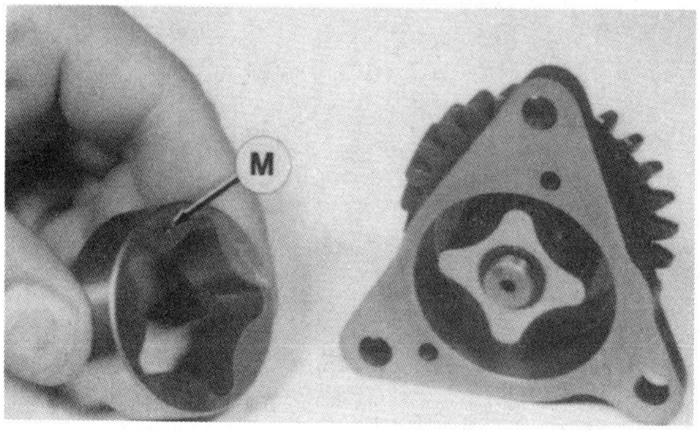

Fig. Y121—Install outer rotor so mark (M) is toward inside of pump.

Remove field coil brushes from brush holder, then remove brush holder (13) from field housing. Separate field coil housing from armature. Remove solenoid retaining screws and withdraw solenoid assembly and armature from drive housing. Push pinion stop (4) downward, remove retaining ring (3) and withdraw clutch assembly (5) from armature shaft.

Inspect all components for wear or damage. Brush length wear limit is 7.7 mm (0.030 in.).

Use an ohmmeter or test light to test armature for grounded or open windings. There should not be continuity between commutator and armature shaft. There should be continuity between two different commutator bars. To test for short circuited winding, use an armature growler.

Use an ohmmeter or test light to check field winding for grounded or open winding. There should not be continuity between field coil brush and field frame. There should be continuity between the two field brushes.

To reassemble starter, reverse the disassembly procedure.

ALTERNATOR. To check alternator unregulated voltage output, disconnect two-pin connector at alternator. Use a voltmeter set at 100 volts ac to measure voltage across alternator connector terminals. With engine running at fast throttle, the alternator should produce a minimum voltage reading of 30 volts ac.

Refer to Fig. Y125 for an exploded view of alternator. Unscrew nut (2) and tap end of shaft screw (13) to separate rotor (11) from housing (5). Press against end of shaft screw to force shaft out of bearing (8) and rotor. Remove harness clamp screw and remove connector from wiring harness (1). Remove stator mounting screws and remove stator (6) from housing.

Renew rotor if magnets are loose or damaged. Check stator for indications of burned wiring or other damage and renew as necessary.

To reassemble, reverse the disassembly procedure. When installing shaft screw, pulley half (12) must bottom against rotor (11) and bearing (8) must bottom against spacer (9). Bearing (4) must be discarded if removed. Install new bearing (4) so it bottoms in housing. Tighten nut (2) to 27 N·m (20 ft.-lbs.).

VOLTAGE REGULATOR/RECTIFIER. The voltage regulator/rectifier should provide 18 amps direct current at 13.5-15 volts at full engine speed. Rectifier/regulator is available only as a unit assembly.

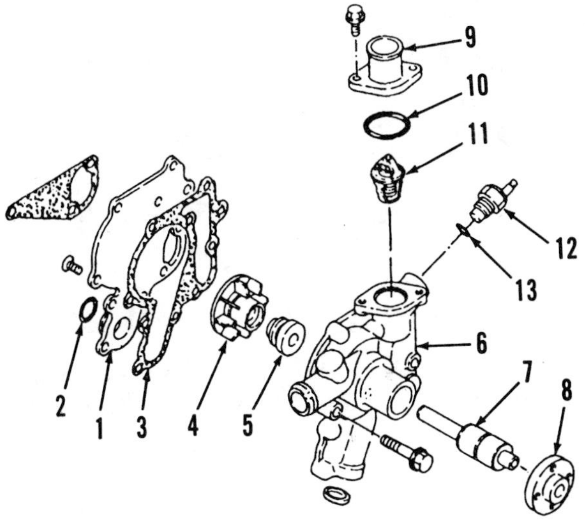

Fig. Y123—Exploded view of water pump.

1. Cover
2. "O" ring
3. Gasket
4. Impeller
5. Seal
6. Housing
7. Shaft & bearing assy.
8. Flange
9. Thermostat cover
10. "O" ring
11. Thermostat
12. Temperature sensor
13. Gasket

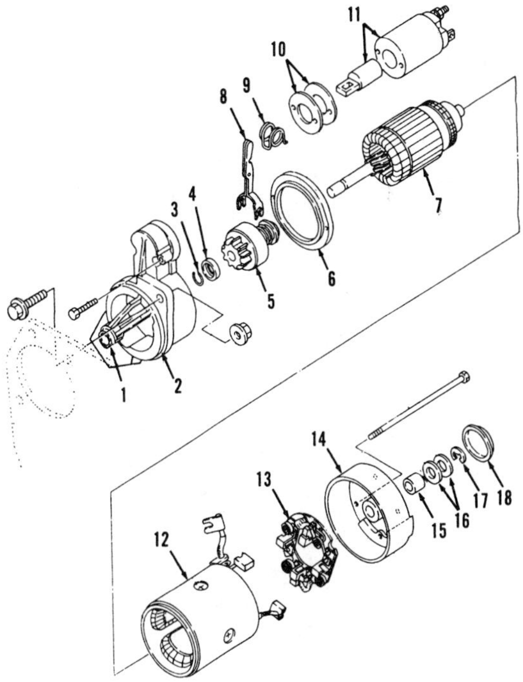

Fig. Y124—Exploded view of starter motor.

1. Sleeve
2. Drive end frame
3. Snap ring
4. Pinion stop
5. Drive pinion & clutch assy.
6. Cover
7. Armature
8. Shift lever
9. Spring
10. Shims
11. Solenoid assy.
12. Field coil assy.
13. Brush holder
14. End cover
15. Sleeve
16. Shims
17. "E" ring
18. Cap

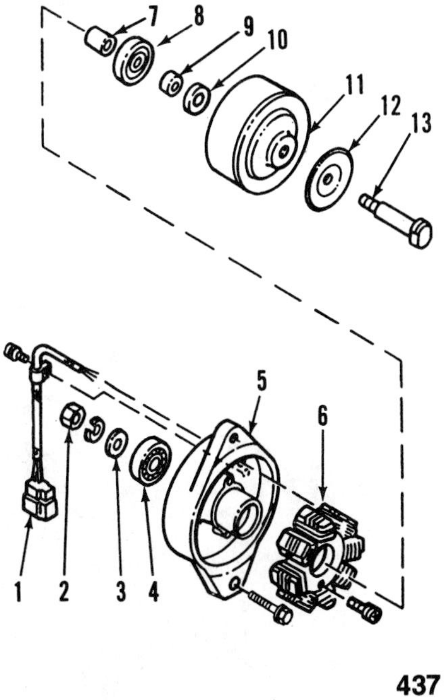

Fig. Y125—Exploded view of alternator.

1. Wiring harness
2. Nut
3. Washer
4. Bearing
5. Housing
6. Stator
7. Spacer
8. Bearing
9. Spacer
10. Washer
11. Rotor
12. Pulley half
13. Shaft

YANMAR

Model	No. Cyls.	Bore	Stroke	Displ.
3TN66UJ	3	66 mm (2.60 in.)	64.2 mm (2.53 in.)	658 cc (40.15 cu.in.)
3TNA72UJ	3	72 mm (2.83 in.)	72 mm (2.83 in.)	879 cc (53.64 cu.in.)

ENGINE IDENTIFICATION

The Yanmar 3TN66UJ and 3TNA72UJ are four-stroke, three-cylinder, water cooled, indirect injection diesel engines. Camshaft is located in cylinder block and valves in head are operated through push rods and rocker arms. Number 1 cylinder is at flywheel end of engine and firing order is 1-3-2. The engine identification plate is located on the rocker cover.

MAINTENANCE

LUBRICATION. Recommended engine lubricant is good quality engine oil with API classification CD. Select oil viscosity depending on expected ambient temperature range as shown on temperature chart in Fig. Y150.

Recommended oil change interval is after every 200 hours of use. Oil filter is a spin-on paper element cartridge. On new and reconditioned engines, lubricating oil and filter should be changed after first 50 hours of operation.

AIR FILTER. The air filter should be cleaned after every 200 hours of operation or more frequently if operating in extremely dusty conditions. Remove dust from filter element by tapping lightly with your hand, not on a hard surface. Compressed air at pressure under 30 psi (210 kPa) may also be directed inside the filter element, but be careful not to damage the element. It is recommended that filter be renewed once a year or as necessary.

WATER PUMP BELT. Adjust belt tension so belt will deflect approximately ½ inch (13 mm) when moderate thumb pressure is applied to belt midway between pump and alternator pulleys. Reposition alternator to adjust belt tension.

COOLING SYSTEM. Drain, flush and refill cooling system annually. Manufacturer recommends using low silicate antifreeze that does not contain stop leak additive. Cooling system capacity is approximately 3.8 L (1 gal.).

A thermostat is used to maintain cooling system temperature. Manufacturer recommends renewing thermostat annually.

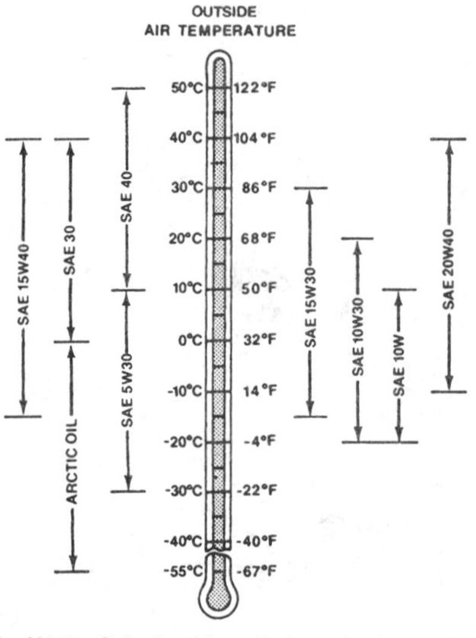

Fig. Y150—Select engine oil viscosity depending upon expected ambient air temperature range.

Model 3TN66UJ

FUEL FILTER. A renewable paper element type fuel filter (1—Fig. Y151) is located between the fuel feed pump and fuel injection pump. Water and sediment should be drained from filter bowl (7) after every 50 hours of operation or as necessary. To drain water, unscrew retaining ring (8) and remove cover and filter assembly. Clean cover and reinstall filter assembly. Loosen bleed screw (3) and turn key switch to ON position to activate fuel feed pump. When fuel flows from bleed screw, tighten screw and turn key switch to OFF position.

Filter element should be renewed annually, or sooner if loss of engine power is evident. Filter element may be removed and installed as outlined above.

Model 3TNA72UJ

FUEL FILTER. A renewable paper element type fuel filter (1—Fig. Y152) is used between the fuel feed pump and fuel injection pump. Water and sediment should be drained from filter after 50 hours of operation or as necessary. To drain water from filter, raise the hood and remove shield from left side of engine. Loosen bleed screw (3) and place end of filter drain hose in a container. Loosen drain screw (7) and allow water and fuel to drain, then tighten drain screw. With bleed screw still loose, turn key switch to ON position to acti-

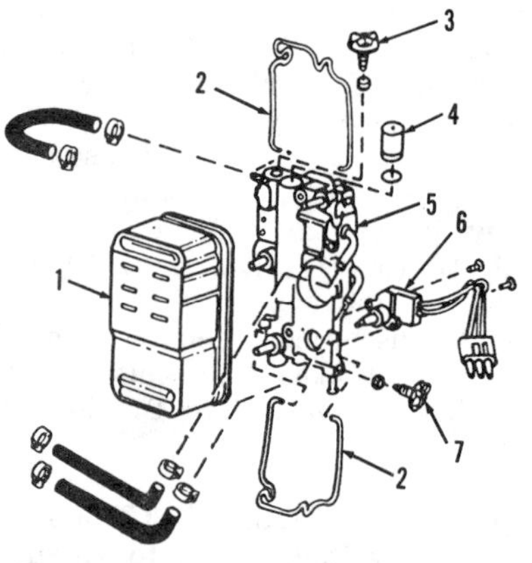

Fig. Y152—Exploded view of fuel filter assembly used on 3TNA72UJ engine.

1. Filter element
2. Clamps
3. Bleed screw
4. Heater core
5. Filter base
6. Water-in-fuel sensor
7. Drain screw

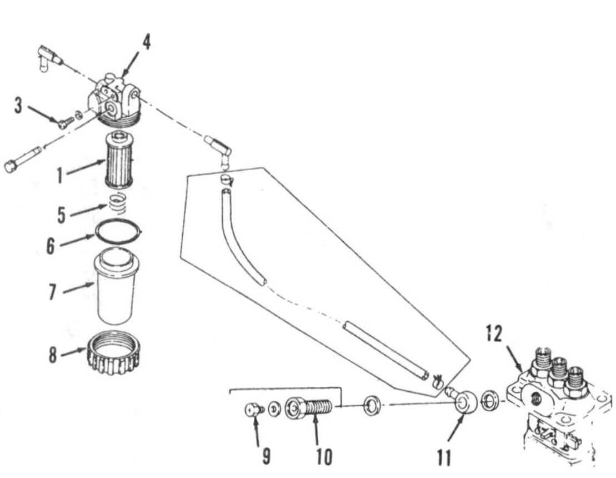

Fig. Y151—Exploded view of fuel filter assembly used on 3TN66UJ engine.

1. Filter element
3. Bleed screw
4. Filter housing
5. Spring
6. Gasket
7. Filter bowl
8. Retaining ring
9. Bleed screw
10. Fuel inlet fitting
11. Fuel inlet line
12. Injection pump

vate fuel feed pump, but do not start engine. When fuel flows from bleed screw, tighten bleed screw and turn key switch to OFF position.

Filter element should be renewed annually, or sooner if loss of power is evident. To renew filter, loosen bleed screw (3) and drain screw (7) and allow all fuel to drain from filter. Release filter clamps (2) and remove filter element (1). Install new element and close drain screw. Turn key switch to ON position until fuel begins to flow from bleed screw, then tighten bleed screw and turn key switch to OFF position.

BLEED FUEL SYSTEM. To bleed air from fuel system, first loosen bleed screw (3—Fig. Y151 or Fig. Y152) on filter housing. Turn key switch to ON position to activate fuel feed pump. When air-free fuel flows from bleed screw, tighten screw. Loosen bleed screw (9—Fig. Y151) on fuel injection pump fuel inlet fitting. When air-free fuel flows from bleed screw, tighten screw and turn key switch to OFF position. If engine will not start, loosen the fuel injector line fittings at the injectors and crank engine until fuel flows from all lines. Tighten the injector line fittings and start engine.

INJECTION PUMP TIMING

Timing of fuel injection pump is accomplished by varying the thickness of shims installed between the injection pump and mounting surface of engine front cover.

To check injection pump timing, thoroughly clean injector, injector line and pump and remove fuel injection line to number 1 cylinder (cylinder at flywheel end of engine). Remove number 1 delivery valve holder, delivery valve spring (1—Fig. Y153) and delivery valve (2), then reinstall holder. Connect JDF-14 spill pipe adapter or other suitable spill pipe to delivery valve holder. Turn flywheel in normal direction until number 1 piston TDC timing mark (TDC—Fig. Y154) on crankshaft pulley or flywheel aligns with timing index mark (A). Turn key switch to RUN position, but do not start engine. Turn flywheel clockwise approximately 90 degrees and note that fuel should be flowing from number 1 open delivery valve. Turn flywheel slowly in normal direction of rotation (counterclockwise) so that number 1 piston is starting up on compression stroke and observe fuel flow from open fuel delivery valve. Fuel flow should change to drops and then stop. Just as fuel stops, stop turning flywheel and check injection timing mark (F—Fig. Y154) on crankshaft pulley or flywheel. Note that if fuel does not stop flowing, number 1 piston is on exhaust stroke instead of compression stroke. Turn flywheel one complete revolution and repeat test.

If timing mark (F) is not aligned with timing index mark (A), repeat check several times to be sure you are observing fuel flow correctly. If injection timing is not correct, remove injection pump and adjust timing as follows: If timing mark has not reached timing index mark when fuel stops flowing, retard injection timing by increasing total thickness of shims under injection pump. If timing mark has passed timing index mark, advance timing by decreasing thickness of pump mounting shims. Changing shim thickness 0.2 mm (0.008 in.) will change position of flywheel timing mark about 4.5 mm (0.180 in.).

Install thickest shim against the injection pump and thinnest shim against timing gear housing. Tighten pump mounting bolts and recheck timing. When injection pump timing is within specifications, install fuel delivery valve and spring using a new bronze gasket. Tighten delivery valve holder to a torque of 42 N·m (31 ft.-lbs.). Install fuel injection line; bleed air from line before tightening fitting at injector.

GOVERNOR

The governor is located within the fuel injection pump camshaft housing on all models. Low idle speed should be 1450 rpm. Engine governed fast idle no-load speed should be 3600-3670 rpm for 3TN66UJ engines and 3400-3450 rpm for 3TNA72UJ engines. Refer to Fig. Y155 for location of maximum governed speed and idle speed adjustment screws on early 3TNA72UJ engines. On late 3TNA72UJ engines and all 3TN66UJ engines, fast idle adjusting screw is located on top of governor housing and slow idle adjusting screw is located on side of governor housing.

Fig. Y155—View showing speed adjusting screws on early (prior to engine serial number 004060) 3TNA72UJ engines.

1. Locknut
2. Cap
3. Fast idle adjustment screw
4. Slow idle adjustment screw
5. Locknut
6. Seal wire

REPAIRS

TIGHTENING TORQUES

Refer to the following special tightening torques:

Cylinder head bolts—
3TN66UJ . . . . 34 N·m (25 ft.-lbs.)
3TNA72UJ . . . . 61 N·m (45 ft.-lbs.)
Valve rocker arm shaft . . . . . . 26 N·m (19 ft.-lbs.)

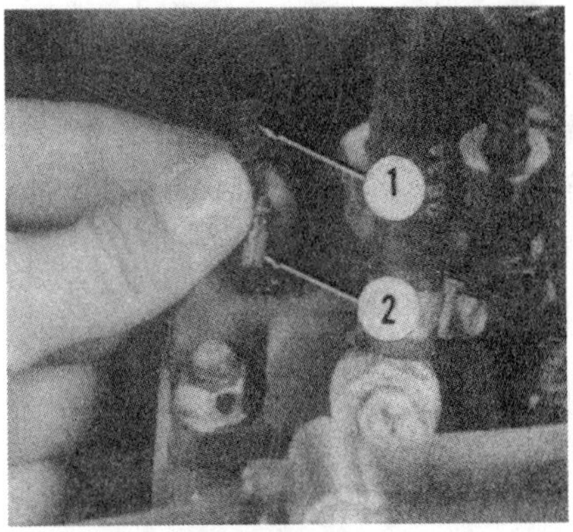

Fig. Y153—To check fuel injection pump timing, remove delivery valve spring (1) and valve (2) from number 1 cylinder fuel injection discharge fitting. Refer to text.

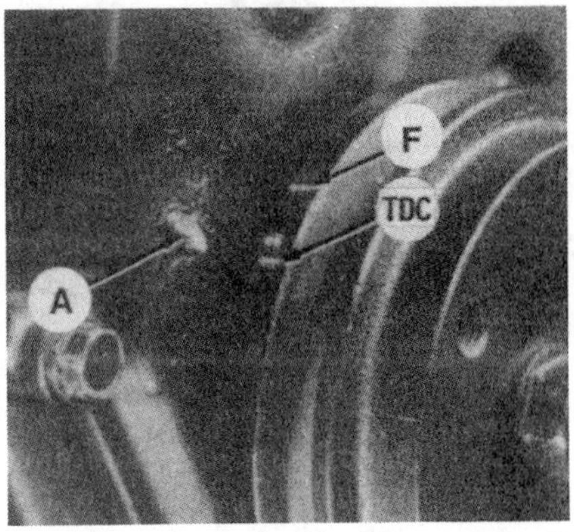

Fig. Y154—Top dead center (TDC) and beginning of injection (F) timing marks for each cylinder are located either on crankshaft pulley or flywheel. Refer to text to check injection timing.

Rocker arm cover nuts . . . . 18 N·m (13 ft.-lbs.)
Exhaust manifold—
3TN66UJ . . . . 11 N·m (8 ft.-lbs.)
3TNA72UJ . . . . 26 N·m (19 ft.-lbs.)
Intake manifold . . . . 11 N·m (8 ft.-lbs.)
Connecting rod bolts—
3TN66UJ . . . . 25 N·m (19 ft.-lbs.)
3TNA72UJ . . . . 22 N·m (16 ft.-lbs.)
Main bearing bolts—
3TN66UJ . . . . 54 N·m (40 ft.-lbs.)
3TNA72UJ . . . . 79 N·m (58 ft.-lbs.)
Crankshaft pulley bolt . . . . . . 115 N·m (85 ft.-lbs.)
Flywheel bolts . . . . 83 N·m (61 ft.-lbs.)
Timing gear housing bolts . . . . . 9 N·m (80 in.-lbs.)
Water pump mounting bolts . . 26 N·m (19 ft.-lbs.)
Injector nozzle nut . . . . 40 N·m (30 ft.-lbs.)
Injector . . . . 50 N·m (37 ft.-lbs.)
Injection pump . . . . 20 N·m (15 ft.-lbs.)

WATER PUMP AND COOLING SYSTEM

To remove water pump, first drain cooling system. Remove engine side panels and grille. Loosen alternator mounting and brace to remove the fan belt. Disconnect coolant hoses from pump. Disconnect coolant temperature sender wire. Unbolt and remove water pump assembly.

To disassemble, unbolt and remove thermostat housing (17—Fig. Y157) and thermostat (16). Remove temperature sender (13). Remove belt pulley (1), then use a suitable puller to remove pulley flange (2). Remove cover plate (11). Press shaft and bearing assembly (3) rearward from pump housing. Press shaft out of impeller (9).

To reassemble, press new bearing and shaft (longer end of shaft first) into pump housing from the front until bearing is flush with front surface of pump housing. Support impeller end of shaft, then press pulley flange onto shaft until flange is flush with end of shaft. Press new shaft seal (8) into housing bore. Install new ceramic insert in impeller, being careful not to touch lapped surface of insert. Press impeller onto shaft until top of impeller is flush with end of shaft. Install cover, pulley, thermostat and housing, and temperature sender.

Reinstall water pump and tighten mounting cap screws to 26 N·m (19 ft.-lbs.). Adjust belt tension so belt deflects about 13 mm (½ in.) when moderate thumb pressure is applied midway between pump and alternator pulleys.

THERMOSTAT

The thermostat is mounted in the water pump housing. To remove thermostat, drain cooling system and unbolt thermostat housing (17—Fig. Y157) from top of water pump housing. Thermostat should start to open at 71° C (160° F) and be fully open at 85° C (185° F).

When installing thermostat, be sure all gasket surfaces are clean and use new gasket and sealing ring. Tighten thermostat housing cap screws to 26 N·m (19 ft.-lbs.).

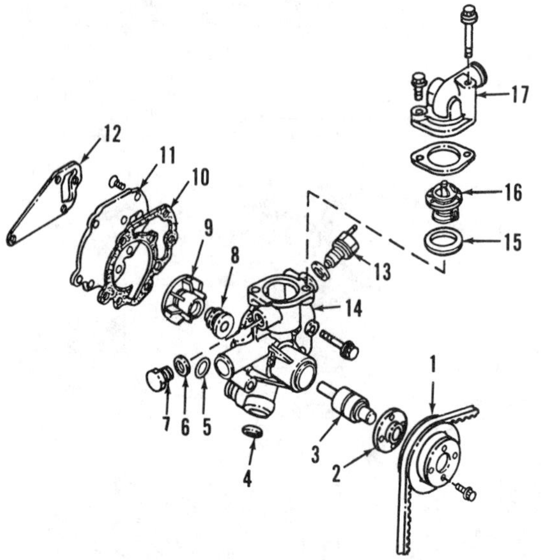

Fig. Y157—Exploded view of water pump assembly used on Model 3TNA72UJ. Water pump used on Model 3TN66UJ is similar.

1. Pulley
2. Flange
3. Shaft & bearing assy.
4. Plug
5. "O" ring
6. Gasket
7. Plug
8. Seal
9. Impeller
10. Gasket
11. Plate
12. Gasket
13. Temperature sender
14. Pump housing
15. Gasket
16. Thermostat
17. Thermostat housing

VALVE ADJUSTMENT

Valve clearance should be adjusted with engine cold. Specified clearance is 0.2 mm (0.008 in.) for both intake and exhaust valves. Adjustment is made by loosening locknut and turning adjusting screw at push rod end of rocker arm until a 0.2 mm (0.008 in.) thick feeler gage is a tight slip fit between end of valve stem and rocker arm pad.

To adjust valves, turn crankshaft in normal direction of rotation so that number 1 piston is at top dead center on compression stroke (number 1 cylinder is at flywheel end of engine). Adjust intake and exhaust valves on number 1 cylinder. Continue to turn crankshaft until next piston in firing order (1-3-2) is at top dead center on compression stroke and adjust valves for that cylinder in same manner. Repeat the procedure for each cylinder in order.

ROCKER ARMS AND PUSH RODS

Rocker arm shaft assembly (Fig. Y158) and push rods can be removed after removing rocker arm cover. Rocker arms for all cylinders pivot on full length shaft that is held by support brackets attached to cylinder head by cap screws. All valve train components should be identified as they are removed so they can be reinstalled in original locations if reused.

Rocker shaft diameter should be 9.97-9.99 mm (0.3925-0.3933 in.) with wear limit of 9.96 mm (0.392 in.) on Model 3TN66UJ. Specified rocker shaft diameter is 11.966-11.984 mm (0.471-0.472 in.) with a wear limit of 11.96 mm (0.471 in.) on Model 3TNA72UJ.

Rocker arm inside diameter should be 10.00-10.02 mm (0.3937-0.3945 in.) with wear limit of 10.09 mm (0.397 inch) for Model 3TN66UJ, and 12.00-12.02 mm (0.4724-0.4732 in.) with a wear limit of 12.09 mm (0.476 in.) on Model 3TNA72UJ. Clearance between rocker arm and shaft should not exceed 0.13 mm (0.005 in.) on any model engine.

Check push rods by rolling on flat plate and renew any that are bent or if ends are galled, worn or chipped. Push rod length is 114-115 mm (4.488-4.528 in.) on Model 3TN66UJ and 141-142 mm (5.551-5.590 in.) on Model 3TNA72UJ.

On all models, loosen rocker arm adjustment screw locknuts and back screws (6) out before installing push rods, rocker arm supports and rocker arms. Tighten rocker arm support bolts to a torque of 26 N·m (19 ft.-lbs.) on all models. Adjust valves as previously outlined and install rocker arm cover.

CYLINDER HEAD

To remove cylinder head, drain engine coolant and proceed as follows: Remove intake and exhaust manifolds, water pump, fuel injectors, glow plugs, rocker arm cover, rocker arm assembly and the push rods. Remove cylinder head bolts and lift cylinder head from engine.

Before installing cylinder head, thoroughly clean head and block gasket surfaces and inspect head for cracks, distortion or other defects. If alignment dowels (roll pins) were removed with head, pull pins from head and install in cylinder block. Check cylinder head gasket surface with straightedge and feeler gage. Head must be reconditioned or renewed if distortion (warp or twist) exceeds 0.15 mm (0.006 in.). A maximum of 0.20 mm (0.008 in.) of material may be removed from cylinder head gasket surface.

Place new gasket on block with trade mark or engine model number up and locate gasket on the two roll pins. Place head on block so that the two roll pins enter matching holes in head. Lubricate bolt threads and install the bolts, tightening in three steps in the sequence shown in Fig. Y159 until final torque of 34 N·m (25 ft.-lbs.) for Model 3TN66UJ or 61 N·m (45 ft.-lbs.) for Model 3TNA72UJ is reached.

Complete installation of cylinder head by reversing removal procedure. Yanmar recommends retorqueing cylinder head bolts after first 50 hours of operation.

VALVE SYSTEM

Valves seat directly in cylinder head on Model 3TN66UJ. Cylinder head on Model 3TNA72UJ is equipped with renewable valve seats. On all engines, valve seat and face angle is 30 degrees for intake valves and 45 degrees for exhaust valves. Recondition exhaust seats using 45, 60 and 30 degree stones and recondition intake seats using 30, 60 and 15 degree stones. Valve seat width specifications are as follows:

Model 3TN66UJ
Intake valve seat—
Desired width . . . . 1.15 mm (0.045 in.)
Maximum width . . . . 1.65 mm (0.065 in.)
Exhaust valve seat—
Desired width . . . . 1.41 mm (0.056 in.)
Maximum width . . . . 1.91 mm (0.075 in.)

Model 3TNA72UJ
Intake valve seat—
Desired width . . . . 1.44 mm (0.057 in.)
Maximum width . . . . 1.98 mm (0.078 in.)
Exhaust valve seat—
Desired width . . . . 1.77 mm (0.070 in.)
Maximum width . . . . 2.27 mm (0.089 in.)

After reconditioning seats, check distance (D—Fig. Y160) valve head is below cylinder head surface. Standard valve head recess distance for Model 3TN66UJ is 0.40 mm (0.016 in.) for intake and 0.85 mm (0.033 in.) for exhaust valve. Model 3TNA72UJ intake valve recess should be 0.50 mm (0.020 in.) and exhaust valve recess should be 0.85 mm (0.034 in.). If valve head recess is 1.0 mm (0.039 in.) or more, install new valves, valve seats (3TNA72UJ) and/or cylinder head.

If new valve seats are installed, grind new seats to proper seat width and check to be sure correct valve recess is maintained.

Refer to the following valve stem specifications:

Model 3TN66UJ
Intake valve
stem OD . . . . 5.460-5.475 mm (0.215-0.216 in.)
Wear limit . . . . 5.40 mm (0.213 in.)
Exhaust valve
stem OD . . . . 5.445-5.460 mm (0.214-0.215 in.)
Wear limit . . . . 5.40 mm (0.213 in.)

Model 3TNA72UJ
Intake valve
stem OD . . . . 6.945-6.960 mm (0.273-0.274 in.)
Wear limit . . . . 6.90 mm (0.272 in.)

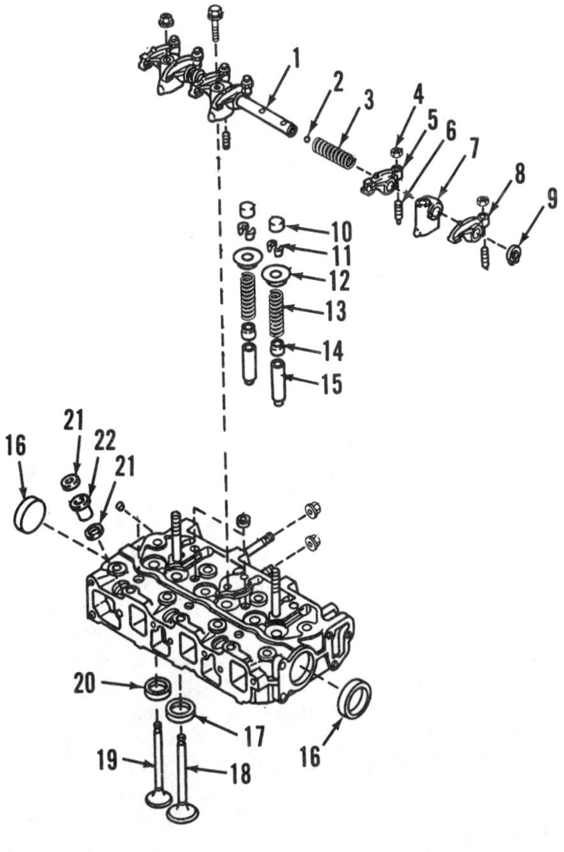

Fig. Y158—Exploded view of cylinder head and valve mechanism used on 3TNA72UJ engine. Cylinder head used on 3TN66UJ engine is similar except valve seat inserts (17 and 20) are not used.

1. Rocker shaft
2. Ball
3. Spring
4. Locknut
5. Rocker arm, exhaust
6. Adjusting screw
7. Support
8. Rocker arm, intake
9. Snap ring
10. Valve caps
11. Split retainer
12. Spring seat
13. Valve spring
14. Seal
15. Valve guide
16. Plugs
17. Valve seat
18. Intake valve
19. Exhaust valve
20. Valve seat
21. Gaskets
22. Teflon seal

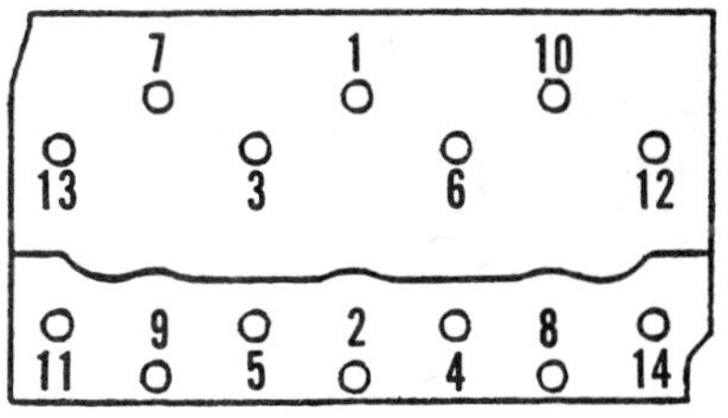

Fig. Y159—Cylinder head bolt tightening sequence for all engines.

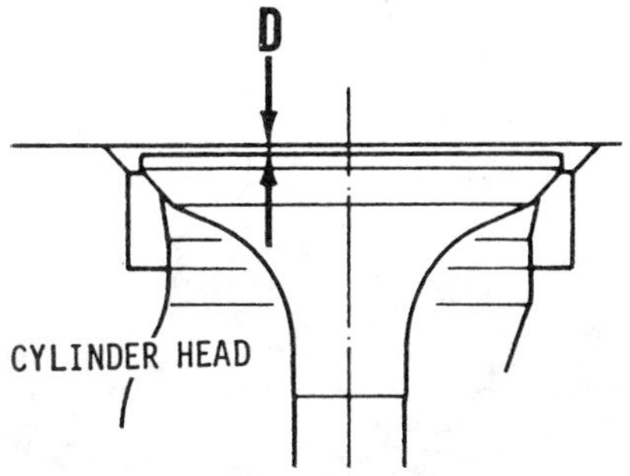

Fig. Y160—Top of valve head should be below cylinder head surface. Refer to text for specifications.

Exhaust valve
stem OD . . . . 6.945-6.960 mm (0.273-0.274 in.)
Wear limit . . . . 6.90 mm (0.272 in.)

Check valve guides against the following specifications and renew any guides with excessive wear. When installing new guides, drive old guide out toward top of head using suitable piloted driver. Install new guide with piloted driver so that distance from spring seat surface of cylinder head to top of guide (H— Fig. Y161) is 7 mm (0.276 in.) for Model 3TN66UJ or 9 mm (0.354 in.) for Model 3TNA72UJ. Refer to valve guide specifications in following table:

Model 3TN66UJ
Guide ID . . . . 5.50-5.51 mm (0.2165-0.2170 in.)

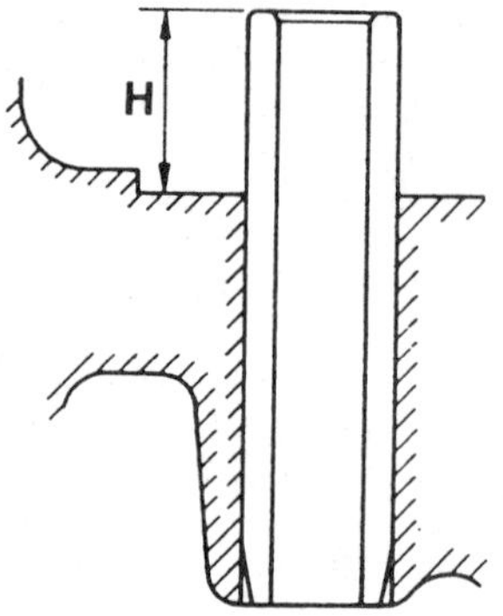

Fig. Y161—Valve guides must be installed to specified height (H) as outlined in text.

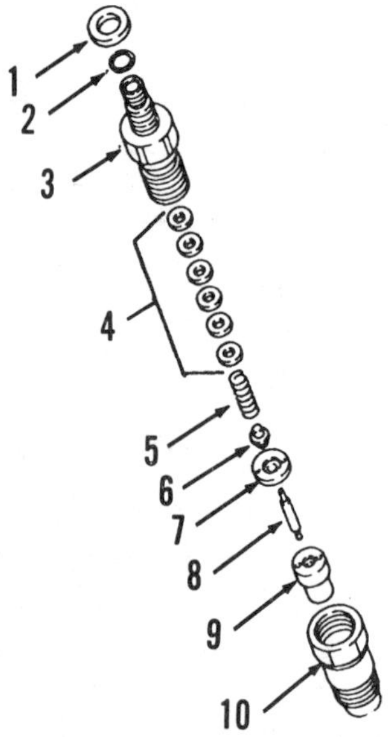

Fig. Y162—Exploded view of throttling pintle type fuel injector used on all engines.

1. Bronze gasket
2. "O" ring
3. Nozzle holder
4. Shims
5. Spring
6. Spring seat
7. Stop plate
8. Nozzle needle
9. Nozzle body
10. Nozzle nut

Wear limit . . . . 5.58 mm (0.220 in.)

Model 3TNA72UJ
Guide ID . . . . 7.005-7.020 mm (0.2758-0.2764 in.)
Wear limit . . . . 7.08 mm (0.279 in.)

After installing guides, measure inside diameter of guide and if necessary, use suitable reamer to obtain proper fit of valve stem to guide.

Intake and exhaust valve springs are alike on all models. Spring free length should be 28 mm (1.102 in.) for Model 3TN66UJ and test length at 125 N (28 lbs.) force is 17 mm (0.591 in.). Spring free length should be 37.4 mm (1.472 in.) for Model 3TNA72UJ and test length at 299 N (67 lbs.) force is 22.5 mm (0.886 in.). Be sure to install springs with closer spaced coils next to cylinder head.

INJECTORS

REMOVE AND REINSTALL. Prior to removing injectors, thoroughly clean injectors, injector lines, fuel injection pump and surrounding area using suitable solvent and compressed air. Remove the fuel return line from top of injectors. Remove the fuel injection lines from injection pump and injectors. Cap all openings to prevent entry of dirt. Unscrew the fuel injectors from cylinder head. Remove gaskets and Teflon seal from bottom of injector bores in cylinder head if they did not come out with the injectors.

When installing fuel injectors, first be sure injector bores in cylinder head are clean and all old seals are removed. Insert new bronze gaskets and the Teflon seals in injector bores and install the injectors. Tighten the injector to a torque of 50 N·m (37 ft.-lbs.). Bleed air from injector lines before tightening high pressure fuel line fittings at injector end.

TESTING. A complete job of testing, cleaning and adjusting the fuel injector requires use of special test equipment. Use only clean approved testing oil in tester tank. Injector should be tested for opening pressure, seat leakage and spray pattern. Before connecting injector to test stand, operate tester lever until oil flows, then attach injector to tester line.

WARNING: Fuel emerges from injector with sufficient force to penetrate the skin. When testing injector, keep yourself clear of nozzle spray.

Close valve to tester gage and operate tester lever a few quick strokes to purge air from injector and to be sure nozzle valve is not stuck.

OPENING PRESSURE. Open the valve to tester gage and operate tester lever slowly while observing gage reading. Opening pressure should be 11,242-12,200 kPa (1630-1770 psi). Opening pressure is adjusted by changing thickness of shim stack (4—Fig. Y162). Increasing shim thickness will increase opening pressure.

SPRAY PATTERN. The spray pattern should be conical, well atomized, and emerging in a straight axis from nozzle tip. If spray is drippy, ragged or to one side, nozzle must be cleaned or renewed.

SEAT LEAKAGE. Wipe nozzle tip dry with clean blotting paper, then slowly actuate tester lever to obtain gage pressure approximately 1960 kPa (285 psi) below opening pressure and hold this tester pressure for 10 seconds. If drop of oil appears at nozzle tip, or oil drips from nozzle, clean or renew nozzle.

OVERHAUL. Hard or sharp tools, emery cloth, grinding compounds or other than approved tools, solvents and lapping compounds, must never be used. Wipe all dirt and loose carbon from injector, then refer to Fig. Y162 and proceed as follows:

Secure injector nozzle in holding fixture or soft jaw vise and loosen nut (10). Unscrew nut and remove internal parts (4 to 9) from injector. Be careful not to lose shims (4). Place all parts in clean diesel fuel or calibrating oil as they are removed, and take care not to mix parts with those from other injector assemblies.

Clean exterior surfaces with brass wire brush to loosen carbon deposits; soak parts in approved carbon solvent if necessary. Rinse parts in clean diesel fuel or calibrating oil after cleaning to neutralize carbon solvent.

Clean nozzle spray hole from inside using pointed hardwood stick. Scrape carbon from nozzle pressure chamber using hooked scraper. Clean valve seat using brass scraper. Inspect nozzle valve (8) and body (9) for scratches or wear. Be sure that nozzle valve slides smoothly in nozzle body bore; nozzle valve should slide down into nozzle body by its own weight. If valve sticks in nozzle body, reclean or renew nozzle assembly.

Reclean all parts by rinsing thoroughly in clean diesel fuel or calibrating oil and assemble injector while immersed in fuel or oil. Be sure shim stack is intact, then install shims (4), spring (5), spring seat (6), stop plate (7), nozzle valve assembly (8 and 9) and the nozzle nut (10). Tighten nozzle nut to a torque of 40 N·m (30 ft.-lbs.). Do not over-tighten nozzle nut in attempt to stop any leakage; disassemble and clean injector if leak is noted. Retest assembled injector.

INJECTION PUMP

Thoroughly clean pump, fuel injectors, lines and surrounding area using suitable solvent and compressed air. Disconnect fuel return line from injectors and remove the high pressure fuel injection lines.

Plug all openings to prevent entry of dirt. Remove the cover and fuel shut-off solenoid from side of fuel injection pump camshaft housing. On Models 3TN66UJ and later 3TNA72UJ, pull out the governor link snap pin and disconnect governor link from pump control rack. On early 3TNA72UJ engines, pump control rack fits into a notch in governor link. On all engines, remove pump retaining screws and lift pump from top of timing gear housing.

Fuel injection pump should be serviced only by a qualified fuel injection repair shop.

To install injection pump, reverse removal procedure using same pump mounting shims as were removed. Do not tighten fuel injection line fittings at injectors until air has been bled from lines. Check injection pump timing as outlined in MAINTENANCE section.

TIMING GEAR COVER

To remove timing gear cover, engine must be removed from the tractor. Remove alternator, fan if equipped and the crankshaft pulley. On early 3TNA72UJ engines, remove cover and fuel shut-off solenoid from side of gear cover. On all engines, unbolt and remove cover from front of engine.

Thoroughly clean timing cover and mounting gasket surfaces. Install new crankshaft oil seal with lip to inside and lubricate seal. Apply a bead of silicone gasket material on front cover mounting surface. Install timing cover on engine, aligning holes in cover with the two hollow alignment dowel pins in front of engine. Tighten retaining cap screws to 9 N·m (80 in.-lbs.). Clean crankshaft pulley hub and apply lubricant on seal contact surface. Install the crankshaft pulley and tighten retaining bolt to a torque of 115 N·m (85 ft.-lbs.). To complete reassembly of engine, reverse removal procedure.

TIMING GEARS

The timing gears can be inspected after removing timing gear cover. Standard backlash between mating timing gears is 0.04-0.12 mm (0.0016-0.0047 in.); maximum allowable backlash is 0.2 mm (0.008 in.). Check gear teeth for excessive wear or tooth damage and renew as necessary. If necessary to renew gears, remove valve cover, rocker arm assembly and the push rods to relieve valve spring pressure on camshaft.

To remove idler gear and shaft, refer to Fig. Y163 and remove snap ring (1), washer (2) and gear (3). On early 3TNA72UJ engines (prior to S.N. 004061), stub shaft is a press fit in cylinder block. To remove shaft, first remove set screw from end of shaft and install a cap screw and washer in end of shaft. Then use a suitable slide hammer puller to grasp cap screw and pull shaft from cylinder block. On later 3TNA72UJ engines and all 3TN66UJ engines, stub shaft (5) is bolted to cylinder block.

Idler stub shaft diameter should be 19.959-19.980 mm (0.786-0.787 in.) and wear limit is 19.93 mm (0.785 in.). Specified bushing (4) to idler shaft oil clearance is 0.020-0.062 mm (0.001-0.002 in.) with maximum allowable clearance of 0.15 mm (0.006 in.). Renew bushing (4) if worn and idler gear (3) is otherwise serviceable.

To remove crankshaft gear, engine camshaft gear or injection pump camshaft gear, it is recommended that the shafts be removed from engine as gears are a press fit on their respective shafts.

When installing timing gears, be sure all timing marks are aligned as shown in Fig. Y165.

INJECTION PUMP CAMSHAFT

The engine must be removed from tractor in order to remove injection pump camshaft (2—Fig. Y164). Remove fuel injection pump and timing gear cover. On early 3TNA72UJ engines, block the timing gears and loosen camshaft gear retaining nut. On 3TN66UJ and later 3TNA72UJ engines, remove governor housing and governor flyweight assembly from rear of camshaft as outlined in GOVERNOR section. On all models, remove the idler gear. Remove screw retaining camshaft front bearing through hole in camshaft gear. Remove camshaft and bearings from engine.

Inspect camshaft and bearings for wear or other damage and renew as necessary. Minimum allowable camshaft

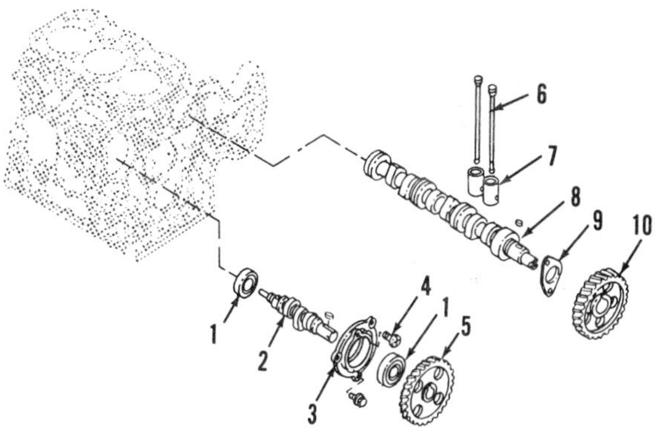

Fig. Y164—Exploded view of engine camshaft and fuel injection pump camshaft assemblies used on late 3TNA72UJ and all 3TN66UJ engines. Early 3TNA72UJ engines are similar except bearing case (3) is not used.

1. Bearing
2. Fuel injection pump camshaft
3. Bearing case
4. Bearing retaining screw
5. Gear
6. Push rods
7. Cam followers
8. Engine camshaft
9. Thrust plate
10. Gear

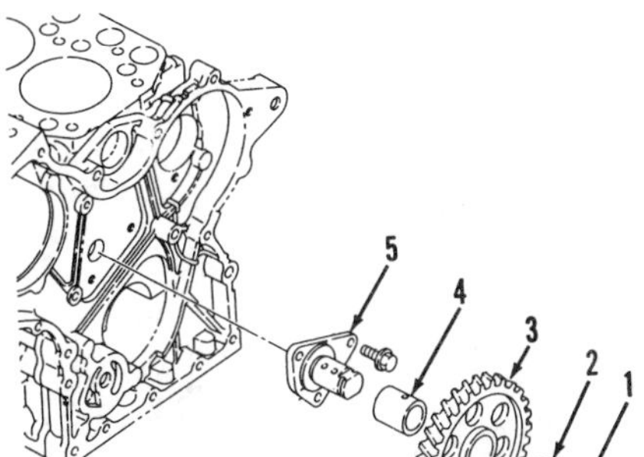

Fig. Y163—Idler gear (3) is mounted on stub shaft (5). On early 3TNA72UJ engines, stub shaft is press fit in timing gear housing.

1. Snap ring
2. Washer
3. Idler gear
4. Idler gear bushing
5. Stub shaft

Fig. Y165—Engine is correctly "timed" when timing marks on idler gear are aligned with marks on injection pump camshaft gear (A), crankshaft gear (B) and engine camshaft gear (C). Timing gears shown are for 3TN66UJ engine; timing gears for 3TNA72UJ engine have single marks instead of double marks.

lobe height is 30.9 mm (1.217 in.) for all engines.

Install camshaft in timing gear housing, being careful not to damage cam lobes or journals. Tighten bearing retaining screw to 20 N·m (15 ft.-lbs.). On early 3TNA72UJ engines, tighten camshaft gear retaining nut to 88 N·m (65 ft.-lbs.). On 3TN66UJ and late 3TNA72UJ engines, install governor flyweight assembly and governor housing. On all engines, install idler gear, aligning timing marks as shown in Fig. Y165. Install timing gear cover and tighten retaining screws to 9 N·m (80 in.-lbs.). Install fuel injection pump, bleed air from fuel system and check and adjust injection timing as outlined in MAINTENANCE section.

CAMSHAFT AND BEARINGS

The engine must be removed from tractor in order to remove engine camshaft. The camshaft (8—Fig. Y164), gear (10) and thrust plate (9) can be removed from cylinder block after removing timing cover, valve cover, rocker arm assembly and push rods. Use magnetic tools to lift cam followers from camshaft or invert engine so that followers fall away from cam. Remove camshaft thrust plate retaining bolts through web holes in camshaft gear and pull camshaft from cylinder block. If necessary to renew camshaft, thrust plate or gear, carefully press shaft from plate and gear.

Camshaft end play in block is controlled by the thrust plate. End play should be 0.05-0.15 mm (0.002-0.006 in.). Renew thrust plate if end play exceeds 0.40 mm (0.016 in.); also check thrust surface wear on timing gear and camshaft.

The camshaft journal at timing gear end of engine rides in a renewable bushing in cylinder block, while the other journals ride directly in unbushed bores of block. Renew camshaft bushing if oil clearance is 0.18 mm (0.007 in.) or more. Camshaft journal to unbushed block bore maximum oil clearance is also 0.18 mm (0.007 in.).

Refer to the following camshaft specifications:

Model 3TN66UJ

Journal OD (gear end)—
Standard . . . . 35.94-35.96 mm (1.415-1.416 in.)
Wear limit . . . . 35.85 mm (1.411 in.)

Journal OD (intermediate and flywheel end)—
Standard . . . . 35.91-35.94 mm (1.414-1.415 in.)
Wear limit . . . . 35.81 mm (1.410 in.)

Bushing ID—
Standard . . . . 36.00-36.065 mm (1.417-1.420 in.)
Wear limit . . . . 36.10 mm (1.421 in.)

Block bores ID—
Standard . . . . 36.00-36.025 mm (1.417-1.418 in.)
Wear limit . . . . 36.10 mm (1.421 in.)

Journal clearance (gear end)—
Standard . . . . 0.040-0.125 mm (0.0016-0.0049 in.)
Wear limit . . . . 0.18 mm (0.007 in.)

Journal clearance (intermediate and flywheel end)—
Standard . . . . 0.065-0.115 mm (0.0026-0.0045 in.)
Wear limit . . . . 0.18 mm (0.007 in.)

Cam lobe height—
Standard . . . . 29.97-30.03 mm (1.180-1.182 in.)
Wear limit . . . . 29.70 mm (1.169 in.)

Model 3TNA72UJ

Journal OD (gear and flywheel ends)—
Standard . . . . 39.94-39.96 mm (1.572-1.573 in.)
Wear limit . . . . 39.85 mm (1.569 in.)

Journal OD (intermediate)—
Standard . . . . 39.910-39.935 mm (1.571-1.572 in.)
Wear limit . . . . 39.85 mm (1.569 in.)

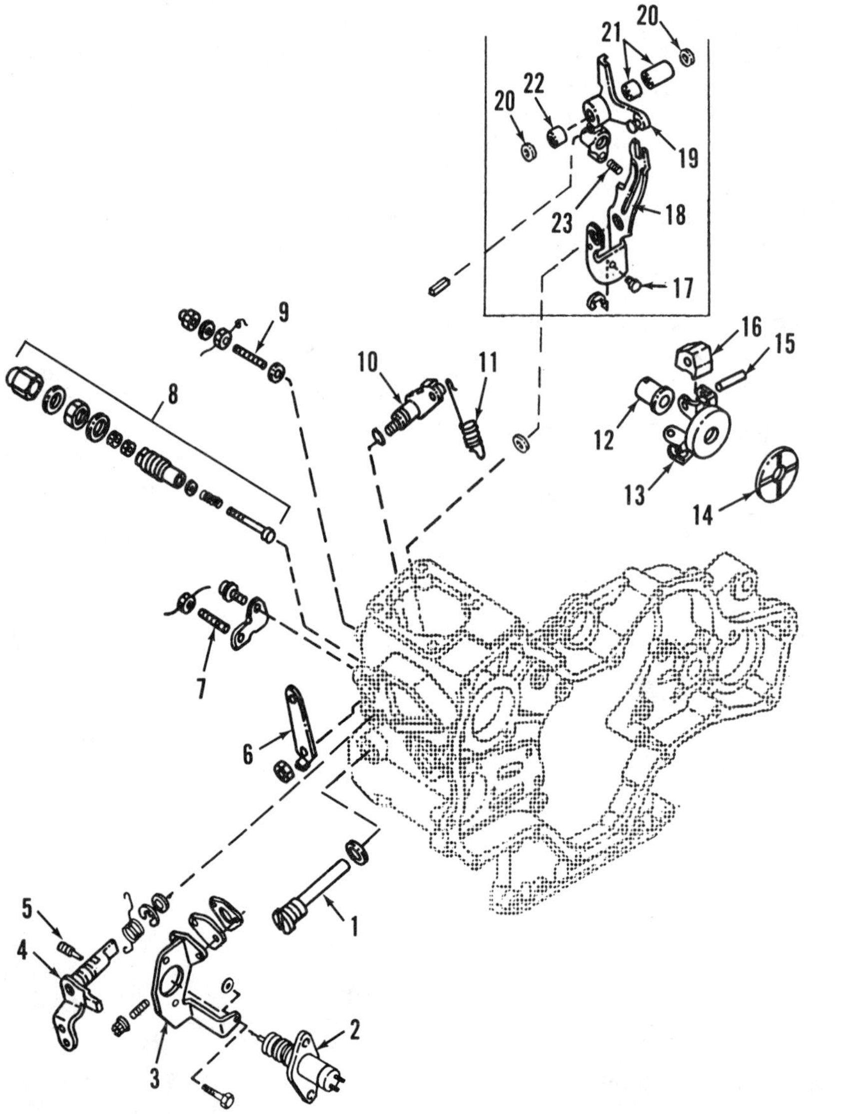

Fig. Y166—Exploded view of governor linkage used on early 3TNA72UJ engines (prior to engine serial number 004060).

1. Governor shaft
2. Fuel shut-off solenoid
3. Solenoid bracket
4. Fuel shut-off lever
5. Set screw
6. Throttle lever
7. Low idle speed adjusting screw
8. Fuel control regulator assy.
9. High idle speed adjusting screw
10. Fuel regulator lever
11. Governor spring
12. Governor sleeve
13. Governor flyweight support
14. Thrust washer
15. Pin
16. Flyweight
17. Pin
18. Governor lever
19. Governor arm
20. Shims
21. Bushings
22. Bushing
23. Spring

Bushing ID—
Standard . . . . 40.00-40.065 mm (1.575-1.577 in.)
Wear limit . . . . 40.10 mm (1.579 in.)
Block bores ID—
Standard . . . . 40.00-40.065 mm (1.575-1.577 in.)
Wear limit . . . . 40.10 mm (1.579 in.)
End journal clearance—
Standard . . . . 0.040-0.085 mm (0.0016-0.0033 in.)
Wear limit . . . . 0.18 mm (0.007 in.)
Intermediate journal clearance—
Standard . . . . 0.065-0.115 mm (0.0026-0.0045 in.)
Wear limit . . . . 0.18 mm (0.007 in.)
Cam lobe height—
Standard . . . . 33.95-34.05 mm (1.337-1.341 in.)
Wear limit . . . . 33.75 mm (1.329 in.)

To remove cam followers (7—Fig. Y164), the cylinder head must first be removed. Be sure to identify each cam follower and its position in cylinder block so they can be reinstalled in same bore from which they were removed. Cam follower standard diameter is 17.95-17.968 mm (0.7067-0.7074 in.) and wear limit is 17.93 mm (0.706 in.) for Model 3TN66UJ. Cam follower standard diameter is 20.927-20.960 mm (0.824-0.825 in.) and wear limit is 20.93 mm (0.824 in.) for Model 3TNA72UJ. Cam follower-to-bore clearance should not exceed 0.10 mm (0.004 in.) for Model 3TN66UJ and 0.15 mm (0.006 in.) for Model 3TNA72UJ. Inspect cam follower face for uneven wear, scuffing, pitting or other damage and renew as necessary. If camshaft is renewed, cam followers should also be renewed.

GOVERNOR

Early Model 3TNA72UJ

To remove governor assembly, remove grille and left side panel. Disconnect throttle cable and fuel shut-off linkage. Remove oil filter and dipstick tube. Remove cover plate from rear of timing gear housing. Remove cover from side of timing gear housing. Disconnect fuel lines from injection pump and remove injection pump. Disconnect governor spring (11—Fig. Y166) from throttle lever. Remove throttle lever retaining nut and withdraw throttle lever (6) and fuel regulator lever (10). Remove set screw (5) from rear of housing and withdraw shut-off lever (4). Remove governor shaft (1), turn governor weights so one weight is toward engine and withdraw governor control linkage (18 and 19) from timing gear housing. Remove governor sleeve (12), flyweight assembly (13, 15 and 16) and thrust washer (14).

Inspect all parts for wear or damage and renew as necessary.

To reassemble, reverse the disassembly procedure. Tighten governor shaft (1) to 49 N·m (36 ft.-lbs.). Check and adjust engine slow and fast idle speeds as outlined in MAINTENANCE section.

Model 3TN66UJ and Later 3TNA72UJ

To remove governor assembly, first remove fuel shut-off solenoid and cover from side of timing gear cover. Remove pin (9—Fig. Y167) and disconnect governor link (10) from injection pump rack. Unbolt and remove governor linkage housing (8) from rear of timing gear housing. Remove governor sleeve (15) and flyweight assembly (17, 18 and 19) from rear of injection pump camshaft. Remove governor linkage from housing as necessary.

Inspect all parts for wear or damage and renew as necessary.

To reassemble, reverse the removal procedure. Check and adjust engine slow and fast idle speeds as outlined in MAINTENANCE section.

OIL PUMP AND RELIEF VALVE

Early Model 3TNA72UJ

The gerotor type oil pump is mounted on front of engine timing gear housing and gear on pump shaft is driven from crankshaft gear.

To remove pump, first remove fuel injection pump, timing gear cover, idler

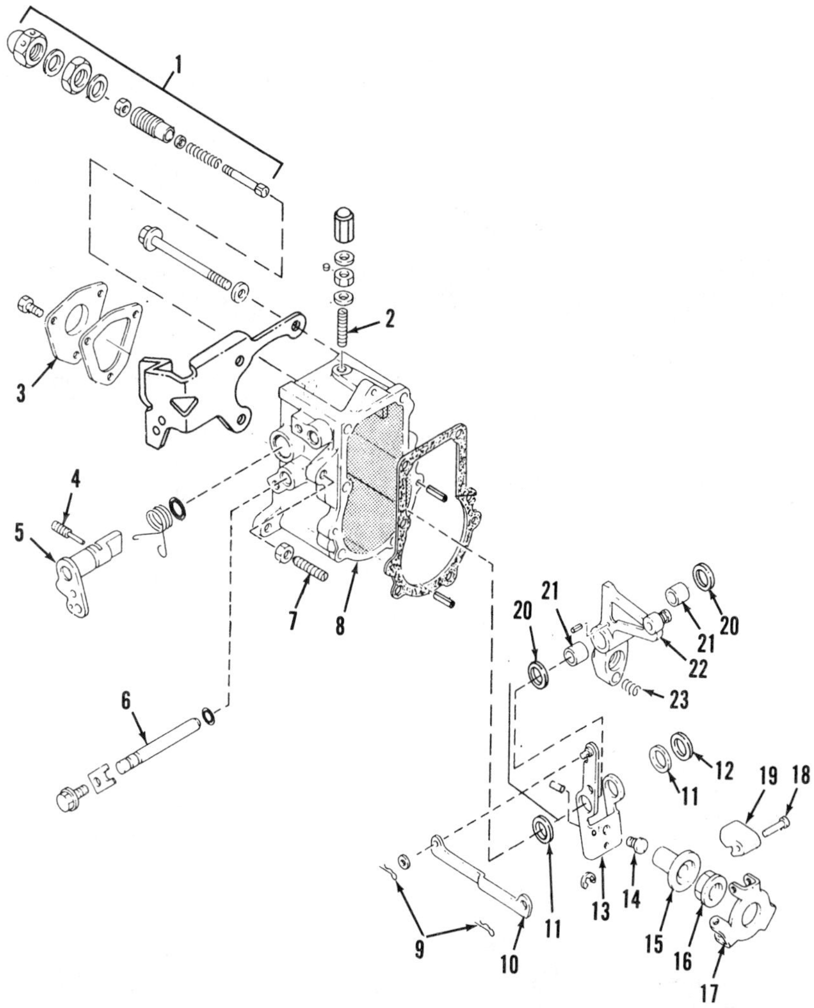

Fig. Y167—Exploded view of governor linkage used on later 3TNA72UJ and all 3TN66UJ engines.

1. Fuel control regulator assy.
2. High idle speed adjusting screw
3. Cover
4. Set screw
5. Fuel shut-off lever
6. Governor shaft
7. Low idle speed adjusting screw
8. Governor housing
9. Pins
10. Governor link
11. Spacer
12. Shim
13. Governor lever
14. Pin
15. Governor sleeve
16. Nut
17. Governor flyweight support
18. Pin
19. Governor flyweight
20. Shims
21. Bushing
22. Governor arm
23. Spring

gear and injection pump camshaft. Remove cover (1—Fig. Y169) from rear of injection pump camshaft housing. Turn slow idle adjusting screw counterclockwise several turns. Remove oil filter and dipstick tube. Remove oil filler/cover (2) and fuel shut-off solenoid and mounting bracket from side of housing. Disconnect governor linkage spring. Remove governor shaft and lift governor levers out of top of housing. Use a suitable puller to remove oil pump drive gear (10). Unbolt and remove oil pump housing (4). Withdraw pump inner and outer rotors (11) from timing gear housing.

Refer to the following specifications and renew pump components as necessary.

Rotor-to-pump body end clearance—
- Specified . . . . 0.08-0.15 mm (0.003-0.006 in.)
- Wear limit . . . . 0.25 mm (0.010 in.)

Outer rotor-to-pump body clearance—
- Specified . . . . 0.10-0.15 mm (0.004-0.006 in.)
- Wear limit . . . . 0.25 mm (0.010 in.)

Inner rotor-to-outer rotor clearance—
- Specified . . . . 0.10-0.15 mm (0.004-0.006 in.)
- Wear limit . . . . 0.25 mm (0.010 in.)

Inner rotor shaft diameter—
- Specified . . . . 12.67-12.685 mm (0.4988-0.499 in.)
- Wear limit . . . . 12.65 mm (0.4980 in.)

Pump bore diameter—
- Specified . . . . 12.70-12.718 mm (0.50-0.5007 in.)
- Wear limit . . . . 12.80 mm (0.504 in.)

Pump shaft-to-bore clearance—
- Specified . . . . 0.015-0.048 mm (0.0006-0.0019 in.)

When reassembling, lubricate all parts with clean engine oil. Be sure that pump inner and outer rotors are installed with punched marked sides toward gear housing and that thrust washers, governor weight assembly and sleeve are positioned correctly on pump shaft. Tighten pump housing cap screws to 9 N·m (80 in.-lbs.). Install governor components and tighten governor shaft to 49 N·m (36 ft.-lbs.). Be sure that timing marks on injection pump camshaft gear, idler gear engine camshaft gear and crankshaft gear are aligned as shown in Fig. Y165. Tighten injection pump camshaft retaining screw to 20 N·m (15 ft.-lbs.), and tighten oil pump gear retaining nut to 88 N·m (65 ft.-lbs.).

Complete installation by reversing removal procedure. Check injection pump timing and adjust slow idle speed as outlined in MAINTENANCE section.

Oil pressure regulating valve (5—Fig. Y169) is located on the oil pump housing and regulates pressure at 296-434 kPa (43-63 psi). Pressure may be adjusted by removing cap (8) and increasing thickness of shims (7). Changing shim thickness 0.2 mm (0.008 in.) will change oil pressure approximately 6 kPa (0.85 psi).

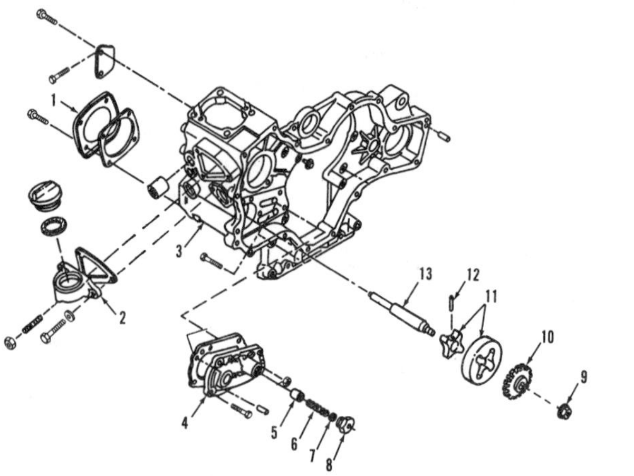

Fig. Y169—Exploded view of engine oil pump used on early 3TNA72UJ engines.

1. Cover
2. Oil filler/cover
3. Timing gear housing
4. Oil pump housing
5. Pressure relief valve
6. Spring
7. Shims
8. Cap
9. Nut
10. Drive gear
11. Inner & outer rotors
12. Drive pin
13. Pump shaft

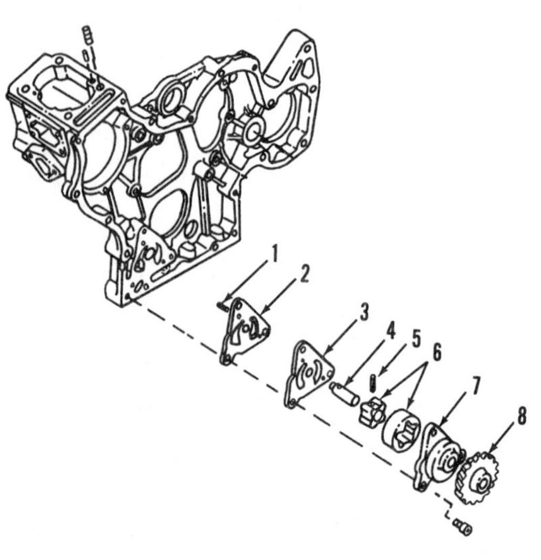

Fig. Y170—Exploded view of engine oil pump used on later 3TNA72UJ and all 3TN66UJ engines.

1. Dowel pin
2. Gasket
3. End cover
4. Pump shaft
5. Drive pin
6. Inner & outer rotors
7. Pump housing
8. Drive gear

Models 3TN66UJ and Later 3TNA72UJ

The gerotor type oil pump (Fig. Y170) is mounted on front of engine timing gear housing and gear on pump shaft is driven from crankshaft gear.

To remove pump, first remove timing gear cover, then unbolt and remove pump from front of engine. The pump end cover (3) can then be removed from pump body to inspect pump. Clearance between outer rotor and pump body should be 0.10-0.17 mm (0.004-0.007 in.) with wear limit of 0.25 mm (0.010 in.). Clearance between lobes of inner and outer rotors should be 0.05-0.1 mm (0.002-0.004 in.) with maximum allowable clearance being 0.25 mm (0.010 in.). Rotor end clearance in pump body should be 0.03-0.09 mm (0.0012-0.0035 in.) and maximum end clearance is 0.25 mm (0.010 in.). Check for looseness of drive shaft in pump body and renew pump if loose or wobbly.

When assembling pump, lubricate all parts with clean engine oil. Be sure that outer rotor is installed with punched marked side toward inside of pump housing (7).

To reinstall pump, reverse the removal procedure. Tighten mounting screws to 25 N·m (18 ft.-lbs.).

The system pressure relief valve is located on the back side of oil filter mounting adapter (4—Fig. Y171) and regulates oil pressure to 295-434 kPa (43-63 psi). Valve pressure setting may be adjusted by removing valve retainer (12) from end of valve and adding shims (11). Increasing shim thickness 1 mm (0.039 in.) will increase oil pressure approximately 13.8 kPa (2.0 psi) on Model 3TN66UJ and 10.9 kPa (1.6 psi) on Model 3TNA72UJ.

PISTON AND ROD UNITS

Engine must be removed from tractor to remove pistons and connecting rods. Piston and rod units are removed from above after removing pto electric clutch, flywheel assembly, oil pan, oil pump pickup tube and the cylinder head. Be sure to remove ridge (if present) from top of cylinders prior to removing pistons. Keep rod bearing cap with matching connecting rod.

To install piston and rod units, be sure that recess in top of piston (1—Fig. Y172) is on same side as identification mark (2) on connecting rod and that marks face toward fuel injection pump side of engine. Lubricate cylinder wall, crankpin, piston and rings and rod bearing. Place ring end gaps at equal spacing at 120 degrees around piston. Install piston and rod unit using suitable ring compressor. Do not reuse connecting rod bolts. Lubricate connecting rod bolt threads with engine oil and tighten bolts to a torque of 25 N·m (19 ft.-lbs.) on 3TN66UJ engine and 22 N·m (16 ft.-lbs.) on Model 3TNA72UJ engine.

PISTONS, PINS, AND RINGS

Pistons are fitted with two compression rings and one oil control ring with a coil type expander under the oil ring. Piston pin is full floating and is retained by a snap ring at each end of pin bore in piston. If necessary to remove piston from connecting rod, remove snap rings and push pin from piston and rod.

Inspect piston skirt for scoring, cracks and excessive wear. Measure piston skirt at right angle to pin bore and at a distance of 5 mm (0.2 in.) from bottom edge of skirt on Model 3TN66UJ, 8 mm (0.3 in.) on Model 3TNA72UJ. Piston skirt standard diameter for Model 3TN66UJ is 65.927-65.957 mm (2.596-2.597 in.) and wear limit is 65.85 mm (2.593 in.). Piston skirt standard diameter for Model 3TNA72UJ is 71.922-71.952 mm (2.832-2.833 in.) and wear limit is 71.81 mm (2.827 in.). Pistons are available in standard size and 0.25 mm (0.010 in.) oversize.

Piston pin to piston pin bore clearance should be zero to 0.017 mm (0-0.0007 in.). Maximum allowable piston pin clearance in piston is 0.045 mm (0.0018 in.). Be sure piston and connecting rod are positioned as shown in Fig. Y172 before installing piston pin. Be sure pin retaining snap rings are securely installed at each end of pin bore in piston.

Install new rings in piston ring grooves and use a feeler gage to measure ring side clearance. Renew piston if side clearance is excessive. Push rings squarely into cylinder bore and measure ring end gap using a feeler gage. Refer to the following specifications:

Model 3TN66UJ
Ring side clearance—
Top ring . . . . 0.065-0.100 mm (0.0026-0.0039 in.)
Second ring . . . . 0.030-0.065 mm (0.0012-0.0026 in.)
Oil ring . . . . 0.020-0.055 mm (0.0008-0.0022 in.)
Wear limit (all rings)...0.20 mm (0.008 in.)
Ring end gap—
Compression rings...0.15-0.30 mm (0.006-0.012 in.)
Oil ring . . . . 0.15-0.35 mm (0.006-0.014 in.)

Model 3TNA72UJ
Ring side clearance—
Top ring . . . . 0.075-0.110 mm (0.0030-0.0043 in.)
Second ring . . . . 0.030-0.065 mm (0.0012-0.0026 in.)
Oil ring . . . . 0.020-0.055 mm (0.0008-0.0022 in.)
Wear limit (all rings) . . . . . . 0.20 mm (0.008 in.)
Ring end gap—
Compression rings . . . . 0.10-0.25 mm (0.004-0.010 in.)
Oil ring . . . . 0.15-0.35 mm (0.006-0.014 in.)

When installing piston rings, place oil ring expander in groove, then install oil ring on top of expander with ring end gap at opposite side of piston from expander ends. Install compression rings with side having identification mark facing up and with end gaps at 120 degree spacing from each other and oil ring end gap.

CYLINDERS

The cylinder block for all engines is sleeveless and the pistons and rings ride directly in the cylinder block bores. Pistons and rings are available in 0.25 mm (0.010 in.) oversize. Cylinders may be rebored and honed to this oversize if worn beyond limit for use of standard size pistons and rings. Check cylinder bores in block for cracks, scoring or excessive wear. Measure cylinder bore at six different places, taking two measurements 90 degrees apart at top, middle and bottom to determine amount of cylinder wear and out-of-round. Cylinder bore specifications are as follows:

Model 3TN66UJ
Standard cylinder
bore . . . . 66.00-66.03 mm (2.599-2.600 in.)
Wear limit . . . . 66.20 mm (2.606 in.)

Model 3TNA72UJ
Standard cylinder
bore . . . . 72.00-72.03 mm (2.835-2.836 in.)
Wear limit . . . . 72.20 mm (2.843 in.)

Cylinder out-of-round condition should be less than 0.01 mm (0.0004 in.); maximum allowable out-of-round is 0.02 mm (0.001 in.). Rebore and hone cylin-

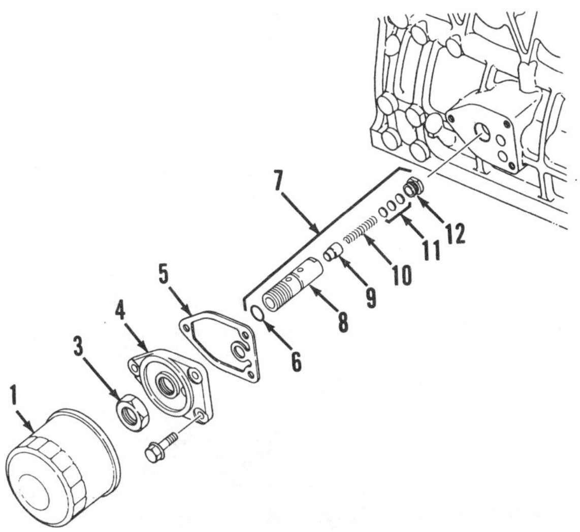

Fig. Y171—Oil pressure relief valve (7) is located on back side of oil filter base on later 3TNA72UJ and all 3TN66UJ engines. If disassembled, be sure valve retainer (12) is securely staked to adapter body (8).

1. Oil filter
3. Hex nut
4. Filter base
5. Gasket
6. "O" ring
7. Relief valve assy.
8. Adapter body
9. Relief valve poppet
10. Spring
11. Adjusting shims
12. Valve retainer

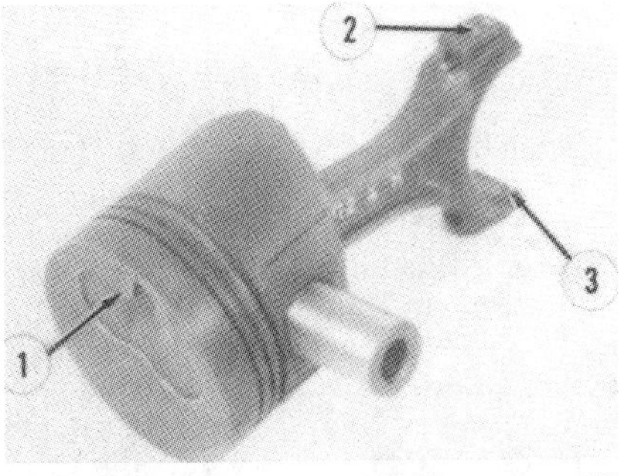

Fig. Y172—Identification mark on piston crown (1) must be on same side as identification marks on connecting rod (2) and face injection pump side of engine when installed in cylinder block.

ders or renew block if out-of-round condition or cylinder wear is excessive.

CONNECTING RODS AND BEARINGS

The connecting rod is fitted with a renewable bushing at piston pin end and a slip-in precision fit bearing at crankpin end. Rod twist or out of parallel condition should be less than 0.05 mm (0.002 in.) per 100 mm (4 in.). Maximum allowable rod distortion is 0.08 mm (0.003 in.) per 100 mm (4 in.).

Crankpin standard diameter is 35.97-35.98 mm (1.416-1.417 in.) for Model 3TN66UJ and 39.97-39.98 mm (1.5736-1.574 in.) for Model 3TNA72UJ. Specified crankpin bearing oil clearance is 0.02-0.07 mm (0.0008-0.0028 in.). Crankpin bearing oil clearance wear limit is 0.15 mm (0.006 in.) for all models.

Rod side clearance on crankpin journal for all models should be 0.2-0.4 mm (0.008-0.016 in.). Maximum allowable side clearance is 0.55 mm (0.022 in.). Renew connecting rod and/or crankshaft if clearance is excessive.

Refer to the following piston pin specifications:

Model 3TN66UJ
Pin OD . . . . 19.99-20.00 mm (0.7870-0.7874 in.)
Wear limit . . . . 19.90 mm (0.786 in.)
Pin bushing ID . . . . 20.025-20.038 mm (0.788-0.789 in.)
Wear limit . . . . 20.10 mm (0.791 in.)
Pin-to-bushing clearance—
Standard . . . . 0.025-0.047 mm (0.001-0.002 in.)
Wear limit . . . . 0.11 mm (0.004 in.)

Model 3TNA72UJ
Pin OD . . . . 20.99-21.00 mm (0.826-0.827 in.)
Wear limit . . . . 20.98 mm (0.826 in.)
Pin bushing ID . . . . 21.025-21.038 mm (0.826-0.827 in.)
Wear limit . . . . 21.10 mm (0.831 in.)
Pin-to-bushing clearance—
Standard . . . . 0.025-0.047 mm (0.001-0.002 in.)
Wear limit . . . . 0.11 mm (0.004 in.)

When reinstalling connecting rod, make certain that same rod cap is installed as was removed and that tangs in cap and rod are on same side. Renew connecting rod cap screws. Lubricate threads of rod cap screws with engine oil and tighten to 25 N·m (19 ft.-lbs.) on Model 3TN66UJ engine and 22 N·m (16 ft.-lbs.) on Model 3TNA72UJ engine.

CRANKSHAFT AND MAIN BEARINGS

The crankshaft is supported in four main bearings. Crankshaft end play is controlled by split thrust washers at each side of number 1 main bearing cap.

To remove crankshaft, first remove engine from tractor. Remove water pump, timing gear cover, fuel injection pump, cylinder head and cam followers, idler gear, engine camshaft, fuel injection pump camshaft, oil pump gear and timing gear housing. Remove piston and rod units, flywheel, flywheel housing and rear bearing seal retainer. Be sure that position number is marked on each main bearing cap, then unbolt and remove main bearing caps and lift crankshaft from cylinder block.

Check all main bearing journals, thrust surfaces at rear journal, and the crankpin journals for scoring or excessive wear. Main bearings and rod bearings are available in standard size and 0.25 mm (0.010 in.) undersize. Refer to CONNECTING RODS AND BEARINGS paragraph for information on crankpin journal specifications. Refer to the following table for main bearing specifications:

Model 3TN66UJ
Main journal OD—
Standard . . . . 39.97-39.98 mm (1.5736-1.5740 in.)
Wear limit . . . . 39.90 mm (1.572 in.)
Main bearing clearance—
Standard . . . . 0.020-0.072 mm (0.0008-0.0028 in.)
Wear limit . . . . 0.15 mm (0.006 in.)

Model 3TNA72UJ
Main journal OD—
Standard . . . . 43.97-43.98 mm (1.731-1.732 in.)
Wear limit . . . . 43.92 mm (1.729 in.)
Main bearing clearance—
Standard . . . . 0.020-0.070 mm (0.0008-0.0028 in.)
Wear limit . . . . 0.15 mm (0.006 in.)

Crankshaft end play should be 0.095-0.270 mm (0.004-0.011 inch). Maximum allowable end play is 0.33 mm (0.013 inch).

If crankshaft gear was removed, heat gear to approximately 150° C (300° F) before pressing gear onto crankshaft. Be sure that all bearing surfaces are clean and free of nicks or burrs. Install bearing shells with oil holes in cylinder block and bearing shells without oil holes in the main bearing caps. Bearing sets for all journals are alike. Lubricate crankshaft and bearings, then set crankshaft in bearings. Lubricate upper thrust washer halves (those without locating projection) and insert thrust washers between block and crankshaft with oil groove side of thrust washers toward crankshaft shoulders. Using light grease, stick lower thrust washer halves in number 1 main bearing cap with tabs on washers aligned with notches in cap and with bearing side out. Lubricate bearing and install cap. Lubricate remaining bearing caps and install in same positions as they were removed. Note that side of caps with embossed letters "FW" and an arrow points toward flywheel end of engine.

Lubricate threads of main bearing cap screws with engine oil. Tighten the main bearing cap screws to 54 N·m (40 ft.-lbs.) on Model 3TN66UJ or 79 N·m (58 ft.-lbs.) on Model 3TNA72UJ. Check to be sure that crankshaft turns freely in bearings before completing reassembly of engine.

CRANKSHAFT FRONT OIL SEAL

The crankshaft front oil seal is located in the timing gear cover and can be renewed after removing crankshaft pulley. Pry seal from timing cover, taking care not to damage the cover. Seal lip rides on hub of crankshaft pulley; inspect pulley hub for excessive wear at point of seal contact. Install new seal with lip to inside using suitable driver. Lubricate seal lip and crankshaft pulley hub, install pulley on crankshaft with hole in pulley hub aligned with dowel pin in crankshaft gear. Tighten pulley retaining bolt to torque of 115 N·m (85 ft.-lbs.).

CRANKSHAFT REAR OIL SEAL

The crankshaft rear oil seal can be renewed after removing flywheel and the seal retainer. Install new oil seal with lip to inside. Press seal into retainer until flush with outer surface of retainer. If crankshaft is grooved at original oil seal contact surface, new seal can be installed 3 mm (1/8 in.) farther into seal retainer. Lubricate seal lip and install retainer with new gasket. Tighten retainer-to-cylinder block cap screws to 11 N·m (95 in.-lbs.) and retainer-to-oil pan cap screws to 9 N·m (80 in.-lbs.).

ELECTRICAL SYSTEM

GLOW PLUGS

All engines utilize glow plugs as an aid in engine starting. The glow plugs are connected in parallel so that if one plug is burned out, the others will continue to operate. The glow plug indicator light will be on while the glow plugs are heating. A timer in the circuit will turn the indicator light off after about 10 seconds; at that time, turn switch to "START" position to start engine.

The glow plugs can be checked with an ohmmeter; renew plug if there is no continuity. Glow plug resistance should be 1.35-1.65 ohms.

To check glow plug controller circuit, check for voltage at glow plug power lead when key switch is turned to ON position. If there is no voltage, remove glow plug controller from connector socket and turn key switch ON. Check for voltage at connector terminal red wire. If there is voltage, renew controller unit.

ELECTRIC STARTER

Engines may be equipped with either a Nippon Denso or Hitachi starter. The Nippon Denso starter (Fig. Y174) uses a gear reduction drive system while the Hitachi starter (Fig. Y175) drives directly from the armature shaft.

On all models, starter current draw under load should be 230 amps or less at 300 rpm (starting rpm). On Nippon Denso starter, no-load current draw should be 90 amps or less at starter speed of about 3000 rpm. On Hitachi starter, no-load current draw should be 60 amps or less at starter speed of about 7000 rpm.

NIPPON DENSO STARTER. To disassemble, disconnect field lead, remove through-bolts and separate starter motor from solenoid housing. Remove end frame (1—Fig. Y174). Pry brush springs away from brushes and disengage field coil brushes from brush holder. Pull negative brushes outward about 6.0 mm (1/4 in.), then release springs to hold brushes in this position. Separate brush holder (2) from field coil housing (3). Withdraw armature (5) from field housing.

To disassemble reduction gears and clutch assembly, remove retaining screws and separate solenoid housing (10) from clutch housing (19). Remove clutch assembly and drive pinion from clutch housing. Push retainer (23) back and remove retaining ring (24), then withdraw components from clutch shaft (16).

Remove cover (7) from solenoid housing and withdraw plunger (8). Remove contact plates (9) if necessary. Note that contact plate located on left side of housing is smaller than contact plate on right side.

Inspect all components for wear or damage. Brush length wear limit is 8.5 mm (0.035 in.).

Use an ohmmeter or test light to test armature for grounded or open windings. There should not be continuity between commutator and armature shaft. There should be continuity between two different commutator bars. To test for short circuited winding, use an armature growler.

Use an ohmmeter or test light to check field winding for grounded or open winding. There should not be continuity between field coil brush and field frame. There should be continuity between the two field brushes.

To reassemble starter, reverse the disassembly procedure while noting the following special instruction: Apply grease to clutch shaft (16) and bearings and to pinion (11), rollers (12) and retain-

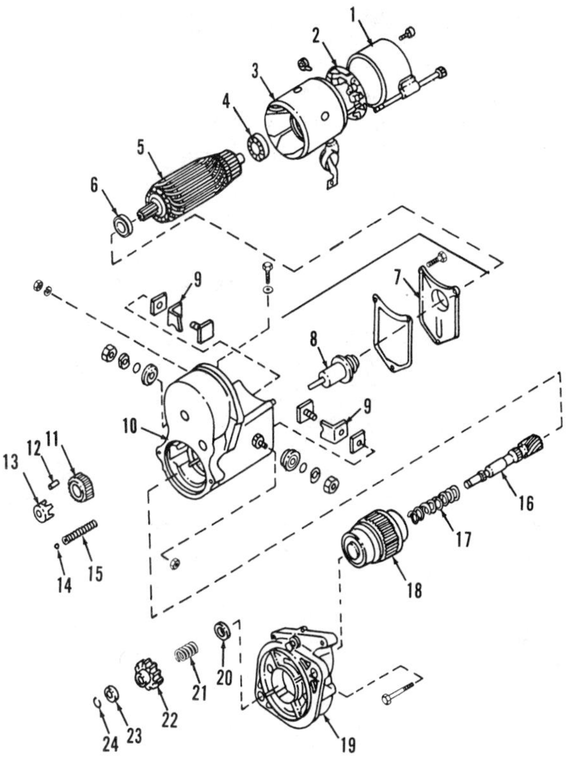

Fig. Y174—Exploded view of Nippon Denso starter motor used on some engines.

1. End frame
2. Brush holder
3. Field coil frame
4. Bearing
5. Armature
6. Bearing
7. Cover
8. Plunger
9. Contact plates
10. Solenoid housing
11. Pinion gear
12. Roller
13. Retainer
14. Ball
15. Spring
16. Clutch shaft
17. Spring
18. Clutch assy.
19. Clutch housing
20. Washer
21. Spring
22. Drive gear
23. Retainer
24. Snap ring

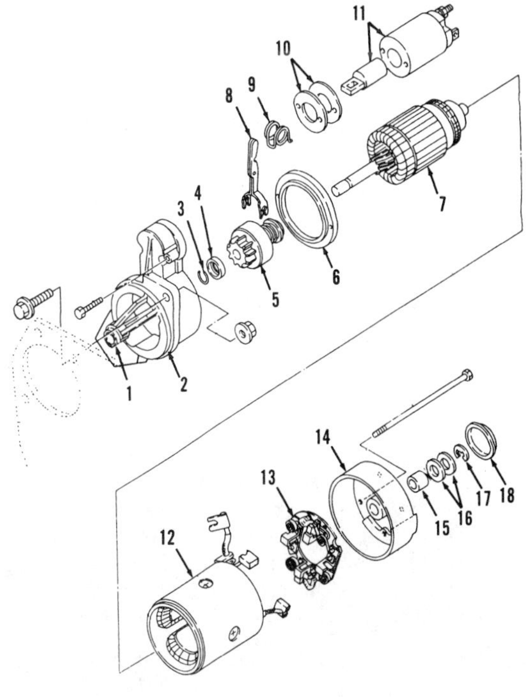

Fig. Y175—Exploded view of Hitachi starter motor used on some engines.

1. Sleeve
2. Drive end frame
3. Snap ring
4. Pinion stop
5. Drive pinion & clutch assy.
6. Cover
7. Armature
8. Shift lever
9. Spring
10. Shims
11. Solenoid assy.
12. Field coil assy.
13. Brush holder
14. End cover
15. Sleeve
16. Shims
17. "E" ring
18. Cap

er (13). Be sure that steel ball (14) is installed in end of clutch shaft. When installing end frame (1), be sure that field coil brush wires do not contact end frame. Tighten starter through-bolt nuts to 88 N·m (65 ft.-lbs.).

HITACHI STARTER. To disassemble starter, disconnect solenoid wire. Remove through-bolts and retaining screws from end cover. Pry plastic cap (18—Fig. Y175) from end cover (14), remove "E" ring (17) shims (16) and end cover. Pry springs away from negative brushes, pull brushes up and release springs to hold brushes in this position. Remove field coil brushes from brush holder, then remove brush holder (13) from field housing. Separate field coil housing from armature. Remove solenoid retaining screws and withdraw solenoid assembly and armature from drive housing. Push pinion stop (4) downward, remove retaining ring (3) and withdraw clutch assembly (5) from armature shaft.

Inspect all components for wear or damage. Brush length wear limit is 7.7 mm (0.030 in.).

Use an ohmmeter or test light to test armature for grounded or open windings. There should not be continuity between commutator and armature shaft. There should be continuity between two different commutator bars. To test for short circuited winding, use an armature growler.

Use an ohmmeter or test light to check field winding for grounded or open winding. There should not be continuity between field coil brush and field frame. There should be continuity between the two field brushes.

To reassemble starter, reverse the disassembly procedure.

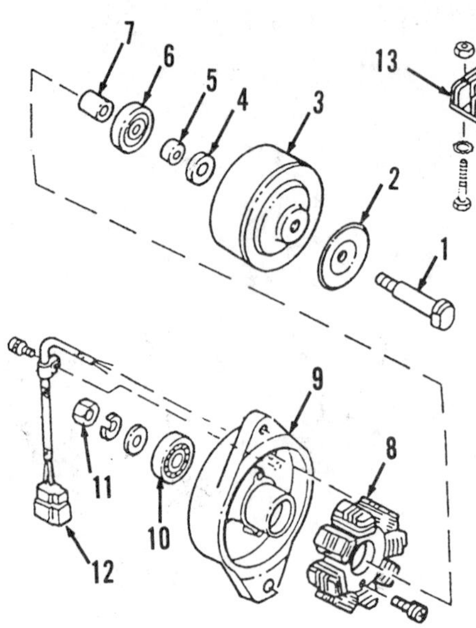

Fig. Y176—Exploded view of Kokosan alternator used on some engines.

1. Shaft
2. Plate
3. Flywheel
4. Washer
5. Spacer
6. Bearing
7. Locking collar
8. Stator
9. End frame
10. Bearing
11. Nut
12. Wiring harness
13. Voltage regulator

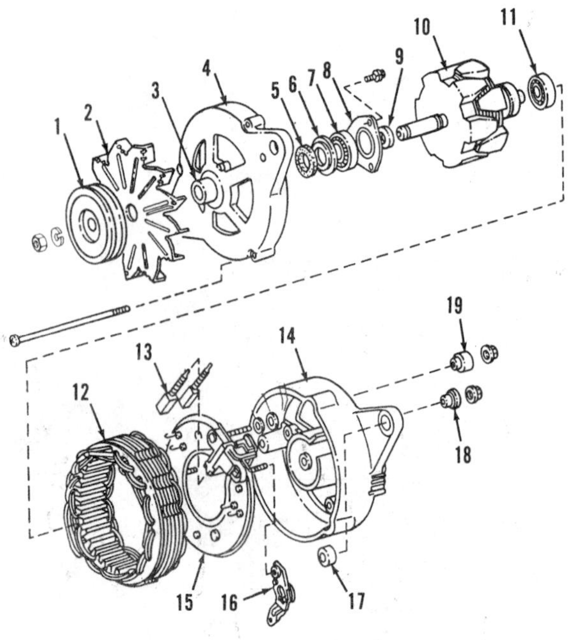

Fig. Y177—Exploded view of Nippon Denso alternator used on some engines.

1. Pulley
2. Fan
3. Locking collar
4. Drive end frame
5. Felt washer
6. Cover
7. Bearing
8. Retainer plate
9. Locking collar
10. Rotor
11. Bearing
12. Stator
13. Brushes
14. Stator housing
15. Rectifier assy.
16. Voltage regulator
17. Bushing
18. Insulator
19. Bushing

ALTERNATOR

Kokosan Alternator

To disassemble alternator, clamp pulley in a vise and remove nut (11—Fig. Y176). Tap end of shaft (1) with a soft hammer to separate flywheel (3) from stator housing (9). Remove harness clamp screw and remove connector from harness leads (12). Remove stator mounting screws and remove stator (8) from housing (9).

Renew flywheel if magnets are loose or damaged. Check stator for indications of burned wiring or other damage and renew as necessary. Check bearings (6 and 10) for roughness or damage and renew as necessary.

To reassemble, reverse the disassembly procedure. Tighten retaining nut (11) to 27 N·m (20 ft.-lbs.).

Nippon Denso Alternator

To disassemble alternator, remove through-bolts and separate drive end frame (4—Fig. Y177) from stator housing (14). Remove pulley (1) and fan (2), then tap rotor (10) from end frame. Remove bearing retainer (8), bearing (7), cover (6) and felt washer (5). Remove nuts and insulators from rear of stator housing, then pry stator from end frame. Unsolder brush leads, voltage regulator leads and stator leads as necessary to remove brushes, voltage regulator and rectifier assembly.

Use an ohmmeter or battery powered test light to check rotor winding for short or open circuit. There should be continuity indicated when tester leads are connected to rotor slip rings. There should not be continuity when tester leads are connected to rotor frame and each of the slip rings.

Renew brushes if exposed length is less than 5.5 mm (0.220 in.). Check for continuity between each of the brushes and ground. Renew rectifier-brush holder assembly if continuity is indicated.

Inspect stator for discoloration or burned odor, which would indicate faulty wiring, and renew as necessary. There should be equal resistance readings between each pair of stator lead wires. There should not be continuity indicated between any of stator leads and stator frame.

There should be continuity indicated between each of the three inner terminals (A—Fig. Y178) of rectifier assembly.

To check rectifier diodes, touch ohmmeter leads to output post (O) and to each of the three outer terminals (B) in turn. Switch ohmmeter test leads and check each terminal again. There should be continuity in only one direction. Repeat the test, touching ohmmeter test leads to ground post (G) and to each of the three outer terminals (B). There should be continuity in only one direction. Renew rectifier assembly if any of the six diodes tests defective.

To reassemble alternator, reverse the disassembly procedure while noting the following special instruction: When resoldering lead wires, use only 60-40 rosin core solder. Do not heat connections longer than necessary as excess heat will damage rectifier assembly. Be sure that stator lead wires do not contact housing frame when assembled. Depress brushes into brush holder and insert a wire through hole in rear of stator housing to hold brushes in place. Remove the wire after drive end frame and stator housing are assembled. Tighten rotor nut to 54 N·m (40 ft.-lbs.).

Hitachi Alternator

To disassemble alternator, remove end cover (15—Fig. Y179). Remove through-bolts and separate drive end frame (2) from stator housing (10). Remove nut from rotor shaft and tap rotor (6) from end frame. Remove retainer plate (5) and bearing (4). Unsolder voltage regulator and brush holder wire leads and separate voltage regulator (12) from brush holder (13). Unsolder stator wire leads and remove rectifier assembly (11).

Use an ohmmeter or battery powered test light to check rotor winding for short or open circuit. There should be continuity indicated when tester leads are connected to rotor slip rings. There should not be continuity when tester leads are connected to rotor frame and each of the slip rings.

Renew brushes (14) if exposed length is less than 5.5 mm (0.220 in.). Check for continuity between each of the brushes and between each brush and ground. Renew brush holder assembly if continuity is indicated.

Inspect stator for discoloration or burned odor, which would indicate faulty wiring, and renew as necessary. There should be equal resistance readings between each pair of stator lead wires. There should not be continuity indicated between any of stator leads and stator frame.

To test rectifier diodes, connect tester leads to ground post (A—Fig. Y180) and each of the three outer terminals (C). Switch ohmmeter test leads and check each terminal again. There should be continuity indicated in only one direction. Connect tester leads to output terminal (B) and to each of the three outer terminals (C), then reverse tester leads and repeat test. There should be continuity indicated in only one direction. Renew rectifier assembly if any of the diodes fails the test.

To reassemble alternator, reverse the disassembly procedure. When resoldering lead wires, use only 60-40 rosin core solder. Do not heat connections longer than necessary as excess heat will damage rectifier assembly. Be sure that stator lead wires do not contact housing frame when assembled. Depress brushes into brush holder and insert a wire through hole in rear cover to hold brushes in place. Remove the wire after alternator is assembled. Tighten rotor nut to 54 N·m (40 ft.-lbs.).

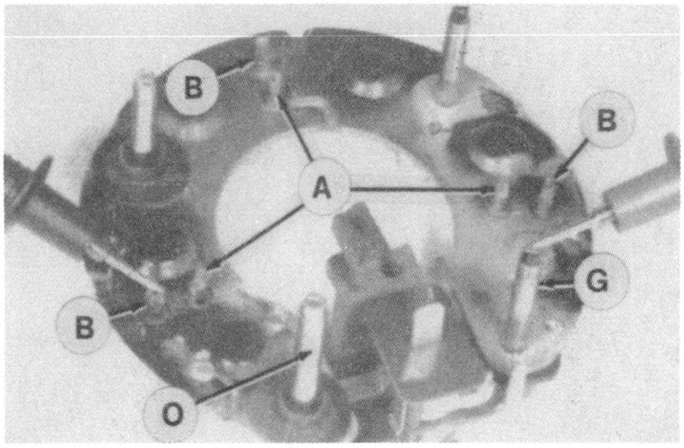

Fig. Y178—Use an ohmmeter or continuity test light to test rectifier diodes. Refer to text for procedure.

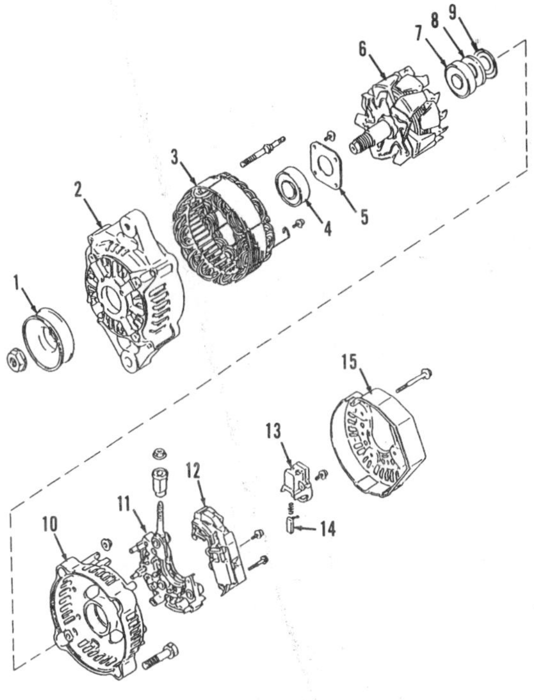

Fig. Y179—Exploded view of Hitachi alternator used on some models.

1. Pulley
2. End frame
3. Stator
4. Bearing
5. Retainer plate
6. Rotor
7. Bearing
8. Thrust washer
9. Cover
10. Housing
11. Rectifier assy.
12. Voltage regulator
13. Brush holder
14. Brush
15. End cover

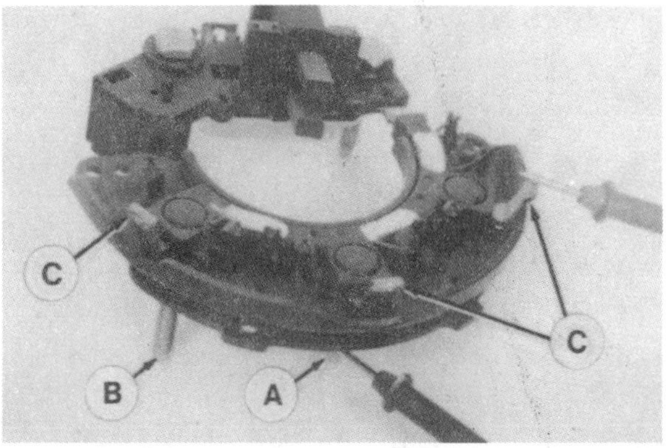

Fig. Y180—Use an ohmmeter or continuity test light to test rectifier diodes. Refer to text for procedure.

NOTES

Genuine John Deere Service Literature

For ordering information, call John Deere at 1-800-522-7448.
All major credit cards accepted.

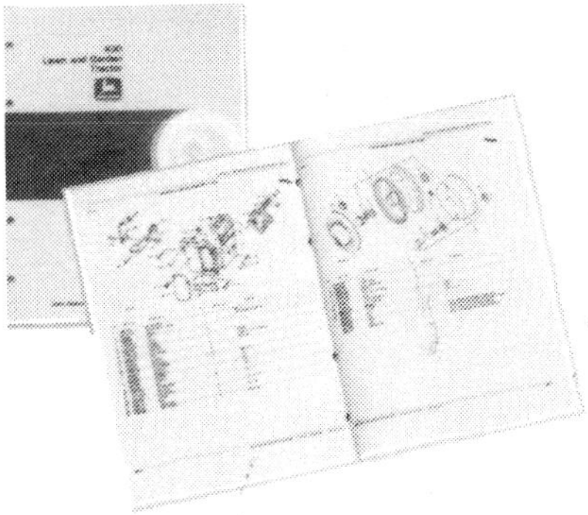

PARTS CATALOG

The parts catalog lists service parts available for your machine with exploded view illustrations to help you identify the correct parts. It is also useful in assembling and disassembling.

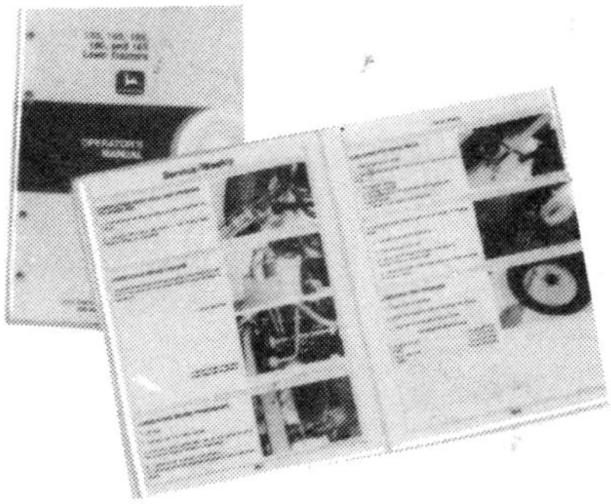

OPERATOR'S MANUAL

The operator's manual provides safety, operating, maintenance, and service information about John Deere machines.

The operator's manual and safety signs on your machine may also be available in other languages.

TECHNICAL AND SERVICE MANUALS

Technical and service manuals are service guides for your machine. Included in the manual are specifications, diagnosis, and adjustments. Also illustrations of assembly and disassembly procedures, hydraulic oil flows, and wiring diagrams.

Component technical manuals are required for some products. These supplemental manuals cover specific components.

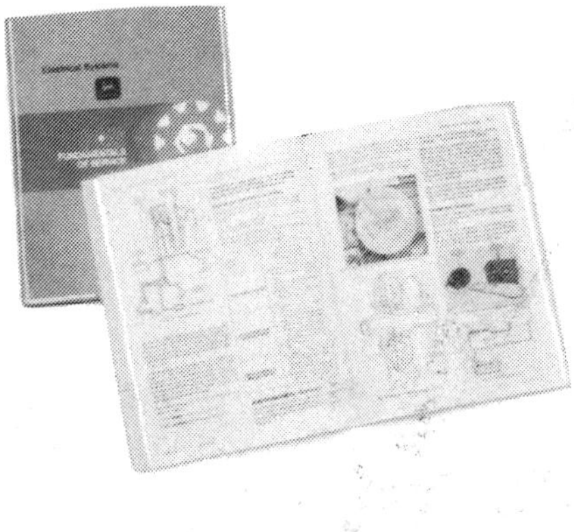

FUNDAMENTALS OF SERVICE MANUALS

er most makes and types of
tell you how to SERVICE
nanual starts with basic theory
colorful diagrams and
hys" and "hows" of adjustments
n this reference library.